W9-BYO-580

WORLD CLIMATE

Essentials of
World Regional Geography

third edition

Publisher: Emily Barrosse and John Vondeling
Acquisitions Editors: Jennifer Bortel and Kelley Tyner
Developmental Editor: Nancy Lubars
Marketing Strategist: Erik Fahlgren
Project Editors: Anne Gibby and Bonnie Boehme
Production Manager: Charlene Catlett Squibb
Art Director: Caroline McGowan
Cover Designer: Ruth Hoover

Cover Image and Credit: China, Yunnan, Yuanyang, mist over terraced
paddy fields, *©Yann Layma/Tony Stone Images*

ESSENTIALS OF WORLD REGIONAL GEOGRAPHY, Third Edition
ISBN 0-03-025984-3
Library of Congress Catalog Card Number: 99-63101

Copyright © 2000, 1998, 1995, 1993 by Harcourt, Inc.

Address for domestic orders:
Harcourt College Publishers, 6277 Sea Harbor Drive,
Orlando, FL 32887-6777
1-800-782-4479

Address for international orders:
International Customer Service, Harcourt, Inc.
6277 Sea Harbor Drive, Orlando FL 32887-6777
(407) 345-3800
Fax (407) 345-4060
e-mail   hbintl@harcourt.com

Address for editorial correspondence:
Harcourt College Publishers, Public Ledger Building, Suite 1250, 150 S.
Independence Mall West, Philadelphia, PA 19106-3412

Web Site Address
http://www.harcourtcollege.com

Printed in the United States of America

12345678 032 10 987654

Essentials of World Regional Geography

third edition

Christopher L. Salter

Joseph J. Hobbs

Jesse H. Wheeler, Jr.

J. Trenton Kostbade

University of Missouri–Columbia

Harcourt College Publishers

Fort Worth Philadelphia San Diego New York Orlando Austin
San Antonio Toronto Montreal London Sydney Tokyo

Contents Overview

This edition of the *Essentials of World Regional Geography* is dedicated to the college students of the United States who are growing up in a world that knows ever more regional conflict. May our words and your own wisdom help you understand the global issues that will be a part of your very lives, whether you want to deal with them or not. We are convinced, as geographers, that it is always better to know what is going on about you, whether at the local or the global scale.

About the Authors

Christopher L. "Kit" Salter

Kit did his undergraduate work at Oberlin College, with a major in Geography and Geology. He spent three years teaching English at a Chinese university in Taiwan immediately following Oberlin. Graduate work for both the M.A. and the Ph.D. was done at the University of California, Berkeley. He taught at UCLA from 1968 to 1987, when he took on full-time employment with the National Geographic Society in Washington with his wife, Cathy. They were both involved in the Society's campaign to bring geography back into the American school system. Kit has been professor and chair of the Department of Geography at the University of Missouri–Columbia since moving to the Heartland in 1988.

Themes that have made Salter glad he chose geography as a life field include landscape study and interpretation in both domestic and foreign settings; landscape and literature in order to show students that geography occurs in all writing, not just in textbooks; and geography education to help learners at all levels see the critical nature of geographic issues. He has written more than 125 articles in various aspects of geography; has traveled to a lot of the places that he writes of in this text; and has been wise enough to have a son in Spain, a daughter in the Bay Area, and a neat wife in the heart of the Heartland—Breakfast Creek, Missouri.

Joseph J. Hobbs

Joe Hobbs received his B.A. at the University of California–Santa Cruz in 1978 and his Ph.D. at the University of Texas–Austin in 1986. He is a professor at the University of Missouri–Columbia, and a geographer of the Middle East with many years of field research on biogeography and Bedouin peoples in the deserts of Egypt. Joe's interests in the region grew from a boyhood lived in Saudi Arabia and India. His research in Egypt has been supported by Fulbright fellowships, the American Council of Learned Societies, the American Research Center in Egypt, and the National Geographic Society Committee for Research and Exploration. He has served as the team leader of the Bedouin Support Program, a component of the St. Katherine National Park project in Egypt's Sinai Peninsula. He is the author of *Bedouin Life in the Egyptian Wilderness* and *Mount Sinai* (both University of Texas Press) and *The Birds of Egypt* (Oxford University Press). He teaches graduate and undergraduate courses in world regional geography, environmental geography, the geography of the Middle East, the geography of caves, and the geography of global current events. In 1994 he received the University of Missouri's highest teaching award, the Kemper Fellowship. Since 1984, he has led "adventure travel" trips to remote areas in Latin America, Africa, the Indian Ocean, Asia, Europe, and the High Arctic. Joe lives in Missouri with his wife Cindy, daughters Katherine and Lily, and a wildlife menagerie of tortoises and Old World chameleons.

Preface

Approach and Scope

The world consists of a continuous interplay between features of place, population, resources, location, and cultural systems. World regional geography is one means of attempting to chart and make sense of those dynamics. In this text, we have made an effort to help you see what aspects of a place give it identity and personality, what features make it newsworthy now and might make it a critical feature of the 21st century. Geographers make no claim for being able to foresee the future, but geographic analysis of past events helps us to realize exactly what characteristics have been most significant in giving a locale the history and geographic development it possesses. Our text is a study in such a vision.

At the beginning of the book, we study the ways in which geographers see the world and then describe the most essential environmental elements necessary for wise analysis. Each of the components of Chapters 1 and 2 occurs again in the particular mix of characteristics and influences that differentiate one region from another. By constructing a story of contemporary concerns and historical themes, with the help of numerous maps and illustrative photographs, the text serves as a guide to understanding the complex and demanding world in which today's students will play out their lives and influence the world of their children's lives as well.

Cooperation and conflict, resource abundance and scarcity, floods and droughts . . . all of these elements are part of the environmental reality that humankind not only has known but has authored through the transformation of the Earth's surface. As students of geography, we would like to believe that everyone thinking about world regional geography will come to grips with the immense power and responsibility associated with our human capacity—and compulsion—to change the Earth. The message of geography must be seen not only as one of understanding what has caused the present to have the shape that it does, but also to better comprehend ways for humankind to enhance Earth's conditions and landscapes of the future.

Features of This Edition

This edition has been updated so that all of the statistical tables reflect the most current information available in early 1999. These data have been added to all tables, as well as to textual comments. In addition, new maps have been added to reflect additional cartographic perspectives either called for by readership of the second edition or selected by the authors to further introduce geographic realities to the students seeking an understanding of the world of the 21st century. There have also been changes and additions to the illustrations in the text, and captions have been edited to try to bring the landscape messages of such photos even more clearly to the consciousness of the students and faculty working with this text.

This edition gives strong attention to the phenomenon of globalization. This perspective has been prompted in large part by the almost worldwide financial crisis that began in Southeast Asia in 1997, swept much of Asia in the following year, and rocked Russia in 1998 and Brazil in 1999. The authors anticipate that social and political unrest, and impacts on the global exchange of ideas, goods, and services prompted by this economic turmoil, will have long-lived effects; and so students will benefit greatly from the currency of this text.

While continuing the changes we introduced into the second edition of *Essentials of World Regional Geography*, we have added some new elements. The features that we feel give personality to this text include:

- **Regional Snapshot** This visual part opener opens each regional section. It consists of the major themes that help to define the region expressed with photographs and maps of representative natural and cultural landscapes.

- **Global View Maps** We have included new maps in this edition that demonstrate the increasingly global nature of virtually all spatial interaction. In these maps we show how the world is interconnected, and give attention to cultural interaction such as in patterns of religion or popular culture or political change. The future will grow ever more interlinked, but at the same time there is likely to be continued growth of landscapes of local diversity, even within the larger global network. Our text continues to provide evidence of both spatial dynamics.

One of the new dimensions of the third edition of *Essentials* is the end of chapter material, which includes the following:

- **Summary with Selected Key Terms** A summary concludes each chapter. In these paragraphs of review, key terms are shown in bold face, and the dominant themes of the chapter are outlined in concise, focused paragraphs.

- **Review Questions** These questions follow the summary; they are drawn from the material in the chapter and may serve either as guidelines for a thoughtful review of the materials or as essay questions for exams.

- **Discussion Questions** The final segment of this new material is built around discussion questions that set the chapter materials in a broader context. These questions may be used to stimulate classroom discussion, or they, too, may serve as exam questions or essay themes for the student's written work.

We think of this new aspect of our text as a user-friendly (whether student or professor) aid to help the reader make still better sense out of the wide range of geographic knowledge associated with the regions and nations of the world.

• **Problem Landscapes** Fourteen chapters incorporate case studies referred to as "problem landscapes." Issues in land use, water rights, boundary disputes, and other conflicts are brought to the fore to help students gain an understanding of how geography gives personality to specific places and regions.

• **Geographic Profiles** Each of the eight world regions discussed in this text is introduced by a chapter that lays out basic dimensions of each world region. In this profile, the themes that provide both a sense of the individuality of the region and its broader commonalities are chronicled.

• **Regional Perspectives** Geographic issues that work their influence across all or part of a region are defined and made relevant in these features. Such perspectives also support the increasingly global perspectives shared by the authors and the general geographic community.

• **Landscape in Literature** Essays, novels, and short stories have been excerpted in an effort to provide an additional perspective on regional identity. One of the most important lessons that a student can carry away from the study of world regional geography is an understanding that people's environmental perception and personal sense of place are shaped by features of both the objective and the subjective world. In the literacy excerpts we try to show the variety of the ways in which the world can be seen.

• **Definitions & Insights** These special boxes bring to center stage critical terms and concepts in geography. These terms have immediate utility where they are found, but they represent geographic concepts that travel all across the face of the Earth.

Organization of the Text

Each of the eight world regions we have built the text around has certain common features that provide for internal coherence. The regional profile opens the study of the region—outlines basic characteristics of that region and describes the sorts of physical and cultural features that create regional unity. Not every region has the same elements of unity, but each region can be identified by similar traits generally found across the region. For example, not all of the people in Middle and South America speak Spanish, but the Spanish language and Spanish cultural influences in Part VIII serve as major features of cohesion.

Chapters 1 and 2 consider how a geographer sees the world and illustrate how critical environmental assessment, modification, and management apply to such a vision. It is in these two chapters—Chapter 2 especially—that a clear link between human–environmental modification and world regional problems

is established. Such a viewpoint is essential not only for comprehending the world in which our students will live but also for crafting a geographic perspective on issues that range from resources to demography.

Chapters 3, 4, and 5 explore Europe. In addition to the profile chapter, the next two chapters divide Europe into the coreland that contains the majority of nations that are part of the European Union (EU). While the EU does not define the full range of geographic traits of northwest Europe, this dynamic 15-nation organization is bringing a unity to economic and even political accomplishments in Europe that is unprecedented during the past century. The realms that lie to the north, the south, and the east of the European coreland are given their own identities so that students can leave Europe with a good sense of the political and historical diversity that characterizes this critical world region.

Chapters 6 and 7 describe and discuss the geography of Russia and the Near Abroad. This is a region in transition; indeed, there is much debate among geographers about whether or not the 15 successor states of the Soviet Union should be handled as a unit. The authors believe that the legacy of Soviet rule still lingers sufficiently strongly, particularly in economic relationships and in cultural geography (witness the populations of Russians in the Baltic states and in Kazakstan), so that at least for a few more years, there will be more benefits than drawbacks in having students introduced to these countries under the same conceptual roof. Long-time users of this text will be pleased that the historical depth that Wheeler and Kostbade gave to this region still remains in this edition. The new edition is thoroughly updated to give considerable attention to the fascinating economic geography of oil in the Caspian Basin.

Because it contains approximately 60 percent of the world's oil, the Middle East has immense relevance to people around the world. Chapters 8 and 9 take students deep into a region that is likely to make headlines throughout their lives. These chapters aspire to lead students to a well-founded understanding of the circumstances underlying strife in the Middle East. Understanding of the Arab-Israeli conflict begins here with comprehension of the importance of sacred places in Judaism and Islam, the exile of the Jews, the Zionist movement, and the establishment of Israel. The subsequent wars are described thoroughly because each is necessary to comprehend the difficult issues confronting would-be peacemakers in the region. The Palestinian-Israeli peace process—which is, above all, a set of geographic problems—is updated through the Wye agreement of 1998. Importantly, these chapters also contain a wealth of material unrelated to the questions of conflict and oil, and students should come away with a strong appreciation of the cultural and environmental features of the nations that comprise the Middle East.

Monsoon Asia is the realm covered in Chapters 10 through 14. This region is home to more than half of the world's population and some of the world's most dynamic—and most troubled—economies. The Asian Tigers were the forces that led to the 21st century being anticipated as The Pacific Century three years ago. These chapters chronicle the changes that have come over the region and point out why such a label is somewhat less likely now.

Even while regional commonalities based on economic development are central to much of our treatment here, these chapters also craft a clear picture of unique national characteristics.

Numerous reviewers of the second edition were pleased by the thorough treatment afforded to the Pacific World. Chapters 15 and 16 have made no sacrifices in this edition. A strong foundation for the environmental setting of this vast region remains, along with several case studies of historical and contemporary problems in environmental management. There is new treatment of the perceived dilemma the majority of white Australians face in deciding whether to forge a new, multi-ethnic Australia oriented to the Pacific Rim and Asia, or to retain ties with the Commonwealth and the British monarchy. As in the previous edition, much attention is paid to the world view unique to Australia's Aborigines.

Africa has long troubled students of geography, who have had difficulty coming to terms with such a vast and complex region. The authors believe strongly that this region can be studied and comprehended as well as any other. We have used Chapter 17 to lay out the major themes that unite and differentiate the sub-regions and countries that make up Africa south of the Sahara. Chapter 18 is a comprehensive and thoroughly updated journey through each of the seven sub-regions.

Chapter 19 is the regional profile chapter for Middle and South America, the world of Latin America. The long chapter on all of Latin America in the second edition has been reorganized into Chapter 20 on Middle America with Mexico, Central America, and Caribbean America. The world south of Panama is now Chapter 21, South America.

Part IX assesses Anglo America, with the profile chapter (Chapter 22) looking at the broad world that lies between the northern border of Mexico and the North Pole. Chapter 23 is devoted to Canada because of the significant role that nation plays in the lives of the major readership of this text, while Chapter 24 focuses exclusively on the United States. Themes as diverse as environmental perception, resource development, urbanization, human mobility, and regional identities are covered in this final chapter.

The progression through the text may be taken from Chapters 1 through 24 in straight sequence, or a class may chart its own course after reading the overview in Chapters 1 and 2. The eight geographic profile chapters (Chapters 3, 6, 8, 10, 15, 17, 19, and 22) call attention to the elements that give character and identity to each of the eight regions in the text. The regional chapters that round out each section are fundamentally self-contained in their presentations. A glossary at the end of the text provides term definitions, should a class cover only a portion of the text in their particular study of world regional geography.

Ancillaries

This text is accompanied by a number of ancillary publications to assist instructors and enhance student learning:

- The Study Guide and Student Resources Workbook aids students in self-evaluation on text and test material.

- A Place/Name Workbook and Map-Pack contain unlabeled maps with exercise questions.

- An Instructor's Manual/Test Bank, written by authors C. L. Salter and J. J. Hobbs, contains both general and curricular suggestions for the teaching of world regional geography as well as multiple-choice classroom tested questions keyed to the text.

- The ExaMaster™ Computerized Test Bank is the software version of the printed Test Bank. Instructors can create a great variety of examinations in a multiple-choice format. A command will also reformat multiple-choice questions into short-answer questions. ExaMaster™ has capabilities for recording and graphing students' grades. Available in Macintosh® and Windows™ formats.

- 100 full-color overhead transparencies of maps and tables.

- World Regional Geography Instructor's Resource CD-ROM has been created to provide instructors with an exciting new tool for classroom presentation. The CD-ROM contains graphic files and images and line art from the textbook, and maps and tables from our collection of overhead transparencies. The World Regional Geography Instructor's Resource CD-ROM files can be opened directly or can be imported into a variety of presentation packages, including Power Point.

- The World Regional Geography Companion Web site includes instructor- and student-specific content and functionality. Instructors can use the Web site as a tool for classroom preparation, taking advantage of the free downloads of images from the World Regional Geography Instructor's Resource CD-ROM, pdf files of the Instructor's Manual, access to a monitored and archived user list, and a rich set of graphic lecture notes.

Students who use the World Regional Geography Companion Web site have access to a whole host of content and self-assessing tools. Students can use the site to review notes prepared by a World Regional Geography instructor. They can self-quiz using our state-of-the-art Quizzing and Testing program. They can use the fully functional, interactive glossary drawn from the text, as well as a rich set of annotated links.

A new addition to the Companion Web site is the Harcourt Atlas Collection: Regions of the World. This online atlas contains an image-rich mapping tutorial, physical and regional maps, an audio pronunciation guide, and much more.

- Users of World Regional Geography also have access to a World Regional Geography Companion WebCT Course. Designed explicitly for online use, the World Regional WebCT Course takes advantage of the robust WebCT Course management and communication tools: Synchronous and Asynchronous Chat, Gradebook, Attendance tracking features, Quizzing and Testing, Glossary, and a rich set of annotated links. The graphics-rich, hyperlinked content has been written specifically for the online environment. Instructors can use the course as a supplement to a traditional course or as a full-blown distance-learning course.

• Our World Regions Interactive CD-ROM is a brand-new interactive multimedia tutorial that encourages students to identify places, as well as to understand the principles and processes of that identification. Students use photographic and textual clues and the GIS model to define geographic regions of their own design. A feedback screen prompts students to reevaluate their region, encouraging critical thinking and trial-and-error modeling.

Harcourt College Publishers may provide qualified adopters with complimentary instructional aids or supplement packages. Please contact your sales representative for more information. If, as an adopter or potential user, you receive supplements you do not need, please return them to your sales representative or send them to:

Attn: Returns Department
Troy Warehouse
465 South Lincoln Drive
Troy, MO 63379

Web site passwords for students and instructors are available through your local representative. Other student ancillaries are sold through your bookstore. Please contact your local sales representative or customer service (800-237-2665) for more information.

Acknowledgments

We wish to acknowledge the long list of reviewers whose comments helped to bring this third edition to completion. We sincerely thank each and every one of these distinguished geographers and outstanding teachers and students.

Reviewers

Thomas F. Baucom, Jacksonville State University
Richard Benfield, Central Connecticut State University
Edward Boyle, Indiana University of Pennsylvania
Robert Brinkman, University of South Florida
Ron Foresta, University of Tennessee
James F. Fryman, University of Northern Iowa
Charles Gritzner, South Dakota State University
James Harlan, University of Missouri–Columbia
John Harmon, Central Connecticut State University
Marcia Holstrom, San Jose State University
David C. Johnson, University of Southwestern Louisiana
Tarek A. Joseph, University of Michigan, Dearborn
Max Lu, Kansas State University
Robin R. Lyons, San Joaquin Delta College
Barbara McDade, University of Florida
Paul Marr, Shippensburg University
William T. Mealor, Jr., University of Memphis

Richard Pillsbury, Georgia State University
Portia Reuben, Macomb College
Jeffery P. Richetto, University of Alabama
Tom Ross, University of North Carolina, Pembroke
Dan Selwa, Coastal Carolina University
Steven Silvern, University of Wisconsin–Oshkosh
Robert Mark Simpson, University of Tennessee, Martin
Joseph P. Stoltman, Western Michigan University
Michael Sullivan, Ball State University

The authors wish to extend special thanks and appreciation to the Saunders College Publishing editorial staff who labored so diligently to help us all meet tight deadlines. Jennifer Bortel was helpful from the very beginning right to the time she was given freedom and transferred to Michigan. Kelley Tyner did a fine job when she came to the project. Nancy Lubars did a lot of heavy lifting during the chapter flow in this edition, as did Mary Beth Smith. We thank them both for their dedication. A special thanks to Anne Gibby, project editor, for carrying the heavy pencil but the light spirit in the crush of publication deadlines. Bonnie Boehme finished up the task with skill and grace.

Kit Salter tosses a special nod to James Harlan, who has managed so much data in his search for the most up-to-date comprehensive and reliable statistics and articles on widely varied geographic themes. Students Hank Rademacher, Denise McElroy-Pass, Andy Wolff, Kelly Thomas, Jerry Busbee, Eric Gates, and associate Daryll McCarthy all can find evidence of their clear thinking in portions of the book. Kit thanks them all for their assistance, and he also thanks Dave's Diner for its great morning staff (Debby, Michele, and Kathryn) and the open counter space. As always, the real balance in a seemingly endless project like this textbook comes from Chloe—writer, gardener, editor, and great spirit.

Joe Hobbs thanks his colleague Kit Salter for asking him along on the journey of writing this text, and for his steady march through this leg of the journey. Also, thanks go to Jim Harlan for his painstaking work on statistics, populations, and end-of-chapter materials. The teaching assistants in the Department of Geography at the University of Missouri—Hillery Hartinger, Alex Oberle, Dina Clot, Scott Campbell, David Fox, Sally Coyne, and Tim Butchart—helped Joe catch some of the errata in the second edition and other important changes in this edition. Most of all, Joe thanks his ever-supportive family, especially Mom, partner and best friend Cindy, and little geographers Katie and Lily.

Kit Salter
Joseph Hobbs
July 1999

Contents

5 *Monsoon Asia*

Chapter 10   A Geographic Profile of Monsoon Asia   278

10.1 Definition and Basic Magnitudes   *279*
10.2 A Broad Range of Environments   *280*
10.3 Climate and Vegetation   *283*
10.4 Cultural Landscapes and Signatures of the Past   *285*
10.5 Distinctive Contemporary Landscapes   *289*
10.6 Population, Settlement, and Economy   *290*

Definitions & Insights: Monsoon   285

Regional Perspectives: The Ming Voyages   286

Definitions & Insights: Economic Growth Rates   293

Chapter 11   Complex and Populous South Asia   296

11.1 The Cultural Foundation   *298*
11.2 Regions and Resources   *299*
11.3 Climate and Water Supply   *303*
11.4 Food and Population   *306*
11.5 Industry, from Bhopal to Booming Bangalore   *307*
11.6 Social and Political Complexities   *309*

Regional Perspective: "Upstream" and "Downstream" Countries   300

Regional Perspective: Is Agricultural Progress Keeping Pace With Population Growth?   305

Definitions & Insights: The Sacred Cow   307

Definitions & Insights: The Caste System   311

Problem Landscape: Nuclear Shock Waves Rumble Across the Subcontinent   313

Chapter 12   The Tigers and Tribulations of Southeast Asia   316

12.1 A Region of Diverse Cultural Influences   *317*

Ray Ellis/Photo Researchers, Inc.

12.2 Area, Population, and Environment   *318*
12.3 The Economic Pattern   *320*
12.4 The Countries   *324*

Regional Perspective: The Economic Meltdown   320

Definitions & Insights: Rubber   322

Definitions & Insights: The Boat People—Sailing Home   328

Problem Landscape: The Spratly Islands, Hot Zone of the South China Sea   334

Chapter 13   China: Enormous Ambition, Enormous Struggle   336

13.1 Area and Population   *337*
13.2 Physical Setting and Major Landscape Elements   *340*
13.3 Continuity and Ongoing Change in China   *349*
13.4 Landscape Legacies From the Maoist Revolution   *352*
13.5 China's Urban Centers   *355*
13.6 China of the 21st Century   *356*
13.7 Taiwan   *358*
13.8 Mongolia   *359*

Regional Perspective: The Silk Road   344

Problem Landscape: "The Pulse of China"   346

Definitions & Insights: Mao Zedong   351

Definitions & Insights: Dazhai: A Model Landscape   352

Chapter 14   Japan and the Koreas: Adversity and Prosperity in the Western Pacific   362

14.1 The Japanese Homeland   *363*
14.2 Historical Background   *370*
14.3 Japan's Postwar "Miracle"   *372*
14.4 Japanese Industry   *374*
14.5 The Social Landscape   *375*
14.6 Divided Korea   *376*

Problem Landscape: Living on the Ring of Fire   366

Definitions & Insights: Japan, Korea, and the Law of the Sea   376

Regional Perspective: Nuclear Power and Nuclear Weapons in the Western Pacific   377

6 *The Pacific World*

Chapter 15   A Geographic Profile of the Pacific World   384

15.1 Melanesia, Micronesia, and Polynesia   *385*
15.2 Land and Life on the Islands   *388*
15.3 Subjugation, Independence, and Development   *393*

Definitions & Insights: Deforestation and the Decline of Easter Island   390

Regional Perspective: Nuclear Weapons Testing in the Pacific   392

7 *Africa South of the Sahara*

Katrinka Ebbe

8 *Latin America*

9 *Anglo America*

List of Maps

1 Perspectives and Issues in World Regional Geography

◀ The palace of the Imam in Wadi Dahr, Yemen.
Joseph J. Hobbs

Chapter 1

Seeing the World Through a Geographer's Eyes

▲ Rio de Janeiro, Brazil, occupies a spectacular setting of bays, peninsulas, islands, low mountains, and world-famous beaches. This landscape not only serves as a hallmark for world tourist trade, but it also demonstrates the smooth blend of physical and cultural landscape features often found in an urban center nested in a dramatic physical setting. Brissaud-Figaro/Gamma Liaison

CHAPTER OUTLINE

1.1 The Geographer's Field of Vision

1.2 The World in Spatial Terms

1.3 Places and Regions

1.4 Physical Systems

1.5 Human Systems

1.6 Environment and Society

1.7 The Uses of Geography

1.8 Careers in Geography: Regional, Systematic, and Technical Specialties

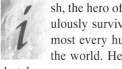 sh, the hero of George Stewart's *Earth Abides*, miraculously survives an unexplained plague that kills almost every human in the United States, and perhaps the world. He slowly adjusts to his solitude and undertakes a cross-country car trip in the hope of finding some other humans still alive. In the passage that follows, all he has found so far is a dog he calls Princess. He has reached a desert region in the American West.

Ish in the Desert

. . . he did not go any further that night. With his new-found sense of freedom he rather enjoyed merely taking additional chances. He pulled the sleeping-bag out. Unrolling it, he lay on the sand under the slight shelter of a mesquite bush. Princess lay beside him, and went soundly to sleep, tired from her run. Once he awoke in the night, and lay calm at last. He had passed through so much, and now he seemed to know a calm which would never pass. Once Princess whimpered in her sleep, and he saw her legs twitch as if she were still chasing the rabbit. Then she lay quiet; he, too, slept away. When he awoke, finally, the dawn was lemon yellow above the desert hills. He was cold, and he found Princess close up against his sleeping bag. He crawled out just as the sun was rising. *This is the desert, the wilderness. It began a long time ago. After a while, men came. They camped at the springs and left chips of stone scattered about in the sand there, and wore faint trails through the lines of the mesquite bushes, but you could hardly tell that they had been there. Still later, they laid down railroads, and strung up wires, and made long straight roads. Still, in comparison with the whole desert, you could hardly tell that man had been there, and*

ten yards aside from the steel rails or the concrete pavement, it was all the same. After a while, the men went away, leaving their works behind them.

There is plenty of time in the desert. A thousand years are as a day. The sand drifts, and in the high winds even the gravel moves, but it is all very slow. Now and then, once in a century, it may be, there comes a cloudburst, and the long-dry streambeds roar with water, rolling boulders. Given ten centuries perhaps, the fissures of the earth will open again and the black lava pour forth.

But as the desert was slow to yield before man, so it will be slow to wipe out his traces. Come back in a thousand years, and you will still see the chips of stone scattered through the sand and the long road stretching off to the gap in the knife-like hills on the horizon. As for the copper wires, they are next to immortal. This is the desert, the wilderness—slow to give, and slow to take away.

For a while the speedometer needle stood at 80, and he drove with the wild joy of freedom, fearless at the thought of a tire blowing out. Later, he slowed down a little, and began to look around with new interest, his trained geographer's mind focusing upon that drama of [humankind's] passing. In this country, he saw little difference.[1]

In the final lines of this passage, Stewart reveals that Ish had been trained as a geographer. As Ish looks out on this desert scene with its minimal signs of earlier human occupance, he begins to learn what the world offers him for survival.

[1]George Stewart, *Earth Abides*. 1949. pp. 50–51. Greenwich, Conn.: A Fawcett Crest Book.

1.1 The Geographer's Field of Vision

Geography is the study of Earth's surface and the spatial distributions and patterns of its physical and human characteristics. Geographers determine what distinct forces of nature and culture have been at work in the creation of the **landscape**—the collection of physical and human geographic features on the Earth's surface—and evaluate the role such a landscape will have on the economic and social development of the local area. Such evaluation involves **environmental perception**—our individual response to environmental features—and **environmental assessment**—our determination of the value and proper management of a given environment. Observation and analysis of the distribution and arrangement of given geographic phenomena lead to better understanding of the landscapes around us, and is at the heart of geography.

Our hope in writing this text is that you, too, will begin to develop a sense of the geographer's mind and learn the richness and usefulness of a geographer's field of vision. The ability to come upon a scene—whether a desert landscape, an urban intersection, a broad, well-farmed valley, a photograph on the front page of a newspaper, or an image on television—and to begin to make sense of the physical and cultural geographic elements that create this image is one of the skills we hope you will achieve with this text.

Geography has many perspectives, and in 1994, our profession created a learning model (Table 1.1) that divides geography into six realms, or the Six Essential Elements (see *Geography for Life: National Geography Standards 1994* for a full discussion of the elements and their associated Geography Standards). The geographer's field of vision draws on each of these elements in distinct ways.

1.2 The World in Spatial Terms

We recognize that all phenomena of the world can be organized in spatial patterns. As random as a landscape might seem at first glance, there is a spatial order in the location of all physical and cultural phenomena (Fig. 1.1). As we open each of this text's chapters on specific world regions, we will give you facts on the area, population, and characteristics of the natural environment of each region. Learning how to see the patterns and distributions of physical and cultural features in these varied landscapes is fundamental to seeing the world in spatial terms.

Maps

To think geographically is to think in terms of spatial distributions, patterns, associations, and interconnections. This inevitably means to think in terms of maps, for maps portray spatial patterns and associations with a clarity beyond the reach of words. Students should refer to maps frequently as they read geographic material, use outline maps for note taking, make their own sketch maps to show important spatial relationships, and try to visualize map relationships when patterns are stated in words in text material. In short, thinking geographically is to think spatially.

A person's life is filled with numerous local spatial realities. The route you take when you walk, ride, or drive to school is built around spatial decisions: which road, which highway, what parking lot or structure? When you come into a classroom—particularly on the first day of a new term—you have a whole raft of spatial decisions to make: Sit by the aisle, sit in the back, sit by friends, sit in new territory, sit alone? In adjusting to a new campus and town, you ask yourself, Where do I eat? What street do I explore to find a bookstore, coffee house, the Salvation Army Thrift Store, a theater? When you have the leisure or the money to seek entertainment, where do you go? Life is quilted together by the sorts of spatial decisions we make, and each of those is part of our personal geography.

One reason for studying geography is to learn where things are in the world, to acquire a framework within which countries, important cities, rivers, mountain ranges, climatic zones, agricultural and industrial areas, and other features can be related to each other. However, it is not enough simply to know the facts of location; one must also develop an understanding and appreciation of the *significance* of location.

Great Britain and New Zealand: An Example of Location Dynamics.   Perhaps an illustration will clarify these remarks. Let us compare the location of Great Britain with that of New Zealand (Fig. 1.2, p. 7). Both are island countries. Westerly winds, which blow off the surrounding seas, bring abundant rain and moderate temperatures throughout the year. Their climates are remarkably similar, even though these areas are in opposite hemispheres and are about as far from each other as it is possible for two places on the Earth to be.

Great Britain is located in the Northern Hemisphere, which contains the bulk of the world's land and most of the principal centers of population and industry; New Zealand is on the other side of the equator, in the Southern Hemisphere. Great Britain is located near the center of the world's land masses and is separated by only a narrow channel from the densely populated industrial areas of western continental Europe; New Zealand is surrounded by vast expanses of ocean. Great Britain is located in the western seaboard area of Europe, where many major ocean routes of the world converge; New Zealand is far away from the centers of world commerce. For more than four centuries, Great Britain has shared in the development of northwestern Europe as a great organizing center for the world's economic and political life; New Zealand, meanwhile, has existed in comparative isolation. Great Britain, in other words, has a central location within the existing frame of human activity on the Earth, whereas New Zealand has a peripheral

Table 1.1   The Six Essential Elements and the Eighteen Geography Standards

I. The World in Spatial Terms

Geography studies the relationships between people, places, and environments by mapping information about them into a spatial context.

The geographically informed person knows and understands

1. how to use maps and other geographic representations, tools, and technologies to acquire, process, and report information from a spatial perspective
2. how to use mental maps to organize information about people, places, and environments in a spatial context
3. how to analyze the spatial organization of people, places, and environments on Earth's surface

II. Places and Regions

The identities and lives of individuals and peoples are rooted in particular places and in those human constructs called regions.

The geographically informed person knows and understands

4. the physical and human characteristics of places
5. that people create regions to interpret Earth's complexity
6. how culture and experience influence people's perceptions of places and regions

III. Physical Systems

Physical processes shape Earth's surface and interact with plant and animal life to create, sustain, and modify ecosystems.

The geographically informed person knows and understands

7. the physical processes that shape the patterns of Earth's surface
8. the characteristics and spatial distribution of ecosystems

IV. Human Systems

People are central to geography; human activities, settlements, and structures help shape Earth's surface, and humans compete for control of Earth's surface.

The geographically informed person knows and understands

9. the characteristics, distribution, and migration of human populations
10. the characteristics, distribution, and complexity of Earth's cultural mosaics
11. the patterns and networks of economic interdependence
12. the processes, patterns, and functions of human settlement
13. how the forces of cooperation and conflict among people influence the division and control of Earth's surface

V. Environment and Society

The physical environment is influenced by the ways in which human societies value and use Earth's natural resources, while at the same time human activities are influenced by Earth's physical features and processes.

The geographically informed person knows and understands

14. how humans modify the physical environment
15. how physical systems affect human systems
16. the changes that occur in the meaning, use, distribution, and importance of resources

VI. Uses of Geography

Knowledge of geography enables people to develop an understanding of the relationships between people, places, and environments over time—that is, of Earth as it was, is, and might be.

The geographically informed person knows and understands

17. how to apply geography to interpret the past
18. how to apply geography to interpret the present and plan for the future

Source: *Geography for Life: National Geography Standards 1994.* National Geographic Research and Exploration, Washington, D.C. Pp.34–35.

Figure 1.1 Geographers are continually interested in determining what role humankind has played in changing the Earth's surface. In this scene from southern Australia, even the narrow presence of a dirt road means everything to people who live at the end of, or along, this road. The geography of connections and networks of interaction are central to human development and play a major role in the geographer's field of vision. *David Doubilet*

location. Centrality of location is a highly important factor to consider in assessing the economic as well as political geography of any country, region, or other place. Location, in other words, is of major significance.

Mental Maps

Our understanding of location, however, is not completely objective, for each of us has our own personal sense of geography—that is, a series of mental maps—which includes relative location. A **mental map** is the collection of personal geographic information that we use to order the images and facts we have about places, both local and distant. The mental map also serves to accommodate changes that one might imagine for a landscape (Fig. 1.3). In this course, we will confront and re-shape your mental map repeatedly as we introduce new images and facts about places that you have possibly never seen but about which you may have some impressions.

The Language of Maps

Because maps are so basic to all geographic understanding, it is important to learn the language of maps. Cartographers are the geographers and skilled draftspeople who create maps through careful design and skillful presentation of geographic information. Increasingly, cartography—the art and science of making maps—utilizes computer manipulation of spatial data, a field increasingly dependent on computers and known as computer cartography. **Cartography** is a graphic portrayal of location. The cartographer shows where things and places on Earth are

located in relation to each other (Fig. 1.4, p. 8). Thus, maps are indispensable tools for discovering, examining, and portraying spatial relationships and associations. Some maps show the distribution and interrelation of things that are fixed in place, such as structures and terrain features, whereas other maps portray the flow of people, goods, and ideas from place to place. A map communicates locational facts and concepts to us by means of a specialized "language" that has to be learned so that the map can be properly understood and used. Major elements of this language are **scale**, **projection**, and **symbolization**.

Definitions & Insights

SPATIAL ANALYSIS

The term **spatial** comes from the noun **space**, and it relates to the ways in which space is organized and patterned. In **spatial analysis**, geographers examine the patterns created in the distribution of phenomena (farms, auto dealerships, port cities, Baptists, etc.). In considering spatial organization, one studies the decision-making processes and outcomes that give such distributions their particular configuration. Important to understanding this term is understanding that all phenomena possess a spatial organization of one sort or another. Even aspects of the world that appear random are generally bound by some system of ordered location and distribution.

6

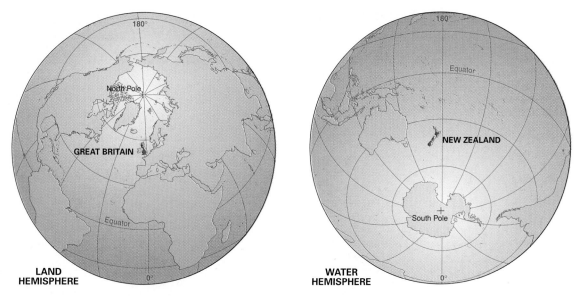

LAND
HEMISPHERE

WATER
HEMISPHERE

Figure 1.2 In the top map, note how the major land masses are grouped around the margins of the Atlantic and Arctic Oceans. The British Isles and the northwestern coast of Europe lie in the center of the "land hemisphere," constitute 80 percent of the world's total land area, and have approximately 91 percent of the world's population. New Zealand lies near the center of the opposite hemisphere, or "water hemisphere," which has only 20 percent of the land and 9 percent of the population.

Scale. A map is a reducer; it enables us to comprehend an extent of Earth-space by reducing it to the size of a single sheet. The amount of reduction appears on the map's scale, which shows the actual distance on Earth as represented by a given linear unit on the map. A common way of denoting scale is to use a representative fraction, such as 1/10,000 or 1/10,000,000.

Figure 1.3 Personal perception of the environment is widely varied. The potential future scenes, for example, envisioned by three people looking at undeveloped land may be quite different. Geography is concerned with understanding where these images come from, what might be necessary to bring such images to reality, and what the environmental consequences of such development might be. *Source: The Key to the National Geography Standards. Illustration by Suzanne Dunaway.*

In other words, one linear unit on the map (for example, an inch or centimeter) equals 10,000 or 10,000,000 such real-world units on the ground. A large-scale map is one with a relatively large representative fraction (for example, 1/10,000 or even 1/100) that portrays a relatively small area in more detail. A "global" or small-scale map has a relatively small representative fraction (for example, 1/1,000,000 or 1/10,000,000) that, in contrast, portrays a relatively large area in more generalized terms (see Fig. 1.4).

Projection. A map projection is a device to minimize distortion in one or more properties of a map (direction, distance, shape, or area). All maps inherently distort because it is not possible to represent the three-dimensional curved surface of Earth with complete accuracy on a two-dimensional flat sheet of paper. If the area represented is very small (for example, a large-scale map), the distortion may be slight enough to be disregarded, but maps that represent larger spatial areas (small-scale maps) may introduce very serious distortions. A classic example is the well-known **Mercator projection** (Fig. 1.5), originally designed to aid navigation. On that projection, every straight line has a constant bearing (direction). To achieve this, both the parallels of latitude and the meridians of longitude are shown as straight lines. A globe can show both geographic shapes and locations without distortion; lines of longitude draw closer together as they go from the equator to the poles where they converge. On a Mercator map these lines remain parallel. For these reasons, a Mercator map greatly exaggerates the east-west dimension of areas near the

Figure 1.4 Map of major world regions that form the basic framework of this text. Some regions overlap others. Russia and the Near Abroad refers to the total expanse occupied by the former Soviet Union (Union of Soviet Socialist Republics) prior to the period of dissolution that began with independence for the Baltic Republics, 1990–1991 (see chapter 6). Names and boundaries of the 15 new independent states in the Russian Realm are shown in Figure 6.3.

poles. It also exaggerates the north-south dimension because the parallels are not spaced evenly between the equator and the poles, as they are on a globe, but gradually draw farther and farther apart with increasing distance from the equator. On a Mercator map, the mathematical, or absolute, location (latitude and longitude) of every place is accurate, but areas near the poles are greatly distorted in size and shape, with the result that Greenland appears bigger than all of South America, although in reality it is only slightly larger than Mexico. There is great variety in the map projections utilized in reproducing the shape of continents on a flat surface (see Fig. 1.5).

Symbolization. Maps enable us to extract certain information from the totality of things, to see patterns of distribution, and to compare these patterns with each other. No map is a complete record of an area; it represents instead a selection of certain details, shown by symbols, that a cartographer records to accomplish a particular purpose. Unprocessed data must be classified in order to provide categories that the symbols will represent. The details may be categories of physical or cultural forms (rivers, roads, settlements, and so on), aggregates such as 100,000 people or 1 million barrels of crude oil, or averages such as population density (the number of persons per square

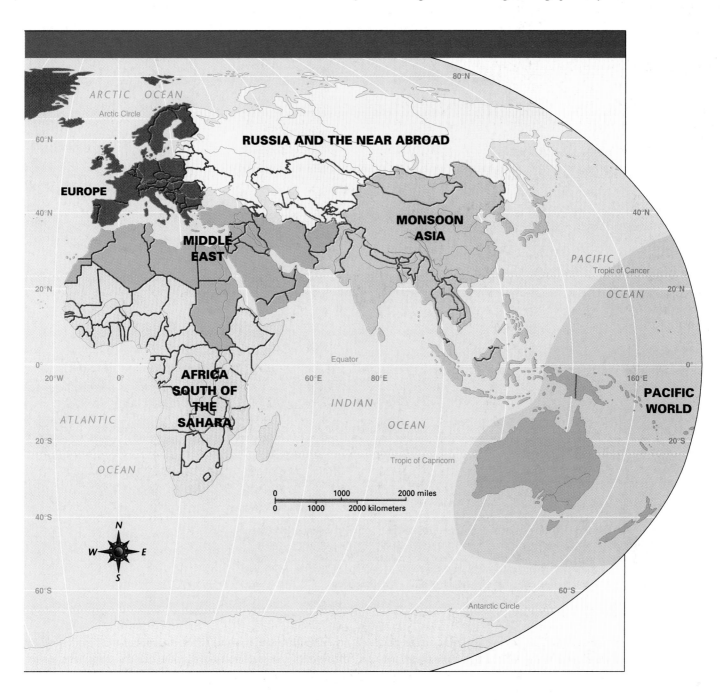

mile within a defined area). Aggregates or averages are frequently ordered into ranked categories in a graded series (for example, population densities of 0–49, 50–99, or 100 or more people per square mile) for portrayal on the map. The categories are not self-evident; they are selected by the cartographer, sometimes by elaborate statistical procedures. To represent varied phenomena, the cartographer has available a wide range of **symbols:** lines, dots, circles, squares, shadings, and others. Color, increasingly, is used in the production of maps.

Dots usually portray quantities, and the **dot map** is one of the most common types of maps used to show distributions of people or things on Earth (see Fig. 2.17). On a dot map of population, for example, each dot represents a stated number of people and is placed as near as possible to the center of the area that these people occupy. In interpreting such a map, we are not greatly concerned with the individual dots but with the way the dots are arranged. We look for the pattern and distribution of a given phenomenon.

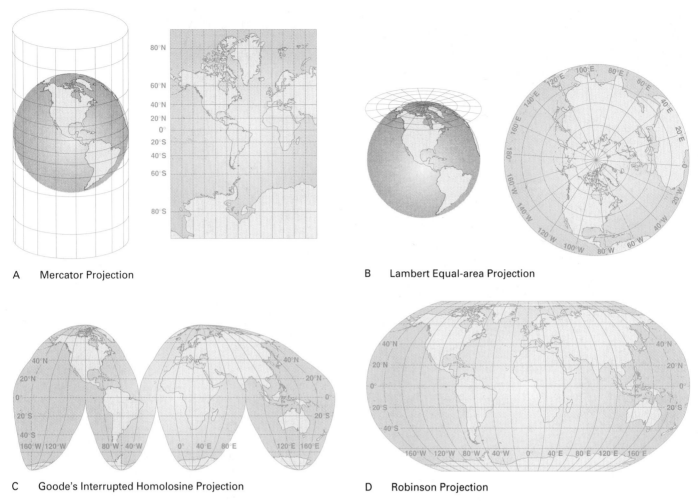

A   Mercator Projection

B   Lambert Equal-area Projection

C   Goode's Interrupted Homolosine Projection

D   Robinson Projection

Figure 1.5 These assorted map projections each attempt to represent the three-dimensional surface of the Earth on a two-dimensional sheet of paper. Note the name of each different projection included in this set of map figures and try to determine what function that projection deals with. Look at the maps at the front of any good college atlas (these are from Goode's 20th edition published by Rand McNally) and you will be able to find examples and a detailed explanation of a variety of projections.

One complexity of cartography is that different scales and ways of showing distributions can give different impressions. For example, methods of symbolization other than dots could be used to show the distribution of world population and might convey rather different ideas from the dot map presented in Figure 2.17. Some alternative methods might include the use of a single symbol for each country, with the size of the symbol proportional to the country's population, or drawing a map on which the area of each country is proportional to the item being shown—such as the number of doctors per 100,000 population. This sort of a proportional map is called a **cartogram** (Fig. 1.6a). As you compare the distribution of any given entity by differing gradients (colors or patterns) in **choropleth maps** (Fig. 1.6b), you can see a graphic distinction from the cartogram. In looking at the two maps in Figure 1.6 you can see

how the different goals of the cartographer—as evidenced by the style of map selected—will influence the maps selected for a text, or for a presentation. Even if symbolization by dots is retained, one map might use one dot per 100,000 people while another uses one dot per 1 million people, also producing a very different look to the two maps.

Once a pattern is perceived, the question arises: Why is this located or distributed the way it is? The search for answers generally involves an attempt to explain relationships between the mapped subject and other phenomena that apparently influence its distribution or people's perceptions of it. In seeing the world in spatial terms, we can see how geographers use maps as tools in understanding patterns of spatial distribution.

An additional aspect of meaning on maps comes from the use of **parallels of latitude** and **meridians of longitude.** The

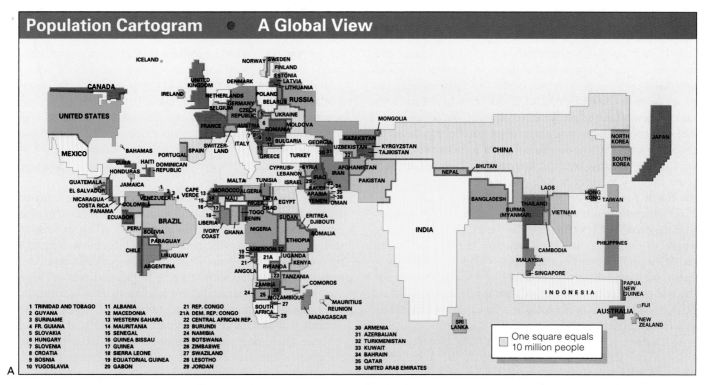

Population Cartogram • **A Global View**

1 TRINIDAD AND TOBAGO
2 GUYANA
3 SURINAME
4 FR. GUIANA
5 SLOVAKIA
6 HUNGARY
7 SLOVENIA
8 CROATIA
9 BOSNIA
10 YUGOSLAVIA

11 ALBANIA
12 MACEDONIA
13 WESTERN SAHARA
14 MAURITANIA
15 SENEGAL
16 GUINEA BISSAU
17 GUINEA
18 SIERRA LEONE
19 EQUATORIAL GUINEA
20 GABON

21 REP. CONGO
21A DEM. REP. CONGO
22 CENTRAL AFRICAN REP.
23 BURUNDI
24 NAMIBIA
25 BOTSWANA
26 ZIMBABWE
27 SWAZILAND
28 LESOTHO
29 JORDAN

30 ARMENIA
31 AZERBAIJAN
32 TURKMENISTAN
33 KUWAIT
34 BAHRAIN
35 QATAR
36 UNITED ARAB EMIRATES

One square equals 10 million people

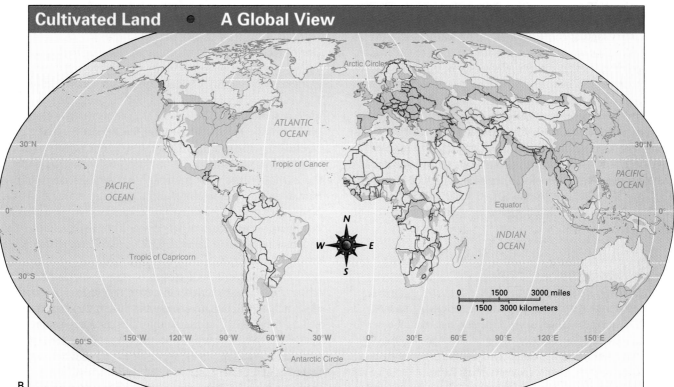

Cultivated Land • **A Global View**

Figure 1.6 (a) This cartogram represents one of the ways cartographers represent different patterns on Earth. In a cartogram such as this one of world population by country, the shape and size of each country named differs in direct relationship to the size of its population. The cartogram provides the viewer a graphic and geometric sense of a world ranking of national population, or of any other item that the cartogram is designed to represent. (b) World map of cultivated land. This is a choropleth map designed to show the approximate distribution of global cropland by a colored pattern. This is a common cartographic convention used to show generalized distribution of everything from hospital beds to religious preferences through distinctive patterns of color or shading.   *Source: U.S. Department of Agriculture.*

term latitude denotes position with respect to the equator and the poles. Latitude is measured in degrees (°), minutes ('), and seconds ("). The equator, which circles the globe midway between the poles, has a latitude of 0°. All other latitudinal lines are parallel to the equator and to each other and therefore are called parallels. Every point on a given parallel has the same latitude. Places north of the equator are in north latitude; places south of the equator, in south latitude. The highest latitude a place can have is 90°N or 90°S latitude. Thus, the latitude of the North Pole is 90°N, and that of the South Pole is 90°S. Places near the equator are said to be in low latitudes; places near the poles, in high latitudes. The Tropic of Cancer and the Tropic of Capricorn, at 23.5°N and 23.5°S respectively, and the Arctic and Antarctic Circles, at 66.5°N and 66.5°S respectively, form convenient and generally realistic boundaries for the low and high latitudes. Places occupying an intermediate position with respect to the poles and the equator are said to be in middle latitudes.

Meridians of longitude are straight lines connecting the poles. Every meridian is drawn due north and south. They converge at the poles and are farthest apart at the equator. Longitude, like latitude, is measured in degrees, minutes, and seconds. The meridian that is most often used as a base (starting point) is the one at the Royal Astronomical Observatory in Greenwich, England. It is known as the meridian of Greenwich, or the prime meridian, and has a longitude of 0°. Places east of the prime meridian are in east longitude; places west of it are in west longitude. The meridian of 180°, exactly halfway around the world from the prime meridian, is the other dividing line between places east and west of Greenwich called the International Date Line. At the equator, 1° of longitude is equivalent to 69.15 statute miles (111.29 km). However, because the meridians converge toward the poles, a degree of longitude at the Arctic Circle is equivalent to only 27.65 miles (44.50 km). The combination of latitude and longitude gives us mathematical or absolute location.

1.3 Places and Regions

Geographers are keenly involved in attempting to understand the nature of place. Given a list of places, one's mind tends to produce an image for each of them, whether or not you have been there. For example, think about your reactions to these place names: New Orleans; Denver; Paris; Paris, Missouri; Surf City; your hometown; your neighborhood. Each of these place words conjures up a mental image in your mind—that is, draws information from your mental map of the world. Those images include both physical and cultural components. They provide a sense of **place identity,** which is an essential aspect of geographic learning.

Recall that in the selection from *Earth Abides* that opens this chapter, Ish is trying to determine what lies before him.

As his story continues, he realizes that, in the desert environment, if the highway has been washed out, he can detour around the problem area and travel across sections of the desert landscape with little inconvenience. Later, when he gets to some large cities, he finds himself awed by the capacity to drive through major urban corridors as the only car in motion—often the only person in town. Place after place provides him with distinct sensations, even with virtually no people to shape his reactions.

Like Ish, you will begin to add places to your own mental map of the world as you become increasingly confident in your study of such spatial relationships—the recurring patterns of things in connection with other things. Places most often take part of their identity from the human influences that have been important in shaping such settings. In this text, we will look continually at the interaction between human influences and the environmental setting. In the geographer's field of vision, phenomena such as settlement, economic activity, lines of transport and communication, and human movement all relate closely to the physical nature of place and to differing human perceptions of even the same places.

Landscape Analysis

One of the things that lends color and interest to geography as a field of study is the opportunity to observe and interpret the extraordinary variety of landscapes that the world affords. This study carries intellectual as well as aesthetic rewards. A great deal of both the human record and nature's works are inscribed in the Earth's surface and both must be subjected to careful mapping and rigorous analysis if one is truly to understand and appreciate what is "written" in the landscape. From the earliest times, the landscape has been a data bank of human-environmental interaction and transformation. The landscapes that humankind has created serve as a primary document—the record of where we have lived and traveled, and of what we have done to try to make Earth more responsive to human needs.

Some landscapes are intricate, irregular, and hard to characterize (Fig. 1.7a), whereas others are regular, sharply defined, and conducive to precise geographic description and characterization (Fig. 1.7b). In the study of landscapes, geographical analysis is most often drawn to the features that are typical—landscape units that repeat themselves and that are subject to mapping and orderly analysis. American small towns, for example, generally share the characteristics of a Main Street, small cafes, corner markets and bars, local churches, and proximate farmland. Yet, even with such similarities, each small town takes pride in elements of individuality and uniqueness. The geographer is engaged by the things that are representative. Many present-day landscapes are described as suffering from visual blight and, as such, are therefore uninteresting (or even menacing) to tourists. But these same landscapes may convey

A

B

Figure 1.7 (a) This aerial shot of Iowa farmland demonstrates the blend of physical features and human landscape transformation. The differing crop patterns, the few stands of trees that remain by streams, and the trees that have been planted by settlements are all signatures of the cultural landscape. All of these elements play a role in the creation of a regional identity. There has often been considerable human attention to the planning and shaping of the landscape, as is so evident in this scene. (b) Farm terraces. This landscape represents one of the most demanding examples of agricultural transformation in the rural world. The effort required to create small level terrace fields, to move the topsoil and then bring it back once the fields are leveled and walled for the sake of water control, is massive, and representative of Asian terraces particularly. The farm population undergoes such labors because a level field makes better use of rainfall, is easier to plow and cultivate, and can be harvested more efficiently. Where water can be captured for plant irrigation on the terraces, this effort is even more productive. In lands where only rainfall farming can be practiced, there is still a higher yield on land that has been terraced. This particular landscape is in eastern China. The rice bowls are awaiting the return of the farm crew that is constructing the terrace walls of local stone.   (a) ©1994 Stephen Graham/Dembinsky Photo Associates; (b) Kit Salter*

significant geographic messages to someone trying to understand the geographic genesis and implications of a given scene.

Regions as Human Constructs

Regions are human constructs and, as such, they are collections of relatively homogeneous characteristics defined and connected through the process of human observation and analysis. Author George Stewart talks about the desert environment as a region in *Earth Abides.* Some of you have probably planned vacation junkets in part because of your perception of distinct regional environmental and geographic characteristics. We have divided the world into regions (see Fig. 1.4) for this text because it is impossible to introduce order and logic to something as massive, complex, and diverse as the Earth's surface without a system and organized framework built around smaller spatial

units. The need for analysis and interaction with such smaller units is the very genesis of the regional concept (Table 1.2).

Think of your own view of the world. You have, in all likelihood, a reasonably well-developed sense of the regional characteristics of your home territory or the region where you have lived the longest. You have, in your own personal mental maps, a feeling for climate, population characteristics, economic activity, and cultural traits, and perhaps even for vegetation and landforms, in your home region. But what of the same considerations for, say, upland Philippines? Libya? Iceland? the Texas-Mexico borderlands? We continually use the regional concept because of the desire we have to bring some sort of order to our mind and its collection of geographic images and information, but this concept is not always expressed in formal and objective ways (Fig. 1.8). The geography of regions and

Table 1.2 The Major World Regions: Basic Data

Region	Area (thousands/ sq mi)	Area (thousands/ sq km)	Estimated Population (millions)	Annual Rate of Increase (%)	Estimated Population Density (sq mi)	Estimated Population Density (sq km)	Infant Mortality Rate (per thousand)	Urban Population (%)	Arable Land (%)	Per Capita GNP ($US)
Europe	1827.2	4732.4	508.7	0.07	278.4	107	7.1	72	25	18,324
Russia and the Near Abroad	8474.5	21,949	291.2	−0.04	34	13	19.3	65	20	1803
Middle East	5787.8	14,990.6	420.9	2.2	72.7	28	57.9	53	8	2463
Monsoon Asia	7725	20,007.9	3278.1	1.4	424.3	164	39.8	32	18	2503
Pacific World	3262.7	7852.3	29.2	1.1	8.7	3	17.3	70.9	5.8	15,435
Africa South of the Sahara	8499.4	22,013.4	595.9	2.6	87.8	34	95.1	26	6.1	514
Latin America	7786.2	13,025.8	499.8	1.8	64.2	25	29.1	72.3	7.3	3687
Anglo America	7094.4	18,374.5	300.8	0.6	42	16	6.7	75	13	27,104
World Regional Totals	**50,457.2**	**122,945.9**	**5924.6**	**1.4**	**120.3**	**46**	**40.1**	**43.7**	**12.3**	**5037**
U.S.A. (for comparison)	3536.3	9159	270.2	0.6	76	29	7	75	20	28,020

Note: The land area of Antarctica is included in the World data and is not considered a part of a Major World Region.

Sources: *World Population Data Sheet*, 1998; United Nations Statistics Division, 1998; *Almanac of Politics and Government,* 1998; *World Factbook,* 1998.

places helps us achieve such an order. While the geographer's field of vision makes continual demands upon our understanding of place, personality, and identity, it also helps us identify arbitrary regional characteristics that we find useful in creating a sense of geographic order in our minds.

We have used an arbitrary but fairly common system of regionalization in this text. We have divided the world up into eight major regions. Each of these regions has varying features that provide internal cohesion and coherence. The opening chapters of the discussions of these eight regions—the Geographic Profile chapters (chapters 3, 6, 8, 10, 15, 17, 19, and 22)—outline basic characteristics and describe the sorts of physical and cultural features that create a regional unity. Not every region has the same elements of homogeneity, but each

Figure 1.8 Regions are human constructs and regional borders are decided upon by human decision-making. For a region to be useful, it has to conform to certain criteria decided upon by the map maker or the person determining the region. In this street corner map photographed in San Francisco, the person selling the used clothing sees the United States divided into four regions and offers clothes from the Cheap region to attract purchasers to the corner. *Kit Salter*

region can be identified by characteristics of language, religion, geographical proximity, or some other blend of features that connect the parts. Not all of Latin America, for example, speaks Spanish, but the Spanish language and Spanish cultural influence serve as a major feature of cohesion in the Latin American world region.

The route by which you explore the world in a textbook is, like regional borders, also arbitrary. In this text we take you from the European world to worlds culturally more distinctive but end up with a return to Anglo America. Once you read chapters 1 and 2, however, you can really chart any route you wish through the text, preferably beginning with the profile chapter each time you go to a new world region and then reading all the subsequent regional chapters before turning to yet another region. As you grow to comprehend the physical and cultural features that develop a regional identity, you are beginning to think like a geographer.

Figure 1.9 Culture defines each group's unique view of itself and others, and includes the material goods, skills, and social behavior passed on to successive generations. It is expressed through art, language, beliefs and institutions, the built environment, and numerous other features seen in the landscape. Such expressions are continually subject to change and to diffusion. *Illustration by Suzanne Dunaway*

1.4 Physical Systems

In the opening of this chapter, Ish looks about the desert region that he is crossing and is most powerfully impressed with the evidence of physical systems that created this arid landscape. In seeing the world through a geographer's eyes he realizes that the human impact upon the desert has been minimal. The dominant force in that scene is the set of physical systems—climate, landforms, vegetation, erosion, and time—that created the soils and land cover he travels through in his initial search for human companionship. Chapter 2 in this text introduces, explains, and illustrates the physical systems that are fundamental to understanding geography.

1.5 Human Systems

Human systems are the powerful mechanisms that people have created to make the Earth more satisfying, more productive, and more clearly reflective of human needs and desires. The driving force in such modification of the environment is the power of culture. Cultural forces shape not only the political interactions of groups but very often also the physical appearance of landscapes from a local to global scale. A group's culture includes the values, beliefs, aspirations, modes of behavior, social institutions, knowledge, and skills that are learned and transmitted within the group. It also includes the material culture—the group's tangible possessions and products that play a role in cultural expression and daily life. People generally perceive their environment in terms of their culture.

Each culture group alters the environment in its own way, thereby creating a distinctive **cultural landscape** (the landscape as transformed by human action). Cultures are dynamic

and constantly changing. They change by learning and adapting traits from the cultures of other groups with whom they come in contact, a process known as **cultural diffusion** (Fig. 1.9). The theory of **environmental determinism,** in which the role of culture was seen as subordinate to the power of the physical environment and its influence upon human actions, had its origin in the early geographies of the Greeks and Romans. It came to its strongest expression in the 19th century through the work of German geographer Friedrich Ratzel (1844–1904) and his American student, Ellen Churchill Semple (1863–1932). **Environmental possibilism**—the belief that humans always had a range of opportunities in any given environment—grew out of the work of the French geographer Paul Vidal de la Blache (1845–1918) and, by the early years of the 20th century, particularly through the work of Carl O. Sauer (1889–1975), culture began to be seen as a powerful engine in the shaping of human landscape through differing social preferences about what development might occur in a specific environment. In this text, we will see that culture and environment are in continual tension simply because there are so many alternative ways to create a setting for human activity.

Because of the complexity of human systems and their central role in shaping cultural landscapes and world regional geography, this element of the National Geography Standards is divided into a series of themes that appear throughout this text. All influences of Human Systems appear in each region, but in differing force and impact.

Human Populations. Central to all geographic analysis in human geography is the nature of population (see Fig. 1.6). The dimensions of growth, mortality, health, education, economic vitality, demographic patterns, migration, and mobility are some of the most frequently used indices of human activity. The next chapter introduces population themes; in your own efforts to make sense of the world, you will find that human population plays a dominant role in the geographic activities of settlement, landscape transformation, and resource utilization.

Cultural Mosaics. Recognizing the broadly differing cultural and architectural mosaics that occur in farming villages and urban settlements (Fig. 1.10), the geographer recognizes landscape variety, ethnic differentiation, and cultural dissimilarity to be part of daily reality. The political and cultural dynamics that are set in motion by these mosaics are part of the geographic patterns that are central to an understanding of the personality and identity of all places and regions you encounter, not only here in this text, but at any scale—your dorm, neighborhood, or distant nations and regions of the world.

Patterns of Economic Interdependence. Were ours a world of abundant resources, small populations, and rigid geographic mountain and desert barriers set between peoples, there would be the potential for imagining worlds of high cultural and economic isolation. Although even today some populations do exist in relative isolation in tropical rain forests, on islands, or in mountain highlands, such isolation is unusual. The regions that this text explores are characterized by profound and increasing interconnectedness. Economic geography is at the heart of this linkage: The creation of raw materials from local resources, the movement of goods and people to centers of manufacturing, and the critical movement of finished goods to markets are integral components both of economic and world regional geography. The geographer sees not only the characteristic elements that can be mapped in the study of this economic interdependence,

but also the human actions that created these specific patterns. Virtually every change associated with the development of human society arises from the interaction of human and physical systems; geography focuses on just such interaction.

Here in the beginnings of the 21st century, we tend to think of economic interdependence as reflecting our own nation's intricate involvement in global economic trade networks. However, economic systems are interdependent constructions that can be as local as a grocery store selling apple cider from a nearby farmer or as global as the same store selling Swiss chocolates. The nature of this geographic interdependence depends upon the level of economic development of the area. This concept of interdependence weaves through every chapter of our text.

Globalization is a term that will become increasingly common as we grow into the new century. When you see basketball shoes being designed in Europe, fabricated in East Asia, marketed in Anglo America, and traded as stock all across the world's markets, you begin to comprehend how absolutely global economic connections are. Add to that reality the nature of political entanglements, human migration streams, and telecommunication linkages and you see how much of our future will be deeply influenced by economic, political, and demographic activities that take place not only halfway across the world, but all across the globe. The concept of globalization is yet another aspect of geography that is beginning to touch all of our lives.

As countries and regions industrialize and networks of geographic interdependence expand in extent and complexity, their occupational structure changes markedly. The number of farmers decreases and, as global population grows, their proportion in the population also steadily decreases. With improving technology, the remaining farmers often produce more. Employ-

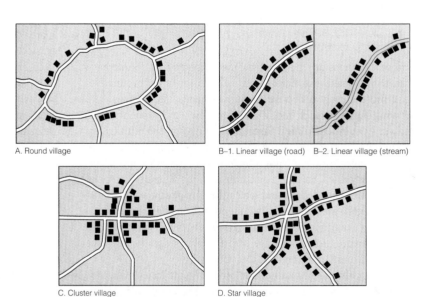

A. Round village   B-1. Linear village (road)   B-2. Linear village (stream)   C. Cluster village   D. Star village

Figure 1.10 Village settlements have long followed fairly standard geometries. These five sketch maps show the most common spatial patterns followed in laying out a village landscape.

ment in manufacturing, trade, and services becomes more dominant. Per-person industrial output and average incomes rise, and industrial and commercial cities grow in size and complexity. Countries or regions begin to specialize in producing and exporting certain types of goods and services. Such items are products of the basic industries that are the main elements of the area's economy. Understanding the geographic and other conditions relative to the growth of such industries (or their failure to develop) casts much light on both the economic and social character of countries and regions.

In recent decades, trends have shown that some industrial societies are becoming **postindustrial;** that is, the industrial labor force begins to decrease in importance and employment in trade and services becomes more dominant. In some countries, industrial output may substantially decrease, but in most industrial societies productivity increases through technological innovations despite a smaller labor force. This has been true in manufacturing and in agriculture. The overall welfare of such a society comes to depend largely on productivity and incomes in the trade and service sectors. Those human activities lead to the exchange of goods and the provision of services such as transportation, banking, and information management. The geographer is always trying to identify and comprehend economic patterns on the landscape and must be continually conscious of the change in the scale from local to global in the study of any of these phenomena. It may help the student of geography to think of countries and regions as either preindustrial, industrializing, industrial, or, to some extent, postindustrial, and to relate other geographic circumstances to the current stage of economic development.

Human Settlement. One of the most important maps you can study is that of **human settlement.** It might be a sketch map given to you by a friend who owns a mountain cabin that you are trying to find in a snowstorm, or it could be a map showing the locations of the Midwest farm villages and towns that were flooded in the Great Flood of '93. Where people settle, where they prosper, where they fail and leave, and where they seem just to persist are all elements of spatial and cultural significance. Settlement points on a map are beginning points in trying to understand a place, a region, a human pattern. In Ish's exploratory trek across the country from California to New York in *Earth Abides,* he has the whole country's landscape to select from in his decision about where to live in the face of such solitude and uncertainty. Millions—in fact, billions—of people have been faced with location decisions about human settlement and, for most of history, rural settlements in different patterns have been selected (see Fig. 1.10). Studies of such decisions and the resulting patterns are central to the discipline of geography and a useful understanding of the world's regions and nations.

Forces of Human Cooperation and Conflict. Humankind seems not to be a peaceful species. From the beginning, there

have been geographic tensions over space. Boundaries, resources, settlements, and population movement and growth are some of the most contentious aspects of the human use of the Earth. However, while the geography of military conflict seems to gain the attention of historians and journalists, there in fact has been a great deal of cooperation in the human use of Earth. To better understand how regions and nations have created and maintained the places and patterns they possess, the significance of cooperation and conflict must be in mind continually.

We live in a world that has been divided into a multitude of political units—**nations** (groups of people who possess common traits and occupy a specific area), **sovereign states** (autonomous political units), subdivisions of these (provinces, states, counties, and so on), and dependent political entities that remain as legacies from a past colonial age. Separated by boundaries drawn on maps and in many cases shown by markers on the ground, these political divisions form a complicated patchwork enclosing most of the land surface of Earth. The political status of any area is an important feature of its geography. In a broad sense, political status includes not only the political organization of the area but also the area's impact on regional and global politics and economics. Although these elements are at times difficult to define precisely, their influence should not be ignored in geographic study.

Just as globalization is a term of importance in the context of economic interdependence, it is also central to understanding world patterns of cooperation and conflict. For example, North Korea's firing of ballistic missiles across the Sea of Japan in 1998 is an event that is not limited in impact to East Asia. Political centers and associated military bases all across the globe take quick note of such action. We are, as will be noted often in our text, all part of a global world. Such reality generates both good and bad outcomes. The challenge to the geographer is to determine what forces are at work at the different levels of human interaction, because virtually all the world's landscapes reflect and exhibit cooperation as well as conflict.

1.6 Environment and Society

The relationship between environment and society represents one of the most critical interfaces examined in world regional geography. The ever-increasing control that humankind has over the environment—not only in technology but also in population growth and patterns of consumption—has meant a rapid and serious reappraisal of human interaction with the environment. At the same time, nature has shown that people have grown too confident of their authority over nature (Fig. 1.11). Almost a century and a half ago, George Perkins Marsh (1801–1882) published *Man and Nature* (1864), setting forth his almost unprecedented view that the human species had been

Figure 1.11 Erosion caused by storm waves and rainfall can undermine whole settlements that have been placed close to the sea because of our human fascination with view and proximity to water. Even though human response to the damage caused in such a setting is one of regret, the odds are that people will rebuild in the same environmental setting after this particular storm and erosional event because of continued attraction to the geographic features of such a setting. *Jack Dermid Photo Researchers, Inc.*

a quite considerable force for change in nature.[2] Technology and population size are the major causes of Marsh's nontraditional perspective. It has been in the last several thousand years—and especially in the most recent centuries—that humankind has had a truly powerful impact on the Earth's surface. In the decades since author Rachel Carson (1907–1964) wrote *Silent Spring* (1962), the world has become ever more concerned about the mounting impact of the search for and depletion of resources, the transfer of water, the relentless human modification of the environment, and our increasing dependence on petrochemical fixes.[3]

1.7 The Uses of Geography

Recall again the opening paragraphs of this chapter. George Stewart uses the expression "the geographer's mind" to refer to the ways in which Ish could and would read the landscape in his search for a base for security and productivity. This perspective is central to this text. Knowledge of geography is not just "state capitals" or "capes and bays" (the two most commonly used casual references to the content of geography). It is understanding the ways in which humankind perceives and interacts with the Earth's surface, its resources, and its peoples—in the present

and over time, both locally and globally. This broad perspective leads to a wide variety of careers and professions within the field of geography (Fig. 1.12).

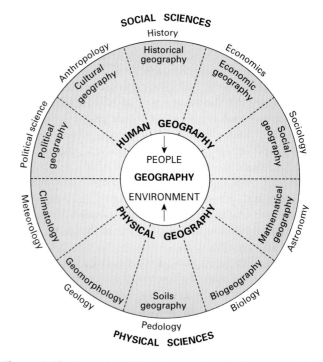

Figure 1.12 Selected subfields of geography. This diagram identifies the main subject-matter areas within human geography and physical geography, and links them with the most closely related disciplines in the social sciences and the physical sciences.

[2]George Perkins Marsh, *Man and Nature.* Ed. David Lowenthal. 1965. Cambridge, Mass.: Harvard University Press.

[3]Rachel Carson, *Silent Spring.* 1962. Boston, Mass.: Houghton Mifflin.

18

1.8 Careers in Geography: Regional, Systematic, and Technical Specialties

This textbook makes use of specialized work by geographers in many regional, systematic (the topical study of various physical or cultural geographic elements), and technical fields. Such specializations have gradually developed as geographic knowledge about world complexities has increased. Today, the Association of American Geographers (AAG) lists more than 100 specialties ("proficiencies") among its more than 7000 members. Regional specialties of the members show a preference for Western rather than non-Western areas. The relative dearth of U.S. geographers with professional expertise in major developing areas is a disadvantage, but it does open opportunities for younger geographers willing to undertake the linguistic and other challenges of studying non-Western areas.

Most U.S. geographers emphasize systematic specialties over regional specialties. Specialists in physical geography study spatial patterns and associations of natural features. Prominent subfields include geomorphology (the study of landforms, and a field in which geography intersects with geology), climatology (climatic processes and patterns), biogeography (study of biotic resources), and soils geography. Closely related to physical geography is the large interdisciplinary field of environmental studies, which is concerned with reciprocal relationships between society and the environment. There is a whole subfield here related to natural hazards because of their impact on the interface of human activity and environmental processes. Another allied specialty is medical geography, which focuses on spatial associations between the environment and human health and on locational aspects of disease and health-care delivery.

Specialists in economic geography study spatial aspects of human livelihood. Major occupations and products are the emphasis, and there is an active theoretical component that intersects with the field of economics. Agricultural geography, manufacturing geography, and transportation geography are important subfields that deal with these major aspects of our transformation of the land and resources. Marketing geography is a relatively small but active applied offshoot of economic geography and is of particular importance in the continuing expansion of residential areas on the edges of cities of all sizes. The study of economic development is an extremely important interdisciplinary field in which many geographers specialize. This text pays much attention to contrasts (expressed or implied) between "developed" and "developing" areas with respect to the economy. Urban geography studies the locational associations, internal spatial organization, and functions of cities.

Geographers specializing in cultural geography concern themselves with places of origin ("culture hearths"), diffusion, interactions, landscape evidence and analysis, and regionalization of culture. Closely allied to cultural geography is the field of cultural ecology, which focuses on the relationship between

Figure 1.13 The use of the computer in geographic analysis and the mapping sciences has been an area of rapid development in the past three decades. There is increasing career potential in Geographic Information Systems (GIS), and many departments of geography have developed computer labs for research and instruction in map design and production, data display and analysis, and for the easier and more immediate updating of various layers of geographic information. *Courtesy Geosystems, Inc.*

culture and environment. Social geography deals with spatial aspects of human social relationships, generally in urban settings; population geography assesses population composition, distribution, migration, and demographic shifts. Political geography studies such topics as spatial organization of geopolitical units, international power relationships, nationalism, boundary issues, military conflicts, and regional separatism within states. Historical geography studies the geography of past periods and the evolution through time of such geographic phenomena as cities, industries, agricultural systems, and rural settlement patterns.

Technical specialties in geography include cartography, computer cartography, remote sensing, quantitative methods (mathematical model building and analysis), air photo interpretation, and Geographic Information Systems (GIS). The last of these—GIS, which involves the computer manipulation and analysis of spatial data—is a very successful interface between the analysis of spatial data and the computer (Fig. 1.13). There is increasing use of this technique in everything from spatial analysis to model building and analysis to sophisticated cartography. Because so much of the data utilized in the management of cities, businesses of all scales, government, and even agriculture is digital data, the computer has become central to all of these fields. With the capacity to array such data in a geographic and cartographic form, an understanding of GIS has become a highly marketable skill.

Increasingly, departments of geography are developing curricula and laboratories to train students in the study of spatial statistics, GIS classes, computer cartography, remote sensing, and digital image processing so that geographers can be active players in this field of professional growth and expanding significance. All of these technique-oriented specialties center on the cartographic (mapped) expression of spatial data, while remote sensing and air photo interpretation utilize various kinds of satellite imagery or photo coverage to assess land use or geographic patterns. The expanding use of the computer offers geographers greater speed, accuracy, and facility in incorporating new and changing data into the creation of maps, graphs, and charts. This has become an area of major professional activity for geographers.

Career lines in geography are found in government agencies at the local, state, regional, and federal level. Firms working with land-use decisions at all scales use geography and geographers—particularly in the areas of GIS, remote sensing, and computer cartography. Retail firms find the problem-solving skills inherent in geographic analysis useful in management, survey design and implementation, estimation of public response to innovations or expansion to new locations, and policy shifts. Issues of environmental perception, management, shifts in utilization, and description all represent career options for geographers. The most essential lesson to learn as one pursues a career or profession that utilizes skills gained in studying geography is that although there may be few classified ads that proclaim "Geographer wanted!", there are worlds of opportunity in those same pages that speak to the very skills and learning represented in the themes we bring you in this world regional geography.

Human history has been shaped by the interplay of geographic influences and human ambitions—whether at the scale of a peasant farmer deciding to irrigate his fields in order to better feed his family, or of an army commander trying to expand his realms of military authority. Knowing about the spatial distribution of resources, mountain passes, ethnic populations, animal herds, and settlements, for example, means that one can achieve a clearer understanding of the past. One of the prime uses of geography, then, is to be able to determine and understand the map of the locale being studied at the time of the event you are attempting to comprehend.

Understanding one place helps prepare you for seeing the meaning of yet other places. In this respect, no place is truly an island, although some locales are more insular than others; nor are places the same in the magnitude, intensity, and reach of the geographic effects they engender. A hurricane in Florida, for example, will have an impact on citrus prices in California. The return of Hong Kong to China in 1997 influenced financial markets all around the world. Such effects may be physical, economic, political, cultural, or social, and they arise from trade, investment, migration, military action (or the threat of it), and other mechanisms of spatial interaction. Especially large and complex effects are generated by such major nations as the United States, Japan, Germany, and Brazil, but many smaller and less powerful places may have an impact disproportionate to their size.

All of these geographic perspectives previously discussed play a role in our understanding of world regional geography. The role of the map, of the distributional patterns of human resources and physical properties, and of the dynamics of interaction between varied and significant players and places in the drama have keen importance to our making sense of the world. Many of the problems, for example, that your generation will face are problems that are evident in today's cultural landscape—that is, the landscape transformed by human action. We want you to see a benefit in studying the world's regions and nations through a geographer's eyes, and developing a geographer's field of vision.

SUMMARY WITH SELECTED KEY TERMS

- **Geography** is the study of the Earth's surface and the analysis of its **spatial distributions and patterns of human and physical characteristics**, brought to light not only through objective analysis, but also through the study of **landscapes in literature** and regional **popular culture**.

- The 1994 National Geography Standards are a recent tool created for **spatial analysis**. These are built around **Six Essential Elements** (including **the world in spatial terms, places and regions, physical systems, human systems, environment and society,** and **the uses of geography**) and **Eighteen Geography Standards**—all of which are designed to illustrate what American students at grades 4, 8, and 12 should know and be able to do in geography.

- Central to such learning are **maps, map projections, map scale, map symbols, mental maps, latitude and longitude, cartograms,** and other aspects of cartography. Such tools are seen as the **language of maps,** and such language is vital to both geographic analysis and the **geographer's field of vision.**

- **Regions are human constructs** and, as such, they have an arbitrary and fluid quality. Critical to their existence is a **clear definition of the criteria** that are utilized to define such regions and the resultant regional maps that are created. For this text, we have created **eight world regions.**

- The **human systems** incorporated in the U.S. Geography Standards include themes of **human population, cultural mosaics, patterns**

of economic interdependence, human settlement, and forces of cooperation and conflict. The interplay of these human realms is constant and operates at all scales in world regional geography.
- Environment and society are themes of human-environmental interaction that focus upon the ways in which humans modify the physical environment, how the physical environment influences human activity, and how the human definition and utilization of resources changes over time.
- Human history has been shaped continually by the nature of our geographic setting. Geography serves as a useful tool not only in understanding past history but also in the geographic analysis of contemporary human activities and in planning for a better-thought-out future development.

- Increasing sophistication in the human capacity to understand, shape, and analyze spatial patterns through digital data has expanded career options for people working in geography. Careers in Geographic Information Systems (GIS) and many other realms of landscape analysis, environmental management, and government work in land management utilizing computers and digital imaging are continually expanding in number and significance. Growing attention to the location, careful stewardship, and conservation of resources at both the local and global level has also provided an expanding base for careers in geography.

REVIEW QUESTIONS

1. Describe the world that Ish explores at the beginning of the chapter.
2. List the Six Essential Elements in the U.S. Geography Standards.
3. What are the geographic features that make Great Britain and New Zealand different?
4. List the eight world regions that are utilized in this text.
5. Define map projection and list four kinds shown in the text.
6. Define and explain the use of latitude and longitude.
7. Define "cartogram" and explain why it has such a different look to it than a world map.
8. Define "mental map" and give an example of a place in your own mental map.

9. List the five themes noted under Human Systems in the Geography Standards.
10. Relate Fig. 1.9 to one or more of the themes asked for in question 9 and suggest a similar scene you know of in your own world.
11. List examples of scenes that might have appeared in Fig. 1.11 that are from your own region in the United States. Why do people live in such threatening environmental settings?
12. Name five professions that do not have the word "geography" in them, but that clearly use some of the skills that are part of the geographer's field of vision.

DISCUSSION QUESTIONS

1. What is author George Stewart's sense of the human use of and impact upon the desert in *Earth Abides*?
2. What difference does it make that Ish is a geographer in the novel that opens this chapter?
3. Talk about the ways in which your daily life is involved with spatial decisions and spatial analysis.
4. Discuss the ways in which the relative isolation of New Zealand compared to Great Britain has been significant for both nations.
5. List and discuss possible world regional systems other than the one used in creating the eight world regions in this text. What might make such alternative regional systems more or less useful than the one used in this text?
6. Paraphrase the three distinctive perception systems illustrated in Figure 1.3. Can you explain what is driving each one? Why are

they competitive? Can you think of other ways in which that landscape might be viewed in that cartoon?
7. Discuss maps as the language of geography. What sorts of things can you say with that language?
8. Select any two of the Six Essential Elements associated with the National Geography Standards and show how landscape reflects their influence.
9. Discuss at least one way you can define and utilize each of the Eighteen U.S. Geography Standards.
10. Looking at the image of Fig. 1.8, discuss the origins and geographic implications of the homemade map.

Chapter 2

Physical and Human Processes That Shape World Regions

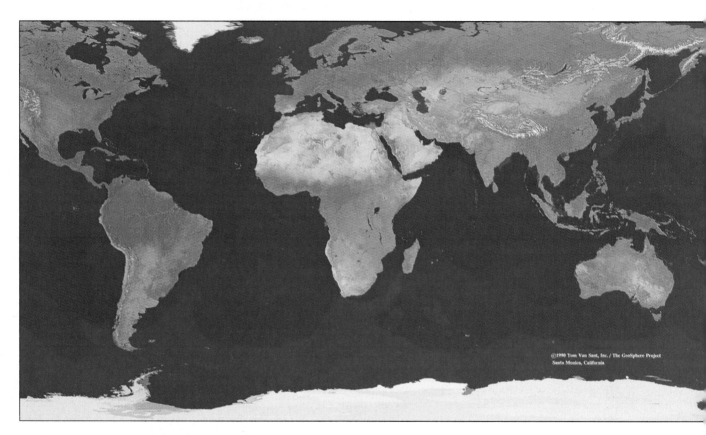

▲ *Earth's environments as seen from space. This composite of satellite images shows forested areas in green, snowcapped mountains and masses of ice in white, largely barren deserts in yellowish-tan, and croplands, grasslands, and tundra in varied shades. Note the prominence of the tropical rain forest of the Amazon Basin in South America, the mass of high mountains in western China and its borderlands, the huge coniferous forest regions of Russia and Canada, and the east-west belt of deserts in North Africa and adjacent Asia.* ©1990 Tom Van Sant, Inc./The Geosphere

*i*n the following chapters you will study the astonishing variety of environments, cultures, and events that characterize the geography of our world today. It is a complex world. You cannot learn what "makes Earth tick" without knowing a great deal about particular places. But details should not overwhelm you; in this chapter you will be introduced to the larger geographic processes and patterns that shed light on the details and make them more meaningful. You will probably want or need to review some of this material as you study the subsequent chapters in the book.

The geographer sees physical and human processes at work around the world. We begin here by introducing the climatic patterns, types of vegetation, and geologic processes which are major parts of the global environmental context for human activities. We introduce modern patterns of people-land associations by viewing them as the products of revolutionary changes in the past: the arrival of agriculture and industrialization. You will see where rich and poor countries are located on the face of the Earth, and understand some of the causes of their prosperity and poverty. You will see where and why populations are increasing and what the implications of that growth are. You will study human impacts on the atmosphere and consider actions to solve some of the most important global problems of our time.

2.1 Climatic Processes and Their Geographic Results

As you experience a warm, dry, cloudless summer day or a cold, wet, overcast winter day, you are encountering the **weather**—the atmospheric conditions prevailing at one time and place. The **climate** is a typical pattern recognizable in the weather of a large region over a long time. Along with surface conditions such as elevation and soil type, climatic patterns have a strong correlation with patterns of natural vegetation and, in turn, with human opportunities and activities on the landscape.

Precipitation

Water is essential for life on Earth, so we begin our study of climates with precipitation. Warm air holds more moisture than cool air, and therefore precipitation is best understood as the result of processes that cool the air to release moisture (Fig. 2.1). Precipitation results when water vapor in the atmosphere cools to the point of condensation, changing from a gaseous to a liquid or solid form. The amount of cooling necessary depends on the original temperature and the amount of water vapor in the air.

For this cooling and precipitation to occur, generally air must rise in one of several ways. In equatorial latitudes or in the high-sun season (summer, when the sun's rays strike Earth's surface more vertically) elsewhere, air heated by intense surface radiation may rise rapidly, cool, and produce a heavy downpour of rain—an event that occurs often in summer over the Great Plains of the United States. Precipitation that originates in this way is called **convectional precipitation** (Fig. 2.2). **Orographic (mountain-associated) precipitation** results when moving air strikes a topographic barrier and is forced upward. Most of the precipitation falls on the windward side of the barrier, and the lee (sheltered) side is likely to be excessively dry. Such dry areas—for example, Nevada, which is on the lee side of the Sierra Nevada range in the western United States—are in a **rain shadow.** Rain shadows are the primary cause of arid and semiarid lands in some regions.

Cyclonic or **frontal precipitation** is generated in traveling **low-pressure cells (cyclones)** that bring air masses with different characteristics of temperature and moisture into contact (Figs. 2.2 and 2.3). A cyclone may overlie hundreds or thousands of square miles of Earth's surface. In the atmosphere, air moves from areas of high pressure to areas of low pressure. Air from masses of different temperature and moisture characteristics is drawn into a cyclone, which has low pressure. One air mass is normally cooler, drier, and more stable than the other. Such masses do not mix readily and tend to retain distinctive characteristics. They come in contact within a boundary zone 3 to 50 miles wide (*c.* 5 to 80 km) called a **front.** A front is named

Figure 2.1 World precipitation map, modified from a map by the United States Department of Agriculture.

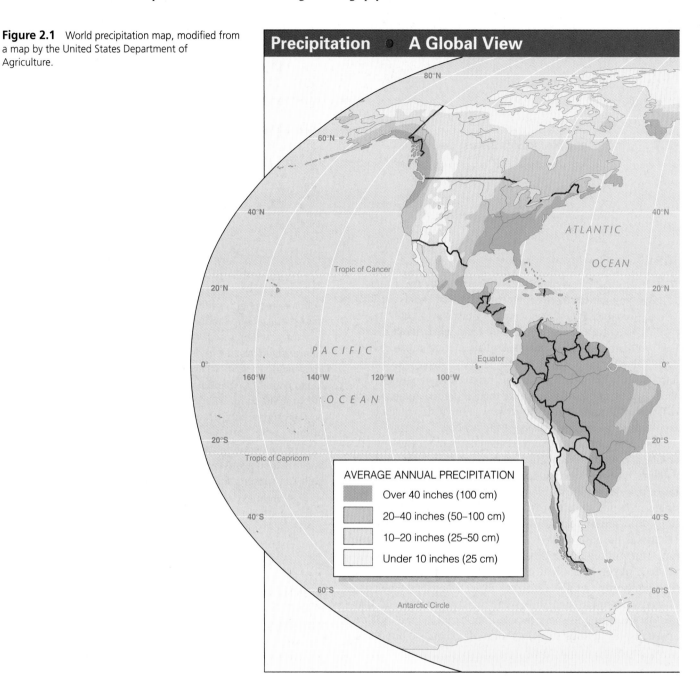

according to which air mass is advancing to overtake the other. In a **cold front,** the colder air wedges under the warmer air, forcing it upward and back. In a **warm front,** the warmer air rides up over the colder air, gradually pushing it back. But whether a warm or cold front, precipitation is likely to result because the warmer air mass rises and condensation takes place.

The middle latitudes, especially in winter, experience a succession of traveling cyclones moving from west to east in an airstream called the **westerly winds** (Fig. 2.4, p. 27). These cyclones rotate slowly, somewhat like whirls and eddies in a stream. Normally there are two fronts as a cyclone passes, first

the warm front and then the cold front. But the cold front moves faster and eventually overtakes the warm front. The cyclone is then said to be **occluded,** and it disappears from the atmospheric pressure map.

Large areas of the Earth receive very small amounts of precipitation (see Fig. 2.1), primarily because of subsiding air masses of high atmospheric pressure. Some parts of the atmosphere generally exhibit high pressure; these are known as semipermanent **high-pressure cells,** or **anticyclones** (see Fig. 2.4). In such a high, the air is descending and thus becomes warmer as it comes under the increased pressure (weight) of the air

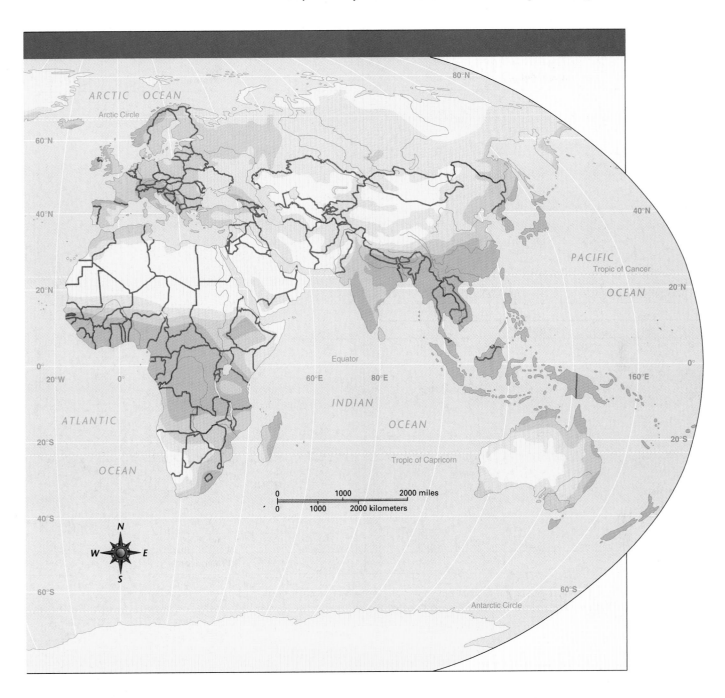

above it. As it warms, its capacity to hold water vapor increases, its relative humidity decreases, and the result is minimal condensation and precipitation. Streams of dry, stable air moving outward from the anticyclones often bring prolonged drought to the areas below their path. Most famous are the **trade winds**—streams of air that originate in semipermanent anticyclones on the margins of the tropics and are attracted equatorward (becoming more moisture-laden as they do so) by a semipermanent low-pressure cell, the **equatorial low.**

Cold ocean waters are responsible for the existence of coastal deserts in some parts of the world, such as the Ata-

cama Desert in Chile and the Namib Desert in southwestern Africa. Here, air moving from sea to land is warmed. Instead of yielding precipitation, its capacity to hold water vapor is increased and its relative humidity is decreased; the result is little precipitation. But many areas of excessively low precipitation result from a combination of influences. The Sahara of northern Africa, for example, seems to be primarily the result of high atmospheric pressure. The rain-shadow effect of the Atlas Mountains and the presence of cold Atlantic Ocean waters along its western coast contribute to the Sahara's dryness.

Types of Precipitation

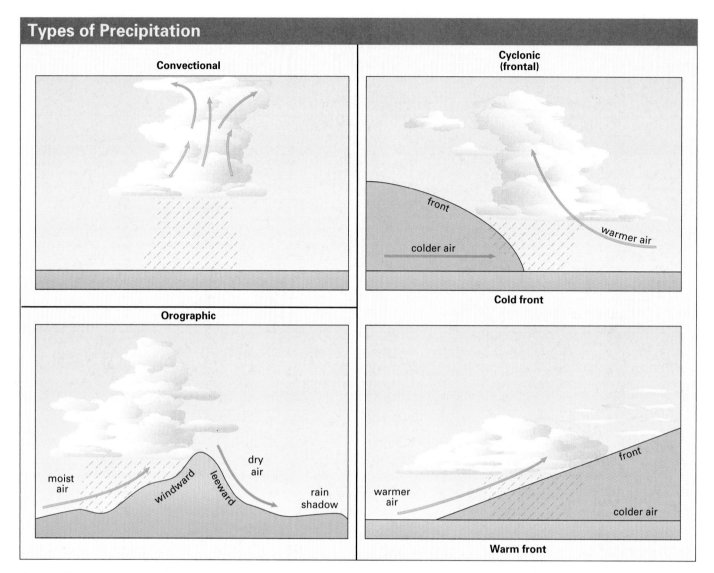

Figure 2.2 Diagrams showing origins of convectional, orographic, and cyclonic precipitation. Note that cyclonic precipitation results from both cold fronts and warm fronts.

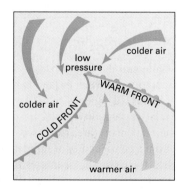

Figure 2.3 Diagram of a cyclone (Northern Hemisphere). The entire system is moving west to east in the airstream of the westerly winds.

Temperature

In the middle and high latitudes, the most significant variable in determining temperatures is seasonality, which is related to the inclination of the Earth's polar axis as the Earth orbits the sun over a period of 365 days (Fig. 2.5, p. 28). On June 21, the first day of summer in the Northern Hemisphere, the northern tip of Earth's axis is inclined toward the sun at an angle of 23½ degrees from a line perpendicular to the plane of the ecliptic (the great circle formed by the intersection of the Earth with the plane of Earth's orbit around the sun). This is the summer solstice in the Northern Hemisphere, and winter solstice in the Southern Hemisphere. A larger portion of the Northern Hemisphere than of the Southern Hemisphere remains in daylight, and warmer temperatures prevail. On September 23, and again

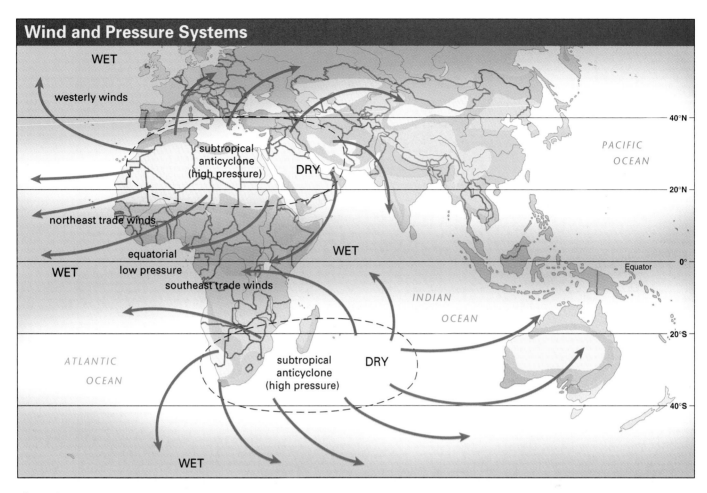

Wind and Pressure Systems

Figure 2.4 Idealized wind and pressure systems. Irregular shading indicates wetter areas.

on March 20, the Earth reaches the equinox position. Its axis does not point toward or away from the sun, so days and nights are of equal length at all latitudes on the Earth. On December 22, the first day of winter in the Northern Hemisphere, the southern tip of Earth's axis is inclined toward the sun at an angle of 23½ degrees from a line perpendicular to the plane of the ecliptic. This is the winter solstice in the Northern Hemisphere, and the summer solstice in the Southern Hemisphere. A larger portion of the Southern Hemisphere than of the Northern Hemisphere remains in daylight, and warmer temperatures prevail.

Most of the sun's visible short-wave energy that reaches the land or ocean surface is absorbed, but some of it returns to the atmosphere in the form of infrared long-wave radiation, which generates heat and is the principal agent that warms the atmosphere. Although the sun is the initial and primary source of the heat, the air is heated mainly by reradiation from the underlying land or water surface.

The Earth varies greatly from place to place in the total amount of energy received annually from the sun; it also varies in the resulting air temperatures. In general, absent the effects of

cloud cover, the lower the latitude of a place, the more solar energy it receives annually. The major reason for this is that on average throughout the year, the sun's rays strike the Earth more nearly vertically at lower latitudes, concentrating a given amount of solar energy on a smaller extent of surface than at higher latitudes (Fig. 2.6, p. 29). A second reason is that when solar rays approach the Earth at a more nearly vertical angle in lower latitudes, they must pass through a smaller thickness of absorbing and reflecting atmosphere before reaching the surface (see Fig. 2.6). However, many humid areas near the equator have a relatively abundant cloud cover that reflects, scatters, or absorbs much of the incoming solar radiation and thus reduces the amount reaching the surface. The result is that the areas of highest annual solar energy actually received at the surface are in the vicinity of 20° to 30° north and 20° to 30° south of the equator. In these subtropical zones, the average angle of the sun's rays is still close to vertical and the cloud cover is much less. The world's greatest deserts are found at these latitudes.

Great differences also exist in the world's annual and seasonal temperatures. In lowlands near the equator, temperatures

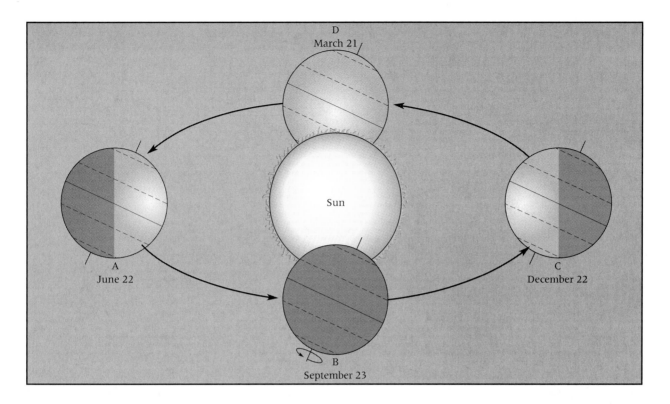

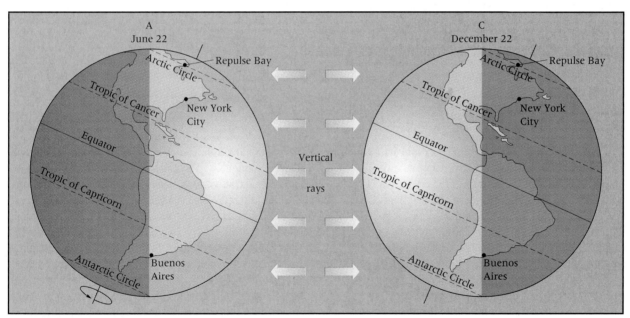

Figure 2.5   Geometric relationships between Earth and the sun at the solstices. Note the differing day lengths at the summer and winter solstices in the Northern and Southern Hemispheres.

remain high throughout the year, while in areas near the poles, temperatures remain low for most of the year. The intermediate (middle) latitudes have well-marked seasonal changes of temperature, with warmer temperatures generally in the summer season of high sun, and cooler temperatures in the winter season of low sun (when sunlight's angle of impact is more oblique and daylight hours are shorter). Intermittent incursions of polar and tropical air masses increase the variability of temperature in these latitudes, bringing unseasonably cold or warm weather.

Also, air at lower elevations is denser and contains more water vapor and dust than air at higher elevations. Thus it ab-

The Earth's Receipt of Solar Energy

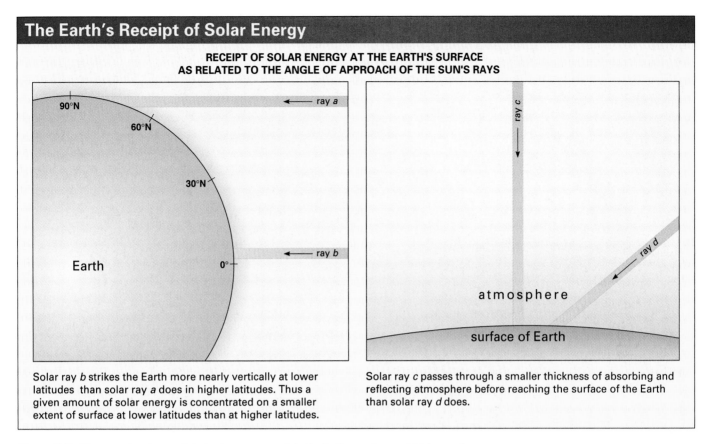

**RECEIPT OF SOLAR ENERGY AT THE EARTH'S SURFACE
AS RELATED TO THE ANGLE OF APPROACH OF THE SUN'S RAYS**

Solar ray *b* strikes the Earth more nearly vertically at lower latitudes than solar ray *a* does in higher latitudes. Thus a given amount of solar energy is concentrated on a smaller extent of surface at lower latitudes than at higher latitudes.

Solar ray *c* passes through a smaller thickness of absorbing and reflecting atmosphere before reaching the surface of the Earth than solar ray *d* does.

Figure 2.6 Diagrams showing receipt of solar energy as related to Earth curvature and thickness of atmosphere.

sorbs more radiation and is warmer. Increasing elevation lowers temperatures on average about 3.6°F (2.0°C) for each increase of 1000 feet (305 m) in elevation. As a result, a place on the equator at an elevation of several thousand feet may have temperatures resembling those of a middle-latitude lowland.

Major Types of Climates and Natural Vegetation

Weather reflects a great variety of local climates. Geographers group the local climates into a limited number of major **climate types** (Fig. 2.7), each of which occurs in more than one part of the world and is associated closely with other types of natural features, particularly vegetation. Geographers recognize 10 to 20 major types of terrestrial **ecosystems,** called **biomes,** which are categorized by dominant type of natural vegetation (Fig. 2.8, pp. 32–33). Climate plays the main role in determining the distribution of biomes, but differing soils and landforms may promote different types of vegetation where the climatic regime is essentially the same. Vegetation and climate types are sufficiently related that many climate types take their names from vegetation types—for example, the **tropical rain forest climate** and the **tundra climate.** You may easily see the geo-

graphic links between climate and vegetation by comparing the maps in Figures 2.7 and 2.8. The spatial distributions of climate and vegetation types do not overlap perfectly, but there is a high degree of correlation.

In the **ice-cap, tundra,** and **subarctic climates,** the dominant feature is a long, severely cold winter, making agriculture difficult or impossible. The summer is very short and cool. There is no vegetation on the ice caps. **Tundra vegetation** is composed of mosses, lichens, shrubs, dwarfed trees, and some grasses. Needleleaf evergreen coniferous trees can stand long periods of freezing weather with attendant lack of water. Thus, **coniferous forest** (often called boreal forests or *taiga,* their Russian name) occupies large areas in the subarctic climate.

In **desert** and **steppe climates,** the dominant feature is aridity or semiaridity. Deserts and steppes occur in both low and middle latitudes. Agriculture in such regions usually requires irrigation. The principal region, sometimes called the "Dry World," extends in a broad band across northern Africa and southwestern and central Asia. The deserts of the middle and low latitudes are generally too dry for either trees or grasslands. They have **desert shrub** vegetation, and some areas have practically no vegetation at all. The bushy desert shrubs are *xerophytic* (literally, "dry plant"), having small leaves, thick bark,

Figure 2.7 Map showing the world distribution of the types of climate discussed in this text.

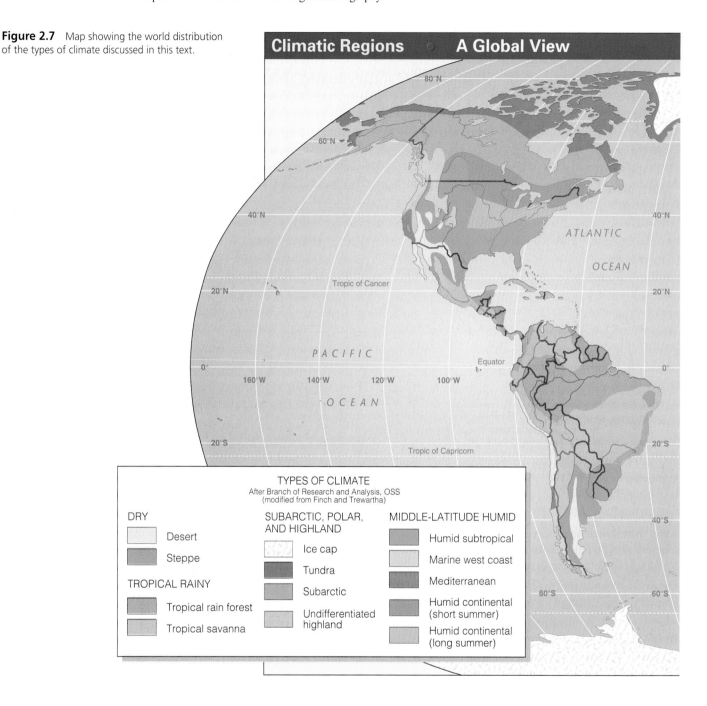

Climatic Regions   A Global View

TYPES OF CLIMATE
After Branch of Research and Analysis, OSS
(modified from Finch and Trewartha)

DRY
- Desert
- Steppe

TROPICAL RAINY
- Tropical rain forest
- Tropical savanna

SUBARCTIC, POLAR, AND HIGHLAND
- Ice cap
- Tundra
- Subarctic
- Undifferentiated highland

MIDDLE-LATITUDE HUMID
- Humid subtropical
- Marine west coast
- Mediterranean
- Humid continental (short summer)
- Humid continental (long summer)

large root systems, and other adaptations to absorb and retain moisture. Grasslands dominate in the more moist steppe climate, a transitional zone between very arid deserts and humid areas. The biome composed mainly of short grasses is also called the **steppe.** The steppe region of the United States and Canada originally supported both tall and short grass vegetation types, known in those countries as *prairies.*

Rainy low-latitude climates include the **tropical rain forest climate** and the **tropical savanna climate.** The critical difference between them is that the tropical savanna type has a pronounced dry season, which is short or absent in the tropical rain forest climate. Heat and moisture are almost always available in the tropical rain forest biome, where broadleaf evergreen trees dominate the vegetation. In tropical areas with a dry season but still enough moisture for tree growth, **tropical deciduous forest** replaces the rain forest. Here the broadleaf trees are not green throughout the year; they lose their leaves and are dormant during the dry season and then add foliage and resume their growth during the wet season. The tropical deciduous forest approaches the luxuriance of tropical rain forest in wetter areas but thins out to low, sparse **scrub and thorn forest** in drier areas. **Savanna** vegetation, which has taller grasses than the steppe, occurs in areas of greater overall rainfall and more pronounced wet and dry seasons.

The humid middle-latitude regions have mild to hot summers and winters ranging from mild to cold, with several types of climate. In the **marine west-coast climate,** occupying the western sides of continents in the higher middle latitudes (for example, in the Pacific Northwest region of the United States), warm ocean currents moderate the winter temperatures, and summers tend to be cool. Coniferous forest dominates some cool, wet areas of marine west-coast climate; examples of growth are the Douglas fir and redwood forests of the western United States.

The **mediterranean (dry-summer subtropical) climate** (named after its most prevalent area of distribution, the lands around the Mediterranean Sea) typically has an intermediate lo-cation between a marine west-coast climate and the lower latitude steppe or desert climate. In the summer high-sun period it lies under high atmospheric pressure and is rainless. In the winter low-sun period it lies in a westerly wind belt and receives cyclonic or orographic precipitation. **Mediterranean scrub forest,** known locally by such names as *maquis* and *chaparral,* characterizes mediterranean climate areas. Because of hot, dry summers, the vegetation consists primarily of xerophytic shrubs.

The **humid subtropical climate** occupies the southeastern margins of continents and is characterized by hot summers, mild to cool winters, and ample precipitation for agriculture. The **humid continental climate** lies poleward of the humid

Figure 2.8   World biomes (natural vegetation) map.

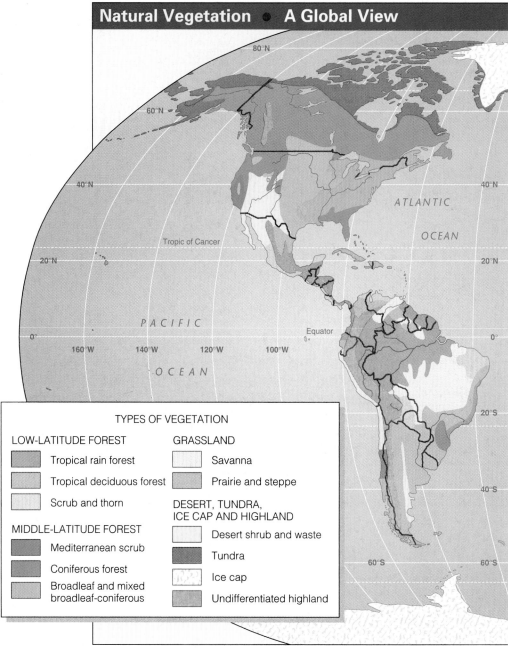

Natural Vegetation ● A Global View

TYPES OF VEGETATION

LOW-LATITUDE FOREST

- Tropical rain forest
- Tropical deciduous forest
- Scrub and thorn

MIDDLE-LATITUDE FOREST

- Mediterranean scrub
- Coniferous forest
- Broadleaf and mixed broadleaf-coniferous

GRASSLAND

- Savanna
- Prairie and steppe

DESERT, TUNDRA, ICE CAP AND HIGHLAND

- Desert shrub and waste
- Tundra
- Ice cap
- Undifferentiated highland

subtropical type; it has cold winters, warm to hot summers, and enough rainfall for agriculture, with the greater part of the precipitation in the summer. In middle-latitude areas with these two climate types, a **broadleaf deciduous forest** or **mixed broadleaf-coniferous forest** (as in northeastern North America) is found. As cold winter temperatures freeze the water within reach of plant roots, broadleaf trees shed their leaves and cease to grow, thus reducing water loss. They then produce new foliage and grow vigorously during the hot, wet summer. Coniferous forests can thrive in some hot and moist locations where porous sandy soil allows water to escape downward, giving conifers (which can withstand drier soil conditions) an ad-

vantage over broadleaf trees. Pine forests on the coastal plains of the southern United States are an example.

Undifferentiated highland climates exhibit a range of conditions according to elevation and exposure to wind and sun. Undifferentiated highland vegetation types differ greatly depending on elevation, degree and direction of slope, and other factors. They are "undifferentiated" in the context of world regional geography because a small mountainous area may contain numerous biomes, and it would be impossible to map them on a small scale. The world's mountain regions have a complex array of natural conditions and opportunities for human use. In climbing from sea level to the summit of a high mountain peak

near the equator (in western Ecuador, for example), a person would experience many of the major climate and biome types to be found in a sea level walk from the equator to the North Pole!

2.2 Biodiversity

Anyone who has admired the nocturnal wonders of the "barren" desert or appreciated the variety of the "monotonous" Arctic can vouch for an astonishing diversity of life even in the biomes people favor least. Geographers and ecologists recognize the ex-

ceptional importance of some biomes because of their **biological diversity**—the number of plant and animal species present and the variety of genetic materials these organisms contain.

The most diverse biome is the tropical rain forest. From a single tree in the Peruvian Amazon region, entomologist Terry Erwin recovered about 10,000 insect species. From another tree several yards away he counted another 10,000, many of which differed from those of the first tree. Alwyn Gentry of the Missouri Botanical Gardens recorded 300 tree species in a single-hectare (2.47-acre) plot of the Peruvian rain forest. Just 25 years ago, scientists calculated that there were 4 to 5 million species of plants and animals on Earth. Now, however, their estimates

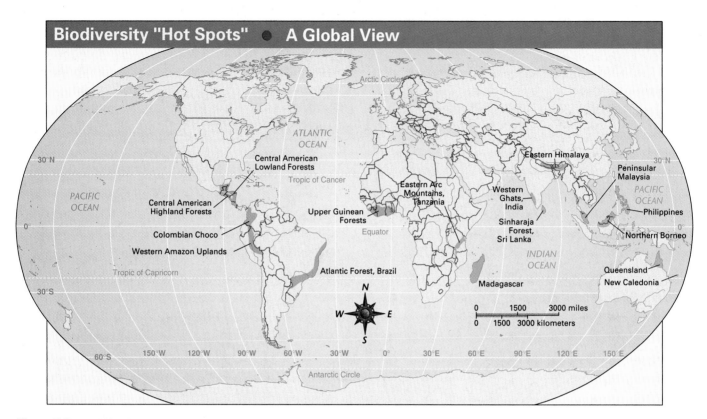

Figure 2.9 World biodiversity hot spots.

are much higher, in the range of 40 to 80 million. This startling revision is based on research, still in its infancy, on species inhabiting the rain forest.

Such diversity is important in its own right, but it also has vital implications for nature's ongoing evolution and for people's lives on Earth. Humankind now relies on a handful of crops as staple foods. In our agricultural systems, the trend in recent decades has been to develop high-yield varieties of grains and to plant them as vast monocultures (single-crop plantings). This trend, which characterizes the **Green Revolution** (see Chapter 10, p. 292), may render agriculture more vulnerable to pests and diseases.

In evolutionary terms, we have reduced the natural diversity of crop varieties that allows nature and farmer to turn to alternatives when adversity strikes. As we remove tropical rain forests and other natural ecosystems to provide ourselves with timber, agriculture, or living space, we may be eliminating the foods, medicines, and raw materials of tomorrow—even before we have collected them and assigned them scientific names. "We are causing the death of birth," laments biologist Norman Myers.[1]

Regions where human activities are rapidly depleting a rich variety of plant and animal life are known as **biodiversity hot spots,** which have been ranked in a list of places scientists believe deserve immediate attention for study and conservation

(Fig. 2.9). These regions are: the Choco region of western Colombia; the uplands of the western Amazon Basin in Ecuador and Peru; Brazil's Atlantic coast; the highland and lowland forest zones of Central America; the entire island of Madagascar; the eastern Himalaya Mountains; the Philippine Islands; peninsular Malaysia; northwestern Borneo; the Queensland province in northeastern Australia; the island of New Caledonia; the upper Guinean forest zone of West Africa; Tanzania's eastern arc mountains; the Western Ghat mountains of southern India; and the Sinharaja forest zone of Sri Lanka. By referring to the map of the Earth's biomes in Figure 2.8, you will see that most of these hot spots are within tropical rain forest areas. Many too are islands which tend to have high biodiversity because species on them have evolved in isolation to fulfill special roles in the ecosystem.

2.3 Landforms, Rock Structures, and Soils

Earth's landforms (Fig. 2.10, p. 36) have been created over geologic time by the **tectonic processes** of warping, folding, faulting, and vulcanism and the **gradational processes** of weathering, erosion, and deposition (see Glossary and the diagrams in Fig. 2.11, p. 38). All four classes of landforms are associated with **igneous, metamorphic, and sedimentary rocks** (see Glossary), with the mix of rock types varying greatly from

[1]Wilson, Edward O. 1989. "Threats to Biodiversity." *Scientific American* 261 (3): 60–66.

one area to another. A **plain** is a relatively level area of slight elevation. Most of the world's people live on plains, especially in China, the Indian subcontinent, Europe, western Russia and Ukraine, and the United States. A **plateau** is an elevated plain, often bounded on one or more sides by an **escarpment.** Plateaus are especially characteristic of Africa, the Arabian Peninsula, western China, western Australia, southeastern Brazil, western United States, Mexico, and eastern Canada. Hill lands are prominent and densely populated in eastern China and southern India, and occur in many other parts of the world. Often they have resulted from long-continued erosion of plateaus and mountains, and many areas of hill lands form strips alongside plateaus and mountains (for example, in western Africa). Mountains are generally higher and more rugged than hill lands, are more of a barrier to movement, and offer fewer possibilities for human settlement and use. Most mountains, particularly the highest ones such as the Himalayas, Alps, and Andes, compose a branching worldwide system of geologically young ranges; those in Europe and Asia radiate in several arms from the Pamir Knot in Asia. This system, characterized by many active volcanoes and frequent earthquakes, is associated with zones of instability where crustal masses known as **plates** are in contact (see Fig. 2.12, p. 39, and **Plate Tectonics** in the Glossary).

Several tectonic processes have been very important in landform creation (see Fig. 2.10). Forces in the Earth warp rock masses upward to form domelike or arched structures, and downward to form structural basins. More intense bending of rock layers **(folding)** produces an accordion-like series of upfolds **(anticlines)** and downfolds **(synclines).** The upfolds often appear as ridges and the downfolds as valleys on the landscape, and erosive processes have produced anticlinal valleys and synclinal ridges in some places. When they are crowded together or pulled apart by the irresistible forces in the Earth's crust, rocks often break and the separate portions move past each other along the line of fracture. This process is known as **faulting,** and the break in the rocks is called a **fault.** Displacements of rock masses upward or downward along faults have often been great enough to create block mountains. Most commonly, the rock mass of the mountain is tilted in such a way as to create a steep face (fault scarp) on the side where the fault occurs and a longer, gentler "back slope" behind the scarp. This is the case in California's Sierra Nevada range, where the fault scarp faces eastward and the back slope faces westward. Some block mountains have been formed between roughly parallel faults and have a scarp face on two sides. A landform of this type is called a **horst,** of which there are many examples in the Basin and Range Country of the United States' Intermountain West. A steep-sided trough known as a **rift valley** or **graben** is formed when a segment of the crust is displaced downward between parallel **tensional faults** (in which the rocks are pulled apart) or when segments of the crust that border it ride upward along parallel **compressional faults** (in which the rocks are crowded together). Both kinds of faulting were involved in the formation of the spectacular Great Rift Valley, which extends for thousands of miles in Africa and southwestern Asia (see p. 422). Both folding and faulting have been major forces shaping the topography of the world's mountainous regions.

Definitions & Insights

THE SECOND LAW OF THERMODYNAMICS

Geographers find it useful to understand how energy flows through ecosystems because this process is central to the question of how many people the Earth and its regions can support and at what levels of resource consumption. Food chains (in which, for example, an antelope eats grass and a leopard eats an antelope) are short, seldom consisting of more than four feeding (trophic) levels, and the collective weight (known as biomass) and absolute number of organisms decline substantially at each successive trophic level (there are fewer leopards than antelopes). The reason is a fundamental rule of nature known as the **second law of thermodynamics,** which states that much of the energy involved in doing work is lost to the surrounding environment. In living and dying, organisms use and lose the high-quality, concentrated energy that green plants produce. An organism loses about 90 percent of its energy in heat and feces as it passes to the next feeding level on the food chain. The amount of animal biomass that can be supported at each successive level thus declines geometrically, and little energy remains to support the top carnivores.

The second law of thermodynamics has important consequences and implications for human use of resources. The higher we feed on the food chain (the more meat we eat), the more energy we use. The question of how many people Earth's resources can support thus depends very much on what we eat. The affluent consumer of meat demands a huge expenditure of food energy, because that energy flows from grain (producer) to livestock animal (primary consumer, with a 90 percent energy loss), and then from livestock animal to human consumer (secondary consumer, with another 90 percent energy loss). Each year the average U.S. citizen eats 100 pounds of beef, 50 pounds of pork, and 45 pounds of poultry. By the time it is slaughtered, a cow has eaten about 10 pounds of grain per pound of its body weight; a pig, 5 pounds; and a chicken, 3 pounds. Thus in a year, a single American consumes the energy captured by two thirds of a ton of grain—1330 pounds—in meat products alone. By eating grain exclusively (feeding lower on the food chain), many people could live on the energy required to sustain just one meat-eater.

Figure 2.10 World map of landforms. The map is generalized to show patterns of plains, plateaus, hill lands, and mountains (modified from a map by the U. S. Department of Agriculture).

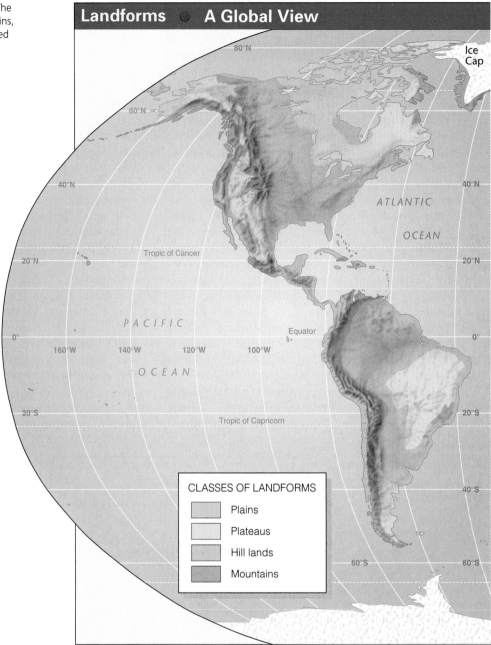

2.4 Two Revolutions That Have Shaped Regions

The geographer's vision is necessarily historical, for it is impossible to understand the current appearance of the landscape without reference to its past. An especially useful perspective on the current spatial patterns of our relationship with Earth is to view these patterns as products of two "revolutions" in the relatively recent past: the **Agricultural** or **Neolithic (New Stone Age) Revolution** that began in the Middle East about 10,000 years ago, and the **Industrial Revolution** that began in 18th-century Europe. Each of these revolutions transformed humanity's relationship with the natural environment. Each increased substantially our capability to consume resources, modify landscapes, grow in number, and spread in distribution.

Hunting and Gathering

Until about 10,000 years ago, our ancestors practiced a **hunting and gathering** livelihood. Joined in small bands consisting of extended family members, they were nomads practicing

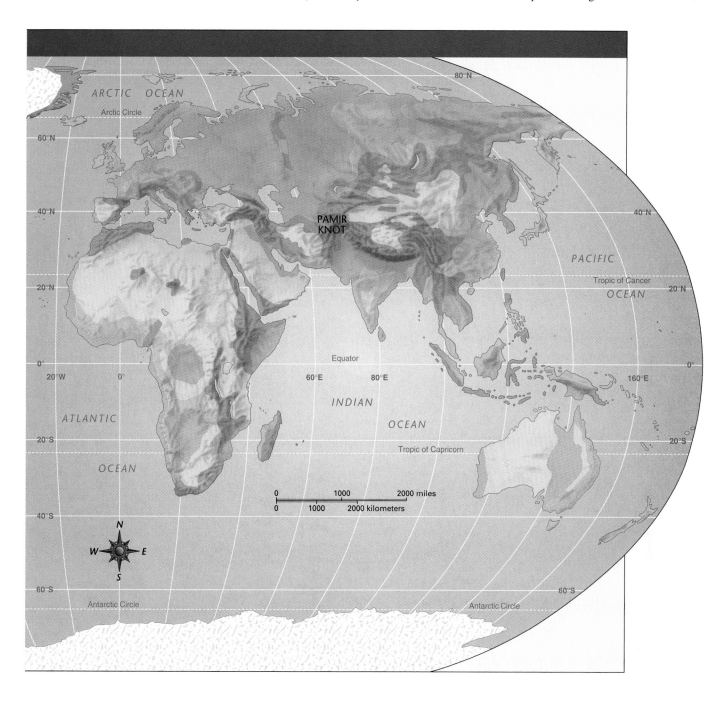

an **extensive land use** in which they covered large areas to locate foods such as seeds, tubers, foliage, fish, and game animals. Moving from place to place in small numbers, they had a relatively limited impact on natural environments compared with human impact today. Many scholars praise these preagricultural people for the apparent harmony they maintained with the natural world in both their economies and their spiritual systems. Hunters and gatherers have even been described as the "original affluent society," because after short periods of work to collect the foods they needed, they enjoyed long stretches of leisure time. Studies of those few hunter-gatherer cultures that lingered into the 20th century, such as the San ("Bushmen") of southern Africa and several Amerindian groups of South America (Fig. 2.13, p. 40), suggest that although their life expectancy was low, they suffered little from the mental illnesses and broken family structures that characterize industrial societies.

Hunters and gatherers were not always at peace with one another or with the natural world, however. With upright posture, stereoscopic vision, opposable thumbs, an especially large brain, and no rutting season, *Homo sapiens* was from its earliest times an **ecologically dominant species**—one that

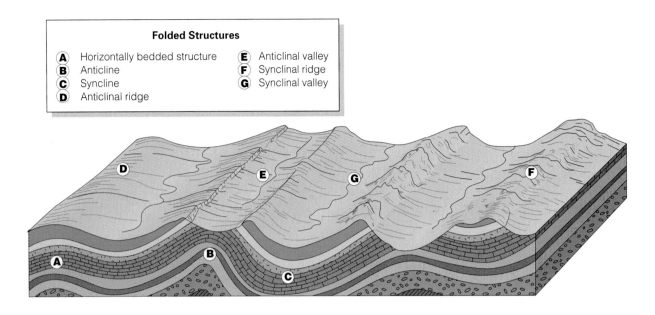

Folded Structures

(A) Horizontally bedded structure
(B) Anticline
(C) Syncline
(D) Anticlinal ridge

(E) Anticlinal valley
(F) Synclinal ridge
(G) Synclinal valley

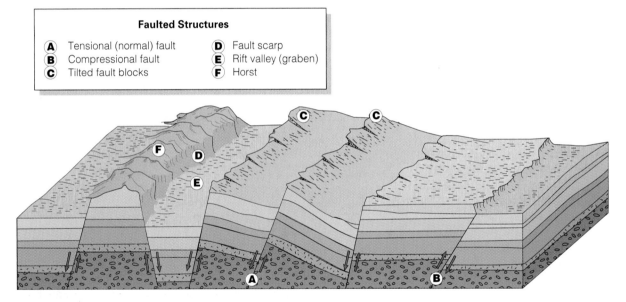

Faulted Structures

(A) Tensional (normal) fault
(B) Compressional fault
(C) Tilted fault blocks

(D) Fault scarp
(E) Rift valley (graben)
(F) Horst

Figure 2.11 Folded and faulted rock structures.

competes more successfully than other organisms for nutrition and other essentials of life, or that exerts a greater influence than other species on the environment. Using fire to flush out game or create new pastures for the herbivores they hunted, preagricultural people, unlike other animals, shaped the face of the land on a vast scale relative to their small numbers. Many of the world's prairies, savannas, and steppes where grasses now prevail originated from hunters and gatherers setting fires repeatedly. These people also overhunted and in some cases eliminated animal species. The controversial **Pleistocene Overkill** hypothesis states that rather than being at harmony with nature, hunters and gatherers of the Pleistocene Era (2 million to

10,000 years ago) hunted many species to extinction, including the elephant-like mastodon of North America.

The Agricultural Revolution

Despite these excesses, the environmental changes that hunters and gatherers could cause was limited. Humankind's power to modify landscapes took a giant step with the domestication (controlled breeding and cultivation) of plants and animals—the **Agricultural Revolution.** Why people began to produce rather than continue to hunt and gather plant and animal foods—first in the Middle East and later in Asia, Europe, Africa,

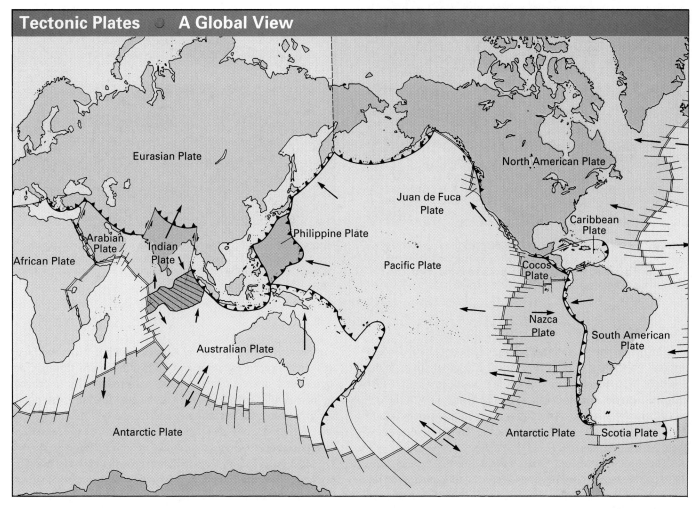

Tectonic Plates   A Global View

Figure 2.12 World map of the major tectonic plates and their general direction of movement. Most tectonic activity occurs along the plate boundaries where plates separate, collide, or slide past one another. *After Gabler et al. 1994:349*

and the Americas—is uncertain. Two theories prevail. Climatic change in the form of increasing drought and reduced plant cover may have forced people and wild plants and animals into smaller areas, where people began to tame wild herbivores and sow wild seeds to produce a more dependable food supply. A more favored theory is that their own growing populations in areas originally rich in wild foods compelled people to find new food sources, so they began sowing cereal grains and breeding animals. The latter process apparently began about 8000 B.C. in the Zagros Mountains of what is now Iran. The culture of domestication spread outward from there but also developed independently in several world regions.

In choosing to breed plants and animals, people gave up the nomadic, extensive land use of hunting and gathering for the sedentary, **intensive land use** of agriculture and animal husbandry. The new system enabled them to create large and reliable surpluses of food. Through **dry farming,** or planting and harvesting according to the seasonal rainfall cycle, population

densities could be 10 to 20 times higher than they were in the hunting and gathering mode. By about 4000 B.C., people along the Tigris, Euphrates, and Nile rivers began **crop irrigation**—bringing water to the land artificially—an innovation that allowed them to grow crops year-round, independently of seasonal rainfall or river flooding (Fig. 2.14). Irrigation technology allowed even more people to make a living off the land; irrigated farming yields about five to six times more food per unit area than dry farming. In ecological terms, the expanding food surpluses of the Agricultural Revolution raised the Earth's **carrying capacity**—that is, the size of a species' population (in this case, humans) that the Earth's ecosystems can support.

This steep increase in food surpluses freed more people from the actual work of producing food, and they undertook a wide range of activities unrelated to subsistence needs. Irrigation and the dependable food supplies it provided thus set the stage for the development of **civilization,** the complex culture of urban life characterized by the appearance of writing, economic

Figure 2.13 A native American of the Coronado tribe in western Ecuador. The red dye in his hair comes from a bean called achiote. His hunting and gathering livelihood is rarely practiced in the world today, but was the exclusive occupation of humanity prior to about 10,000 B.C. *Joseph J. Hobbs*

Figure 2.14 The Tigris River in southeastern Turkey. After 4000 B.C., highly productive irrigated agriculture in Southwest Asia and North Africa sparked rapid population growth and the emergence of the world's first cities. *Joseph J. Hobbs*

specialization, social stratification, and high population concentrations. By 2500 B.C., for example, 50,000 people lived in the southern Mesopotamian city of Ur, in what is now Iraq. Other **culture hearths**—regions where civilization followed the domestication of plants and animals—emerged between 8000 and 2500 B.C. in China, Southeast Asia, the Indus River Valley, West Africa, Mesoamerica, and the Andes.

The agriculture-based urban way of life that spread from these culture hearths had larger and more lasting impacts on the natural environment than either hunting and gathering or early agriculture. Acting as agents of humankind, domesticated plants and animals proliferated at the expense of the wild varieties people came to regard as pests and competitors. Agriculture's permanent and site-specific nature magnified the human imprint on the land, while the pace and distribution of that impact increased with growing numbers of people.

The Industrial Revolution

The human capacity to transform natural landscapes took another giant leap with the **Industrial Revolution,** which began in about 1700 A.D. This new pattern of human-land relations was based on breakthroughs in technology that several factors made possible (Fig. 2.15). First, western Europe had the economic capital necessary for experimentation, innovation, and risk. Much of this money was derived from the lucrative trade in gold and slaves undertaken initially in the Spanish and Portuguese empires after 1400. Second, in Europe prior to 1500, there had been significant improvements in agricultural productivity, particularly with new tools such as the heavy plow, and with more intensive and sustainable use of farmland. Crop yields increased, and human populations grew correspondingly. A third factor was population growth itself. More people freed from

work in the fields represented a greater pool of talent and labor in which experimentation and innovation could flourish. As agricultural innovations and industrial productivity improved, an increasingly large proportion of the growing European population was freed from farming, and for the first time in history there were more city-dwellers than rural folk. The process of industrialization continues to promote urbanization today.

Most geographers see population growth today as a drain on resources, but the Industrial Revolution illustrates that, given the right conditions, more people do create more resources. Inventions such as the steam engine tapped the vast energy of

Figure 2.15 A tweed mill in Stornoway on Scotland's Lewis Island. The process of industrialization which began in 18th-century Europe has rapidly transformed the face of the Earth. *Joseph J. Hobbs*

40

fossil fuels—initially coal and later, oil and natural gas. This energy, the photosynthetic product of ancient ecosystems, allowed Earth's carrying capacity for humankind to be raised again—this time into the billions.

As they began to deplete their local supplies of resources needed for industrial production, Europeans began to look for these materials abroad. As early as their **Age of Exploration,** which began in the fifteenth century, Europeans probed ecosystems across the globe to feed a growing appetite for innovation, economic growth, and political power. Thus, the process of European **colonization** was linked directly to the Industrial Revolution. Mines and plantations from faraway central Africa and India supplied the copper and cotton that fueled economic growth in Belgium and England.

No longer dependent upon the foods and raw materials they could procure within their own political and ecosystem boundaries, European vanguards of the Industrial Revolution had an impact on the natural environment that was far more extensive and permanent than that of any other people in history. There are many measures of the unprecedented changes that the Industrial Revolution and its wake have wrought on Earth's landscapes. Between 1700 and 1990, the total forested area on Earth declined by 20 percent. During the same period, total cropland grew 460 percent, with more expansion in the period from 1950 to 1990 than in the 150 years from 1700 to 1850. Human use of energy increased more than a hundredfold from 1700 to 1990. Today the single species of humankind uses about 40 percent of Earth's land-based photosynthetic productivity in the production of crops and exploitation of forests and other natural vegetation. Of particular interest to geographers is how the costs and benefits of such expansion are distributed in unequal patterns across the Earth.

2.5 The Geography of Development

One very notable characteristic of human life on Earth is the large disparity between wealthy and poor people, both within and among countries. At a high level of generalization, the world's countries can be divided into "haves" and "have-nots" (Table 2.1 and Fig. 2.16). Writers refer to these distinctions variously as "developed" and "underdeveloped," "developed" and "developing," "more developed" and "less developed," "industrialized" and "nonindustrialized," and "north" and "south," based on the general geography of the wealthier countries occupying the middle latitudes of the Northern Hemisphere. This text uses the distinction between **more developed countries (MDCs)** and **less developed countries (LDCs)** as a framework to help explain the world's complexities. It must be emphasized, however, that as an introductory tool, this division cannot account for the tremendous variations and ongoing changes in economic and social welfare that characterize the world today.

Table 2.1	Characteristics of More Developed Countries (MDCs) and Less Developed Countries (LDCs)	
Characteristic	**MDC**	**LDC**
Per capita GNP and income	High	Low
Percent in middle class	High	Low
Percent in manufacturing	High	Low
Energy use	High	Low
Percent urban	High	Low
Percent rural	Low	High
Birth rate	Low	High
Death rate	Low	Low
Population growth rate	Low	High
Percent under age 15	Low	High
Percent literate	High	Low
Leisure time available	High	Low
Life expectancy	High	Low

Some countries, such as the "Asian tigers," are best described as **newly industrializing countries (NICs)** since they do not fit either the MDC or LDC idealized type. We will describe these cases in the relevant discussions of the Earth's regions.

Measures of Development

It is customary to distinguish MDCs from LDCs by examining annual **gross national product (GNP)**—the total output of goods and services a country produces in a year—or by per capita gross national product, which is the GNP divided by the country's population. The gulf between the world's richest and poorest countries is startling (Table 2.2). The average per capita GNP in the MDCs is 17 times greater than in the LDCs. In 1998, with a per capita GNP of $45,360, Luxembourg was the world's richest country, while Mozambique was the poorest with just $80. These raw numbers suggest that economic productivity and income alone characterize **development,** which, according to a common definition, is a process of improvement in the material conditions of people through diffusion of knowledge and technology.

Such economic definitions reveal little about measures of well-being such as income distribution, gender equality, literacy, and life expectancy. Recognizing the shortcomings of strictly economic definitions, the United Nations Development Programme created the **Human Development Index (HDI),** a scale that considers these attributes of quality of life. According to this index, Canada in 1998 was "the world's best place to live," although it is seventeenth in per capita GNP ranking.

Figure 2.16 Wealth and poverty by country. Note the concentration of wealth in the middle latitudes of the Northern Hemisphere.

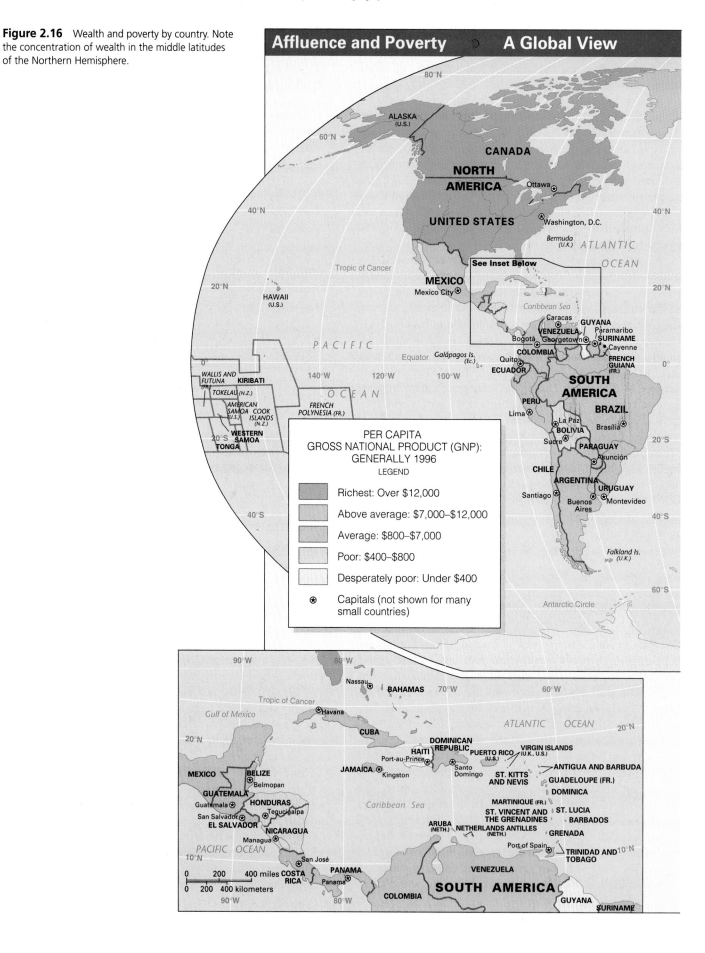

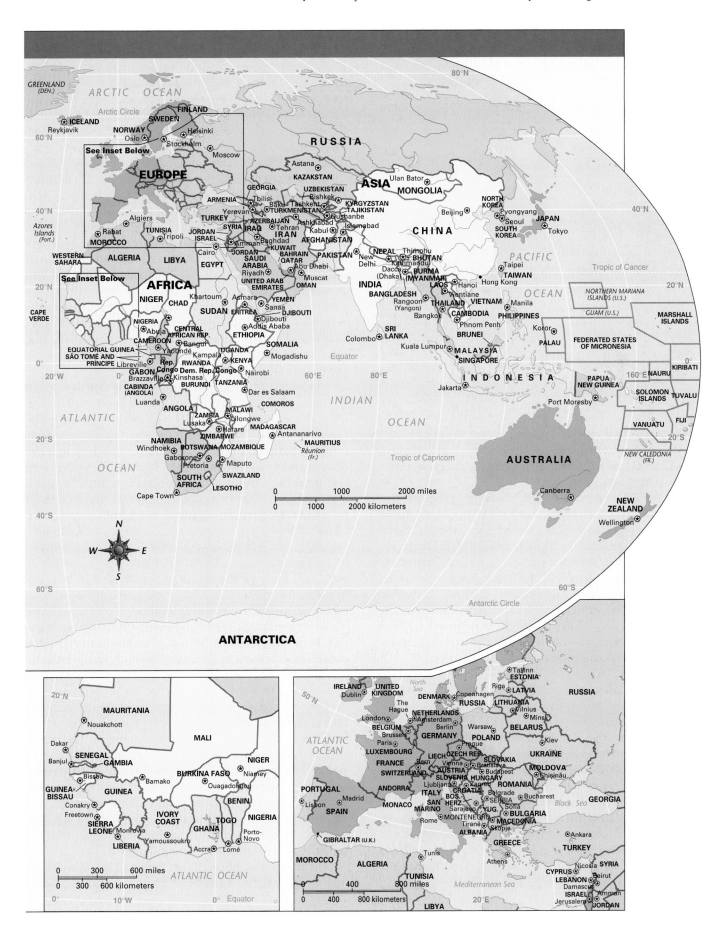

Table 2.2   Top 10 and Bottom 10 Countries in Per Capita GNP (US$, 1998)

Top Ten		Bottom Ten	
Country	GNP	Country	GNP
1. Luxembourg	45,360	1. Mozambique	80
2. Switzerland	44,350	2. Ethiopia	100
3. Japan	40,940	3. Dem. Rep. of Congo	130
4. Norway	34,510	4. Chad	160
5. Denmark	32,100	5. Burundi	170
6. Germany	28,870	6. Tanzania	170
7. Austria	28,110	7. Malawi	180
8. U.S.A.	28,020	8. Niger	200
9. Iceland	26,580	9. Sierra Leone	200
10. Belgium	26,440	10. Nepal	210

Source: Population Reference Bureau, World Population Data Sheet, 1998.

Notes:
- If included, Singapore would be No. 6 in the Top Ten with a GNP of $30,550.
- GNPs for the oil-rich Mid-East countries were not listed by the PRB.

Following Canada, in descending order, are France, Norway, the United States, Iceland, the Netherlands, Japan, Finland, New Zealand, and Sweden. In HDI terms, Sierra Leone in West Africa is the world's worst place to live.

On the basis of per capita GNP, 1.2 billion, or 20 percent, of the world's 6.0 billion people inhabit the MDCs. Most citizens of these countries, such as the United States, Canada, Japan, Australia, New Zealand, and the nations of Western Europe, enjoy an affluent lifestyle with freedom from hunger. Employed in industries or services, most of the people live in cities rather than in rural areas. Disposable income, or money that people can spend on goods beyond their subsistence needs, is generally high. There is a large middle class. Population growth is low as a result of low birth rates and low death rates. Life expectancy is long, and the literacy rate is high.

Life for the planet's other 80 percent, or about 4.8 billion people, is very different. In the LDCs, including most countries in Latin America, Africa, and Asia, poverty and often hunger prevail. Most people engage in subsistence agriculture, and the industrial base is small. The middle class tends to be small, with an enormous gulf between the vast majority of poor and a small wealthy elite, which owns most of the private landholdings. With high birth rates and falling death rates, population growth is high. Life expectancy is low. The literacy rate is low.

With four-fifths of the world's people living in the poorer, less-developed countries, it is important to understand the root causes of underdevelopment and to appreciate how wealth and poverty affect the global environment in very different but equally profound ways.

Causes of Disparities

Many theories attempt to explain the disparities between MDCs and LDCs. The most widely debated, and also the most embraced in the LDCs, is **dependency theory,** which argues that the worldwide economic pattern established by the Industrial Revolution and the attendant process of colonialism continues today. In his book *Ecological Imperialism* (New York: Cambridge University Press, 1986), historian and geographer Alfred Crosby explains the rich-poor divide and dependency theory by idealizing two very different patterns by which European powers utilized foreign lands during the Industrial Revolution. In the pattern of **settler colonization,** Europeans sought to create new Europes, or "neo-Europes," in lands much like their own: temperate midlatitude zones with moderate rainfall and rich soils where they could raise wheat and cattle. Thus, between 1630 and 1930, more than 50 million Europeans emigrated from their homelands to create European-style settlements in what are now Canada, the United States, Argentina, Uruguay, Brazil, South Africa, Australia, and New Zealand. These lands were destined to become some of the world's wealthier regions and countries.

In contrast to their preference to settle familiar middle latitude environments, Europeans viewed the world's tropical lands as sources of raw materials and markets for their manufactured goods. The environment was too different from home to make settlement attractive. In establishing a pattern of **mercantile colonialism,** Crosby explains, Europeans were less inhabitants than conquering occupiers of the colonies, overseeing indigenous peoples and resettled slaves in the production of primary or unfinished products: sugar in the Caribbean; rubber in Latin America, West Africa, and Southeast Asia; and gold and copper in southern Africa, for example. Colonialism required huge migrations of people to extract the Earth's resources, including 30 million slaves and contracted workers from Africa, India, and China to work mines and plantations around the globe.

In the mercantile system, the colony provided raw materials to the ruling country in return for finished goods; thus people in India might purchase clothing made in Great Britain from the cotton they had harvested. The relationship was always most advantageous to the colonizer. Great Britain, for example, would not allow its colony India to purchase finished goods from any country but Britain. It prohibited India from producing any raw materials the empire already had in abundance, such as salt (India's Mohandas Gandhi defiantly violated this prohibition in his famous "March to the Sea"). Finished products (or "value-added" products) are worth much more than the raw materials they are made from, so focusing manufacturing in the ruling country concentrated wealth there while discouraging industrial and economic development in the colony.

The colony was obliged to contribute to, but was prohibited from competing with, the economy of the ruling country—a relationship that dependency theorists insist continues today. Dependency theory states that to participate in the world economy, the former colonies but now-independent countries continue to depend upon exports of raw materials to, and purchases of finished goods from, their former colonizers, and this disadvantageous position keeps them poor. With independence, the former colonies needed revenue. To earn that money, they continued to produce the goods for which markets already existed—generally the same unprocessed primary products they supplied in colonial times. Dependency theorists argue that when former colonies try to break their dependency by becoming manufacturers, the former colonizers impose trade barriers and quotas to preclude that step. To support their argument, they point out that countries like Thailand and South Korea, which the Europeans never colonized, and which never developed these dependencies, are among the most prosperous LDCs or are in that select group of NICs.

Many geographers, however, view dependency theory as too simplistic and politically charged. Rather, they consider a wider and more complex set of factors, including culture, location, and natural environment, to explain why some countries are wealthy and others poor. For example, because it is situated close to a great mainland with which to trade, the island of Great Britain enjoys a central location favorable for economic development. Japan has a similar location relative to the Asian landmass. In contrast, landlocked nations such as Bolivia in South America and numerous nations in Africa have locations unfavorable for trade and economic development, and they have not overcome this disadvantage.

A superabundance of one particular resource (such as oil in the Persian/Arabian Gulf states) or a diversity of natural resources has helped some countries to become more developed than others. The former Soviet Union and the United States developed superpower status in the 20th century in large part based on the enormous natural resources of both countries. In other cases, human industriousness has helped to compensate for resource limitations and promote development. Japan, for example, has a rather small territory with few natural resources (including almost no petroleum); yet, in the second half of the 20th century, it became an industrial powerhouse largely because the Japanese people united in common purpose to rebuild from wartime devastation, placing priorities on education, technical training, and seaborne trade from their advantageous island location. Conversely, cultural or political problems like corruption and ethnic factionalism can hinder development in a resource-rich nation, as in the mineral-wealthy Democratic Republic of Congo.

Whether because of neocolonialism (as many dependency theorists argue) or a more complex array of variables, many developing countries continue to rely heavily on income from the export of a handful of raw materials. This makes them vulnerable to the whims of nature and the world economy. The economy of a country heavily dependent on rubber exports, for example, may suffer if an insect pest wipes out the crop or if a foreign laboratory develops a synthetic substitute. When demand for rubber rises, that country may actually harm itself trying to increase its market share by producing more rubber, because in the process it drives down the price. If the country withholds production to shore up rubber's price, it provides consuming countries with an incentive to look for substitutes and alternative sources. The developing country is in a dependent and disadvantaged position.

Environmental Impacts of Underdevelopment

Of particular interest to geographers are the environmental impacts of relations between MDCs and LDCs, especially the impacts on the poorer countries. LDCs generally lack the financial resources needed to build roads, dams, energy grids, and the other infrastructure assets they perceive as critical to development. They turn to the World Bank, International Monetary Fund (IMF), and other institutions of the MDCs to borrow funds for these projects. Many borrowers are unable to pay even the interest on these loans. When lender institutions threaten to sever assistance, borrowing countries often try to quickly raise cash to avoid this prospect. One method is to dedicate more quality land to the production of **cash (commercial) crops,** luxuries such as coffee, tea, sugar, coconuts, and bananas exported to the MDCs. Governments or foreign corporations often displace or "marginalize" subsistence farmers in the search for new lands on which to grow these commercial crops. In the process of **marginalization,** poor subsistence farmers are pushed onto fragile, inferior, or marginal lands which cannot support crops for long and which are degraded by cultivation. In Brazil's Amazon Basin, for example, peasant migrants arrive from Atlantic coastal regions where government and wealthy private landowners cultivate the best soils for sugarcane and other cash crops. The newcomers to Amazonia slash and burn the rain forest to grow rice and other crops that exhaust the soil's limited fertility in a few years. They move on to cultivate new lands, and in their wake come cattle ranchers whose land use further degrades the soil.

National decision makers often face a difficult choice between using the environment to produce more immediate or more long-term economic rewards. In most cases they feel compelled to take short-term profits, and by cash cropping and other strategies initiate a sequence having sometimes tragic environmental consequences. Most LDCs have resource-based economies that rely not on industrial productivity but upon stocks of soils, forests, and waters. The long-term economic health of these countries could be assured by the perpetuation of these natural assets, but—to pay off international debts and meet other needs—the LDCs generally draw on their ecological capital faster than nature can replace it. In ecosystem terms, they exceed **sustainable yield** or the **natural replacement rate,** the highest rate at which a renewable resource can be used without decreasing its potential for renewal.

Perspective

The Dark Side of Globalization

One of the most remarkable international trends of the late twentieth century was *globalization*—the spread of free trade, free markets, and investments across borders—and the political and cultural adjustments that accompany this diffusion (see Chapter 1, p. 16). Much of this process has been fueled by multinational companies (also called transnational companies, or companies having operations outside their home countries) as they increased their investments abroad. Most of the companies are based in the MDCs, but multinational corporations also have grown in LDCs such as Mexico. After 1990, investment by MDC multinational firms became the largest source of outside financing for the LDCs—larger even than foreign aid. Individual and corporate investors in the MDCs also invested heavily in stock market securities of the LDCs, reasoning that these developing nations had much greater economic growth potential than the mature economies of the MDCs. The opportunities appeared particularly great in the former communist countries, including Russia, whose doors had been opened to free enterprise and whose long-deprived citizens had a strong appetite for consumer goods.

Throughout the 1990s there was debate about the pros and cons of globalization. With the economies of many LDCs—known to foreign investors as "emerging markets"—booming, the cheering rose above the skeptics' mutterings. Advocates of globalization argued that a new global economy would bring increased prosperity to all the world. Innovations in one country could be instantly transferred to another country, productivity would increase, and standards of living would improve. With once torpid economies like that of Malaysia registering rapid economic growth, and with the Russian stock market rising over 70 percent in periods of less than six months, the optimists' views seem justified. Other analysts were more cautious, worrying that globalization might actually increase the gap between rich and poor countries; a selected few developing nations might prosper from increased foreign investment and resulting industrialization, but the hoped-for "global" wealth would bypass other countries altogether. And the increasing interdependence of the world economies would make all the players more vulnerable to economic and political instability.

Events in 1997 and 1998 focused attention on the drawbacks of globalization. A wave of currency devaluations in East and Southeast Asia in 1997 led to deep recessions in Thailand, Malaysia, Indonesia, South Korea, Hong Kong, and Japan (see Chapter 12, p. 320). There were fears of an "Asian contagion" that would harm the economies of the MDCs in the West as Asian demand for manufactured goods fell. But for more than a year the Asian financial crisis failed to shatter the confidence of consumers and investors abroad. Stock markets in the United States reached historical highs in mid-July 1998.

In August 1998, a global economic crisis was triggered by an unexpected source: Russia, a country accounting for only 1 percent of the world's trade. With its reserves severely depleted by a hope-lessly unproductive economy (see Chapter 6, p. 187), the Central Bank of Russia gave up trying to defend Russia's currency, the ruble, on the international market. Currency exchange ceased, making it virtually impossible for Russians to buy the U.S. dollars that had become the *de facto* currency for much of the economy. The value of the ruble fell by more than half. During the last week of August, institutions such as foreign banks that had bought the country's debt in the form of bonds saw the value of their holdings fall to around 10 percent of their worth. On August 27, trading was halted on shares of the Republic National Bank in New York as its owner announced huge losses in Russian government bonds.

Fear gripped Wall Street as it confronted what it fears most: uncertainty. Republic was a bank with conservative lending practices, so investors worried about what might happen to the shares of more aggressive lending institutions. And what would happen in Latin America and other emerging markets if, as in Russia, there were to be a collapse in the prices of emerging market bonds? Such a collapse would make it much more expensive for the LDCs to borrow to meet their financial needs. With borrowing becoming more expensive, these countries might not be able to pay off their debts and cover their high trade deficits. That would force these countries to raise interest rates to defend the exchange value of their currencies. High interest rates tend to slow economic growth and cause prices to fall—a phenomenon known as **deflation.**

In tropical biomes, people cut down ten trees for every one they plant. In Africa, that ratio is 29 to 1. In 1950, about 30 percent of Ethiopia's land surface was forested. Today, less than 1 percent is in forest. Such countries are ecologically bankrupt; they have exhausted their environmental capital.

Many political and social crises result from this bankruptcy: the wars, refugee migrations, and famines of the Horn of Africa, for example, have underlying environmental causes. Such problems are becoming more central to the national interests of the United States and other powerful MDCs. The

The devaluations of currencies, first in East Asia, then in Russia, had already given rise to serious deflation, with marked consequences within those countries and throughout the international marketplace. For example, while the devalued ruble means that goods priced in rubles become increasingly expensive for Russians (that is, **inflation**), non-Russians can buy exported goods with dollars or other "hard" currencies at cheap prices. This leads to a downward spiral in prices that eventually affects all manufacturing economies, as companies are forced to reduce prices of their goods to remain competitive. Lower earnings, compounded by the slump in consumer confidence, leads to job losses and economic recession.

This was the prospect hanging over the world's financial markets on August 27. Fearing that Russia's crisis would spread to Latin America and beyond, investors in the MDCs rushed money out of stocks and into the refuge of cash and money market funds. On that single day, stock markets in the United States, Canada, France, Germany, and Italy fell more than 6 percent. Venezuela, Brazil, and Argentina each lost about 10 percent (Fig. 2.A). On top of its already massive declines, Russia's stock market fell another 17 percent. Globalization was showing its ugly side.

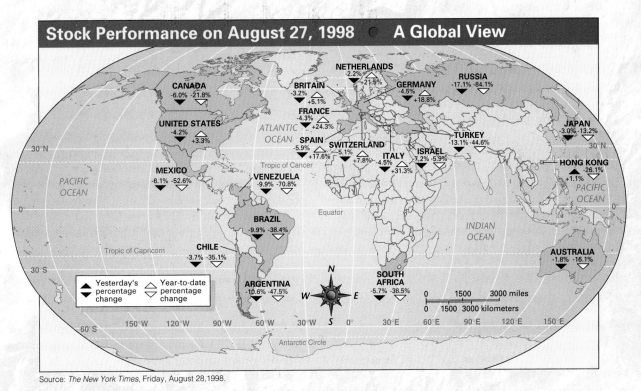

Source: *The New York Times*, Friday, August 28, 1998.

Figure 2.A Global stock market falls on August 27, 1998; 1998 year-to-date market gains or losses.

U.S. Central Intelligence Agency, for example, is increasingly using Geographic Information Systems (GIS) and other geographic techniques to analyze the natural phenomena and human misuses of environment that are root causes of war and threats to global security.

People in the LDCs feel the impacts of environmental degradation more directly than do people in the MDCs. The world's poor tend to drink directly from untreated water supplies and to cook their meals with fuelwood rather than fossil fuels. They are more dependent upon nature's abilities to replenish and cleanse itself, and thus they suffer more when those abilities are diminished (see Definitions and Insights, page 49).

Two Types of Overpopulation

High rates of human population growth intensify the environmental problems characteristic of the LDCs. More people cut more trees, a phenomenon that suggests there is a problem of **people overpopulation** in the poorer countries. Many persons, each using a small quantity of natural resources daily to sustain life, may add up to too many people for the environment to support.

There is another type of overpopulation, **consumption overpopulation,** that is characteristic of the MDCs. In the wealthier countries a few persons, each using a large quantity of natural resources from ecosystems across the world, may also add up to too many people for the environment to support. With just five percent of the world's population, the United States accounts for about a third of the world's annual energy consumption. Americans use more fossil fuels, iron, and copper than do the inhabitants of Latin America, Africa, and Asia combined. One analyst calculated that in his or her lifetime, the average U.S. citizen will consume more than 250 times as many goods as the average person in Bangladesh. Such disparities suggest that, just as Bangladesh is "underdeveloped," the United States is "overdeveloped." Much of this overconsumption is unnecessary for sustenance or well-being and impacts markedly the global environment.

If the vast majority of the world's population were to consume resources at the rate that U.S. citizens do, the environmental results might be ruinous. Even if consumption levels in the LDCs do not rise substantially, the sheer increase in numbers of people in those countries suggests that degradation of the environment will accelerate in the coming decades. The basic attributes of the world's population geography provide an insight into this dilemma.

2.6 Population Geography

Earth's human population 10,000 years ago, before the Agricultural Revolution, was probably about 5.3 million. By 1 A.D. it was probably between 250 and 300 million, or about the population of the United States today. The first billion was reached about 1800. Then a staggering population explosion occurred as a result of the Industrial Revolution. The second billion came in 1930, the fourth in 1975, and the sixth in 1999. As one measure of ecological dominance, humankind is now by far the most populous large mammal on Earth and has succeeded where no other animal has in extending its range to the world's farthest corners. As Figure 2.17 reveals, some world regions have especially dense populations. Notable are China and India, where productive agriculture and a long time span of occupation are the main reasons for high density, and western

Europe and the northeastern United States, where industrial productivity is the main reason.

Figure 2.18, p. 52, illustrates the **population explosion** as seen in the classic J-shaped curve of exponential growth in the human population. Most striking is the upward curve of growth after the dawn of the Industrial Revolution, with the greatest increase in the population growth rate taking place since 1950. At the current rate of 1.4 percent annual growth, our population would double in 49 years.

Dense settlement and growing numbers pose a fundamental question: Will we exceed Earth's carrying capacity for our species, and what will happen if we do? Early in the Industrial Revolution, an English clergyman named Thomas Malthus (1766–1834) postulated that human populations, which can grow geometrically or exponentially, would exceed food supplies, which usually grow only arithmetically or linearly. He predicted a catastrophic human die-off as a result of this irreconcilable equation. He could not have foreseen that the exploitation of new lands and resources, including tapping into the energy of fossil fuels, would permit food production to keep pace with or even outpace population growth for at least the next two centuries. However, the **Malthusian scenario** of the race between food supplies and mouths to feed remains a source of constant and important debate today.

Population Concepts and Characteristics

Two primary variables determine population change in a given village, city, country, or over the entire Earth: birth rate and death rate. The **birth rate** is the annual number of live births per thousand people in the population. The **death rate** is the annual number of deaths among that same thousand people. The **population change rate** is the birth rate minus the death rate in that population. On a worldwide average in 1998, the birth rate was 23 per thousand and the death rate 9 per thousand. By year's end, among the 1000 people, 23 babies had been born but 9 people of varying ages had died, resulting in a net growth of 14. That figure, 14 per thousand, or 1.4 percent, represents the 1998 population change rate (population growth rate) for the world. As no one leaves or enters Earth, there is no need to consider a third variable, **migration,** at this scale. At the scale of a village, city, or country, migration does affect the population. Compulsory and voluntary migration are strong forces in the world's regions and countries, and this text cites frequent examples of both.

Many factors affect birth rates. Better educated and wealthier people have fewer children. Conversely, poorer, less educated people generally want and have more children. Poor parents view additional children as an economic asset rather than a burden, because they represent additional labor to work in fields or factories and will care for them in their old age. People in cities tend to have fewer children than those in rural areas. Those who marry earlier have more children. Couples with access to and understanding of contraception may have

Definitions & Insights

THE FUELWOOD CRISIS

The removal of tree cover in excess of sustainable yield (as in highland Nepal, for example) illustrates the many detrimental effects that a single human activity can have in the LDCs. These impacts comprise the complex phenomenon, widespread in the world's LDCs, known as the **fuelwood crisis.** As people remove trees to use as fuel, for construction, or to make room for crops, their existing crop fields lose protection against the erosive force of wind. Less water is available for crops because, in the absence of tree roots to funnel water downward into the soil, it runs off quickly. Increased salinity (salt content) generally accompanies increased runoff, so that the quality of irrigation and drinking water downstream declines. Eroded topsoil can choke irrigation channels, reduce water delivery to crops, raise

floodplain levels, and increase the chance of floods destroying fields and settlements. As reservoirs fill with silt, hydroelectric generation and therefore industrial production is diminished. Upstream, where the problem began, fewer trees are available to use as fuel.

People must now change their behavior. Women, most often the fuelwood collectors, must walk farther to gather fuel. Turning to animal dung and crop residues as sources of fuel deprives the soil of the fertilizers these poor people most often use when farming. Reduced food output is the result. A family may eventually give up one cooked meal a day or tolerate colder temperatures in their homes because fuel is lacking—measures that negatively impact the family's health.

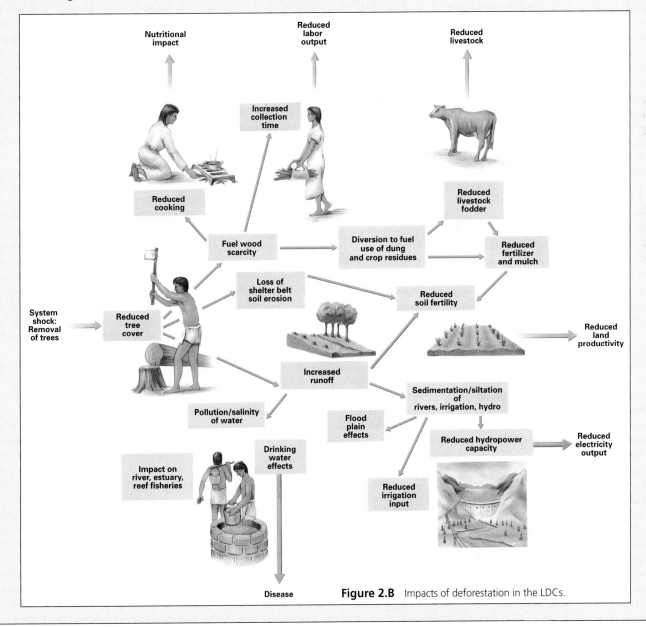

Figure 2.B Impacts of deforestation in the LDCs.

Figure 2.17 World population map. Each dot represents 100,000 people. The French geographer Jean Brunhes insisted that the two most significant world maps are those of population and rainfall. There are striking concentrations of people in the intensively and long-cultivated farming regions of southern and eastern Asia, and in the industrialized regions of Europe, the eastern United States, and Japan. Note the tendency of people to congregate on the margins of continents, near the sea. Populations are notably sparse in the more challenging environments of the tropical rain forest, deserts, subarctic taiga, and arctic tundra.

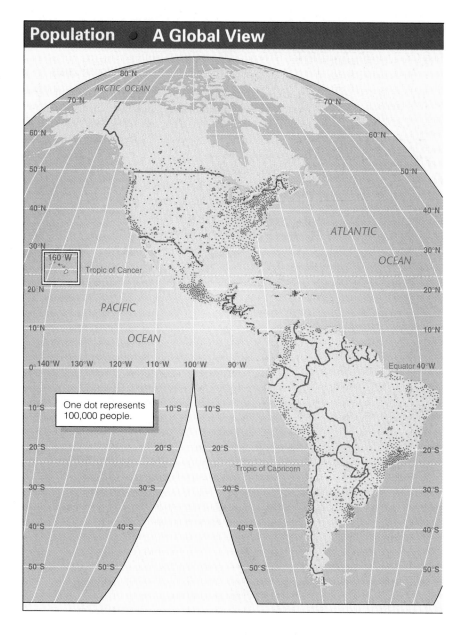

fewer children. Cultural norms are important, because even where contraception is available, for religious or social reasons a couple may decide not to interfere with what they perceive as God's will or the social status associated with a larger family.

Death rates are affected mainly by health factors. Improvements in food production and distribution help reduce death rates. Better sanitation, hygiene, and drinking water eliminate infant diarrhea, a common cause of infant mortality in LDCs. The availability of antibiotics, immunizations, insecticides, and other improvements in medical and public health technology have a marked correlation with declining death rates.

The explosive growth in world population since the beginning of the Industrial Revolution is the result not of a rise in birth rates, but of a dramatic decline in death rates, particularly in the LDCs. The death rate has fallen as improvements in agri-

cultural and medical technologies have diffused, or spread, from the MDCs to the LDCs. Until recently, however, there were no strong incentives for people in LDCs to have fewer children. With birth rates remaining high and death rates falling quickly, the population has grown sharply.

With about 9 out of 10 babies worldwide born in the LDCs, the current and projected rates of population growth are distributed quite unevenly between the MDCs and LDCs. This phenomenon is apparent in the age-structure diagrams typical of these countries. An **age-structure profile** (often called a "population pyramid") classifies a population by gender and by five-year age increments (Figs. 2.19 and 2.20). One important index these profiles show is the percentage of a population under age 15. The very high proportion typical of LDCs, about 35 percent, is remarkable, for it means that populations will continue to

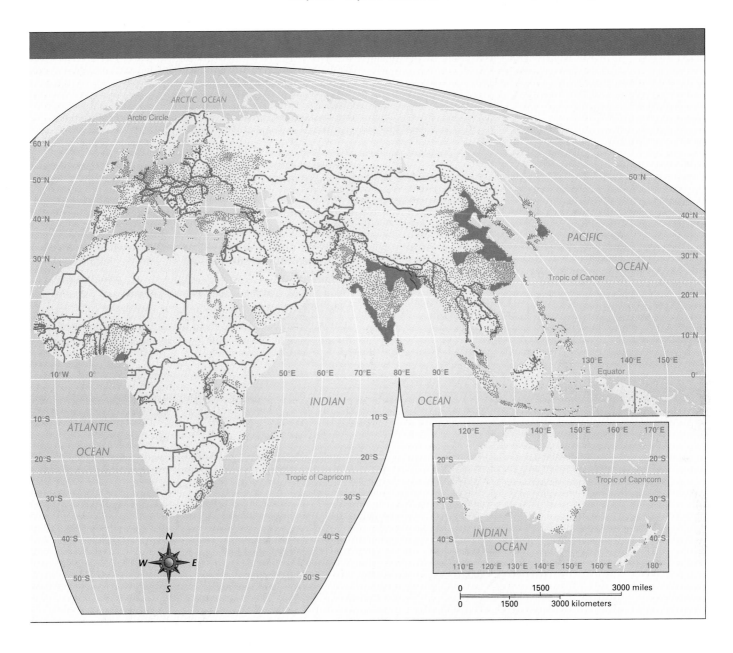

grow in these poor countries as these children enter their reproductive years. The bottom-heavy, pyramid-shaped age-structure diagram of LDCs contrasts markedly with the more rectangular structure of the MDCs. The wealthier countries have a much more even distribution of population through age groups, with a modest share of 19 percent under age 15. Such profiles suggest that, when migration is not considered, their population growth will be low in the near future.

The Demographic Transition: Will the LDCs Complete It?

The history of population change of the MDCs provides a model whose usefulness in predicting the population future for LDCs is still unknown. Known as the **demographic transition,**

this model depicts the change from high birth rates and high death rates to low birth rates and low death rates that accompanied economic growth in the MDCs. Four stages comprise the model (Fig. 2.21, p. 54). In the first or preindustrial stage, birth rates and death rates were high and population growth negligible. In the second or transitional stage, birth rates remained high but death rates dropped sharply after about 1800 with the medical and other innovations of the Industrial Revolution. In the third or industrial stage, beginning around 1875, birth rates began to fall as affluence spread. Finally, after about 1975, some of the industrialized countries entered the fourth or postindustrial stage, with both low birth rates and low death rates, and therefore (once again, as in stage one) low population growth. Some MDCs, including Sweden, Greece, Italy, Portugal, Spain, and Slovenia, officially register **zero population growth** (ZPG),

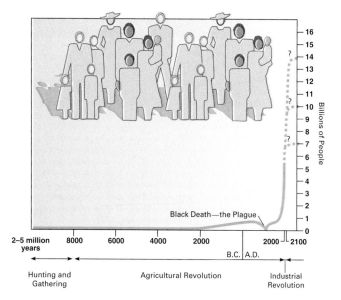

Figure 2.18 The J-shaped curve tracing exponential growth of the human population.

a rate of equal birth rates and death rates, and a few (in 1998, Estonia, Latvia, Lithuania, Germany, Belarus, Bulgaria, the Czech Republic, Hungary, Romania, Russia, and Ukraine) are experiencing a negative growth rate with more deaths than births.

Viewed as a group, the LDCs have proceeded from stage one to stage two of the demographic transition. With falling or already low death rates and only slightly declining birth rates, many LDCs appear to have stalled in this stage of very high population growth. There are two very distinct points of view on where these LDCs will proceed from this stage. Some observers believe that the poorer countries will follow the wealthier ones through the transition, achieving prosperity and a stabilized population. Most recognizable among the optimists are the so-called **technocentrists** or **cornucopians,** who argue that human history provides insight into the future: Through their technological ingenuity, people always have been able to conquer food shortages and other problems, and therefore always will. The late Julian Simon, a University of Maryland economist, argued that far from being a drain on resources, additional people create additional resources. Technocentrists

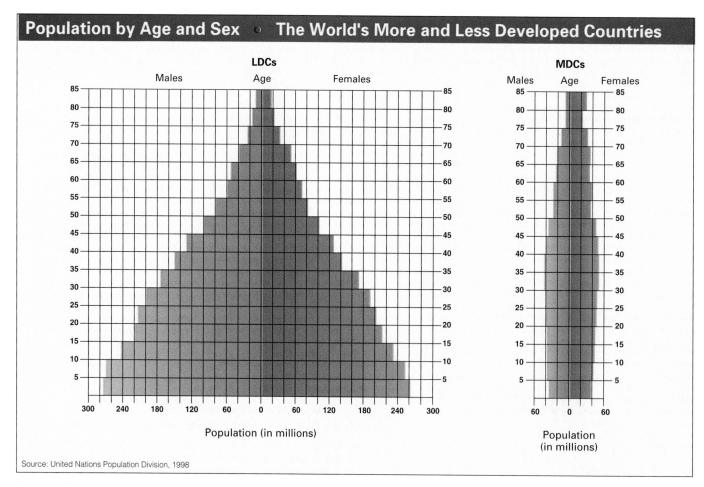

Figure 2.19 Age-structure profiles of the LDCs and MDCs in 1998. *Source: Population Reference Bureau.*

thus insist that people can raise the Earth's carrying capacity indefinitely and that the die-off that Malthus predicted will always be averted. Our more numerous descendants will instead enjoy more prosperity than we do.

In contrast, the **neo-Malthusians** argue that although successful so far, we cannot indefinitely increase Earth's carrying capacity. There is an upper limit beyond which growth cannot occur, with calculations ranging from 8 to 40 billion. Neo-Malthusians insist that LDCs cannot remain indefinitely in the transitional stage. Either they must intentionally bring birth rates down and make it successfully through the demographic transition, or they must unwillingly experience nature's solution, a catastrophic increase in death rates. By either the "birth rate solution" or "death rate solution," the neo-Malthusians argue, the less developed world will confront a population crisis.

Whereas technocentrists view the equation between people and resources passively, insisting that no corrective action is needed, neo-Malthusians tend to be activists who describe dire scenarios of a death rate solution to motivate people to adopt the birth rate solution. Biologist Paul Ehrlich thinks that we are increasingly vulnerable to a Malthusian catastrophe, particularly as HIV and other viruses diffuse around the globe with unprecedented speed. "The only big question that remains," wrote Ehrlich, "is whether civilization will end with the bang of an all-out nuclear war, or the whimper of famine, pestilence and ecological collapse."[2] Such dire warnings have earned many neo-Malthusians the reputation of being "gloom and doom pessimists."

Lifeboat Ethics

Neo-Malthusian ecologist Garrett Hardin offers the **ethics of a lifeboat** in an essay so distressing but challenging that we are obliged to consider and respond to it. "People turn to me," writes Hardin, "and say 'my children are starving. It's up to you to keep them alive.' And I say, 'The hell it is. I didn't have those children.'"[3] He describes our world not as a single "spaceship earth" or "global village" with a single carrying capacity, as many environmentalists do, but as a number of distinct "lifeboats," each occupied by the citizens of single countries and each having its own carrying capacity. Each rich nation is a lifeboat comfortably seating a few people. The world's poor are in lifeboats so overcrowded that many fall overboard. They swim to the rich lifeboats and beg to be brought aboard. What should the passengers of the rich lifeboat do? The choices pose a dilemma.

[2]Ehrlich, Paul R. 1988. "Populations of People and Other Living Things." In *Earth '88: Changing Geographic Perspectives*, Harm De Blij, ed. pp. 302–315. Washington, D.C.: National Geographic Society.

[3]Hardin, Garrett. 1968. "The Tragedy of the Commons." *Science* 162: 1243–1248.

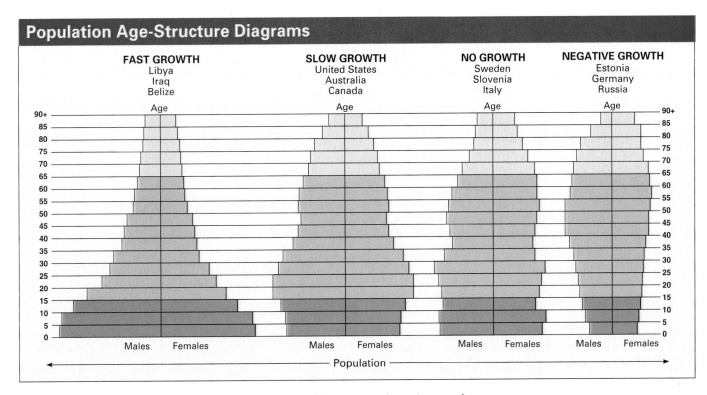

Figure 2.20 Age-structure profiles typical of countries with rapid, slow, zero, and negative rates of population growth. *Source: Population Reference Bureau.*

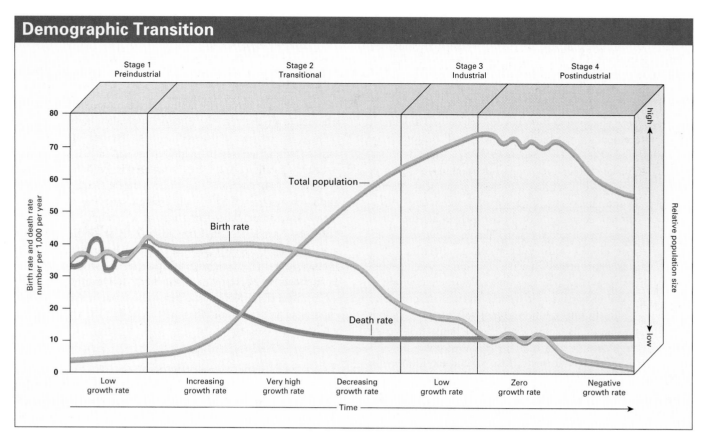

Demographic Transition

Figure 2.21　The demographic transition model. In the MDCs, economic development has resulted in a transition from high birth rates and high death rates to low birth rates and low death rates. Population growth rates at stages 1 and 4 are low because birth rates and death rates are nearly equal.

Hardin proposes the following: There are 50 rich passengers in a boat with a capacity of 60. Around them are 100 poor swimmers who want to come aboard. The rich boaters have three choices. First, they could take in all the swimmers, capsizing the boat with "complete justice, complete catastrophe." Second, as they enjoy an unused excess capacity of 10, they could admit just 10 from the water. But which 10? And what about the margin of comfort that excess capacity allows them? Finally, the rich could prevent any of the doomed from coming aboard, ensuring their own safety, comfort, and survival.

Translating this metaphor into reality, as an occupant of the rich lifeboat United States, for example, what would you do for the drowning refugees from lifeboat Haiti?

Hardin's choice is the third: drowning. To preserve their own standard of living and ensure the planet's safety, the wealthy countries must cease to extend food and other aid to the poor, and must close their doors to immigrants from poor countries. "Every life saved this year in a poor country diminishes the quality of life for subsequent generations," Hardin concludes. "For the foreseeable future, survival demands that we govern our actions by the ethics of a lifeboat."[4]

[4]*Ibid.*

In the United States and other Western countries there is now a growing sense of **donor fatigue.** Wearied by constant images of people in need, and feeling their contributions might be only marginally useful, people who can afford to give are simply tired of thinking about giving. The neo-Malthusian view that Hardin represents would see donor fatigue as beneficial to the biosphere.

2.7　Challenges of a New Millennium

Hardin assumes that people overpopulation is the greatest threat to Earth and does not consider the dangers posed by consumption overpopulation. He foresees a static situation in which the wealthy countries alone decide Earth's fate. But changes are ongoing in the economic status, political power, populations, and resource uses of poor countries. The oil embargo and energy crisis of 1973–74, for example, illustrated the ability of some LDCs to change the course of world events. Decisions about the future must be based on a more thorough understanding of the distributions and uses of global resources.

Birth rates in the LDCs are now falling, and so consequently is the world's rate of population growth (however, because the population base is so large, the actual number of people added to our population—about 80 million per year—is still greater than at any time in the past). In view of declining birth rates worldwide, the United Nations in 1997 revised its projection for future population growth downward. The agency predicted 9.4 billion by 2050 (half a billion fewer than that projected just two years earlier) and a stabilization at 10.7 billion in 2120 (the earlier projection was 12.6 billion). This revision prompted a rash of popular and academic articles proclaiming "the population explosion is over" and even some essays arguing there would soon be too few people on Earth, particularly in the MDCs. Other sources cautioned that it was too soon to declare the population bomb defused. "In 1997, world population growth turned a little slower," a Population Institute report argued. "The difference, however, is comparable to a tidal wave surging toward one of our coastal cities. Whether the tidal wave is 80 feet or 100 feet high, the impact will be similar."[5]

Considering the population projections, the geographer asks whether, how, and, above all, where global resources can support 10.7 billion people. About 11 percent of Earth's land surface is now in crops, 25 percent in pasture and rangelands, and 31 percent in forests and savanna; 33 percent is "little productive" (including deserts, ice caps, and other harsh biomes, and paved and built-up areas). The proportion dedicated to crops grew from the Neolithic until 1981, when the total area of farmland began to decrease. The other productive areas of pasture, rangeland, forests, and savanna are also decreasing. Only the "little productive" category is growing. After a long period of growth, crop yields on existing farmlands are also diminishing. Worldwide, irrigated area per capita has been decreasing since 1978, and per capita grain production has been decreasing since 1984. In sum, just as our demands for more food and other photosynthetic products are growing, supplies are dwindling.

Confronting Climate Change

While attempting to improve the technical means to meet the growing demand for food, we must also be concerned with the long-term effects of technology on the atmospheric processes identified at the start of this chapter. Climate change as a result of human activity could reverse efforts to improve life on the planet. This problem may become the most significant environmental challenge of the new millennium.

There is evidence that the Earth's atmosphere is warming, probably because of the **greenhouse effect** (Fig. 2.22). Formulated in 1827, the theory of the greenhouse effect begins with an observation that the Earth's atmosphere acts like the transparent glass cover of a greenhouse (Fig. 2.23) (for modern purposes, like a windows of a car). Visible sunlight passes through the glass to strike the Earth's surface. Ocean and land (the car

[5]Holmes, 1997: A7.

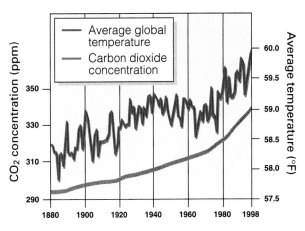

Figure 2.22 Industrialization and burning of tropical forests have produced a steady increase in carbon dioxide emissions. Most scientists believe these increased emissions explain the corresponding steady increase in the global mean temperature.

upholstery) reradiate the incoming solar energy as invisible infrared radiation (heat). Acting like the greenhouse glass or car window, the Earth's atmosphere traps some of that heat.

In the atmosphere, naturally occurring greenhouse gases such as carbon dioxide (CO_2) and water vapor make Earth habitable by trapping heat from sunlight. Concern over global warming focuses on human-made artificial sources of greenhouse gases, which trap abnormal amounts of heat. Carbon

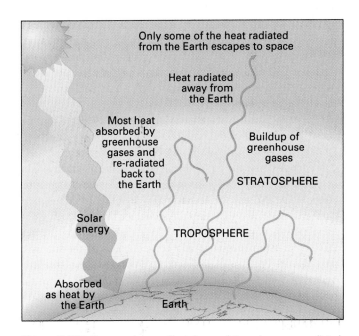

Figure 2.23 The greenhouse effect. Some of the solar energy radiated as heat (infrared radiation) from the Earth's surface escapes into space, while greenhouse gases trap the rest. Naturally occurring greenhouse gases make Earth habitable, but carbon dioxide and greenhouse gases emitted by human activities accentuate the greenhouse effect, making Earth unnaturally warmer.

dioxide released into the atmosphere from burning coal, oil, and natural gas is the greatest source of concern, but methane (from rice paddies and the guts of ruminating animals like cattle), nitrous oxide (from the breakdown of nitrogen fertilizers), and chlorofluorocarbons, or CFCs (from cooling and blowing agents), are also human-made greenhouse gases. Continued production of these gases may have the effect of rolling up the car windows on a sunny day—increased temperatures and physical problems for the occupants result.

While formal records of meteorological observations began just over a century ago, past climates left evidence in the form of marine fossils, corals, glacial ice, fossilized pollen, and annual growth rings in trees. These sources, along with formal records, indicate that the 20th century was the warmest century in the last 600 years, and the warmest years in that 600-year period were 1990, 1995, 1997 and 1998. There are indications that global warming is accelerating; the ten hottest years since record-keeping began in 1880 occurred in the twelve-year span between 1986 and 1998.

Until 1995, many scientists believed it was premature to state firmly that global climatic change, and in particular global warming, was now taking place as a result of human activities. Experts on climate change held out the possibility that the recent periods of warm years may have been natural variations. But in 1995, breakthroughs in computerized models of climate change prompted the United Nations–sponsored Intergovernmental Panel on Climate Change, the world's leading organization concerned with this issue, to conclude that the warming of the last century "is unlikely to be entirely due to natural causes" and that "a pattern of climatic response to human activities is identifiable in the climatological record." Increasing numbers of former skeptics now hold the view that people are responsible for the increasing warmth of Earth's atmosphere. The theory of the greenhouse effect is the most widely adopted explanation for global warming.

In 1896, Swedish scientist Svante Arrhenius feared that Europe's growing industrial pollution would eventually double the amount of carbon dioxide in the Earth's atmosphere and as a consequence raise the global mean temperature as much as nine degrees Fahrenheit (5°C). Computer-based climate change modellers today also use the scenario of a doubling of atmospheric carbon dioxide, and they similarly conclude that the mean global temperatures might warm from 3 to 9° Fahrenheit by the year 2030. The estimated range and consequences of the increases vary widely because of different assumptions about the little-understood roles of oceans and clouds. Most models concur that because of the slowness with which the world's oceans respond to changes in temperature, the effects of this rise will be delayed.

Geographers are most concerned with where the anticipated climatic changes will occur. Computer models conclude that there will not be a uniform temperature increase across the entire globe but that the increases will vary spatially and seasonally. Differing models produce contradictory results about the timing

and impacts of warming. Two teams of scientists publishing their results in the journal *Nature,* for example, first assumed a doubling of CO_2. Both then predicted rises in winter precipitation (due to increased evaporation) and in temperature, but assuming different seasonal combinations, they reached different conclusions about the effects on agriculture in the United States. One assumed warmer, drier summers and forecast that crop productivity would decline by 20 percent. The other team assumed warmer winters and forecast a 10 percent increase in crop yields.

There is general agreement that with global warming the distribution of climatic conditions typical of biomes will shift poleward. Many animal species will be able to migrate to keep pace with changing temperatures, but plants, being stationary organisms, will not. Conservationists who have struggled to maintain islands of habitat as protected areas are particularly concerned about rapidly changing climatic conditions.

There is also general agreement that if global temperatures rise, so will sea levels as polar and glacial ice melts, and as seawater warms and thus occupies more volume. Decision makers in coastal cities throughout the world, in island countries, and in those with important lowland areas adjacent to the sea are worried about the implications of rising sea levels (see Chapter 15, p. 391). The coastal lowland countries of the Netherlands and Bangladesh have for many years been outspoken advocates of reductions in greenhouse gases. President Gayoom of the Maldives, an Indian Ocean country comprised entirely of low islands, pleaded, "We are an endangered nation!"

There is growing international political consensus to take the strong measures that may be necessary to prevent or greatly reduce global warming. Present efforts to reduce the production of greenhouse gases focus on the MDCs, where about 85 percent of the emissions originate. The United States alone now accounts for 25 percent of all emissions, China for 9 percent, and more than 70 LDCs together for only about 5 percent. However, this balance is likely to change as the MDCs adopt stricter emissions standards and as coal-wealthy China matures into a major industrial power, possibly with less stringent standards. In 1997, 160 countries signed the **Kyoto Protocol,** a landmark international treaty on climate change. The agreement requires 38 MDCs (known as the "Annex I" countries) to reduce carbon dioxide emissions to more than 5 percent below their 1990 levels by the year 2012. The United States pledged to cut its emissions to 7 percent below their 1990 levels by that date. The European Union made a promise of 8 percent, and Japan 6 percent.

Meeting these pledges will require substantial legislative, economic, and behavioral changes in these more affluent countries. Transportation and other technologies that use fossil fuels will have to become more energy efficient, making these technologies at least temporarily more expensive. Gasoline prices will rise, so consumers will feel the pinch. Advocates of the protocol argue that the initial sacrifices will soon be rewarded by a more efficient and competitive economy powered by cleaner, cheaper sources of energy derived from the sun. Opponents, however, feel that higher fuel prices will be too costly for the United States and other industrialized economies to bear. Their position will make it difficult for many signatories of the Kyoto Protocol to ratify and then implement the treaty. In the United States a long and difficult debate will ensue as the Senate confronts the task of ratifying the agreement. One of the main obstacles to ratification in the United States and other MDCs is that China, along with all the world's LDCs, is not required by the Kyoto Protocol to take any steps to reduce greenhouse gas emissions.

The LDCs which rely largely on fuelwood rather than fossil fuels release relatively little CO_2 into the atmosphere, but they do contribute to the greenhouse effect. When burned, trees not only release the CO_2 they stored in the process of photosynthesis, but they cease to exist as organisms which remove CO_2 from the atmosphere. However, because these countries produce relatively little CO_2, policy-makers are considering innovative ways to reward them and encourage them to develop clean industrial technologies. One system being considered in the wake of the Kyoto Protocol is that of "tradable permits," in which each country would be assigned a right to emit a certain quantity of CO_2, according to its population size, setting the total at an acceptable global standard. The United States, already an overproducer by this scale, could purchase emission rights from a populous country like China, which currently "underproduces" CO_2. China would be obliged to use the income to invest in energy-efficient and nonpolluting technologies. Most notable about this system is that it views global warming as a global problem.

Concrete steps toward reducing another global threat, the **"ozone hole,"** have already occurred. CFCs destroy the stratospheric ozone, a layer of gas about 10–30 miles above the Earth that absorbs much of the ultraviolet radiation from the sun and thus protects plants, animals, and people from the sun's harmful effects. A rise in human skin cancer in Australia recorded since 1985, and a growing incidence of blindness in sheep in Chile, are among the indications that destruction of ozone has immediate and negative consequences. Due to the Earth's rotation and climatic patterns, depletion of stratospheric ozone is concentrated at the Earth's poles, particularly over Antarctica, as "holes" in the ozone layer. Breaking up in the polar spring season, these zones elongate and drift equatorward. In the mid 1990s, scientists discovered that, as anticipated, ozone holes were beginning to creep over populated areas of northern Europe, posing a serious threat to large numbers of people, livestock, and crops.

In contrast to the still-controversial geography of global warming, the geography of future ozone depletion has for many years been almost unanimously regarded as a major threat. The world's 37 major CFC-producing countries took action in 1989 with a treaty known as the **Montreal Protocol,** in which they agreed to cease production of these chemicals between the years 2000 and 2010. The MDCs which initiated the treaty agreed to cut their production sooner than the LDCs would have to. The United States ceased production in 1996. Many countries applauded its leadership but also reminded the United States that it should lead the world in a more vigorous attack on the problem of CO_2 emissions.

Sustainable Development

Actions against the potential consequences of global warming and ozone depletion are part of a growing agenda aimed at reversing destructive global patterns of resource and landscape use. Within the last two decades, new concepts and tools for managing the Earth in an effective, long-term way have emerged. Known collectively as **sustainable development** or **ecodevelopment,** these ideas and techniques consider what both MDCs and LDCs can do to avert the possible Malthusian dilemma and improve life on the planet.

The World Conservation Union defined sustainable development as "improving the quality of human life while living within the carrying capacity of supporting ecosystems." Sustainable development refutes what its proponents perceive as the current pattern of unsustainable development, whereby GNP and other measures of economic growth are based in large part on the destruction of global resources. According to these measures, a country that plunders its resource base for short-term profits gained through deforestation increases its GNP and appears to be more "developed" than a country that protects its forests for long-term harvesting of sustainable yield. Deforestation appears to be beneficial to a country because it raises GNP through the production of pulp, paper, furniture, and charcoal. However, GNP does not measure the negative impacts of deforestation, such as erosion, flooding, siltation, and malnutrition.

Sustainable development is a complex assortment of theories and activities, but its proponents call for eight essential changes in the way people perceive and use their environments:

1. People must change their worldviews and value systems, recognizing the finiteness of resources and reducing their expectations to a level more in keeping with Earth's environmental capabilities. Proponents of sustainable development argue that this change in perspective is needed especially in the MDCs, where, rather than trying to "keep up with the Joneses," people should try to enjoy life through more social instead of material pursuits.
2. People should recognize that development and environmental protection are compatible. Rather than viewing environmental conservation as a drain on economies, we should see it as the best guarantor of future economic well-being. This is especially important in the LDCs, with their resource-based economies.
3. People all over the world should consider the needs of future generations more than we do now. Much of the wealth we generate is in effect borrowed or stolen from our descendants. Our economic system values current environmental

benefits and costs far more than future benefits and costs, and thus we try to improve our standard of living today without regard to tomorrow.

4. Communities and countries should strive for self-reliance, particularly through the use of appropriate technologies. Villages, for example, could rely increasingly on solar power for electricity rather than be linked into national grids of coal-burning plants.

5. LDCs need to limit population growth as a means of avoiding the destructive impacts of "people overpopulation." Advances in the status of women, improvements in education and social services, and effective family planning technologies can help limit population growth.

6. Governments need to institute land reform, particularly in the LDCs. Poverty is often not the result of too many people on too little total land area, but of a small, wealthy minority holding a disproportionally high share of quality land. To avoid the environmental and economic consequences of marginalization, a more equitable distribution of land is needed.

7. Economic growth in the MDCs should be slowed to reduce the effects of "consumption overpopulation." If economic growth, understood as the result of consumption of natural resources, continues at its present rate in excess of sustainable yield, the Earth's environmental "capital" will continue to diminish rapidly.

8. Wealth should be redistributed among the MDCs and LDCs. Because poverty is such a fundamental cause of environmental degradation, the spread of a reasonable level of prosperity and security to the LDCs is essential. Proponents argue that this does not mean that rich countries should give cash outright to poor countries. Instead, the lending institutions of MDCs could forgive some existing debts owed by LDCs, or use such innovations as **debt-for-nature swaps,** in which a certain portion of debt is forgiven in return for the borrower's pledge to invest that amount in national parks or other conservation programs. Reducing or eliminating trade barriers that MDCs impose against industries in LDCs would also help redistribute global wealth.

Some geographers and other scientists believe that sustainable development will be the "third revolution," a shift in our ways of interacting with Earth so dramatic that it will be compared with the origins of agriculture and industry. The formidable changes called for in sustainable development are attracting increasing attention, perhaps because at the present there are no comprehensive alternative strategies for dealing with some of the most critical issues of our time.

SUMMARY WITH SELECTED KEY TERMS

- **Weather** is the atmospheric conditions prevailing at one time and place while **climate** is a typical pattern recognizable in the weather of a region over a long period of time. Climatic patterns have a strong correlation with patterns of vegetation and, in turn, with human opportunities and activities in relation to the landscape.

- Warm air holds more moisture than cool air; therefore **precipitation** is best understood as the result of processes that cool the air to release moisture. **Convectional** precipitation is the result of air being heated by intense surface radiation and rising rapidly to cool and produce rain. **Orographic** precipitation results when moving air confronts a topographic barrier and is forced upward to cool and produce rain. **Cyclonic (frontal)** precipitation is generated in **cyclones** or traveling low-pressure cells that bring air masses of different characteristics of temperature and moisture into contact.

- Most of the sun's visible short-wave energy that reaches the Earth is absorbed, but some of it returns to the atmosphere in the form of **infrared long-wave radiation,** which generates heat and helps to warm the atmosphere. Although the sun is the primary source of heat, the air is mainly heated by **re-radiation** from the land and water.

- In general, the lower the latitude of the place, the more solar energy it receives annually. However, many humid areas near the equator have such abundant cloud cover that the amount of solar radiation reaching the surface is greatly reduced. The result is that **subtropical zones** or areas in the vicinity of 20 to 30 degrees north and south of the equator receive the highest annual solar energy.

- Geographers group local climates into major **climate types,** each of which occurs in more than one part of the world and is associated with other natural features, particularly vegetation. Geographers recognize 10 to 20 major types of **ecosystems** or **biomes,** which are categorized by the type of natural vegetation. Vegetation and climate types are so sufficiently related that many climate types are named from the vegetation types—for example, the **tropical rain forest climate** and the **tundra climate.**

- Geographers and ecologists recognize the importance of some biomes because of their **biological diversity**—the number of plant and animal species and the variety of genetic materials these organisms contain. Regions where human activities are rapidly depleting a rich variety of plant and animal life are known as **biodiversity hot spots,** a ranked list of places scientists believe deserve immediate attention for study and conservation.

- Earth's landforms have been created by **tectonic processes** of warping, folding, faulting, and vulcanism and the **gradational processes** of weathering, erosion, and deposition. All four classes of landforms—**plains, plateaus, hill lands, and mountains**—are associated with **igneous, metamorphic, and sedimentary rocks.**

- A useful perspective on the current spatial patterns of our relationship with Earth is to view these patterns as products of two "revolutions" in the relatively recent past: the **Agricultural** or **Neolithic (New Stone Age) Revolution** and the **Industrial Revolution.** Each of these transformed humanity's relationship with the natural environment.

- High rates of human population growth intensify the environmental problems characteristic of the **LDCs.** More people use more re-sources, a phenomenon that suggests there is a problem of **people overpopulation** in the poorer countries. **Consumption overpopulation** is more characteristic of the **MDCs** where fewer people use large quantities of natural resources from across the world.

- Considering population projections, the geographer must ask how and where global resources can support 10.7 billion people by the year 2120. Our demands for more food and other photosynthetic products are growing while supplies are dwindling.

- Computer-based climate change models use the scenario of a doubling of atmospheric carbon dioxide and indicate that the mean global temperature might warm from 3 to 9 degrees by the year 2030. Thus the theory of the **greenhouse effect** is the most widely-adopted explanation for global warming. The general agreement is that with global warming the distribution of climatic conditions typical of biomes will shift poleward.

- Within the last 2 decades, new concepts and tools for managing the Earth in an effective, long-term way have emerged. Known collectively as **sustainable development** or **ecodevelopment,** these ideas and techniques consider what humanity can do to avert the **Malthusian dilemma** and improve life on the planet.

REVIEW QUESTIONS

1. List the major climate types and their associated natural vegetation. Identify the general locations of these climate types on maps (Figs. 2.7, 2.8).
2. List Earth's biodiversity hot spots and identify their locations in association with Earth's biomes.
3. What are some of the prevailing theories concerning the Agricultural Revolution? What factors made the Industrial Revolution possible?
4. List and define the measures of development used to distinguish MDCs from LDCs.

5. According to the "dependency theory," what are the causes of disparity between MDCs and LDCs?
6. What are the primary environmental impacts of underdevelopment?
7. List and define the two types of overpopulation.
8. List and define the four stages of demographic transition.
9. Explain Garret Hardin's "lifeboat analogy."
10. What long-term effects on Earth's atmospheric processes can be attributed to modern technology?

DISCUSSION QUESTIONS

1. Explain the high degree of spatial correlation between the distributions of Earth's climate and vegetation types.
2. What are some of the possibly negative ramifications of the **Green Revolution** on agriculture and the Earth's biodiversity?
3. Explain how *Homo sapiens* has always been an ecologically dominant species.

4. What characteristics other than productivity and income are considered in the **Human Development Index (HDI)** and why?
5. Debate Garrett Hardin's "lifeboat" ethics. Do you agree or disagree with his view, and why?

2 Europe

This map shows the shifting borders so definitive of East Central Europe this century. Fragmentation and boundary change have long characterized the Balkan region and beyond.

Europe has grand tradition written all across its landscape. Historical patterns of urban land use are under constant pressure from new buildings and increasing urban densities, but old structures generally maintain their elegance and ceremonial importance. *Guido Alberto Rossi / The Image Bank*

European capital cities often drape themselves across major waterways. Such centers are dominant, often putting other smaller cities in their shadows. *Colour Library Books*

Europe is alive with blends of waterways, mountains, and vegetation that have economic as well as cultural significance. These settings are popular tourist destinations, giving Europeans an effective incentive to maintain the beauty of their environment. *©1995 Dembinsky Photo Associates*

Regional Snapshot

The very dynamics that helped make the mid-18th century Industrial Revolution such a global success have been changing steadily. The labor force in mining, heavy manufacturing, and farming has been losing demographic and economic importance all across western Europe. *Jake Sutton / Gamma Liaison*

Agricultural productivity from Scandinavia to the Mediterranean rim has long been a dominant characteristic of this region. Fields push right up to urban settlements; rural households are interspersed with village settlements. Farmland and agricultural activity represent a continuing lifestyle in Europe. *Jay Fries / The Image Bank*

Ever since the collapse of the USSR and the 1990 tearing down of the Berlin Wall, Europe has been reinventing its regional political shape and identity. The political force newly realized in popular uprisings has expressed itself in many ways across the political landscape of the region. *AP / Wide World Photos*

Chapter 3

A Geographic Profile of Europe

▲ Europe from space. This view of Europe is a mosaic composed of a number of National Oceanic Atmospheric Administration (NOAA) weather satellite images taken on cloud-free days. The colors approximate natural tones and are a good indication of different ground cover. In western Europe, the light and midgreen colors indicate pasture and cultivated land. The bluer greens of eastern Europe show greater aridity and a less lush vegetation. This mosaic is composed of summer images so little snow is evident. Most evident in this graphic is the geographic fact that Europe is more a peninsula of the massive Eurasian landmass than a free-standing continent, even though convention marks Europe as one of the world's continents. However, in terms of political, economic, and social significance, it is very much a continent. (NOAA)

CHAPTER OUTLINE

3.1 Definition and Basic Magnitudes

3.2 The Physical Setting: Its Significance and Transformation

3.3 Climate Patterns

3.4 The Importance of Rivers and Waterways

3.5 Agriculture and Fisheries

3.6 European Patterns of Language and Religion

3.7 The Impact of Europe on the World in the Colonial Age

3.8 Europe's Rise to Global Centrality— and Potential Decline

3.9 Europe's Highly Developed and Diversified Economy

3.10 Problems of Europe's Environmental Pollution

3.11 Europe's Population Dynamics

3.12 The Drive for New Patterns of Regional Cooperation

Many regions of the globe have contributed to shaping today's world but none more impressively than Europe. The influence of Europe on modern nation building, the development of science and technology, the rise of advanced capitalist economies, and expansion overseas is central to world history, even though there are both positive and negative perspectives to these areas of influence. Today, the region continues to play a vital role in the world's economic, political, cultural, and intellectual life while it attempts to craft an ever closer regional compact through the expanding map of the **European Union (EU).**

3.1 Definition and Basic Magnitudes

In this book, we define Europe as the countries of Eurasia lying west of Russia and the Near Abroad and Turkey. Thus defined, Europe is a great peninsula, fringed by lesser peninsulas and islands, and bounded by the Arctic and Atlantic Oceans, the Mediterranean Sea, Turkey, the Black Sea, and Russia and the Near Abroad (Fig. 3.1). Europe achieves the label of continent largely because fundamental geographic standards for such language were determined and defined by Europeans when they had enormous global authority.

Understanding the geographic reality of Europe requires a grasp of some basic magnitudes that are very different from those of the United States. For example, all of Europe has an area only about two-thirds as large as that of the 48 contiguous states of the United States. This means that the areas of Europe's many countries tend to be rough equivalents of those of U.S. states. One of the striking things about Europe is the large impact that societies nourished on some of these small pieces of land have had on the world.

A different set of comparative magnitudes emerges when comparing the populations of the European countries with those of the individual United States. California—with its pop-

ulation of nearly 33 million—would be a rather populous country if it were European, but even California is considerably exceeded in population by six European countries (Table 3.1). Four major countries—Germany, the United Kingdom, France, and Italy—far outsize all others in Europe in population. Their respective populations range from about 82 million for Germany to 59 million each for France, Italy, and the United Kingdom. Together the four countries represent about one-half of Europe's population, or nearly as many people as there are in the entire United States.

Another useful comparison between European countries and the United States is by density of population. Even excluding great disparities among small units such as Malta and the **microstates** of Andorra and Monaco, Europe exhibits an extreme range of population density—from over 1197 persons per square mile (over 430 per sq km) in the intensively developed Netherlands to about 7 per square mile (3 per sq km) in Iceland (Table 3.1 and Fig. 3.2). Comparisons with the United States reveal about the same range, although in general the European countries are more densely populated. Table 3.1 provides a set of data that will allow a comparison within Europe and with other world regions and nations as well.

This is a region worthy of some detailed study, for many of the cultural characteristics and patterns that give image, personality, and tension to the world today derive from the influences of **Western civilization** in its many manifestations.

3.2 The Physical Setting: Its Significance and Transformation

There is a broad range of environmental settings in Europe. The application of human creativity to this varied environment is an important feature of Europe's long record of struggle and accomplishment. The roots of this development within Europe far antedate the Age of Discovery (see Chapter 1). Later, profits

Europe ● Location

Estimated population in millions

- Over 80
- 60–80
- 30–60
- 10–30
- Under 10

⊛ Capital and largest city
★ Capital of country
● Largest city of country

Figure 3.1 Europe showing political units as of early 1999. Note that in nearly all European countries the political capital is also the largest city (metropolitan area).

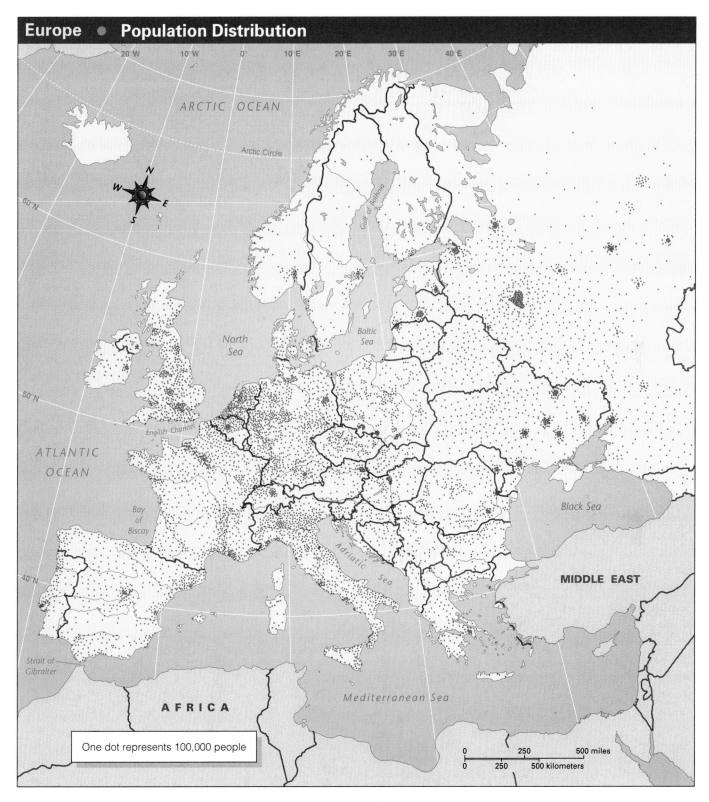

Europe • **Population Distribution**

One dot represents 100,000 people

Figure 3.2 The distribution of the human population is one of the most significant patterns to be found in geography. In this map of European population distribution, the strong role of rivers, coasts, and lowlands is evidenced. In addition, the long-time concentration of urban industrial population on the island of Great Britain, along the coastlines of northwest Europe, on the North European Plain, and along both margins of the Italian peninsula is apparent. The dot map clearly distinguishes high population areas from areas that are relatively less densely populated.

Table 3.1   Europe: Basic Data

Political Unit	Area (thousand/ sq mi)	Area (thousand/ sq km)	Estimated Population (millions)	Estimated Annual Rate of Increase (%)	Estimated Population Density (sq mi)	Estimated Population Density (sq km)	Infant Mortality Rate	Urban Population (%)	Arable Land (% of total area)	Per Capita GNP ($US)
British Isles										
United Kingdom	93.3	241.6	59.1	0.2	634	245	6.1	90	29	19,600
Ireland	26.6	68.9	3.7	0.5	138	53	5.5	57	14	17,110
Total	**119.9**	**310.5**	**62.8**	**0.2**	**524**	**202**	**5.8**	**88**	**26**	**18,864**
Western Central Europe										
France	212.4	550.1	58.8	0.3	277	107	5.1	74	32	26,270
Germany	137.8	356.9	82.3	−0.1	610	236	4.9	85	34	28,870
Belgium	11.8	30.5	10.2	0.1	866	334	5.8	97	24	26,440
Netherlands	13.1	33.9	15.7	0.3	1197	461	5.7	61	26	25,940
Luxembourg	1	2.6	0.4	0.4	427	165	4.9	86	24	45,360
Switzerland	15.3	39.6	7.1	0.3	466	180	4.7	68	10	44,350
Austria	31.9	82.6	8.1	0.1	253	38	4.8	65	17	28,110
Total	**423.3**	**1096.3**	**182.6**	**0.1**	**431.4**	**166**	**5.1**	**79**	**30**	**28,249**
Northern Europe										
Denmark	16.4	42.5	5.3	0.2	324	125	5.8	85	61	32,100
Norway	118.5	306.9	4.4	0.3	37	14	4	74	3	34,510
Sweden	158.9	411.6	8.9	0	56	22	3.9	83	7	25,710
Finland	117.6	304.6	5.2	0.2	44	17	3.5	65	8	23,240
Iceland	38.7	100.2	0.3	0.9	7	3	5.5	92	1	26,580
Total	**450.1**	**1165.8**	**24.1**	**0.2**	**53.5**	**21**	**4.5**	**78**	**8**	**28,200**
Southern Europe										
Italy	113.5	294	57.5	0	508	196	5.8	67	32	19,880
Spain	192.8	499.4	39.4	0	204	79	4.7	64	31	14,350
Portugal	35.6	92.1	10	0	280	108	6.9	48	32	10,160
Greece	49.8	129	10.5	0	211	81	8.1	59	23	11,460
Malta	0.1	0.3	0.4	0.6	3035	1172	10.7	89	38	7600
Total	**391.8**	**1014.8**	**117.8**	**0**	**300.7**	**116**	**7.2**	**64**	**30**	**16,413**
East Central Europe										
Poland	117.5	304.3	38.7	0.1	329	127	12.2	62	46	3230
Czech Republic	29.8	77.2	10.3	−0.2	345	133	5.9	77	*	4740
Slovakia	18.6	48.2	5.4	0.2	291	112	10.2	57	*	3410
Hungary	35.7	92.5	10.1	−0.4	284	110	10	63	51	4340
Romania	88.9	230.3	22.5	−0.2	253	38	22.6	55	43	1600
Bulgaria	42.7	110.6	8.3	−0.5	194	75	15.6	68	34	1190
Albania	10.6	27.5	3.3	1.2	312	120	20.4	37	21	820
Yugoslavia	39.4	102	10.6	0.2	270	104	14	51	30	3000
Bosnia-Herzegovina	19.7	51	4	0.6	203	78	11.6	40	20	3200
Croatia	21.6	55.9	4.2	0.1	193	75	8	54	32	3800
Macedonia	9.8	25.3	2	0.8	208	80	16.4	60	24	990
Slovenia	7.8	20.3	2	0	255	98	4.7	50	10	9240
Total	**442.1**	**1145.1**	**121.4**	**0.01**	**274.6**	**106**	**12.6**	**59**	**34**	**3012**
Summary Total	**1827.2**	**4732.4**	**508.7**	**0.07**	**278.4**	**106**	**7.1**	**72**	**25**	**18,324**

Sources: *World Population Data Sheet,* Population Reference Bureau, 1998; United Nations Statistics Division, 1998; *Almanac of Politics and Government,* 1998; *World Factbook,* CIA, 1998.

Definitions & Insights

WESTERN CIVILIZATION

Western civilization and related terms, such as **the West** and **Westernization,** are indispensable to a study of world geography because they connote innumerable traits that give geographic distinctiveness to places. In general, Western civilization refers to the sum of values, practices, and achievements that had roots in ancient Greece, Rome, Mesopotamia, and Palestine; that subsequently flowered in Europe; and which are still being developed and modified there and in other areas to which they have been diffused. The term incorporates a set of languages, religious practices (associated with "Western" Christian churches including the Roman Catholic Church and the Protestant churches but not the Orthodox Eastern churches), systems of law, and systems of social, economic, and political organization. Among the core concepts of modern Western civilization are strong commitments (not always effective) to education, experimental science, technological progress, economic development, democratic representative government, and explicit protection of individual rights and liberties. A strong reliance on private capitalism developed prior to and during the Industrial Revolution, although this has been modified by experiments in communism, socialism, and fascism during the 20th century. **Westernization** refers to the process whereby non-Western societies acquire Western traits, adopted with varying degrees of completeness. As used today, "the West" refers primarily to Europe, Anglo America, Australia, and New Zealand, although the concept embraces other countries—such as the Latin American countries, Israel, and South Africa—to a lesser degree.

from exploitation of a worldwide colonial realm funded technological advances and major economic growth. It should not be forgotten, however, that Europeans also exploited each other: The unbelievably primitive working conditions of early English coal mines are only one of many possible illustrations. The marshaling of human energy to extract resources and transform the environment over centuries by the Europeans has created landscapes of considerable significance.

Physical Attributes

Irregular Outline

A noticeable characteristic of Europe is its extremely irregular coastal outline. The main peninsula of Europe is fringed by numerous smaller ones, most notably the Scandinavian, Iberian, Italian, and Balkan peninsulas (see Fig. 3.4). Offshore are numerous islands, including Great Britain, Ireland, Iceland, Sicily, Sardinia, Corsica, and Crete. Around the indented shores of Eu-

rope, arms of the sea penetrate the land in the form of significant **estuaries** (the tidal mouth of a river), and countless harbors offer protection for shipping. This complex mingling of land and water provides many opportunities for maritime activity, and much of Europe's history has focused on maritime trade, sea fisheries, and sea power.

Northerly Location

Another striking environmental characteristic of Europe is its northerly location: Much of Europe lies north of the 48 conterminous states (Fig. 3.3). Despite their moderate climate, the British Isles are at the same latitude as Hudson Bay in Canada. Athens, Greece, is only slightly farther south than St. Louis, Missouri. One effect of a northerly latitude that visitors to such cities as Berlin and Stockholm notice is the long duration of daylight during summer (giving a boost to the flow of tourist dollars into the economy) and the brevity of daylight in the winter.

Temperate Climate

The overall climate of Europe is more temperate than its northerly location would suggest. Winter temperatures, in particular, are mild for the latitude. For example, London has approximately the same average temperature in January as Richmond, Virginia, which is 950 miles (*c.* 1500 km) farther south. These anomalies of temperature are caused by relatively warm currents of water (the **Gulf Stream** and the **North Atlantic Drift**) which originate in tropical western parts of the Atlantic Ocean, drift to the north and east, and make the waters around Europe much warmer in winter than the latitude would warrant (Fig. 3.4). This influence is carried as far north as the Scandinavian Peninsula and into the Barents Sea, allowing Murmansk, Russia, to be an ice-free port even though it lies within the Arctic Circle.

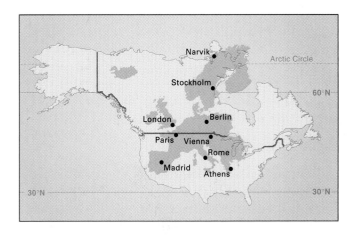

Figure 3.3 Europe in terms of latitude and area compared with the United States and Canada. Most islands have been omitted.

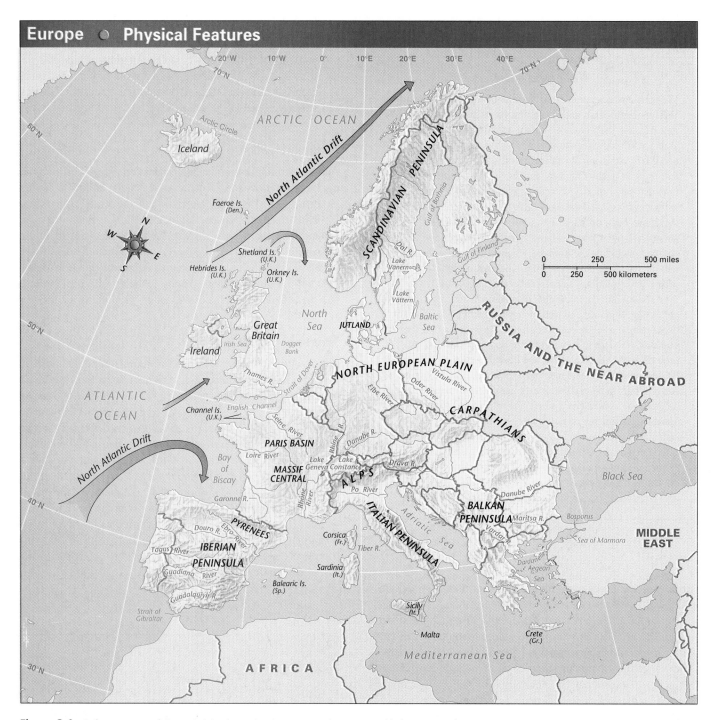

Europe ○ Physical Features

Figure 3.4 Reference map of the main islands, peninsulas, seas, straits, topographic features, and rivers of Europe. A few lakes are also shown.

The same winds that bring warmth in winter and coolness in summer also bring abundant moisture. Most of this falls as rain, although the higher mountains and more northerly areas have considerable snow. Ample, well-distributed, and relatively dependable moisture has always been one of Europe's major assets. Actually, the average precipitation in most places in the European lowlands is only 20 to 40 inches (c. 50 to 100 cm) per year (Fig. 3.5). In some parts of the world nearer the equator, this amount would be distinctly marginal for agriculture. But in most of Europe, the moisture is sufficient for a wide range of crops because mild temperatures and high atmospheric humidity lessen the rate of evapotranspiration (Fig. 3.6).

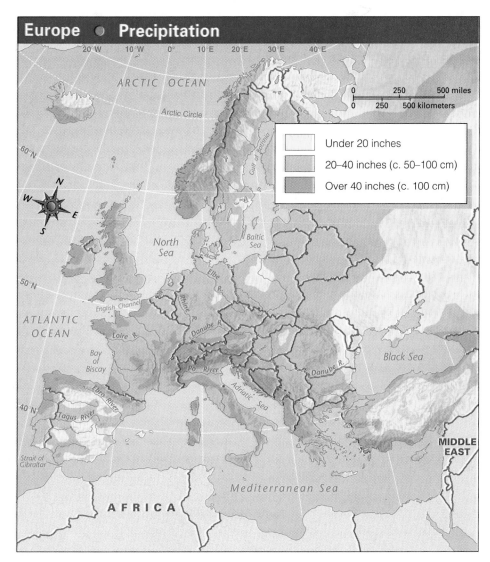

Figure 3.5 Average annual precipitation in Europe. Most parts of Europe receive sufficient total precipitation for crop production, although in areas that have a mediterranean climate, the concentration of rain occurs in the colder half of the year, thereby reducing its usefulness. Highlands along windward coasts often receive heavy orographic precipitation. International boundaries are not shown for areas outside Europe as defined in this text.

Advantages and Limitations of a Varied Topography

Europe's topographic features are highly diversified. Each class of features—plains, plateaus, hill lands, mountains, and water bodies—is well represented (Fig. 3.7). The whole physical assemblage, overspread with a varied plant life and enriched by the human associations and constructions of an eventful history, presents a series of distinctive and often highly scenic landscapes (Fig. 3.8).

One of the most prominent surface features of Europe is a plain that extends without a break from the Pyrenees Mountains at the French-Spanish border, across western and northern France, central and northern Belgium, the Netherlands, Denmark, northern Germany, and Poland, and far into Russia. Known as the **North European Plain** (see Fig. 3.7), it has outliers in Great Britain, the southern part of the Scandinavian peninsula, and southern Finland. For the most part, the plain is undulating or rolling. Flat stretches are generally on alluvial land. The North European Plain contains the greater part of Europe's cultivated land, and it is underlaid in some places by deposits of coal, iron ore, potash (used in the production of soaps, glass, and the manufacture of other potassium compounds), and other minerals that have been important in the region's significant industrial development. Many of the largest European cities—including London, Paris, and Berlin—developed on the plain. From northeast France eastward, a band of

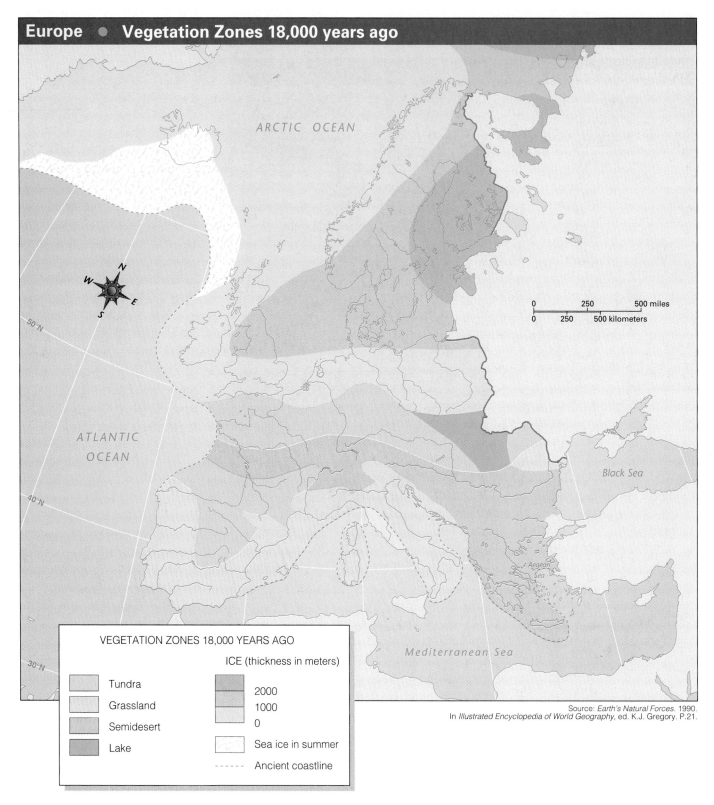

ARCTIC OCEAN

ATLANTIC OCEAN

Black Sea

Aegean Sea

Mediterranean Sea

0 — 250 — 500 miles
0 — 250 — 500 kilometers

VEGETATION ZONES 18,000 YEARS AGO

ICE (thickness in meters)

Tundra
Grassland
Semidesert
Lake

2000
1000
0

Sea ice in summer
Ancient coastline

Source: *Earth's Natural Forces.* 1990.
In *Illustrated Encyclopedia of World Geography*, ed. K.J. Gregory. P.21.

Figure 3.6 Europe was deeply influenced by the southward sweep of major continental glaciation. This map shows how Europe was most profoundly shaped by the presence and processes of the advancing and retreating ice ages. The whole of Scandinavia was under ice, and the soils of the North European Plain and British Isles were enriched by the depositions associated with this movement.

Legend:
- Humid mid-latitude plains
- Humid mid-latitude hill lands (including small areas of low mountains and of plains)
- Mediterranean (dry-summer) subtropical hills, tablelands, and small plains
- Arctic tundra
- Glacially scoured subarctic plains and hills forested in conifers
- Mountains
- Approximate boundaries between marine west coast climate (west of line) and humid continental climate (east of line) between the Alps and the Baltic Sea

Selected cities are shown as reference points.

Figure 3.7 Natural regions. The "natural regions" of the map are attempts to show natural features as composite habitats. Each color symbol represents a distinctive association of landforms and climate, with natural vegetation also stated or implied. The result is a broad-scale view of the *natural settings* for human activity in Europe.

Figure 3.8 The European capacity to blend water, mountains, flowers, and small-scale human settlements has been a hallmark for centuries. This scene from Lake Thun and Spiez in Switzerland represents such a landscape. This pattern has promoted tourism for decades and decades. © 1995 Dembinsky Photo Associates

especially dense population extends along the southern edge of the plain; this concentration apparently has been present since prehistoric times. It coincides with (1) extraordinarily fertile soils formed from deposits of **loess** (a windblown soil of generally high fertility), (2) an important natural transportation route skirting the highlands to the south, and (3) a large share of the mineral deposits of the region.

South of this northern plain, Europe is predominantly mountainous or hilly, although the hills and mountains enclose many plains, valleys, and plateaus. The hill lands and some low mountains are geologically old. Exposed to erosion for a long time, these uplands are often rather smooth and round in outline. But many mountains in southern Europe are geologically young, and they are often high and ruggedly spectacular, with jagged peaks and snowcapped summits. They reach their peak of height and grandeur in the Alps.

Glaciation has played a major role in the shaping of the landscape of Europe as well. During the periods of ice advance in the Great Pleistocene Age (c. 2,000,000 to 10,000 years B.P.), continental ice sheets formed over Europe, beginning on the Scandinavian Peninsula and Scotland. Glacial evidence suggests that there were four major periods of widespread glacial coverage. Today, landscapes of **glacial scouring**—the erosive action of ice masses in motion—characterize most of Norway and Finland, much of Sweden, parts of the British Isles, and Iceland. These changes have often created favorable sites for hydroelectric installations. **Glacial deposition**—the process of offloading rock and soil in glacial retreat or lateral movement—had the predominant effect on the present landscape. Glacial deposits of varying thickness were left behind on most of the North European Plain. Most of the deposits still reflect their glacial origins, although careful farming and use of fertilizers has made them productive. Loess soils, created in part by

glacial action and carried south by wind, play an important role in contemporary European farming.

The North European Plain is bordered by glaciated lowlands and hill lands in Finland and eastern Sweden and by rugged, ice-scoured mountains and fjords in western Sweden and most of Norway. The British Isles also have considerable areas of glacially scoured hill country and low mountains, along with lowlands where glacial deposition occurred.

3.3 Climate Patterns

Climate patterns tend to be fairly uniform over wide areas of the Earth. This allows classification of the Earth's surface into climatic regions, each characterized by a particular type of climate. Certain ranges and combinations of temperature and precipitation conditions define the climate types. Associated with and strongly influenced by each type are certain vegetation and soil conditions. One common classification is shown in this text on the map of world climate types (pp. 30–31). An examination of this map reveals that Europe, despite its modest dimensions, has remarkable climatic diversity.

Marine West Coast Climate

This is the type of climate in which Atlantic influences are dominant. It extends from the coast of Norway to northern Spain and inland to west central Europe. The main characteristics are mild winters, cool summers, and ample rainfall, with many drizzly, cloudy, and foggy days. Throughout the year, changes of weather follow each other in rapid succession as different air masses temporarily dominate or collide with each other along weather fronts. Most precipitation is frontal in origin or results from a combination of frontal and orographic (highland) influences (see Chapter 2). In lowlands, winter snowfall is light, and the ground is seldom covered for more than a few days at a time. Summer days are longer, brighter, and more pleasant than the short, cloudy days of winter, but even in summer there are many chilly and overcast days. The frost-free season of 175 to 250 days is long enough for most crops grown in the middle latitudes to mature, although most areas have summers that are too cool for heat-loving crops such as corn (maize) to ripen.

Humid Continental Climates

Inland from the coast, in western and central Europe, the marine climate gradually changes. Winters become colder and summers hotter; cloudiness and annual precipitation decrease. Influences of maritime air masses from the Atlantic diminish and are modified by continental air masses from inner Asia. At a considerable distance inland, conditions become sufficiently different that two new climate types are apparent: the humid continental short-summer climate in the north (principally in Poland, Slovakia, and the Czech Republic) and the humid continental long-summer climate in the warmer south (principally in Hungary, Romania, Serbia, and northern Bulgaria). The natural vegetation is mostly forest, and soils vary greatly in quality. Among the best soils are those formed from alluvium and loess along the Danube River.

Mediterranean Climate

Southernmost Europe has a distinctive climate—the mediterranean (dry summer–wet winter) subtropical climate. This pattern of precipitation results from a seasonal shifting of atmospheric belts. In winter, the belt of westerly winds shifts southward, bringing precipitation at a time of relatively low evaporation. In summer, the belt of subsiding high atmospheric pressure over the Sahara shifts northward, bringing desert conditions. Mediterranean summers are warm to hot, and little precipitation occurs during the summer months when temperatures are most advantageous for crop growth. Winters are mild; frosts are few. Drought-resistant trees originally covered mediterranean lands, but little of the forest remains. It has been replaced by the wild scrub that the French call *maquis* (called *chaparral* in the United States and Spain), or by cultivated fields, orchards, or vineyards. Much of the land consists of rugged, rocky, and badly eroded slopes where thousands of years of deforestation, overgrazing, and excessive cultivation have taken their toll. The subtropical temperatures make possible a great variety of crops, but irrigation is necessary to counteract the dry summers. However, with irrigation, an enormous variety of crops can be raised productively. In addition, this climate is a boon for summer travelers and the money made from its exploitation for tourism is of immense economic importance in Mediterranean Europe.

Subarctic and Tundra Climates

Some northerly sections of Europe experience the harsh conditions associated with subarctic and tundra climates. The subarctic climate, characterized by long, severe winters and short, rather cool summers, covers most of Finland, the greater part of Sweden, and some of Norway. A short frost-free season, coupled with thin, highly leached, acidic soils, handicaps agriculture. Human settlement is scanty, and a forest of needle-leaf conifers, such as spruce and fir, covers most of the land. In the tundra climate of northernmost Norway and much of Iceland, cold winters combine with brief, cool summers and strong winds to create conditions hostile to tree growth. An open, windswept landscape results, covered with lichens, mosses, grass, low bushes, dwarf trees, and wildflowers. Wildlife is present, but human inhabitants are few. Agriculture in a normal sense is not feasible. In the taiga region, south of the tundra, a slight increase in moisture and temperature allows some forest growth.

Highland and Ice-Cap Climates

The higher mountains of Europe, like high mountains in other parts of the world, have an undifferentiated highland climate varying with elevation and differential exposure to sun, wind, and precipitation. Given enough height, the variety can be startling. The Italian slope of the Alps, for instance, ascends from subtropical conditions at the base of the mountains to tundra and ice-cap climates at the highest elevations. The ice-cap climate experiences temperatures that average below freezing every month in the year, generally enabling ice fields and glaciers to be preserved.

3.4 The Importance of Rivers and Waterways

As one might expect in such a humid area, Europe has numerous river systems that are very important economically for transport and water supply, the generation of electricity, and the creation of regional images. Geographic discussion of these rivers requires an initial understanding of certain terms. A **river system** is a river together with its tributaries, and a **river basin** is the whole area drained by a river system. The **source** of a river is its place of origin; the **mouth** is where it empties into another body of water, often forming an estuary. As they near the sea, many rivers become sluggish, depositing great quantities of sediment to form **deltas** and often dividing into a number of separate channels, known as **distributaries** (see Fig. 3.4). The management of a river requires a high level of cooperation throughout the river system because major change upstream—taking water out for irrigation or putting in industrial pollutants, for example—will have significant impact on downstream populations and their water use.

Rivers were an important part of Europe's transport system at least as far back as Roman times and probably before; today they are still central to the transporting of large quantities of bulk cargo at low cost, mostly in motorized (self-propelled) barges. The more important rivers for transportation are largely those of the highly industrialized areas in northwestern Europe. An extensive system of canals connects and supplements the rivers. During the time of the Roman Empire, canals were built throughout northern Europe and Britain. These were used primarily for military transport, although some systems in southeastern Britain were also used to help field drainage.

The Dutch developed the pound (pond) lock for canals in the late 14th century, and this led to continued expansion of regional connections of waterways through canals, as well as the use of canals to link farmlands with the coasts. In the late 1700s, Britain became especially active in such construction as it was moving raw materials toward factory locations in the early years of the Industrial Revolution. This period of active growth continued until the middle of the 19th century, when the expanding influence of railroads and the steam engine provided real competition for canals. The longest canal constructed in this period of British investment in canals was the 140 mile (230 km) long Leeds-Liverpool Canal that was begun in 1770 and completed in 1816.

Important seaports have developed along the lower courses of many rivers, and some have become major cities. London on the Thames, Antwerp on the Scheldt, Rotterdam in the delta of the Rhine, and Hamburg on the Elbe are outstanding examples. Often the river mouths are wide and deep, allowing ocean ships to travel a considerable distance upstream and inland. This is true even of short rivers such as the Thames and Scheldt.

The Rhine and the Danube are European rivers of particular importance, both touching or crossing the territory of many countries (see Fig. 3.4). In the case of the Rhine, these countries include Switzerland—where the river rises in the Alps—Liechtenstein, Austria, France, Germany, and the Netherlands. Highly scenic for much of its course, the Rhine River is Europe's most important inland waterway. Along or near it, a striking axis of intense urban-industrial development and extreme population density has developed. At its North Sea end, the Rhine axis connects to world commerce by the world's most active seaport, Rotterdam, in the Netherlands.

The Danube (German: *Donau*) touches or crosses more countries than does any other river in the world. From its source in the Black Forest of southwestern Germany, not far from Lake Constance in Switzerland, it flows eastward through Austria, with Vienna on its banks, and serves as the border between Slovakia on the north and south through Hungary, with Budapest on its banks. Further south it serves as the border between Croatia on the west and Yugoslavia on the east. It then turns east, passing by Belgrade, and then becomes the border between Yugoslavia and Romania on the north and, further east, the border between Bulgaria on the south and Romania on the north. Continuing further east, the Danube enters Romania, turns north, and makes one more sharp turn to the east before pouring into the Black Sea. Within Europe, the Danube is also unusual in its southeasterly direction of flow (see Fig. 3.4).

3.5 Agriculture and Fisheries

Agriculture was the original foundation of Europe's economy, and it is still a very important component. Food provided by the region's agriculture allowed Europe to become a relatively well-populated area at an early time. After about 1500, a period of steady agricultural improvement began. Introduction of important new crops such as the potato played a part (see Table 3.2 on crop diffusion), but so did such practical improvements as new systems of crop rotation and scientific advances that produced a better knowledge of the chemistry of fertilizers. The continuing expansion of industrial cities provided growing markets for European farmers, who received some protection through tariffs or direct subsidies to encourage production and support rural incomes. Both of these governmental benefits are now enjoyed by farmers throughout the European Union (EU). Large surpluses of many products have emerged as a result of the EU's **Common Agricultural Policy (CAP).** These include grains, butter, cheese, olive oil, and table wines.

Table 3.2 Diffusion of Foodstuffs Between New World and Old World: The Columbian Exchange	
This table shows some of the most significant foodstuffs diffused from the New World to the Old World after 1492, as well as the major New World foodstuffs taken into world trade after 1492 by the ships and crews that came to the newly found lands of North, Middle, and South America.	
Diffused from Old World to New World	**Diffused from New World to Old World**
wheat	maize
grapes	potatoes
olives	cassava (manioc)
onions	tomatoes
melons	beans (lima, string, navy, kidney, etc.)
lettuce	pumpkins
rice	squash
soybeans	sweet potatoes
coffee	peanuts
bananas	cacao
yams	

As a region, Europe is by no means agriculturally self-sufficient. Imports of supplemental agricultural commodities are necessary. Increasingly, policies dealing with agricultural goods are taken under consideration by the EU.

Throughout history, fishing has been an important part of the European food economy, and the coasts are still thickly dotted with fishing ports. At times, control of fishing grounds has been a major commercial and political objective of nations, even resulting in warfare. Fisheries are particularly important in shallow seas that are rich in the small organisms known as plankton. These organisms are the principal food for schools of herring, cod, and other fish of commercial value. The Dogger Bank (to the east of Great Britain) in the North Sea and the waters off Norway and Iceland are major fishing areas, and Norway and Denmark are Europe's leading nations in total catch. Iceland, which has few other resources and few people, ranks fourth in total catch and depends economically mainly on exports of fish and fish products, which represented 75 percent of all Iceland's exports by value in 1998.

3.6 European Patterns of Language and Religion

Patterns of language and religion in Europe helped establish the foundation for Europe's dominance in a large part of the world's economic, cultural, and political development during the past five centuries.

Language

Language is one of the most highly identifiable elements in a culture group's personality. In learning a language, one also learns a great deal about tradition, history, cultural mores, belief systems, and the spirit of a people. Language also plays a major role in the tensions frequently at work in the geography of borders and ethnicity in Europe (Fig. 3.9).

Europe emerged from prehistory as the homeland of many different peoples. In ancient and medieval times, certain of

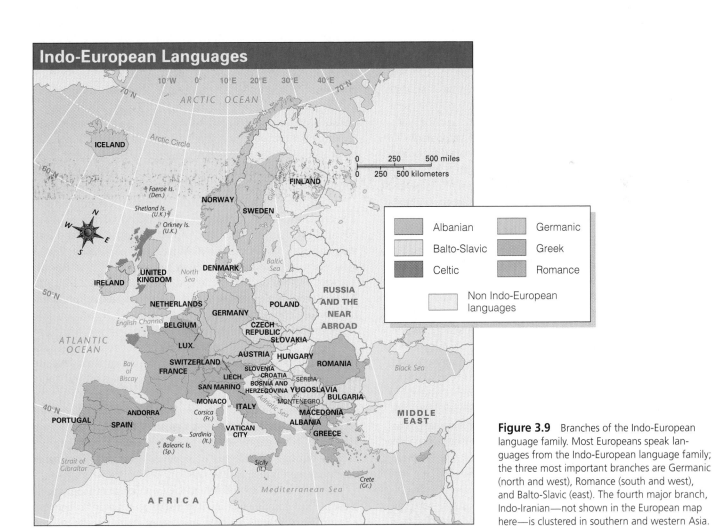

Figure 3.9 Branches of the Indo-European language family. Most Europeans speak languages from the Indo-European language family; the three most important branches are Germanic (north and west), Romance (south and west), and Balto-Slavic (east). The fourth major branch, Indo-Iranian—not shown in the European map here—is clustered in southern and western Asia.

these peoples experienced periods of vigorous expansion, and their languages and cultures became widely diffused. First came the expansion of the Greek and Celtic peoples, and later that of the Romans, the Germanic (Teutonic) peoples, and the Slavic peoples. As each expansion occurred, traditional languages persisted in some areas but were displaced in others. Each of the important languages eventually developed many local dialects. With the rise of centralized nation-states in early modern times, particular dialects became the bases for standard national languages.

The first millennium B.C. witnessed a great expansion of the Greek and Celtic peoples. In peninsulas and islands bordering the Aegean and Ionian Seas, the early Greeks evolved a civilization that reached unsurpassed heights of philosophical inquiry and literary and artistic expression. Greek adventurers, traders, and colonists used the Mediterranean Sea as their highway to spread classical Greek civilization and its language along much of the Mediterranean shoreline, although there was already a strong Phoenician influence along the shores of the eastern Mediterranean before the Greeks arrived. Evidence of the early geographic range and subsequent influence of the Greek language and culture is apparent in the many Greek elements in modern European languages. But over time, the use of Greek in most areas disappeared as new peoples and languages were introduced, expanded, and gained authority.

Europe's Celtic languages expanded at roughly the same time as Greek, and, like Greek, are represented today only by remnants. The expansion of preliterate Celtic-speaking tribes radiated from **hearth areas**—regions of original development—in what is now southern Germany and Austria, eventually occupying much of continental Europe and even the British Isles. Conquest and cultural influence by later arrivals, mainly Romans and Germans, eventually eliminated the Celtic languages except for a few traces that survive today.

In present-day Europe, the overwhelming majority of the people speak **Romance, Germanic,** or **Slavic languages.** The Romance languages evolved from **Latin,** originally the language of ancient Rome and a small district around it. In the few centuries before and just after the birth of Christ, the Romans subdued territories extending through western Europe as far as Great Britain, and the use of Latin spread to this large empire. Latin had by far the greatest impact in the less developed and less populous western parts of the empire in southwest Asia. Over a long period, extending well beyond the collapse of this western part of the empire in the 5th century, regional dialects of Latin survived and evolved into Italian, French, Spanish, Catalan, Portuguese, Romanian, and other Romance languages of today. It's worth noting here that language frontiers do not always coincide exactly with political frontiers. For example, the French language extends to western Switzerland and also to southern Belgium, where it is known as Walloon.

In the middle centuries of the first millennium A.D., the power of Rome declined and a prolonged expansion by Germanic and Slavic peoples began. Germanic peoples first appear in history as a group of tribes inhabiting the coasts of Germany and much of Scandinavia. They subsequently expanded southward into Celtic lands east of the Rhine. Roman attempts at conquest were repelled, and the Latin language had little impact in Germany. In the 5th and 6th centuries A.D., Germanic incursions overran the western Roman Empire, but in many areas the conquerors eventually were absorbed into the culture and language of their Latinized subjects.

However, the German language expanded into, and remains, the language of present-day Germany, Austria, Luxembourg, Liechtenstein, the greater part of Switzerland, the previously Latinized part of Germany west of the Rhine, and parts of easternmost France (Alsace and part of Lorraine). The Germanic languages of Europe include many languages other than German itself. In the Netherlands, Dutch developed as a language closely related to dialects of northern Germany, and Flemish—almost identical to Dutch—became the language of northern Belgium. The present languages of Denmark, Norway, Sweden, and Iceland descended from the same ancient Germanic tongue, though Finnish did not.

English is basically a Germanic language, although it has many words and expressions derived from French, Latin, Greek, and other languages. Originally, English was the language of the Germanic tribes known as Angles and Saxons who invaded England in the 5th and 6th centuries A.D. The Norman conquest of England in the 11th century established French for a time as the language of the English court and the upper classes. Modern English retains the Anglo-Saxon grammatical structure but borrows great numbers of words from French and other languages. English is now the principal language in most parts of the British Isles, having been imposed by conquest or spread by cultural diffusion to areas outside of England.

Slavic languages are dominant in most of eastern Europe. The main ones today include the Russian and Ukrainian of the adjoining Russia and the Near Abroad; Polish; Czech and Slovak; and Serbian, Croatian, and Bulgarian in the Balkan Peninsula. They apparently originated in eastern Europe and Russia, and were spread and differentiated from each other during the Middle Ages as a consequence of migrations and cultural contacts involving various peoples.

A few languages in present-day Europe are not related to any of the groups just discussed. Some are ancient languages that have persisted from prehistoric times in isolated (usually mountainous) locales. Two outstanding examples are Albanian (Indo-European in its origin) in the Balkan Peninsula and Basque, spoken in or near the western Pyrenees Mountains of Spain and France. Some languages unrelated to others in Europe do have relatives in Russia. They reached their present locales through migrations of peoples westward. The prime examples are Finnish and Hungarian, also called Magyar.

While it is easy to understand the importance of language as a medium of communication, it is just as essential to realize that a language represents a whole universe of culture, tradition, and history. The kaleidoscope of languages in Europe has contributed

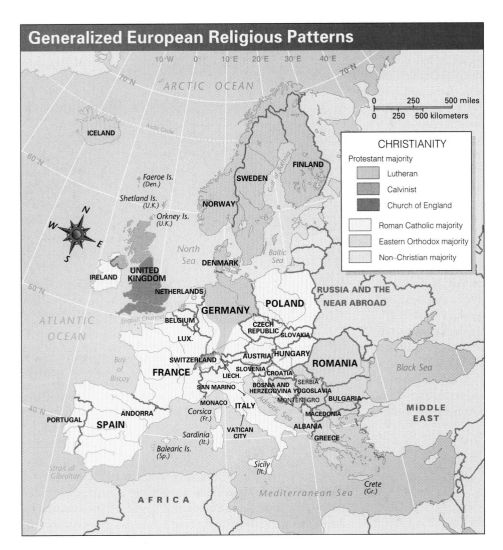

Generalized European Religious Patterns

Figure 3.10 The religious patterns of Europe are largely expressions of distinct sects of Christianity. There are also significant areas of non-Christian populations. They are shown by the symbol of the "Non-Christian majority" category in the map.

to the patterns of cooperation, affiliation, tension, and distrust that characterize this region today. Language plays this same role in all other parts of the world, especially where a great number of cultures and languages are resident in a relatively small area. The current tensions that are so significant in the former Yugoslavia area derive in part from language differences—and their associated religious and other cultural distinctions—within the region.

Religious Belief Systems

Europe's diverse patterns of religious belief systems are powerful indicators of culture, badges of nationality, and repositories of achievement. They—along with language—stand very high among the factors that differentiate Europe geographically (Fig. 3.10). During the long course of European history, both these cultural systems have played an important role in political conflict, including warfare, repression, discrimination, and terrorism.

In the Roman Empire, Christianity arose, spread, survived persecution, and became a dominant institution in the late stages of the empire. It survived the empire's fall and continued to spread, first within Europe and then to other parts of the world. In total number of adherents, the Roman Catholic Church, headed by the Pope in the Vatican, is Europe's largest religious division, as it has been since the Christian Church was first established. Today, it remains the principal faith of a highly secularized Europe. The main areas that are predominantly Roman Catholic include: Italy, Spain, Portugal, France, Belgium, the Republic of Ireland, large parts of the Netherlands, Germany, Switzerland, Austria, Poland, Hungary, the Czech Republic, Slovakia, Croatia, and Slovenia.

During the Middle Ages, a center of Christianity evolved in Constantinople (now Istanbul, Turkey) as a rival to Rome. From that seat of power, the Orthodox Eastern Church spread to become, and remains, dominant in Greece, Bulgaria, Romania, Serbia, and Macedonia, as well as in the Ukraine and Russia to the east.

In the 16th century, the Protestant Reformation took root in various parts of Europe, and a subsequent series of bloody religious wars, persecutions, and counterpersecutions left Protestantism dominant in Great Britain, northern Germany, the Netherlands, Denmark, Norway, Sweden, Iceland, and Finland. Except for the Netherlands, where a higher Catholic birth rate has since reversed the balance, these areas are still mainly Protestant.

The Islamic (Muslim) faith, which is the principal religion of Albania and is an important faith in other parts of the Balkan Peninsula, plays a major influence in the contemporary struggle of European states in south-central Europe to achieve both political and religious stability. Islam was once widespread on the Balkan Peninsula, where it was established by the Ottoman Turks during their period of rule from the 14th to the 19th centuries. It was also the religion of the Moors, a powerful cultural presence on the Iberian Peninsula from the 8th century until the end of the 15th century. Its presence has served as a dominant factor in the ethnic unrest in the former Yugoslavia since 1991, especially in the Kosovo conflict in Yugoslavia. Religious differences continue to surface as a source of discord in Bosnia and Herzegovina (Fig. 3.11).

Under the Roman Empire, Jewish minorities spread from Palestine into Europe, where they have since persisted as a significant minority population in many countries despite recurrent persecution.

Figure 3.11 Although media attention in recent years has focused upon Bosnia in the former Yugoslavia as the setting for Muslim influences in southern and eastern Europe, the Moorish influence is more geographically widespread in Europe. This religious structure near Oberammergau, Germany, demonstrates the breadth of such diffusion. © 1996 Howard Garrett/Dembinsky Photo Associates

3.7 The Impact of Europe on the World in the Colonial Age

There is a global imprint of European influence expressed through language, custom, economic systems, and technology diffusion. Far more than any other region, Europe has shaped the human geography of the modern world. Before the late 15th century, Europe played a minor role in world trade patterns. Western Europe saw goods moving from the southern coast of France and northern Italy northwest to Britain. A more active European trade connected Italy with the margins of the Mediterranean Sea, and beyond to southwest Asia. However, major trade routes had centers much further east and south (Fig. 3.12). The longest trade link was the centuries-old Silk Route that moved goods overland from Peking (Beijing) in China to Venice. The route was more than 5000 miles (9000 km) long.

From the beginning of the Age of Discovery in the 15th century, European seamen, missionaries, traders, soldiers, and colonists burst upon the world scene. By the end of the 19th century, Europeans had created a world in which they and their descendants were culturally dominant. This influence is seen in everything from landscape to philosophical mind-set.

The process of exploration and discovery by which Europeans filled in the world map began with 15th-century Portuguese expeditions down the west coast of Africa. In 1488, a Portuguese expedition headed by Bartholomew Diaz rounded the Cape of Good Hope at the southern tip of Africa and opened the way for subsequent European voyages eastward into the Indian Ocean. Then, in 1492, North America was brought into contact with Europe when a Spanish expedition commanded by a Genoese (Italian), Christopher Columbus (Cristóbal Colón), crossed the Atlantic to the Caribbean Sea. Less than half a century later, these early feats of exploration culminated in the first circumnavigation of the globe by a Spanish expedition commanded initially by a Portuguese, Ferdinand Magellan—who named the Pacific Ocean. Worldwide exploration, both coastal and inland, continued apace for centuries, with leadership being wrested from the early Spanish and Portuguese by the ascending Dutch, French, and English.

The discovery of new peoples and places by the Europeans generally meant that missionaries, soldiers, and traders were seldom far behind the explorers, often traveling with the first ships. Trading posts, Christian missions, and military garrisons were established along many coasts or, in some cases, inland. Inland posts were especially notable in Latin America, where they were often a response to the highland location of Indians who could be Christianized, robbed of precious metals, and put to work in mines or on the land. Many European outposts came to dominate the areas where they were established and eventually became bases from which complete colonial control was often extended.

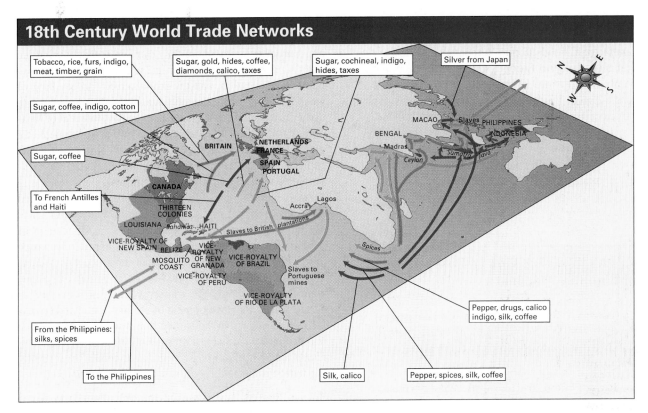

Figure 3.12 In the 18th century the global flow of goods, slaves, and information was carried all around the world by European merchant fleets. From the small countries in Europe came the decisions, the marine power, the mercantile systems, and the military force to promote the global movement of sugar, spices, and human cargo. The details provided in the items along the sides of the map show how many goods flowed to and from how many places. *Source:* World Economy. *1993. In* The Illustrated Encyclopedia of World Geography, *ed. Stuart Corbridge. New York: Oxford University Press. PP. 14–15.*

This process of discovery inevitably set in motion a whole ensemble of changes, for both the people discovered and those doing the discovering. As researchers look more at the lives of common people rather than only at the social patterns and **material culture**—the artifacts of a group—of the wealthy and politically powerful, it is increasingly clear that being "discovered" extracts a high cultural cost. Every exploring, discovering, exploiting, and conquering people has had its own explanations about why such geographic acts were meant to bring benefits to those being found. However, the recipients of such discovery have long complained about the ruinous impact such "discovery" had on their lives.

Subsequent settlement by European colonists, on differing scales and at different times, took place in a great many non-European areas. In well-populated or environmentally difficult areas, European populations remained small minorities, but in other circumstances local versions of European societies took such firm root that they became numerically dominant. Today the descendants of transplanted Europeans make up most of the population in the United States, Canada, and parts of Latin America, Australia, New Zealand, and they are the financially dominant group in South Africa.

In carrying on their various overseas enterprises, Europeans not only migrated themselves but often also transferred non-Europeans from place to place. The most monumental instance of this was the massive forced migration of Africans to the New World that began in the 16th century. Such transfers greatly influenced the ultimate racial and cultural makeup of sizable areas.

Also of great importance in shaping the world's geography was the transfer of plants and animals from one place to another. A few major examples are the introduction of hogs and cattle and the reintroduction of horses to the Americas from Europe; of tobacco and corn (maize) from the Americas to Europe and other parts of the world; of rubber from South America to Asia; of the potato from the Americas to Europe, where it became a major food; and of coffee from Africa to Latin America, where it became the principal basis of a number of national economies.

A worldwide system of trade, with Europe at its core, was a major outcome of European expansion. The system principally involved the movement of raw materials and food products from the rest of the world to Europe and the return sale of European manufactured goods to non-European areas. In general, this exchange favored Europe, and wealth from the exchange mainly accumulated there, particularly in Seville, Spain, during the Age of Discovery. The consequent infusions of money enabled certain European cities, particularly London, to grow and become centers of world finance. Today other centers of commerce and finance—notably in the United States and in Japan and Hong Kong in East Asia—have risen to challenge Europe's dominance.

3.8 Europe's Rise to Global Centrality— and Potential Decline

What factors account for the rise of this Eurasian peninsula as the center of world trade and industrialization? Many interpretations are possible, with various scholars emphasizing such factors as the following:

1. *Capitalism.* Parts of Europe had already developed capitalist institutions by the end of the Middle Ages, and the colonial age saw further rapid development of capitalism in the region. Profit became an acceptable motive, and the relative freedom of action afforded by the capitalistic system provided opportunities for gaining wealth by taking risks. Energetic entrepreneurs found it possible to mobilize capital into large companies and to exploit both European and overseas labor (Fig. 3.13).

2. *Technology.* By the end of the Middle Ages, Europeans had reached a level of technology generally superior to that of non-Europeans with whom they came in contact during the early Age of Discovery. In particular, achievements in shipbuilding, navigation, and the manufacture and handling of weapons gave them decided advantages. Nations that had earlier made significant progress in the development of technology—such as China—were disinclined to compete at a global level during the Age of Discovery. This gave Europe a virtual free rein in energetic expansion and control of technological innovation for several centuries.

3. *Science.* Europe's scientific prowess must not be overlooked as an explanation for the region's rise to world dominance. Technology is partly applied science, and the foundations of modern science were constructed almost entirely in Europe during the centuries when European influence was becoming paramount. Until recently, the great names of science were overwhelmingly the names of Europeans. As late as World War II, the development of atomic fission and associated space science in the United States was carried out by a team composed largely of prominent refugee European scientists, and Nobel Laureates continue to reflect the scientific capacities of the peoples of, and from, this region.

In the 20th century, however, the preeminence in world trade and industry of leading European countries diminished. What caused the relative decline in its fortunes? Some important reasons and circumstances include:

1. *War dislocation.* Europe suffered enormous casualties and damage in World Wars I and II, which were initiated in Europe and fought mainly there. Recovery was eventually achieved with U.S. aid, but the region's altered position

Figure 3.13 Paris, France, has been one of the most innovative metropolitan centers in terms of urban showcase architecture. These public housing towers surround one side of the impressive new Grand Arch La Defence, a new structure that puts Paris in a competition with Berlin for having some of the most striking urban structures of the late 20th century. *BKA/Network Aspen*

could not be reversed. The wars destroyed Europe's ability to maintain its predominance in the face of such trends as anticolonialism and thereby accelerated the capacity of the United States and Russia and the Near Abroad to rise to world power.

2. *New nationalism.* Rising **nationalism**—the quest of a nationality to possess its own homeland—in the colonial world during the 20th century resulted in the virtual end of the European colonial empires within a surprisingly short time after World War II. Taking advantage of a weakened Europe and a mounting disapproval of colonialism in the world at large, one European colony after another gained independence quickly in the decades following the war. Opposition to continued European control was often spearheaded by colonial leaders who had been educated in Europe and had absorbed nationalistic ideas there; nationalism is a sentiment that has had its most pervasive expression in Europe. This same drive received an additional boost with the collapse of the Soviet Union in 1991. Much of the bloodshed and warfare in the former Yugoslavia and in the Balkan Peninsula is fueled by a rising sense of Serb and non-Serb nationalism.

3. *Ascendence of the United States and Russia.* Europe's predominance was seriously eroded by the rising economic and political stature of the United States and, to some degree, Russia and its neighboring states. These enormous countries, each far larger than any European country, outstripped Europe in military power, economic resources, and world influence, particularly in the rebuilding years of 1950–1970. With the collapse of Communism and the USSR, Russian influence diminished considerably.

4. *Shift in world manufacturing patterns.* Europe once enjoyed nearly a monopoly in exports of manufactured goods, but in the past three decades, manufacturing has developed in many countries outside of Europe. Japan, China, South Korea, Hong Kong (now part of China), and Taiwan (a contested province of China) are prime examples just in East Asia alone. In the decades following World War II—although Japan began its transformation in the 1870s—these Asian countries emerged from more traditional rural societies to become industrial forces capable of competing with Europe in markets all over the world. The United States began its industrialization earlier than Japan, but it too reached industrial maturity in the 20th century and also became a vigorous competitor of Europe.

5. *Energy factors.* Europe's ability to assert itself in world affairs has been weakened by the region's new dependence on outside sources of energy. The region's traditionally significant coal fields (Fig. 3.14) have become increasingly costly to exploit, and despite the recent development of North Sea oil and gas resources (which benefit primarily Great Britain and Norway), Europe is now very dependent on Middle East oil and other imported sources of energy.

Figure 3.14 The mining of coal has been an important European occupation for centuries. Seen here are miners coming off shift at a British colliery. The children of these miners, however, will not find their fathers' jobs as easy to secure as in the past. Industrial changes and the ever-diminishing role played by coal in Europe—especially western Europe—make this a vanishing occupation. *Jake Sutton/Gamma Liaison*

The speed of technology transfer and the complex global interdependence of capital flow makes it difficult to pinpoint current, much less future, centers of manufacturing. However, current patterns of industrial activity in Europe show that the coalescence of resources, skilled labor, cultures that promote entrepreneurship, and spatial networks of transport and communication are all still in place in this region. It is certain that Europe will continue to play a major role in manufacturing and global trade for a long time to come, even though the very urban centers that began these patterns are losing relative vitality as new manufacturing areas emerge both within and beyond Europe.

3.9 Europe's Highly Developed and Diversified Economy

Diversity is a keynote of the European economy, just as it is of the European environment. The economic vitality that the region has been characterized by for the past two centuries plays a major role in the current global centrality of Europe. Numerous forms of economic activity are highly developed, and much variety exists from one country and region to another. Most European countries rank high in productivity and affluence compared with other countries of the world. Western Europe was the first world region to evolve from an agricultural society into an industrial economy. The key series of events, commonly called the **Industrial Revolution** (see further on), were followed by vast advances in productive (and destructive) knowledge, technology, and environmental transformation. Even military devastation did not stop the economic dynamism of Europe, which capitalized upon U.S. aid after World War II—primarily the Marshall Plan that began in 1947—to rapidly reorganize, rebuild, and attain new heights of production, affluence, and global influence.

One of the most significant facts of global geography is the high proportion of the world's manufacturing capacity and output that is European. Today, this proportion is decreasing, but approximately one quarter to one third of the world output of most major industrial products originates in Europe. Industrialization goes far toward explaining the highly urban character of most European countries as well as their high standing in overall productivity and average income.

Europe was the place where a large-scale manufacturing industry, using machines driven by inanimate power, first arose. Notable for its surge of inventiveness, the Industrial Revolution began to be felt in the first half of the 18th century. Industrial innovations, many of which British inventors developed, made water power—and then coal-fueled steam power—increasingly available to turn machines and drive gears in the new factories. The invention of a practical steam engine by James Watt in 1769 in Great Britain suddenly made coal a major resource and greatly increased the amount of power available. New processes and equipment—including the invention of **coke** (coal with most of its volatile constituents burned off)—made iron smelting relatively cheaper and, hence, more abundant. The invention of new industrial machinery made of iron and driven by steam engines multiplied the output of manufactured products, principally textiles at first. Then, in the 19th century, British inventors developed processes that allowed steel, a metal superior to iron in strength and versatility, to be made on a large scale and cheaply for the first time. Such developments led the world into the modern age of massive, mechanized industrial production. At the same time, this shift to new industrial activity also stimulated significant human migration from the countryside into the cities. Europe was the primary hearth for all of these changes.

Europe has one of the highest regional averages of urban population in the world (see Fig. 3.2). In the United Kingdom and Belgium, more than 90 percent of the population is urban, with Belgium reaching 97 percent urban in 1998. More than 85 percent of the population of Germany, Denmark, and Luxembourg resides in cities, while Sweden is 83 percent urban. These high levels of urbanization complement the industrial nature of these nations' economies, but they also have an influence on levels of population growth. These nations generally have the lowest levels of crude birth rate (the total number of live births per 1000 population per year) and often the highest per capita **GNP (Gross National Product)** levels.

However, as later chapters will show, the high rates of industrialization and urbanization are far from uniform in Europe. From the western and northwestern subregions toward eastern

and central Europe, there is a gradient toward lower levels of industrial activity, higher ratios of nonurban population, and lower per capita GNP levels.

Moving Toward a Postindustrial Society

Scholars have traced a slowly rising trend of European technical inventiveness since the Middle Ages, involving such things as clocks, gears, waterwheels, windmills, the invention of printing in 15th-century Germany, improvements in shipbuilding, drainage of mines by pumping to permit mining at deeper levels, improvements in artillery and other armaments, and many others. The practical improvements leading to greater production were furthered by mathematical and scientific discoveries and by the development of a scientific viewpoint toward nature and the further possibilities for its ever more productive use. At the same time, still prior to the 18th century, capitalist values, practices, and institutions were evolving and becoming prominent—for example, private ownership of property; favorable attitudes toward profit making, including interest from loans; commercial banking, insurance, and credit institutions; double-entry bookkeeping; and corporate forms of business organization. Increased law and order imposed by newly centralized states—especially in England, the Netherlands, France, and Spain—protected the rising commercial and industrial interests which profited steadily from overseas trade.

The developments spawned by the Industrial Revolution continue to the present time. Today, industrial technology has advanced from the crude steam engines and textile machines of 18th-century Britain to modern electronics, computer-controlled factory robots, and the human exploration of space. As it has advanced, it has diffused geographically. Early coal-based and steam-powered industry first spread from Britain to Belgium, and then more widely in Europe and the eastern United States. It began to expand in influence to some non-European areas like Japan in the 1870s. Today's more modern technology is continuing its expansion to Monsoon Asia, Latin America, and parts of Africa with a rising tempo.

Much evidence exists that the older and more advanced industrial societies are in the process of becoming "postindustrial"; that is, they will eventually be societies in which industrial workers are nearly as uncommon as farmers are now in most "industrial" societies. The introduction and improvement of **computer-controlled robots** (machines programmed to do specific jobs without further direction) on industrial production lines is moving this possibility along. The recent occurrence of serious **technological unemployment** (the supplanting of workers by technology) in Western Europe is, in part, related to this trend. Hence, the more advanced and prosperous European countries are beset by such concerns as (1) where new jobs will come from (most likely from the "service" sectors of the economy), (2) how to make the transition from manufacturing to service roles without too much social dislocation, and (3) how to allocate income between a few highly productive industrial workers who tend robots and a predominant mass of service employees whose output per worker will often be less than a robot's, and whose basic employment is increasingly at risk.

While this process is moving across northwestern Europe at the present, it is a major shift in employment and manufacturing that characterizes the entire developed world. It is significant not only as an economic transformation, but also as a societal change.

The Role of Energy in Europe's Development

Energy resources are central to any program of industrialization. When Europe began to industrialize, the energy resource situation was quite favorable. There were fewer people, the scale of production was much smaller, and raw materials were generally sufficient for the simpler requirements of the time. A relatively primitive technology was served by extensive forests, numerous small-scale sources of water power, and a large variety of minerals in small deposits found in widespread locations. When coal became the main source of power and heat in the 18th century, Europe—especially England—was favored by its many coal fields. These fields proved to be the main localizing factor for much of Europe's industrial and urban development: During the early period of industrialization, electricity was still in the future and coal was expensive to transport; consequently, the new factories were generally built on or near the coal fields. The working population, in turn, clustered near the factories and coal mines (see fig. 3.14) and developed a group of industrial districts that incorporated mines, factories, and urban areas.

These older districts are still very prominent on the map of Europe today. Notable examples include most major cities in Great Britain, although not London; the east-west line of cities strung across Belgium south of Brussels; the huge urban agglomeration in western Germany called the Ruhr; and the Polish industrial cities near the Czech border. Economic depression and urban blight are increasingly common in these places, which had the advantage of an early start but now suffer from steadily changing industrial circumstances, including increasing competition from non-European industries; competition from newer products made elsewhere (for example, new synthetic fibers competing with the cottons and woolens of old textile centers); and decreasing requirements for labor due to the increasing automation of factories.

Industrial Resources and Landscapes Today: The Impact of New Patterns

A few countries—notably Sweden, Finland, Norway, and Austria—currently have surpluses of wood. Hydroelectricity has been developed in the Alps, the Scandinavian Peninsula, and along the Rhone and Danube rivers. Most deposits of metallic ores are too depleted or too small to be important today. But the most far-reaching change is the altered situation of coal.

Figure 3.15 The discovery of major oil resources in the North Sea in the last three decades has both changed the appearance of the seascape and given Norway, for one, a major economic independence. This photo shows how these oil production platforms have become a much more commonplace feature of the North Sea. *Mark A. Leman/Tony Stone Images*

Worked long and intensively, most European coal fields are now expensive to mine and yield less coal. Many were phased out as Europe shifted to cheaper imported oil for power after World War II, and as it became apparent that coal had a major negative impact on the quality of city air. Indeed, new environmental demands have further increased the cost of coal as a fuel source because of its implication in acid precipitation as well as in the generation of smog. North Sea oil production started only in the 1970s; prior to that much of the oil used in Europe after World War II was imported from the Middle East or Russia. In a broader sense, Europe, which was once relatively self-sufficient in energy resources, must now buy and import energy and raw materials from a global base, and most European countries now enter world trade actively to acquire energy resources.

This shift away from relative energy self-sufficiency has had implications for economic well-being, factory employment levels, and the vitality of the economic landscape in Europe. The transition is an ongoing example of the globalization of manufacturing, trade, and political interdependency. What Europe has achieved in developing a much wider network of raw material imports and the export of high value-added manufactured products is a pattern that is occurring in all world regions. Because of its colonial legacy and earlier worldwide trade dominance, Europe has more experience in such global trade patterns than most of the world's manufacturing centers.

A number of European countries do have mineral resources of considerable importance. As already mentioned, in the 1970s, the United Kingdom became a large producer and net exporter of oil and also a large producer of natural gas when its offshore fields in the North Sea were brought into production (Fig. 3.15). Norway's oil and gas fields in the North Sea likewise yield a large production. For several decades, the Netherlands has been a major producer and exporter of natural gas.

83

Large iron ore fields are worked in eastern France and northern Sweden. Coal from southern Poland is a mainstay of that country's economy, and Germany is still a sizable coal producer.

3.10 Problems of Europe's Environmental Pollution

A much-publicized problem in some European areas is **environmental pollution.** For example, pollutants pour into the Mediterranean Sea from urban-industrial districts in Europe, Africa, and Southwest Asia. Wastes discharged into this most historical of seas damage both the coastal tourist industry and fisheries. The Mediterranean, almost totally enclosed by land, cannot cleanse itself by flushing wastes into the ocean. By contrast, the North Sea is far more able to interchange water with the ocean, thanks to strong currents that move in and out through the broad opening to the Atlantic between Scotland and Norway. But even the North Sea has become alarmingly contaminated with pollutants discharged by great industrial metropolises, the undersea oil industry, and a heavy volume of shipping.

Still another famous water feature, the Rhine River, is under stress from untreated wastes. Stringent legislation requiring municipalities and industries to maintain waste-treatment plants has lessened the pollution, but the Rhine, from Basel, Switzerland, to the North Sea, is still so dangerous to human health that swimming is generally forbidden. Aquatic life in the Rhine continues to be damaged, not only by toxic wastes but also by excessive warmth when water withdrawn for industrial use is returned to the river in a heated state. Nuclear stations, for example, use large volumes of Rhine water to cool radioactive cores and produce steam to generate electricity. The return of this water heats the river. Similarly, steel mills and coke works withdraw water to cool red-hot metal or to quench flaming coal in the coking process, and this water is still warm when it comes back to the river.

Atmospheric pollution is another widespread environmental affliction in Europe, as it is in North America and various other parts of the world where there are massive metropolises and industrial concentrations. One form of this pollution is known as **acid rain** (sometimes called **acid precipitation**). This is precipitation that has interacted with airborne industrial pollutants, causing the moisture to become highly acidic. Especially publicized have been the ravages of pollutants that have killed great numbers of trees in some of Europe's finest forests. Emissions containing sulfur, when combined with precipitation, produce weak sulfuric acid, and those containing nitrogen yield weak nitric acid. Both can damage trees, soils, and aquatic life in lakes that receive acid precipitation and its runoff. There has been much alarm about lakes that are "dying." Germany and Scandinavia are among the European areas particularly affected.

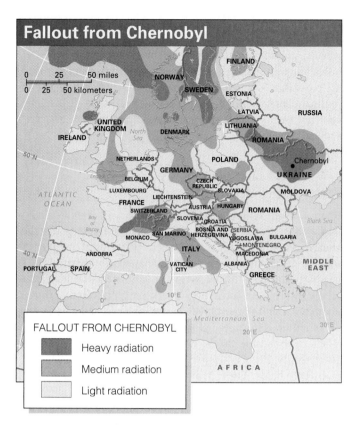

Figure 3.16 Extent of radiation dispersal seven days after the April 26, 1986, partial meltdown and fire at Chernobyl. The release of radioactive material lasted 10 days in all, and its distribution varied with changing wind patterns. Initially, southeasterly winds blew the radiation over Poland, Ukraine, Belarus, Latvia, Lithuania, Finland, and across Norway and Sweden. The winds then changed, bringing fallout to most of Europe. The extent of the fallout was determined by whether or not rain washed the contamination from the atmosphere. Restrictions on livestock movement and slaughter were introduced in the United Kingdom and the Nordic countries.

Another form of atmospheric pollution caused worldwide alarm in 1986 when significant **radioactive fallout** from a nuclear disaster—a partial meltdown and resulting fire—at the **Chernobyl** nuclear power plant in the Ukraine (western part of Russia and the Near Abroad) was deposited on parts of Europe as well as on a large area of Russia and the Near Abroad itself (Fig. 3.16). Sections of Scandinavia and Finland were particularly affected; in fact, some northerly areas used by Lapp reindeer herders became so radioactive that the wild forage could no longer be used and the animals that had grazed on it had to be destroyed.

Many other examples of environmental damage can be drawn from Europe. For instance, old mining and industrial regions contain numerous waste heaps, sometimes centuries old, that are full of poisonous substances. Seepage of these poisons into groundwater may cause cancers and a variety of health problems. Today there is a very active environmental consciousness in Europe, as in many other parts of the world, but many problems deriving from environmental neglect are so deep-seated that solutions will be slow and very costly. This is particularly true of former Communist bloc areas in eastern Europe, where high rates of unemployment make it particularly difficult to close factories for environmental reasons.

3.11 Europe's Population Dynamics

A major reason for Europe's importance in the global economy lies in the region's large and technically skilled population. Europeans represent one of the great masses of world population (see Table 3.1). Europe's approximately 509 million people in 1998 was nearly twice that of the United States. One out of every 11 people in the world live in a space half the size of the United States. Birth rates are low in this region of relative affluence and high urbanization, and the whole region is increasing in numbers much more slowly than most of the world.

Major Population Belts and Their Significance

Except for some northern, rugged, or infertile areas, European population density is everywhere greater than the world average (120/sq mi; 46/sq km). However, the greatest densities appear along two belts of industrialization and urbanization near coal or hydroelectric power sources. One belt extends north-south from Great Britain to Italy (see Fig. 3.2). In addition to large parts of Great Britain, it includes extreme northeastern France, most of Belgium and the Netherlands, Germany's Rhineland, northern Switzerland, and—across the Alps—northern Italy. The second belt of dense population extends west-east from Great Britain to southern Poland and continues into Ukraine. It corresponds to the first belt as far east as the Rhineland, where it forms a relatively narrow strip eastward across Germany, western Czechoslovakia, and Poland. Although the two belts represent a minor part of Europe's total extent, they contain more large cities and a greater value of industrial output than the rest of Europe combined. Only in eastern North America and Japan are there urban-industrial belts of the same order of importance and complexity.

The European belts coincide with major routeways that were in use very early on for migration, trade, and military movement. Many cities along these routes are very old. With the rise of modern industry, coal fields along the west-east belts became important, and in the southern part of the north-south belt, the age of electricity saw the development of many industries based on the hydroelectric resources of the Alps. Both belts benefit from the relatively good soil they encompass.

An issue of growing concern in Europe is the tension caused by high rates of domestic unemployment in many EU nations and the steady increase in rates of legal and illegal immigration. From 1990–1995, Germany alone absorbed 2,400,000 foreigners, most of them from Central and Eastern Europe, during the bloodshed of the Bosnian chaos in the breakup of the former Yugoslavia. By 1996, 8.8 percent of Germany's population were foreigners; in France, 6.3 percent; Italy, 6 percent; and the United Kingdom, 3.4 percent. Switzerland was 19.5 percent foreign, and the United States was 7.9 percent foreign born. The EU has eliminated border controls as one aspect of the new openness of flow among its member nations. This means, in essence, that Italy—with its 5000 miles (9000 km) of coastline plays the gatekeeper role for much of the annual migrant flow into western Europe. The EU's open borders, however, are intended for legal and official citizens of the EU nations. As it is, a significant migrant flow into Europe comes either in the form of migrants seeking asylum, or as illegal immigrants hoping to keep a low profile long enough to be able to take advantage of periodic amnesties granted by most EU nations. Such an amnesty leads generally to legal citizenship.

During periods of more rapid economic growth a decade ago, the immigrant pool (which often meant needed labor resources) was more quickly welcomed. In fact, major flows of **guest workers**—workers encouraged to come to industrial and agricultural centers that were not able to meet labor needs with domestic populations—were a part of steady immigration to Germany and France and other nations as well (Fig. 3.17). But current high levels of unemployment (Italy, 13 percent; France 12 percent; Germany, 12 percent; and the EU average, 11 percent) have changed the social climate toward immigrants—both legal and illegal—considerably.

Another factor that has influenced this demographic equation has been the governmental strain resulting from the increasingly large welfare benefits being paid out by European governments. As the European population gets older, lives longer, and expects a continuation of the generous welfare programs that have evolved since the end of World War II, governments are faced with a significant budget requirement. Such a mind-set makes the vision of an immigrant population standing in need of benefits a particularly unsettling one, even for this region that has such a long history of active human migration.

3.12 The Drive for New Patterns of Regional Cooperation

Since World War II, the countries of Western Europe have moved toward greater economic, military, and political cooperation. Searching for development and security, the countries have banded together in a series of organizations designed to foster unity. In eastern Europe there were efforts to create

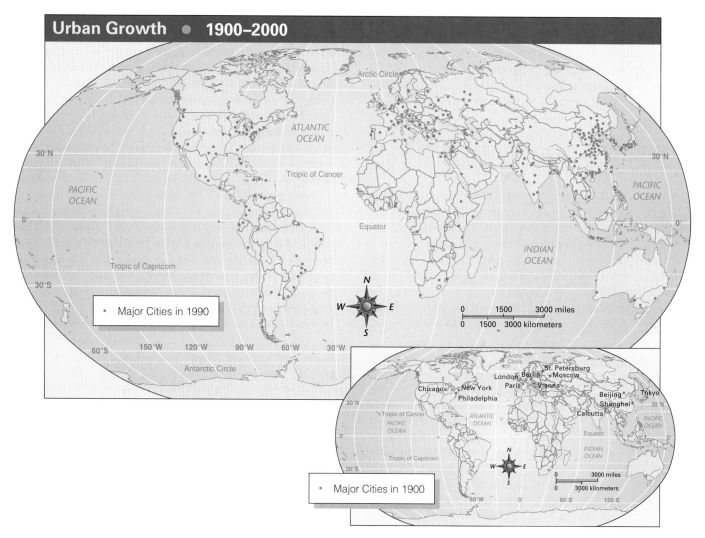

Urban Growth ● **1900–2000**

Major Cities in 1990

Major Cities in 1900

Figure 3.17   In the past century, the growth of cities all across the globe has been a true urban explosion. A century ago, there were 13 cities that were larger than 1,000,000 inhabitants. A study of this map shows the increase in the number of large cities. Europe has gained much population growth overall, and in city growth the increase in "million cities" has touched all parts of the region. The steady inflow of guest workers during the decades since the conclusion of World War II has been a significant factor in such population shift and growth. *Source:* Human Settlement. *1992. In* The Illustrated Encyclopedia of World Geography, *ed. John R. Short. New York: Oxford University Press. P. 26.*

economic and security pacts under the former Soviet leadership, but the collapse of Communism a decade ago in that region aborted such cooperative arrangements.

What is responsible for the new age of cooperation in a region too well known for national quarrels and strife? Important among the many motivations are: a long-standing ideal of European unity dating from the Roman Empire and the medieval Catholic Church; the experience of two World Wars (1914–1918; 1939–1945); perceived military and political threats from the then-USSR after World War II; and a search for economic betterment through the concerted action of nations working in an elusive unity. After World War II, economic recovery was slow, and there were fears in western Europe of military aggression by Russia and the Near Abroad and/or a political takeover by powerful Communist parties in Italy and France. In the meantime, Russia and the Near Abroad and the Communist governments of the satellite countries east of the Iron Curtain (the term Winston Churchill devised for the western boundary of the Soviet bloc in a 1956 speech in Fulton, Missouri) saw U.S. military power in Europe as a major threat to be countered by economic and military cooperation under Soviet leadership. Thus the desire for economic recovery and military security in both the western and eastern nations set the stage for a remarkable period of supranational organization.

The alliances developed in this era of regional cooperation have played a major role in both the economic and political development of the region since the end of World War II. The collapse of the Soviet Union and the Communist satellite nations in the early 1990s has changed the political fabric of the region, and opened new possibilities—and new demands—for broader regional cooperation.

In 1948, a major step was taken in this direction when the **Organization for European Economic Cooperation (OEEC)** was formed by many European states to coordinate the use of U.S. aid proffered under the 1947 **Marshall Plan** to bolster Europe against Communism. The Marshall Plan was terminated in 1952, but the OEEC continued to function. In 1960 the name was changed to the **Organization for Economic Cooperation and Development (OECD).** The membership now includes the United States and several other countries outside Europe.

But the most significant economic organization among Europe's western nations was the European Economic Community (EEC), also called the Common Market, formed in 1957 and now officially called the European Union or EU (since November 1, 1993). It was initially established by six countries: France, the former West Germany, Italy, Belgium, Luxembourg, and the Netherlands. In 1967, this group was reorganized into the European Community. These countries had already eliminated barriers to trade in coal and steel and now looked toward removal of all economic barriers. Later, Great Britain, Denmark, Ireland, Greece, Portugal, and Spain became members. Cyprus, Malta, and Turkey are currently associate members. As of 1998, the fifteen members of the European Union included the United Kingdom, Ireland, Portugal, Spain, France, Luxembourg, Germany, Italy, Greece, Austria, Belgium, the Netherlands, Denmark, Sweden, and Finland. These fifteen countries have a total population of more than 375 million people and produce two-fifths of the world's global exports. Nations that are requesting admission to the EU include Bulgaria, Cyprus, the Czech Republic, Estonia, Hungary, Latvia, Lithuania, Poland, Romania, Slovakia, Slovenia, and Turkey.

The EEC was initially designed to secure the benefits of large-scale production by pooling the resources—natural, human, and financial—and markets of its members. Hence, tariffs were eliminated on goods moving from one member state to another, and restrictions on the movement of labor and capital between member states were greatly reduced. Monopolistic trusts and cartels that formerly restricted competition were discouraged. Meanwhile, a common set of external tariffs was established for the EEC's entire area to regulate imports from the outside world, and a common system of price supports for agriculture replaced the individual systems of member states. The founders of the EEC anticipated that free trade within such a populous and highly developed bloc of countries would (1) stimulate investment in mass-production enterprises, which could, wherever they were located, sell freely into all EEC countries, and (2) encourage a productive geographical specialization, with each part of the Community expanding lines of production for which it was best suited. Thus, each country might achieve greater production, larger exports, lower costs to

consumers, higher wages, and a higher level of living than it could achieve on its own.

It is possible that this success would have been attained without the formal organization of an international economic community. For example, both Switzerland, which is a close neighbor of the European Union but not a member, and Norway, which voted in 1995 not to join the EU, have among the highest GNP and per capita incomes in the world. Sweden, which joined the EU in 1995, also ranks near the top of the world scale in affluence. However, Switzerland, Norway, and Sweden have all depended heavily on trade and other dealings with the Common Market, and thus their prosperity has been closely tied to that of the Market. Six nations that were not initially enrolled in the EEC—Iceland, Norway, Sweden, Switzerland, Finland, and Austria—belonged for many years to a separate trading organization called the **European Free Trade Association (EFTA).** It has worked closely with the Common Market and both are increasingly integrated within a single market area, particularly for nonagricultural products. The Common Market made preferential trade and economic aid treaties with many other countries—mainly ex-colonies of Common Market members—in areas outside of Europe. By late 1996, however, three of the EFTA nations (Sweden, Finland, and Austria) had joined the EU.

The European Union, under provisions of the **Maastricht Treaty of European Union** that went into force on November 1, 1993, is now trying to achieve major new steps in unification (Fig. 3.18). These steps include the removal of nontariff trade barriers, such as "quality standards," that still exist between member countries. There has been major energy focused upon the implementation of a single EU currency—the **euro**—in 1999. Progress has been slow because of questions concerning national sovereignty and occasional protest by populations through the EU who do not want to lose the autonomy of having their own unique currency. On January 1, 1999, the euro became the currency of record, but it will not be until January 1, 2001, that euro notes and coins will be introduced. On July 1, 2002, national coins and currency will cease to serve as legal tender. Part of the significance of this monetary innovation are the fiscal criteria that were determined upon in the December, 1991, meetings in Maastricht, Netherlands. At the 1999 implementation of the euro, the United Kingdom, Sweden, and Denmark elected not to participate, and Greece was required to wait until 2001 to join because it had not yet met the budgetary criteria established for membership in the European Monetary Union. This means that eleven of the EU member nations began electronic use of the euro on January 1, 1999. The euro and this move toward an ever more closely linked European Union community of nations represents a major shift in the body politic of the dominant economic and cultural forces of Europe.

In the sphere of military defense, the key international body in non-Communist Europe became the **North Atlantic Treaty Organization (NATO).** This military alliance was formed in 1949 and now includes the United States, Canada, a majority of

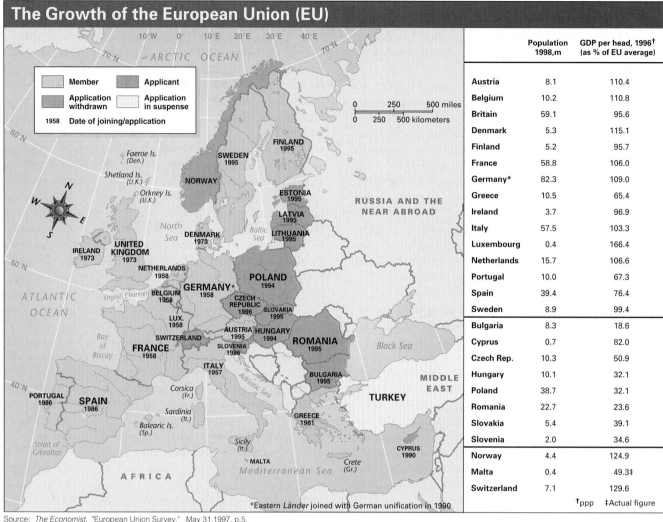

The Growth of the European Union (EU)

	Population 1998,m	GDP per head, 1996[†] (as % of EU average)
Austria	8.1	110.4
Belgium	10.2	110.8
Britain	59.1	95.6
Denmark	5.3	115.1
Finland	5.2	95.7
France	58.8	106.0
Germany*	82.3	109.0
Greece	10.5	65.4
Ireland	3.7	96.9
Italy	57.5	103.3
Luxembourg	0.4	166.4
Netherlands	15.7	106.6
Portugal	10.0	67.3
Spain	39.4	76.4
Sweden	8.9	99.4
Bulgaria	8.3	18.6
Cyprus	0.7	82.0
Czech Rep.	10.3	50.9
Hungary	10.1	32.1
Poland	38.7	32.1
Romania	22.7	23.6
Slovakia	5.4	39.1
Slovenia	2.0	34.6
Norway	4.4	124.9
Malta	0.4	49.3‡
Switzerland	7.1	129.6

[†]ppp   ‡Actual figure

*Eastern *Länder* joined with German unification in 1990

Source: *The Economist,* "European Union Survey." May 31,1997, p.5.

Figure 3.18 The European Union has grown steadily from six nations in 1952 to nine in 1973, 10 in 1981, 12 in 1986, and 15 in 1997, with another dozen interested in becoming part of this extraordinary collection of diverse but economically ambitious European countries. The table to the right of the map provides population data and comparative economic efficiency.

the European nations west of the former Iron Curtain, and Turkey. In 1997, NATO membership was extended to Poland, Hungary, and the Czech Republic. The accession process to full membership was intended to take two years. The NATO membership offer to three of the countries that had been part of the Union of Soviet Socialist Republics for nearly half a century led to considerable anger from Russia. It saw this eastward expansion of NATO as particularly threatening. After considerable negotiation, Russia finally agreed to accept this act because of its own deeper involvement in the NATO-sponsored Partnership for Peace program. Russia also played a role in the NATO-led Stabilization Force (SFOR) that enforced the terms of the 1995 Dayton Accords between Bosnia-Herzegovina and Serbia.

NATO member countries pledged to settle disputes among themselves peacefully, to keep their individual and joint de-fense capacities in good order, to consider an attack upon any of them in Europe or Anglo America, e.g., the Falkland Islands, as an attack upon all, and to come to the aid of the country or countries being attacked. Questions arise today as to whether NATO is necessary any longer and about what its role should be in view of the dissolution of the Soviet Union and the retreat of Soviet power from eastern Europe. The Partnership for Peace program of military and political cooperation (including joint military exercises and peacekeeping operations) with Russia continues to be a response to such questions. Eighteen ex-Soviet Iron Curtain allies joined the program on June 22, 1994.

The response of the former Communist East to the formation of the Common Market and NATO was the creation in 1955 of an international economic organization called the **Council for Mutual Economic Assistance (COMECON)** and

International Organizations of Europe, January 1999

Legend:
- European Union (EU) members
- Countries being considered for next EU enlargement
- North Atlantic Treaty Organization (NATO) members
- (E) European Free Trade Association (EFTA) members
- (O) Organization for Economic Cooperation and Development (OECD) members

*Note: Turkey was granted a customs union treaty with the EU in 1996.

Sources: EUROSTAT (Statistical Office of the European Communities) (1996). CIA, *The World Fact Book* (1995).

Figure 3.19 Some of the key organizations in Europe's intricate framework of political, strategic, and economic relationships. The sequence of the creation and elaboration of the various steps leading to the contemporary European Union (EU) is chronicled in the text. The critical fact to remember is that this region has a long history of conflicts that have been political, religious, and economic. The EU marks a very momentous innovation in European regional cooperation.

a military alliance called the **Warsaw Pact.** The Former Soviet Region dominated the Communist "satellite" states in these organizations. The new eastern Europe still needs freely entered agencies of economic cooperation and collective security, but, since COMECON and the Warsaw Pact were not freely entered, they thus have been terminated. Currently, NATO and the EU are working with a variety of potentially useful and productive collaboratives with the eastern European nations. For example, in the former Yugoslavia—once an east European nation—warring factions have thrown NATO into a new role as a peacekeeping force. The 1995 Dayton Accords tried to bring stability to a European region already known for centuries of discord. In the winter of 1999, NATO played a military role in attempting to hold back Serbian military advances against ethnic Albanians in the Kosovo region of the former Yugoslavia.

These examples of virtually unprecedented collaboration and economic and political cooperation among European nations reflect the changing geopolitical nature of the world today. Patterns of traditional geographic isolation, independence, and relative autonomy are being modified by these new regional and increasingly global linkages. Because of the enormous economic strength of the giant nations of Europe, and the market associated with those countries with the largest populations, the new profile created by the EU sets the stage for still additional innovations in regional efforts to unify the historically individualistic nations of Europe (Fig. 3.19).

SUMMARY WITH SELECTED KEY TERMS

- Europe is a **Eurasian peninsula,** called a continent because of the political and economic authority held in Europe two centuries ago. It has a series of other smaller peninsulas, rivers, mountain systems, and plains, which are generally much more densely settled than the United States. The European population is about twice the size of the U.S. population. **Western civilization** has its origins in this region, and its influences have reached all areas of the world.

- **Northerly location, temperate climate, the North Atlantic Drift, and varied topography** are all factors in the productivity and long history of productive settlement of this region. The **North European Plain** is a major belt of settlement and agricultural productivity. The **marine west coast climate** and the **mediterranean climate** are the two most characteristic climates of Europe. Major rivers flow out from mountain highlands in west central Europe and spill into the North Atlantic, the Mediterranean, the Baltic, the Black Sea, and other proximate seas. The **Rhine** and the **Danube** are the two most important rivers.

- European languages derive primarily from **Indo-European** roots and are expressed primarily in **Romance, Germanic,** and **Slavic** languages. English is a Germanic language and, like most European languages, it is enriched by many other tongues. The languages of Europe reflect a long history of **human migrations** from east to west and a pattern of continual mixing. The dominant religious influence in Europe has been Christianity, with the major three religions being Protestant, Roman Catholic, and Eastern Orthodox. There are significant pockets of Islam and Judaism as well as many regional variants on all of these **belief systems.**

- Before the **Age of Discovery,** Europe played only a minor role in world trade networks. From the beginning of the 16th century until late in the 19th century, Europe was at the center of global patterns of **colonization** and foreign settlement, **long-distance trade, slave trade,** and **agricultural innovation.** One of the major outcomes of the maritime trade and travel such patterns engendered was the **Columbian Exchange,** or the diffusion of crops and animals from the Old World of Europe and Asia to the New World of the Western Hemisphere. Major crops, including **maize, coffee, cacao, rice, grapes,** and **potatoes,** were taken to new locales and became major food crops.

- The **Industrial Revolution** originated in Europe, with energy coming from coal and factory technology focused on textiles and iron manufacture. **Rivers, canals,** and **railways** all played a role in making Europe strong, and a shipbuilding industry and skill in maritime endeavors from exploration to navigation gave Europe a dominance in the 18th and 19th centuries that it has only slowly given up.

- Europe rose to dominance in world trade and politics because of factors including **capitalism, technology,** and **science.** However, recent decades have seen a global change in the patterns of power, and factors such as **war dislocation, rising nationalism,** the **ascendance of the United States,** a **shift in world manufacturing patterns,** and **new energy sources** have combined to diminish Europe's global centrality. Europe continues to be a dominant force in world economic, political, and social realms, but there is a steady increase in competition from many other world regions.

- Europe has a history of human migrations coming from the east and going to all parts of not only Europe, but the world beyond. **Contemporary migration streams** from eastern Europe to west and northwestern Europe, and from northern Africa to southern and southwestern Europe, are aspects of migration that shape the current scene. **Guest workers,** very important in the rapid economic growth of Europe in the 1970s and 1980s, have recently become a demographic problem, just as **technological unemployment** has. Population growth rates vary a great deal from country to country and region to region in Europe, but overall it has the lowest fertility rates of any world region.

- The **European Union (EU)** is a collection of fifteen nations that has grown from six member states who came together to gain economic strength four decades ago. Now the EU is leading the way in the creation of a Europe characterized by **open borders, limited or no tariffs,** and, in time, a common currency called the **euro.** The **North Atlantic Treaty Organization (NATO)** is another example of a supranational organization that attempts to create a military and political force that gives Europe a more powerful voice than it would achieve through its individual states.

REVIEW QUESTIONS

1. What defines the continent of Europe?
2. What are its major physical and environmental characteristics?
3. Define the major types of climates and differentiate between the two dominant ones.
4. List four major rivers in Europe and explain their historical and contemporary importance.
5. What are the dominant languages and religions in Europe and what have the sources been for these cultural features?
6. Define the factors that led to Europe's central role in the Age of Discovery and the subsequent patterns of colonization and world trade patterns of the 18th and 19th centuries.
7. Explain the origins and significance of the Industrial Revolution, including what role resources had in that revolution.

8. List some of the environmental problems that Europe is faced with; describe how they came about and explain why they are difficult to fix.
9. Describe patterns of human migration in the history and in the contemporary world of Europe. Make a sketch map that will show these spatial patterns.
10. Recall the Six Essential Elements in Chapter 1 and describe a scene from the European scene that would be suitable as an example for each Element.
11. Describe the efforts Europe has been undertaking since World War II to build a stronger regional Europe.

DISCUSSION QUESTIONS

1. What is the impact of the North Atlantic Drift (Gulf Stream) on European patterns of agriculture, settlement, and economy?

2. Try to explain the reasons for the relatively low population growth rates for European countries, taking care to point out the significant differences from country to country.

3. What are the factors that made Europe the center of the world for three centuries after the beginnings of European settlement of North, South, and Middle America?

4. Considering the importance of spatial analysis, how would you describe and explain the patterns of European cultural influences around the world today?

5. What are the factors that may lead to a steady decline in the relative importance of Europe in the global economy in the next two decades? Are there other realms in which Europe is likely to lose relative importance?

6. Consider the factors that make up an agricultural scene in Europe. Use a photograph or recall a trip you or someone you know has taken, and explore the features of that scene that are natural versus the features that have been modified by human effort. From that observation, make some comments about agriculture in Europe today.

7. What are the factors working against the success of the EU in Europe today? Are they likely to be overcome?

8. If you were a major Russian leader, describe how you would have responded when NATO invited Poland, Hungary, and the Czech Republic to join NATO. What could have changed your mind, and changed your reaction?

Chapter 4

West Central Europe's Growing Global Strength

▲ The Potsdamer Platz in Berlin, Germany, continues to be one of the most ambitious and massive construction sites in all of Europe. With the German decision to remove the national capital from its Cold War placement in Bonn in the west to Berlin in the east, an enormous investment in construction and image enhancement has been undertaken. The red structure in Potsdamer Platz is called the Daimler Benz "Red Info Box." Messerschmidt, Leo de Wys, Inc.

*t*his chapter looks at the process and ramifications of the Industrial Revolution, current post-industrial dynamics, and the regional landscapes in this economic hub of Europe. This monumental process of regional economic change led to major demographic shifts, social change, and, ultimately, European concerns about the environmental deterioration that seems to travel with industrial activity. West Central Europe is now undergoing a transformation toward more service industries and less traditional manufacturing. Fewer jobs are opening in smokestack-laden industrial zones and more people are bringing computer skills to their employment. Carry these images with you as we learn in the pages that follow of the incremental abandonment of 19th- and early-20th-century industrial landscapes in the coreland of northwest Europe and their partial replacement with modern information-centered office complexes.

countries differ sharply in their human geographies and national perspectives, and the whole area is differentiated in a mosaic of regions and localities exhibiting great physical and cultural variety.

In the 19th and early 20th centuries, the largest coal and iron ore deposits in Europe provided bases for large-scale industrial development (see Figs. 4.1 and 4.2). Consequently, a series of major industrial concentrations now stretches from the United Kingdom to Poland along the main belt of mineral deposits. Many individual industrial cities lie outside the main coal and steel axis, and some of these—such as Paris and London—are sufficiently important to be considered major industrial concentrations in their own right. In the typical industrial agglomeration, coal mining and iron and steel production have been economic foundations, with chemicals, textiles, and heavy engineering also common.

4.1 The Geography Behind West Central Europe's Economic Strength

Three of Europe's major countries and a series of smaller nations and microstates occupy the British Isles and the west central portions of the European mainland. The United Kingdom and the Republic of Ireland share what is called the British Isles. Germany; France; their smaller neighbors of the Netherlands, Belgium, Luxembourg, Switzerland, and Austria; and the microstates of Liechtenstein, Monaco, and Andorra may be conveniently grouped as the countries of West Central Europe. Considered as a whole, these units are highly developed (see chapter opener photo), and they form Europe's economic and political core. Development is far more intensive in some areas than in others, but the overwhelming majority of people in the coreland live in areas that are among the world's most intensively developed cultural landscapes (see Fig. 4.1 and Table 3.1).

Even given the complexity of Europe, the geography of the northwestern coreland is unusually intricate. Spatial variations are great, layers of history appear at every turn, and functional linkages are extraordinarily complicated. The individual

4.2 Forces of Change in the British Isles

Located off the northwest coast of Europe, two countries occupy the British Isles: the Republic of Ireland, with its capital at Dublin, and the larger United Kingdom of Great Britain and Northern Ireland, with its capital at London. The United Kingdom incorporates the island of Great Britain, consisting of England, Scotland, and Wales, plus the northeastern corner of Ireland and most of the smaller out-islands (Fig. 4.3). Altogether, it encompasses about four-fifths of the area and well over nine-tenths of the population of the British Isles (see Table 3.1).

Political Subdivisions and World Importance of the United Kingdom

England, which lies only 21 miles (34 km) from France at the closest point (see Fig. 4.1), is the largest of the four main subdivisions of the United Kingdom (the others being Scotland, Wales, and Northern Ireland [see Fig. 4.3]). These all were originally independent political units. England conquered

West Central Europe ● Industrial Concentrations

Figure 4.1 Industrial concentrations and cities, seaports, internal waterways, and highlands in West Central Europe. Older industries such as coal mining, heavy metallurgy, heavy chemicals, and textiles cluster around highland margins in the congested districts shown as "major industrial concentrations." Local coal deposits provided fuel for the Industrial Revolution in most of these districts, which have shifted increasingly to newer forms of industry as older industries have declined. But the new and more diversified industries have also proliferated in metropolitan London, Paris, and smaller industrial cities away from the old "major industrial concentrations." Relative sizes of dots and circles indicate rough groupings of industrial cities and seaports according to economic importance. Rivers and canals shown on the map are navigable by barges and, in some instances, by ships. The Channel Tunnel (Eurotunnel) will change forever the sense of proximity or isolation perceived by the political units on the map.

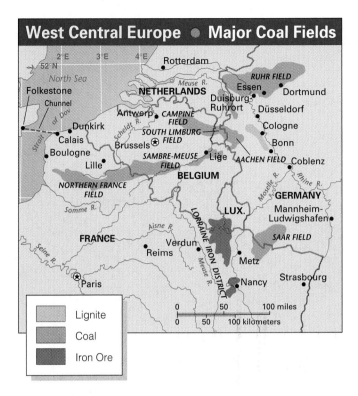

Figure 4.2 Major coal fields of West Central Europe. Note the line of fields stretching across western Germany, the Low Countries, and northeastern France. The Northern France and South Limburg coal fields have been closed. The early importance of these fields as centers of 19th-century industrialization is being eclipsed now by the increasing importance of imported natural gas and petroleum, but the footprint of significant West Central European industrial origins is shown in this map. Stars show national capitals. *After a map by the U.S. Geological Survey*

Wales in the Middle Ages but preserved some cultural distinctiveness associated with the Welsh language.

The international influence once attained by the United Kingdom and the imprint it left on the world are a legacy of the 19th century. When British power was at its peak in the century, between the defeat of Napoleonic France and the outbreak of World War I (1815 to 1914), the United Kingdom was generally considered the world's greatest power. Its overseas empire eventually covered a quarter of the Earth, and the influence of English law, education, and culture spread still farther. Until the late 19th century, the United Kingdom was the world's foremost manufacturing and trading nation. The Royal Navy dominated the seas, and the British merchant marine carried half or more of the world's ocean trade. London became the center of a free trade and financial system that invested its profits from industry and commerce around the world.

From the late 19th century onward, the rate of economic growth in the United Kingdom tended to be slower than that of many other industrialized or industrializing countries. Thus, a relative decline in the country's economic importance set in. The free trade and financial system based in London was damaged by World War I, and decline accelerated after World War

II, although London remains a major world financial center. Relative economic decline has been accompanied by a drop in political stature. The large colonial countries once included in the British Empire have all gained independence, and Britain's remaining overseas possessions are scattered and small. The United Kingdom, however, is still a country of consequence in many fields. It plays a major role in the European Union and is associated with many of its former colonies in the worldwide Commonwealth of Nations.

Regional Variations on the Island of Great Britain

The physical base of the United Kingdom is Great Britain, an island about 600 miles (somewhat under 1000 km) long by 50 to 300 miles (80 to nearly 500 km) wide. For its small size, this island provides a homeland with a notable variety of landscapes and modes of life. Some of this variety can be attributed to a broad fundamental division of the island into two contrasting areas: Highland Britain and Lowland Britain (see Fig. 4.3).

Highland Britain

The principal highlands include the Scottish Highlands; the Southern Uplands of Scotland, the Pennine mountains and the Lake District of northern England; the mountains that occupy most of Wales, the uplands of Cornwall and Devon in southwestern England; and the mountainous and hilly rim of Ireland. The mountains are low—Ben Nevis in the Scottish Highlands is the highest at 4406 feet (1343 m)—but many slopes are steep, and the mountains often look high because they rise precipitously from a base at or near sea level. Highland Britain includes two important low-lying areas: (1) the boggy, agricultural Central Plain of Ireland, and (2) the Scottish Lowlands, a densely populated industrialized valley separating the rugged Scottish Highlands from the gentler and more fertile Southern Uplands of Scotland. The Scottish Lowlands compose only about one-fifth of Scotland's area, but they incorporate more than four-fifths of its population and its two main cities: Glasgow (with a population of 1.8 million in the metropolitan area, including Glasgow city proper [689,000] plus suburban and satellite districts nearby) on the west, and Edinburgh, Scotland's political capital (with a population of 630,000 in the metropolitan area and 443,600 in the city proper), on the east.[1] In

---

[1] Unless otherwise noted, city populations in this text are rounded approximations for metropolitan areas as of the late 1990s. They are only crudely comparable from one country to another because of variations in the accuracy of statistics and the actual definition of "metropolitan area." Such figures, based on standard sources such as *Goode's World Atlas* (Rand McNally), the *New International Atlas* (Rand McNally), the *Statesman's Year-Book* (St. Martin's Press), the *CIA Fact Book*, and the *Britannica Book of the Year,* do enable students to broadly group urban areas by orders of magnitude—an activity far more meaningful than attempts to memorize populations city by city.

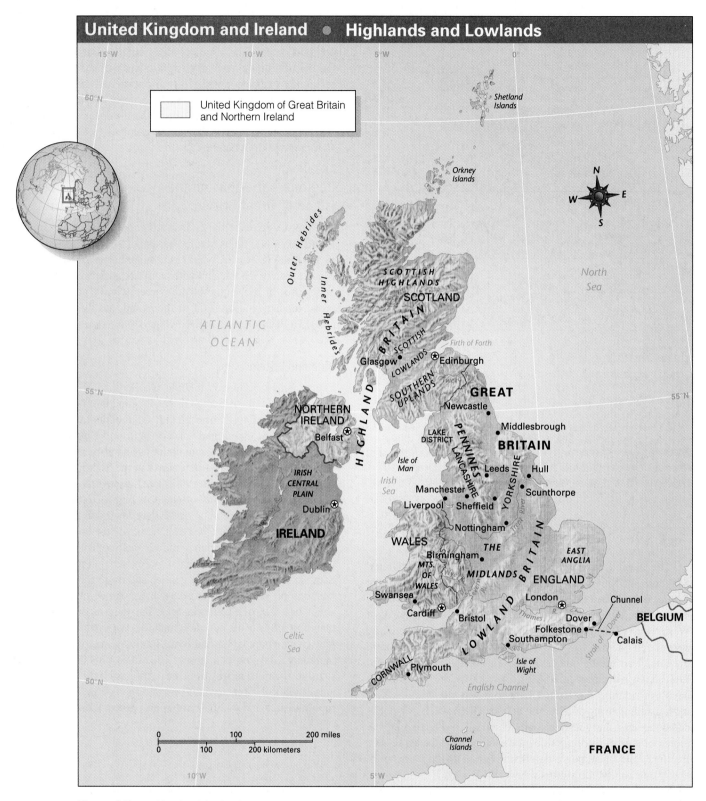

Figure 4.3 Highland and Lowland Britain.

the summer of 1997, the Scots voted overwhelmingly to create their own parliament after a 290-year union with England. This new body politic will give Scotland more independence in its domestic matters.

Lowland Britain

Practically all of Lowland Britain lies in England and it is often called the English Lowland. It has better soils than the highlands and a gentler topography (see Fig. 4.4). Most of it exhibits a mosaic of well-kept pastures, meadows, and crop fields fenced by hedgerows and punctuated by closely spaced villages, market towns, and industrial cities. The largest industrial and urban districts, aside from London, lie in the midlands and the north, around the margins of the Pennine Chain. Here, clusters of manufacturing cities rise in the midst of coal fields.

Geographic Elements and Impact of Britain's Industrial Revolution

Development in the Industrial Revolution

Four industries were of major significance in Britain's industrial rise: *coal mining, iron and steel, cotton textiles,* and *shipbuilding*. The advantage of an early start in these enterprises gave the United Kingdom industrial preeminence through most of the 19th century. Exported surpluses paid for massive food and raw material imports, and made Britain the hub of world trade and very wealthy. Current difficulties in these same industries—due to growing international competition, changing patterns of demand, and shifting urban populations—are fundamental to many of the problems that Britain faces in the 21st century.

The British Coal Industry

Coal became a major industrial resource between 1700 and 1800, when it fueled the blast furnace and the steam engine. James Watt's improvements in 1765 on Thomas Newcowmen's earlier design of a steam engine led to a steady expansion of the British role in the Industrial Revolution. Coal was coked for use in blast furnaces, and coke replaced the charcoal previously used for smelting iron ore. With Britain's development of an economically practical steam engine, coal supplanted human muscle and running water as the principal source of industrial energy. Coal production rose in England and Wales until World War I. Then, between 1914 and the 1990s, output fell by about two-thirds, the number of jobs in coal mining was cut by more than four-fifths, and practically all of the coal export trade disappeared. Major factors contributing to this decline included (1) increasing competition from foreign coal in world markets, (2) depletion of Britain's own coal, which was easiest and cheapest to mine (although large reserves of coal still remain), (3) the environmental unattractiveness of coal, and (4) increas-

Figure 4.4 Fields in the English Lowland normally are fenced by hedgerows, seen here in southwestern England near the mouth of the River Severn. Hedgerow trees blend in the distance to give a misleading impression of forest. This landscape was created centuries ago by the Enclosure Movement, in which large landowners appropriated local "commons," or unfenced land tilled by villagers. Hedges were planted so that sheep could be raised under a system of controlled breeding. This closure of lands that had historically provided a safety valve for marginal peasant farmers helped promote the migration of peasant youth to the newly industrializing English cities. *Jesse H. Wheeler, Jr.*

ing substitution of other energy sources for coal—both in overseas countries and in Britain itself. In the late 1960s and the 1970s, large-scale development of newly discovered oil and gas fields underneath the North Sea revolutionized Britain's energy situation. By the 1980s, once-dominant coal's principal remaining role in Britain had been reduced to supplying fuel for generating domestic electricity, and even that is under steady competition by natural gas and oil.

Iron and Steel

Steel was a scarce and expensive metal before the Industrial Revolution. Iron, not too cheap or plentiful itself, was more commonly used. It was made in small blast furnaces by heating iron ore over a charcoal fire, with temperatures raised by an air blast from a primitive bellows. Supplies of iron ore in Great Britain were ample for the needs of the time, but the island was largely deforested by the 18th century and charcoal was in short supply. By 1740 Britain was importing nearly two-thirds of its iron from countries with better charcoal supplies, notably the American colonies, Sweden, and Russia.

This heavy import dependence then changed when 18th-century British ironmasters—led in the late 1750s by Abraham Darby when he was able to replace charcoal with coal and coke—made revolutionary changes in iron production. There was improvement of the blast mechanism; invention and use of a refining process called puddling to make the iron more malleable; and adoption of the rolling mill in place of the hammer

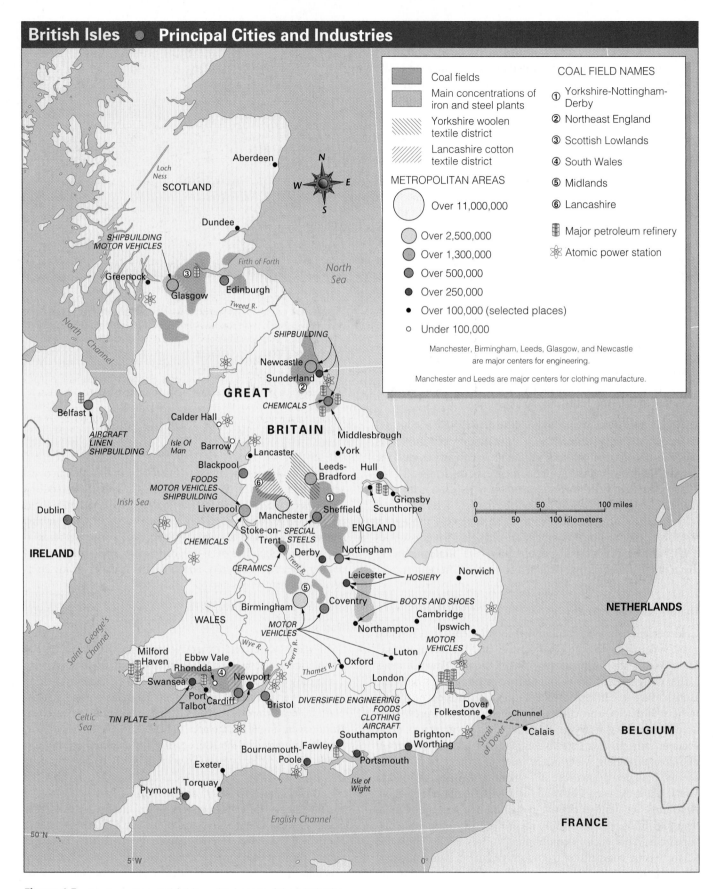

Figure 4.5 Principal cities, coal fields, and industries of the British Isles.

for final processing. By 1855, Britain was producing about one-half the world's iron, much of which was exported. Further advances in production resulted when British inventors developed the Bessemer converter and the open-hearth furnace in the 1850s. Steel, a metal superior in strength and versatility to iron, could now be made on a large scale and cheaply for the first time.

The early iron industry in Great Britain used domestic ores, but these eventually became inadequate, and Britain now relies almost entirely on high-grade imported ores. And although British inventors and entrepreneurs led the world into the age of cheap, mass-produced, and extensively used iron and steel, Britain soon lost its leadership in steel production. The United States and Germany surpassed the British output before 1900. By 1998, the EU produced a total of 161 million tons of steel, with Germany producing 44.7 million tons—the fourth highest global output. China led world production with 114 million tons; the United States had 98 million tons, and Japan produced 93.5 million tons in 1998. England is now in the next tier.

Cotton Milling in Lancashire

During the period of Britain's industrial and commercial ascendancy, cotton textiles were its leading export by a wide margin, amounting to almost one quarter of all exports by value in 1913. The enormous cotton industry eventually became concentrated in Lancashire, in northwestern England, with Manchester (population: 2.8 million, metropolitan area; 438,000, city proper) the principal commercial center (see Fig. 4.5). Imported raw cotton was the basis for this industry, and manufactured textiles were exported through the Mersey River port of Liverpool (population: 1.5 million, metropolitan area; 539,000, city proper), although Manchester itself became a supplementary port after completion of a ship canal to the Mersey Estuary in 1894.

A hand-labor textile industry in Lancashire dated from the Middle Ages and found itself involved in the Industrial Revolution in several lines of development. First, a series of inventors, in good part Lancashire men, mechanized and speeded up the spinning and weaving processes. Then, power was applied to drive enlarged and improved machines—first using falling water from Pennine streams, then steam fueled by coal from underlying Lancashire fields. In 1793, the invention of the cotton gin by American Eli Whitney provided a machine which separated seeds from raw cotton fibers economically, making cotton a relatively cheap material from which inexpensive cloth could be manufactured. Industrial growth acted like a magnet to bring farm youth to the cities in quest of factory labor. These new factory complexes that shone in the soot of England's expanding cities were difficult environments. They did, however, provide alternatives to unemployment and underemployment in the countryside, and they were seen by many youth as the only means of breaking patterns with the past and getting into the city.

For more than a century, until World War I, Lancashire dominated world trade in cotton and cotton textiles. But 1913 saw the peak of British production; after that, the decline of the industry was nearly as spectacular as its earlier rise. The root of the trouble was increased foreign competition. Throughout the 20th century, country after country surpassed the United Kingdom in cotton textile production and exports, with a resulting slide in employment and great economic distress in Lancashire. Lancashire's future as an industrial area lies mainly with other

Definitions & Insights

ENCLOSURE

One of the rural changes that played a role in the promotion of rural youth migration to the city was a shift in rural land use called **enclosure** (also spelled **inclosure**). This was the establishment of fences, hedges, ditches, or other barriers around the perimeters of farm fields (see Fig. 4.4). The process began in the 12th century in England, but it became widespread in the mid-18th and early 19th centuries. Landowners realized that, with enclosure, they could make more money on their animal herds—especially sheep—by controlling breeding and, hence, producing better wools for the growing textile trade. Enclosure also provided abundant, cheap labor for new city industries because this act eliminated or severely restricted peasant access to lands traditionally held in **commons** for those farmers who had little or no land for themselves. The commons had served traditionally as unfenced lands that all farmers could run animals on, or even plant. For the farmers who had counted on access to the commons, enclosure meant that an already marginal agricultural operation became even more untenable. It was this marginal peasant population that became so important to the growing urban textile centers. These rural youth—particularly the youngest children of land-poor families—came to the factory centers and found themselves pressed into factory work in difficult situations. Although the full set of influences on this critical demographic shift is quite complex, the act of enclosing the rural commons did lead to a major new **migration stream**—the patterns of migration between the countryside and the city—in Britain as farm youth moved to the cities. It also led to higher levels of agricultural productivity because the enclosed fields were farmed more thoroughly and intensively. Enclosure, in various forms, spread to other nations of the European coreland.

industries such as engineering (a general name for a wide range of industries that produce machinery, vehicles, and tools), chemicals, and clothing.

Shipbuilding on the Tyne and Clyde

When wood was the principal material used in building ships, Great Britain, as a major seafaring and shipbuilding nation, eventually created serious difficulties for itself due to deforestation. Large quantities of timber were imported from the Baltic area and from North America, and the importation of completed ships from the American colonies, principally New England, was so great that an estimated one-third of the British merchant marine was American-built by the time of the American Revolution. During the first half of the 19th century, American wooden ships posed a serious threat to Britain's commercial dominance on the seas.

In the later 19th century, iron and steel ships propelled by steam transformed ocean transportation. By the end of the century, Britain led the way and built four-fifths of the world's seagoing tonnage. Two districts developed as the world's greatest centers of shipbuilding: the Northeast England district, with a great concentration of shipyards along the lower Tyne River between Newcastle (metropolitan population: 1.3 million) and the North Sea; and the western Scottish Lowlands, with shipyards lining the River Clyde for miles downstream from Glasgow. Both districts had a seafaring tradition (Newcastle in fishing and the early coal trade, and Glasgow in trade with America), both were immediately adjacent to centers of the iron and steel industry, and both had suitable waterways for location of the shipyards. Their ships were built mainly for the United Kingdom's own merchant marine, as they still are, but even the minority built for foreign shipowners were an important item in Britain's exports.

The Industrial Revolution and the Growth of Cities

The Industrial Revolution gave a powerful impetus to the growth of cities at the same time that Britain's total population was increasing rapidly. In 1851, the United Kingdom became the world's first predominantly urban nation, and it remains so today. Most of the large cities, with the notable exception of London, are on or near the coal fields that supplied the power for early industrial growth. The most striking cluster is composed of cities associated with the coal fields that flank the Pennines to the east, south, and west—that is, the industrial districts of Yorkshire, the Midlands, and Lancashire respectively. In contrast, the largest metropolitan areas that lie at a considerable distance from the coal fields—aside from London—are generally smaller and more widely spaced. Most of them are in southern England. For many years, these cities have tended to grow more rapidly and be more prosperous than the coal-field

industrial cities, which tend to be physically unattractive and too dependent on old and failing industries. The transportability of electricity, natural gas, and oil has made it unnecessary for industries to locate near coal, and recent development tends to favor the relatively pleasant environments of southern England.

Urban Location Dynamics: Two Perspectives on London

Perspective I: The Specific Location of London. Cities are almost always where they are because of the presence of some particular geographical characteristic(s). These might include a ford across a river, a strategic pass, the location of a resource, the intersection of significant migration paths, or even a religious or historic site. London, as one of the most dominant of the European cities, is characterized by a complex set of urban location dynamics that are worth some detailed consideration. While what follows is the story of London, it is important to realize that many aspects of this development would apply to Paris or Brussels or Berlin. Use this chronicle as a set of specifics that has broad implications for urban positioning across the globe.

The London metropolitan area, with more than 11 million people and 6,574,000 in the city proper, sprawls over a considerable part of southeastern England (Fig. 4.6). Its historical background is unique in that it was already a major city before the Industrial Revolution, and its economy is not based so much on industry. Although the metropolitan area contains a great deal of manufacturing, this sector does not dominate. Instead, it shares leadership with several other major elements, some of which developed before the Industrial Revolution and have greatly expanded since then. These include commerce, government, finance and insurance, and corporate administration. Basic elements more recently developed include communications media (publishing, television, films) and tourism.

Its function as a seaport and commercial center laid the original foundation for London's development. Trade generated population growth, profits, and financial operations, including insurance (which began with marine insurance covering risks to ships and cargoes). Population, trade, and wealth attracted government. Finance, trade, and government attracted corporate offices. The large, growing, and increasingly prosperous population provided both a labor force and a large market for manufacturing.

London became an important seaport as early as the Roman occupation of England in the 1st century A.D. and continues to be one today. Certain aspects of London's location and evolution as a port not only account for the city's own development but also shed some light on port cities in general. Such factors include:

1. Location relatively near major trading partners. London is located in the corner of Britain nearest the European continent, which was the most important area with which

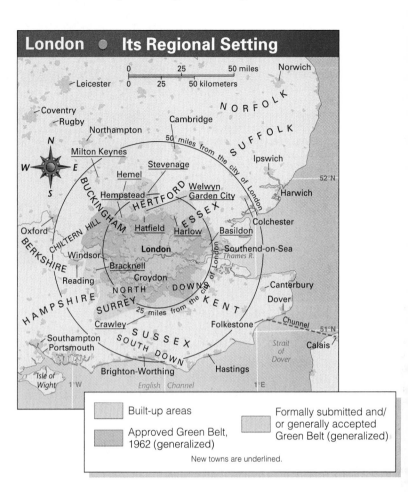

Figure 4.6 The maps of London illustrate the structure and setting of a major world metropolis. They may be used together with the New York maps panel in Chapter 24 to examine general concepts of urban geography, such as the Sector Model and the Concentric Zone Model (see Glossary). The City of London, built originally by Roman conquerors of Britain, contains one of the world's greatest financial districts. The Green Belt limits further urban sprawl and the New Towns are planned communities built to provide residential areas and employment for migrants from London proper in order to lessen congestion there. The West End is the governmental, hotel, and entertainment center. The Docks, built as artificial anchorages for shipping, have been closed but are finding new life as tourist venues. There is an ingenious river management construction called the Thames Barrier that lies across the Thames just upstream from Tilbury Docks. These gates enable engineers to close the river to rising flood waters upon command, a task increasingly significant because of subsidence in the Thames Estuary. *Various details on the maps are generalized from the* Atlas of Britain and Northern Ireland *(1963), by permission of Clarendon Press, Oxford; from a government map of the Green Belt on an Ordnance Survey base, Crown copyright; and from maps in J. T. Coppock and High C. Prince (eds.),* Greater London, *Faber and Faber, 1964.*

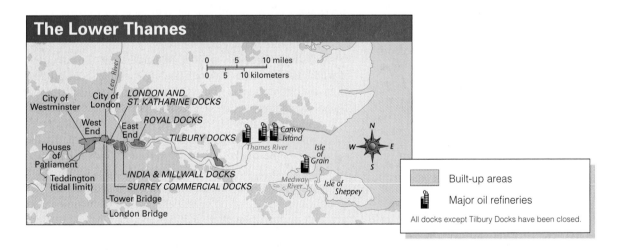

England traded before the Age of Discovery and which is still of dominant importance to British trade today.

2. *Location in a productive area.* London is located in the fertile English Lowland, which was the most productive and populous part of Great Britain in the preindustrial period, supplying most of the wool that was Britain's main export and providing most of the British market for imports.

3. *Location facilitating transport connections with the surrounding region.* From London, a major route to the interior, with branches, was provided by the Thames River and its tributaries. Before the age of the railroad and the truck, water transport had greater advantages over land transport than it has today, and these natural inland waterways were extensively used for transporting goods to and from London.

Figure 4.7 London on the Thames River in southeastern England is a major node in the worldwide system of places that geographers study. Tower Bridge is at the lower left, with new London Bridge next upstream. The financial district (City of London) is at the right center, adjacent to London Bridge. *Colour Library Books*

During the Industrial Revolution, their utility was enhanced by canals (little used today) that connected the Thames system with other British rivers. Furthermore, London became a major junction for roads as early as Roman times, and again the city's location provided advantages. Tidal marshes originally fringed the lower Thames, forming a major barrier to transportation, but London's site provided firm ground that penetrated these marshes and approached the river bank on both sides. Highways could avoid the marshes by converging on London and crossing the river there.

4. *Location on a river (and estuary) permitting relatively deep land penetration of ocean ships.* A seaport set inland in such a manner can serve more trading territory in the surrounding countryside, within a given radius, than can a port on a comparatively straight stretch of coast or on a promontory. Dense urban development surrounds the port of London on all margins of the Thames (Fig. 4.7). Not only does the river provide good channel access for oceangoing ships, but London's network of highway connections to the rest of Britain facilitates port activities that link the island with global trade. The inland location of London's port tends to maximize the distance goods can be carried toward many destinations by relatively cheap water transportation (ocean vessels) and by minimizing the more expensive land transport.

Although London, like many other ports, originally developed well inland, later growth of the port has been downstream toward the ocean. This has been necessary to accommodate the growth in size (and hence the efficiency) of ships, because large ships cannot go as far up the river. Although ships have been growing larger for centuries, in the industrial age they have been growing at an accelerated pace, especially with the advent of oil supertankers and other very large ships in the last few decades. In addition, more efficient loading and unloading methods, such as containerization, require large areas of relatively open ground along the waterfront for tracking and mov-

ing containers, and such areas cannot generally be found in the congested older ports built up for earlier ships and methods. Instead, they must be sought in the areas downstream from the original port. Thus, a large share of the facilities in the older part of the London port have closed (Fig. 4.8), with much loss of employment, whereas new port areas, using modern methods that require fewer people, are being developed far out on the Thames estuary.

Perspective II: London As an Exemplar of Industrial Shift. As in the case of urban location dynamics, the city of London serves well as an example for another common economic and geographic dynamic characterizing the past four decades: industrial and commercial relocation.

Industrial difficulties in older industrial areas and countries, of which Britain is the most outstanding example, are a product of many circumstances, but the most vital—at least for the United Kingdom—is the increasing degree of international industrial competition that developed in the 20th century, especially since World War II. Intensified competition is partly the logical outcome of the continued spread of industrialization around the world, accentuated in recent decades by the rapid improvement of communications and transportation. This brought a precipitous drop in the costs both of maintaining contact with distant production facilities and markets and of shipping materials, subassemblies, and finished products for long distances. Distance from competitors no longer offers much protection, and distance from markets is no longer much of a handicap. Competition has also been accentuated by the systematic and scientific development of new products with the potential to compete with and replace older ones. A 20th-century example is the replacement of older means of transport—particularly rail connections—by the automobile and truck, and the attendant replacement of coal by oil. Another is

102

the development of synthetic fibers, made from wood or petroleum, to compete with cotton and wool.

In the strenuous competition for international markets, older industrial areas tend to lose to newer ones. Some major reasons include:

1. *Loss of a resource advantage.* A common example is the lessening of the significance of local coal. Location on a coal field was the main basis for the development of many older districts, but that advantage has been lost because of (a) depletion of the local coal, (b) the rise of lower-cost coal fields elsewhere, or (c) replacement of the coal by oil or natural gas.
2. *Cheaper labor in newer areas.* In older districts, there is time for unionization to occur and to increase wages, whereas a new area is likely to have an unorganized workforce to whom even low wages seem a great improvement from urban unemployment or rural underemployment. With lower wages, prices can be lower.
3. *Technological advantages of newer areas.* Such areas can be developed with the newest and most efficient technology. In older areas, there is a large investment in older plants and equipment, which is difficult to write off, scrap, and replace with the latest technology. In some cases, strong unions may oppose the new technology because it provides fewer jobs.
4. *Unemployment resulting from the efforts of older industries to survive.* Even if an industry in an older area survives the competitive struggle, unemployment and hardship in the community often result because the industry is likely to survive only through increased mechanization, automation, and the use of robots. Production may be maintained or even expanded by this tactic, but workers will be fewer in number.
5. *Conservatism of management.* After years or decades of success, there is often a tendency for the management as well as the workers in an older industrial area to be understandably resistant to necessary changes. There is a natural tendency to be complacent in the belief that procedures that worked in the past will continue to succeed in the future. Many charges of this nature have been leveled against established management teams not only in Britain, but anywhere that dislocation occurs.

As the process of dislocation takes place in one sector of the economy, there is often a parallel expanding employment sector. This is currently taking place in the *service sector*—a term embracing all workers not directly involved in producing material goods. In industrialized nations, the pattern that is ever more significant is the reduction of a manufacturing base and the expansion of the service base. Apparently this trend must continue if an increase in employment is to be enjoyed: Agriculture has fallen to low levels of employment, and industrial jobs are being lost to competition and to automation. Increasingly, Britain—along with other industrial nations—is a country of white-collar workers. The expanding service sector that employs them is highly varied in its makeup and includes gov-

Figure 4.8 This derelict 19th-century factory building radiates the disuse and abandonment that characterize urban landscapes that have been left in the wake of new suburban development. Some of these buildings, however, are renovated by people who believe that their architecture and urban centrality make refurbishing worth the investment of capital and considerable human energy. Such an effort can promote a full-scale neighborhood transformation—both in terms of appearance and social characteristics. *David Hoffman/Tony Stone Images*

ernment, finance, transportation, insurance, health services, management, sales, education, research, personal services, law, tourism, and communications media.

There are many unanswered questions about this service sector–oriented, "postindustrial" society toward which the industrialized Western nations are moving. These questions are focused on:

1. *The role of exports.* Will this society be able to maintain itself industrially with lightly manned, highly automated manufacturing industries?
2. *Capacity of the service sector.* Can the service industries provide an adequate supply of exports needed in international trade? The United Kingdom long supported itself in part by "invisible exports" such as capital, financial services, insurance services, ocean shipping, airline services, and tourism, and other Western nations are now involved in such activities. But a productive industrial sector remains important to thriving economies. Just as important, can modern service jobs—what are sometimes described as knowledge-worker jobs—be expanded rapidly enough to supply high employment in the face of declining industrial employment?
3. *Income potential in the service sector.* Can service jobs be made productive enough to provide high incomes to those holding them? There is reason to doubt the ability of many parts of the service sector to increase productivity at rates comparable to those of large-scale and highly mechanized

agricultural or manufacturing industries. If such doubts prove justified, the countries with a high dependence on services are likely to experience relatively slow economic growth unless they are able to create a service sector that is based more on information management and less on fast food.

4. *Location of the service sector.* Will new service employment locate in the afflicted old industrial areas where it is especially needed, or will it grow mainly elsewhere? In Britain, the primary growth of such employment, as with employment in newer forms of manufacturing, has taken place in London and surrounding parts of southeast England rather than in the depressed coal-field industrial areas. This has accentuated the sharp contrast between the relatively prosperous London region and the rest of the country.

The answers to the foregoing questions will go far toward determining the future of Western countries both within and outside of Europe. The United Kingdom will be a key country to watch because the process of change is relatively far along there and is proceeding rapidly.

Two Examples of British Efforts to Change Urban Decline

All through the heart of industrial Europe, there has been classic urban decline. Industrial centers that played major roles in early industrial growth over the past two centuries found themselves generally involved, as well, in the war efforts associated with both World War I and World War II. However, after World War II, there were major rebuilding campaigns that brought new structures, new technologies, and new urban development to western Europe. The two cases cited here are examples of cities that got left behind in the 1950s and 1960s and have now made major attempts to redefine themselves and deal economically and culturally with the late 1990s and beyond.

Glasgow. The history of Glasgow, Scotland, is a history of steelmaking, shipbuilding, international trade, and industrial invention and innovation. The commercial strength of Glasgow came from the union of Scotland with England in 1707. This port city soon became a major trading center for American tobacco and expanded to play a central role in cotton and sugar trade as well. In the early 19th century, iron and steel became dominant.

Overall, this urban center played a role in all of the major industrial and trade patterns of western Europe associated with the Industrial Revolution. But like so many such cities, the shift of urban populations and activities after World War II left Glasgow darkened and faced with powerful economic decline. In the place of sooty factories and stone monuments to early industrial power now stands a city increasingly given over to high-tech businesses, tourist trade, and regional shopping activities. In the center of the city, classic old bank buildings have been converted

Figure 4.9 Muslim women on the street in Glasgow, Scotland, reflecting a steady demographic shift taking place in most old industrial cities in the British Isles. *Joseph J. Hobbs*

to clubs, restaurants, and outdoor cafes. In addition, many of the Victorian structures that defined Glasgow's city center have been spared the wrecker's ball and there is now an approximate square mile of ritzy shops, outdoor cafes, monumental buildings with new functions, and an elegant Gallery of Modern Art—all in the Royal Exchange Square in the city center.

It will always be difficult for an urban center to sustain itself on shopping alone, but Glasgow is counting on the expanding role of discretionary travel among the peoples of the EU and beyond, and its rebuilding reflects this in an effective mode. Glasgow was proclaimed Britain's "City of Architecture and Design" for 1999 and tourist trade and a rapidly growing service sector have brought a strong measure of success to the city in its efforts to reshape itself for the next century.

A classic problem plaguing Glasgow and other cities that have lost major industrial prominence is the loss of a middle-class urban population. Many of the families that were central to the initial industrial and fiscal strength of the city have moved out of the city to newer suburban areas for environmental and employment reasons, taking a major segment of the tax base the city had traditionally relied on for support. Such a demographic shift means a continuing struggle to acquire capital adequate to the tasks of continuing the development of the new business, tourist, and shopping centers (Fig. 4.9). The numbers now offered by the Glasgow government are that some

"2,000,000 people make use of the facilities of a city in which fewer than 200,000 pay council tax." A geographic footnote on the case of Glasgow is that the Scottish Parliament is housed in Edinburgh, Scotland's capital and Glasgow's perennial rival—making continued governmental funding of Glasgow's new trajectory of development even more uncertain.

Birmingham. Birmingham is England's second largest city (after London) and its growth and change in the past half century had a distinct catalyst from the Glasgow experience. It began in 1962 when a decision was made in Pakistan to construct a major dam in Azad Kashmir. This dam—the Mangla dam—caused the displacement of more than 100,000 people. Using funds from Pakistan's governmental payout to displaced families, thousands of Pakistanis chose to migrate to England, or more specifically, to Birmingham, then the industrial heartland of Britain. By the early 1990s, Birmingham was more than 20 percent nonwhite, and it is estimated that by the year 2000, more than half of all the school children in its schools will be nonwhite. This will make Birmingham England's first majority "minority" city.

This major demographic shift has been one of greater smoothness than was anticipated. There were some periods of protest, but there has been a steady movement of the children of these migrant families into professional roles, and there has also been associated immigration of other Asian peoples who have added major business and commercial vitality to Birmingham.

An additional factor that has given this city a distinct, late-20th-century pattern of change has been its decision to seek a major role as a convention city. A National Exhibition Centre opened in 1976; an international convention center opened in the city's downtown in 1991, and Birmingham is now constructing its tallest skyscraper near the downtown convention center. There does continue to be, however, the problem of continuing high unemployment for ethnic minorities, but slowly, new Asian immigrant business people are expanding everything from small family-owned businesses to textile businesses with hundreds of employees. As service industries increase in association with convention business and tourism, and as Asian immigrant youth move through the educational system and grow into more professional roles, Birmingham takes on the look of a more global city that is reflective of western Europe's growing role as a global migration destination.

Food and Agriculture in Britain

A major characteristic of the United Kingdom's agriculture is its high concentration on grass farming for milk and meat production. Meat and, to an even greater degree, fresh milk are traditionally more difficult and expensive to transport than grains and many other vegetable products; consequently, they afford local producers a degree of advantage over competitors located at a greater distance from the market.

Some 2 percent of the working population of Great Britain is employed in agriculture, forestry, and fishing. A little more than a quarter of the total land area is given over to crops, with nearly half being used for pasture and rough grazing. Even with such a broad arable base, more than 60 percent of Britain's food supply is imported. Land in crops is used primarily to produce barley or other supplementary feeds for dairy cattle, beef cattle, and sheep.

Total food production has increased sharply since World War II, mainly an outcome of various government policies aimed at reducing imports. A large increase in wheat and barley production has been fostered by subsidies to farmers and has been expensive. However, despite the high cost of subsidies, Britain's membership in the Common Market—and now the European Union—probably will continue to foster the tendency to expand British agricultural production. Up to the late 1990s, the agricultural policy of the EU emphasized subsidies to agriculture, including generally high import tariffs on farm products from non-EU countries (see Problem Landscape on p. 108). As long as this policy holds, British farmers will continue to receive incentives in the form of relatively high and protected prices, and the British people, like the other peoples in the European Union, will eat more expensive food, produced to a greater degree within the home country or at least within the EU. This agricultural aspect of EU policies will continue to foment political tension as other countries continue to try to compete with EU-produced farm products.

Great Britain's fishing industry has been in steady decline since the 1960s. EU policies have led to fishing restrictions and a decline in catch, but even with this, Britain possesses the largest fishing fleet in the EU. Major ports still central to fisheries include Hull, Grimsby, Fleetwood, and Plymouth in England and Peterhead in Scotland.

The English Channel Tunnel. The 31-mile three-tube underwater tunnel that links southern Britain with Calais, France, is also called the **Chunnel** or **Eurotunnel**. It opened for commercial travel in 1994 (a year and a half later than its scheduled opening), but discussion about the possibility of a tunnel link between the British Isles and the European continent began 140 years earlier during the reign of Britain's Queen Victoria. The estimated cost for this structure, which houses two commercial rail lines and a service shaft, was set at $7.7 billion in the initial 1987 plans. The final cost came to more than $15 billion in 1994, making it the most expensive single building project in human history. It is now possible to board a train in London and arrive in central Paris in three hours, traveling at speeds that can reach 200 miles per hour. In 1997 design started on the 16-mile (26-km) twin-tube tunneling that will complete the high-speed connection between Paris and London.

The Channel Tunnel has opened a whole new era of speculation about the impact of reducing the insularity of Britain—a geographic characteristic that the British have long seen as

The Channel Tunnel

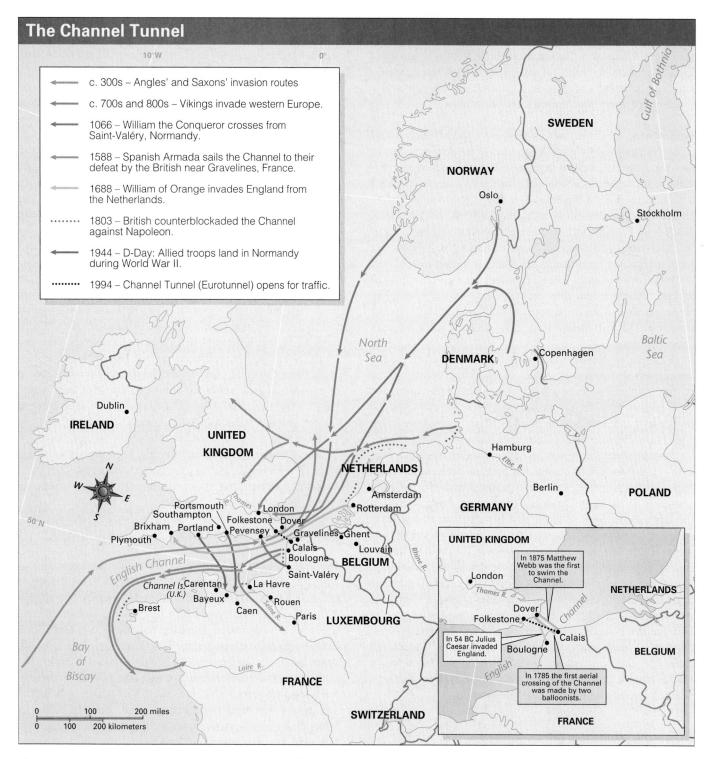

- c. 300s – Angles' and Saxons' invasion routes
- c. 700s and 800s – Vikings invade western Europe.
- 1066 – William the Conqueror crosses from Saint-Valéry, Normandy.
- 1588 – Spanish Armada sails the Channel to their defeat by the British near Gravelines, France.
- 1688 – William of Orange invades England from the Netherlands.
- 1803 – British counterblockaded the Channel against Napoleon.
- 1944 – D-Day: Allied troops land in Normandy during World War II.
- 1994 – Channel Tunnel (Eurotunnel) opens for traffic.

In 1875 Matthew Webb was the first to swim the Channel.

In 54 BC Julius Caesar invaded England.

In 1785 the first aerial crossing of the Channel was made by two balloonists.

Figure 4.10 History in the English Channel. This map shows the great variety of ways in which the English Channel has been challenged and (generally) overcome across the centuries. The current rail traffic through the Channel Tunnel (also called Eurotunnel and Chunnel) adds a new dimension to this map of interaction between the continent and the island of Great Britain.

central to their traditional ability to chart a history somewhat independent of western Europe. However, this opening to the vital landscapes of France, Germany, and the Low Countries has the potential of significant economic expansion not only for London, but for Birmingham and other British manufacturing (and tourist) centers as well (Fig. 4.10).

4.3 Ireland: New Directions, New Potentials

Ireland is a land of hills and lakes, marshes and peat bogs, cool dampness and verdant grassland (Fig. 4.11). The island consists of a central plain surrounded on the north, south, and west by hills and low, rounded mountains. The Republic of Ireland, which occupies a little over four-fifths of the island—and is included in the definition of the term "British Isles"—is in the midst of a transition from an agricultural to an industrial economy. Manufacturing is now a larger employer than agriculture and supplies more exports by value. But the transition is far from complete. Agriculture is still more important as an employment sector in Ireland than in most countries in western Europe, and the Republic is a considerable exporter of agricultural commodities and an importer of manufactures. Another indicator of the transitional nature of this still industrializing economy is the relatively low income level of the population. In per capita income, Ireland has ranked near the bottom of western Europe's countries for the past decades, but recent economic growth has begun to change this pattern (see Table 3.1).

Both nature and history contributed to the long delay in Irish industrialization. Some of the delay was due to an almost total lack of mineral resources that played such a major role in early industrial development on the island of Great Britain. But Ireland's retarded development was also a consequence of its long subordinate relationship to England. England completed an effective conquest of Ireland in the 17th century and subsequently made the island a formal part of the United Kingdom. Ireland's status, in reality, was virtually colonial. Most of the land was expropriated and divided among large estates held by English landlords with Irish peasants functionally tenants on these estates. Whatever income was gained in this rule generally found its way to Britain and Ireland became a land of deep poverty, sullen hostility, and periodic violence.

Into this setting has come a major surge of industrial development. It has been accomplished under a program designed to attract foreign-owned plants by a combination of cheap labor, tax concessions, and help in financing plant construction. Hundreds of plants have been attracted, although most of them are rather small. They are owned mainly by U.S., British, and German companies, and their major products are processed foods, textiles, office machinery, organic chemicals, and clothing. Most of the new plants have been built in and around the two main cities—Dublin (population: 1.2 million, metropolitan area; 533,929 city proper) and Cork (population: 174,400)—and in the vicinity of Shannon International Airport in western Ireland. The development of new high-tech industries has been a critical aspect of Ireland's rapid economic growth in the 1990s. Partly because of a highly developed work force, some 25 percent of the country's exports in the late 1990s came from electronic products. In terms of overall growth, Ireland increased its manufacturing output during the 1990s on a par with the strongest global growth patterns. In terms of west central Europe, it became the showcase of growth as its GDP increased above the levels of more fully developed nations.

Figure 4.11 The combination of abundant rainfall, rich peaty soils, and a continuing maintenance of family farms gives Ireland a lush pastoral landscape that is both agriculturally productive and attractive to tourists. This scene is near Lough Derravaragh County, Westmeath, Ireland. *Dembinsky Photo Associates*

PROBLEM LANDSCAPE

France in the Dock as Farmers Wage Fruit War

... Every April at the start of the strawberry season the war against cheaper Spanish produce begins in France. Since 1993, French farmers have attacked more than 60 Spanish lorries transporting cheaper fruit and vegetables to markets within the European Union.[1]

ONE OF THE MOST IMPRESSIVE GEOPOlitical shifts of the past half century in Europe has been the evolution of the European Union from the 1957 Treaty of Rome. The initial European Economic Community (EEC) became the European Community (EC), and now the functioning unit of great consequence in Europe is the fifteen-member European Union (EU).

The EU has had an impact on all facets of the European economy. For agriculture it is an influence in the various support programs and marketing arrangements for member nations. One of the most vocal populations—in occasional support, but more frequently in a role of dissent—has been the French farmers. In the decades since the 1957 Treaty of Rome, they, like farmers throughout Europe, have undergone profound change in farm organization, levels of mechanization, size of farms, and dependence upon foreign markets. This article about the fruit war deals with some of those elements.

The EU makes a continual effort to treat the regions embraced by EU membership as one open market area. Early in European efforts to stabilize and optimize

farm output, the European Community devised a Community Common Agricultural Policy (CCAP) which was intended to maximize agricultural productivity and efficiency. The benefits of disregarding national boundaries and individual support programs within the EU focused upon maximizing agricultural production in regions best suited to a particular crop or suite of crops. In broad terms, there has been a major shift in the expansion of wheat, barley, and maize (corn) production in Europe and especially in France. In vegetable production there has been a major expansion from subsistence farming and market gardening in the Paris basin to a much wider region, now linked to the Paris market by development of both highways and rail systems.

The French farmers referred to in the article above are representative of the recently vocal farmers who find the support systems of CCAP to work sometimes to their benefit, but sometimes to their perceived detriment. In this case, French farmers found that the elimination—required by the EU—of the French tariffs put local farmers—especially those near the Spanish border—at an economic disadvantage. While the importation of Spanish vegetables is not of major overall economic significance, the influence of EU policies overall has caused the French to rise in protest because of the perception of cultural, as well as financial, hardship being perpetrated by the EU. Because France delivers almost a quarter of the EU's agricultural production annually, there is considerable interest in what these farmers say and do.

The cultural impact that the French fear most is the further loss of the French

family farm. Since 1995, more than 4 million farm jobs have been lost in France, and the country is now approximately 80 percent urban. The average French farm size has grown to nearly 30 hectares (75 acres) through the remaining farmers expanding to absorb the unworked farmland. Though this total is about one eighth the size of a United States farm, it does mean that it takes increasing capital to be able to stay afloat in such an operation. For the French, to see all farming becoming capital intensive, export-market–oriented, and controlled by the EU is quite unacceptable. The French continue to view themselves as an agricultural nation, holding the customs of the family farm central to their identity and their moral well-being. The economic growth resulting from EU membership has extracted its cost by increasing the migration of French youth away from family farms toward Paris and other cities. Meanwhile, the aging farm population goes further into debt purchasing new farm machinery that might enable them to increase productivity and stay competitive in the domestic, EU, and global marketplaces.

The importation of Spanish vegetables is only one symptom of a larger irritation that has led French farmers to periodic protests in Paris as well as along the highways where the lorries carry produce from competing farmers.

With shifting farm technologies and national demographics, and increasing global agricultural competition, the problem landscape for the French farmer is a specter overshadowing family farms all across the globe.

[1] *The European*, 2–8 May, 1996, p. 1–2.

With now only 9 percent of its labor force on farms, the Republic of Ireland is loosening its traditional dependency on agriculture to a marked degree. Although the large estates of the period of English rule are gone and the Irish farmer now tends to be a landowner, there continues to be classic rural poverty. Farms tend to be both small and under-mechanized, but even they have taken on some additional value as Ireland has become a major tourist destination. Most important, though, is the increase in high-tech industrial activity that has reduced Ireland's traditionally high unemployment rate to under 10 percent.

Northern Ireland

The Protestant North of Ireland—now known as Northern Ireland—is a part of the United Kingdom. Shortly after the end of World War I, the island of Ireland was divided into a largely Protestant Northern Ireland, and the Republic of Ireland, which was nearly entirely Catholic. The Protestants were the descendants of Scottish Presbyterians and other British Protestants who were brought to the north of Ireland in the 17th century under the auspices of the British crown. They were intended to form a nucleus of loyal population in a hostile, conquered, largely Roman Catholic country. Protestant entrepreneurs were able to develop shipbuilding and linen textile industries in and near the North's main city, Belfast (population: 685,000, metropolitan area, 297,000 city proper). It was in Belfast that the English White Star Line had its luxurious ocean liner, *Titanic,* constructed and outfitted. These industries have declined recently, but Northern Ireland has been effective in attracting new industries seeking cheap labor despite recent political violence.

Ever since the 1960s, the religious tensions in Northern Ireland—also known as Ulster—have been manifest in car bombings, the shooting of British police and soldiers, and street bloodshed through clandestine bombings. In the spring of 1998, former senator George Mitchell of the United States was central to an intense effort to move the Northern Ireland peace talks to a new level of productive interaction between the Catholics and Protestants on this corner of the island of Ireland. A document was signed by the antagonistic factions on Good Friday 1998. This document was given a majority vote of support by both IRA (the Irish Republican Army) and Ulster populations, leading people to think that perhaps the disquiet that had been so characteristic of Northern Ireland since its 1922 inception might be coming to an end. However, such hopes were premature. A bloody bombing in a crowded marketplace in Omagh (60 miles due west of Belfast) in August 1998 was, in fact, the event that most completely set the stage for cautious talks between the two sides split by long-time religious enmity. Both sides declared a cease-fire after the Omagh bombing and political progress toward open discussions and resolution of the problem of clandestine arms and violence began in earnest. While there is no certainty yet that these issues of religion and tradition and political autonomy will be cleanly resolved, in late 1998 there was greater optimism about such an outcome than had been felt for decades.

4.4 Realities and Images of France

France has long been one of the world's more significant countries. It is the largest country by area in Europe as defined in our text (see Table 3.1), and is Europe's leading nation in value of agricultural output. The country is one of the world's major industrial and trading nations and a nuclear power. France has lost its former significance as a major colonial power, but it still maintains special relations with a large group of former French colonies. More than a dozen are now Departments of France and send representatives to the Assembly in Paris. Still another important foundation for France's international importance is the high prestige of French culture in many parts of the world.

Topography and Frontiers

France lies mostly within an irregular hexagon framed on five sides by seas and mountains (Fig. 4.12). In the south, the Mediterranean Sea and Pyrenees Mountains form two sides of the hexagon; in the west and southwest, the Bay of Biscay and the English Channel form two sides; in the southeast and east, a fifth side is formed by the Alps and the Jura Mountains, and, farther north, by the Vosges Mountains, with the Rhine River a short distance to the east. Part of the sixth or northeastern side of the hexagon is formed by the low Ardennes Upland, which lies mostly in Belgium and Luxembourg, but which has broad lowland passageways leading into France from Germany both north and south of the Ardennes.

Aside from the large Massif Central in south central France, all the country's major highlands are peripheral. Even the Massif Central, a hilly upland with low mountains in the east, is skirted by lowland corridors that connect the extensive plains of the north and west with the Mediterranean littoral. These corridors include the Rhône-Saône Valley between the Massif to the west and the Alps and Juras to the east, and the Carcassonne Gap between the Massif and the Pyrenees. Thus, France is a country that has no serious internal barriers to movement.

The state that emerged as modern France was to a considerable degree a natural fortress, as two of its three land frontiers lay in rugged mountain areas. The Pyrenees are a formidable barrier, lowest in the west, where the fiercely independent Basques—neither French nor Spanish in language—occupy an area extending into both France and Spain. In southeastern France, the Alps and Jura Mountains follow France's boundaries with Italy and Switzerland. The sparsely populated Alpine frontier between France and Italy is even higher than that of the

France ● Index Map

URBAN AREAS

- ✪ Over 10,000,000 (national capital)
- ◉ Over 1,000,000
- ● Over 600,000
- ○ Over 275,000
- ○ Over 100,000 (selected places)
- • Selected smaller places

City-size symbols are based on metropolitan area population estimates.

LANDFORMS

- Lowlands
- Hills and uplands
- Low mountains
- Swiss Plateau
- High mountains

Figure 4.12 General location map of France.

Pyrenees: Mont Blanc, the highest summit, reaches 15,771 feet (4807 m). The number of important routes through the mountains is limited, and the Alps have been an important defensive rampart throughout France's history. The Jura Mountains, on the French-Swiss frontier, are lower but also difficult to cross, as they are arranged in long ridges separated by deep valleys.

France's Flourishing Agriculture

Agriculture is a more important part of the economy in France than it is in most highly developed countries. Since World War II, rapid economic change has been accompanied by a massive migration from farms to cities. The number of farmers has fallen to the point that only about 5 percent of the country's labor force is now in agriculture (as of 1998). But this is still higher than it is in many developed countries, and, although their numbers have fallen, France's farmers have become more efficient. Hence, agricultural output has increased greatly, which is a common experience in industrializing countries.

Despite its relatively small size on a world scale, France is a major producer and exporter of agricultural products. Two products—wheat and wine—are of primary importance in the export trade, but France also has large exports of barley, cereal products, corn, dairy products, fruits, and vegetables. For a country with only four-fifths the area of Texas to attain such a large volume of farm exports, some human and physical advantages in agriculture must be present. One advantage often cited is the country's relatively small population compared to its land area. But one must remember that a density of 277 people per square mile is low only by western European standards; it is almost four times the density of the United States (76 per sq mi). Also, France's membership in the European Union is a major advantage. This gives French farmers free access to a huge and affluent consuming market protected by tariffs from non-EU producers. In addition, the European Union budget provides generous price supports at rates above world prices for many agricultural products, with French farmers the leading beneficiaries.

France's physical advantages include superior topographic, climatic, and soil conditions. The main topographic advantage is the high proportion of plains. Large areas, especially in the north and west, are level enough for cultivation. Climatically, most of the country has the moderate temperatures and year-round moisture of the marine west coast climate. The wettest lowland areas are in the northern part of the country, and it is here that crop and dairy production is especially intense. Southern France often experiences notably warmer temperatures, allowing a sizable production of corn in the southwestern plains. Productive vineyards are also concentrated in the south and southwest, although important vineyard areas are found as far north as Champagne, east of Paris. Finally, northeastern France from the frontier to beyond Paris has fine soils developed from loess. As in much of Europe, these exceptionally fertile soils are used principally to produce wheat and sugar beets. This wheat belt in northeastern France is the country's chief breadbasket; the beet production from the same area provides not only sugar but also livestock feed (beet residues from processing) for meat and milk production. Wheat and beets tend to occupy the best soils because both crops are unusually sensitive to soil fertility.

The family farm in France is much more than just an agricultural unit. It is a cultural institution and carries its tradition with an energy that is reflected both in the productivity of the fields and in the politics that surround French agriculture.

French Cities and Industries

The Primacy of Paris

Paris is the greatest urban and industrial center of France, completely overshadowing all other cities in both population and manufacturing development. With an estimated metropolitan population of about 10 million (and 2,152,329 in the city proper), it is by far the largest city on the mainland of Europe (excluding Moscow in Russia). Paris is located at a strategic point relative to natural lines of transportation, but it is especially the product of the growth and centralization of the French government and of the transportation system created by that government. Like London, it has no major natural resources for the industry in its immediate vicinity, yet it is the greatest industrial center of its country.

The Primate City: The Example of Paris

By formal definition, a **primate city** is one which is many times larger (in population) than any other city in the country. In a more common usage, an urban center is the primate city when its demographic, economic, political, and cultural importance far outshines all other cities in a country (Fig. 4.13). Such a city is considered primate by the **hierarchal rule**—meaning that its cultural and economic influence overshadows all other urban centers. In the rank order of urban-center size in France (using city-proper population figures and not the metropolitan conurbation in which these cities are central), Paris leads with 2,152,329; then Marseilles with 800,309; Lyon, 417,479; Nice, 342,903; and Toulouse, 258,598.

The concept of the primate city is, however, a more culturally defined term, and in that sense, there is a singularity in the authority and dominance of Paris, making it an excellent example of a contemporary and popular primate city (Fig. 4.14). The unusual aspect of the Paris example is that a primate city is more often a geographic characteristic in *developing* nations, where the dominant city is the center of everything—industry, politics, culture, and government. In France, the lesser cities have considerable industrial and cultural strength, but they all continue to exist in the shadow of Paris politically.

Figure 4.13 Notre Dame rests on a small island in the Seine River in Paris, standing majestic on its site as a major landscape signature of this most famous European city. The Seine functions as a major transportation artery for the city and the region, but it is better known as a landscape hallmark of the French capital. *Francois Dumont/ Tony Stone Worldwide*

Paris began on an island in the Seine River that offered a defensible site and facilitated crossing of the river. Its early growth, which dates from Roman times if not before, was furthered by its location in a highly productive agricultural area and amidst navigable streams as well as land routes crossing the Seine. The rivers that join in the vicinity of Paris include the Seine, which comes from the southeast and flows northwestward from Paris to the English Channel; the Marne, which flows from the east to join the Seine; and the Oise, which joins the Seine from the northeast (see Fig. 4.13). In recent centuries, these rivers were modified and canals were provided where necessary to give Paris waterway connections with seaports on the lower Seine, with the coal fields of northeastern France, with Lorraine, and with the Saône and Loire rivers (see Fig. 4.1).

In the Middle Ages, Paris became the capital of the kings who gradually extended their effective control over all of France. As the rule of the French monarchs became progressively more absolute and centralized, their capital, housing the administrative bureaucracy and the court, grew in size and came to dominate the cultural as well as the political life of France. When national road and rail systems were built in relatively recent times, their trunk lines were laid out to connect Paris with the various outlying sections of the country. The result was a radial pattern, with Paris at the hub. As the city grew in population and wealth, it became an increasingly large and rich market for goods.

The local market, plus transportation advantages and proximity to the government, provided the foundations upon which a huge industrial complex developed. Speaking broadly, this development involves two major classes of industries. On the one hand, Paris is the principal producer of the high-quality luxury items—fashions, perfumes, cosmetics, jewelry, and so

on—for which France has long been famous. The trades that produce these items are very old. Their growth before the Revolution of 1789 was based in considerable part on the market provided by the royal court; since the Revolution, the production of specialty goods has been favored by the continued concentration of wealth in the city. On the other hand, Paris is now the country's leading center of engineering industries, secondary metal manufacturing, and diversified light industries. These industries are concentrated in a ring of industrial suburbs that sprang up in the 19th and 20th centuries. Automobile manufacturing is the most important single industry, but a great variety of other goods is produced. In addition, the city's economic base includes an endless variety of services for a local, regional, national, and worldwide clientele.

With the more recent growth of Marseilles, Lyon, Arras, and Lille, the dominance of Paris has been steadily modified and diminished. Two ports near the mouth of the Seine handle much of the overseas trade of Paris: Rouen (population: 380,000 metro, 102,722 city), the medieval capital of Normandy, is located at the head of navigation for smaller ocean vessels using the Seine; however, with the increasing size of ships in recent centuries, more and more of Rouen's port functions have been taken over by Le Havre (population: 253,675 metro, 195,932 city), located at the entrance to the wide estuary of the Seine.

French Urban and Industrial Districts Adjoining Belgium

France's second-ranking urban and industrial area, located in the north near the Belgian border, consists of cities clustered on or near the country's once-leading coal field (now closed). A bit north of the field, several cities, of which Lille is the largest, form a metropolitan agglomeration of more than two million people. This region, called the Nord, is both an important contributor to, and a major problem in, the economy of France. In the 19th and early 20th centuries, it developed a rather typical concentration of the coal-based industries of that period, with close juxtaposition of coal mining, steel production, textile plants, coal-based chemical production, and heavy engineering. Such operations were not among France's growth industries in the second half of the 20th century, and the region suffers now with extraordinary employment problems, including classic decline in heavy industries and only modest labor replacement by other industries and service employment.

Urban and Industrial Development in Southern France

France's third largest urban-industrial cluster centers on the metropolis of Lyon (metropolitan area population: about 1.3 million) in southeastern France. The valleys of the Rhône and Saône rivers join here, providing routes through the mountains to the Mediterranean, northern France, and Switzerland. Since prehistoric times, the valley has been a major connection between the North European Plain and the Mediterranean. Today, it forms the leading transportation artery in France, providing links between Paris and the coast of the Mediterranean via superhighways, rails that carry some of the fastest trains in the world, and a barge waterway formed by the Rhône, Saône, and connecting rivers and canals to the north.

In addition, the Rhône-Saône Valley corridor between the Alps and Jura Mountains on the east and the Massif Central on the west also plays a prominent role in French energy production. Recently, France has stressed hydroelectric and nuclear energy to compensate for the inadequacy of its fossil fuel supplies, and the Rhône Valley plays a major role in both. A series of massive dams on the Rhône and its tributaries produce about one tenth of France's electricity, and further power comes from several large nuclear plants in the Rhône Valley using uranium from the nearby Massif Central and foreign sources.

France's principal Mediterranean cities and metropolitan areas—Marseilles, Toulon, and Nice—are strung along the indented coast east of the Rhône delta. Marseilles (population of metropolitan area: 1,231,000) ranks in a class with Lyon and Lille in metropolitan population and is France's leading seaport. It is located on a natural harbor far enough from the mouth of the Rhône to be free of the silt deposits that have clogged and closed the harbors of ports closer to the river mouth. Recently, rapid growth of port traffic and manufacturing in the Marseilles area has been particularly associated with increasingly large oil imports and related development of the petrochemical industry. Toulon (population: 437,825), southeast of Marseilles, is France's main Mediterranean naval base, and Nice (metropolitan population: 517,291), near the Italian border, is the principal city of the French Riviera, probably the most famous resort dis-

Figure 4.14 Bastille Day—July 14th—is a day of international celebration. The French mark this date as the birth of the French Revolution and take great pride in the changes associated with that political upheaval in 1789. It is a festive and wildly popular day in France and among the French all over the world. *Martine Mouchy/Tony Stone Images*

trict in Europe. Along this easternmost section of the French coast, the Alps come down to the sea to provide a spectacular shoreline dotted with beaches. Near Nice, a string of resort towns and cities along the coast includes the tiny principality of Monaco, which is nominally independent but closely related to France economically and administratively.

The Dispersed Cities and Industries of Western France and the Massif Central

Western France and the Massif Central are considerably less populous, urban, and industrial than the parts of France to the northeast and east. The largest urban center, Bordeaux, has only about 700,000 people in its metropolitan area, and the next two larger cities, Toulouse and Nantes, have only about 650,000 and 495,229 respectively. Bordeaux is a seaport on the Garonne River, which discharges into the Gironde Estuary. The city passed its name to the wines from the surrounding region, exported through Bordeaux for centuries. Nantes is the corresponding seaport where the other main river of western France, the Loire, reaches the Atlantic. Toulouse, a very old and historic city with modern machinery, chemical, and aircraft plants, developed on the Garonne well upstream from Bordeaux, at the point where the river comes closest to the Carcassonne Gap. Lying between the Pyrenees and the Massif Central, the Gap provides a lowland route from France's Southwestern Lowland to the Mediterranean.

Recent Dynamism in French Urban and Industrial Development

France shows striking contrasts in the nature of its development before and since World War II. Since the Industrial Revolution until recent decades, the country tended to fall behind its European neighbors, except for Spain and Italy, in industrialization and modernization. It changed enough to become a considerable industrial power while remaining more rural and less urban and industrial. In the decades following World War II, France's old patterns were rapidly modified with new development.

Population growth was encouraged after World War II by government subsidies for children. Inadequate coal resources became less of a handicap as other energy sources increased in availability and importance. Protection of inefficient producers from competition was weakened or abolished by France's membership in the European Union. The French government promoted the combination of companies into larger units capable of higher efficiency and international competitiveness. France was well placed, in both site and situation, to move into newly developing industries and technologies. It had a large pool of labor available for transfer from an over-manned agriculture sector, a relatively modest share of its manpower and capital tied up in older industries, and an excellent educational system. Not the least of its advantages was the adoption of a future-oriented national economic planning system following the conclusion of World War II.

Tourism.   Tourist flows of the late 1990s have now made France the world's number one tourist destination. In addition, Paris is the most visited city. In the late 1990s, annual income from tourist activity in France amounted to more than $30 billion. Major tourist destinations included the traditional Paris landmarks as well as the redone Disneyland Paris, the Cote d'Azur in the south of France, and the Cannes film festival. The real impact of tourism in France, however, can be seen in the small-scale ways in which it has reached the Mediterranean coast, the mountain flank towns and villages, and the urban sidewalks and ceremonial spaces all through the country. The widely popular image of romance and conversation at outdoor cafes, and of leisurely meals with fine wine over which memorable bicycle, pedestrian, train, barge, and car trips are recounted, is evidence of the major impact of tourism on the well-being of the French economy. These landscape elements that may seem so Parisian or French in their origin have now been widely disseminated all across the globe. Tourism is a truly global activity and generates significant economic impact wherever it flourishes.

French National and Regional Planning

The new dynamism of France has been achieved with deliberate planning. Since 1947, the country has engaged in national and regional planning with the stated objectives of improving national output and living standards, reducing the considerable differences in prosperity of different sections of the country, and making France a pleasant place to live in a technological age. This type of government planning is envisaged as complementary to the activities of private enterprise, with the state using various means to influence the decisions of private companies while at the same time cooperating with them.

In addition to fostering overall national growth and improvement, French planning is directed toward reducing imbalances of long standing among the regions of France. In the mid-1960s, the government called particular attention to the contrasting levels of development among four major regions—Paris, the Northeast, the Southeast, and the West (Fig. 4.15). The most significant contrasts were (1) between the Paris region and the rest of the country and (2) between the less developed and poorer West and the rest of the country. Because modern economic growth is concentrated strongly in cities, the French planners are attempting to divert development from Paris to the main cities of other regions. They use restrictions and penalties on various types of new development in the Paris region, a varied array of financial rewards for firms locating facilities in regions of greater need, and direct location or relocation of government-owned facilities in the regional centers.

An outstanding success story in the promotion of regional centers is the rapid growth of Toulouse in southwestern France as a major European center for the aircraft and aerospace industries. Within the Paris region itself, a similar policy of economic and population decentralization continues, with functions such as wholesale markets and universities moving from the old city to the suburbs and several large new planned cities being constructed in the suburbs. Paris nonetheless maintains its economic and cultural primacy through an increasing importance in high-tech locations and administrative function in the capital.

Although there has been a real decentralization of growth in France during the planning period, some analysts believe that decentralization might well have occurred without any planning policy. They maintain that after manufacturing industries reach a certain stage of development, the work in them becomes so standardized and routine that little skill is required from the workforce, and such businesses then tend to relocate from industrial cities to previously nonindustrial places where lower wages can be paid to unskilled and nonunionized labor. This point of view sees French economic decentralization as a normal result of the evolution of industries, and it questions much

Figure 4.15 France, with its four regions of contrasting levels of economic development: Paris is at the highest level, with declining development from the Northeast to the Southeast to the West.

of the planning process as an expensive irrelevancy. However valid such criticisms may be, some decentralization of jobs and prosperity in France is essential for political stability and national cohesion.

4.5 The Pivotal Role of Germany

Germany reappeared on the map of Europe as a unified country in 1990. Between 1949 and 1990, there had been two Germanies: West Germany (German Federal Republic), formed from the occupation zones of Britain, France, and the United States after Germany's defeat in World War II in 1945; and Communist East Germany (German Democratic Republic), in territory occupied by the former Soviet Union. The urban landscape and political climate of West Berlin was a part of West Germany for those decades although the city of Berlin overall was entirely surrounded by East Germany. The withdrawal of Soviet support for East Germany in 1989 led to the rapid collapse of its Communist government in 1990, the reunification of Germany, and the de facto inclusion of East Germany in the EU.

The new Germany is, by a wide margin, the most dominant country of Europe. Its population of 81 million is much greater than that of any other nation in the region. Economic consider-

ations are even more important. The former West Germany alone, with 61 million people, was Europe's leading industrial and trading state and was the principal focus of the economy of western Europe. The former East Germany was an advanced country by Soviet bloc standards but an economic disaster by West German standards. Many billions of dollars and much time will be required to rehabilitate Germany's East, but this area can be expected to contribute increasingly to overall German dominance in the European and global economy, even though there is a continuing irritation felt by the (once) West Germans because of the tax costs they are having to pay for the national efforts to bring (once) East Germany into fuller employment and more productive economic levels of activity.

The Significance of the German Environment

Germany arrived at its favorable economic position by skillfully exploiting the advantages of its centrality within Europe, together with certain key features of a diverse environment (Fig. 4.16). Broadly conceived, the terrain of Germany can be divided into a low-lying, undulating plain in the north and higher country to the south (Fig. 4.17). The lowland of northern Germany is a part of the much larger North European Plain. The central and southern countryside is much less uniform. To the extreme south are the moderately high German (Bavarian) Alps, fringed by a flat to rolling piedmont or foreland that slopes gradually down to the east-flowing upper Danube River. Between the Danube and the northern plain is a complex series of uplands, highlands, and depressions. The higher lands are predominantly composed of rounded, forested hills or low mountains. Interspersed with these are agricultural lowlands draining to the Rhine, Danube, Weser, or Elbe Rivers.

The climate of Germany is maritime in the northwest and increasingly continental toward the east and south. The annual precipitation is adequate but not excessive, with most places in the lowlands receiving an average of 20 to 30 inches (c. 50 to 75 cm) per year. Farmers, however, must contend with soils that generally are rather infertile. Two areas are major exceptions:

1. The southern margin of the North German Plain forms a belt with soils of extraordinary fertility formed from deposits of **loess.** This belt continues beyond Germany both east and west.
2. The long, nearly level alluvial Upper Rhine Plain, extending from Mainz to the Swiss border and mostly enclosed by highlands on both sides, is also an area with fertile agricultural soils.

Locational Relationships of German Cities

Within the environmental framework just described, Germany's major cities tend to be located in the most economically

Regional Perspective

The Geography of Discontent

With the steady expansion of the European Union over the past decade, and the number of nations that have already asked to be considered for membership in the EU, it seems strange that there is a simultaneous devolution of European political units from larger to smaller. Yet, looking at the tensions in the fall of 1998, one can see that political and ethnic forces are at work seeking new political autonomy for all sizes of places. In Germany, for example, there has been such a focus on not letting too much political power become centered in one locale again that there has been major power given to the sixteen German *Lander* (states). This has enabled these separate political units to play significant roles in modifying or even blocking reform or federal projects. This **devolution**—the dispersal of political power and autonomy to smaller spatial units—of the power structure of the federal government has generally been coupled with democratic reforms.

In the fall of 1997, Scotland voted 3 to 1 in favor of establishing their own parliament, making a move toward greater independence. Scotland is similar in size and population to Denmark, Finland, or Ireland, so the success of an in-dependent Scotland will have increasing attraction to those who are seeking more freedom from the governments that have controlled their economies for decades, even centuries.

Ever since the 1975 death of Generalissimo Franco, Spain has given more latitude to 17 new regions, including those of the Basques and the Catalans—who have been famous for seeking such authority at the cost of bloodshed and continuing belligerence. France has been following the same theme with its own expansion of independence to 22 new mainland regions outside of Paris. It is hoped by these central governments—not just in Spain and France, but elsewhere in Europe as well—that this devolution, accomplished in an ordered and peaceful manner, will quell the strong separatist movements that have sparked fights and bad blood through much of Europe.

Another aspect of this process took place in the former Yugoslavia for most of the 1990s. In the winter of 1999, Kosovo—a subsection of Serbia in the former Yugoslavia—made world news after the North Atlantic Treaty Association (NATO) undertook to bomb Slav forces out of the countryside of Kosovo, where the population is 90 percent ethnic Albanian. This devolution through efforts to develop new political units bonding similar peoples to other similar peoples —called ethnic cleansing in some contexts—has the potential in Europe of continuing as a subtheme of political geography for the indefinite future.

This potential for atomization of places in Europe into ever smaller, more homogeneous units is in stark—but not unrealistic—contrast to the continuing fiscal union being sought by the member nations of the EU, particularly as the euro currency became an electronic currency in early 1999, and coin and bill in 2002. In some situations it is even felt that the two forces—as contradictory as they seem in terms of simple geography —may even support each other. For example, some feel that the EU's economic growth for economic reasons may help nudge Ireland and Northern Ireland closer together, even though religion has kept them distant and unwelcoming to each other for most of the 20th century.

The map in Figure 4.A shows the hot spots for the geography of discontent.

advantageous areas and those most strategic with respect to transportation. A string of major metropolitan areas is located in the western part of Germany, along and near the Rhine River (see Fig. 4.16). This waterway has been a major route since prehistoric times. Four of the eight largest German metropolises—Düsseldorf, Cologne, Wiesbaden-Mainz, and Mannheim-Ludwigshafen-Heidelberg—are on the Rhine itself. The other four—Essen, Wuppertal, Frankfurt, and Stuttgart—are on important Rhine tributaries. This north-south belt of cities also includes several other large metropolitan centers, such as Dortmund, Duisburg, and Aachen, plus numerous smaller places. The concentration of seven of the forenamed cities at or just north of the boundary between the Rhine Uplands and the North German Plain reflects additional conditions (Fig. 4.18).

Historically, some of the seven cities profited from their location on the agriculturally excellent loess soil just north of the Uplands. This concentration of large and medium-sized urban centers differentiates the German landscape from the French world, in which Paris plays such a profoundly singular and dominant role.

Other German cities developed as early industrial centers before the Industrial Revolution, drawing on the Uplands for ores, waterpower, wood for charcoal, and surplus labor. Then, in the 19th and early 20th centuries, all seven cities grew in connection with the development of both the Ruhr coal field, located at the south edge of the Plain and east of the Rhine, and the smaller Aachen field. Except for Aachen and Cologne, the cities became part of "the Ruhr," Europe's greatest concentration of

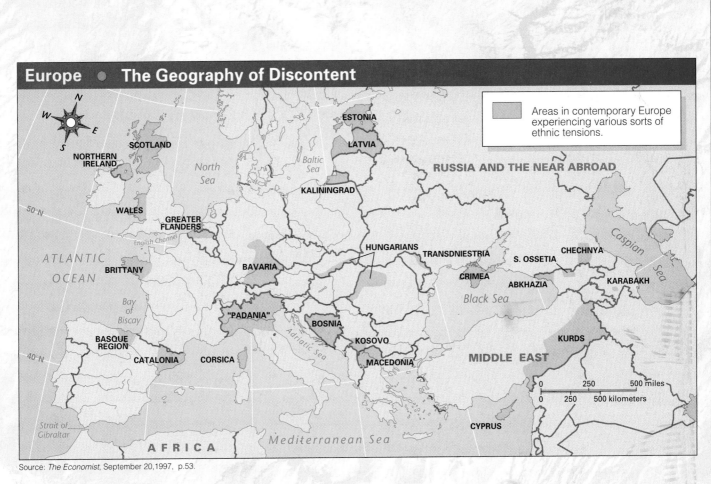

Europe • The Geography of Discontent

Source: *The Economist*, September 20, 1997, p.53.

Figure 4.A   Areas in contemporary Europe experiencing various sorts of ethnic tensions.

coal mines and heavy industries within a network of active river transport. Still other important German cities lie apart from the Rhine and Ruhr. These include (1) Hanover, Leipzig, and Dresden, spread along the loess belt across Germany; (2) the North Sea ports of Hamburg, located well inland on the Elbe estuary, and Bremen, up the Weser estuary; (3) two cities in the southern uplands—Munich, north of a pass route across the Alps, and Nuremberg; and (4) Berlin, whose large size belies its unlikely location in a sandy and infertile area of the northern plain. Berlin is mainly an artificial product of the central governments—first of Prussia and then of Germany—which developed it as their capital. Berlin's metropolitan area population is 4,150,000 and Hamburg's is 2,385,000; the remaining cities previously named vary between 500,000 and 2 million.

Geographic Evolution and Problems of Germany's Industrial Economy

Germany's development into one of the world's greatest industrial areas has roots in its distant past. Certain areas, such as the Harz Mountains, were exceptionally important centers of European mining in the Middle Ages, and many urban handicraft industries were also well established in medieval times. By about 1800, there were three German regions with unusually high concentrations of small-scale industries. One was the Rhine Uplands, in which industry spread along both sides of the river north of Wiesbaden and also away from the river in both directions, but especially toward the east. A second was Saxony, north of and including the Erzgebirge (Ore Mountains) on the

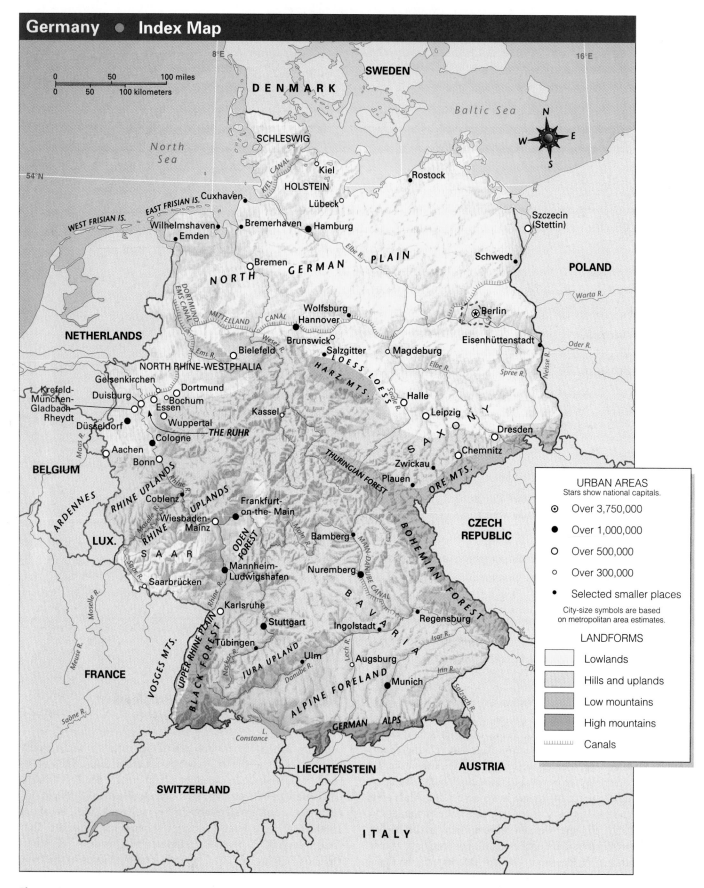

Figure 4.16 General location map of Germany.

German-Czech border. The third was Silesia (now in Poland), consisting of the plains along the upper Oder River—which contained the major mineral resources of Upper (southeastern) Silesia—and the Sudeten Mountains adjoining the plains. All three regions mined metallic ores, smelted the ores with charcoal, used the waterpower of mountain streams, and produced textiles as well as metals and metal goods.

In the 19th and early 20th centuries, each of the foregoing regions developed into a major center of coal-based industry. As the Industrial Revolution spread into Germany in the early 19th century, each area discovered coal resources that were conveniently located and, in the case of the Rhine area and Upper Silesia, extremely large. The Ruhr coal field, located at the north edge of the Rhine Uplands (see Fig. 4.18), is the greatest coal field in Europe, and the Upper Silesia field is also of major importance. It is in this region that the firms with the names of Krupp, Thyssen, and Stinnes had their origins. These same firms continue to be the global players in the mergers and the industrial expansion so characteristic of the globalization of the European economy in the late 20th century. In both areas, the abundance of coal favored the growth of large iron and steel industries, which had become increasingly dependent on imported ores. Secondary development—such as metal goods, machinery, and coal-based chemicals—evolved much more in the Ruhr than in Silesia, perhaps because of the Ruhr area's much better market position. Today, "the Ruhr" denotes an area of somewhat indefinite extent beyond the coal field proper. The "Inner Ruhr" (total population: about 5 million) denotes the more concentrated urban-industrial zone incorporating Essen, Duisburg, and Dortmund, whereas the "Outer Ruhr" includes Düsseldorf and other more widely spaced cities.

In Saxony, coal resources were much poorer. Brown coal and lignite fields were eventually developed, but the heat produced by these resources is only about one quarter of that produced by bituminous coal, which yields coke for the iron and steel industry. Hence, an iron and steel complex did not arise in Saxony, although factory-type industries grew rapidly and promoted the growth of Leipzig, Dresden, and other cities. Machinery, precision instruments, textiles, and chemicals were (and are) the main emphases. Saxony's industrial area is roughly triangular and is often called the Saxon Triangle (see Fig. 4.1).

Much of Germany's present industry, however, is not within the Ruhr or Saxony. A wide scatter of industry and of urban centers is characteristic. Many cities are large, but none is as gigantic and nationally dominant as London or Paris. The number and scatter of fairly large industrial cities apparently resulted in good part from two circumstances:

1. Germany was divided for a long time into petty states, a number of which eventually became internal states within German federal structures. These units often promoted the growth of their own capitals, and many of the large cities of today were once the capitals of independent German states.
2. German cities were able to secure power from a distance

Figure 4.17 The meadows that flank the Wetterstein Range in Bavaria in south Germany frame the images that make this farming area so popular for hikers and tourists. The Bavarian Alps that rise in the background are part of the European mountain system that has played a major role in borders and battles for centuries in this region. *Josef Beck/FPG International*

during the age of industrialization. In the sequence of German development, railways generally reached major cities before coal became practically an industrial necessity. Thus, when coal was needed, the cities away from the fields could generally get it and continue to industrialize and grow. Waterways also helped many cities get coal; power grids then began to provide electricity from a distance.

Today, there is a major tendency in former West Germany, as in many other advanced industrial countries, for factory industry to shift from the cities to more suburban locations and to rural areas. The attractions appear to be lower wages, lower site costs, and environmental amenities. The shift is allowed by the wide availability of power in the form of electricity, access by superhighways, and the ability and increasing willingness of labor to commute long distances by car or rail.

The former West Germany recovered rapidly from World War II. The United States Marshall Plan (officially called the European Recovery Program, but better known for its namesake, U.S. general George Marshall) was authorized in 1948. From 1948 until 1952 the United States funneled more than $13 billion in capital goods, as well as management expertise and political stability, into West Germany in an effort to keep Communism from spreading west into that wartorn region. A new

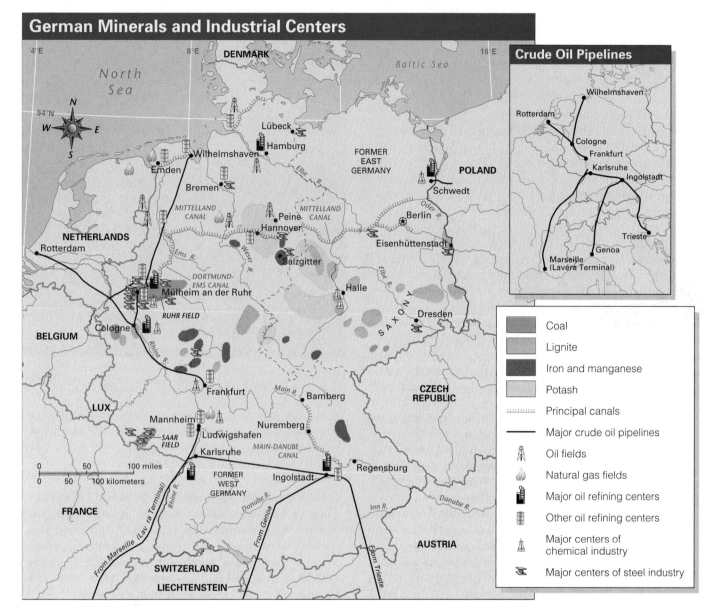

German Minerals and Industrial Centers

Crude Oil Pipelines

Legend:
- Coal
- Lignite
- Iron and manganese
- Potash
- Principal canals
- Major crude oil pipelines
- Oil fields
- Natural gas fields
- Major oil refining centers
- Other oil refining centers
- Major centers of chemical industry
- Major centers of steel industry

Figure 4.18 Major mineral deposits and iron-and-steel manufacturing centers, chemical-manufacturing centers, and oil-refining centers in Germany. Dashes show the boundary between former East Germany and former West Germany.

currency was provided in 1948 and a government empowered to act was formed in 1949. Former production levels were regained and surpassed. German cities were badly bombed in World War II (Fig. 4.19), but industrial capacity was not as badly damaged as is sometimes suggested. Various circumstances facilitated the process of economic rebuilding, not the least of which were the rather freewheeling free-enterprise approach followed by the government and the massive aid received from the United States in those early years. The skills of the population were still there, and labor was cheap for a time, as workers were abundant and glad to be working again.

Markets were hungry for industrial products. By the 1960s the pace of growth slowed, but it was still fast enough to provide jobs for large numbers of foreign workers (sometimes known as **guest workers**) who came in to take jobs vacated by upwardly mobile German workers.

Some analysts cite a further factor as an important contributor to cheap and abundant labor for a time—a rapid and large decrease in the farm population, releasing large numbers of workers to nonfarm work, which usually proved more remunerative. Germany had traditionally maintained a relatively large proportion of its people on the land by restricting imports and

Figure 4.19 The resurgence of Germany from the ruins of World War II is dramatically illustrated by this photo of Frankfurt-am-Main compared with the photo of the shattered city in 1949.   *Left: H.P. Merten/TheStock Market. Right: AP/Wide World.*

raising the prices of foreign food products competitive with German products. The situation then changed when the European Common Market deprived the West German farmer of protection from European competitors at the same time that industrial expansion opened new jobs. The resulting exodus of farmers from the land was so great that the West German farm labor force declined from 5.2 million to 1.7 million in the period between 1950 and 1976, freeing millions of workers for industrial and service employment. Even this decline, with the accompanying consolidation of farms, left most farms too small to provide, by West German standards, a good living for their operators.

The pace of development slowed during the 1970s and 1980s, although West Germany continued to be a very wealthy and productive country. West German coal became increasingly unable to compete with the imported petroleum which now replaces it as Germany's main energy source. New plants to generate electricity with nuclear energy have sparked violent antinuclear protests, and the need for cheap energy led West Germany to quarrel with the United States in the early 1980s in order to complete a deal for Soviet natural gas from a new Russia-to-Europe pipeline. The export of German industrial products to world markets became more difficult as Japanese and other foreign competition stiffened. However, despite the various difficulties of the former West Germany's "economic miracle," it ranked with the United States and Japan as one of the world's three largest exporters of goods at the time of the 1990 reunification.

The former East Germany (1949–1990) was Communist-ruled and Soviet-dominated, but it was still German. Even Soviet exploitation and the inefficiencies of bureaucrats attempting to plan and operate a "command economy" were unable to destroy completely a historical and traditional German capacity to produce. East Germany was the most productive of the Soviet-launched and closely managed countries that lay between Russia and western Europe. When East Germany collapsed and extensive outside investigation became possible, it was revealed to be a poor and inefficient country by Western standards. Images of life in West Germany, and the West in general, led to an upwelling of desire in its population to leave East Germany (Fig. 4.20), stimulating major migration streams from East to West Germany after the 1989 opening of the Wall. Many billions of dollars of investment and considerable economic pain over a number of years will be needed to bring the economy of what is now the eastern part of reunified Germany closer to the levels of productivity and living standards in the western part. Huge costs are also involved in the cleanup of severe environmental damage incurred under the East's Communist rule.

Germany as a Migrant Destination. In June 1997, the EU countries signed the Amsterdam Treaty, which committed the European Union to creating an "area of freedom, justice, and security" within five years. This was intended to mean that frontiers among the 15 members of the EU would be open and allow free movement. However, increases in migration from

Figure 4.20   Berlin citizens from both sides of the Berlin Wall watch its demolition in June 1990. This structure had served as one of the most powerful icons of the Cold War since its construction in 1961.   *AP/Wide World Photos*

Turkey—which was turned down for EU membership in 1998—and from Albania, Bosnia, and other Balkan areas caught in ethnic warfare have made Germany particularly uneasy about the open border policy. The current pattern in the EU is to accept migrants who seek political asylum; if they are turned down, they are given two weeks to leave the EU country. Since Italy and Austria are the gateway EU countries for major migrant streams, when immigrants are turned away there, they often are officially "lost" and then tend to move north and west, into the heart of the EU nations.

This knocking on the door has been positive in one sense because of the relatively low (or even negative) population growth in most of the EU nations, particularly in Germany and France. And immigration was important to the dramatic economic growth of the 1960s and 1970s. With the slowdown and the high rates of domestic unemployment in the 1990s, however, Germany has found the migration scenario to be more troublesome. Table 4.1 shows the sources of current immigration in Germany and the changing magnitude of the Turkish migration stream.

The Political and Spatial Shift from Bonn to Berlin.   When Bonn was made the capital of West Germany in 1949, it was thought that this was only a temporary shift, one that was necessitated by the enclosure of the traditional capital city of Berlin by hostile Soviet control (see Fig. 4.20). In 1991, after the political reunification of Germany, it was decided to move the national capital from Bonn back to Berlin. This process is slated to be completed by the end of 1999 so that the year 2000 will see Germany governed, again, from Berlin. In this move, Bonn will be left with the *Bundesrat* (federal council) as well as eight federal ministries.

Part of this political shift is being set in motion by Europe's largest building project: the Potzdamer Platz (see chapter opener photo). In late 1998 it was said that there were six Berlins visible in the 17-acre construction site in the heart of old Berlin: (1) the baroque city that was fashioned in the last years of the 18th century by Frederick the Great; (2) the imperial Berlin of Bismarck and Kaiser Wilhelm II (spanning the creation of modern Germany in the 1870s through World War I); (3) the architectural remnants of the hyperinflated Weimar Republic of the 1920s; (4) Nazi Berlin of 1933–1945; (5) the postwar Berlin that was divided into its western enclave and the eastern city run by Moscow until 1989; and (6) the unified Berlin that has been preparing itself to stand again as the capital of a unified Germany. In terms more easily grasped by American readers, Potzdamer Platz is often called "old Berlin's 'Times Square.'"

The Berlin that is promoting the massive rebuilding project with its profound images of a new European unity ("We are the laboratory of unity," said Berlin's mayor, Eberhard Diepgen, in 1988) has gained a visibility that only a few anticipated. The city has created a vivid red observation platform overlooking the Potzdamer Platz site, which has received more than two million visitors, who, the mayor claims, are drawn by "construction site tourism" (see photo on p. 92).

Berlin has more than 4,150,000 people in its metropolitan area and 3,472,000 in the city proper. Bonn has a population of 575,000 in its metropolitan area and 293,000 in the city proper. The shift of the capital from west to east, from the present to the past (and future), is a geographic shift of major political and cultural magnitude.

4.6 Benelux: Trade Networks as Resources

Belgium, the Netherlands, and Luxembourg have been closely associated throughout their long histories. Today, they are often referred to collectively as "the Benelux Countries." An older name often applied to the three is "the Low Countries." In its

Table 4.1 Recent Migrants in Germany

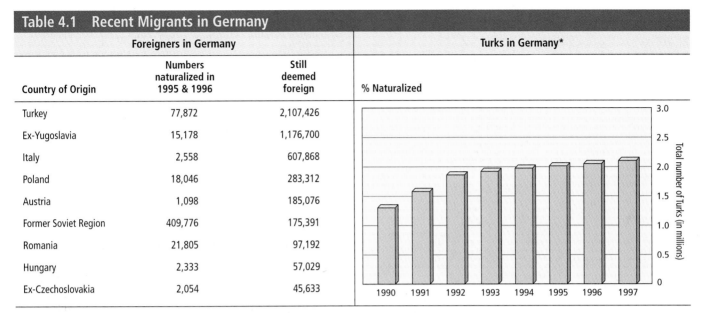

	Foreigners in Germany		Turks in Germany*
Country of Origin	Numbers naturalized in 1995 & 1996	Still deemed foreign	% Naturalized
Turkey	77,872	2,107,426	
Ex-Yugoslavia	15,178	1,176,700	
Italy	2,558	607,868	
Poland	18,046	283,312	
Austria	1,098	185,076	
Former Soviet Region	409,776	175,391	
Romania	21,805	97,192	
Hungary	2,333	57,029	
Ex-Czechoslovakia	2,054	45,633	

Sources: German Interior Ministry; *The Economist*, Vol. 348, July 4, 1998.　　　　* Estimate

strictest sense, however, this term is properly applied only to Belgium and the Netherlands, because approximately the northern two-thirds of the former and practically all of the latter consist of a very low plain facing the North Sea (Fig. 4.21). This plain is the narrowest section of the North European Plain (see Fig. 4.16). Large sections near the coast, especially in the Netherlands, are indeed below sea level and have been made into major farming and urban landscapes through the engineering of **polders**—land that lies below sea level but that has been protected from the sea by dikes and canal systems (Fig. 4.22).

The economic life of the Benelux countries is characterized by an intensive and interrelated development of industry, agriculture, and trade. An especially distinctive feature of their economies is a remarkably high development of, and dependence on, international trade.

Ports of the Netherlands and Belgium

Exploitation of commercial opportunities by the Low Countries is reflected in the presence of three of the world's major port cities within a distance of about 80 miles (*c.* 130 km): Rotterdam and Amsterdam in the Netherlands and Antwerp in Belgium. Amsterdam (population: 1.2 million metro) is the largest metropolis and the constitutional capital of the Netherlands, although the government is actually located at The Hague (population: 773,000). The city was the main port of the Netherlands during the 17th and 18th centuries and for most of the 19th century, while the Dutch colonial empire was expanding. The principal element in this empire was the Netherlands East Indies (now Indonesia), where the Royal Dutch Shell Company originated. Until the loss of Indonesia after World War II, Amster-

dam built a growing trade in imported tropical specialties. Many of these imports were re-exported, often after processing, to other European countries. It also became, and remains, an important manufacturing center. Profits accumulated during these centuries supplied funds for the large overseas investments of the Netherlands and made Amsterdam an important financial center, which it continues to be.

Amsterdam also remains an important port, but it has lost much of its relative position in this respect during the past century. The original sea approach via the Zuider Zee—a landlocked inlet of the North Sea—proved too shallow for the larger ships of the 19th century. The opening in 1876 of the North Sea Canal solved this problem for a time, but even successive expansions of the canal did not give access to the sea comparable to that of Rotterdam. And the Dutch loss of Indonesia in 1949 greatly decreased the port's importance as an entrepôt, although continuing ties with its former colony have given Amsterdam a large Indonesian community that imparts a distinctive ethnic and cultural flavor to the city.

Rotterdam (population: 1.1 million, metropolitan area; 598,236, city proper), the world's largest port in tonnage handled, is better situated for Rhine shipping than either Amsterdam or Antwerp, being located directly on one of the navigable distributaries of the river rather than to one side (see Fig. 4.21). Accordingly, Rotterdam controls and profits from the major portion of the river's transit trade, receiving goods by sea and dispatching them upstream by barge, and receiving goods downstream by barge and dispatching them to sea. Two developments are mainly responsible for the port's tremendous expansion. First was the opening in 1872 of the New Waterway, an artificial channel to the sea far superior to the shallow and

Figure 4.21 General location map of the Benelux countries: Belgium, the Netherlands, and Luxembourg. Note the extensive area of polder land reclaimed from the sea.

treacherous natural mouths of the Rhine and the sea connections of either Amsterdam or Antwerp. Second, and more fundamental, is the increasing industrialization of areas near the Rhine (particularly the Ruhr district), for which Rotterdam has long been the main sea outlet. A large addition to Rotterdam's facilities known as Europoort lies along the New Waterway at the North Sea entrance to the Rotterdam port area and is specially designed to handle supertankers and other oversized container carriers (Fig. 4.23).

Antwerp (population: 700,000), located about 50 miles (80 km) up the Scheldt River, is primarily a port for Belgium itself but also accounts for an important share of the Rhine transit trade. Belgium's coast is straight and its rivers shallow, so that the deep estuary of the Scheldt gives Antwerp the best harbor in the country, even though it must be reached through the Netherlands. Transit trade is facilitated by the fact that Antwerp lies slightly closer to the Ruhr than does either Rotterdam or Amsterdam.

Figure 4.22 Windmills are a common feature in the polder landscape of the Low Countries. This scene from Alkmaar, Hoorn, in the Netherlands shows the fertile farmland resulting from the poldering process. *Tony Stone Images*

Fig. 4.23 The industrial and commercial complex of Rotterdam and Europoort serve to illustrate the continuing significance of water transportation in all of Europe, especially northwest Europe. This break-of-bulk point serves a major role for trade coming from and going to a wide world network. This continues to be the largest port on the Rhine River and in Europe. *Sam. C. Pierson, Jr./Photo Researchers, Inc.*

Industrial Patterns in Benelux

All the Benelux countries are highly industrialized. Before major discoveries of natural gas were made in the northeastern Netherlands in the 1950s, Belgium and Luxembourg were better provided with mineral resources. This helped them become somewhat more industrialized than the Netherlands. Now, however, the Netherlands, with its advantageous trade position and natural gas, is the leading country within Benelux in total value of industrial output.

Heavy Industry in the Sambre-Meuse District and Luxembourg

One of Europe's major coal fields crosses Belgium in a narrow east-west belt about a hundred miles (161 km) long (see Fig. 4.2). It follows roughly the valleys of the Sambre and Meuse Rivers and extends into France on the west and Germany on the east. Liège (population: 750,000) and smaller industrial cities along this Sambre-Meuse field account for most of Belgium's metallurgical, chemical, and other heavy industrial production. A sizable iron and steel industry developed here well before the Industrial Revolution, using local iron ore and charcoal. Subsequently, local coal and the early adoption of British techniques made Liège the first city in continental Europe to develop modern large-scale iron and steel manufacturing. The Sambre-Meuse district also specializes in smelting imported nonferrous ores, with much of the metal being exported, though it now suffers from the customary problems of old, coal-based heavy industrial districts. Coal production has declined drastically in the face of competition from imported oil and even coal imported from better and less depleted fields in North America and else-

where. International competition has also adversely affected the iron and steel industry. Newer industrial plants have generally located in northern Belgium, where unions are weaker and wages lower. Belgium's capital and largest city, Brussels (population: 2,390,000, metropolitan area), is the leading center of a breed of new, less energy-dependent cities, where industry is able to utilize less skilled, nonunion employees.

Luxembourg's most important exports come from the iron and steel industry, carried on in several small centers near the southern border where the Lorraine iron ore deposits of France overlap into Luxembourg. Production is on a large scale, and the very small home market both necessitates and permits export of most of the metal. Luxembourg is attempting, with some success, to follow the European pattern in diversifying its manufacturing industries and exports.

Industries in the Netherlands

Until after World War II, the Netherlands had a smaller development of industry than might have been expected of a country in the heart of western Europe. It had no internal power resources except one small coal field (now closed) in the southernmost province called Limburg. The country's dense population provided cheap labor, and its location allowed cheap import of fuel and cheap export of products. So the Netherlands developed no heavy industrial region but had a rather scattered development of light industries such as textiles and electrical equipment. This situation changed after World War II as imported oil increasingly supplanted coal in the industries of western Europe. More oil imports now enter Europe via Rotterdam-Europoort than any other port, with some arriving as crude oil and some refined further in the Rotterdam area before forwarding. This activity has generated one of the world's largest concentrations of refineries. Petrochemical plants near the refineries represent the main branch of the sizable Dutch chemical industry. Port locations and dependence on port functions also characterize many other important Dutch industries and activities, particularly the operation of a good-sized merchant fleet. Port activities reflect not only the country's present emphasis on trade but also a long maritime tradition. The Dutch played a prominent part in European discoveries and colonization overseas, and for a brief time in the 17th century were probably the world's greatest maritime power. In the early part of the 20th century they were able to still capitalize on their earlier dominance when Indonesia became a colony of the Dutch in 1910. Major petroleum reserves were discovered and exploited during the following four decades, and when Indonesia gained independence in 1949, the Dutch had significant oil holdings in the East Indies.

Certain other industries of consequence are in the major ports but also in smaller Dutch cities. Among them are the food processing, electrical machinery, textile, and apparel industries. Some food industries process imported foods for European distribution, but more of them handle products of the remarkably productive Dutch agriculture. Certain industries tend toward interior locations having relatively cheap, productive labor and good transport connections. Among the best known are the electrical and electronics industries.

The Distinctive Agricultural Patterns of the Low Countries

Both the Netherlands and Belgium are very productive agriculturally, and are characterized particularly by extremely high yields per unit of land. In fact, yields are so high that the total output of farm products is surprisingly great for such small countries. Such productivity results mainly from the following circumstances:

1. *Fertile soils.* Parts of each country have very fertile soils, the largest such areas being the polders (diked and drained lands) in the Netherlands and the loess lands of Belgium.
2. *Farm labor.* Despite rapid decreases of farm population in recent times, there is still an intensive use of labor on the small farms that characterize these countries.
3. *Highly modified landscape.* There is an intensive application of capital in such forms as water control, agricultural machines, knowledge, technical expertise, and especially fertilizer. Both countries are export producers of nitrogen fertilizers made from oil, natural gas, or coal, and both countries have developed specialty crops and very productive greenhouse agriculture.
4. *Role of the EU.* Agriculture in the Low Countries has profited from Common Market tariff protection and subsidies for agricultural producers. These have stimulated spectacular increases in output, to the extent that overproduction of certain products such as milk and butter has become a more worrisome problem than food supply, even in such crowded countries as these.
5. *Trade networks.* Belgium and the Netherlands profit agriculturally from trading relationships that enable the two countries to specialize in farm commodities particularly suited to their conditions. The dense and affluent populations in and near the Low Countries provide a large market for such specialties. Requirements for other commodities such as grains are met increasingly by imports.

Definitions & Insights

THE POLDER AND THE CULTURAL LANDSCAPE

One of the most definitive elements of the Low Countries landscape is the presence of land reclaimed from the North Atlantic. This pattern exists in other parts of the world, but the leading example of this expansion of the cultural landscape is the **polder land** in the Netherlands. These are lands that have been surrounded by dikes and artificially drained. The process of turning former swamps, lakes, and shallow seas into agricultural land has been going on for more than seven centuries. An individual polder, of which there are a great many of various sizes, is an area enclosed within dikes and kept dry by constant pumping into the drainage canals that surround it. About 50 percent of the Netherlands now consists of an intricate patchwork of polders and canals, and Belgium has a narrow strip of such lands behind its coastal sand dunes. Even though the polders are the best agricultural lands in the Netherlands and production from them is the heart of Dutch farming, this land has become increasingly important

in accommodating the continuing growth of the urban landscape. These highly manipulated landscapes—perfect examples of a combination of tradition and modern engineering to expand a limited farmland resource—represent the conflict created by the modification of a natural landscape—the shallow seascape—into a more productive cultural landscape. In a major 1953 flood, more than 1800 people drowned, providing further evidence of nature's capacity to remind engineers of the potential danger in massive landscape transformation. Polders not only underlie much of the extremely productive agriculture of the Low Countries, but they are also central to tourist images of Holland: windmills, dikes, and gardenlike farmscapes. The entire scene, however, is always subject to threats from the sea at times of storm. Like all modified and artificial landscapes, polders require strict human maintenance to continue the unusually high levels of productivity that have been wrested from nature.

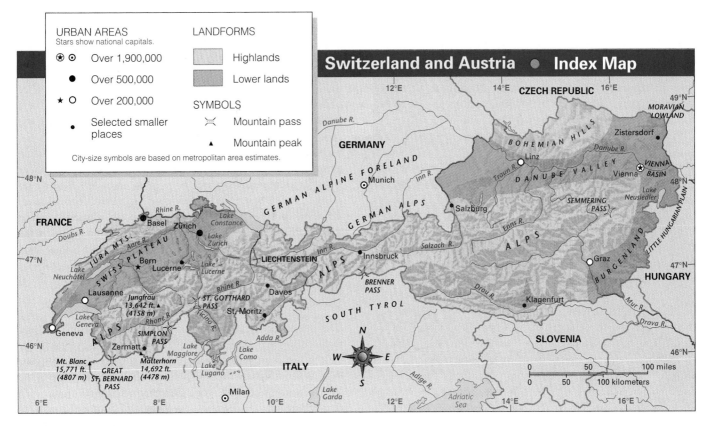

Figure 4.24 General location map of Switzerland and Austria.

Population and Planning Problems

Countries as intensively developed as the Netherlands and Belgium, with high standards of living and high aspirations, must consider how development can be guided to provide an environment offering pleasant living conditions for their crowded populations. In the Netherlands, such planning problems are especially acute as this country has a population density higher than virtually any country of Europe. It also has a rapidly growing population. In this situation, the Dutch are trying to limit urban sprawl, provide more outdoor and waterfront recreational areas, and, in general, maintain a pleasant and varied environment. They are especially concerned with maintaining the open space that is semiencircled by Amsterdam, Utrecht, Rotterdam, The Hague, and other cities. These urban areas are tending to coalesce into what is referred to as Randstad Holland—the "Ring City" of Holland.

4.7 Switzerland and Austria

The two small countries of Switzerland and Austria, located in the heart of Europe, have many environmental and cultural similarities but are remarkably different in their historical development. Switzerland represents perhaps the world's foremost example of the economic and political success of a small neutral nation, whereas Austria has experienced great difficulties in

determining its role in modern Europe, although these have lessened markedly in recent years.

With respect to physical environment, Switzerland and Austria have much in common. More than half of each country is occupied by the high and rugged Alps (Fig. 4.24). North of the Alps, both countries include strips of the rolling morainal foreland of the mountains. The Swiss section of the Alpine foreland, often called the Swiss Plateau, extends between Lake Geneva on the French border and Lake Constance (German: *Bodensee*) on the German border. The Austrian section of the foreland, slightly lower in elevation, lies between the Alps and the Danube River from Salzburg on the west to Vienna on the east. The Swiss and Austrian sections are separated from each other by a third portion of the foreland in southern Germany. On their northern edges, both Switzerland and Austria include mountains that are much lower and smaller in extent than the Alps. These are the Jura Mountains, on the border between Switzerland and France, and the Bohemian Hills of Austria, on the border with the Czech Republic.

The Disparity Between Resources and Economic Success

Historical assessment of economic success has often suggested a close link between such success and a generous resource base. In a comparison of Austria and Switzerland, however, it can be seen that natural resources and raw materials play only a partial role in their distinct levels of economic achievements. Based on

127

natural resources alone, one might expect Austria to be the more successful country economically. It has more arable land than Switzerland and more per person, primarily as a result of Austria's greater proportion of nonmountainous terrain. It also has more forested land than Switzerland, and more per person. This results mainly from the fact that the Austrian Alps have somewhat lower elevations than the Swiss Alps and hence a smaller proportion of land is above the tree line. The same contrast holds true for minerals. Austria has no mineral resources of truly major size, but it does have a varied output of minerals that do possess value collectively. Switzerland, on the other hand, is almost devoid of significant mineral resources. Both countries depend heavily on abundant hydroelectric potential, but Austria's resource base is greater.

In economic success, however, the advantage lies with Switzerland, which in 1997 had a per capita GNP of $44,350, while Austria, in contrast, had a per capita GNP of $28,110. But this must be seen in perspective, as Austria is quite well off today, largely as a result of economic growth since World War II.

Historical developments that favored Switzerland far outweighed Austria's natural resource advantage. The primary reason seems to be the long period of peace enjoyed by Switzerland. Except for some minor internal disturbances in the 19th century, Switzerland has been at peace inside stable boundaries since 1815. The basic factors underlying this long period of peace seem to be (1) Switzerland's position as a buffer between larger powers, (2) the comparative defensibility of much of the country's terrain, (3) the relatively small value of Swiss economic production to an aggressive state, (4) the country's historic and political value as an intermediary between belligerents in wartime and as a refuge for people and money, and (5) Switzerland's policy of strict and heavily armed neutrality.

Switzerland

During well over a century and a half of peace, the Swiss have had the opportunity to develop an economy finely adjusted to the country's potentials and opportunities. In particular, they have skillfully exploited the fields of banking, tourism, manufacturing, and agriculture. Favored by Swiss law, the country's banks enjoy a worldwide reputation for security, discretion, and service. The result has been a massive inflow of capital—including some of dubious origins—to Swiss banks. Switzerland's largest city, Zürich (population: 921,446, metropolitan area; 342,804, city proper), is one of the major centers of international finance, and other Swiss cities are also heavily involved.

Switzerland has scenic resources in abundance, and the country has long been a major tourist destination. Development of winter sports makes snow an important resource, and Alpine ski resorts such as Zermatt, Davos, and St. Moritz are world famous. Less publicized is Switzerland's development of the manufacturing of delicate timepieces and high-technology instrumentation, based primarily on hydroelectricity from moun-

tain streams and the skills of Swiss workers and management. Because most of its raw materials and fuels must be imported, industrial specialties in Switzerland minimize the importance of bulky materials and derive value from skilled design and workmanship. Reliance on hydroelectricity has facilitated the development of many small industrial centers, mostly in the Swiss Plateau. Five of the country's six largest cities are strung along the Plateau. Except for Zürich, these cities are under 500,000 in metropolitan area population, although Geneva approaches that figure. The remaining city, Basel (population: 405,079), lies beyond the Juras at the point where the Rhine River turns north between Germany and France. Basel is the head of navigation for Rhine barges that connect Switzerland with ocean ports and other destinations via river and canal.

The National Unity of Switzerland

The internal political organization of Switzerland expresses and makes allowances for ethnic diversity. The country was originally a loose alliance of small sovereign units known as cantons. When a stronger central authority became desirable in the 19th century, not only were the customary civil rights of a democracy guaranteed, but governmental autonomy was retained by the cantons except for limited functions specifically assigned to the central government. Although the functions allotted the central government have tended to increase with the passing of time, each of the local units (now 26 in number) has preserved a large measure of authority. Local autonomy is supplemented by the extremely democratic nature of the central government. In no country are the initiative and the referendum more widely used to submit important legislation directly to the people. Thus, guarantees of fundamental rights, local autonomy, and close governmental responsiveness to the will of the people are used to foster national unity despite the potential handicaps of ethnic diversity and local particularities.

The central government, in turn, has pushed economic development vigorously. Two outstanding accomplishments have been the construction and operation of Switzerland's railroads (Fig. 4.25) and the development of the hydroelectric power generating system. Despite the difficult terrain, the Swiss government has built a railway network of great density, which carries enough international transit traffic to earn important revenues. The hydroelectric system, utilizing the many torrential streams of the mountains, yields about 55 percent of the country's electric output and places Switzerland very high among the world's nations in hydroelectricity produced per capita.

Austria: Problems and Readjustments

Austria emerged in its present form as a remnant of the Austro-Hungarian Empire when that empire disintegrated in 1918 as a consequence of defeat in World War I. Austria's population is essentially German in language and culture—and the countries have been closely linked historically—but a unification with Germany was forbidden by the victors. The economy was seriously disoriented by the loss of its empire, and the country limped through the interwar period until it was absorbed by Nazi Germany in 1938. In 1945, after defeat in World War II, Austria was reconstituted as a separate state but was divided into occupation zones administered, respectively, by the United States, the United Kingdom, France, and the Soviet Union. During 10 years of occupation, the Austrian Soviet Zone in the east (which included Vienna)—already badly damaged in the war—was subjected to extensive eastward removal of industrial equipment for reparations. It was only in 1955 that this occupation was ended by agreement among the four powers and Austria became an independent, neutral state.

As the core area of the earlier Austro-Hungarian (Hapsburg) empire in southeastern Europe, Austria developed a diversified industrial economy prior to 1914. Its industries depended heavily on resources and markets provided by the broader empire that covered much of the North European Plain. Austrian iron ore was smelted mainly with coal drawn from Bohemia and Moravia, now in the Czech Republic. Austria's textile industry specialized to a considerable extent on spinning, leaving much of the weaving to be done in Bohemia. Austrian industry was not outstandingly efficient, but it had the benefit of a large protected market in which one currency was in use. In turn, the producers of foodstuffs and raw materials also had a market within the empire protected by external tariffs.

When the empire disintegrated at the end of World War I, the areas that had formed Austria's protected markets were incorporated into independent states. These states, motivated by a desire to develop industries of their own, began to erect tariff barriers, and other industrialized nations began to compete with Austria in their markets. The resulting decrease in Austria's ability to export made it more difficult for the country to secure the imports of food and raw materials that its unbalanced economy required. Such difficulties were further increased after World War II by the absorption of East Central Europe into the Soviet Communist sphere. The necessary 1990s reorientation of Austrian industries toward new markets, principally in western Europe, was not easy because Austria did not join the EU until 1995. As a consequence, Austrian exports of goods were consistently inadequate to pay for imports. However, the country has been able to compensate for this deficit by revenues from a growing tourist industry.

Austria made notable progress after World War I, and has made especially rapid progress since World War II, in building a successful economy suited to its status as a small independent state. Major elements in the economy include forest industries, tourism, some iron and steel production based on Austrian iron ore and imported ore and fuel, increased hydroelectric development, traditional manufacturing of clothing and crafts, and larger-scale engineering and textile industries. The latter industries contribute quite importantly to exports, but development is

Figure 4.25 The great achievement of the Swiss government in creating transportation routes through the rugged Alps is graphically illustrated by this view of bridges in precipitous terrain. *Swiss National Tourist Office*

not sufficient to satisfy the internal market, and Austria is now a net importer of machinery and textiles.

Since the end of the empire, Austrian agriculture has moved toward greater specialization in dairy and livestock farming. Such production is a logical use of the large upland areas of Austria that are best adapted to pasture. An enlarged acreage devoted to feed crops such as barley and corn accompanies the increased emphasis on animal products. But Austria is still far behind western Europe's leading countries in agricultural efficiency. Many small and poor farms still exist, and 8 percent of the labor force is still in agriculture (as of 1997)—a high proportion by western European standards. Meanwhile, substantial imports of food and feed remain necessary.

A significant aspect of Austrian readjustment since World War I has been the lessened importance of the famous Austrian capital, Vienna (population: 1.9 million). Although Vienna is still by far the largest city in Austria, there has been a pronounced tendency for population and industry to shift away from the capital toward the smaller cities and mountain districts. From a position as the capital of a great empire, Vienna has regressed to the capital of a small country. Its present metropolitan area population is less than the population of the city proper in 1918.

Vienna's importance, however, has persisted since Roman times and is based on more than purely Austrian circumstances. The city is located at the crossing of two of the European continent's major natural routes: the Danube Valley route through the highlands separating Germany from the Hungarian plains and southeastern Europe, and the route from Silesia and the North European Plain to the head of the Adriatic Sea. The latter follows the lowland of Moravia to the north and makes use of the passes of the eastern Alps, especially the Semmering Pass,

to the south. Thus, Vienna has long been a major focus of transportation and trade and also a major strategic objective in time of war. In the geography of the early 21st century, Vienna finds itself particularly well-positioned for the potential expansion and development of the Eastern European nations and the western tiers of the states emerging from the former Soviet Region.

SUMMARY WITH SELECTED KEY TERMS

- This chapter gives attention to the **Industrial Revolution,** particularly the forces that led to its genesis and development in Great Britain. **Coal, steam power,** and the **railroad** all took new roles in an assemblage of industrial changes, **demographic shifts,** and growing urbanism in **England** and **Scotland** in the middle 17th century as part of this revolution.

- The nations that are considered part of West Central Europe in this chapter include the **United Kingdom, France, Germany, the Low Countries (the Netherlands, Belgium, Luxembourg), Austria,** and **Switzerland,** as well as a number of smaller nations and microstates. The giant nations in this region, however, are considered to be Germany, the United Kingdom, and France. They are the primary engines for the **economic momentum** in this part of the world.

- The United Kingdom is made up of **England, Scotland, Wales,** and **Northern Ireland.** It forms the greater part of two major islands: **Great Britain** and **Ireland,** which are called the **British Isles.** The island of Great Britain has a highlands and a lowlands with distinctive landscapes and mineral resources. Most of the 18th- and 19th-century industrial and urban development of Britain related to the location of **coal** and **iron**—the two most dominant resources of the Industrial Revolution. It was later that the location of London on the **Thames Estuary** enabled that city to rise to vital importance with neither coal nor iron as a **local resource.**

- The four major factors that led to Britain's importance in the genesis of the Industrial Revolution were **coal, iron and steel, cotton textiles,** and **shipbuilding.** The process of **enclosure** also helped promote a demographic shift that provided urban factory workers for the new textile mills. London is a major example of a dominant urban center that reflects the forces of innovation and change that characterize this era of industrial growth and development. Central to London's development were **location on the Thames, proximity to regional trading partners, the high level of productivity of the London region (even without coal or iron ore),** and **the significant role of English shipping** for almost all of the 18th and 19th centuries. London is also used to illustrate late-20th-century **industrial relocation.** Central to this locational shift are a **loss of resource advantage, cheaper labor in new areas, technological and employment changes,** and **managerial conservatism.**

- The island of **Ireland** has long experienced a tense relationship between **Roman Catholics,** who make up the major population on the island, and Protestants, who came from Scotland and Britain in the 17th century and settled in **Ulster,** the locale now called **Northern Ireland.** Economically, Ireland has encouraged **high-tech industries,** as well as tourism, to play an important role in its recent development.

- **France** is the largest of the West Central European countries. The nation has an **internal coherence** to it and reasonably clear national

borders made up of rivers, mountains, and seacoasts. **Agriculture** is more important in France than in any other West Central European country. It plays a role not only in **economic** but also in **social terms.** Markets for French agricultural products are not only domestic and regional, but global—especially for **wines, olives,** and **wheat.** Paris, an example of a **primate city,** is the most dominant city in all of France and more singular than are other national capitals in this region. The growth of Paris from a small settlement on an island in the Seine River into a world capital is an example of many geographic features coming to play in a single location. These factors include **location, transportation, human migration, markets, cultural uniqueness,** and **colonialism.** France is made up of four regions, each less economically vibrant than the prior one. The regions, in order, are the **Paris region, the Northeast, the Southeast,** and **the West.** Tourism continues to be a strong factor in the economic health of France, with Paris the most popular **tourist city destination** in the world.

- **Germany,** re-formed when **West Germany reunified in 1990 with East Germany** upon the collapse of the former Soviet Union and its collection of satellite nations lying between Russia and West Central Europe, has the **largest economy and population of all Europe.** German **urban centers** have a much **broader distribution** than do the industrial cities of France, and even of Britain. These cities, generally, have been located in **direct response to availability of coal, river systems, iron ore,** and **the soils of the North European Plain.** The German adoption of the Industrial Revolution occurred much later than the British or Belgian experience, but by the beginning of the 19th century, Germany had developed major industrial zones in the **Rhine uplands, Saxony,** and **Silesia** (now in Poland). All three of these centers had **coal, iron ore, hydroelectric power, charcoal, labor,** and the **technology** for **textile manufacturing** and production of **metal goods.** The geography of German manufacturing is more widespread than is common in West Central Europe. There has been considerable immigration of **guest workers or foreign laborers.** Germany has also been a target destination for migrating peoples from the Balkan Peninsula, Turkey, Russia, and various countries neighboring Russia. At the end of the 1990s, Germany was in the process of **moving its federal capital** from **Bonn,** where it had been since 1949, to **Berlin,** as a result of reunification.

- The **Netherlands, Belgium,** and **Luxembourg** are called **the Low Countries,** or **Benelux.** They have developed global trading networks, busy river systems, and active expansion of a land base through **polders,** and continue to derive significant income from agriculture. Three major port cities in Benelux are **Rotterdam** and **Amsterdam** in the Netherlands, and **Antwerp** in Belgium.

- Rotterdam, in the Netherlands, serves as an example of the ways that the river systems of West Central Europe have played and continue to play a central role in the economic growth and importance of port cities on the North Sea. The reasons for the design, engineering, and construction of **polders** in the Low Countries derive from some of the same influences that make Rotterdam and other port cities so vital to current economic health.

- **Switzerland** and **Austria** represent two nations with somewhat similar geographic features that have had very distinct **historical geo-graphies.** They are both in the **Alps,** have played the **buffer state** role in the heart of Europe, and they both are significant **tourist destinations.** Neither has particularly abundant **natural resources,** but scenery, **hydroelectricity,** and a strong **cultural tradition** have all played a role in their development. Switzerland has gained more stability with its **traditional neutrality** and financial management, while Austria has had its fate tied more closely to Germany and battles on the **North European Plain.**

REVIEW QUESTIONS

1. List the major nations of West Central Europe and locate the most important beds of coal and iron ore, both in the 18th century and now.
2. What are the four industries that were central to the Industrial Revolution? Discuss each of them in relation to the development of industrial growth in the United Kingdom.
3. Discuss the role of agriculture in France, explaining why it is so much more central a role in that country than in the United Kingdom.
4. Using maps and the text, and thinking about Thinking of the World in Spatial Terms from Table 1.1, Chapter 1, why is Paris the city that it is? What is it that not only makes it central to France and the French population, but enables it to be the most popular tourist city in the world?
5. Explain why West Germany went to the great expense in 1990 of bringing East Germany back into the single state of Germany.
6. Compare and contrast the maps of industrial centers in the United Kingdom, France, and Germany. What role do resources play in those distinct maps?
7. Looking at the population map of Europe in Chapter 3, suggest reasons for the active migration from eastern Europe and Russia into West Central Europe, and especially Germany.
8. What are the reasons for—and possibly against—the move of the capital of Germany from Bonn in the west to Berlin in the eastern part of Germany?
9. What is it that has made Rotterdam the most active and prosperous port in West Central Europe? What are the geographic features that have led to this and what are the cultural features?
10. Explain the concept of the polder. Where is it important and how do you think its construction cost and maintenance costs have been justified?
11. What role has geography had in the distinct histories of Switzerland and Austria?

DISCUSSION QUESTIONS

1. What is the impact of the Eurotunnel (the Chunnel or the English Channel Tunnel) on Britain? On France? On West Central Europe?
2. Discuss the locales that have been caught up in the Geography of Discontent (see Regional Perspective, pp. 116–117) that is cited in this chapter. Where are they? How would you characterize the features that put them in this category?
3. Consider the growth of the German economy and the return of the federal capital from Bonn to Berlin. What are the various conversations that such a move might raise? Give a French, a Dutch, an English, a German, and a Russian perspective in such conversations.
4. Discuss the importance of European rivers and of European coal resources shown on the maps in this chapter. What can you say about the role of geographic resources and landscapes?
5. Discuss how the National Geography Standards' Six Essential Elements that deal with Places and Regions would be helpful in making an analysis of the geographic features that are talked about in question 4 above.
6. What is the pattern of population growth in Europe? What is the significance of the strong regional differences in growth rates?
7. Think of your mental maps of European cities. Which cities most quickly come to mind as places you would like to visit? How many of them are in West Central Europe? What are the qualities that make them so appealing to you? Talk about how your mental maps of these cities have been shaped. How might they be changed?
8. Discuss your migration decision-making process if you are living in the Balkan Peninsula. What forces might lead you to leave? What news and images affect the decisions you make about where you want to go?
9. If you were asked to create a Disneyworld complex or theme park in Berlin, would you have it be the same as the one outside Paris? Why or why not?
10. Discuss the likelihood of the United Kingdom, Germany, and France being able to bring old industrial cities into the 21st century as economically viable urban centers. Give examples of where this has worked or not worked.

Chapter 5

Europe: The North, the South, the East

▲ A classic Mediterranean landscape in Tuscany, north central Italy, near Florence. The stone town of San Gimignano on the hilltop overlooks small farm homes and vegetation characteristic of the Mediterranean world. Note the close-packed rows of grapevines in the vineyard (deep green) to the immediate left of the stone houses in the center of the view. *Jay Fries/The Image Bank*

CHAPTER OUTLINE

5.1 The North: Cohesive, Independent, and Prosperous

5.2 The South: Mediterranean Patterns and Problems

5.3 The East: Legacies of Empire and Uncertainties of the Future

*P*resent-day Europe's northwestern coreland—the focus of Chapter 4—is ringed by three outer groups of countries: Northern Europe (sometimes called Norden), Southern Europe (best known as Mediterranean Europe), and East Central Europe. Although these countries all possess distinctive traits and unique histories, today they tend to play a more minor role than do the coreland nations in influencing the other regions and nations of the world. We explore in this chapter the geographic personalities and economic and political dynamics of these nations outside the European center. Area and population data can be found in Table 3.1.

5.1 The North: Cohesive, Independent, and Prosperous

Denmark, Norway, Sweden, Finland, and Iceland are the countries of Northern Europe (Fig. 5.1). The peoples of these lands recognize their close relationships with each other and group their countries geographically under the regional term Norden, or "the North." A more common term to describe these countries is **Scandinavia**, or the Scandinavian countries. But this regional name is ambiguous: It occasionally refers only to the two countries that occupy the Scandinavian Peninsula, Norway and Sweden; more often it includes these countries plus Denmark, and sometimes it includes these three plus Iceland. When Finland is included in the group, the term **Fennoscandia**, or Fennoscandian countries, is used, thus acknowledging the cultural and historical differences between the Finns and the other northern nations.

Dominant Traits of Northern Europe As a Region

"The North" is an accurate descriptive term for these lands. No other highly developed countries have their principal populated areas so near the North Pole (see Fig. 3.2). Located in the general latitude of Alaska, the countries of Northern Europe represent the northernmost concentration of advanced industrial societies in the world.

Westerly winds from the Atlantic, warmed in winter by the North Atlantic Drift (see p. 68), moderate the climatic effects of this northern location in some areas. Temperatures average above freezing in winter over most of Denmark and along the coast of Norway. But, away from the direct influence of the west winds, winter temperatures average below freezing and are particularly severe at elevated, interior, and northern locations. In summer, the ocean tends to be a cooling rather than a warming influence, and most of Northern Europe's Fahrenheit temperatures in July average no higher than the 50s or low 60s ($10°$ to $18°$C). The populations of the various countries tend to cluster in the southern sections, and all countries except Denmark have considerable areas of sparsely populated terrain where the problems of development are largely those of overcoming the rigors of a northern environment.

Historical interconnections and cultural similarities are important factors in the regional unity of Northern Europe. Historically, each country has been more closely related to others of the group than to any outside power. In the past, warlike relations often prevailed, and for considerable periods some countries ruled others of the group. But since the early 19th century, relations among the Northern European peoples have been peaceful, and have come to be expressed in close international cooperation among the five countries.

The languages of Denmark, Norway, and Sweden descend from the same ancient tongue and are mutually intelligible. Icelandic, although a branch of the same root, is more difficult for the other peoples because it has changed less from the original Germanic language and has borrowed less from other languages. Only Finnish, which belongs to a different language family, is entirely distinct from the others. Even in Finland, however, about six percent of the population speaks Swedish as a native tongue, and Swedish is recognized as a second official language.

Among all these countries, cultural unity is also embodied in a common religion. The Evangelical Lutheran Church is the

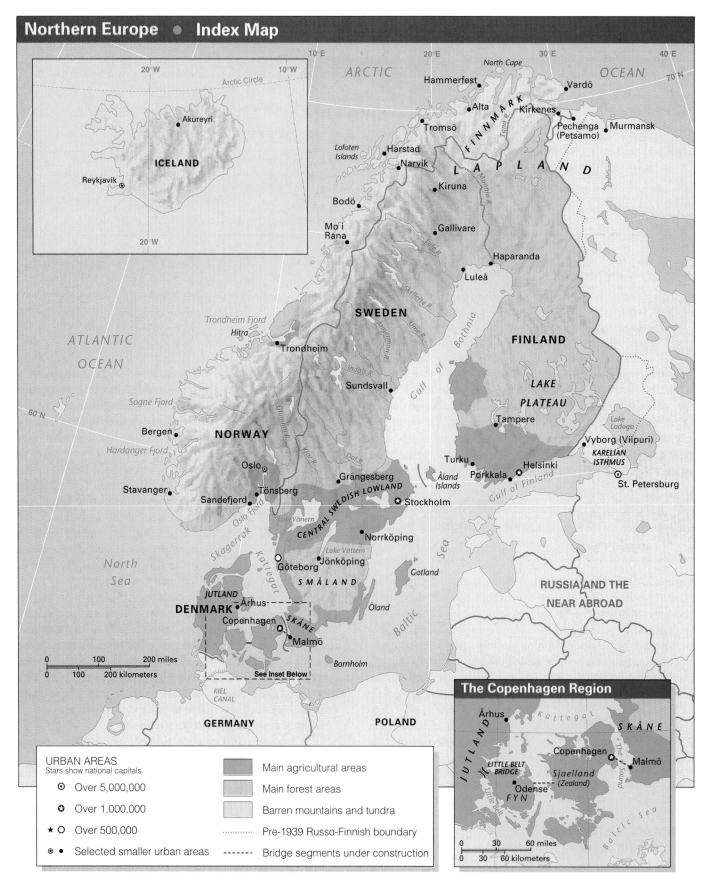

URBAN AREAS
Stars show national capitals.

⊙ Over 5,000,000

✪ Over 1,000,000

★ ○ Over 500,000

⊛ ● Selected smaller urban areas

Main agricultural areas

Main forest areas

Barren mountains and tundra

.... Pre-1939 Russo-Finnish boundary

-------- Bridge segments under construction

Figure 5.1 Index map of the North. Small tracts of agricultural land are scattered through the areas of forest, mountain, and tundra, and smaller forest tracts occur within the agricultural areas.

Figure 5.2 Farmland in the North often comes down to the edges of the fjords that provide both transport for farm surpluses and a connection with the more densely settled world. This photo of Aurlandfjord, Norway, shows how little of the mountain flank landscape has been opened for farming   *Frank Clarkson/Gamma Liaison*

church of approximately 90 percent of the population, at least nominally. It is a state church, supported by taxes, and is probably the most all-embracing organization outside of the state. The countries of Northern Europe also exhibit basic similarities with respect to law and political institutions. They all have a long tradition of individual rights, broad political participation, limited governmental powers, and democratic control. Today, they are recognized as strongholds of democratic institutions, not just within Europe, but globally.

Small size and resource limitations made it necessary for each of the countries of Northern Europe to build a highly specialized economy in pursuit of a high standard of living. Landscape images of the North often revolve around small, picturesque villages nestled on the flanks of mountains rising out

of placid fjords (Fig. 5.2). Success of settlements in such environments has been so marked that these countries are probably known as much for high living standards as for any other characteristic. In all five, high standards of health, education, security for the individual, and creative achievement are characteristic.

Denmark

Denmark has the distinction of possessing the largest city in Northern Europe, while, at the same time, it is the most dependent on agriculture of any country in the region. The Danish capital of Copenhagen (population: 1,353,333) has one-third of Denmark's population in its metropolitan area, allowing it—like Paris—to play the role of a primate city. Denmark has a far greater density of population than the other countries of Northern Europe, a fact accounted for by the presence of Copenhagen, the greater productivity of the land, and the lack of any sparsely populated zone of frontier settlement.

Copenhagen and the Danish Straits

Copenhagen lies on the island of Sjaelland (Zealand) at the extreme eastern edge of Denmark (see Fig. 5.1). Sweden lies only 12 miles (*c.* 20 km) away across the Sound, the main passage between the Baltic and the North Seas. The city grew beside a natural harbor well-placed to control traffic through the Sound. Before the 17th century, Denmark controlled adjacent southern Sweden as well as the less favored alternative channels to the Sound—the Great Belt and the Little Belt. Tolls were levied on all shipping passing to and from the Baltic. Although the days of levying tolls are now long past, Copenhagen still does a large transit and entrepôt business in North Sea–Baltic Sea trade. Under construction is a major bridge and tunnel project that will truly change the linkage between Scandinavia and the heart of Western Europe. The Oresoundbridge will be 16 km in length, and include a 2 km tunnel. The eastern anchor will rest in Malmo, Sweden. The western terminus will lead to the Copenhagen (Denmark) airport. This project will establish the first surface link between the North and Denmark and Germany.

Trade led to industry, and Copenhagen is the principal industrial center of Denmark as well as its chief port and capital. Outstanding industries include the processing of both exported and imported foods, the manufacture of chemicals, brewing, and engineering.

Factors in Denmark's Agricultural Success

About three-fifths of Denmark is cultivated—the largest proportion of any European country. It is fortunate that so much of the nation is arable, because the country is greatly lacking in other natural resources. Danish agriculture is so efficient that only 5 percent of the working population is employed on the land. Agriculture, however, is basic to the country's economy. Many additional workers engage in processing and marketing

Figure 5.3   Norway was heavily glaciated during the Pleistocene. This tongue of active glacier flow is the Briksdal Glacier in western Norway. *Joseph J. Hobbs*

the country's agricultural products and in supplying the needs of the farms. Approximately one-fifth of Denmark's exports relate to agriculture.

Few countries or areas that depend so heavily on agriculture are as materially successful as Denmark. Danish agriculture is based on the highly specialized and carefully fostered development of animal husbandry, which was emphasized when the competition of cheap grain from overseas brought ruin to the previous Danish system of grain farming in the latter part of the 19th century. Until recently, Danish animal husbandry focused on dairy farming, but in recent years the production and sale of animals for meat has become more important than dairying.

A number of factors have been responsible for the general success the Danes have had in agriculture. Membership in the EU and a location adjacent to the largest country in western Europe (Germany) have provided a ready market for Danish agricultural products. Two other factors have had a major impact as well: The Danes have developed agricultural cooperatives to a high level of efficiency and have given keen attention to adult education in addition to traditional schooling, leading to a skilled labor force in agriculture as well as a high level of overall literacy.

Norway

Norway stretches for well over a thousand miles (*c.* 1600 km) along its west side and around the north end of the Scandinavian peninsula. The peninsula, as well as Finland, occupies part of the Fennoscandian Shield, a block of ancient and very stable metamorphic rocks. The western margins of the block in Norway are extremely rugged. Most areas have little or no soil to

cover the rock surface, which was scraped bare by the continental glaciation that centered in Scandinavia. About 70 percent of Norway is classified as wasteland, 3 percent as arable or pasture land, and the remainder as forest (see Fig. 5.2).

The nature of the terrain hinders not only agriculture but also transportation. Glaciers deepened the valleys of streams flowing from Norway's mountains into the sea, and when the ice melted, the sea invaded the valleys. Thus were created the famous **fjords**—long, narrow, and deep extensions of the sea into the land, usually edged by steep valley walls. It is difficult to build highways and railroads parallel to such a coast. The fjords provide some of the finest harbors in the world, but most of them have practically no **hinterland** or productive rural landscape dependent upon the central places established on the fjords. Much of Norway's population is scattered along these waterways in many relatively isolated clusters, most of which are small and only moderately interconnected.

More than half of the population of Norway lives in the southeastern core region, which centers on the capital, Oslo (population: 758,949). Located at the head of Oslo Fjord, where several valleys converge, Oslo is the principal seaport and industrial, commercial, and cultural center of Norway as well as its largest city. In this region valleys are wider and the land is less rugged. The most extensive agricultural lands and the largest forests in the country are located here. Hydroelectricity powers sawmilling, pulp and paper production, metallurgy, electrochemical industries, and industries that produce various consumer goods for Norwegian consumption.

Apart from its scanty agricultural base, Norway has rich natural resources relative to the size of its population, which is

Regional Perspective

Norway and "No" to the European Union

In November 1994, Norwegians voted, by a margin of 53 to 47 percent, *not* to join the European Union (EU). This negative vote came about largely because recent petroleum discoveries in Norway's sector of the North Sea have given it an economic stability and the capacity to decline the interference of the EU as well as the benefits of membership. Sweden, during the same time period, voted *to* join the EU by the reverse percentages. In the European Economic Area (EEA) this means that only Iceland, Liechtenstein, and Norway are not members of the increasingly powerful EU. As the regional influence of the EU continues to expand, Norway's decision becomes even more significant.

In the case of Norway, the situation has dimensions that are unusual for Europe. Because Norway's economy is so strong and stable, there was a reluctance to allow the EU to interfere. For example, in 1995, Norway was second only to Saudi Arabia in the export of petroleum worldwide. In addition, Norway gains more than 99 percent of its electricity from hydroelectricity, so that its oil wealth is almost entirely an exportable resource. In economic growth, Norway experienced a rate of 4.5 percent in 1995, one of the highest in Europe. Norway's 5 percent unemployment rate is one of the lowest in Europe. And in the effort to get the EU nations' financial houses in order for the implementation of a common EU currency, Norway is one of the few nations that already meets the essential EU criteria—a balanced budget, modest national debt levels, currency stability, and low inflation—required for participation in the new monetary unit. Norway's status improved when its government moved into a time of budget surplus and it became a net lender in 1996.

The forces that led the battle against membership in the EU—at a time when even most of the Eastern European nations are petitioning for EU membership —were coalitions of environmentalists, farmers, fishermen, and nationalists. Their call was to preserve the relative independence of Norway and not slip into the situation where uniquely Norwegian situations would be tampered with by bureaucrats in Brussels—the administrative "capital" of the EU. On the other hand, the people nervous about the "no" vote warned that Norway's petroleum income will surely decline in the future, and that the aging of the population will change the budget picture profoundly as ambitious promises for retirement and medical benefits become more costly. However, the farmers and fishermen wanted to retain their Norwegian state subsidies, which are even more generous than those offered by the EU. In the end, it was Norwegian suspicions of transferring political power to Brussels that resulted in the unusual "no" vote.

small and increasing at a very slow rate. Key resources include spectacular scenery, a vast potential for the generation of hydroelectricity, forests, and a variety of minerals (Fig. 5.3). The country's resource position was greatly enhanced by the discoveries and development, in the 1960s and 1970s, of large oil and natural-gas deposits far offshore in the Norwegian sector of the North Sea floor. In just a few years, oil and gas exports grew from nothing to the country's largest by value, and Norway advanced from being reasonably well-off to being one of Europe's richest countries as measured by per capita GNP (see Table 3.1).

Meanwhile, older economic emphases remain important. Manufacturing is a large employer and provides exports of goods related to Norway's resources. And Norway has had a major fishing industry for centuries. An ancient seafaring tradition going back to the Vikings is expressed in one of the world's largest merchant fleets.

In the late 1990s Norway began an ambitious investment in transportation infrastructure. Bridges, tunnels, and highways have received much attention, especially in the more remote parts of the country. The government is also trying to find a way to stem the increasing migration stream from isolated and remote settlements to small towns and cities in the south. The thought is that if, for example, a tunnel to Hitra Island (see Fig. 5.1), some 200 miles northwest of Oslo, can make the 4,100 inhabitants feel more closely connected to the mainland, it will reduce the number of youth who leave the island in quest of jobs and a more urban lifestyle. Hundreds of millions of dollars were spent in 1998 in just such capital construction projects—even though relatively small numbers of people were directly affected by such work.

Sweden

Sweden is the largest in area and population (8.9 million) and the most diversified of the countries of Northern Europe. In the northwest, it shares the mountains of the Scandinavian peninsula with Norway; in the south, it has rolling, fertile farmlands like those of Denmark; in the central area of the great lakes lies another block of good farmland. To the north, between the mountains and the shores of the Baltic Sea and Gulf of Bothnia, Sweden consists mainly of ice-scoured, forested uplands. A smaller area of this type occurs south of Lake Vättern.

Economy and Resources

The high development of engineering and metallurgical industries is the feature that most distinguishes the Swedish economy from that of the other countries of Northern Europe. These industries have evolved out of a centuries-old tradition of mining and smelting. The iron ores of the Bergslagen region, just north of the Central Swedish Lowland, have been made into iron and steel since medieval times. Originally, the fuel used to smelt the ore was charcoal made from the surrounding forest. When coal and coke began to replace charcoal as fuel in Europe's iron industries in the late 17th century, Sweden was handicapped by its scarcity of coal. But, in the age of electricity, the Swedes responded to this problem by a heavy reliance on electric furnaces using current from hydropower plants; more recently, this kind of smelting has been supplemented by a Swedish-invented oxygen process. This electric process is expensive but gives Swedes an extraordinary control of quality, and "Swedish steel" has long been a synonym for steel of the highest grade. The modern Swedish steel industry still centers in many small towns, but output supplemented by production from some large conventional plants built on the Baltic and Bothnian coasts which use imported coal and coke. And although Sweden now produces large quantities of ordinary steel, the ultrafine steels and the products made from them still impart a special character to Swedish metallurgy.

Among the specialties of the Swedish steel industry that yield a large volume and value of exports are automobiles (Volvo and Saab), industrial machinery (including robots), office machinery, telephone equipment, medical instruments and machinery, and electric transmission equipment. The country's reputation for skill in design and quality of product also extends to items such as ball bearings, cutlery, tools, surgical instruments, and the glassware and furniture crafted in Småland, the infertile and rather sparsely populated plateau south of the Central Lowland.

The introduction of new technologies in the forestry industries has reduced job opportunities in this most traditional occupation in the country's remote lands. Recent promotions, however, have touted remote villages in the Swedish north as excellent settings for computerized telephone call-centers. Because of a tradition of well-educated, keyboard-literate youth, even in the remote reaches of the country, the Swedish government is attempting to lure telecommunication firms to these places. The hope is also that the far north and upland will seem less distant and remote if a young population will come there for good jobs and global computer linkages. Thus, Sweden—like Norway with its transportation investments—is hoping to

slow down the steady migration of young populations from villages to the cities in the south.

Overall, Swedish manufacturing occurs principally in numerous centers in the Central Lowland. These places, ranging in population from about 125,000 to mere villages, have grown up in an area favorably situated with respect to minerals, forests, water power, labor, food supplies, and trading possibilities. The Central Lowland is the historic core of Sweden and has long maintained important agricultural development and a relatively dense population.

Unlike Denmark, Sweden has not developed a specialized export-oriented agriculture; and unlike Norway, Sweden commands enough good land to supply all but a minor share of its small population's food requirements. More than 92 percent of Sweden is nonagricultural, but there is some good land, largely in Skåne at the southern tip of the peninsula and in the Central Lowland. Skåne, with relatively fertile soils like those of adjacent Denmark, and with Sweden's mildest climate, is the prime agricultural area. This fertile agricultural area is the source for the term "Scandinavia."

Stockholm and Göteborg

Stockholm (population: 880,096) is the second largest city in Northern Europe (Fig. 5.4). The location of the capital reflects the role of the Central Lowland as the center of the Swedish state and the early orientation of that state toward the Baltic and trans-Baltic lands. Stockholm is the principal administrative, financial, and cultural center of the country and shares in many of the manufacturing activities typical of the Central Lowland.

Figure 5.4   Stockholm, Sweden's capital, has the low skyline typical of European cities. The oldest part of the city and the main harbor are in the center of the photograph.   *Peter Jordan/Gamma Liaison*

But in the past century, as Sweden has traded more via the North Sea, Stockholm has been displaced by Göteborg (population: 700,000), located at the other end of the Central Lowland, as the leading port of the country and, in fact, of all Northern Europe. In addition to its North Sea location, Göteborg has a harbor that is ice-free year round, whereas icebreakers are needed to keep Stockholm's harbor open in winter.

The Swedish Policy of Neutrality and Preparedness

In the Middle Ages and early modern times, Sweden was a powerful and imperialistic country. In the 17th century, the Baltic Sea functioned almost like a Swedish lake. During the 18th century, however, the rising power of Russia and Prussia put an end to Swedish imperialism, aside from a brief campaign in 1814 through which Sweden won control of Norway from Denmark. Since 1814, Sweden has never engaged in a war, and it has become known as one of Europe's most successful neutrals. The long period of peace has undoubtedly been partially responsible for the country's success in attaining a high level of economic welfare and a reputation for social advancement. However, Sweden stays heavily armed while maintaining its traditional policy of neutrality.

Finland

Conquered and Christianized by the Swedes in the 12th and 13th centuries, Finland was ceded by Sweden to Russia in 1809 and was controlled by the latter nation until 1917. Under both the Swedes and the Russians, the country's people developed their own culture and feelings of nationality. This independence has been also promoted by the fact that the Finnish language is entirely distinct from the other Scandinavian languages. The opportunity for independence provided by the collapse of Tsarist Russia in 1917 was eagerly seized.

Most of Finland is a sparsely populated, glacially scoured, subarctic wilderness of coniferous forest, ancient igneous and metamorphic rocks, thousands upon thousands of lakes, and numerous swamps. Despite this environment, Finland was primarily an agricultural country until very recently, and the majority of the population resides in the relatively fertile and warmer lowland districts scattered through the southern half of the country. Hay, oats, and barley are the main crops of an agriculture in which livestock production, especially dairying, predominates.

To pay for a wide variety of imports, the nation depends heavily on exports of forest products. About two-thirds of Finland is forested, mainly in pine or spruce, and about two-fifths of the country's exports in 1996 consisted of wood products—paper, pulp, and timber. Forest production is especially concentrated in the south central part of the country, often referred to as the Lake Plateau. A poor and rocky soil discourages agriculture here, but a multitude of lakes, connected by streams, aid in transporting the logs.

Helsinki (population: 1,016,291), located on the coast of the Gulf of Finland, is Finland's capital, largest city, main seaport, and principal commercial and cultural center. It is also the country's most important and diversified industrial center. Hydroelectricity originally powered Finland's industries, but this source has now been surpassed by both thermal and nuclear power. The thermal stations use imported fossil fuels.

Problems of a Buffer State

For centuries, Finland has been a buffer between Russia, Scandinavia, and the former eastward reach of German power. In 1939, Finland refused to accede to Soviet demands for the cession of certain strategic frontier areas and was overwhelmed by the Former Soviet Region in the "Winter War" of 1939–1940. Then, in an attempt to regain what had been lost, Finland fought with Germany against the Soviet Union from 1941 to 1944 during World War II and again was defeated. Following the war, the economy of Finland was heavily burdened by (1) the necessity for rebuilding the northern third of the country, devastated by retreating German soldiers after Finland surrendered to the Soviets, (2) the necessity of resettling one-tenth of the total population of the country after they fled as refugees from border areas ceded to the Soviets, (3) the loss of the ceded areas themselves, some portions of which were of considerable importance in the prewar economy, and (4) the necessity of making large reparations to the Soviet Union.

In spite of these difficulties, however, the Finns were able to make a rapid recovery. The economy drastically changed in some respects. One of the most striking changes was the rise of metalworking industries. This was made necessary when the Soviets required that a large part of the reparations be paid in metal goods. Several small steel plants and other metalworking establishments were built to meet this demand, and they continue to operate to the present day. Finland currently gains some 40 percent of its export capital from machinery, and another 25 percent from paper and forestry products.

Iceland

Iceland is a fairly large (39,699 sq mi/102,819 sq km), mountainous island in the Atlantic Ocean just south of the Arctic Circle. Its rugged surface shows the effects of intense glaciation and vulcanism. Some upland glaciers and many active volcanoes and hot springs remain. The vegetation consists mostly of tundra, with considerable grass in some coastal areas and valleys. Trees are few, their growth discouraged by summer temperatures averaging about 52°F (11°C) or below, as well as by the prevalence of strong winds. The cool summers are also a great handicap to agriculture. Mineral resources are almost nonexistent.

Despite the deficiencies of its environment, however, Iceland has been continuously inhabited at least since the 9th century and is now the home of a progressive and democratic republic with a population of about 300,000. Practically the entire population lives in coastal settlements, with the largest concentration in and near the capital, Reykjavik, which has a metropolitan population of just over 153,000. Because of the proximity of the relatively warm North Atlantic Drift, the coasts of Iceland have winter temperatures that are mild for the latitude. Reykjavik has an average January temperature that is practically the same as New York City's.

The economy of Iceland depends basically on fishing, farming, and the processing of their products. Agriculture focuses on cattle and sheep raising and a limited production of potatoes and hardy vegetables. The real backbone of the economy is fishing, which supplied more than three-quarters of all exports by value in 1996.

For centuries, Iceland was a colony of Denmark. In 1918, it became an independent country under the same king as Denmark and in 1944 declared itself a republic. Iceland has strategic importance because of its position along major sea and air

routes across the North Atlantic; the country is also a member of the North Atlantic Treaty Organization (NATO).

Greenland, the Faeroes, and Svalbard

Denmark and Norway possess outlying islands of some significance. Greenland, earlier a colony of Denmark, became a Danish province in 1953, and in 1979 was granted home rule. It is the world's largest island (839,00 sq mi / 2,175,600 sq km) and lies off the coast of North America. Greenland and the Faeroe Islands—between Norway and Iceland—are considered integral, although self-governing, parts of Denmark, and their peoples have the rights of Danish citizens. Although Greenland has nearly one-fourth the area of the United States, about 85 percent is covered by an ice cap, and the population, mainly distributed along the southern coast, amounts to only about 57,000. Nearly all of this population is of mixed Eskimo and Scandinavian descent. The principal means of livelihood are fishing, hunting, trapping, sheep grazing, and the mining of zinc and lead. Under the auspices of NATO, joint Danish-American air bases are maintained there, and the United States has a giant radar installation at Thule in the remote northwest.

The Faeroes are a group of treeless islands where some 40,000 people of Norwegian descent make a living by fishing and grazing sheep. Norway controls the island group of Svalbard, located in the Arctic Ocean and commonly known as Spitsbergen, which is also the name of its main island. Although largely covered by ice, the main island has the only substantial deposits of high-grade coal in Northern Europe. Mining is carried on by Norwegian and Russian companies.

Definitions & Insights

HYDROELECTRICITY

In the generation of electricity, there is no means more environmentally benign and geographically sensitive than water power. Water that is ponded—or captured behind dams—to develop **water head,** or the volume of water necessary to turn turbines, possesses the potential for **multiple use.** Once the water moves through a complex of turbines and dynamos—generating power with virtually no waste product of any sort—it returns to a stream system that exists downstream from the dam. In small systems, power is generated by diverting flowing streams into small generating units and then returning water directly to the stream system. Thus, water is a **flow resource** or a **renewable resource.**

In hydroelectric complexes that require the construction of dams, there is the inevitable loss of some settlement land as rivers are stopped in their flow, ponded to depths of sometimes scores or hundreds of feet, and then released through major turbine and penstock networks. This system of power generation works best where there is little or only modest silt carried in the water being utilized; a heavier silt load in the reservoir behind the dam tends to settle, reduces the water head obtainable at that site, and thus shortens the life of the hydroelectric project. In the larger projects built around massive river systems, major flood control components are commonly designed into the project. This has great potential benefit to downstream farming and urban populations. Currently, Norway gets 99 percent and Iceland 94 percent of their power from hydroelectricity. It is a perfect resource in the upland areas of the North, and an important resource in this region particularly.

5.2 The South: Mediterranean Patterns and Problems

On the south, Europe is separated from Africa by the Mediterranean Sea, into which three large peninsulas extend (Fig. 5.5). To the west, south of the Pyrenees Mountains, is the Iberian Peninsula, unequally divided between Spain and Portugal. In the center, south of the Alps, is Italy and its southern offshoot, the island of Sicily. To the east, between the Adriatic and Black Seas, is the Balkan Peninsula, from which the Greek subpeninsula extends still farther south between the Ionian and Aegean Seas. The four main countries of Southern Europe, as considered in this text, are Portugal, Spain, Italy, and Greece. But Southern Europe also includes the small island country of Malta, the microstates of Vatican City (enclosed within Rome) and San Marino, and the British colony of Gibraltar. Three of the peninsular countries include islands in the Mediterranean, the largest of which are Italy's Sicily and Sardinia, the Greek island of Crete, and the Spanish Balearic Islands. Some Mediterranean islands lie outside these countries, the most notable being Corsica, which is a part of France, and Cyprus.

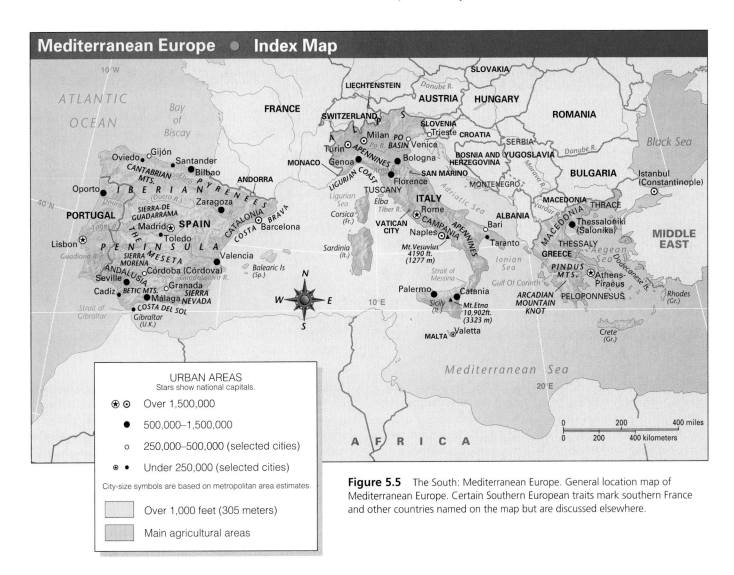

Mediterranean Europe ● Index Map

URBAN AREAS
Stars show national capitals.

⊛ ⊙ Over 1,500,000

● 500,000–1,500,000

○ 250,000–500,000 (selected cities)

⊗ • Under 250,000 (selected cities)

City-size symbols are based on metropolitan area estimates.

Over 1,000 feet (305 meters)

Main agricultural areas

Figure 5.5 The South: Mediterranean Europe. General location map of Mediterranean Europe. Certain Southern European traits mark southern France and other countries named on the map but are discussed elsewhere.

Topography and Population Distribution

In terrain and population distribution, as well as in climate and agriculture, the countries of Southern Europe present various points of similarity. Rugged terrain predominates in all four major countries. Plains tend to be small and face the sea, separated from each other by mountainous territory. Population distribution corresponds in a general way to topography, with the lowland plains being densely populated and the mountainous areas much less so, although a number of rough areas attain surprisingly high densities.

The Ecology of Mediterranean Agriculture

A distinctive natural characteristic of Southern Europe is its **mediterranean** (dry-summer subtropical) climate, which combines mild, rainy winters with hot, dry summers. Characteristics associated with this climate become increasingly pronounced toward the south. The northern extremities of both Italy and

Spain have atypical climatic characteristics. Except for a strip along the Mediterranean Sea, northern Spain has a marine climate like that of northwestern Europe—cooler and wetter in summer than the typically mediterranean climate areas—and the basin of the Po River in northern Italy is distinguished by a relatively wet summer and cold-month temperatures in the lowlands averaging just above freezing. Much of the high interior plateau of Spain, called the Meseta, is cut off from rain-bearing winds and has somewhat less precipitation and colder winters than is typical of the mediterranean climate, although the seasonal regime of precipitation with dry summers is characteristically a mediterranean climate trait. As such, the Meseta is enormously productive agriculturally (Fig. 5.6), and the agriculture is based principally on crops that are naturally adapted to winter rainfall and summer drought.

Although all agricultural patterns create a mosaic of crops and land-use patterns in **agricultural geography**, winter wheat deserves special note here. Although it is a common grain crop in much more than the Mediterranean world, it has great importance

Figure 5.6 The Mediterranean world is ablaze with the colors of its flowers, fruit crops, and sun. This market scene on Portugal's Madeira Island shows the color and variety of local farm products, as well as the potential tourist interest in such a setting. *Hugh Sitton/Tony Stone Images*

in this region, and has for thousands of years. This grain crop is generally planted at the end of the harvest of a summer crop. In September or as late as October, farmland is cleared of crops that were planted in spring, cultivated through the summer, and harvested before the first frost. Wheat is planted as fall temperatures drop and daylight shortens. Winter wheat is an unusually hardy grain, for it has to endure any snow cover that comes, although this is not common in the South. Winter wheat tends to give a higher yield than does **spring wheat,** which is planted in late spring and harvested in early fall. Tillers (grain heads) form on winter wheat, and when cold weather and snow come, the plant goes dormant. In the spring rains and/or snow melt, this wheat begins growing again, taking advantage of the moisture, the return of the longer hours of daylight, and the new warmth. In the spring, before any other crops have grown long enough—or have even been planted—to bring the farmer a harvest, winter wheat comes to fruition. Such timing produces a very popular and usable crop at the outset of the busy summer farming season.

Barley, a more adaptable grain, tends to supplant wheat in some areas that are particularly dry or infertile. Other typical crops in the South are olives, grapes, citrus, and vegetables. The olive tree and the grapevine have extensive root systems and certain other adaptations that allow them to survive the summer droughts. Olive oil is a major source of fat in the typical Mediterranean diet, and virtually all of the world supply is produced in countries that touch or lie near the Mediterranean Sea, or in a mediterranean climate zone. The principal use of grapes is for wine, a standard household beverage in Southern Europe and a major export product. Where irrigation is lacking, vegetables are grown which will mature during the wetter winter or in the spring. The most important are several kinds of beans and peas.

These are a source of protein in an area where meat animals make only a limited contribution to the food supply; feedstuffs are not available to fatten large numbers of animals, and parched summer pastures further inhibit the meat supply. Extensive areas too rough for cultivation are used for grazing, but the carrying capacity of such lands is relatively low. Sheep and goats, which can survive on a sparser pasturage than cattle, are favored. They are kept only partially for meat, and the total amount they supply is relatively small. In many places, grazing depends on **transhumance**—utilizing lowland pastures during the wetter winter and mountain pastures during the summer. In some areas, nonfood crops supplement the basic Mediterranean products. An example is tobacco, which is an export of Greece.

Mediterranean agriculture comes to its peak of intensity and productivity in areas where the land is irrigated. While irrigation on a small scale is found in many parts of Southern Europe, a few irrigated areas stand out from the rest in size and importance. Among these are northern Portugal, the Mediterranean coast of Spain, the northern coast of Sicily, and the Italian coastal areas near Naples. The largest of all, the plain of the Po River in northern Italy, uses irrigation to supplement year-round rainfall. In northern Portugal, irrigated corn (maize) replaces wheat as the major grain, and some cattle are raised on irrigated meadows. Grapes are the other agricultural mainstay. In the Mediterranean coastal regions of Spain, irrigation makes possible the development of extensive orchards. Oranges are the most important product, with growth concentrated around the city of Valencia, which has given its name to a type of orange. Lemons are particularly important on the island of Sicily. The district around Naples, known as Campania, is more intensively farmed, productive, and densely populated than any other agricultural area of comparable size in Southern Europe.

Irrigation water applied to exceptionally fertile soils formed of volcanic debris from Mount Vesuvius supports a remarkable variety of production that includes almost every crop grown in Southern Europe. Despite the intensive cultivation and productivity of many locales in this area, the overcrowded population on its tiny farms is notably poor. Fruits and vegetables, a large proportion of which are often destined for export, tend to supplement, and sometimes largely to displace, other types of agricultural products.

Spain

On the Iberian Peninsula, the greater part of the land consists of a plateau—the Meseta—with a surface lying between 2000 and 3000 feet (*c.* 600–900 m) in elevation. The plateau surface is interrupted by deep river valleys and ranges of mountains. Population density is generally restricted, mainly by lack of rainfall, to between 25 and 100 per square mile (*c.* 10–40 per sq km), compared to the European average of approximately 290 per square mile (*c.* 120 per sq km). The plateau edges are mostly steep and rugged (see photo, p.145). The Pyrenees and the Cantabrian Mountains border the Meseta on the north, and the Betic Mountains, culminating in the Sierra Nevada, border it on the south. Most of the population of Iberia is distributed peripherally on discontinuous coastal lowlands.

In Spain, the three largest cities are the capital, Madrid (population: 4.7 million), and the two Mediterranean ports of Barcelona (population: 4.1 million) and Valencia (population: 1.3 million). Madrid was deliberately chosen as the capital of Spain in the 16th century because of its location near the geographic center of the Iberian Peninsula, approximately equidistant from the various peripheral areas of dense population and of sometimes separatist political tendencies. Located in a poor region, Madrid has had little economic excuse for existence during most of its history. But its position as the capital has made it the center of the Spanish road and rail networks and thus has given it nodal business advantages. In Spain's recent economic boom, the city has begun to attract industry as well as tourists on a considerable scale.

Barcelona is Spain's major port and the center of a diversified industrial district. The main elements of Barcelona's industrialization have been (1) the early formation of a commercial class that supplied financing, (2) the availability of hydroelectric power, mainly from the Pyrenees, (3) cheap and sufficiently skilled labor, and (4) imported raw materials.

Valencia, Spain's third city in population, is the business center and port for an unusually productive section of coastal Spain. An extensive irrigation system in the area was developed by the Moors in the Middle Ages. Today, the density of population on irrigated land resembles that of Italy's Campania, but here, much of the agricultural effort goes into producing oranges, which are Spain's largest single agricultural export.

The main period of Spanish power and influence began in 1492, when a centuries-long struggle—the Reconquista—

Figure 5.7 A doorway to La Mezquita, or the Great Mosque of Cordoba, Spain. This Moorish temple was built in the 8th century and turned into a Christian cathedral in 1236. It exists today in southern Spain as one of the most visually stunning examples of the blending of Moorish and Christian sacred space. *Kit Salter*

ended with Spain's defeat of the Kingdom of Granada, the last part of Iberia held by the Muslim Moors (Fig. 5.7). In that same year, Columbus, a Genoese Italian navigator in the service of Spain, landed in the Americas in search of a sea route to India and East Asia. For a century thereafter, Spain was the greatest power in Europe and probably in the world. It rapidly built an empire in areas as widespread as Italy, the Low Countries, Latin America, Africa, and the Philippines. These possessions were then lost at various times during a long, slow decline of Spanish power over more than three centuries. The Canary Islands, off the Atlantic coast of Morocco, are still Spanish-held and governed as an integral part of Spain.

Portugal

Portugal also played a part in expelling the Moors from Iberia and in the 15th century took the lead in seeking a sea route around Africa to the Orient. The first Portuguese expedition to succeed in the voyage, headed by Vasco da Gama, returned

Landscape in Literature

The Walking Drum

This excerpt from Louis L'Amour's The Walking Drum *is presented here to provide a sense of how big a role trade and markets played in European life almost a thousand years ago. Although set in France, the images and the lore captured in this scene spill over into the rest of Mediterranean Europe and other regions as well. As you read these paragraphs, think about the routes that had to be traveled to bring all of these goods and people together in this market setting. Think about the role of change agent played by the vendors and the soldiers who traveled with the merchant caravans from the distant source points for many of these goods.*

The fair at Provins was one of the largest in France during the twelfth century. There was a fair in May, but the most important was that in September. Now the unseasonably cold, wet weather had disappeared, and the days were warm and sunny.

Long sheds without walls covered the display of goods. Silks, woolens, armor, weapons, leather goods, hides, pottery, furs, and every conceivable object or style of goods could be found there.

Around the outer edge of the market where the great merchants had their displays were the peasants, each with some small thing for sale. Grain, hides, veg-etables, fruit, goats, pigs, and chickens, as well as handicrafts of various kinds.

Always there was entertainment, for the fairs attracted magicians, troupes of acrobats, fire-eaters, sword-swallowers, jugglers, and mountebanks of every kind and description.

The merchants usually bought and sold by the gross; hence, they were called grossers, a word that eventually came to be spelled grocer. Dealing in smaller amounts allowed too little chance for profit, and too great a quantity risked being left with odds and ends of merchandise. The White Company had come from Spain with silk and added woolens from Flanders. Our preferred trade was for lace, easy to transport and valued wherever we might go.

Merchants were looked upon with disdain by the nobles, but they were jealous of the increasing wealth and power of such men [as the merchants in this chapter], or a dozen of others among us.

The wealth of nobles came from loot or ransoms gained in war or the sale of produce from land worked by serfs, and there were times when this amounted to very little. The merchants, however, nearly always found a market for their goods.

At the Provins fair there were all manner of men and costumes: Franks, Goths, Saxons, Englanders, Normans, Lombards, Moors, Armenians, Jews, and Greeks. Although this trade was less than a century old, changes were coming into being. Some merchants were finding it profitable to settle down in a desirable location and import their goods from the nearest seaport or buy from the caravans.

Artisans had for some time been moving away from the castles and settling in towns to sell their goods to whoever passed. Cobblers, weavers, coppers, and armorers had begun to set up shops rather than doing piecework on order. The merchant-adventurers were merely distributors of such goods. . . .

The Church looked upon the merchants with disfavor, for trade was considered a form of usury, and every form of speculation considered a sin. Moreover, they were suspicious of the far-traveling merchants as purveyors of freethinking.

Change was in the air, but to the merchant to whom change was usual, any kind of permanence seemed unlikely. The doubts and superstitions of the peasants and nobles seemed childish to these men who had wandered far and seen much, exposed to many ideas and ways of living. Yet often the merchant who found a good market kept the information for his own use, bewailing his experience and telling of the dangers en route, anything to keep others from finding his market or his sources of cheap raw material.[1]

[1]Louis L'Amour. *The Walking Drum.* Toronto: Bantam Books, 1984. 256–258.

from India in 1499. For the better part of a century thereafter, Portugal dominated European trade with the East and built an empire there and across the Atlantic in Brazil. Then there was a rapid decline in Portuguese fortunes, although they controlled Brazil until the 19th century, when it gained independence, and they also held large African territories until the 1970s.

The two large cities of Portugal are both seaports on the lower courses of rivers that cross the Meseta and reach the At-lantic through Portugal. Lisbon (population: 2,561,225), on a magnificent natural harbor at the mouth of the Tagus River, is both the leading seaport of the country and its capital. The smaller city of Oporto (population: 1,174,461), at the mouth of the Douro River, is the regional capital of northern Portugal and the commercial center for trade in Portugal's famous port wine, which comes from terraced vineyards along the hills overlooking the Douro Valley.

Italy

In Italy, the Alps and the Apennines are the principal mountain ranges (Fig. 5.8). Northern Italy includes the greater part of the southern slopes of the Alps. The Apennines, lower but often rugged, form the north-south trending backbone of the peninsula, extending from their junction with the southwestern end of the Alps to the toe of the Italian boot, appearing again across the Strait of Messina in Sicily, where Mount Etna, a volcanic cone, reaches 10,902 feet (3323 m). Near Naples, Mount Vesuvius, at 4190 feet (1277 m), is one of the world's most famous volcanoes. West of the Apennines, most of the land between Florence on the north and Naples on the south is occupied by a tangled mass of lower hills and mountains, often of volcanic origin. Both Sicily and Sardinia, the two largest Italian islands, are predominantly mountainous or hilly.

Parts of the Italian highlands have population densities of more than 200 per square mile (*c.* 75 per sq km), but even these areas are much less densely populated than most Italian lowlands. The largest lowland, the Po Plain, has about two-fifths of the entire Italian population, with nonmetropolitan densities frequently reaching 500–700 per square mile (*c.* 190–310 per sq km). Italy overall has a population density of nearly 508 per square mile, more than twice the density of any other large Mediterranean nation. Some other lowland areas with extremely high densities include the narrow Ligurian Coast centering on Genoa, the plain of the Arno River inland to Pisa and Florence, the lower valley of the Tiber River including Rome, and Campania around Naples.

The Po Basin

The Po Basin is the economic heart of modern Italy and the most highly developed and productive part of Southern Europe.

Comprising just under a fifth of Italy's area, this slightly more moist mediterranean climate zone contains about two-fifths of Italy's population (who are more prosperous by far than the rest) and accounts for approximately one-half of its agricultural production and two-thirds of its industrial output.

A very rapid expansion and diversification of Po Basin industry, previously dominated by textiles, occurred after World War II. Prominent components of this expansion were (1) the growth of engineering and chemical industries, (2) the development of a sizable iron-and-steel industry, mainly in northern Italy but partly in peninsular Italy, and (3) a shift to domestic natural gas and imported oil as the predominant sources of power—with hydroelectric energy continuing to play an important role.

Many cities in the basin share in the region's industry, but the leading ones are Milan (population: 1,371,008), Turin (population: 961,916), and Genoa (population: 675,596). Milan is Italy's industrial capital, largest city, and main center of finance, business administration, and railway transportation. It is advantageously located between the port of Genoa and important passes, now shortened by railway tunnels, through the Alps to Switzerland and the North Sea countries. Turin is the leading center of Italy's automobile industry. It is particularly associated with FIAT SpA, which is by far the largest Italian auto manufacturer. The port of Genoa functions as a part of the Po Basin industrial complex, although it is not actually in the basin. It is reached from the Milan-Turin area by passes across the narrow but rugged northwestern end of the Apennines. The city dominates the overseas trade of the Po area and is Italy's leading seaport. Genoa is now much larger and economically more important than its old medieval commercial and political rival, Venice (population: 425,000), whose location on islands at the eastern edge of the Po Plain is relatively unfavorable for serving present-day industrial centers.

Figure 5.8 Rugged mountains in northern Italy where the Alps merge with the Apennines. Genoa, Italy's main seaport, is relatively near. The landscape exhibits a spectacular display of quarries producing the famous Carrara marble.
Pierre Boulat/Woodfin Camp & Associates

Figure 5.9 Panorama of Rome. The tree-lined Tiber River winds through the view. The low skyline and uniform architecture are characteristic of older European cities. The gleaming white structure in the left foreground is the Victor Emmanuel Monument, celebrating the 19th-century unification of Italy. The ancient Roman Colosseum is at the upper right. *Guido Alberto Rossi/The Image Bank*

In Italy, the largest cities outside of the northern industrial area are the capital, Rome (population: 2.6 million), and Naples (population: 1.05 million). Rome (Fig. 5.9) has a location that is roughly central within Italy and is predominantly a governmental, religious, and tourist center. Naples, located farther south on the west coast of the peninsula, is the port for the populous and productive, but very poor, Campanian agricultural region described earlier. It is also the main urban center of one of Italy's major tourist regions, with attractions such as Mount Vesuvius, the ruins of Pompeii, and the island of Capri in the vicinity.

Italy's main period of preeminence occupied approximately five centuries, during which the Roman Empire ruled over the whole Mediterranean basin and some lands beyond. Within the Empire, Christianity began and spread, and by the time the Empire collapsed in the 5th century A.D., Rome had become the seat of the Popes and the Catholic Church. During the later Middle Ages, a number of Italian cities—notably Venice, Genoa, Milan, Bologna, and Florence—became powerful independent states. Venice and Genoa controlled maritime empires within the Mediterranean region, and all the cities profited as traders between the eastern Mediterranean and Europe north of the Alps. After unification in the 19th century, Italy attempted to emulate past glories, but the empire that it acquired in northern Africa—on the eastern Mediterranean and Ethiopia—was lost when Italy was defeated in World War II.

Domestic Imbalances in Italy's Economic Growth—and Implications for the EU

There has long been an imbalance in the pace and scale of growth between northern Italy (Milan, Turin, the Po Valley) and southern Italy (Rome, Naples, and islands in the Mediter-

ranean). In the 1990s, northern Italy had labor shortages while the south had approximately 20 percent unemployment. The productivity of labor and industry in the south is approximately 80 percent that of the north. Per capita income in the south is less than three-fifths that of the north. Even with considerable aid from Rome, this regional imbalance continues apace.

Shortly after World War II, the Italian government began efforts to develop the southern part of its peninsula. Agricultural development, land reform, and the introduction of subsidized industries were among the methods used to stimulate development. But the differential between north and south remains and is widely perceived in Italy as a major national problem. This was accentuated by the 1996 efforts of the Po Valley region to draw attention to the possibility of secession and the creation of a new country, to be called "Padania," made up of the northern third of the Italian peninsula. This region is home to most of Italy's heavy industry and considerable agricultural and resource wealth. Given the history of political flux in Europe, even seemingly facetious suggestions like this have to be given some consideration.

Italy's domestic imbalance is of particular geographic concern because one of the prime goals of the European Union—with its new euro currency—is to bring all of the member nations of the EU into a greater common prosperity. It is hoped that with the full achievement of **open borders**—a treaty-guaranteed goal of the EU—and with a common currency built around the euro, the wealth of some of the member nations will transcend historic borders and regional differences and expand prosperity regionwide. However, if a single EU country cannot achieve this prosperity, what will be the impact on the future scenarios of the EU? In the case of Italy, it is felt that this regional differentiation is both historic and not particularly unusual. The pattern of relative wealth in highly urbanized regions with a concentration of industrial activity, good harbors, networks of transportation, and early geographic centrality in the beginnings of major industrial growth occurs in many parts of Europe. The lesson to be learned from the Italian example is that the broad expansion of wealth across the full realm of the European Union will not happen nearly as quickly as promised—if it happens at all.

Greece

In Greece, most of the peninsula north of the Gulf of Corinth is occupied by the Pindus Mountains and the ranges that branch from them. Extensions of these ranges form islands in the Ionian and Aegean Seas. Greece south of the Gulf of Corinth, commonly known as the Peloponnesus, is composed mainly of the Arcadian mountain knot. Along the coasts of Greece, many small lowlands face the sea between mountain spurs and contain the majority of the people. A particularly famous lowland, although far from the largest, is the Attica Plain, still dominated as in ancient times by Athens and its seaport, Piraeus. Centuries before the Christian era, Greek artists, architects, authors, philosophers, and scholars produced many of the ideas and works that laid the foundation for Western civilization. Between about 600 and 300 B.C., the sailors, traders, warriors, and colonists of

Figure 5.10 The small island of Mykonos in the Aegean islands of Greece is one of many famous tourist sites in the Mediterranean world. The light-colored architecture, the fishing boats, and the deforested—and fully settled—slopes are characteristically Mediterranean. Vegetation on the slope in the background has turned brown in the regular midsummer drought. Both the island and the town are named Mykonos. *Toshi Chatelin*

the Greek city-states spread their culture throughout the Mediterranean area. A second period of Greek power and influence occurred in the Middle Ages, when Constantinople (modern Istanbul, Turkey) was the capital of the large Byzantine Empire. This Greek empire developed and diffused the Eastern Orthodox branch of Christianity.

Athens is the largest city in Greece by an overwhelming margin. With a metropolitan population of 3,072,922 (including the port of Piraeus), representing three-tenths of the population of Greece and growing rapidly, Athens has no national rival. It is the capital, main port, and main industrial center. Its development is based on imported fuels, and materials are used to manufacture products that are sold almost entirely in a domestic market whose main center is Athens itself.

Tourism is the mainstay of the Greek economy, with nearly $4 billion a year coming from this trade. Greece has not only the beauty of the Mediterranean to bring people to its shores, but also the ruins of early Greek culture and ongoing festivals in art and drama—all combining to make this an economic and cultural environment of major significance (Fig. 5.10).

Geographic Factors Influencing Southern Europe's Uneven Development

Today, Southern Europe is, as it has long been, one of the relatively poor parts of Europe, along with East Central Europe (see Table 3.1). The main factor contributing to Southern European poverty seems to have been a lagging development of industry related to unfavorable social and resource conditions during the 19th and early 20th centuries, when many European countries were industrializing rapidly. Lack of industrial employment has left too many people in crowded small-farm areas where output per person (although not always per acre) is low. Among the inimical social conditions in Southern Europe were:

1. *Trade deficiency.* The absence of large-scale trade and therefore of capital accumulation to invest in manufacturing created a steady lack of development capital.
2. *Low educational levels.* The preponderance of poor and uneducated populations offered little in the way of skilled labor or dynamic markets and, in combination with the trade deficiency, worked to slow economic growth.
3. *Land tenure patterns.* The control of land and often government by wealthy landowners whose interests lay in maintaining the agrarian societies that they dominated meant that there was little drive for a significant change in the way that society and the economy were organized.
4. *Ineffectual government assistance.* The prevalence of unstable, undemocratic, and ineffective governments little oriented toward economic development served as a continuing impediment to real economic growth.
5. *Resource deficiencies.* Resource deficiencies included a lack of good coal and iron ore deposits, except in northern Spain;

deficient wood supplies due to dryness and centuries of deforestation; and lack of water power in the areas of mediterranean climate, where streams often dry up in summer.

6. *Transportation difficulties.* The rugged topography of much of the region made construction of adequate rail and road systems difficult and expensive.

A variety of these factors worked in different combinations as major impediments to economic development in the region. Only three sizable industrial areas developed in Southern Europe before World War II. The most important was in the Po Plain of northern Italy. Here, imported coal and raw materials, hydroelectric power from the surrounding mountains, and, cheap labor formed the basis for an area specializing in textiles, but including some heavy industry, automobile production, an aircraft industry, and engineering. A second area very similar in its foundations and specialties, but developed on a much smaller scale, came into existence in the northeastern corner of Spain, in and around Barcelona. A third area, focused on heavy industry, developed in the coal- and iron-mining section of northern coastal Spain. All three areas lie near the foot of mountains with humid conditions that provide the basis for hydroelectric power development.

Economic Progress in the Last Half Century

In the later 20th century, the economies of Southern Europe began to show a new dynamism. Although the South is still relatively poor compared to the North and West Central Europe, recent growth has been shaped by both industrial modernization and an expanding role of the EU. Industrial and service activities led the way, and there was a sharp decline in the proportion of the population still employed in agriculture. A number of factors played a part in this revitalization:

1. *Transportation changes.* Improved transportation, particularly in ocean transport, made the importation of fuels and materials for manufacturing more economical.

2. *Energy shift.* Coal ceased to be the dominant source of power in Southern Europe, replaced by oil or, in Italy, by natural gas and oil. The countries in Southern Europe became major importers of oil, as did their industrial competitors. Thus, Southern Europe moved to a more even footing in securing power than when it had to depend on large imports of coal. In Italy, the power situation also improved with the discovery of large natural gas deposits, principally under the Po Plain, as well as development of an active geothermal program.

3. *Market expansion.* Market conditions for Southern European exports improved greatly in the decades after World War II, largely as a result of the rising prosperity of northwestern Europe and the expansion of European markets. These exports included goods such as subtropical fruits and vegetables, Italian-built automobiles, and, less obvi-

Figure 5.11 The western Po valley around Milan, Italy, is a hub of varied manufacturing in this most industrial part of Italy. This Turin-based firm specializes in the fabrication of full-scale, plastic decoy tanks and fighter planes. Such products as these may have been used in the Gulf War by Iraq in an effort to have the United Nations forces expend missiles on these plastic shells at considerable financial cost to the United Nations war effort. *Enzo Signorelli/Gamma Liaison International*

ously, labor. Streams of temporary migrants from Southern Europe found jobs in booming, labor-short industries to the north. Absence of the migrants relieved unemployment, and their remittances added purchasing power in their home areas.

4. *Capital infusion.* There were other infusions of foreign money into the Southern European countries. Each country received American economic and military aid, and favorable economic circumstances and government policies attracted a considerable amount of foreign investment in industrial and other facilities (Fig. 5. 11). Southern Europe also saw a tremendous boom in tourism. To increasingly prosperous Europeans and Americans, Southern Europe offers a combination of spectacular scenery (see Fig. 5.10), historical depth exceeding that of even the other parts of Europe, a sunny subtropical climate, the sea, and the beaches. The ancient cities, now modern and bustling with activity around their monuments of the past, supplemented attractions with new or greatly expanded resort areas such as Spain's Costa Brava and Costa del Sol, the Balearic Islands, the Italian Riviera, and various other places.

Economic and Demographic Trends: The Shift from Agriculture to Industry

Since World War II, Italy, Spain, Portugal, and Greece have all evolved from agricultural to industrial and service-oriented countries. Agricultural exports are still important, but manufactured exports have become more valuable. Large tourist industries "marketing" sun, scenery, historic sites, and beaches supplement exports with an increasing economic significance.

5.3 The East: Legacies of Empire and Uncertainties of the Future

As defined in this chapter and text, the East consists of East Central European countries that are attempting to recover from more than four decades of Communist rule. They are Poland, the Czech Republic, Slovakia, Hungary, Romania, Bulgaria, Slovenia, Croatia, Bosnia-Herzegovina, Yugoslavia, Albania, and Macedonia, occupying eastern Europe between the Baltic, Adriatic, and Black Seas (Fig. 5.12). Prior to German reunification in 1990, the former East Germany also was Communist-controlled, and it will be brought into the present discussion where appropriate. Many of these countries (including East Germany) became parts of the former Soviet empire during World War II, when Soviet troops entered them in pursuit of retreating German armies in 1944 and 1945. Exceptions were Albania and the former Yugoslavia, where national Communist resistance forces took power on their own as German and Italian power collapsed. In the others, Soviet occupation forces supported local Communists, and all had Communist "satellite" governments subservient to the Soviets by 1948.

Following their takeover by Communism, the nations of East Central Europe (and East Germany) were reshaped along Communist lines, with one-party dictatorial governments; national economies planned and directed by organs of the state; abolition of private ownership (with some exceptions) in the fields of manufacturing, mining, transportation, commerce, and services; abolition of independent trade unions; and varying degrees of socialization of agriculture. Attempts by East Germany in 1953, Hungary in 1956, and Czechoslovakia in 1968 to break away from Soviet control were crushed by military force.

In 1989 and 1990, Soviet backing for the region's Communist order was abruptly withdrawn because of the inability of the Russian government to continue financial support and because of rapidly expanding calls for greater autonomy by the former satellite nations. Several Communist regimes (including East Germany's) quickly collapsed and were replaced by democratically elected non-Communist governments. However, "reformed" and renamed Communist parties and ex-Communist officials still play a significant role in the region. In late 1995, a Communist was elected president of Poland, and in that same year, the Communists gained more than 20 percent of the common vote—the largest of any party—in a parliamentary election in Russia. Governments are now struggling to build democracies with capitalist economies out of the economic wreckage of frequently nonproductive, nonprofitable, noncompetitive, state-owned enterprises left as a Communist legacy. The struggle to make Western free-market economies work in these nations of East Central Europe has been a contest of a varied and often serious nature.

Political Geography of the "Shatter Belt"

The extension of Soviet power into East Central Europe at the end of World War II was in keeping with the history of the region. In the Middle Ages, several peoples in the region—the Poles, Czechs, Magyars, Bulgarians, and Serbs—enjoyed political independence for long periods and at times controlled extensive territories outside their homelands. Their situation then deteriorated as stronger powers—Germans, Austrians, Ottoman Turks, and Russians—pushed into East Central Europe and carved out empires. These empires frequently collided, and the local peoples were caught in numerous wars that devastated great areas, often resulted in a change of authority, and sometimes brought about large transfers of population from one area to another.

Origins of the Present Pattern of Tensions

The present pattern of countries in East Central Europe resulted from the disintegration of the Ottoman Turkish, Austro-Hungarian, German, and Tsarist Russian Empires in the 19th and early 20th centuries. World War I hastened this process, in which Germany, Austria-Hungary, and Turkey were on the losing side. Russia, originally allied with the victors, suffered disastrous defeats and then withdrew from the war following the Bolshevik Revolution of 1917. It was not represented at the Paris Peace Conference of 1919, where the victorious Western powers rearranged parts of the political map of East Central Europe in an attempt to satisfy the aspirations of its various nationalities. Poland, which had been

Definitions & Insights

SHATTER BELT

In regions characterized by frequent or even continual shifting of boundaries due to warfare, rapidly changing roles of authority, or frequent friction between two major nations, there tends to be a fluidity in the alignment and independence of the adjacent smaller nations. The result is a **shatter belt**. Historically the term has referred to the East Central European nations from Poland in the north to Albania in the south. In broader use, however, this term connotes any region that experiences continual demographic shifts, through ongoing migrations, efforts at **ethnic cleansing**—the often bloody removal of ethnic populations from a state or region controlled by different and more powerful ethnic groups—and continual internecine warfare and border disputes. This pattern of geopolitical splintering is at the base of enormous instability, tension, ongoing enmity, and, very often, local warfare. In the case of the former Yugoslavia, the shatter-belt characteristic has a history of leading to much broader wars, and it carries that same potential even today.

East Central Europe • **Index Map**

URBAN AREAS
Stars show national capitals

⊛ ⊙   Over 3,500,000

✪ ●   1,000,000–3,500,000

★ ○   500,000–1,000,000

☆ ∘   300,000–500,000

⊛ •   Selected smaller places

The independent successor nations of former Yugoslavia are underlined.

City-size symbols are based on metropolitan area estimates.

Mountainous areas

Major industrial concentrations

Main territorial changes since 1938

Main agricultural areas

Figure 5.12   The "mountainous areas" are broadly generalized. For identification of cities shown by letter in the major industrial concentrations, see Figure 3.1. City-size symbols are based on metropolitan area estimates.

partitioned among three empires in the 1700s, was reconstituted as an independent country. Czechoslovakia—the homeland of the closely related Czech and Slovak peoples who divided in 1993 by their own peaceful preference—was in 1919 carved out of the Austro-Hungarian Empire as an entirely new country. Hungary, greatly reduced in size, was severed from Austria. The Kingdom of Serbia, which had won independence from Turkey in the 19th century, was joined with the small independent state of Montenegro and several regions taken from Austria-Hungary to form the new Kingdom of the Serbs, Croats, and Slovenes, later known as Yugoslavia. Romania, which had been independent of Turkish control since the mid-19th century, was enlarged by territories taken from Austria-Hungary, Russia, and Bulgaria. Bulgaria had already achieved independence from Turkey in the second half of the 19th century, as had Albania immediately before the outbreak of World War I.

This longer view of regional historical geography reminds us that the civil war of the 1990s fought in the former Yugoslavia has long been characteristic of this shatter belt of Eastern Europe. There have been large population transfers in the region since the beginning of World War II. During that war and the immediate postwar years, transfers involving millions of Germans, Poles, Hungarians, Italians, and others "simplified" the ethnic distribution patterns but at enormous human cost. People often were uprooted without notice, losing all their possessions, and were dumped as refugees in a so-called "homeland" that many had never seen. The prewar populations of Germans in Poland and Czechoslovakia were expelled at the end of World War II and forced into East and West Germany. Many died in this process of forced migration. Millions of Jews were systematically killed throughout East Central Europe during the German occupation in World War II. Ethnic minorities now constitute only 2 percent of the population in East Central Europe's largest country, Poland, which transferred most of its German population to Germany and whose territory containing Lithuanians, Russians, and Ukrainians was taken away.

A number of countries still have sizable minorities—for example, Hungarian minorities in the Slovak Republic, Vojvodina in Serbia, and the mountainous area called Transylvania in Romania. However, while Communist control put a quieting blanket over most of East Central Europe, its collapse beginning in 1989 was accompanied by a rapid reemergence of some old ethnic quarrels. For example, fighting among diverse ethnic elements has fractured the former federal state of Yugoslavia (Fig. 5.13).

The Slavic Realm of Europe

The expulsion of Germans from East Central Europe and the extermination or flight of more than six million Jews in total

during the period of Nazi control in the 1930s and '40s intensified the Slavic character of the region. Practically all of Europe's Slavic people live there, and they form a large majority within the region.

The Major Slavic Groups

The Slavic peoples are often grouped into three large divisions: (1) East Slavs of the Former Soviet Region, (2) West Slavs, including Poles, Czechs, and Slovaks, and (3) South Slavs, including Serbs, Croats, Slovenes, Bulgarians, and Macedonians. Another group is the Romanians.

Although the various Slavic peoples speak related languages, such languages may not be mutually intelligible. For example, the Polish language is not easily understood by a Czech, although Poles and Czechs are customarily grouped as West Slavs. Other significant differences also exist. Serbian and Croatian, for example, are essentially one language in spoken form, but the Serbs, like the Bulgarians and most of the East Slavs, use the Cyrillic alphabet (based on the Greek language), whereas the Croats use the Latin alphabet, as do the Slovenes and the West Slavs. The Romanian language is classed with the Romance languages derived from Latin; however, the contemporary Romanian language contains many Slavic words and expressions. Romanians are generally classified as non-Slavs and are an additional example of the ethnic variety of Eastern Europe.

A religious division exists between the West Slavs, Croats, and Slovenes, who are Roman Catholics for the most part, and the East Slavs, Romanians, Bulgarians, Serbs, and Macedonians, who adhere principally to various branches of the Orthodox Christian Church.

Non-Slavic Peoples

The principal non-Slavic elements in East Central Europe are the Hungarians and the Albanians. The Hungarians, also known as Magyars, are the descendants of nomads from the east who settled in Hungary in the 9th century. They are distantly related to the Finns and speak a language entirely distinct from the Slavic languages. Roman Catholicism is the dominant religious faith in Hungary, although substantial Calvinist and Lutheran groups have long existed there. The Albanians speak an ancient language that is not related to the Slavic languages except in a very distant sense. Excluding Turkey, Albania is the only European state in which Muslims form a majority of the population, although Slavic Muslims (Moslems) are the largest ethnic group in Bosnia and Herzegovina (in the former Yugoslavia). Serbia, Macedonia, Montenegro, and Bulgaria have many Albanian or Turkish Muslims. These groups are legacies from former Turkish rule.

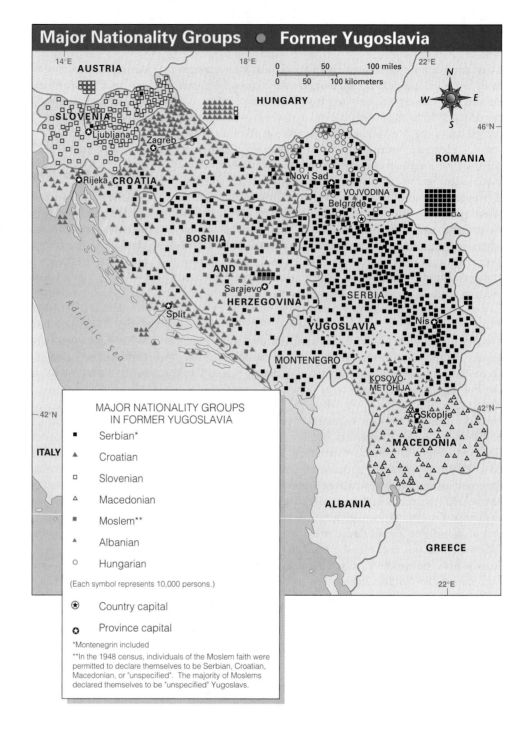

Major Nationality Groups ● Former Yugoslavia

Figure 5.13 This map has been redrawn from a map in George W. Hoffman and Fred Warner Neal, *Yugoslavia and the New Communism* (New York: Twentieth Century Fund, 1962), p. 30, by permission of the authors and the Twentieth Century Fund. Names in the largest type denote the six republics that constituted the former Socialist Federal Republic of Yugoslavia. The Vojvodina and Kosovo-Metohija areas are former "autonomous provinces" brought under tight control by Serbia in 1988. The future status of the political divisions and boundaries within former Yugoslavia continued to be very uncertain in early 1999.

MAJOR NATIONALITY GROUPS IN FORMER YUGOSLAVIA

- ■ Serbian*
- ▲ Croatian
- □ Slovenian
- △ Macedonian
- ▪ Moslem**
- ▴ Albanian
- ○ Hungarian

(Each symbol represents 10,000 persons.)

- ✪ Country capital
- ✪ Province capital

*Montenegrin included
**In the 1948 census, individuals of the Moslem faith were permitted to declare themselves to be Serbian, Croatian, Macedonian, or "unspecified". The majority of Moslems declared themselves to be "unspecified" Yugoslavs.

Physical Geography: A Checkerboard of Habitats

A glance at Figure 5.12 shows that East Central Europe forms a kind of crude checkerboard of mountains and plains. The *mountains*—located in the center and south—come in a wide range of elevations and degrees of ruggedness. The *plains*—concentrated in the north and in the center along the Danube—exhibit varying degrees of flatness, fertility, warmth, and humidity, and agriculturally they range from some of Europe's better farming areas to some of the poorest. A handful of Europe's largest rivers carry the drainage of the region to the Black Sea, the Baltic Sea, or the North Sea.

The Northern Plain

Most of Poland lies in a large northern plain between the Carpathian and Sudeten mountains on the south and the Baltic Sea on the north. The plain is a segment of the North European Plain, which extends westward from Poland into Germany and eastward into the Ukraine in the Former Soviet Region. Central and northern Poland exhibit land that is rather sandy and infertile, with many swamps, marshes, and lakes. At the south, the plain rises to low uplands. Here, where Poland touches Europe's loess belt, are the country's most fertile soils.

Poland's largest rivers, the Vistula (Polish: *Wisla*) and Oder (Polish: *Odra*), wind across the northern plain from Upper Silesia to the Baltic. The Oder is the main internal waterway, carrying some barge traffic between the important Upper Silesian industrial area (with which it is connected by a canal) and the Baltic. Poland's main seaports—Gdansk (formerly *Danzig*), Gdynia, and Szczecin (formerly *Stettin*)—are along or near the lower courses of these rivers. Gdansk and Gdynia form a metropolitan area of about 900,000 people, and Szczecin has about 450,000.

The Central Mountain Zone

A central mountain zone is formed by the Carpathian Mountains and lower ranges farther west along the Czech frontiers. The Carpathians, considerably lower and less rugged than the Alps, extend in a giant arc for about a thousand miles (*c.* 1600 km) from Slovakia and southern Poland to south central Romania. West of the Carpathians, lower mountains enclose the hilly and highly industrialized Bohemian basin in the Czech Republic. On the north, the Sudeten Mountains and Ore Mountains separate this republic from Poland and the former East Germany. Between these ranges, at the Saxon Gate, the valley of the Elbe River provides a lowland connection and a navigable waterway from Bohemia to the industrial region of Saxony in Germany and, farther north, to the German seaport of Hamburg and the North Sea. To the southwest, the Bohemian forest occupies the frontier zone between Bohemia and former West Germany.

Between mountain-rimmed Bohemia and the Carpathians of Slovakia, a busy passageway is provided by the lowland corridor of Moravia in the Czech Republic. Through this corridor run major routes connecting Vienna and the Danube valley with the plains of Poland. Near the Polish frontier, the corridor narrows at the Moravian Gate between the Sudeten Mountains and the Carpathians. Just beyond this gateway, East Central Europe's most important concentration of coal mines and iron and steel plants developed in the Upper Silesian–Moravian coal field of Poland and the Czech Republic.

The Danubian Plains

Two major lowlands, bordered by mountains and drained by the Danube River and its tributaries, compose the Danubian plains. One of these, the Great Hungarian Plain, occupies two-thirds of Hungary and smaller adjoining portions of Romania, Serbia, and Croatia. It is very level in most places and contains much poorly drained land near its rivers. The second major lowland is composed of the plains of Walachia and Moldavia in Romania, together with the northern fringe of Bulgaria. The Danubian plains contain the most fertile large agricultural regions in East Central Europe.

The Danube River, which supplies a vital navigable water connection between these lowlands and the outside world, rises in the Black Forest of southwestern Germany and follows a winding course of some 1750 miles (*c.* 2800 km) to the Black Sea. Upstream from Vienna, the river is swift and hard to navigate, although large barges go upstream as far as Regensburg, Germany. Below Vienna, the Danube flows leisurely across the Hungarian and Serbian plains past Budapest (Fig. 5.14) and Belgrade. In the border zone between Serbia and Romania, the river follows a series of gorges through a belt of mountains about 80 miles (*c.* 130 km) wide. The easternmost gorge is the Iron Gate gorge. Beyond it, the Danube forms the boundary between the plains of southern Romania and northern Bulgaria. The river then turns northward into Romania and enters the Black Sea through a marshy delta. Traffic on the Danube is modest compared with the tonnage carried by barges on the Rhine or by road and rail in East Central Europe. The river flows mostly through agrarian areas, and heavy industries—the most important generators of barge traffic—are found in only a few places along its banks.

The Southern Mountain Zone

East Central Europe's southern mountain zone occupies the greater part of the Balkan Peninsula. Bulgaria, the Yugoslav successor states, and Albania, which share this zone, are very mountainous, although important lowlands exist in some

Figure 5.14 The Danube River at Budapest. Government buildings line the waterfront beyond the Chain Bridge. The domed structure is Hungary's national Parliament. The view was taken from Buda on the higher west bank, looking toward Pest across the river. *H. Todd Stradford, Jr.*

places. In Bulgaria, the principal mountains are the Balkans, extending east-west across the center of the country, and the Rhodope Mountains in the southwest. These are rugged mountains that attain heights of over 9000 feet (c. 2800 m) in a few places. Between the Rhodope Mountains and the Balkan Mountains is the productive valley of the Maritsa River, constituting, together with the adjoining Sofia Basin, the economic core of Bulgaria. North of the Balkans, upland plains, covered with loess and cut by deep river valleys, slope to the Danube.

In southern Serbia and neighboring areas to the west and south, a tangled mass of hills and mountains constitutes a major barrier to travel. Through this difficult region, a historic lowland passage connecting the Danube valley with the Aegean Sea follows the trough of the Morava and Vardar rivers. At the Aegean end of the passage is the Greek seaport of Salonika (Thessaloniki). The Morava-Vardar corridor is linked with the Maritsa Valley by an east-west route that leads through the high basin in which Sofia, the capital of Bulgaria, is located. Along the rugged, island-fringed Dalmatian Coast, the Dinaric Alps rise steeply from the Adriatic Sea. The principal ranges run parallel to the coast and are crossed by only a few significant passes.

The Varied Range of Climates

In most of East Central Europe, winters are colder and summers are warmer than in western Europe. In the Danubian plains, the humid continental long-summer climate is comparable to that of the corn and soybean region (Corn Belt) in the United States' Midwest, but the plains of Poland have a less favorable humid continental short-summer climate, comparable to that of the American Great Lakes region. The most exceptional climatic area is the Dalmatian Coast. Here, temperatures are subtropical and precipitation, concentrated in the winter, reaches 180 inches (about 260 cm) per year on some slopes facing the Adriatic. This is Europe's greatest precipitation. The climate of the Dalmatian Coast is classed as mediterranean (dry-summer subtropical). Together with spectacular scenery, this environmental attribute has fostered a sizable tourist industry. All through this region tourist dollars are not only becoming more important than ever before —and are seen as a sign of welcomed economic change—but that same upswing places Western cultural influences right in the center of these countries. The capital cities are the most frequently sought tourist destinations, and it is from those same centers that political and cultural influences radiate.

Patterns of Erratic Agricultural Development Under Communism

Following World War II, the new Communist governments liquidated the remaining large private holdings in East Central Europe, and programs of collectivized agriculture on the Soviet model were introduced. Some farmland was placed in large state-owned farms on which the workers were paid wages, but most land was organized into collective farms owned and worked jointly by peasant families who shared the proceeds after operating expenses of the collective had been met. **Collectivization**—the bringing together of individual land holdings into a government-organized and government-controlled agricultural unit—met with strong resistance, and it was discontinued in the 1950s in former Yugoslavia and Poland. Today, all but a minor share of the cultivated land in Poland is privately owned. The remaining countries are pursuing programs to reprivatize their farmland, most of which was still in collective ownership in mid-1996. Conversion to private farming after four decades of collectivization presents painful obstacles, and progress is slow, even though images of this transformation continue to fuel hopes of the region's small farmers.

Despite recent attempts to diversify and intensify agriculture, farming in East Central Europe remains primarily a crop-growing enterprise based on corn and wheat—the leading crops in the region from Hungary and Romania south. In the Czech and Slovak republics and Poland, with their cooler climates, corn is unimportant, but wheat is a major crop as far north as southern Poland's belt of loess soils. Most of Poland lies north of the loess belt and has relatively poor sandy soils; here, rye, beets, and potatoes become the main crops. Livestock raising is a prominent secondary part of agriculture throughout East Central Europe. The region's meat supply is much less abundant than in western Europe and comes primarily from

Figure 5.15   Brasov is Romania's second largest city and an important economic, cultural, and tourist center. The center of the city dates back to medieval times.   *Bill Bachman/Photo Reseachers, Inc.*

pigs. The more southerly areas contain poor uplands that pasture millions of sheep, the main source of meat in Bulgaria, Albania, and parts of former Yugoslavia.

Urban and Industrial Development in East Central Europe

East Central Europe has few cities of major size, although small regional centers, often old and historic, are widespread (Fig. 5.15). Only two areas—the Upper Silesian–Moravian coal field of Poland and the Czech Republic, and the industrialized Bohemian basin—exhibit a closely knit web of urban places. Even the agglomeration of mining and metallurgical centers in Upper Silesia–Moravia—a Ruhr in miniature—contains only about 3.5 million people. The Polish part of Upper Silesia–Moravia is somewhat larger than Poland's capital, Warsaw (population: 1.64 million). In all the other countries, the largest cities are national capitals. None of the capitals, including Warsaw, ranks in Europe's top ten cities in metropolitan population. The largest, Budapest, has about 2.6 million people; Bucharest has 2.3 million; Prague, Sofia, and Belgrade each have between 1 and 1.7 million; Zagreb, Skopje, and Sarajevo each have over 500,000; and both Ljubljana and Tirane have 250,000 to 500,000. In every instance, the capital is its country's leading center of diversified industries and serves also as a primate city (Fig. 5.16).

Industrial Trends in the Communist Era

The ascension of Communism in East Central Europe following World War II inaugurated a new industrial and urban era in this region. With minor exceptions, the existing industries were taken out of private hands, and national economic plans were developed. Communist planners did not aim at a balanced development of all types of industry. Instead, they stressed the types that were deemed most essential to industrial development as a whole. Investment was channeled heavily into a few fields: mining, iron and steel, machinery, chemicals, construction materials, and electric power. These were favored at the expense of consumer-type industries and agriculture.

The planned expansion of mining and industry in East Central Europe under Communism produced notable increases in the total output of minerals, manufactured goods, and power. But it also produced inefficient, overmanned industries ineffective in their later competition for the world markets of the post-Communist period. The central planning agencies of the Communist governments maintained a rigid control over individual industries. Plant managers were directed to produce certain goods in quantities determined by state governmental planners, and the success of a plant was judged by its ability to meet production targets rather than its ability to sell its products competitively and at a profit. Political reliability and conformity to the central plan were qualities much desired in plant managers.

Figure 5.16 Industrial development in Eastern Europe has lagged behind the post–World War II buildup that characterized so much of Western Europe. In this scene of the Nowa Huta steelworks near Cracow, Poland, can be seen an industrial plant that was intended to replicate some of the industrial power that Pittsburgh, Pennsylvania, developed nearly a century ago. Plants such as this become difficult to maintain because of the competition of steel from new mills, but to close these older plants creates a social dislocation that is always significant. *Simon Fraser/Science Photo Library/Photo Researchers, Inc.*

This system resulted in shoddy goods that were often in short supply or in mountainous oversupply because production was not being driven or keyed by buyer demand and satisfaction.

A key element in East Central Europe's industrialization in the Communist era was its heavy dependence on the Soviets for vital minerals. These included iron ore, coal, oil, natural gas, and other minerals, which were imported in great quantities and over long distances by rail or pipeline from Soviet mines and wells. In return, the Soviet Union was a large importer of East Central Europe's industrial products.

During the 1970s and 1980s, however, the region's trade relationship with the Soviet Union weakened. Trade and financial relations with Western countries, companies, and banks expanded rapidly. With relatively little to export, the East Central European countries borrowed massively from Western governments and banks to pay for imports from the West. This course was followed particularly by Poland and Romania. New industrial plants and equipment acquired in this way were supposed to be paid for by goods that would be produced for export to the West by industries that had learned efficient production and marketing techniques from the West. But then the Western economies began to slump, lowering the demand for imports.

Figure 5.17 This 1989 anti-Communist rally of shipyard workers in Gdansk, Poland, reflects the surge of protest that rapidly overthrew the Communist order in most of East Central Europe and in the former East Germany as well. This meeting was addressed by former United States president George Bush and "Solidarity" leader Lech Walesa. *Chick Harrity/U.S. News and World Report*

This led to a situation by the 1980s in which large debts to Western governments and banks needed to be repaid if countries were to maintain any credit at all. However, with the help of mismanagement by Communist bureaucrats, exports with which to pay were not being produced or could not be sold. The result was a severe impediment to economic expansion, falling standards of living as the governments squeezed out the needed money from their people, and, in Poland, social action through the formation of an independent trade union called **Solidarity,** which paralyzed the country and threatened to upset Communist dominance (Fig. 5.17). These conditions, along with the USSR's own growing distress, set the stage for the withdrawal of Soviet control and de-Communization in 1989 and 1990.

The Political Revolutions of 1989 and the Recession of Communism

As late as the early summer of 1989, the former East Germany and the East Central European countries seemed firmly in the control of totalitarian Communist governments. Then, in the late summer and autumn, the Communist order began to crumble. Public demands for freedom, democracy, and a better life gathered momentum in country after country. Communist dictators who had ruled for many years were forced out, and reformist governments took charge. This liberalizing process continued into the 1990s. By mid-1991, democratic multiparty elections had been held in all countries.

In a process that has had bold implications for the political as well as the economic landscape of Eastern Europe, the whole geography of economic activity underwent profound change in the early 1990s. Ten highlights of the economic restructuring of the region included:

1. *privatization*—the privatizing of state-owned enterprises;
2. *increasing the efficiency of state-run enterprises* by allowing noncompetitive enterprises to fail and removing diseconomic support of poor performers;
3. *encouraging new private enterprises* to develop, in many cases with foreign capital and management playing a role in the growth of these new units;
4. *developing new institutions* required by a market-oriented economy (banks, insurance companies, stock exchanges, accounting firms, and so on);
5. *fostering joint enterprises* between state-owned firms and foreign firms (Fig. 5.18);
6. *ending price controls* and allowing prices to reflect competition in the market—a shift that has been particularly difficult for the firms that paid little attention to markets, the need for their product, or quality;
7. *eliminating bureaucratic restrictions* on the private sector;
8. *making currencies internationally convertible* to increase trade and thus increase competition;
9. *converting farmland* from state and collective ownership to private ownership; and
10. *expanding access to international communications media* of all sorts to help local entrepreneurs learn from foreign

Figure 5.18 One of the most popular and widespread signatures of a U.S. presence abroad is the nearly universal Marlboro Man. This Moscow scene, with the ten-story-high image of the Big Sky–country cowboy associated with Marlboro cigarettes, is another sign of the rapid expansion and flow of Western, particularly American, cultural icons and influences into Europe and beyond and, in this case, into Russia. There are parallels to this image, and others, of numerous United States exports all through Europe and, more recently, East Central Europe as well. *Swersey/ Gamma Liaison*

examples—such as encouraging travel of potential foreign investors to plant sites to assess the prospects for success in joint ventures.

This economic transformation has been highly varied in its success in East Central Europe. The period from 1989 to the mid-1990s saw only spasmodic economic progress in the region. Economic hardship, uncertainty, and uneasiness were widespread. Loans by the International Monetary Fund (IMF) and other outside sources were helpful but modest in scale. Unemployment was high in all countries. Freeing of prices as a move toward a market economy often resulted in more and better goods in stores, but at prices few local customers could afford. The increased prices created great hardship for pensioners and others living on low fixed incomes. Meanwhile, a high proportion of all industries remained state-owned despite various efforts to dispose of them to private owners. Such privatization was slowed not only by the decrepitude of many properties but also by resistance to change on the part of workers and managers whose jobs, security, and power were at stake. Similarly, the privatization of state-owned farmland moved very slowly. Most prospective farmers were reluctant to exchange the relative security of employment on collectivized farms for the potential hazards of private farming. The uncertainty generated by all these changes was further promoted by the success of Communist party candidates in the region in elections in the mid-1990s.

Poland

Poland lies in the northeast corner of Europe as defined in our text. It is nearly 50 percent arable, sited as it is on the North European Plain (although most of Poland lies north of the fer-

tile loess belt), with another 13 percent in meadow and some 28 percent in forest. It has served as a migration corridor for millennia, and even over the past century has had its borders changed a number of times in response to the political effects of World Wars I and II (see Fig. 5.12). Currently Poland has a favored position among the countries of the post-Soviet break-up because of the catalyst role it played in the early 1980s with Solidarity's first rising in the Gdansk shipyards (see Fig. 5.17). The Gdansk Accords gave Polish workers the right to strike and to form free trade unions. This, in essence, began the end of Soviet control of Eastern Europe. Warsaw, Poland's capital and primate city, has given much attention to the reconstruction of the structures that survived heavy bombing toward the end of World War II.

Like other nations in the East that are still experimenting with new freedoms and challenges following the collapse of Communism, Poland is still experiencing political shifting. The elections of 1995 saw a significant return of authority to re-elected Communist party members. However, in 1997 Poland was offered membership in NATO, and accepted. Poland also suffers from continued environmental problems spawned by the Soviet promotion of economic development and industrial growth between the 1950s and the end of the 1980s. By the end of the 1980s, 75 percent of Poland's rivers had been declared biologically dead.

The Czech Republic

The economically diversified Czech Republic, where a long tradition of economic accomplishment was undercut by Communism, is also enjoying more than usual success in breaking out of the Communist mold, although, like Hungary, it is still a poor area by broader European standards. It is 43 percent arable and one-third forested. It was the economic disparity between the Czech and Slovak sectors of the former Czech and Slovak Federal Republic (Czechoslovakia) that led to its dissolution on January 1, 1993, and the creation of the new independent countries of the Czech Republic and Slovakia. Many Czechs felt that less industrialized and less prosperous Slovakia was a drag on the federal economy; Slovaks, in turn, resented "dictation" by the Czech majority. This process of disengagement was called the Velvet Revolution, which began in 1989, led to the 1993 break and, finally, resulted in the 1996 delineation of new borders between the two independent countries. Prague, like Warsaw, is both the capital and the primate city of the Czech Republic. Proximity to Germany and European tourists plays a major role in its economic growth, with more than $2,600,000 in tourist dollars earned in 1996.

Slovakia

After the 1993 creation of Slovakia, Bratislava (on the Danube River) became its national capital. Slovakia is one-third arable and two-fifths forested, and is somewhat more ethnically diverse than either the Czech Republic or Poland. As a result, Slovakia has a potential problem of political separatism within its minority community of about 600,000 ethnic Hungarians. This tension has been exacerbated by the 1993 completion of the Gabcikovo Nagymaros Dam on the Danube River. This dam draws out water upstream from Hungary and has led to Hungarian downstream complaints about dire environmental and economic consequences.

Slovakia still suffers from the fact that many professionals, and much potential foreign investment, stayed with the Czech Republic after the 1993 split. In the late 1990s Slovakia had an unemployment rate of 15 percent, but there is hope that the upswing in European economic growth might enlarge the tourist flow to landscapes along the Danube, to the Tatra Mountains in the east, and to the country's many caves and thermal spas.

Hungary

Hungary has been a relatively bright spot economically in East Central Europe. Its move toward free-market capitalism began early and has been aided by Western and Japanese investments larger than those received by any other East Central European country in recent decades. A relatively small but growing class of "newly rich" private entrepreneurs has emerged, often as participants in joint business ventures with foreign companies. This new "business bourgeoisie" contrasts, however, with a far larger class of "newly poor" persons living near or below the poverty line.

Romania

In 1993 Romania applied for membership in the EU and was turned down. It also was not included in the first set of Eastern European countries asked to join NATO. Part of the reason that the country has fared so poorly in its quest for greater recognition from western Europe and the world beyond is because of its poor economic performance in the collection of nations taken over by the Soviet Union after World War II, even though it had a strong resource base in oil and coal. Currently the country has to import oil and gas in order to provide its own energy needs.

The shadows of Communist control of Romania, and especially the government of Nicolae Ceausescu from 1965 until his death in 1989, have been particularly persistent in their influence on Romania. With a long eastern coastline on the Black Sea, and hundreds of miles of Danube River landscape, Romania has hoped for an influx of foreign exchange from the tourist trade. But the tourist traffic has largely been from the east—mostly Russia—which recently has grown so small that no foreign exchange is gained from the trade. Industry accounts for one-third and agriculture for one-fifth of Romania's annual Gross Domestic Product (GDP).

Bulgaria

Bulgaria has the Danube River and its flood plain as its northern border, the Balkan Mountains coursing from east to west through the middle of the country, and mountains separating it from Macedonia and Greece in the south and southwest. To the southeast lies the lowlands that separate Bulgaria from Turkey. Some 34 percent of the country is arable, while another 30 percent is forested and 16 percent meadows and pastures. Tradiionally the population has been largely agricultural, but in recent decades—especially under the USSR after World War II—there has been expanded effort in industrialization. Currently Bulgaria is 68 percent urban.

One of the most important economic resources the country possesses is the rail route that links Instanbul, Turkey, with points in central and western Europe. The Maritsa River that flows from west to east and drains into the Aegean Sea also provides an important travel corridor. Manufacturing and mining are responsible for more than 40 percent of Bulgaria's GDP, with machinery and chemicals responsible for 25 percent of the industrial sector. Agricultural products now account for only 13 percent of the total annual GDP.

The Breakup of Yugoslavia

Economic progress in Yugoslavia was disrupted by ethnic warfare following the dissolution of the Yugoslav federal state in 1991. Previously, under the Communist regime instituted by Marshal Josip Broz Tito during World War II, Yugoslavia had been organized into six Socialist People's Republics: **Serbia** (composition: 63 percent Serb in 1997); **Croatia** (Croats 78 percent, Serbs 12 percent); **Slovenia** (Slovenes 91 percent); **Bosnia and Herzegovina** (also spelled Hercegovina; Slavic Muslims 38 percent, Serbs 40 percent, Croats 22 percent); **Montenegro** (Montenegrins 69 percent, Slavic Muslims 13 percent, Albanians 6 percent); and **Macedonia** (Macedonians 65 percent, Albanians 22 percent, Turks 5 percent). In addition, two "autonomous provinces" were created: Kosovo (Albanians 77 percent, Serbs 13 percent) and Vojvodina (Serbs 54 percent, Hungarians 19 percent).

In 1988, Serbia took direct control of Kosovo and Vojvodina as part of Serbia's push (under an elected Communist president, Slobodan Milosevic) for greater influence within federal Yugoslavia. This push was challenged by all the other republics except Montenegro; they demanded a looser federation with more autonomy for each republic, and in 1991 they all declared independence. Slovenia, Croatia, and Bosnia and Herzegovina were admitted to the United Nations in 1992.

The secessions left federal Yugoslavia a rump state composed of Serbia and Montenegro. The Dayton Accords of 1995 attempted to institute a means of establishing a more viable series of ethnic and political units in the lands of the former Yugoslavia. In June 1991, Slovenia and Croatia declared their independence from Yugoslavia. That political move to segment the already patchwork nation of Yugoslavia into smaller, more homogeneous political units was simply a renewed attempt toward **"balkanization,"** or the generally "bloody" breakup of larger political units into smaller ethnic units. As soon as Croatia made its declaration in 1991, the Croatian Serbs took over one-third of the new country and began an **"ethnic cleansing,"** the forced emigration—or killing—of peoples different from the ethnic group in power. As a result of this act, the tensions between the Serb peoples and the Muslim peoples of the former Yugoslavia have become a major focus of the world press.

U.N. troops have been deployed in Bosnia since late 1995 as part of the Dayton Accords signed in December 1995 by the presidents of Bosnia, Croatia, and Serbia on behalf of Serbia and the Bosnian Serb Republic. The Accords and the positioning in this region of 60,000 NATO troops brought a slowdown to the Serb military's attempts to continue ethnic cleansing, and in a sense, political efforts have replaced military efforts. The NATO presence has initiated some effort to reunite ethnically diverse populations, to bring Muslims back to their homes in cities that now are controlled by Serbs, and to stimulate government cooperation at all levels in this effort.

In 1914, the assassination of Archduke Francis Ferdinand of Austria-Hungary by a Serbian nationalist in Sarajevo was the primary catalyst for World War I. The assassination was part of an effort to change the political pattern of this region. The same forces of ethnic cleansing and assassination continue to break up and realign the struggling populations of the Balkan Peninsula today, and it is this part of central and eastern Europe that is breaking up into ever smaller and smaller ethnic, religious, and political groups—all generally demanding autonomy, political independence, and widespread political recognition. The warfare in Kosovo in 1997–1999 is a variant of this same theme of Serb desire for more homogeneous nation-states, and for more autonomy in the realignment of the tortured landscapes that were held together by Tito for decades before his 1980 death. The efforts of the Kosovo Liberation army to gain full autonomy and independence is more of the continuing Balkan—perhaps global—effort to turn an ethnic realm into a nation state.

Albania

This small country has gained most of its visibility since the end of World War II by linking itself economically with the People's Republic of China. For China, Albania became a unique locus to show off its political capacity to relate to a non-Asian, distant country. The Albanians had been on the bottom rungs of economic and cultural development for so long that even such a distant ally seemed very beneficial. However, even that linkage did not enable the Communist government of Albania to design and achieve a pattern of productive economic growth in the years following Italy's wartime occupation.

In 1944, Communist Enver Hoxha took charge of Albania and, as the dominant leader for four decades, he oversaw first the Soviet linkage, then the China connection, then harsh isolation. With Hoxha's ouster in 1984 and the opening up of eastern Europe to new political, economic, and social winds, there was a possibility that Albania would grow into a position of greater strength. This has largely been unrealized; Albania is one of the few countries in the East in which a majority of its population continues to live in the rural sector (63 percent). There is the hope that tourism might be able to capitalize on the more than 100 miles of Albanian coastline on the Adriatic Sea, but there is no infrastructure yet established to begin to highlight that locale. The coast has been more important as a point of departure for Albanians and others trying to gain access to the eastern shores of Italy as part of the growing refugee flow from the Balkan Peninsula.

Macedonia

Macedonia is another of the new political units carved—perhaps ruptured—from the Yugoslavia that had been held together so forcefully by Josip Tito for more than three decades. With the breakup of Yugoslavia and the warfare in Bosnia and, later, Kosovo, Macedonians also began to search for more independence. However, like the whole southwestern corner of the Balkan Peninsula, Macedonia is home to a number of ethnic and religious groups—and the history of cooperative living in this difficult environment has not been peaceful. The demographic

PROBLEM LANDSCAPE

"Kosovo Refugees: Pawns in a NATO-Serb Clash?"

"RIBAR, SERBIA. AUG. 23. WHAT turned the Baftiu family into refugees early this morning was the artillery shell that blew the wall of their upstairs bedroom into the hallway and down the staircase. In their terror the family, and the 15 refugees from Government attacks last week who were taking shelter with them, fled in such haste that they left behind the provisions they had in place to grab on the way out if Government forces approached. The thin plastic sacks packed with clothes and the bags of flour still stood by the front door."

In this lead column from a *New York Times* article in August 1998 (by Mike O'Connor, August 24, 1998), the dynamics of yet another Balkan drama are outlined. The government that is cited is the Serbian government that continues as one of the two republics remaining in the political space called Yugoslavia. The two provinces of Serbia are Montenegro and Kosovo. Kosovo is made up of 90 percent ethnic Albanians (the country immediately south and west of Yugoslavia), and therein lies the key to this particular Problem Landscape.

Yugoslavia—the land of the southern Slavs—has long been a political patch-work quilt of differing religions (Roman Catholics, Eastern Orthodox, Muslims, and some others), differing ethic backgrounds (Slavs and non-Slavs are the most critical pairing in this region), and differing major powers on the outside looking in on Yugoslav unrest (Russia from the east; the EU and NATO from the west). This combination of geopolitical stresses is not unique to this area, but the history of such tensions in the Balkan Peninsula is splattered with warfare, bloodshed, refugee flows, and ethnic cleansing. Kosovo has it all.

Similar shellings and Serb efforts to force Muslims out of Bosnia and Croatia in the early 1990s led to a short period of NATO bombing in 1995 and a set of shaky peace talks outside of Dayton, Ohio. The 1995 Dayton Peace Accords launched 60,000 NATO troops into the northwestern sectors of the former Yugoslavia, and what was to be a multi-month presence developed into a multi-year task. It did, however, almost eliminate the armed harassment of non-Serb villagers in efforts to get them to move to other locations.

In Kosovo, the equation has been made more complex by the fact that the ethnic Albanians of this southwestern province of Serbia have taken up arms themselves and created the Kosovo Liberation Army (KLA). In early 1999, blood flowed on both sides as the KLA attempted to secure territory while the Serbs killed many men, women, and children in villages. As a result, NATO efforts to suppress the firepower of the Serbians in order to resettle the harried ethnic Albanians have been undermined again and again by KLA attacks on Serb police and military posts. In the eyes of the KLA, only full independence and autonomy are seen as an acceptable outcome of these months of tension and bloodshed.

For NATO, the idea of having Kosovo gain such independence is nearly as unattractive as such an outcome is to the Serbian government in Yugoslavia. There are many small, well-armed, and restless populations all through Europe and in the nations on its margins. Most of them would take the granting of full autonomy to Kosovo as a green light to increase their own struggle for greater independence. Neither NATO nor the individual governments of Spain, Germany, the United Kingdom, or Greece—to name just a few—would be able to support such an outcome.

Former U.S. president George Bush spoke of the potential for a "new world order" when the former Soviet Union collapsed, the Berlin Wall came down, and Germany was unified. In images of such a new order, there will have to be a zone for bloody and seemingly endless local efforts to redesign the politics of space and borders. Kosovo is a template for such horrors at the present.

element leading to the most difficulty is the tension between ethnic Albanians (with a large proportion of that group Muslim) and the Macedonians.

Macedonia is 24 percent arable and another quarter of its land is meadow and pasture. It is landlocked and the political need to maintain positive relations with Greece—from whom it is partially cleaved—in order to have predictable access to the Greek port of Thessaloniki further complicates the autonomy of Macedonia. In 1995 it brokered a treaty with Greece over the use of the name Macedonia—which Greece has claimed historically. In the late 1990s, the ethnic Albanians in the country also pursued the formation of an independent nation within Macedonia, in another Balkan move parallel to the Kosovo difficulty discussed in the Problem Landscape above (Fig. 5.19).

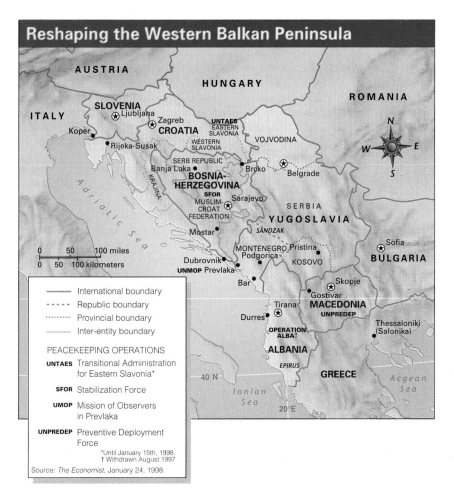

Reshaping the Western Balkan Peninsula

Figure 5.19 Ever since the 1980 death of Josip Tito—the fiercely independent leader who had held Yugoslavia together for more than four decades in the face of both German and Soviet aggression—Yugoslavia has been the victim of powerful forces of fragmentation. This map shows evidence of both boundary and demographic shifts in the past two decades. It also shows evidence of continuing international efforts to keep open warfare to a minimum. As this book goes to press, the North Atlantic Treaty Organization is just beginning its third month of the bombing of Yugoslavia in an effort to retard or eliminate the process of "ethnic cleansing" that has caused the removal of hundreds of thousands of ethnic Albanians from Kosovo at the hands of Slavic forces. The fragmentation of this region of the Balkan Peninsula looks as though it will continue into the new century with certainty.

SUMMARY WITH SELECTED KEY TERMS

- Present-day Europe is ringed with three sets of nations called **the North, the South,** and **the East.** These represent nations that are more peripheral to the **European Union** nations that lie in the core of Europe. The North includes the **Scandinavian** countries of **Norway, Sweden,** generally **Denmark,** and sometimes **Iceland.** If **Finland** is added, this cluster is called **Fennoscandia,** or the **Fennoscandian** countries.

- The North is the world's most northerly highly settled, economically developed region. This is because of **North Atlantic Drift,** a northward extension of **the Gulf Stream** that comes from the Caribbean Sea and the Gulf of Mexico, bringing unusual warmth to **North Sea** and even the **Baltic Sea,** and allowing Murmansk, Russia, to be an **ice-free port,** even though it is within the **Arctic Circle.** The winter climate of the whole of western Europe is much more moderate than places in North America at the same latitude.

- Countries in Fennoscandia have close connections through language (except for Finland) and religion. The **Evangelical Lutheran Church** is the church of 90 percent of the people. Finnish has its origin—unlike the other languages of the North—in languages of Central Asia. Other qualities that link countries in the North are set-

tlements in the southern reaches of the countries, **very sparse populations** in the north and in the mountains of all the countries (except Denmark, which has no mountains to speak of), **high levels of literacy and education,** and major economic importance in **forestry, fishing, skilled manufacturing,** and—in Denmark and part of Sweden—very **productive agriculture. Norway** has also very significant **petroleum reserves** in the North Sea. **Hydroelectricity** is also a major resource in all of the North except Denmark.

- **The South** is the term used for the countries on the northern margin of the **Mediterranean Sea** (excluding France) including **Portugal, Spain, Italy, and Greece,** as well as a number of major islands and some microstates including **San Marino, Vatican City,** the British colony of **Gibraltar,** and the small state of **Malta.** The countries on the eastern edge of the Aegean Sea are also part of **the East.** The **Iberian Peninsula** and the **Italian Peninsula** are of major importance to the South, and the **Balkan Peninsula** is seen as part of the East.

- **The South** is much influenced by the **mediterranean climate,** which is characterized by **summer drought** and **winter precipitation. Irrigation** has been an important factor in the South's agricultural history. Important crops are **winter wheat, olives, citrus, grapes** and associated wine making, and **vegetables.**

- **The Moors and the Islamic/Moorish influence** have been an important part of Spanish history. Also central was **Christopher Columbus** because of his voyage to America in 1492. **Portugal** had great fame through the voyages of **Vasco da Gama** and with the control of **Brazil. Italy** had some colonial influence, and is currently the most industrial and economically significant nation of the South. The **Po Valley** is central to industrial northern Italy, and houses two-fifths of the nation's population. There is a classic and costly economic imbalance between **northern Italy** and **southern Italy.** "Padania" is the name suggested for a new country that some residents hope might be created out of the prosperous north of Italy.

- **Diffusion, migration, the Roman Catholic Church, merchants,** and **trade** are themes that provide images of 12th-century **Mediterranean Europe** (and beyond) in *The Walking Drum*. This **literature** excerpt (see p. 144) deals with **periodic markets** and major **agricultural fairs.** The variety of languages, customs, and trade and **travel patterns** help provide images of medieval Europe.

- The South has lagged behind West Central Europe in development because of **trade deficiency, low educational levels, land tenure patterns,** and **resource deficiencies.** These traditional characteristics have changed somewhat in recent decades by **transportation changes, energy shifts, market expansion,** and **capital infusion.**

- **The East** is made up of **Poland, the Czech Republic, Slovakia, Hungary, Romania, Bulgaria, Slovenia, Croatia, Bosnia-Herzegovina, Yugoslavia, Albania,** and **Macedonia.** These countries have been shaped powerfully by shifting human migrations and power struggles between various empires over the past millennium. Most critical has been the **Communist control of the Soviet Union** after the end of World War II until the **collapse of the USSR in 1989.** The term **"shatter belt"** is useful in explaining the shifting political affiliations of these nations.

- **Slavs** have been a dominant population group in the East, especially in the **Balkan Peninsula. East Slavs** of the former Soviet Union; **South Slavs** including **Serbs, Croats, Slovenes, Bulgarians,** and **Macedonians;** and **West Slavs** including **Poles, Czechs,** and **Slo-**

vaks are the three divisions that still operate in the political tensions of this region. The current **geopolitics** of the region still are responding to this Slavic population and its relationship to Russia and to other, non-Slavic local populations.

- The environment of the East includes the **Northern Plain,** with Poland dominant; the **Central Mountain Zone,** with the Czechs dominant; the **Danubian Plains,** including major landscapes in Hungary, Romania, Serbia, and Croatia; and the **Southern Mountain Zone,** with Bulgaria, part of Yugoslavia, and Albania dominant. Varied climates exist.

- **Collectivization** under Communist rule during the Soviet era from 1947 until nearly 1990 was the primary agricultural motif. Corn and wheat are the primary crops and farmland has now been largely privatized but is still less productive than croplands to the west in Europe. **Livestock raising** and **sheep** flocks in the uplands of the East are part of the farming pattern. There is a steady **rural-to-urban** migration all through the region.

- **Mining, iron and steel, machinery production, construction materials,** and **electrical power** were the highlights of the industrial effort during the Soviet years, with the majority of **raw materials** coming from Russia. The shift to the market economy has left many firms uncompetitive and many abandoned. There has been a major effort to make the industrial base of the East more productive, including **increasing the efficiency of state-run enterprises** (not everything has been privatized), **encouragement of new private enterprises, fostering joint enterprises, elimination of price controls and bureaucratic restrictions, expanding international connections,** and **increasing the quality** of the region's manufactured products.

- Politically, the disintegration of the former Yugoslavia has been the most disabling phenomenon of the post-Soviet era, leading to the **Bosnian conflict,** the **Dayton Peace Accords of 1995,** and, in 1998, the emergence of the **Kosovo Liberation Army** and its active military action against the Serbian forces controlling Yugoslavia. These various campaigns of **ethnic cleansing** have led to **NATO involvement,** including bombing and the stationing of forces in Bosnia. In 1998 NATO extended an invitation to Poland, Hungary, and the Czech Republic to join NATO.

REVIEW QUESTIONS

1. List the countries of the North and outline their similarities and their distinctive differences.

2. Which of the countries in the North has a language fully unlike the other four, and explain the historical geography that explains this distinction. What are other examples of a highly distinctive national image characteristic of the North?

3. Describe the climate patterns of the North, comparing this region with North America at the same latitudes. Explain what geographic feature is most responsible for the regional warmth of the North. What economic impact does this lead to?

4. Reconsidering the National Geography Standards of Chapter 1, explore your mental map of the European North and try to determine the source of your images. Do the same for the South and the East. Consider the urban images and their relative strength compared to environmental or agricultural images.

5. List the countries of the South, and the islands related to these countries.

6. What is the ecology of Mediterranean agriculture, and outline the climate and soil patterns that relate to it, explaining why such agriculture is so historically significant. Give some detail to the role of winter wheat in your discussion.

7. Inventory the major urban centers that appear from west to east in the South, noting the comparative roles of industry, governmental function, tourism, and agriculture in defining their major images and geographic characteristics.

8. What are the factors that have influenced the South's relatively uneven economic development, and what changes have been realized in these patterns since World War II?

9. List the nations that make up the East as defined in this text, and relate them to broader patterns of European historical geography.

10. Define "shatter belt" and cite examples of such a phenomenon both in the East and in the North. Where—in what regions, and in what country or countries—does the concept play the most significant role?

11. Outline the breakup of the former Yugoslavia, and use the aspects of change in the last decade there to discuss aspects of the Geography Standards, especially relating to cultural mosaics, forces of cooperation and conflict, and human settlement.

12. What has been the Slavic presence in the East and what role does that play in the stability of the region today? Use Kosovo and other nations carved from the former Yugoslavia to define your observations.

13. Using river systems, topography, and soils, define the physical geography of the East, linking such characteristics to history where possible.

14. Utilizing "Human Systems" from Table 1.1, show the ways that Essential Elements are reflected in the landscape of the East in this chapter.

DISCUSSION QUESTIONS

1. What physical and cultural characteristics have blended to create the lifestyle of the North?

2. Looking at the larger map of Europe and western Russia, discuss the ways in which basic physical geography has had a strong role in historical development of the region.

3. Try to imagine the North without the North Atlantic Drift. How would the reality of that region change, and what significance would such changes have?

4. Think about growing up in rural or isolated lands of the North. What would be the factors that pulled you toward other regions? What can governments do to reduce the steady rural-to-urban migration in that region?

5. Consider the tourist images of the South. What are the geographic elements in such images? What is the economic importance of such images, and how are they promoted?

6. Why do you think countries in the South played such a significant role in cultural development two millennia ago, but play such a relatively modest role now? What influence have geographic factors had in this shift?

7. Discuss efforts in the north of Italy to secede from the rest of the country. Why do such thoughts emerge? What political and economic and cultural significance might they have?

8. Discuss the influence of fifty years of Soviet control on the nations of the East. What might have been patterns of post–World War II development had the Cold War not been so strongly developed?

9. If you had money to invest in a place that was going through a political change from collectivization to privatization, where—in the East—would you invest your money? What sort of industry would you promote? What sort of enterprise would you avoid?

10. Discuss the patterns of ethnic cleansing that have emerged in the past decade in the Balkan Peninsula. Are there any other parts of Europe where there is a potential for similar conflict? What dynamics have made this process so characteristically Balkan? What might lead to a change in such patterns?

3 Russia and the Near Abroad

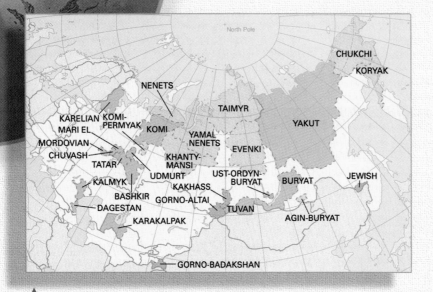

The Soviet Union was, and Russia is, a "land empire" built when a single group imposed its political will and established economic sovereignty over disparate peoples in a vast hinterland. Ethnic diversity made the long-term survival of the Soviet Union an unlikely prospect, and now threatens the unity of the Russian Federation.

Russians and other ethnic groups in the region have contributed enormously to the arts, sciences, and technology, particularly in the 19th and 20th centuries. *Sovfoto / East Foto*

The world's largest country, Russia, is endowed with natural resources but, paradoxically, is handicapped by huge transportation distances, high elevations, and natural hazards in its most productive areas. Jesse Wheeler described Siberia as a "bleak storehouse of resources." *Tass / Sovfoto*

Regional Snapshot

The prospects for Russia, so recently the core of a superpower state, are uncertain. Russia appears to be a "misdeveloped" country torn between conflicting economic, political, and cultural choices. *Sovfoto / East Foto*

The Soviet Union experienced unimaginable suffering and loss of life in the two world wars of the 20th century. The stamina of the Russians and other peoples in the region has been tested repeatedly in troubled times, most recently with the financial crisis that erupted in 1998. *Sovfoto*

Seen from Russian eyes, the 14 other successor states of the Soviet Union comprise the "Near Abroad," a strategically vital perimeter retaining important economic and ethnic ties with Russia. While this is a region in transition, it remains valid and useful to approach Russia and the Near Abroad as a unit of study. *Sovfoto*

Chapter 6

A Geographic Profile of Russia and the Near Abroad

▲ *The Caspian Sea and mouth of the Volga River viewed from space. The Volga mouth is in the upper right portion of the image. Long known as the source of most of the world's caviar, the Caspian Sea is emerging as one of the world's great petroleum-producing regions. Once marshalled exclusively by the Soviet Union, these resources are now divided among five independent nations.* NASA

CHAPTER OUTLINE

6.1 Area and Population

6.2 The Natural Environment

6.3 Russian Territorial and Cultural Evolution

6.4 The Soviet Union

6.5 The Second Russian Revolution

6.6 Russia, the Near Abroad, and the Wider World

6.7 Russia, the Misdeveloped Country

*f*rom 1917 to 1991, the huge area described in this text as Russia and the Near Abroad and in other sources as the Former Soviet Union comprised a single Communist-controlled country called (from 1922) the Union of Soviet Socialist Republics (USSR). After World War II, the government based in the Kremlin fortress in Moscow exercised control over Communist "satellite" countries in eastern Europe and maintained strong influence in other Marxist states, particularly in LDCs not aligned with the capitalist West. For four decades, the Cold War between the Soviet bloc of nations and the Western bloc led by the United States dominated world politics. The perspectives and actions of the Soviet Union became major determinants of world events.

Suddenly, in 1991, the country split into 15 independent nations (see Table 6.1 and Definitions and Insights, p. 171). Russia is by far the largest in area, population, and political and economic influence. From the Russian perspective, the remaining 14 countries comprise the "Near Abroad," and because of Russia's prominence the region is known in this text as "Russia and the Near Abroad." Twelve of the units are associated loosely in the Commonwealth of Independent States (CIS), but tendencies toward further political fragmentation and decentralization are still great. Without the USSR's Red Army to impose order, long-simmering ethnic conflicts have boiled over. Political and economic redevelopment is under way, but the future of this huge territory is uncertain.

This chapter offers geographic and historical perspectives on a major world region in which momentous changes are taking place. Placing the 15 former republics of the Soviet Union, now 15 independent countries, in a single region is somewhat awkward but still necessary at the outset of the 21st century. These countries have often uncomfortable strategic and economic associations. For example, the Soviet-era oil refinery and pipeline system still links Russia with Kazakstan, Azerbaijan, Ukraine, Belarus, and other countries. Since the breakup of the Soviet Union, Russia has periodically shut off the pipelines supplying natural gas to Ukraine, Belarus, and Moldova to obtain political concessions from those nations. The needs of these countries to buy and sell oil could provide a strong impetus to hold the Commonwealth of Independent States together or to create new political and economic blocs.

Although the Soviet government failed in its vision of creating a single, unified "Soviet people" from the hundreds of ethnic groups living within the boundaries of the USSR, it did succeed in transforming the human geography of the vast Soviet empire. For many years to come, the 15 countries of that crumbled empire will be attempting to sever or reestablish ties with the imperial hub of Moscow. For its part, the Russian government in Moscow will try vigorously to maintain influence over the former republics. Some analysts believe that Russia may even attempt to reexert control over some of the recently independent nations, especially those vital to the economic and political security of the Russian Federation.

6.1 Area and Population

With an area (including inland waters) of 8.5 million square miles (21.9 million sq km), the region of Russia and the Near Abroad is smaller than Africa but larger in area than any other major world region recognized in this text (see Table 1.2, p. 14). A good indication of its staggering size is the fact that this region (and, notably, Russia alone) spans eleven time zones; by comparison, the United States from Maine to Alaska spans seven. However, so much of this vast region is sparsely populated that its estimated population of 291 million in mid-1998 ranked it only seventh among the eight world regions. The average population density of 34 per square mile (13 per sq km) is only half that of the United States. Great stretches of economically unproductive terrain separate many outlying populated areas from other population centers.

The eventful geopolitical history of Russia and the Near Abroad has given this region land frontiers with 12 countries in Eurasia (Fig. 6.1). Between the Black Sea and the Pacific, the region directly borders Turkey, Iran, Afghanistan, China,

Stars show political capitals.

- ⊛ Over 8,000,000
- ⊙ Over 5,000,000
- ✪ ● Over 1,500,000
- ★ ○ 750,000–1,500,000
- ☆ ○ 350,000–750,000 (selected places)
- ⊛ • Selected smaller places

Figure 6.1 General reference map of the 15 independent successor states that emerged from the dissolution of the Soviet Union. The inset map shows subordinate ethnic political units that did not have Union Republic status in the Soviet era (one unit, Karelia, was downgraded from that status). Names of the titular nationalities are shown.

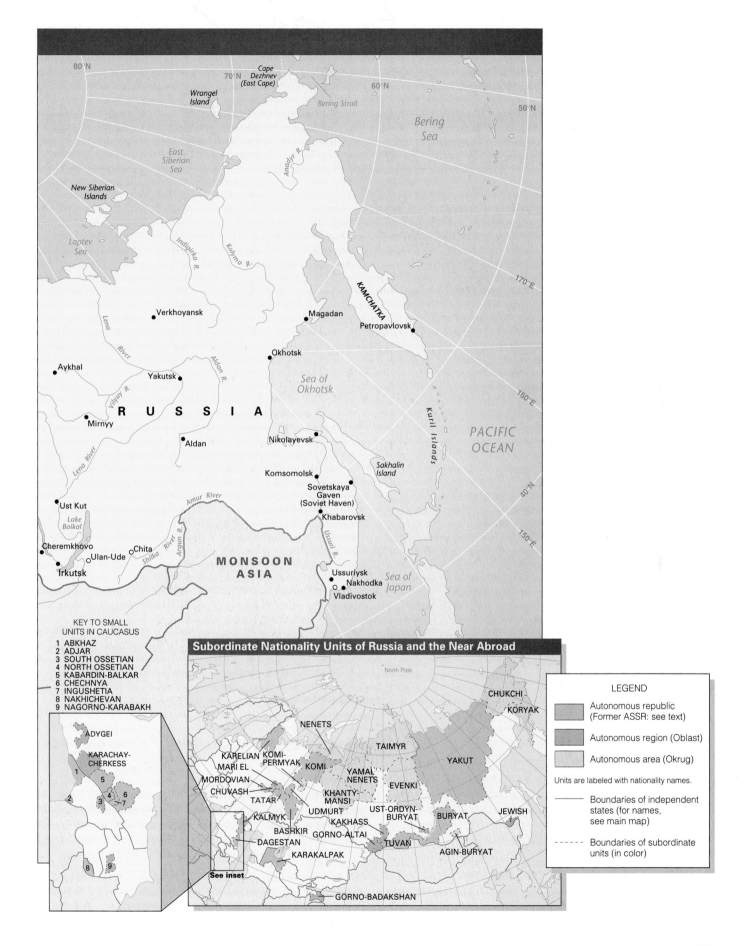

RUSSIA

Cape Dezhnev (East Cape)
Wrangel Island
Bering Strait
Bering Sea
East Siberian Sea
New Siberian Islands
Laptev Sea
Indigirka R.
Kolyma R.
KAMCHATKA
Verkhoyansk
Magadan
Petropavlovsk
Okhotsk
Sea of Okhotsk
Lena River
Aldan R.
Aykhal
Yakutsk
Vilyuy R.
R U S S I A
Kuril Islands
PACIFIC OCEAN
Mirnyy
Aldan
Nikolayevsk
Lena River
Komsomolsk
Sakhalin Island
Sovetskaya Gaven (Soviet Haven)
Amur River
Khabarovsk
Ust Kut
Lake Baikal
Ussuri R.
Cheremkhovo
Chita
Shilka River
Argun R.
MONSOON ASIA
Ulan-Ude
Irkutsk
Ussuriysk
Nakhodka
Sea of Japan
Vladivostok

KEY TO SMALL UNITS IN CAUCASUS

1 ABKHAZ
2 ADJAR
3 SOUTH OSSETIAN
4 NORTH OSSETIAN
5 KABARDIN-BALKAR
6 CHECHNYA
7 INGUSHETIA
8 NAKHICHEVAN
9 NAGORNO-KARABAKH

ADYGEI
KARACHAY-CHERKESS

Subordinate Nationality Units of Russia and the Near Abroad

North Pole
CHUKCHI
KORYAK
NENETS
TAIMYR
KARELIAN
KOMI-PERMYAK
MARI EL
KOMI
YAKUT
MORDOVIAN
YAMAL NENETS
CHUVASH
KHANTY-MANSI
EVENKI
TATAR
UDMURT
UST-ORDYN-BURYAT
BURYAT
KALMYK
KAKHASS
JEWISH
BASHKIR
GORNO-ALTAI
TUVAN
DAGESTAN
KARAKALPAK
AGIN-BURYAT
See inset
GORNO-BADAKSHAN

LEGEND

Autonomous republic (Former ASSR: see text)

Autonomous region (Oblast)

Autonomous area (Okrug)

Units are labeled with nationality names.

——— Boundaries of independent states (for names, see main map)

----- Boundaries of subordinate units (in color)

169

Table 6.1 Russia and The Near Abroad: Basic Data

Political Unit	Area (thousand/ sq mi)	Area (thousand/ sq km)	Estimated Population (millions)	Estimated Annual Rate of Increase (%)	Estimated Population Density (sq mi)	Infant Mortality Rate	Urban Population (%)	Arable Land (% of total area)	Per Capita GNP ($US)
Slavic States and Moldova									
Russia	6520.6	16,888.4	146.9	−0.5	23	17	73	*	2410
Ukraine	223.7	579.4	50.3	−0.6	225	14	68	56	1200
Belarus	80.1	207.5	10.2	−0.4	127	12	69	29	2070
Moldova	12.7	32.9	4.2	0.1	330	20	46	50	590
Total	**6837.1**	**17,708.2**	**211.6**	**20.5**	**31**	**15.8**	**71**	**49**	**2070**
Transcaucasia									
Georgia	26.9	69.7	5.4	0.4	202	17	56	*	850
Armenia	10.9	28.2	3.8	0.6	349	14	67	29	630
Azerbaijan	33.4	86.5	7.7	1.1	230	19	52	18	480
Total	**71.2**	**184.4**	**16.9**	**0.8**	**237**	**16.7**	**57**	**21**	**632**
Central Asia									
Kazakstan	1031.2	2670.8	15.6	0.5	15	25	56	15	1350
Uzbekistan	159.9	414.1	24.1	2	151	26	38	10	1010
Turkmenistan	181.4	469.8	4.7	1.7	26	42	45	3	940
Kyrgystan	74.1	191.9	4.7	1.5	63	28	34	*	550
Tajikistan	54.3	140.6	6.1	1.7	113	42	45	6	940
Total	**1500.9**	**3887.2**	**55.2**	**1.5**	**37**	**32.6**	**44**	**13**	**1053**
Baltic States									
Estonia	16.3	42.2	1.4	−0.4	88	10	70	22	3080
Latvia	24	62.2	2.4	−0.6	102	16	69	27	2300
Lithuania	25	64.8	3.7	−0.1	148	10	68	49	2280
Total	**65.3**	**169.2**	**7.5**	**20.3**	**115**	**12**	**69**	**34**	**2436**
Summary Total	**8474.5**	**21,949**	**291.2**	**−0.04**	**34**	**19.3**	**65**	**20**	**1803**

Sources: *World Population Data Sheet,* Population Reference Bureau, 1998; United Nations Statistical Division, 1998; *Almanac of Politics and Government,* 1998; *World Factbook,* CIA, 1997.

Mongolia, and North Korea. Pakistan and India also lie close by. In the Pacific Ocean, narrow water passages separate the Russian-held islands of Sakhalin and the Kurils from Japan. If India, Pakistan, and Japan are considered, the Asian part of Russia and the Near Abroad is a near neighbor of about half the world's people.

In the west, the region has frontiers with Romania, Hungary, Slovakia, Poland, Finland, and Norway. These differ greatly from the sparsely populated, arid, and mountainous frontiers in Asia. On the frontier between the Black and Baltic Seas, international boundaries pass through well-populated lowlands that have long been disputed territory between the Russian state and other

countries. During World War II (1939–1945), the Soviet Union expanded its national territory westward in several stages (see Chapters 5 and 7).

Stretching nearly halfway around the globe in northern Eurasia, the region of Russia and the Near Abroad has formidable problems associated with climate, terrain, and distance. Most of it is handicapped economically by excessively cold, infertile soils, marshy terrain, aridity, and ruggedness. Natural conditions are more similar to those of Canada than to those of the United States. Interaction with a complex and demanding environment was a major theme in Russian and Soviet expansion and development. Nature provided large assets but also posed great problems.

Regional Perspective

Areal Names of Russia and the Near Abroad

Geographic study of Russia and the Near Abroad requires an understanding of some areal names. The region corresponds in extent with the former Union of Soviet Socialist Republics, often shortened to Soviet Union or USSR. This state came into existence in 1922 following the overthrow of the last Romanov tsar in the Russian Revolution of 1917 and the subsequent civil war. Prerevolutionary Russia is often called Old Russia, Tsarist Russia, Imperial Russia, or the Russian Empire.

During the Communist era from 1917 to 1991, Westerners often called the country Soviet Russia. In the past, the name Russia has been used loosely to refer to the entire country either before or after the revolution, but it is now properly restricted to the huge Russian Federation, which was the largest of the 15 Soviet Socialist Republics (Union Republics or SSRs) that made up the Soviet Union. The full name of the Russian Federation during the Communist period was Russian Soviet Federated Socialist Republic (RSFSR). The loosely aligned Commonwealth of Independent States (CIS) was formed by 12 of the 15 former Union Republics late in 1991. Estonia, Latvia, and Lithuania are not members of this organization.

The area west of the Ural Mountains and north of the Caucasus Mountains has been known historically as European Russia. Similarly, the Caucasus and the area east of the Urals has been called Asiatic Russia, Soviet Asia, or the eastern regions. Here, Transcaucasia is the region in and south of the Caucasus Mountains (see Fig. 7.1, p. 192); Siberia is a general name for the area between the Urals and the Pacific; and Central Asia is a general name for the arid area occupied by five states with large Muslim populations immediately east and north of the Caspian Sea.

6.2 The Natural Environment

The Role of the Climatic and Biotic Environment

Russian and Soviet expansion and development went forward in a harsh climatic setting (for a map of the region's climatic zones, see Fig. 2.7). Severe winter cold, short growing seasons, drought, and desiccating summer winds that shrivel crops in the steppes are major handicaps. Associated with these rigorous conditions are significant advantages in the form of tillable soils, the world's largest forests, natural pastures for livestock, and a diverse wild fauna. However, many of these resources are only marginally useful because of unfavorable climatic factors.

Most parts of the region display continental climatic influences characterized by long cold winters, warm to hot summers, and low to moderate precipitation. The severe winters are the result of the northerly continental location coupled with mountain barriers on the south and east. Four-fifths of the total area is farther north than any point in the conterminous United States (Fig. 6.2). The region lies mainly in the higher middle and lower high latitudes of the Northern Hemisphere, where landmasses reach their greatest extent relative to the bordering oceans. The climatic influence of the land overpowers that of the sea in continental interiors, and the most extreme continental climates of the world prevail. The lowest temperature ever recorded in the Northern Hemisphere ($-90°F$) occurred in the Siberian settlement of Verkhoyansk (67°N, 135°E).

Continuous chains of mountains and high plateaus occupying the southern and eastern margins of the region act as a screen against moderating influences from the Indian and Pacific oceans. In addition, most of the region is a zone of high atmospheric pressure in winter, with outflowing winds. Westerly winds from the Atlantic moderate the winter temperatures somewhat in the west, but their effects become steadily weaker toward the east, where it is much colder.

Aside from the interior of eastern Siberia, most parts of the region of Russia and the Near Abroad experience winter temperatures that do not differ greatly from the temperatures of places located at comparable latitudes and elevations in the interior of North America. Indeed, owing to moderating influences from the Atlantic, many places in the region's west and south are actually warmer in midwinter than comparable interior places in North America. Nevertheless, the region as a whole is definitely handicapped by the winter climate.

Huge areas in the tundra and subarctic climatic zones are permanently frozen at a depth of a few feet. This condition of **permafrost** makes construction difficult. Heat generated by buildings melts the upper layers of permafrost and causes foundations and walls to sink and tilt. Most buildings are elevated on pilings to reduce this risk (Fig. 6.3). Pipelines carrying crude oil, which is very hot when it comes from the ground, may likewise melt the permafrost and sink unless they are heavily insulated or built on elevated supports.

The average frost-free season of 150 days or less in most areas except the extreme south and west is too short for a wide range of crops to mature. Most places are relatively warm dur-

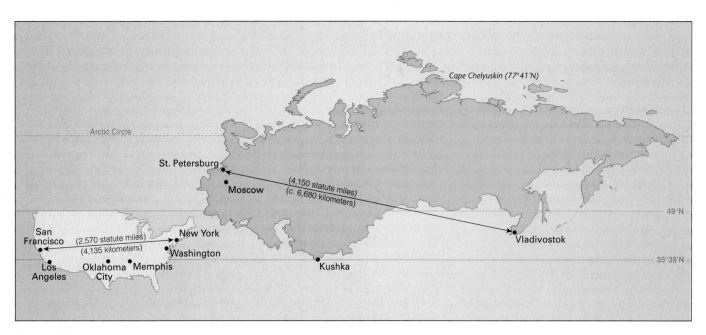

Figure 6.2 Russia and the Near Abroad compared in latitude and area with the conterminous United States.

ing the brief summer, and the southern steppes and deserts are hot. A factor partially offsetting the brevity of the summer is the long summer daylight period that encourages plant growth.

From an agricultural standpoint, a general shortage of moisture and the periodic occurrences of drought are even greater handicaps than low temperatures. The annual average precipitation is less than 20 inches (under 50 cm) nearly everywhere except in the extreme west, along the eastern coast of the Black Sea and the Pacific littoral north of Vladivostok, and in some of the higher mountains. North of the Arctic Circle and in the southern deserts, the average precipitation is less than 10 inches (25 cm). Most precipitation is derived from the Atlantic Ocean, so the total amount tends to decrease from west to east. Precipitation also decreases toward the south, with 10 to 20 inches in the black-earth belt (see Fig. 7.1, p. 192) but only 5 to 10 inches or even less in the dry steppes and deserts farther south. The black-earth belt—the region's most important grain-producing region—is subject to severe droughts and hot, desiccating winds *(sukhovey)* that may greatly damage or destroy a crop. Areas north of this belt in the western part of the region have somewhat greater and more dependable rainfall, coupled with a lower rate of evaporation resulting from lower temperatures. For these reasons, they are better supplied with moisture than the black-earth belt. However, this advantage is offset by their poorer soils, cooler summers, shorter growing seasons, and larger proportion of poorly drained land.

Figure 6.3 In the High Arctic, buildings must be erected on pilings so that they do not melt the permafrost below. This is the Russian coal-mining settlement of Barentsburg in Norway's Svalbard (Spitsbergen) Archipelago. *Joseph J. Hobbs*

There are five main climatic belts in the region of Russia and the Near Abroad: tundra, subarctic, humid continental, steppe, and desert. In this order these belts, each with its associated vegetation and soils, succeed each other from north to south (see Figs. 2.1, 2.7, and 2.8). Smaller scattered areas of subtropical, mediterranean, and undifferentiated highland climates also exist.

The most productive human activities take place in the subarctic, humid continental, and steppe climatic zones. The subarctic climate is essentially coextensive with the Russian taiga, or northern coniferous forest—the largest continuous area of forest on Earth (Fig. 6.4). The main trees are spruce, fir, larch, and pine, which are useful for pulpwood and firewood but frequently are too small, twisted, or knotty to make good lumber. Nevertheless, large reserves of timber suitable for lumber exist

Figure 6.4 The taiga in Russia is Earth's largest continuous forest biome. This scene is in northwestern Karelia, near Russia's border with Finland. *Joseph J. Hobbs*

in parts of the taiga, and this is one of the major areas of lumbering in the world. Many observers fear that Russia, in its drive to advance the economy, will deplete the taiga. There is some agricultural settlement in the taiga, especially toward the south. However, farming is marginal, handicapped by short summers, winter cold, unseasonable frosts, the marshy and swampy character of much of the land, and the prevalence of poor soils. The dominant soils are podzols (Russian for "ashes underneath"), which have a grayish, bleached appearance when plowed, lack well-decomposed organic matter, and are low in natural fertility. Their acidity is unfavorable for most crops and also for bacteria, earthworms, and other soil-improving organisms.

A humid continental climate occupies a triangular area south of the subarctic climate, narrowing eastward from the western border of the region of Russia and the Near Abroad to the vicinity of Novosibirsk. This area has the short-summer subtype of humid continental climate, comparable to that of the Great Lakes region and the northern Great Plains of the United States and adjacent parts of Canada. Mixed or broadleaf deciduous forest supplants the evergreen taiga forest; oak, ash, maple, elm, and other broadleaf trees alternate with conifers. To the south, in the "wooded steppe" or "forest steppe"—transitional between forest and open steppe grassland—clumps of trees intermix with grassy areas. Both climate and soil are more favorable for agriculture in the humid continental climate than in the subarctic. Soils developed under broadleaf deciduous or mixed forest are normally more fertile than podzols, although less so than grassland soils.

The steppe climate characterizes the grassy plains south of the forest in Russia and is the main climate of Ukraine, Moldova, and Kazakstan. The average annual precipitation is 10 to 20 inches (c. 25–50 cm), an amount barely sufficient for unirrigated crops. Recurrent droughts add to the hazards of farming. Nevertheless, this is the most important area of crop and livestock production in the region of Russia and the Near Abroad; the low and variable rainfall is partially offset by the fertility of soils in the famous black-earth belt. The main soils of this belt are the **chernozems**—meaning "black earth." Among the best soils to be found anywhere, these are very thick, productive, and durable. A similar belt occupies the east-ern Great Plains of North America. Their great fertility is due largely to an abundance of humus in the topsoil, the presence of sufficient lime to neutralize excessive acidity, and less leaching of nutrients by rainfall.

The steppe includes extensive areas of chestnut soils in the areas of lighter rainfall. Chestnut soils are a lighter color than chernozems and lack their superb fertility, but are among the world's better soils. In North America, a belt of chestnut soils lies west of the chernozems in the Great Plains. The most characteristic vegetation of the steppe is short grass, forming a carpet that is nearly continuous. Pastoral nomads grazed their herds from an early time on the treeless steppe grasslands, which stretched over a vast area between the forest and the southern mountains and deserts. Today, however, much of this area is cultivated, with wheat the main crop (Fig. 6.5). The steppe also is a major producer of sugar beets, sunflowers grown for vegetable oil, various other crops, and all the major types of livestock.

Crop growing is difficult in the desert climate areas east and immediately north of the Caspian Sea in the Central Asian

Figure 6.5 Wheat harvest in the Russian steppe just west of the Ural Mountains. This location is in the non-Slavic ethnic area of Bashkiria. *Tass/Sovfoto*

173

states. Cotton is the most important crop where irrigation water is available, as in oases watered by the Syr Darya, Amu Darya, and other rivers that originate in high mountains to the south and east.

The Role of Rivers

The Moscow region lies on a low upland from which a number of large rivers radiate like spokes of a wheel (Fig. 6.6). The longest ones lead southward: the Volga to the landlocked Caspian Sea, the Dnieper (Dnepr) to the Black Sea, and the Don to the Sea of Azov, which connects with the Black Sea through a narrow strait. Shorter rivers lead north and northwest to the Arctic Ocean and Baltic Sea. These river systems are accessible to each other by easy portages. In the early history of Russia, the rivers formed natural passageways for trade, conquest, and colonization. They were especially crucial in Siberia. Half again as large as the United States, Siberia is drained by some of the greatest rivers on Earth: the Ob, Yenisey, Lena, and Kolyma, all flowing northward to the Arctic Ocean; and the Amur, flowing northward to the Pacific. By following these

rivers and their lateral tributaries, the Russians advanced from the Urals to the Pacific in less than a century.

Alteration of rivers for power, navigation, and irrigation became an important and much publicized aspect of economic development under the Soviets. The main element in the program was the construction of large dams along such major rivers as the Dnieper, Don, Kama, Irtysh, Ob, Yenisey, Angara, and Volga. Of these, the Volga is perhaps the most important; Russians call it *Matushka*—"Mother." It rises about 200 miles (320 km) southwest of St. Petersburg and flows 2300 miles (3680 km) to the Caspian Sea. Some geographers regard it as the dividing line between Europe and Asia, or as Europe's largest and longest river.

Before railroads supplemented river traffic, the Volga was Russia's premier commercial artery. Traditionally, wheat, coal, and pig iron from the Ukraine, fish from the Caspian Sea, salt from the lower Volga, and oil from Baku, on the western shore of the Caspian, travelled upriver toward Moscow and the Urals. Timber and finished products moved downriver to the lower Volga and the Ukraine (Fig. 6.7). Boatmen towed barges upstream, requiring 70 days to go from Astrakhan, near the Volga

Figure 6.6 Map of landforms and water features in the region of Russia and the Near Abroad.

Figure 6.7 Timber moving on the Volga, a major transportation artery in Russia. This scene is near Saratov, on the lower Volga; note the typical steppe vegetation on the banks. *Joseph J. Hobbs*

mouth, to Kazan, on the middle Volga. The steamboat arrived in the late 1800s, bringing an end to the way of life still remembered in the famous Russian song, "The Volga Boatmen." The ethnic mosaic of the Volga region nevertheless remains very complex (see inset map of Fig. 6.1). In the autonomous regions of Tatarstan, Chuvashia, and Mari-El on the middle Volga, non-Russian groups of Turkish and Finnish origin make up the majority of the population.

The Volga was difficult to navigate until the latter half of the 20th century. There were shoals and shallows, especially in very dry summers, and ice still closes the waterway each winter for 120–160 days. In the 1800s, construction began on dams and reservoirs which would impound and conserve spring floodwaters for the dry summer months. At the same time, canals were begun which would increase the navigable length of the Volga and associated waterways.

Within the past 60 years, the "Great Volga Scheme" has transformed the river. The goal has been to control completely the flow of the river with a stairway of huge reservoirs, each of which reaches upstream to the dam forming the next reservoir, thus assuring complete navigability during the six months when the river is ice-free, and supplying hydroelectric power and water for irrigation. Behind the dams that are the backbone of the

Great Volga Scheme, the reservoirs are so vast that they are generally called "seas." With these improvements in navigation, the Volga secured its place as Russia's most important internal waterway. Large numbers of barges and log rafts, as well as a fleet of passenger vessels that carry tourists, use it.

A major link in the inland waterway system is the Volga-Don Canal, opened in 1952 to tie together the two rivers where they approach each other in the vicinity of Volgograd (see Figs. 6.1, 6.8). The completion of the canal meant that the White Sea and Baltic Sea in the north were linked to the Black Sea and Caspian Sea in the south in a single water transport system. The accomplishment was no small task. The Don River, which empties into the Black Sea via the Sea of Azov, is about 150 feet (45 m) higher than the Volga. The ground that separates these two rivers rises to almost 300 feet (90 m). Engineers solved this discrepancy with 13 locks, each with a lift of about 30 feet (9 m), which carry water 145 feet (44 m) above the Don and drop it 290 feet (87 m) to the Volga. To provide water for the canal and the lower Don, they erected a seven-mile-long dam across the Don at Tsimlyansk. The 1000-square-mile (260 sq km) reservoir which formed behind it was for a time the world's largest artificial lake. The Tsimlyansk Sea looks vast, but is very shallow.

Figure 6.8 A lock in the Volga-Don Canal, a vital link for trade between the Volga watershed and the Black Sea. *Joseph J. Hobbs*

The Role of Topography

Most of the important rivers of Russia and the Near Abroad wind slowly for hundreds or thousands of miles across large plains. Such plains, including low hills, compose nearly all the terrain from the Yenisey River to the western border of the region. The only mountains in this lowland are the Urals, a low and narrow range (average elevation: less than 2000 feet/ 610 m) located about midway between the western frontier of Russia and the Yenisey. The Urals trend almost due north and south but do not occupy the full width of the lowland. A wide lowland gap between the southern end of the mountains and the Caspian Sea permits uninterrupted east-west movement by land. Cut by river valleys offering easy passageways, the Urals do not constitute a serious barrier to transportation.

Between the Urals and the Yenisey River, the great lowland of western Siberia (West Siberian Plain) is one of the flattest areas on Earth. Immense wetlands, through which the Ob River and its tributaries slowly wind their way, cover much of the plain. This waterlogged country, underlain by permafrost that blocks drainage, is a major barrier to land transport and is extremely uninviting to settlement. In the spring, tremendous floods occur when the breakup of ice in the upper basin of the Ob releases great quantities of water while the river channels farther north are still frozen, acting as natural dams.

The area between the Yenisey and Lena Rivers is occupied by the hilly Central Siberian Uplands, which have a general elevation of 1000 to 1500 feet (c. 300–450 m). Mountains dominate the landscape east of the Lena River and Lake Baikal. Extreme northeastern Siberia is an especially bleak and difficult country for human settlement. High mountains rim the region of Russia and the Near Abroad on the south from the Black Sea to Lake Baikal and lower mountains rim the region from Lake Baikal to the Pacific. Peaks rise to over 15,000 feet (c. 4500 m) in the Caucasus Mountains between the Black and Caspian Seas and in the Pamir, Tien Shan, and Altai mountains east of the Caspian Sea. From the feet of the Pamir, Tien Shan, and Altai ranges, and the lower ranges between the Pamirs and the Caspian Sea, arid and semiarid plains and low uplands extend northward and gradually merge with the West Siberian Plain and the broad plains and low hills west of the Urals.

6.3 Russian Territorial and Cultural Evolution

Russian political and cultural origins reach more than a thousand years into the past. The central figures were the Slavic peoples who colonized Russia from the west, interacted culturally with many other peoples, stood off or outlasted invaders, and conquered in a piecemeal fashion the giant territory that would become the Soviet Union.

Early Scandinavian and Byzantine Influences

Slavic peoples have inhabited Russia since the early centuries of the Christian era. During the Middle Ages, Slavic tribes living in the forested regions of western Russia came under the influence of Viking adventurers from Scandinavia known as Rus or Varangians. The warlike newcomers carved out trade routes, planted settlements, and organized principalities along rivers and portages connecting the Baltic and Black Seas. In the 9th century, the principality of Kiev, ruled by a mixed Scandinavian and Slavic nobility, achieved mastery over the others and became a powerful state. The culture it developed was the foundation on which the Russian, Ukrainian, and Belarusian cultures later arose.

Contacts with Constantinople (now Istanbul, Turkey) much affected Kievian Russia. This great city, located on the straits connecting the Black Sea with the Mediterranean, was the capital of the Eastern Roman or Byzantine Empire, which endured for nearly 1000 years after the collapse of the Western Roman Empire in the 5th century A.D. Constantinople became an important magnet for Russian trade, and the Russians borrowed heavily from it culturally. In 988 the ruler of Kiev, Grand Duke Vladimir I, formally accepted the Orthodox Christian faith from the Byzantines. He had his subjects baptized, and Orthodox Christianity became a permanent feature of Russian life and culture. Moscow came to be known as the "Third Rome" (after Rome itself and Constantinople) for its importance in Christian affairs. The Bolshevik Revolution in 1917 began a 75-year period of official repression and neglect of the Church. Since the breakup of the Soviet Union, Orthodox Christianity has experienced a rebirth; in Moscow alone, the number of parishes has more than quadrupled since 1990 (Fig. 6.9).

The Tatar Invasion

Yet another cultural influence reached the Russians from the heart of Asia. The steppe grasslands of southern Russia had long been the habitat of nomadic horsemen of Asian origin. During the later days of the Roman Empire and in the Middle Ages, these grassy plains, stretching far into Asia, provided the Huns, Bulgars, and other nomads a passageway into Europe. In the 13th century, the Tatars took this route. Many steppe and desert peoples were in their ranks, but Mongols led them. In 1237, Batu Khan ("Batu the Splendid"), grandson of Genghis Khan, launched a devastating invasion that brought all the Russian principalities except the northern one of Novgorod under Tatar rule.

The Tatars collected taxes and tributes from the Russian principalities but generally allowed the rulers of these units latitude in local government. Even the princes of Novgorod, who were not technically under Tatar control, paid tribute in order to avoid trouble. The Tatars established the Khanates (governmental units) of Kazan, Astrakhan, and Crimea. Russians called the Kazan Tatars "the Golden Horde," after the brightly colored tents in which they lived. When Tatar power declined in the 15th century, the rulers of the Moscow principality were able to begin a process of territorial expansion that resulted in the formation of present-day Russia. Today, oil-rich Tatarstan, with its capital at Kazan, is one of Russia's most important "autonomous" political units (see Chapter 7, p. 209).

The Empire of the Tsars

The Russian monarchy reached outward from Muscovy, its original domain in the Moscow region. From the 15th century until the 20th, the tsars built an immense Russian empire by accretion around this small nuclear core. This imperialism by land (see p. 179), in the era when the maritime powers of western Europe were expanding by sea, brought a host of alien peoples under tsarist control. The motivations were diverse. Quelling raids by troublesome neighbors, particularly nomadic Muslim peoples in the southern steppes and deserts, was one objective. Many Russian cities, such as Volgograd (originally Tsaritsyn and then Stalingrad) on the Volga River, were founded as fortified outposts on the steppe frontier. In the wilderness of Siberia, the search for valuable furs and minerals—especially gold—stimulated early expansion, and the missionary impulse of Orthodox priests also played a role. Land hunger and a desire to escape serfdom and taxation led to the flight of many peasants into the fertile black-earth belt of southwestern Siberia, and many landlords also moved there with their serfs.

The initial outward thrust from Muscovy under Ivan the Great (who reigned in 1462–1505) was mainly northward and eastward. The rival principality of Novgorod was annexed and a domain secured that extended northward to the Arctic Ocean and eastward to the Ural Mountains. Ivan the Terrible (who reigned in 1533–1584) added large new territories to Russia by conquering the Tatar khanates of Kazan and Astrakhan, thus giving the Russian state control over the entire Volga River. Later tsars pushed the frontiers of Russia westward toward Poland and southward toward the Black Sea. Peter the Great (reigned in 1682–1725) defeated the Swedes under Charles XII to gain a foothold on the Baltic Sea, where he established St. Petersburg as Russia's capital and its "Window on the West." Catherine the Great (reigned in 1762–1796) secured a frontage on the Black Sea at the expense of Turkey.

Meanwhile, the conquest of Siberia proceeded rapidly. At the end of Ivan the Terrible's reign, traders and Cossack military pioneers were already penetrating this wild and lonely territory, where there were only scattered indigenous inhabitants. The Cossacks were peasant-soldiers in the steppes, formed originally of runaway serfs and others fleeing from tsardom.

Figure 6.9 Services in a Russian Orthodox church in Moscow. Since the breakup of the USSR there has been a resurgence of organized worship throughout the region. *Sovfoto/Eastfoto*

They eventually gained special privileges as military communities serving the tsars.

In 1639, a Cossack expedition reached the Pacific. Russian expansion toward the east did not stop at the Bering Strait but continued down the west coast of North America as far as northern California, where the Russian trading post of Fort Ross was active between 1812 and 1841. In 1867, however, Russia sold Alaska to the United States and withdrew from North America. Russia's selling price was two cents per acre. In the United States, the transaction was widely derided as "Seward's Folly," after William Henry Seward, the American secretary of state who negotiated it.

The relentless quest of Russian fur traders for sable, sea otter, and other valuable furs brought great cultural changes to aboriginal peoples in Siberia and Pacific North America. Such effects became increasingly pronounced, especially in the last decades of the Russian empire when millions of Russians moved into Siberia.

In the Amur River region near the Pacific, the Russians were not able to consolidate their hold for nearly two centuries as a result of opposition by strong Manchu emperors who claimed this territory for China. They were thus barred from the east Siberian area best suited to grow food for the fur-trading enterprise and best endowed with good harbors for maritime expansion. This situation changed in 1858–1860 following the victory of European sea powers over China in the Opium Wars. Russia was able to add the Amur region to its earlier gains in the Ob, Yenisey, and Lena basins. Tsarist Russia conquered most of the southern zone of high mountains and dry lands only in the 19th century. This annexation involved a long series of military actions in the Caucasus region and Turkestan, the arid and semiarid area east of the Caspian Sea.

6.4 The Soviet Union

War, Revolution, and More War

Russia has often triumphed over powerful invaders and inflicted shattering losses on them. Outstanding examples include the defeats of invading Swedish forces led by the warrior-king Charles XII in 1709; of French and associated European forces under Napoleon I in 1812; and of German and associated European forces sent by Hitler into the Soviet Union in World War II. In each case, the Russians lost early battles and much territory, but eventually inflicted a crushing and decisive defeat on the invaders. The success of Old Russia and the Soviet Union in withstanding invasions by such formidable armies was due in part to environmental rigors (particularly the brutal Russian winter) invaders faced, the overwhelming distances of a huge country with poor roads, and the defenders' love of their homeland. It also was due to talented Russian military leadership and to the willingness of good and poor generals alike to lose great numbers of men in combat. Finally, the successful defense used the **scorched earth** tactic to protect the motherland; rather than leave Russian railways, crops, and other resources to fall into the invaders' hands, the defenders destroyed them.

The Russian Revolution of 1917, which set the stage for the formation of the Soviet Union, was really two revolutions that occurred against the backdrop of World War I. In 1914, when a Serb in Sarajevo assassinated the heir to the throne of Austria-Hungary, a complicated series of alliances required Tsar Nicholas II to commit Russian troops to fight with Serbia, France, and Great Britain against Austria-Hungary and Germany in World War I. The first revolution in Russia began early in 1917 as a general protest against the terrible sacrifices of Russian forces on the Eastern Front during this war. That revolt overthrew Nicholas II, the last of the Romanov tsars. Later in the year came the Bolshevik Revolution, often called the October Revolution (according to the old tsarist calendar) or November Revolution (by the newer Soviet calendar). Led by Vladimir Ilyich Lenin (1870–1924), the Bolshevik faction of the Communist Party seized control of the government. The new regime made a separate peace with Germany and its allies and survived a difficult period of civil war and foreign intervention between 1917 and 1921. Lenin presided over the establishment of Russia's successor state, the Soviet Union, in 1922.

World War II found Russians again allied with France and Britain in a far more ferocious war against Germany. The Soviet Union's success in withstanding the German onslaught that began as "Operation Barbarossa" in June 1941 surprised many outside observers who had predicted that the Soviets would prove too weak and disunited to resist for more than a few weeks or months. The crucially important cities of Leningrad (now St. Petersburg) and Moscow held out through brutal sieges. Late in 1942 the German push eastward was decisively checked along the Volga River in the huge Battle of Stalingrad (see p. 180).

This engagement was the turning point in the war; thereafter, Soviet and Allied forces reversed the Nazi advance. By the end of the war in 1945, Soviet armies had taken Berlin and linked up with U.S. forces at the Elbe River in Germany.

The failure of powerful Germany to conquer the Soviet Union was a clear indication that the strength of the Soviet Union had been underrated and that the country's power would henceforth be a major feature of world affairs. But the war's impact on the USSR would linger. The German invasion took an estimated 20 million or more Soviet lives and caused the relocation of millions of people. It did enormous damage to settlements, factories, and livestock. As a strategic precaution during the war, Lenin's successor, Josef Stalin, directed major Soviet industries to relocate eastward away from the front, a move which has had a lasting imprint on the region's economic geography.

Russians and Other Nationalities

The region of Russia and the Near Abroad is a mosaic of peoples speaking over 100 languages, many organized politically into independent states and subordinate units based on nationality. By far the largest and most powerful state is the Russian Federation (Russia). Like most of the smaller countries of the Former Soviet Region, it inherited the Soviet system of administrative areal subdivision into *oblasts* (or similar units called *krays*), often comparable in area and population to the states of the United States. An oblast or kray is an economic area focusing on its largest city, which serves as the political capital and for which the unit is named. The most populous are Moscow Oblast and St. Petersburg Oblast, centered on Russia's two main cities. Moscow is the capital of the Russian Federation.

Many smaller nationalities are organized into politically subordinate "autonomous" units (see Fig. 6.1, inset). The Soviets had established 16 Autonomous Soviet Socialist Republics (ASSRs) as homelands for ethnic minorities, in theory endowing them with limited self-governing powers. Until the breakup of the Soviet Union, however, Moscow governed these ASSRs like all other regions. The other Soviet-era autonomies included 5 "autonomous regions" (autonomous oblasts) and 10 "autonomous areas" (autonomous okrugs). Soviet labelling of these units as "autonomous" came back to haunt the nation of Russia in the 1990s (see the section on "The Future of the Russian Federation" in Chapter 7, p. 208). As a result of the devolution of power since 1991, within the Russian Federation there are now 89 "subjects" divided into four categories: 52 oblasts (regions), 6 krays (territories), 21 republics, and 10 autonomous okrugs (districts which are ethnic subdivisions of oblasts or krays).

The policy of the Soviet Union permitted the different nationalities to retain their own languages and other elements of their traditional cultures. Alphabets were created for nationalities that previously had no written language. In spite of such cultural concessions to non-Russians, however, the Russian-dominated regime implemented a deliberate policy of **Russification.** This was an effort to implant Russian culture in non-Russian regions, and it began even before the 1917 revolution. As the tsarist empire expanded from the original small state of Muscovy around Moscow, non-Russian peoples, especially those of the east and south, were not absorbed on a basis of equality with Russians. Instead, Russian governors ruled them as colonials. Russian officials, soldiers, traders, priests, and immigrant farmers carried Russian culture throughout the empire but rarely attempted to assimilate the peoples they regarded as "alien" or "backward."

Favoritism toward Russians and "Russianness" carried over from Old Russia to the Soviet era. A strong means of Russification was the migration of large numbers of Russians to cities, industries, and (in Kazakstan) state farms in non-Russian republics. Russians were very prominent in positions of responsibility even within the non-Russian republics; throughout the country they held the overwhelming majority of top posts in the Communist Party, the government, and the military.

Russian ways had relatively little impact on many culturally complex and cohesive peoples such as the Georgians and the Armenians. Even Slavic kinspeople of the Russians, such as the Ukrainians, insisted on retaining their cultural autonomy. In general, the policy of Russification was a failure because of strong nationalist sentiments throughout the Soviet Union.

The Communist Economic System

The Communist economic system that played such a central role in Soviet life following the Bolshevik Revolution was an attempt to put into practice the economic and social ideas of the 19th-century German philosopher Karl Marx (who actually spent most of his adult life in England and is buried in London). According to Marx, the central theme of modern history is a struggle between the capitalist class (bourgeoisie) and the industrial working class (proletariat). He forecast that exploitation of workers by greedy capitalists would lead the workers to revolt, overthrow the capitalists, and turn over ownership and management of the means of production to new workers' states. In the classless societies of these states, there would be social harmony and justice, with little need for formal government.

Marx's utopian vision has not materialized anywhere, but it did promote revolutions and provide guidelines for communist political systems in many countries. The ideas of Marx as Lenin interpreted and implemented them ("Marxism-Leninism") provided the philosophical basis for the Soviet Union's centrally planned "command economy"—an economy that is now being restructured with great difficulty. This economy developed as a mixture of advanced technology and an economic bureaucracy inherited from Tsarist Russia. Beginning in 1928, the Soviet command economy operated according to Five-Year Plans that prescribed goals of production for the nation—for example: types and quantities of minerals, manufactured goods, and agricultural commodities to be produced; factories, transportation links, and dams to be built or improved; and locations of new residential areas to house industrial workers.

The revolutionary leaders had essentially two aims: They meant to abolish the old aristocratic and capitalist institutions of Tsarist Russia and to develop a strong socialist state able to stand on an equal footing with the major industrial nations of the West. For the first decade, they consolidated their hold on the country and put a limited part of their program into effect. Large-scale industry, banking, and foreign trade were nationalized, while the New Economic Policy (NEP), announced in 1921, permitted some private trading, together with private ownership of small industries and agricultural land.

Lenin, the chief architect of the revolution, died in 1924. It was mainly his successor, Josef Stalin, who forged the Soviet system. Abandoning the privatization of the New Economic

Regional Perspective

The USSR and Other Land Empires

Some geographers have characterized the Soviet Union as a **land empire.** Rather than establishing its colonies overseas as did imperial powers such as Spain and Great Britain, Russia founded its colonies in its own vast continental hinterland. Many of the colonized peoples had little in common with the Russian ethnic majority which ruled from faraway Moscow.

Like former colonies of overseas empires such as Great Britain, non-Russian regions of the Soviet Union's periphery were drawn into a relationship of economic dependency upon the imperial Russian center. The Central Asian republics, for example, followed Moscow's demands to grow cotton, which was shipped to Moscow to be manufactured into expensive clothing marketed in Central Asia and elsewhere in the empire. This pattern contributed to the growth of the USSR's economy but often inhibited local development. Other entities characterized historically as land empires include the United States, China, Brazil, and the Ottoman, Moghul, Aztec, and Inca Empires.

STALINGRAD—SACRED SPACE

Sacred space or **sacred place** may be defined as any locale which people hold in reverence. The most obvious sacred places in the realms of ordinary experience are places of worship such as synagogues, churches, and mosques. Cemeteries, too, evoke a special code of behavior in visitors, and are managed as sacred sites. Places on Earth where large numbers of people have lost their lives are among the most significant sacred spaces—the Gettysburg Battlefield and the now-barren site of the federal building in Oklahoma City are examples in the United States. Russia may have the world's most extensive network of sacred places associated with the loss of life.

Perhaps in no other country is the memory of World War II etched so actively and vividly in the national consciousness as it is in Russia. Even today, millions of pilgrims annually visit a large number of war monuments and cemeteries in an effort to heal deep emotional wounds and to keep alive the memory of Russia's costly wartime resistance. These rites of visitation and commemoration pass to each new generation; immediately after their wedding ceremony, for example, a newlywed couple places flowers at the local tomb of the unknown soldier. Contemporary Russian pride and nationalism have strong roots in wartime sacrifice, and the experiences of those who survived the war continue to influence Russian politics and international relations.

During the war, superiors urged Russian soldiers to fight to the death for the Motherland: Russia was sacred ground to be defended at any cost. Today, the battlegrounds where those soldiers fell are sacred places. They are kept hallowed by the continuous ritual visitation of veterans, war widows, and two generations of descendants of war survivors.

Stalingrad (now Volgograd) is the greatest of all these sites. The visitor to Volgograd cannot help but feel the pain of war that has lingered more than 50 years. All over the city are monuments to remind the living of the dead. The devastated shell of a mill stands as the only physical artifact of the past, but the memorials built after the war rekindle the emotional losses most strongly. Stalingrad's central memorial is the complex on Mamayev Hill, a mecca for Russians

(Fig. 6.A). Old soldiers, still wearing their medals, look war-weary even now as they shuffle through. Grandmothers lead small children to place flowers at the feet of statues. Over the mass grave of an estimated 300,000 Soviet and German soldiers, a huge hand raises an eternal torch. The honor guard changes in goosesteps hourly. This sacred place is dominated by the world's largest statue: a female sword-wielding figure called the Russian Motherland, 279 feet (85 m) high and weighing 8000 tons.

Located on the border of the steppe and semidesert where the Volga and Don rivers are closest together, Volgograd is situated strategically. The city became an important grain and livestock center after a railway connected it to the Don valley. Later, the rail connected the city to the Caucasus region, between the Caspian and

Figure 6A Mamayev Hill monument to the memory of the defenders of Stalingrad in World War II. *Joseph J. Hobbs*

Policy, he and his long despotic regime embarked on full-scale Communism and used terror to sweep aside real and suspected opposition. Stalin's government launched a ruthless drive for comprehensive planning, forced socialization of the economy, and massive industrialization. The motivation was partly ideological and partly defensive. Insisting that because the world socialist revolution had not yet occurred, and because the USSR was surrounded by hostile capitalistic states such as Germany, Stalin resolved to "build socialism in one country." After

his death in 1953, Stalin's successors abolished and modified many rigidities and cruelties of the Stalin era and were able to raise the Soviet standard of living somewhat. Such changes, however, did not revolutionize the country's basic structure and ideology, and economic conditions gradually worsened.

Under the Communists, an agency in Moscow called Gosplan (Committee for State Planning) formulated national plans for the Soviet Union. Once approved at the highest level, the plans were transmitted downward through a huge administrative

Black Seas. Early in World War II, Hitler resolved to capture the oil fields of the Caucasus. Rostov-on-Don and Stalingrad stood in his way. Though strategically vital, Stalingrad was to Hitler as much a symbolic as a military prize: Since it bore the name of Russia's leader, its fall would be of great propaganda value to the German war effort. Equally, its salvation from the invader was a goal of nearly religious significance for the Stalin regime.

The German offensive on the city began in August 1942. On September 7, from a bunker in Moscow's Kremlin, Stalin ordered: "Not another step backward." He issued this directive:

> Comrades and citizens of Stalingrad: Each of us must apply ourselves to the task of defending our beloved town, our homes, and our families. Let us barricade every street, transform every district, every block, every house, into an impregnable fortress.

Russian resistance at Stalingrad is an extraordinary chapter in the history of warfare. The invaders were unprepared for the resolve of Russian fighters, as these passages from the diary of a German soldier reveal:

> September 11 [1942]. Our battalion is fighting in the suburbs of Stalingrad. Firing is going on all the time. Wherever you look, fire and flame. Russian cannon and machine guns are firing out of the burning city. Fanatics!
>
> September 16. Our battalion, plus tanks, is attacking the grain elevator. The battalion is suffering heavy losses. The elevator is occupied not by men, but by devils, that no bullets or flames can destroy.
>
> September 18. Fighting is going on inside the elevator. If all the buildings of Stalingrad are defended like this, then none of our soldiers will get back to Germany!
>
> September 26. We don't see them at all. They've established themselves in houses, and cellars; they're firing from all sides, including from our rear. Barbarians! They use gangster methods!

As the fighting persisted, a Russian soldier wrote of this scene:

> Stalingrad is no longer a town. By day, it is an enormous cloud of burning, blinding smoke; it is a vast furnace, lit by the reflection of the flames. And when night arrives, one of those very hot, noisy, bloody nights, the dogs plunge into the Volga and swim desperately to gain the other bank. The nights of Stalingrad are a terror for them. Animals flee from this hell. The hardest stones cannot bear it for long. Only men endure.

On November 8, when his soldiers held nine-tenths of the city, Hitler said in an after-dinner speech in Munich:

> I wanted to get to the Volga, at a particular point where stands a certain town—bears the name of Stalin himself. I wanted to take the place, and you know, we've done it. We've got it really, except for a few enemy positions still holding out. Now, people say, "Why don't they finish the job more quickly?" Well, I prefer to do the job with quite small assault troops. Time is of no consequence at all.

Hitler was mistaken, of course. The severe Russian winter had begun, fierce Russian resistance continued in the city, and time was very much against his forces. The Russians struck back on November 19, and soon confined the German Sixth Army within a pincer. The invaders were now on the defensive. Radio Moscow broadcast (in German) this demoralizing Christmas Day message to them:

> Every seven seconds, a German soldier dies in Russia. Stalingrad is a mass grave.

The final Russian assault began on January 11, 1943. Two weeks later, Hitler denied General Von Paulus's request to surrender in order to save German lives. On January 31, Hitler made Von Paulus a field marshal, reminding him that no field marshal had ever been taken alive by the enemy. On the day of his promotion, Field Marshal Von Paulus surrendered his forces to the Russians.

In Germany, news of the Russian victory at Stalingrad was announced on February 3, 1943. The communique noted that German forces had succumbed to superior enemy strength and to "unfavorable circumstances." It was a devastating defeat for the Germans: Two armies consisting of 24 generals, 2000 officers, and 90,000 soldiers were taken prisoner. Enough materiel was lost to equip one-fourth of the German army. Two years earlier, the Germans could not have imagined such a defeat. But the victory for the Soviet Union exacted an unimaginable cost. After the five-month battle, Stalingrad lay in ruins; Soviet authorities dubbed it a "city without an address" and, in honor of its defenders, a "hero-town." In 1993, fifty years after the battle, Russian military authorities finally released figures on the number of dead. In this sacred ground lay the bodies of three and a half million soldiers and civilians, of whom 2.7 million were Soviets.

---

bureaucracy (apparat) until they finally reached the operating level of individual factories, farms, and other enterprises. This unwieldy process generated inefficiency in a number of ways:

1. Fear of offending superiors made persons at lower levels reluctant to suggest ways to improve efficiency.
2. It was hard for planners in Moscow to manage an area larger than North America as though it were one gigantic corporation. Trying to coordinate such a huge and diverse body of enterprises, materials, labor, and consumer demand from one central point was too great a task.
3. In freer economies, the market—the desires and abilities of consumers and businesses to make and buy things—largely determines what will be produced. The Soviet planning bureaucracy had no free market to guide it, so often goods were produced that people would not buy, or goods were not produced even though people would have liked to buy them.
4. Although Soviet decision-makers knew that more free

181

enterprise might result in more efficient operation of plants and farms, they did not widely grant such freedoms. They were afraid of the uncertainties of free markets and unwilling to surrender their decision-making powers.

5. Gosplan stated production targets in quantitative rather than qualitative terms, often resulting in shoddy workmanship and unsalable goods. Enterprise managers often lobbied to keep their production targets as low as possible.

Communist planners changed the country's spatial organization, altered the interaction of people and nature, and added many new elements to the cultural landscape. They constructed new cities and transport links, enlarged older cities, carried out a massive expansion of mining and manufacturing, increased the amount of cultivated land, and reorganized the countryside by collectivized agriculture. These are still significant features on the landscapes of Russia and the Near Abroad.

The Soviet Union emphasized some grandiose economic projects. Such enterprises harnessed the energies and resources of the whole country to achieve specific objectives. The government called on the people to sacrifice in order to make the country strong and provide a better life in the future; the term "Hero Project" was meant to incite enthusiasm. A few examples of such projects include the construction of large tractor plants in the 1920s and 1930s at Kharkov in Ukraine, Stalingrad on the Volga, and Chelyabinsk in the Urals; the "Virgin and Idle Lands" program of the 1950s to expand grain acreage east of the Volga; and the "Project of the Century"—the construction of the Baikal-Amur Mainline (BAM) Railroad during the 1970s and 1980s, to provide an alternate link to the Pacific north of the Trans-Siberian Railroad in the Far East. These accomplishments were impressive, but they failed to give the Soviet people the higher standard of living they desired.

Collectivized Agriculture

Current efforts to privatize industry and agriculture throughout the region of Russia and the Near Abroad are slowed by the difficulties inherent in a complete overhaul of the often inefficient state-run system. A successful future for agriculture in particular will require that the legacy of past mismanagement be reversed.

Farmers and farming had troubled careers under the Soviet system of collectivized agriculture. Between 1929 and 1933, about two-thirds of all peasant households in the Soviet Union were collectivized; their landholdings were confiscated and reorganized into two types of large farm units (Fig. 6.10): the collective farm *(kolkhoz)* and the factory-type state farm *(sovkhoz)*.

As originally conceived by the Soviet state, the system of collectivized agriculture was to result in the following major advantages:

1. The old arrangement of small, fragmented individual holdings separated by uncultivated boundary strips would be replaced by larger fields incorporating the boundary strips,

Figure 6.10 A scene on a collective farm in Moldova. Advanced technology keeps company with holdovers from Old Russia in the economies of Russia and the Near Abroad. *Sovfoto/Eastfoto*

increasing the amount of cultivated land and promoting mechanized farming.

2. Increased mechanization would release surplus farm labor for employment in factories and mines, promoting industrialization and creating the large urban working class looked to as the principal support for the Communist system.

3. Mechanization, improved methods of farming, and reclamation of new land under state supervision would result in greater overall production.

4. Increased production, plus easier collection of surpluses from a smaller number of farm units, would result in larger and more dependable food supplies for growing urban populations and greater tax revenues to use in building industry.

5. Liquidation of individual peasant farming would remove the most important capitalist element still remaining in the USSR.

6. Consolidation of individual farmsteads and villages into fewer but larger communities on the collective farms would permit the government to more efficiently and cheaply administer, monitor, and indoctrinate the rural population, and provide them with services including education, health care, and electricity.

The rural people resisted collectivization fiercely. In their own version of scorched earth, peasants and nomads slaughtered millions of livestock and burned crops to avoid turning them over to the "socialized sector." Government reprisals followed, including wholesale imprisonments and executions, together with confiscation of food at gunpoint (often including the peasants' own food reserves and seed). The more prosperous private farmers, known as *kulaks,* were killed, exiled, sent to labor camps, or forced to starve. Famines took millions of lives. Soviet leaders disregarded these costs, and collectivization was virtually complete by 1940.

The drive to increase the national supply of farm products also demanded an enlargement of cultivated area. In 1954, the

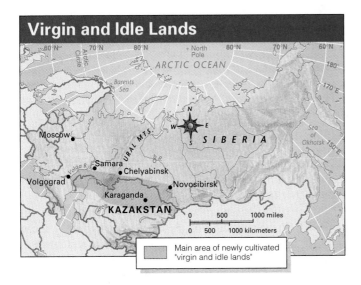

Figure 6.11 The area shown in green encloses most of the new lands brought under cultivation in the period 1954–1957. Modern international boundaries are shown.

Soviet Union instituted a program to increase the amount of grain (mainly spring wheat and spring barley) produced by bringing tens of millions of acres of **virgin and idle lands** or **"new lands"** into production in the steppes of northern Kazakstan and adjoining sections of western Siberia and the Volga region (Fig. 6.11). Hundreds of new state farms were organized. Between 1954 and 1960, the cultivated area of the Soviet Union was enlarged by over 90 million acres (36 million hectares), with most of the increase taking place in 1954–1956 in Kazakstan, Siberia, and the Volga region. Despite bad weather in some years, with crop failures and low production, the Virgin and Idle Lands Scheme added to Soviet grain output and moved the center of grain farming eastward. This lessened the impact of bad weather in a given year, as a poor winter wheat crop in the west could be offset by a good spring wheat crop in the east, and vice versa. However, there have been problems in maintaining satisfactory production on a long-term basis. The "new lands" have low precipitation and require careful management to conserve soil moisture and prevent wind erosion. Some of the most marginal land is no longer productive.

The Superpower

The principal goal of Soviet national planning after 1928 was a large increase in industrial output, with emphasis on heavy machinery and other capital goods, minerals, electric power, better transportation, and military hardware. The drive to industrialize had far more success than the planned expansion of agriculture. Masses of peasants were converted into factory workers, new industrial centers were created, and old ones were enlarged. The increase in urban population was phenomenal: In 1926 only 18 percent of the Soviet Union's population

lived in cities, but by 1998 the average figure for the 15 countries of Russia and the Near Abroad had risen to an estimated 65 percent (the world average in 1998 was 44 percent; in the United States, it is 75 percent).

The prodigious drive to remake the USSR industrially unleashed a large expansion of railways and ocean shipping, the mass production of millions of new apartments, space flight, and a huge military machine. The Soviet Union became the military power strong enough to survive Germany's onslaught in World War II. After the war, the Soviets maintained large armed forces and accumulated a huge arsenal of conventional and nuclear weapons in the "arms race" against the world's only other superpower, the United States. At the height of the Cold War, 15 to 20 percent of the country's GNP was dedicated to the military (in contrast to less than 10 percent in the United States), representing a considerable diversion of investment away from the country's overall economic development. While Russian arms sales abroad continue to grow today, economic reformers are grappling with the chore of converting a large share of the military economy to production for civilian use (Fig. 6.12).

To meet the needs of its people, the USSR operated as a collectivized welfare state, with the government providing guaranteed employment, low-cost housing, free education and medical care, and old-age pensions. Social services were often minimal but in some sectors were quite successful; the literacy rate, for example, rose from 40 percent in 1926 to 98.5 percent in 1959. However, military-industrial superpower status was generally achieved at the expense of Soviet consumers, whose needs were slighted in favor of heavy metallurgy and the manufacture of machinery, power-generating and transportation equipment, and industrial chemicals. Lines of consumers waited in stores to purchase scarce items of clothing and everyday conveniences. The exasperation of shoppers confronted by long lines and empty shelves was an important factor generating dissatisfaction with

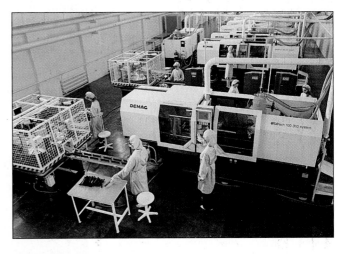

Figure 6.12 The RONIS factory in the Russian city of Rostov-on-Don is one of Europe's largest producers of videocassettes and compact disks. *Sovfoto/Eastfoto*

the economic system and leading to demands that the system be redesigned. Shortages of food and industrial products available to urban consumers were largely due to producers withholding goods in the hope of higher prices and to large-scale bartering of food from state farms and collectives in exchange for needed goods from large industrial enterprises.

6.5 The Second Russian Revolution

Internal freedoms and prosperity did not accompany geopolitical power in the Soviet Union. Near paralysis gripped the flow of goods and services in the late 1980s, when large demonstrations and strikes underscored public anger at a political and economic system that was sliding rapidly downhill. The Communist system came under open challenge on the grounds that it stifled democracy, failed to provide a good living for most people, and thwarted the ambitions of the country's many ethnic groups for a greater voice in running their affairs. The outpouring of dissent was unprecedented in Soviet history.

Worsening economic conditions led to official calls for fundamental reform in the mid-1980s. The revamping of the economic system became an urgent priority during the regime of Mikhail Gorbachev that began in 1985. Gorbachev initiated new policies of *glasnost* ("openness") and *perestroika* ("restructuring") to facilitate a more democratic political system, more freedom of expression, and a more productive economy with a market orientation (Fig. 6.13). The government was reorganized, and the Communist Party's control over political and economic affairs eventually ended.

At the same time, there were increasing problems of disunity as the different republics and the ethnic groups organized by the Soviet system into nominally "autonomous" units took advantage of new freedoms to resurrect old quarrels and demand that their units be given more autonomy. Fighting among ethnic groups erupted in several republics. Meanwhile, in all of the republics, declarations of sovereignty and in some cases outright independence challenged the authority of the central government. In 1991, the USSR and most other nations recognized the independence of the three Baltic republics of Estonia, Latvia, and Lithuania. The process of the empire's disintegration had begun.

At the center of the crumbling empire, having failed to reverse the downward slide of the economy, Gorbachev faced growing sentiment to scrap the command system and move as rapidly as possible to a market-oriented economy. Early in 1991, strikes in coal fields vital to the national fuel supply compounded the country's economic troubles. Laboring in some of the world's most blighted industrial districts, the miners called for decentralization of political and economic control as a means of achieving a better life.

The miners' cause was championed by Boris Yeltsin, a leading advocate of rapid movement toward a market-oriented economy and greater control by individual republics over their own resources, taxation, and affairs. Yeltsin's political stature grew when, in June 1991, the citizens of Russia chose him president of the giant Russian Federation in the Soviet Union's first open democratic election. Gorbachev, himself elected president of the USSR by a Congress of People's Deputies rather than by the people as a whole, opposed Yeltsin's proposals for radical reform. He advocated more gradual movement toward free-market orientation within the state-controlled planned economy

Figure 6.13 Freedom of expression erupted all over the USSR during the Gorbachev era. This was an artist's kiosk in Leningrad (now St. Petersburg) in 1991. *Joseph J. Hobbs*

and insisted on the need for unity and continued political centralization within the Soviet state. Even Gorbachev's gradualist approach was too radical for a third body of opinion, held by hardliners in the Communist Party—the government bureaucracy, the military, and the police—who called for strong measures to enforce greater efficiency within the existing system.

Matters came to a head in August 1991, when an attempted coup by hardliners failed. Yeltsin gained enhanced stature from his defiant opposition to the coup. Gorbachev resigned his position as head of the Communist Party and began to work with Yeltsin to reconstruct the political and economic order. His efforts to preserve the Soviet Union and a modified form of the Communist economic system failed. His program of *perestroika* was supposed to rebuild a sense of individual involvement in the Russian economy, but Gorbachev seesawed on his reform program and ended up fighting the forces of liberalization he had unleashed. During the autumn of 1991, the Communist Party was disbanded and the individual republics seized its property. On December 25, 1991, Gorbachev resigned the presidency, and the national parliament formally voted the Soviet Union out of existence on the following day. A powerful empire had quickly and quietly faded away, to be replaced by 15 independent countries.

6.6 Russia, the Near Abroad, and the Wider World

With the collapse of the Soviet Union, the Russian Federation took over the property of the former government within the Federation's borders (including Moscow's Kremlin; see Fig. 6.14). Russia also took custody of the international functions of the USSR, including its seat at the United Nations. In early December 1991, the republics of Russia, Ukraine, and Belarus (formerly Belorussia) formed a loose political and economic organization called the Commonwealth of Independent States (CIS) headquartered in Minsk, Belarus. Except for Estonia, Latvia, and Lithuania, the other republics eventually joined the CIS, in which Russia took a strong leadership role. Disputes erupted, notably between Russia and Ukraine, over such questions as ownership of property inherited from the defunct USSR, particularly the Black Sea naval fleet based in Ukraine's Crimea region, and disposition and control of nuclear weapons.

Since the Soviet Union dissolved in 1991, many important links between Russia and the other 14 successor states have persisted. Economic, ethnic, and strategic links are especially crucial. For example, one major economic link is the vital flow of Russian oil and gas to other states such as Ukraine, Belarus, and the Baltic states, which are highly dependent on this supply of energy (see Chapter 7, p. 193). Ethnicity links approximately 25

million Russians who resided in the 14 smaller states at the time of independence. Although a considerable number have migrated to Russia since 1991, many have had difficulties finding housing and employment, and the great majority still live in the other countries. In certain areas, notably eastern Ukraine and Crimea, many ethnic Russians are causing political instability because they want to secede and form their own state or join Russia. Russians in the 14 non-Russian nations complain that governments and peoples discriminate against them.

In response, the Russian government has said that it has a right and a duty to protect Russian minorities in the other countries. Russia's perspective on the other 14 former members of the USSR is that they constitute a special foreign policy region, the country's "near abroad," in which Russia's special interests and influence must be recognized. Some observers contend that Russia's new activism in this border region is an effort to reestablish the countries of the former socialist camp as buffers between Russia and the "far abroad."

In this national security interest, Russia forcefully asserts claims to a special sphere of influence in the Near Abroad. Along the outer frontiers of the cordon of successor states, Russia maintains a chain of military bases where, in 1998, an estimated 100,000 Russian soldiers were stationed. Within the Near Abroad, notably in Moldova, Tajikistan, Georgia, Armenia, and Azerbaijan, Russian troops have become engaged frequently as "peacekeepers" in local ethnic conflicts. There are fears in the West and among Russia's neighbors that "peacekeeping" is merely a euphemism for a restoration of Russian imperial rule. Some observers contend that Russia wants access to important resources and economic assets in its former republics, such as uranium in Tajikistan, aviation plants in Georgia, military plants in Moldova, and the Black Sea coast and naval fleet in Ukraine's Crimea. Noting Moscow's reluctance to recall Russian troops from the Near Abroad, some analysts fear that Russia's reasoning may be that Moscow's empire will again be expanded, and it is not worth withdrawing troops and dismantling bases only to redeploy and rebuild them later.

Within Russia there is a body of public opinion favoring reassertion of Russian control over its former empire, but the strength of this feeling is unknown. Many Western analysts anticipate that Russia will attempt increasingly to "Finlandize" its Near Abroad, referring to the Soviet Union's historic influence over its Nordic neighbor. This would involve Russia's insistance on authority over the foreign policy and security policy of such nations as Ukraine, Estonia, Latvia, and Lithuania. With the recent election of pro-Russian governments in Ukraine and Belarus, it looks increasingly likely that there will emerge a Slavic political and economic *troika* consisting of Russia, Ukraine, and Belarus, possibly dominated by Russia. And while a significant reintegration around Russia may occur in the Slavic nations, most of the Central Asian countries may gravitate toward neighboring Islamic nations such as Turkey or Iran (see Chapter 7, p. 217).

Figure 6.14 Moscow's Red Square, seat of power for the USSR and now the Russian Federation. The Lenin Mausoleum, just to the right of the line of conifers on the left side of this image, backs up to the Kremlin wall. *Joseph J. Hobbs*

6.7 Russia, the Misdeveloped Country

Despite heightened public expectations for a better life in Russia, criticism of the triumphant President Yeltsin and his policy of economic reform toward freer markets increased throughout the course of his two terms in office. In September 1993, reacting to the parliament's efforts to slow the pace of economic reforms, Yeltsin dissolved the parliament and set a December date for a referendum on the draft constitution and election of a new two-chamber parliament: a lower house, or State Duma, of 450 deputies, and an upper house, or Federation Council, composed of representatives from 88 subdivisions of the Russian Federation. Then, in October 1993, Russian army units quickly and violently suppressed an armed uprising against the president in Moscow by members of the old parliament and their supporters. In December, Russia's electorate approved the draft constitution (which gave the president broad powers), but the parliamentary elections did not yield a clear majority favoring rapid market reforms. The new legislature was generally hostile to Yeltsin, and was dominated by Communists and ultranationalists like Vladimir Zhirinovsky.

In the 1995 parliamentary elections, continued public dissatisfaction with the new Russia gave once-disgraced Communists nearly a quarter of the legislature's seats. The Communists advanced a candidate for the prize of Russia's presidency in 1996. Although Boris Yeltsin's prospects of winning appeared dim in advance of the election, especially because most Russians expressed dissatisfaction with the state of the economy, the majority preferred an uncertain future under Yeltsin rather than a return to the Communist past, and they elected him president for a second term concluding in 2000.

Early in his first term, President Yeltsin introduced a program of rapid economic reform—known widely as economic "shock therapy"—designed to replace the Communist system with a free-market economy. It removed price controls and encouraged privatization of businesses. Products and services from the private sector had already been indispensable during the Communist era, but now they were to be the centerpiece of the economy. Expansion of private-sector goods and services was swift; by the time the first phase of the privatization process officially ended in 1994, two-thirds of Russian industry was in private hands and 40 million Russians owned shares in newly privatized firms.

Results of "shock therapy" in Russia were initially mixed. Former employees of the Communist Party resented the loss of their jobs and privileges. Printing of money to meet state obligations caused inflation. Poor people living on fixed incomes struggled to survive high prices for basic necessities. A new consumer-oriented society developed along class lines (Fig. 6.15). There was growing unemployment and homelessness, and a widening gap between rich and poor. Access to choice goods and services was no longer forbidden to ordinary citizens as it was during the Communist era, but prices were so high that most people could not afford to pay them.

One of the major components that has emerged from Russia's new economic geography is the **underground economy,** also known as the "countereconomy" or "second economy" (or economy *na levo*—"on the left"). Widespread barter has resulted from the declining value of the ruble (see Definitions and Insights, p. 187). Many people resort to selling personal possessions to buy high-priced food and other necessities. Rampant black-marketeering has developed since prices were decontrolled. Much of this traffic is still officially illegal, but there

Figure 6.15 Free enterprise thrives throughout what was once the Soviet Union. Customers are lined up to buy food from a street vendor in St. Petersburg. *Joseph J. Hobbs*

is a general tendency to overlook such transactions because they are so essential to the economy. Russia's new private entrepreneurs include a large criminal "mafia" that preys on government, business, and individuals, raising public fears and prompting some people to call for a return to the relative stability and security of the Soviet state. The unprecedented experiment to transform the colossal Communist state into a bastion of free-market democracy has produced strange and often unfortunate results, leading some observers to classify Russia not as a "less developed" or "more developed" country, but as a "misdeveloped" country.

With its bizarre mix of legitimate and black-market enterprises, Russia rumbled along. Then came August 1998. Depleted of cash and therefore unable to defend Russia's currency on the international market, the country's Central Bank devalued the ruble and suspended its exchange for foreign currency. An international financial crisis ensued (see Chapter 2, p. 46). The long-suffering Russian consumer faced an unprecedented explosion in the prices of food and other goods and services. The financial chaos prompted a political crisis. President Yeltsin sacked the reformist prime minister Sergei Kiriyenko, whom he had appointed only five months earlier, and replaced him with the less reform-minded Viktor Chernomyrdin— whom Yeltsin had fired to bring in Kiriyenko. The Duma refused to confirm Chernomyrdin, however, forcing Yeltsin to nominate a prime minister more acceptable to the opposition: Yevgeny Primakov, an important Soviet-era bureaucrat still enjoying good relations with Russia's Communists.

In this unstable environment, the United States, its Western allies, and the International Monetary Fund were loathe to pledge any new financial support or implement financial bailout packages for Russia. Since Yeltsin had come to power, far less of the aid that was pledged from these sources actually reached its targets. The lending agencies complained that what support

did reach Russia disappeared with few apparent results, often apparently into the pockets of the tycoons and the mafia that control so much of the economy. Revered in the heady days of newfound freedoms in Russia, Americans and Westerners in general rapidly lost popularity among Russians. Many Russians perceived Yeltsin to be "selling out" to the West, and resented their growing dependence upon its institutions, exports, and media. The Russian poet Yevgeny Yevtushenko lamented what he called "the McDonaldization of Russia."

Definitions & Insights

BARTER

Barter is the exchange of goods or services in the absence of cash. Bartering was a common practice in the USSR, and has increased in importance since Russia became independent; it now accounts for about half the transactions in Russian industries and 40 percent of national tax payments. Barter is even a medium of foreign exchange (see discussion of Belarus in Chapter 7, p. 194). When companies do not have cash to buy equipment, pay utility bills, pay workers, or pay taxes, they resort to barter. For example, a Novgorod-based company which mainly manufactures valves also produces railroad-car brake pads, which it uses to pay rail cargo fees, which have been kept artificially high as a means of subsidizing passenger travel.

Companies that owe unpaid taxes also like to use barter as a means of avoiding state seizure of their cash assets. While critics argue that barter corrupts free enterprise by distorting prices and making a mockery of tax collection, others emphasize that it allows Russian individuals and businesses some insulation against severe economic turmoil.

SUMMARY WITH SELECTED KEY TERMS

- It is still necessary to place the **15 former republics of the Soviet Union** into a single region. These 15 countries will be attempting to sever or reestablish ties to Moscow for many years to come. Some analysts believe that Russia may even attempt to reexert control over some of the recently independent nations.
- The area west of the Ural Mountains and north of the Caucasus Mountains has been known as **European Russia.** Similarly, the Caucasus and the area east of the Urals has been called **Asiatic Russia. Siberia** is the general name for the area between the Urals and the Pacific; **Central Asia** is the general name for the arid area occupied by five states immediately east and north of the Caspian Sea.
- Russia and the Near Abroad span 11 time zones. Stretching nearly halfway around the globe, the region has formidable problems associated with **climate, terrain, and distance.**
- The most productive human activities take place in the **subarctic, humid continental, and steppe climatic zones.**
- The most productive agricultural area is located in the steppe climate characterized by the **grassy plains** south of the forest in Russia and is the main climate of Ukraine, Moldova, and Kazakstan. Here the low and variable rainfall is partially offset by the fertility of the **chernozems ("black earth")** which are among the best soils to be found anywhere.
- In early Russia, the **rivers formed natural passageways** for trade, conquest, and colonization. Especially important in Siberia, the Russians followed tributaries of some of Earth's greatest rivers to advance from the Urals to the Pacific in less than a century. More recently, **alteration of rivers** for power, navigation, and irrigation became an important aspect of economic development under the Soviet regime.
- **Large plains** dominate the terrain from the Yenisey River to the western border of the country. The area between the Yenisey and Lena rivers is occupied by the hilly **Central Siberian Uplands.** In the southern and eastern parts of the region mountains, such as the **Caucasus Mountains,** and the **Pamir, Tien Shan,** and **Altai** ranges, dominate the landscape.
- **Slavic and Scandinavian** peoples were greatly influential in early cultural development which was the foundation for the Russian, Ukrainian, and Belarusian cultures.

- Early contacts with the **Eastern Roman** or **Byzantine Empire** led to the establishment of **Orthodox Christianity** as a feature of Russian life and culture. Although it was repressed and neglected after the **Bolshevik Revolution,** this religion has been experiencing a rebirth since the breakup of the Soviet Union.
- From the 15th century until the 20th, the **tsars** built an immense Russian empire around the small nuclear core of **Muscovy.** This imperialism brought great areas of land and hosts of alien peoples under tsarist control. **Cossack** expeditions extended to the Pacific, across the **Bering Strait,** and on down the west coast of North America as far as northern California.
- Russia has often triumphed over powerful foreign invaders. These achievements have been due in part to the environmental rigors the invaders faced, the overwhelming distances involved, and the defenders' willingness to accept large numbers of casualties and implement a **scorched earth** tactic to protect the motherland.
- The policy of the Soviet Union permitted the dfferent nationalities to retain their own languages and other elements of traditional cultures. However, the regime implemented a deliberate policy of **Russification** in an effort to implant Russian culture in non-Russian regions.
- To meet the needs of its people, the **USSR** became a **collective welfare state,** with the government providing guaranteed employment, low-cost housing, free education and medical care, and old-age pensions. However, **military-industrial superpower** status was generally achieved at the expense of the ordinary consumer.
- Current efforts to privatize industry and agriculture are slowed by the difficulties inherent in a complete overhaul of the old inefficient state-run systems such as the **collectivized agriculture** of the Communists.
- Many important links between Russia and the other 14 successor states—**the Near Abroad**—have persisted. Economic, ethnic, and strategic links are especially crucial and the Russian government has said that it has the right and duty to protect Russian minorities in the other countries. Some observers contend that Russia is attempting to reestablish the "near abroad" as buffers between Russia and the "far abroad."

REVIEW QUESTIONS

1. List the five climatic belts of this region and briefly describe their attributes.
2. Using maps and the text, locate and list the major river systems.
3. What wars have been significant to the development of this region?
4. List the reasons why Russia has so often won the conflicts with foreign invaders.
5. What were the perceived advantages of "collectivized agriculture"?
6. Name and describe some Soviet Communist projects that changed the cultural landscape of this region.

7. List some of the successful elements of the drive to industrialize the Soviet Union.
8. List some of the negative results of Boris Yeltsin's "shock therapy" economic reform.
9. List the areal names that have been used for this region.
10. List the ethnic, cultural, and political external forces that have helped shape this region.

DISCUSSION QUESTIONS

1. Examine and discuss the significance of the area and population of Russia and the Near Abroad.
2. What roles have the natural environment played in this region's history and development?
3. What other cultures were important in Russia's early years and what were the major contributions of these cultures?
4. Examine and discuss how warfare has been such a major influence on this region.
5. What were the foundations for the Communist economic system and what is this system's legacy for today's economy?
6. Why did the "Second Russian Revolution" occur? What are some of the difficulties that the present government is facing?
7. Locate, examine, and discuss the important physical features of this region.
8. Locate, examine, and discuss the important rivers and water bodies of this region.

Chapter 7

Fragmentation and Redevelopment in Russia and the Near Abroad

▲ *Children at a day-care center in Ulyanovsk, Russia. They are growing up in the new and uncertain world that follows decades of Soviet rule.* Joseph J. Hobbs

CHAPTER OUTLINE

7.1 Peoples and Nations of the Fertile Triangle

7.2 Agriculture in the Coreland

7.3 The Industrial Potential of the Russian Coreland

7.4 The Russian Far East

7.5 The Northern Lands of Russia

7.6 The Future of the Russian Federation

7.7 The Caucasus

7.8 The Central Asian States

*t*he region of Russia and the Near Abroad has an uneven spatial structure of population and development. Nearly three-fourths of the people, and an even larger share of the cities, industries, and cultivated land of this immense region, are packed into a triangular **Fertile Triangle** composing roughly one-fifth of the region's total area (Fig. 7.1). Lying mainly west of the Ural Mountains but narrowing eastward into Asia, the triangle is also called the **Agricultural Triangle** and the **Slavic Coreland**. The rest of the region of Russia and the Near Abroad lies mostly in Asia and consists of land where settlement is extremely spotty, handicapped by nonagricultural environments except in limited areas. Most of the non-Slavic population of the region lives there (Fig. 7.2), but many immigrant Slavs also reside there, generally in cities. These areas outside the coreland are a storehouse of minerals, timber, and waterpower, and governments have made major investments to develop these resources. This has resulted in a series of discrete production nodes widely separated by taiga, tundra, mountains, deserts, wetlands, and frozen seas. The nodes are scattered through the Caucasus region, Central Asia, the Far East, and the Northern Lands.

7.1 Peoples and Nations of the Fertile Triangle

Slavic peoples are the dominant ethnic groups in most of the region of Russia and the Near Abroad, in both numbers and political and economic power. The major groups are the Russians (whose cultural geography is described in Chapter 6), Ukrainians, and Belarusians (Byelorussians). Most of this Slavic population lives in the triangular coreland extending from the Black and Baltic Seas to the neighborhood of Novosibirsk in Siberia. The greatest part of the coreland is in Russia, but Ukraine, Belarus, Moldova, Estonia, Latvia, Lithuania, and northern Kazakstan also fall within it. The triangle has about one-half the area and more than four-fifths the population of the United States. Lowlands predominate—the only mountains are the Urals, a small segment of the Carpathians, and a minor range in the Crimean Peninsula. The original vegetation was mixed coniferous and deciduous forest in the northern part and steppe in the south. Moscow (population: 8.6 million, city proper; 13.15 million, metropolitan area) and St. Petersburg (formerly Leningrad; population: 4.27 million, city proper; 5.53 million, metropolitan area) are by far the largest cities. Moscow, capital of Russia before 1713 and of the Soviet Union from 1918 to 1991, remains the most important manufacturing city, transportation hub, and cultural, educational, and scientific center in the region (see Fig. 6.14). It is the political capital of the Russian Federation.

Ukraine

Russians are by far the largest ethnic group in the coreland, followed by Ukrainians. Although closely related to the Russians in language and culture, Ukrainians are a distinct national group. The name Ukraine translates as "at the border" or "borderland." In this area, armies of the Russian tsars fought for centuries against nomadic steppe peoples, Poles, Lithuanians, and Turks before the Russian Empire absorbed Ukraine in the 18th century. For three centuries Ukrainians were subordinate to Moscow, first as part of the Russian Empire and then as a Soviet republic. Ukraine's industrial and agricultural assets were always vital to the Soviet Union; the Bolshevik leader Lenin once declared, "If we lose the Ukraine, we lose our head." Since achieving independence from Moscow in 1991, Ukraine has restored the Ukrainian language to its educational system. However, a majority of Ukrainians still want Russian to be a second official language, along with Ukrainian.

Today, Ukraine is one of the most densely populated and productive areas in the region. With about 50 million people as of 1998, the republic is not far behind such countries as the United Kingdom and France in population.

Ukraine lies partly in the forest zone and partly in the steppe. On the border between these biomes is the historic city

Regional Divisions of Russia and the Near Abroad

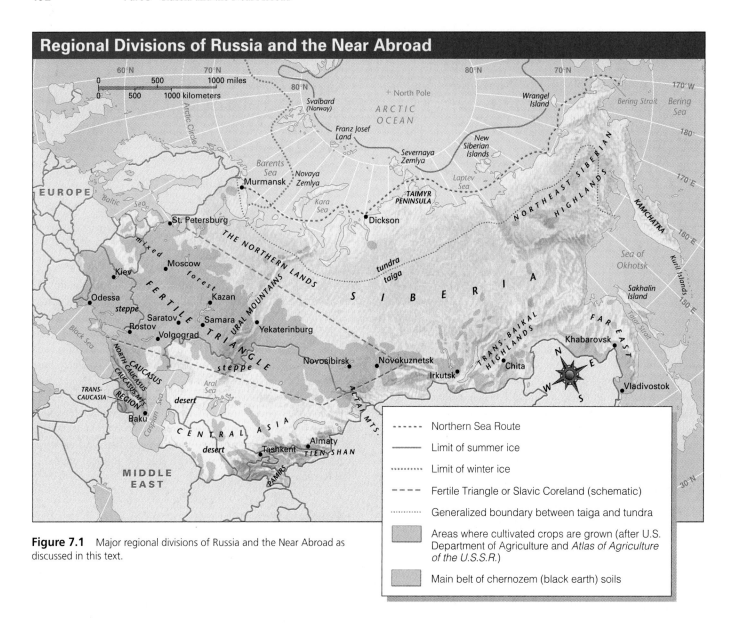

Figure 7.1   Major regional divisions of Russia and the Near Abroad as discussed in this text.

of Kiev (population: 2.64 million, city proper; 3.25 million, metropolitan area), Ukraine's political capital and a major industrial and transportation center (Fig. 7.3). The country is an important producer of wheat, barley, livestock, vegetable oil (mainly from sunflower seeds), beet sugar, and many other products, grown mainly in the steppe region south of Kiev. North of Kiev is Chernobyl, where an explosion at a nuclear power station in 1986 rendered parts of northern Ukraine and adjoining Belarus incapable of safe agricultural production for years to come.

Ukraine has large deposits of coal, iron, manganese, salt, and natural gas which contribute to heavy industry (see Fig. 7.5). The coal comes from the Donets Basin (Donbas) coal field near the industrial center of Donetsk (population: 1.12 million, city proper; 2.13 million, metropolitan area), where most of

Ukraine's iron and steel plants are also located. Iron ore comes by rail from central Ukraine in the vicinity of Krivoy Rog (population: 737,000, city proper), the most important iron-mining district in the region of Russia and the Near Abroad. Most of the ore is used in Ukrainian iron and steel plants, but some is exported to East Central European countries and Russia. This area is also an important center of chemical manufacturing. Surrounding the inner core of mining and heavy-metallurgical districts in Ukraine is an outer ring of large industrial cities that carry on metal fabricating and other types of manufacturing. These cities include the Ukrainian capital of Kiev on the Dnieper River; the machine-building and rail center of Kharkov (population: 1.62 million, city proper; 2.05 million, metropolitan area), about 250 miles (400 km) east of Kiev; and the seaport and diversified industrial center of Odessa (population:

Ethnic Map of Russia and the Near Abroad

KEY TO
ETHNIC GROUPS

1 Russians	11 Abkhazians	19 Karakalpaks
2 Ukrainians	12 Estonians	20 Uzbeks
3 Belarusians	13 Karelians, Komi	21 Turkmen
4 Poles	and other	22 Azeris
5 Bulgarians	Finno-Ugrian	23 Yakuts
6 Lithuanians	14 Chuvash	24 Buryats
7 Latvians	15 Tatars	25 Armenians
8 Moldovans	16 Bashkirs	26 Tuvinians
9 Tajiks	17 Kazaks	27 N. Caucasians
10 Georgians	18 Kyrgyz	

Sparsely populated areas in Central Asia

Northern minority peoples

- - - - Pre-1991 international boundary

——— International boundary

Figure 7.2 The Soviet Union had difficulties trying to hold together such a vast collection of ethnic groups. The Russian Federation faces many of the same challenges.

1.10 million, city proper; 1.19 million, metropolitan area), on the Black Sea.

Since independence, Ukraine has taken important steps on the world stage. In 1994, the Ukrainian parliament voted overwhelmingly to become a nuclear-free nation. By electing to sign the Nuclear Nonproliferation Treaty, Ukraine committed itself to disposing of the 1800 nuclear warheads which had made it the world's third largest nuclear weapons power. The United States rewarded this move with a much-increased foreign aid package to Ukraine, which now is the third largest recipient of U.S. economic assistance after Israel and Egypt.

At the same time, Ukraine has improved relations with Russia that had soured since the collapse of the USSR. The 1994

election of President Leonid Kuchma, who favored economic union and closer political ties with Russia, began to correct some of the problems that followed Ukraine's efforts to distance itself from Russia following independence. The Ukrainian economy had faltered, especially because the country remained extremely dependent upon Russia for oil, gasoline, natural gas, and uranium for its nuclear power reactors. With independence, Ukraine had to pay world market prices for these sources rather than the low prices the Soviets previously subsidized.

Following Ukraine's independence in 1991, one source of trouble with Russia had been the Crimean Peninsula, which Russia's Catherine the Great annexed in 1783 but which Soviet leader Nikita Khrushchev returned to Ukraine in 1954. In 1991,

Figure 7.3 The heart of Kiev, on the banks of the Dnieper River. Kiev is the third largest metropolis in the region of Russia and the Near Abroad, and is the mother city of the Ukrainian and Russian branches of Orthodox Christianity. Kiev was devastated by the military actions of World War II, and massive reconstruction of buildings took place after the war. *Jerry Alexander/Tony Stone Images*

Ukraine gave the Crimea special status as an autonomous republic. Crimeans then elected their own president, a pro-Russian politician who advocated the Crimea's reunification with Russia. (Ethnic Russians make up 70 percent of Crimea's population and about 22 percent of Ukraine's total population.) Other residents wanted complete independence for Crimea. In 1997 diplomacy triumphed as Kuchma and Russia's President Yeltsin signed a "Friendship Treaty" which allowed Russia to operate its Black Sea fleet in the Ukrainian port of Sevastopol. Because of its accommodating climate, the Crimean Peninsula has itself often been a venue for diplomacy. Sheltered by the Yaila Mountains, Crimea's south coast has the relatively mild temperatures and minimal summer rainfall associated with the mediterranean or dry-summer subtropical climate. It is a picturesque area of orchards, vineyards, and resorts, of which Yalta is the most famous.

Belarus

The much smaller state of Belarus (formerly called Belorussia and "White Russia"; population: 10.2 million) adjoins Ukraine on the north. It developed slowly in the past, mainly because of a lack of fuels other than peat. Recently, however, oil and gas pipelines from other areas such as Siberia have provided raw material for new oil-refining and petrochemical industries. Some oil production has begun in Belarus itself. Under Soviet rule, Belarus became an industrial powerhouse, shipping televisions, farming equipment, automobiles, textiles, and machinery

to both Moscow and its eastern European satellites. Belarus's foreign trade is now almost exclusively with Russia, much of it in the form of barter. In 1998, for example, Belarus paid off a debt to Russia's natural gas monopoly (Gazprom) by supplying it with $300 million worth of refrigerators, tractors, and other hard goods. Unlike Russian industries, most manufacturing enterprises in Belarus are still state-run.

Minsk (population: 1.66 million, city proper; 1.69 million, metropolitan area) is the political capital and main industrial city. Belarus is too cool, damp, and infertile to be prime agricultural country, but it has a substantial output of small grains, hay, potatoes, flax, and livestock products. Southern Belarus contains the greater part of the Pripyat (Pripet) Marshes, large sections of which have been drained for agriculture.

Like Ukraine, Belarus has agreed to eliminate the nuclear weapons it has possessed since the breakup of the Soviet Union. Also like Ukraine, Belarus is reestablishing close ties with Moscow. In 1994, Belarus and Russia agreed to unify their monetary systems. In the agreement, Belarus gave up sovereignty over its currency and banking system in exchange for preferential access to Russian energy and other resources. In a 1995 referendum, an overwhelming majority of Belarusians voted for closer economic integration with Russia and for the adoption of Russian as their official language. In 1997, Belarus and Russia formed a "union state" to more closely link the economies, political systems, and cultures of these predominantly Slavic countries. During its subsequent financial crisis, Russia evaluated the costs of absorbing some of the financial liabilities of Belarus, and backed away from implementing these linkages.

Moldova

Tiny Moldova (formerly Moldavia; population: 4.2 million), made up largely of territory that the Soviet Union took from Romania in 1940, adjoins Ukraine at the southwest. Moldovans are a people with many Slavic cultural characteristics who speak a dialect of Rumanian. The area, however, has many problems of ethnic antagonism and political separatism within its borders. In the eastern part of the country, in a sliver of territory along the Dniestr River known as Transdniestria, Ukrainians and Russians have been agitating violently to form a separate state. Elsewhere the minority of nationalistic Moldovans, who claim that Moldovans are Romanians, want Moldova to join Romania.

Industry has increased markedly in Moldova since World War II, especially around the capital of Chisinau (also known as Kishinev; population: 667,100, city proper). Mainly a fertile, black-earth steppe upland, however, Moldova is primarily agricultural. An important specialty is the growing of vegetables and fruits, especially grapes for winemaking.

The Baltic States and Kaliningrad

Estonia, Latvia, and Lithuania, known as the Baltic states or the Baltics, lie between Belarus and the Baltic Sea. These countries share many environmental, economic, and historical traits with the other nations of the Fertile Triangle. However, the indigenous inhabitants of the Baltics are not Slavs, and they have long tried zealously to safeguard their national identities against Russian encroachment. They were part of the Russian Empire before World War I but became independent following the Russian Revolution. The Soviet Union reabsorbed them in 1940.

Today, they are again independent states, but they are still linked closely to the remainder of the Former Soviet Union in various ways. Russia uses Baltic ports to transport goods abroad and exports fossil fuels and a variety of other goods and services to the Baltic nations. But Estonia, Latvia, and Lithuania want to minimize their ties with their giant neighbor to the east. They have refused to become members of the Commonwealth of Independent States. They would like to become members of the European Union (EU; see p. 87), but only Estonia has been accepted for "fast track" EU incorporation. The Baltics would also like to become members of the North Atlantic Treaty Organization (NATO; see p. 87), but Russia considers that option a serious threat to its security, and resists it. So, while the Baltic governments prefer to portray themselves as European, as many outside observers do (the French Geological Institute recently calculated that Lithuania is geographically the heart of Europe), vital and difficult links with Russia

suggest that these nations will in many ways belong in a sphere of "former Soviet" countries for some years to come.

In these countries, dairy farms alternate with forests in a hilly landscape. Mineral resources are scarce; the most notable are deposits of oil shale in Estonia, mined for use as fuel in electric power plants. Like Belarus, the Baltic states benefit from natural gas and petroleum brought from Russian fields located hundreds and even thousands of miles away. Expansion of industry after World War II helped make the area one of the more prosperous and technically advanced parts of the former USSR. Its largest city and manufacturing center is the port of Riga (population: 848,000, city proper), in Latvia (population: 2.4 million).

About 660,000 people, or 85 percent of Latvia's non-Latvians, are Russian-speaking—mostly retired Soviet military officers and their offspring. While independent Lithuania and Estonia established procedures early on to grant citizenship to ethnic Russians, Latvia delayed taking such measures until late 1994 and gave in only under pressure from Russia and the West. The Latvian government feared that enfranchised Russians would destabilize the country in future elections or even try to reannex Latvia to Russia. But Moscow's economic leverage prevailed; it threatened to cut off the shipment of its oil through Latvia (the transit fees generate important revenue for Latvia) and to stop buying Latvian agricultural exports. Latvia relented, in part by agreeing to allow all children born on Latvian soil since 1991 to become Latvian citizens, and by allowing more non-Latvians to apply for citizenship.

The presence of Russian troops in the Baltic countries was a serious foreign relations problem after the countries gained independence. With independence, only Lithuania (population: 3.7 million; capital, Vilnius—population: 581,500, city proper) successfully negotiated the immediate withdrawal of Russian troops. A large number of these Russian forces resettled in neighboring Kaliningrad, an anomalous Russian enclave two countries away from Russia. As the northern half of the former German East Prussia, the territory was transferred to the USSR at the end of World War II to become the Kaliningrad Oblast. Soviet authorities expelled nearly all of Kaliningrad's Germans to Germany, and Russians now make up more than three-fourths of the population. Kaliningrad remains a massive military and naval establishment, with defense workers and their dependents and retired military families comprising most of the population of about one million.

Unwanted Russian troops remained in Latvia and Estonia until late 1994. Before then, Russia had argued that it could not pull out its troops because Latvians and Estonians were violating the rights of 300,000 ethnic Russians in Estonia and a large minority in Latvia. Since independence, Estonia (population: 1.4 million; capital, Tallinn—population: 447,700, city proper; Fig. 7.4) has developed close economic ties with Finland, which quickly replaced Russia as the country's dominant trading partner.

Figure 7.4 Estonia's capital, Tallinn, is a vibrant and colorful city. *Joseph J. Hobbs*

7.2 Agriculture in the Coreland

The Fertile Triangle is the agricultural core of Russia and the Near Abroad, and is by far the leading area of production for all the major crops except cotton and for all the major types of livestock. There are two main agricultural zones within the Triangle: a black-soil zone in the southern steppes, and a non-black-soil zone corresponding roughly to the region of mixed forest. The black-soil zone includes chernozem soils and associated chestnut and other grassland soils. Wheat does well on these soils, making the region one of the major wheat-growing areas of the world. The black-soil zone is also the principal producing area for sugar beets, sunflowers, hemp (grown mainly for oil), barley, and corn. Irrigation waters reach the area from the Dnieper, Don, Volga, and other rivers. In the **non-black-soil zone**, with its cooler and more humid climate and poorer soils, rye has been traditionally the main grain crop for human consumption. Agriculture in this zone also emphasizes dairy production, potatoes, oats, and hay.

Some countries in the region of Russia and the Near Abroad, notably Russia and Ukraine, are global-scale producers of such farm commodities as wheat, barley, oats, rye, potatoes, sugar beets, flax, sunflower seeds, cotton, milk, butter, and mutton. The overall high output, however, does not reflect high agricultural productivity per unit of land and labor. The independent nations of the region will be struggling for many years to overcome the agricultural inefficiencies of the state-operated and collective farms that long dominated the agricultural sector (see

Chapter 6, p. 182). On the state-operated farms, workers received cash wages in the same manner as industrial workers. Bonuses for extra performance were paid. Workers on collective farms received shares of the income after obligations of the collective had been met. As on the state farms, there were bonuses for superior output. These methods, however, failed to provide enough incentives for highly productive agriculture. Farm machinery stood idle because of improper maintenance; since no individual owned the machine, the incentive to repair it was diminished. There were also shortages of spare parts. Poor storage, transportation, and distribution facilities, plus wholesale pilfering, caused alarming losses after the harvest. Younger people were deserting the farms for urban work, often living with the family in the countryside but commuting to their city jobs. With the shortage of younger male workers, women, children, and the elderly did a large share of the farm work.

In an effort to reverse such disturbing trends, all of the countries are promoting land reform by privatizing collective and state farms and developing more independent or "peasant" farming. Privatization of the collective and state farms occurred rapidly in the Baltic states and Armenia. Agricultural reform began in Russia, but then ground to a halt. After 1991, the Russian government held about 30 percent of the country's farms in reserve as state farms *(sovkhozes)* and technically allowed the remaining 70 percent to be privatized. However, arguing that only rich land barons would be able to purchase land available on auction, Communist opposition in Russia's legislature (Duma) has largely prevented the privatization of the collective farms *(kolkhozes)* that make up half of Russia's approximately 50,000 farms.

As of 1999, the failure to reform Russia's inefficient, still largely collectivized agricultural system was exacting a high cost. Russia's 1998 grain harvest totalled only 52 million tons, down from 88 million tons in 1997, representing the worst harvest since 1953. About 80 percent of Russia's farms were in debt, and the country was importing more than one-fourth of its food. However, individual initiative remained strong; backyard gardens and livestock accounted for perhaps one-third of the meat and half the vegetables produced in Russia.

7.3 The Industrial Potential of the Russian Coreland

The Soviet Union's emphasis on heavy industry and armaments required the use of huge quantities of minerals, most of which existed in the Soviet Union. From the beginning of its Five-Year Plans, the Soviet Union stressed the need for heavy industry, which it developed in three principal areas of mineral abundance: Ukraine, the Urals, and the Kuznetsk Basin (Fig. 7.5). One of the most problematic legacies of the breakup of the Soviet Union is that production of industrial commodities is not evenly spread across the region. The main production of each commodity tends to come from only a few areas, such as textiles from the Moscow region; steel from Ukraine, the Urals, and areas immediately south and north of Moscow; automobiles from the Volga and Moscow areas; and raw cotton from Central Asia. The manufacture of machinery is widespread, but a given city specializes only in particular kinds of machinery, such as grain harvesters at Rostov-on-Don and textile machinery near Moscow.

This regional specialization, which the 15 successor countries inherited from Communist planning, has made continued cooperation essential among the now-independent states. For example, a single factory in Belarus produces all the electric motors required by certain types of appliance factories throughout the region of Russia and the Near Abroad. For the short term at least, it is in the interest of the producers and the consumers in the former republics to maintain trading relations.

In the case of Russia, the industrial resources are there, the country has a large and highly skilled workforce, and the markets both within and outside its borders are vast. Sadly, however, Russia has squandered its industrial promise in recent years. The process of privatizing formerly state-controlled industries has been corrupted by the Russian mafia and by "cronyism," the deal-making and favoritism among Russia's new tycoons (known as *oligarchs* in Russia) that thwart the emergence of a genuine free market. The largest industries are extremely inefficient and continue the Soviet practice of churning out products that no one needs. A single steel factory typically employs more than 10,000 workers—far more than are needed—in great contrast to the lean and efficient plants of the West. Such inefficient enterprises are propped up by the barter system and by a round-robin cycle in which everyone owes money to someone but no one ever pays up. "The effect," wrote one analyst, "is to keep alive concerns that chew up $1.50 worth of resources in order to turn out a product that is worth only $1 to consumers. Economists call this 'negative value added.' Ordinary folks call it economic suicide."[1] Russia's gross domestic product (GDP) shrunk by more than 40 percent between 1991 and 1998. This fall in production is the largest any industrialized country has ever experienced in peacetime.

In its persistent conflict between capitalist reformers and old-school communists, Russia's political leadership has been powerless to prevent the deterioration of the country's economy. Suffering under an enormous debt burden, the Russian government is trying to keep spending down in part by not paying salaries to federal employees; the debt in unpaid wages amounts to an estimated 25 percent of the country's gross domestic product. Many privately owned industries are too unproductive to pay wages to their employees. Unpaid, workers cannot pay taxes back to the government. The government has therefore turned to other sources for the tax revenues it desperately needs—for example, by cracking down on the street vendors who have been among the vanguards of free-market enterprise in Russia. The widespread failure of economic reforms in Russia, in both the agricultural and industrial sectors, points toward a likely reversal of the reform process—toward more state control of the economy. While the country is in flux, it is worthwhile to study its considerable potential, boding perhaps a brighter future for Russia.

Industries of the Moscow and St. Petersburg Regions

The industrialized area surrounding Moscow is known as the Central Industrial Region, Old Industrial Region, or Moscow-Tula-Nizhniy Novgorod Region. These names indicate important characteristics of the region. It has a central location physically within the populous western plains and is functionally the most important area of Russia and the Near Abroad. It lies at the center of rail and air networks reaching Transcaucasia, Central Asia, and the Pacific, and is connected by river and canal to the Baltic, White, Azov, Black, and Caspian Seas (Moscow calls itself the "Port of Five Seas").

This region ranks with Ukraine as one of the two most important industrial areas in the region of Russia and the Near Abroad. South of Moscow lies the metallurgical and machine-building center of Tula (population: 530,300, city proper), and to the east is the diversified industrial center of Nizhniy Novgorod

---

[1]Michael M. Weinstein. August 30, 1998. "Russia Is Not Poland, and That's Too Bad." *New York Times*, p. 5.

Industrial Areas and Railroads of Russia and the Near Abroad

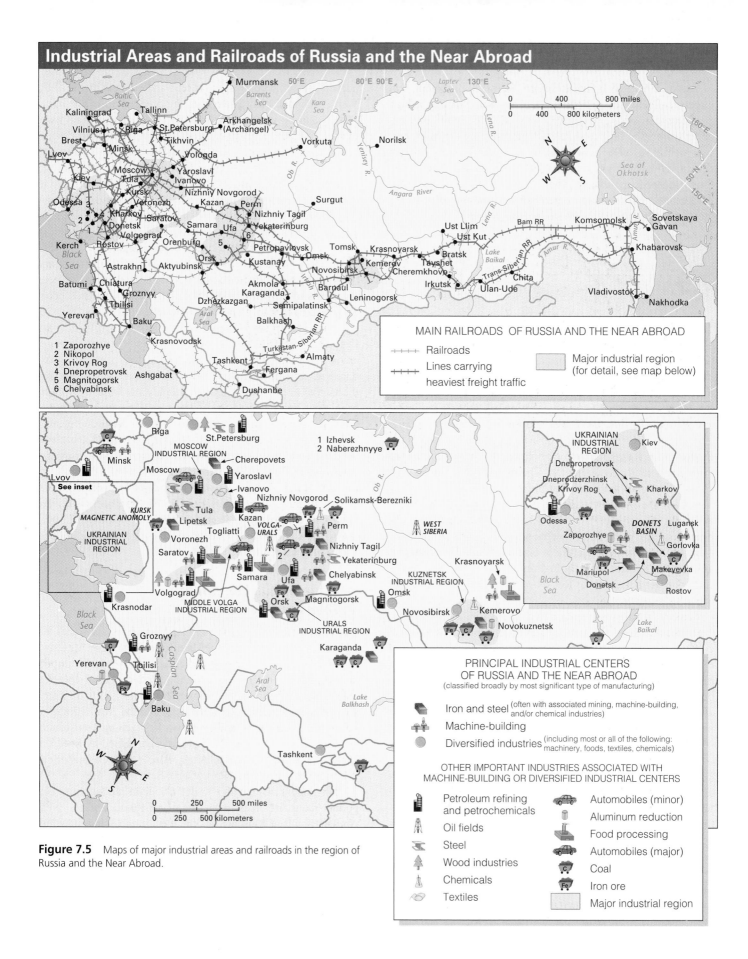

Figure 7.5 Maps of major industrial areas and railroads in the region of Russia and the Near Abroad.

Figure 7.6 St. Petersburg, established on the Baltic Sea at the mouth of the Neva River to serve as Russia's "Window on the West." *Joseph J. Hobbs*

(formerly Gorkiy; population: 1.38 million, city proper; 2.4 million, metropolitan area), on the Volga. Textile milling, mainly using imported U.S. cotton and Russian flax, was the earliest form of large-scale manufacturing to be developed in the Moscow region. The area still has the main concentration of textile plants (cottons, woolens, linens, and synthetics) in the region of Russia and the Near Abroad, with Moscow and Ivanovo the leading centers. The breakup of the Soviet Union has done much damage to this textile industry, however. The price of Central Asian cotton quadrupled with independence, forcing mills to close and consumers to turn to cheap clothing imports.

The Moscow region achieved industrial eminence despite a notable lack of mineral resources. The area's well-developed railway connections, partly a product of political centralization, provide good facilities for an inflow of minerals, materials, and foods, and for a return outflow of finished products to all parts of Russia and the Near Abroad.

The St. Petersburg area (Fig. 7.6) is not as great an industrial center as the Moscow region, Ukraine, or the Urals, but is still a significant presence in industrial development. As in the case of Moscow, local minerals have not formed the basis of this industrialization. Instead, metals from outside the area provide material for the metal-fabricating industries that are the leaders in St. Petersburg's diversified industrial structure. Supported by university and technological-institute research workers, the city's highly skilled labor force played an extremely significant role in early Soviet industrialization. They pioneered the development of many complex industrial products, such as

power-generating equipment and synthetic rubber, and supplied groups of experienced workers and technicians to establish new industries in other areas. Industrial innovation is now centered in Moscow, but St. Petersburg is still preeminent in some specialized industries, such as the making of large turbines for hydroelectric stations.

Industries of the Urals

The Ural Mountains contain an extraordinarily varied collection of useful minerals, including iron, copper, nickel, chromium, manganese, bauxite, asbestos, magnesium, potash, and industrial salt. Low-grade bituminous coal, lignite, and anthracite are mined. The important Volga-Urals oil fields lie partly in the western foothills of the Urals, and a major gas field lies at the southern end of the mountains.

The former Communist regime fostered the development of the Urals as an industrial region well removed from the exposed western frontier of the Soviet Union. The major industrial activities are: heavy metallurgy, emphasizing iron and steel and the smelting of nonferrous ores; the manufacture of heavy chemicals, based on some of the world's largest deposits of potassium and magnesium salt; and the manufacture of machinery and other metal-fabricating activities.

The Soviets modernized and expanded tsarist-era metallurgical and machine-building plants in the Urals and constructed several immense new plants. Their most famous and spectacular creation was the Soviet Union's largest iron-and-steel center at Magnitogorsk (population: 425,500, city proper) in the southern Urals. Located near a reserve of exceptionally high-grade iron ore, this place was not even a village prior to 1931. Construction of a huge plant and a city to house the workers began that year. Other giant Soviet-built iron-and-steel mills are scattered through the Urals at Chelyabinsk, Nizhniy Tagil, and Orsk.

The years that saw the creation of Magnitogorsk also saw a large expansion of coal mining in the Kuznetsk Basin (Kuzbas), located in southern Siberia more than 1000 miles (c. 1600 km) to the east of Magnitogorsk. A railway shuttle developed, with Kuznetsk coal and coke moving to Magnitogorsk and other industrial centers in the Urals, and Urals iron ore (primarily from Magnitogorsk) moving to a new iron-and-steel plant in the Kuznetsk Basin. Thus was created the famous Urals-Kuznetsk Combine (*Combinat* in Russian). Each end of the Combine, however, soon became partially independent of the other as coal was exploited closer to the Urals and iron closer to the Kuznetsk Basin. Much iron ore now reaches the Urals from mines that have been developed in a huge ore formation between the Moscow region and Ukraine called the Kursk Magnetic Anomaly, or KMA (so named because of its disturbing effect on compass needles).

Yekaterinburg (formerly Sverdlovsk; population: 1.38 million, city proper; 1.62 million, metropolitan area), located at the eastern edge of the Ural Mountains, is the largest city of the Urals and the region's preeminent economic, cultural, and transportation center. The second most important center is Chelyabinsk (population: 1.09 million, city proper; 1.11 million, metropolitan area), located 120 miles (193 km) to the south. This grimy steel town was infamous in Old Russia as the point of departure for exiles sent to Siberia. Today, areas in the vicinity are highly polluted with radioactive waste from Soviet nuclear operations. In western Siberia, between the Urals and the large industrial and trading center of Omsk (population: 1.17 million, city proper; 1.19 million, metropolitan area), rail lines from Yekaterinburg and Chelyabinsk join to form the Trans-Siberian Railroad, the main artery linking the Far East with the coreland (see railroad map, Fig. 7.5). Omsk is a major metropolitan base for Siberian oil and gas development. It is also Siberia's most important center of oil refining and petrochemical manufacturing.

Industries of the Kuznetsk Region

From Omsk, the Trans-Siberian Railroad leads eastward to the Kuznetsk industrial region, the most important concentration of manufacturing east of the Urals. The principal localizing factor for industry here is the enormous reserve of coal mentioned above. The manufacture of iron and steel is also a major industrial activity of the Kuznetsk region, whose main center is Novokuznetsk (population: 574,900, city proper). The industry draws its iron ore from various sources among Russia's Central Asian neighbors.

The largest urban center of the Kuznetsk region is Novosibirsk, a diversified industrial, trading, and transportation center located on the Ob River at the junction of the Trans-Siberian and Turkestan-Siberian (Turk-Sib) railroads. Sometimes called the "Chicago of Siberia," Novosibirsk (population: 1.37 million, city proper; 1.4 million, metropolitan area) has developed from a town of a few thousand at the turn of the century. Hundreds of factories produce mining, power-generating, and agricultural machinery, tractors, machine tools, and a wide range of other products.

The city of Akademgorodok ("Academy Town" or "Science Town"; population: 100,000, city proper), Siberia's main center of scientific research, is an outlying satellite community of Novosibirsk. The city was established in 1957 as a parklike haven where some of the Soviet Union's greatest scientific minds could research and develop both civilian and military industrial innovations. Reflecting a general crisis in the post-Soviet military-industrial establishment, Akademgorodok's funding has shrunk dramatically, forcing many prominent scientists to seek menial jobs. Internationally there are fears that underpaid Russian nuclear scientists will sell their know-how or materials to governments or terrorist organizations seeking to develop atomic weapons. There have already been hundreds of thefts of radioactive substances at nuclear and industrial institutions in the former USSR. The destination points of these materials remain largely unknown. Former Soviet nuclear experts have applied for work or are already working in Iran, Iraq, Algeria, India, Libya, and Brazil.

Industries of the Volga

Numerous industrial cities outside the main industrial concentrations are scattered through the region. The most notable are cities along the Volga River from Kazan southward, in the region Russians know as the Povolzhye. The largest are Samara (formerly Kuybyshev; population: 1.19 million, city proper; 1.22 million, metropolitan area), Volgograd (formerly Stalingrad; population: 997,400, city proper; 1.03 million, metropolitan area), Saratov (population: 894,600, city proper; 901,600, metropolitan area; heart of the Volga German Autonomous Republic until World War II), and Kazan (population: 1.08 million, city proper; 1.17 million, metropolitan area). These four cities in the middle Volga industrial region have diversified machinery, chemical, and food-processing plants. Under the Soviets, Samara was one of the military-industrial "closed cities" (closed to foreigners because of security concerns) whose industries Stalin commanded to be hastily built as a defensive measure against German attack in World War II. Like Nizhniy Novgorod upriver on the Volga, the city is undergoing a process known officially as "conversion"—that is, the retooling of military enterprises for civilian-oriented manufacturing. Kazan has a major university whose alumni include Vladimir Lenin and Leo Tolstoy. Maxim Gorki wrote about Kazan in *My Universities*; down along the Volga wharves, he found "a whirling world

Figure 7.7 Renowned for its drab functionality, Russian mass housing, like these apartment blocks for auto workers in the factories at Togliatti, also offers some amenities, such as these fountains. *Joseph J. Hobbs*

where men's instincts were coarse and their greed was naked and unashamed."

In recent decades, the Volga cities have seen an upsurge of industrial activity, particularly the construction of dams and hydroelectric stations along the Volga River and its large tributary, the Kama. A second major development has been the rapid rise of the Volga region to leadership in the automobile industry. Togliatti (Fig. 7.7; population: 630,000, city proper), on the Volga, is by far the leading center for the production of passenger cars in the region of Russia and the Near Abroad. The cars come from the Volga Automobile Plant, which was built and equipped for the Soviet government by Italy's Fiat Company; the city itself takes its incongruous name from a former leader of Italy's Communist Party.

Large-scale exploitation of petroleum in the nearby Volga-Urals fields has contributed to the industrial rise of the Volga cities. Prior to the opening of fields in western Siberia during the 1970s, the Soviet Union's most important area of oil production was the Volga-Urals fields, stretching from the Volga River to the western foothills of the Ural Mountains. Ufa (population: 1.09 million, city proper) and Samara are the region's principal oil-refining and petrochemical centers, but all of the larger cities along the Volga and Kama rivers—such as Perm on the Kama (population: 1.03 million, city proper; 1.04 million, metropolitan area)—as well as some smaller cities in the oil fields have a share of these industries.

7.4 The Russian Far East

The Far East is Russia's mountainous Pacific edge. Most of it is a thinly populated wilderness in which the only settlements are fishing ports, lumber and mining camps, and the villages and camps of aboriginal peoples. Port functions, fisheries, and forest industries provide the main support for most Far Eastern communities, and the output of coal, oil, and a few other minerals is small. Most of the Russians and Ukrainians who make up the majority of its people live in a narrow strip of lowland behind the coastal mountains in the southern part of the region. This lowland, drained by the Amur River and its tributary, the Ussuri, is the region's main axis of industry, agriculture, transportation, and urban development. Several small-to-medium-sized cities form a north-south line along two important arteries of transportation: the Trans-Siberian Railroad and the lower Amur River. At the south on the Sea of Japan is the port of Vladivostok (population: 631,500, city proper), which is kept open throughout the winter by icebreakers. About 50 miles (80 km) east of the city, the main commercial seaport area of the Far East has developed at Nakhodka (population: 163,000, city proper; Fig. 7.8) and nearby Vostochnyy (East Port). Both ports are nearly ice-free.

Figure 7.8 Rail yard on the electrified Trans-Siberian Railroad at the busy port of Nakhodka on the Pacific. The view shows passenger cars, timber for export, and a trainload of containers. Much container traffic moves between Pacific countries and European countries via this Trans-Siberian "bridge." *Tass/Sovfoto*

North of Vladivostok lies a small district that is the most important center of the Far East's meager agriculture, producing cereals, soybeans, sugar beets, and milk for Far Eastern consumption. The Far East is far from self-sufficient in food and consumer goods; large shipments from the Russian coreland and from overseas supplement local production.

The diversified industrial and transportation center of Khabarovsk (population: 601,000, city proper) is located at the confluence of the Amur and Ussuri Rivers, where the main line of the Trans-Siberian Railroad turns south to Vladivostok and Nakhodka and the Amur River turns north to the Sea of Okhotsk.

Before World War II, the Soviet Union and Japan held the northern and southern halves, respectively, of the large island of Sakhalin, which today has important forest and fishing industries and some coal, petroleum, and natural gas reserves. At the end of the war, the USSR annexed southern Sakhalin and the Kuril Islands and repatriated the Japanese population. Russian control of these former Japanese territories continues to be a major problem in relations between the two countries (see Chapter 14).

The Kurils are small volcanic islands that screen the Sea of Okhotsk from the Pacific. Fishing is the main economic activity. At the north, the Kurils approach the mountainous peninsula of Kamchatka (total population is 450,000, with 320,000 living in two cities). Like the islands, Kamchatka is located on the Pacific Ring of Fire—the geologically active perimeter of the Pacific Ocean—and has 23 active volcanoes (Fig. 7.9). Soviet authorities protected Kamchatka for decades as a military area, and no significant development activities occurred there. A land of rushing rivers, extensive coniferous forests, and gurgling hot springs, Kamchatka is now one of the last great wilderness areas on Earth, resembling the U.S. Pacific Northwest landscape of a century ago. A struggle is now underway between those who want to develop Kamchatka's gold and oil resources and those who wish to set the land aside in national parks and reserves where ecotourism would be the only significant source of revenue.

7.5 The Northern Lands of Russia

North and east of the Fertile Triangle and west of (and partially including) the Pacific littoral lie enormous stretches of coniferous forest (taiga) and tundra extending from the Finnish and Norwegian borders to the Pacific. These outlying wilderness areas in Russia may for convenience be designated the Northern Lands, although parts of the Siberian taiga extend to Russia's southern border.

These difficult lands form one of the world's most sparsely populated large regions. The climate largely prevents ordinary types of agriculture, but hardy vegetables, potatoes, hay, and barley are grown in scattered localities, and there is some dairy

Figure 7.9   Kamchatka is a land of fire and ice. The 11,364-foot (3464-m) Koryak volcano towers over the city of Petropavlovsk. *Sovfoto/Eastfoto*

farming. A few primary activities, including logging, mining, reindeer herding, fishing, hunting, and trapping support most people. Significant towns and cities are limited to a small number of sawmilling, mining, transportation, and industrial centers, generally along the Arctic coast to the west of the Urals, along the major rivers, and along the Trans-Siberian Railroad between the Slavic Triangle and Khabarovsk.

The oldest city on the Arctic coast is Arkhangelsk (Archangel; population: 376,200, city proper), located on the Northern Dvina River inland from the White Sea. Tsar Ivan the Terrible established the city in 1584 for the purpose of opening seaborne trade with England. Today, the city is the most important sawmilling and lumber-shipping center in the region of Russia and the Near Abroad. Despite its restricted navigation season from late spring to late autumn, it is one of the more important seaports.

Another Arctic port, Murmansk (population: 410,234, city proper; Fig. 7.10), is located on a fjord along the north shore of the Kola Peninsula west of Arkhangelsk, and is the headquarters for important fishing trawler fleets that operate in the Barents Sea and North Atlantic. Murmansk also has a major naval base and cargo port, and is home port to the icebreakers that escort cargo vessels and carry Western tourists to the North Pole and other High Arctic destinations (Fig. 7.11). It is connected to the coreland by rail. The harbor is open to shipping all year,

Figure 7.10 The ice-free port of Murmansk is situated on a fjord at the northern end of the Kola Peninsula. *Joseph J. Hobbs*

thanks to the warming influence of the North Atlantic Drift, an extension of the Gulf Stream. Murmansk and Arkhangelsk played a vital role in World War II, continuing to receive supplies by sea from the Soviet Union's Western allies after Nazi forces captured and closed off the other ports of the western Soviet Union.

Murmansk and Arkhangelsk are western termini of the Northern Sea Route (NSR), a waterway the Soviets developed to provide a connection with the Far East via the Arctic Ocean (see Fig. 7.1). Navigation along the whole length of the route is possible for only up to four months per year, despite the use of powerful icebreakers (including nine nuclear-powered vessels; see Problem Landscape, p. 204) to lead convoys of ships. Areas along the route provide such cargoes as the timber of Igarka and the metals and ores of Norilsk. Some supplies are shipped north on Siberia's rivers from cities along the Trans-Siberian Railroad and then loaded onto ships plying the Northern Sea Route for delivery to settlements along the Arctic coast. The Northern Sea Route is now open to international transit shipping, and, if not for the capricious sea ice, it would be an attractive route. The distance between Hong Kong and all European ports north of London is shorter via the Northern Sea Route than through the Suez Canal. Russia is anxious to develop the Northern Sea Route more for its own exports, since Russian exports by land to western Europe must now pass over the territories of Ukraine, Belarus, and the Baltic countries.

The railroad that connects Murmansk with the coreland serves important mining districts in the interior of the Kola Peninsula and logging areas in Karelia, which adjoins Finland. The Kola Peninsula is a diversified mining area, producing nickel, copper, iron, aluminum, and other metals, as well as phosphate for fertilizer. The peninsula and Karelia are physically an extension of Fennoscandia, located on the same ancient, glacially scoured, granitic shield that underlies most of the Scandinavian peninsula and Finland. Karelia, with its thousands of lakes, short and swift streams, extensive softwood forests, timber industries, and hydroelectric power stations, bears a close resemblance to nearby parts of Finland. In fact, Finland annexed some of this area during World War II.

Important energy resources, including good coking coal, petroleum, and natural gas, are present in the basin of the Pechora River, east of Arkhangelsk. The principal coal-mining

Figure 7.11 The Russian nuclear-powered icebreaker *Yamal* at the North Pole in August 1996. The former Soviet icebreaker fleet still escorts commercial ships on the Northern Sea Route but during the summer carries Western tourists such as these to the High Arctic. *Joseph J. Hobbs*

PROBLEM LANDSCAPE

Nuclear Waste Simmers in Russia's High Arctic

ON THE MAP, THE RUSSIAN ISLANDS OF Novaya Zemlya ("New Land"), separating the Barents and Kara Seas in the high Arctic, appear remote, wild, and untouched. These ice- and tundra-covered islands are indeed an isolated wilderness, which is precisely why Soviet authorities perceived them as prized grounds for nuclear weapons testing and nuclear waste dumping.

Soon after the breakup of the Soviet Union in 1991, a former Soviet radiation engineer stepped forward with disturbing information: Soviet authorities had ordered the dumping of highly radioactive materials off the coast of Novaya Zemlya for the past 30 years [Fig. 7.A]. From the 1950s through 1991, the Soviet navy and icebreaker fleet secretly and illegally used the shallow waters off the eastern shore of Novaya Zemlya as a nuclear dumping ground. At least twelve nuclear reactors removed from submarines and warships —six of which still contained their highly radioactive nuclear fuel—were sent to the

shallow seabed. Some of these individual reactors contain roughly seven times the nuclear material contained in the Chernobyl reactor which exploded in 1986. Between 1964 and 1990, 11,000 to 17,000 containers of solid radioactive waste were also dumped here at shallow depths between 200 and 1000 feet. Soviet seamen reportedly cut holes in those "sealed" containers which otherwise would not have sunk. This dumping occurred even though the Soviet Union joined other nations in a 1972 treaty which allowed only low-level radioactive waste to be dumped at sea, and then only at depths greater than 12,000 feet.

The danger posed by the nuclear material is not limited to the islands' shores but extends through the entire ecosystem of Russia's Arctic seas, and possibly beyond. Contamination can begin at the base of the food chain (the phytoplankton which "bloom" in the area each spring) and pass upward to higher trophic levels, where fish, seals, walrus, polar bears, and people feed. Alaskan authorities are worried, and Norwegians are particularly concerned. Norway's prime minister described the dumping as a "security risk to people and to the natural biology of northern waters." Western European and North American

markets for Norwegian and Russian fish, which now represent significant exports for these countries, may refuse purchases if radiation levels rise.

Due to the long half-life of the nuclear materials, the dangers posed by 30 years of haphazard dumping near Novaya Zemlya may last thousands of years. The threat can be reduced greatly by retrieving the waste and disposing of it in terrestrial sites deep underground, which are considered safer. (The Russian military supports a proposal to convert some of the shafts used in 1950s atomic weapons tests to permanent radioactive waste repositories.) The major obstacle to this step is the extreme costliness of such an operation— estimated to be in the hundreds of billions of dollars—which is well beyond Russia's economic means. An arms-agreement obligation for Russia to decommission about 80 nuclear submarines based in Murmansk is also costly and, because of the radioactive materials involved, potentially dangerous. Confessing and confronting the nuclear legacy of the Soviet Union, Russia has asked the United States and other nations to help pay for an environment with fewer nuclear hazards. The United States, in turn, is asking Russia not to export its nuclear power technologies to Iran

center is Vorkuta (population: 106,900, city proper), infamous for the labor camp where great numbers of political prisoners died during the Stalin regime. U.S. and Norwegian oil companies are working together to explore for, and possibly develop, oil for export from the Pechora region.

East of the Urals, the swampy plain of western Siberia north of the Trans-Siberian Railroad has seen a surge of petroleum and natural gas production since the 1960s. The West Siberian fields are the largest producers in the region of Russia and the Near Abroad. In a broader perspective, these are the largest oil and gas fields in Russia's considerable share of world reserves; Russia's 50 billion barrels of proven oil reserves represent 5 percent of the world total, and its 50 trillion cubic meters of

natural gas a third of proven global reserves. By the late 1990s, even though oil and gas production in postindependence Russia had experienced a marked decline, exports of these fossil fuels accounted for nearly half of Russia's hard-currency export earnings. Extraction and shipment of oil and gas take place in West Siberia under frightful difficulties caused by severe winters, permafrost, and swampy terrain. Huge amounts of steel pipe (Fig. 7.12) and pumping equipment have been required to connect the remote wells with markets in the coreland and Europe.

Still farther east, the town of Igarka, located about 425 miles (c. 680 km) inland on the deep Yenisey River, is an important sawmilling center accessible to ocean shipping during

and other countries where the West fears they may be used to develop weapons.

The predicament of Novaya Zemlya and the future of the vast wilderness in Russia's Far East reflect important questions about environment and development all across Russia and the Near Abroad. Like the United States and other industrialized countries, the Soviet Union experienced problems of environmental pollution caused by rapid economic growth. Russia and the other countries must now deal with the Soviet legacy of decades of environmental neglect. Environmental cleanup is hindered by the massive cost of repairing past damage and a reluctance to bear the expenses of stringent enforcement of pollution-abatement measures. Some Russian sources estimate that 80 percent of all industrial enterprises in Russia would go bankrupt if forced to comply with environmental laws. A desperate search for hard currency and widespread corruption have led to other environmental problems, including the wide-scale selling off of natural resources like timber, and to poaching of animals. With the breakup of the USSR, many environmentalists had predicted more protection for nature. They have been disappointed.

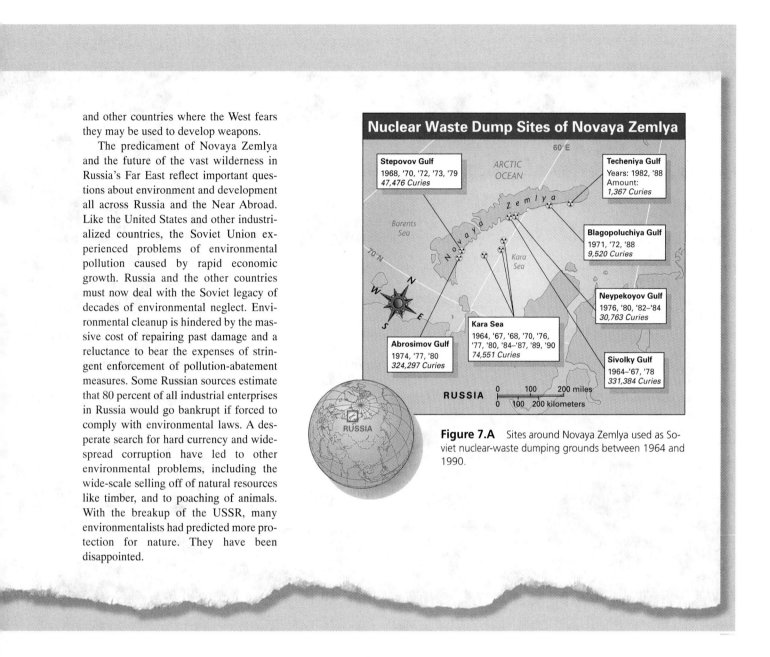

Figure 7.A Sites around Novaya Zemlya used as Soviet nuclear-waste dumping grounds between 1964 and 1990.

the summer and autumn navigation season. Northeast of Igarka lies Norilsk (population: 166,100, city proper), the most northerly mining center of its size on the globe. Rich ores yielding nickel, copper, platinum, and cobalt have justified the region's large investment in this Arctic city, including a railway connecting the mines with the port of Dudinka on the Yenisey downstream from Igarka.

In central and eastern Siberia, the taiga extends southward beyond the Trans-Siberian Railroad. A vast but sparsely settled hinterland is served by a few cities spaced at wide intervals along the railroad: Krasnoyarsk (population: 868,600, city proper) on the Yenisey River, Irkutsk (population: 579,900, city proper) on the Angara tributary of the Yenisey near Lake

Baikal, and others. From east of Krasnoyarsk, a branch line of the Trans-Siberian leads eastward to the Lena River. Years of construction in the 1970s and 1980s under very difficult conditions resulted in continuation of this line to the Pacific under the name Baikal-Amur Mainline, or BAM (Fig. 7.13). The chief purposes of the railroad were to open important mineralized areas east of Lake Baikal and to lessen the strategic vulnerability caused by the close proximity of the Trans-Siberian Railroad to the Chinese frontier. The Pacific terminus is located on the Gulf of Tatary, which connects the Sea of Okhotsk with the Sea of Japan (see Fig. 6.1, p. 169; and Fig. 7.5).

The BAM railway passes through Bratsk (population: 259,100, city proper), site of a huge dam and hydroelectric

Figure 7.12 Laying an oil pipeline in Russia's Northern Lands. Immense quantities of steel have been required for thousands of miles of oil and gas pipes that have been laid in Russia and other parts of Russia and the Near Abroad since World War II. *Tass/Sovfoto*

power station on the Angara River. Lake Baikal, which the Angara drains, lies in a mountain-rimmed rift valley and is the deepest body of fresh water in the world at more than a mile deep in places. Lake Baikal is at the center of an environmental controversy over the estimated 8.8 million cubic feet of wastewater which pour into it daily from two large paper and pulp mills on its shores. The waste jeopardizes Baikal's unique natural history as the lake contains an estimated 1800 endemic plant and animal species (see Landscape in Literature, p. 207).

Siberia's development is based in large part on the hydroelectric plants located along the major rivers (Fig. 7.14). The largest hydropower stations are two on the Yenisey (at Krasnoyarsk and in the Sayan Mountains) and two on Angara, the

Yenisey's large tributary (at Bratsk and Ust Ilim). These power stations and dams are among the largest in the world. They supply inexpensive electricity, consumed mainly in mechanized industries that use power voraciously. These include cellulose plants that process Siberian timber and aluminum plants making use of aluminum-bearing material (alumina) derived from ores in Siberia, the Urals, and other areas and imported from abroad via the Black Sea.

Scattered throughout central and eastern Siberia are important gold-mining centers, mainly in the basins of the Kolyma and Aldan Rivers and in other areas in northeastern Siberia. (In Stalin's time, the Kolyma mines, located in one of the world's harshest climates, developed an evil reputation as death camps for great numbers of political prisoners forced to work there.) Large coal deposits are mined along the Trans-Siberian Railroad at various points between Irkutsk and the Kuznetsk Basin.

Aside from irregularly distributed centers of logging, mining, transportation, and industry peopled mostly by Slavs, the Northern Lands are home mainly to non-Slavic peoples who make a living by reindeer herding, trapping, fishing, and, in the more favored areas of the taiga, by cattle raising and precarious forms of cultivation (Fig. 7.15). The domesticated reindeer is especially valuable to the tundra peoples, providing meat, milk, hides for clothing and tents, and draft power. The Yakuts, a Turkic people living in the basin of the Lena River, are among the most prominent of the non-Slavic ethnic groups in the Northern Lands. Their political unit, Sakha (formerly Yakutia), is larger in area than any former Union Republic except the Russian Federation, within which it lies. Sakha has only about 1 million people inhabiting an area of 1.2 million square miles (3.1 million sq km). The capital city of Yakutsk (population: 193,900, city proper) on the middle Lena River has road connections with the BAM and Trans-Siberian railroads and with the Sea of Okhotsk. Riverboats on the broad and deep Lena provide a connection with the Northern Sea Route and the southern rail system during

Figure 7.13 Linkup of tracklayers from east and west on the Baikal-Amur (BAM) railroad, September 19, 1984. The two ends of the line met in the mountainous taiga between Lake Baikal and the Pacific. The laying of a symbolic "golden rail" marked the completion of ten years of work under difficult climatic and terrain conditions. The parallel to the driving of the "golden spike" near Logan, Utah, completing the first transcontinental rail line in the United States, is striking. *Novosti/Sovfoto*

Landscape in Literature

Baikal

Valentin Rasputin, a native and still a resident of Irkutsk in Siberia, is one of Russia's leading writers and most outspoken environmental activists. He is associated with a Russian literary tradition called "rural prose" or "village prose," which presents a realistic view of the often difficult lives of people in Russia's vast hinterland. In this passage, Rasputin argues that Lake Baikal (Fig. 7.B) is a spiritual resource—a marked contrast to the economic cornucopia that most Soviet (and now Russian) planners have perceived it to be.

"The sacred sea," "the sacred lake," "the sacred water"—that is what native inhabitants have called Baikal from the beginning of time. So have Russians, who had already arrived on its shores by the seventeenth century, as well as travelers from abroad, admiring its majestic, supernatural mystery and beauty. The reverence for Baikal held by uncivilized people and also by those considered enlightened for their time was equally complete and captivating, even though it touched mainly the mystical feelings in the one and the aesthetic and scientific impulses in the other. The sight of Baikal would dumbfound them every time because it did not fit their conceptions either of spirit or of matter: Baikal was located where something like that should have been impossible; it was not the sort of thing that should have been possible here or anywhere else, and it did not have the same effect on the soul that "indifferent" nature usually does. This was something uncommon, special and "wrought by God."

Baikal was measured and studied in due course, even, in recent years, with the aid of deep-sea instruments. It acquired definite dimensions and became subject to comparison, alternatively likened to Lake Tanganyika and to the Caspian Sea.

They've calculated that it holds one fifth of all fresh water on our planet, they've explained its origin, and they've conjectured as to how species of plants, animals and fish existing nowhere else could originate here and how species found only in other parts of the world many thousands of miles away managed to end up here. Baikal is not so simple that it could be deprived of its mystery and enigma that easily, but based on its physical properties it has, nevertheless, been assigned a fitting place alongside other great wonders that have already been discovered and described, as well it should. And it stands alongside them solely because Baikal itself, alive, majestic, and not created by human hands, not comparable to anything and not repeated anywhere, is aware of its own primordial place and its own life force.

Nature as the sole creator of everything still has its favorites, those for which it expends special effort in construction, to which it adds finishing touches with a spe-

cial zeal, and which it endows with special power. Baikal, without a doubt, is one of these. Not for nothing is it called the pearl of Siberia. We will not discuss its natural resources at present, for that is a separate issue. Baikal is renowned and sacred for a different reason—for its miraculous, life-giving force and for its spirit, which is a spirit not of olden times, of the past, as with many things today, but of the present, a spirit not subject to time and transformations, a spirit of age-old grandeur and power preserved intact, of irresistible ordeals and inborn will.

From Valentin Rasputin, "Baikal." In *Siberia on Fire: Stories and Essays by Valentin Rasputin.* DeKalb: Northern Illinois University Press, 1989: 188–189.

Figure 7.B Fishermen at work in Lake Baikal, Earth's deepest lake. *Tass/Sovfoto*

207

Figure 7.15 Yakut reindeer herders in the northeast Siberian region of Sakha. *Sovfoto/Eastfoto*

Figure 7.14 Siberia's hydroelectric power potential is vast. This river in extreme northeastern Siberia is one of the abundant northward-flowing drainages in the region. *Joseph J. Hobbs*

the warm season. Sakha will figure prominently in future mineral and timber development, with its large reserves of diamonds, natural gas, coking coal, other minerals, and timber.

7.6 The Future of the Russian Federation

Russia's internal republic of Sakha is the largest in area of many autonomies (nationality-based republics and lesser units) in the Russian Federation. They are successors to the autonomous ASSRs, autonomous oblasts, and autonomous okrugs of the Soviet era (see inset map, Fig. 6.1). Scattered autonomies are located in other independent successor states of the former USSR, but the majority are in the Russian Federation, where they occupy over two-fifths of the total area and represent somewhat under one-fifth of the total population.

Nearly half of the population in Russia's autonomies is made up of ethnic Russians, who actually compose a majority in many units. Employment in new industries and mining attracted many of these Russians during the Communist period. Titular nationalities of the autonomies are ethnically diverse

(see Fig. 7.2): Some, such as the Karelians, Mordvins, and Komi, are Finnic (related to the Finns and Magyars); others are Turkic (Tatars, Bashkirs, and Yakuts, for example), Mongol (Buryats near Lake Baikal, for example), and members of many other ethnic groups. Russian expansion brought these peoples into the Russian Empire at different times over a period of several centuries, and the Communist rulers of the USSR organized them into "autonomous" units ranked on a nationalities ladder with the former SSRs (Union Republics) at the top. Some of the lesser units eventually climbed the ladder to become Union Republics.

The largest areal units form a nearly solid band stretching across the northern part of the Northern Lands from Karelia, bordering Finland, to the Bering Strait. The Karelian, Komi, and Sakha (Yakut) republics are in this group. This band also includes several large but thinly inhabited units of aboriginal peoples that the Soviets designated as "autonomous okrugs." Pastoralism generally predominates over agriculture in the autonomies. Large cities are few; the only "million cities" are Kazan in Tatarstan, Ufa in Bashkiria, and Perm in Komi-Permyak. Manufacturing is poorly developed except in a handful of cities, such as Naberezhnyye Chelny in Tatarstan, where Russia's largest truck plant is located. However, minerals and mining are of major importance in some units. Oil and gas abound in the Volga-Urals fields in Tatarstan and Bashkiria, and there are high-grade coal deposits at Vorkuta in the Komi Republic. The diamonds mined in Sakha represent one-fourth of the world's production.

Such resources have great potential significance in the geopolitical realm. Many of the former Soviet ASSRs, recalling Moscow's long-standing promises of self-rule for them, have issued declarations of sovereignty asserting their right to greater self-direction of their internal affairs and greater control over their own resources. Moscow has relented to some of them. The Sakha (Yakut) Republic, for example, in 1992 won the right to keep 45 percent of hard currency earnings from foreign sales of Sakha diamonds, compared with only a small fraction in the Soviet era. In turn, Sakha must now pay for the government subsidies which Moscow previously paid to local industries. Without credits from Moscow, Sakha no longer pays taxes to Moscow. Salaries and other indices of living standards in the Sakha Republic have risen since this agreement was implemented.

Such regional demands for greater self-rule threaten the unity of the Russian Federation. Eighteen out of 20 of Russia's republics (the 16 Soviet-era ASSRs plus 4 autonomous regions upgraded to republic status) signed a 1992 Federation Treaty which grants them considerable autonomy. The treaty calls for the devolution of power centralized in Moscow, and for more cooperation between regional and federal governments. Each republic of the Russian Federation is legally entitled to have its own constitution, president, budget, tax laws and other legislation, and foreign and domestic economic partnerships. But many republics, and the oblasts and krays that have had their own elected governors since 1997, are complaining that Moscow is not honoring the treaty, and they are looking for regional solutions to their economic and other problems. Thus far, eight regional "economic associations" have emerged as a result (Fig. 7.16).

Chechnya (then joined with Ingushetia) and Tatarstan did not sign the 1992 Russian Federation Treaty and insisted on independence from Moscow. Oil-rich Tatarstan now has its own constitution, parliament, flag, and official language, and has

Figure 7.16 Russia's regional associations. *Source:* The Economist, *January 25, 1997, p. 47.*

since signed the treaty. Other autonomous regions will be watching to see what happens in Tatarstan and Chechnya. The oil-rich, Turkic-speaking Bashkirs and the Chuvash might push for more independence. This could begin a process that would virtually cut Russia in half. While some analysts argue that this process of devolution in Russia may actually bring more stability and prosperity to the country, the government in Moscow worries that what happened to the USSR might happen to Russia itself. The situation in Chechnya (population: 1.2 million) is of particular concern.

The Chechens, who are Sunni Muslims (see Chapter 8, p. 232), have periodically resisted Russian rule, and Moscow has punished them. Russian troops attempted but failed to seize control over Chechnya soon after its 1991 declaration of independence. At a cost of an estimated 30,000–80,000 lives, Russian troops succeeded in exerting physical control over most of Chechnya.

Russia's involvement in Chechnya was very unpopular among the Russian people. It revealed deep divisions in Russian political and military leadership and greatly weakened support for Boris Yeltsin. In August 1996, Chechen forces recaptured the capital city of Groznyy (population: 401,000 prior to the fighting in 1994–1995), renaming it Jokhar-Gala. The prospect of further Russian losses led President Yeltsin's security chief, Alexander Lebed, to negotiate a peace agreement with the Chechens in September. The pact required an immediate withdrawal of Russian forces from Chechnya and deferred the question of Chechnya's permanent political status for five years— beyond the expiration of Yeltsin's second term in office. Before the end of the five-year period, the Chechen people are to hold a referendum on the republic's future. In 1997 they elected a president (Aslan Maskhadov) they believe will be the father of their nation. Later that year Maskhadov and President Yeltsin signed a formal peace treaty, which normalized relations between Russia and Chechnya but left the future uncertain. Moscow continues to insist that Chechnya is part of the Russian Federation, and Chechens continue to demand complete independence for the country they would call Ichkeria. The future of Chechnya and the Russian Federation remain uncertain. Thus far, none of the autonomies that have declared independence from Russia has officially secured it. Moscow and the world watch nervously.

7.7 The Caucasus

The far southern Caucasus region borders Russia between the Black and Caspian Seas. It includes the rugged Caucasus Mountains, a fringe of foothills and level steppes to the north, and the area to the south known as Transcaucasia. The Greater Caucasus Range forms an almost solid wall from the Black Sea to the Caspian (Fig. 7.17). It is similar in age and character to the Alps but is much higher, with the highest point in Europe at

Figure 7.17   Mt. Kazbek (15,812 ft, 5033 m) in the Greater Caucasus Range towers over a small settlement in northern Georgia.   *Sovfoto*

Mt. Elbrus (18,510 feet, 5642 m), on the Russia/Georgia border. In southern Transcaucasia is the mountainous, volcanic Armenian Plateau. Between the Greater Caucasus Range and the Armenian Plateau are subtropical valleys and coastal plains where the majority of people in Transcaucasia live. A humid subtropical climate prevails in coastal lowlands and valleys of Georgia, south of the high Caucasus Mountains. Mild winters and warm to hot summers combine with the heaviest precipitation to be found anywhere in the region of Russia and the Near Abroad— 50 to 100 inches (*c.* 125–250 cm) a year. The lowland area bordering the Black Sea in Georgia is the most densely populated part of Transcaucasia. Such specialty crops as tea, tung oil, tobacco, silk, and wine grapes, together with some citrus fruits, grow there. There is too much freezing weather for the citrus industry to flourish; "subtropical" areas in Transcaucasia have many frosts. The lowlands of eastern Transcaucasia, bordering the Caspian Sea, receive so little precipitation in most places that they are classified as subtropical steppe.

Russians and Ukrainians predominate in the North Caucasus, but non-Slavic groups form a large majority in Transcaucasia. The Caucasian isthmus between the Black and Caspian Seas has been an important north-south passageway for thousands of years, and the population includes many different peoples who have migrated into this region at various times. Dozens of nationalities can be distinguished, with most of these small cultures confined to mountain areas that became their refuges in past times. The republic of Dagestan, within Russia on its southeastern border with Georgia and Azerbaijan, is home to 34 different ethnic groups. Some mountain villages, particularly in southeastern Azerbaijan, have achieved international fame as the abodes of the world's longest-lived people— some up to 168 years! In addition to Russians and Ukrainians, most of whom live north of the Greater Caucasus Range, the most important nationalities are the Georgians (Fig. 7.18), Armenians, and Azerbaijanis, each represented by an independent country. Throughout history, all have stubbornly maintained their cultures in the face of pressure by stronger intruders.

These nationalities have ethnic characteristics and cultural traditions that are primarily Asian and Mediterranean in origin. Their religions differ. The Azerbaijanis (also known as Azeri Turks) are Muslims. The Georgians belong to one of the Eastern Orthodox churches. The Armenian Apostolic Church is a very ancient, independent Christian body that is an offshoot of Eastern Orthodoxy; Armenia became the world's first Christian country in 301 A.D. (Fig. 7.19). There has been a history of animosity between the Armenians and Azeri Turks (including a war between them in 1905), growing out of the Turks' persecution of Armenians in the Ottoman Empire prior to and during World War I. Turkey and Azerbaijan still have not accepted responsibility for their roles in the Armenian Genocide, in which an estimated 1.5 million Armenians died between 1915 and 1918. The descendants of Armenians who fled the genocide have established themselves in many different nodes around the world, and today outnumber Armenians living within the country.

Early in the 1990s, historical animosity flared into massive violence between Armenians and Azeris over the question of Nagorno-Karabakh, a predominantly Armenian enclave within Azerbaijan, governed by Azerbaijan, but claimed by Armenia. From 1992 until 1994, when Armenian forces finally secured the region, fighting over Nagorno-Karabakh took thousands of lives and created nearly a million refugees. Large numbers of Armenians fled from Azerbaijan to Armenia as refugees. There was also a refugee flight of Azeris from Armenia to Azerbaijan. Tens of thousands of Azeris also fled into Iran. Iranian Azeris, who make up a third of Iran's population, feel that Iran has leaned too far toward Armenia as a de facto ally against Turkish influence in the Caucasus and Central Asia.

Russia negotiated a cease-fire between Armenia and Azerbaijan in March 1994, but the countries are still technically at war. Armenians are angry that Moscow did not restore to Armenia the enclave of Nagorno-Karabakh, which the Bolsheviks had put under Azeri control. But the Armenians also embrace Russia as a strategic ally against their historic enemy, the Turks, who virtually surround them.

In Georgia, there also have been recent episodes of ethnic and political violence, particularly between the Georgians and the primarily Muslim South Ossetians, who would like to free themselves from Georgian control and establish an Ossetian nation. The North Ossetians, whom they wish to join, live adjacent to them in Russia.

Another Muslim people, the Abkhazians, who make up 1.8 percent of Georgia's population and are concentrated in the

Figure 7.18 Ethnic Georgians selling produce in a market at Tbilisi, the main city and capital of Georgia. *Katrinka Ebbe*

province known as Abkhazia, also seek freedom from the Georgians. In 1993, Abkhazian separatists captured the Georgian Black Sea port of Sukhumi, in Abkhazia. Russian forces initially aided them, both to regain access to Black Sea resorts and to take revenge on Georgian president Eduard Shevardnadze, the former Soviet foreign minister whom many Russians hold partly responsible for the breakup of the USSR. Russia was also putting pressure on Georgia to rejoin the CIS. Shevardnadze agreed in October 1993 that Georgia would join the CIS, and agreed to allow Russian bases on Georgian soil and Russian troops on the border with Turkey. Russian forces then put a stop to the Abkhazian offensive, but did not drive the rebels from the territory they had captured. Georgians see Russia as trying to use the Abzkakian problem as a means of retaining leverage in

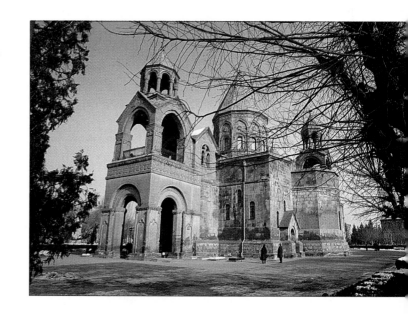

Figure 7.19 The Holy Echmiadzin Cathedral in Echmiadzin, Armenia. With foundations dating to 301 A.D., this is reputed to be the oldest Christian church in the world. *Bruce Brander/Photo Researchers, Inc.*

Regional Perspective

Oil in the Caspian Basin

Estimated oil reserves beneath and adjacent to the Caspian Sea total 200 billion barrels, or about one-tenth of the world's total. Reserves in Azerbaijan, Kazakstan, and Turkmenistan represent the world's third most important oil region, behind the Persian/Arabian Gulf and Siberia. The potential wealth this resource represents will be tapped only through a difficult and delicate resolution of several geographical and political problems

(Fig. 7.C). One of the first to be dealt with was the question of which countries owned the fossil fuels lying under the Caspian waters. With relatively small reserves of oil lying adjacent to their Caspian shorelines, Iran and for awhile Russia argued that the lakebed resources should be treated as common property, with shares divided equally among the five surrounding states. Eventually Russia conceded that the basin should have

five separate sovereign territorial waters, with each country having exclusive access to the reserves in its domain. Iran unhappily accepted this outcome, which gives Azerbaijan the largest share of Caspian oil.

Azerbaijan's oil situation epitomizes the economic prospects and obstacles facing the southern countries of the region of Russia and the Near Abroad. The Azeri resource is large; there are huge re-

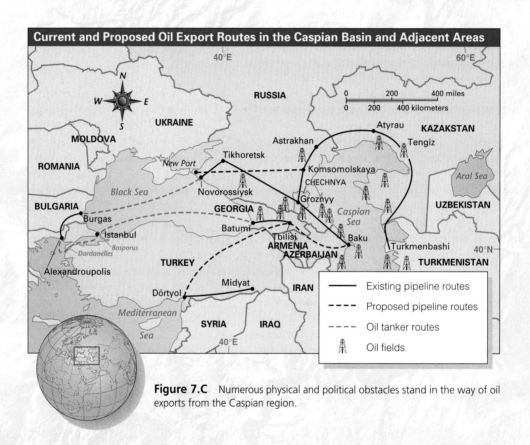

Figure 7.C Numerous physical and political obstacles stand in the way of oil exports from the Caspian region.

the strategically vital Caucasus region. For its part, Georgia wants to loosen ties with Russia in favor of better relations with the West and its allies; most road and rail links between the countries have been severed, and Turkey has already replaced Russia as Georgia's main trading partner. Georgia also wants to reestablish its historically important position in overland trade with the Central Asian countries, particularly by serving as an

outlet for the export of oil from the Caspian basin (see Regional Perspective above).

The Caspian Sea, which is actually the world's largest freshwater lake in area, is an important fishery, especially for its sturgeon, which supply over 90 percent of the world's caviar (sturgeon eggs). When only the USSR and Iran bordered the Caspian, this resource was managed sustainably. Now, how-

serves of oil and gas under the bed of the Caspian Sea. Planners study Azerbaijan's oil with a difficult question in mind: What is the best way to get this oil to market? They are considering a variety of export pipeline routes, each of which has a different set of geopolitical, technical, and ecological problems. Since the Caspian is essentially a large lake, Azerbaijan is effectively landlocked, and Azeri oil must cross the territories of other nations to reach market. These territories, however, are embroiled in wars and political unrest left in the wake of the collapse of the USSR. Russia wants to retain strong influence over Azerbaijan in part by insisting that Azeri oil reach export terminals on the Black Sea by passing through Russian territory, as it now does, through a pipeline to the Russian port of Novorossiysk. Turkey opposes this strongly, on the grounds that oil-laden ships navigating the narrow Bosporus and Dardanelles straits imperil Turkey's environment. In response, Iran has offered a swap with Azerbaijan: Iran would import Azeri oil for Iran's domestic needs and would export equivalent amounts of its own oil from established ports on the Persian Gulf. Iran also argues that its territory offers the shortest and safest pipeline route for fossil fuel exports from Turkmenistan, Kazakstan, and Azerbaijan. (Iran is also in an excellent geographical position to facilitate transportation of other goods to and from the Central Asian nations, using its Gulf of Oman and Persian/Arabian Gulf ports. To stimulate this trade, Turkmenistan and Iran began linking their rail systems in 1996.)

Azerbaijan's leaders, however, want more ties with the West (which is generally hostile to Iran) and independence from Moscow. The United States and its Western allies anticipate that by about 2005 Azerbaijan and Kazakstan will become major counterweights to the volatile oil-producing states of the Middle East, and they want to be able to obtain Caspian Sea oil without relying on Russia's or Iran's goodwill. Azerbaijan, therefore, in 1995 signed an agreement with a U.S.-led Western oil consortium to export oil through a combination of new and old pipelines to the Georgian port of Batumi on the Black Sea. To avoid a direct confrontation with Russia over this sensitive issue, some of the initial production passed through the existing Russian pipeline to Novorossiysk.

This compromise applied only to the first stage of production, through 1997. As this book went to press, a permanent route had yet to be established, with both Turkey and Russia vying for the pipeline to pass through their territories. Russia insists that the environmental hazard of navigating the Turkish straits can be averted by offloading oil at the Bulgarian Black Sea port of Burgas, shipping it overland through a new pipeline to the Greek Aegean Sea port of Alexandroupolis, and reloading it there. The Turkish route would be far from the straits, skirting hostile Armenia to run across Anatolia to the Mediterranean port of Ceyhan. The Western powers prefer this route. So do the oil-producing countries, if only because the Russian route would lead through the breakaway republic of Chechnya—perhaps exposing the pipeline to a greater risk of sabotage. Proponents of the Turkish route also have security risks in mind: the pipeline might have to take a detour to skirt the volatile Kurdish-dominated region of southeastern Turkey.

Other Caspian Basin countries face similar questions about how best to export their resources. Turkmenistan, on the lake's southeast shore, has large oil deposits and the fourth-largest natural gas reserves in the world. In 1997, Turkmenistan began exporting natural gas through a pipeline into northern Iran. The country's leadership, however, is in favor of the chief export route promoted by the United States, with oil and gas travelling westward through pipelines on the bed of the Caspian Sea to Baku, then through Georgian and Turkish pipelines to the Mediterranean Sea. Turkmenistan is also considering the construction of a pipeline for natural gas export from eastern Turkmenistan through Afghanistan to Pakistan.

Of the five Central Asian countries, Kazakstan has the largest fossil fuel endowment. In the late 1990s it began exporting its oil from fields around Tengiz to join the Baku-to-Novorossiysk pipeline at Komsomolskaya. The Kazaks are working with Omani, Russian, and American firms to complete a pipeline bypass from Komsomolskaya to a new port facility that would be built just north of Novorossiysk. Kazakstan is also studying the possibility of building a pipeline to export oil eastward into China.

ever, the recently independent countries bordering the Caspian are exerting tremendous pressure on this resource in an attempt to boost their market shares in competition with Russia and Iran. The result has been a rapid drop in the sturgeon population and soaring prices of caviar on the world market.

Oil and gas are the most valuable minerals of the Caucasus region. The main oil fields are along and underneath the Caspian Sea around Azerbaijan's capital, Baku (population: 1.15 million, city proper; 2.02 million, metropolitan area), and north of the mountains at Groznyy in Chechnya. Dry and windswept Baku, the largest city of the Caucasus region, was the leading center of petroleum production in Old Russia and, at the dawn of the 20th century, was the world's leading center of oil production. Under the Soviets it continued its leadership until it was decisively surpassed in the 1950s by the Volga-Urals fields and again in the 1970s by the West Siberian fields. However, petroleum development is now reestablishing the Caucasus as a vital world region.

Diverse manufacturing industries have developed in Transcaucasia on a small scale. Most factories are located in and near the political capitals: Baku; Tbilisi (population: 1.27 million, city proper; 1.46 million, metropolitan area) in Georgia; and Yerevan (population: 1.20 million, city proper) in Armenia. Oil-refining and petrochemical industries are significant in the areas that produce oil and gas. Despite these industries, there are important tourist resorts along the Black Sea coast and in the mountains of Georgia.

7.8 The Central Asian States

Across the Caspian Sea from the Caucasus region lie large deserts and dry grasslands, bounded on the south and east by high mountains. Like Transcaucasia, this Central Asian region has a variety of non-Slavic nationalities who outnumber Slavs in all the countries. Those with the largest populations are four Muslim peoples speaking closely related Turkic languages—the Uzbeks, Kazaks, Kyrgyz, and Turkmen—and a Muslim people of Iranian origins, the Tajiks (Fig. 7.20). Each has its own independent nation (see Table 6.1).

Division and Conquest

The peoples of this region are heirs of ancient oasis civilizations and of the diverse cultures of nomadic peoples. Small-scale social and political units such as clans, tribes, and petty autocracies were associated with particular oases. Larger empires controlled some of these at various times. Conquerors have repeatedly possessed this area, which lies on the famed Silk Road (see p. 344), an ancient route across Asia from China to the Mediterranean. Among them were Alexander the Great, the Arabs who brought the Muslim faith in the 8th century, the Mongols, and the Turks. By the time of the Russian conquests in the 19th century, Turkestan (as it was then known) was contested between feuding, tradition-bound Muslim khanates with political ties to China, Persia, Ottoman Turkey, and British India. Cultural and political fragmentation continued up to the Russian Revolution and into the period of civil war.

Nationalist identities and elites existed in Central Asia, but nations did not exist as large communities bound together by common interests and allegiance. During World War I, Central Asian peoples revolted against the tsarist government, which had begun to draft the Muslims for menial labor at the front. This generated widespread disorder and bloodshed until the Bolshevik government gained control in the early 1920s and organized the area into nationality-based units. The Central

Figure 7.20   The elders of this community in Tajikistan are about to partake in the traditional meal celebrating Novruz, the Muslim New Year. *Sovfoto*

Asian "nationalities" are themselves to a large degree artificial creations of the Bolsheviks.

Most of the 20th century saw a large incursion of Russians and Ukrainians into Central Asia, mainly into the cities. They included political dissidents banished by the Communists, administrative and managerial personnel, engineers, technicians, factory workers, and, in the north of Kazakstan, a sizable number of farmers in the "new lands" wheat region. Most of the Slavic newcomers live apart from the local Muslims. With the breakup of the Russian-dominated Soviet Union, their destiny is uncertain.

Kazakstan provides a good example of the dilemma facing ethnic Russians in Central Asia. At the time it became independent in 1991, Kazakstan's population of 17 million was roughly 40 percent Kazak and 39 percent Russian, with the rest a mix of nationalities. Many ethnic Kazaks, who believe they were treated as second-class citizens until independence, have been seeking to rectify their status. They are increasingly predominant in government and business. Kazak is now the official language and Russian is the language of "interethnic communication." By the end of 1995, the non-Kazak residents were told they had to decide if they wanted Kazak citizenship. With rising ethnic tensions between 1991 and 1998, about one million Russians, along with a half million ethnic Germans, emigrated from Kazakstan, and the exodus is continuing.

Environment and Agriculture

The five Central Asian states are composed predominantly of plains and low uplands, except for Tajikistan and Kyrgyzstan, which are extremely mountainous and contain the highest summits in the region of Russia and the Near Abroad, in the Pamir and Tien Shan ranges (see Fig. 6.6). The Kazak Upland in east

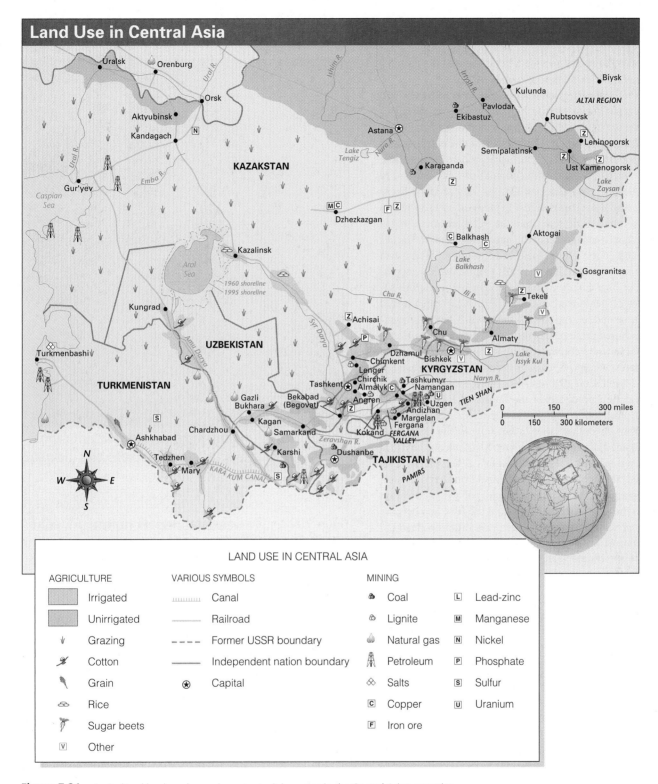

Land Use in Central Asia

LAND USE IN CENTRAL ASIA

AGRICULTURE
- Irrigated
- Unirrigated
- Grazing
- Cotton
- Grain
- Rice
- Sugar beets
- [V] Other

VARIOUS SYMBOLS
- Canal
- Railroad
- Former USSR boundary
- Independent nation boundary
- Capital

MINING
- Coal
- Lignite
- Natural gas
- Petroleum
- Salts
- [C] Copper
- [F] Iron ore
- [L] Lead-zinc
- [M] Manganese
- [N] Nickel
- [P] Phosphate
- [S] Sulfur
- [U] Uranium

Figure 7.21 Agricultural lands and some important mining areas in the Central Asian countries.

central Kazakhstan is a hilly area with occasional ranges of low mountains. Central Asia is almost entirely a region of interior drainage. Only the waters of the Irtysh, a tributary of the Ob, reach the ocean; all the other streams either drain to enclosed lakes and seas or gradually lose water and disappear in the Central Asian deserts.

Most people live in irrigated valleys at the base of the southern mountains (Fig. 7.21). There, most soils (many formed of loess) are fertile, the growing season is long, and rivers flowing from mountains provide irrigation water. The principal rivers in the heart of the region are the Amu Darya (Oxus) and Syr Darya, both of which empty into the enclosed Aral Sea. Their

215

tributaries and other rivers provide a large share of the region's irrigation waters, which numerous canals carry.

Most of the larger irrigated districts are in Uzbekistan. Especially important are the fertile Fergana Valley on the upper Syr Darya, almost enclosed by high mountains, and the oases around Tashkent (capital of Uzbekistan and Central Asia's largest city; population: 2.09 million, city proper; 2.33 million, metropolitan area), Samarkand (population: 366,000, city proper), and Bukhara (population: 228,000, city proper). These old cities became powerful political centers on caravan routes connecting southwestern Asia and the Mediterranean basin with eastern Asia. Samarkand is the successor to many earlier cities that occupied the site on loess hills beside the Zeravshan River. It was the capital of Tamerlane's empire, which encompassed much of Asia and, like Bukhara today, has numerous superb examples of Islamic architecture (Fig. 7.22). Streams fed by melting snow and ice in some of the world's highest mountains have sustained these places through their long histories.

Cotton began to be cultivated here during the Civil War in the United States, when U.S. exports abroad virtually ceased. It remains the region's major crop. This area grows over nine-tenths of the cotton produced in the region of Russia and the Near Abroad. Production supplies textile mills in the Moscow region, Tashkent, and other cities in far-flung locations, including many outside the region. However, a significant environmental price was paid as a result of cotton's development here. Driven by the need for foreign exchange from export sales, Communist planners pushed an expansion of irrigated cotton production to the point that the rivers and the Aral Sea became highly polluted with agricultural chemicals draining from the fields. Diversion of water from the rivers caused the Aral Sea to shrink rapidly in volume and area, virtually destroying a vast natural ecosystem and an important regional fishery (Figs. 7.21 and 7.23).

Mulberry trees to feed silkworms are grown around the margins of many irrigated fields and along irrigation canals in the oases. Irrigated rice and other grains, sugar beets, vegetables, vineyards, and orchards of temperate fruits are also important components of this region's production. Commercial orchard farming is prominent around Almaty ("City of Apples"; population: 1.17 million, city proper; 1.19 million, metropolitan area), the former capital of Kazakstan.

Most of Central Asia outside the oases is too dry for cultivation. However, ethnic Russians and Ukrainians practice non-irrigated grain farming in northern Kazakstan, which extends into the black-earth belt. Huge acreages, primarily on chestnut soils, were planted in wheat under the "virgin and idle lands" scheme of 1954–1956 (see Chapter 6, p. 183).

Many of the region's people previously were pastoral nomads who grazed their herds on the natural forage of the steppes and in mountain pastures in the Altai and Tien Shan ranges. Over the centuries, there was a slow drift away from nomadism. The Soviet government accelerated this process by collectivizing the remaining nomads—often by harsh measures

Figure 7.22 A 16th-century madrasa (Islamic school) on the Registan Square in the heart of the storied ancient city of Samarkand, Uzbekistan. *Karen Sherlock/Tony Stone Images*

in the face of strong resistance—and settling them in permanent villages. Livestock raising became a form of ranching in the process, but herdsmen must still accompany the grazing animals from one range to another. Sheep and cattle are the principal livestock animals, followed by goats, horses, donkeys, and camels.

Forces of Modernization

In recent decades, parts of Central Asia have become increasingly important in the production of minerals, particularly natural gas, coal, oil, iron ore, nonferrous metals, and ferroalloys. Kazakstan (population: 15.6 million) has the greatest natural resource wealth, and political stability under a nominally democratic government since independence has made it the largest target for Western investment in Central Asia. Large reserves of bituminous coal are mined at Karaganda (population: 596,000) in central Kazakstan. The Karaganda coal basin has become a very important producer, and Karaganda itself has experienced spectacular growth. Much of the coal is shipped to metallurgical works in the Urals.

Despite Kazakstan's coal wealth, equipment and management problems have resulted in declining production in the country since the breakup of the Soviet Union. In 1996, the country's new capital was chosen to be established at Astana (formerly Akmola and Tselinograd; population: 287,000), northwest of Karaganda, as part of a larger plan to redevelop Kazakstan's industrial heartland.

Coal from Karaganda and other Central Asian locations, as well as from the Kuznetsk Basin, provides fuel for large-scale metallurgical production in the region. Of primary importance is the smelting of nonferrous metals, especially copper, lead, and zinc. In the mid-1990s, prospecting began in Kazakstan in what may be the world's third largest gold deposit. This rich and diversified mineralized region of Central Asia also holds impressive reserves of chrome and nickel in the Ural foothills of northern Kazakstan, natural gas near the Amu Darya, petroleum in fields bordering the Caspian Sea, and large iron ore deposits in northern Kazakstan.

Turkmenistan (population: 4.7 million; capital, Ashkhabad —population, 407,000) still has a Communist government. Ethnic Russians comprise about 10 percent of the population and Uzbeks another 10 percent. Natural gas exports by pipeline to western Europe are the country's main source of hard currency.

Uzbekistan, Central Asia's most populous country (24.1 million), does not have access to the Caspian oil fields. The country is largely agricultural, with cotton exports the most important source of hard currency. Like other LDCs, Uzbekistan faces the challenge of transforming itself from an exporter of raw materials to an exporter of manufactured goods. The country is also trying to reverse the ecological damage done by years of Soviet emphasis on cotton production heedless of the damages done by pesticides and overirrigation. Since becoming independent, Uzbekistan has dramatically cut the area planted in cotton and increased production of food crops.

Kyrgyzstan (population: 4.7 million), with its capital at Bishkek (population: 583,900, city proper), is, like the other Central Asian countries, the successor to a republic which Stalin carved artificially out of Russian Turkestan. Searching for identity in the post-Soviet period, the people and government of Kyrgyzstan have resurrected the oral epics which tell the story of Manas, a great warrior who, a thousand years ago, fought off the enemies of the Kyrgyz nomadic tribes and unified the tribes for the first time. As it restores these ancient roots, Kyrgyzstan is modernizing. Most of the Soviet-era state-owned businesses are now privatized, and the country has its own stock exchange.

Central Asia and the Wider World

Central Asia's future is of outstanding international importance. In an attempt to create a common market and integrate their economies more effectively, the five countries formed a regional organization called the "United States of Central Asia"

Figure 7.23 Rusting vessels on former sea bottom testify to the shrinkage of Central Asia's Aral Sea. The major cause of the catastrophic water loss is diversion of the Amu Darya and Syr Darya for irrigation of cotton. *G. Pinkhassov/Magnum*

in 1993. Although primarily intended to meet regional economic needs, the group may become preoccupied with Central Asia's external relations. The region borders post-revolutionary Iran, war-torn Afghanistan, and the great power of China. Central Asia's population of over 50 million is growing rapidly. Many kinship relations exist between the Central Asian Muslims and their neighbors on the other side of the international frontiers. Along with internal ethnic conflicts and dissatisfaction over living conditions, these international affiliations may make the region a major political problem area.

Both Iran and Turkey are vying for increased influence in the region. Support for Iran is largely lacking, in part because the majority of Central Asians practice Sunni rather than Shiite Islam. Ethnicity is also important; the four Turkic states are inclined to orient with Turkey, with which they have already established cultural ties such as educational exchanges and shared media. Turkey also has extended economic assistance and established many small business ventures and large construction contracts in Central Asia. Turkmenistan has good relations with both Turkey and Iran and has recently completed a rail link with Iran. Turkmenistan also tries to maintain good ties with Russia.

The region is also becoming a focus for rivalry between the historical enemies Turkey and Russia. Russians are now concerned that Turkey is winning too much influence in the new Muslim states of Central Asia. For their part, many Turks believe Russia will try to take over the Transcaucasus region again and then turn to Central Asia. Some Turks uphold a dream of Pan-Turkism, uniting all the Turkic peoples of Asia from Istanbul to the Sakha Republic in Russia's Siberia region—a prospect that Moscow does not like.

Kazakstan is developing strong ties with the West, and, like Ukraine and Belarus, has agreed to eliminate its nuclear weapons. In return, Kazakstan wants the West to help guarantee the security of its borders with Russia, China, the Caspian Sea, and three other Central Asian nations. The country also favors close coordination with Russia, and in 1996 joined Russia and Kyrgyzstan in an agreement calling for the creation of a common market to ensure the free flow of goods, services, and economic capital between them and to promote coordinated industrial and agricultural policies.

International concern about the region's stability focuses on Tajikistan (population: 6.1 million; capital, Dushanbe—population: 528,600). Russian diplomats have a "domino theory" about the region stemming from a fear that Islamic fundamentalism might spread from Tajikistan to neighboring Central Asian countries. Tajikistan is still a Communist country, but its government has the backing of Washington and Moscow because they see the alternative as a radical Islamic government. At issue is whether or not, with support from Afghan guerrillas and Middle Eastern activists, Tajikistan's Islamists can gain control of the country and spread their influence into neighboring states like resource-rich Kazakstan. After the country gained independence, a coalition of prodemocracy and Islamic rebels, comprised largely of members of the Garm and Badakhshan ethnic groups, who have little economic and political power in the country, lost a bloody civil war against the Russian-backed government. More than 50,000 people died in the fighting, and an estimated 20 percent of the country's population fled, mostly into neighboring Afghanistan. In 1997, as part of a United Nations–brokered peace agreement between the rebels and the government, a former rebel leader was installed as Tajikistan's deputy premier.

Russia remained skeptical of Tajikistan's truce with its Islamist opposition, and in 1999 still had 25,000 troops in Tajikistan to back the government. Russia wants Tajikistan to be a buffer so that political violence does not spread into neighboring former republics on Russia's borders. President Yeltsin declared that Tajikistan's borders "are effectively Russia's"— its main barrier against the infiltration from Afghanistan of Islamic militancy, revolution, guns, and drugs. Despite Tajikistan's formal independence, it is now in fact a "client state" of Russia. Moscow currently funds 70 percent of the Tajik national budget. Ethnic Russians make up 12 percent of the country's population and Uzbeks another 23 percent. Secular Turkey also fears the specter of a rising tide of fundamentalism originating in Tajikistan and neighboring Afghanistan, as forces of both tradition and change advance into Central Asia from the heart of the Middle East.

SUMMARY WITH SELECTED KEY TERMS

- Nearly three-fourths of the people, and an even larger share of the cities, industries, and cultivated land of the region, are packed into a triangular **Fertile Triangle** composing one-fifth of the total area. This triangle is also called the **Agricultural Triangle** and the **Slavic Coreland.** The rest lies mostly in Asia and consists of land where settlement is spotty and handicapped by the environment.

- There are two main agricultural zones within the Fertile Triangle: a **black-soil zone** in the southern steppes, and a **non-black-soil zone** corresponding roughly to the region of mixed forest. The black-soil zone is the principal producer of wheat and the non-black-soil zone has traditionally produced rye in its cooler and poorer soils.

- Situated partly in the forest zone and partly in the steppe, **Ukraine** translates as "at the border" or "borderland." Today, Ukraine is one of the most densely populated and productive areas of the region. Ukraine's industrial and agricultural assets have always been vital to this region.

- Under Soviet rule, **Belarus** became an industrial power, shipping commodities to both Moscow and its eastern European satellites. Belarus' trade is now almost exclusively with Russia, much of it in the form of barter. In 1997, Belarus and Russia formed a **"union state"** to more closely link these predominantly Slavic countries.

- **Moldova** is made up of territory that the Soviet Union took from Romania in 1940. A culturally diverse nation, Moldova has serious ethnic and political problems within. While industry is improving, agriculture remains the economic base.

- The **Baltic States** share many traits with the other nations of the Fertile Triangle. However, the indigenous people of the Baltics are not **Slavs,** and they have long tried to safeguard their identities against Russian encroachment.

- In an effort to reverse the inefficiencies left by the state-run system, all of the countries are promoting land reform by privatizing collective and state farms and developing more independent or **"peasant"** farming.

- The production of industrial commodities is not evenly spread across the region. This has made continued cooperation among now-independent states essential.

- The industrialized area around Moscow is known as the **Central Industrial Region,** Old Industrial Region, or Moscow-Tula-Nizhniy Novgorod Region. It has a **central location** within the populous western plains and is functionally the most important area of Russia and the Near Abroad.

- The **Ural Mountains** and the **Kuznetsk Basin** contain an extraordinarily varied collection of valuable minerals. This led to these areas combining into an industrial region, well removed from the exposed western frontier of the Soviet Union, called the **Urals-Kuznetsk Combine.**

- Large-scale exploitation of petroleum in the **Volga-Urals** fields has contributed to the industrial rise of the **Volga cities.** Prior to the opening of the fields in **western Siberia,** the Soviet Union's most important area of oil production was the Volga-Urals fields.

- Most of the **Russian Far East** is a thinly populated wilderness in which the only settlements are fishing ports, lumber and mining camps, and villages of aboriginal peoples. Several small- to medium-sized cities form a north-south line along the **Trans-Siber-**

ian Railroad and the lower **Amur River,** the two main transportation arteries.
- **The Northern Lands** comprise one of the world's most sparsely populated areas. Here lie enormous stretches of coniferous forest (**taiga**) and **tundra** extending from the Finnish and Norwegian borders to the Pacific.
- **Siberia's** development is based in large part on **hydroelectric plants** located on the major rivers. These power stations and dams are among the largest in the world. They supply inexpensive electricity, consumed mainly in mechanized industries that use power voraciously.
- The far southern **Caucasus** region borders Russia between the Black and Caspian Seas. It includes the Caucasus Mountains, a fringe of foothills and level steppes to the north, and the area south known as **Transcaucasia.** Russians and Ukrainians are the majority north

of the Greater Caucasus Range, while south the important nationalities are the **Georgians, Armenians,** and **Azerbaijanis,** each represented by an independent country and stubbornly maintaining their unique cultures.
- **The Central Asian states** lie immediately east and north of the Caspian Sea and are populated in the majority by a variety of non-Slavic peoples. Those with the largest populations are the **Uzbeks, Kazaks, Kyrgyz, and Turkmen**—all speaking closely related Turkic languages—and a people of Iranian origins, the **Tajiks.** Most of these people live in irrigated valleys at the base of the southern mountains which lie on the path of the famed **Silk Road,** an ancient route from China to the Mediterranean.

REVIEW QUESTIONS

1. Using maps and the text, locate the Fertile Triangle.
2. Using maps and the text, locate the black-soil and the non-black-soil zones.
3. Using maps and the text, locate the major industrial centers and list the major industries of each.
4. List the five major seas that are connected to Moscow by rivers and canals.
5. List the nations of the Caucasus and briefly describe their ethnic compositions.

6. List some of the outside forces that have occupied the Central Asian states.
7. Briefly describe the physical aspects of the Caucasus.
8. Briefly describe the physical aspects of the Central Asian states.
9. Using maps and the text, locate the taiga and the tundra and briefly describe each.
10. Briefly describe the physical aspects of the Far East.
11. Name and locate the two important transportation arteries of the Far East.

DISCUSSION QUESTIONS

1. Discuss the significance of the Fertile Triangle in this region.
2. Examine the attributes of the Russian Far East. Why are the Trans-Siberian Railroad and the Amur River so important here?
3. What two distinct climate and vegetative zones dominate the Northern Lands? What are the main attributes of the populace in the Northern Lands?
4. Examine and discuss some of the difficulties—present, past, and future—for the Russian Federation.

5. Describe the physical aspects, the population, and the economy of the Caucasus.
6. Why has the area of the Central Asian states been subject to such repeated division and conquest? What present regional and world political roles do these states play?
7. List the major centers of industry and their products.
8. Why is cooperation among the major centers of industry so important in this region?

4 The Middle East

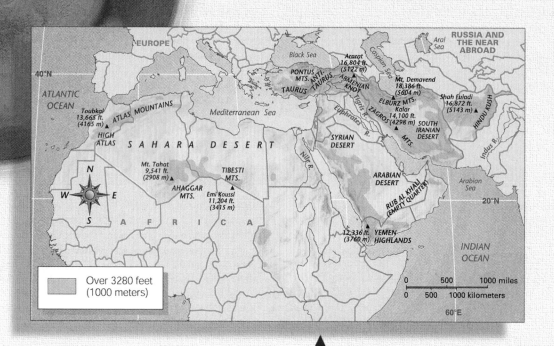

The Middle East is a pivotal global crossroads linking Asia, Europe, Africa, the Mediterranean Sea, and the Indian Ocean.

Many of the world's earliest and most accomplished civilizations emerged in the Middle East. The three great monotheistic faiths of Judaism, Christianity, and Islam were born here. *Mike Barlow*
©1992 / Dembinsky Photo Associates

Generally the region is arid, but great river systems and freshwater aquifers sustain large human populations.
Ray Ellis / Photo Researchers, Inc.

Regional Snapshot

Wars exacted a costly toll on the Middle East in the 20th century. Israel and its Arab neighbors are engaged in an unprecedented search for lasting peace.
Earth Observation Satellite Company, Lanham, MD / Science Photo Library / Photo Researchers, Inc.

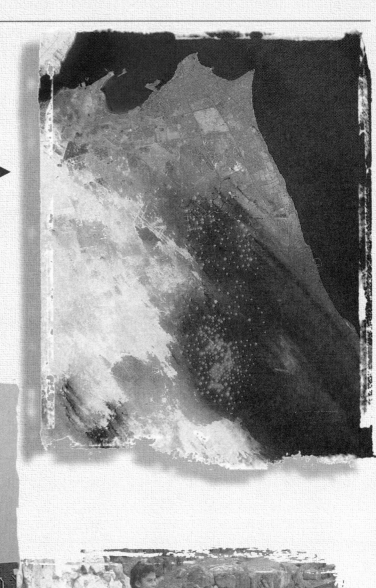

Regional and global powers have frequently contested for—and controlled strategically and economically—important areas of the Middle East. More than two-thirds of the world's proven oil reserves are in the Middle East.
Paul Lowe–Network / Matrix

Middle Easterners include Jews, Arabs, Turks, Persians, Berbers, and other ethnic groups who practice a wide variety of ancient and modern livelihoods.
Joseph J. Hobbs

Chapter 8

A Geographic Profile of the Middle East

▲ *A photograph looking northeastward from the U.S. Space Shuttle* Atlantis *in April 1991. In the left foreground is the Nile Valley. The Gulf of Suez and Gulf of Aqaba frame the triangle of Egypt's Sinai Peninsula. The political boundary between Egypt and Israel is clearly visible; agricultural development on the Israeli side gives Israel a shade of green. The rift valley follows the Gulf of Aqaba northward to the Dead Sea and beyond, up the Jordan River.* NASA/Mark Martin/Photo Researchers, Inc.

CHAPTER OUTLINE

8.1 Environmental Setting

8.2 World Importance of the Middle East

8.3 The Heartland of Islam

8.4 The Middle Eastern Ecological Trilogy: Villager, Pastoral Nomad, and Urbanite

What and where is the Middle East? The term itself is Eurocentric, created by Europeans and their American allies who placed themselves in the figurative center of the world. They began to use the term prior to the outbreak of World War I, when the "Near East" referred to the territories of the Ottoman Empire in the Eastern Mediterranean region, the "East" to India, and the "Far East" to China, Japan, and the western Pacific rim. With "Middle East" they designated as a separate region the countries around the Persian Gulf (known to Arabs as the "Arabian Gulf," and in this text as the "Persian/Arabian Gulf" and simply "The Gulf"). Gradually, the perceived boundaries of the region grew. Today, depending upon what source you consult, you might find that the Middle East includes only the countries clustered around the Arabian Peninsula, or that (as in this text) it spans a vast 6000 miles (9700 km) west to east from Morocco in northwest Africa to Afghanistan in central Asia, and 3000 miles (4800 km) north to south from Turkey, on Europe's southeastern corner, to Sudan, which adjoins East Africa (Figs. 8.1 and 8.2). Thus defined, the region incorporates 22 countries and the disputed Western Sahara region (Table 8.1), occupying 5.79 million square miles (14.99 million sq km) and inhabited by about 421 million people as of mid-1998.

The region's occupants themselves now use the term "Middle East" to describe their region. It is a fitting designation for their location and their cultures, for they are literally in the middle. The region is a physical crossroads, where the continents of Africa, Asia, and Europe meet and the waters of the Mediterranean Sea and the Indian Ocean mingle. Its peoples—Arab, Jew, Persian, Turk, Kurd, Berber, and others—express in their cultures and ethnicities the coming together of these diverse influences. Occupying as they do this strategic location, the nations of the Middle East have throughout history been unwilling hosts to occupiers and empires originating far beyond their borders. They have also bestowed upon humankind a rich legacy which includes the ancient civilizations of Egypt and Mesopotamia and the world's three great monotheistic faiths: Judaism, Christianity, and Islam. People outside the region tend to forget about such contributions as they associate the Middle East with military conflict and terrorism. The region may even be on the eve of a new era of peace and reconciliation. The following two chapters attempt to make the complex, often destructive, but sometimes hopeful events in the Middle East more intelligible by presenting the geographic context within which they occur.

8.1 Environmental Setting

The margins of the Middle East are occupied mainly by oceans, seas, high mountains, and deserts: To the west lies the Atlantic Ocean; to the south, the Sahara Desert and the highlands of East Africa; to the north, the Mediterranean, Black, and Caspian seas, together with mountains and deserts lining the southern land frontiers of Russia and the Near Abroad; and to the east, the Hindu Kush mountains on the Afghanistan-Pakistan frontier and the Baluchistan Desert straddling Iran and Pakistan. The land is composed mainly of arid and semiarid plains and plateaus, together with considerable areas of rugged mountains and isolated "seas" of sand.

Heat and Aridity

Aridity dominates the Middle East (see Fig. 8.3, the world precipitation map in Fig. 2.1, the climate map in Fig. 2.7, and the world vegetation map in Fig. 2.8). Most of the region is part of what geographers call the Dry World—a belt of deserts and dry grasslands extending across Africa and Asia from the Atlantic Ocean nearly to the Pacific. At least three-fourths of the Middle East has an average yearly precipitation of less than 10 inches (25 cm), an amount too small for most types of unirrigated agriculture under the prevailing temperature conditions. Sometimes, however, localized cloudbursts release moisture that allows plants, animals, and even small populations of people— the Bedouins, Tuaregs, and other pastoral nomads—to live in

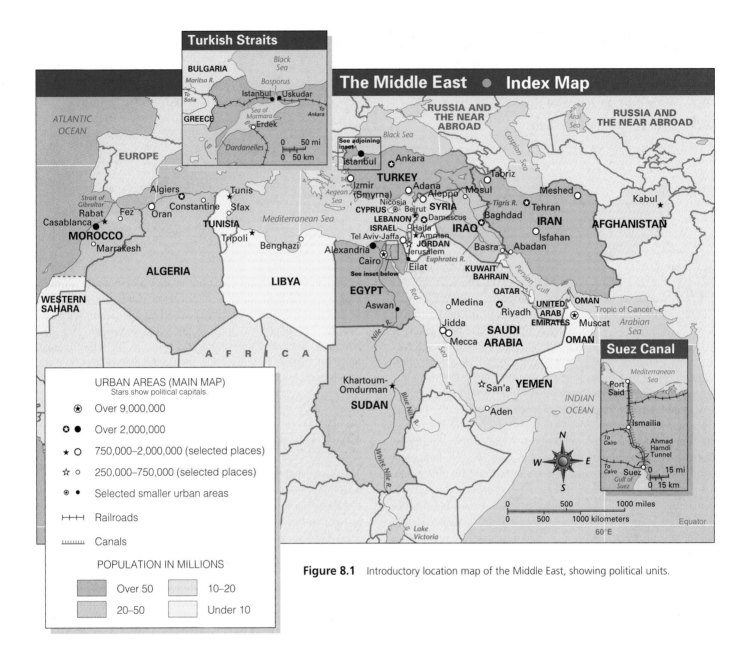

Figure 8.1 Introductory location map of the Middle East, showing political units.

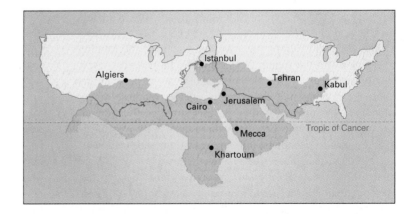

Figure 8.2 The Middle East compared in latitude and area with the conterminous United States.

Table 8.1 Middle East: Basic Data

Political Unit	Area (thousand/ sq mi)	Area (thousand/ sq km)	Estimated Population (millions)	Estimated Annual Rate of Increase (%)	Estimated Population Density (sq mi)	Infant Mortality Rate	Urban Population (%)	Arable Land (% of total area)	Per Capita GNP ($US)
Arab League States									
Egypt	384.3	995.3	65.5	2.2	171	63	43	3	1080
Sudan	917.4	2376.1	28.5	2.1	31	70	27	6	184
Tunisia	60	155.4	9.5	1.9	158	35	61	32	1930
Algeria	919.6	2381.8	30.2	2.4	33	44	56	3	1520
Morocco	172.3	446.3	27.7	1.8	161	62	52	22	1290
Lebanon	4	10.4	4.1	1.6	1030	34	87	30	2970
Syria	71	183.9	15.6	2.8	220	35	51	32	1160
Jordan	34.3	88.8	4.6	2.5	134	34	78	5	1650
Iraq	168.9	437.5	21.8	2.8	129	127	70	13	2000 (GDP)
Saudi Arabia	830	2149.7	20.2	3.1	24	35	51	2	10,600 (GDP)
Kuwait	6.9	17.9	1.9	2.3	271	10	100	0.3	16,700 (GDP)
Bahrain	0.3	0.8	0.6	2	266	14	88	3	13,000 (GDP)
Qatar	4.2	10.9	0.5	1.7	125	12	91	1	21,300 (GDP)
UAE	32.3	83.7	2.7	2.2	84	11	82	1	23,800 (GDP)
Yemen	203.8	5278.4	15.8	3.3	77	77	25	3	380
Oman	4.2	10.9	2.5	3.9	31	27	72	0.3	9500 (GDP)
Total	**3813.5**	**9877.2**	**251.7**	**2.4**	**66**	**60.3**	**49**	**6**	**2446**
Other Units									
Libya	679.4	1759.6	5.7	3.7	8	60	86	1	5800
Turkey	297.1	769.5	64.8	1.6	218	42	64	36	2830
Cyprus	3.6	9.3	0.7	0.7	210	8	68	47	7550
Iran	631.7	1636.1	64.1	1.8	102	35	61	11	1500
Afghanistan	251.8	652.2	24.8	2.5	98	150	18	12	800 (GDP)
Israel	8	20.7	6	1.5	129	6.7	90	21	15,870
Western Sahara	102.7	266	0.2	2.9	2	150	*	0	300
Gaza	*	*	1.1	4.6	*	33	*	*	590
West Bank	*	*	1.8	3.4	*	27	*	*	1200
Total	**1974.3**	**5113.4**	**169.2**	**1.9**	**85.7**	**54.3**	**58**	**11**	**2576**
Summary Total	**5787.8**	**14,990.6**	**420.9**	**2.2**	**72.7**	**57.9**	**53**	**8**	**2463**

Sources: *World Population Data Sheet,* 1998; United Nations Statistics Division, 1998; *Almanac of Politics and Government,* 1998; *World Factbook,* 1997.

the desert. Even the vast Sahara, the world's largest desert, which outsiders perceive as utterly desolate and barren, supports a surprising diversity and abundance of life. Plants, animals and even people have developed strategies of **drought avoidance** and **drought endurance** to live in this harsh biome. Plants either avoid drought by completing their life cycle quickly wherever rain has fallen, or endure drought by using

their extensive root systems, small leaves, and other adaptations to take advantage of subsurface or atmospheric moisture. Animals endure drought by calling upon extraordinary physical abilities—a camel can sweat away a third of its body weight and still live, for example—or avoiding the worst conditions by being active only at night, or by migrating from one moist place to another. (Migration to avoid drought is the strategy pastoral

Figure 8.3 Principal highlands and deserts of the Middle East.

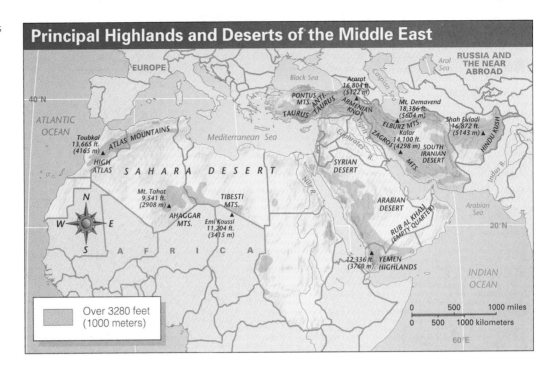

nomads use.) Populations of people, plants, and animals are all but nonexistent in the region's vast sand seas, including the Great Sand Sea of western Egypt (Fig. 8.4) and the Empty Quarter of the Arabian Peninsula.

Middle Eastern climates exhibit the comparatively large diurnal (daily) and seasonal ranges of temperature that are characteristic of dry lands. Summers in the lowlands are very hot almost everywhere. The hottest shade temperature ever recorded on Earth, 136°F (58°C), occurred in Libya in September 1922. Many places regularly experience daily maximum temperatures over 100°F (38°C) for weeks at a time. Day after day a baking sun assails the parched land from a cloudless sky. Night brings some relief, when the lack of cloud cover and low humidity allows heat to radiate back into space. Human settlements located near the sand seas often experience the unpleasant combination of high temperatures and hot, sand-laden winds, creating the sandstorms known locally by such names as *simuum* ("poison") and *scirocco*. Only in mountainous sections and in some places near the sea do higher elevations or sea breezes temper the intense midsummer heat. The population of Alexandria explodes in summer as Egyptians flee from Cairo and other hot inland locations. In Saudi Arabia, the government relocates from Riyadh to the highland summer capital of Taif to escape the heat.

Moisture from Seas and Mountains

Lower winter temperatures bring relief from the summer heat, and the more favored places receive enough precipitation to grow winter wheat or barley and some other cool-season crops.

In general, Middle Eastern winters may be characterized as cool to mild. But very cold winters and slight-to-moderate snowfalls are experienced in the high interior basins and plateaus of Iran, Afghanistan, and Turkey. These locales generally have a steppe climate. Only in the southernmost reaches of the region, such as Sudan's upper Nile Basin, do temperatures remain consistently high throughout the year. A savanna climate and biome prevail there.

Most areas bordering the Mediterranean Sea have 15 to 40 inches of precipitation a year, falling almost exclusively in

Figure 8.4   Sand "seas" cover large areas in Saudi Arabia, Iran, and parts of the Sahara Desert. This is the edge of the Great Sand Sea near Siwa Oasis in Egypt's Western Desert.   *Joseph J. Hobbs*

Figure 8.5 "The Treasury," a temple carved in red sandstone, probably in the 1st century A.D., by the Nabateans at their capital of Petra in southern Jordan. *Joseph J. Hobbs*

Figure 8.6 Rainfall is heavy on the Black Sea side of Turkey's Pontic Mountains. Note that roofs are pitched to shed precipitation. *Joseph J. Hobbs*

winter, while the summer is dry and warm—a typical Mediterranean climate pattern. Throughout history, people without access to perennial streams have stored this moisture to make it available later for growing those crops that require the higher temperatures of the summer months. The Nabateans, for example, who were contemporaries of the Romans in what is now Jordan, had a sophisticated network of limestone cisterns and irrigation channels (Fig. 8.5). Rainfall sufficient for unirrigated summer cropping (dry farming) is concentrated in areas along the southern and northern margins of the region. The Black Sea slope of Turkey's Pontic Mountains is lush and moist in the summer, and tea grows well there (Fig. 8.6). In the southwestern Arabian Peninsula a monsoonal climate brings summer rainfall and autumn harvests to Yemen and Oman, probably accounting for the Roman name for the area: *Arabia Felix,* or "Happy Arabia" (Fig. 8.7).

Mountainous areas in the Middle East, like the river valleys and the margins of the Mediterranean, play a vital role in supporting human populations and national economies. Due to orographic or elevation-induced precipitation, the mountains tend to receive much more rainfall than surrounding lowland areas.

There are three principal mountainous regions of the Middle East (see Fig. 8.3). In northwestern Africa between the Mediterranean Sea and the Sahara Desert, the Atlas Mountains of Morocco, Algeria, and Tunisia reach over 13,000 feet (3965 m) in elevation. Mountains also rise on both sides of the Red Sea, with peaks up to 12,336 feet (3760 m) in Yemen. A larger area of mountains, including the highest peaks in the Middle East, stretches across Turkey, Iran, and Afghanistan. On the eastern border with Pakistan, the Hindu Kush range has peaks of over 25,000 feet (7600 m). The loftiest and best-known mountain ranges in Turkey are the Taurus and Anti-

Figure 8.7 The mountainous landscape near Sana'a, Yemen's capital. The only crop under cultivation in this scene is the stimulant known as *qat (Cathya edulis).   Joseph J. Hobbs*

Taurus, and in Iran, the Elburz and Zagros Mountains. These chains radiate outward from the rugged Armenian Knot in the tangled border country where Turkey, Iran, and the countries of the Caucasus meet. One of the world's most dramatic and culturally important mountains, Mount Ararat, is an extinct, glacier-covered volcano of 16,804 feet (5122 m) towering over the border region between Turkey and Armenia (Fig. 8.8). Many Biblical scholars think the ark of Noah lies high on the mountain, for in the book of Genesis this boat was said to have come to rest "in the mountains of Ararat."

Where Forests Stood

Extensive forests existed in early historical times in the Middle East, particularly in these mountainous areas, but overcutting and overgrazing have almost eliminated them. Since the dawn of civilization in this area, at around 3000 B.C., people have cut timber for construction and fuel faster than nature could replace it. Egyptian King Tutankhamen's funerary shrines and Solomon's Temple in Jerusalem were built of cedar of Lebanon. So prized has this wood been through the millennia that only a few isolated groves of cedar remain in Lebanon. Described in ancient times as "an oasis of green with running creeks" and "a vast forest whose branches hide the sky," Lebanon is now largely barren (Fig. 8.9). Lumber is still harvested commercially in a few mountain areas such as the Atlas region of Morocco and Algeria, the Taurus Mountains of Turkey, and the Elburz Mountains of Iran, but the total supply falls far short of demand.

Mineral Wealth and Shortages

A shortage of mineral resources, especially those useful for industrialization, also handicaps many parts of the Middle East. Good deposits of coal are rare, except in Turkey. The region is rich in petroleum and natural gas, but the largest deposits are confined to a few countries bordering the Persian/Arabian Gulf (Saudi Arabia, Iraq, Kuwait, Iran, United Arab Emirates, Oman, Qatar), plus Libya, Algeria, and Egypt in North Africa. By coincidence, the countries rich in oil tend to have relatively small populations, while the most populous nations have few oil reserves (Iran is an exception). Although scattered deposits of metals occur, only a few are important on a global scale. Large salt deposits are common, and phosphate rock, useful as a chemical and fertilizer material, is mined commercially in Morocco, Tunisia, and Jordan. Israel and Jordan extract potash,

Figure 8.8 Behind this cart, the mighty volcano of Mt. Ararat rises to 16,804 feet (5122 m) to dominate the frontier regions of Turkey, Armenia, and Iran.   *Joseph J. Hobbs*

Figure 8.9 One of the twelve groves of cedars remaining in Lebanon. People of the Mediterranean Basin harvested this valuable resource for about five thousand years, nearly depleting it. This symbol of Lebanon now has complete protection.   *Leonard Wolfe/Photo Researchers, Inc.*

another chemical and fertilizer material, from the briny waters of the Dead Sea.

Aside from oil and gas, which currently account for all but a tiny percentage of Middle Eastern mineral production by value, the outlook for mineral extraction is poor. The region is generally handicapped by an inadequate natural resource base. Thus, despite the very high national GNP and per capita GNP that characterize some of the Middle Eastern nations, most countries in the region are clearly recognizable as LDCs.

8.2 World Importance of the Middle East

Throughout history, the sparsely populated deserts and mountains of the Middle East, separating the humid lands of Europe, Africa, and Asia, have been a hindrance to overland travel between those regions. Nevertheless, circulation of people, goods, and ideas has taken place along certain favorable routes, and the scattered population centers of the region have had a history of vigorous interaction with the outside world and with each other. Cultures of this region have made many fundamental contributions to humanity. Many of the plants and animals upon which the world's agriculture is based were first domesticated in the Middle East between 5000 and 10,000 years ago, in the course of the Agricultural Revolution. The list includes wheat, barley, sheep, goats, cattle, and pigs, whose wild ancestors people processed, manipulated, and bred until their physical makeup and behavior changed to suit human needs. The interaction between people and the wild plants and animals they eventually domesticated took place mainly in the **Fertile Crescent** stretching from Israel to western Iran (see p. 230).

By about 6000 years ago, people sought higher yields by irrigating crops in the rich but often dry soils of the Tigris, Euphrates, and Nile River valleys. Their efforts produced the enormous crop surpluses that allowed **civilization**—a cultural complex based on an urban way of life—to emerge in Mesopotamia (literally, "the land between the rivers" Tigris and Euphrates) and Egypt. Accomplishments in science, technology, art, architecture, language, mathematics, and other areas diffused outward from these centers of civilization. Egypt and Mesopotamia are thus among the world's great **culture hearths** (see Chapter 2, p. 40).

Most of the Old World's great empires have included portions of the Middle East. Some were indigenous, including the Egyptian, Babylonian, Phoenician, and Assyrian Empires. Foreign emperors and kings of Greece, Rome, and Christian Byzantium ruled the Middle East before the coming of Islam. Following the death of Islam's Prophet Muhammad in Arabia in 632, a great Islamic Empire arose and grew with such vigor that just a century later it stretched from Spain to Central Asia, with Damascus serving as its capital and cultural center. The most recent Islamic Empire collapsed in the early 20th century when

World War I brought an end to the Ottoman Empire that had been based in Istanbul beginning in 1453. Foreigners once again asserted control over the region after the war, when victorious Great Britain and France divided the remains of the Ottoman Empire between them, laying the groundwork for what is known today as the "Arab-Israeli Conflict" (see Chapter 9). After World War II, these great powers withdrew from the region and left newly independent nations with the task of reconciling often irreconcilable problems, such as those created when Britain and France made conflicting promises to both Arabs and Jews and drew national boundaries that bore little relationship to the distributions of local ethnic and religious groups.

Since World War II, several international crises have been precipitated by events in the Middle East. Although the colonial age has ended, strong outside powers heavily dependent upon Middle Eastern oil for their current and future industrial needs have competed for political influence and supplied their proxy armies. In strategic terms, the area is thus known as a **shatter belt**—a large, strategically located region composed of conflicting states caught between the conflicting interests of great powers. For example, in the 1967 and 1973 Arab-Israeli wars, Soviet-backed Syrian forces fought U.S.-backed Israeli troops.

The United States has always maintained a precarious relationship with the key players in the Middle East arena. On the one hand, the United States has pledged unwavering support for Israel, but on the other it has courted Israel's traditional enemies such as oil-rich Saudi Arabia. That Arab kingdom and its neighbors around the Persian/Arabian Gulf possess more than 60 percent of the world's proven oil reserves and are thus vital to the long-term economic security of the Western industrial powers and Japan. The United States and its Western allies made it clear they would not tolerate any disruption of access to this supply when Iraqi troops directed by Iraqi president Saddam Hussein occupied Kuwait on August 2, 1990. U.S. president George Bush drew a "line in the sand," proclaiming "we cannot permit a resource so vital to be dominated by one so ruthless—and we won't."

The region's strategic crossroads location also often has made it a cauldron of conflict. From very early times, overland caravan routes, including the famous "Silk Road" (see p. 344), crossed the Middle East with highly prized commodities traded between Europe and Asia. The security of these routes was vital, and countries on either end could not tolerate threats to them. Then, in 1869, one of the world's most important waterways opened in the Middle East. Slicing 107 miles (172 km) through the narrow Isthmus of Suez, the British- and French-owned Suez Canal linked the Mediterranean Sea with the Indian Ocean, saving cargo, military, and passenger ships a journey of many thousands of miles around the southern tip of Africa (see Figs. 8.1 and 8.10). Egyptian president Gamal Abdel Nasser's nationalization of the canal in 1956 led immediately to a British, French, and Israeli invasion of Egypt and a conflict known as both the "Suez Crisis" and the "1956 Arab-Israeli War."

Definitions & Insights

THE FERTILE CRESCENT

The Fertile Crescent is an arc-shaped area of relatively high precipitation stretching from Iran's Zagros Mountains westward through northern Iraq, southern Turkey, Syria, Lebanon, Israel, and western Jordan. Since about 10,000 years ago, relatively abundant rainfall (over 12 inches or 30 cm annually) provided excellent habitat for diverse plant and animal life. This resource wealth attracted human settlers, whose growing populations may have led by necessity to the earliest known domestications of plants and animals (see Chapter 2, p. 39). The productive seasonal agriculture supported by good water supplies and soils allowed Neolithic (New Stone Age) cultures to flourish throughout the Fertile Crescent until about 3000 B.C. After that time, far more productive irrigated agriculture—largely outside the Fertile Crescent and along the river valleys in Mesopotamia and Egypt—made those regions the main centers of cultural development.

Figure 8.A   The Fertile Crescent was a focal region for the domestication of plants and animals, leading to the emergence of urban cultures.

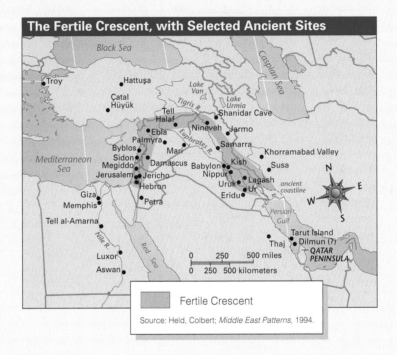

The Fertile Crescent, with Selected Ancient Sites

Source: Held, Colbert; *Middle East Patterns*, 1994.

Figure 8.10   The strategically vital Suez Canal zone saw bitter fighting in the Middle East wars of 1956, 1967, and 1973.   *Joseph J. Hobbs*

Because of its crossroads location, the Middle East has many important airports serving intercontinental routes, most notably in Cairo, Tel Aviv, and the United Arab Emirates cities of Dubai and Abu Dhabi. At the same time, Middle East infrastructure suffers a general lack of long-distance rail lines and highways.

8.3 The Heartland of Islam

The Middle East is so varied in its ways of living that generalizations about the culture of the entire region can be misleading. Nevertheless, a surprising amount of similarity can be discerned in modes of life prevailing in places as far apart as Morocco and Afghanistan. Because of Islam's powerful influence, not merely as a set of religious practices but as a total way of life, it is possible to distinguish an Islamic Middle Eastern culture that is characteristic of almost the entire region. In the Middle East, only Israel, within its pre-1967 borders, and Cyprus have non-Muslim majority populations (a Muslim is a person who practices the faith of Islam). Christians are minorities in all nations of the Middle East except Cyprus.

There is a tendency for outsiders to associate Islam and the Middle East exclusively with Arabs (people whose native language is Arabic). However, while most of the region's occupants are Arabs, it is important to note that two of the region's most populous states—Turkish Turkey and Persian Iran—are mostly non-Arab. The languages of Turkish (in the Altaic language family) and Persian (in the Indo-European language family) are entirely unrelated to Arabic. About 20 million Kurds—a people living in Turkey, Iraq, Iran, and Syria—speak Kurdish (in the Indo-European language family). Many people in North Africa speak Berber (in the Afro-Asiatic language family). It is also important to recognize that although political circumstances made them enemies in the 20th century, Arabs and Jews lived in peace for centuries and share many cultural traits. For example, both speak Semitic languages (in the Afro-Asiatic language family) and recognize Abraham as their patriarch. Finally, it is important to note that the world's most populous Islamic nation, Indonesia, is neither Arab nor Middle Eastern.

Roots of the Faith

An understanding of the religious tenets, culture, and diffusion of Islam is vital for appreciating the cultural geography of the Middle East. Islam is a monotheistic faith built upon the foundations of the region's earliest monotheistic faith, Judaism, and its offspring, Christianity. Indeed, Muslims call Jews and Christians "People of the Book," and their faith obliges them to be tolerant of these special peoples. Muslims believe that their Prophet Muhammad was the very last in a series of prophets who brought the Word to humankind. Thus they perceive the Bible as incomplete but not entirely wrong—Jews and Christians merely missed receiving the entire message. Muslims disagree with the Christian concept of the divine trinity, and regard Christ as a prophet rather than as God.

Muhammad was born in 570 A.D. to a poor family in the western Arabian (now Saudi Arabia) city of Mecca. Located on an important north-south caravan route linking the frankincense-producing area of southern Arabia (now Yemen and Oman) with markets in Palestine (now Israel) and Syria, Mecca was a prosperous city at the time. It was also a pilgrimage destination, because more than 300 deities were venerated in a shrine there called the *Ka'aba* (the Cube; Fig. 8.11). Muhammad married into a wealthy family and worked in the caravan trade. Muslim tradition holds that when he was about 40 years old, Muhammad was meditating in a cave outside Mecca when the Angel Gabriel appeared to him and ordered him to repeat the words of

Figure 8.11 The black-shrouded shrine known as the Ka'aba (just right of center) in Mecca's Great Mosque is the object toward which all Muslims face when they pray, and is the centerpiece of the pilgrimage to Mecca required of all able Muslims. *Nabeel Turner/Tony Stone Images*

God which the angel would recite to him. Over the next 22 years, the prophet related these words of God (Allah) to scribes who wrote them down as the Qur'an (or Koran), the Holy Book of Islam.

During this time, Muhammad began preaching the new message, "There is no god but God," which the polytheistic people of Mecca viewed as heresy. As much of their income depended upon pilgrimage traffic to the Ka'aba, they also viewed Muhammad and his small band of followers as an economic threat. They forced the Muslims to flee from Mecca and take refuge in Yathrib (modern Medina), where a largely Jewish population had invited them to settle. There were subsequent skirmishes between the Meccans and Muslims, but in 630 the Muslims prevailed and peacefully occupied Mecca. The Muslims destroyed the idols enshrined in the Ka'aba, which became a pilgrimage center for their one God.

The Diffusion of Islam

After Muhammad's death in 632, Arabian armies carried the new faith far, and quickly. The two decaying empires that then prevailed in the Middle East—the Byzantine or Eastern Roman Empire, based in Constantinople (now Istanbul), and the Sassanian Empire, based in Persia (now Iran) and adjacent Mesopotamia (now Iraq)—put up only limited military resistance to the Muslim armies before capitulating. Local inhabitants generally welcomed the new faith, in part because administrators of the previous empires had not treated them well, while the Muslims promised tolerance. Soon the Syrian city of Damascus became the center of a Muslim empire. Baghdad assumed this role in 750 A.D.

Arab science and civilization flourished in the Baghdad immortalized in the legends of the Thousand and One Nights. There were important accomplishments and discoveries in mathematics, astronomy, and geography. Scholars translated the Greek and Roman classics and, if not for their efforts, many of these works would never have survived to become part of the modern European legacy. It was an age of exploration, when Arab merchants and voyagers visited China and the remote lands of southern Africa. Many important discoveries by the Arab geographers were recorded in Arabic, a language unfamiliar to contemporary Europeans, and had to be rediscovered centuries later by the Portuguese and Spaniards.

The Five Pillars of Islam

Despite their differences, Sunni and Shi'ite Muslims (see Definitions and Insights) are united in support of the five fundamental precepts, or **pillars,** of Islam. The first of these is the profession of faith: "There is no god but God, and Muhammad is His Messenger." This expression is often on the lips of the devout Muslim, both in prayer and as a prelude to everyday activities. The second pillar is prayer, required five times daily at prescribed intervals. Two of these prayers mark dawn and sunset. Business comes to a halt as the faithful prostrate themselves before God. Muslims may pray anywhere, but wherever they are they must turn toward Mecca. There also is a congregational prayer at noon on Friday, the Muslim sabbath.

The third pillar is almsgiving. In earlier times, Muslims were required to give a fixed proportion of their income as charity, similar to the concept of the tithe in the early Christian church. Today the donations are voluntary. Even Muslims of very modest means give what they can to the more needy.

The fourth pillar is fasting during Ramadan, the ninth month of the Muslim lunar calendar. During Ramadan, Muslims are required to abstain from food, liquids, smoking, and sexual activity from sunrise to sunset. The lunar month of Ramadan occurs earlier each year in the solar calendar and thus periodically falls in summer. In the torrid Middle East, that timing imposes special hardships on the faithful, who, even if they are performing manual labor, must resist the urge to drink water during the long, hot days.

The final pillar is the pilgrimage *(hajj)* to Mecca, Islam's holiest city. Every Muslim who is physically and financially capable is required to make the journey once in his or her lifetime. A lesser pilgrimage may be performed at any time, but the prescribed season is the twelfth month of the Muslim calendar. Those days witness one of Earth's greatest annual migrations, as about two million Muslims from all over the world converge on Mecca (Fig. 8.11). Hosting these throngs is an obligation the government of Saudi Arabia fulfills proudly and at considerable expense, but with some trepidation in recent years because of the security threat foreign visitors may pose to the host country and because accidents such as tent city fires have cost many lives. Many pilgrims also visit the nearby city of Medina, where Muhammad is buried. Most Muslims regard the *hajj* as one of the most significant events of their lifetimes. All are required to wear simple seamless garments and, for a few days, the barriers

Definitions & Insights

SHI'ITE AND SUNNI MUSLIMS

A serious schism developed very early within Islam, and it persists today, continuing to have an important impact on relations among the Muslim nations and between some Muslim countries and the West. The split developed because the Prophet Muhammad had named no successor to take his place as the leader (caliph) of all Muslims. Some of his followers argued that the person with the strongest leadership skills and greatest piety was best qualified to assume this role. These followers became known as **Sunni,** or orthodox, Muslims. Others argued that only direct descendants of Muhammad, specifically through descent from his cousin and son-in-law Ali, could qualify as leaders. They became known as **Shi'a,** or Shi'ite, Muslims. The military forces of the two camps engaged in battle south of Baghdad at Karbala in 680 A.D., and in the encounter, Sunni troops caught and brutally murdered Hussein, a son of Ali. Thereafter, the rift was deep and permanent. The martyrdom of Hussein became an important symbol for Shi'ites, who still today regard themselves as oppressed peoples struggling against cruel tyrants—including some Sunni Muslims.

Today only two Muslim countries, Iran and Iraq, have Shi'ite majority populations. Significant minority populations of Shi'ites are in Syria, Lebanon, Yemen, and the Arab states of the Persian/Arabian Gulf. Shi'ite Iran challenged the United States in an infamous hostage ordeal in 1980 that began after Muslim clerics assumed control of Iran's government. After Israel's invasion of Lebanon in 1982, Lebanese Shi'ite Muslims sympathetic to Iran abducted several U.S. and other Western citizens and killed hundreds more. The Shi'ite Muslim government of Iran continues to be an outspoken critic of Saudi Arabia (a largely Sunni nation), causing constant apprehension on the part of the oil kingdom's Western allies. The West's fear of a Shi'ite rebellion leading to the emergence of a new Shi'ite state in southern Iraq contributed to the allies' decision to halt their assault on Iraq after Iraqi forces withdrew from Kuwait in 1991.

separating groups by income, ethnicity, and nationality are broken. Pilgrims return home with the new stature and title of "hajj" but also with humility and renewed devotion.

Interpretations of the Faith

While all Muslims share the five pillars and other tenets, they vary widely in other cultural practices related to their faith, depending upon what country they live in, whether they are from the desert, village, or city, and how much education and income they have. The governments and associated clerical authorities in Saudi Arabia and Iran insist upon strict application of Islamic law *(shari'a)* to civil life; in effect, there is no separation between church and state. The Qur'an does not state

Regional Perspective

Islamic Fundamentalism

A wave of Islamic "fundamentalism" has recently swept the Islamic world. Arguing that "Islam is the solution," Islamists (as they are more accurately known) reject what they view as the materialism and moral corruption of Western countries and the political and military support these countries lend to Israel. Both Sunni and Shi'ite Muslims have advanced a wide range of Islamic movements, notably in Iran, Lebanon, Egypt, Afghanistan, Sudan, and Algeria.

Although nominally religious, the more radical of these movements have political and cultural aims, particularly the destabilization or removal of U.S. and Israeli interests in the Middle East and abroad. In 1993, followers of the radical Egyptian cleric Shaykh Umar Abdel-Rahman bombed New York City's World Trade Center as a protest against American support of Israel and Egypt's pro-Western government. In 1998, supporters of Osama bin Laden, a former Saudi businessman living in exile in Afghanistan, bombed U.S. embassies in Kenya and Tanzania as part of an avowed worldwide struggle against American imperialism and immorality. And in an attempt to destabilize and replace Egypt's government, Islamists of the al-Jami'at al-Islamiyya movement attacked and killed foreign tourists in Egypt in the 1990s. In the 1980s, members of the pro-Iranian Hizbullah, or "Party of God," in Lebanon kidnapped foreigners as bargaining chips for the release of comrades jailed in other Middle Eastern countries. Within Israel and the autonomous Palestinian territories of the West Bank and Gaza Strip, Palestinian members of HAMAS (an Arabic acronym for the Islamic Resistance Movement) have carried out terrorist attacks on Israeli civilians and soldiers in an effort to derail implementation of the peace agreements reached between the Israeli government and the Palestine Liberation Organization (PLO). In Algeria, years of bloodshed have followed the government's annulment of 1991 election results which would have given the Islamic Salvation Front (FIS) majority control in the parliament. Muslim sympathizers have carried the battle to France, bombing civilian targets in protest against the French government's support for the Algerian regime.

In all of these situations, a tiny minority of Muslims carried out terrorist actions which the great majority condemned. Mainstream Islamic movements have distinguished themselves through public service to the needy, and through encouragement of stronger moral and family values. For most Muslims, the growing Islamist trend means a re-embrace of traditional values like piety, generosity, care for others, and Islamic legal systems which have proven effective for centuries—values which now pose a reasonable alternative to Western cultural influences and often repressive political and administrative systems. For many people outside the region, however, "Muslim" and "terrorist" have become synonymous—an erroneous association which can be overcome in part by careful study of the complex Middle East.

that women are required to wear veils, but it does urge them to be modest, and it portrays their roles as different from those of men. Clerics in Saudi Arabia insist that women wear floor-length, long-sleeved black robes and black veils in public, that they not travel unaccompanied by a male member of their families, and that they not drive cars. In Egypt, by contrast, Muslim women are free to appear in public unveiled if they choose. However, in most Muslim countries, conservative ideas about the role of women are still very strong: They should be modest, retiring, good mothers, and keepers of the home. The Qur'an portrays women as equal to men in the sight of God and, in principle, Islamic teachings guarantee the right of women to hold and inherit property.

Most Muslim women argue that what others often see as "backward" cultural practices are in fact very progressive. For example, their modest dress compels men to evaluate them on the basis of their character and performance, not their attractiveness. Segregation of the sexes in the classroom makes it easier for both women and men to develop their confidence and skills. Sexual assault is rare. A married woman retains her maiden name. These apparent advantages can be weighed against the drawbacks that women are generally subordinate to men in public affairs and have fewer opportunities for education and for work outside the home.

8.4 The Middle Eastern Ecological Trilogy: Villager, Pastoral Nomad, and Urbanite

In the 1960s, the American geographer Paul English developed a useful model for understanding relationships between the three ancient ways of life that still prevail in the Middle East today: villager, pastoral nomad, and urbanite (see Fig. 8.12). Villagers are the subsistence farmers of rural agricultural areas; pastoral nomads are the desert peoples who migrate with their livestock, following patterns of rainfall and vegetation; and urbanites are the inhabitants of the large towns and cities. Describing each of these as a component of the **Middle Eastern ecological trilogy,** English explained how each of them has

A

B

C

Figure 8.12   Faces of the Middle Eastern ecological trilogy: (a) villagers at a melon harvest in central Turkey; (b) Ma'aza Bedouin nomads with sheep in the Eastern Desert, Egypt; and (c) Turkish shoppers in the bazaar of Izmir.   *Joseph J. Hobbs*

an important, usually mutually beneficial, pattern of interaction with the other two.

The peasant farmers of Middle Eastern villages (the villagers) represent the cornerstone of the trilogy. They grow the staple crops such as wheat and barley which feed both the city-dweller and the pastoral nomad of the desert. Neither urbanite nor nomad could live without these. The village also unwillingly provides the city with tax revenue, soldiers, and workers. And before the mid–20th century, villages provided pastoral nomads with plunder as the desert-dwellers raided their settlements and caravan supply lines. Generally, however, the exchange is beneficial: The nomads provide villagers with livestock products including live animals, meat, milk, cheese, hides, and wool, and with desert herbs and medicines, while educated and progressive urbanites provide technological innovations, religious instruction, and cultural amenities.

There is little direct interaction between urbanites and pastoral nomads, although some manufactured goods such as clothing travel from city to desert, and some desert-grown folk medicines pass from desert to city. Historically the exchange has been violent, as urban-based governments have sought to control the movements and military capabilities of the elusive and sometimes hostile nomads. Pastoral nomads once plundered rich caravans plying the major overland trade routes of the Middle East. Governments did not tolerate such activities and often cracked down hard on those nomads they were able to catch.

In the 1970s, Paul English wrote an article marking the "passing of the ecological trilogy." He noted that cities were encroaching on villages, villagers were migrating into cities and giving some neighborhoods a rural aspect, and pastoral nomads were settling down—thus, the trilogy no longer existed. In reality, while the makeup and interactions of its parts have

changed somewhat, the trilogy model is still valid and useful as an introduction to the major lifeways of the Middle East. It is especially significant that a given man or woman in the Middle East strongly identifies himself or herself as either a villager, a pastoral nomad, or an urbanite. This perception of self has an important bearing on how these people of very different backgrounds interact, even when they live in close proximity. Urban officials may work in rural village areas, but they remain at heart and in their perspectives city people, and usually live apart from farmers. Similarly, extended families of pastoral nomads may settle down and become farmers but continue to identify themselves by affiliation with the nomadic tribe. Many continue to harvest desert resources on a seasonal basis, and retain marriage and other ties with desert-dwelling kinsmen.

The Village Way of Life

Agricultural villagers historically represented by far the majority in Middle Eastern populations; only within recent decades have urbanites begun to outnumber them. In this generally dry environment, the villages are located near a reliable water source, with cultivable land near by. They tend to be composed of closely related family groups, with the land of the village often owned by an absentee landlord. Most often, the villagers live in closely spaced flat-roofed houses made of adobe, mud brick, or concrete blocks. Production and consumption focus on a staple grain. As land for growing fodder is often in short supply, villagers keep only a limited number of sheep and goats, and rely in part on nomads for pastoral produce. Residents of a given village usually share common ties of kinship, religion, ritual, and custom, and the changing demands of agricultural seasons regulate their patterns of activity.

Village life has been increasingly exposed to outside influences since the mid-18th century. Contacts with European colonialism brought significant economic changes, including the introduction of cash crops and modern facilities to ship them. Improved and expanded irrigation, financed initially with capital from the West, brought more land under cultivation. Recent agents of change have been the countries' own government doctors, government teachers, and land reform officers. Products of modern technology such as radios, television, sewing machines, and motorcycles have modified old patterns of living. The young and more ambitious have been drawn to urban areas. Improved roads and communications in turn have carried urban influences to villages, prompting villagers to become more integrated into national societies.

The Pastoral Nomadic Way of Life

Pastoral nomadism emerged as an offshoot of the village agricultural way of life not long after plants and animals were first domesticated in the Middle East. Rainfall and the wild fodder it produces, although scattered, are sufficient resources to support small, mobile groups of people who migrate with their sheep, goats, and camels (and in some locales, cattle) to take advantage of this changing resource base. In mountainous areas they follow a pattern of **vertical migration** or **transhumance,** moving with their flocks from lowland winter to highland summer pastures. In the flatter expanses which comprise most of the region, the nomads have a pattern of **horizontal migration** over much larger areas where rainfall is typically far less reliable than in the mountains. In addition to selling or trading livestock in order to obtain foods, tea, sugar, clothing, and other essentials from settled communities, pastoral nomads also hunt, gather, work for wages, and, where possible, grow crops. Their many-faceted livelihood has been described as a strategy of "risk minimization," based on the exploitation of multiple resources so that some will support them if others fail.

Although renowned in Middle Eastern legends and in popular Western films like *Lawrence of Arabia,* pastoral nomads have been described as being "more glamorous than numerous." It is still impossible to obtain adequate census figures on the number still living in the deserts of the Middle East, though estimates range from 5 to 13 million. The late 20th century witnessed the rapid and progressive settling down, or **sedentarization,** of the nomads—a process attributed to a variety of reasons. In some cases prolonged drought virtually eliminated the resource base on which the nomads depended. Traditionally they were able to migrate far enough to find new pastures, but modern national boundaries now prohibit such movements. Some have returned with the rains to their desert homelands, while others have chosen to remain as farmers or wage laborers in villages and towns. On the Arabian Peninsula, the prosperity and technological changes prompted by oil revenues made rapid inroads into the material culture—and then the livelihood preferences—of the desert people; many preferred the comforts of settled life. Some governments, notably those of Israel and prerevolutionary Iran—unable to count, tax, conscript, and control a sizable migrant population—compelled nomads to settle.

Pastoral nomads of the Middle East identify themselves primarily not by their nationality but by their tribe. The major ethnic groups from which these tribes draw are the Arabic-speaking Bedouins of the Arabian Peninsula and adjacent lands; the Berber and Tuareg of North Africa; the Kababish and Bisharin of Sudan; the Yoruk and Kurds of Turkey; the Qashai and Bakhtiari of Iran; and the Pashtun of Afghanistan. Members of a tribe claim common descent from a single male ancestor who lived countless generations ago; their kinship organization is thus called a **patrilineal descent system.** It is also a **segmentary kinship system,** so called because there are smaller subsections of the tribe, known as clans and lineages, which are functionally important in daily life. Members of most closely related families comprising the lineage, for example, share livestock, wells, trees, and other resources. Both the larger clans, made up of numerous lineages, and the tribes possess territories. Members of a clan or tribe typically allow members of another clan or tribe to use the resources within its territory on the basis of "usufruct," or nondestructive mutual use.

Although some detractors have depicted pastoral nomads as the "fathers" rather than "sons" of the desert, blaming them for wanton destruction of game animals and vegetation, there are numerous examples of pastoral nomadic groups who have developed indigenous and very effective systems of resource conservation. Most of these practices depend upon the kinship groups of family, lineage, clan, and tribe to assume responsibility for protecting plants and animals.

The Urban Way of Life

The city was the final component to emerge in the ecological trilogy, beginning in about 4000 B.C. in Mesopotamia and 3000 B.C. in Egypt. Unlike the villages they resembled in many ways, the early cities were distinguished by their larger populations (more than 5000 people), use of written languages, and presence of monumental temples and other ceremonial centers. The early Mesopotamian city and, after the seventh century, the classic Islamic city, called the **medina,** had several structural elements in common (Fig. 8.13). The medina was characterized

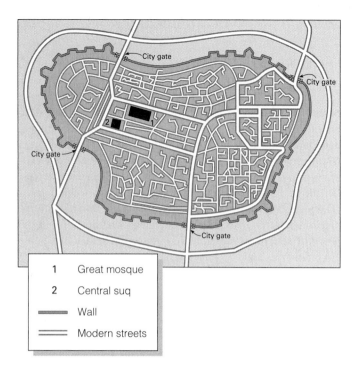

1	Great mosque
2	Central suq
▬	Wall
═	Modern streets

Figure 8.13   An idealized model of the classic *medina* or Muslim Middle Eastern city.

Figure 8.14   Cairo's City of the Dead. The domes mark tombs of Egypt's 15th-century ruling class. Left of center, the individual posts rising from the ground are individual graves from the 20th century. Until 1967, these graves and the monumental tombs were the only features on this landscape on the eastern edge of Cairo. But population growth and refugee movements in the last three decades have led to massive settlement of this cemetery. Single- and multiple-family dwellings, such as those on the right side of this photograph, now cover most of the graves; the living dwell with the dead.   *Joseph J. Hobbs*

by a high surrounding wall built for defensive purposes. The city center was dominated by the congregational mosque and often an attached administrative and educational complex. Although Islam is often characterized as a faith of the desert, religious life has always been focused in, and diffused from, Middle Eastern cities. The importance of the city's congregational mosque in religious and everyday life is often emphasized by its large size and outstanding artistic execution.

A large commercial zone, known as a *bazaar* in Persian and *suq* in Arabic, and recognizable as the ancestor of the modern shopping mall, typically adjoined the ceremonial and administrative heart of the city. Merchants and craftsmen of different commodities occupied separate spatial areas within this complex, and visitors to an old medina today can still expect to find streets where spices, carpets, gold, silver, traditional medicines, and other goods are sold exclusively. Smaller clusters of shops and workshops were located at the city gates.

Residential areas were differentiated as quarters not by income group but by ethnicity; the medina of Jerusalem, for example, still has distinct Jewish, Arab, Armenian, and non-Armenian Christian quarters. Homes tended to face inward toward a quiet central courtyard, buffering the occupants from the noise and bustle of the street. The narrow, winding streets of the medina were intended for foot traffic and small animal-drawn carts, not for large motorized vehicles, a fact which accounts for the traffic jams in some old Middle Eastern cities today and for the wholesale destruction of the medina in others.

Those medinas that survive today are gently decaying vestiges of a forgotten urban pattern. Periods of European colonialism and subsequent nationalism changed the face and orientation of the Middle Eastern city. During the colonial age,

resident Europeans preferred to live in more spacious settings at the outer edges of the city, and later the national elite followed this pattern. In recent times, independent governments have adopted Western building styles, with broad traffic arteries cutting through the old quarters and large central squares near government buildings. This opening up of the cityscape has scattered commercial activity along the wide avenues, thus diluting the prime importance of the central bazaar as the focus of trade.

Rural-urban migration and the city's own internal growth contribute to a rapid rate of urbanization which puts enormous pressure on services in the poorer Middle Eastern nations. Governments often build high-rise public housing to accommodate the growing population, contributing to a cycle in

Figure 8.15   The port of Jubail in Saudi Arabia's Eastern Province is an entirely modern, meticulously planned city, founded upon the kingdom's oil wealth.   *Robert Azzi/Woodfin Camp & Associates*

236

which the urban poor move into the new dwellings only to leave their old quarters as a vacuum to draw in still more rural migrants. In Cairo, millions of former villagers now live in the "City of the Dead," an extraordinary urban landscape composed of multistory dwellings erected above graves—a last resort for the poor who have no other place to go (Fig. 8.14). The overwhelmingly largest, or primate, city (see p. 111), so characteristic of Middle Eastern capitals, thus grows at the expense of the smaller city. Much of the rural-urban migration and subsequent urban gridlock and squalor could probably be avoided if governments invested more in the development of villages and smaller cities. The oil-rich countries with relatively small populations generally enjoy an urban standard of living equaling that of affluent Western countries. Modern industrial cities such as Jubail in Saudi Arabia and others founded on oil wealth were built virtually overnight, providing fascinating contrast to the region's colorful, complex ancient cities (Fig. 8.15).

SUMMARY WITH SELECTED KEY TERMS

- Events in the Middle East profoundly affect the daily lives of people around the world, yet this region is **often misunderstood.** Misleading stereotypes about its environment and people are common and people outside the region often associate it with military conflict and terrorism.
- The Middle East has served as a pivotal global crossroads, linking Asia, Europe, Africa, and the Mediterranean Sea with the Indian Ocean. These nations have historically been unwilling hosts to **occupiers and empires** originating far beyond their borders.
- The margins of this region are occupied by oceans, high mountains, and deserts. The land is composed mainly of **arid and semiarid plains and plateaus,** together with considerable area of **rugged mountains and isolated "seas" of sand.**
- The region has bestowed upon humanity a rich legacy of **ancient civilizations,** including Egypt and Mesopotamia, and the **three great monotheistic faiths of Judaism, Christianity, and Islam.**
- **Aridity dominates** the environment with at least three-fourths of the region receiving less than ten inches of yearly precipitation. Plants, animals, and people have developed strategies of **drought avoid-**

ance and **drought endurance** to live here. In addition, **great river systems and freshwater aquifers** have sustained large human populations.
- Many of the plants and animals upon which the world's agriculture depends were first domesticated in the Middle East in the course of the **Agricultural Revolution.** The interaction between people and the wild plants and animals that were domesticated took place mainly in the **Fertile Crescent.**
- Since World War II, several **international crises and wars** have been precipitated by events in the Middle East. **Strong outside powers heavily depend upon this region** for their current and future industrial needs.
- An **Islamic Middle Eastern culture** is characteristic of almost the entire region with only Israel and Cyprus having majority non-Muslim populations.
- Middle Easterners include **Jews, Arabs, Turks, Persians, Berbers, and other ethnic groups** who practice a wide variety of ancient and modern livelihoods.

REVIEW QUESTIONS

1. Using maps and the text, list and locate the nations of the Middle East.
2. What are the major climatic patterns of the Middle East?
3. Explain how residents have culturally adapted to heat and aridity in the Middle East.
4. Where were the extensive forests of the Middle East found during early historical times? What nation supplied the cedar for King Tutankhamen's funerary shrines?

5. List some of the non-Arab nations of the Middle East and define their cultural composition.
6. Using maps and the text, locate the two great culture hearths of Mesopotamia and Egypt.
7. Using maps and the text, explain why the Middle East has held such a strategic importance in international conflict.
8. Define vertical migration (transhumance) and horizontal migration in relation to the Middle East.

DISCUSSION QUESTIONS

1. Explain the origin of the term "Middle East."
2. Explain how heat and aridity have helped shape human culture in the Middle East.
3. Why is this region considered such an important **culture hearth** for the rest of the world?
4. Islam is often considered more a way of life than just a religion. Why?
5. Examine and discuss the **Shi'ite and Sunni** differences. List the **Five Pillars of Islam.**

6. List the **non-Arab** nations of the Middle East and describe their peoples.
7. The **ecological trilogy** asserts that three ancient ways of life prevail in the Middle East. What are these?
8. Examine and discuss the important land features of the Middle East.
9. Examine and discuss the important water bodies in and around the Middle East.

Chapter 9

The Middle Eastern Countries: Modern Struggles in an Ancient Land

▲ *Faces of the rural Middle East. These men are from a village in Upper (southern) Egypt.*
Joseph J. Hobbs

onflict in the Middle East has been one of the most persistent and dangerous problems in global affairs in recent decades. This chapter introduces the key players and issues in conflict and discusses the central features of land and life in the individual countries of the Middle East. Hostilities in this region generally focus on questions of who owns land and water and are of great interest to geographers. The ongoing peace process represents an opportunity to resolve these issues and provide peace "dividends"—not just to the region's inhabitants but to the international community. The consequences of a breakdown in the peace process would likewise affect the wider world. The challenges posed in the Middle East are enormous, and so are the opportunities.

9.1 The Arab World

The Arab world, stretching east-west from the Indian Ocean to Morocco and southward from the Mediterranean to the fringes of tropical Africa, includes almost all of the Middle East. The peoples who inhabit it are ethnically and culturally diverse, but an overwhelming majority speak dialects of Arabic and are therefore known as Arabs. Originally, the Arabs were inhabitants of the Arabian Peninsula, but conquests after their conversion to Islam spread their language and associated Islamic culture widely.

Despite strong economic and political contrasts between Arab nations, there is also a sense of Arab unity. This sense of Arab community has several important roots—namely, a common language, cultural heritage, and majority belief in Islam, coupled with pride in past achievements and the realization of a need for united efforts to improve social and economic conditions throughout the Arab world. In recent decades, regional unity has been enhanced by opposition to perceived economic and political intervention by foreigners—a feeling most sharply focused on the state of Israel, which was founded under foreign auspices on a territory inhabited mostly by Arabs.

9.2 Israel, Jordan, Lebanon, and Syria: Uneasy Neighbors

The Promised Land

Depending upon one's perspective, the Jewish connection with the geographical region known as Palestine—essentially the area now composed of Israel, the West Bank, and the Gaza Strip—is most significant on a time scale of either about 4000 years or about 100 years. According to the Bible, around 2000 B.C. God commanded Abraham and his kinspeople, known as Hebrews (later as Jews), to leave their home in what is now southern Iraq and settle in Canaan. God told Abraham that this land of Canaan—geographical Palestine—would belong to the Hebrews after a long period of persecution. The Bible says that the Hebrews did settle in Canaan, until famine struck that land. At the command of Abraham's grandson Jacob, the Hebrews—known then as Israelites—relocated to Egypt, where grain was plentiful. That began the long sojourn of the Israelites in Egypt, which, according to the Bible, ended in about 1200 B.C. when Moses led them out (the Exodus).

According to Jewish history, the prophecy of Abraham was first fulfilled when the Israelites settled once again in their "promised land" of Canaan. The Jewish king Saul unified the twelve tribes who were Jacob's descendants into the first united Kingdom of Israel in about 1020 B.C. In about 950 B.C. in Jerusalem—the capital of a kingdom enlarged by Saul's successor David—King Solomon built Judaism's First Temple. He located it atop a great rock known to the Jews as Even HaShetiyah, the "Foundation Stone," plucked from beneath the throne of God to become the center of the world and the core from which all the world was created.

The united Kingdom of Israel lasted only about 200 years before splitting into the states of Israel and Judah. Empires based in Mesopotamia destroyed these states: The Assyrians attacked Israel in 721 B.C. and the Babylonians sacked Judah in 586 B.C. The Babylonians destroyed the First Temple and exiled the Jewish people to Mesopotamia, where they remained until conquering

Landscape in Literature

The Book of Job

One who has "the patience of Job" has the stamina of this personality of the Jewish Bible, the Christian Old Testament, and the Muslim Qur'an. His story, dating to the first millennium B.C., is that of a wealthy man who loses every material thing as Satan tests his faith. Job's devotion to God remains unshaken. In this passage, Job speaks of how efficient people are in exploiting the Earth to suit their needs, but says the search for wisdom is much more elusive; wisdom dwells with God. In its detailed inventory of the riches of the Middle East at this early date, and in its insightful perspective on how people change the face of the Earth, this is a remarkable piece of geographic prose.

Surely there is a mine for silver,
And a place where they refine gold.
Iron is taken from the dust,
And from rock copper is smelted.
Man puts an end to darkness,

And to the farthest limit he searches out
The rock in gloom and deep shadow.
He sinks a shaft far from habitation,
Forgotten by the foot;
They hang and swing to and fro far from
    men.
The earth, from it comes food,
And underneath it is turned up as fire.
Its rocks are the source of sapphires,
And its dust contains gold.
The path no bird of prey knows,
Nor has the falcon's eye caught sight
    of it.
The proud beasts have not trodden it,
Nor has the fierce lion passed over it.
He puts his hand on the flint;
He overturns the mountains at the base.
He hews out channels through the rocks;
And his eye sees anything precious.
He dams up the streams from flowing;
And what is hidden he brings out into the
    light.
But where can wisdom be found?
And where is the place of understanding?
Man does not know its value,
Nor is it found in the land of the living.

The deep says, "It is not in me;"
And the sea says, "It is not with me."
Pure gold cannot be given in exchange
    for it,
Nor can silver be weighed as its price.
It cannot be valued in the gold of Ophir,
In precious onyx, or sapphire.
Gold or glass cannot equal it,
Nor can it be exchanged for articles
    of fine gold.
Coral and crystal are not mentioned;
And the acquisition of wisdom is above
    that of pearls.
The topaz of Ethiopia cannot equal it,
Nor can it be valued in pure gold.
Where then does wisdom come from?
And where is the place of understanding?
Thus it is hidden from the eyes of
    all living,
And concealed from the birds of
    the sky.

—Job 28:1–21 New American
    Standard Bible

Persians allowed them to return to their homeland. In about 520 B.C. the Jews who returned to Judah, the land from which they take their name, rebuilt the Temple (the "Second Temple") on its original site. A succession of foreign empires came to rule the Jews and Arabs of Palestine: Persian, Macedonian, Ptolemaic, Seleucid, and, around the time of Christ, Roman. Herod, the Jewish king who ruled under Roman authority and was a contemporary of Christ, greatly enlarged the Temple complex.

The Jewish Diaspora

The Jews of Palestine revolted against Roman rule three times between 64 and 135 A.D. The Romans quashed these rebellions in a series of famous sieges, including those of Masada and Jerusalem. The Romans destroyed the Second Temple, and a third has never been built. All that remains of the Second Temple complex is a portion of the surrounding wall built by Herod.

Today this Western Wall, known to non-Jews as the "Wailing Wall," is the most sacred site in the world accessible to Jews (Fig. 9.1). Religious tradition prohibits Jews from ascending the Temple Mount above—the area where the Temple actually stood. After the temple's destruction, that site was occupied by a Roman temple, and then in 691 by the Muslim shrine called the Dome of the Rock, which still stands today. The mostly Muslim Arabs know the Temple Mount as *Haraam ash-Shariif,* meaning "The Noble Sanctuary."

The victorious Romans scattered the defeated Jews to the far corners of the Roman world. Thus began the Jewish exile or **diaspora.** In their exile the Jews never forgot their attachment to the promised land. The Passover prayer ends with the words "Next year in Jerusalem!" In Europe, where their numbers were greatest, Jews were subjected to systematic discrimination and persecution, and were forbidden to own land or engage in a number of professions. Known as **anti-Semitism,** the hatred of

Figure 9.1   Sacred places in Judaism and Islam. Below and to the right of the golden dome is the Western Wall, part of the retaining wall which supported Judaism's holiest site, the Temple in Jerusalem. Near the bottom-left corner of the wall is the ancient tunnel whose other end was opened in 1996. The tunnel opening sparked widespread unrest among Muslim Palestinians, who perceived it as an incursion against one of Islam's holiest sites, the Dome of the Rock, whose golden dome rises above the Western Wall. The Dome of the Rock is widely believed to occupy the site where the Jewish Temple's Holy of Holies stood, and where the Ark of the Covenant rested for many centuries. The minaret at the upper far right belongs to the al-Aqsa Mosque, another holy site to Muslims. The discarded palm fronds in the foreground are remainders from a recent celebration of the Jewish festival known as Succoth.   *Joseph J. Hobbs*

Jews developed deep roots in Europe. This sentiment in part grew out of the perception that Jews were responsible for the murder of Christ, and the fact that Christian Europeans were prohibited from practicing usury, or moneylending. They assigned this role to Jews but then resented paying debts to the despised moneylenders.

In the 1930s anti-Semitism became state policy in Germany under the Nazis, led by Adolf Hitler. Many German Jews—including Albert Einstein—fled to the United States, and others emigrated to Palestine in support of the **Zionist movement,** which aimed at establishing a Jewish homeland in Palestine with Zion (a synonym for Jerusalem) as its capital. Most Jews were not as fortunate as the emigrants. Within the boundaries of the Nazi empire that dominated most of continental Europe during World War II, Hitler's regime executed its "final solution" to the Jewish "problem." Nazi Germans and their allies killed an estimated six million Jews, along with other "inferior" minorities, including Gypsies and homosexuals. It was this **Holocaust** that prompted the victorious allies of World War II, from their powerful position in the newly formed United Nations, to create a permanent homeland for the Jewish people in Palestine.

The Birth of Israel

The modern state of Israel was carved from lands whose fate had been undetermined since the end of World War I. The Ottoman Empire, based in what is now Turkey, had ruled Palestine and surrounding lands in the eastern Mediterranean since the 16th century. After the British and French defeated the Turks in World War I and destroyed their empire, they divided the region between them. The British received the mandate for Palestine, Transjordan (modern Jordan), and Iraq, and the French received the mandate for Syria (now Syria and Lebanon). British administrators of Palestine made conflicting promises to Jews and Arabs, on the one hand promoting Jewish immigration to Palestine with an eye to the eventual establishment of a Jewish state there, and on the other imposing limits on this immigration and implying that any Jewish state would not be established in the heart of predominantly Arab Palestine. Placing themselves in a no-win position, in 1947 the British decided to withdraw from Palestine and leave the young United Nations with the task of determining the region's future.

The United Nations Partition Plan of 1947 implemented a "two-state solution" to the problem of Palestine. It established an Arab state and a Jewish state. The plan was deeply flawed. The states' territories were long, narrow, and almost fragmented, giving each side a sense of vulnerability and insecurity (Fig. 9.2). War broke out in May 1948 between newborn Israel and the armies of the neighboring Arab countries of Transjordan, Egypt, Iraq, Syria, and Lebanon. The smaller but better-organized and more highly motivated Israeli army defeated the Arab armies, and Israel acquired its "pre-1967" borders (see Fig. 9.2). Israel gained control of about 77 percent of the territory of the Palestine mandate, including the coastal areas, the northern hill country of Galilee, the dry and thinly populated southern triangle called the Negev, the western portion of the city of Jerusalem in the Judean hills, and a corridor leading to the city from the coastal plain. Egypt occupied the Gaza Strip, a piece of land on the Mediterranean shore adjacent to Egypt's Sinai peninsula. Inhabited mostly by Palestinian Arabs, the Gaza Strip was part of the independent Palestine the U.N. had envisioned. Transjordan occupied the West Bank—the predominantly Arab hilly region of central Palestine on the west side of the Jordan River—and the entire old city of Jerusalem, including the Western Wall and Temple Mount. The Palestinian Arab state envisioned in the U.N. Partition Plan was thus stillborn.

In keeping with national legislation known as the **Law of Return,** the Jewish state of Israel has always granted citizenship to any Jew who wishes to live there. Immigrants from Europe (Ashkenazi Jews) and from the Middle East (Sephardic Jews) have populated Israel since its founding. New waves of Jewish immigration followed the dissolution of the Soviet Union and a change of government in Ethiopia, where an ancient Jewish group called Falashas lived in isolation for thousands of years. Since 1985, Israel has accommodated about 500,000 Russian Jews and 30,000 Ethiopian Jews. Members of both groups often complain that other Israelis discriminate against them.

In 1946, Jews made up 31 percent of the population of the mandate of Palestine (up from 11 percent in 1922 and 16 percent in 1931). In 1998, they made up 82 percent of the population of 5.8 million living within Israel's pre-1967 borders.

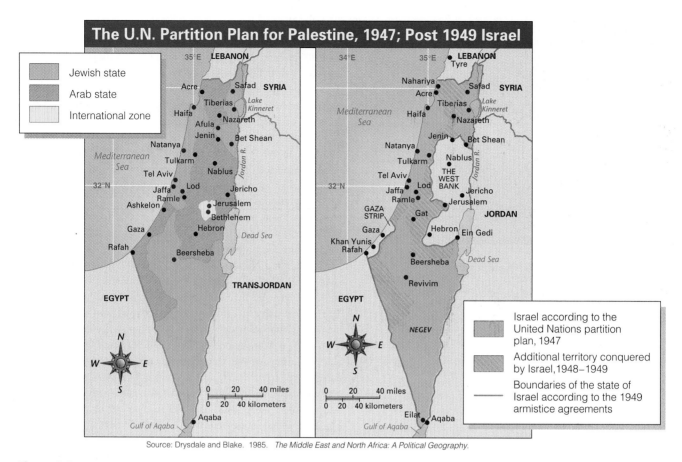

Source: Drysdale and Blake. 1985. *The Middle East and North Africa: A Political Geography.*

Figure 9.2　Maps of the 1947 U.N. Partition Plan for Palestine and of Israel's pre-1967 borders. The war, which began as soon as Britain withdrew from Palestine and Israel proclaimed its existence, aborted the U.N. Partition Plan and created a tense new political landscape in the region.

Palestinian Arabs with and without Israeli citizenship, and a variety of other ethnic groups, comprise the balance of the population. Another 2.3 million people live in the West Bank, Gaza Strip, and Golan Heights—areas occupied by Israel in the wars of 1967 and 1973. About 145,000 of those are Jewish settlers, and most of the rest are Palestinian Arabs (in the West Bank and Gaza Strip) and Druze and other Arabs (in the Golan Heights). Another 110,000 Jewish settlers live in East Jerusalem, which is also part of the territories Israel conquered in 1967.

A Geographic Sketch of Israel

Aided by large amounts of outside capital, primarily from the United States, the Israelis have developed their republic as a modern Westernized state. Still far from self-sufficient economically, however, Israel has difficulty in financing both a large military budget and extensive social programs, including those which service the recent immigrants from Russia and Ethiopia.

Despite many obstacles, one of Israel's greatest achievements since independence has been the expansion and intensification of agriculture (see Fig. 9.3). Both cultivated land and irrigation in Israel proper have greatly increased. Production

concentrates on citrus fruits (including the famous Jaffa oranges), which provide export revenue, and on dairy, beef, and poultry (eggs and meat) products, as well as flowers, vegetables, and animal feedstuff. Israel's intensive, mechanized agriculture, oriented to nearby urban markets, resembles in many ways the agriculture of densely populated areas in western Europe. Collectivized settlements called *kibbutzim* (singular, *kibbutz*) are a distinctively Israeli feature on the agricultural, and cultural, landscape. Many of these lie near the frontiers and have defensive as well as agricultural and industrial functions. Far more numerous and important in Israel's agricultural economy are other types of villages, including the small farmers' cooperatives called *moshavim* (singular, *moshav*) and villages of private farmers.

Agricultural development is concentrated in the northern half of the country, with its heavier rainfall (concentrated in the winter half-year in this Mediterranean climate) and more ample supplies of surface and underground water for irrigation. Annual precipitation averages 20 inches (50 cm) or more in most sections north of the city of Tel Aviv and in the interior hills where Jerusalem lies. Due to the rain shadow effect, very little rain falls in the deep rift valley of the Jordan River. From Tel Aviv southward to Beersheba, annual rainfall decreases to

Figure 9.3 Israel and its neighbors, with the occupied territories. Most highland areas are mountains except in Israel and Jordan, where hills predominate. International boundaries show the status prior to the 1967 Arab-Israeli War.

about 8 inches (20 cm), and in the central and southern Negev it drops to 4 inches or less. In the northern half of the country, irrigation water from varied sources (streams, lakes, springs, wells, artificial reservoirs to catch runoff, and reclaimed sewage water) helps intensify agriculture and overcome the summer drought. Surface water is scarce in the semiarid and arid south, and where underground water is found, it is often too saline for most crops. To expand agriculture along the coastal plain and in the northern Negev, Israel transfers large quantities of water from the north by the aqueduct and pipeline network of the National Water Carrier. The largest source for this network is Lake Kinneret (Lake Tiberias, or Sea of Galilee), which is fed by the upper Jordan River (see also Regional Perspective, p. 244).

Israel's main hope for continued support of its growing population lies in expanding industry, trade, and services, including tourism. Industry has increased greatly since independence, although—rather like Japan—little of Israel's industrialization can be based on its own mineral resources. Metal-bearing ores and mineral fuels are scarce, although potash, bromine, and other materials extracted from the Dead Sea have provided an important basis for expanded chemical manufacturing. Immi-

grants and the country's own advanced educational system have provided the technical and scientific skills required for diverse industries, including many high-technology enterprises. Israel's diamond-cutting industry is its most internationally well-known enterprise.

Israel's proclaimed capital of Jerusalem (population: 573,000, excluding East Jerusalem in occupied territory) is the country's largest city. Tel Aviv–Jaffa (population: 357,100, city proper; 1.88 million, metropolitan area) on the Mediterranean coast is Israel's greatest industrial center. The leading seaport and center of heavy industry (including steel milling and oil refining) is Haifa (population: 248,200) in northern Israel. The ancient town of Beersheba (population: 144,700) is the northern gateway, administrative center, and main industrial center of the Negev frontier region. To the south lies Israel's small Red Sea port and resort town of Eilat. Thanks to successful peace accords with Egypt in 1979 and Jordan in 1994, open doors now link Eilat with the neighboring resort towns of Taba in the Egyptian Sinai and Aqaba in Jordan. Tourism is booming, and planners envision a "Red Sea Riviera" that will grow on the three countries' Gulf of Aqaba shores to meet the demands of European sun-worshippers.

Regional Perspective

Water in the Middle East

In the arid Middle East, where most water is available either from rivers or from underground aquifers that cross national boundaries, control over water is an especially difficult and potentially explosive issue. The ongoing peace process could settle some of the more contentious disputes. In its September 1995 accord with the PLO, for example, Israel pledged to increase Palestinian access to the fresh-water aquifers underneath the West Bank, which supply about 40 percent of Israel's water. In 1998, however, a newer Israeli government insisted that water resources on the West Bank must remain under strict Israeli control, and Israel's infrastructure minister called for large swaths of territory in the West Bank to be held by Israel to secure the aquifers below. Jordan and Israel promised in their 1994 peace treaty to reach an agreement on sharing waters from the Jordan River (which forms a portion of their common border) and its tributary, the Yarmuk River. They are discussing a joint venture to build the "Dead-Red Canal," which would connect the Gulf of Aqaba with the Dead Sea. The gravity flow of seawater to the Dead Sea would spin turbines and run generators to produce electricity to be shared between the two nations.

Still to be addressed is the sharing of water between Israel and Syria. In occupying the Golan Heights, Israel controls an important watershed and some of the northern bank of the Yarmuk River on the border with Jordan. For many years, Israel indicated it would never allow Syria and Jordan to construct the "Unity Dam" which would store waters of the Yarmuk River to be shared between those countries. Israel thus implied it would bomb the dam rather than allow it to deprive Israel of Jordan River water.

As a "downstream" riverine, or **riparian,** nation, completely dependent upon water originating outside the country, Egypt considers relations with upstream countries vital to its long-term national security. Egypt and Sudan are signatories of a Nile Waters Agreement that apportions water between them, but Egypt does not have similar agreements with other countries. Ethiopia's diversion of water from vital Nile headstreams is a potential future source of friction with Egypt and Sudan, but at present Ethiopia has little money to finance the dams, canals, and pumps that large-scale diversions would require. Meanwhile, Egypt's demands on Nile waters are increasing. Egypt is now excavating a multi-billion-

dollar canal that will transport water from Lake Nasser over a distance of 300 miles (500 km) to the Kharga Oasis of the Western Desert. Proponents of the scheme insist it will result in the cultivation of nearly 1.5 million acres (600,000 hectares) of "new" land and provide a living for hundreds of thousands of people. Critics argue that it is a waste of money, and that salinization and evaporation will take a huge toll on the cultivated land and the country's water supply.

Turkey is an "upstream" state and is the source of four-fifths of Syria's water and two-thirds of Iraq's. These downstream neighbors are distraught by the diminished flow and quality of water resulting from Turkey's comprehensive Southeast Anatolia Project. When completed, the project is expected to reduce Syria's share of the Euphrates waters by 40 percent and Iraq's by 60 percent. Also increasing the likelihood of serious future tension is a history of strained relations among Turkey, Syria, and Iraq, accompanied by the fact that no commonly accepted body-of-water law governs the allocation of water in such international situations.

The Arab-Israeli Conflict

Israel's relations with the Arab world have always cast a cloud on this young country. Central issues—all of which have geographic underpinnings—include (1) Israel's right to exist, (2) Israel's possession of occupied territories taken from Arab countries in warfare, (3) the rights of the Arab Palestinians who fled as refugees in wartime or were overrun by Israeli occupying forces, (4) who should control Jerusalem, and (5) Jordan River water appropriation. Poor relations between Israel and its Arab neighbors have exploded into full-scale war on five occasions since Israel's declaration of independence on May 14, 1948. These conflicts have profoundly affected the geography of the Middle East and have had a marked impact on international economies and political relations. In order to understand these effects and the issues of the current Middle East peace process begun in the 1990s, it is essential to be familiar with these five

wars (including the 1948 war, described above), particularly as they have rearranged the boundaries of nations and territories.

The 1956 War.   Israeli forces invaded Egypt's Sinai peninsula shortly before a British and French invasion of the Suez Canal Zone in the autumn of 1956. Egypt's President Gamal Abdel Nasser had nationalized the Suez Canal earlier in the year, ostensibly as a means of paying for the construction of the Aswan High Dam. This antagonized the British and French, who had built and still owned the canal, and viewed it as a vital asset to their national economic interests. At the same time, Israel wanted to destroy Soviet-made arms which Egypt had positioned in the Sinai, and to open the Gulf of Aqaba to Israeli shipping. Pressure from both the United States and the Soviet Union quickly forced all of the invaders to withdraw from Egypt. It was a sensational victory for the firebrand nationalist

Nasser. Israel won the right of navigation through the Gulf of Aqaba. Great Britain and France were humiliated.

The 1967 War ("The Six Day War"). This struggle was precipitated in part when, by positioning arms at the Strait of Tiran **chokepoint** (see Figs. 9.4 and 9.26, and Definitions and Insights, p. 246), Egypt again closed the Gulf of Aqaba to Israeli shipping. President Nasser and his Arab allies took several other belligerent but nonviolent steps toward a war they were ill-prepared to fight. Israel elected to make a preemptive strike on its Arab neighbors, virtually destroying the Egyptian and Syrian air forces on the ground. Israel gave Jordan's King Hussein an opportunity to stay out of the conflict. However, Jordan went to war and quickly lost the entire West Bank and the historic Old City of Jerusalem. The entire nation of Israel was transfixed by the news that Jewish soldiers were praying at the Western Wall. The Israeli army (Israeli Defense Forces, or IDF) also seized the Gaza Strip, Egypt's Sinai Peninsula (thus closing the Suez Canal), and the strategic Golan Heights section of Syria overlooking Israel's Huleh Valley, in which the upper Jordan River flows. Israel had tripled its territory in six days of fighting (see Fig. 9.3).

The 1973 War. On October 6, 1973, the Jewish holy day of Yom Kippur, Egypt and Syria launched a surprise attack on Israel with the hope of rearranging the humiliating, stalemated political map of the Middle East. Egypt's army initially showed surprising strength, penetrating deep into Sinai and overturning Israel's image of invulnerability. Israeli troops soon surrounded Egypt's army, but in the ensuing disengagement talks Egypt won back the eastern side of the Suez Canal and by 1975 was able to reopen it to commercial traffic. Syrian forces won back a small sliver of the Israeli-occupied Golan Heights.

The United States supported Israel during the conflict. Fearing a move by the Soviet Union, Syria's ally, President Richard M. Nixon placed U.S. forces on an alert status that presumed a

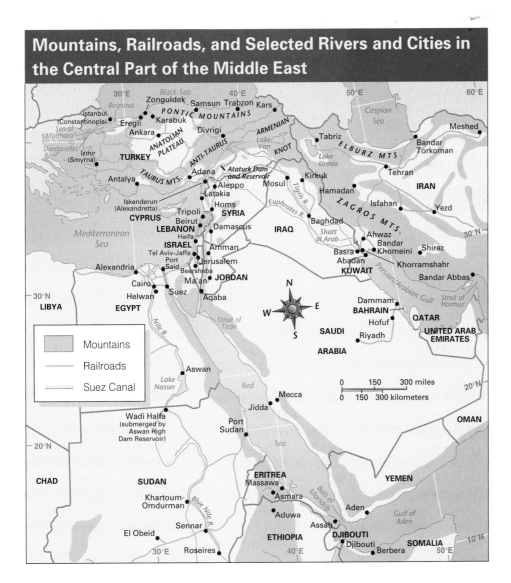

Figure 9.4 Mountains, railroads, and selected rivers and cities in the central part of the Middle East. Mountainous areas are generalized.

Definitions & Insights

CHOKEPOINTS

The Strait of Tiran at the mouth of the Gulf of Aqaba is one of the Middle East's many **chokepoints,** or strategic narrow passageways on land or sea that may be easily closed off by use of force—or even the threat of force. Other notable chokepoints in the area are the Strait of Gibraltar at the mouth of the Mediterranean Sea; the Dardanelles and Bosporus Straits (together known as the Turkish Straits; see Fig. 9.4) linking the Black Sea and Aegean Sea; the Bab el-Mandeb at the mouth of the Red Sea; the Strait of Hormuz at the mouth of the Persian/Arabian Gulf; and the Suez Canal. Notable events in military history and the formation of foreign policy in the Middle East focus on these strategic places. For example, Iran's plans to station Chinese-made silkworm missiles on the Strait of Hormuz and thus threaten international oil shipments led to a new level of United States involvement late in the Iran-Iraq War of 1980–88; and Egyptian closure of the Strait of Tiran chokepoint to Israeli shipping in 1967 helped precipitate the Six Day War.

possible nuclear exchange. And, led by Saudi Arabia, furious Arab allies of Egypt and Syria imposed an **oil embargo,** halting petroleum sales to the United States for eight months beginning in October 1973, precipitating the country's **energy crisis,** when oil supplies plummeted and prices soared.

In 1979, Israel returned Sinai to Egypt under the U.S.-sponsored Camp David Accords, and Egypt recognized Israel's rights as a sovereign state. The Gaza Strip, West Bank, and Golan Heights continued to be held by Israel and came to be known as the "occupied territories."

The 1982 War.   This conflict was one facet of a larger war in Lebanon involving both civil strife and international intervention. Lebanon had become the main stronghold of the Palestine Liberation Organization (PLO), which used the country as a base for guerrilla operations against Israel. Under the leadership of Yasir Arafat, the PLO had been established in 1964 with the ambition of establishing an independent Palestinian state either in the occupied territories or in Israel itself. In June 1982, Israel invaded Lebanon with the avowed intention of smashing the PLO and establishing security for the northern Israeli frontier. Syria, whose forces had previously occupied Lebanon's Bekaa Valley on an alleged peacekeeping mission for the Arab League, resisted the Israeli advance and gave support to the PLO. Israeli troops routed and inflicted heavy casualties on both the Syrians and the PLO.

The Israeli push continued northward to the suburbs and vicinity of Beirut, Lebanon's capital, culminating in a ferocious aerial and artillery bombardment of the city itself that ended only when U.S. president Ronald Reagan exerted pres-

sure on Israel to cease fire. Under supervision by U.S. and other international peacekeeping troops, the PLO withdrew from Beirut to Tunisia. After the multinational troops withdrew, alleged Muslim assailants killed Lebanon's Christian president Bashir Gemayal. Angry Christian forces then massacred hundreds of Palestinian and other civilians in refugee camps south of Beirut. U.S. and other multinational forces were again deployed to Beirut, and American ships used firepower to assist the new Christian president in widening his area of authority.

Lebanese Shi'ite Muslim guerrillas, seeking to avenge United States attacks on Lebanon and its support of Israel, soon bombed American and French targets in Beirut, killing over 200 U.S. marines in a barracks near Beirut's airport. The multinational forces withdrew, but Syrian troops remained in Lebanon and allowed Iranian forces and Iran's Shi'ite Muslim allies in Lebanon to establish a stronghold in the Bekaa Valley town of Baalbek. Subjected to persistent guerrilla attacks, Israeli troops withdrew from all of Lebanon except a self-declared "security zone" in the south, inhabited by both hostile Shi'ite Muslims and Israel's de facto Christian Lebanese allies. Israel continues to exert control over this zone and engages in regular skirmishes with Hizbullah, a Shi'ite Muslim organization sympathetic to Iran and hostile to the peace process between Israel and its Arab neighbors.

The Palestinian Question

The creation of Israel, the subsequent wars, and the Camp David Accords left the future of the Palestinians unresolved. The Palestinians are Arabs who historically have lived in the geographic region of Palestine (which is roughly the area now made up of Israel and the occupied territories) and who recognize themselves as a distinct ethnic group. About 700,000 Palestinians are now Israeli citizens residing within the pre-1967 boundaries of Israel. Another five million Palestinians live outside Israel's pre-1967 borders, in the West Bank, the Gaza Strip, and abroad. Prior to and during the fighting of 1948, approximately 800,000 Palestinian Arabs were forced, or chose, to flee from the new state of Israel to neighboring Arab countries (Fig. 9.5). The United Nations established refugee camps for these displaced persons in Jordan, Egypt, Lebanon, and Syria. Little was done to resettle them in permanent homes, and both the Arab governments and the refugees themselves continued to insist on the return of the refugees to Israel and the restoration of their properties there.

The refugee situation became more complex as a result of the 1967 war, in which Israel took over the areas where most of the camps were located. Many persons in the camps again took flight, and Palestinians fleeing villages and towns in the newly occupied territories joined them as refugees (see Fig. 9.5). Some later returned to their homes, but an estimated 116,000 either did not attempt to return or were denied permission by Israel authorities to do so.

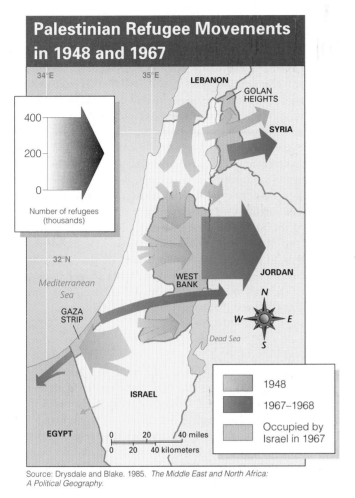

Palestinian Refugee Movements in 1948 and 1967

Source: Drysdale and Blake. 1985. *The Middle East and North Africa: A Political Geography.*

Figure 9.5 Palestinian refugee movements in 1948 and 1967. Many who fled in the first conflict relocated again in the second.

Figure 9.6 A Jewish settlement in the West Bank, near Jerusalem. The pitched roofs indicate that these dwellings have piped-in supplies of water. Most Palestinian homes in the West Bank lack plumbing and so have flat roofs on which water can be collected and stored. In the foreground is a Palestinian Arab shepherd with his flock. *R. Norman Matheny/ Christian Science Monitor*

After 1967, and particularly after the election of its conservative Likud Party to power in 1977, Israel moved to strengthen its grip on the occupied territories through security measures and the government-sponsored establishment of new Jewish settlements, most of which were, and are still, spread through the West Bank (Fig. 9.A). These are generally not frontier farming settlements but communities inhabited by relatively prosperous middle-class Jews who commute to jobs in Israel. They are not built as temporary encampments but as permanent fixtures meant to create what have been described as "facts on the ground"—an Israeli presence so entrenched that its withdrawal would be almost inconceivable (Fig. 9.6). Some of the motivation for this settlement is ideological, as the West Bank is composed of historical Judea and Samaria, which many devout Jews regard as part of Israel's historic and inalienable homeland.

The proliferation of new settlements, which persisted until Israel's liberal Labor Party came to office in 1992, increased the bitterness between the Arab world and Israel and drove a wedge into the Jewish population of Israel itself. Many Israelis strongly opposed further settlement in the occupied territories and annexation of these territories to the Jewish state, since, if the territories were annexed, Israel would become a country whose population would be only approximately 60 percent Jewish. With an Arab minority that has a higher birth rate than that of the Jews, Jews could eventually become the minority population in their own country.

In 1987, Palestinians instigated a popular uprising in the occupied territories known as the Intifada (Arabic for "the shaking"). Initially they relied on rocks and bottles to engage Israeli troops, who answered with rubber and metal bullets. International media coverage of a Palestinian "David" fighting an Israeli "Goliath" did much damage to Israel's image abroad, and many Israelis came to question the justice and importance to security of Israel's continued occupation. Popular opinion and, by 1992, the new Labor government began to consider the previously unthinkable: giving the Palestinians at least some control over land and internal affairs in the occupied territories.

Jordan, Between Iraq and Israel

Jordan (1998 population: 4.2 million), bordering Israel on the east, is mostly desert and semidesert. The main agricultural areas and settlement centers, including the largest city and capital, Amman (population: 965,000, city proper; 1.63 million, metropolitan area), are in the northwest. Prior to the 1967 war with Israel, northwestern Jordan included two upland areas—the West Bank, or Palestine Hills, on the west, and the East Bank, or Transjordanian Plateau, on the east—separated by the deep rift valley occupied by the Jordan River, Lake Kinneret, and the Dead Sea (see Fig. 9.3). The valley is the deepest

Figure 9.7 At 1286 feet (385 m) below sea level, the surface of the Dead Sea is the lowest point on Earth. This lake occupies a portion of the rift valley system that extends from northern Lebanon into southern Africa. This is a view from the Israeli shore of the Dead Sea, looking eastward to the escarpment on the Jordanian side. Note the large palm plantation in the foreground. *Joseph J. Hobbs*

depression on the Earth's land surface, lying about 600 feet (183 m) below sea level at Lake Kinneret and nearly 1300 feet (400 m) below sea level at the Dead Sea (the lowest point on Earth; Fig. 9.7). Although annual precipitation in the valley increases northward to an average of more than 12 inches (30 cm) in the vicinity of Lake Kinneret, most of the valley bottom lies in deep rain shadow and receives less than 4 inches (10 cm) a year. The bordering uplands, rising to more than 2000 feet (610 m) above sea level and more and lying in the path of moisture-bearing winds from the Mediterranean, receive precipitation during the cool season that averages 20 inches (50 cm) or more annually within two north-south strips. It was within these belts, one in the West Bank (occupied by Israel in 1967) and the other in the hilly East Bank, that most of Jordan's pre-1967 population resided. The largest numbers were villagers supporting themselves by cultivating winter wheat and barley, vegetables, and fruits, especially olives and grapes.

Jordan is a poor country. Today only about 5 percent of Jordan is cultivated, although much larger areas are used to graze sheep and goats. Rainfall is so scarce that only small areas can be cultivated without irrigation, and water available for irrigation is very limited. Some food has to be imported, although irrigated vegetables (especially tomatoes), together with olives, citrus fruits, and bananas, provide substantial exports. Only limited manufacturing is carried on in Jordan, and the contrast with the large and technologically advanced industries of neighboring Israel is extreme. Natural resources are limited, although rock phosphate and potash extracted from the Dead Sea provide major exports. They are shipped through Jordan's small Red Sea port of Aqaba, located on the Gulf of Aqaba and adjoining Eilat in Israel. Aqaba also handles much transit traffic, mostly by truck, for Iraq and was especially

important to Iraq during its war against Kuwait and the Western coalition.

There is an enormous potential for growth in Jordan's tourism industry, especially now that the country has established peaceful relations with Israel. Traditionally constrained within their tiny country, Israelis are now enjoying the freedom of traveling in neighboring Arab states and enjoying such sites as Jordan's Nabatean city of Petra (see Fig. 8.5) and Egypt's Sinai beaches.

Jordan's economic and political situation has always been precarious. Throughout his long reign, which ended with his death from cancer in 1999, King Hussein withstood various crises resulting from the entry of great numbers of Palestinians, including the PLO. For a time, Jordan was the chief operating base of the PLO, but in 1970 and 1971, the king expelled the armed Palestinian forces, who relocated to Lebanon. Palestinians still make up about 60 percent of Jordan's population, not including those in the Israeli-controlled West Bank that Jordan annexed in 1950 and that Israel took in the 1967 war.

Lebanon and Syria

Israel's immediate neighbors on the north are Lebanon and Syria (see Fig. 9.3). Their physical and climatic features are similar to those of Israel and Jordan. As in Israel, narrow coastal plains backed by highlands front the sea. The highlands of Lebanon and Syria are loftier than those of Israel and Jordan, Mount Hermon, and the Lebanon and Anti-Lebanon ranges reach above 10,000 feet (*c.* 3050 m) in the Lebanon range. Cyclonic precipitation brought by air masses off the Mediterranean is supplemented by orographic rain and snow induced by the mountain chains that parallel the coast and block the path of the prevailing westerly winds.

Lebanon and Syria as a result have agricultural patterns that conform to the mediterranean climatic pattern of winter rain and summer drought. The coastal plains and seaward-facing mountain slopes are settled by villagers growing Mediterranean crops such as winter grains, vegetables, and drought-resistant trees and vines. Precipitation decreases and pastoralism increases toward the east. The Anti-Lebanon range and Mount Hermon are drier than the Lebanon range, which screens out moisture they would otherwise receive. These mountains, and the eastern slopes of the Lebanon range as well, are used primarily for grazing.

Lebanon's Bekaa Valley, drained in the south by the Litani River, lies between the Lebanon and Anti-Lebanon ranges and is a northward continuation of the Jordan rift valley and the African Great Rift Valley. Baalbek, its chief city and the site of an important ancient Roman temple, became a Syrian military stronghold in 1976 and a center of pro-Iranian Hizbullah activities against Israel after 1982. The political and military volatility of the Bekaa was conducive to the emergence of a flourishing drug industry—producing both hashish (from marijuana plants) and opium (from poppies)—from the mid-1970s until the early 1990s.

A semiarid zone covers central and northern Syria east of the mountains. In southeastern Syria, it trends gradually to the Syrian Desert, most of which lies in Iraq and Jordan. The semiarid region, much of which has good soils, is Syria's main producer of cotton, wheat, and barley. Cotton is grown as an irrigated summer crop, whereas wheat and barley, primarily nonirrigated, are winter crops.

Syria is more industrialized than Lebanon, but neither country compares with Israel industrially. Textile milling, agricultural processing, and other light industries predominate. Both countries are deficient in minerals, although Syria has an oil field of moderate importance in the east.

Three large cities of long historic importance are in Syria and Lebanon. Damascus (population: 1.55 million), Syria's capital, is in a large irrigated district in the southwest at the eastern foot of the Anti-Lebanon range. Its Umayyad Mosque, originally a Byzantine cathedral, is one of the great monuments of the Middle East. The main city of northern Syria, Aleppo (population: 1.54 million), is an ancient caravan center located in the "Syrian Saddle" between the Euphrates valley and the Mediterranean. And the old Phoenician city of Beirut (population: 509,000, city proper; 1.68 million, metropolitan area), on the Mediterranean, is Lebanon's capital, largest city, and main seaport. Equipped with modern harbor facilities and an important international airport, Beirut before 1975 was one of the busiest centers of transportation, commerce, finance, and tourism in the Middle East. Indeed, for both its prosperity and its natural beauty, pre-1975 Lebanon had earned the name "Switzerland of the Middle East." Then a precarious political balance broke down and the country was plunged into civil war (Fig. 9.8). Large parts of the capital city were devastated, and only now is Beirut rising from the ashes in an ambitious, expensive reconstruction program.

The origins of the civil war of 1975–1976 were rooted in Lebanon's mixture of religious communities that entered at various times and became strongly localized in particular areas.

Currently, seventeen separate ethnic and religious groups, known as "confessions," are recognized officially (Fig. 9.9). Maronite Christians prevail in the Lebanon range north of Beirut, but close to the northern border with Syria Sunni Muslims predominate. Shi'ite Muslims are the majority in the Bekaa Valley and along Lebanon's southern border with Israel. The Shouf Mountains, a portion of the Lebanon range south of Beirut, are home to the Druze, who practice an ancient and distinct form of Shi'ite Islam. Greek Orthodox, Greek Catholic, and nearly a dozen other Christian sects are also present in Lebanon. Beirut itself is divided (most strictly, in the war years of the 1970s and 1980s, by the "Green Line") between predominantly Muslim West and Christian East. Conflict between the communities broke out from time to time, and foreign powers that gained influence with one or another group heightened the antagonisms.

What the French carved out of greater Syria and put together as Lebanon in 1920 was a collection of small geographic areas, each dominated by one or a few minorities. A 1932 census revealed a majority of Christians in Lebanon, and the National Pact, or constitution, drafted for an independent Lebanon in 1946 distributed power according to those census figures. Maronite

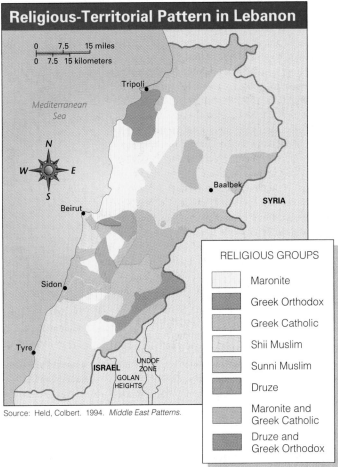

Religious-Territorial Pattern in Lebanon

RELIGIOUS GROUPS

- Maronite
- Greek Orthodox
- Greek Catholic
- Shii Muslim
- Sunni Muslim
- Druze
- Maronite and Greek Catholic
- Druze and Greek Orthodox

Source: Held, Colbert. 1994. *Middle East Patterns.*

Figure 9.8 The Lebanese Civil War of 1975–1976 devastated the capital city of Beirut. This is the lobby of the Holiday Inn, where a fierce tank battle had taken place a few days earlier. *Joseph J. Hobbs*

Figure 9.9 Generalized map of the distribution of religious groups in Lebanon.

PROBLEM LANDSCAPE

On the Brink of Peace

PRESIDENT ANWAR SADAT OF EGYPT SET the precedent in 1979: Arabs could make peace with Israel. Egypt's Arab neighbors scorned this populous and influential Arab country for a decade for its peacemaking, and Sadat was assassinated because of it. However, in the process, Israel restored the Sinai to Egyptian control and the United States rewarded Egypt handsomely with about $2 billion in aid annually. Along with about $3 billion awarded Israel annually, these countries have been the two largest recipients of U.S. foreign aid.

Israel and its remaining Arab adversaries stand to gain much more through the implementation of a comprehensive framework for peace. The United States initiated a sequence of events early in the 1990s that makes that prospect more likely than ever before. After Iraq invaded Kuwait in 1990, the United States and its European allies needed to build a firm coalition of Arab military and political will to oppose Iraq. Backed also by a sympathetic Soviet Union, the United States obtained this support in part by promising to pursue Middle East peace aggressively once Iraq was defeated. The presidents of both countries (George Bush in the United States and Mikhail Gorbachev of the Soviet Union) kept their pledges, and, in Madrid in October 1991, they sponsored the historic opening of talks between Israel and four Middle Eastern delegations: Egyptian, Syrian, Lebanese, and—because Israel did not recognize the PLO—a joint Jordanian-Palestinian delegation that excluded PLO members.

Little progress was made in the on-again, off-again peace talks until the Labor Party in Israel, under Prime Minister Yitzhak Rabin, came to power in 1992. Throughout the summer of 1993, Rabin's ministers undertook secret talks in Oslo,

Norway, with members of the PLO. In early September came the historic announcement that Israel recognized the legitimacy of the PLO as the sole representative of the Palestinians. The PLO in turn recognized Israel's right to exist and renounced terrorism. On September 13, 1993, on the White House lawn, U.S. president Bill Clinton orchestrated a historic handshake between PLO leader Yasir Arafat and Israeli prime minister Rabin. The leaders signed an agreement known as the "Gaza-Jericho" Accord, which established a framework for peace with the following major points:

1. A five-year interim period of limited self-rule (autonomy) for Palestinians in the occupied territories, beginning in 1993 in the Gaza Strip and the West Bank town of Jericho. Self-rule was to cover education, culture, health, taxation, and tourism. By the third year (1996), negotiations were to begin on the permanent status of the territories, and by May 4, 1999, the permanent status issues were to be resolved.
2. Withdrawal of Israeli troops from Palestinian areas in the Gaza Strip, Jericho, and Palestinian towns in the West Bank.
3. Elections of a Palestinian Authority to oversee self-rule.
4. Formation of a Palestinian police force.
5. Continued Israeli protection of Jewish settlers in the Gaza Strip and West Bank.
6. Discussion of the possible repatriation of Palestinian refugees from the 1967 war.
7. Discussion of the status of Jerusalem, claimed by both Israelis and Palestinians as their "eternal" capital.

Israeli forces did withdraw from Jericho and from Palestinian areas of the Gaza Strip. Yasir Arafat took up residence in the Gaza Strip—another historic, previously

unimaginable event. However, Israeli and Palestinian public approval of the peace process has periodically waxed and waned since September 1993. Arafat's PLO has been forced to deal with a series of violent attacks on Jewish civilians and soldiers by the radical Palestinian organization HAMAS. While these terrorist attacks caused many Israelis to reject the peace process, subsequent PLO detention and prosecution of Palestinian extremists also hurt Palestinian support for the PLO, which many Palestinians now see as an agent for Israeli interests. Many Palestinians turned against the peace process after February 1994, when a right-wing Israeli extremist opposed to Israel's negotiations with the Palestinians shot and killed Muslim Palestinian worshippers in the Tomb of the Patriarchs, a holy place in the West Bank town of Hebron where the prophet Abraham is said to be buried.

Nevertheless, the peace process remained on track. In September 1995, Israel and the PLO signed another agreement to implement the withdrawal of Israeli troops from Palestinian cities in the West Bank. Palestinians then acquired self-rule over about 30 percent of the West Bank, with the rest, including all Jewish settlements, remaining in Israeli hands. The agreement reaffirmed that the final status of the territories would be negotiated beginning in 1996.

Many of the approximately 136,000 Jewish settlers living in the West Bank (where about one million Palestinians also live) are vehemently opposed to the peace process. Regarding Jewish presence in the area as a God-given right, they fear the peace process will result in an independent Palestinian state and the forced withdrawal of Israeli Jews. On November 4, 1995, after a peace rally in Tel Aviv, an Israeli Jew who held this view assassinated Prime Minister Yitzhak Rabin. The assassin proclaimed that his purpose was to destroy the peace process.

The short-term impact of Rabin's murder was a surge in popular Israeli support for his legacy of the peace process. This support was quickly reversed by a series of bloody suicide bombings carried out by HAMAS militants against civilian targets in Israel during early 1996. These attacks contributed to the defeat of Rabin's successor, Shimon Peres, and the victory of the conservative Likud Party's Benjamin Netanyahu, in Israel's 1996 general election.

As a candidate for prime minister, Netanyahu had vowed to step up Jewish settlement in the West Bank and to slow the pace of the peace process in the interest of Israel's security. He also promised that he would not meet with Palestinian leader Yasir Arafat, whom he labelled a terrorist. Under much international pressure to fulfill Israel's treaty obligations to its Arab neighbors, however, Prime Minister Netanyahu, after late 1996, held face-to-face talks with Arafat. Under Netanyahu, Israel refused to abide by his predecessors' negotiated timetable for Israeli withdrawals from the West Bank; he argued that the PLO must first demonstrate its commitment and ability to fight Palestinian terrorism against Israel. Violence on both sides continued, with HAMAS periodically attacking Jewish civilian targets, and Jewish settlers sometimes killing Arabs. Late in 1996, Israeli authorities opened a new entrance to an ancient tunnel near the Western Wall and the Dome of the Rock. Some Palestinian Muslims perceived this action as an incursion against their sacred space, and mounted demonstrations against Israel. These quickly escalated into widespread unrest, with armed Palestinian police and unarmed Palestinian civilians alike engaging Israeli troops. Casualties were high

(Box continued on following page)

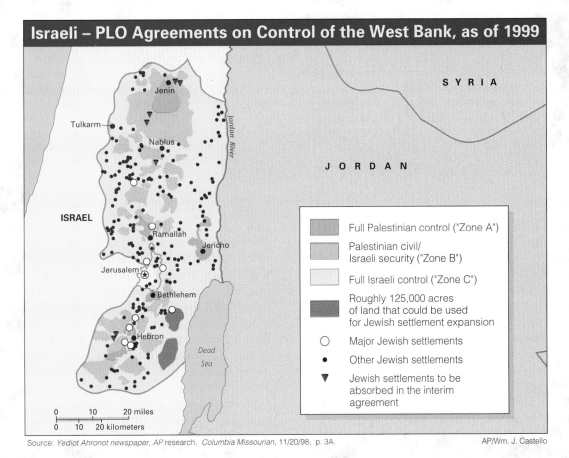

Source: *Yediot Ahronot newspaper, AP* research. *Columbia Missourian*, 11/20/98, p. 3A.   AP/Wm. J. Castello

Figure 9.A   Israeli and Palestinian areas of control in the West Bank in 1999.

On the Brink
of Peace *(continued)*

on both sides, and the disturbances threatened to bring down the delicate peace process. However, early in 1997, Israel and the PLO signed an agreement providing for further Israeli troop withdrawal from Palestinian areas in the West Bank, including the city of Hebron.

In October 1998, after more than a week of intensive negotiations in Maryland, Prime Minister Netanyahu, Chairman Arafat, Jordan's King Hussein, and President Clinton concluded an agreement, known as the "Wye River Memorandum," promising the transfer of more West Bank lands from Israeli to Palestinian control, and promising increased Palestinian efforts to crack down on Palestinian terrorists and so guarantee the security of Israelis in the Palestinian territories and in Israel. Most critically, Israel pledged to transfer an additional 13 percent of the territory it controlled in the West Bank from the area designated in the peace process as "Zone C"

(complete Israeli control) to "Zone B" (Palestinian civilian control, Israeli security control), and pledged to transfer 14 percent of the West Bank from "Zone B" (Palestinian civilian control, Israel security control) to "Zone A" (complete Palestinian control). That brought the total of West Bank lands under complete Palestinian control to 17 percent, leaving 57 percent completely in Israeli hands, and 26 percent under joint control (see Figure 9.A). Israel also released some Palestinian prisoners from Israeli jails, allowed the opening of an international Palestinian airport in Gaza, and accepted the establishment of two corridors of safe passage for Palestinians between the West Bank and the Gaza Strip. However, insisting that the PLO was still not ensuring Israel's security, Netanyahu refused to implement the territorial transfers promised in the Wye Agreement. This freeze in the peace process contributed to Netanyahu's defeat in the 1999 election, when the Labor Party's Ehud Barak became Prime Minister.

The peace process has both progressed and faltered on other fronts as well. In

October 1994, following the PLO's lead (and again under President Clinton's eye), Jordan signed a treaty with Israel. It called for an end to hostility between the two nations; establishment of full diplomatic relations; the opening of exchanges in trade, tourism, science, and culture; and an end to economic boycotts. Meanwhile, Syria and Israel have failed to reach a peace agreement because of a deadlock on the issues of recognition of Israel and return of occupied land: Israel refuses to return the Golan Heights to Syria until Syria recognizes the sovereignty and legitimacy of Israel, and Syria refuses to recognize the sovereignty and legitimacy of Israel until Israel returns the Golan Heights. Syrian president Hafez al-Assad may be waiting to see how successful the implementation of Israel's existing peace treaties is before concluding his own with Israel. And because Syria is the dominant power in Lebanon, with tens of thousands of troops stationed in that tiny, politically weak country, peace between Israel and Lebanon is improbable unless Israel and Syria first come to terms.

Christians held a majority of seats in Parliament, followed by Sunni Muslims and Shi'ite Muslims. Top leadership positions were allocated by a formula still followed today, with the president a Maronite Christian, the prime minister a Sunni Muslim, and the speaker of Parliament a Shi'ite Muslim.

There was never an effort to unify the disparate geographic and ethnic enclaves of this mountainous land. Meanwhile, Muslim populations grew. Eventually they came to outnumber the Christians, who were unwilling to relinquish their political and economic privileges. Pushed from Jordan, PLO guerrillas established a virtual "state within a state" in the southern slums of Beirut and in Shi'ite-controlled areas of southern Lebanon from which they launched attacks on Israel.

The volatile mix of Muslims versus Christians exploded in 1975, complicated by shifting alliances with outside powers, including Israel and Syria. By the end of the war in 1976, as many as 100,000 people, most of them civilians, had been killed. Lebanon effectively had become a number of geographically distinct microstates or fiefdoms—a politically unstable situation that set the stage for the next war in 1982.

Lebanon is now on a steady course of reconciliation and reconstruction. A peace accord known as the Taif Agreement has redistributed power to reflect the new demographic composition in which Muslims outnumber Christians. Once dominated by Christians, the Lebanese army is now comprised of soldiers from all of Lebanon's ethnic confessions. There are numerous social and education programs underway which mix children from different ethnic groups, with the hope they will learn tolerance. The Taif Agreement called for the withdrawal of Syrian troops, but the government in Damascus has never accepted the 1920 division of greater Syria, which gave a portion of Syria to Lebanon, and as of 1999 tens of thousands of Syrian soldiers remained in Lebanon. Surprisingly perhaps, many Lebanese belonging to all the major ethnic groups want the Syrians to remain to keep the peace while the country's former enemies continue the long, slow healing process.

Syria, like Lebanon, is a country beset by religious factionalism. In Syria, however, Muslims outnumber Christians nine to one. In 1970, a coup overthrew the ruling Sunni Muslim establishment and brought to power Hafez al-Assad, a member of the Alawite sect of Shi'ite Islam. Under his authoritarian rule as president, dominance of the government quickly passed to the Alawites. The police and army crushed opposition within the other communities. President Assad is a skilled politician who plays Middle Eastern affairs cautiously. His regime survived the economic and political setbacks accompanying the downfall of its benefactor the Soviet Union, and has slowly adopted a more accommodating relationship with the United States and the West.

9.3 Egypt: The Gift of the Nile

At the southwest, Israel borders Egypt (officially the Arab Republic of Egypt), the most populous Arab country. This ancient land, strategically situated at Africa's northeastern corner between the Mediterranean and Red Seas, is utterly dependent on a single river. Not only does the Nile supply the water that enables 66 million Egyptians to exist, but it created the alluvial floodplain and triangular delta on which more than 95 percent of all Egyptians live (see map, Fig. 9.10). The Greek historian and geographer Herodotus called Egypt "an acquired country—the gift of the Nile." The river's bountiful waters and fertile silt helped Egypt become a culture hearth that produced remarkable achievements in technology and cultural expression over a period of almost 3000 years until about 500 B.C., when a succession of foreign empires came to rule the land.

The Nile Valley may appropriately be described as a "river oasis," for stark, almost waterless desert borders this lush ribbon (Fig. 9.11). Only 3 percent of Egypt is cultivated, and nearly all this land lies along and is watered by the great river. The conversion of the original papyrus marshes and other wetlands along the Nile to the thickly settled, irrigated landscape of today is a process that has been unfolding for more than fifty centuries. It is difficult to imagine that, in the time of the pharaohs, crocodiles and hippopotami swam in the Nile and game animals typical of East Africa roamed the nearby plateaus.

Egyptian Agriculture

Ancient Egyptian civilization was based on a system of **basin irrigation,** in which the people captured and stored in built-up embankments the flood waters which spread over the Nile floodplain each September. Egyptian farmers allowed the water to stand for several weeks in the basins, where it supersaturated the soil and left a beneficial deposit of fertile volcanic silt washed down by the Blue Nile from Ethiopia. They drained the excess water back into the Nile and planted their winter crops in the muddy fields. They were able to harvest one crop a year with this method. They were able to produce a second or third crop on limited areas of land using water-lifting devices such as a weighted lever or *shaduf* (a bucket on a counterweighted pole, manually operated to raise water from canal to field) and an

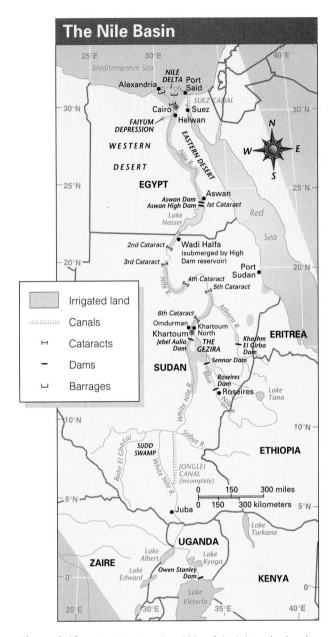

Figure 9.10 The Nile Basin. The widths of the irrigated strips along the Nile are exaggerated.

Archimedes' screw (a cylinder containing a screw, also manually turned to raise the water).

However, achieving this so-called **perennial irrigation** on a vast scale required the construction of barrages and dams—innovations of the much later 19th and 20th centuries (see Fig. 9.10). It was during this time—beginning late in the 19th century—that French and British colonial occupiers of Egypt, anxious to boost Egypt's exports of cotton (a summer crop demanding perennial irrigation), began Egypt's conversion from basin to perennial irrigation by constructing a number of barrages, or low barriers designed to raise the level of the river

Figure 9.11 The Nile Valley in southern Egypt. The Nile River snakes through the left side of the cultivated area. Few other places on Earth exhibit such a stark contrast between productive, populated landscapes and those virtually devoid of water, people, and vegetation. *Joseph J. Hobbs*

Figure 9.12 Egypt's controversial Aswan High Dam. On the left is the dam's hydroelectric station, which generates much-needed and pollution-free electricity for Egypt's industries. *Joseph J. Hobbs*

high enough that the water flows by gravity into irrigation canals. The barrages permit a constant flow of water to irrigated land even in periods of low water levels. They are not designed to store large amounts of water—a function now performed by two dams in Upper Egypt (the old Aswan Dam, completed in 1902, and the huge Aswan High Dam, completed in 1970) and several others along the Nile and its tributaries in Sudan and Uganda.

The enormous Aswan High Dam (Fig. 9.12) is located about five miles (eight km) upstream from the older Aswan Dam, on the stretch of the Nile known as the First Cataract (the Nile's many cataracts are areas where the valley narrows and rapids form in the river). Its reservoir, Lake Nasser, stretches for more than 300 miles (480 km) and reaches into northern Sudan.

Like all giant hydrological schemes, the Aswan High Dam has generated benefits and liabilities, and its construction was controversial. On the plus side, by storing water in years of high flood for use in years of low volume, the High Dam provides a perennial water supply to Egypt's preexisting 6 million irrigated acres (2.4 million hectares), thus extending the cultivation period. Egypt weathered a two-year drought in the late 1980s thanks to the excess waters stored in Lake Nasser. The dam made possible the reclamation of considerable amounts of additional farmland, particularly on the desert fringe adjoining the western Nile delta. Navigation of the Nile downstream from (north of) Aswan is much easier, and the country's industries rely heavily on the hydroelectricity produced at the dam.

On the other hand, the dam has caused the water table to rise, making it harder for irrigated soil to drain properly (When farmers use too much water, standing water evaporates and leaves a deposit of salt so that the once-fertile soil loses its productivity— a problem known as **salinization**). Also, perennially available canal waters are ideal breeding grounds for the snail which hosts the parasite that causes schistosomiasis or bilharzia, a debilitating disease affecting a large proportion of Egypt's rural popula-

tion. Mediterranean sardine populations—deprived of the rich silt that nurtured their feeding grounds—have plummeted off the Nile delta, and the sardine industry has diminished. And without the free fertilizer the silt offered, Egypt's bill for artificial fertilizers has increased. Generally, however, Egyptians are very proud of the High Dam, and it has increased food output dramatically for Egypt's mushrooming population.

Egypt's food crops include corn, wheat, barley, rice, millet, fruit, vegetables, and sugarcane. Berseem (clover), a legume, is grown for livestock feed and to enrich the soil with nitrogen. Long-staple Egyptian cotton has long been the main commercial export crop for the hard currency Egypt needs desperately. Egypt was once a great food exporter as well, but the country's population has grown so explosively that food now comprises almost 20 percent of Egypt's import expenditures. With the rural-urban migration of recent decades, the proportion of Egypt's labor force employed in agriculture has declined substantially, but the actual number of people living and working on the land has increased. Rural population densities already are among the highest in the world, and the crowding is growing worse despite a rapid migration to Egypt's main metropolis, Cairo (population: officially 6.8 million, city proper; 9.3 million, metropolitan area—but probably much higher)—the largest metropolis in Africa and the Middle East—and other cities. So far, the government's efforts to lure urban dwellers and factories out of the Nile Valley to desert "satellite cities" have borne little fruit.

Population Growth and the Challenge to Industrialize

Efforts of the Egyptian government to improve the rural situation during recent decades have had some success. Land reform has been carried out, cooperative farming fostered, water supply improved, high-yielding grain varieties introduced, and public health programs instituted. Birth rates, however, have continued to be high, and Egypt is confronted with rapid population increases (2.2 percent in 1998) that threaten to outrun the food supply and bring to naught all attempts to raise the average standard of living. Perhaps nowhere on Earth is the Malthusian scenario—of population growth outpacing availability of food and other resources—more imaginable than in the confines of the Nile Valley.

Egypt has been industrializing in a modest way for a long time. Cotton textiles, food processing, clothing manufacture, and cement production are major industries, and there is a sizable output of chemicals, such as nitrate and phosphate fertilizers. The manufacture of automobiles and other consumer durables involves primarily the assembling of imported components. The main industrial centers are, by an overwhelming margin, Cairo, the capital, and the main port of Alexandria (population: 3.38 million, city proper), located, respectively, at the apex and western seaward edge of the Nile delta.

Egypt lacks the mineral wealth to support extensive industrialization, however. The country is self-sufficient in some vital minerals, including oil; located mainly along the Gulf of Suez and including many offshore wells, oil production provides an important part of the country's exports. But Egypt's oil and other exports are far too small to pay for a great diversity of needed imports, and the country has had to depend heavily on foreign loans and grants, especially from the United States and the European Union. Income from tourism, particularly to the outstanding temples and tombs of ancient Egypt and to the world-famous dive sites on the Red Sea and Gulf of Aqaba, helps redress the unfavorable trade balance, as do transit fees from ships using the Suez Canal. However, tourism is a very vulnerable resource for the country. After radical Islamists bent on destabilizing the government massacred foreign visitors at an ancient temple in November 1997, tourism to Egypt plummeted.

9.4 Sudan: Bridge Between the Middle East and Africa

Egypt is bordered on the south by Africa's largest country, the vast and sparsely populated tropical republic of Sudan (population: 29 million). Formerly controlled by Great Britain and Egypt, Sudan received its independence in 1956. This southern neighbor is vitally important to Egypt, for it is from Sudan that the Nile River brings the water that sustains the Egyptian people. The Nile receives all its major tributaries in Sudan (see Fig. 9.10), and storage reservoirs exist there on the Blue Nile, the White Nile, and the Atbara River, all of which benefit Egypt as well as Sudan itself (see Regional Perspective, p. 244).

The Nile's main branches—the White Nile and Blue Nile—originate respectively at the outlets of Lake Victoria in Uganda and Lake Tana in Ethiopia and then flow eventually into Sudan. The Blue Nile, Atbara, and Sobat Rivers, flowing from seasonally rainy highlands in Ethiopia, are primarily responsible

for the annual summer floods of the main river. In contrast to the Ethiopian rivers, however, the White Nile, fed by one of the world's largest lakes, maintains a fairly constant flow through the year. The Owen Stanley Dam in Uganda generates hydroelectricity and increases the storage capacity of Lake Victoria and thus facilitates control of the White Nile. In southern Sudan, the river passes through a vast wetlands called the Sudd, where much water is lost by evaporation. An artificial channel to straighten the river's course and increase the volume and velocity of its flow through the Sudd was under construction for a long time. Called the Jonglei Canal, its future is uncertain. Civil war in Sudan has completely halted its progress since 1985. Predicting the death of the Sudd wetlands if the canal is completed, many environmentalists would be pleased if work were not resumed.

Approximately 6 percent of Sudan is cultivated, but only about 10 to 20 percent of the cultivated land is irrigated. The largest block of irrigated land is found in the Gezira (Arabic for "island"), between the Blue and White Niles; water flows by gravity to the fields through canals from Sennar Dam and Roseires Dam on the Blue Nile. The most important cash crop and export of Sudan is irrigated long-staple Egyptian-type cotton, although many Sudanese support themselves by raising cattle, camels, sheep, and goats. The country's industries are relatively meager and there are few mineral reserves, with the exception of some oil deposits being developed in southern Sudan.

Most of Sudan is an immense plain, broken frequently by hills and in a few places by mountains. Climatically and culturally, the area is transitional between the Middle East and Africa South of the Sahara. A contrast exists between Sudan's arid Saharan North, peopled mainly by Arabic-speaking Muslims, and the seasonally rainy equatorial South, with its grassy savannas, papyrus wetlands, and non-Arabic-speaking peoples of many tribal and ethnic affiliations who practice both Christian and animistic faiths. About three-fourths of Sudan's total population is Muslim.

There has been much political friction between the two sections. An all-out civil war has raged between northern and southern Sudan since 1983, with southern factions of the Sudanese People's Liberation Army (SPLA) engaging government army troops from the north. Some casualty estimates exceed one million people, mostly civilians. Among the reasons for this conflict have been historical antagonisms and economic disparities between the more developed North and the less developed and poorer South, along with efforts by the Arab-dominated government to impose Islamic law *(shari'a)* and the Arabic language on the South and to exploit the South economically. Warfare and prolonged drought have devastated southern Sudan and adjacent Ethiopia during recent years, and Sudan in particular has hovered perilously close to catastrophic famine. Huge tides of refugees have flowed back and forth between the two countries. The military regime has meanwhile been accused of gross human rights violations, including "ethnic cleansing" of the Nuba, Nuer, Dinka, and other tribespeople in the South. International relief agencies have been allowed to provide food and medical aid in only a small portion of the war zone, and there are no statistics on the immense suffering and death suspected of occurring over the wider region. In 1998, the United Nations estimated that more than half the children in the 12 rebel-held areas were malnourished as a result of the Khartoum government's refusal to allow in relief supplies.

Sudan's foreign relations have changed markedly since 1989, when an Islamic military regime led by General Omar al-Bashir overthrew an elected government. The new rulers allied themselves with militant Iran. In 1993, after obtaining information linking Sudanese officials with the bombing of New York City's World Trade Center, the United States branded Sudan a "terrorist" nation and halted economic aid. Following the analysis of intelligence information about bombings of U.S. embassies in Kenya and Tanzania, in August 1998 American cruise missiles obliterated a pharmaceutical factory—alleged to be a chemical weapons factory—in Sudan's capital.

Sudan is overwhelmingly rural. The largest urban district, formed by the capital, Khartoum (population: 924,505, city proper; 1.45 million, metropolitan area), and the adjoining cities of Omdurman and Khartoum North, is located toward the center of the country at the junction of the Blue and White Niles. Transportation within Sudan as a whole is poorly devel-

oped. Several cataracts (see Figure 9.10) prohibit navigation on the Nile north of Khartoum, but the White Nile is continuously navigable at all seasons from Khartoum to Juba in the deep South. Most of the country has no railways or good highways, although there is widespread airplane service and much automobile traffic despite the lack of surfaced roads. Khartoum itself has rail connections with several widely separated sections (see Fig. 9.4). The main route for rail freight connects Khartoum and the Gezira with Port Sudan (population: 305,385, city proper), a modern, well-equipped port that handles most of Sudan's seaborne trade.

9.5 Libya: Deserts, Oil, and a Defiant Survivor

Egypt's neighbor to the west, Libya—formerly an Italian colony—became an independent kingdom in 1951. In 1969, Libya became a republic after Colonel Muammar al-Qaddafi and other army officers led a coup against the monarchy. Some 97 percent of the country's people are Muslim Arabs and Berbers.

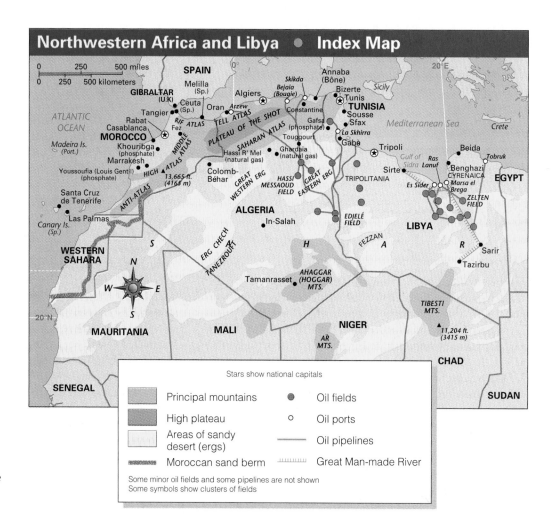

Figure 9.13 General reference map of the Maghreb countries, Western Sahara, and Libya.

Figure 9.14 By tapping into "fossil waters" in aquifers beneath the Sahara Desert, Libyan farmers are able to use center-pivot irrigation to produce a surprisingly abundant yield of wheat and other crops. *Tony Stone Images/Derek Bayes*

Libya is made up of three major areal divisions: Tripolitania in the northwest, Cyrenaica in the east, and the Fezzan in the southwest (Fig. 9.13). The principal cities are the capital, Tripoli (population: 551,477, city proper), in Tripolitania, and Benghazi (population: 282,192, city proper), in Cyrenaica.

Most of Libya lies in the Sahara and is too dry to support cultivation except at scattered oases watered by wells and springs. The average population density is only about 8 per square mile (3 per sq km). However, this figure has little meaning since most of Libya is uninhabited; its 5.7 million people are heavily concentrated in coastal lowlands and low highlands along the Mediterranean, where limited winter rains and irrigation with groundwater allow small-scale agriculture emphasizing crops associated with the Mediterranean region. In southern Libya, center-pivot irrigation of wheat is carried on, in which deep wells tap **fossil waters** that accumulated in deep limestone aquifers in ancient times when rains fell abundantly in what is now the Sahara Desert (Fig. 9.14). And in an ongoing project called the "Great Man-made River," these same fossil waters are diverted from 1155-feet (347 m) deep wells around Sarir and Tazirbu in east-central Libya by a pipeline 13 feet (4 m) in diameter stretching 560 miles (900 km) to Sirte and Benghazi on the Mediterranean (see Fig. 9.13). The goal of this effort, which Qaddafi calls "the world's largest civil engineering project," is agricultural self-sufficiency through the irrigation of fruits, vegetables, and wheat in northern Libya. However, some Western military analysts believe the huge pipeline far exceeds what would be needed for the transfer of water, and might be designed to conceal the movements of Libyan troops and materiel in some future conflict.

The Libyan economy operated at a deficit for many years, but, in 1961, when the first Libyan oil was exported, the country's financial situation took a sharp turn for the better. Intensive prospecting by numerous Western oil companies revealed large oil deposits, particularly in Cyrenaica. Crude oil represents nearly all of Libya's exports (primarily to western Europe), and returns from oil sales have generated governmental revenues that have increased enormously as a result of the steep jump in the price of crude oil in 1973 and after.

Libya has spent large amounts of its oil revenue on public works, education, housing, aid to agriculture, and other projects for social and economic development. The country has also spent large sums on weapons, reportedly including a chemical weapons factory that U.S. military authorities have hinted they will destroy before it becomes operational.

Libya's revolutionary government, still headed by Colonel Qaddafi, has long been involved in an exchange of hostile rhetoric and sometimes hostile action with the United States, its Western allies, and Israel. In 1986, American warplanes bombed Tripoli and Benghazi in reprisal for alleged terrorist acts against American citizens instigated by Libya, which is on the U.S. list of "terrorist" states. An apparent target of the attack was Colonel Qaddafi himself, whose daughter was killed in the raid. After Western intelligence authorities linked two Libyans to the 1988 bombing of Pan American Airlines' flight 103 over Lockerbie, Scotland, Libya was subjected to an international air embargo. The embargo, which was lifted in 1999, stipulated that until Libya allowed extradition of the two suspects, other nations would not permit flights of Libyan aircraft over their airspace or allow their national carriers to serve Libya.

9.6 Northwestern Africa: The "Maghreb"

Arabs know the northwestern fringes of Africa as the "Maghreb" (meaning "Place of the West") of the Arab world. Despite its peripheral location, this area is Middle Eastern in many essential characteristics. Today, it includes four main political units: Morocco, Algeria, Tunisia, and the disputed Western Sahara (see Fig. 9.13).

Morocco is geographically and culturally the closest to Europe; only the 11-mile-wide (18 km) Strait of Gibraltar separates it from Spain. Islamic influences crossed the strait beginning early in the 8th century, and prevailed in southern Spain until 1492, when Spain's monarchs ordered both Jews and the Muslim Moors to be expelled from Spain. France became the dominant power in the Maghreb during the 19th and early 20th centuries. France invested large sums to develop mines, industries, irrigation works, power stations, railroads, highways, and port facilities, but French rule was very unpopular. In 1956, after long agitation by local nationalists, Tunisia and Morocco secured independence. In Algeria, independence in 1962 came only after a bitter civil war lasting eight years.

Muslim Arabs form the majority in the Maghreb. The ethnically distinct Berbers, who are most numerous in Morocco and Algeria, converted to Islam after the Arabs brought the new faith into the region in the 7th century A.D. Many Berbers now speak Arabic, and most have adopted Arab customs. In colonial times, nearly two million Europeans, primarily French but also of Spanish and Italian origin, lived in the Maghreb. About two-thirds were in Algeria—mostly in Algiers (population: 1.52 million, city proper; 1.74 million, metropolitan area) and Oran (population: 490,788, city proper). Others settled in Casablanca (population: 2.14 million, city proper; 2.94 million, metropolitan area), Tunis (population: 596,654, city proper; 1.23 million, metropolitan area), and other cities, where they formed a class of business and professional people, administrators, office workers, and skilled artisans. In Algeria, agriculture supported about 25,000 Europeans. Characteristic of a colonial economy, their farms, worked by indigenous laborers and tenants, were much larger and more mechanized than Muslim-owned farms, and they included the largest share of Algeria's best agricultural land. Smaller numbers of European farmers also lived in Morocco and Tunisia.

Independence for the Maghreb countries was followed by a large exodus of Europeans, mainly to France. Most European-held farmland was expropriated and redistributed to private Muslim farmers and cooperatives or placed in large units run by the three governments. The new governments also redistributed some land held by Muslim landowners. All three countries today show a contrast between the small and generally poor farms of the "traditional sector," in which the bulk of the agricultural population still lives, and the larger and more mechanized and productive units in the "modern sector" inherited from the former European settlers.

Most people in the Maghreb live in the coastal belt of mediterranean climate (called "The Tell" in Algeria), with its cool-season rains and hot, rainless summers. Here, in a landscape of hills, low mountains, small plains, and scrubby mediterranean vegetation, people grow the familiar products associated with the mediterranean climate (Fig. 9.15).

Inland from the coastal belt of the Maghreb, the Atlas Mountains extend in relatively continuous chains from southern Morocco to northwestern Tunisia, with the highest peaks in the High Atlas Mountains of Morocco. They block the path of moisture-bearing winds from the Atlantic and the Mediterranean, and some mountain areas receive 40 to 50 inches (c. 100–130 cm) or more of precipitation annually. The precipitation nourishes forests of cork oak or cedar in some places. In Algeria, the mountains form two east-west chains, the Tell Atlas nearer the coast and the Saharan Atlas (see Fig. 9.13) further south.

All the countries reach the Sahara to the south. Some pastoral nomadic tribesmen still migrate through the Sahara with their livestock of camels, sheep, and goats. Clusters of oases exist in a few places, such as in the dramatic sandstone Ahaggar Mountains of southern Algeria, which rise high enough to catch moisture from the passing winds. A prominent line of oases fed by springs, wells, mountain streams, and *foggaras* (tunnels from mountain water sources; known as *qanats* in Iran) lies along or near the southern base of the Atlas Mountains. Several large areas of sandy desert (*ergs*) and a barren gravel plain, the Tanezrouft, occupy portions of the Algerian Sahara.

Figure 9.15   The western Maghreb has breathtaking physical landscapes, from the Mediterranean regions in the north to the Atlas Mountains and Sahara Desert in the south. Traditional village and urban architecture lends yet more unique character to the human environment. This is Valle du Dades in Morocco, with the Atlas Mountains in the background.   *Tony Stone Images/Nicholas DeVore*

The Maghreb economies benefit from a valuable endowment of minerals. Algeria's oil, located in the Sahara, is the region's greatest asset (see Fig. 9.13). The oil flows through pipelines to shipping points on the Mediterranean in Algeria and Tunisia. Algeria also extracts natural gas from the Sahara, some of which is exported in liquefied form. Algeria's National Liberation Front (FLN), which ruled the country for three decades after independence, built a few state-owned heavy industries that refine oil, liquefy natural gas, make petrochemicals, and manufacture small quantities of iron and steel. Petroleum products, crude oil, and gas make up well over nine-tenths of Algeria's exports by value. Morocco's exports also include some minerals, notably phosphate, but agricultural exports and clothing make up a much larger total. Consumer industries, including plants that assemble foreign-made components, are—after oil—the leading industries in the Maghreb countries.

Algeria's oil boom ended in the 1980s, but population growth has continued, fueling an economic crisis and shortages of food and consumer goods that contributed to widespread unrest in the 1990s. Algeria also has built up a staggering foreign debt that is contributing to instability. Some economic relief might be provided by a pipeline that will soon carry Algerian natural gas to Europe. Meanwhile, political instability in Algeria is reverberating through the entire Maghreb and across the Mediterranean to the region's former colonial power, France. Since 1992, when Algeria's military government canceled the results of parliamentary elections that would have given majority seats to the Islamic Salvation Front (FIS), civil strife has claimed an estimated 80,000 lives within Algeria. Violent sympathizers of the FIS, such as the Armed Islamic Group, have carried their insurgency to France, bombing government and civilian targets to protest France's support of the military government in Algiers. Four million Muslim immigrants, mostly

Algerians, already lived in France prior to 1992. The civil strife in Algeria has given rise to a new tide of immigrants to France, and to heightened French hostility toward the Arabs.

Morocco and Tunisia view events in neighboring Algeria with anxiety. Citing the violence in Algeria, Tunisia banned Islamist political parties in advance of multiparty parliamentary elections in 1994. Tunisia's government has promoted family planning and women's rights, and is often cited as an "oasis" of openness and progressive change in the Middle East. In Morocco, King Muhammad VI rules over a constitutional monarchy that promises political accommodation of Islamic groups, in contrast with Algeria's experience. The country has shown exceptional success in bringing down birth rates; while the average woman had seven children in 1980, she now has four.

The Western Sahara

Morocco and Algeria have often been at odds over the future of Spain's former dependency known as Western Sahara. This sparsely populated coastal desert (population 200,000), which contains rich phosphate deposits, adjoins Morocco on the south and is claimed by that country. When Spain relinquished control in 1975, the territory was partitioned between Morocco and Mauritania, but in 1979 Mauritania withdrew its presence. Algeria objected to Moroccan control and, together with other outside parties, gave assistance to an armed resistance movement in Western Sahara against Morocco called the Polisario Front. Meanwhile, the Organization of African Unity (OAU), composed of all of Africa's sovereign states except South Africa, called for a referendum whereby the Western Saharans could determine their own future, and at a 1984 summit meeting, the OAU gave a seat to a Western Saharan government-in-exile, the Saharan Arab Democratic Republic (SADR). Morocco then withdrew from the organization in protest.

Morocco has aggressively asserted its claim to Western Sahara. Morocco has sold Western Sahara's offshore fishing rights to European countries. To stave off Algerian interests, Moroccan technicians have also erected a 700-mile-long sand barrier parallel to Western Sahara's border with Algeria, reinforcing it with land mines and 150,000 troops (see Fig. 9.13). West of this berm, in Morocco-held territory, is the world's largest reserve of phosphate.

By 1989, Morocco and the Polisario Front had agreed to a referendum on independence, and a peace treaty was signed in 1991. To influence the outcome of the referendum in its favor, Morocco relocated 40,000 Moroccans into the region in 1991; they remain there today under Moroccan guard. The Moroccan government has also used tax breaks and other subsidies to attract new Moroccan settlers there. Such questionable actions led the United Nations to repeatedly suspend the referendum. While the status of Western Sahara remains unresolved, more than 150,000 Saharawis (as the indigenous people of Western

Sahara are known) live in refugee camps in the Algerian desert, as they have for decades.

9.7 The Gulf Oil Region

Petroleum is vital to the industrialized world as a source of fuels, lubricants, and chemical raw materials. So essential is this greasy mixture of hydrocarbons that world production of crude oil grew from an average of 7 million barrels per day in 1945 to about 63 million barrels (2.5 billion gallons) per day in the all-time peak year of 1979. A crucial feature of world geography is the concentration of approximately two-thirds of the world's proven petroleum reserves in a few countries that ring the Persian/Arabian Gulf.

All but about 1 percent of the Persian/Arabian Gulf oil region's proven reserves of crude oil are located in Saudi Arabia, Iraq, Kuwait, Iran, and the United Arab Emirates (UAE), with smaller reserves in Oman and Qatar. Saudi Arabia, by far the world leader in reserves, has about 26 percent of the approximately one trillion barrels of proven crude oil on the globe. Iraq has an additional 10 percent, Kuwait 9, UAE 7, and Iran 6. By comparison, Venezuela has 6 percent of world reserves, Russia another 6, Mexico 5, and the United States only 2.

Most oil deposits in the Gulf countries lie along, underneath, and near the Gulf itself (Fig. 9.16). The oil is marketed largely outside the region, although the larger Gulf states require considerable amounts for their internal needs (Fig. 9.17). Exports of goods from all the states that touch the Gulf are composed overwhelmingly of crude oil and/or refined products. Such exports have long yielded most of the vital foreign exchange used to purchase food, equipment, materials, and technology for development and defense. Oil from the Gulf is sold to many countries, but most of it is marketed in western Europe and Japan. The United States also imports large amounts of Gulf oil but has a much smaller relative dependence on this source than do Japan and Europe. However, the Gulf region is very important to the United States because of (1) the heavy dependence of close American allies on Gulf oil, (2) the importance of the oil as a future reserve, (3) the great involvement of American companies in oil operations and oil-financed development in Gulf countries, and (4) the large economic impact of Gulf "petrodollars" spent, banked, and invested in the United States. Maintaining a secure supply of Gulf oil has therefore been one of the longstanding pillars of U.S. policy in the Middle East.

Most oil produced in the Gulf region is shipped to market by oceangoing tankers, generally the huge vessels called supertankers. These ships load oil at terminals spaced along the Gulf. Some crude oil moves by pipelines from oil fields in the Gulf countries to tanker terminals on the Mediterranean and Red Sea coasts. A growing proportion of extracted crude oil is refined in the Gulf states for local use and export (Fig. 9.18). Petrochem-

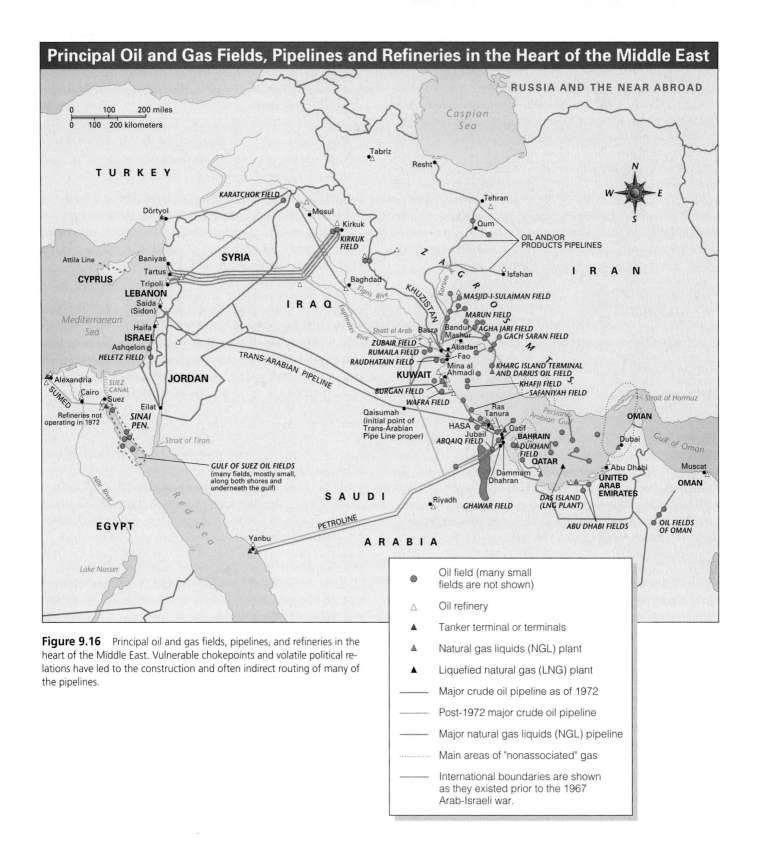

Principal Oil and Gas Fields, Pipelines and Refineries in the Heart of the Middle East

Legend:
- Oil field (many small fields are not shown)
- Oil refinery
- Tanker terminal or terminals
- Natural gas liquids (NGL) plant
- Liquefied natural gas (LNG) plant
- Major crude oil pipeline as of 1972
- Post-1972 major crude oil pipeline
- Major natural gas liquids (NGL) pipeline
- Main areas of "nonassociated" gas
- International boundaries are shown as they existed prior to the 1967 Arab-Israeli war.

Figure 9.16 Principal oil and gas fields, pipelines, and refineries in the heart of the Middle East. Vulnerable chokepoints and volatile political relations have led to the construction and often indirect routing of many of the pipelines.

ical industries exist in most of the producing countries, with the largest development in Saudi Arabia. The industries use local natural gas as their primary source of energy. Huge gas reserves are associated with oil deposits, and other major reserves exist apart from the oil fields, in "nonassociated" deposits. Growing amounts of gas are exported in liquefied form; Saudi Arabia is the world's largest exporter.

Foreign companies, principally British and American, originally developed the oil industry in the Gulf. These firms made huge profits from their oil concessions, but, with a few excep-

260

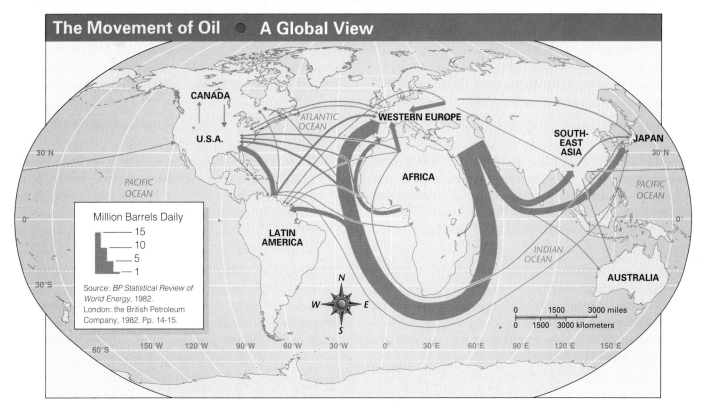

The Movement of Oil ● A Global View

Million Barrels Daily

15
10
5
1

Source: *BP Statistical Review of World Energy*, 1982. London: the British Petroleum Company, 1982. Pp. 14-15.

Figure 9.17 The global importance of the Middle East as a source of energy is expressed profoundly in this depiction of the movement of oil worldwide.

tions, after 1972 the companies saw their properties nationalized by the Gulf governments. However, many outside companies still work closely with national oil companies in the Gulf states, providing them with technical and managerial expertise, skilled workers, equipment, marketing channels, and varied services.

In addition to taking over the ownership and management of their oil industries, all of the Gulf states except Oman and Bahrain banded together with some other oil-exporting countries of the world to form the Organization of Petroleum Exporting Countries (OPEC), with the aim of joint action to maximize returns from oil for the participating countries. Formed in 1960, OPEC was relatively obscure until 1973, when world oil prices took a jump as a result of the Arab-Israeli "October War." OPEC subsequently coordinated a long series of price rises, which peaked in 1981 when the organization's price for "Arab Light Crude" reached $34 (U.S.) per barrel, compared with $2 a barrel in early 1973.

These events had enormous repercussions for the world economy. Immense wealth was transferred from the MDCs to the OPEC countries to pay for indispensable oil supplies. The skyrocketing cost of gasoline and other oil products helped cause serious inflation in the United States and many other countries, and contributed to the 1973 **energy crisis** in the United States. Desperately poor LDCs found that high oil prices not only hindered the development of their industries and transportation, but also reduced food production because of high prices for fertilizer made from oil and natural gas. For the MDCs, the transfer of wealth to OPEC had some return bene-

fits, as most of the money found its way back to the developed world in payments for goods and services and in the form of bank deposits and investments. In the Gulf countries, the oil bonanza produced a wave of spending for military hardware, showy buildings, luxuries for the elite, and ambitious development projects of many kinds. Per capita benefits to the general populace were greatest in the oil states of the Arabian Peninsula, where small populations and immense inflows of oil

Figure 9.18 Symbols of modernization and tradition in Saudi Arabia: A coastal oil refinery and a mosque are compatible features on the landscape of the Middle East. *Paul Lowe-Network/Matrix*

261

money made possible the abolition of taxes, the development of comprehensive social programs, and heavily subsidized amenities such as low-cost housing and utilities, including water distilled from the Gulf by desalting plants.

Then, at the beginning of the 1980s, the era of continually expanding oil production, sales, and profits by the OPEC states came to an end. After 1973, the high price of oil stimulated oil development in countries outside of OPEC. Oil-conservation measures such as a shift to more fuel-efficient vehicles and furnaces were instituted. Substitution of cheaper fuels for oil increased. Coal replaced oil in many electricity-generating stations. Oil refineries were converted to make gasoline from cheaper "heavy" oils rather than the more expensive "light" oils previously used. Meanwhile, the world entered a period of economic recession, due in part to high oil prices. Decreased business activity reduced the demand for oil. Profits of the world oil industry (and taxes paid to governments) were severely cut, large numbers of refineries had to close, and much of the world tanker fleet was idled. Oil prices rose temporarily in 1990–91 when the flow of Iraqi and Kuwaiti oil was cut off following Iraq's military takeover of Kuwait in August 1990, but the prices soon fell again. In the 1990s, Saudi Arabia and other oil-rich Gulf states implemented internal economic austerity measures for the first time. However, the immense oil and gas reserves still in the ground seem to guarantee that the Gulf region will continue to have a major long-term impact on the world, and will remain prosperous as long as these finite resources are in demand in the MDCs.

Saudi Arabia, the Oil Colossus

With an area of roughly 830,000 square miles (2.1 million sq. km), Saudi Arabia occupies the greater part of the Arabian Peninsula, the homeland of Arab civilization and the faith of Islam. The interior of the country is an arid plateau, fronted on the west by mountains that rise steeply from a narrow coastal plain bordering the Red Sea. Most of the plateau lies at an elevation of 2000 to 4000 feet (*c.* 600–1200 m), with the lowest elevations near the Gulf and in the eastern part of the gigantic sea of sand called the Empty Quarter (*ar Rub al Khali;* see Fig. 8.3, p. 226). Mountains (Arabic: *jebels*) rise locally from the plateau, and the plateau surface is trenched by many valleys (Arabic: *wadis*) that carry water whenever it happens to rain. Most interior sections average less than 4 inches (10 cm) of precipitation annually, but years may pass at a given locale with no precipitation at all. Rainfall is greatest along the mountainous western margin of the peninsula, especially in the southwestern Asir region, near the border with Yemen, which benefits from summer monsoon rains. There is very little nonirrigated agriculture in the desert kingdom; about 95 percent of Saudi Arabia's present crop production depends on irrigation. Cultivated land makes up less than 1 percent of the country's area, yet surprisingly, Saudi Arabia is self-sufficient in many agricultural staples

and is an exporter of wheat, irrigated mostly in the central part of the country by gigantic center-pivot sprinkler systems.

The estimated population of Saudi Arabia was about 20 million in 1998. About one quarter of that population consists of foreign workers drawn to the kingdom by jobs ranging from the purely menial to the highly technical. Saudi Arabia's best-known city is Mecca (population: 550,000, city proper), the birthplace of the Prophet Muhammad and the principal holy city of the Islamic world. The interior city of Riyadh (population: 1.3 million, city proper) is Saudi Arabia's capital. Here, futuristic structures and traffic-clogged thruways have sprouted like magic from the dusty, bedraggled, mud-brick town where King Abdul Aziz al Saud established the government of the newly consolidated Saudi state in 1932. The port city of Jidda (population: 1.3 million, city proper) on the Red Sea, some 40 miles (64 km) from Mecca, is the main point of entry for foreigners who make the pilgrimage to Mecca (see Fig. 8.11), most arriving at the city's major international airport.

The first oil discovery in Saudi Arabia in commercial quantities was made at Dammam on the shore of the Gulf in 1938, but large-scale production was delayed until after World War II. Oil production was further developed by the Arabian American Oil Company (ARAMCO), which was originally owned by four United States companies but which was nationalized in the late 1970s. Most of Saudi Arabia's oil fields, including the world's largest, the Ghawar field, are located in the eastern margins of the country, with much extracted from deposits underlying the Gulf. The great thickness of Saudi Arabia's oil-bearing strata and the high reservoir pressures have made it possible to secure an immense amount of oil from a relatively small number of wells. Such advantages are widespread in the Gulf oil states. The productivity of each well makes each barrel inexpensive to extract and makes it simple to increase or reduce production quickly in response to world market conditions.

Some Saudi Arabian oil is processed in refineries within the country and in nearby Bahrain, but most of the production leaves Saudi Arabia in crude form via tankers, using loading terminals that jut into both the Gulf (at Ras Tanura) and the Red Sea. However, the proportion of Saudi Arabia's national income derived from overseas sales of refined products and petrochemicals is expected to increase in the future. To meet the anticipated demand, huge new refining and petrochemical complexes, with associated industries of various types, have been developed at Jubail on the Gulf and Yanbu on the Red Sea (see Fig. 8.15). Their petrochemical industries depend mainly on the use of inexpensive natural gas associated with Saudi Arabia's oil fields.

Industrial development has also been pushed, but Saudi Arabia is far from being a major industrial power. Most factories produce and assemble consumer goods for domestic consumption. The country's transition to a standard of living comparable to the MDCs is not complete. Despite the new oil wealth, Saudi Arabia still bears various marks of an LDC. For example, infant mortality remains high (see Table 8.1, p. 225).

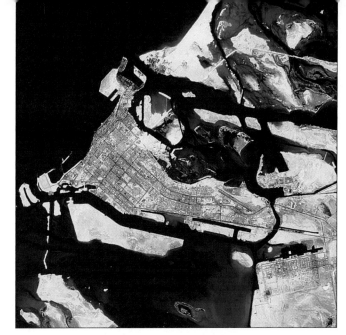

Figure 9.19 These extraordinary before-and-after satellite images reveal how oil wealth has transformed some Middle Eastern landscapes. The first shows Abu Dhabi, capital of the UAE, in 1972. The city occupies much of the natural island in the center. The second image of the same area, recorded in 1984, reveals the outcome of huge investments, including humanmade islands and deepwater port facilities. *NRSC Ltd./Science Photo Library/Photo Researchers, Inc.*

And there are striking discrepancies in income; while the sharing of oil wealth has brought once undreamed-of prosperity to most of the population, the Shi'ite Muslim minority in the Eastern Province remains poor.

With the oil boom, a new consumerism replaced the spartan lifestyle of earlier times. The new emphasis on material consumption in Saudi Arabia offers a strong contrast to the austere puritanism of the Wahhabi Muslims, whose political leader, King Abdul Aziz al Saud (1888–1953), consolidated many tribal and regional units into the Saudi kingdom before the advent of oil. However, many ancient Muslim traditions continue, and such links with the past are fostered by the conservative Saudi family that rules this absolutist monarchy. Restrictions on the freedoms of women (who are prohibited from driving automobiles, for example) are a very publicized facet of the Saudi attempt to preserve tradition while the tide of modernity rushes in. Far more threatening to the stability of the oil-rich kingdom are recent political acts of violence, such as the bombings of U.S. military facilities there. These attacks were probably carried out by Islamist organizations intent on purging the region of secular and "imperialist" Western interests and replacing the monarchy with an Islamic government.

Bahrain, Qatar, and the UAE

The sheikdom of Bahrain (ruled by a monarch called a Shaykh or emir; population: 600,000) occupies a small archipelago in the Gulf between the Qatar peninsula and Saudi Arabia. A modern four-lane causeway connects the main island to nearby Saudi Arabia. In addition to being an oil producer, Bahrain is a center of oil refining, seaborne trade, ship repair, air transportation, business administration, international banking, and aluminum manufacturing using imported alumina and local natural gas. Like Saudi Arabia, Bahrain is an oil-financed welfare state. However, in contrast to its giant neighbor, Bahrain's petroleum reserves are expected to be exhausted within two decades.

East of Bahrain along the Gulf lie the small emirates of Qatar (population: 500,000) and the United Arab Emirates (UAE) (population: 2.7 million). The UAE is a loose confederation of seven units, only three of which produce oil; only Abu Dhabi and Dubai produce it in major quantities. Qatar and the UAE are extremely arid and, prior to the oil era, supported only a meager subsistence economy based on fishing, seafaring, herding, and a little agriculture, augmented by smuggling and piracy. Today the UAE is capitalizing on its central location between Asia, Africa, and Europe with the construction of sophisticated air and seaport facilities. Like nearby Abu Dhabi's facilities, Dubai's new seaport of Jebel Ali is a colossal excavation easily visible to astronauts and satellites in space (Fig. 9.19). Unlike its more conservative neighbors, the UAE has promoted itself as a destination for sun-worshipping international tourists and shoppers. Until their economic crisis occurred in 1998, Russians comprised a substantial proportion of the foreign visitors.

The Southern Margins of Arabia

Known to the ancient Romans as Arabia Felix, or "Happy Arabia," because of their verdant landscapes, Yemen (population: 15.8 million) and Oman (population: 2.5 million) occupy, respectively, the mountainous southwestern and southeastern corners of the Arabian Peninsula.

The present Republic of Yemen was created in 1990 by the union of the former North Yemen and South Yemen. Its political capital and largest city is San'a (population: 450,000, city proper; 926,595, metropolitan area), located in the former North Yemen. Until the 1960s, North Yemen was an autocratic monarchy, extremely isolated and little known. In the mid-1960s, a republican government supported by Egypt and opposed by Saudi Arabia took control of large sections, at a cost of much bloodshed.

The former South Yemen, known until 1967 as Aden, a British colony and protectorate, lies along the Indian Ocean between the former North Yemen and Oman. A Marxist state supported by the Soviet Union and other East-bloc nations, South Yemen had hostile relations with neighboring countries and the Western world. It served as an East-bloc foothold on the Arabian Peninsula near the strategic chokepoint of Bab el-Mandeb connecting the Red Sea with the Indian Ocean. The main city, Aden (population: 350,000, city proper; 400,783, metropolitan area), is located on a fine harbor near the straits.

Civil war erupted between the north and the south in May 1994, with the south proclaiming its secession from the four-year-old union. Northern forces besieged Aden and overran most of the south. Today, an uneasy union holds the two Yemens together in a single official nation.

Yemen receives the heaviest rainfall in the Arabian Peninsula, enjoying a summer rainfall maximum derived from the same monsoonal wind system that brings summer rain to the Horn of Africa and the Indian subcontinent. Its main subsistence crops are spring-sown millet and sorghums. Most of the country's best agricultural land consists of terraces and is dedicated to the cultivation of *qat (Cathya edulis),* a stimulant shrub chewed by virtually all adults in the country on a daily basis (see Fig. 8.7, p. 228). Most of the country's people live in the rainier highlands, while the coastal plain is extremely arid and sparsely populated.

Yemen's meager agricultural output and lack of industrial production has been compensated for during recent decades by earnings sent home by Yemeni workers in the oil fields and services of the wealthy Gulf countries. The support of Yemen's government for Saddam Hussein's regime in Iraq during the 1991 Gulf War prompted Saudi Arabia to expel millions of these expatriate Yemeni workers, leading to a precipitous decline in Yemen's economy. Oil production began in Yemen itself only in the 1980s. The fields are located in a contested border area between Yemen and Saudi Arabia and subjected periodically to raids and kidnappings by groups opposed to Yemen's government, so production has been slow to increase.

Most of the Sultanate of Oman lies along the Indian Ocean and the southern shore of the Arabian Sea, with a small detached part bordering the Strait of Hormuz and the Gulf. As in Yemen, the coastal region is very arid (although verdant in irrigated spots), and most people live in interior highlands where mountains induce a fair amount of rain. For about 2000 years, beginning around 1000 B.C., Oman was the world's

Figure 9.20 The frankincense tree *(Boswellia sacra)* of Oman. The prized frankincense is the tree's gum, which in ancient times was harvested, like rubber, by cutting and "bleeding" the tree. *Joseph J. Hobbs*

only producer of frankincense, a prized commodity which made the region a hub of important oceanic and continental trade routes (Figure 9.20). Now, crude oil from fields in the interior and piped to the coast furnishes practically all of Oman's exports.

The country's strategic position on the Strait of Hormuz—through which much of the oil produced by the Gulf countries is transported—has made it an object of concern in industrial nations, particularly Great Britain, with which it has historic ties, and the United States. These Western powers have rights of access to base facilities in Oman in the event of any military emergency that threatens the continued flow of oil from the Gulf region. Allied forces used the bases after Oman joined the United Nations coalition against Iraq in the Gulf War of 1991.

Kuwait and the Gulf War of 1991

With 9 percent of the world's proven oil reserves within its borders, tiny Kuwait (6900 sq mi, 17,900 sq km—about the size of the U.S. state of New Jersey) has the world's third largest store of proven petroleum resources (after Saudi Arabia and Iraq). Associated with the oil is much natural gas, which Kuwait uses for fuel within the country and also liquifies and makes into petrochemicals for export. Oil production is greatest in the Burgan field south of Kuwait City. Oil revenues have changed Kuwait from an impoverished country of boatbuilders, sailors, small traders, and pearl fishermen to an extraordinarily prosperous welfare state. Generous benefits have flowed to Kuwaiti citizens, but not to the Asian and other non-Kuwaiti expatriates drawn by employment opportunities to this emirate. (By 1990, foreign workers actually outnumbered Kuwaiti citizens in the country's total population of 2,143,000.) The country's total population stood at 1.9 million in 1998, as the government has tried to reduce reliance on foreign workers.

In August 1990, Kuwait took center stage in world affairs when invading military forces at the command of Iraqi president Saddam Hussein overran the country. Claiming that Kuwait was historically part of Iraq and calling it Iraq's "nineteenth province," the Baghdad regime attempted to annex Kuwait. Saudi Arabia and other Gulf oil states seemed in danger of invasion. Strenuous worldwide opposition to Iraq's threats and aggression resulted in a sweeping embargo on Iraqi foreign trade imposed by the Security Council of the United Nations. Led by the United States, a coalition of 28 U.N. members (including several Arab states) staged a rapid buildup of armed forces in the Gulf region.

Beginning in January 1991, the coalition's crushing air war drove the Iraqi air force from the skies, massively damaged Iraq's infrastructure, and hammered the Iraqi forces in Kuwait and adjacent southern Iraq with intensive bombardment. In the ensuing ground war, the Iraqis were driven from Kuwait within 100 hours. Overall, an estimated 110,000 Iraqi soldiers and tens of thousands of Iraqi civilians were killed, while the technologically superior coalition forces suffered only 340 combat deaths, of which 148 were American.

During the conflict, which came to be known as the Gulf War, the Iraqis also made war on the environment, setting fire to hundreds of oil wells in Kuwait (Fig. 9.21) and allowing damaged wells to discharge huge amounts of crude oil into the Gulf. Ecological damage was severe.

A formal cease-fire came on April 6, 1991, with harsh conditions imposed on Iraq by the U.N. Security Council. Particularly important was the abolition of Iraqi chemical, biological, and nuclear "weapons of mass destruction" capabilities. Iraq agreed to pay reparations for the damage its forces had caused, and it renounced its claims to Kuwait. To guarantee the boundary between Iraq and Kuwait, the United Nations sent a multinational U.N. force to police a security zone along the border. A rapid withdrawal of coalition forces from the war zone then took place. Today, U.N.-sponsored economic sanctions against Iraq continue, with shortages of food and medical supplies bringing great hardship to the country's population. Only in late 1996 did the United Nations agree to allow Iraq to sell limited quantities of oil abroad, with the proceeds limited to purchasing food and humanitarian supplies.

After the Gulf War cease-fire, Saddam Hussein's surviving forces mercilessly put down rebellions by Shi'ite Arabs in southern Iraq and by Kurds (members of a non-Arab ethnic group) in northern Iraq. Great numbers of terror-stricken Kurds fled to the mountainous borderland where Iraq, Turkey, Iran, and Syria meet. An international effort to aid them slowly gained momentum, hindered by rugged terrain, lack of roads, and the sheer number of people requiring food, water, medicine, and protection from Saddam Hussein's soldiers. To deter Iraq's military and provide a safe haven for the Kurds in the north and the Shi'ites in the south, the United Nations established "no-fly zones" north of the 36th parallel and south of the 32nd (later, 33rd) parallel, where Iraqi aircraft were prohibited from operating. Following these actions, and with U.N. monitoring, repair of the devastation throughout Kuwait and Iraq

Figure 9.21 Iraqi troops set more than 600 oil wells ablaze before retreating from Kuwait in 1991. This image shows some of the fires on April 28, 1991. Kuwait City occupies much of the peninsula in the top center. *Earth Observation Satellite Company, Lanham, Maryland/Science Photo Library/Photo Researchers, Inc.*

began, and large numbers of refugee Kurds moved back to their home areas.

Throughout the 1990s, the regime of Saddam Hussein played a "cat and mouse" game with United Nations inspectors whose duty was to eliminate Iraq's ability to produce weapons of mass destruction. United States, British, and other forces were repeatedly deployed to strike against Iraq unless the country allowed unimpeded access by the U.N. inspectors, and in each case until 1998 the Baghdad regime relented at the last moment and averted war.

After being denied access to some sites in December 1998, the U.N. inspectors left Iraq in protest. Almost immediately, in an operation dubbed "Desert Fox," military forces of the United States and Britain launched cruise missiles and dropped bombs on scores of military and political targets across Iraq. The Baghdad regime vowed that U.N. inspectors would never again set foot in the country, began flaunting the "no-fly zones," and restated its wartime position that Kuwait was an illegitimate country. The United States articulated a new policy to support opposition groups that might replace the regime of Saddam Hussein, and the prospect for long-term conflict with Iraq continued.

After the Gulf War, the emir of Kuwait returned to his country from Saudi Arabia, where he had sought refuge from the Iraqi invasion. He faced a restive Kuwaiti population intent on securing more democracy and personal freedoms than had been available under the emir's authoritarian rule in the past. Meanwhile, Kuwait continues to live nervously in Iraq's shadow. It has excavated a trench on its border with Iraq to slow any invasion, and has continued to practice war games with U.S. military troops in anticipation of future incursions by its northern neighbor.

Iraq, Troubled Oil State of Mesopotamia

The Republic of Iraq occupies one of the world's most famous centers of early civilization and imperial power. This riverine land of Mesopotamia has had a complex geopolitical history, having been a seat of empires, a target of conquerors, and, in the 20th century, a focus of oil development and political and military contention involving many other nations.

Interaction with the wider world began very early. The ancient Babylonians and Assyrians (occupying lands later known as Iraq) overcame neighboring peoples and built a succession of empires reaching from the Gulf to the Mediterranean; the earliest such empire, the Akkadian, dates to 2350 B.C. Irrigated land on the Tigris-Euphrates plain gave empire-builders a dependable food base, and the Fertile Crescent corridor between mountains to the north and deserts to the south was a passageway for military movement and trade. Much later, during the centuries of Arab ascendancy after the death of the prophet Muhammad, Baghdad on the Tigris became the capital of an imposing Arab and Muslim empire.

Over thousands of years, such periods of imperial triumph alternated with times of internal disorder and weakness when outsiders such as the Persians and the Romans subjugated Iraq. In the 13th century, Mongol invaders devastated the area, which subsequently became a part of the Turkish Ottoman Empire. British forces defeated the Turks in Iraq in World War I, and after the war Great Britain took over the country as a mandated territory under the League of Nations until it gained its independence in 1932.

Iraq's government was a pro-Western monarchy from 1923 to 1958, when a revolution engineered by army officers made the country a republic. Over the next two decades, Iraq's political instability was reflected in many coups and attempted coups as various factions struggled for power. By the middle 1970s, Iraq had become an authoritarian one-party state ruled by President Saddam Hussein in the name of the Arab Baath Socialist Party. Aspiring to leadership in the Arab community, this Sunni Muslim-based regime asserted its dedication to Arab solidarity, opposition to Israel, freedom from Western imperialism, and strong internal development for a self-dependent, secularized, and socialist Iraq. In 1972, the Iraqi government signed a treaty of friendship with the Soviet Union, and Iraq received large Soviet arms shipments and equipment and technology for

many new factories. Iraq also nationalized its oil in 1972; prior to then, as was typical of the early oil industry in the Middle East, Great Britain and other Western nations produced and marketed Iraqi petroleum. After 1973, increasing oil revenues from the country's huge reserves (about 10 percent of the world's total) underwrote new social programs, large construction projects, and the creation of many new businesses under state, private, and joint owner-ship. New high-rise buildings, busy expressways, and an active commercial and social life gave metropolitan luster to ancient Baghdad.

In 1980, perceiving internal weakness in the neighboring fledgling Islamic Republic of Iran, as well as an opportunity to boost his stature in the Arab world and a means of securing land and mineral resources on the Iranian side of the Shatt al-Arab waterway ("Shore of the Arabs," formed by the confluence of the Tigris and Euphrates rivers), President Hussein launched an invasion of his more populous neighbor. Thus began an eight-year war that proved enormously costly to both sides, with an estimated 500,000 people killed or wounded. Iran mounted a far better defense than Iraq had anticipated, expending its greater manpower in bloody "human wave" assaults to counter Iraq's greater firepower from tanks, artillery, machine guns, and warplanes.

Although the fighting was localized mainly along the international border, many noncombatant nations became involved in various ways. For example, Saudi Arabia and other wealthy Arab Gulf states provided large-scale financing and arms to Iraq because they feared the destabilizing effects of an Iranian victory, particularly concerning what other Arab territories Iran might overrun, or what Arab populations might become receptive to Iran's revolutionary message that political Islam should replace secular governments and monarchies. The United States and the Soviet Union pursued their respective geopolitical interests in ways that generally benefitted Iraq, although neither of them declared formal support for either of the warring countries. Turkey, sharing a border with each, greatly expanded its trade with both. France sold Iraq sophisticated missiles and aircraft. A complex web of geopolitical relationships thus made the Iran-Iraq War a storm center in the political world. The war ended in stalemate in 1988, with neither side gaining significant territory or other assets.

Iraq occupies a broad, irrigated plain drained by the Tigris and Euphrates rivers (see Fig. 9.4), together with fringing highlands in the north and deserts in the west. In recent decades, Iraq's government has expanded irrigation on the plain, partly through the construction of dams on the Tigris and Euphrates and their tributaries for multiple uses of flood control, hydroelectric development, and irrigation storage. An estimated 70 percent of Iraq's 22 million people—about 60 percent of whom are Shi'ite Muslims—are urban dwellers—a much higher percentage than in most Middle Eastern countries. Most of the rest are villagers depending on agriculture for a living. Small bands of nomadic Bedouins still raise camels, sheep, and goats in the western deserts. Some of the Kurdish tribesmen in the moun-

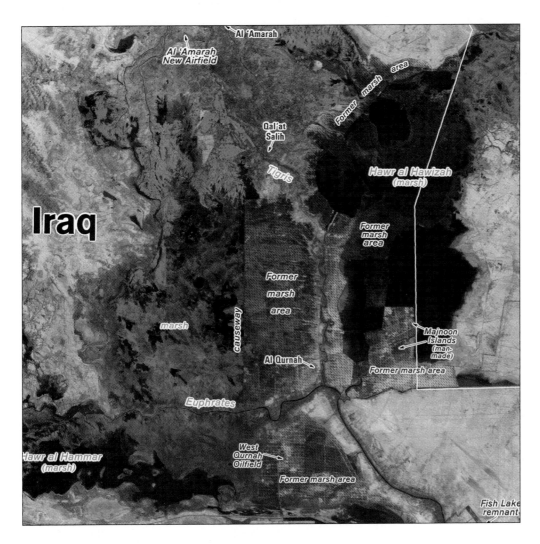

Figure 9.22 Southeastern Iraq in August 1992, after the draining of the marshes near the Tigris River was well underway. In order to drive out Shi'ite inhabitants, Iraqi engineers erected causeways to divert water away from and drain the wetland areas labeled former marsh areas here. The destruction of the southern marshes widened after this satellite image was taken. *Central Intelligence Agency,* "Iraq: A Map Folio."

tain foothills of the north are seminomadic graziers, although most Iraqi Kurds are sedentary farmers.

A Sunni Muslim people of Indo-European origins, the Kurds are by far the largest of several non-Arab minorities in Iraq. They have revolted against the central authority on numerous occasions, most recently during the Gulf War of 1991. The Kurdish question has international implications, because the area occupied by the Kurds also extends into Iran, Turkey, Syria, and the southern Caucasus region. European powers promised an independent state of Kurdistan at the end of World War I, but the governments concerned never took steps to create it. A book about the Kurds is appropriately entitled *No Friends But the Mountains.*[1]

Baghdad (population: 3.8 million, city proper; 5.5 million, metropolitan area), Iraq's political capital, largest city, and main industrial center, is located strategically on the Tigris near its closest approach to the Euphrates River. Like other metropolises in the LDCs, Baghdad has received a huge influx of villagers seeking a better life. The Iraqi preference for single-family homes finds expression in an endless spread of low-rise houses with garden plots. This mass of housing stretches outward for many miles from the cluster of office buildings and Western-style hotels in the downtown core of the city. The country invested heavily in rapid reconstruction of its capital city after the 1980–88 war, and again after the 1991 conflict.

Basra (population: 406,296, city proper; 900,000, metropolitan area) is Iraq's main seaport, though it is located upriver on the Shatt al-Arab. It suffered extensive destruction in the Iran-Iraq War and again in the Gulf War. Located in the predominantly Shi'ite southern portion of Iraq, Basra has long been the center of smoldering rebellion against the repressive Sunni-dominated government in Baghdad. One of the great environmental and cultural tragedies of the late 20th century has been Baghdad's systematic draining after 1991 of the marshes adjoining the southern floodplains of the Tigris and Euphrates rivers. Perceiving the ancient Shi'ite culture of the Maadan, or "Marsh Arabs," inhabiting the wetlands as enemies of the state, the Iraqi military built massive embankments and canals to drain the marshes. Authorities in Baghdad told international observers this was an "agricultural reclamation scheme," but its real purpose was to put an end to the agriculture and fisheries of the local people and to force them to flee into neighboring Iran and into settlements within Iraq where they could be controlled (Fig. 9.22).

[1]Bullock, John. 1993. *No Friends But the Mountains.* London: Oxford University Press.

Oil and Upheaval in Iran

Iran, formerly Persia, was the earliest Middle Eastern country to produce oil in large quantities. The Anglo-Persian Oil Company, now the British Petroleum Company (BP), began commercial production in 1912. The main oil fields form a line along the foothills of the Zagros Mountains in southwestern Iran. British-built pipelines linked the large refinery at Abadan with the inland oil fields. Abadan, which was heavily damaged by Iraqi shelling in the 1980–88 Iran-Iraq War, lies on the Shatt al-Arab waterway about 50 miles (80 km) from the Gulf. It is accessible only to small tankers, so nearly all exports of crude oil from Iran pass through a newer terminal at Kharg Island (see Figure 9.16), which allows supertankers to load in deep water.

Revenues from the oil industry (fully nationalized in 1973) are crucially important to Iran, a country where a dry and rugged habitat makes it hard to provide a good living for its rapidly growing population of 64.1 million. Arid plateaus and basins, bordered by high, rugged mountains, are Iran's characteristic landforms (Fig. 9.23). Encircling mountain ranges prevent moisture-bearing air from reaching the interior, creating a classic rain-shadow desert. Only an estimated 11 percent of Iran is cultivated. Enough rain falls in the northwest during the winter half-year to permit unirrigated cropping (dry farming), but rainfall sufficient for intensive agriculture is largely confined to a densely populated lowland strip between the high, volcanic Elburz Mountains and the Caspian Sea. The western half of this lowland receives more than 40 inches (*c.* 100 cm) annually, including both summer and winter rainfall, and produces a

variety of subtropical crops such as wet rice, cotton, oranges, tobacco, silk, and tea.

In the other regions of Iran, agriculture depends mainly or exclusively on irrigation. Many irrigated districts lie on gently sloping alluvial fans (triangular deposits of stream-laid sediments) at the foot of the Elburz and Zagros Mountains; the capital, Tehran (population: 6.75 million, city proper), is in one of these areas south of the Elburz. *Qanats* (underground sloping tunnels) furnish the water supply for a large share of the country's irrigated acreage. However, they are expensive to build and maintain, and an increasing part of Iran's water supply during recent times has come from stream diversions or from deep wells equipped with diesel pumps. Wheat is the most widespread and important crop, as it generally is in Middle Eastern areas having winter precipitation and summer drought. Cattle are raised in areas where there is enough moisture to grow forage for them, but drier areas support sheep and goats. Many livestock belong to seminomadic mountain peoples—the Qashqai (Kashgai), Baktiari, Lurs, Kurds, and others—whose independent ways have long been a source of friction between these tribesmen and the central government. Before the Islamic revolution in Iran in 1979, it was government policy to sedentarize Iran's nomads so that they did not threaten the state.

Prior to World War I, Iran was an exceedingly poor and undeveloped country, in marked contrast to its imperial grandeur when Persian kings ruled a great empire from Persepolis. In the 1920s, a military officer of peasant origins, Reza Khan, seized control of the government and began a program to modernize Iran and free it from foreign domination. He had himself crowned as Reza Shah Pahlavi, the founder of the new Pahlavi dynasty. Influenced by the modernizing efforts of Mustafa Kemal Ataturk in neighboring Turkey, the new Shah (king) introduced social and economic reforms. During World War II, Reza Shah displayed such pro-German leanings that British and Russian troops invaded his country in 1941 to secure a "back door" through which Allied supplies could reach the Russian front. Reza Shah was forced to abdicate, and in 1941 his young son took the throne as Mohammed Reza Shah Pahlavi.

After the war, the new Shah continued and expanded the program of modernization and Westernization begun by his father. Progress was relatively slow until the 1970s, when Iran played a leading role in raising world oil prices. Mounting oil revenues underwrote explosive industrial, urban, and social development. New port facilities, hundreds of new factories, and surfaced highways were built. Extraction of coal, metals, and other minerals was stepped up. The new industrial development centered in Tehran, but Isfahan (population: 1,220,595, city proper) received a large steel mill, and Tabriz (population:

Figure 9.23 Iran's rugged, arid Zagros Mountains, with the coast of the Gulf; part of the Strait of Hormuz is visible in the lower right. Plants and animals may have first been domesticated in this region. This is a mosaic of 12 Landsat images. *NASA*

1,166,203, city proper) in the northwest became a center of machine-tool production to support the nation's industries.

In the countryside, the Shah's government attempted to upgrade agricultural productivity through land reform and the introduction of better methods, tools, seeds, fertilizers, and animal varieties. Iran made major efforts to increase the supply of irrigation water and electricity by building large storage dams and hydropower stations along mountain rivers fed by melting snows. The largest river-control scheme involved the Karun River and smaller rivers in the southern oil-bearing region called Khuzistan, which adjoins Iraq, where there continues to be an emphasis on such cash crops as sugarcane, sugar beets, and citrus fruits. Under the Shah, reforms in the agricultural sector were largely unrewarding, and poverty-stricken families from the countryside poured into Tehran and other cities. From this devoutly Muslim group of new urbanites came much of the support for the revolution that ousted the Shah in 1978–1979.

Oil money funded Iran's development under the Pahlavi dynasty, but the vast sums the regime spent on the military undercut the benefits to Iran's people. Arms came primarily from the United States, which allied itself with Iran as one of the "Twin Pillars" of American interests in the oil-rich Gulf region (the other pillar was Saudi Arabia). The 1978–79 revolution that overthrew the Shah and abruptly took Iran out of the American orbit shattered these arrangements. The country became an Islamic republic governed by Shi'ite clerics, who included a handful of revered and powerful ayatollahs ("signs of God") at the head of the religious establishment, and an estimated 180,000 priests called mullahs. These religious leaders continue to supervise all aspects of Iranian life and perform many functions allotted to civil servants in most countries. They base their authority on Shi'ite interpretations of the Islamic faith. (About 93 percent of Iran's population is Shi'ite.)

The central figure among the revolutionaries who overthrew the Shah was the Ayatollah Ruhollah Khomeini, an elderly critic of the regime who was forced into exile by the Shah in 1964. Living in Iraq until 1978, when President Hussein expelled him, and then in Paris, Khomeini sent repeated messages to Iran's Shi'ites that helped spark the revolution. During 1978, the country experienced many riots and strikes, which eventually caused the Shah to flee the country and abdicate the throne. (After hospitalization in the United States and life in exile in Panama and Egypt, the Shah died in Cairo in 1981 and is buried there; Fig. 9.24.) The Ayatollah Khomeini returned in triumph to Tehran in 1979, where he soon was able to take control of the government. Meanwhile, his followers held 52 U.S. diplomatic personnel as hostages in Tehran for 444 days—an enormously publicized event that helped defeat U.S. president Jimmy Carter in his bid to be reelected in 1980; Carter had failed through diplomatic and military means to free the hostages.

The revolution was the product of overwhelming opposition to the Shah and his regime among most elements of Iranian society. Opponents ranged from Communists to Westernized

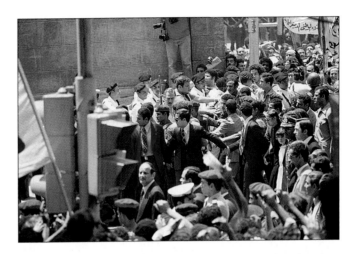

Figure 9.24 The late 1970s and early 1980s were turbulent times in the Middle East. The exiled Shah of Iran was welcomed only in the United States and in a single sympathetic country in the region: Egypt. Here a line of dignitaries, besieged by crowds, is following the casket of the Shah to its resting place in a Cairo mosque in July 1981. At the far right center, wearing a military uniform, is Egypt's president Anwar Sadat, who was assassinated by Islamist militants three months later. *Joseph J. Hobbs*

liberals to Shi'ite clerics and their followers. Grievances against the Shah included widespread corruption within the ruling circles, whereby the Shah, his numerous relatives, members of the government, and court favorites enriched themselves. Many people objected to the huge sums spent on arms, the exaggerated pomp of the Shah's court, the heavyhanded control of dissent by the secret police (SAVAK), and the Shah's harassment of the influential merchants in the bazaars. Landowners (including the mullahs) whose holdings had been diminished or amalgamated by the Shah's land reforms were resentful. But the driving force of the revolution was a furious tide of Shi'ite religious sentiment directed against modernization, Westernization, Western imperialism, communism, the exploitation of underprivileged people in Iran and elsewhere, and the institution of monarchy, which Shi'ite beliefs regard as illegitimate.

What has emerged since the revolution is a theocratic state, guided by edicts of the Ayatollah Khomeini (who died in 1989) and his successors, who have undertaken to create a society governed by Islamic law (*shari'a*). Political and religious dissidence has been crushed. In contrast to their status during the Shah's reign, when they were encouraged to have stronger public roles, women in postrevolutionary Iran have lost many freedoms. Under the new regime, "polluting" influences of Western thought and media have been purged; however, the late 1990s saw a general softening of restrictions on personal freedom of expression. While the government continues to be run by the Ayatollah's adherents, Western analysts alternately perceive it to be either developing a more accommodating attitude toward the international community, or exporting ever more dangerous notions of revolutionary Islam to currently pro-Western countries in the Middle East. A new detente opened between Iran and the United States and between Iran and Great Britain in 1998. Iranian and American scholars and athletes visited each other's country, and Iran's lifting of the death threat against the British author Salman Rushdie for his allegedly blasphemous book *The Satanic Verses* led to the restoration of diplomatic relations between Iran and the United Kingdom.

9.8 Turkey, Where East Meets West

Once a Muslim state at the center of a disintegrating empire, Turkey transformed itself in the 20th century into a secular national unit with fewer ties to traditional Islam and with a marked Western orientation. The Turks who organized the Ottoman Empire beginning in the 14th century were pastoral nomads from interior Asia (major Turkic elements still exist in Russia and the Near Abroad and China), and from the 16th to the 19th century their empire was an important power, centering on the large Anatolian Peninsula between the Mediterranean, Aegean, and Black Seas which forms most of the present Republic of Turkey (see Fig. 9.4).

Modern Turkey was created from the wreckage of the old empire after World War I. Its founder, Mustafa Kemal Ataturk, was determined to Westernize the country, raise its standard of living, and make it a strong and respected national state. He inaugurated social and political reforms designed to break the hold of traditional Islam and open the way for modernization and Turkish nationalism. Islam had been the state religion under the Ottoman Empire, but Ataturk separated church and state, and to this day Turkey is the only Muslim Middle Eastern country to officially cleave them. Wearing the red cap called the *fez*, an important symbolic act under the Ottoman caliphs, was prohibited, and state-supported secular schools replaced the religious schools that had monopolized education. To facilitate public education and remove further traces of Muslim dominance, Latin characters replaced the Arabic script of the Qur'an. Slavery and polygamy were outlawed, and women were given full citizenship. Legal codes based on those of Western nations replaced Islamic law, and forms of democratic representative government were instituted, although Ataturk ruled in dictatorial fashion. Turkey has continued to have trouble establishing a fully democratic system, and there have been frequent periods of military rule. After decades of secularization, there was, for two years (1996–1997), a democratically elected government coalition comprised of avowedly secular politicians and Islamist leaders, including the prime minister (Necmettin Erbakan), who wanted to see Turkey reestablished on Islamic foundations. In 1998, however, Turkey banned government participation by the pro-Islamic Welfare (Refah) Party which had put the prime minister in office. Backed by the army, the new government reaffirmed Ataturk's vow that Turkey remain secular. The regime's perception that the Islamist tendencies threaten the state represents a setback for democracy and human rights in Turkey—there are even new restrictions against women wearing Islamic-style head scarves—and raises the spectre of violent dissent.

From a physical standpoint, Turkey is composed of two units: the Anatolian Plateau and associated mountains occupying the interior of the country, and the coastal regions of hills, mountains, valleys, and small plains bordering the Black, Aegean, and Mediterranean Seas. The Anatolian Plateau, an area of wheat and barley fields and grazing lands, is bordered on the north by the Pontic Mountains and on the south by the Taurus Mountains (Fig. 9.25). The annual precipitation, concentrated in the winter half-year, is barely sufficient for grain. The plateau ranges from 2000 to 6000 feet (*c.* 600–1800 m), and is highest in the east where it adjoins the high mountains of the Armenian Knot. Its surface is rolling and windswept, hot and dry in summer and cold and snowy in winter, with a natural vegetation of short steppe grasses and shrubs. Production of cereals (especially barley, wheat, and corn) and livestock (especially sheep, goats, and cattle), employing both traditional and mechanized means, prevails in Anatolia.

Coastal plains and valleys along and near the Aegean Sea, Sea of Marmara, and Black Sea are generally Turkey's most densely populated and productive areas. Here are grown such cash crops as hazelnuts, tobacco, grapes for sultana raisins, and figs. Irrigated sugar beets are important in the small European section of Turkey known as Thrace. The Black Sea coastlands differ from the rest of Turkey in that they experience summer as well as winter rain and have much greater precipitation than other parts of the country (see Fig. 8.6). Turkey in general has a physical presence quite unlike the popular image of the Middle East as a desert region—it is, in fact, the only country in the region that has no desert, and cultivated land occupies about a third of Turkey's land area.

Figure 9.25   The Taurus Mountains of the Anatolian Plateau, in south-central Turkey. *Joseph J. Hobbs*

Figure 9.26 Istanbul straddles the strategic chokepoint of the Bosporus Strait, which separates Europe (left) from Asia (right). *Joseph J. Hobbs*

Relative poverty compared with the nearby European countries is a striking characteristic of present-day Turkey. There remains a very strong rural and agricultural component in Turkish life, with about one-third of the population still classified as rural. However, the country has embarked on a course of change that promises to modernize its agriculture, expand industry, and raise its general standard of living. Early in the 1990s Turkey began an impressive agriculture scheme called the Greater Anatolia Project (GAP) or the Southeast Anatolia Project (SEAP). Its aim is to convert the semiarid southeast quarter of the country into the "Breadbasket of the Middle East." The centerpiece of this massive irrigation project is the Ataturk Dam, whose sluice gates closed down on the waters of the Euphrates River in 1991. About 20 other dams on the Turkish Euphrates and Tigris are also part of the project, which will provide hydroelectricity as well as water. The GAP is controversial, particularly because it is being implemented in the part of the country where a restive Kurdish population is seeking recognition, autonomy, and, among some factions, independence from Turkey. In addition, Turkey's Tigris and Euphrates waters flow downstream into neighboring states, raising serious questions about downstream water allocation and quality (see Regional Perspective, p. 244).

The value of output from manufacturing in Turkey is already greater than that from agriculture. Turkey's predominant industries are characteristic of LDCs: textiles, agricultural processing, cement manufacturing, simple metal industries, and assembly of vehicles from imported components. The country is still very dependent on imports for much of its machinery as well as for many other types of manufactured goods. An iron and steel mill in the northern interior, based on Turkish coal and iron ore, is the largest heavy-industrial establishment. Turkey does produce some oil and has valuable deposits of chromium and other metallic and nonmetallic minerals, and its transportation facilities are more adequate than those of most Middle Eastern coun-

tries. This rapidly growing nation of 65 million is a member of the North Atlantic Treaty Organization (NATO) and an associate member of the European Union. Turkey is thus a kind of "in-between" country: Economically, it is well below the level of MDCs but above most of the world's LDCs; culturally, it is between traditional Islamic and secular European ways of living.

Istanbul (population: 7.77 million, city proper), formerly Constantinople and Byzantium, is Turkey's main metropolis, industrial center, and port. One of the world's most historic and cosmopolitan cities, Istanbul was for many centuries the capital of the Eastern Roman (Byzantine) Empire. It became the capital of the Ottoman Empire when it fell to the Turks in 1453. However, the capital of the Turkish Republic was established in 1923 at the more centrally located and more purely Turkish city of Ankara (population: 2.84 million) on the Anatolian Plateau. Both Istanbul and Ankara have shantytowns that house migrants from impoverished rural Turkey. The same is true of Turkey's third largest city, the seaport of Izmir (population: 2.02 million, city proper; 2.41 million, metropolitan area) on the Aegean Sea.

Istanbul is located at the southern entrance to the Bosporus Strait, the northernmost of the three water passages (the Dardanelles, Sea of Marmara, and Bosporus) that connect the Mediterranean and Black Seas and are known as the Turkish Straits (see inset map, Fig. 8.1, p. 224, and Fig. 9.26). The straits have long been a focus of contention between Turkey and Russia. In recent years, relations between the Turks and their neighbors to the north have been relatively tranquil, although the Turks, who have fought many wars against the Russians, maintain a high level of military preparedness within the NATO alliance. Some of Turkey's military posture has also been shaped by fragile relations with its NATO neighbor, Greece. The main questions recently at issue have been the status of Cyprus and conflicting claims to the oil and gas resources of the Aegean seabed.

9.9 Ethnic Separation in Cyprus

The large Mediterranean island of Cyprus, located near southeastern Turkey (see Fig. 9.4), came under British control in 1878 after centuries of Ottoman Turkish rule. In 1960, it gained independence as the Republic of Cyprus. An all-important problem on the island is the division between the Greek Cypriots, who are Greek Orthodox Christians comprising about three-fourths of the estimated population of 700,000, and the Turkish Cypriots, who are Muslims comprising about one-fourth. Agitation by the Greek majority for union *(enosis)* with Greece was prominent after World War II and led in the 1950s to widespread terrorism and guerrilla warfare by Greek Cypriots against the occupying British. Violence also erupted between Greek advocates of *enosis* and the Turkish Cypriots, who greatly feared a transfer from British to Greek sovereignty.

In 1974, a major national crisis erupted when a short-lived coup by Greek Cypriots, led mainly by military officers from Greece, temporarily overthrew Cyprus' President Makarios, who had followed a conciliatory policy toward the Turkish minority. Turkey then launched a military invasion that overran the northern part of the island. Cyprus was soon partitioned between the Turkish north and the Greek south. A buffer zone (the "Attila Line" or "Green Line") sealed off the two sectors from each other, and even the main city of Nicosia was divided. A separate government was established in the north, and in 1983 an independent "Turkish Republic of Northern Cyprus" was proclaimed. Only Turkey has recognized this state. Meanwhile, the internationally recognized Republic of Cyprus functions in the Greek-Cypriot sector, which comprises somewhat more than three-fifths of the island's land. Both republics have their capitals in Nicosia (population: 188,800, metropolitan area), which was the capital before partition.

Prior to the partitioning of Cyprus, the north had dominated the economy, but since then the north has had severe economic difficulties while Greek Cyprus has prospered. The Turkish sector was seriously weakened during the 1974 crisis by the flight of an estimated 200,000 Greek Cypriots to the south as refugees. There was a return flow of Turkish Cypriots entering the north, and thousands of immigrants from Turkey settled in the north, but this immigration has not compensated for the almost total loss of Greek entrepreneurs, farmers, skilled workers, and consumers. Most outside nations refused to trade directly with Turkish Cyprus after the invasion, and the trade and economic assistance offered by an economically weak Turkey were insufficient to provide much momentum. The economically depressed north remains tied to the struggling economy of the Turkish mainland and is exceedingly dependent on aid from Turkey.

The Greek sector, by contrast, was able to make effective use of economic aid from Greece, Britain, the United States, and the United Nations. A construction program provided new housing, commercial buildings, roads, and port facilities. Tourism based on both beach and mountain resorts was greatly expanded. An efficient telecommunications system gave the south new links to all parts of the world; hundreds of new businesses were attracted there by such factors as favorable tax policies, modern facilities, dependable overseas communications, an educated and reasonably priced labor force, the relative security provided by the island location, a government friendly to foreign business, and amenities provided by the tourism industry. A duty-free zone allows goods to be landed and transshipped without payment of customs duties. Greek Cyprus is expected to become a member of the European Union, pending the outcome of negotiations concerning some of its social and military policies. Of great concern is Greek Cyprus' purchase of antiaircraft missiles from Russia in 1998. Turkey has vowed to destroy them, and since Greece has a defense pact with Greek Cyprus, this could set off a war between Turkey and Greece—two NATO allies.

9.10 Rugged, Strategic, Devastated Afghanistan

High and rugged mountains dominate Afghanistan, the only landlocked nation in the Middle East (Fig. 9.27). Historically, it has occupied an important strategic location between India and the Middle East. Major caravan routes crossed it, and a string of empire-builders sought control of its passes. Today it has limited resources, poor internal transportation, and little foreign trade; Afghanistan is one of the poorest of the world's LDCs.

Afghanistan's population is estimated at about 22 million, only about one-third the size of neighboring Iran's. Most people live in irrigated valleys around the fringes of a mass of high mountains occupying a large part of the country. The country's second most populous area is the foothills and steppes on the northern side of the central mountains. Most of its inhabitants live in oases forming an east-west belt along the base of the mountains. Northern Afghanistan borders three of the five Central Asian countries, and millions of people on the Afghan side are related to peoples of those countries.

The most heavily populated section is the northeast, particularly the fertile valley of the Kabul River, where the capital and largest city, Kabul (population: 1.42 million, city proper; elevation: 6200 ft/1890 m) is located. Most of the inhabitants of the

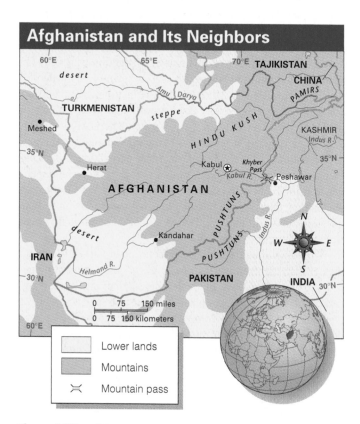

Figure 9.27   Afghanistan in its regional setting.

southeast are Pushtuns (also known as Pashtuns or Pathans). Their language, Pushtu, is related to Farsi (Persian), which is the main language of administration and commerce in Afghanistan. The Pushtuns are the largest and most influential of the numerous ethnic groups that make up the Afghan state. The independent-minded tribal Pushtun people have always been slow to recognize the authority of central governments. In the days of Great Britain's Indian Empire, the area was rife with warfare among tribes, tribal raids on British-controlled areas, and British punitive expeditions against the tribes. In those days, Peshawar—on the Indian (now Pakistan) side of the Khyber Pass into Afghanistan—became a noted garrison town for British and Indian troops.

Since Britain's withdrawal from India in 1947, the border region has continued to be a source of friction between Afghanistan and Pakistan. The friction has been partially due to border incidents but has also grown out of proposals that the Pushtun-inhabited areas of Pakistan be incorporated in a separate state ("Pushtunistan"), either independent or affiliated with Afghanistan. Pakistan has firmly opposed such proposals.

The only minerals currently extracted are minor quantities of natural gas, coal, and a few others; Afghanistan is overwhelmingly a rural agricultural and pastoral country. The land exhibits a wide range of climatic conditions corresponding to differences in elevation, but is generally so mountainous and arid that only an estimated 12 percent is cultivated. Enough rain falls in the main populated areas during the winter half-year to permit dry farming of winter grains. A variety of cultivated crops, fruits, and nuts are important locally. Raising of livestock on a seminomadic and nomadic basis is widespread. In general, Afghanistan's agriculture bears many of the customary Middle Eastern earmarks: traditional methods, simple tools, limited fertilizer, and low yields.

Through most of the 20th century, this highland country was remote from the main currents of world affairs. After the Islamic revolution in Iran in 1979, however, Afghanistan's location next to that oil-rich country made it once again the target of foreign interests. The Soviet military intervention of 1979 and the ensuing devastation catapulted the country into world prominence. In 1973, the Afghan monarchy had been overthrown by a military coup, and a communist government oriented to the Soviet Union took power. Widespread rebellions followed, and in 1979 the Afghan government called on the USSR for assistance. In a surprise move, the Soviets responded with sizable military force, and a long internal war began. Widespread killing and maiming of civilians, sowing of land mines over vast areas, destruction of villages, burning of crop fields, killing of livestock, and pollution or destruction of irrigation systems by Soviet ground and air forces caused several million Afghan refugees to flee into neighboring Pakistan and Iran (Fig. 9.28). Meanwhile, arms from various foreign sources filtered into the hands of the *mujahadiin,* the anti-Soviet rebel bands who kept up resistance in the face of heavy odds. The United States was one of the powers supporting the rebels, and in that sense waged a proxy war against the Soviet Union in Afghanistan.

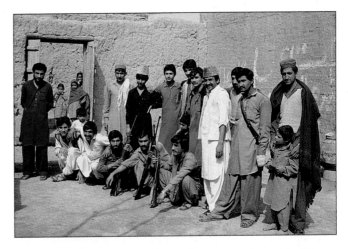

Figure 9.28 When this photo was taken in 1987, the *mujahadiin* fighters pictured here in an Afghan refugee camp outside Peshawar, Pakistan, were regularly involved in firefights with Soviet troops across the border in Afghanistan. More recently, many of the various guerrilla factions that made up the anti-Soviet resistance have been involved in conflicts with one another. *Joseph J. Hobbs*

Support lent to the *mujahadiin* and sympathetic Arab fighters by the United States and "moderate" Arab states such as Egypt and Saudi Arabia came back to haunt those countries in the 1990s. Intelligence analysts credit some of the groups armed and trained by these countries with some spectacular acts of terrorism, including an assassination attempt on Egyptian president Hosni Mubarak, the bombing of U.S. military barracks in Saudi Arabia, and the bombing of U.S. embassies in Kenya and Tanzania. The East African bombings in 1998 led within days to a cruise missile strike by the United States on six guerrilla camps near Khost, Afghanistan.

The USSR's motives for its invasion of Afghanistan may have included a desire to prevent by force the spread of Iranian-style Islamic fundamentalism into that country (which is 84 percent Sunni Muslim and 15 percent Shi'ite Muslim) or into the Central Asian Soviet republics deemed vital to the superpower's security. For its part, the United States warned the Soviet Union that it would not tolerate further Soviet expansionism.

Only the Soviet Union's desire, under Mikhail Gorbachev, to put a more humanitarian face on the Communist giant, and the realization that the USSR could not establish a lasting presence in Afghanistan and win its "Vietnam War," led to a withdrawal of Soviet troops in 1988 and 1989. It is estimated that, of the 15.5 million people who lived in Afghanistan when it was invaded in 1979, one million died, two million were displaced from their homes to other places within the country, and six million fled as refugees into Pakistan and Iran. As of 1999, fewer than half of these refugees had returned.

The *mujahadiin* forces succeeded in overthrowing the Communist government of Afghanistan in 1992, but since then rival factions among the formerly united rebels have been engaged in civil warfare. By late 1996 one of the factions, the Taliban (backed by Saudi Arabia and Pakistan), controlled two-thirds of the country's provinces, including the capital of Kabul.

Definitions & Insights

THE LASTING LEGACY OF LAND MINES

Land mines are antitank and antipersonnel explosives, usually hidden in the ground, designed to kill or severely wound the enemy. There are many kinds: Soviet troops planted at least nine varieties in Afghanistan, including seismic mines, which are triggered by vibrations created by passing horses or people; mines with devices that pop up and explode when a person approaches; "butterfly" mines meant to maim rather than kill; and explosives disguised as toys, cigarette packs, and pens.

There are an estimated 10 million land mines in Afghanistan, or 40 per square mile (15 per sq km). Even after the fighting ends, reconstruction of this agricultural country cannot begin in earnest until the mines are cleared. The obstacles are enormous in rugged Afghanistan. Mines typically remain active for decades after they are planted, and tend to last longer in arid places. The mountainous terrain prevents systematic clearance of the explosives. Many mines were scattered indiscriminately from the air and will be difficult to locate. Heavy rains such as those which fell on Afghanistan in the spring of 1996 move the mines downhill and away from known locations. And it is expensive to clear mines; Kuwait spent $800 million to rid its landscape of them after the Gulf War. So far, Western countries have donated $12 million toward the expense of clearing Afghanistan's land mines; Russia has contributed nothing.

Land mines are a global problem. Up to 70 million mines contaminate the land in 64 countries, and they kill or maim more than 25,000 civilians each year. The numbers and concentrations in some countries are staggering. The leader in numbers is Egypt, with 23 million mines; in density of mines, Bosnia-Herzegovina leads with 152 per square mile (58 per sq km). The United Nations estimates that with current technology, it would take $33 billion and 1100 years to clear the world's land mines.

In 1997 several events occurred which may portend an eventual end to the land mine scourge. The world's most visible critic of land mines, Princess Diana of Wales, was killed in an automobile accident in Paris in August. In her wake there were renewed calls for a moratorium on the manufacture and use of antipersonnel land mines. Two months after Diana's death, an anti–land mine activist named Jody Williams won the 1997 Nobel Peace Prize. Then in December, representatives from 125 nations met in Ottawa, Canada, to sign a treaty strictly banning the use, production, storage, and transfer of antipersonnel land mines. Many of the largest producers, users, and exporters of these explosives—including the United States, Russia, China, India, Pakistan, Iran, Iraq, and Israel—refused to sign. The United States argued that land mines were necessary to protect South Korea against an invasion from North Korea, and failed to win an exemption on this count that would have allowed it to sign the treaty.

Too many countries regard these as legitimate weapons to make the prospect of a mine-free world likely. Costing as little as two dollars each, they are very affordable weapons for cash-strapped warring LDCs (and so have been dubbed the "Saturday-night specials of warfare"). For now, these hidden horrors continue to be planted and to be cleared, as one de-mining specialist put it, "one arm and one leg at a time."

Proclaiming itself the sole legitimate government of Afghanistan, the Taliban immediately imposed a strict code of Islamic law in the regions under its control and gained international notoriety for its austere administration. The Taliban removed almost all women from the country's workforce, forbade education of girls, and outlawed "un-Islamic" practices such as dancing, kite-flying, television, bird-keeping, and beard-trimming. The Taliban continued to make advances against its opponents inside Afghanistan, and by 1999 controlled more than three quarters of the country's territory. The neighboring Central Asian countries and Russia began to fear the spread of the Taliban's extreme interpretation of Islam into their nations, and Russian military advisors were sent to train Uzbek forces along the Afghan frontier. Late in 1998, Iran amassed troops on its border with Afghanistan and stepped up its rhetoric against the Taliban regime. Iran's hostility toward the regime in Kabul stemmed from the Taliban's strident interpretation of Sunni Islam; its apparent associations with the United States, Saudi Arabia, and Pakistan; alleged Taliban murders of Iranian diplomats; and alleged Taliban complicity in the smuggling of opium and heroin from Afghanistan into Iran (Afghanistan is the world's second largest producer of opium, after Burma).

The scars of war will last a very long time in this country. When peace does come, Afghanistan, like Kuwait, will have to confront another problem: clearing the millions of land mines sown across the landscape during wartime.

SUMMARY WITH SELECTED KEY TERMS

- **The Arab World** stretches from Morocco to the Indian Ocean. While an ethnically diverse region, most inhabitants speak dialects of Arabic and are known as Arabs. The wide distribution of these peoples is due primarily to the rise and spread of **Islam and its empire.**
- **The Promised Land** is a reference to the land promised to Abraham and his kin by God. Historically known as **Canaan** and geographically as **Palestine,** it is now mostly inside the nation of **Israel.**
- The Roman Empire scattered the Jews into exile, known as the **diaspora,** after the destruction of the Second Temple. Hebrew teachings explain how the Jewish people will eventually be reunited in Jerusalem (**Zion**).
- The **United Nations Partition Plan of 1947** attempted a **"two state solution"** to the Arab/Jewish conflict. However, it produced almost fragmented states, making both sides vulnerable to the other and complicating the issue further. The Arab/Israeli conflict has continued from that time through numerous **wars, crises,** and **terrorist actions.**

- The famous **Bekaa Valley,** shared by Lebanon and Syria, has been a center of political activity since before Roman times, and continues to be a center for pro-Iranian **Hizbullah** activities today.
- Lebanon and Syria are the home of the historic and modern urban centers of **Damascus, Aleppo, and Beirut.** Before **Lebanon's Civil War,** Beirut was known as the **"Play Area"** of the Middle East.
- Sudan is a nation of marked environmental diversity and is considered a **transition zone.** This diversity has led to long-term military and political conflict between the northern and southern peoples.
- Most of Libya lies in the **Sahara** and its primary natural resource is **oil.** The revenues generated from the exported oil are used to support social and economic projects along with large purchases of weapons and weapons development. Muammar al-Qaddafi's brand of socialism/communism and his support for international terrorism have earned the nation a dubious reputation.
- The **"Maghreb" (Place of the West)** is composed of Morocco, Algeria, Tunisia, and the disputed Western Sahara. This area is closer, geographically and culturally, to Europe than is any other of the Middle East.
- **Two-thirds of the world's proven petroleum reserves** are concentrated in a few countries that ring the **Persian/Arabian Gulf.** All but 1 percent of the Gulf's reserves are located in Saudi Arabia, Iraq, Kuwait, Iran, and the United Arab Emirates (UAE). Saudi Arabia controls approximately 26 percent of the world's oil production.
- **Iraq,** the region of **Mesopotamia ("Land between the rivers"),** lies on the eastern end of the Fertile Crescent. It is the home of the **first known civilization** and **culture hearth.** While the **Baath (Arab Socialist) Party** under **Saddam Hussein** has driven for Arab solidarity, freedom from the West, and economic development, it has also caused the Iraqis to suffer heavy casualties in the long

Iran/Iraq War and a political, economic, and military thrashing in the **Gulf War** for its attempt at expansion (for oil and seaports) into Kuwait and other areas of the Persian Gulf.
- **Iran (formerly known as Persia)** was the earliest Middle Eastern country to produce oil in large quantities. The oil revenues were used to modernize and Westernize the nation during the **Pahlavi Dynasty.** The rapid changes in the culture precipitated a revolutionary movement, which forced the **Shah** to abdicate and flee. The new **theocratic state,** since the death of the **Ayatollah Khomeini,** is alternately perceived to be either developing a more accommodating attitude toward the international community, or exporting ever more dangerous notions of revolutionary Islam to the region.
- **Turkey** was formerly the seat of power for the **Islamic Ottoman Empire.** It is now an independent, secular nation with fewer ties to traditional Islam. Turkey has become a **rather unique Middle Eastern nation** based upon such attributes as: official separation of church and state, democratic government, membership in NATO and full citizenship for women.
- **Cyprus** has a history of foreign rule by the Greeks, Romans, Ottoman Turks, and the British. Though independent today, the ethnic composition of Cyprus is **three-fourths Greek** and **one-fourth Turkish.** Despite hopes for **enosis (unification),** several internal conflicts have occurred in efforts to gain control of this large island near southeast Turkey.
- **Afghanistan** is one of the world's poorest nations, being **landlocked** as well as having **limited resources, poor internal transportation,** and **little foreign trade.** Its **historic location** along the ancient caravan routes made it the target of empires. The present situation next to oil-rich and revolutionary Iran has again made it a **strategic target.**

REVIEW QUESTIONS

1. List and discuss places sacred to Judaism, Christianity, and Islam.
2. List and describe the major conflicts that have occurred between Arabs and Israelis.
3. Using maps and the text, identify the major physical and cultural features of Israel, Jordan, and Syria.
4. List the effects of dam construction on the Nile.
5. In what ways is Sudan's north different from the south?
6. What are some of the major factors driving U.S. interests in the Gulf Oil Region?

7. Using maps and the text, locate the countries of the Gulf Oil Region. Which of these are more important in petroleum production?
8. List some of the unique aspects of Turkey in relation to most of the other countries of the Middle East.
9. What have been the primary goals of the Baath regime in Iraq? List some of the Baath regime's results.
10. What were some of the grievances against the Pahlavi dynasty that led to revolution in Iran?

DISCUSSION QUESTIONS

1. Where are the "chokepoints" of the Middle East?
2. Examine and discuss the **birth of Israel** and its lasting legacy for the rest of the world.
3. What is the origin of the phrase **"Gift of the Nile"**? How does this relate to modern Egypt?
4. Examine and discuss the political problems associated with Sudan and Libya and the rest of the world.
5. Explain how the **"Maghreb"** is closely associated with Europe.
6. What regions or nations of the world are primarily dependent upon the **Gulf Oil Region** for petroleum? What is the U.S. interest?

7. Why did the United States and other nations feel driven to engage Iraq in the **Gulf War**?
8. Examine and discuss Iran's complicated struggle between traditional Islamic culture and modern society.
9. Why has Turkey, once the home of the **Ottoman Empire,** evolved as a comparatively Westernized nation in the Middle East?
10. Examine and discuss the significance of Afghanistan's **location** in the Middle East.

5 Monsoon Asia

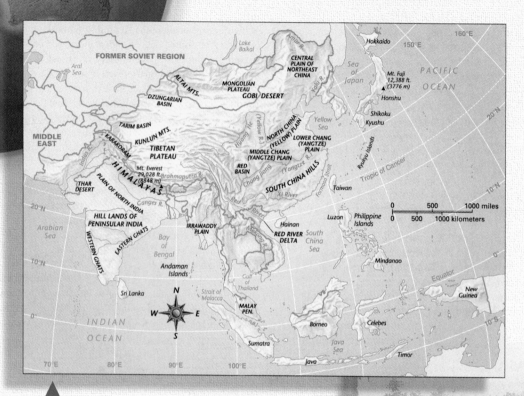

Monsoon Asia expresses a great crescent from the margins of the Middle East around to the very edge of Eurasia. Massive river networks flow down from mountain highlands to the edges of varied seas, and in these river systems, life and civilization have been nurtured and harried for thousands of years.

The city is gaining strength and importance all through Monsoon Asia. It is the gateway to western influences, and the setting for the most rapid social change and experimentation. It represents both energy and uncertainty throughout all of Monsoon Asia.
Jeff Greenburg / Photo Researchers, Inc.

A long cultural tradition exists in virtually all of Monsoon Asia. Patterns of architecture, family structure, social organization, religion, and attitudes toward the environment have their roots in history that goes back thousands of years and hundreds of generations. Such tradition generally stays vital even in the face of rapid economic change. *Bruce Burkhardt / Westlight*

Regional Snapshot

Population is a theme of continual fascination and concern to all of Monsoon Asia. The young ages of the peoples, the increasing longevity of its seniors, and the rural majority of most nations combine to make the fact that nearly three-fifths of the world's population lives in this one region a reality of major consequence.
L. Rebman / Photo Researchers, Inc.

From Japan in an Asian arc around to India, high technology has been introduced into manufacturing as part of ambitious regional programs of economic development. The reputation of the "Asian Tigers" comes in part from such investment decisions and resultant productivity. *Tadanori Saito / PPS / Photo Researchers, Inc.*

Monsoon Asia is still predominately an agricultural world. For millennia, peasant farmers have made and remade small, nearly garden-size farming plots for a \rariety of crops. Irrigation has been added when possible and such farms manifest very close management of the rural landscape. *Brian Brake / Photo Researchers, Inc.*

Religion runs through the life patterns of virtually all peoples in Monsoon Asia. Sacred waters run in major rivers; sacred mountains stand guard over ceremonial landscapes; and people associate their surroundings with everything from personal karma to major political flashpoints of conflict. *Joel Simon / Tony Stone Images*

Chapter 10

A Geographic Profile of Monsoon Asia

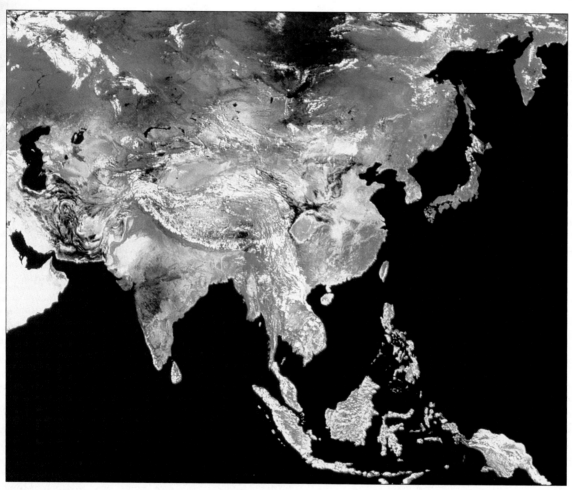

▲ This composite satellite image is from the National Oceanic and Atmospheric Administration (NOAA) and shows the full landscape of Monsoon Asia, with its great mountain highlands and plateaus; its desert expanses; ranges of borderland mountains; verdant river valleys; and the plains and river systems with dense accumulations of farming and urban populations. Image colors in red denote vegetation; in grey, semi-desert; and in white, desert sands. The mottled red and white in Southeast Asia is tropical rain forest.   Earth Satellite Corporation/Science Photo Library/Photo Researchers, Inc.

CHAPTER OUTLINE

10.1 Definition and Basic Magnitudes

10.2 A Broad Range of Environments

10.3 Climate and Vegetation

10.4 Cultural Landscapes and Signatures of the Past

10.5 Distinctive Contemporary Landscapes

10.6 Population, Settlement, and Economy

*a*sia extends, in a broad, sweeping crescent, from Pakistan and the Indus River through an arc of mountains, rivers, plains, islands, and seas to the northern island of Japan. In this landmass of more than 8 million square miles (20,761,000 sq km) lives more than half of the world's population. The cultural makeup of these 3.3 billion people is as diverse and expressive as are the physical environments they inhabit. The world's highest mountain peaks, some of the longest rivers, and Earth's most highly transformed and densely settled river plains are the setting for some of our oldest civilizations and our most modern developing economies. Just as this region has played a major role in societal development over the past millennia, so, too, will it be a major force in defining the world of the 21st century.

"The Orient" is the term traditionally used to refer to the countries occupying the southeastern quarter of Eurasia. The term "orient" comes from the Latin meaning "to rise," or "to face the east." "Occident" means "to set" (as in the sun—in the west). Early on, the point from which this daily orbit of the sun was described was Western Europe. Hence, the Orient and the Occident became European terms for spatial regional identification. In this text we use the term Monsoon Asia rather than the Orient for this region because of the significant role the seasonal shift of wind patterns plays all the way from Pakistan in the west to Japan in the east (Fig. 10.1).

Monsoon Asia encompasses the following regions: East Asia, which embraces Japan, North and South Korea, China, Taiwan, Macao, and countless nearshore islands; South Asia, which includes Pakistan, India, Sri Lanka, Bangladesh, and the mountain nations of Bhutan, Nepal, and offshore islands; Southeast Asia, which is the term used for the dominant peninsula jutting out from the southeast corner of the Asian continent and includes the countries of Burma (Myanmar), Thailand, Laos, Cambodia, Vietnam, Malaysia, and Singapore; and the island world that rings this peninsula, which includes the countries of Indonesia, the Philippines, Brunei, and East Timor.

This crescent of Monsoon Asia has been home to some of the most important cultural developments of humankind—in landscape transformation, settlement patterns, religion, art, and political innovations. It is not the magnitude of population alone that makes understanding Monsoon Asia central to our study of world regional geography; it is also the pivotal role of past and present Asian cultural innovations and a significant global economic role that make this region a major building block in the process of better understanding the nature of the world today.

10.1 Definition and Basic Magnitudes

Table 10.1 portrays the demographic and geographic magnitude of Monsoon Asia. Within it lie the major subregions of the Indian subcontinent, Southeast Asia, the Chinese realm, and Japan and Korea. There is profound geographic variety in this region that encompasses less than one-eighth of Earth's surface, but which is home to approximately 60 percent of the world's population. The birth rate ranges from a 0.32 annual rate of increase in Japan to nearly 3.0 in parts of Southeast Asia. In Hong Kong, Macao, and Singapore, some of the world's highest urban densities are found, while in Mongolia and Nepal, population densities are extremely low. Singapore is one of the few nations to claim a 100 percent urban population, while 92 percent of the Nepalese population is still rural and nonurban. Only 2 percent of Bhutan is arable, while farmers in India and Bangladesh both cultivate more than 50 percent of their land.

To gain a feeling for the diversity of this region, it is important to see it today as a collection of distinctive approaches to achieving economic growth and, in the face of the severe economic decline of 1997–1998, sustaining such growth. This challenge is set against a backdrop of widely varying distinctive histories, particularly in the development of technology, farming, literature, and religion. At the same time, images in the West of prosperous Japanese and Hong Kong tourists festooned with cameras and sophisticated electronic gadgetry must be tempered with the reality of a large Asian peasantry still

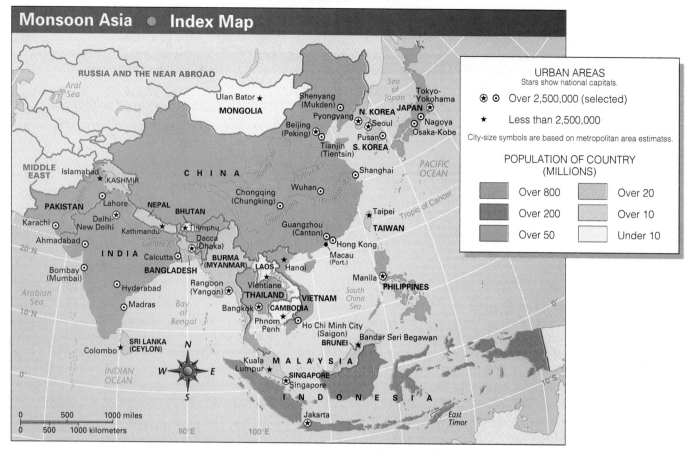

Monsoon Asia • Index Map

Figure 10.1 Introductory location map of Monsoon Asia, showing political units as of early 1999. East Timor is a former Portuguese colony occupied by Indonesia, whose claim is not recognized by the United Nations. Macao is scheduled to be returned to China late in 1999.

working the land with traditional tools and only modest evidence of recent change in agricultural technology. Monsoon Asia, to a greater extent than other regions chronicled in our text, is a restless amalgam of tradition and innovation, rural industriousness and urban experimentation, and societal continuity and tense assimilation of Western popular culture. This restlessness is expressed with particular force in the varied ways in which the Asian nations—especially those in East and Southeast Asia—are attempting to reduce and distance themselves from the financial upset that began in Thailand in 1997 and rippled through much of Southeast and East Asia these last several years. The region is in many ways an evocative microcosm of the forces and elements of contemporary world regional geography.

10.2 A Broad Range of Environments

The setting in which the Asian drama is being enacted is a complex intermingling of many topographies (Fig. 10.2). However, a certain broad order appears if the surface features are conceived of as three concentric arcs, or crescents, of land: an inner western arc of high mountains, plateaus, and basins; a middle arc of lower mountains, hill lands, river plains, and shallow basins; and an outer eastern arc of islands and seas.

The Inner Highland

The inner highland is composed of the world's highest mountain ranges, interspersed with plateaus and basins. At the south of the continent, the great wall of the Himalaya, Karakoram, and Hindu Kush mountains overlooks the north of the Indian subcontinent. At the north, the Altai, Tien Shan, Pamir, and other towering ranges separate this region of Asia from the countries of the former Soviet Union. Between these mountain walls lie the sparsely inhabited Tibetan Plateau, at over 15,000 feet (c. 4500 m) in average elevation, and the dry, thinly populated basins and plateaus of Xinjiang (Sinkiang) and Mongolia (Fig. 10.3).

River Plains and Hill Lands

The area between the inner highland and the sea is principally occupied by river floodplains and deltas bordered and separated by hills and relatively low mountains. Major components are:

280

Table 10.1 Monsoon Asia: Basic Data

Political Unit	Area (thousand/ sq mi)	Area (thousand/ sq km)	Estimated Population (millions)	Annual Rate of Increase (%)	Estimated Population Density (sq mi)	Estimated Population Density (sq km)	Infant Mortality Rate	Urban Population (%)	Arable Land (%)	Per Capita GNP ($US)
Indian Subcontinent										
India	1148	2973.3	988.7	1.9	861	332	72	26	55	380
Pakistan	297.6	770.8	141.9	2.8	477	184	91	28	23	480
Bangladesh	50.3	130.3	123.4	1.8	2454	947	82	17	67	260
Nepal	52.8	136.8	23.7	2.2	449	173	79	10	17	210
Bhutan	18.1	47	0.8	3.1	44	17	71	13	2	390
Sri Lanka	25	64.8	18.9	1.3	757	292	16.5	22	16	740
Total	**1591.8**	**4123**	**1297.4**	**2**	**815**	**315**	**68.6**	**25**	**47**	**382**
Southeast Asia										
Myanmar (Burma)	253.9	657.6	47.1	2	185	71	83	25	15	660
Thailand	197.3	511	61.1	1.1	310	120	25	31	34	2960
Vietnam	125.7	325.6	78.5	1.2	625	241	38	20	22	290
Cambodia	68.2	176.6	10.8	2.4	158	61	116	6	16	300
Laos	89.1	230.8	5.3	2.8	59	23	97	19	4	400
Malaysia	126.9	328.7	22.2	2.1	175	68	10	57	3	4370
Singapore	0.236	0.611	3.9	1.1	16,415	6336	3.8	100	4	30,550
Brunei	2	4.5	0.3	2.2	155	60	24.7	67	1	15,800
Indonesia	705.2	1826.5	207.4	1.5	294	113	66	37	8	1080
Philippines	115.1	298.1	75.3	2.3	654	252	34	47	26	1160
Total	**1683.6**	**4360**	**511.9**	**1.6**	**304**	**117**	**49.8**	**35**	**14**	**1509**
Chinese Realm										
China (PRC)	3601	9327	1249.2	1	345	133	31	30	10	750
Macao	0.008	0.02	0.5	1	59,700	23,044	5	97	0	13,600
Taiwan	13.9	36	21.7	1	1555	600	6.7	75	24	13,970
Mongolia	604.8	1566.4	2.4	1.6	4	2	49	57	1	360
Total	**4219.7**	**10,929.4**	**1273.8**	**1**	**302**	**117**	**22.9**	**31**	**9**	**980**
Japan and Korea										
Japan	145.3	376.3	126.4	0.2	869	335	3.8	78	13	40,940
North Korea	46.5	120.5	22.2	0.9	477	184	39	59	18	900
South Korea	38.1	98.6	46.4	1	1218	470	11	79	21	10,610
Total	**229.9**	**595.4**	**195**	**0.5**	**848**	**327**	**17.9**	**76**	**15**	**29,165**
Summary Total	**7725**	**20,007.9**	**3278.1**	**1.4**	**424**	**164**	**39.8**	**32**	**18**	**2503**

Sources: *World Population Data Sheet*, 1998. United Nations Statistics Division, 1998. *Almanac of Politics and Government*, 1998. *World Factbook*, 1998.

1. the immense alluvial plain of northern India, built up through ages of meandering and deposition by the Indus, Ganges, and Brahmaputra rivers (Fig. 10.4);
2. the uplands of peninsular India, geologically an ancient plateau but largely hilly in aspect;

3. the plains of the Irrawaddy, Chao Praya (Menam), Mekong, and Red rivers in peninsular Southeast Asia, together with bordering hills and mountains;
4. the uplands and densely settled small alluvial plains of southern China;

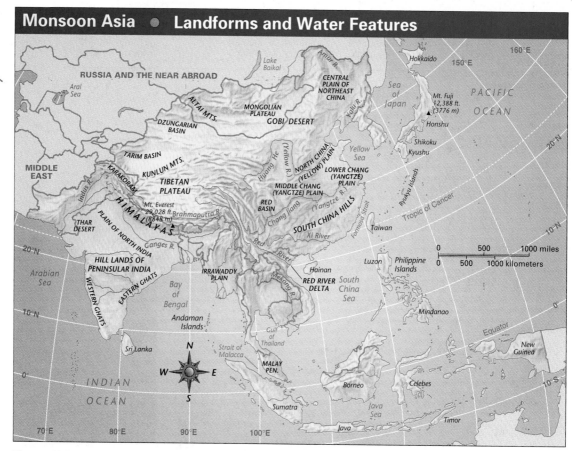

Figure 10.2   Major landforms and water features of Monsoon Asia.

Figure 10.3   Dunhuang—in China's western Gansu province—is an ancient stop on the Silk Route that linked northern China with trading ports as far west as Baghdad and various Mediterranean ports. This century a whole series of grottoes was discovered carved into the cliffsides of the river valley seen in the distance. These caves were used for ceremonial and religious displays by travelers either relieved to have successfully crossed the deserts of western China or nervous about now beginning such a traditionally difficult crossing. In the foreground can be seen small ceremonial structures built by the farmers of this wheat-growing and carpet-weaving stop on this ancient route between China and the West.   *Kit Salter*

Figure 10.4 In India, the great rivers play a role not only in the flow of agricultural life but also in religious life. Here the holy act of washing in the waters of the Ganges River is both a sacred and a social act. *Cathlyn Melloan Tony Stone Worldwide, Ltd.*

5. the broad alluvial plains along the middle and lower Chang Jiang (Yangtze River) in central China and the mountain-girded Red Basin on the upper Chang Jiang;
6. the large delta plain of the Huang He (Yellow River) and its tributaries in North China, backed by loess-covered hilly uplands; and
7. the broad central plain of Northeast China (Manchuria), almost completely enclosed by mountains.

Offshore Islands and Seas

Offshore, a fringe of thousands of islands, mostly grouped in great **archipelagoes**—clusters of islands—borders the mainland. On these islands, high interior mountains with many volcanic peaks are flanked by coastal plains where most of the people live. Three major archipelagoes incorporate most of the islands—the East Indies, the Philippines, and Japan. Sri Lanka, Taiwan, and Hainan are large, densely populated islands outside these archipelagoes.

Between the archipelagoes and the mainland lie the China Seas, and, to the north, the Sea of Japan. At the southwest, the Indian peninsula projects southward between two immense arms of the Indian Ocean—the Bay of Bengal and the Arabian Sea.

10.3 Climate and Vegetation

The climatic pattern of the region, like the physiographic, is one of almost endless variety. Two unifying elements, however, are present throughout those parts of Asia inhabited by considerable numbers of people. These are (1) the dominance

of warm climates, and (2) a monsoonal regime of precipitation (Fig. 10.5).

Temperature and Precipitation

In the most populated parts of Monsoon Asia, temperatures are tropical or subtropical. The principal exceptions exist in northern sections of China, Korea, and Japan. Here, summers are warm to hot in the lowlands, but the growing season is relatively short and winters are cold. The arid, sparsely populated basins and plateaus of Xinjiang and Mongolia also have warm summers and sometimes bitterly cold winters. The higher mountain areas and the Tibetan Plateau have undifferentiated highland climates varying with elevation, aspect, and latitude. Permanent snow fields and glaciers occur at the higher elevations.

Annual precipitation varies from near zero in parts of the Chinese desert, the Takla Makan, and parts of Xinjiang to more than 400 inches (1000 cm) in parts of the Khasi Hills of northeastern India. A monsoon climate, or at least a climate with monsoonal tendencies, prevails nearly everywhere in the populous middle arc of plains and hills all across Asia, and in many parts of the islands as well (see Fig. 10.5).

Types of Climate

Monsoon Asia is characterized for the most part by a warm, well-watered climate. There are seven main types of climate customarily recognized in Asian climatic classifications: **(1) tropical rain forest, (2) tropical savanna, (3) humid subtropical, (4) humid continental, (5) steppe, (6) desert, and (7) undifferentiated highland.**

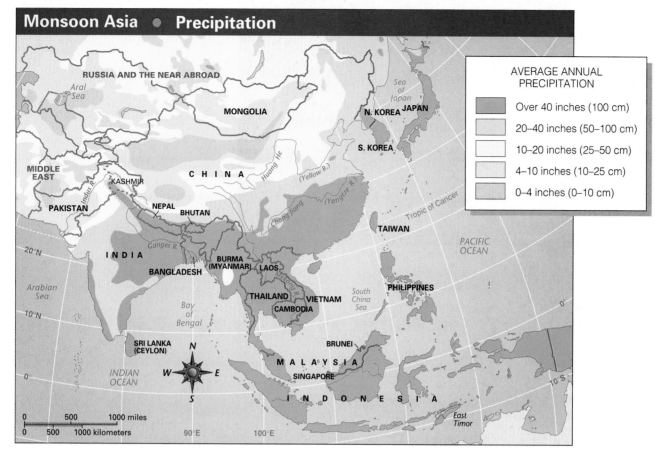

AVERAGE ANNUAL PRECIPITATION

- Over 40 inches (100 cm)
- 20–40 inches (50–100 cm)
- 10–20 inches (25–50 cm)
- 4–10 inches (10–25 cm)
- 0–4 inches (0–10 cm)

Figure 10.5 The pattern of precipitation in Monsoon Asia is defined in part by orographic influences caused by the numerous mountain ranges that lie toward the interior. With air masses full of moisture from their origins in the seas on the margins of the east and south of Monsoon Asia, much precipitation is coaxed out of the air masses that flow toward the Asian interior, giving maximum levels of precipitation toward the coast and diminishing toward the interior. The true interior of this region is arid and in some cases complete desert.

Most rainy tropical climates (rain forest and savanna) are generally found between the equator and the Tropics of Cancer and Capricorn, with major regions closer to the equator. Consequently, high temperatures are experienced throughout the year in the lowlands. Based on temperature alone, the year-round growing season would seem to offer maximum possibilities for agriculture. However, the high temperatures and heavy rains promote rapid leaching of mineral nutrients and destruction of organic matter, with the result that most soils are relatively infertile despite the surprisingly thick cover of trees and grasses they often support. A brief description of the Asian climate zones follows.

Tropical rain forest. This climate zone is typically found within 5 or 10 degrees of the equator. Precipitation is spread throughout the year so that each month has considerable rain. The amount of precipitation is at least 30 to 40 inches (c. 75–100 cm), often 100 inches (c. 250 cm). Average temperatures vary only slightly from month to month; Singapore, for example, exhibits a difference of only 3°F between the warmest and coolest months. Monotonous heat prevails year-round,

although some relief is afforded by a drop of 10 to 25°F (6–14°C) in the temperature at night, and sea breezes refresh coastal areas. In such climates as this, it is said that "nighttime is the winter" of the region.

Tropical rain forest climate and vegetation are characteristic of most parts of the East Indies, the Philippines, and the Malay Peninsula. Rain forest vegetation is also found in certain other areas (generally along coasts) that experience a dry season but in which the precipitation of the rainy season is sufficiently heavy to promote a thick growth of trees.

Tropical savanna. The savanna climate, like the tropical rain forest climate, is characterized by high temperatures year-round but is customarily found in areas farther from the equator, and the average temperatures vary somewhat more from month to month. However, the most important difference between the two climates is that the savanna has a well-defined dry season, lasting in some areas as long as six or eight months each year, creating a problem for agriculture.

The main Asian areas of tropical savanna climate occur in southern and central India, in the greater part of the Indochinese

Definitions & Insights

MONSOON

A monsoon is the seasonal wind shift with associated rain that is so characteristic of much of Monsoon Asia. Because of enormous differentials in the way in which the sea and land absorb insolation (heat from the sun), air masses are kept unstable. If coastal waters and the adjacent coastline receive approximately equal amounts of warmth from the sun, the land will get warm or hot much more quickly than the sea water. As a result, air over the land begins to get unstable and rise. Such ascending air creates a target for the air masses over the water, and these marine winds begin to blow toward the land. In Asia, the summer monsoon is characterized by winds blowing from the sea to the land and bringing—as you would expect—high humidity, moist air, and generally predictable rains. Because of the moisture carried by such air masses, these wind shifts are the sources of major rainfall in the late spring, summer, and early fall seasons. Agriculture and patterns of human activity all key on these incoming rains. If there is a little bit of elevation in the landscape—hills or mountain flanks—the moist air is driven higher, cools, and releases even more rain.

In the winter, the land loses its relative warmth, while the sea and coastal waters all maintain their warmth longer. As a result, the wind shifts and air masses begin to flow from the inland areas of Monsoon Asia toward the sea. Monsoon flow goes in the opposite direction. However, because there is little moisture in the source areas over land for the winter monsoon, there is much less rain. A monsoon climate, therefore, is generally characterized by spring and summer precipitation and a long dry season in the low sun (winter) cycle. Japan, however, experiences a relatively unusual pattern of winter precipitation related to the monsoon wind shifts. The winter winds blow out of China and east Asia, wash across the Sea of Japan, and pick up moisture and drop heavy, wet snow on the west coast of the Japanese islands.

The monsoon has different local traits all across Monsoon Asia, but the general pattern of major air flows from sea to land in the warm season and from land to sea in the cooler seasons is quite common. The power of these patterns shows up not only in farming patterns, but in literature, myth, and general environmental perception of this region by inhabitants as well as travelers to these well-watered landscapes that arc from Pakistan to Japan. These traits are particularly evident in the coastal plains and lowlands that are found in South Asia, the peninsula and islands of Southeast Asia, and the eastern one-third of China.

The monsoon has a dominant role in the cultures because agricultural success is tied so closely to the arrival of spring and summer rains. This pattern of monsoon rains and air movement shapes not only the nature of farming, but the personality of place as well.

---

peninsula, and in eastern Java and the smaller islands to the east. In Asia, the characteristic natural vegetation associated with this climate is a deciduous forest of trees smaller than those of the tropical rain forest. Tall, coarse grasses, like bamboo—a very common vegetation form in African and Latin American savannas—are found only in limited areas, and even they are thought to have been produced by repeated burning of forests.

Humid subtropical. This climate zone occurs in southern China, the southern half of Japan, much of northern India, and a number of other countries. It is characterized by warm to hot summers, mild or cool winters with some frost, and a frost-free season lasting 200 days or longer. The annual precipitation of 30 to 50 inches (*c.* 75–125 cm) or more is fairly well-distributed throughout the year, although monsoonal tendencies produce a dry season in some areas. The natural vegetation is a mixture of evergreen hardwoods, deciduous hardwoods, and conifers.

Humid continental. This climate characterizes the northern part of eastern China, most of Korea, and northern Japan. It is marked by warm to hot summers, cold winters with considerable snow, a frost-free period of 100 to 200 days, and less precipitation than the humid subtropical climate. Most areas experience a dry season in winter. The predominant natural vegetation is a mixture of broadleaf deciduous trees and conifers, although prairie grasses are thought to have formed the original cover in parts of northern China.

Steppe and desert. These climates, whose characteristics have been previously described, are found in Xinjiang and Mongolia, and in parts of western India and Pakistan (see Fig. 10.3).

Undifferentiated highland. The undifferentiated highland climate is most extensive in the Tibetan Highlands and adjoining mountain areas. It is made up of a broad range of montane microclimates that provide little base for plant growth and vary from place to place according to elevation, aspect, and latitude.

10.4 Cultural Landscapes and Signatures of the Past

A cultural landscape can be described as a **palimpsest**—a term which describes an artist's canvas that has been sketched on, erased, and sketched on again, but with markings left from the earlier drawings. In the same sense, the cultural landscape has tangible and visible evidence of earlier human efforts to settle, to farm, and to create landscapes that satisfy human needs (and desires). Even though such needs change over time from cul-

Regional Perspective

The Ming Voyages

We often think of the Age of Discovery as simply the chronicle of what the West did after Columbus launched his maritime effort to get to Asia by sea in 1492. The map of the world might have a very different look if the Ming Voyages—from an earlier age of Chinese discovery—had been dealt with in a different way by the Chinese and other Asian nations. From 1405–1433 a Chinese Muslim admiral, Zheng He (Cheng Ho), commissioned by the Ming Court in China, led seven long-range maritime expeditions. He sailed from a port in southeast China to large overseas Chinese communities in Indonesia and then on to India, the Arabian Sea, and even to Mombasa on the East African coast. There were 63 ocean-going junks (a Chinese sailing vessel) in Zheng He's fleet, the largest of which—at 440 feet long and 180 feet wide, with a crew of more than 400 men—was six times the size of the Niña, the Pinta, or the Santa Maria. The Chinese certainly had the maritime power and the manpower to colonize new lands in the early 15th century, and were it not for the death of the sponsoring Chinese emperor in 1433 and the general Chinese inclination to stay focused on the world of East Asia, the Earth might have had a very different Age of Discovery. After 1433, these voyages ended when the new Ming Court prohibited any Chinese ships from leaving traditional coastal shipping lanes.

tural diffusion, the markings of earlier efforts stand in mute testimony to the variety of ways people interact with their landscapes.

In learning to "read" a cultural landscape, there are features that have such distinctive associations with place that they serve to mark that locale in a specific and unique way. The Great Wall in China and the Taj Mahal in India are two universally known features of human engineering (Fig. 10.6). These features are called **landscape signatures.** They are generally features created by human modification of the environment, although sometimes physical features may be so distinctive—the mouth of the Ganges River in Bangladesh, for example—that they play the same role. Landscape signatures are useful to know as a means of assigning landscape and regional identity to specific places.

ancient heritage, coupled with resentment of indignities and repression suffered under colonial administrations, contributed to rising nationalism and, consequently, the end of Western colonial control after World War II. But the geographic heritage from the colonial past remains pervasive, and some of its major aspects are summarized here as essential background for the discussions that follow (Fig. 10.7).

European penetration of this part of Asia began at the end of the 15th century. The early explorers (from Portugal, Spain, and the Netherlands) gradually extended political control over some islands and limited areas near the coast. In the 18th and 19th centuries, the pace of colonization and economic control quickened, and large areas came under European sway. By the end of the 19th century, Great Britain was supreme in India, Burma

European Signatures of Colonialism

Like much of the modern world, Asia is an area where many traditional geographic patterns were created or reshaped by Western colonialism. Today, the region is postcolonial and made up of many independent nations, some of which contain areas in which very early civilizations developed. Pride in this

Figure 10.6 The Great Wall is one of China's most remarkable landscape features. As it exists today it represents the collection of independently constructed wall segments going back to before the second century B.C. The segment of wall that is most commonly seen is a bus ride from Beijing and is now surrounded by tourist shops and even a set of modern-day *yurts* (the Mongol traditional skin tents) used for tourist accommodations. The segment of wall in this photo lies west of the mile and a half of wall that is in travel posters and slide sets. This view shows the state of collapse that is more genuinely representative of the wall segments that still stand. *Kit Salter*

286

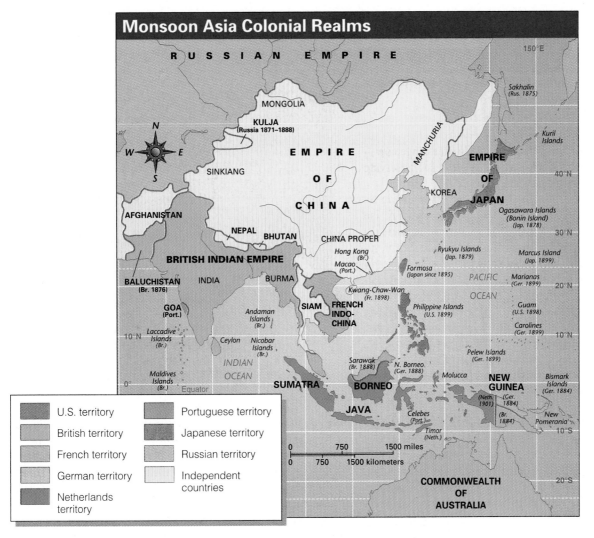

Monsoon Asia Colonial Realms

Legend:
- U.S. territory
- British territory
- French territory
- German territory
- Netherlands territory
- Portuguese territory
- Japanese territory
- Russian territory
- Independent countries

Figure 10.7 This map shows the general distribution of colonial realms—as well as independent countries—in Monsoon Asia at the beginning of the 20th century. The British influence is seen as the most dominant and the one that continued to the middle of the century, finally eclipsed almost totally with the 1997 return of Hong Kong to China. China had small regions controlled by Germany, Russia, Japan, and Great Britain, but never fell under full control of any colonizing power. It was at the beginning of this century that the United States also began to play a role in the development of some Pacific territories, with the Philippines and Hawaii as the most significant. *From "The World in 1900" in R.R. Palmer, Atlas of World History. Chicago: Rand McNally, 1957. Pp. 169–171.*

(Myanmar), Ceylon (now Sri Lanka), Malaya, and northern Borneo; the Netherlands possessed most of the East Indies (Indonesia); and France had acquired Indochina (now Vietnam, Laos, and Cambodia) and small holdings around the coast of India, and Portugal had Goa in India, and Macao in China. China, although retaining a semblance of territorial integrity, was forced to yield possession of strategic Hong Kong to Britain in the 1842 Treaty of Nanking and to grant special trading concessions and extraterritorial rights to various European nations and the United States through the latter half of the 19th century. In 1898, the Philippines, held by Spain since the 1500s, came into American control.

The few Asian countries to escape domination by the Western powers during the colonial age were (1) Thailand, which formed a buffer between British and French colonial spheres in Southeast Asia; (2) Japan, which withdrew into almost complete seclusion in the mid-17th century but which emerged in the later 19th century as the first modern, industrialized Asian nation and soon acquired a colonial empire of its own; and (3) Korea, which also followed a policy of isolation from foreign influences until 1876, when a trade treaty was forced on it by Japan.

During the age of European expansion, Asia constituted an extraordinarily rich colonial area. Western nations extracted vast quantities of tropical agricultural commodities and, in turn, found large markets for their manufactured goods. Westerners also found a fertile field for investment in plantations, factories, mines, transportation, communication, and electric-power facilities. Some of the region's most important cities—for

287

Figure 10.8 Buddhist monks in their saffron robes in Bangkok, Thailand. Buddhism is the main religion in several Southeast Asian countries and also plays a role in other countries in Monsoon Asia. *Lorina Ebbe*

example, Shanghai in China, Calcutta in India, and Singapore—were developed largely by Western capital as seaports serving Western colonial enterprise.

Western dominance of Asia ended in the 20th century by circumstances that included (1) the weakening effect of the conflicts among the Western nations in the two world wars; (2) the rise of Japan to great-power status and its successful, although temporary, military challenge to the West in the early stages of World War II; and (3) the rise of anticolonial movements in areas subject to European control. After World War II, virtually all colonial possessions in Asia gained independence, except for Hong Kong (returned to China in mid-1997) and Macao (slated for transfer from Portuguese to Chinese sovereignty in 1999). Largely as a consequence of these changes, 20th-century Asia has been a region marked by revolution, war, and considerable turmoil.

Religion

Beyond temple architecture and sacred groves, religion influences societal organization, patterns of settlement and mobility, and even a people's inclination or disinclination to adopt foreign cultural practices. For these reasons, religion assumes a profound role in our efforts to read the Asian landscape.

The variety of religions in Monsoon Asia is broad. Hinduism is dominant in India, although many other religions are practiced there as well. The Islamic faith prevails in Pakistan, Bangladesh, most parts of the East Indies, parts of the southern Philippines, parts of western China, and among the native Malays of the Malay Peninsula. Various forms of Buddhism are dominant in Burma, Thailand, Cambodia, Laos, Tibet, and Mongolia. The people of Sri Lanka and Nepal are uneasily divided between Buddhism and Hinduism. The religious patterns of the Chinese and Vietnamese are somewhat more difficult to describe, even if effects of Communist rule are disregarded. Among them, Buddhism, Confucianism, and Taoism have all exerted an important influence, often in the same household. The same general situation has prevailed in both Koreas. In Japan, religious affiliations often overlap, with an estimated 84 percent of the Japanese population sharing the strongly nationalistic religions of Shintoism and Buddhism. The Philippines, with a large Roman Catholic majority, is the only Christian nation in Monsoon Asia, although Christian groups are found in various other areas.

The religions just named are indigenous to Asia except for Christianity and Islam, both of which originated in the Middle East. Other customs include the veneration of ancestors which is especially prominent among Chinese, Vietnamese, and Kore-

ans. Often elaborate rituals for everyday living are followed. Asian hill tribes—particularly in insular Southeast Asia—are largely animists, believing that natural processes and objects possess souls. And, because of their particular cultural significance and widespread adoption in Asia, Buddhism and Confucianism are worthy of some additional consideration. (Hinduism also plays a major role in the religious geography of Monsoon Asia, and it is discussed in more detail in Chapter 11.)

Buddhism

Buddhism had its origin in India and was diffused to China from Central Asia by the first century A.D. along the major Silk Road trade routes, to Korea by the fourth century, and from there to Japan by 552. It was proclaimed the state religion of Japan in 593 and it continues to be a significant religious influence there today. It reached Tibet in the middle of the eighth century, and its arrival was followed by decades of contest there between Chinese and Indian Buddhists (Fig. 10.8). The Chinese were defeated in this struggle and expelled from Tibet at the end of the eighth century.

In Japan, contemporary evolution of this religion has focused upon the *Soka Gakkai, the Value Creation Society* of the Nichiren sect of Buddhism. In this Japanese context, the belief system now promotes the use of mass media and effective organization to expand material benefit and contemporary happiness for its followers. While it has evolved in many distinct ways since its inception, the Nichiren sect has evolved the furthest from the initial Buddhist abandonment of material goods and the self-discovery and pilgrimages of Siddhartha Gautama in fifth-century B.C. India.

Figure 10.9 The terraced gardens and small plots of the Asian farmer remind travelers of how energetically the Asian peasant farmer has remade the landscape in order to gain better and more predictable productivity. Step terraces are designed to allow water to flow by gravity through all the fields, generally reentering a stream at a lower level. This scene is from Sichuan Province in China. *Kit Salter*

Confucianism

Confucianism is another religious belief system that has a continuing presence in Asia. While it is a belief system primarily associated with Chinese populations (both inside and outside of China proper), the broad spatial distribution of Chinese outside Asia means that it appears as a cultural influence in many countries and among many non-Chinese as well. Confucianism comes from the writings of Confucius (c. 551–479 B.C.), collected primarily in his *Analects*. Mencius (c. 371–288 B.C.) later became a major force in the widespread diffusion of this belief system throughout China. The major tenets of Confucianism are embodied in a system of ethical precepts for the proper management of society. Confucius never proclaimed his beliefs to constitute a religion; he was simply attempting to create a social contract between different classes central to Chinese government and society. He viewed his philosophy as secular, and its diffusion to Korea, Japan, and Vietnam has been part of the cultural innovation and baggage taken abroad by a steady stream of Chinese migrants to those places.

10.5 Distinctive Contemporary Landscapes

The current Asian landscape and culture are universally expressed in the several landscape features highlighted here—garden agriculture and use of water—which have an influence on social and visual patterns all across this crescent of Asia.

Garden Agriculture

Garden agriculture has a long tradition in Monsoon Asia. Because of the huge population in virtually all Asian nations, a shortage of land has always faced farmers. Rural families have traditionally relied on additional children as a source of inexpensive and significant labor, in the belief that more hands could do more farming. Whether you are traveling in the coastal lowlands of southeast China, the uplands of the island of Luzon in the Philippines, the hill lands of Burma (Myanmar), or the broad plain of northern India, there is a world of small garden plots that ring the cities and extend out to towns and villages as well. During the period of intense promotion of the merits of agriculture by the Chinese Maoist government in the early 1970s, romantic images of the beauty of simple farming were widely shown in popular magazines. Although farm mechanization in the past three decades has introduced significant change in Asian agriculture, there continues to be the rich landscape signature of delicate, small, and generally privately worked garden plots.

"Teaching Water"

The Chinese describe irrigation practices as the process of "teaching water" how to do what the farmer needs it to do to make agriculture more productive. In the landscapes of Asia, water has been "taught" to flow in canals, to pond in artificial reservoirs, to warm in slow-moving sheets atop terraced rice fields, to flow in and out of different levels of farmland with gravity as the major force of movement, and to be the center of attention in many agricultural ceremonies (Fig. 10.9). This talent of the Asian farmer and engineer to transform the Earth so that water can be better controlled, or better "taught," comes from the realization that if labor is abundant and precipitation relatively scarce or unpredictable, useful landscape modifications allowing irrigation can be designed and created by manual labor. The benefit of such hydraulic control has traditionally led to higher yields, more certainty of crop success, and labor opportunities for marginally unemployed farm populations.

10.6 Population, Settlement, and Economy

Approximately three-fifths of the world's people live in Monsoon Asia (see Table 10.1). They range from small tribal groups with locally distinct cultures in the hills and uplands of Southwest Asia and southwest China, to major and expansive culture groups like the Chinese and the Indians. Mongoloid peoples form a majority in China, Japan, Korea, Burma, Thailand, Cambodia, Vietnam, and Laos. But the majority of the people in India, although darker skinned than Europeans, are considered in many classification systems to belong to the Caucasian race; and similar peoples form a majority of the native inhabitants of the Malay Peninsula, the East Indies, and the Philippines.

Distribution of Population

The densest Asian populations are found on river and coastal plains, although surprisingly high densities occur in some hilly or mountainous areas. Higher mountains, steppes, deserts, and some areas of tropical rain forest are very sparsely inhabited.

Most of the countries have experienced large increases in population during recent centuries, especially since the beginning of the 19th century, and most of the additional people have accumulated in areas that were already the most crowded. Food production levels remain precarious, although such factors as increased grain supplies from new high-yielding varieties, the ability of Thailand and Burma to continue to export rice, and the availability of surplus grain from overseas countries are generally making it possible to maintain adequate levels of nutrition. Through irrigation, farmland expansion, and increasing use of chemical fertilizers and new seed stock, Asian agriculture has grown to be much more productive than it has been historically. How long this can continue in the face of continuing population increases—and, just as importantly, in the face of changing diet patterns in China that are moving away from rice and toward increased meat and wheat consumption—remains to be seen. While Japan has lowered its rate of population increase to below that of the United States and Canada, and while China has considerably reduced its growth rate in the past decade by stringent birth-control measures built around a limit of one child per couple, many Asian countries continue to have high fertility rates characteristic of the "developing" world.

Dominance of Village Settlement

Although an estimated 900 million people in Monsoon Asia live in urban settlements, the main unit of Asian settlement is the village. About two-thirds of the region's people are residents of an estimated 1.9 million villages. Highly urbanized Japan, Singapore, Macao, and Brunei do not fit this pattern, nor, in lesser measure, do Taiwan, South Korea, and Mongolia. But in all the other countries, the typical inhabitant is a villager. The Asian village is essentially a grouping of farm homes, although some villages house other occupational groups, such as miners or fishermen. Clusters of houses bunched tightly together are typical, and cheap and simple structures—often made of local clays and other building materials—are characteristic. Piped water and indoor plumbing continue to be somewhat exceptional in village homes, although the availability of electricity has been steadily expanding. Details of village life vary according to culture and place; for instance, Hindu villages have segregated quarters for different castes, even though the caste system has been officially outlawed.

As in all corners of the agricultural world, the original siting of villages was closely adapted to natural conditions. For example, villages are slightly elevated in floodplains located on natural levees (raised river banks built up by deposition of sediments during floods), dikes, or raised mounds. Early villages in Indonesia were often built in defensible mountain sites, although the Dutch colonial administration gradually required the building of villages along main roads and trails in the lowlands to make it easier to exercise control, collect taxes, and draft soldiers or laborers for road work or other projects.

One of the landscape signatures associated with village settlement in Monsoon Asia is the continuing tradition of slash-and-burn agriculture, particularly in the island world of Southeast Asia. In 1997 and 1998 Indonesia suffered through two seasons of the worst forest fires in two decades, and perhaps of the century. At least a portion of the fires were caused by runaway fires associated with the traditional slash-and-burn swidden farming. The other major cause was a very severe drought, influenced by El Niño. An additional factor was the steadily expanding forest clearance for commercial forestry and the creation of new plantations for the oil palm. Such clearing often results in attendant piles of forest debris that can serve as crisp fuel for forest fires. Estimates of the amount of land burned in the Indonesian fires of 1997–1998 run as high as 2 million hectares.

Rural-to-Urban Population Shift

Even though the village is central to Asian demography, this era of unprecedentedly rapid cultural diffusion and change since World War II has rapidly accelerated migration. Rural peoples have been leaving the farming life and going to the cities at a pace never before experienced.

In the late 1940s, approximately 85 percent of China's population lived in the countryside. Although there were massive cities such as Shanghai, Beijing, and Guangzhou (Canton), the great bulk of the Chinese population was still deeply involved in rural activity and living in relatively small villages. However, in the 1990s, China underwent a demographic shift that has taken it to a nearly 30 percent urban population, in large part due to economic growth in the country's urban (especially coastal) regions. India also defined just under 30 percent of its

Figure 10.10 In the design of the lobby of the Shanghai Hilton Hotel, careful attention has been given to artistic motifs that reflect Chinese culture, while plants and open space are mixed to assure comfortable surroundings for social as well as business conversations. There are even young musicians who play above the restaurant, creating an environment that seems wonderfully tranquil in comparison with the crowds and activity outside the hotel. Hotels such as these are big draws for tourists from all over the world. *Kit Salter*

population as urban in 1998. This demographic shift has meant enormous problems for city management as the rural migrants seek space, jobs, services, food, and goods.

The magnitude of these demographic shifts in China and India is so great that such moves can mean the transfer of hundreds of millions of people within a decade in those two Asian countries alone. Those left behind in the countryside, consequently, have been prompted to introduce more agricultural mechanization in order to fill the resulting labor gaps. At the same time, cities have been energetically absorbing this labor force into everything from small sweat-shop activities to major industrial enterprises.

The lure of the city—so powerfully expressed via today's widely available television and movies—has never played its siren song so successfully before. The whole demography of Monsoon Asia is shifting from a traditional rural past to an evolving urban present (Fig. 10.10).

Means of Livelihood

Japan was the first Asian country to develop modern cities and modern types of manufacturing on a large scale, even while maintaining a broad-based village structure (Fig. 10.11). For more than a century it has been a major industrial power. But China and India also have important and dynamically expanding industrial bases. These two nations are the largest in the world in population, and they are much better supplied with mineral resources than Japan. They also have cheaper labor. But Japan's labor force is more skilled, and Japan has shifted increasingly to types of industry requiring such labor and advanced technology. Not only does it lead Asia in **high value-**

added manufacturing—the process of refining and fabricating more valuable goods from raw or semiprocessed materials —but Japan continues to be the leader in the provision of financial services and other activities essential to steady economic growth.

The remaining Asian countries present highly varied patterns of urbanism and associated development of manufacturing and services (see Table 10.1). The urban centers of Hong Kong and Singapore are outstandingly productive in proportion to their size. South Korea and Taiwan also have a relatively high level of development of skilled labor. This pattern of developing major urban sectors skilled in finance, services, and global networking has become a very significant hallmark of the East Asian nations of Monsoon Asia, and an indicator of aggressive economic development (Table 10.2).

In Monsoon Asia as a whole, agriculture remains the major source of livelihood. In most countries—Japan is a conspicuous exception—the majority of the people continue to be farmers. Two major types of agriculture—plantation agriculture and shifting cultivation—are discussed in the chapter on Southeast Asia, the subregion in which these forms of farming are the most prominent. In the steppes and deserts, nomadic or seminomadic herding and oasis farming are practiced. Over large sections of Asia, most farmers make a living by cultivation of small rain-fed or irrigated plots worked by family labor. In Communist-held areas—particularly in China—the past four decades have seen a variety of experiments in agricultural organization. The current pattern has largely reverted to family farming operations with markets controlled—at least in part— by local and provincial governments, or, increasingly, by an expanding free market system.

Figure 10.11 This winter scene from Japan serves to modify the general images that so often represent Japan only by scenes of Tokyo traffic and crowding. This village scene from Tsumago in northern Honshu reflects the traditional architecture and street scale that is still representative of most of rural and small-town Japan, particularly in areas distant from the environs of Tokyo. *Art Wolfe/Tony Stone Images*

Table 10.2 The Asian Economic "Flu"

	GDP per head 1996,$	GDP growth			
		% annual average 1970–96	Estimate 1997	Forecast 1998	1999
China	3,120	9.1	8.9	6.3	7.5
Hong Kong	25,400	7.5	5.1	1.8	3.8
Indonesia	4,280	6.8	5.4	−5.2	2.9
Malaysia	9,703	7.4	7.4	1.6	1.8
Philippines	3,060	3.6	4.8	1.9	4.0
Singapore	25,650	8.2	7.6	2.7	5.0
South Korea	12,410	8.4	5.6	−2.5	1.7
Taiwan	17,720	8.3	6.3	5.0	5.7
Thailand	8,370	7.5	−0.7	−4.0	3.7
Rich industrial countries	22,700	2.7	2.8	2.6	2.6

Patterns of economic growth in East and Southeast Asia were particularly impressive between 1970 and 1996. The near economic collapse in Thailand in mid-1997 caused a major shift in such growth. All the figures in Table 10.2 are compared with the composite growth of a block of more developed industrial nations. Source: *The Economist,* March 7, 1998, Survey p. 5.

Although there has been some farm mechanization in Asia, the general practice (outside Japan) continues to be characterized by the steady input of large amounts of very arduous hand labor. Production is often of a semisubsistence character. This type of agriculture, which is often referred to as intensive subsistence agriculture, is built around the growing of cereals. Where natural conditions are not suitable for irrigated rice, grains such as wheat, barley, soybeans, millet, sorghums, or corn (maize) are raised. However, irrigated rice yields the largest amount of food per unit of area where conditions are favorable for its growth, and this crop is the agricultural mainstay in the areas inhabited by a large majority of Asian farmers. Because of the role of rice in the traditional Asian diet, it is the crop of choice, other things being equal.

In response, in part, to these distinctive landscapes and growing cities, tourism has also grown enormously in Asia (see Fig. 10.10). In Hong Kong, Thailand, South Korea, and Japan, for example, tourist income has increased five- to seven-fold in the past decade. The draw to touring Asia has long been fueled by the classic and traditional landscape treasures of art, architecture, and shrines and gardens. The upswing in these recent years has been stimulated by the construction of world-class hotels in major cities all across Asia, and by the continuing skill the region has in extending services and welcome to foreign visitors. This inflow of tourist dollars has not only been an important source of investment capital for the countries, but tourist traffic has led to the upkeep and significant refurbishing

of traditional landscape features—such as temples, sacred burial sites, and monumental structures—that might have otherwise been given less attention in the current rush to modernize and, in many cases, Westernize.

The Green Revolution

As a step toward eliminating hunger by using science to increase yields of rice—the single most important crop in Monsoon Asia—the **International Rice Research Institute (IRRI)** was founded in the Philippines in 1962. The institute is one facet of a worldwide research effort involving use of new grains in association with a modification of traditional farming practices. Governments also have invested increased capital in building better roads and bridges to enable farmers to get their surpluses to markets, and to bring new farming tools and materials to the countryside more efficiently.

Notable success has been achieved in breeding the new high-yielding varieties of seed stock, and there has been a large upsurge of production in certain areas where the new strains have been widely introduced. Peasant farmers have changed their **crop calendars**—the dates for planting, cultivating, and harvesting crops—to accommodate an increasing dependence on chemical fertilizers as well. The entire effort has come to be known as the **Green Revolution.**

For Asian farmers to capitalize fully on the Green Revolution, they must overcome many obstacles. For example, success requires levels of capital that are often beyond the present means or inclination of peasant farmers, landlords, and governments to provide. Such expenditures are needed for water-supply facilities (for example, the tubewells that have burgeoned by hundreds of thousands in the Indo-Gangetic Plain of the northern Indian subcontinent), chemical fertilizers, and chemicals to control weeds, pests, and diseases. And as agriculture becomes more mechanized, considerable increases in the costs of machinery, fertilizers, and fuel have to be borne by farmers who have, in many cases, had little experience with the cash economy needed to achieve a positive return on their investment. Governments have to improve transportation so that the large quantities of fertilizer required by the new seed varieties can be delivered in a timely fashion. Not only must there be these associated infrastructural changes to support this "revolution," but the crop calendar of the farmer becomes much less forgiving as many of the newest seed grains demand more precise water, fertilizer, and cultivation requirements than traditional grains. Grain-storage facilities, now subject to plundering by rats, have to be improved.

Overcoming the financial obstacles is rendered more difficult by the widespread system of share tenancy. If a farmer is a share tenant, he generally has no security of tenure on the land, and thus he cannot be sure that money he invests in the Green Revolution will actually benefit him in the future. He may not wish to assume any additional risk, even if credit on reasonable terms is available. Even if he remains on his holding, the land-

Definitions & Insights

ECONOMIC GROWTH RATES

The unprecedented pace of economic growth engineered by Monsoon Asia between 1970 and 1996 is shown in Table 10.2. While these numbers are indeed impressive, it is important to note the way in which those growth rates impact the landscapes and the societies of these nations. Associated, invariably, with such rapid growth rates are increasing rural-to-urban migration, increased consumption of consumer goods, expanding urban sprawl, and growing dependence upon both the sale of export goods and the consumption of domestically produced and import goods. While such phenomena may seem more like economics than geography, every one of these elements has an impact on village settlement and farming patterns, on urban growth and levels of urban employment, on goods put into the export stream, and on domestic labor patterns. Strong declines—such as those shown in Table 10.2 for the years following 1996—will have, as well, an influence on international migration patterns as Asian youth come home from education abroad to help the family maintain their domestic levels of well-being during the relatively hard times of slow growth. Thus, even though the numbers on the table may seem like pure economic statistics, every figure reflects a landscape and social impact.

As noted, the pattern of steady economic growth in the industrial sector of many Monsoon Asia nations—especially the ones in East Asia—has been changing drastically since 1997. In the summer of that year, Thailand's currency began to plummet, bringing down domestic financial structures and banks. Upon that collapse, nations in this region began to experience the same sort of weakening of their own currencies and associated declines of stock prices and stock market values. In the late 1990s nations not only in Monsoon Asia but in Latin America, Europe, and North America were attempting to stabilize their economies in such a way as to avoid the "Asian flu"—the term given to the weakening of the currency and markets in Thailand and adjacent nations. This regional financial deterioration has been particularly significant in Indonesia, the Philippines, Malaysia, South Korea, and—perhaps most significantly—Japan. This nation that served as the engine of East Asian economic prosperity in the early part of the decade was severely hit with the flu. China, Taiwan, and other nations in Monsoon Asia and beyond have also been influenced by subsequent decline in their economic growth rates.

---

lord may take up to half of the increased crop while bearing little or none of the additional expense. Landlords, in their turn, may be content to collect their customary rents without expending the additional capital necessary in this new mode of farming, or they may endeavor to turn tenants off the land in order to create larger spatial units that they themselves can farm more profitably with machinery and hired labor. In fact, landowners with large holdings often become the chief beneficiaries of the new technology, with many smaller farmers becoming a class of landless workers hired for low wages on a seasonal basis or migrants in the rural to urban migration stream.

Other problems associated with the Green Revolution include damage to ecosystems by large infusions of agricultural chemicals, and the economic dislocations that result when rice-importing countries become more self-sufficient, thus causing hardships for rice exporters. The Asian experiment with the Green Revolution provides yet additional evidence of the widespread ramifications that are set in motion when any major technological shift—either agricultural or industrial—is introduced into societies. Indian or Filipino farmers, for example, who had developed patterns of reasonable self-sufficiency through traditional agricultural practices, can find their lives turned upside down as they try to adopt some of the standards of the Green Revolution.

The scale of culture change that must accompany the pattern shifts in a farmer's involvement in the Green Revolution

may seem acceptable to an urbanite—whether from Asia or the world beyond—but to the traditional farmer, such changes are not only financially costly but serve to rupture long-held traditions. As you explore the various nations and landscapes of Monsoon Asia, keep in mind how much impact the global flow of information, fads, and images has on human ambitions for an improved livelihood and lifestyle. The ways in which peoples of Monsoon Asia have responded to these pulsating influences of change—from sophisticated mechanization (Fig. 10.12) to maintenance of traditional customs and costumes (see Fig. 10.8) and growth—are significant to your understanding of world regional geography.

Impact of Natural Events

Even with Monsoon Asia actively involved in the experiments, technology transfer, and diffusion of new agricultural practices as part of the Green Revolution, there are parts of the region that are virtually untouched by all this change. The most significant example is North Korea, where its population of some 20 million experienced four years of major food deficiency in the 1990s. The country experienced a **100-year flood**—a flood of such a magnitude that it is expected to occur only once in a century—in 1995–1996, and it suffered through a 1997 drought that was partially caused by El Niño. In 1995–1996, the food crop yield in North Korea was less than 2,800,000 tons, or two

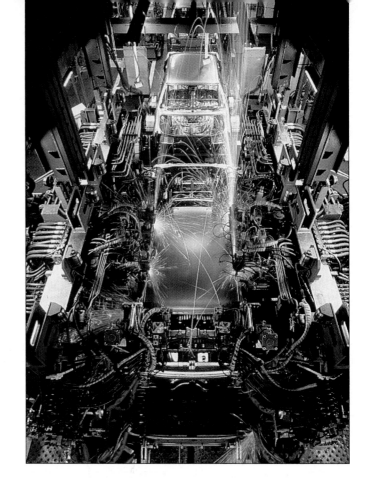

Figure 10.12   This high-tech manufacturing conveyor belt in a Honda automotive plant in Japan represents the sophistication that Japan has brought to its manufacturing. The spot-welding robot is called "General Welder" by the employees who keep the line running smoothly.
Tadanori Saito/PPS/Photo Researchers

million tons less than was necessary to feed their population at even a minimal level. Yields and total output have been down dramatically ever since mid-1990, leading to the unprecedented North Korean acceptance of grain from South Korea.

The major flooding of 1995–1996 also had other impacts on North Korea. The country's major mining area was flooded, meaning that mining itself was disrupted significantly. This led not only to major industrial unemployment in mining and associated industries, but also to a major drop in the generation of electricity. That, then, pulled the plug on other manufacturing. This combination led to a reduction in winter heating, overall electricity availability, and general industrial activity. At the same time, agriculture had dropped its output to about half of the normal productivity. So, even while parts of Monsoon Asia are experiencing steady increases in agricultural yields through the Green Revolution, other locales are either slow or unable to adopt the changes, often harmed by unpredictable weather conditions.

SUMMARY WITH SELECTED KEY TERMS

- **Monsoon Asia** includes the countries of **Japan, North and South Korea, China, Taiwan, Macao, Pakistan, India, Sri Lanka, Bangladesh, Bhutan, Nepal, Burma (Myanmar), Laos, Thailand, Cambodia, Vietnam, Malaysia, Singapore, Indonesia, the Philippines, Brunei, East Timor,** and numerous islands scattered along the edges of this major continental block. The adjective "monsoon" is used because of the central role this precipitation pattern plays all through this region.

- **Three concentric arcs** make up the broad physiography of the region. The arcs include the highest mountain ranges of the **Himalaya, Karakoram,** and **Hindu Kush.** The next arc, going eastward, is made up of major floodplains, deltas, and relatively low mountains. These include the **Indus, Ganges, and Brahmaputra Rivers** in South Asia; the **Irrawaddy, Chao Praya (Menam), Mekong, and Red Rivers** in Southeast Asia; and the **Chiang Jiang (Yangtze) and Huang He** in East Asia.

- In areas of dense settlement, **climates are warm** and **monsoonal.** Varied types include **tropical rain forest, savanna, humid subtropical, humid continental, steppe, desert,** and **undifferentiated highland.** Although there is significant desert landscape, there has been a long cultural history of caravans trekking across it, connecting East Asia with Europe.

- **Landscape signatures** of many sorts exist in Monsoon Asia, including indigenous ones (like **the Great Wall** in China and the **Taj Mahal** in India and the intensely worked **garden plots** supporting peasant households), but there are also signatures of the **colonial pe-**riod. Great Britain, the Netherlands, France, and Portugal were the most important colonial powers in this region. Most of these domains were relinquished by the middle of the 20th century, with the British return of **Hong Kong** and the Portuguese return of **Macao** (both to China) fundamentally closing the colonial period. **The Chinese Ming Voyages** in the early 15th century illustrate another potential for colonialism, but one that did not materialize.

- Major **religions** include **Hinduism, Islam, Buddhism, Confucianism, Taoism,** and **Christianity.** There continue to be tensions between populations of the different belief systems, especially in the case of Pakistan (Islam) and India (Hinduism). Wherever Buddhism and Confucianism are predominant there tends to be less tension because of the capacity of those religions to accommodate the existence of other belief systems.

- **Garden agriculture, education,** and **"teaching water" (irrigation)** are three very significant cultural features of Monsoon Asia. The combination of concern for **intensive farming** and **water management** have created distinctive landscapes both in mountains and on the plains. The importance of education has enabled Asians to adapt productively to **technology transfer.**

- **Three-fifths of the world's population lives in Monsoon Asia.** Partly because of the importance of rural settlement and associated farm labor, and partly because of cultural support of large families, China and India have enormous populations. These two countries and Indonesia are three of the four most populated nations in the world. **Birth control programs and family planning** have been successful in reducing **fertility rates,** and increasing **migration from villages to cities** has also been important in reducing population growth rates. **China** has been the most efficient in wide-scale reduction of population growth, and India will possibly become the most populous nation in the world by the middle of the 21st century or even sooner.

- **Villages** are the most common settlement form in Monsoon Asia, although the percentage of people resident in urban places has been steadily increasing all through the region. Japan was the first Monsoon Asian nation to adopt **Western industrialization.** In the past three decades, there has been widespread economic development, especially in East Asia and Southeast Asia (Taiwan, South Korea, Hong Kong [before it was returned to China], Singapore, Indonesia, Thailand, and Malaysia), but economic decline since 1997 has been rapid and significant, particularly in Japan. China and Taiwan have been particularly diligent in avoiding the decline.

- **The Green Revolution** is the name of a broad effort to increase agricultural productivity in dominant crops through expanded use of **chemical fertilizers, new seed stock, a more exacting crop calendar,** and better access to good **market roads, markets, and credit.** While there has been some success with the innovations associated with the Green Revolution, it has been difficult to overcome the entrenchment of strong traditional agricultural practices. **North Korea** represents one nation that has had virtually no productive encounter with these innovations and has, in addition, been plagued by both floods and drought, bringing the nation to the end of the 20th century with a great deal of failure to mark its half century of experimentation with Communism.

REVIEW QUESTIONS

1. What are the nations that make up Monsoon Asia?
2. Name the largest plains and river valleys in Monsoon Asia, and comment on their role in human settlement.
3. What are the characteristics of the Monsoon climate? How does agriculture relate to such patterns?
4. Where does the term "the Orient" come from? The "Far East"? What is the problem in using such names now?
5. What percent of the world's population lives in Monsoon Asia? Describe the range of population growth rates in the region. What areas have had the greatest success in reducing fertility rates?
6. Discuss the reasons that rural-to-urban migration has an impact on population growth rates. Give examples in Monsoon Asia where this trend is significant.
7. List the seven different climate classifications that are found in Monsoon Asia. Take three and discuss the types of landscapes and settlements they support.
8. Define a "palimpsest" and explain how it is used in geography. What examples might you find in Monsoon Asia to support your definition and explanation?

9. Using some of the maps in Chapter 10, talk about the colonial period and explain where and why colonial influences are the most pronounced.
10. What are the primary religions in Monsoon Asia? What landscape features would they be likely to create or influence?
11. Explain the significance of garden agriculture, "teaching water," and education in Monsoon Asia. In what way are these three things part of geographic perspective? What three Geography Standards would reflect and support your answer?
12. Look at the world population dot map in Chapter 2 and discuss the population settlement patterns that are evident in Monsoon Asia.
13. Discuss tourism, industrialization, and rural-to-urban migration in Monsoon Asia. How are these distinct processes and activities related? Where, in this region, would you see the clearest presence of any or all of the three?

DISCUSSION QUESTIONS

1. Discuss the role that the natural environment has had in the patterns of settlement in Monsoon Asia.
2. Discuss the ways in which patterns of human settlement relate to climate patterns. Use various maps to support your perception and answers.
3. What are the hallmarks of colonialism in Monsoon Asia? Name the good and bad influences that this process has had on the region.
4. Discuss the Ming Voyages and talk about the ways in which that quarter-century might have had a very different impact on world history and geography.

5. What is the consequence of the explosion of nuclear devices by India and Pakistan in May 1998? What sort of impact will this act have beyond the region?
6. Name the countries that have not been colonized in Monsoon Asia. In what ways have their histories been distinct from those of the nations that underwent colonization?
7. Discuss the sorts of changes that take place in migrating from the countryside to the city in Monsoon Asia. What economic and demographic changes does such a migration set in motion?

Chapter 11

Complex and Populous South Asia

▲ *The human fabric of South Asia is extremely rich and varied. This man is from Peshawar, Pakistan.* Joseph J. Hobbs

CHAPTER OUTLINE

11.1 The Cultural Foundation

11.2 Regions and Resources

11.3 Climate and Water Supply

11.4 Food and Population

11.5 Industry, from Bhopal to Booming Bangalore

11.6 Social and Political Complexities

a triangular peninsula that thrusts southward a thousand miles (*c.* 1600 km) from the main mass of Asia splits the northern Indian Ocean into the Bay of Bengal and the Arabian Sea. The peninsula is bordered on the north by the alluvial plain of the Indus and Ganges (Ganga) Rivers, north of which rise the highest mountains on Earth. The entire unit—peninsula, plain, and fringing mountains—is often called the Indian subcontinent. It contains the five countries of India, Bangladesh, Pakistan, Nepal, and Bhutan (Fig. 11.1) in an area a little more than half the size of the 48 conterminous United States (Fig. 11.2). India outranks all of the other South Asian countries in both area and population. Off the southern tip of India, across the narrow Palk Strait, lies the island nation of Sri Lanka, which shares many physical and cultural traits with the subcontinent.

Mountains enclose the subcontinent on its landward borders (Fig. 11.3). Its northern boundary lies in the Himalaya Mountains and the Karakoram Range. Nepal and Bhutan, on the southern flank of the Himalaya, are small, rugged, and remote. They are buffers between India and China, which have engaged in sometimes violent border disputes since the early 1950s. From each end of this massive wall, lower mountain ranges trend southward to the sea. Until 1947, the entire area, except the small Himalayan states and Sri Lanka, was referred to as "India." It was for well over a century the most important unit in the British colonial empire—the "jewel in the crown." In 1947 it gained freedom but in the process became divided along religious lines into two countries—the secular but predominantly Hindu nation of India and the Muslim nation of Pakistan. Pakistan had two parts, West Pakistan and East Pakistan, separated by Indian territory. In 1971, an Indian-supported revolt in East Pakistan led to the birth there of the new independent country of Bangladesh.

India is one of the most important nations to gain independence since World War II. It is the world's largest democracy, a demographic giant in which 989 million people (as of mid-1998), or nearly one-sixth of the human population, live. Its area of 1.1 million square miles (3 million sq km) is exceeded

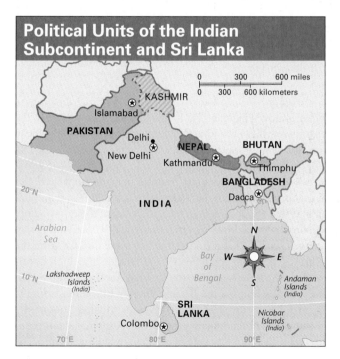

Figure 11.1 Political units of the Indian subcontinent and Sri Lanka. The future status of Kashmir, disputed between India and Pakistan, with some parts occupied by China, remains unresolved.

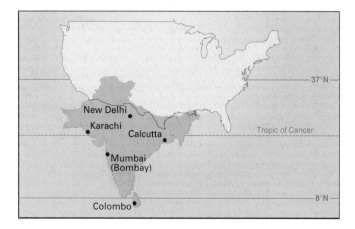

Figure 11.2 The Indian subcontinent and Sri Lanka, compared in latitude and area with the conterminous United States.

297

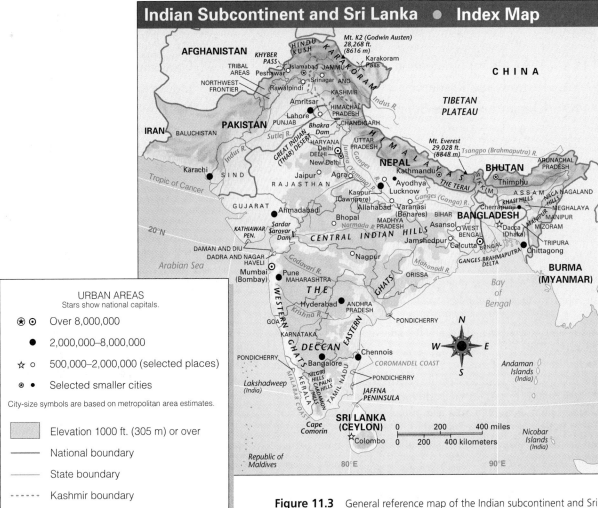

Figure 11.3   General reference map of the Indian subcontinent and Sri Lanka. Note how mountainous areas frame the northwestern, northern, and northeastern portions of this region, while ocean waters set the boundaries in the south.

by that of only a few other countries. Victory in the 1971 war with Pakistan confirmed India as the leading power in south Asia, although the development of nuclear weapons in both India and Pakistan has added a frightening new dimension to the conflict between these rivals.

Pakistan and Bangladesh are much smaller than India in area and population, but both are among the world's ten largest countries in population. Altogether, about 1.3 billion people, more than one out of every five on Earth, lived on the subcontinent and Sri Lanka in 1998.

11.1 The Cultural Foundation

The Indian subcontinent is one of the world's culture hearths. Among the animals first domesticated in the region were zebu cattle, which today are important livestock in many world regions. Before 3000 B.C., some of the world's earliest cities (notably Mohenjo Daro) developed on the banks of the Indus River

in what is now Pakistan. Their inhabitants were probably the ancestors of the Dravidians, a dark-skinned people whose stronghold is now southern India.

Between 2000 and 1000 B.C. came the Aryans (or Indo-Aryans), tribal cultivators and pastoral nomads from the inner Asian steppes who were the source of many ethnic and cultural attributes that are dominant in the subcontinent today—notably the Caucasoid racial majority, the numerically dominant Aryan languages such as Hindi (most of which were derived from the Sanskrit used by the conquerors), and Hinduism, which developed from the beliefs and customs of the Aryans. In the eighth century, Arabs brought the Islamic religion into India from the west, to what is now Pakistan. Waves of Muslim conquerors who came through mountain passes from Afghanistan and central Asia brought a wider spread of Islam in later centuries. The last great invasion, by a Mongol-Turkish dynasty called the Moguls (Mughals), began in the early 16th century, soon after the first Europeans came to India by sea in the Age of Discovery. Before it succumbed to rebellious local Hindus and Sikhs and to the British Empire, the Mogul empire ruled nearly all of

Figure 11.4 For Indians and foreigners alike, the Taj Mahal in Agra symbolizes the cultural achievements of the Indian subcontinent. *Joseph J. Hobbs*

the subcontinent from its capital at Delhi. Some of the region's most spectacular artistic and scientific achievements date to the peak of Mogul power in the 16th and 17th centuries, including the Red Fort in Delhi and the tomb of Mogul emperor Shah Jahan and empress Mumtaz Mahal (the world-famous Taj Mahal) in Agra (Fig. 11.4). Muslims now make up about 12 percent of India's population—seemingly a small number except that it equals more than 100 million people, ranking India fourth among the world's countries with the largest Muslim populations, behind Indonesia, Pakistan, and Bangladesh.

British colonialism left an indelible mark on the subcontinent's cultural landscape. To serve the needs of their empire, the British developed the great port cities of Karachi, Bombay (now called Mumbai), Madras (now called Chennois), and Calcutta, and India's current inland capital of New Delhi. The existence of English as the de facto national language of both India and Pakistan, and the presence in Great Britain today of large numbers of immigrants from the subcontinent, bear witness to the close relations between these countries.

This simple sketch of Dravidian, Aryan, Arab, and British influences can only begin to shed light on the extreme social diversity of the Indian subcontinent. There is an enormous range of ethnic groups, social hierarchies, languages, and religions among regions of the subcontinent, and even within single settlements. The subcontinent is the most culturally complex area of its size on Earth. Some of the problems related to this complexity are discussed later in the chapter.

11.2 Regions and Resources

Physical conditions are extremely important to the overwhelmingly rural, resource-based economies of the Indian subcontinent and Sri Lanka. With relatively little foreign aid reaching the region, people's survival depends quite directly on local agricultural resources, and future industrialization will have to

draw largely on available natural resources. Fortunately, the subcontinent has some generous endowments of natural resources. It also has colossal natural hazards, however, including floods, droughts, landslides, and earthquakes.

The subcontinent can be roughly subdivided into three natural areas: the outer mountain wall, the northern plain, and peninsular India.

The Outer Mountain Wall

A series of parallel mountain ranges forms an inverted "U" around the north of the subcontinent (see Fig. 11.3). In the west, the mountains extend northeastward from the Arabian Sea to Kashmir. From here, the longest section of the wall trends southeast and then eastward to the subcontinent's northeast corner in the Indian state of Arunachal Pradesh. Finally, a short eastern leg of the "U" reaches southward along the border with Burma (Myanmar) to the Bay of Bengal.

Pakistan's borders with Iran and Afghanistan traverse the western section of the mountain wall. Here the mountains are rugged almost everywhere, and in most places they extend well into the bordering countries. Desert and steppe climates reach upward to high elevations. These mountains provide only a few crossing places, the most famous of which is the Khyber Pass through the Hindu Kush mountains on the border between Pakistan and Afghanistan. Such passes played a fateful role for thousands of years as gateways into India for conquerors from the outer world, including the Aryans and the Arabs.

The towering northern segment of the mountain wall has been much less passable. Across the north of the subcontinent, the Himalaya extend about 1500 miles (*c.* 2400 km) from Kashmir to the northeastern corner of India (Fig. 11.5). Paralleling the Himalaya on the northwest, and separated from them by the deep gorge of the upper Indus River, lies another exceedingly high range called the Karakoram. Beyond the

Figure 11.5 The Himalaya dwarf most of the world's other great mountain ranges. These peaks tower above the Gokyo Valley, near Mount Everest in eastern Nepal. Potatoes and barley are grown in the fields seen here, at about 17,000 feet (5414 m) above sea level. *Joseph J. Hobbs*

Regional Perspective

"Upstream" and "Downstream" Countries

The region of the subcontinent called the Punjab, meaning "Five Waters" (five Indus River tributaries), is split between two countries, with its rivers flowing through Indian territory before entering Pakistan (see Fig. 11.3). In geographical and geopolitical terms, India is an "upstream country" and Pakistan a "downstream country" where Punjab waters are concerned. There are no internationally accepted laws regarding water-sharing between upstream and downstream states, and the upstream country often exercises its geographical advantage to siphon off what the downstream country views as too much water. Riparian (river-owning) countries often come close to, or actually go to, war over the problem. In many cases, formal treaties have averted what could have become

major international conflicts (see also Chapter 9, p. 244).

After 1947, relations between India and newly-independent Pakistan eroded in large part because of the disputed waters of the Indus and the five Punjab rivers. For many years the parties could not agree on how much of the water India could divert and how much should remain for Pakistan. The countries finally reached a satisfactory resolution in the 1960 Indus Waters Treaty. The agreement allocates the water of the three eastern rivers of the Punjab to India, which in return allows unrestricted and undiminished flow of the two others, and the Indus River itself, into Pakistan. The two countries have abided by the agreement ever since, despite their periodic wars and constant state of tension.

Another landmark water-sharing agreement was signed between India and Bangladesh in 1996. For two decades these nations had a bitter disagreement about the allocation of water from the Ganges River. In the 1996 accord, which is meant to be in effect for 30 years, upstream India guarantees a larger share of water to downstream Bangladesh. This achievement may open the way to the integrated development of what is known as the Ganges-Brahmaputra-Barak (GBB) river basin, a watershed that is home to 500 million people in Bangladesh, Bhutan, China, India, and Nepal. Cooperation to settle the vital issue of water allocation among these countries could promote peace and reduce poverty in the region.

Himalaya and the Karakoram is the rugged and very high Tibetan Plateau, controlled by China. The two ranges contain 92 of the about 100 world peaks whose elevations are 24,000 feet (7315 m) or higher, culminating in the Earth's two loftiest summits: Mt. Everest (29,028 ft/8848 m) in the Himalaya on the border between Nepal and China, and Mt. K2, or Godwin Austen (28,268 ft/8616 m), in the Karakoram within the Pakistani-controlled part of Kashmir. Geographical surveys recently quieted a brewing debate over whether K2 might actually be higher than Everest: The Himalayan peak is the highest. Both great ranges, however, are young, dynamic, and still growing, so the two peaks may still be in competition. Although most passes through these ranges are higher than any peak in the Alps of Europe, a trickle of trade and cultural exchange has crossed these mountains for millennia. However, neither commerce nor military forces have ever crossed this barrier on any large scale.

The kingdoms of Nepal (capital, Kathmandu; population 235,160; Fig. 11.6) and Bhutan (capital, Thimphu; population 12,000) are situated on this mountain wall. Their landlocked locations and difficult topographies certainly contribute to their slow economic and social development. Both countries have productive agricultural regions in the "middle ranges," the bands of foothills between the vast lowland of the subcontinent and the towering Himalayan peaks. In Nepal the capacity of this productive area has been saturated, and poor peasants "marginalized" by rapid population growth have sought new ground on

steeper slopes in nearby Bhutan and in Nepal's lowland region of the Terai. Once a malaria-plagued region of dense forest and savanna, the Terai helps serve as a "safety valve" for Nepal's overpopulation. Control of malaria began in the 1950s and people began migrating there to clear and cultivate the land.

Figure 11.6 Nepal's capital, Kathmandu, sprawls across a plain in the country's "middle" mountain range. In the foreground is the Royal Palace. *Joseph J. Hobbs*

Nepal's Himalayan region is also a popular destination for tourists. An estimated 120,000 trekkers visit annually, bound especially for the Everest region in the east and the Annapurna area of the west. They bring in much-needed foreign currency to this poor country but also create problems by littering and inducing social change in traditional cultures. Bhutan has been far more careful in opening its doors to tourism, placing restrictions on the numbers permitted in and on the routes they may follow.

The mountains along and near the subcontinent's border with Burma (Myanmar) are much lower than the Himalayan ranges but are also nearly impenetrable. They are rugged, cloaked with dense vegetation, and soaked with rain. Cherrapunji in the Khasi Hills of India has an average annual rainfall of 432 inches (1097 cm), of which more than nine-tenths falls in the summer half-year. It ranks with a station on Kauai in the Hawaiian Islands as one of the two spots with the greatest annual rainfall ever recorded on Earth.

The Plains of India, Pakistan, and Bangladesh

Just inside the outer mountain wall lies the subcontinent's northern plain. This large alluvial expanse contains the core areas of the three major countries. In the west, the plain is split between Pakistan and India. The Indus River and its tributaries and distributaries cross the portion lying in Pakistan. The climate here is desert and steppe; the dry region straddling the two countries is known as the Thar Desert or the Great Indian Desert. Irrigation water from the rivers is essential to the many millions of people living here.

East of the Punjab lies the part of the northern plain traversed by the Ganges (Ganga) River and its tributaries, especially the Jumna (Yamuna). The western portion of this region is known as the Doab, "the land between the two rivers." The northern plain is India's core region, containing more than two-fifths of the country's population as well as its capital, numerous other major cities, and many places of great significance in the religions of Hinduism, Buddhism, and Jainism.

This region of India is relatively well-watered, especially toward the east, where an agriculture based primarily on irrigated rice supports population densities that are unusually high even for the subcontinent. At the narrow western end of the Ganges-Jumna section of the plain, the old fortified capital of Delhi arose as a bastion against invaders from beyond the western mountains (Fig. 11.7). The present (and adjoining) capital of New Delhi was founded after 1912 as the capital of British-controlled India. Today, the metropolitan population of Delhi–New Delhi is an estimated 8.4 million (7.2 million, city proper), making it India's third largest metropolis after Mumbai and Calcutta.

The southeastern part of the northern plain lies in the region of Bengal, which is essentially the huge delta formed by the combined flow of the Ganges and Brahmaputra. Bengal, which is divided between India and Bangladesh, is extremely densely

Figure 11.7 The Mogul emperor Shah Jahan established Delhi as India's capital early in the 17th century. His monumental Red Fort still dominates this old city. Adjacent New Delhi is India's capital today. *Frederica Georgia/Photo Researchers*

populated. Another magnet for migrants from Bangladesh and from rural India is the English-founded port of Calcutta in the Indian state of West Bengal. A city of 11.02 million in its metropolitan area (4.4 million, city proper), Calcutta has a worldwide reputation for teeming slums and dire poverty.

The Bangladesh portion of the delta could yield many million more emigrants as it ranks, along with the Indonesian island of Java, as one of the two most crowded agricultural areas on Earth. Its problems increase sharply during the many periods when floods strike (Fig. 11.8). Tropical cyclones from the Bay of Bengal cause some of the flooding. Deforestation upstream on the steep slopes of the Himalaya has increased water runoff and sediment load in the Ganges, also contributing to Bangladesh's flood problems. Bangladesh must also keep a wary eye on the potential for the sea level to rise if the world's temperatures increase according to many "greenhouse effect" models.

Figure 11.8 Bangladesh is both a low-lying coastal land and a downstream country of several large watersheds. As a result of storms blowing in from the Bay of Bengal and of accelerated runoff from rainfall and snowmelt on deforested mountains upstream, it is subjected to devastating floods. This is an aerial view following the cyclone (hurricane) which struck Bangladesh on April 29, 1991, leaving more than 120,000 people dead and inundating most of the country's precious farmlands. *Bruce Brander/Photo Researchers, Inc.*

Peninsular India

The southern peninsular portion of the subcontinent, entirely within India, consists mainly of a large volcanic plateau called the Deccan. It is relatively low, generally less than 2000 feet (*c.* 600 m) above sea level. Its rivers run in valleys cut well below the plateau surface. This topographic problem, together with the rivers' seasonal flow regime, makes it difficult to use their waters for irrigation of the upland surface without large inputs of capital and technology. Wells and the rain-catchment ponds called tanks are used, but water is in shorter supply and population density is much lower than in the better-watered parts of the northern plain (Fig. 11.9).

Hills and mountains outline the edges of the Deccan. In the north, belts of hills (the "Central Indian Hills" in Fig. 11.3) separate the plateau from the northern plain. On both east and west the plateau is edged by ranges of low mountains called Ghats ("steps"), which overlook low-lying alluvial plains along the coasts of the Bay of Bengal and the Arabian Sea. The low and discontinuous eastern mountains are known as the Eastern Ghats. The western mountains, which are higher and which present a relatively continuous west-facing escarpment, are the Western Ghats (Figure 11.10). The two ranges merge near the southern tip of the peninsula in clusters known locally as the Nilgiri, Palni, and Cardamon "hills." They are really rather high mountains, with some peaks rising above 8000 feet (*c.* 2400 m). The colonizing British founded several "hill stations" in them for recreation and administration during summer months, when the climate on the subcontinent's lowlands is stifling. They are now popular tourist destinations for Indians during that season.

The coastal plains between the Ghats and the sea are occupied by ribbons of extremely dense population, with the density

Figure 11.9 Water is at a premium in the dry period preceding the summer monsoon in South Asia. It is usually the task of women to draw water from communal supplies such as this one in a small settlement near Mumbai. *Joseph J. Hobbs*

reaching a peak in the far south along the Malabar Coast of western India and with a slightly lower density on the corresponding Coromandel coast in the east. The largest metropolis of the far south, Chennois (formerly Madras; population: 3.84 million, city proper; 5.42 million, metropolitan area) is a seaport on the Coromandel coast. The main metropolises of the interior Deccan are Hyderabad (population: 3.06 million, city proper; 4.34 million, metropolitan area) and Bangalore (population: 2.66 million, city proper; 4.13 million, metropolitan area).

Sri Lanka

Sri Lanka (population: 18.9 million), formerly called Ceylon, is a tropical island country. It is a land apart; despite its proximity to the Indian subcontinent, some geographers place it in the Southeast Asian realm. Its main physical and cultural characteristics are linked to the subcontinent, however. Its name conveys its physical beauty: Sri Lanka means "Resplendent Isle" in the native Sinhalese language, and medieval Arabs knew it as

Figure 11.10 Southwest India's mountains, known as the Western Ghats, have been widely cleared of native forest by logging and agricultural expansion. This is a tea plantation in eastern Kerala state. *Joseph J. Hobbs*

Figure 11.11 Verdant landscapes are typical of all but northernmost Sri Lanka. This view is near Sigiriya in the south-central part of the country. *Joseph J. Hobbs*

"Serendip," the island of serendipity (Fig. 11.11). The island consists of a coastal plain surrounding a knot of mountains and hill lands (see Fig. 11.3). Most people live either in the wetter southwestern portion of the plain, in the south-central hilly areas, and in the drier Jaffna Peninsula of the north. Coconuts and rice are the major crops of the low southwestern coast and Jaffna Peninsula, while tea and rubber plantations dominate the economy of the uplands. Colombo (population: 615,000, city proper; 2.05 million, metropolitan area), in the southwest, is the capital, chief port, and only large city.

Sri Lanka's economy is highly commercialized. Three cash crops—rubber, coconuts, and the world-famous Ceylon tea—occupy the greater part of the agricultural land and supply about 30 percent of export earnings. Clothing manufacture in the Colombo area has grown into an export industry. Expansion of rice production has been so successful that near self-sufficiency has replaced the need for major rice imports. This development has been helped by declining rates of population growth. However, with its limited land area, Sri Lanka has a high population density of 757 per square mile (272 per sq km), just below that of India.

External influences on this attractive island have diversified its population and sown seeds of unrest (see p. 314). Centuries of recurrent invasion from India were followed by Portuguese domination in the early 16th century, Dutch in the 17th, and British from 1795 until the country was granted independence in 1948 as a member of the British Commonwealth. This eventful history, plus the longstanding commercial importance of the sea route around southern Asia, has given Sri Lanka a polyglot population which includes Burghers (descendants of Portuguese and Dutch settlers) and Arabs.

The two major ethnic groups, distinguished from each other by language and religion, are the predominantly Buddhist Sinhalese, making up about 82 percent of the population, and the Hindu Tamils, constituting about 9 percent. The light-skinned

Sinhalese are an Indo-Aryan people who settled in Sri Lanka about 2000 years ago. The dark-skinned Tamils, whose main area of settlement is the Jaffna Peninsula and adjoining areas in the north, are descendants of early invaders and more recent imported laborers from southern India.

11.3 Climate and Water Supply

Climatic conditions in the subcontinent and Sri Lanka vary between remarkable extremes, from Himalayan ice and snow fields to the year-round tropical heat of peninsular India, and from some of the world's driest climates to some of the wettest. Climatic and biotic types in the subcontinent include undifferentiated highland climates in the northern mountains; desert and steppe in Pakistan and adjacent India; humid subtropical climate in the northern plains; tropical savanna in the peninsula, except for a patch of tropical steppe in the rain shadow of the Western Ghats; and rain forest on the seaward slopes of the Western Ghats and the coastal plain at their base, and in parts of the Ganges-Brahmaputra Delta and the eastern mountains near Burma (see Fig. 2.7).

Heat is nearly constant except in the mountains. The tropical peninsula is hot all year. The subtropical north has stifling heat before the "break" of the wet monsoon, and warm conditions even in the winter. Some of the warmest temperatures on the globe occur in the plains of Pakistan and northwestern India. Delhi, for instance, averages 94°F (34°C) in both May and June. The highest temperatures over most of the subcontinent occur in May and June, just before monsoonal rainfall brings relief.

Another near-constant is the extreme seasonality of rainfall, offering a stark contrast between a short, wet summer and arid or semiarid conditions the rest of the year. The "humid" parts of

Figure 11.12   The torrential monsoon rains are a regular and welcome feature of land and life in South Asia. This is a flooded road in south-central Sri Lanka.   *Joseph J. Hobbs*

Figure 11.13   Aridity and deforestation are among the urgent environmental problems faced by Pakistan. Foraging by goats and tree-cutting for fuel have created a depauperate landscape in the country's Karokoram Mountain region.   *Joseph J. Hobbs*

the subcontinent are rather dry for most of the year, while floods are often a threat or a reality during the short rainy season that corresponds to a season of snow melt along upper river courses in the Himalaya. These striking conditions are caused by seasonally-reversing winds known as monsoons—with a wet monsoon in the summer and a generally dry monsoon in the winter.

The **southwest** or **wet monsoon** is at its height from June to September, and most parts of the subcontinent receive the bulk of their annual rainfall during those months (Fig. 11.12). There are two main arms of this monsoon. One arm, approaching from the west off the Arabian Sea, strikes the Western Ghats and produces heavy rainfall on these mountains and the coastal plain. The amount of rain diminishes sharply in the interior Deccan rain-shadow region to the east of the mountains. Here, the annual precipitation over a large area is barely sufficient for dry farming and in some years is so low that serious crop failures occur.

The second major arm of the wet monsoon, approaching from the Bay of Bengal, brings moderate amounts of rain to the eastern coastal areas of the peninsula and heavy precipitation to the Ganges-Brahmaputra Delta region and northeastern India. Wet monsoon winds pass up the Ganges Valley to drop moisture that diminishes in quantity from east to west. Both arms of the monsoon bring some rain to Pakistan, but the total is so small that semiarid or desert conditions prevail in most areas (Fig. 11.13).

The **dry monsoon** of the winter half-year is also called the **northeast monsoon**, because it often blows from the northeast over the peninsula. It also frequently blows from the northwest over much of the northern plain. Blowing mainly from land to sea, this monsoon brings dry weather to most parts of the subcontinent, with occasional light rains in areas outside the tropics. An exception occurs in the far south of the peninsula. Here, the heaviest rainfall of the year falls along the eastern coast and in adjacent uplands during the four-month period from October through January. In addition to widespread drought, the winter monsoon brings cooler weather to the subcontinent, especially the north.

With precipitation so concentrated within a short part of the year, and with the extreme pressure of so many people on so little land, a crop season extended by irrigation is vital. People in

Regional Perspective

Is Agricultural Progress Keeping Pace With Population Growth?

Poverty and marginal human health are already problems in the Indian subcontinent and Sri Lanka, and the prospect of continued high population growth raises the question of whether growth in food supplies can avert an eventual Malthusian crisis.

In India, a few people are very wealthy, and there is an emerging middle class of as many as 200 million people (Fig.11.A). Most people, however, are even poorer than the averages indicate. The prosperous minority is surrounded by a sea of overwhelming poverty. India and its neighbors are countries where "people overpopulation" is a problem: Crowding is a relatively recent phenomenon in the region, resulting from the declining death rates that accompanied the spread of modern public health and medical techniques to the subcontinent. Population growth has accelerated rapidly since independence as birth rates have remained relatively high while death rates have continued to fall. Given increases in urbanization and industrialization, and the dividends from India's strenuous family planning programs, future population growth in the region is expected to slow. The population base is already so vast, however, that even modest growth will add huge numbers. The issue of population growth is especially critical for Pakistan, which has never mounted an effective family planning program,

mainly because of opposition from Muslim religious authorities, who regard birth control as an intervention against God's will (Pakistan's birth rate in 1998 was 39 per thousand, compared with 27 per thousand in India).

Agricultural output in South Asia has been increasing rapidly since independence. Most notably, despite its huge and rapidly growing population and irregular distribution of food supplies, India has managed to remain self-sufficient in food production, and has large strategic reserves of staple grains. The successes of agriculture in the subcontinent have been due mainly to the increased use of artificial fertilizers, the introduction of new high-yield varieties of wheat and rice associated with the Green Revolution (see Chapter 10, p. 292), the application of more labor from the growing rural population, increased irrigation, the spread of education (Fig. 11.B), and the development of government extension institutions to aid farmers.

A large increase in the use of artificial fertilizers began in South Asia after the mid-1960s, especially in conjunction with the use of new strains of wheat and rice as part of the Green Revolution. The new developments first took root in the Punjab area of India and Pakistan and are continuing to spread from there. They have been accompanied by increasing irrigation.

Social conditions and services in rural areas are improving in some parts of South Asia, notably in India, suggesting that development is progressing and that agricultural production has so far been sufficient to keep up with population growth—signs that a Malthusian catastrophe long predicted for the region is not imminent.

One of the challenges confronting India is raising the status of women. Despite a 1961 ban on dowries (the money and gifts given by a bride's parents to the groom), the practice is continuing. So is the rate of the killing of brides who do not provide enough dowry. The burden placed on the bride's family has also prompted parents to abort female fetuses, which they now can often detect with ultrasound technology.

Technology is thus a mixed blessing in this traditional society. Generally it plays a constructive role, since, where face-to-face efforts are not possible, radio and television can encourage positive change in the most remote villages. India now has its own space program, and has put several communications satellites into orbit. Such accomplishments, with the possibility for much further improvement, indicate that a process of development is unfolding in South Asia. The challenge will be for technology and human resources to keep agricultural growth apace with or ahead of population growth.

Figure 11.A   This Indian family visiting the Red Fort in Agra is drawn from the growing ranks of the subcontinent's middle class. *Joseph J. Hobbs*

Figure 11.B   Education and agricultural production are advancing in South Asia despite great odds. These schoolboys are having a lesson in a mosque in Peshawar, Pakistan.   *Joseph J. Hobbs*

Figure 11.14 These meticulously crafted and maintained terraces produce a high yield of rice in Pakistan's northern Punjab. *Joseph J. Hobbs*

the subcontinent therefore expend an enormous amount of labor to get additional water onto the land. Modern technology does the job in places where huge dams impound rivers and where networks of canals distribute water from these reservoirs.

The main locales for such large-scale development are in the steppe and desert areas of the Punjab and the lower Indus Valley (Sind Province) in Pakistan. Since the 19th century, when the British began irrigation works in these areas, rapid colonization has changed dry lands from sparsely populated to densely populated areas. Both the Indian and Pakistani parts of the Punjab still normally produce surpluses of wheat. India's East and West Punjab and Pakistan's Sind and Punjab Provinces produce surplus rice, including the Basmati variety, which is an important export (Fig. 11.14).

A massive irrigation and hydroelectric power project is now underway in the north-central part of peninsular India, where the Sardar Sarovar Dam is being constructed on the Narmada River (see Fig. 11.3). Largely because it is expected to displace 320,000 tribal and rural people and inundate 28,000 acres of cropland and 32,000 acres of forest, the dam is opposed by environmentalists and human rights advocates both in India and abroad. The World Bank withdrew its financial support of the project in 1993 in response to these protests, but, asserting that its benefits will greatly outweigh its costs, India has pressed ahead with the dam's construction. It is the centerpiece of the massive Narmada Valley Development Project, where 30 dams will ultimately displace about one million people.

11.4 Food and Population

The major staple food crops of the subcontinent, along with population densities, correlate spatially with the amount of water available. Rice is the basis of life in the wetter areas and, with its high caloric yields per acre, is associated with the highest population densities (see world population map in Chapter 2, p. 51). Rice dominates in the delta area of Bengal (in both India and Bangladesh), in the adjacent lower Ganges valley, and in the coastal plains of the peninsula. Irrigated wheat is the staple crop and food in the drier upper Ganges valley of India and in the dry Punjab of India and Pakistan. Here, the caloric yield and population densities are intermediate. Unirrigated sorghums and millets are dominant over most of the Deccan plateau and in other areas where low rainfall cannot be supplemented much by irrigation. Caloric yields from these crops are low, and so are population densities, although the "low" densities in some places are more than 200 people per square mile (c. 80 per sq km). Maize is an important grain in the lower regions of the Himalaya. However, rice is the preferred food almost everywhere, and where enough water is available, there are patches of irrigated rice. A host of minor crops supplements the staple grains.

Industrial and Export Crops

Cotton, jute, tea, and rice are the main crops grown in the subcontinent for industrial use and export. Irrigated cotton in the Punjab and lower Indus Valley makes Pakistan a cotton producer exceeded among the world's states only by China, the United States, and India. Textiles, yarn, and raw cotton contribute 60 percent of Pakistan's exports. In India, cotton is grown mainly in the interior Deccan on soils that hold moisture well and produce a crop without irrigation. India's large textile industry absorbs most of the production.

The Ganges-Brahmaputra Delta is the world's greatest producing area for jute, the principal material for burlap. The delta lies in the former province of Bengal, which was partitioned between India and (East) Pakistan when British rule ended. The subcontinent's independence from Britain and the subsequent division of India presented a dilemma for the jute industry. Most of the jute-growing areas lay on the Pakistani side (now Bangladesh), while the jute mills, which had developed in the Calcutta area, went to India. Since independence, India has increased raw jute production on its side of the border, and Bangladesh has developed jute mills of its own.

Of the four major commercial crops in the subcontinent, only tea is exported on a sizable scale. India is the world's greatest exporter of tea. Unlike cotton, jute, and most other crops of the subcontinent, tea is principally a plantation crop. The plantations were developed, and are still largely owned, by British interests. Production is greatest in the northeastern state of Assam, with a secondary center in the mountainous far south

Definitions & Insights

THE SACRED COW

Many attitudes, beliefs, and practices associated with cattle make up the world-famous but often poorly understood "sacred cow" concept of India. There are nearly 200 million cattle in India, representing about 15 percent of the world total and the largest concentration of domesticated animals anywhere on Earth. India's dominant religion of Hinduism forbids the slaughter of cows but allows male cattle and both male and female water buffalo to be killed. Reverence for the cow is well founded. Indians favor cow's milk, ghee (clarified butter), and yoghurt over dairy products from water buffalo. They value cows as producers of male offspring, which serve as India's principal draft animal. Both cows and bullocks provide dung, an almost universal fuel and fertilizer in rural India.

Reverence for the cow in particular and cattle in general pervades the Hindu religion and mythology. The bull Nandi is associated with the Hindu god Shiva. People allow cattle to freely roam the streets of Indian cities. As a symbol of fertility, cows are associated with (but not worshipped as) several deities. The mother of all cows, Surabhi, was one of the treasures churned from the cosmic ocean. Hindus honor cows at several special festivals. They use cow's milk in temple rituals. They believe the "five products of the cow"—milk, curds, ghee, urine, and dung—have unique magical and medicinal properties, particularly when combined. All over India there are *goshalas*, or "old folks' homes," for aged and infirm cattle.

Remarkably, however, the subcontinent countries of Pakistan, Bangladesh, and India are major exporters of leather and leather goods. In India the leather industry is mainly in the hands of Muslims, who do not share Hinduism's prohibitions against killing or eating cattle. Muslims are forbidden to eat pork; therefore, pigs, a major food resource in many developing countries, are of little importance here. They are eaten mainly by Christians, very low-caste Hindus (some of whom are pig breeders), and tribal peoples.

of peninsular India (see Fig. 11.10). Coffee, rubber, and coconuts are minor plantation crops, and India exports products made from coir (coconut fiber).

Opium is a major illegal export crop of Pakistan. Cultivation is restricted to the rugged mountain region of the Northwest Frontier Province, inhabited mainly by Pushtun tribespeople, along the frontier with Afghanistan. Processing of the raw opium into heroin is also a significant industry in the region and in adjoining areas of Afghanistan. The government of Pakistan is under pressure from the United States and other heroin-consuming nations to crack down on the illegal trade.

11.5 Industry, from Bhopal to Booming Bangalore

In the partition following independence from Britain, India received almost all of the subcontinent's modern industries and most of its mineral wealth and energy supplies. Pakistan, including the present Bangladesh, was left with a much smaller industrial infrastructure and fewer natural resources. However, in global terms, the subcontinent has a small but growing industrial base. Handicraft workers still exceed factory workers by many millions, and handicrafts supply many everyday needs for the population. Most visitors to the subcontinent are astounded to see human hands performing, on a vast scale, those tasks routinely done by machines in the MDCs (Figure 11.15).

India's leading factory industries are cotton textiles, jute products, food processing, and iron and steel. Modern cotton mills are concentrated mainly in the metropolitan area of Mumbai (formerly Bombay; population: 9.9 million, city proper; 12.6 million, metropolitan area) and neighboring areas along and near the west coast. Of basic importance to cotton manufacturing are the large home market, cheap labor, hydroelectric power from stations in the Western Ghats, and the large domestic production of cotton. As a major world producer of cotton

Figure 11.15 This is the central laundry for the city of Karachi. All the washing is done by hand. *Joseph J. Hobbs*

goods, India supplies cloth to enormous numbers of its impoverished citizens and is regularly a leading exporter of textiles and clothing.

The Bengal is an important industrial region. As previously discussed, India's jute industry is concentrated in and near Calcutta. While jute is only a minor export for India, in Bangladesh both raw jute and jute products are major exports. India's iron and steel industry is concentrated in the Bengal and in adjacent parts of the northeastern portion of the peninsular uplands. The most valuable assemblage of mineral resources in India—including iron ore, coal, manganese, chromium, and tungsten—is found in and near this area. These resources have led to the rise of small- to medium-sized industrial and mining cities in a region that in other respects is one of the least developed parts of the country. The largest center is Jamshedpur (population: 670,000, city proper; 850,000, metropolitan area), 150 miles (c. 240 km) west of Calcutta.

Until quite recently India has been intent on keeping out most foreign manufacturing, and has sought industrial self-sufficiency. With this policy, India's industries have expanded rapidly and its output of manufactured goods has become diversified and sophisticated, but inefficiency in some sectors has led recently to a greater openness to imports. Smaller industries such as engineering and chemicals are growing at a faster rate than older sectors like the textile industries.

Today, there are very few items among the great variety of machines used in a modern society that are not produced somewhere in India. There are several automobile and truck assembly plants in the country. Motorcycles are manufactured and bicycles, produced in many small factories, have become a significant export. More recent additions to the array of manufacturing include tractors, bulldozers, data-processing equipment, silicon chips for electronics, and computer software. Bangalore is known as India's "silicon valley," and its space-age character presents a startling contrast to the largely rural and timeless image that many outsiders have of India. Software developers in Bangalore were among the pioneers attempting to tackle the so-called "Year 2000 Bug," which would have confused computers worldwide as they interpreted "00" to mean 1900 rather than 2000.

Mumbai is India's "Wall Street" and also its "Hollywood"—the center of a prolific and accomplished film industry (and the world's largest) that produces movies of international stature as well as the regular fare, which a huge domestic market consumes passionately (Fig. 11.16).

India produces a wide range of chemical products, from vitamins to fertilizers. Inadequate factory safety measures set the stage for a horrendous accident in the chemical industry on December 2, 1984. At a Union Carbide factory in the north-central city of Bhopal, the cooling system failed in a tank containing highly toxic gas used in the manufacture of pesticides. The ensuing explosion and the release of toxic fumes killed nearly 15,000 people and seriously injured more than 200,000, the majority of whom were poor people living in the plant's shadow. The multinational corporation paid a $470 million settlement

Figure 11.16 Mumbai produces the romance and action films which are the staples of Indian cinema. *Joseph J. Hobbs*

for an accident that cost India an estimated $4.1 billion in economic damage alone.

Pakistan and Bangladesh are still far behind India in total industrial output. However, Pakistan has made rapid progress in some areas since independence. Its principal success has been the development of a cotton-textile industry. Pakistan's main industrial centers are its two largest metropolitan areas: the seaport of Karachi (population: 4.9 million, city proper; 5.3 million, metropolitan area) at the western edge of the Indus River delta and the cultural center of Lahore (population: 2.7 million, city proper; 3.03 million, metropolitan area; Fig. 11.17) in the Punjab. The capital city of Islamabad, the "City of Islam" (population: 204,000, city proper) is a very modern metropolis mainly of administrative significance. In Bangladesh, industrial development centers in Dacca (Dhaka; population: 3.64 million, city proper; 6.54 million, metropolitan area), the capital and largest city, and Chittagong (population: 1.57 million, city proper; 2.34 million, metropolitan area), the main seaport.

Inadequate energy resources are likely to be a major problem confronting further industrialization in the subcontinent. The major domestic sources of commercial energy are India's coal deposits and the hydropower generated in all the countries but lowland Bangladesh. All countries in the subcontinent are heavily dependent on imported energy, especially oil, and would probably be more so if their levels of industrial development and consumption were higher. India has a modest but

Figure 11.17 Lahore's 17th-century Badshahi Mosque, built by the Mogul emperor Aurangzeb. *Joseph J. Hobbs*

Figure 11.18 The central rail station in New Delhi, India, is one of the hubs in a vast network dating to the British colonial period. *Joseph J. Hobbs*

rapidly growing oil output from fields in Assam, Gujarat, and offshore of Mumbai. Natural gas from domestic wells is especially important to Pakistan, and smaller quantities are in India and Bangladesh. India derives about 2 percent of its electricity from nuclear power plants.

One factor working strongly in favor of industrial development in India, Pakistan, and Bangladesh is the existence of a superior rail transportation network (Fig. 11.18). Railways are universally regarded as one of the positive legacies of British colonialism in the subcontinent. The vast network created by the British has been expanded and improved upon by the independent nations. In addition to transporting goods to internal markets and to ports for export abroad, this system carries millions of passengers daily.

11.6 Social and Political Complexities

On a physical map, the Indian subcontinent looks like a discrete unit, marked off from the rest of the world by its mountain borders and seacoasts. But the social complexity of this area is so great that it has never been unified politically except during a relatively brief period in the 19th century and the first half of the 20th century, when the outside power of Great Britain imposed this unity. Even then, a variety of political units under indigenous rulers retained varying degrees of autonomy, although all of these "princely states" were ultimately under British control.

As soon as British power withdrew, the divisive force of conflicting social groups asserted itself, splitting the subcontinent into Muslim and Hindu countries. Within a quarter century, social and geographic divisiveness split Pakistan into two countries. All countries of the subcontinent and Sri Lanka have serious problems in internal or international relations, and sometimes in both.

Religion and the Problems of Sectarianism

The major social divisions within the subcontinent center on religion and language. Religious differences are particularly troubling (Fig. 11.19). The most serious division has been between the two major religious groups, the Hindus and the Muslims. Indigenous Hinduism was the dominant religion at the time Islam made its appearance in the region. It continues to have the largest number of adherents: an estimated 83 percent of the population of India, 11 percent in Bangladesh, and 2 percent in Pakistan.

The faith of Islam is described in Chapter 8. It is much more difficult to characterize the complex and regionally varied faith of Hinduism. It lacks a definite creed or theology. It is very absorptive, encompassing an unlimited pantheon of deities and an infinite range of types of permissible worship. Briefly, it has the following notable elements distinguishing it from the region's other religions: Most Hindus recognize the social hierarchy of the **caste system**, deferring authority to the highest (Brahmin) caste. They practice rituals to honor one or another of five principal deities, attending temples to worship them in the form of sanctified icons in which the gods' presence resides. They believe in reincarnation and the transmigration of souls. They are supposed to be tolerant of other religions and ideas. They participate in folk festivals to commemorate legendary heroes and gods. To earn religious merit and to struggle toward liberation from the bondage of repeated death and rebirth, they make pilgrimages to sacred mountains and rivers.

The Ganges is a particularly sacred river to Hindus, who believe it springs from the matted hair of the god Shiva, and who make pilgrimages to its city of Varanasi (Benares) in the state of Uttar Pradesh (Fig. 11.20). Many elderly people go to die in this city and to be cremated where their ashes may be thrown into the holy waters. The Indian government is now attempting to clean up the Ganges, polluted in part by incompletely cremated corpses; many of the faithful poor cannot afford to buy the fuel

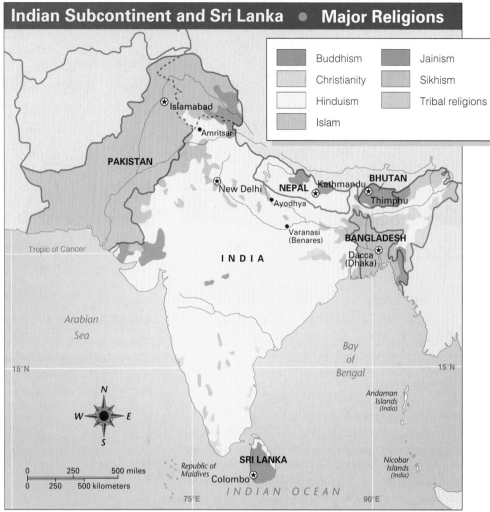

Indian Subcontinent and Sri Lanka ● Major Religions

Legend:
- Buddhism
- Christianity
- Hinduism
- Islam
- Jainism
- Sikhism
- Tribal religions

Figure 11.19 Major religions of the Indian subcontinent and Sri Lanka. If not for the massive migrations of Muslims to West Pakistan and East Pakistan (now Pakistan and Bangladesh) in 1947, there would be many more Muslim areas in India. Note the markedly localized distributions of Buddhists, Jains, and Sikhs.

Source: Stansfield and Zimolzak. 1982. *Global Perspectives: A World Regional Geography.*

needed for thorough immolation. Officials in Varanasi have released scavenging water turtles to dispose of the cadavers.

Seldom in history have two large groups with such differing beliefs lived in such close association with each other as have Hindus and Muslims in the subcontinent. Islam holds to an uncompromising monotheism and prescribes uniformity in religious beliefs and practices. Hinduism is monotheistic for some believers but polytheistic for others, and asserts that a variety of religious observances is consistent with the differing natures and social roles of humans. The exuberant and noisy celebrations of the Hindu faith are a striking contrast to the austere ceremonies of Islam. Islam has a mission to convert others to the true religion, while most Hindus regard proselytizing as essentially useless and wrong. Islam's concept of the essential equality of all believers is a total contrast to the inequalities of the caste system endorsed in Hinduism. Islam's use of the cow for food is anathema to Hinduism (see Definitions and Insights, p. 307).

Practical differences have sometimes added to doctrinal and cultural differences to produce outright conflict between the groups. During the British occupation, the formerly subordinate Hindus came to dominate the civil service and most businesses. Many Muslims feared the results of being incorporated into a state with a Hindu majority, and their demands for political separation led to the creation of the two independent states (India and Pakistan) from the British colony.

Immediately preceding and following partition in 1947, violence broke out between the two peoples on a huge scale, and hundreds of thousands of lives were lost in wholesale massacres. Mass migrations between the two countries involved more than 15 million people. Pakistan today has a large population known as Mohajirs or "migrants," the Muslim immigrants and their descendants who poured into the region from India when partition occurred. Particularly in Karachi and elsewhere in the southern province of Sind, violent conflict frequently occurs among rival factions among the Mohajirs, and between the Mohajirs and other ethnic groups including Pathans and Biharis. In 1971, the events that created Bangladesh—revolt, repression, and Indian intervention against Pakistan—again brought huge casualties and millions of refugees.

Figure 11.20   The Hindu faithful pray and bathe on the banks of the sacred Ganges River in Varanasi.   *Joel Simon/Tony Stone Images*

Since 1947 India and Pakistan have been in persistent conflict over the status of Kashmir (called Jammu and Kashmir in India), a disputed province straddling their northern border. Before independence, Kashmir was a princely state administered by a Hindu maharajah. About three-fourths of Kashmir's estimated population of 12 million is Muslim, which is the basis of Pakistan's claim to the territory. But, under the partition arrangements, the ruler of each princely state was to have the right to join either India or Pakistan, as he chose. Kashmir's Hindu ruler chose India, which is the legal basis of India's claim.

After partition, fighting between India and Pakistan led to a cease-fire line leaving eastern Kashmir, with most of the state's population, in India, and the more rugged western Kashmir in Pakistan. Beginning in the mid-1950s, China pressed its own claims to remote northern mountain areas of Kashmir and occupied some of the border territories. Pakistan ceded part of the occupied territory to China and established friendly relations with its giant northern neighbor, while India rejected China's claims. Until now, India's recurrent small-scale military actions have failed to dislodge Chinese forces. India now holds about 55 percent of the old state of Kashmir, Pakistan 30 percent, and China 15 percent.

Beginning with the conflict that accompanied partition and independence, three wars between India and Pakistan have effected little change in Kashmir. In 1965, conflict began in Kashmir, spread to the Punjab, and escalated to a brief but indecisive

Definitions & Insights

THE CASTE SYSTEM

Serious problems of religious division exist not only between Hindus and minority groups in India, but also within Hinduism. A fundamental feature of Hinduism has been the division of its adherents into the most elaborate caste system ever known. Traditionally, every Hindu is born into a particular caste. Caste membership is inherited and cannot be changed. Particular castes are associated with particular religious emphases, and their members are expected to follow traditional caste occupations. With certain exceptions, marriage outside the caste is forbidden, and meals may be taken only with fellow caste members or those from higher castes. Castes form a hierarchy that determines a person's social rank, with the Brahmin caste at the top (comprising about 5 percent of all Hindus) and three others (Kshatriya, Vaisya, and Sudra) below.

At the bottom of the social ladder are the *dalits* or "untouchables" (about 26 percent of all Hindus). They are not part of the caste system but are literally "outcast" because, according to Hindu belief, they are not twice-born. The untouchables are so called because they traditionally performed the worst jobs, such as handling of corpses and garbage, and therefore their touch would defile caste Hindus.

Forces of modernization are challenging this ancient, rigid social system. Brahmin privileges are being increasingly challenged and in most areas have been restricted by law. Indian leaders of high caste have championed the untouchables' cause for both moral and practical reasons. Indian law now forbids recognition of untouchables as a separate social group. Disintegration of the caste system has been especially rapid in the cities, where it has been hastened by the close intermingling of different castes in factories, public eating places, and public transportation. In India's vast rural areas the caste system is much more entrenched. Politicians use caste divisions to their advantage to bring out voters, promising if elected to act in the interests of their respective large social blocs. Confronting the fact that upper castes account for less than one-fifth of India's population but command more than half of the best government jobs, the Indian government has instituted an affirmative action program to allocate more jobs to members of lower castes. In 1997, an untouchable was elected president of India for the first time in the country's history.

full-scale war involving tanks, airborne forces, and widespread air raids. Renewed hostilities in Kashmir in 1971, as part of the war in which India supported the revolt of Bangladesh against Pakistan, again did not alter the political landscape of Kashmir. During that conflict, the Siachen Glacier in the Karokoram—at about 20,000 feet (2800 m)—earned the title of "world's highest battlefield."

In 1989, Kashmir's Muslim majority escalated the campaign for secession from India, and the strife has continued ever since. India has stationed more than 700,000 troops on its side of the "line of control" in an effort to quell Muslim insurgency. Pakistan has made repeated pleas, so far unsuccessfully, for foreign diplomats to mediate the conflict, hoping that Kashmir might gain the stature of such international concerns as Northern Ireland and Cyprus. As of early 1999, it appears that the two countries will themselves engage in unprecedented diplomatic efforts to solve the crisis. There would be economic benefits for both India and Pakistan if peace returned to the region. The stunning snow-crowned peaks and flower-laden valleys of this western Himalayan region once lured millions of Indian tourists and ranks of Western trekkers each year, and the tourism development potential is enormous.

Pakistan has internal problems centering on religious divisions, mainly between the majority Sunni Muslims (70 percent of the country's population) and Shi'ite Muslims (30 percent). Shi'ite Muslims opposed the government's efforts in the 1980s to establish an avowedly Sunni Islamic state, and there was violent conflict between members of the two sects. Pakistani factions who preferred a Western-style secular democracy also rioted in protest against the Islamization of the government. Democratic elections in the 1980s and 1990s twice installed Benazir Bhutto, a liberal, Western-educated woman, as the country's prime minister. Her regimes were so riddled with corruption that she was twice voted out of office to be replaced by Nawaz Sharif, who promised to cleanse the country of corruption and embrace the traditional Islamic system of law known as *shari'a*.

In India, sectarian violence (known in the region as "communal" violence) between Hindus and Muslims is a constant threat to the country's social and political fabric. Contention over places sacred to both faiths sometimes ignites widespread violence. A 16th-century mosque in the city of Ayodhya in Uttar Pradesh was a place of prayer for Muslims and also revered by the Hindus as the birthplace of the god-king Ram. Backed by Hindu nationalists in the provincial government, a mob of about 250,000 Hindu fundamentalists demolished the mosque in December 1992 with the intention of building a Hindu temple on the site. The ensuing communal violence between Hindus and Muslims throughout India, notably including the cosmopolitan and usually tolerant urbanites of Mumbai, left thousands dead in the weeks which followed.

In addition to the Muslims who make up 12 percent of the population in India, there are other significant religious minorities in that country. About 17 million Sikhs are concentrated mainly in the unusually prosperous state of Punjab. Their faith emerged in the 15th century as a religious reform movement intent on narrowing the differences between Punjab's Hindus and Muslims. It is now India's fourth largest religion, after Hinduism, Islam, and Christianity. Sikh men are recognizable by their turbans, which contain their never-shorn hair, and by a bracelet called a "bangle" worn on the right arm. They regard men and women as equals, and disavow the caste system recognized by Hindus.

In India's Punjab, where Sikhs make up 60 percent of the population and Hindus 36 percent, the 1980s and early 1990s were violent years during which about 20,000 people died in armed clashes. Sikh factions intent on establishing their own homeland of Khalistan—a movement prompted in part by New Delhi's plans to divert water from the Punjab—challenged Indian authority and turned Amritsar's Golden Temple, the holiest site in the Sikh faith, into their military stronghold (Fig. 11.21). In a controversial move to quell the revolt, Prime Minister Indira Gandhi ordered troops to storm the Golden Temple in June 1984. The resulting deaths and desecration led directly to Mrs. Gandhi's assassination by her own Sikh bodyguards later that year. Indian authorities accused Pakistan of arming the Sikh militants.

There are about 22 million Christians in India, most of whom live in the south of the peninsula; Buddhists, Jains, Parsis, and members of a variety of tribal religions make up the numerous smaller remaining religious minorities. Most of the estimated 7 million Buddhists are recent converts from among India's lowest castes. Numbering perhaps 200,000 and concentrated mainly in Mumbai, the famously entrepreneurial Parsis have attained wealth and economic power far out of proportion to their number. Their religion is the ancient pre-Islamic Persian faith of Zoroastrianism, known mainly for its reverence of fire.

India's 5 million Jains also have influence beyond what their numbers suggest, as they control a significant share of India's business. Their faith, founded upon the teachings of Vardhamana Mahavira (a contemporary of Buddha, *c.* 540–468 B.C.), is renowned for its respect for geographical features and animal life. Jains believe that souls are in people, plants, animals, and nonliving natural entities such as rocks and rivers. Jainism has taken the principle of nonviolence (*ahimsa*, also present in Hinduism) to mean they should not even harm microbes. Jain worshippers therefore often wear masks to prevent inhalation of microscopic organisms, and Jains cannot be farmers because they would have to destroy plant life and living organisms in the soil (Fig. 11.22).

Religious and political troubles are also rife in the mountain country of Bhutan. The country is a monarchy ruled from the capital of Thimphu by a Buddhist king of the Drukpa tribe, to

Figure 11.21   The Golden Temple (established in 1577, rebuilt in 1764) in Amritsar is the holiest shrine of the Sikh religion.   *Paolo Koch/Photo Researchers*

PROBLEM LANDSCAPE

Nuclear Shock Waves Rumble Across the Subcontinent

SINCE INDEPENDENCE JUST OVER 50 YEARS ago, India has avowedly been a secular democratic state. An explictly sectarian political party captured the highest number of the country's parliamentary seats for the first time in the national election in 1998. This was the Bharatiya Janata Party, or BJP, a Hindu nationalist party that promised India's Hindu majority the strongest say in the country's affairs. The BJP platform included a pledge to build a Hindu temple atop the ruins of the mosque at Ayodhya and the adoption of a code of civil law that would strip Muslims of separate laws in matters of marriage, divorce, and property rights. However, in order to attract the support of non-Hindu parties and thereby gain enough parliamentary seats to ensure his place as prime minister, the BJP leader Atal Vajpayee dropped many of the Hindu-oriented elements of his agenda. Putting forward a charter he called a "document of unity," he pledged to eradicate hunger in India within five years, increase spending on education, deliver potable water to all villages, encourage economic self-reliance, weed out corruption, and "promote peaceful relationships with all neighbors"—a promise of reconciliation with Pakistan. But another, more troubling message to Pakistan was Prime Minister Vajpayee's declaration that India would reevaluate its nuclear arms policy "and exercise the option to induct nuclear weapons."

On May 11, 1998, much to the surprise of the U.S. Central Intelligence Agency and other Western intelligence agencies, India conducted three underground nuclear tests in the Thar desert. With the blasts, India's government seemed to be trying to stake India's claim as a great world power, exhibit its military muscle to Pakistan and China, and garner enough political support for the BJP to form an outright parliamentary majority in the future. Initial reactions among India's vast populace were highly favorable; 91 percent of those polled within three days of the event supported the tests. Outside India there was alarm. India had defied an informal worldwide moratorium on nuclear testing that went into effect in 1996, when 149 nations (not including India and Pakistan) signed the Comprehensive Test Ban Treaty (known as the Nuclear Non-Proliferation Treaty), which prohibits all nuclear tests. Obliged by a 1994 law to deter potential nuclear weapons powers from crossing the threshold, the United States immediately passed a series of economic sanctions against India, cutting off military and economic aid and using its clout to prevent the World Bank and International Monetary Fund from extending loans to India. Japan (India's biggest source of aid), Germany, and Australia also announced sanctions. "India will not be cowed by threats and punitive steps," Prime Minister Vajpayee answered.[1]

The world's eyes quickly turned to India's neighbor to the west, where a senior Pakistani diplomat announced, "India's actions, which pose an immediate and grave threat to Pakistan's security, will not go unanswered."[2] Governments around the world pleaded with Pakistan to refrain from answering India with nuclear tests of its own, arguing that Pakistan would have a public relations triumph by using restraint: It could appear to be a mature and responsible power while India revealed itself as a dangerous rogue state. They also argued that if Pakistan tested its bombs, the United States and other powers would have to levy crushing economic sanctions against the country.

Prime Minister Sharif of Pakistan chose to ignore these admonitions in favor of appeasing his populace, which demanded a tit-for-tat response to India's blasts. Pakistan conducted six nuclear tests of its own, to match India's total, by the end of May. The action triggered economic sanctions against Pakistan, which suffered more immediately and severely than India's much larger economy did. Foreign investment virtually disappeared in Pakistan, and by the end of 1998 the country faced a deficit of $3.5 billion on its import bills and interest payments. The government imposed economic austerity measures on the population but, paradoxically, announced its intention to go ahead with the construction of a large dam on the northern Indus River. The same crowds who had cheered the tests at the end of May were back out on the streets of Pakistan three months later to protest the government's incompetence and the economic crisis it precipitated. His regime endangered, Prime Minister Sharif soon pledged to sign the Comprehensive Test Ban Treaty within a year if international economic sanctions were lifted. India quickly matched Pakistan's offer.

India and Pakistan have joined only five other nations—the United States, Russia, China, Great Britain, and France—in acknowledging that they possess nuclear weapons. There is considerable disagreement about what this new regional and global shift in the balance of power means. Some analysts fear an escalating nuclear arms race that could result in a border skirmish in Kashmir—leading to the mutual assured destruction of Pakistan and India, or perhaps China and India. Others, particularly within South Asia, argue that the weapons represent the best deterrent against conflict, as they did for decades between the countries of NATO and the Warsaw Pact. Few people anywhere argue with the contention that the nuclear rivalry between India and Pakistan has taken an enormous economic toll in two nations that need to wage war on poverty.

[1] McGeary, Johanna. May 25, 1998. "Nukes . . . They're Back." *Time,* pp. 34–42.

[2] Kinzer, Stephen. May 15, 1998. "Pakistan Is Under Growing Pressure Not to Respond to India with Atom Test." *The New York Times,* p. A5.

Figure 11.22 A priest in a Jain temple in Mumbai wears a mask so as not to breathe in and thereby harm microbes.   *Joseph J. Hobbs*

which most of the government ministers also belong. The government forbids political parties, and is being challenged by a number of outlawed organizations, including the Bhutan People's Party (BPP). The BPP represents ethnic Nepalis, who are Hindus and who for decades have been migrating out of their own overcrowded Himalayan kingdom to work the productive soils of Bhutan. Ostensibly as a means of promoting national unity, Bhutan's king has outlawed the use of the Nepali language in Bhutan's schools and insisted that all inhabitants of Bhutan wear the national dress and hairstyle of the Drukpa tribe. Violent clashes ensued between ethnic Nepalis and Drukpas during the 1990s.

Nepal itself has suffered violent consequences of one-party rule. The country was an absolute monarchy until 1991, when prodemocracy riots and demonstrations forced King Birendra to accept a new role as constitutional monarch and permit elections for a national parliament.

In Sri Lanka, internal violence reflects antagonisms between ethnic groups and discontent with economic and political conditions, especially among the Tamils. Since the mid-1980s, more than 50,000 people have been killed and many more rendered homeless as the Tamil majority in the north has fought for independence from Sri Lanka's overall majority Sinhalese government. Tamil fighters want to establish their own homeland, Tamil Eelam, in the north and west of the country. Guerrilla warfare in 1987 by the organization of Tamil separatists called the Tamil Tigers, or the Liberation Tigers of Tamil Eelam (LTTE), led to military action by Indian troops invited in to help restore order. India withdrew in 1990, but the troubles have continued. The 1991 assassination in southern India of Indian prime minister Rajiv Gandhi, son of the slain Indira Gandhi,

was linked to Indian sympathizers of the Tamil separatist movement in Sri Lanka. In December 1995, Sri Lankan government troops succeeded in retaking the Tamils' geographical stronghold on the Jaffna Peninsula. The Tamil guerrillas retreated to the wild forests just south of the Jaffna and continued their insurgency with the sinking of Sri Lankan naval ships, a devastating car bombing of downtown Colombo, and even an attack on the Sinhalese Buddhists' holiest site, the Temple of the Tooth in Kandy. With that desecration, Sri Lanka's government effectively withdrew its pledge to find a peaceful settlement with the Tamil rebels.

Problems of Language

Language is another strongly divisive factor in India and Pakistan. The languages of the subcontinent fall into two principal families: in the north, the Aryan languages of the Indo-European language family, and in roughly the southern third of peninsular India, the languages of the Dravidian language family. The various languages within each of these groups are fairly closely related to one another. Languages of major importance, each with at least tens of millions of speakers, include four languages of the Dravidian group and eight Aryan languages. Hundreds of other languages and dialects are also spoken. Many people speak two or even three languages, but overall literacy rates are low and communication is often a problem.

Since British colonial times, English has been the *lingua franca* and the de facto official language of the subcontinent. However, only a very small percentage of the population, mainly the educated, are literate in English. This fact, along with nationalist sentiments, have prompted calls for indigenous languages to be adopted as official languages. After independence and partition, disputes arose in each country over which language should be chosen. India decided on Hindi, an Aryan language spoken by the largest linguistic group, and made it the country's national language in 1965. This government decision resulted in strong protests, especially in the Dravidian south. The result has been a proliferation of official languages to satisfy huge populations; India now has 15, all of which appear on the country's paper currency. Bloody rioting between speakers of different languages has occurred at times in areas of India where such peoples are mixed, and demands by language groups have been major factors in reshaping the boundaries of a number of India's internal political units.

In Pakistan, Urdu, a language similar to Hindi but written in Perso-Arabic script, is the official language. As in India, English is widely used in Pakistan's business and government. In Bangladesh, Bengali is the predominant and official language.

The subcontinent's accomplishments in recent decades have been remarkable. India has lowered its birth rate and achieved considerable growth in agriculture and industry. More uniquely among the world's LDCs, it has done so while maintaining, despite all its internal divisions and frictions, a Western-style representative democracy. But sectarian violence between Hindus and Muslims, a growing Hindu nationalist movement which threatens India's remarkable diversity, Sikh desire for autonomy, and the problem of Kashmir all pose long-term difficulties for this South Asian giant. Pakistan and Nepal must confront the problems of rapid population growth, deforestation, and land degradation where the balance between people and resources is already precarious. Bangladesh faces the huge challenge of dealing with the environmental consequences of upstream deforestation and repeated flooding assaults from the

Bay of Bengal. Sri Lanka's future may be brightened either by a forced reunification of the country, or by the government's willingness to tolerate a de facto partition of the country into Sinhalese and Tamil realms, so long as the Tamils do not push for independence. And all of these nations struggle with the chronic problem of poverty, which ultimately is the source of many of the region's most severe environmental and political dilemmas.

SUMMARY WITH SELECTED KEY TERMS

- South Asia is mostly a **triangular peninsula** that thrusts from the main mass of Asia and splits the northern **Indian Ocean** into the **Bay of Bengal** and the **Arabian Sea.** The entire unit is often called the **Indian subcontinent.**
- South Asia contains the five countries of **India, Bangladesh, Pakistan, Nepal, and Bhutan** in an area a little more than one-half of the 48 conterminous United States. The island nation of **Sri Lanka** lies off the southern tip of India and shares many physical and cultural traits with the subcontinent.
- Mountains enclose the subcontinent on its northern borders. There lie the **Himalaya Mountains** and the **Karakoram Range.** The countries of **Nepal and Bhutan,** on the southern flank of the Himalaya, are small, rugged, and remote.
- Until 1947, the entire area, with the exception of the small Himalayan states and Sri Lanka, was referred to as "India." For over a century it was the most important unit in the British colonial empire—the **"jewel in the crown."**
- The Indian subcontinent is **one of the world's culture hearths.** Some of the world's earliest cities developed on the banks of the Indus River before 3000 B.C. The **Aryans** arrived between 2000 and 3000 B.C. and were the source of many ethnic and cultural attributes of the subcontinent today.
- The subcontinent's **northern plain** contains the core areas of the three major countries of India, Bangladesh, and Pakistan. This plain is watered by three great rivers: the **Indus River,** the **Ganges River,**

and the **Brahmaputra River.** While mostly desert and steppe in Pakistan, the plain is relatively well-watered in India where it supports unusually **high population densities.**
- The southern portion of the subcontinent is mainly a large volcanic plateau called the **Deccan.** Generally low in elevation, it is rimmed by highlands: the **Western Ghats,** the northern **Central Indian Hills,** and the low-lying **Eastern Ghats.**
- Climates vary with **undifferentiated highland** in the northern mountains to **desert and steppe** in Pakistan and western India; **humid subtropical** occurring in the northern plain; **tropical savanna** in the peninsula with the exceptions of the **rain shadow** in the lee of the Western Ghats; and the **rain forests** of the Western Ghats' seaward slopes, the coastal plain to the south, and part of Bangladesh.
- The **southwest** or **wet monsoon** is at its height from June to September, blowing in **from the Arabian Sea** and **from the Bay of Bengal.** The **dry** or **northeast monsoon** occurs in the winter, with dry, cooler winds from the northeast coming from high-pressure systems over the continent.
- The subcontinent is home to many different faiths including: **Hinduism, Sikhism, Christianity, Islam, Buddhism, Zoroastrianism, and Jainism.** While **hundreds of languages and dialects** are spoken, they fall into two principal families of **Aryan and Dravidian. English** is spoken often for business but only a small percentage of the people, generally the well educated, is literate in this language.

REVIEW QUESTIONS

1. List the four groups of peoples that have been most significant in forming the cultural foundation of this region.
2. Using maps and the text, locate the Outer Mountain Wall, the Northern Plain, Peninsular India, and Sri Lanka. List the major resources and attributes of these areas.
3. Using maps and the text, locate the main agricultural production areas of rice, irrigated wheat, sorghums, millets, jute, and tea.
4. List the main industrial and export crops of this region.
5. List the major non-agricultural industries of this region.

6. Using maps and the text, locate the areas of differing climates.
7. Identify the areas of high, intermediate, and low population densities and relate these to climate.
8. Locate Sri Lanka and discuss the reasons for its diversified population and internal unrest.
9. Using maps and the text, list and locate the three major rivers of the Northern Plain.
10. Using maps and the text, list and locate the major industrial regions along with their primary industries.

DISCUSSION QUESTIONS

1. Explain why India could be considered one of the most important nations to have gained independence since World War II.
2. Why is the Indian subcontinent considered the most culturally complex area of its size on Earth?
3. Discuss the impact that the Outer Mountain Wall has had on the subcontinent's development.
4. Why are the wet and dry monsoons so significant to this region?
5. What legacy was left to this region by the British Empire?

6. What disputes center on Kashmir and the Punjab?
7. Discuss how language and religion have been divisive to this region.
8. What are the major contributors to Bangladesh's flood problems?
9. What have been the results of India's attempt at industrial self-sufficiency?
10. What are the major points of contention in the nuclear rivalry between India and Pakistan?

Chapter 12

The Tigers and Tribulations of Southeast Asia

▲ *The peoples of Southeast Asia are modernizing rapidly, but not sacrificing cherished traditions. This young dancer is from Phnom Penh, Cambodia.* Paula Bronstein/ Tony Stone Images

CHAPTER OUTLINE

12.1 A Region of Diverse Cultural Influences

12.2 Area, Population, and Environment

12.3 The Economic Pattern

12.4 The Countries

Southeast Asia, from Burma (Myanmar) in the west to the Philippines in the east, is a fragmented region of peninsulas, islands, and intervening seas. East of India and south of China, between the Bay of Bengal and the South China Sea, the large Indochinese Peninsula projects southward from the continental mass of Asia (Fig. 12.1). From it, the long, narrow Malay Peninsula extends another 900 miles (c. 1450 km) toward the equator. Ringing the south and east of this continental projection are thousands of islands, among which Sumatra, Java, Borneo, Celebes (Sulawesi), Mindanao, and Luzon are outstanding in size. Another large island, New Guinea, east of Celebes, is culturally a part of the Melanesian archipelagoes of the Pacific World, but its western half, held by Indonesia, is politically part of Southeast Asia.

Southeast Asia is composed politically of ten states: Burma, Thailand, Laos, Cambodia, Vietnam, Malaysia, Singapore, Indonesia, Brunei, and the Philippines. With the exception of Thailand, which was never a colony, all of these states have become independent from their colonial powers since 1946. These ten nations together occupy a land area less than one-half that of China, and are fragmented both politically and topographically. There is also a high degree of cultural fragmentation, which has often led to contention and warfare between differing cultural groups. Outside intervention has generally complicated and worsened local discord, producing enormous suffering in the region and long-lasting traumatic effects among the foreign soldiers who fought losing battles against the determined inhabitants of Southeast Asia.

12.1 A Region of Diverse Cultural Influences

Southeast Asia is one of the world's culture hearths, having contributed domesticated plants (including rice) and animals (including chickens) and the achievements of civilizations to a wider world. Cultural innovations and accomplishments have been diffused from several centers which themselves were built upon many influences reaching Southeast Asia from abroad. Indian and Chinese traits were especially strong in shaping the region's cultural geography. Hinduism came from India, and Hinayana Buddhism from India through Ceylon (Sri Lanka). The Chinese cultural influences of Confucianism, Taoism, and Mahayana Buddhism were particularly strong in Vietnam. China periodically demanded and received tribute from states in Burma (Myanmar), Sumatra, and Java. The faith of Islam came from the west in the 14th and 15th centuries and took permanent root; Southeast Asia is the eastern margin of the world of Islam, and Indonesia has more Muslims than any other country in the world today.

The historic and contemporary monumental architectures of Southeast Asia reflect these diverse origins. In the ninth century, the advanced Sailendra culture of the Indonesian island of Java built Borobudur, the world's largest Buddhist *stupa* (shrine) and still the largest monument in the Southern Hemisphere. In the 12th century, a Hindu king built the extraordinary complex of Angkor Wat as the capital of the Khmer empire

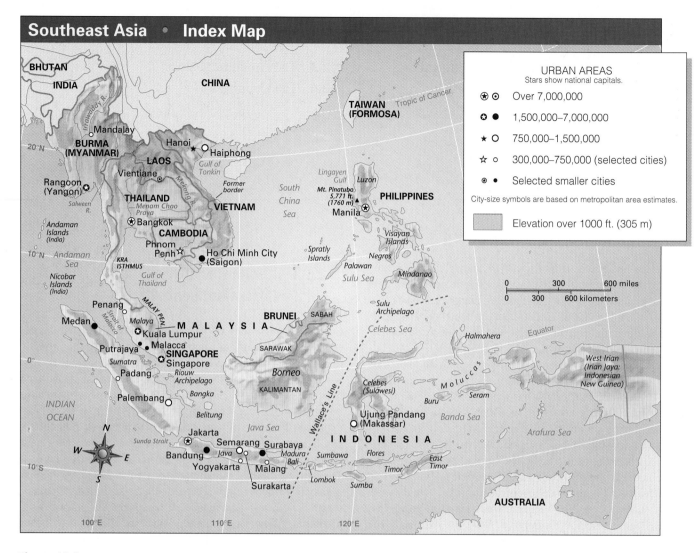

Figure 12.1   Introductory reference map of Southeast Asia.

based in Cambodia (Fig. 12.2a). The first independent Thai kingdom emerged in the 13th century, and in the 14th century Thai kings ruling from the great city of Ayuthaya (near modern Bangkok) extinguished the Khmer empire and extended their own control into the Malay Peninsula and Burma. The world's largest mosque stands in modern Brunei, a tiny oil-rich country on the northern tip of the island of Borneo, and before an economic crisis shook Malaysia in 1997, this hopeful country erected the world's tallest buildings (1462 feet/446 m)—the Petronas towers in Kuala Lumpur (see Fig. 12.2b).

12.2 Area, Population, and Environment

It is approximately 4000 miles (c. 6400 km) from western Burma to central New Guinea and 2500 miles (c. 4000 km)

from northern Burma to southern Indonesia (Figs. 12.1 and 12.3). Despite these great distances, the total land area of Southeast Asia is only about 1.7 million square miles (4.4 million sq km), or approximately half the size of the United States (including Alaska). The estimated population of the region was 512 million in mid-1998, giving it an overall density of 304 per square mile (117 per sq km)—an average that embraces very great extremes within the region (see Table 10.1). The average population density of Southeast Asia is high compared to that of much of the world but low compared to most other areas on the seaward margins of east Asia. For example, although Southeast Asia has nearly four times the overall density of the United States, its density is well under half that of the Indian subcontinent and China proper (the humid eastern part of China south of the Great Wall).

Southeast Asia has been less populous than India and China proper throughout history. The entire region probably contained only about 10 million people around the year 1800 and lacked

A

B

Figure 12.2 (a) Angkor Wat, the 12th-century capital of the Khmer empire, is a symbol of the rich cultural history of Southeast Asia. (b) Modernity and prosperity characterize some nations in Southeast Asia today. The twin Petronas Towers in Kuala Lumpur are the world's tallest buildings. *(a) Jerry Alexander/Tony Stone Images, (b) Robin Moyer/Gamma Liaison*

widespread dense populations. Environmental difficulties probably had much to do with the historical slowness of population growth in Southeast Asia. By land the region is relatively isolated, since the northern Indochinese Peninsula is an area of high and rugged mountains. Over many centuries, the ancestors of most of the present inhabitants entered the area as recurrent thin trickles of population, crossing this mountain barrier as refugees driven from previous homelands to the north.

There are formidable environmental challenges within the region itself. Southeast Asia is truly tropical, with continuous heat in the lowlands, torrential rains, a prolific vegetation difficult to clear and keep cleared, soils that are generally leached and poor, and a high incidence of disease. A four- to six-month dry season on the Indochinese Peninsula and on scattered smaller areas in the islands has unfavorable agricultural effects, compounded by the high evaporation rates produced by tropical heat. Despite the lush vegetation, most soils have little fertility. The nutrients of the natural ecosystem are largely retained in the natural vegetation itself; when people clear the natural vegetation and plant crops, soil quality deteriorates rapidly. In mountainous areas, deforested slopes erode rapidly; therefore, during the rainy season, rivers carry enormous volumes of mud and silt.

More spectacular environmental difficulties result from the location of part of the region on the Pacific "ring of fire" (see Chapter 14), subjecting Indonesia and the Philippines espe-

cially to earthquakes, the tidal waves or tsunamis created by earthquakes, and volcanic eruptions. Violent wind-and-rain storms known locally as typhoons (known elsewhere as hurricanes and cyclones) often strike coastal Vietnam and the northern Philippines.

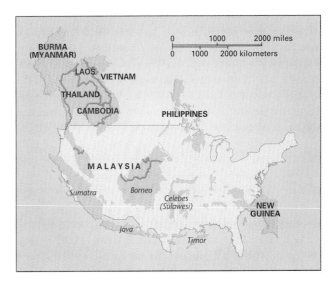

Figure 12.3 Southeast Asia compared in area (but not latitude) with the conterminous United States.

Regional Perspective

The Economic Meltdown

"In Southeast Asia, the economic miracle is over."[1] Thus declared a respected economist at the Massachusetts Insitutute of Technology in November 1997. Following years of spectacular growth, in which economies across the region grew at rates up to 8 percent annually, 1997 saw a stunning reversal of fortunes in Southeast Asia. Up to 1997, the region's economic "Tiger Cubs" were praised for high rates of savings, balanced budgets, and low inflation. Export-driven growth rates were climbing, particularly with increasing manufacturing of computers and other electronics. Gradually, however, the appetite of the main comsumer for the products, the United States, began to

wane, and trade deficits developed in the producing countries. At the same time, the emerging economies pursued expensive public works projects, such as Malaysia's new capital of Putrajaya. In some countries, corruption added to the potential for economic crisis as insolvent companies were propped up with infusions of cash from sympathetic politicians (in a phenomenon known as "crony capitalism"). Banks extended risky loans for both private and public businesses and construction projects. The loans were generally made in U.S. dollars. When local currencies rapidly lost their values (see discussion of Thailand on p. 326), the borrowers were unable to repay these

loans. Economies melted down. Stock markets, which had seen explosive growth, imploded; Malaysia's market, for example, lost 80 percent of its value between September 1996 and September 1998. Among the region's stronger economies (excluding Vietnam, Cambodia, Laos, and Burma, which had far less to lose), Malaysia, Thailand, and Indonesia fell into economic recession—characterized by a relatively short fall in economic output and a moderate rise in unemployment—and, as the crisis dragged on, threatened to tip into outright depression. Only the Philippines and Singapore escaped the bloodletting.

[1] Switow, Michael. "Asia Sputters Into Low Gear." *Christian Science Monitor*, December 1, 1997. p. B6.

As is typical in LDCs throughout the world, the recent introduction of modern medical and other technologies into the region have reduced the effects of these environmental difficulties and have lowered death rates, producing a tremendous population increase in the last century and a half. This increase, which accelerated after World War II, is now slowing but is still rapid. Between 1973 and mid-1998, population grew from an estimated 316 million to 512 million, a 25-year increase of 196 million people, or 62 percent. During recent times, the race between population growth and food supply has seen food supply winning in some countries but losing to population growth in others. From the 1970s through the 1990s, large increases in food production significantly outpaced population growth in Malaysia and Indonesia. At the other end of the scale were Cambodia, Laos, and Vietnam, in which war, repression, and inefficient communalization of agriculture thwarted significant progress.

12.3 The Economic Pattern

Southeast Asia as a whole is among the world's poorer regions. As of 1996, the per capita gross national product (GNP) of six of its ten countries was below the average for the world's LDCs. In contrast, Singapore, Malaysia, and Thailand were

above average for LDCs, and because of their high rates of industrial productivity and economic growth, they have often been described as "Asian Tigers," or at least "tiger cubs" (along with Hong Kong, South Korea, and Taiwan). In fact, the highly successful commercial-industrial city-state of Singapore has a per capita GNP that places it in the category of MDCs (see Table 10.1). Beginning in 1997, however, a series of events precipitated an economic crisis that roared through the region and threatened to declaw these emerging economies (see Regional Perspective, above).

One means by which countries of Southeast Asia have sought to strengthen their economic and security interests has been to form a regional organization. Known as the Association of Southeast Asian Nations (ASEAN), the coalition was formed in 1967 to promote regional trade and minimize alignment with the two world superpowers, the United States and the Soviet Union. Singapore, Malaysia, Thailand, Indonesia, Brunei, and the Philippines were founding members. Vietnam joined in 1995 with the promise of strengthening regional ties and adopting the more open economic system the Vietnamese call "doi moi," or renovation. Burma and Laos joined in 1997, and Cambodia finally joined in 1999. After United States troops withdrew from bases in the Philippines in 1991, ASEAN members grew concerned that they lacked a security "umbrella" and that China or possibly even Japan might become more aggressive in the region. Several ASEAN members, notably Indonesia,

Malaysia, and Singapore, therefore strengthened ties with the U.S. military to deter other powers from attempts to dominate the region. The organization thus aspired to become both a large trading bloc and a substantial security counterweight to China.

Despite rapid urbanization and industrialization in recent years, the majority of the people in Southeast Asia are still farmers (see Table 10.1), and manufacturing is less developed than in Japan, China, and India. Petroleum and minerals contribute significantly to the incomes of a few countries.

Shifting Cultivation

Permanent agriculture has been difficult to establish in most parts of this mountainous tropical realm. Much land is used only for a shifting, subsistence form of cultivation, in which a migratory farmer clears and uses fields for a few years and then allows them to revert to secondary growth vegetation while he moves on to clear a new patch. Since about 15 years are necessary for an abandoned field to be restored to forest that can again be cleared for farming, only a small proportion of the land can be cultivated at any one time. This extensive form of land use can support only sparse populations, with farmers often growing crops using only simple digging sticks and little additional cultivation. Unirrigated rice is the principal crop, and corn, beans, and root crops such as yams and cassava are also grown. Agriculture is supplemented by hunting and gathering in forested areas. These migratory subsistence farmers often differ ethnically from adjacent settled populations, having historically been driven to refuge in the backcountry by stronger invaders.

Sedentary Agriculture

Most of Southeast Asia's people live, often in extremely dense clusters, in scattered areas of permanent sedentary agriculture. These core regions stand in striking contrast to the relatively empty spaces of the adjoining districts. Superior soil fertility in these areas appears to have been the main locational factor in most instances. A few areas of Southeast Asia exhibit fairly dense populations without any corresponding soil superiority. Such areas are ordinarily characterized by plantation agriculture of rubber and other crops (see p. 322).

The typical inhabitant of the areas of permanent sedentary agriculture is a subsistence farmer whose main crop is wet rice, grown with the aid of natural flooding or by irrigation (Fig. 12.4). In some drier areas, unirrigated millet or corn replaces rice. The major crop is supplemented by secondary crops that can be grown on land not suited for rice and other grains. Prominent secondary food crops include coconuts, yams, cassava, beans, and garden vegetables. Some farmers grow a secondary cash crop such as tobacco, coffee, or rubber. In addition, fishing is important both along the coasts and along inland streams and lakes. Because of their importance in the food sup-

Figure 12.4 A farmer plowing and hoeing a rice field in Bali, Indonesia, in preparation for planting. Rice seedlings grown in a seedbed will be transplanted by hand to the flooded field. The soil must have a creamy consistency when the young plants are set out. Water buffaloes (upper left) are the main work animals used in Indonesian rice farming. *CARE*

ply, many fish are harvested from artificial fish ponds or from flooded rice fields, in which they are grown as a supplementary "crop."

Plantation Agriculture

Subsistence agriculture in Southeast Asia exists alongside tropical plantation cash crops grown for export. The region is one of the world's major supply areas for such crops. This highly commercial type of production is a legacy of Western colonialism, which began in the 16th century. Until the 19th century, however, the newcomers were interested mainly in the region's location on the route to China. They were generally content to control patches of land along the coasts and to trade for some goods produced by local people. During the 19th century, the Industrial Revolution in Europe and Anglo America greatly enlarged the demand for tropical products, and as a result, Europeans extended their control over almost all of Southeast Asia. European colonists used their capital and knowledge, along with indigenous and imported labor, to bring about a rapid increase in production for export. The colonizers gave emphasis to certain local commodities, such as copra (coconut meat) and spices, and introduced a number of entirely new crops.

The usual method of introducing commercial production was to establish large estates or plantations managed by Europeans but worked by indigenous labor or labor imported from other parts of Monsoon Asia (see Chapter 2, p. 44). Development of these enterprises was aided by a favorable climate, the abundance of land, and the availability of cheap transportation by water. The major difficulty was the recruitment of adequate labor from a population already fully engaged in food production and not poor enough to be drawn away from village life into a wage-labor economy. The solution in many areas was large-scale importation of contract labor from India and China. These massive migrations further complicated an already complex ethnic mixture, and account in part for the unusual distribution of ethnic groups in the region today.

Plantation activity came to have widespread repercussions for the economic life of local inhabitants. Many of them learned by example and entered commercial production on a small scale on their own. Today, small landowners command an important share of the export production of most "plantation" crops in Southeast Asia.

Many types of plantation cash crops have been produced in Southeast Asia. Wide variations have occurred over the years in the crops grown, the centers of production, the amounts exported, and the prosperity of the producers. Factors such as fluctuating world demands, regional and world competition, changing political conditions, and the occasional ravages of plant diseases have caused these shifts. The major plantation cash crops now exported from Southeast Asia are:

1. *Rubber.* Over four-fifths of the world's natural rubber is produced in Southeast Asia, primarily in Malaysia, Indonesia, and Thailand. However, synthetic rubber, most of which is made from petroleum, now supplies the greater part of the world's rubber consumption.
2. *Oil palm and coconut palm products.* Palm products consist primarily of palm oil, coconut oil, and copra (dried coconut meat from which oil is pressed). Malaysia is the world leader in palm oil production, and the Philippines and Indonesia dominate the world output of coconut palm products.
3. *Tea.* Indonesia is the region's leading producer of tea, and is the world's second largest.

Many nations in Southeast Asia produce other export crops. These include coffee from Indonesia, cane sugar from the Philippines and Indonesia, pineapples (of which the Philippines and Thailand have become the world's leading producers and exporters), and many others. However, the countries of Southeast Asia are generally coming to depend less heavily on their agricultural exports as mining and manufacturing activities develop.

Commercial Rice Farming

One significant impact of Western colonialism on the economy of Southeast Asia was the stimulation of commercial rice farming in areas that had formerly been unproductive. The development of plantation agriculture, mining, and trade in Southeast Asia provided a market for rice by creating a large class of people who worked for wages and had to buy—rather than produce—their food. During the same period, Western economic, medical, and sanitary innovations helped decrease death rates and bring about an enormous increase in population and a growing demand for food. Western technology facilitated the

Definitions & Insights

RUBBER

The most notable cash crop introduced into Southeast Asia was rubber, in the 1870s. The Englishman Sir Joseph Priestley had given the name "rubber" to the latex of the *Hevea brasiliensis* tree in 1770, when he discovered he could use it to rub errors off the written page. Extensive exploitation of this tree in its native Brazil gave rise to the "rubber boom" of the late 19th century, when Brazilian rubber traders were so wealthy that they used bank notes to light cigars and sent their shirts to be laundered in Europe. In 1876 an Englishman named Henry Wickham smuggled 70,000 rubber seeds out of Brazil. Delighted botanists at London's Kew Gardens cultivated them and shipped them on to Ceylon (Sri Lanka) and Southeast Asia for experimental commercial planting. Rubber trees proved exceptionally well-adapted to the climate and soils of South, and later Southeast, Asia (Fig. 12.A). Back in Brazil, however, a fungus known as leaf blight wiped out the rubber plantations, and by 1910, bats and lizards came to inhabit the mansions of Brazil's rubber barons. Fortunately, the Kew Gardens stock was free of the fungus; by 1940, 90 percent of the world's natural rubber came from Asian plantations.

Figure 12.A The rubber tree has been more commercially successful in Southeast Asia than in its native Brazil. *Thomas D.W. Friedman/Photo Researchers*

bulk processing and movement of rice and aided in the development of drainage, irrigation, and flood-control facilities needed to produce it on a large scale in previously undeveloped areas. As a consequence, an Asian pioneer movement into areas capable of expanded rice production took place.

Three of these areas gained prominence in commercial rice growing—the deltas of the Irrawaddy, Chao Praya, and Mekong rivers, located in Burma, Thailand, and Vietnam and Cambodia respectively. Almost impenetrable wetlands and uncontrolled floods had kept these deltas thinly settled, but incentives and methods for settlement have turned them into densely populated areas within the past century. The farms there are larger than most in east Asia, and the surpluses of rice produced on them until the 1960s represented nearly one-third of the world's total rice exports. War and economic dislocation then ended the ability of Vietnam and, in the 1970s, Cambodia to generate this surplus production. Expansion of production in Thailand partially replaced the loss of the Mekong Delta as a surplus rice-producing area; also in Thailand, corn (maize) became a major food export. In the 1990s, Thailand, Burma, and the resurgent Vietnam had become the region's leading rice exporters. Since most Southeast Asian countries are also major grain importers, there is a substantial regional market for rice.

Agricultural Growth and Deforestation

Near areas of dense settlement in Southeast Asia are large areas still relatively undeveloped, despite the fact that they apparently are capable of supporting large populations. Clearing of forests, draining of wetlands, and construction of irrigation systems proceed when population growth demands such expansion. Many environmentalists view the attendant destruction of tropical rain forest here as an international ecological problem. Southeast Asia's forests were already largely destroyed by the early 1990s; for example, only 15 percent of Thailand's forests remained at that time, along with 14–20 percent of those in the Philippines and 16–19 percent of those in Vietnam. Deforestation has moved into Malaysia, where 47 percent of the land area is still forested, and to Indonesia, where 54 percent is forested. Malaysia is now the world's largest exporter of tropical hardwoods, and, at current rates of deforestation, it will have logged all of its principal forest reserves in Sarawak on the island of Borneo by the year 2005.

Although Southeast Asia's tropical forests are smaller in total area than those of central Africa and the Amazon Basin, they are being destroyed at a much faster rate, most significantly by commercial logging for Japanese markets rather than by subsistence farmers. Environmentalists fear the irretrievable loss of plant and animal species and the potential contribution to global warming that deforestation in this region may cause. In 1995, both Malaysia and Indonesia moved to slow the destruction, with Indonesia taking the aggressive step of banning the use of fire to clear forests. However, enforcing such legislation proved impossible. Hundreds of Indonesian and Malaysian companies,

most of them large agricultural concerns (such as palm oil plantations and pulp-and-paper companies) with close ties to the government, continued to use fire as a cheap and illegal method of clearing forests. In 1997, a widespread drought attibuted to the El Niño warming of Pacific Ocean waters turned the annual July–October burning season into a manmade holocaust. Deliberately set fires raced out of control over large areas of Sumatra and Borneo, resulting in a choking haze that shut down airports, closed schools, deterred tourists, and caused respiratory distress to millions of people in Indonesia, Malaysia, Singapore, the Philippines, Brunei, and Thailand.

Many of the species that find habitats in these forests are **endemic** (found nowhere else on Earth). Indonesia, which contains 10 percent of the world's tropical rain forests, is known, after Brazil, as the world's second most important "megadiversity" country—with about 11 percent of all the world's plant species, 12 percent of all mammal species, and 17 percent of all bird species within its borders. The Southeast Asian region is also particularly significant in biogeographical terms because of the so-called "Wallace's Line"—named for its discoverer, English naturalist Alfred Russel Wallace—which divides it (see Fig. 12.1). Nowhere else on Earth is there such a striking local change in the composition of plant and animal species in such a small area on either side of this divide. East of the line (separating the Indonesian islands of Bali and Lombok, for example), marsupials are the predominant mammals, while west of the line, placental mammals prevail. Similarly, bird populations are remarkably different on either side of the line.

Production and Reserves of Minerals

Petroleum is the most important mineral resource in Southeast Asia, although the region's reserves and production are not impressive on a world scale. Seven countries—Indonesia, Malaysia, Brunei, Burma, Thailand, Vietnam, and the Philippines—produce oil. Their combined output is about 5 percent of world oil production, with Indonesia the leading producer. Indonesia's main oil fields are in Sumatra and Indonesian Borneo (known as Kalimantan).

Although it does not loom large on the world scene, Southeast Asian oil is very important to the countries that own and produce it. Oil and associated natural gas provide nearly all the exports of Brunei and make that tiny Muslim country one of the world's wealthiest nations as measured by per capita GNP (Fig. 12.5). Oil, gas, and refined products supply almost one-fourth of Indonesia's exports and about one-tenth of Malaysia's. Singapore profits as a processor of oil. The principal customer for Southeast Asian oil and natural gas (shipped in liquid form) is Japan, which is the leading trade partner of several Southeast Asian countries.

Considerable mineral wealth other than oil and gas exists in Southeast Asia. A wide variety of metal-bearing ores are extracted in many locations, and many unexploited reserves remain. Tin is the most important ore currently mined, with about

Figure 12.5 Brunei is a tiny, prosperous Muslim nation. This mosque of Omar Ali Saifuddin was designed by an Italian architect and cost $5 million to build. *Paul Harris / Tony Stone Images.*

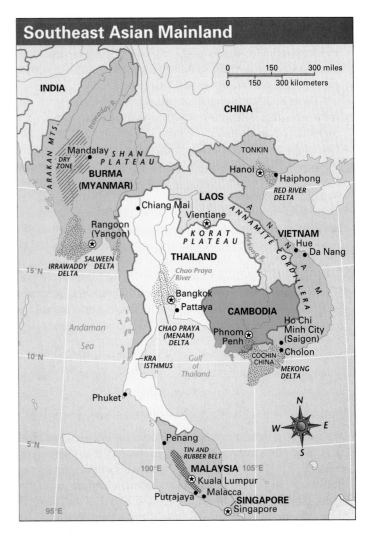

Figure 12.6 Index map of the Southeast Asian mainland. Stars show political capitals and the independent state of Singapore. Area symbols show river deltas, the Dry Zone of Burma (Myanmar), and the Tin and Rubber Belt of Malaysia.

one-fifth of world output from Malaysia and another fifth from Thailand and Indonesia combined. The Philippines produces the greatest variety of metals—notably copper, chromite, nickel, silver, and gold—in significant amounts. The most serious mineral deficiency in the region is the near absence of high-grade coal. However, there are sizable hydropower potentials that could supply additional energy.

12.4 The Countries

Although the countries of Southeast Asia have many broad similarities, each has its own distinctive qualities, exhibiting a different combination of environmental features, indigenous and immigrant peoples, economic activities, and culture traits. Many of the characteristics and problems of these countries are the outcome of European imperialism, but in no two countries have the results of colonialism been the same. Following are portraits of the major countries.

Burma (Myanmar)

Burma (renamed Myanmar by its military government in 1989; 1998 population: 47.1 million) is centered in the basin of the Irrawaddy River and includes surrounding uplands and mountains. Within the basin are two distinct areas of dense population: the Dry Zone, around and south of Mandalay (population: about 533,000, city proper), and the Irrawaddy Delta,

which includes the capital and major seaport of Rangoon (Yangon); population: 2.5 million, city proper; 2.8 million, metropolitan area; (Fig. 12.6). The Dry Zone has been the historical nucleus of the country. The annual rainfall of this area—34 inches (87 cm) at Mandalay—is exceptionally low for Southeast Asia, and there is a dry season of about six months. The people are supported by mixed subsistence and commercial farming, with millet, rice, and cotton as major crops. During the past century, the Dry Zone has been surpassed in population by the Delta, where a commercial rice-farming economy now provides one of the country's two leading exports (forest products is the other). The Irrawaddy forms a major artery of transportation uniting the two core areas.

The indigenous Burman people, most of whom live in the Irrawaddy Basin, number an estimated 69 percent of the country's population. About 90 percent of Burma's citizens are Buddhists. A minority of around one million mainly Hindu Indians lived here prior to World War II. Between the end of World War II and 1970, however, most of these Indians were ejected from this area through expropriations of property and other oppressive measures resulting from the nationalization of the economy, including land ownership. A variety of hill-dwelling tribes live in the Arakan Mountains of the west and the northern highlands. The Shan Plateau of eastern Burma is inhabited by the ethnic Shans to the north and ethnic Karens to the south and in the delta of the Salween River. They account, respectively, for about 9 and 7 percent of the country's population. The Shan Plateau in northeastern Burma is part of the "Golden Triangle"—the world's leading source area for opium. As the world's largest producer, Burma annually harvests about 60 percent of the world's opium supply.

Great Britain conquered Burma in three wars between 1824 and 1885. It became an independent republic outside the British Commonwealth in 1948. There has been almost constant civil war since independence. At one time no fewer than eight different rebellions were in progress. Both communism and ethnic separatism have supplied major motivations for rebel forces drawn principally from the Karens, the Shans, a number of less numerous hill peoples, and Burmese Muslims who fear the creation of a pure Buddhist state. The economy has been badly damaged by the fighting, and for years was mismanaged under a form of rigid state control promoted as "The Burmese Way to Socialism." Burma became the poorest noncommunist country in Southeast Asia, but, since the late 1980s, it has begun to abandon socialism in favor of a free-market economy and has enjoyed considerable economic growth. China is Burma's main trading partner and military ally.

The repressive military government that seized power in Burma in 1988 has yielded little to popular pressure for democratization. In 1995 the government did release the opposition leader and Nobel Peace Prize laureate Aung San Suu Kyi, who had been under house arrest since leading her antimilitary political party to victory in the 1990 elections (the results of which the military government nullified). Since her release "The Lady," as Ms. Suu Kyi is popularly known, has called for dialogue with the military junta, and has appealed for the parliament that was elected in 1990 to be convened. Her pleas so far have been ignored. She has been more successful in soliciting support abroad for boycotts of companies doing business with Burma. Many U.S. corporations, including Pepsi and Phillips Petroleum, have responded by withdrawing from Burma. There is pressure on another American firm, Unocal, to suspend its contract to join a French firm in developing Burma's considerable offshore natural gas reserves in the Yadana Field of the Andaman Sea. The two companies have already built a pipeline from the field to a port west of Bangkok, in Thailand, from where the gas is exported abroad. The pipeline has to be well defended, in view of the Burmese government's many opponents who might try to destroy it as a means of weakening the regime.

Thailand

Thailand (formerly Siam; 1998 population: 61.1 million) is located in the delta of the Chao Praya River, known to Thais as the Menam ("The River"). Annual floods of this river irrigate the rice that is the most important crop in this Southeast Asian land of relative abundance and progress. Thailand is now the world's largest exporter of rice.

On the lower Chao Praya is Bangkok (population: 5.9 million, metropolitan area), the capital, main port, and only large city. It is a notoriously crowded and polluted city; air pollution is responsible for as many as 400 deaths yearly. But it is also a city of stunning architectural marvels, particularly its Buddhist temples. Bangkok's network of canals (*klongs*), thronged with both commercial and residential activities, give it a distinctively amphibious character; especially renowned is its "floating market" (Fig. 12.7).

Areas outside the river's delta are more sparsely populated and include the mountainous territories in the west and north— inhabited mainly by ethnic Karens and a variety of mountain

Figure 12.7 The floating market of Damnern Saduak near Bangkok, Thailand, typifies the amphibious nature of life in Southeast Asia. Water transportation is vitally important in the river deltas and islands where most Southeast Asians live. *George Chan/Photo Researchers*

tribes—and the dry Korat Plateau to the east, populated by the Thai, related Laotian, and Cambodian peoples. To the south in the Kra Isthmus, a part of the Malay Peninsula, live more than 2 million Malays. Most Malays are Muslims, but an estimated 95 percent of the people of Thailand are Buddhists whose faith includes elements borrowed from Hinduism and local spiritualism. Spectacular Buddhist temples are a prime attraction for one of Southeast Asia's largest tourist industries, as are ceremonies celebrating Thailand's monarchy and other historical traditions.

Governed today as a democracy in which military influence is strong and in which the monarchy has largely symbolic functions, Thailand was the only Southeast Asian country to preserve its independence throughout the period of colonialism. This peculiar status was due to then Siam's position as a buffer between British and French colonial spheres. A number of border territories were lost, however, and the country has exhibited irredentist (separatist) tendencies for many years. Thailand's attention recently has focused more on combating insurgencies within its own outlying territories than on attempting to regain lost territories.

The main ethnic minority of Thailand is Chinese. An estimated 14 percent of the population, or 8.6 million ethnic Chinese, reside in the country and are claimed as citizens by both Thailand and China. They control much of the country's business, a situation that has roused ill feeling on the part of the Thais and makes the loyalty of the Chinese a matter of importance. Thailand has enjoyed more internal tranquility than most of Southeast Asia, but since the late 1960s, guerrilla insurrections motivated both by communist aims and by ethnic separatism have persisted in the northeast and south. In the late 1970s, the situation was further complicated by a mass movement of refugees from adjoining Cambodia. In 1993, Malay guerrillas seeking independence for Thailand's Muslim south burned schools and ambushed government troops in that area.

In spite of conflict and refugee problems, Thailand has had one of the strongest economies in Southeast Asia. From 1993–96, the economy grew a robust 8 percent annually. The country shifted away from its traditional exports of garments, textiles, cut gems, and seafood in favor of "medium-tech" industries such as computer assembly. Then came 1997, when Thailand was the first "domino" to fall in the Asian economic crisis. Thailand's currency, the *baht*, was "pegged" to the U.S. dollar, meaning that its value fluctuated in a narrow range relative to the dollar. In 1996 and 1997, the currencies of Thailand's main competitors, Japan and China, fell, putting Thailand at an increasingly competitive disadvantage as the U.S. dollar rose in value. As the costs of Thailand's exports grew, its economy weakened and its banks, which had funneled billions of dollars into questionable investments, began to look vulnerable. Currency speculators, who profit by taking advantage of declining currencies, bet that the *baht* was due for a steep decline. The government exhausted its reserves of foreign currency and raised interest rates in a failed effort to prop up the currency. The *baht* fell precipitously when its peg to the dollar was re-

moved, and the increase in interest rates put a virtual halt to the country's economic development activities. While the currency crisis and subsequent economic recession spread from Thailand throughout the region, Thailand's emerging consumer society suddenly found itself bereft. Economic growth had attracted tens of thousands of the rural poor to the factories of Bangkok, where they made garments for such major U.S. retailers as the Gap, Nike, and London Fog. Now they were laid off and joined the swelling ranks of Thailand's poor. Even before the slump, Thailand had a huge gap between the wealthy and the poor, with more than 50 percent of the country's wealth in the hands of the richest 10 percent of its population.

Through the mid-1990s, tourism was Thailand's leading source of foreign exchange. Thailand's assets and attractions are diverse. A liberal social climate has given rise to the unique phenomenon of sex tourism, with package tours catering to an international clientele, particularly Japanese men. Thailand's overall annual income from prostitution is estimated to be between $22 and $27 billion, and prostitutes send as much as $300 million each year to relatives in the villages from which they migrated. However, the spread of HIV infection among prostitutes in the red-light districts of Bangkok and the once-popular beach resorts of Pattaya and Phuket now threaten both the country's public health and its lucrative tourist trade. Thailand's government has responded with an aggressive and increasingly successful anti-AIDS public awareness campaign, which has resulted in a decrease in the number of new HIV infections every year since 1991.

Other Southeast Asian countries lag far behind Thailand in attacking the AIDS scourge. The government of the Philippines has been unsuccessful in mounting an anti-AIDS campaign because of opposition from the country's powerful and anticontraception Catholic clergy. Church officials have denounced the government's anti-AIDS program as "intrinsically evil," and have set boxes of condoms on fire at antigovernment demonstrations. Similarly, in the conservative and largely Islamic nations of Indonesia and Malaysia, Muslim clerics denounce their governments' anti-AIDS campaigns as efforts to encourage promiscuity.

Vietnam, Cambodia, and Laos

Vietnam, Cambodia, and Laos are the three states that emerged from the former French colony of Indochina. Vietnam is the largest of the three and is geographically the most complex. Three major regions have long been recognized in Vietnam: Tonkin, the northern region, consists of the very densely populated delta of the Red River and a surrounding frame of sparsely populated mountains; Annam, the central region, includes the sparsely populated Annamite Cordillera and numerous small, densely populated pockets of lowland along its seaward edge; and Cochin China, the southern region, lies mainly in the delta of the Mekong River and has a fairly high population density. Almost nine-tenths of Vietnam's people are ethnic Vietnamese, who are related closely to the Chinese in many cultural respects, including language and the shared religious elements of Confucianism, Buddhism, and ancestor veneration. A Catholic minority reflects the French occupation, and there are minor native religions. Many other minority groups complicate the region's ethnic makeup. Lao people form a minority in the mountains of northern Vietnam, and Cambodians are a minority in southern Vietnam. There is a large Chinese minority in the major cities of the three countries.

Cambodia and Laos occupy interior areas except for Cambodia's short coast on the Gulf of Thailand. They are very different countries physically, and a particular culture dominates each. Cambodia is mostly plains along and to the west of the lower Mekong River, although it has mountainous fringes to the northeast and southwest. Almost all the 11 million people of Cambodia are Cambodians, also known as Khmer people. Their language is Khmer, and their predominant religion is Buddhism.

Laos is extremely mountainous, sparsely populated, and landlocked. It has 68 identified ethnic groups, the largest of which is the ethnic Lao, who are linguistically and culturally related to the Thais, and whose dominant religion is Buddhism. Their largest numbers are in the lowlands, especially in and near the capital, Vientiane. Most of the ethnic minorities of Laos live in the mountain regions, including the country's second largest group, the Hmong.

The Vietnam War

France conquered Indochina between 1858 and 1907. The French first extinguished the Vietnamese Empire, which covered approximately the territory of today's Vietnam, and defeated the Chinese whom the Vietnamese called upon for help. Later, the French took Laos (in 1893) and western Cambodia (in 1907) from Thailand. France's administration of Indochina, centered in Hanoi (population: 1.07 million, city proper; 3.06 million, metropolitan area) and Saigon (now officially Ho Chi Minh City, but still known popularly as Saigon; population: 3.02 million, city proper; 3.92 million, metropolitan area), left a considerable cultural imprint. Its major economic success lay in opening the lower Mekong River area to commercial rice production, thus converting both Vietnam and Cambodia into important exporters of rice.

Japanese forces overran French Indochina in 1941, initiating five decades of warfare in the area. In World War II, a communist-led resistance movement was formed to carry on guerrilla warfare against the Japanese. When French forces attempted to reoccupy the area after the end of World War II, these forces fought to expel them. After eight years of warfare, the Vietnamese, with Chinese materiel support, destroyed a French army in 1954 at Dien Bien Phu in the mountains of Tonkin, and France withdrew, to end its century of colonial occupation.

Four states came into existence with France's departure. Laos and Cambodia became independent noncommunist states. North Vietnam, where resistance to France had centered,

emerged as an independent communist state. South Vietnam gained independence as a noncommunist state that received many anticommunist refugees from the north. Vietnamese Catholics were prominent among those who fled southward. This partition of Vietnam into two states (see Fig. 12.1) was largely the work of United States power and diplomacy.

From the outset, South Vietnam was a client state of the United States, while North Vietnam became allied first with communist China and then with the Soviet Union. Meanwhile, warfare in Vietnam continued, eventually to a different conclusion. Between 1954 and 1965, an insurrectionist communist force supported by North Vietnam and known as the Viet Cong achieved increasing successes in South Vietnam in an attempt to reunify the country on its own terms. A similar situation existed in Laos. Increasing intervention by the United States in South Vietnam escalated from a small scale to a large scale in 1965 and eventually involved the full-scale commitment of half a million U.S. military personnel (Fig. 12.8a). North Vietnam

A

B

Figure 12.8 (a) U.S. troops in combat in a Vietnamese village in 1972. The war inflicted enormous suffering on all sides. (b) Vietnam opened its doors to capitalism and Western goods in the mid-1990s. *(a) Johner Vie Retro Guerre DuVietnam/Gamma Liaison, (b) Noboru Komine/Photo Researchers*

responded by committing regular army forces against American and South Vietnamese troops. The United States never invaded North Vietnam, although American air strikes there were devastating. American bombs, napalm, and defoliants such as Agent Orange caused enormous damage to the natural and agricultural systems of Vietnam, destroying an estimated 5.4 million acres (2.1 million hectares) of forest and farmland in an effort described in military parlance as "denying the countryside to the enemy."

The United States avoided leading an invasion of the North in part because of the risks posed by Chinese and Soviet support of the North. In limiting the theater of ground warfare, however, the United States found itself unable to expel from the South a North Vietnamese army that was determined, skillfully commanded, increasingly better equipped by its allies, accomplished in guerrilla tactics, and willing to bear heavy losses. About one million Vietnamese died in the conflict, and about 58,000 U.S. soldiers and support staff perished. In 1973 almost all American forces were withdrawn from the costly war, which was extremely divisive at home. There are still 2238 Americans listed as "missing in action" in Indochina.

North Vietnam completed its conquest of South Vietnam in 1975. Saigon was renamed Ho Chi Minh City, after the original leader of Vietnam's Communist Party. Vietnam entered a period of relative internal peace; but there were ongoing conflicts with Vietnam's 1979 conquest of Cambodia, continued warfare against Cambodian and Laotian resistance groups, and a brief border war with China to resist Chinese aggression in 1979. Within its borders, reunited Vietnam unleashed a repression strong enough to cause a massive outpouring of refugees. This situation intensified, especially for the country's ethnic Chinese, after China invaded Vietnam in 1979.

Vietnam Today

Vietnam is now restoring its war-torn landscape through large-scale reforestation, agricultural reclamation, and nature conservation programs, aided by international organizations such as the World Wildlife Fund. Animal species new to science have been discovered in the Vu Quang Nature Reserve in the rugged Annamite Cordillera region on the border with Laos. These include two hoofed mammals—the Vu Quang ox and the giant muntjac, a new species of barking deer. Such natural treasures, along with Vietnam's tigers, rhinoceroses, and elephants, are helping to lure international ecotourists and provide much-needed revenue.

Tourism is already booming, especially in the south, augmenting a generally positive picture of Vietnam's economy (Fig. 12.8b). The communist government halted collectivized farming in 1988, turning control of land over to small farmers under 20-year lease agreements. The results have been impressive; Vietnam since 1988 has become a major exporter of rice. In addition to joining ASEAN, Vietnam in 1995 restored full diplomatic relations with the United States, and U.S. businesses

Definitions & Insights

THE BOAT PEOPLE—SAILING HOME

The desperation of many people in Vietnam and its neighboring war-torn countries created the phenomenon of the "boat people"—refugees who fled onto the open ocean on large rafts, with chances of survival variously estimated at 40 to 70 percent. After 1974, more than 1.5 million Vietnamese fled Vietnam.

The United States provided permanent refuge for more than half of those who resettled, and is now home to more than one million persons of Vietnamese descent. France, Canada, and Australia also received sizable numbers. Large numbers of Vietnamese and Laotian refugees occupy camps in China near its borders with those countries. Boat people also filled refugee camps in Hong Kong, Malaysia, Indonesia, Thailand, and the Philippines (Fig. 12.B).

The United States and other Western governments have withdrawn the welcome mat for the refugees, and the Asian countries are already undertaking or are considering measures to repatriate the Vietnamese boat people within their borders to Vietnam. Just as these destinations are closing their doors, Vietnam is undertaking political and economic reforms that are providing incentives for the expatriates to return home, and for resident Vietnamese to stay home. Since 1990, more than 800,000 overseas Vietnamese, mostly from the United States and France, have returned to Vietnam. Their repatriation has pumped welcome economic capital into Vietnam's fledgling capitalist economy. Many Vietnamese who aspired to leave their home country have heard accounts of the difficulties refugees have faced in adjusting to their adopted countries, and have chosen to remain in the hopeful nation of Vietnam.

Figure 12.B Warfare and poverty in Vietnam prompted an exodus of boat people. While some have been accommodated in new lands, others have languished in refugee camps like this one in Hong Kong. *Alan Evrard/Photo Researchers*

began scrambling for consumers in this promising market of 79 million people. The dual personalities of north and south are being perpetuated in the new economic climate, with most investment and infrastructural improvement focused in the south. Along with a growing gap between the rich and the poor, this trend could slow overall development for the country. In an effort to better integrate the nation, Vietnam has begun a costly and ambitious effort to build a new road which will run the entire length of the country.

Like that of many LDCs, however, Vietnam's progress is also threatened by the specter of uncontrolled population growth. Since the war with the United States ended in 1975, Vietnam's population has increased by more than 60 percent. It will probably reach 168 million by 2025, a staggering figure in view of the country's limited size and resource base. The government in Hanoi has mounted a family planning campaign in an effort to reduce birth rates, but so far most Vietnamese have ignored it and continue to prefer large families.

Cambodia's Killing Fields

Cambodia escaped the kinds of ills known to Vietnam until relatively late, but then suffered even worse catastrophes. After five years of fighting in the country, a communist insurgency overcame the American-backed military government in 1975. Known as the Khmer Rouge and led by a man with the *nom de guerre* of Pol Pot, this communist organization ruled the country for four years with a savagery almost unparalleled. In 1973, the population had been estimated at 7.5 million, but by 1979 it was estimated at only about 5 million. Mass murders between 1975 and 1979 eliminated nearly a third of the population. Some intended victims managed to escape as refugees from Cambodia's "killing fields" to adjoining Thailand. The Cambodian communists' stated policy was to build a "new kind of socialism" by eradicating the educated and the rich, emptying the cities, and breaking up the family. The population of Phnom Penh, the capital and by far the largest city, was expelled on a death march to the countryside. The city's population dropped from over one million in 1975 to an estimated 40,000–100,000 in 1976. By 1998, it had recovered to an estimated 300,000.

In 1979, the communist/nationalist imperialism of Vietnam came to dominate Cambodia's political landscape. There had been escalating border conflicts between the two countries, and in 1979 Vietnamese troops quickly conquered the greater part of Cambodia and installed a puppet communist government. Then the Khmer Rouge, supported by China (Vietnam had by then become a close ally of China's rival, the Soviet Union), began a guerrilla resistance against the Vietnamese. Two noncommunist Cambodian guerrilla forces also took the field. These three resistance forces often quarreled among themselves, but all maintained the conflict with the Cambodian/Vietnamese army of the Phnom Penh government, which periodically pursued them into Thailand and engaged Thai border military units. At first, the Vietnamese rejected international food aid for

Cambodia and followed a policy of systematically starving the country, apparently to weaken resistance. Later they attempted to promote recovery under their puppet government. Peace negotiations among Cambodia's contending factions led, in 1991, to an agreement resulting in the restoration of democratic government under United Nations supervision. A new constitution with new leaders took effect in 1993, and Cambodia's former king Norodom Sihanouk returned to the throne. In 1997, Hun Sen, a former Khmer Rouge leader, ousted the popularly elected royalist leader Prince Norodom Ranariddh in a violent coup. Under much international pressure, Hun Sen permitted elections to be held in 1998. The vote, which most international observers judged to be fair, left Hun Sen's political party with a majority of the parliament's seats but not the two-thirds needed to form a government. Even while Hun Sen's political opponents took to the streets to declare the elections unfair, Hun Sen began an effort to form a coalition government that would include members of the opposition. Finally succeeding in forming a government legitimately and peacefully, he secured Cambodia's membership in ASEAN and regained Cambodia's seat at the United Nations.

Even in this relatively peaceful time, the people of Cambodia must deal with a persistent scourge of war: land mines. Sown by hostile forces during years of warfare, the antipersonnel explosives remain active indefinitely. More than 35,000 Cambodians have been maimed by mines, and 200–300 more casualties occur each month. People sow new mines, imported mainly from China and Singapore, to protect their property in Cambodia.

The Lao Way

Laotians also suffer from ordnance dating to the Vietnam war. Between 1964 and 1973, U.S. warplanes dropped more than two million tons of bombs—more than the United States dropped on Germany in World War II—on the Laotian frontier with Vietnam in an effort to disrupt communist supply lines on the nearby Ho Chi Minh trail and to prevent communist troops from entering Laotian cities. Today millions of unexploded American cluster bombs remain in the region. Smaller than a tennis ball, the cluster bomb contains more than 100 steel ball bearings. Many curious Laotian children who play with them are killed or dismembered. About half of the territory of Laos may contain unexploded ordnance, according to United Nations studies, creating a serious deterrent to farming in a country with inadequate food supplies.

The administrative capital of Laos is Vientiane (with just 132,000 people), located on the Mekong River where it borders Thailand. In 1975, when Laos came under Vietnamese occupation and the Pathet Lao organization overthrew the Laotian monarchy and established a new "People's Democratic Republic," Vientiane replaced the royal capital of Luang Prabang ("Royal Holy Image"), located upriver on the Mekong. The communist dictatorship now rules Laos in a benign fashion, encouraging a market economy and foreign investment. The government calls its conservative approach to economic reform and social change "the Lao Way." To slow the potential flood of foreign world views and materials into the country, Laos has built only one bridge to Thailand across the Mekong River, and has resisted appeals for the construction of a second bridge.

Landlocked Laos is now one of the world's least developed countries, ranking 133 out of the 173 countries on the United Nations Human Development Index. About 85 percent of its people are peasant farmers. Exports are limited to timber, hydroelectric power (to Thailand), and garments and motorcycles assembled by cheap Laotian labor from imported materials. Without significant industries, trade, or a stock exchange, Laos is already too poor to have felt the severe impact of the economic crisis that shook Asia beginning in 1997. However, the crisis in neighboring Thailand has prevented Laos from embarking on projects to build more than 20 new dams which would have produced electricity for sale to Thailand. Like those of Vietnam and Cambodia, the Laotian economy has also suffered from the **brain drain** prompted by years of warfare, political turmoil, and underdevelopment—conditions which caused the best-educated and most talented of the country's inhabitants to flee. After the 1975 takeover, 343,000 Laotians, mostly members of the Hmong tribe, fled to neighboring Thailand. Many Hmong refugees subsequently emigrated to the United States.

Given the conditions of warfare, political turmoil, oppression, and inefficient economic systems, it is not surprising that the three Indochinese countries are, with Burma, the poorest in Southeast Asia. However, their outlook seems to be improving as tensions with the outer world relax, privatization of communist economies increases, the supply of goods and services improves, and the number of returning refugees exceeds emigrants.

Malaysia and Singapore

The small island of Singapore (Fig. 12.9), narrowly separated from the southern tip of the mountainous Malay Peninsula, lies at the eastern end of the Strait of Malacca, which is the major passageway through which sea traffic funnels between the Indian Ocean and the South China Sea. For centuries, European sea powers contested for control of this passageway. British control over the strait became continuous from 1824 and it was exercised from the port of Singapore, founded on the southern side of the island five years earlier. Under the British, Singapore, with its relatively central position among the islands and peninsulas of Southeast Asia, developed into not only a major naval base but also the region's major **entrepôt**. Singapore ranks with Rotterdam as the world's largest ports in tonnage of goods shipped.

From Singapore, the British gradually extended their political hold over the adjacent southern end of the Malay Peninsula. This expansion gave Britain control over the part of the penin-

Figure 12.9 Singapore, located at a crossroads where the Pacific and Indian Oceans meet, exhibits new skyscrapers of a rising business center, older buildings once occupied by the former British colonial administration (foreground), and a spacious harbor (background). *Ron McMillan/Gamma Liaison*

sula which is now included, along with northwestern Borneo (except Brunei), in the independent country of Malaysia. This southern end of the peninsula is known as Malaya, Peninsular Malaysia, or West Malaysia. The Borneo section is composed of the units (former colonies) of Sarawak and Sabah, which together are known as East Malaysia.

During the British period, Malaya developed a highly commercialized and relatively prosperous economy. Tin and rubber became the major commercial products, with oil palms and coconuts of secondary importance. Chinese and British companies pioneered the tin-mining industry in the late 19th century. The building of railroads gave access to a line of rich tin deposits along the western foothills of the mountains. These rail lines also provided transportation for rubber when new electrical and automotive industries stimulated a greatly increased demand for rubber during the early 20th century. A densely populated belt of tin mines and rubber plantations (see Fig. 12.6) developed in the foothills between Malacca and the hinterland of Penang (George Town; population: 219,000, city proper; 500,000, metropolitan area). Within this belt, the inland city of Kuala Lumpur (population: 1.15 million, city proper; 1.48 million, metropolitan area) became the leading commercial center and capital of the country (see Fig. 12.2b). The country's future capital, Putrajaya, is under construction about 25 miles (40 km) south of Kuala Lumpur.

The development of a commercial economy gave Malaya an ethnically mixed population. The native Muslim Malays played only a minor role in this development, often preferring to remain subsistence rice farmers in small coastal deltas. Chinese and Indian immigrants and their descendants became the principal farmers, wage workers, and businessmen of the tin and rubber belt. So heavy was the immigration that 26 percent of Malaysia's population today is Chinese and 7 percent is Indian. The growth of Singapore involved even heavier immigration from overseas, so that over three-fourths of Singapore's present population of 3.9 million is Chinese.

Ethnic antagonisms between Malays and Chinese have been fundamental in shaping the political geography of Malaysia and Singapore. Malaya accepted independence in 1957 only on the condition that Singapore not be included. This was to ensure that the Malays would have a majority in the new state. In 1963, however, the Federation of Malaysia was formed, composed of Malaya, Singapore, and the former British possessions of Sarawak and Sabah in sparsely populated northern Borneo. The non-Chinese majorities in Borneo were counted on to counterbalance the admission of Singapore's Chinese. This experiment in union lasted only until 1965, when Singapore was expelled from the federation and left to go its own way as an independent state.

Malaysia and Singapore are among the more economically successful of the world's formerly colonial states that have received independence since World War II. Singapore has become one of the outstanding centers of manufacturing, finance, and trade along the Pacific Rim. Its industries include oil refining, machine building, and many others. The tourists it attracts each year far outnumber its resident population. It has now reached a level of income higher than that of many of the poorer countries in Europe. Singapore suffered less than any of the other Southeast Asian nations during the region's economic crisis of the late 1990s. This was due in part to a geographical advantage: its ability to conduct trade with a vast regional hinterland without having to attend to the struggles of a large and poor rural population of its own. Its prosperity is accompanied by a strict sense of propriety. For example, chewing gum is banned and graffiti artists are flogged.

Malaysia's dependence on exports of rubber and tin has greatly lessened, although it still leads the world as an exporter of both commodities. A more diversified export pattern now includes such items as crude oil (from northern Borneo), timber, palm oil, and electronic components. Malaysia is the world's largest exporter of semiconductors and refrigerators, and has recently begun manufacturing automobiles for export. Malaysia's annual per capita income is far below Singapore's but is quite high among the LDCs and exceeds that of any other state in Southeast Asia except Brunei and Singapore (see Table 10.1).

Throughout the country the slogan "Wawasan 2000" ("Vision 2000") is emblazoned on buses, billboards, and pamphlets; it refers to the year by which national economic planners had intended Malaysia to be a fully developed country. The slogan has rung hollow since 1997, however, when, after a decade in which the economy grew at a robust 8 percent annually, Malaysia began falling into recession. The country has abandoned its ambitious and expensive development projects and

adopted economic austerity measures, including even an appeal to individuals to cut down on the number of sugar lumps added to their tea. Like the other "emerging markets," Malaysia had been benefitting from huge and often speculative foreign capital investments. And, like the others, Malaysia suffered in 1997 and 1998 when worried foreign investors rapidly withdrew their capital (see "The Dark Side of Globalization" on pp. 46–47). In 1998 Malaysia effectively pulled out of the orbit of globalization and took drastic actions to protect itself from the whims of foreign investment. It required foreign investors to wait a year before repatriating their funds. Rejecting the notion that free markets would continue to benefit his country, Malaysia's prime minister imposed strict currency controls, pegging the value of the Malaysian *ringgit* to that of the dollar. This made Malaysian money worthless overseas but ensured that Malaysian money abroad would return home. And the fixed rate of the *ringgit* made it easier for businesses to plan for the future, since they knew the value of the goods they exported and imported would not be subjected to wide fluctuations. Malaysian politicians blamed the International Monetary Fund and other institutions in the West for Asia's financial troubles, and in turn these institutions condemned Malaysia for retreating from the principles of the global free market.

Indonesia

Indonesia is by far the largest and most populous country of the region (1998 population: 207 million), and is the fourth most populous country on Earth. Its 13,600 islands comprise over two-fifths of Southeast Asia's land area and contain about two-fifths of its population. The large population of Indonesia results from an enormous concentration of people on the island of Java. About 120 million people lived there in 1998, representing nearly 60 percent of Indonesia's population and about 25 percent of that of all Southeast Asia. The island's population density is about 2193 per square mile (850 per sq km). In contrast, the remainder of Indonesia, which is about 13 times larger than Java in land area, has an average density of only about 117 per square mile (*c.* 45 per sq km).

This extraordinary concentration of population on one island is owed in part to the superior fertility of Java's soils, the best of which have been derived from materials poured out of its many volcanic peaks. However, it also has cultural and historical roots, particularly resulting from the concentration of Dutch colonial activities in Java (see following paragraph). Other islands of Indonesia (with the notable exception of Borneo) have areas of fertile volcanic soil, but such areas, although generally more densely populated than adjoining nonvolcanic areas, seldom attain the extremely high population densities found on Java. The economic development of some islands has been handicapped by their unfriendly coastlines, along which coral reefs, cliffs, and wetlands create difficulties of access.

The Dutch East India Company, after lengthy hostilities with Portuguese and English rivals and with indigenous states, secured effective control of most of Java in the 18th century. Large sections of the remaining islands, however, were not brought under colonial control until the 19th century, and only in 1904 was the conquest of northern Sumatra completed. The Netherlands undertook strenuous efforts to exploit the natural wealth of Java with the introduction of the **Culture System** in 1830. Under this system, Dutch colonizers required Javanese farmers to contribute land and labor for the production of export crops under Dutch supervision. From the Dutch point of view, this harsh system was successful. The Culture System was abolished in 1870, but indigenous commercial agriculture continued to expand along with plantation production to support increasing numbers of people. The introduction of the Culture System thus appears to have set off the enormous increase in Java's population, which is more than 20 times larger than it was a century and a half ago. Development of the other main islands began later and has been less intensive. Nevertheless, the eastern coastal plain of Sumatra, inland from the great fringing swamp, has now surpassed Java in agricultural exports.

Java's concentration of people and production and Sumatra's importance in export production are evident in the distribution of Indonesia's larger cities. Of twelve cities having estimated metropolitan populations of more than 500,000, eight are on Java. These include four "million-cities," headed by Jakarta, the country's capital and main seaport, and the largest metropolis in Southeast Asia (population: 9.16 million, city proper; 10.2 million, metropolitan area). Java's second largest city is Surabaya (population: 2.7 million, city proper). Of the remaining Indonesian cities with more than half a million people, three are in Sumatra and one (Ujung Pandang) is in Celebes (Sulawesi).

The Indonesian state has had a turbulent career. Increasing nationalism during the Japanese occupation of World War II led to a bitter struggle for independence from the Netherlands following the war. This struggle finally succeeded in 1949. Another confrontation with the Netherlands in 1962 brought western New Guinea (West Irian) out of Dutch control and into the Indonesian state. Being Papuan rather than Indonesian in culture, West Irian had not originally been included in Indonesia. However, a United Nations plebiscite in 1969 ratified Indonesian control, and relatively friendly relations have been reestablished with the Netherlands.

The presence of some 300 different ethnic groups, combined with physical fragmentation and economic problems, has made it difficult to attain peace, order, and unity in Indonesia. Although 87 percent of the population is at least nominally Muslim (Indonesia has a larger total population of Muslims than any other nation in the world), and despite the presence of an official national language (Bahasa Indonesian), great cultural diversity is reflected in the fact that over 200 languages and dialects are in use. Various groups in the outer islands have resented the dominance of the Javanese, and such animosities have escalated at times to armed insurrections. The government carried on a particularly long and bitter struggle against an in-

dependence movement led by the Catholics of East Timor—a former Portuguese possession that Indonesia occupied after the collapse of the Portuguese colonial empire in the 1970s. In 1998, Indonesia and Portugal finally reached an autonomy agreement that would give the Timorese the right to local self-government and control of educational and cultural affairs, while Indonesia would retain control over foreign and military matters and some fiscal policies. By 1999, Indonesia appeared to be considering outright independence for East Timor.

In 1965 ("The Year of Living Dangerously," as a well-known American film was titled), an attempted communist coup against the longstanding regime of Indonesian president Sukarno was unsuccessful and resulted in the massacre of about 300,000 communists and their supporters by the Indonesian army and by Islamic and nationalist groups. It also brought to power a government in which army influence was predominant and which still controls the country. For more than three decades this government's president was an Army general named Suharto, whose regime was regularly criticized for human rights abuses.

Suharto ruled Indonesia undemocratically and often repressively, but argued that he was a benevolent figure acting on behalf of his people. Indonesians generally tolerated his authority because the country's economy was advancing rapidly. The economic boom of the 1980s and early 1990s centered in the oil industry (in the hands of the state firm Pertamina) and in food production. Oil fields located mainly in Sumatra and Kalimantan produce 2 to 3 percent of the world's oil, and their production has been expanding. Crude oil and liquefied natural gas account for about 25 percent of Indonesia's exports and have tied its economy very closely to that of Japan, which is the main market for these products. Indonesia is a member of OPEC, the Organization of Petroleum Exporting Countries. In addition to oil and gas, clothing, wood products, and various tropical agricultural products are important contributions to the export trade. Tourism has also brought welcome foreign exchange earnings to Indonesia. The country's principal attraction by far is the legendary island of Bali, home of Indonesia's only remnant of the India-born Hindu culture that permeated the islands from the 4th to 16th centuries (Fig. 12.10). Bali is world-famous for its unique culture and forms of art and dance.

Indonesia's economy collapsed under the force of the spreading Asian economic crisis of the late 90s. The *rupiah*, Indonesia's currency, lost 80 percent of its value in less than a year. The Suharto govenment removed expensive subsidies on gasoline and kerosene, and the resulting protests quickly grew into vast riots. The economic troubles were compounded by a prolonged drought attributed to El Niño's warming Pacific waters, which caused significant declines in the country's rice harvest. Skyrocketing prices for rice and widespread rice shortages helped fuel widening popular discontent, which focused on two targets: First were the ethnic Chinese who represent 3 percent of Indonesia's population but control half the economy, and second was President Suharto. Rioters looted and destroyed

Figure 12.10 The cultural arts of Bali's Hindu people are world-renowned, making this Indonesian island a popular tourist destination. *George Hunter/Tony Stone Images*

hundreds of Chinese-owned shops in Indonesia's major cities, forcing thousands of ethnic Chinese to flee the country. Since the Chinese had dominated the food production business in Indonesia, food shortages intensified in the wake of the disturbances. Protestors and rioters also called increasingly for the resignation of President Suharto, on whom they also blamed their economic troubles. Bowing to the relentless pressure, and averting a potential bloodbath, Suharto resigned from office in May 1998, to be succeeded by his vice president, B. J. Habibie.

The scale and pace of Indonesia's economic decline are dizzying. The country's economic productivity has declined substantially. Millions of people have slipped below the poverty line. "People who had moved out of mud huts and into wooden shacks as the economy surged have now been thrown back into the fields," one observer wrote.[2] The nation's unemployment rate was expected to increase to as much as 40 percent in 1999. The turmoil has created a new generation of boat people, economic refugees who are fleeing Indonesia for what they perceive as better lands, notably Malaysia. When Malaysia boomed, it had welcomed Indonesians to do much of the country's work, but with its own recession, Malaysia is now turning the unwelcomed masses back.

Rapid population growth in Indonesia has compounded the effects of the economic crisis. Indonesia's population grew by 78 million, or more than 60 percent, in the 23 years from 1975 to 1998. The government lacks an aggressive family planning program and instead is exporting its "surplus" population by promoting emigration from Java and other densely populated islands to some of the sparsely inhabited outer islands of the vast Indonesian archipelago.

[2]WuDunn, Sheryl. "Asia's Tigers Lick Wounds and Worry." *New York Times*, September 16, 1998. p. C6.

The Philippines

The Philippine archipelago includes over 7000 generally mountainous islands. The two largest islands, Luzon and Mindanao—almost equal in size—account for two-thirds of the total area.

The Philippines were a Spanish colonial possession governed from Mexico from the late 16th century until 1898 when the United States achieved victory in the Spanish-American War and gained control of the Philippines. With the exception of the Japanese occupation of 1942–1944, the Philippines were controlled by the United States from 1898 until 1946, when independence was granted.

Spanish missionary activity in the Philippines succeeded in creating the only Christian nation in Asia. The country's Spanish legacy is still important. Ever since its founding in 1571, the Spanish-oriented capital of Manila (population: 1.73 million, city proper; 8.59 million, metropolitan area) has been the major metropolis and only large city of the islands. The society created by Spain was composed of a small upper class of Hispanicized Filipino landowners and a great mass of landless peasants. Problems created by this uneven distribution of agricultural land have remained as a major source of difficulty for the Philippines. Discontented peasants gave much support to a communist-led revolt after World War II. It was eventually suppressed, in large part through granting land (generally on sparsely populated Mindanao) to surrendered rebels. Revolt flared up again in the 1960s, however, and continued on a reduced scale in the 1990s.

American involvement began with the suppression of a Philippine independence movement but eventually ended with Philippine independence, along with economic development and tutelage in democracy. Preferential treatment in the U.S. market stimulated the growth of major Philippine export industries—coconut products, sugarcane, and abaca. Education in English grew so that even now English is spoken by about 45 percent of the people and serves as a bridge between many of the diverse linguistic groups of the population. After independence, Tagalog, one of the most widely used indigenous languages, became the principal base for the official national language of Pilipino, spoken by 55 percent of the populace. English also remains an official language. Both Tagalog and English are second languages for the majority of Spanish speakers.

The Philippine economy has expanded considerably since independence. Until the 1970s, the most striking growth was in agricultural exports such as coconut products, sugar, bananas, and pineapples, and in copper and other metal exports. More recently, there has been growth in manufacturing and food production. Labor-intensive manufacturing, such as that of electronic devices and clothing, now accounts for about 40 percent of all exports. Japanese and American capital, together with American agricultural science and technology, have been very important in Philippine economic development. The United States and Japan buy more than half of all Philippine exports.

PROBLEM LANDSCAPE

The Spratly Islands, Hot Zone of the South China Sea

About 60 islands make up the Spratly Island chain, which lies in the South China Sea between Vietnam and the Philippines (see Figure 12.1). They are an idyllic tourist destination, where divers can hire luxury boats to explore the coral reefs and palm-lined beaches of remote atolls. However, the islands are much more significant for their strategic location between the Pacific and Indian Oceans. During World War II, Japan used the islands as a base for attacking the Philippines and Southeast Asia. Still more significantly, as much as one trillion dollars in oil and gas may lie beneath the seabed around the Spratlys.

Not surprisingly, many nations covet control of the Spratlys. Six nations claim some or all of the islands: China, Vietnam, Taiwan, Malaysia, Brunei, and the Philippines. During the Cold War, the competing claimants felt it was too hazardous to push their claims on the islands. As the Cold War drew to a close, however, the situation became more volatile. All the contenders except Brunei placed soldiers, airstrips, and ships on the islands. In 1988 the Chinese Navy invaded seven of the islands occupied by Vietnam, killing about 70 Vietnamese soldiers. In 1992, China again landed troops in the islands and began exploring for oil in a section of the seabed claimed by Vietnam. In 1995, China moved to expand its territorial claims on islands claimed already by the Philippines. Indonesia, which has no claims on the Spratlys, is sponsoring unofficial workshops on joint efforts in oil exploration among the six claimants in an effort to defuse the emerging crisis. There are fears that as the countries' petroleum needs grow, they will seek more aggressively to gain control of the Spratlys. An incident in these remote islands could trigger a much wider and more serious conflict in Asia.

Despite encouraging progress in recent decades, the Philippine republic remains poor and potentially explosive politically, and has not joined the ranks of "Asian Tigers" or "Tiger Cubs." The country has taken a conservative approach to its development, and has borrowed money carefully to invest in modest projects, avoiding the aircraft industry, the building of a new national capital, and other such showy and expensive enterprises undertaken in other Southeast Asian nations in the 1990s. Since the mid-1970s, the Philippines has also adhered closely to the terms of loans and financial reforms imposed by the International Monetary Fund. The country did not boom, but neither did it bust in the economic crisis that rolled across Asia in 1997 and 1998.

Economic expansion in the Philippines has been hard-pressed to stay ahead of the more than 70 percent growth in population during the 23 years between 1975 and 1998. The predominant Roman Catholic culture (Catholics make up 83 percent of the population, with 11 percent adhering to other Christian denominations) encourages large families, and the Philippines has a high annual population growth rate of 2.3 percent (while food production has been increasing at only about 1 percent annually in the 1990s). Church officials have labeled the government's current promotion of family planning as "demographic imperialism" masterminded by the United States.

Compounding the problem, land and wealth continue to be extremely unevenly distributed, as Philippine society continues to be dominated by just a few hundred wealthy families. On the island of Mindanao, rebels of the Moro Islamic Liberation Front have been fighting the government in an effort to establish a separate Islamic state. Their rebellion is rooted in economic as well as political grounds, with claims that Muslims have not benefitted from the region's economic growth. Other defiant rebel populations became more moderate and entered negotiations with the government after 1986, when the country's dictator, Ferdinand Marcos, was ousted in a popular uprising and replaced by a democratically elected president,

Figure 12.11 After lying dormant for six centuries, Mount Pinatubo in the Philippines exploded into life in 1991 with a violence that caused over 350 deaths and enormous property damage. Both the Philippines and Indonesia lie in the volcanic and earthquake-prone ring of fire around the Pacific rim. *Alberto Garcia/SABA*

Corazón Aquino. Marcos and his wife, Imelda, had looted the country of billions of dollars over a period of twenty years and headed a corrupt and abusive regime virtually ignorant of the needs of the ordinary Filipino.

The United States, which has maintained friendly ties with its former colony and long depended on important military bases there, has watched with much concern the country's struggle to resolve its difficult problems. During the 1980s many Filipino legislators strongly resisted the prospect of renewing the leases under which the United States held two mil-itary bases on Luzon, Subic Bay and Clark Field. Then, in 1991, the eruption of volcanic Mount Pinatubo put an end to the debate over the fate of these two American outposts (Fig. 12.11); volcanic ash forced their closure. Pinatubo also ejected enough ash high into the Earth's atmosphere to cool temperatures globally over the following several years, at least temporarily confounding efforts to track the trend in global warming due to the greenhouse effect. Meanwhile, the U.S. base at Subic Bay was converted into a thriving free-trade port that the Filipinos hope will become Asia's "New Hong Kong."

SUMMARY WITH SELECTED KEY TERMS

- Southeast Asia is composed politically of ten states: **Burma, Thailand, Laos, Cambodia, Vietnam, Malaysia, Singapore, Indonesia, Brunei,** and the **Philippines.** With the exception of Thailand, all of these states were formerly **colonies of foreign powers.**
- Southeast Asia is **one of the world's culture hearths,** having contributed **domesticated plants** (including rice) and **animals** (including chickens) and the **achievements of civilization** to a wider world.
- This region poses **numerous environmental challenges** for its inhabitants. These include: **tropical climates, poor soils, disease, long dry seasons, deforestation, heavy erosion, tectonic processes,** and **Monsoon climate patterns.**
- The majority of the populace of this region are still **farmers,** with **manufacturing less developed.** The land is often used for **shifting, subsistence cultivation.** The densest populations exist where there are better-than-average soils.

- **Plantation cash crops** are grown for export and are the legacy of **Western colonialism.** The climate, cheap transportation, and abundance of land aided in the development of plantation agriculture.
- Southeast Asia's **tropical forests** are being destroyed at a faster rate than others of the world, mostly for **commercial logging** for Japanese markets.
- Southeast Asia as a whole is **among the world's poorer regions.** However, Singapore, Malaysia, and Thailand, because of their high rates of industrial productivity and economic growth, have often been described as **"Asian Tigers" or "Tiger Cubs."**
- **Considerable mineral wealth** including oil, gas, and a wide variety of metal-bearing ores are extracted in many locations, and many **unexploited reserves** remain. The most serious deficiency is the **near absence of high-grade coal.**

REVIEW QUESTIONS

1. Using maps and the text, locate the countries of this region and identify the major physiographic features.
2. List the major Indian and Chinese traits that have been especially strong in shaping this region.
3. Using maps and the text, explain why this region has been considered relatively isolated.
4. List the members of ASEAN and explain why the coalition was formed.
5. Using the maps and text, identify the areas of densest population for this region and relate those to agriculture.

6. List the typical plantation cash crops and locate where these crops are produced.
7. Using the maps and text, list and locate the prominent commercial rice production areas.
8. List the major mineral resources of this region and locate their production areas.
9. Locate and explain the significance of the "Golden Triangle."
10. Explain how colonialism has had such a lasting effect on this region.

DISCUSSION QUESTIONS

1. What are the two distinctly populated areas of Burma? What is the ethnic population makeup of Burma? What is the political atmosphere like in Burma?
2. What is the ethnic population makeup of Thailand? What is the political atmosphere like in Thailand? Why is tourism such a strong economic draw in Thailand?
3. What are the three major regions of Vietnam? What is its present political atmosphere like? What is the ethnic population makeup of Vietnam?
4. What is the ethnic population makeup of Cambodia? What is the political atmosphere like in Cambodia?

5. What is the ethnic population makeup of Laos? What is the political atmosphere like in Laos? What problems, most resulting from warfare, are still being suffered in Laos?
6. Why are Malaysia and Singapore considered to be more successful than most formerly colonial states?
7. What are the most distinctive physical, cultural, and political attributes of Malaysia?
8. What are some of the foreign countries that have been influential in the Phillipines' development?
9. Where and how did the Asian economic crisis begin?
10. Which countries have been most and least affected by the region's economic turmoil?

Chapter 13

China: Enormous Ambition, Enormous Struggle

▲ This urban scene from near the center of Beijing possesses all the architectural and functional elements that characterize China's headlong rush into a tourist-welcoming, globally connected, and architecturally varied new century. Ever since Deng Xiaoping's welcome to Western trade, technology, and elements of popular culture, China—particularly Beijing and Shanghai—has been a billboard for East Asian showcasing of the Asian capacity to demonstrate the international diffusion of urban change elements. The scene shows the tri-wheeled pedi-cab that evolved from the traditional rickshaw, the ubiquitous bicycle, the tourist bus, and the increasingly present private automobile.   Cameraman International, Ltd.

CHAPTER OUTLINE

13.1 Area and Population

13.2 Physical Setting and Major Landscape Elements

13.3 Continuity and Ongoing Change in China

13.4 Landscape Legacies from the Maoist Revolution

13.5 China's Urban Centers

13.6 China of the 21st Century

13.7 Taiwan

13.8 Mongolia

So often images of China are built almost entirely around population issues. While there is real dimension to having the world's largest population—as China does—the real drama of China relates to the ways in which its people have made their landscape accommodate such populations. From the earliest dynasties dating back to the second millennium B.C.—and the monumental construction of China's Great Wall in the third century B.C.—to the present, the story of China has been one of steady manipulation of a not particularly fertile landscape. China is a nation that has maintained a distinctly Chinese perspective for nearly 4000 years, despite major foreign infusion through Mongol, Manchu, and Western control of all or part of the country.

Outsiders have long had a fascination with China. The West has sent missionaries, businesspeople, young scholars hopeful of understanding this giant of the East, and—now—countless tourists. There is a grandeur to China that we have long appreciated. It has long been on the horizon of interest and opportunity for people of the West. In the past decades it has had a special importance for merchants of the United States.

China currently engages in active industrial and economic development in an effort to pull away from its past as a monumental agricultural nation. At the same time, the need to feed its population and raise fiber for its enormous textile base forces it to stay attentive to growth in the rural sector as well. With the reclaiming of Hong Kong in 1997 after a century and a half of British control, and with relations with the United States ever uncertain, China faces a broad array of demands on its social and economic fabric. Not only is China continually contested by the forces of nature, such as the annual rise and fall of the Chang Jiang (river) and the typhoons hugging the face of the East Asian coast, but it must deal with the insurgencies that have played an ongoing role in fomenting instability in this enormous nation. The unfolding Chinese drama represents an effort to resolve the tension between enormous ambition and enormous struggle. And, particularly in light of the recent Asian economic slowdown and potential for serious recession, China has continued to make monumental efforts to guarantee its continuing economic growth and market development—on both domestic and foreign fronts.

13.1 Area and Population

China has the world's largest total national population, estimated at 1.25 billion (excluding Taiwan) in 1997—over one-fifth of the world's people—and increasing by 13 million per year. Population is a very serious matter for an industrializing but underdeveloped country whose area of about 3.6 million square miles (9.3 million sq km) is only slightly larger than that of the United States (Fig. 13.1), but whose inhabitants outnumber the United States population nearly 4.62 to one. This population lives in a state officially called the People's Republic of China (PRC), which plays a major role in world affairs (Fig. 13.2). It is, nonetheless, still relatively poor and largely agricul-

Figure 13.1 China compared in latitude and area with the conterminous United States.

337

China and Mongolia • Index Map

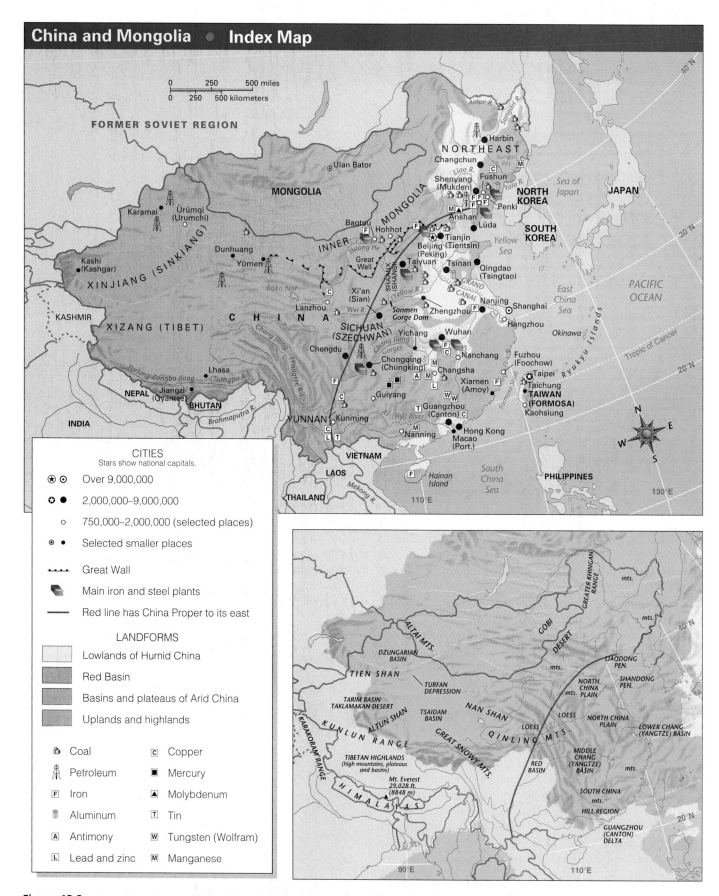

Figure 13.2 General location map of the Chinese realm.

tural despite major industrial development and economic growth under the Communist regime that came to power in 1949 (see Table 10.1).

Over 870 million Chinese (about 70 percent of China's population) live in rural settlements and are largely engaged in and supported by agriculture. This accumulation of peasant humanity must wrest food for itself and for the ever expanding urban population from a landscape where nine-tenths of the land is not in cultivation because of steep slopes, dryness, or short growing seasons. China's arable area, concentrated in the plains and river valleys of humid China, is about half the arable area of the United States, but, through **multiple-cropping** practices (growing successive crops on a given field throughout the year), the sown acreage may actually be larger than the area farmed in the United States. China's arable area provides, on the average, about one-third of an acre of arable land per person to the agricultural population, and only one-fourth of an acre per person to the total population.

In the past decade, China has grown at an average of 9 percent annually in terms of economic development. It has now engineered the world's third largest economy. This growth has a major urban component, but it is also built around 25 million rural industrial enterprises with more than 125 million employees. This base amounts to some 40 percent of China's industrial labor force. These rural-based, small-scale industries generate some 50 percent of China's annual industrial output, 25 percent of its total economic value, and more than one-third of China's exports. The magnitude of this growth is shown by the fact that in 1979, trade between the United States and China amounted to $2.4 billion while in 1998 it had grown to $49.7 billion. With numbers like this—even with the regional downswing of economic health in the late 1990s—China has a possibility of becoming the world's largest economy sometime after 2010. China is a nation of major economic and demographic magnitude.

Population Policy

China's success in reducing population growth in the past decade has been one of the global success stories in population management (Fig. 13.3), though not without controversy and serious social impact. The drive for smaller families—begun seriously in the late 1970s—was motivated by the fact that even with China's impressive recovery from war in the 1950s, there was only a modest increase in China's per capita food output for two decades. Increases in agricultural output were largely matched by rapid population growth during the time when the Mao Zedong regime made little or no effort to curb the birth rate (Fig. 13.4).

In the early years of the revolution, Mao felt that every new pair of hands born to China could be a productive addition to the country's economic infrastructure. Then, toward the end of Mao's life in 1976, the regime instituted one of the most stringent programs of birth control in the world. This has culminated

Figure 13.3 The Chinese government has been promoting "one couple, one child" for over two decades in an effort to slow down the annual population increase. This poster exhorts young couples to observe this constraint with the phrase "For the sake of a prosperous today and a beautiful tomorrow, limit your family to one child." *Forrest Anderson / Gamma Liaison*

in the "one child campaign," which aims to limit the number of children per married couple to one. Accordingly, the government continually exhorts its young families to limit family size to a single child. It takes note of individual family birthing patterns—especially in the cities—and maintains surveillance through local authorities. It dispenses free birth-control devices, free sterilization operations, free hospital care in delivery, free medical care for the child, free education for the child,

Figure 13.4 This 10-feet-high painting of Chairman Mao Zedong (Mao Tse-tung) maintains its position atop the central gateway to the massive complex of The Forbidden City in Beijing. Mao appears to look out on Tiananmen Square, nearly 100 acres of open space cleared by the removal of makeshift housing in the years following the victory of the Communists over the Kuomintang in 1949. It was in this square in June 1989 that China defined its disinclination to allow democracy to grow by the massacre of unarmed students who were protesting governmental restrictions on public assembly. *Kit Salter*

an extra month's salary each year for the parents, and other favors and preferences to induce compliance with the one-child-per-family norm. Those who violate the norm are subjected to constant social and political pressure, denial of the privileges accorded one-child families, pay cuts, and fines. Women who become pregnant without permission are pressured to have a free state-supplied abortion, with a paid vacation provided.

Under such programs, China's birth rate declined by 1995 to 1 percent, the lowest in any major less developed country (LDC), and its rate of natural population increase dropped to about 60 percent of the average for other LDCs. However, China's base population is so large that even the relatively low birth rate achieved by 1995 still meant a net increase of approximately one million new mouths a month, or a gain of approximately the urban population of New York City annually! This is in a country where the death rate has also been drastically lowered by better food availability, better transportation, and improved and more readily available public health facilities. The regime hopes to continue to decrease childbearing to the extent that population in the 21st century will stabilize at 1.3 billion people, but some population experts are projecting a total of 1.4 billion by the year 2010, or 1.5 billion by 2025. The higher number may be more likely since, all through the 1990s, government officials were reporting widespread disregard of the "one couple, one child" policy. This was particularly true in the countryside as peasants became wealthier and hence more ready to pay the fines in a continuing personal quest for larger families, or to produce a son if their one child was a female—or simply because governmental policies seem never to be enforced as closely in the countryside as in the city.

Urban populations have been more responsive to the continual exhortation for couples to have only one child. At the same time, however, the rapid growth of the urban population in China and the increasing number of women in the urban work force have led to decreasing birth rates in the cities because of the increasing importance of material goods and new—for China—lifestyle patterns that do not necessarily revolve around parenting.

There is also a new class of child called "the little emperor," who is a single child—most often male—much loved, and often spoiled. Historically there were usually many siblings to diffuse parental care across the lives of a number of children. Now, with few siblings and an improvement in the standard of living, China is fearful of the power and consequence of having so many children growing up without the social benefit of an extended family and sibling interaction.

There is, as well, the growing social concern about who will care for the aging parents who may not have a family of children to support them in their old age. As family size diminishes, and as the one couple, one child policy is ever more fully followed in urban China, the traditional safety net of extended families and children to care for seniors vanishes. If, for example, a young urban couple has the allowed one child, and that child is female, the likelihood is quite good that she will marry and become—by Chinese tradition—a part of the husband's family to such an extent that she (or they) may not be available to care for her parents as they get older. Since China has traditionally used the family as the accommodation of choice for senior citizens, either a major shift in social services and associated taxation will be necessary, or a full generation of seniors will find no easy setting for their final years. By 1998, China already had more than 115 million people over the age of 60, and that population segment was growing annually as the health of China's people improved steadily.

These unexpected social outcomes of a successful family-planning campaign serve to remind people that virtually every change in one social domain has the capacity to send ripples of influence across all or at least many other areas of social concern. Population planning is particularly likely to have such varied impacts. Figure 13.5 is a map that shows the spatial distribution of the Chinese population.

13.2 Physical Setting and Major Landscape Elements

China is vast in size but, like all extremely large countries, contains a great deal of unproductive and thinly settled land. In China, this land lies mainly in the western half of the country. Here are huge outlying areas principally composed of high mountains and plateaus, together with arid or semiarid plains, where rainfall is generally insufficient for agriculture. This dry, sparsely settled country probably contains about 5 percent of China's population (see Fig. 13.5)

The arid interior area is in marked contrast to the better-watered, more densely settled eastern core of the country. Thus, China is divided approximately into a western half, or **Arid China**, and an eastern core region, or **Humid China**, in which the country's population, developed resources, and productive capacity are heavily concentrated. A rough boundary between the two major divisions is an arc drawn from the northeastern corner of India to the northern tip of China's Northeast. This line corresponds in a general way to the stretch of land whose average annual rainfall is 20 inches (c. 50 cm), with Arid China to the west and Humid China to the east.

Arid China

The principal regions of Arid China are the Tibetan Highlands, Xinjiang (Sinkiang), and Nei Mongol (Inner Mongolia). All three are identified historically with non-Han populations, although now—after four decades of politically promoted domestic migration away from the east coast urban centers toward the borders with Russia—Han Chinese comprise the great majority in Nei Mongol and may become a majority in Xinjiang.

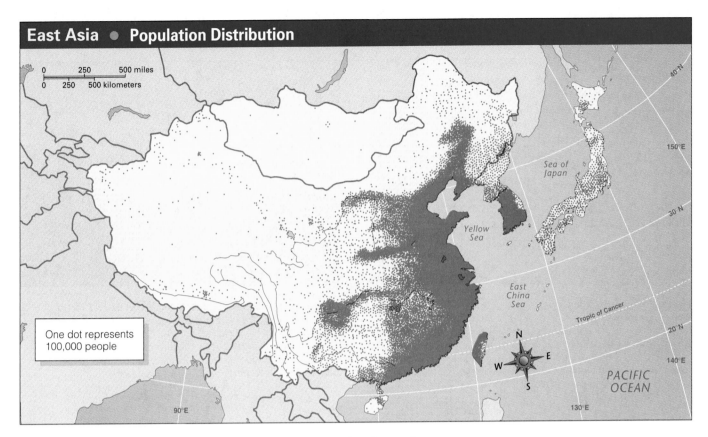

One dot represents 100,000 people

Figure 13.5 This dot map of population distribution in East Asia shows the influence of waterways, coastlines, and mountain systems in the shaping of human activities. Coastal areas with open access to the adjacent seas get washed with monsoonal precipitation, as do mountain flanks on the windward side. These areas of relatively high precipitation have long favored settlement and agricultural activity. River systems have historically provided transportation, fishing resources, and water resources for settlement and irrigation. Upland and interior areas are often too arid for dense settlement but such locales may play pivotal roles as market places, transportation hubs, and sometimes religious sites. Population distribution in China and East Asia reflects these very basic geographic realities all across the map. *After Nelson et al.* Human Geography. *1995. p. 280.*

Tibetan Highlands

The very thinly inhabited Tibetan Highlands (see Fig. 13.5) occupy about one-fourth of China's area. Included are Tibet itself and fringes of adjoining provinces. Most of this vast area is a very high, barren, and mountainous plateau averaging nearly three miles (*c.* 5000 m) in elevation. To the northwest, high basins of internal drainage are common, some containing large salt lakes. To the southeast, the plateau is cut into ridge and canyon country by the upper courses of great rivers such as the Tsangpo (the Brahmaputra of India), the Mekong, and China's own Chang Jiang (Yangtze Kiang or Yangtze River) and Huang He (Hwang Ho or Yellow River). It is along the Chang Jiang's flow through the eastern edge of this plateau—in movement toward the well-settled middle Chang Jiang basins of central China—that the Chinese have begun to focus their plans to capture and control the power of this greatest river of China, which travels more than 3400 miles in its descent from the Tibetan Highlands through the Three Gorges Project landscape, to its delta in the Shanghai area in the east (see Problem Landscape, p. 346).

Lower elevations, warmer temperatures, and greater precipitation in parts of the southeastern plateau result in some extensive grasslands and stands of conifers. Around the edges of the Tibetan Highlands are huge mountain ranges: the Himalayas on the south; the Karakoram and Kunlun Shan to the north and northwest; and the Qinling Shan and others to the east.

The people of the Highlands, numbering approximately 4.6 million (2.1 million in the political area called the Tibetan Autonomous Region) are predominantly sedentary farmers and animal herders (Fig. 13.6). A few hardy crops, especially barley and root crops, are basic to agriculture in this restrictive environment. In grasslands at higher elevations, a nomadic minority graze their flocks of yaks, sheep, and goats. Yaks are particularly important not only as durable beasts of burden in these high elevations, but also as providers of meat, milk and butter, leather, and hair traditionally woven into cloth.

The Tibetan Highlands are under Chinese political control, and Lhasa, the capital, now has more than 310,000 Han Chinese residents. This province's population—beyond the capital—is comprised predominantly of Tibetans, who are distinguished by

341

Figure 13.6 Yaks continue to play both a ceremonial and a utilitarian role in the farmlands of Tibet. They do better in the high plateaus of this region than any of the traditional lowland farm animals. The banners are a reminder of the traditionally significant role these animals have played in this land of harsh farming conditions. Even though China has made some effort to introduce small-scale farm machinery to its agriculture, traditional patterns still persist—especially in the distant upland regions. *D.E. Cox/Tony Stone Images*

Figure 13.7 This is a scene from Dunhuang in the arid interior of China in western Gansu Province. This oasis setting was a major stop on the Silk Road that has connected northern China with the Mediterranean region for more than 2000 years. Dunhuang has now become an important tourist destination because of the discovery of scores of caves with delicate religious paintings on their walls. *Kit Salter*

their own language and by their adherence to Lamaism, the Tibetan variant of Buddhism. Prior to the Communist era, China had sometimes exerted loose control over Tibet, but Chinese authority vanished with the overthrow of the Manchus and the Qing Dynasty in 1911. From 1912 until the Chinese Communists' conquest of 1951, Tibet existed as an independent state, with its capital and main religious center at Lhasa. After the Chinese completed roads to Tibet in late 1954, an increase in restrictive measures by the Communist government contributed to the rise of guerrilla warfare. This culminated in a large-scale Tibetan revolt in 1959 and the flight of the Dalai Lama (the spiritual and political head of Lamaism) and many other refugees to India. Subsequently, the Chinese drove most of the monks from their monasteries, expropriated the large monastic landholdings, and made institutional changes similar to those designed to implement socialism and prohibit organized religion in the rest of China. Tibetan resistance to Chinese rule and aggressive repressions by the Chinese continue.

Xinjiang (Sinkiang)

Xinjiang adjoins the Tibetan Highlands on the north. It has an area of roughly 635,000 square miles (*c.* 1.6 million sq km) and a population of 16.2 million (as of 1998). It consists of two great basins—the Tarim Basin to the south and the Dzungarian Basin to the north (see Fig. 13. 2). These basins are separated by the lofty Tien Shan range. The Tarim Basin is rimmed to the south by the mountains bordering Tibet, and the Dzungarian Basin is enclosed on the north by the Altai Shan and other

ranges along the southern border of Russia and Mongolia. Both basins are arid or semiarid. The Tarim Basin is particularly dry, since it is almost completely enclosed by high mountains that block off rain-bearing winds. This basin includes the Taklamakan Desert—which, translated, means "Once you get in, you'll never get out"—perhaps the driest region in Asia. The basin varies in elevation from 2000 to 6000 feet (about 600–1800 m) above sea level; the smaller adjoining Turfan Depression drops to 928 feet (283 m) below sea level.

The great majority of Xinjiang's population is concentrated in oases, located mainly at points around the edges of the basins where streams from the mountains enter the basin floors. For many centuries, these oases were stations on caravan routes crossing central Asia from Humid China toward the Middle East and Europe on the historic Silk Road (Fig. 13.7). Under the People's Republic, the ancient routes have been superseded by modern transportation. A railroad connects Lanzhou (Lanchow; population: 1.6 million), a major industrial center and supply base in northwestern China proper, with Urumqi (Urumchi; population: 1.2 million), the capital of Xinjiang; several major roads link the oasis cities, and several major airfields have been built in the region.

The transport links are key elements in a drive to expand the economic significance of this remote part of China and bring it under firmer political control. Early in the Communist era, demobilized soldiers and urban youths from eastern China were organized into quasi-military production and construction divisions to push expansion of irrigated land on state-operated farms in this arid landscape. Now cotton is a major crop. The region also has deposits of coal, petroleum, iron ore, and other minerals, and a considerable expansion of mining and manufacturing has occurred under the auspices of the People's Republic.

This region has been valuable for both its isolation and its resources. It was in Xinjiang that the Chinese exploded their first nuclear bomb in 1964. It continues to be the most mineral-rich province in China and has the potential for considerable political as well as economic activity in the decades to come.

Nei Mongol

North and northwest of the Great Wall, rolling uplands, barren mountains, and parched basins stretch into the arid interior of Asia. Here lie slightly more than a million square miles (*c.* 2.6 million sq km) of dry, thinly grassed terrain that is divided about equally between Nei Mongol (Inner Mongolia), the area nearest the Great Wall, and the now-independent country called Mongolia. An overwhelming majority of this Autonomous Region's present population (23 million in 1998) is Han Chinese. This population is concentrated in irrigated areas along and near the great bend of the Huang He. By far the greater amount of territory of Nei Mongol is still the habitat of a very sparse population of Mongol herdsmen. It is from these severe, arid landscapes that the Mongols emerged to control China and an expanse of land that ranged from the Korean peninsula in the east to the margins of Poland in the west in the 13th century. The desert landscape displays the shadows of an extraordinary history of Asia, through its remnants of the Chinese Great Wall, caravan oases and routes, and temples hidden in desert caves at the end (or the beginning) of the long treks to and from distant markets (Fig. 13.A, p. 344).

The traditional picture of nomadic Mongol tribesmen herding their flocks of sheep and goats and using camels and horses for riding and pack purposes is fast disappearing. The economic core of Nei Mongol today is composed of the agricultural areas near the Huang. The irrigated areas, which have been expanded by the People's Republic, produce mainly oats and spring wheat, while unirrigated fields are used mainly for drought-resistant millet and kaoliang—a sorghum grain used for feed and liquor. The main city of the area is Baotou (Paotow; population: 1.2 million), an expanding industrial center on the Huang. One of China's major iron and steel works was opened at Baotou with Soviet help in the 1950s, using nearby resources of coal and iron ore.

Humid China

Humid China, sometimes referred to as Eastern or **Monsoon China**—the core region of the country—includes the densely settled parts of China south of the Great Wall and in the Northeast. The ancient provinces south of the Wall are often referred to by outsiders as **"China proper,"** although the Chinese seem not to employ the term (see Fig. 13.2). China proper includes

two major divisions, North China and South China, which differ from each other in various physical, economic, and cultural respects. North China and South China are realms of major cultural and physical differentiation. As in the United States, there is a strong sense of regional identity that radiates from both of these worlds.

North China

Affected mightily by the monsoon wind patterns that characterize so much of this Asian region, North China has hot, humid, and rain-filled summers and correspondingly cold and dry winters. When the winter monsoon blows out from interior Asia, long, dark, cold winter days dominate North China. Because of the historic disinclination to have interior heating in homes in this region, winters are particularly bitter as people attempt to break the cold with just the heat from the cooking stove or, traditionally, the heated bed called the *kang*.

Although the region receives between 15 and 21 inches of precipitation annually, it is plagued with relatively high variability from year to year (Fig. 13.8). Such a rainfall pattern has led to a history of droughts and floods, reminding one again of the burden of being a densely settled area of low precipitation and high variability. The droughts that have swept across North China have been instrumental in the initiation of the massive Chang Jiang Water Transfer Project, which is designed to bring surplus water from the Three Gorges region of the Chang Jiang through aqueducts to the heavily farmed but water-deficient North China Plain.

Floods in North China almost all relate to the continual problem of managing the Huang He (Yellow River). It is around this 2500-mile-long river and the Wei He tributary (east of Xi'an) that Chinese civilization was founded. The Huang drops down from its source on the Tibetan Plateau and courses for more than a thousand miles through the loess soils of North China. These airborne soils blown from the Gobi Desert to the west and northwest are very poorly structured and they collapse easily into the river system. The river carries this yellow-colored loess soil as silt to the Yellow Sea in the east. It has been this physical process that has built up the broad, fertile, and relatively level North China Plain. Chinese recorded history chronicles 26 major channel changes and different discharge points to the Yellow Sea in the past 1500 years.

Floods are particularly significant on the Huang He. The river carries an enormous quantity of silt that is deposited along the final 500–600 miles of the river's flow through the Tai Hang Mountains on the western edge of the North China Plain to its delta. This sedimentation has elevated the actual stream channel of the river *above* the surrounding landscape. The Huang flows within a critical diking system all the way from the Tai Hang Mountains to the current delta north of the Shandong peninsula. When the dikes break during times of flood, the river pours out of the channel and down to the lower, densely settled

Regional Perspective

The Silk Road

In the early Han Dynasty (from 206 B.C. to about 8 A.D.) the Chinese began a steady exchange of goods for items from the Mediterranean trading ports. From China went silk (not raised in the West until the sixth century), and to China came wool, gold, silver, and glass. The overland route for this exchange—called the Silk Road—was approximately 4000 miles (6000 km) long, going from X'ian (the early Chinese capital of Chang'an)

to Damascus along the Great Wall and the Taklamakan Desert, over the Pamir Mountains, and through Baghdad to the Mediterranean Sea. From there goods were shipped to various European ports. It was this route that played a role in the diffusion of Buddhism into China in the first century A.D. Almost no one ever followed the whole route, for travel was dangerous and every caravan load had to pass numerous checkpoints and pay con-

siderable tolls. Venetian explorer **Marco Polo** is sometimes identified as the first person to travel the complete route, and it was, in part, the power of his description of this grand overland route of trade, diffusion, and cultural exchange that made Christopher Columbus so determined to find a sea route to replace it. The route is currently being considered as a possible base for a United Nations trans-Asian international highway (see Fig. 13.A).

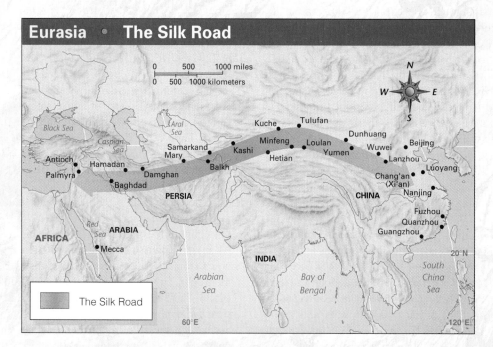

Figure 13.A   The Silk Road (sometimes called The Silk Route) is the nearly 4000-mile link that connected China with European markets and civilization several centuries before the time of Christ. This series of partial trails and routes was almost never traveled by a single person or caravan, but rather was a trade route for many caravan groups each crossing specific terrain. Silks, paper, art goods, and spices went west from China while gold and silver bullion, glass goods, and metalware were major goods going east. The Silk Road has been a major diffusion route of religions, language, and customs as well.

farmlands and cities of the North China Plain. Even after the ruptured dike is repaired and the river is contained once again, the great flood of water that issued forth from the elevated stream bed has nowhere to go. Such widespread ponding and floods have been associated with the Huang He and the North China Plain for centuries. It is for this reason that the Huang He is also called "China's Sorrow."

The major crops in North China are winter wheat, millet, and kaoliang, with some land given over to summer rice crops and corn and a wide variety of kitchen vegetables. North China has exported wheat to South China since the seventh century A.D. Historically much trade has been carried along the Grand Canal, a civil engineering project of the Sui Dynasty (581–618 A.D.). This 1050-mile-long (1700 km) inland waterway linked

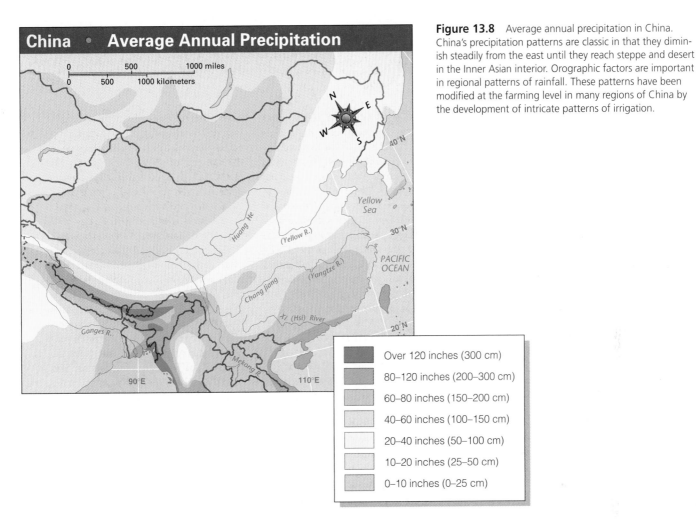

Figure 13.8 Average annual precipitation in China. China's precipitation patterns are classic in that they diminish steadily from the east until they reach steppe and desert in the Inner Asian interior. Orographic factors are important in regional patterns of rainfall. These patterns have been modified at the farming level in many regions of China by the development of intricate patterns of irrigation.

Hangzhou in the Chang Jiang delta region to just southwest of Beijing. Wheat and coal were barged southward and rice was carried north. Although the Grand Canal has never attained a status equal to the image of the Great Wall, it has come much closer to achieving its initial design goals than has the much more widely known Great Wall.

South China

The Qinling Mountains, at approximately 10,000 feet (3048 m), have historically served as the accepted dividing line between North and South China. These mountains run east to west and divide stream drainage between the Huang He to the north and the Chang Jiang to the south. The agricultural dominance of South China comes from the 3400-mile-long (c. 5440 km) Chang Jiang that flows from its origin in the Tibetan Plateau to the delta just north of Shanghai on the east. The more than 700,000 square miles (1,813,000 sq km) are drained by this largest of Chinese rivers. Historically, the Chang Jiang has served as a major east-west transit corridor for China. It can accommodate freighters that draw ten feet all the way from the East China Sea to Wuhan in the central basin. From Wuhan, the

river can be traveled by smaller freighters to Yichang. At that point there historically has been a classic **break-of-bulk** point for further passage through the Chang Jiang gorges: Freight and passengers were off-loaded and put on flat-bottomed junks and sampans which were drawn upstream by human trackers who pulled and rowed the junks through the Three Gorges (see Problem Landscape, p. 346) to Wanxian in Sichuan Province. Today there is steady traffic of tourist ferries and small freighters going up and down the Chang Jiang through the gorges.

The two major basins of the Chang Jiang east of the gorges are broad, densely settled areas with agricultural populations, small towns, and some expanding industrial cities (Wuhan is the most significant of them). The Chang Jiang has long presented the Chinese with difficult settlement decisions, however. Because of the fertile soils and the irrigation benefits associated with residence in the great river floodplains, these lowlands have long been settled and farmed. Such geographic benefits of soil and river transportation, however, always have another side to them—just as parallel environments do in the United States and other nations with heavily settled floodplains. The Chang Jiang Water Transfer Project (Three Gorges Dam) is the latest in Chinese efforts to harness the enormous strength of this

PROBLEM LANDSCAPE

"The Pulse of China"

"THE YANGTZE IS A RAW, UNCONTROLLED and sometimes destructive river sweeping through China, bringing riches to many, but drowning others in its floods. And like the river, change is flooding through the nation. Fortunes are being made in private businesses, but millions of workers who depended totally on their jobs with state enterprises are being laid off. More millions are roaming the country trying to find work. Corruption has become malignant, and environmental damage is off the end of the scale." (From Terry McCarthy, "The Pulse of China." *Time,* June 29, 1998. 31–35.)

It is fitting that *Time* journalist Terry McCarthy uses the Yangtze metaphor to describe the great flow of change that has been washing over China. The river is central to China's sense of identity and has been for well over two thousand years. The "Great River"—the Chang Jiang—has long been central to a Chinese sense of place. This river drains more than 700,000 square miles and has seasonal shifts in river depth of well over 100 feet in the gorges that lie between Sichuan Province in the west and the broad agricultural basins of central China to the east. The current Chang Jiang Water Transfer Project—called the Three Gorges Dam

for short—is the latest and largest effort made by the Chinese to control the enormous power and potential of that river.

In 1919, Sun Yat-sen (sometimes called the George Washington of China) suggested that a major dam be built to help control the Chang Jiang. One goal of the contemporary project is to build a series of dams so strong and high that the flow of the river water through the heavily populated provinces east of the gorges can be evened out, reducing or eliminating flood threats to a region that has suffered major inundations for thousands of years. There is also the goal to create a hydroelectric project that would produce more power than many of China's dams combined. The Three Gorges Dam will be 610 feet high, 1.3 miles (2.1 km) long and the reservoir will extend upstream for 385 miles (620 km). The boldness and scale of this landscape modification project make it akin to the building of the Great Wall and the Grand Canal in the eyes of the Chinese. There has been much disapproval—both domestic and foreign—of the plan because of its potential for environmental disturbance to a whole network of water and plant ecosystems that rely upon the Chang Jiang. The World Bank, after expending $8.7 million for a feasibility study, decided not to fund any part of the project. In May 1996, the U.S.

Export-Import Bank also refused to provide any funding for the megadam project. In view of China's major flooding in the summer of 1998, there is also the claim that the major project has already taken so much capital and administrative energy that ordinary projects that would have helped to diminish the impact of these recent floods were not undertaken.

The project is also disfavored because of the number of villages and archaeological sites that will be drowned by the reservoir that will develop behind this tallest of all China dams (Fig. 13B). It is estimated that completion of the project will displace some 1.2 to 1.4 million people and submerge 13 cities, including the lower half of Wanxian (a city of more than 314,000), 350 to 700 villages, and more than 115,000 acres of arable farmland. The Three Gorges Dam project must be seen as a political statement—as well as an engineering hallmark—further manifesting China's wish to reshape its present and to create a more productive future, with electric power and flood control as a major agent in such transformation—regardless of the potential governmental or environmental cost. The ultimate accounting of the costs and the benefits of this water project will stretch far into the 21st century.

waterway and turn it into a hydroelectric power source and an irrigation and settlement resource, not only for the middle and eastern basins of the Chang Jiang, but also possibly for the North China Plain through an elaborate canal project.

The Red Basin—at the western end of the passage through the gorges of the Chang Jiang—contains more than 110.8 million people. Nature has provided little level land in the basin, but rice and other crops are grown on enormous numbers of small fields in narrow ribbons of valley land and on laboriously terraced hillsides. The Red Basin served as the center of the Chinese government during World War II because of its isola-

tion and distance from Japanese bombers. It continues today to seem almost like an independent nation because of its environmental resources, distance from central government, and long history of relative self-sufficiency.

South of the Chang Jiang basins is the real cultural heartland of South China: the delta region at the mouth of the Xijiang in the Guangdong lowlands. Its central features are the city of Guangzhou (Canton; population: 3.92 million, city proper) and the numerous small, densely farmed and settled lowlands on the Xijiang and its tributaries and **distributaries** (the smaller channels by which a river takes its silt and water

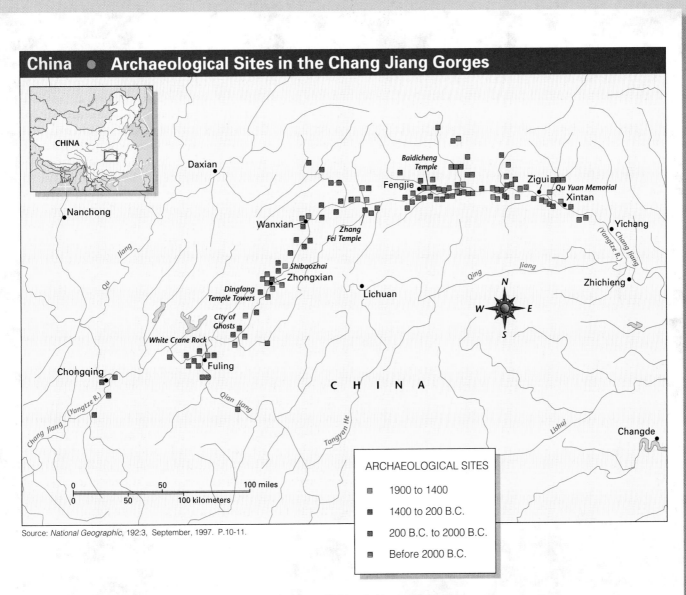

China ● Archaeological Sites in the Chang Jiang Gorges

Source: *National Geographic*, 192:3, September, 1997. P.10-11.

ARCHAEOLOGICAL SITES

- 1900 to 1400
- 1400 to 200 B.C.
- 200 B.C. to 2000 B.C.
- Before 2000 B.C.

Figure 13.B The Chinese damming of the Chiang Jiang (Yangtze River) is called by some China's new Great Wall. The scale and landscape implications of the multiple dam project along hundreds of miles of the river as it winds through the mountains west of the Middle Basin of the Yangtze make it far overshadow any water project undertaken in China. This map shows the extent of the project and the variety of archaeologic sites to be lost from the ponding of the river's water behind the world's tallest dam in Xiling Gorge. The planning, initial construction, budget, and environmental ramifications of the Three Gorges Dam Project have been hotly contested for more than a decade—and some would say for nearly a century. However, China's government continues to stick to its plan to bring the Yangtze under more complete control, to increase China's capacity to generate electricity, and to possibly develop a water transfer scheme that might move some of the Yangtze flow to the water-hungry North China Plain. The project will not be completed until well into the 21st century.

out to sea). This whole area was not brought under Chinese control until the Qin and Han dynasties (*c.* 200 B.C.), or nearly 1000 years after China proper had developed into a viable political state.

The rivers are short, swift, and unnavigable, and each basin constitutes a unit isolated except on the east. Farming is restricted to small and scattered valley lands and adjacent mountain flanks. The coast is dotted with fishing hamlets. This is the only section of China in which the people have taken much interest in seafaring.

The other dimension of South China is the mountainous landscape that lies west of the Guangzhou lowlands. These hills—mostly in Yunnan and Kweizhou provinces—are part of a deeply dissected 6000-foot-high limestone plateau. It is here that some of the world's most dramatic **karst** landscape features (limestone hills with caves and sinkholes) are found (Fig. 13.9). The area is also home to many of China's small minority populations who continue to sustain themselves with an intensive self-sufficiency in hillside farming, one aspect of which, because of the isolation of the region and its distance from the seat of national authority, has been active cultivation of the opium poppy.

China's Northeast

The Northeast—better known to the West as Manchuria—is the part of Humid China lying north and east of the Great Wall.

Figure 13.9 This is a classic "karst landscape." This geographic term, meaning *a limestone landscape with numerous caves and sinks and sharply sided mountains and hills,* is Serbo-Croat in its origin, but this region in China is one of the most dramatic examples of these physical landscape features anywhere in the world. The flatlands are rich in lime but the broken topography caused by the sinks and depressions presents problems to farmers and settlements. In southwest China, karst landscapes have been important as tourist draws, especially in the region shown here in Guilin, Guangxi-Zhuang Autonomous Region.   *Kit Salter*

It is made up of the three provinces of Liaoning, Jilin, and Heilungjiang. It has an area of about 310,000 square miles (*c.* 800,000 sq km) and a population of approximately 105 million (1996). To the west are the Mongolian steppes and deserts; to the east, the area is separated from the Sea of Japan by the Korean peninsula and by Russian territory.

The Manchu emperors of Qing Dynasty China (1644–1911) kept their homeland relatively closed to the Chinese until the very end of their rule. The political instability of East Asia in the early 20th century led the Manchus to encourage Chinese to migrate to the relatively open lands of the Northeast. Millions of Chinese farmers moved to these provinces, pioneered its agricultural frontier, and gave it an overwhelming majority of ethnic Chinese at the outset of the Sino-Japanese War in 1937. During the 20th century, this region became a subject of contention and an area of conflict among China, Russia, and Japan, all of which have controlled it at one time or another. Russian interest has focused on the Chinese Eastern Railway—a shortcut across the northern two provinces between Vladivostok and the Trans-Siberian Railroad east of Lake Baikal—and on the naval and port facilities of Lüda, which the Russians have used at various times.

Japan has long been interested in developing the Northeast's impressive industrial resources and its potential for surplus food production. After the Russo-Japanese War of 1904–1905, fought mainly in Liaoning, Japan became increasingly active in the area and took over actual control in 1931 with the establishment of the puppet state of Manchukuo. In 1945, during the final days of World War II, Soviet forces took it from the Japanese. Eventually, the Chinese Communists incorporated the Northeast into the People's Republic, beginning this region's role as the center of broadly expanded industrial development. Although significant factory plant and equipment was uprooted and shipped back to industrial centers in the former USSR by the Russians at the very end of World War II, the industrial and transportation infrastructure created by the Japanese in the 1930s and 1940s during their occupation was an important asset for the PRC in its later economic development of this region after the ouster of the Chinese Nationalists in 1949.

It was the northward movement—toward Northeast China —of the United Nations troops on the Korean peninsula in 1950 that brought the Chinese army into the Korean Conflict. The North Korean army was being pushed back nearly to the Yalu River—the border between Northeast China and North Korea—when the Chinese decided to step in and stop this military advance on their newly developing industrial heartland in Liaoning. The Communists' geopolitical and military decision is not surprising when you understand the industrial importance the Northeast played for revolutionary China.

The Northeast consists of a broad, rolling central plain surrounded by a frame of mountains that seldom rise higher than 6000 feet (a little over 1800 m). The mountains on the east and north contain much valuable hardwood and softwood timber. The central plain, oriented northeast-southwest, is approximately 600 miles long (966 km) by 200 to 400 miles wide. The soils are very fertile, and the summer rainfall is generally sufficient for the crops that are grown, although spring drought is often troublesome. Conditions are not favorable for irrigated rice, and the winters are generally too severe for winter grains. However, soybeans, kaoliang, millet, corn, and spring wheat do well during the relatively short but warm frost-free season of 150 to 180 days. There is a much greater possibility for future mechanization of agriculture here than in the parts of Humid China south of the Great Wall.

The industries of the region are centered mainly in or near the largest city, Shenyang, in Liaoning. The largest center of iron and steel production is Anshan (population: 2.5 million, city proper), south of Shenyang. The industry is based on deposits of iron ore in a belt that crosses the southern part of the region. Substantial deposits of coal also exist, although the total reserves are far smaller than those of China south of the Great Wall, and only a small fraction of Manchurian coal is suitable for coking. The Daqing oil fields are located northwest of Shenyang in Heilongjiang Province. These fields have been developed into one of China's richest oil production centers since their discovery and energetic development in the 1960s.

13.3 Continuity and Ongoing Change in China

Scholars estimate that the Chinese had already become a major political entity by the time of the Shang Dynasty (*c.* 1523–1027 B.C.). At this early time, Chinese culture had already developed an agricultural system able to support large numbers of people on small areas of intensely worked arable land. During the Shang Dynasty the great alluvial plain of the Huang He in North China already was occupied by Chinese peasants with a garden type of agriculture based on the hoe, irrigated crops, varied fertilizers including night soil (human wastes), intensive use of flat alluvial lands, terracing of fertile hillsides to create workable farm plots, a high degree of control over water, private ownership of land, use of family labor (Fig. 13. 10), and siting of villages on less fertile land. Little attempt was made to farm land not fertile enough to yield a living to families equipped only with hoes. From its origins in the Huang He basin, this system spread throughout the humid eastern part of the country by a process of diffusion that included migrations and conquests over a period of many centuries.

In East Asia, China historically has been as dominant politically as it has been culturally, but this political dominance has often alternated with political disunity and weakness. Unification of the Chinese people into one empire was first accomplished by the Qin Dynasty in the third century B.C.; after that time, periods of strong central government, generally occurring soon after the beginning of a new dynasty, alternated with peri-

ods of weak central government and internal warfare. During periods of unity and strong central power, China generally extended its authority widely, both by direct conquest and by accepting the fealty and tribute of lesser powers around it. Under the Han Dynasty (206 B.C.–220 A.D.), for example, Chinese armies pushed far westward across central Asia and seem to have barely missed contact with the Romans in the vicinity of the Caspian Sea. There was also the steady presence of the Chinese who traveled with the caravans and Chinese goods on the Silk Route. For millennia China has been a presence in Europe because of the great prices paid for goods from the Middle Kingdom—the name that China has given to its country, the middle realm between heaven and earth (see Fig. 13.A).

Under the Yuan (Mongol) Dynasty (1280–1367), the Chinese empire controlled Burma, Indochina, Korea, Manchuria, and Tibet. Under the Manchu (Qing) Dynasty (1644–1911), the Chinese held large areas that eventually were added to the Asian part of the Russian empire. But strong dynasties tended to weaken with time, and then internal disorder, partition, and civil war generally followed. During such periods of weakness, parts of China often were conquered by nomadic invaders (such as Mongols and Manchus) from the north and northwest. Nevertheless, through all the rise and fall of dynasties, a distinctive Chinese civilization persisted and intensified.

China's Push to Industrialize

Mass production of industrial goods did not begin in China until the beginning of the 20th century. Prior to that time, Imperial China's handicraft industries produced high-quality goods that

Figure 13.10 The treadmill used here to pump water from an irrigation canal to the fields is one example of the arduous manual labor still applied in massive amounts to Chinese agriculture. China has learned the value of labor-intensive agriculture partly because of its need to support and utilize such a large population. *Paolo Koch/Photo Researchers*

attracted Western traders in spite of China's official disinterest in such trade. To this day, handicraft and shop-scale industries continue to supply many of China's simpler needs.

China's early factory industries were largely foreign-owned and were attracted to China by the huge market for cheap goods, by cheap labor, and sometimes by certain natural resources. As in so many countries, the first large-scale factory industry to develop was the manufacture of cotton textiles. Japanese and other foreign firms, together with some native Chinese companies, developed a major cotton industry by the 1930s. This industry, and others, developed in such coastal cities as Shanghai and Tianjin. Meanwhile, the Japanese used the minerals of the Northeast (in their puppet state of Manchukuo) to make that area the first center of heavy industry in China in the 1930s.

In their efforts to further industrialize, Communist China directed national economic development in four distinct national drives:

1. **Communization and repair of wartime damage, 1949–1952.** The industrial structure, which had been badly hurt by invasion and civil war from 1937 to 1949, was brought back to prewar production levels under the new system of state ownership. This was done even while China was expending considerable capital and manpower in its involvement in the Korean Conflict.

2. **Heavy-industry development using the Soviet pattern, 1952–1960.** Aided by the Soviet Union, Chinese planners gave priority to coal, oil, iron and steel, electricity, cement, certain machine-building industries, and armaments. China's resources for these industries are considerable, and the country's centrally **planned economy**—a system in which major economic activity is initiated by government planners—achieved major successes in developing them, as has been the case in many Communist countries. There was also a spatial component to this drive, for the Soviets advised the Chinese to diminish the coastal concentration of industrial plants. This led to expansion into new industrial landscapes at Wuhan on the Chang Jiang, at Baotou in Nei Mongol, and even into Inner Asia near the [then] Soviet border in Xinjiang. Pre-Communist-era coal production had peaked at 66 million metric tons per year under Japanese military control in 1942, but by 1958, China's output was 270 million tons and the country ranked third in the world after the USSR and the United States. Subsequently, output increased to 1116 million tons in 1992 (with the United States at 947 million tons), making China the world's largest coal producer. Meanwhile, a maximum pre-Communist-era steel production of 900,000 tons per year was multiplied by a factor of nine, to 8 million tons, by 1958. By 1992, steel output had reached 81 million tons, placing China ahead of Russia (58 million) for third position in the world, but behind Japan (98 million) and the United States (93 million).

Progress in heavy industry has been built on a rich resource base. China has coal resources estimated at about one-tenth of the world total. They are widely spread, but the major producing fields lie under the North China Plain and its bordering hills and in Liaoning Province in the Northeast. Iron-ore resources are less abundant, but there are workable deposits in a number of locations, with the largest in Liaoning. The southern region of the Northeast has been the country's leading heavy-industrial region since modern industries were developed there by the Japanese, but widespread coal and ore deposits have facilitated Chinese Communist development or expansion of these industries at other scattered centers in China proper, in Nei Mongol, and along the Chang Jiang. China's large production of many alloys and a number of other metals has further facilitated development. The discovery and development of major new oil fields in the Northeast, Gansu, Sichuan, and in the western reaches of Xinjiang province have significantly enriched China's resource base in the past four decades. China has gone from being a petroleum importer in the early 1960s to now serving as a major Asian exporter of petroleum products.

3. **Costly political innovation under late Maoism, 1960–1978.** This phase of Communist China's industrial development began with the country's rift with the Soviet Union in 1958–1960 and continued until new economic policies were introduced after the death of Mao in the late 1970s and early 1980s. Progress continued in heavy industry, but the country's general industrial advance slowed in pace, technical development, quality of output, and modernization as a result of (a) near isolation from more advanced industrial countries, (b) the "Cultural Revolution" begun by Mao in 1966 and carried on into the early 1970s, and (c) the customary difficulties that nonmarket, centrally planned economies on the Soviet model have had in developing consumer-oriented industries and advancing technology.

The Cultural Revolution (1966–1972) was Mao's attempt to maintain an ongoing "permanent revolution" in China. Politically the movement was driven by Mao in his effort to throw a newly established urban and intellectual elite off balance by promoting purges of virtually anyone who had offended Communist Party functionaries. Local youth who donned the red armbands of the youthful Red Guards became the change agents in this campaign to unsettle the new bureaucracies. The Red Guards were loosed on the populace to harass those who were accused of putting learning, skill, expertise, or personal matters above revolutionary enthusiasm or Marxist/Maoist goals and principles. Government oppression in this campaign is thought to have led to a million assassinations, executions, or suicides, and to have provided lifetime jobs for those deemed most loyal to the Maoist revolution, regardless of their job performance. Reforms initiated in the late 1970s finally began the process of removing such people from often unproductive positions, but people whose lives were turned upside down by the Cultural Revolution continue to speak of themselves as a lost generation.

Geographically the Cultural Revolution was a revolution in mobility as well. For the first time in the lives of many Chinese, it was suddenly acceptable to travel randomly about the country, taking advantage of free rides on buses, trains, and trucks for the purpose of "making revolution." This meant that Chinese who had lived within known worlds of very small scale were now making journeys all across China in an effort to root out the libraries and lifestyles of people who had wandered, or who were accused of wandering, from the purist revolutionary route that Chairman Mao had outlined.

4. **The impact of the Deng Xiaoping reform era, 1978–1999.** After a short period of internal confusion after the 1976 death of Mao Zedong, another Long March (see Definitions and Insights, opposite) veteran—Deng Xiaoping—ascended to supreme power in China. Under his leadership, China introduced the **Four Modernizations** that focused the nation's energy on accelerated development of agriculture, industry, science and technology, and national defense. Since the late 1980s, China has done a complete about-face in its interaction with the West. China has emerged as one of the world leaders in making the economic transition from a LDC to a rapidly developing economy. Even though the per capita GNP of the Chinese is still relatively low, the overall magnitude of economic activity in China has taken the nation to the third most productive economy in the world.

These four major shifts in governmental control launched China on a pathway of economic change that has brought significant benefit to rural and city folk alike. For example, before the Deng Xiaoping reforms, China had six washing machines for every 100 families. The number now is more than 90. Before Deng, there was one refrigerator per hundred households; now more than half of the Chinese households have one. Not only has the life of the average Chinese been transformed, but light industrial capacity in China is expanding at an unprecedented rate to satisfy both domestic markets and expanding export needs.

An enormous array of manufactured goods is produced in China, including some relatively sophisticated products such as those needed to maintain the country's nuclear arsenal. But China's trade pattern is still that of a country in the early stages of industrialization. It emphasizes export of labor-intensive manufactured goods along with resources or related products such as textiles and clothing, oil, and oil products. Total exports in 1996 had a value of $151 billion, with Hong Kong, Japan, the United States, and the EU receiving more than 75 percent of that goods flow. The total value of China's exports increased to $20 billion in 1982, and $80 billion in 1993. In 1994, exports finally began to approach the $150 billion total. The scattered oil and gas deposits in various parts of the country now account for approximately 5 percent of world oil production, which is enough to supply the country's own low per capita usage and

Definitions & Insights

MAO ZEDONG

Much of the history of China in this century has been shaped by the influence and leadership of Mao Zedong (Mao Tse-tung). He was one of the founders of the Chinese Communist Party (CCP) and in the 1920s was actively involved in efforts to organize and improve life for China's peasants. In 1927 the CCP split from the Kuomintang (Nationalists) and China began a bloody civil war that was to last until 1949. Mao was the major leader in the 1934–35 **Long March** (a more than 3000-mile flight on foot of nearly 100,000 Communists being pursued by Nationalist soldiers) and worked steadily to have China's military fight the Japanese rather than fight among themselves during the Second Sino-Japanese War (1937–1945). With the defeat of Japan in World War II, the CCP and the Nationalists returned to their civil war and, in 1949, two million Chinese Nationalists under Chiang Kai-shek fled to Taiwan and Mao and the CCP claimed victory. From then until Mao's death in 1976, Mao led the People's Republic of China through a demanding and often counterproductive series of experiments in social (re)organization in the Chinese countryside: In 1957 he initiated the Great Leap Forward and the Chinese Commune Movement, and in the mid-1960s he stimulated the Cultural Revolution, all the while making efforts to expand and increase the productivity of China's industries as well. By the time he died, China had regained a global prominence that Mao felt it had been denied for a century and a half under Western colonialism and interference in the nation's domestic affairs. Two years after Mao's death, Deng Xiaoping gained the role of major authority in China and, for the first time since 1949, China opened its doors to the West and began to tailor its economic efforts, devoting more attention to personal incentives for agricultural and industrial success. With this came a new attention to the production of consumer goods and a relaxation of Maoist domestic migration control. Nevertheless, even though many of Mao's favorite programs have been disbanded, he continues to be a figurehead of major significance in contemporary China.

also provide a major export. The enormous potential of both the Chinese market and China's labor force has caused foreign investors to endure high office-space costs, a complex and largely unyielding bureaucracy, and wide governmental swings in political attitudes toward joint ventures. The country's main trading partners and its primary sources of foreign investment and technology are Japan and the United States, with Hong Kong continuing to play the role of an effective middleman between China and the outside world. The $139 billion worth of imports that entered China in 1996 come largely (about two-thirds) from Japan, the EU, Taiwan, the United States, and South Korea.

13.4 Landscape Legacies from the Maoist Revolution

Politically, China, like virtually all nations, has alternated between periods of strength and periods of weakness. But, despite these fluctuations, it has continued to be the home of the dominant culture of East Asia since very ancient times. Surrounding countries and regions show evidence of the long-standing influence of Chinese culture on their languages and writing, religion, agriculture, arts, crafts, institutions, ways of thinking, and histories. Chinese creativity is attested to by many early pioneering Chinese inventions and discoveries, such as gunpowder, paper, the wheelbarrow, the magnetic compass, and their system of writing. It is also attested to by the organizational and engineering skills that were necessary for the building of the Great Wall (much of which was completed before the end of the first millennium B.C.) and the subsequent seventh-century construction of the Grand Canal to link the Chang Jiang valley with the populations and resources of the North China Plain.

Landscape modification and intensification of existing agricultural techniques. China has attacked the problem of agricultural production by intensifying the use of the existing agricultural base in order to raise yields per unit of land. Irrigation and flood-control measures, when applied to land already farmed, may be seen as one phase of this effort. Another approach lies in the application of scientific knowledge and experimentation to develop better seed, equipment, and farming techniques, together with the development of administrative and propaganda machinery to promote their adoption. A third approach lies in increased fertilization. Chinese farmers have long fertilized their fields with animal and human manures, pond mud, plant residues and ashes, oilseed cakes, and green manure crops. Although these traditional fertilizers are still by far the most important source of plant nutrients, the current regime also stresses the use of chemical fertilizers. China's relatively small chemical-fertilizer industry has been rapidly expanded by the purchase of entire plants from abroad. However, the amounts used by Chinese farmers are still relatively small.

Land reclamation. One thing that has often impressed observers in China is the small use that is made of low-yielding second-rate land, even in places where population is excessively crowded on adjacent good land. For instance, in South China, narrow valleys often teem with an overcrowded and struggling population, but broader adjacent uplands are little used. Apparently the Chinese farmer and his family, unequipped with machinery, cannot till a sufficient amount of second-rate land by hand labor to make a living.

However, in the early 1950s, the Communists placed considerable stress on land reclamation and on increasing the amount of land available for potential cultivation. Small gains in the amount of cultivated land were made through reclama-

tion of poorly drained land and extension of irrigation into deserts. However, there has actually been a decrease in the amount of land under cultivation, owing largely to the expansion of the urban base in China. This decrease in farmland suggests the difficulties of the land-reclamation approach, and in recent years the Chinese have scaled down their estimates of reclaimable land.

Mechanization, crop breeding, control of diseases and pests, and water control. Other methods to increase farm output are not yet of first importance but may be crucial in the future. Mechanization has developed slowly. The use of tractors is growing, but primarily in the Northeast and in the west where more extensive farm fields are located. Thus, there has been progress compared to the situation existing in 1949, but, by and large, the rural scene still is dominated by human and animal power and simple hand tools.

Progress also has been reported in crop breeding and in the control of insects and plant diseases, but little definitive information is available. The lack of trained agricultural scientists and technicians has been a hindrance. There is evidence that insecticides and herbicides are coming into wider use. Pest control was greatly publicized during the early years of the People's Republic, with great campaigns to eradicate rats, mice, flies, sparrows, and other pests. Observers were amazed at the sight of Chinese peasants fanatically hunting down pests, propelled by the exhortation of the government often through ubiquitous radio broadcasts into farm fields from speakers attached to telephone poles connecting Chinese villages to the larger world. A major objective of these campaigns was to reduce food losses caused by the pests, which had been credited with the destruction of many millions of tons of grain each year. Antipest campaigns against flies, mosquitoes, and rats have had, in addition, an obvious public health benefit. The evidence suggests that these programs have had at least a limited success. They also were politically important as they demonstrated the government's capacity to marshal enormous numbers of people in a wide variety of national campaigns—many of which were focused upon modification of the environment.

Projects to control and conserve water and to expand the irrigated acreage have been actively promoted. Floods and droughts have always periodically afflicted Chinese crop production, especially in North China. In the past, they have caused millions of deaths from flood and famine. Thus, a major objective is to prevent water from running off in floods and to conserve it in reservoirs for times when it is needed. Three specific types of projects have been and are being carried out to achieve this: (1) construction of large numbers of small-scale dams, ponds, canals, and dikes by local communities; (2) building of large-scale dams and associated structures to control larger waterways, several projects of which are complete or under way (see Problem Landscape, p. 346); and (3) reforestation, which also relates closely to water control. Only 14 percent of China is forested, and large areas are without significant vegetative cover. Most forest growth in China proper has long since been removed, primarily by peasants for use as fuel.

A major objective of water conservancy measures has been to increase the amount of land under irrigation and thereby increase output. China claims to have increased the amount of arable land under irrigation by 200 percent since 1949. A considerable increase has been achieved, many of the existing irrigation works have been improved, and use of power equipment has increased the efficiency of irrigation in some areas. However, in arid areas there continues to be a problem with salinization because of the inability to gain enough water to not only nourish the crops but to wash the salts from the irrigated land as well.

The overall significance of this problem of the arable base in China is highlighted by two statistics. In 1950, at the outset of the Maoist/Marxist Revolution, the per capita amount of arable was 2.7 *mu* (a *mu* is one-fifteenth of a hectare). By 1995 that had been reduced to 1.2 *mu*, even with all of China's efforts to expand the country's arable base. Yet, at the same time, the average per capita grain production had grown from 209 kilo-

grams per year to 380 kilograms per year through increases in agricultural productivity. The current plan of the Chinese government is to open an average of 500 million *mu* of new farmland annually between 1990 and 2000 to compensate for the continuing transformation of productive farmland into industrial and even residential lands as China's urban landscape grows steadily outward from its contemporary cities.

Improvement of transportation. To integrate China's widespread resources and widely scattered producing and consuming centers into a functioning whole, the government has placed great importance on improving the transportation system. Although trucks now play an important role, a good deal of internal commerce is still distributed by human porters, bicycles, pack animals, wheelbarrows, or carts drawn by animals or drawn and pushed by humans. The country has a growing highway system, now several times the length of the serviceable roads existing in 1949. However, much of it consists of earth roads repeatedly lost to heavy seasonal rains. Strategic new roads have been built in western China—for example, to Tibet, the Sino-Indian border, and Nepal. China had extended the 1949 total of 50,000 miles in the national road system more than 700,000 miles (1,120,000 km) by the mid-1990s.

Water transportation was the main long-distance mover of Chinese goods in past times. The Chang Jiang and its many tributaries provided a massive, branching system reaching into many sections of the country. The Grand Canal linked the agriculturally rich lands along the Chang to the locus of political power in North China. But only one-third of the waterway network is in the industrially developed North and Northeast, and not more than one-fourth of the total length of waterways can be utilized by cargo vessels other than small junks and sampans.

The major effort of the Communist government thus far has been directed to the improvement and extension of railways. Prior to 1949, the Northeast had a fairly extensive rail system, developed by Tsarist Russia and expanded by Japan. At the end of World War II, China's North and Northeast contained about 75 percent of China's rail mileage. Most of China south of the Chang Jiang floodplain and all of the western provinces were without rail lines. But the Communists, as a part of their extensive program of railway construction, have made South China accessible by building the first three bridges ever to cross the Chang Jiang. Meanwhile, rail lines have been pushed not only into South China and the Red Basin but also into Xinjiang and across Mongolia to connect with the Trans-Siberian Railroad. Since 1949, the length of China's rail network has nearly tripled, yet that total still is only about one-half of the network of China's navigable waterways (see Fig. 13.11).

Institutional and organizational innovations. Since the Communist Party gained control of China in 1949, it has mandated a series of reorganizations of Chinese agriculture in successive attempts to solve the agricultural problem within the limits of Communist ideology and control.

Transportation of China

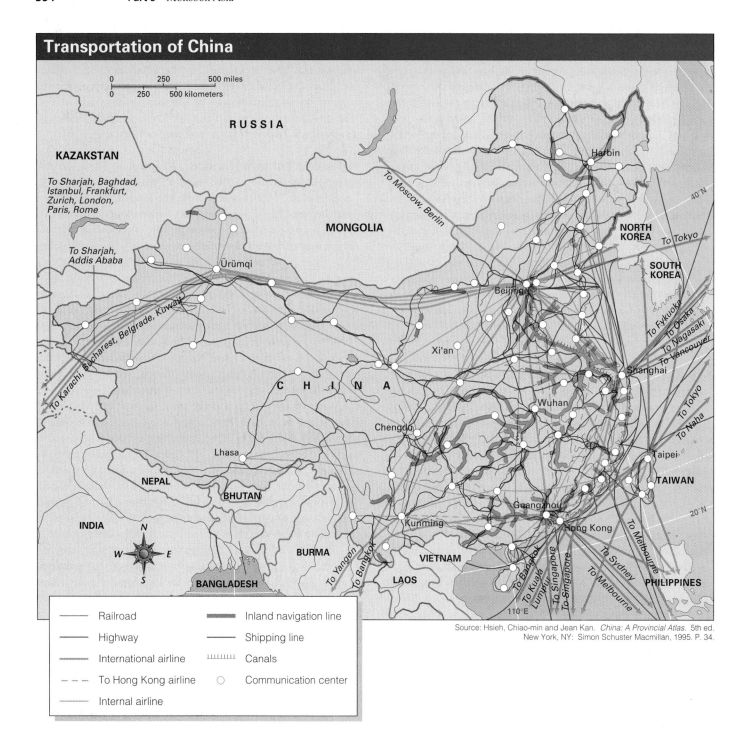

Source: Hsieh, Chiao-min and Jean Kan. *China: A Provincial Atlas.* 5th ed. New York, NY: Simon Schuster Macmillan, 1995. P. 34.

Figure 13.11   One of China's most ambitious efforts in expanding its infrastructural base for continuing economic development has been its transportation system. This 1995 map shows a much-expanded highway, airline, and railroad network. China has long had a well-developed internal waterways and canal system, but even it has been extended through dam construction. The capacity to move goods, raw materials, and people continues to be at the heart of effective economic development. This is a lesson that China is taking very seriously.

With the communes and the other efforts already noted, changes in production failed to achieve the momentum desired by government planners. After the 1978 return of Deng Xiaoping to the major leadership role in the Communist Party, free-market reforms began to take place in China. The results of this experiment with a freer semicapitalist system in agriculture have been good. Chinese agricultural output expanded by approximately 50 percent during the 1980s and 1990s, and rapidly enough—when coupled with the steady decrease in birth rate—to generate an increase of 20 percent in the per capita food supply.

However, bloody military suppression of demonstrations in Beijing in 1989 (centering in Tiananmen Square in Beijing) raised questions about the future. The demonstrators demanded political democracy in addition to the increased economic freedoms. The violent reaction from the Chinese government raised the possibility of China's moving backward into a less free economy in order to preserve the power of the Communist Party and leadership. Since Tiananmen, the Chinese leadership has grown more belligerent about Western efforts to promote democracy and human rights in China. Even with the continued expansion of Western economic influences in China's growing economy, there continues to be a steady tension between the People's Republic and the Western nations that annually increase their import of Chinese goods and the amount of capital they invest in the development of Chinese joint ventures. This serves as yet another example of the tension between ambition and struggle in China.

13.5 China's Urban Centers

Besides an enormous rural population inhabiting an estimated one million villages and hamlets, China has many large cities. China's approximately 375 million urban residents exceeds the total population of the United States and presents a resource and a liability for Chinese economic development and planning. The locations of the country's 12 largest metropolitan cities in geographic clusters illuminates the spatial structure of a major component of China's economy. Table 13.1 shows the enormous growth in urban settlement in China between 1980 and the late 1990s. The growth of mid-sized cities of under 500,000 is particularly remarkable.

Chang Jiang Cities

The Chang Jiang valley city-group is a string of cities along and near the river. This group includes five of China's 12 largest cities, a fact that reflects the collective importance of the Chang Jiang basins as a major axis of the national economy. At the western end of the axis in the Red Basin is Chengdu (population: 3.5 million), a provincial capital and node for transport lines connecting the axis with Arid China. It lies away from the

Table 13.1 Growth in Urban Settlement in China		
Number of cities and towns	**1980**	**1995**
Cities total, of which:	223	640
population over 1 m	15	32
0.5–1.0 m	30	43
0.2–0.5 m	70	192
under 0.2 m	108	373
Towns	2,870	16,992

Source: China News Analysis; *The Economist,* March 8, 1997. P. 18.

river. On the river is another Red Basin metropolis, Chongqing (population: 3.12 million). Downstream lie Wuhan (population: 3.86 million) in the Middle Basin—an important manufacturing, transportation, and agricultural marketing center—and Nanjing (population: 2.6 million) in the Lower Basin. Nanjing has served as national capital for China and is a major agricultural marketing center as well. On a navigable tributary (the Huang Po Creek) near the mouth of the Chang in the Lower Basin is the port of Shanghai (population: 8.2 million; Fig. 13.12). With its location at the seaward end of this urban axis, it serves the whole Chang valley and other areas. Built from a fishing village by foreign enterprise in the 19th century, Shanghai is the largest city and port in China. Fifteen percent of China's tax revenues derive from Shanghai and its industrial and financial sectors. This is the city that reflected the good and the bad of the Western influences in China in the 19th and early 20th centuries. Even today, it is a showcase for impressive joint ventures in Chinese urban architecture and design. Thus, the Chang valley city-group includes a metropolis for each productive basin, plus an overseas global connecting point as well as a transfer and connecting point for interior China.

North China Cities

The cities of North China include the national capital of Beijing (population: 7.36 million), the port city of Tianjin (Tientsin; population: 5.8 million), and far to the southwest, the very ancient city of Xi'an (Sian; population: 2.87 million). Each is located at an entryway to the North China Plain: Tianjin, by the sea; Beijing, near the end of a pass leading through the uplands to the north toward Mongolia and the Great Wall; and Xi'an, in the Wei River valley, which leads upstream to interior China and downstream to the Huang valley and the North China Plain. Beijing is central to all administrative matters in China, being both the national capital and the center of major governmental offices and decision making. A recent addition to the cluster of cities in North China is Zibo, lying east of the Huang He in its delta region. It has grown to a population of 2.48 million (city proper), nearly doubling in size in the past decade.

Figure 13.12 Shanghai, China's biggest city and one of the largest in the world, is in steady growth in an effort to accommodate all of the domestic Chinese who have decided to leave the countryside and find work and life in this port city. There is also an increasing pressure on the city by foreign businesspeople who want to locate in this city that lies between the administrative capital in the north and the commercial centers on the southeast coast in China. Old settlement features like one- and two-story buildings as seen here are continually threatened by new skyscraper construction.   *Kit Salter*

Cities of the Northeast

The Northeast includes another three of China's 12 largest cities: the port of Lüda (formerly Dairen–Port Arthur, with a population of 2.28 million); Shenyang (Mukden; population: 4.65 million), the main metropolis of a group of heavy-industry centers; and Harbin (population: 3.6 million), a rail center with connections to Russia and Korea. Since the Communists took over in 1949 and expanded China's industrial base, these cities of the Northeast have lost some of their relative industrial importance, but they continue to be very significant centers of manufacturing, chemical production, and—increasingly—agricultural marketing and processing.

Cities South of the Chang Jiang

This region south of the Chang Jiang has only one of the 12 largest cities, the port of Guangzhou (Canton; population: 3.92 million, city proper) in the Xi River delta. The lack of major cities in South China reflects in part the regional dominance of Hong Kong (population: 6.2 million) in the region's trade and manufacturing, and partly the relatively small size of the dense lowland rural regions. However, Guangzhou and its vicinity now form a rising center of new industries, particularly around the site of Shenzhen, a new city and Special Economic Zone (population: 875,176), with much financing from Hong Kong and beyond (Fig. 13.13).

13.6 China of the 21st Century

China has undergone extraordinary shifts in economic activity, politics, and demography in the past five decades. From the end of a very costly and divisive civil war in 1949, to the bloody Tiananmen Square massacre in 1989, to the achievement of a decade of unprecedented economic growth by the late 1990s, China has kept a steady hand on economic activity, urban growth, agricultural innovations, and China's central role in what—until the strong economic downturn in East Asia in 1997–1998—was being called the forthcoming Pacific Century (Fig. 13.14). Just as it played a dominant role in the development of an emergent Asian identity and cultural singularity three millennia ago, China now hopes to play a parallel role in the next century's global attention on East Asia.

China has focused great energy on the productive reabsorption of Hong Kong into its national fabric, and it is conscious that Portugal's much smaller colony of Macao will rejoin China in December 1999, adding another 550,000 people, nearly all of Chinese origin, to the rich and rapidly developing coastline near the famous Hong Kong harbor. Macao supports itself largely by transit trade with China, fishing, and tourism, including gambling. It remains to be seen what role it will play upon completion of the return at the very end of the century.

The problem of regional imbalance, however, continues to play a steady subtheme in Beijing's management of the nation's energetic economic growth. The classic 14th-century novel *Romance of the Three Kingdoms* contains this central theme: that "the empire, long united, must divide; long divided, must unite. Thus it has always been." This steady, undulating shift of power and organization in China has not been only a literary motif; it reflects reality right down to the present moment.

Most of the country's industrial development and urban growth is strung along its east coast, near the traditional port and river cities that were strongly influenced by Western economic interests in China from the mid–19th century on. The other areas of prime growth are the rural regions in China's humid east, which are reasonably well-linked to expanding markets and prospering cities. However, western regions of China,

Figure 13.13 The Special Economic Zones (SEZs) of southeast China were organized in the late 1970s to showcase China's ability to deal with the trading, manufacturing, and urban demands of the Western world. This map shows their concentration along China's southeast coast, the traditional area of China's main interaction with trading nations from the West. These SEZs today have become some of the most rapidly growing urban centers in all of China.

mountain villages, and landscapes lacking in surface transportation in the southwest and parts of upland China are all increasingly frustrated. They have seen relatively little evidence of China's touted economic prosperity.

Besides geographic tension, the Chinese central government is having difficulty enforcing the laws of Beijing as the law of the land. Thus, rapid economic change, monumental foreign interest—from both private and government sectors—in joint-venturing in China, and the gains to be captured by working private deals outside the domain of the central or even provincial government are all leading to the promotion of decentralization.

For the geographer, this all demonstrates the persistent influence of a diverse landscape with historically poor interior communications for most of its development. Although new highway and rail systems have begun to work some modification of the dominance of river and canal transportation, the historic pattern in which most of the Chinese people live and work within small communities has created a population with a high potential for strong local regional identities and weak federalist inclinations. Beijing will have to fight the specter of a nation

Figure 13.14 Hong Kong has elevated shopping to an entertainment matched by few cities in Monsoon Asia. This shopping center has created three floors of shopping variety built around an elaborate centerpiece that is intended to make the shopping experience as much social as it is commercial. *Kit Salter*

splintering into small economic worlds that have their own foreign linkages, even if there still would be an alleged political allegiance to the nation that calls itself the Middle Kingdom.

Finally, China continues to struggle with another geographic facet of development—the power of the environment. By the end of 1998, China admitted to problems of major air pollution that, for example, gave Beijing a smog index five times as high as that of Los Angeles. The primary culprit in this is the nearly universal use of soft coal in cooking, home heating, and a great deal of industrial activity. Coal is one of China's most abundant resources, and traditional patterns have made this widespread fossil fuel the resource of choice for the great majority of the population. And with the increasing popularity of the private automobile in China, the potential for even more horrendous air pollution in city and country alike is very real. So, as the peoples of China stride into the 21st century, they will face abundant challenges and anxieties as they transform their environment more than ever, attempting to manage both China's landscape and its population in productive, satisfying, and effective ways while achieving a balance between enormous ambition and enormous struggle. If successful in meeting these challenges, there will be great benefit felt by hundreds of millions, even billions, of people. If China fails, the scale of such failure will be unmatched, in either Asia or the broader world.

13.7 Taiwan

When it was driven from the mainland in 1949, the Chinese Nationalist government fled, with remnants of its armed forces and many of its more prominent adherents, to the island of Taiwan. There were nearly two million Chinese involved in this migration across the 100-mile Taiwan Strait. Here, protected by American sea power, the government reestablished itself with its capital at Taipei (population: 2.5 million, city proper). It continues to hold the island, along with other small neighboring islands, and continues voicing its claim to be the home of the legitimate government of China. Meanwhile, the People's Republic of China claims Taiwan as part of its own historic—and contemporary—territory. The United States backed the Nationalist claim for a time, but, during the 1970s, America and the Communist mainland developed closer relations and, in 1979, the United States withdrew its official recognition of Taiwan and its seat in the United Nations to meet the demands of the People's Republic. However, the United States opposes the annexation of Taiwan to China by force, just as it has opposed Taiwan's call for a Nationalist effort to retake the Chinese mainland during periods of internal weakness in China. It continues to supply Taiwan with weapons and aid.

Taiwan, known by Westerners for centuries as Formosa (Portuguese for "beautiful"), is separated from South China by the Taiwan or Formosa Strait (see Fig. 13.2). From 1895 to 1945 the island was governed by the Japanese, who developed its

economy as an adjunct to that of Japan and called the island by its Chinese name of Taiwan. The island is nearly 12,500 square miles (c. 32,000 sq km) in area and in 1998 had an estimated 21.5 million people, largely Chinese in origin. High mountains rise steeply from the sea on the eastern side of the island but slope on the west to a broad coastal plain. Taiwan lies on the northern margin of the tropics, and its temperatures are further moderated by the warm, northward-flowing Kuroshio, or Japanese Current. Irrigated rice, the main food crop, is favored by the hot, humid summers in the lowlands and by abundant irrigation water from streams originating in the mountains. Sugar and pineapple are also of considerable agricultural importance.

The Chinese Nationalists, operating an authoritarian regime with increasing elements of political democracy, have been strikingly successful in fostering capitalist economic development on the island of Taiwan, a mere 90 miles (145 km) from the Chinese mainland (see Fig. 13.2). A major handicap has been lack of energy resources. Some coal and natural gas exist, and some hydropower has been harnessed, but the island depends heavily on oil imports. However, the island is not without geographic advantages. Temperatures and water supply permit many irrigated rice fields to be double cropped and various tropical crops to be produced, and the native rain forest supplies both soft and hard woods in export quantities. The Japanese had already done much to develop the island as a sugar-producing colony, and American aid and technical help—plus monumental Chinese and Taiwanese entrepreneurship and economic drive—have seen the island develop in this last decade at an economic pace that is unmatched in a region of aggressively developing nations.

On these foundations, the Nationalists have been able to unite cheap Taiwanese labor with foreign capital to build one of Asia's first urban-industrial countries. Its major exports include clothing and textiles, electronic equipment, other machinery, and sugar. Its per capita income is approximately $13,970, or nearly five times that of the People's Republic. The Taiwan population is 75 percent urban, its infant mortality rate is the third lowest in the world (lower than that of the United States), and its crude birth rate has been lowered to .96 annually, one of the lowest rates in Asia. Although the United States, Japan, and Hong Kong account for over one-half of its foreign trade (1996), Taiwan trades with over 150 countries, most of which do not officially recognize its independent existence or the legitimacy of its government. It continues to resist mainland China's overtures for reunion, which include promises of relatively broad autonomy. But pressures on the island are growing. An unpredictable element in the situation lies in the fact that only about 15 percent of the island's people descend from the Nationalist refugees. Except for a relatively few aborigines, the rest are Chinese in culture, but are less China-oriented in political outlook, having arrived mostly in earlier Chinese migrations from the mainland in the mid–17th century. These earlier migrants have been discriminated against by the Nationalists in the past half century, but their power is increasing.

The relationship between Taiwan and China continues to be a curious mix of political alarm and economic interaction. China is Taiwan's major target for its investor capital, now providing more capital than any other source in the world. The island and the mainland had their first formal discussions about their political linkage late in 1998, and although nothing was put on a calendar, it remains clear that the mainland continues to see Taiwan as a renegade province that has, unfortunately for China, become very prosperous and quite attracted to the economic and political systems evolving on the island.

Taiwan has now climbed to the rank of the 20th (out of 192) most prosperous political unit in the world, with an annual per capita GNP now totaling more than $10,000. Unemployment is under 2 percent. The island has shown enormous capacity for managing population growth and becoming an active player in the production of consumer electronics; it is the world's largest producer of, in succession, televisions, watches, personal computers, and track shoes. Small firms have the capacity to respond quickly to market shifts but so far lack the scale to have a major impact on world pricing patterns.

There is no doubt that China and Taiwan will both be intensely concerned with the evolution of each locale's politics, economy, and international standing during the next decade. For Taiwan, there will have to be extraordinary overtures by the mainland to make reunification attractive enough to forsake the independence and maverick role that the island has been playing for the past decade. For China, any appreciable further move by Taiwan toward real independence will be a source of profound alarm. This is a setting to watch carefully as we move into the 21st century.

13.8 Mongolia

Mongolia (until recently, the Mongolian People's Republic, and often referred to as Outer Mongolia in the past) was once part of the Chinese Empire but became a separate Communist country in 1924. Relations with the former Soviet Union were close, both politically and economically. Mongolia has the distinction of being the first Asian country to abandon Communism (1990). It has an area of about 600,000 square miles, with a population of about 2.4 million (in 1998). Some 90 percent of the population consists of various Mongol groups.

The country contains large desert plains in the south and east, locally termed gobis (hence, the "Gobi Desert"); mountain ranges in the west; and grassy valleys and wooded hills and mountains to the north (see Fig. 10.2). Vast herds of livestock, primarily sheep and goats, are grazed. Much of the population still derives its livelihood from animal husbandry, although the Communist government made major efforts to further agriculture. The total amount of cultivated land, however, amounts to no more than 1 percent of Mongolia's area. Sixty percent of Mongolia's annual export income comes from mineral products and another 24 percent from textiles.

More than three-fifths of the population is now classed as urban. Ulan Bator (population: 575,000, city proper) is the capital and major urban center. Political and economic changes reflecting those of the Former Soviet Region are in progress. In 1990, the first multiparty elections ever held resulted in a coalition government committed to changing the Soviet-style "command economy" to a market economy. New private businesses are being established, and many state-owned enterprises and properties are being privatized. Commercial banks, business schools, and a stock market have been opened, and most Russian technical and military personnel have left. Much economic hardship is accompanying the changeover, and public discontent led to a sweeping reformed-Communist victory when national parliamentary elections were held in 1992. However, the reformed-Communist candidate (a hard-line ideologist) was defeated in the presidential election of 1993. There continues to be a tension between the Russian and Chinese peoples who have stayed on in Mongolia in the past decade, with commerce and privatization often giving the Chinese especially a commercial role that the Mongolians only reluctantly accept.

SUMMARY WITH SELECTED KEY TERMS

- China is home to approximately **22 percent of the world's population on about 7 percent of the world's land.** With **3.7 million square miles (9.6 million sq km) in area,** it is the world's third largest nation, and is the largest nation in East Asia, historically central to the development of culture, language, and technological development in the region.
- Population policy in China has been focused on **one child per couple,** taking China to a late 1990s birth rate of **1 percent, lowest of any major less-developed country (LDC).** Chinese planners think that national population will stabilize at a total of 1.4 or 1.5 billion by 2025. Declining family size has led to concern about caring for senior citizens.

- The broad division of China into the **arid west** and the **humid east** is a wide physical division. **Population distribution** is highly asymmetrical, and the **Han** and **non-Han realms** are associated with the **arid (non-Han) west** and the **humid (Han) east.** To separate the arid from the humid regions, an **arc can be drawn from Kunming in Yunnan north-northeast to Beijing.** The land to the east is Humid China, sometimes called **China Proper.** The land to the west and in the **Tibetan uplands** is Arid China, and is much more thinly settled. The **source areas** of the **Huang He (Yellow River)** and the **Chiang Jiang (Yangtze River)** are both in these highlands.
- Humid China—China Proper—is densely settled, focused on **irrigated rice production,** and highly dependent upon the **Chang**

Jiang and other river systems. This region has long been settled and **double-crops** its fields, giving it a high **multiple cropping index.** The **South China** pattern is centered upon rice. **North China, which** is drier than South China, is more dependent upon **winter wheat, spring wheat, corn, soybeans, and sorghum.** The **Huang He (Yellow River)** is major to this region both because of the silt it has deposited upon the North China Plain and because of regular problems with flooding. The Huang is also called **"China's Sorrow."**
- The **Three Gorges Project** is taking place on the Chang Jiang and is engineered to deal with **flood control, electricity generation, and possibly water transfer.** The Chang Jiang has a long history of flooding and very high differences in low and high season flow. The Project has been both celebrated and lamented because of the scale of **environmental change** necessary for its completion.
- The **Northeast**—historically called **Manchuria** by the West—was the early industrial heartland of China. It has **coal, iron ore, petroleum,** and well-developed **transportation systems.** It was the military threat to this area in the **Korean Conflict** that caused the Chinese to take up arms in support of North Korea. This area is likely to grow in population as China experiences more **internal migration.**
- **China's history** gives it a long-time importance in East Asia. **Language, literature, government, agriculture, and religion** have all been major cultural **traits diffused** from China to other parts of East Asia during the last several millennia. The **British presence in Hong Kong** from 1842 on, and the **Japanese efficiency in adopting Western technology,** combined to make China slip into a relatively minor role in the 19th and early 20th centuries. **Mao Zedong and the Communist Party and the Maoist/Marxist Revolution** did a great deal to change that from 1949 on.
- **Revolutionary campaigns** in the first three decades of government after the 1949 defeat of the Chinese Nationalists led to experiments in **communal agriculture, migration control, and backyard industrialization**—all of which were less significant than China's **opening to Western free-market economy** after the death of Mao

in 1976. This change was introduced by **Deng Xiaoping,** who became China's major political figure a few years after Mao's death. For the past two decades, China has modeled much of its most dynamic economic growth after capitalist, rather than Communist/ Marxist/Maoist, economic and political models.
- **Landscape change in China** in the past four decades has given much attention to **expansion of farmland, increase in irrigated land, opening new land in areas not before farmed,** and the **use of more chemical fertilizers, new plant stock, and slowly increasing agricultural mechanization.** There has also been much capital spent in increasing the **railroad network and highway systems,** and in upgrading the continued use of **canal and river systems.**
- **China's urban centers** have continued to grow, with China now between **30 and 40 percent urban.** Cities continue to absorb **migrants coming in from the countryside** in hope of getting closer to the more city-centered economic development. China's aggressive migration management policies of the 1960s and 1970s have relaxed. Considerable economic activity has taken place in the countryside, but it is not evenly distributed. There continue to be tens, perhaps many tens, of millions of **urban poor who are underemployed and unemployed.**
- **Regional imbalances in economic activity and opportunity** continue to plague China. The most prosperous areas are in the **southeast, east coast, and farmlands near large cities,** while regions in the **west, southwest, uplands, and arid China** have much less evidence of economic development.
- **Special Economic Zones (SEZs) and Taiwan** are two special landscapes that have China's attention. The SEZs have been developed to showcase China's ability to absorb and make productive use of Western-style urban growth and development. Taiwan, in its **ever-increasing prosperity,** is being closely watched as it becomes **more democratic, more prosperous, and more inclined to seek a full political and economic independence from China.**

REVIEW QUESTIONS

1. List the area and population of China and explain the importance of both those statistics.
2. What are the characteristics of China's population growth and what governmental policy is at the heart of its contemporary birth control programs?
3. What is the meaning of "the little emperor," and what implications does that phrase and the associated concept have for future generations in China?
4. Define and find on the map Arid China. Discuss some of the associated landscapes.
5. Define and find on the map Humid China. Discuss some of the associated landscapes.
6. What is the economic and political significance of the western highlands in China? Of the Northeast?

7. Explain the importance of the Great Wall, the Grand Canal, and the Three Gorges Project. Differentiate their economic importance from their historical and political significance.
8. Who was Mao Zedong and what role did he play in China in the 20th century? What campaigns is his government associated with? What role do his ideas play now in China?
9. Outline the different eras of development in China between 1949 and 1999. Which of these eras seem to still have importance?
10. Who was Deng Xiaoping? What contemporary programs is he associated with? What is their contemporary importance?
11. What is the relationship between China and Taiwan? Explain potential future scenarios between those two places.

DISCUSSION QUESTIONS

1. What are the geographic features that give China such a monumental image in East Asia? In all of Asia? In world affairs?
2. Discuss the magnitude and significance of the Chinese population and of the government's efforts to control its growth.
3. What are the features of Arid China that support or diminish settlement and economic development?
4. Compare North China and South China in terms of physical features, demographic patterns, and cultural significance. Looking at Essential Element 2 in Chapter 1, what are the regional forces that make the two regions so distinct?
5. How much of China's future is linked with China Proper and how much will come from the world of Western China?

6. What is the cause and the solution for the imbalanced economic development in China? How old is this pattern?
7. In China's efforts to industrialize, what role has been played by Mao Zedong, and what role by Deng Xiaoping?
8. What was the attraction of the Dazhai village model in the 1960s and 1970s? Could such a model occur again in China? Under what conditions?
9. Discuss the origins, the expected outcomes, and the environmental ramifications of the Three Gorges Project.
10. What are the geographic patterns of urban distribution in China? What do such patterns reflect in terms of China's physical landscape and demographic history?

Chapter 14

Japan and the Koreas: Adversity and Prosperity in the Western Pacific

▲ *Ancient and modern ways walk in step in Japan.*   Charles Gupton/Tony Stone Worldwide

CHAPTER OUTLINE

14.1 The Japanese Homeland

14.2 Historical Background

14.3 Japan's Postwar "Miracle"

14.4 Japanese Industry

14.5 The Social Landscape

14.6 Divided Korea

On a cluster of islands off the eastern coast of Asia, the 126 million people of Japan operate an economy larger than that of any nation except the United States. The economic prodigies of the Japanese have been achieved on mountainous, resource-poor, and crowded islands a little smaller in total area than the state of California, but containing nearly half as many people as the entire United States. Japan's dramatic rise from the ashes of World War II has drastically changed the economic and political geography of the world.

And, although the Korean peninsula occupies a seemingly marginal position on the Pacific Rim and the Eurasian landmass, it too has been central in the geopolitical affairs of the second half of the 20th century, primarily because of its strategic location between Japan, China, and the former Soviet Union (now Russia and the Near Abroad), and because of its emerging nuclear weapons capability. The problems of a divided Korea and the shifting economic fortunes of Japan continue to raise questions about the region's stability.

14.1 The Japanese Homeland

There are four main islands in the Japanese archipelago (Fig. 14.1). On the west, the main islands are separated from Russia and Korea by the Sea of Japan and the Korea Strait. On the east, Japan faces the open Pacific. Hokkaido, the northernmost island, is narrowly separated from Russian-controlled islands to its north. Honshu, the largest island, is separated from Shikoku and Kyushu by the scenic and busy Inland Sea. South of Kyushu, the smaller Ryukyu Islands extend southwestward almost to Taiwan.

All of the main islands consist largely of mountains, with the higher peaks generally between 5000 and 9000 feet ($c.$ 1500–2750 m) in elevation. The islands are volcanic (Fig. 14.2) and are subject to frequent and sometimes very severe earthquakes (see Problem Landscape, p. 366). Located west of Tokyo, the volcano Mt. Fuji, whose name means "Fire God-

dess," last erupted in 1707 and is Japan's highest mountain (12,288 ft, 3776 m). Fuji was sacred to the aboriginal Ainu people of Japan as long as 2000 years ago. It is still a holy mountain in the national Shinto faith, and about 300,000 pilgrims and tourists visit its lofty summit each July and August, when it is free of snow. The journey to the peak of this most prominent symbol of Japan is a difficult one, as suggested in the Japanese proverb, "It is as foolish not to climb Fuji-san as to climb it twice in a lifetime."

Climate and Resources

Japan's location—off the east side of the Eurasian landmass and at latitudes comparable to those from South Carolina to Maine (Fig. 14.3)—strongly influences the climate of the islands (see Fig. 2.7). A humid subtropical climate extends from southernmost Kyushu to north of Tokyo, a humid continental long-summer climate characterizes northern Honshu, and Hokkaido has a humid continental short-summer climate. In lowland locations, where most of the people live, the average temperatures for the coldest month range from about 20°F (-7°C) in northern Hokkaido to about 40°F (4°C) in the vicinity of Tokyo and about 44°F (7°C) in southern Kyushu. Averages for the hottest month are only about 64°F (18°C) in northern Hokkaido but rise as you go southward, to about 80°F (27°C) around Tokyo and above 80°F farther south. Conditions are cooler in the mountains, with heavy snowfalls common in winter in Honshu and Hokkaido.

Japan is a humid country. Annual precipitation of more than 60 inches ($c.$ 150 cm) is common from central Honshu south, and 40 inches ($c.$ 100 cm) or more falls in northern Honshu and Hokkaido. Their location off the east coast of Eurasia puts the islands in the path of the inward-blowing summer monsoon from the Pacific, and summer is the main rainy season. But the islands are also subject to the seaward-blowing winter monsoon after it has crossed the Sea of Japan, so that in Japan a good deal of precipitation also occurs in the winter half-year; precipitation is thus more evenly spread through the year than is common in monsoonal climates. Japan is also subjected in summer and fall

Figure 14.1 Index map of Japan and Korea.

to powerful typhoons, the great low-pressure storm systems known as hurricanes in the western Atlantic region. These add to the woes of this hazard-prone country.

Japan's natural resource base is not large enough to meet the country's growing needs. Up to the middle of the 19th century, the country was nearly self-sufficient economically, although the standard of living was low. Japan was already crowded with about 30 million people, but disease, starvation, abortion, and infanticide slowed the rate of natural increase. Rapid population growth in the second half of the 19th century and later was made possible by industrialization, which has continued on an ever-increasing scale except in the years during and just after World War II. Population growth has recently almost ceased, as

delayed marriages and the rising costs of child-rearing have lowered birth rates. Meanwhile, affluence has increased dramatically. The material demands of so many people enjoying a high standard of living in a country with limited natural resources require Japan to import most of its raw materials and energy supplies, and a large share of its food.

Forests and Wood

Forests, which now cover about 67 percent of the land surface, illustrate the progressive inadequacy of Japan's natural resources. In a preindustrial and early industrial society, with a much smaller population, the nation's heavily forested moun-

Figure 14.2 Japan has numerous volcanoes—some active, others intermittently active, and still others extinct. In 1991, Mount Unzen, near Nagasaki in Kyushu, erupted for the first time since 1792. At least 35 people were killed. The photo shows firemen fleeing for their lives as smoking red-hot lava hurtles down the side of the volcano. *The Yomiuru Shimbu*

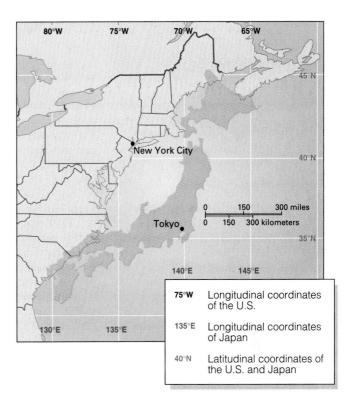

Figure 14.3 Japan's relative size and latitude. Japan compared with the eastern United States in latitude and area. *After a map by E. O. Reischauer.*

tainsides supplied the needed wood, but if today's demand for forest products were met with domestic reserves, the country would be rapidly deforested. The varied local environments of Japan support a wide range of valuable trees, mainly broadleaf types toward the south and needleleaf conifers toward the north. Many mountainsides are quilted with rectangular plots of different kinds of trees planted in rows. Most first-growth timber is long since gone, but some virgin stands are still protected in preserves. Forests are important in Japanese culture, and this affluent country does not wish to lose them. With its enormous demand for wood products, Japan therefore is the world's largest importer of wood, mostly hardwoods from the tropical rain forests of southeast Asia.

Kerosene stoves have largely replaced the charcoal used formerly for household cooking and heating, but Japanese people still use wood for many purposes (Fig. 14.4). Single-family dwellings continue to be built largely of wood, which provides more safety from earthquake shock waves than more rigid construction. However, wood makes the dwellings more vulnerable to the fires that typically sweep through Japanese cities after earthquakes rupture gas lines. Paper made from wood is used to cover the sliding partitions between rooms in homes and apartments. Japan's large publishing industry requires a great deal of paper, and other industrial and handicraft uses are large.

Minerals, Water, Agriculture, and Fisheries

Small deposits of many different minerals, notably coal, iron ore, sulfur, silver, zinc, copper, tungsten, and manganese, supported early industrialization in Japan. The supplies of all of these minerals are inadequate for Japanese needs today, so, like most other raw materials, they must be imported. Japan's mountain streams provided much of the power for early industrialization, but they are so swift, shallow, and rocky that they are of little use for navigation. Most of the water used for irrigation in Japan comes from these streams, which emerge from the mountains, divide into distributaries, and then cross the plains in beds elevated above the level of the cropland. These elevated beds make it possible to get much of the water to the land by gravity.

Living on the Ring of Fire

THE PEOPLE OF JAPAN MUST COPE WITH one of the most hazard-prone regions on Earth. By far the greatest threats are related to Japan's location on the Pacific "ring of fire," the name given to the boundaries of the Pacific plate of the Earth's crust where tectonic activity—including earthquakes and volcanic eruptions—are commonplace and often catastrophic (Fig. 14.A). Japan's coasts must be on guard even for the effects of earthquakes that occur on the other side of the Pacific, in North America, where a strong quake can produce a *tsunami* (commonly called a "tidal wave") that may travel thousands of miles to strike and flood coastal Japan.

On January 17, 1995, a devastating earthquake released the tension that had built up on a transverse fault near historic Kobe, a port city of 1.5 million people. Killing 6400 people and injuring thousands more, damaging 190,000 buildings, displacing 300,000 people, and causing at least $100 billion worth of damage, this was the worst disaster to hit Japan since World War II. The Japanese soon labeled the 6.9-magnitude episode the "Great Hanshin Earthquake," establishing its place in the annals of Japanese disasters, alongside such events as the Great Kanto Earthquake, which killed 143,000 people in and around Tokyo in 1923.

The Great Hanshin quake had many repercussions far from its epicenter. Kobe is Japan's most important heavy-cargo port, and the damage to its facilities resulted in an immediate but short-term 5-percent drop in Japan's exports. Opposition party members in Japan's parliament complained about the government's slow response to the disaster, contributing to an already strong sense of political uncertainty in the country. Kobe's remarkable lack of disaster preparedness and relief focused attention on a common Japanese complaint: that the government invests too much in commerce-oriented development and not enough in the welfare of the Japanese people. Many observers believed that the quake and its aftermath signaled the final blow in a five-year period of decline in Japan's economic and cultural self-confidence.

Japanese authorities did not predict the Great Hanshin Earthquake. After the fact, there were reports that large schools of fish swam uncharacteristically close to the surface of the sea off Kobe in the days preceding the quake. Crows and pigeons reportedly were noisier than usual and flew erratically in the hours before the ground moved. In the weeks preceding the quake, levels of radon gas rose in a water aquifer below the city and then declined sharply after January 9. Animal behavior, water level and composition, and many other variables are often-debated but still-unproven "barometers" of an impending earthquake. Since 1965, Japan has spent more than $1.3 billion in an effort to develop means of predicting earthquakes. During the 1990s, more than $100 million yearly went to this effort. But by 1998, no earthquake had been successfully predicted, and Japan suspended this expensive research program. Its critics argue that earthquakes are chaotic events and will always be impossible to predict. They also point out the hazards of issuing an earthquake warning that is not followed up by a real quake. A false alarm in the Tokyo region would cost an estimated $7 billion a day in lost business and other economic disruptions.

With the prediction program scrubbed, Japan has redoubled its efforts to prepare for future "big ones." New structures must meet a tough safety code which aims at "earthquake proof" construction. The government has resolved to build, by 2010, a new national capital inland, where mountain bedrock would lessen the devastating impacts of seismic shock waves. Much of the devastation in Kobe resulted from the city's location on soft coastal soils and reclaimed lands, which also underlie great parts of Tokyo. Japan's struggle against the seismically inevitable is a national priority and a symbol of the nation's determination to maintain economic success despite environmental adversity.

Figure 14.A For now, Japanese scientists have conceded defeat in their struggle to predict earthquakes, but they continue to improve ways of living with the consequences of movements in the Earth's crust.

Only about one-eighth of this small country's area is arable; Japan is not one of the world's major crop-growing nations. Most cultivable land is in mountain basins and in small plains along the coast. Terraced fields on mountainsides augment this level land. Even on flat land, great numbers of Japan's fields are terraced for the growing of irrigated rice, the country's basic food and most important crop (Fig. 14.5). Irrigated rice is nearly everywhere on arable land, even in most parts of the northern island of Hokkaido. Production of this grain is heavily subsidized by the Japanese government and promoted by a powerful farm lobby, which aims at rice self-sufficiency, with other foods being imported as needed. The country's farmers usually achieve a net export of rice by intensive use of scarce cropland. However, a scarcity of rice throughout east Asia in

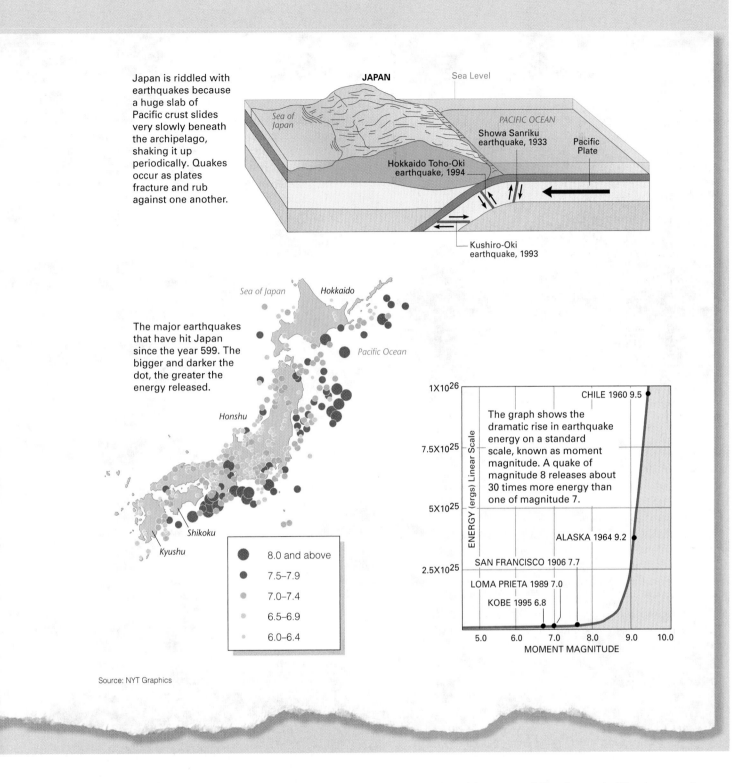

Japan is riddled with earthquakes because a huge slab of Pacific crust slides very slowly beneath the archipelago, shaking it up periodically. Quakes occur as plates fracture and rub against one another.

JAPAN

Sea Level

Sea of Japan

PACIFIC OCEAN

Showa Sanriku earthquake, 1933

Pacific Plate

Hokkaido Toho-Oki earthquake, 1994

Kushiro-Oki earthquake, 1993

Sea of Japan

Hokkaido

Pacific Ocean

The major earthquakes that have hit Japan since the year 599. The bigger and darker the dot, the greater the energy released.

Honshu

Shikoku

Kyushu

- 8.0 and above
- 7.5–7.9
- 7.0–7.4
- 6.5–6.9
- 6.0–6.4

The graph shows the dramatic rise in earthquake energy on a standard scale, known as moment magnitude. A quake of magnitude 8 releases about 30 times more energy than one of magnitude 7.

CHILE 1960 9.5

ALASKA 1964 9.2

SAN FRANCISCO 1906 7.7

LOMA PRIETA 1989 7.0

KOBE 1995 6.8

ENERGY (ergs) Linear Scale

1×10^{26}

7.5×10^{25}

5×10^{25}

2.5×10^{25}

5.0   6.0   7.0   8.0   9.0   10.0

MOMENT MAGNITUDE

Source: NYT Graphics

the early 1990s, and pressure on Japan to reduce its massive trade surplus with the United States, prompted Japan in 1995 to open its rice market to foreign imports for the first time. Rice consumption per person has nevertheless been declining in recent years as the Japanese diet has become more diverse.

Mild winters and ample moisture permit double cropping—the growing of two crops a year on the same field—on most ir-

rigated land from central Honshu south. Rice is grown in summer, and wheat, barley, or some other winter crop is planted after the rice harvest. Overall, about one-third of Japan's irrigated rice fields are sown to a second crop, and more than half of the unirrigated fields are double cropped. **Intertillage**—the growing of two or more crops simultaneously in alternate rows—is also common. Dry farming also produces a variety of crops:

Figure 14.4   Wood is a vital resource in Japan. This wooden home is in Shirakawa, north of Tokyo on the island of Honshu.   *Jean Kugler/FPG International*

sugar beets in Hokkaido; all sorts of temperate-zone fruits and vegetables; potatoes, peas, and beans, including soybeans; and, in the south, tea, citrus fruits, sugarcane, tobacco, peanuts, and mulberry trees to feed silkworms.

Changes in Japan's agricultural sector have accompanied a surge in urban-industrial employment and general affluence in recent decades. Farmers have been moving to towns and the number of agricultural workers has dropped dramatically (from about 25 percent of the labor force in 1962 to 7 percent in 1999). A huge increase in farm mechanization has made it possible to till the land with far less human labor. Millions of small tractors and other types of mechanized equipment are now used on Japan's tiny farms. With this technology and heavy applications of fertilizers (both chemical and processed from human waste), plus improved crop varieties, Japanese agriculture has achieved the world's highest yields per unit of land.

The island people of Japan view the ocean as an important resource and promising frontier. The country has an ambitious "inner space" program of deep-sea exploration to mine the sea's living and mineral riches. A variety of food fish are present in the waters surrounding Japan, and Japanese fishermen range widely through the world's major ocean fishing grounds. Japan is the leading nation in ocean fisheries and the Japanese enjoy the world's highest per capita consumption of fish, including *sashimi*, or raw fish, and other marine foods (Fig. 14.6). The Japanese have also historically enjoyed whale meat, but pressure from international environmentalists compelled Japan in

1985 to support a worldwide moratorium on commercial whaling. Early in the 1990s, along with Iceland and Norway, Japan lodged reservations against the ban and has since resumed small-scale whaling of minke whales for what Japan insists are "scientific research" purposes. This loophole allows whale meat to be widely sold.

In recent years the Japanese have begun to eat less fish and rice and to adopt a more Western diet. While sea fish continue to be the main source of animal protein in the Japanese diet, increased consumption of meat, milk, and eggs has been reflected in a growing number of cattle, pigs, and poultry on Japanese farms. By 1993, livestock surpassed rice as the country's most valuable agricultural product. The animals require heavy imports of feed, and the increasing variety in people's tastes means many more food imports. The United States is Japan's largest supplier of imported feedstuffs and food.

Japan's Core Area

Most of Japan's economic activities and a large majority of its people are packed into a corridor about 700 miles (c. 1100 km) long. This megalopolis extends from Tokyo, Yokohama, and the surrounding Kanto Plain on the island of Honshu in the east, through northern Shikoku, to northern Kyushu in the west (see Fig. 14.1). This highly urbanized core area contains more people than live in the Boston-to-Washington, D.C., "megalopolis" on the United States' east coast. All of Japan's greatest cities and many lesser ones have developed here, on and near harbors along the Pacific and the Inland Sea. They are sited typically on small, agriculturally productive alluvial plains between the mountains and the sea.

The largest urban complex in the world—Tokyo-Yokohama and surrounding suburbs (population: 30.5 million by one estimate; other figures differ according to how the metropolitan area is defined)—has developed on the Kanto Plain, Japan's

Figure 14.5   Hanging rice on racks to dry on the island of Kyushu, Japan. *Todd Stradford, Jr.*

Figure 14.6 Ocean fish are a major element in the Japanese diet. This is the Tsukiji fish market in Tokyo. *Sylvain Grandadam/Photo Researchers, Inc.*

most extensive lowland. Tokyo grew as Japan's political capital and Yokohama as the area's main seaport. This urban complex functions today as Japan's national capital, one of its two main seaport areas (the other is Osaka-Kobe), its leading industrial center, and the commercial center for northern Japan (Fig. 14.7).

Beyond Tokyo, in the entire northern half of Japan there is only one major city—Sapporo (population: 1.74 million, city proper; 2.2 million, metropolitan area), on Hokkaido. About 200 miles (*c.* 320 km) west of Tokyo lies Japan's second largest urban agglomeration, in the Kinki District at the head of the Inland Sea. Here, three cities—Osaka, Kobe, and Kyoto—form a metropolis, together with their suburbs, of nearly 20 million people.

The inland city of Kyoto was the country's capital from the eighth century A.D. until 1869 and is now preserved as a shrine city where modern industrial disfigurement is prohibited (Fig. 14.8). Some U.S. military planners wanted to target Kyoto with a nuclear weapon in August 1945, but the opinion prevailed that it should be avoided because of its historical importance. Kyoto continues to be the destination of millions of pilgrims and tourists annually.

Osaka grew mainly as an industrial center, and Kobe originally developed as the district's deepwater port. Both cities now combine port and industrial functions. Another large metropolis, Nagoya (population: 2.15 million, city proper; 4.8 million, metropolitan area), is situated directly between Tokyo and Osaka, on the Nobi Plain at the head of a bay. It is both a major industrial center and a seaport.

West of the Kinki District, smaller metropolitan cities spot the coreland. The largest are metropolises of over 1.5 million people: the resurrected Hiroshima on the Honshu side of the Inland Sea, Kitakyushu (a collective name for several cities) on the Strait of Shimonoseki between Honshu and Kyushu, and Fukuoka on Kyushu. Hiroshima was destroyed on August 6, 1945, when the United States dropped the first atomic bomb ever used in warfare, detonating it directly over the center of the city. With no topographic barriers to deter the effects of the explosion, the city was obliterated, and an estimated 140,000 died by the end of 1945, when radiation sickness had taken its greatest toll. The city has since rebuilt itself.

Among the several smaller cities of northern Kyushu is Nagasaki, which, on August 9, 1945, was hit by a second, and

Figure 14.7 Tokyo, the capital of an industrious and affluent nation, is the world's largest and one of its most vibrant cities. This is rush hour at the Shinjuko Station. *Paul Chesley/ Tony Stone Worldwide.*

369

Figure 14.8 Kyoto has a unique status in Japanese history and religious life. This is the 14th-century Kinkakuji Temple, also known as the Golden Pavilion. *Bruce Burkhardt/Westlight*

where some still live, principally in Hokkaido. The oldest surviving Japanese written records date from the eighth century A.D. These depict a society strongly influenced by China, often through cultural and ethnic traits reaching Japan by way of Korea. A distinct Japanese culture gradually evolved, and today it is linguistically and in many other ways different from its Chinese and Korean antecedents. Legendary Japanese traditions extend back to the reign of Jimmu, the first emperor, whose accession is ascribed to the year 660 B.C., but which probably took place centuries later. These traditions are important in the Japanese worldview, which says that the islands and their emperor have a divine origin. Jimmu was said to have descended from the Sun Goddess, and the Japanese have known their homeland as the "Land of the Gods." The Japanese concept of Japan as unique and invincible developed early on, shaped by legends such as that of the *kamikaze,* the "divine wind" that repelled a Mongol attack on Japan in the 13th century.

The early emperors gradually extended control over their island realm. A society organized into warring clans emerged. By about the 12th century, powerful military leaders called *shoguns,* who actually controlled the country, diminished the emperor's role. Meanwhile, the provinces were ruled by nobles, or *daimyo,* whose power rested on the military prowess of their retainers, the *samurai.* This structure resembled the European feudal system of medieval times.

final, U.S. nuclear bomb. Nagasaki was chosen during the flight mission after clouds obscured the primary target, Kokura. The bomb detonated over the suburbs rather than the city center, and hills helped to diminish the explosion's impact, but even so, an estimated 70,000 died by year's end. The bombing of Nagasaki effectively brought World War II to its end.

14.2 Historical Background

Japan's eventful history has seen successive periods of isolationism and expansionism, and of economic and military accomplishment and defeat. Japan today is near the top of most indicators of prosperity in the MDCs. The remarkable success story of Japan can best be appreciated by considering its turbulent history and the difficult home environment in which the Japanese have always lived.

Early Japan

The Japanese are descended from a number of primarily Mongoloid peoples who reached Japan from other parts of eastern Asia at various times in the distant past. An earlier non-Mongoloid people, the Ainu, were driven into outlying areas

Early Contacts with Europe

Adventurous Europeans reached Japan early in the 16th century, beginning with the Portuguese in the 1540s. Most of these early arrivals were merchants and Roman Catholic missionaries. Japanese administrators allowed the merchants to set up trading establishments and open commerce, and allowed the missionaries to preach freely. There were an estimated 300,000 Japanese Christians by the year 1600.

These contacts were short-lived. Japan had entered a period when a series of military leaders imposed central authority on the disorderly feudal structure. After winning the battle of Seikigahara in 1603 with arms purchased from the Portuguese, warlord Ieyasu Tokugawa proclaimed himself "Nihon Koku Taikun"—Tycoon of All Japan—and became the first shogun to rule the entire country. During the era of the **Tokugawa Shogunate,** from 1600 to 1868, the Tokugawa family acquired absolute power and shaped Japan according to its will.

To maintain power and stability, the early Tokugawa shoguns wanted to eliminate all disturbing social influences, including foreign traders and missionaries. They feared that the missionaries were the forerunners of attempted conquest by Europeans, especially the Spanish, who held the Philippines. The shoguns drove the traders out and nearly eliminated Christianity in persecutions during the early 17th century. After 1641, a few Dutch traders were the only Westerners allowed in Japan. Their Japanese hosts segregated them on a small island in the harbor of Nagasaki, even supplying them with a brothel so that they would not be tempted to venture out and pollute Japan's ethnic integrity. Under Tokugawa rule, Japan settled into two centuries of isolation, peace, and stagnation.

Westernization and Expansion

When foreigners made a serious attempt to reopen the country to trade two centuries later, they were not thwarted by a Japan that had fallen far behind in technology. Visits in 1853 and 1854 by American naval squadrons of "Black Ships" under Commodore Matthew Perry (who wanted to establish refueling stations in Japan for American whaling ships) resulted in treaties opening Japan to trade with the United States; this was called "gunboat diplomacy." The major European powers were soon able to obtain similar privileges. Some of the great feudal authorities in southwestern Japan opposed accommodation with the West. United States, British, French, and Dutch ships responded in 1863 and 1864 by bombarding coastal areas under the control of these authorities. These events so weakened the faltering prestige of the Tokugawa Shogunate that a rebellion overthrew the ruling shogun in 1868. The revolutionary leaders restored the sovereignty of the emperor, who took the name Meiji, or "Enlightened Rule." The revolution of 1868 is known as the **Meiji Restoration.**

The men who came to power in 1868 aimed at a complete transformation of Japan's society and economy. They perceived that if Japan were to avoid falling under the control of Western nations, its military impotence would have to be remedied. These leaders saw that this would require a reconstruction of the Japanese economy and of many aspects of the social order. They approached these tasks with energy and intelligence and abolished feudalism, but only after a bloody revolt in 1877. The Meiji leaders cemented the power of a strong central government that would remodel and modernize the country while resisting foreign encroachment. Under this style of government, which lasted until 1945, democracy was instituted but was strictly limited, so that Japan was generally ruled by small groups of powerful men manipulating the machinery of government and the prestige of the emperor. Military leaders were very prominent in this power structure.

The new government pressed its people to learn and apply the knowledge and techniques that Western countries had accumulated during the centuries of Japan's isolation. Foreign scholars were brought to Japan, and Japanese students were sent abroad in large numbers. Japan adopted a constitution modeled after that of imperial Germany. The legal system was reformed to be more in line with Western systems. The government used its financial power, which it obtained from oppressive land taxes, to foster industry. New developments included railroads, telegraph lines, a merchant marine, light and heavy industries, banks, and other financial institutions. Wherever private interests lacked the capital for economic development, the government provided subsidies to companies, or built and operated plants until private concerns could acquire them. Rapid ur-

banization accompanied this process of rapid industrialization.

So spectacular were the results of the Meiji Restoration that within 40 years Japan had become the first Asian nation in modern times to attain the status of a world power. Outsiders often spoke of the Japanese in derogatory terms as mere imitators. The Japanese, however, knew which elements of Western technology they wanted, and they adapted them successfully to Japanese needs.

The Prewar Empire

Japan emerged as an imperial power after abandoning the isolationism of the Tokugawa period. With only brief interruptions, the country pursued an expansionist policy between the early 1870s and World War II. By 1941 Japan controlled one of the world's most imposing empires. The extent and dates of the empire's acquisitions are shown in Figure 14.9. Between 1875 and 1879, Japan absorbed the strategic outlying archipelagoes of the Kuril, Bonin, and Ryukyu Islands. Japan won its first acquisitions on the Asian mainland by defeating China in the Sino-Japanese War of 1894–1895, which broke out over disputes in Korea. In the humiliating 1895 treaty of Shimonoseki, China ceded the island of Taiwan (Formosa) and Manchuria's Liaotung peninsula to Japan, and recognized the independence of Korea. Feeling threatened by Japan's emerging strength so close to its vital port of Vladivostok, Russia demanded and won a Japanese withdrawal from the Liaotung peninsula, and leased the strategic peninsula from China. Japan went to war with Russia in 1904 by attacking the peninsula's main settlement of Port Arthur. Russia lost the ensuing battle of Mukden, which involved more than a half-million soldiers, making it history's largest battle to that date. Russia also lost the Russo-Japanese War of 1904, representing the first defeat of a European power by a non-European one. This victory restored the Liaotung peninsula to Japan, gave Japan control over southern Sakhalin Island, and established Korea as a Japanese protectorate until Japan annexed it in 1910.

As a result of World War I, the Caroline, Mariana, and Marshall Islands were transferred from defeated Germany to Japan as a mandated territory. Encroachments on China in Manchuria followed in the 1930s, and an all-out attack that overran the most populous parts of China began in 1937. Japanese troops committed many atrocities in the assault on China; the most notorious was the "Rape of Nanking" (now Nanjing) in 1937, when the invaders took control of China's temporary capital, killing an estimated 300,000 Chinese civilians and soldiers, and raping about 20,000 Chinese women. In 1940, Japan seized French Indochina after Germany crushed France, and after Japan's carrier-based air attack crippled the American Pacific Fleet at Pearl Harbor, Hawaii, on December 7, 1941, Japanese forces rapidly overran Southeast Asia as far west as Burma, as well as much of New Guinea and many smaller Pacific islands.

The motives for Japanese expansionism were mixed. They included a perception of national superiority and "manifest

Figure 14.9   Map showing overseas areas held by Japan prior to 1937 and the line of maximum Japanese advance in World War II.

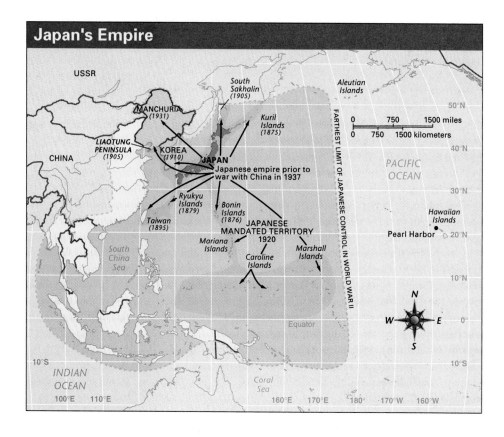

destiny," a desire for security and recognition of the great powers, the wishes of military leaders to aggrandize themselves and gain control of the Japanese government, and an ambition on the part of industrialists in Japan to gain sources of raw materials (such as Manchurian and Korean coal) and markets for Japanese industries. Although the methods of gaining them were often illegitimate, the desire for materials and markets was based solidly on need. Expansion of industry and population on an inadequate base of domestic natural resources had made Japan dependent on sales of industrial products outside the homeland. Such sales provided, as they do now, the principal funds with which the country purchased the imported foods, fuels, and materials that it required. As worldwide depression hurt the trade-dependent prewar Japanese economy, Japan chose to create a Japanese-controlled Asian realm which would insulate the Japanese economy from the vicissitudes of the world economy and elevate the nation to the rank of a leading world power. In 1938, Japan therefore proclaimed its "New Order in East Asia," and in 1940 and 1941, this widened into the "Greater East Asia Co-Prosperity Sphere," a euphemism for Japanese political and economic control over China and Southeast Asia.

Japan's militaristic and colonial enterprises proved to be disastrous at home and abroad. Japanese soldiers inflicted great suffering on civilians throughout the western Pacific. Even now nations are demanding formal apologies from Japan. In August 1995, on the fiftieth anniversary of Japan's defeat in World War II, Japan's prime minister Tomiichi Murayama did issue a for-

mal general apology for his nation's role in the war. Critics say Japan has not fully atoned for its war crimes and are trying, among other efforts, to block Japan's quest to gain a permanent seat on the United Nations Security Council.

August 1945 indeed found Japan completely defeated. Its overseas territories, acquired during nearly 70 years of successful imperialism, were lost. Its great cities were in ruin. In addition to 1.8 million deaths among its armed services, Japan had suffered some 8 million civilian casualties from the American bombing of the home islands; most of its major cities were over half destroyed. Tokyo's population had fallen from nearly 7 million people to about 3 million, most of whom were living in shacks. In 1950, five years after the beginning of the U.S. military occupation of Japan, national production still stood at only about one-third of its 1931 level, and annual per capita income was $32.

14.3   Japan's Postwar "Miracle"

Without benefit of colonies or spheres of influence, Japan has become an economic superpower since World War II. The nation's explosive economic growth after its defeat has been one of the most startling and significant developments of the late 20th century; it is widely known as the Japanese "miracle."

Japan's postwar relations with the United States centered mainly on the American military occupation of the country. The occupation of 1945 to 1952 instituted a series of significant reforms, several of which contributed to Japan's subsequent economic successes. The divine status of the emperor was officially abolished; indeed, when Emperor Hirohito announced Japan's surrender on the radio, it was the first time the stunned Japanese public had ever heard the voice of this mythic figure. A new U.S.–written constitution made Japan a constitutional monarchy with an elective parliamentary government. With the rejection of divine monarchy, the strongly nationalistic Shinto faith lost its status as Japan's official religion, although worship at Shinto shrines was allowed to continue; it still dominates, along with Buddhism, Japan's spiritual life. There was extensive land reform in the countryside to do away with a near-feudal landlord system. Some major Japanese companies were broken up to reduce their monopolistic hold on the economy. Women were enfranchised. A democratic trade union movement was established. The formerly dominant military officer corps was purged. Japan was forbidden to rearm, except for small "self-defense" forces.

When the occupation ended in 1952, leaving U.S. bases in Japan but returning control of Japanese affairs to the Japanese, Japan was placed under American military protection. The U.S. military umbrella over Japan remains controversial today. Many Japanese regard it as an outmoded vestige of colonialism and the war. And in 1995, the trial and conviction of three U.S. servicemen accused of raping a local child in Okinawa (where three-quarters of the U.S. bases and more than half of the U.S. troops in Japan are stationed) intensified Japanese public attention to this issue. Public protests and diplomatic appeals led the United States early in 1996 to agree to return control of a major airbase in Okinawa to Japan, and to relocate some U.S. troops from Okinawa to mainland Japan. On behalf of the American people, U.S. president Clinton also publicly apologized for the rape. In a subsequent referendum, the people of Okinawa voted overwhelmingly in favor of a further U.S. military withdrawal from the island.

Explaining the Miracle

Observers of Japan cite different reasons for the country's economic success. Proponents of dependency theory (see Chapter 2, p. 44) argue that Japan, never having been colonized, escaped many of the debilitating relationships with Western powers that crippled many potentially wealthy countries. Others emphasize Japan's association with the United States following World War II. Some of the postwar U.S.–imposed reforms worked well economically, and relations between the two countries since the war have also generally stimulated Japan's industrial productivity. Without large military expenditures, much of Japan's capital was freed to invest in economic development.

Rapid recovery from the postwar depths began when the United States called on Japanese production to support American forces in the Korean War of 1950–1953. By 1954, Japanese steel production was the largest in the country's history. The country's textile and clothing industries supplied American military needs, then expanded exports so rapidly that United States industries were asking for protection by 1956. Military procurement for American forces continued to support the Japanese recovery through the late 1950s, and U.S. economic aid continued to flow to Japan. The United States permitted many Japanese products to have free access to the United States market for long periods and allowed Japan to continue to protect its economy from imports. Japanese firms had relatively free access to American technology, which they frequently improved on, often producing and introducing the improvements ahead of American firms.

By the 1970s, Japan had become an industrial giant with a GNP far exceeding that of any country except the United States and the Soviet Union. The United States and European nations found themselves at a strong disadvantage to Japanese competitors in many industries. They busily sought ways to match this competition or to protect themselves from it without doing too much damage to their exporting firms, consumers, and overall trade relations. By the 1980s the United States and Japan were at odds over Japan's massive trade surplus with the United States.

Observers often point to several unique features of Japanese management and employment to explain the country's meteoric postwar gains. One is recruitment through an extremely challenging (some say brutal) educational system in which technical training has strong emphasis. A rigorous and stressful testing system controls admission to higher education. Japanese management strategies emphasize benevolence toward employees, encouragement of employee loyalty, and participation of workers in decision making. About 20 percent of Japanese workers enjoy guarantees of lifetime employment in their firms, and large Japanese companies are generally active in providing housing and recreational facilities for their employees.

One essential factor in Japan's postwar economic growth was a high level of investment in new and generally very efficient industrial plants. The very low level of military expenditure and the ability of companies to operate at low profit margins helped boost this investment. Investment capital has also been freed by government policies which cut expenditures on amenities and services such as roads, antipollution measures, parks, housing, and even higher education. Japanese industrial cities have grown explosively and have also become very highly polluted areas of dense and inadequate housing, with few public amenities and snarled transportation. The money "saved" has been available for more direct investment in industrial growth, but the costs to Japanese society have been high.

Also crucial to the high level of industrial investment has been a remarkably high rate of savings by the Japanese people; the average household had $116,000 in savings in 1998. Employees commonly receive only about two-thirds of their earnings each year as regular wages or salary, with the other third in the form of one or two lump-sum payments. Most Japanese workers prefer this because it automatically and at least temporarily saves one-third of their income. In addition, buying on credit is less common in Japan than in the United States. There are few mechanisms for lending, so most Japanese consumers must save money to purchase a big-ticket item. Workers must also save to take care of their children and themselves, because Japanese higher education is expensive and because the country's welfare system, pensions, and social security are quite inadequate.

Some analysts cite elements of Japan's political culture to explain the country's economic successes. One political party, the Liberal Democratic Party, has been repeatedly reelected to power since the time Japan regained its sovereignty in 1952. It is a conservative and strongly business-oriented and business-connected organization. The party has promoted Japanese exports with financial policies that have tended to keep the yen (the Japanese unit of currency)—and thus Japanese goods—inexpensive. The party has also cooperated very closely with Japanese business in the development of new products and new industries.

Some analysts believe that the country's postwar economic miracle grew from an intense spirit of achievement and enterprise among the Japanese. Notably, many Japanese attribute this industrious spirit to Japan's geography as a resource-poor island nation. In order to overcome the constraints nature has placed on them, the Japanese people feel they must work harder.

Despite all of these factors in Japan's economic favor, the Japanese "miracle" did not last. The peak of Japan's postwar success came in the 1980s. A powerful economic boom led to speculative rises in stock prices and land prices. At the end of the 1980s, the total value of all land in Japan was four times greater than that of the United States. The Tokyo Stock Exchange was the world's largest, based on the market value of Japanese shares. Then the "bubble economy" burst, and real estate and stock prices fell by more than 50 percent. Japan had near-zero economic growth through most of the 1990s, and the Japanese industries whose growth had seemed unstoppable experienced an increasing loss of market share to U.S. and European producers.

14.4 Japanese Industry

The gradual evolution of industry over more than three centuries which characterized Europe and the United States has been compressed into little more than a century in Japan. Early iron and steel development was concentrated in northeast Kyushu, on and near the country's most significant coal resources. The product went largely into a new railway system and both commercial and naval shipping (Fig. 14.10). Japanese coal, supplemented increasingly by imports, powered the country's factories, railways, and ships. Hydroelectricity was devel-

Figure 14.10   The port of Tokyo. The island nation of Japan thrives on seaborne trade.   *Karen Kasmauski/Matrix*

oped on many mountain streams and became an important element in the energy economy.

The status of Japan's petrochemical and other oil-based industries reflects its great dependency upon imports. Almost devoid of petroleum reserves, the country obtains almost all of its oil and gas as imports, mostly from the Persian/Arabian Gulf region. Japan therefore has carefully avoided alignment with the West in most Middle Eastern disputes, and only after much pressure from the United States and its Western European allies did Japan contribute financially to the multinational Desert Shield/Desert Storm operations which ousted Iraqi troops from Kuwait in 1991.

The phenomenal advance of industry since the early 1950s has been marked by a series of overlapping booms, along with some declines, in various sectors. In the 1960s, electronics, cameras and optical equipment, petrochemicals, synthetic fibers, and automobiles (notably those produced by Mazda, Honda, Toyota, Mitsubishi, and Nissan) became boom industries. In the 1970s and 1980s, Japan became deeply involved in the manufacture of computers and robots. Supported by the government, Japanese companies attempted to take over the industry's leadership from the United States' computer companies. By the early 1980s, Japan established a clear lead over all other countries in the robotization of industry, but the United States took and continues to hold the lead in computer manufacturing.

For the first time in the postwar era, Japanese industry in the 1990s was not characterized by world supremacy in any single category of manufacturing. While Japanese manufacturing evolved with exceptional speed from relatively simple industries based on domestic natural resources and inexpensive labor to highly sophisticated and futuristic industries based on a highly skilled and educated workforce, Japanese industry must now look for new directions in which to establish leadership. As always, in order to overcome the severe limitations imposed by a small, resource-poor island, the Japanese must live by their wits.

14.5 The Social Landscape

Japan has one of the most homogeneous populations (99.5% Japanese) on Earth. Historically, except for a long-standing community of ethnic Koreans (who have often suffered discrimination in Japan), few non-Japanese have settled in the country. Only with the country's booming prosperity and growing labor shortages of the 1980s did Japan open its doors to a few unskilled and low-skilled immigrants from such countries as Pakistan, Bangladesh, Thailand, Peru, and Brazil. This shortage of racial and ethnic diversity has had mixed results for Japan. Many observers believe that it has helped the country achieve a sense of unity of purpose, allowing the Japanese to persist through periods of adversity—especially the postwar years of reconstruction. However, the Japanese have also earned a reputation for intolerance of ethnic minorities.

Capitalist Japan has a remarkably egalitarian society, with about 80 percent of the population comprising its middle class. The richest third of the population has a total income just three times as great as that of the poorest third (compared with five times as much in the United States). The Japanese have historically embraced the concept of *wa,* or harmony, based in part on the principle of economic equality. The recession of the 1990s, however, resulted in growing joblessness and homelessness in Japan, and began to tarnish the country's self-satisfied image. The country's official unemployment rate crept past 4 percent in 1998. During the Asian financial crisis of the late 1990s, there were loud international appeals for Japan to reform its economic system by allowing inefficient businesses to collapse (rather than prop them up with expensive subsidies) and to resist the temptation to satisfy marginal rural populations with costly education and public works projects. The government generally balked at these suggestions because of the threat they posed to a socially harmonious Japan.

The legendary Japanese work ethic has had its advantages and drawbacks: It has helped the Japanese create a prosperous country, but it has also created a nation of workaholics beset with the same problems—stress, suicide, depression, and alcoholism—experienced by workers in the world's other MDCs. Women have not achieved parity with men in the workplace, and they complain increasingly of discrimination and sexism. In struggling to increase their footing in Japanese society, growing numbers of Japanese women are marrying later, contributing to Japan's remarkably low birth rate (10 per thousand annually, one of the lowest in the world). College graduates who are not hired during the annual recruiting season face difficult obstacles to entering the job market. And for those it affects, the practice of lifetime employment makes it difficult to change jobs. One Japanese critic declared that workers in his country "are owned like pets by their companies and housing is offered as the equivalent of dog food."

Crowding and overdevelopment, accompanied by a striking lack of amenities, are inevitable facts of life in Japan. Officially designated parks comprise 14 percent of Japan's land area, but many of these areas are heavily developed with roads, houses, golf courses, and resorts. There has been enormous growth in the popularity of winter sports destinations in northern Honshu and Hokkaido, and of hot springs and mineral bath resorts throughout the country. Reflecting dissatisfaction with too much growth and development, the conservation ethic is already strong and growing in Japan. However, it is difficult for the Japanese to truly "get away from it all" at home.

Daily life for the Japanese is devoid of many of the benefits associated with the MDCs. Apartments and homes are generally very small. Only about half of the country's homes are connected to modern sewage systems. Smog from automobile exhaust and industrial smokestacks is a persistent health hazard. The ever-present traffic jams are known locally as traffic "wars." Japan's much-vaunted rail system, including its high-speed *Shinkansen,* or "bullet," trains, which carry traffic between northern Honshu and Kyushu at speeds averaging 106 miles (170 km) per hour, is chronically overburdened with passengers (Fig. 14.11). Summarizing these living conditions, the Japanese politician Ichiro Ozawa recently described Japan as having "an ostensibly high-income society with a meager lifestyle."

Japan faces a troubled period as it deals with its economic recession. Most analysts expect the country to recover, perhaps turning current adversity into opportunity as it has in the past. The country's highly educated and skilled workforce is a major resource. Although natural assets are lacking, Japan's favorable location should also assist in its recovery. The western Pacific region may well resume dynamic economic growth through the

Figure 14.11 A bullet train glides through the Ginza district of Tokyo. Japan has an efficient, high-speed rail network. *Pete Seaward/ Tony Stone Images*

first decade of the 21st century, and Japan is well-positioned to benefit from trade with the countries of this region. Japan continues to be an important member of the international community; even during recession it has maintained its status as the world's largest donor of foreign aid.

14.6 Divided Korea

Japan is closest to the Asian mainland at its extreme west, where the peninsula of Korea lies only about 110 miles (177 km) away (see Fig. 14.1). Korea's political history during the 20th century, first as a colony of Japan and then as a divided land composed of the separate countries of North and South Korea, has tended to obscure its distinctive culture and contributions to the world. Although the Koreans have been influenced by Chinese culture, and to a lesser extent by Japanese culture, they are ethnically and linguistically a separate people.

Liabilities of Korea's Location

The two Koreas occupy an unfortunate geographical location. These small countries are near larger and more powerful neighbors (China and Russia) that have frequently been at odds with one another and with the Koreans. North Korea adjoins China along a frontier that follows the Yalu and Tumen Rivers; it faces

Japan across the Korea Strait; and, in the extreme northeast, it borders Russia for a short distance. For many centuries, the Korean peninsula has served as a bridge between Japan and the Asian mainland. From an early time, both China and Japan have been interested in controlling this bridge, and Korea was often a subject or vassal state of one or the other. However, from the late 7th century to the mid-20th century, Korea was a unified state, sometimes invaded and forced to pay tribute, but never destroyed as a political entity. The decline of Chinese power in the 19th century was accompanied by the rise of modern Japan, whose influence grew in Korea, and from 1905 until 1945 Korea was firmly under Japanese control. In 1910 it was formally annexed to the Japanese Empire. The legacy of hostility and occupation continues to cast a shadow over relations between the Koreas and Japan, and disputes periodically emerge between them over such issues as control of islands and fishing rights (see Definitions and Insights, below). In an effort to begin mending these rifts, Japanese prime minister Keizo Obuchi used the occasion of a 1998 visit by South Korea's head of state to apologize for Japan's wartime treatment of the Koreans.

Japan lost Korea at the end of World War II. In accordance with the victorious allies' agreements, Soviet forces occupied Korea north of the 38th parallel and United States forces occupied Korea south of that line. Korea was to have become a unified and independent country, but the occupying powers could not reach agreement on the formation of a government. The occupying powers therefore set up separate governments: the democratic Republic of South Korea in the south, under the

Definitions & Insights

JAPAN, KOREA, AND THE LAW OF THE SEA

Some 90 miles (145 km) between the shores of Japan and South Korea in the Sea of Japan lie two small, inhospitable islands known to Japanese as the Takeshima Islands and to Koreans as the Tokdo Islands (see Fig 14.1). South Korea has actually controlled the islands since 1956, but with only one Korean couple and some Coast Guard personnel living on the islands, Japan has periodically asserted its right to them. In 1996 they became the focus of a dispute between Japan and Korea, not because of any riches they contain, but because of a 1970s United Nations treaty known as the **Convention on the Law of the Sea,** which would permit their sovereign power to have greater access to surrounding marine resources.

The Law of the Sea was initiated in an effort to apportion ocean resources as equitably as possible and to avoid precisely the kind of conflict brewing between Japan and South Korea. The treaty gives a coastal nation mineral rights to its own continental shelf, a territorial water limit of 12 miles offshore, and the right to establish an exclusive economic zone (EEZ) of up to 200 miles offshore (in which, for

example, only fishing boats of that country may fish). The power that controls offshore islands such as the Takeshima/Tokdo Islands can extend the area of its exclusive economic zone even further.

By early 1996, 85 nations had ratified the treaty. South Korea ratified the treaty late in 1995. Japan was preparing to ratify it in 1996 when news reached Tokyo that South Korea had plans to build a wharf on the islands. To avoid provoking Japan, North Korea, and China, South Korea avoided declaring an exclusive economic zone off its waters which would include the islands. But fears that South Korea may build facilities and station more people on the islands as a step toward establishing an EEZ—and thereby excluding Japanese fisherman from the area—caused Japan to restate its claim to the islands in 1996. South Korean officials answered with military exercises near the islands, and South Korean civilians staged loud demonstrations outside Japan's embassy in Seoul. The issue may have to be settled in the same international legal arena in which it originated.

Regional Perspective

Nuclear Power and Nuclear Weapons in the Western Pacific

In 1995, on the fiftieth anniversary of the bombing of Hiroshima and Nagasaki, Americans and Japanese did much soul-searching about the use of nuclear weapons. The surviving decision makers generally continued to insist that the use of these weapons spared many thousands of lives, both American and Japanese, that would have been lost if the United States had instead undertaken an invasion of the Japanese homeland. For their part, most Japanese continued to condemn the decision, and it remains official policy that Japan will never develop or use atomic weapons.

However, Japan has come to rely on nuclear power for about 33 percent of the country's electricity needs. Hydropower still supplies 12 percent, but, with fossil fuels almost absent, the critical lack of energy has compelled the nation to develop the world's most energy-efficient economy and to adopt a technology to which many Japanese are opposed. A fire and explosion at a Japanese nuclear waste reprocessing plant in 1997 released small amounts of radiation into the atmosphere, prompting new domestic concerns about the safety of nuclear energy. Japan imports large quantities of plutonium for use in its nuclear power industry, and the long-distance ocean shipment from Europe of this very hazardous material has contributed to Japan's poor reputation in environmental affairs. International protests against the plutonium shipments have caused Japan to postpone the construction of a series of nuclear breeder reactors, which use and create recyclable plutonium.

Japan continues to worry about the potential nuclear threat from three adversaries, all of which possess or have had programs to develop nuclear weapons: Russia, China, and North Korea. The Japanese feel that the West dismissed such potential threats from Russia too readily when the USSR dissolved. Japan still has territorial disputes with Russia, particularly involving the four Kuril islands of Kunashiri, Etorofu, Shikotan, and Habomai, which the government of Josef Stalin seized at the end of World War II and which are just off the coast of northeast Hokkaido (see Fig. 14.9 and Fig. 6.1). Japan and China also have a territorial dispute concerning the East China Sea, and if, as anticipated, oil is discovered there, relations between the two countries could deteriorate. China continued to test nuclear weapons through 1996, adding to tensions in Japan. Finally, Japan fears reunification in Korea, which, as a former Japanese colony from 1910–1945, has a particular historic dislike of Japan. There is speculation that Japan may do the unthinkable—develop nuclear weapons—to counter the perceived threat of North Korea's nuclear weapons program.

The world also continues to look nervously at the troubled relations between North and South Korea and between North Korea and the West. A crisis flared in the spring of 1994 when heavily-armed North Korea refused to permit full inspection of its nuclear facilities by the International Atomic Energy Agency. The country was suspected of separating plutonium that could be used in making nuclear bombs. Some analysts believed that North Korea had already manufactured one or two nuclear weapons. In 1993, the North Koreans had successfully tested a new medium-range ballistic missile capable of carrying a nuclear warhead to most areas of Japan. Unofficial diplomatic talks between former U.S. president Carter and North Korean leader Kim Il Sung helped scale down the tension level. The United States, Japan, and South Korea agreed to assist North Korea in the construction of two nuclear power plants in exchange for a freeze on North Korea's nuclear weapons program. The United States also promised to deliver 500,000 tons of fuel oil yearly to help meet North Korea's winter heating needs. Because of Congressional refusal to appropriate full funding for these shipments, the United States fell behind on its deliveries in 1998. Meanwhile South Korea, suddenly impoverished by its own economic crisis, found itself unable to advance the funds it pledged to help build the nuclear reactors in the North. North Korea responded to these setbacks by proposing to resume its own nuclear power program, a veiled threat to renew work on nuclear weapons.

---

auspices of the United Nations, and the People's Republic of Korea in the north, a Chinese communist satellite. The occupying powers withdrew most of their forces in 1948 and 1949.

Then, in 1950, North Korea attacked South Korea. United Nations units, made up mostly of U.S. forces, entered the peninsula to repel the communist advance. Late in 1950, when the North Koreans had been driven back almost to the Manchurian border, China entered the war and drove the United Nations forces south. A stalemate then developed just north of the 38th parallel until an armistice was arranged in 1953. The border between the two Koreas, often the most tense boundary on Earth, still follows this armistice line. The 151-mile-long (243 km), 2.5-mile-wide (4 km) demilitarized zone is a virtual no-man's-land of mines, barbed wire, tank traps, and underground tunnels (Fig. 14.12). The world's largest concentration of hostile troops faces off on either side of it. Remarkably, with

Figure 14.12 A United Nations observer peers across the demilitarized zone to North Korea. This is one of the most tense borders on Earth. *Mary Beth Camp/Matrix*

the near absence of human activity, rare birds like the Manchurian crane and other endangered animal species have taken refuge in this narrow strip.

Few lands have ever been more devastated than Korea after years of warfare covering the length and breadth of the peninsula. The tragedy has been all the greater in that Korea is not a poor land by nature. During their period of control, the Japanese developed transportation, agriculture, and industry based on a sizable reserve of mineral and power resources. The Korean people received few benefits, however, because the increased production was put mainly to Japanese uses. The division between north and south handicapped the economy after 1945, and after 1950, the enormous physical destruction of the Korean War set the countries back again. The physical scars of the conflict have since been erased, however, and the economies of both the North and South Korean states have been expanded with the help of considerable outside aid.

A state of cold war and periodic border incidents have nevertheless persisted between North and South Korea. Relations between the two countries thawed encouragingly in 1995 when the South sent emergency stocks of rice to the North, but they soon iced over again in 1996, when a North Korean spy submarine ran aground in South Korea. North Korea apologized for the incident, and began diplomatic talks with South Korea, the United States, and China in 1997. In 1998, for the first time since the war, South Korean tourists were permitted to visit the North, but only on strictly monitored cruises; no Northerners were permitted to travel to the South. Family members stranded on each side of the border are still prohibited from visiting one another.

The border between the two Koreas is unlikely to be the tripwire for a third World War, as was feared in the days of the U.S.–Soviet Union Cold War. Any incident along the border could, however, plunge the two Koreas into war and lead inevitably to the intervention of United States troops—beginning with the 37,000 American soldiers now stationed there—and possibly even to the intervention of Chinese forces. Some analysts fear that Korea may yet become a nuclear battlefield.

Contrasts Between the Two Koreas

The Korean armistice line divides one people into two very different countries (see Fig. 14.13 and Table 10.1). North Korea has an area of about 46,500 square miles (120,500 sq km) inhabited by about 22 million people (1998), with a density approaching 500 per square mile (*c.* 185 per sq km). In 1998, South Korea had about 46 million people in only 38,000 square miles, averaging 1218 per square mile (*c.* 470 per sq km). South Korea is a republic that has fluctuated between attempts at democracy and a repressive military dictatorship. It has a capitalist economy heavily dependent on relationships with the United States and Japan. North Korea is a rigid and very tightly controlled communist state which historically vacillated in its principal ties between China and the Soviet Union; however, with the collapse of the USSR, China has become North Korea's main ally and benefactor.

From World War II until his death in 1994, the dictator Kim Il Sung—whom North Koreans called "The Great Leader"—governed the nation. His son, Kim Jong Il, is now in power and to date has not changed the country's militaristic character. North Korea spends a staggering 26 percent of its gross domestic product on its military, compared with South Korea's expenditure of only 3.4 percent, China's at 2 percent, and Japan's at 1 percent.

There are marked physical contrasts between the two countries. Although both are predominantly mountainous or hilly, North Korea is the more rugged. Relatively level lowland is found mainly along the western side of the peninsula in both countries, but South Korea also has some extensive lowland areas in the southeast. In both countries, mountains rise toward the eastern side of the peninsula and drop abruptly into the Sea of Japan with little or no coastal plain. They are highest and most rugged in northeastern North Korea adjoining Manchuria.

North Korea has a humid continental long-summer climate, with hot summers and quite cold winters. South Korea is mostly a humid subtropical area with much shorter and milder winters. Both are strongly monsoonal, with precipitation concentrated in the summer.

Rice is the staple food in both countries. Rice-growing conditions are best in South Korea and diminish northward. Unlike the North, South Korea is able to double-crop irrigated rice fields by growing a dry-field winter crop such as barley after the rice has been harvested. In the North, the main supplementary crop is corn, which must be grown in the summer and therefore cannot occupy the same fields as rice. Beginning in 1995, successive waves of flood and drought brought famine to North

A

B

Figure 14.13 The two Koreas are like night and day. Affluent shoppers enjoy the fruits of free enterprise in South Korea (a), while military security preoccupies austere, Stalinist North Korea (b). *(a) Alain Evrard/Photo Researchers, Inc. (b) Jeffrey Aaronson/Network Aspen*

Korea. The country's almost impenetrable veil of secrecy has made it difficult to calculate the losses, but estimates of numbers of people who died in the famine through early 1999 range from 900,000 to 2,400,000, or up to 10 percent of the country's pre-famine population. As one crop after another failed, North Korea gradually and reluctantly sought food aid from abroad. It coaxed emergency supplies of wheat from the United States in part by threatening to withdraw from its 1994 agreement with the United States to halt its nuclear weapons program (see Regional Perspective, p. 377).

Another physical contrast between the two Koreas is that most of the nonagricultural natural resources are in North Korea. All of the peninsula's major resources—coal, iron ore, some lesser metallic ores, considerable hydropower potential, and forests—are more abundant in the North than in the South. North Korea was thus originally the more industrialized state, featuring a typical communist emphasis on mining and heavy industry, together with hydroelectric production and timber products. However, while the North remained mired in this stage of development, the South took over industrial leadership in the 1970s and 1980s by very rapidly developing a dynamic and diversified capitalist industrial economy. In an effort to begin to close the gap, North Korea opened the Rajin-Sonbong Free Economic and Trade Zone in the mid-1990s. Centered at the extreme northwest corner of North Korea around the twin cities of Rajin and Sonbong (combined population: 140,000),

and ringed by barbed wire, the 288-square-mile (720 sq km) trade zone is intended to attract foreign investment and industries that will capitalize on this site's favorable location near the borders of Russia and China.

In the late 1990s, South Korea's main exports were electrical and electronic equipment and appliances, automobiles, textiles, shoes, iron and steel (from a huge plant on the southeast coast north of Pusan), and ships from a major shipbuilding industry. Large investments from Japan and the United States, as well as access to markets in those countries, were fundamental to the explosive development of these industries after the Korean War. Also important were the skills of South Koreans receiving higher education abroad and at home (the country has the world's highest number of Ph.D.'s per capita), the availability of inexpensive and increasingly skilled Korean labor, and the vigorous support of industrialization by strong and sometimes ruthless governments.

The burgeoning cities of South Korea are visible evidence of the country's resolve to industrialize. The capital city of Seoul has reached a metropolitan area population of 11.8 million (10.08 million, city proper), ranking it last among the top ten most populous cities in the world. The port of Pusan has 3.8 million people (metropolitan area) and the inland industrial center of Taegu has more than 2 million. By contrast, the only large city in North Korea is the capital, Pyongyang, with a metropolitan population of 2.36 million.

South Korea is one of Asia's success stories, an "Asian Tiger" that enjoyed a rocketing economic growth rate of up to 10 percent annually in the mid-1990s. The development of its high-tech industries has been causing concern to competitors in Japan and the United States; for example, the Samsung electronics firm has established an increasing market share in an industry dominated by Japanese firms. South Korea is also poised for especially strong growth in the automobile (notably Hyundai cars and trucks), semiconductor, information processing, telecommunications, and nuclear energy industries.

South Korea's breathless development stopped and then reversed in 1998 as the shock waves of Asia's spreading economic crisis hit the country. To prevent South Korea from slipping into depression, the International Monetary Fund (IMF) extended the offer of a huge loan package conditional upon the implementation of several reforms, including a restructuring of the way business is conducted. Most of South Korea's wealth is in the hands of about 15 gigantic, interlocking conglomerates called *chaebols*. These firms employ the majority of South Korea's working population and own most of the banks. During the country's explosive period of growth, the banks lent money back to their parent companies for risky investments all over the world, particularly throughout Asia. As the Asian economic crisis spread, those projects were halted, and the borrowers defaulted on their loans. Foreign investors rushed to pull their capital out of South Korea, whose national currency and foreign reserves deflated. The United States saw its fifth-largest trading partner at the brink of economic collapse and—in reaction to a new economic "domino theory" that suggested that if South Korea fell, so too would Japan, and perhaps the United States—stepped in with a large pledge to the IMF bailout fund. Such is the importance of the western Pacific Rim to the global economic system.

SUMMARY WITH SELECTED KEY TERMS

- The people of **Japan** operate an **economy larger than that of any nation except the United States. The Korean peninsula** has also been **central in the geopolitical affairs** of the second half of the 20th century because of its **strategic location.**
- There are four main islands in the Japanese archipelago: **Hokkaido, Honshu, Shikoku, and Kyushu.** The smaller **Ryukyu Islands** extend almost to Taiwan.
- All of the main islands of Japan consist largely of **volcanic mountains** and are subject to **frequent and sometimes very severe earthquakes.**
- **Affluence has increased dramatically in Japan** to the point that material demands require the nation to import most of its raw materials, energy supplies, and a large share of its food.
- The people of Japan view the ocean as an important resource and promising frontier. Japanese have the **world's highest per capita consumption of fish and other marine foods.**
- Most of Japan's economic activities and a majority of its people are packed into a corridor about 700 miles long. This **megalopolis** extends from **Tokyo, Yokohama,** and the surrounding **Kanto Plain** on the island of Honshu, through northern Shikoku, to northern Kyushu.
- Korea's political history of the 20th century, as a **colony of Japan** and then as a **divided land composed of North and South Korea,** has obscured its distinctive culture and contributions to the world.
- Although the Koreans have been influenced by **Chinese culture** and to a lesser extent by **Japanese culture,** they are **ethnically and linguistically a separate people.**
- The two Koreas occupy an **unfortunate geographical location.** They are near larger and more **powerful neighbors** that have frequently been at odds with one another and the Koreans.
- There are **marked physical and economic contrasts between North and South Korea** with North Korea being much more rugged. **North Korea has a humid continental long-summer climate,** with hot summers and cold winters, while **South Korea is mostly a humid subtropical area** with much shorter and milder winters. Capitalist South Korea is far more prosperous than communist North Korea.

REVIEW QUESTIONS

1. Using maps and the text, identify the four main islands and the Ryukyu Islands of Japan.
2. Using maps and the text, identify the areas of Japan's climatic zones of humid subtropical, humid continental long-summer, and humid continental short-summer.
3. List the notable natural resource assets of Japan.
4. Using maps and the text, identify Japan's long megalopolis corridor.
5. Why is Japan such a large importer of wood products while 76 percent of the land cover is forest?
6. What are the major theories for Japan's great economic success?

7. What goals propelled Japan's expansionist drive from 1875 to the end of World War II?
8. What three national adversaries are perceived as major military threats by Japan today?
9. Using maps and the text, explain why the Korean peninsula could be viewed as a "land bridge" between Japan and the rest of Asia.
10. List the physical, political, and economic contrasts between North and South Korea.

DISCUSSION QUESTIONS

1. What have been the consequences of Japan's existence as an archipelago of fragmented islands?
2. Of what significance is Japan's location on the edge of the Pacific Ring of Fire?
3. How has Japanese agricultural production been expanded beyond the limits of its small arable land area?
4. Why would the Japanese believe the ocean or "inner space" to be of such importance?
5. Explain the major reasons for the location of Japan's core area or megalopolis.
6. How did Japan become the first Asian nation to attain the status of a world power?
7. Discuss the significance of Japan's historical background in relation to its status in today's world.
8. Explain why Japan has one of the most homogeneous populations on Earth.
9. Why has the development of the Korean peninsula been so influenced by its proximity to China, Russia, and Japan?
10. What precipitated the division of the Korean peninsula into North and South Korea?

6 | The Pacific World

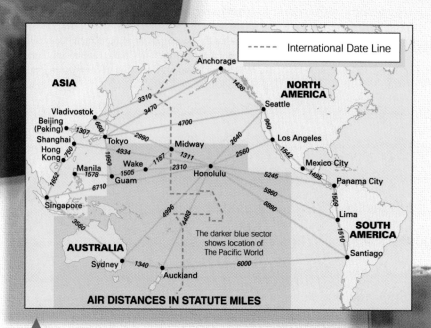

International Date Line

ASIA · NORTH AMERICA · SOUTH AMERICA · AUSTRALIA

Anchorage · Vladivostok · Beijing (Peking) · Shanghai · Hong Kong · Tokyo · Manila · Wake · Guam · Midway · Honolulu · Singapore · Seattle · Los Angeles · Mexico City · Panama City · Lima · Santiago · Sydney · Auckland

3310 · 3470 · 660 · 1307 · 750 · 2990 · 4934 · 1990 · 1578 · 1505 · 1187 · 1311 · 2310 · 1652 · 6710 · 3550 · 4996 · 14083 · 1438 · 4700 · 960 · 2640 · 2560 · 1542 · 1495 · 5245 · 5960 · 6880 · 1509 · 1510 · 1340 · 6000

The darker blue sector shows location of The Pacific World

AIR DISTANCES IN STATUTE MILES

▲

Vast distances separate generally small islands, and a single island continent, across the world's largest body of water.

Micronesia, Melanesia, and Polynesia are geographical devices useful in classifying the environmental and cultural subregions of the Pacific World. *George Holton / Photo Researchers, Inc.*

▼

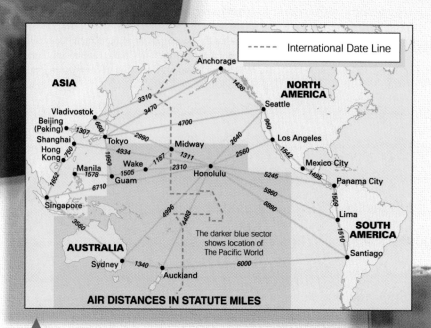

▲

People from the affluent, industrialized West have long stereotyped and subjugated the inhabitants of the Pacific World. The process of decolonization is nearly complete, but some Western powers maintain military interest in the region. *Nik Wheeler / Westlight*

Regional Snapshot

Karl & Jill Wallin / FPG International

Volcanic high islands and continental islands contain most of the region's economic resources and human populations. The low islands are poorer, less populated, and are vulnerable to anticipated rises in sea level. A high degree of endemism is characteristic of island flora and fauna. *Paul Chesley / Tony Stone Images*

The economic and strategic orientations of Australia and New Zealand have been changing, generally away from Britain and toward East Asia and the United States. The economic crisis in Asia may dim the prospects for the long-anticipated emergence of the 21st century as the "Pacific Century." *David Austen / Tony Stone Images*

Indigenous peoples of the Pacific, including the Aborigines of Australia and the Maori of New Zealand, have unique ethnogeographical perspectives that conflict with the world views of the Europeans who have come to dominate their homelands. *Joseph J. Hobbs*

Chapter 15

A Geographic Profile of the Pacific World

▲ *Earth's largest ocean, the vast Pacific. This is a composite of several thousand satellite images.* Tom Van Sant/Geosphere Project, Santa Monica/Science Photo Library/Photo Researchers, Inc.

CHAPTER OUTLINE

15.1 Melanesia, Micronesia, and Polynesia

15.2 Land and Life on the Islands

15.3 Subjugation, Independence, and Development

Covering fully one-third of the Earth's surface, the Pacific World is mostly water. The Pacific itself, the world's largest ocean, is bigger than all the Earth's continents and islands combined. Before World War II, the Western world spawned legends about this ocean and its islands as a kind of utopia. Some islands still do have an idyllic quality for European and American visitors. However, there has long been trouble in this paradise. On many islands, traders, whaling crews, labor agents, and other opportunists exploited the indigenous peoples, reducing their numbers and disrupting their cultures. The military battles of World War II shattered whatever idyllic quality remained in many islands. Yet today, the Pacific mystique, which has been perpetuated in the works of many noted writers, forms part of the allure for the massive development of tourism in such places as the U.S. state of Hawaii, the island of Tahiti, and other islands of the South Pacific.

The Pacific World region (often called Oceania) encompasses Australia, New Zealand, and the islands of the mid-Pacific lying mostly between the Tropics (Table 15.1, p. 388). The Pacific islands nearer the mainlands of east Asia, Russia, and the Americas are excluded here on the basis of their close ties with the nearby continents. Large areas of the eastern and northern Pacific that contain few islands are also largely discounted. Because Australia and New Zealand are sufficiently different from the tropical island realms to the north, they could be considered a separate world region, but they are included here in the Pacific World because of their strong political and economic interests in the tropical islands, their similar insular character, and the ethnic affiliations of their original inhabitants with the peoples of those islands.

15.1 Melanesia, Micronesia, and Polynesia

The Pacific islands are commonly divided into three principal regions: Melanesia, Micronesia, and Polynesia (Fig. 15.1). The islands of **Melanesia** (Greek: "black islands"), bordering Australia on the northeast, are relatively large. New Guinea, the largest, is about 1500 miles long (c. 2400 km) and 400 miles across (c. 650 km) at the broadest point. In general, these islands are hot, damp, mountainous, and carpeted with dense vegetation. **Micronesia** (Greek: "tiny islands") includes thousands of small and scattered islands in the central and western Pacific, mostly north of the equator. Before World War II, Japan held Micronesia under a mandate from the League of Nations, except for the Gilbert Islands, which were a British possession, and Guam, a United States possession. After the war, the islands held by Japan became the Trust Territory of the Pacific Islands, administered by the United States. **Polynesia** (Greek: "many islands") occupies a greater expanse of ocean than does either Melanesia or Micronesia. It is shaped like a rough triangle whose corners are at New Zealand, the Hawaiian Islands, and remote Easter Island.

Diverse peoples inhabit the huge Pacific realm. Four distinct geographical races call the region home. The dark-skinned but non-Negroid Australian Aborigines are indigenous to Australia. The Melanesians, who are also dark-skinned, inhabit New Guinea and islands to the east. The dark-skinned Micronesians, who live west of Polynesia and north of Melanesia, probably originated from the mixing of Melanesians and Southeast Asians. The lighter-skinned Polynesians are relative latecomers to the region; they descended mainly from a culture which originated in the interior of Southeast Asia about 5000 years ago, and then advanced southward into the Malay Peninsula. From there, over a period of several thousand years, people of this "Malayo-Polynesian" culture embarked on voyages to Madagascar, New Zealand, Easter Island, and finally Hawaii, "island-hopping" all the way and accomplishing extraordinary feats of navigation in their outrigger canoes.

Hundreds of distinct cultures emerged from the complex origins of the Pacific peoples. New Guinea and the Solomon Islands alone are home to 800–900 different languages, making them—on a per capita basis—the most linguistically diverse region in the world.

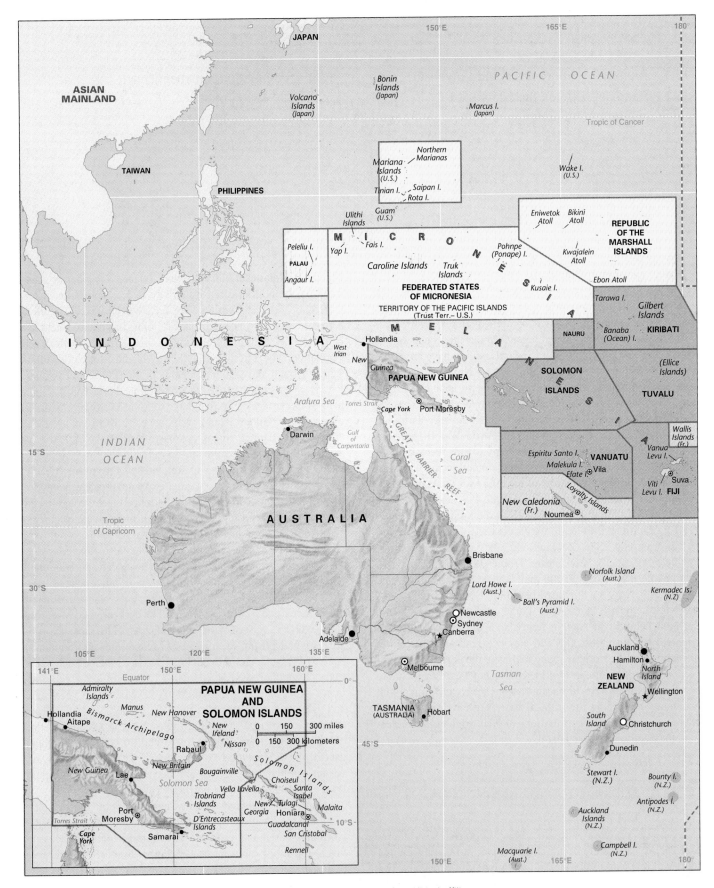

Figure 15.1   General reference map of the Pacific World. In some cases, the tints for political affiliation do not show precise political boundaries. Note the peculiar jog in the international date line near Kiribati. In a bid to attract more tourists by being the first country in the world to greet the new millennium, Kiribati unilaterally decided in 1997 to shift the date line eastward by more than 2000 miles.

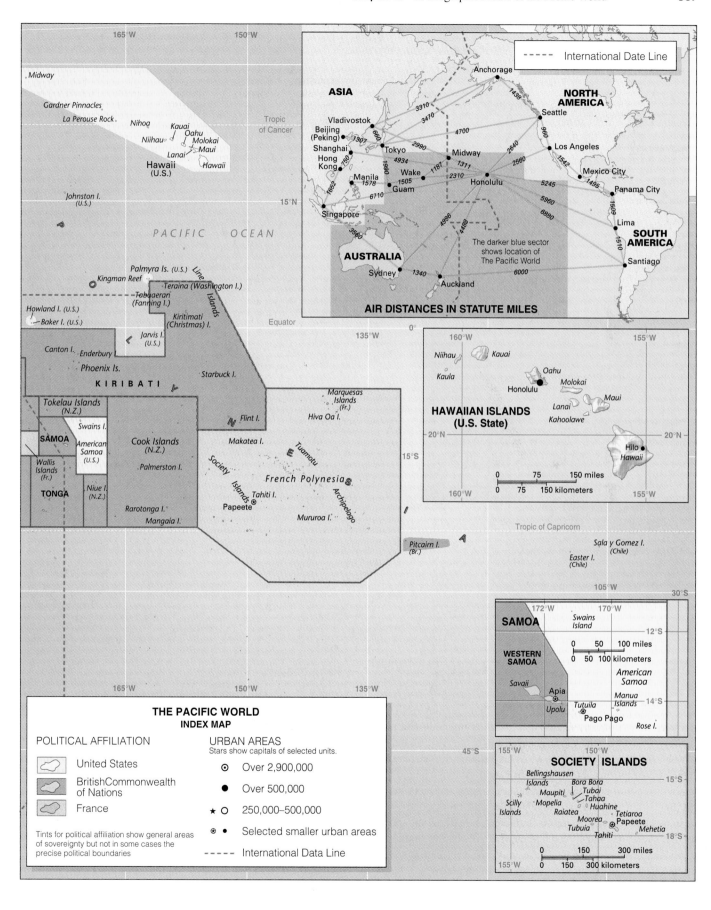

THE PACIFIC WORLD
INDEX MAP

POLITICAL AFFILIATION

United States

BritishCommonwealth of Nations

France

Tints for political affiliation show general areas of sovereignty but not in some cases the precise political boundaries

URBAN AREAS
Stars show capitals of selected units.

⊙ Over 2,900,000

● Over 500,000

★ ○ 250,000–500,000

⊛ ● Selected smaller urban areas

----- International Data Line

AIR DISTANCES IN STATUTE MILES

HAWAIIAN ISLANDS (U.S. State)

SAMOA

SOCIETY ISLANDS

Table 15.1 The Pacific World: Basic Data

Political Unit	Area		Estimated Population (millions)	Annual Rate of Increase (%)	Estimated Population Density (sq mi)	Infant Mortality Rate (per thousand)	Urban Population (%)	Arable Land (%)	Per Capita GNP ($US)	
	(thousand/ sq mi)	(thousand/ sq km)								
Australia and New Zealand										
Australia	2951.1	7644.4	18.7	0.7	6	5.3	85	6	20,090	
New Zealand	103.5	268.1	3.8	0.8	37	6.7	85	9	15,720	
Total	**3055**	**7312.5**	**22.5**	**0.7**	**7.1**	**5.5**	**85**	**6.1**	**19,352**	
Other										
Fed. States of Micronesia	0.3	0.8	0.1	2.6	422	46	27	0	2070	
Fiji	7	18.1	0.8	1.8	114	17	46	10	2470	
French Polynesia	1.4	3.6	0.2	1.8	162	11	54	1	8000	(GDP)
Guam	0.2	0.5	0.2	2.4	700	9.1	38	11	19,000	(GDP)
Marshall Islands	0.07	0.2	0.1	3.6	901	26	65	0	1890	
New Caledonia	7.1	18.4	0.2	1.7	29	8	71	0	8000	(GDP)
Palau	0.2	0.5	0.02	1.6	90	25	69	0	5000	(GDP)
Papua New Guinea	174.8	452.7	4.3	2.4	25	77	15	0.1	1150	
Solomon Islands	10.8	30	0.4	3.2	38	28	13	1	900	
Vanuatu	4.7	12.2	0.2	2.8	38	41	18	2	1290	
Western Samoa	1.1	2.8	0.2	2.4	166	21	21	19	1170	
Total	**207.7**	**539.8**	**6.7**	**2.4**	**32.3**	**57**	**23.5**	**0.6**	**2279**	
Summary Total	**3262.7**	**7852.3**	**29.2**	**1.1**	**8.7**	**17.3**	**70.9**	**5.8**	**15,435**	

Sources: *World Population Data Sheet,* 1998. United Nations Statistics Division, 1998. *Almanac of Politics and Government,* 1998. *World Factbook,* 1997.

15.2 Land and Life on the Islands

The landscapes, seascapes, and cultural geographies of the almost countless islands scattered across the tropical Pacific vary widely. However, there are generally two types of islands—"high" and "low" (Fig. 15.2). Most high islands are the result of volcanic eruptions, but high continental islands (islands attached to continents before sea level changes and tectonic activities isolated them) are much larger and more complex in origin. The low islands are made of coral, a material composed of the skeletons and living bodies of small marine organisms that inhabit tropical seas. These very different settings offer quite different opportunities for human livelihoods.

High Islands

Some of the volcanic high islands of the Pacific comprise **island chains.** These are formed by the oceanic crust sliding over a stationary "**hot spot**" in the Earth's mantle where molten magma is relatively close to the crust. As the crust slides over the hot spot, magma rises through the crust to form new volcanic islands. The Hawaiian Island chain is an excellent example. The big island of Hawaii is now situated over the hot spot and is still active; it is one of the chain's youngest members. The Pacific Plate of the Earth's crust is moving northwestward here; therefore, older volcanic islands which were born over the hot spot have moved northwestward, where they have become inactive. Still older islands farther northwest in the

A

B

Figure 15.2 Characteristic high and low islands of the Pacific. The high island of Bora Bora (a) in French Polynesia is an atoll in the making, roughly in the middle stage of the sequence pictured in Figure 15.4. Clouds are forming over the higher parts of the island. The photograph was taken at low tide; note the extensive sandbars. In the foreground, note the airstrip which has facilitated tourist access to this lovely, world-famous island. With less diverse attractions and resources, low islands (b) are most associated with the coconut civilizations of the Pacific. Low islands are extremely vulnerable to the dangers which would be posed by global warming. *(a) Tony Stone Images/Paul Chesley, (b) FPG International/Karl & Jill Wallin*

chain have submerged to become **seamounts,** or underwater volcanic mountains.

Many of the high volcanic islands are spectacularly scenic (Fig. 15.2a). Steep slopes predominate, and some islands have peaks thousands of feet high. However, there are great variations in elevation, slope, soil, rainfall, and plant life from island to island and even on a given island. Among islands of this type, the Hawaiian, Samoan, and Society groups, all located in Polynesia, are especially well known.

The main cities and seaports in the tropical Pacific islands are in the volcanic high islands and the continental islands. Valuable minerals are scarce, but the soil is generally fertile. Varied physical conditions in the volcanic islands allow a diversity of tropical crops. The islands' original peoples depended mainly on starchy foods such as breadfruit, plantains, taro, sweet potatoes, and yams. Pigs and sea fish supplied animal protein. Europeans, North Americans, and East Asians subsequently introduced other crops and animals, including arrowroot and cassava; bananas; tropical fruits such as mangoes, pineapples, papayas, and citrus; coffee and cacao; sugarcane; and cattle, goats, and poultry.

The newcomers established sugar plantations on some islands, most prominently in the Hawaiian Islands, Fiji Islands (Fig. 15.3), and Saipan in the Marianas. They also introduced laborers from outside the region because the local peoples generally proved too few in number or refused to work in the cane fields. Successive infusions of Chinese, Portuguese, Japanese, and Filipinos brought a polyglot character to Hawaii.

British plantation owners imported mainly Hindu Indian laborers to Fiji, where indigenous Fijians now outnumber ethnic Indians only slightly. Ethnic tensions in Fiji after independence (1970) led to a military takeover of the government by native Fijians in 1987. In Saipan, Japanese interests using Japanese labor developed the island's sugar plantations, but the sugar industry did not survive the devastating U.S. military invasion of

Figure 15.3 Sugarcane is a major commercial crop in a few high islands of the Pacific World. The small locomotive in the view is hauling a load of cane on a plantation in the Fiji Islands. *William A. Noble*

Definitions & Insights

DEFORESTATION AND THE DECLINE OF EASTER ISLAND

"Easter Island is Earth writ small," wrote traveler Jared Diamond. Recent archeological and paleobotanical studies suggest that the civilization that built the island's famous monolithic stone statues destroyed itself through overpopulation and abuse of natural resources (Fig. 15.A). Deforestation, or the removal of trees by people or their livestock, is well known to have serious repercussions wherever it occurs, but perhaps nowhere are these impacts so apparent as they are on Easter Island.

When the first colonists from eastern Polynesia reached remote Easter Island in about 400 A.D., they found the island cloaked in subtropical forest. Plant foods, especially from the Easter Island palm, and animal foods, notably porpoises and seabirds, were abundant. The human population grew rapidly in this prolific habitat. The complex, stratified society that emerged on the island grew to an estimated 7000 to 20,000 people between 1200 and 1500 A.D., when most of the famous statues were built. Apparently in association with their religious beliefs, Easter Islanders erected more than 200 statues, some weighing up to 82 tons and reaching 33 feet (10 m) in height, on gigantic stone platforms. At least 700 more statues were abandoned in their quarry sites and along roads leading to their would-be destinations, "as if the carvers and moving crews had thrown down their tools and walked off the job," Diamond observed.[1]

"Its wasted appearance could give no other impression than of a singular poverty and barrenness,"[2] the Dutch explorer Jacob Roggeveen wrote of the island on the day he discovered it—Easter Sunday, 1722. Not a single tree stood on the island. The depauperate landscape bears testimony to the fate of the energetic culture which built the great statues. By 800 A.D., people were already exerting considerable pressure on the island's forests for fuel, construction, and ceremonial needs. By 1400, people and the rats they introduced to the island caused the local extinction of the valuable Easter Island palm.

Continued deforestation to make room for garden plots and to supply wood to build canoes and to transport and erect the giant statues probably eliminated all of the island's forests by the 15th century. Also by then people had hunted to extinction many terrestrial animal species and could no longer hunt porpoises because they lacked the wood needed to build seagoing canoes. Crop yields declined because deforestation led to widespread soil erosion. There is evidence that in the ensuing shortages, people turned on each other as a source of food. By about 1700 A.D., the population began a precipitous decline to only 10–25 percent of the number who once lived on this isolated Eden.

Figure 15.A   The Polynesian people who erected these *moai* statues between 1200 and 1500 A.D. deforested most of Easter Island—in part to supply levers and rollers for the statues—and thus precipitated the demise of their civilization.   *George Holton/Photo Researchers*

[1]Diamond, Jared. 1995. "Easter's End." *Discover* 16(8): page 64.

[2]Ibid.

Saipan in World War II. Sugar is still important in Hawaii and Fiji, however.

Some high islands to the north and northeast of Australia are continental islands, including New Guinea, New Britain and New Ireland in the Bismarcks, New Caledonia, Bougainville and smaller islands in the Solomons, the two main islands of Fiji, and a few others. Mountains over 16,000 feet (4877 m) high rise in New Guinea, and lower but still imposing mountains exist on various other islands. Valuable minerals are extracted on a few of these islands, including the nickel of the French-held island of New Caledonia (the world's third largest

producer of nickel, after Russia and Canada) and the copper of Bougainville in Papua New Guinea.

Mineral wealth has sometimes brought unrest to these islands, notably New Caledonia and Bougainville. Separatists demanding independence for nickel-rich New Caledonia clashed with French police in 1988. An accord signed after the confrontation guaranteed a 1998 referendum for New Caledonia's people (known as Kanaks), who would then choose to become independent or remain under French rule. In 1998, however, apparently to maintain its interests in the colony's nickel, France postponed the vote for at least 15 years.

In 1988 a crisis also shook Bougainville, a former Australian colony which in 1975 became part of newly independent Papua New Guinea at Australia's insistence. Angry landowners calling themselves the "Black Rambos" destroyed mining equipment and power lines to force the closure of the world's third largest copper mine. These generally younger landholders were receiving none of the royalties and compensation payments made by New Guinean and Australian mining interests to an older generation of the island's inhabitants. Residents of Bougainville concerned about the ruinous environmental effects of copper mining and disturbed that most mine laborers were imported from mainland Papua New Guinea—with little mining revenue remaining on the island—joined the opposition. The growing crisis drove world copper prices to record high levels. Opposition to the Australian-backed mining interests of Papua New Guinea in Bougainville grew after 1988 into a full-fledged independence movement led by the Bougainville Revolutionary Army (BRA). Late in 1989, war broke out when the BRA leader declared independence from Papua New Guinea; most Bougainvilleans wanted to be united again with the Solomon Islands, whose people they are culturally and ethnically related to. Papua New Guinea responded with an economic blockade around Bougainville to prevent supplies from reaching the rebels, who salvaged World War II–era weapons and ammunition to fight superior troops. In a decade of fighting, Bougainville's 200,000 people suffered huge losses, with nearly 20,000 killed and 40,000 made homeless. A cease-fire in 1998 finally brought to an end the longest conflict in the Pacific since World War II.

Low Islands

The low islands are generally smaller than the volcanic high islands and lack the resources to support dense populations. They are typically fringed by the waving coconut palms that are a mainstay of life and the trademark of the "South Sea isles." The islands of the republic of Kiribati (pronounced KIH-rih-bus) are typical of the picturesque low islands idealized by Hollywood and travel brochures (Fig. 15.2b). These islands generally pose several natural hazards to human habitation. The lime-rich soils are often so dry and infertile that trees will not grow, and a shortage of drinking water often prevents permanent settlement. Their low elevation above sea level provides little defense against huge storm waves and tsunamis, the waves generated by earthquakes. There is increasing concern that these islands will be the world's first victims of the rise in sea level due to global warming. Sea levels have risen in recent years at a rate of 2 millimeters (.08 inches) annually, and there are reports of unprecedented tidal surges on Pacific island shores. If the trend continues, the first Pacific islands to be totally submerged would be Kiribati, the Marshall Islands, and Tuvalu, while Tonga, Palau, Nauru, Niue, and the Federated States of Micronesia would lose much of their territories to the sea. The island nations are among the 39 countries comprising the Alliance of Small Island States, which argued forcefully but unsuccessfully at the 1997 Kyoto Climate Change Conference that, by 2005, global greenhouse gas emissions should be reduced to 20 percent below their 1990 levels.

The low islands are made of coral and usually form an irregular ring around a lagoon. Such an island is called an atoll. Generally the coral ring is broken into many pieces, separated by channels leading into the lagoon, but the whole circular group is commonly considered one island. Charles Darwin devised a still widely accepted explanation for the three-stage formation of atolls (Fig. 15.4). First, coral builds a fringing reef around a volcanic island. Then, as the island slowly sinks, the coral reef builds upward and forms a barrier reef separated from the shore by a lagoon. Finally, the volcanic island sinks out of sight and a lagoon occupies the former land area whose outline is reflected in the roughly oval form of the atoll.

(A) Volcanic island with fringing reef

(B) Slight subsidence barrier reef

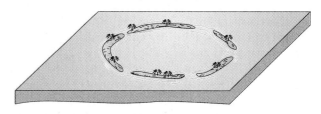

(C) An atoll

Figure 15.4 Charles Darwin's explanation of the development of an atoll. First (top), a fringing coral reef is attached to the volcanic island's shore. Then, as the island subsides (center), a barrier reef forms. With continued subsidence, the coral builds upward, and the volcanic center of the island finally becomes completely submerged, forming an atoll (bottom). *From Gabler et al. 1987. Essentials of Physical Geography, p. 562.*

Regional Perspective

Nuclear Weapons Testing in the Pacific

The remote locales and nonexistent or small human populations of some Pacific islands have made them irresistible venues for weapons tests by the great powers controlling them. In the 1940s and 1950s the United States used the Bikini atoll in the Marshall Islands as one of its chief testing grounds (the island's fame at the time led French designers to name a new two-piece bathing suit after it). In 1946, U.S. government authorities relocated the indigenous inhabitants of Bikini to another island. Over the next 12 years came 23 nuclear blasts. One detonation had the power of a thousand Hiroshima bombs and produced radioactive dust that fell downwind on the Rongelap atoll, where children played in the dust "as though it were snow," one observer wrote.[3] The Marshallese today suffer a legacy of cancers and deformities linked to the weapons tests on Bikini and Eniwetok atolls. So far, the United States has provided more than $63 million to compensate Marshall Islanders made ill by the blasts.

The Bikinians have never accepted their exile. But returning to Bikini is a hazardous prospect given its long legacy of radioactive soil. A repatriation of the Bikinians in the 1960s was found to be premature, as radiation persisted, and the nuclear nomads moved again. Recent tests have determined that Bikini has little ambient radioactivity and is essentially habitable, except for the cesium-137 that permeates the soil and becomes concentrated in fruits and coconuts. Bikinians do not want to go home to stay until they have assurances from the United States that the problem of poisoned produce can be resolved and that the island is completely safe.

However, now that Bikini is at least safe to visit, some Bikinians are capitalizing on its formerly off-limits status. The island's lack of human inhabitants for more than 50 years has had the remarkable effect of rendering Bikini a true marine wilderness. In 1997 it opened as a tourist destination, with naval ships sunk by the atomic blasts a highlight for many divers. The dive masters and tour guides are Bikinians.

Like the United States, France has had a nuclear stake in the Pacific. In 1995, after a three-year hiatus, France resumed underground testing of nuclear weapons on the Mururoa ("Place of Deep Secrets") atoll in French Polynesia. There was an immediate backlash from environmental and other groups. Crew members of the *Rainbow Warrior 2,* the flagship of the environmental organization Greenpeace, confronted French Navy ships near Mururoa in July 1995. French commandos used tear gas to overwhelm the crew and seize the ship. Greenpeace managed to televise the event, inflicting enormous public relations damage to France. The event was the continuation of a long-running feud between Greenpeace and France which had first come to a head in 1985 when French agents attacked and sank the original *Rainbow Warrior* at anchor in New Zealand's port of Auckland as it was preparing to sail to the Mururoa test site. The revelation that the French government was behind the attack was a serious embarrassment for France and boosted antinuclear sentiment throughout the region.

Throughout the Pacific there were strong negative reactions to the 1995

[3]Kristof, Nicholas D. March 5, 1997. "An Atomic Age Eden (but Don't Eat the Coconuts)." *The New York Times,* p. A4.

The Vulnerable Island Ecosystem

Island ecosystems in the Pacific region, like those across the globe, are typically inhabited by **endemic** plant and animal species—those found nowhere else in the world. They result from a process in which ancestral species colonize the islands from distant continents, and, over a long period of isolation and successful adaptation to the new environments, evolve to become new species. Gigantic and flightless animals, like the moa bird of New Zealand and the giant tortoises of the Galapagos and Seychelles islands, are typical of endemic island species.

Generally developing in the absence of natural predators, and inhabiting relatively small areas, island species have proved to be especially vulnerable to the activities of humankind. People purposely or accidentally introduce to the island new plant and animal species (called **exotic** species) which often prey upon or overtake the endemic species. Habitat destruction and deliberate hunting also lead to extinction. For example, New Zealand's moas and the dodo birds of the Indian Ocean island of Mauritius are now "dead as dodos." Indigenous inhabitants of New Zealand set fires to hunt the giant, ostrichlike moa, and had already killed off most of the birds by the time the Maori people colonized New Zealand around 1350 A.D. By 1800, the moa was extinct.

From one end of the Pacific World to the other, there is heightened concern about such human-induced environmental changes. The U.S. Hawaiian Islands have become known to ecologists as the "extinction capital" of the world because of the irreversible impact that exotic species, population growth,

weapons tests. Strongly antinuclear New Zealand protested most loudly. In Australia, customers boycotted imported French goods and local French restaurants, and postal employees refused to deliver mail to the French embassy and consulates. Antinuclear protestors firebombed the French consulate in Perth. Australia joined New Zealand in filing a case at the World Court in The Hague aimed at halting French nuclear tests. Worldwide, informal boycotts diminished sales of French wines. In French Polynesia itself, activists for independence from France used the issue to highlight international attention to their demands. Over several days in September 1995, hundreds of anti-French demonstrators in Tahiti rampaged through French Polynesia's capital city of Papeete, looting and setting parts of the city on fire and causing $11 million worth of damage to the island's international airport, a vital touristic hub.

Throughout the controversy, French authorities insisted that the tests and the atoll itself were safe. About 1500 French military personnel live on Mururoa, subsisting largely on supplies flown in from Tahiti but also on fish caught in local waters. French spokesmen swam in Mururoa's lagoon to show press photographers the safety of the water above the test site, which was located deep in the basalt basement rock where the blasts had been detonated since 1981.

Despite the nuclear tests, the majority of inhabitants of French Polynesia have long favored continued French rule. Tourism and other local businesses contribute only about 25 percent of the territory's revenue; the remainder comes from French economic assistance, largely in the military sector. Many locals, however, whose parents and grandparents gave up subsistence fishing and farming for jobs in the military and its service establishments, are worried about their future. France signed the Comprehensive Test Ban Treaty and ceased its nuclear tests in 1996—raising fears about the potential economic impact if France withdraws its investments and aid from the region.

Similar fears stalk Palau, the Marshall Islands, and the Federated States of Micronesia (FSM), three independent nations carved from the former U.S. Trust Territory of the Pacific. The United Nations awarded that territory, which includes the Bikini and Eniwetok atolls, to the United States in 1946 as the world's first and only political-military zone. The United States still leases some islands to conduct military tests—for example, on Strategic Defense Initiative ("Star Wars") technology—and provides large cash grants and development aid to the three countries. But with the Cold War over and the U.S. treaty arrangements with these nations due for review in 2001, many islanders fear the bottoms of their economies will fall out. The people of the Marshall Islands and FSM are especially worried, since more than half of the gross domestic product of each country is comprised of economic aid from the United States.

and development have had on indigenous wildlife there. Environmentalists are also concerned about the impact of commercial logging on the island of New Guinea. About 22,000 plant species grow there, fully 90 percent of them endemic. And ecologists consider nearby New Caledonia to be one of the world's biodiversity "hot spots" because of ongoing human impact on its unique flora and fauna (see Chapter 2, p. 34).

Human-induced extinctions complement a long list of natural hazards to island species. For example, island animals are particularly prone to natural catastrophes such as typhoons (hurricanes) and volcanic eruptions. Most animals find it difficult or impossible to evacuate in the event of a natural catastrophe, and the island's distance from other lands may hinder or prevent recolonization. So, while island habitats have a higher rate of endemism than similar climates on mainlands, environmental difficulties often cause the number of different kinds of species (the species diversity) to be lower.

15.3 Subjugation, Independence, and Development

Europeans began to visit the Pacific islands early in their Age of Exploration. Spanish and Portuguese voyagers were followed by Dutch, English, French, American, and German explorers. Many famous names are connected with Pacific exploration, including Magellan, Tasman, Bougainville, La Perouse, and Cook. By the end of the 18th century, Europeans knew virtually all of the important islands. For a long period, the European governments exercised only nominal control. European whaling crews, sandalwood traders, indentured labor contractors, and other adventurers abused the indigenous peoples. On island after island, European penetration decimated the islanders and disrupted their cultures. The intruders introduced venereal and other infectious diseases, alcohol, opium, forced labor, and firearms, which greatly increased the slaughter in tribal wars. Four and a half centuries of turbulent Western influence in the Pacific islands do not reflect well on the outsiders.

In recent times, plantation agriculture, mining, fishing, military facilities and activities, and an expansion of tourism have attracted Western personnel and capital to the islands. These activities are widespread except for mining, which is confined largely to New Caledonia, Papua New Guinea, and Nauru. The new surge of Western interest has accompanied a steady process of decolonization. Since the end of World War II, the United States, Britain, Australia, and New Zealand have abandoned most of their colonies in the region. Only France has insisted on holding onto all of its colonies.

The Pacific islands are now midstream between colonialism and independence. Once entirely colonial, the region is a mixture of units still affiliated politically with outside countries and states that have become fully independent. Evolution of the islands away from colonial status began with New Zealand's granting of independence to Western Samoa in 1962. The "typical" Pacific island country has about 100,000 to 150,000 people in an area of 250 to 1000 square miles (c. 650–2600 sq km); consists of a number of islands; is quite poor economically; is an ex-colony of Britain, New Zealand, or Australia; and depends heavily on foreign economic aid. As of 1999, nine fully independent states had emerged, with populations ranging from 4.3 million in Papua New Guinea—which is exceptionally large in population and area for this region—to 10,500 on tiny Nauru island.

In some cases, colonial powers have delayed independence to would-be island nations. They explain that such territories are too small and isolated, or are not economically viable enough, for independence. Some islands remain dependent because they confer unique military or economic advantages on the governing power. For example, French Polynesia has been the locale for French atomic testing (see Regional Perspective, p. 392), and Guam and American Samoa are useful to the United States for military purposes. Guam, which has been a U.S. possession since 1898 and a U.S. territory since 1950 (its residents are American citizens), is especially vital in U.S. strategic thinking. It was the jumping-off point for American B-52 jets on bombing missions to Iraq in the 1990s, and would be used in any U.S. military action in Korea or elsewhere in East Asia. (In the event of such hostilities, the United States would need Japan's permission to operate from bases in Japan, while no such approval would be necessary to act from Guam.)

Another characteristic of the Pacific islands is a general lack of industrial development, combined in many cases with high population densities. The poverty typical of LDCs prevails in the region. The major exception has been tiny Nauru, whose earnings from phosphate—the product of thousands of years of accumulation of seabird excrement, or **guano**—have provided a high average income. Nauru's economy has depended totally on annual exports of two million tons of phosphate and on revenues earned from overseas investments of profits from phosphate sales (the tallest building in Melbourne, Australia, is a product of Nauruan investments and is known affectionately as "Birdshit Tower"). This resource is approaching total depletion, however, and because the mining of phosphate has stripped the island of soil and vegetation, the people of Nauru are now looking for a new island home (Fig. 15.5).

Nauru is an extreme example of how, like LDCs elsewhere, the Pacific nations tend to rely on a few primary products for major exports. Before the intrusion of outsiders, the island economies were based heavily on subsistence agriculture, gathering, and fishing, with coconuts a major element in the food supply. Many countries now rely heavily on modern commercialized versions of the coconut and fishing economy but still continue their dependence on coconuts and fish as major elements in a widespread pattern of subsistence activities. Indeed, the coconut is so important in these islands that they have been said to have a "coconut civilization." The coconut provides both food and drink for the islanders, and the dried meat, known as copra, is the only significant export from innumerable islands. The husks and shells of the nuts have many uses, as do the trunks and leaves of the coconut palms. People make baskets and thatching from the leaves, for example, and use timber from the trunk for construction and making furniture.

Figure 15.5 Phosphate mining has stripped most of the soil and vegetation cover from the island of Nauru, whose inhabitants are now looking for a new home. In the foreground is a plant for processing raw phosphate into exportable fertilizer; behind is a bare stockpile from which the phosphate has been mined. *William E. Ferguson/William E. Ferguson Photography*

Many of the island political units listed in Table 15.1 rely mainly on exports other than coconut products or fish. These units fall into four major categories:

1. *Countries and colonies in which mineral extraction has replaced subsistence economies.* New Caledonia, Nauru, and Bougainville are outstanding examples. Another important producer is Papua New Guinea, where petroleum, gold, and copper provided 72 percent of the country's exports in 1998.
2. *Colonies whose economies are supported and shaped by military activities,* such as Guam, American Samoa, and French Polynesia.
3. *Fiji,* which developed under British rule as a source of plantation sugar using labor largely imported from India. Sugar is still the largest export, but Fiji also exports gold, fish, and timber. Another Pacific locale where sugar is a valuable source of income is the U.S. state of Hawaii.

4. *The Cook Islands,* a self-governing territory in free association with New Zealand which is developing an industrial economy. Labor-intensive clothing manufacturing is the leading export.

In sum, the Pacific islands' economic picture is one of nonindustrial economies (1) exporting products that were important in their precolonial subsistence activities, (2) exporting plantation crops introduced by Westerners, (3) exporting minerals desired by industrial nations, and (4) deriving income from activities connected with the military needs of occupying powers. The most dynamic new element is the rapid growth of tourism as a basic industry, with the United States and Japan the main sources of visitors. The Pacific islands continue to exert an irresistible appeal to people awash in the conveniences and congestion of the industrialized world.

SUMMARY WITH SELECTED KEY TERMS

- **The Pacific World region** (often called **Oceania**) encompasses **Australia, New Zealand, and the islands of the mid-Pacific lying mostly between the Tropics.**
- The Pacific islands are commonly divided into three principal regions: **Melanesia, Micronesia, and Polynesia.**
- Though countless islands are scattered across the Pacific Ocean, there are two general types—**"high islands" and "low islands."** Most high islands are **volcanic or continental** in nature. Low islands are typically made of **coral,** a material composed of the skeletons and living bodies of small marine organisms that inhabit tropical seas.
- The island ecosystems of the Pacific region are typically inhabited by **endemic** plant and animal species—species found nowhere else in the world. Island species are especially vulnerable to the activities of humankind such as habitat destruction, deliberate hunting, or the introduction of **exotic** plant and animal species.

- Europeans began to visit and **colonize** the Pacific islands early in their **Age of Exploration.** However, a steady process of **decolonization** has accompanied a recent surge of Western interest and investment in the region.
- Aside from a few notable exceptions such as **Nauru,** the poverty typical of **LDCs** prevails throughout most of the Pacific region.
- In general, the Pacific islands' economic picture is one of **nonindustrial economies.** Typical economic activities include: **plantation agriculture, mining, and income derived from activities connected with the military needs of occupying powers.**
- During the 1940s and 1950s the United States used the **Bikini Atoll in the Marshall Islands** as one of its chief testing grounds for its **nuclear weapons.** In recent years, strong negative reaction arose throughout the Pacific region as the French resumed **underground testing of nuclear weapons on the Mururoa atoll in French Polynesia.**

REVIEW QUESTIONS

1. Using maps and the text, identify the three principal island groups that comprise the Pacific islands.
2. Describe the major differences between high and low islands.
3. What is a "hot spot" and how does it relate to Pacific islands?
4. What is an "atoll"? Explain how atolls form.

5. How has Nauru been able to avoid the poverty typical of most Pacific islands? And how long will Nauru be able to enjoy its relative prosperity?
6. Describe the major economic activities that occur in the Pacific islands.

DISCUSSION QUESTIONS

1. Describe some of the natural hazards associated with life on the Pacific islands. How will global warming potentially affect life on the islands?
2. Describe the cultural diversity of the Pacific realm.
3. Explain the processes which are responsible for the creation of endemic species. Why are so many endemic species found on islands such as those of the Pacific world?

4. Describe the positive and negative effects of colonialism on the Pacific islands.
5. Describe the worldwide reaction to French nuclear weapons testing in the Pacific world. How did various countries show their dissatisfaction with the tests?

Chapter 16

Prosperous, Remote Australia and New Zealand

▲ *These Aborigines of Australia's northwest Kimberley coast represent the country's indigenous inhabitants.* Joseph J. Hobbs

CHAPTER OUTLINE

16.1 Peoples and Populations

16.2 The Australian Environment

16.3 Australia's Natural-Resource-Based Economy

16.4 New Zealand: Pastoral and Urban

*a*ustralia and New Zealand are unique. Far removed from Europe, their dominant cultures are nevertheless European. In the underdeveloped Pacific World, they are prosperous. Distinctive plants and animals inhabit the often odd landscapes of these islands, stirring a sense of wonder among indigenous people, European settlers, and foreign tourists alike. Writer Alan Moorehead described the reactions of early European visitors: "Everything was the wrong way about. Midwinter fell in July, and in January, summer was at its height. In the bush there were giant birds that never flew and queer, antediluvian animals that hopped instead of walked or sat munching mutely in the trees. Even the constellations in the sky were upside down."[1]

16.1 Peoples and Populations

There is a basic kinship between these two unusual countries, derived from similarities in population, cultural heritage, political problems and orientation, type of economy, and location. Australia and New Zealand are among the world's minority of prosperous countries. Australia's per capita gross national product of $20,000 is comparable to that of Britain, although well below that of the United States and the most prosperous European countries. Australia has been dubbed "The Lucky Country," and a recent World Bank study ranked Australia as the world's wealthiest country based on its natural resource assets divided by the total population. New Zealand is less affluent, but it is still quite prosperous compared with most nations of the world. In both countries there are relatively few people to spread the prosperity among; Australia's 18.7 million people plus New Zealand's 3.8 million (as of 1998) amount to only about 70 percent of the people living in California.

[1]Moorehead, Alan. 1963. *Cooper's Creek.* P. 7.

Both countries trace their prosperity to the wholesale transplanting of a culture—from the industrializing Great Britain—to the remote Pacific beginning in the late 18th century. Australia (established originally as a penal colony for British convicts) and New Zealand are products of British colonization and strongly reflect the British heritage in the ethnic composition and culture of their majority populations. The Australians and New Zealanders speak English, live under British-style parliamentary forms of government, acknowledge the British sovereign as their own (as of 1999; this may change), and attend schools patterned after those of Britain. Only very recent decades of immigration from continental Europe and Asia have given Australia a more cosmopolitan population.

Although both countries are fully independent nations, loyalty to Britain has long been an outstanding characteristic of Australia and New Zealand. They both still belong to the British Commonwealth of Nations. In both world wars, the two countries immediately came to the support of Britain and lost large numbers of troops on battlefields far from home. In World War II, United States forces helped to frustrate a threatened Japanese assault on their homelands. (Since that time, the two countries have sought closer relations with the United States, and British influence has declined sharply.) Their small populations and remote insular locations account for the two countries' historical tendency to seek ties with strong allies, especially naval powers such as Great Britain and the United States.

Both Australia and New Zealand have minorities of indigenous inhabitants. The native Australian people are known as "Aborigines." Colonizing whites slaughtered many of their ancestors and drove the majority into marginal areas of the continent (see Problem Landscape, p. 398). The Aboriginal population now numbers only about 260,000. Living mainly in the tropical north, this minority exists mostly on the fringes of white society. Some carry on a more traditional way of life in the dry Australian wilderness.

In New Zealand, the dominant indigenous peoples are the Maori, a group of Polynesian descent concentrated mainly on

397

PROBLEM LANDSCAPE

Australia's Indigenous People Reclaim Rights to the Land

AS IN THE UNITED STATES, NEWCOMERS to Australia forged a new nation by dispossessing the ancient inhabitants of the land. And as in the United States, in Australia there is a long history of white racism, discrimination, and abuse against people of color, who in Australia's case are the original inhabitants. Aborigines were not even counted in the national census until 1967. The most disadvantaged group in Australian society, Aborigines suffer from high infant mortality (four times that of whites in Australia), low life expectancy (15 to 20 years lower than that of whites), and high unemployment (36 percent, versus 9 percent for whites). Like Native Americans, a disproportionately large share of Aborigines fall prey to the economic and social costs of alcoholism. Statistics indicate they are 29 times more likely than non-Aborigines to end up in jail. Once in jail, they are more likely than non-Aborigines to commit suicide or be killed by guards. The Australian government has begun a program to improve the justice system for native Australians.

One of the largest issues of contention between Australia's indigenous people and the white majority is land rights. As in the United States, European newcomers pushed the native people off productive ranching, farming, and mining lands into special reservations on inferior lands. The Europeans used a legal doc-

trine called *terra nullius,* meaning "the land was unoccupied," to lay claim to the continent. Under the 1976 Land Rights Law, Aborigines were permitted to seek title to vacant state land ("Crown land") that they could prove a historical relationship to, but few succeeded. Following new legislation in 1992, the government began to return titles on a parcel-by-parcel basis—generally in very marginal lands—to Aborigines who had argued their rights successfully. However, most Aborigines were unhappy with the terms of the returned titles, which allowed the native title to coexist with, but not supersede, the established crown title. The government retained mineral rights to the land, and could therefore lease native land to mining companies.

In 1993, Aboriginal leaders expressed their objections to the legislation in a declaration known as the Eva Valley statement, which made four demands:

- Aborigines should have veto rights over mining and pastoral leases on native-title land.
- Mining and pastoral leases cannot extinguish native title to the land.
- There should be no access by developers without Aboriginal permission.
- The federal government, not the states, should control native title issues.

Prime Minister Paul Keating rejected the Eva Valley statement, arguing that the government had already made enough concessions. Aboriginal leaders took their case to the United Nations in Geneva, where they argued that existing legislation to validate mineral leases on native land

was a breach of international human rights conventions, which Australia had signed.

Late in 1993, after the longest senate debate in the country's history, Foreign Minister Gareth Evans succeeded in pushing the Native Title Bill through Australia's parliament. The new legislation addressed the Aborigines' major objections, providing the following concessions:

- Aborigines have the right to claim land leased to mining concerns once the lease expires.
- Aborigines have the right to negotiate with mineral leaseholders over development of their land. However, they do not have the right of veto they asked for.
- Where their native title has been extinguished, Aborigines are entitled to compensation paid by the government.
- With the help of a Land Acquisitions Fund, impoverished Aborigines can buy land to which they have proven native title.

The Native Title Bill is likely to be of great consequence in the states of South and Western Australia, where there is much vacant Crown land and many Aborigines able to file claims on it. Those states are also heavily dependent upon mining. Members of Australia's Mining Industry Council are unhappy with the new legislation and worry that their mine leases will run out before the minerals do, which will require them to negotiate with the Aboriginal titleholders.

Potential Aboriginal claims to land expanded vastly in 1996 with another Australian High Court ruling on what is known as the "Wik case" (named after the indigenous people of north Queensland who

North Island. Whites (known as *Paheka* to the Maori) broke the Maori hold on the land in a bloody war between 1860 and 1870. The Maori, who in 1900 seemed destined for extinction, have rebounded to make up almost 15 percent of New Zealand's population. Although their socioeconomic standing is still

depressed, their situation is much better than that of the Australian Aborigines. The Maori are being increasingly integrated, both socially and racially, with New Zealand's white population. At the same time, a Maori cultural and political movement is resisting the loss of traditional culture which this integration

Figure 16.A Sacred to Aborigines and revered by tourists, Uluru (Ayers Rock) in central Australia is a 1100-foot-high (335 m) sandstone monolith. *Joseph J. Hobbs*

initiated it). The court concluded that Aborigines could claim title to public lands held by farmers and ranchers under long-term "pastoral leases" granted by state governments. These lands comprise about 42 percent of Australia's territory. There is bitter debate about what Aboriginal claims to these lands would mean. Technically, in light of this court decision, Aborigines have a "right to negotiate," meaning they have a decision-making voice in how non-Aboriginal leaseholders use the land. They also have a right to a share of the profits those leaseholders earn from their land uses. White farmers and ranchers protest that they would be ruined economically if they had to share profits, especially with exports of wheat and beef reduced as a result of economic turmoil in Asia. Generally, however, Aborigines have insisted their title rights would not mean running whites out of business, but would simply confer nominal recognition of their title, and access, to the land, especially for visitation to sacred sites. Under pressure from white farmers, ranchers, and miners, Australia's conservative government under Prime Minister John Howard (elected in 1996) reacted to the Wik case with a ten-point plan that would make it much more difficult for Aborigines to make native title claims and to challenge farming, ranching, and mining activities.

In the Northern Territory, which has its own land-rights legislation, vast tracts of land have already been given back to Aborigines. Aborigines there make up 24 percent of the population, and now control 32 percent of the land. Aborigines also hold title to Australia's two greatest national parks, Uluru (Ayers Rock; Fig. 16.A) and Kakadu, both in the Northern Territory. In an arrangement that has become a model worldwide for management of national parks where indigenous people reside, the Aboriginal owners of the reserves have leased them back to the Australian government park system, which comanages them with the Aborigines. Tourism programs in Uluru and Kakadu now highlight the cultural resources of the parks, in addition to their natural wonders. At both parks visitors can enjoy interpretive natural history walks that also emphasize Aboriginal culture and that are conducted by the land's oldest and most knowledgeable inhabitants.

Although there has been remarkable progress in recent years in this area and in others involving the Aborigines, the issue of rights to land has not been resolved to the satisfaction of all. "The handover of Ayers Rock is a turning point in Australia's race relations," declared Charles Perkins, then the country's top Aboriginal civil servant. "It's a recognition that Aboriginal people were the original owners of this country." But Aboriginal activist Galarrwuy Yunupingu declared, "I'd only call it a victory if the governor-general came and gave us title to the whole of Australia, and we leased *that* back to the Commonwealth."[2]

[2]Foster, Catherine. August 3, 1993. "Australia's Ayer's Rock: It's Not Just for Climbing Anymore." *The Christian Science Monitor,* p. 10.

may cause. Like the Aborigines, the Maori are increasingly filing claims against the government to recover what they regard as traditional tribal lands. They assert that New Zealand has violated the spirit and letter of the Treaty of Waitangi, signed in 1840 by Maori tribes and the British government. In the agreement, the Maori gave up political autonomy in exchange for "full and undisturbed" rights to lands, forests, fisheries, and "national treasures" of New Zealand. In one recent concession to such rights, the government has made it legal for Maori to fish without a license.

Landscape in Literature

Aboriginal Views of Earth

"The Aborigine clings to his native soil with every fiber of his being," wrote the ethnographer Carl Strehlow. "Mountains and creeks and springs and water holes are to the Aborigine not merely interesting or beautiful scenic features. They are the handiwork of ancestors from whom he himself has descended. The whole country is his living, age-old family tree."[3]

According to Aborigines, Earth and all things on it were created by the "dreams" of humankind's ancestors during the period of the "Dreamtime." These ancestral beings sang out the names of things, literally "singing the world into existence." The Aborigines call the paths that the beings followed the "Footprints of the Ancestors" or the "Way of the Law"; whites know them as "Songlines." In a ritual journey called "Walkabout," the Aboriginal boy on the verge of adulthood follows the pathway of his creator-ancestors.

In the following passage, Jack Sullivan, whose mother was Aboriginal and father was European, tells the anthropologist Bruce Shaw some accounts of the creation of people, places, and animals. Sullivan, who was born in 1901, "came over to the white side" in his late twenties and took on the life of a white cattle drover, moving in search of work between the cattle stations of northwest Australia. The accounts suggest that, like many of Aboriginal descent today, Sullivan has forgotten many of the details of the Dreamtime stories passed down orally through the generations. However, the words Sullivan spoke into Shaw's tape recorder convey the essential Aboriginal senses of wonder and reverence for places and living things on the landscape. They also carry a note of sadness that Aboriginal people today are forgetting the story of how places came to be.

I don't know how the first lot of blackfellers [Aborigines] were bred from the earlier generation. There are different words about how the world began, the way things went on before humans like you and I came out. It was the same as they reckon among the white men before we came out. There was some sort of man who may have been taller or shorter, different, until a big earthquake sort of washed up and wiped them all out. A big water came over and they were all drowned. White and black tell you the same story, just like you hear in another country or see in the paper where a volcano bursts up or there is an earthquake. Perhaps the land gets broken and the water comes up there. A lot of whites used to tell me this too. I suppose you have heard of or seen those things in the science papers or books? An old Chinaman once told me it happened like that in China, that they dug up this 12 foot man who lived before the Chinamen, together with a lot of early day jars full of gold that had been planted. They took up a little bit of an acre and fiddled around digging up everything, and when they opened the lid there was gold. Well those people all died out and then we came out all over this world. It happened around Jerusalem too. China was a very old country and they reckon that Jerusalem was too. And they reckon with that China Wall now, how was that done before the machines? They reckon they were big blooming stones and how did they carry them there? Just like in Jerusalem, they reckon there was a cold chisel made from copper to cut the iron like a wood chisel, and that there was a mosquito net made from gold. There you are. That was before all the machines. How they made that nobody knows.

I heard once from Duncan and Mandi that in this Hidden Valley they had sort of an early day turnout further up where nobody should go, right in the centre where there was a sort of round cliff and a big long tree. I don't know whether it was a tree or a stone but it was a big long thing standing right in the centre up that creek where it finished. That was where everything must have rushed in the early days, beginning too late and they all ended up there. The first lot were birds and goannas [monitor lizards]

[3]Quoted in Geoffrey Blainey, *Triumph of the Nomads*. Melbourne: Sun Books, 1987, p. 181.

16.2 The Australian Environment

Australia is truly a "world apart." Located in Earth's Southern Hemisphere (Fig. 16.1) where landmasses are few and far between, it is both an island and a continent. Its natural wonders and curiosities have provided endless inspiration—and sometimes consternation—to its inhabitants and to travelers and visiting writers. On arriving in Botany Bay—now Sydney Harbor—in 1788, British major Robert Ross declared, "I do not scruple to pronounce that in the whole world there is not a worse country than what we have yet seen of this. All that is contiguous to us is so very barren and forbidding that it may with truth be said here that nature is reversed; and if not so, she is nearly worn out."[4] Mark Twain marveled at the "Land Down

[4]From Evans, Howard Ensign, and Evans, Mary Alice. *Australia: A Natural History*. Washington, D.C.: Smithsonian Institute Press, 1983. P.12.

400

Figure 16.B Australian Aborigines performing a corroboree (ceremonial dance). *Joseph J. Hobbs*

Auvergne. His country was only from Keep River and Newry this side of Auvergne to Carlton, Ningbing and Ivanhoe. That was his beat. He can show you the places, tell you where this has been and where that goes and where this goes, all the years and corners about the Law [the Aboriginal scheme of customary behavior] and corroborees [ceremonial dances; Fig. 16.B] that went through. And he can figure out where there was a blackfeller in the early days who may have been a turtle turned out of rock down the Keep River from Newry somewhere. That was where they all came in and finished up after a bit of a row according to the big corroboree. Some turned out of birds and some into rocks and trees. [Waddi pointed a place out to me once where] a little mob of ducks had been and ended up in stone streaking along the river. And other places he said were turtles, birds, fish and crocodiles all waterlogged; then the water sort of dried off and left them. It was one of those places they went to every few years to have a corroboree and sing so that all the animals will come out and run over the land. That little place where they finished was where the birds went up from the rocks and shells. Nobody ever mentioned it before Waddi said it. Some fellers knew but forgot it. Waddi still had it in his mind.

From Sullivan, Jack. *Banggaiyerri: The Story of Jack Sullivan As Told to Bruce Shaw.* Canberra: Australian Institute of Aboriginal Studies, 1983, pp. 139–141.

and other animals. Those animals were human beings too before the present blackfellers, according to the old people who used to show us the caves.

Old Waddi told me a good tale once when we were sitting in my camp about all the times when they were travelling around. He talks man to man sensibly about all turnouts right from the jump to the finish. You would get a lot of information from him about these olden days. He knows this country inside out although he never went much over

Under" in his book *Following the Equator:* "To my mind the exterior aspects and character of Australia are fascinating things to look at and think about. They are so strange, so weird, so new, so uncommonplace, such a startling and interesting contrast to the other sections of the planet."[5]

[5]Twain, Mark. *Following the Equator: A Journey Around the World.* Hartford, Conn.: American Publishing Company, 1897.

Including the offshore island of Tasmania, Australia has an area of nearly 3 million square miles (7.8 million sq km), approximately equal to the area of the contiguous United States. However, most of the continent is very sparsely populated, and in total population Australia is a relatively small country. The sparseness of Australia's population and its concentration into a small part of the total land area are closely related to the continent's physical characteristics, among which aridity and low average elevation are outstanding. Australia is the world's oldest

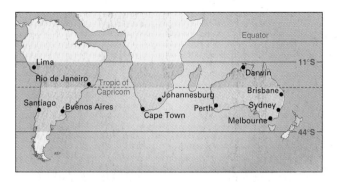

Figure 16.1 Australia and New Zealand compared in latitude and area with southern Africa and South America. The distances separating the continents have been compressed in this figure.

continent. This has given the elements time to wear down the surface of the land, so that it is also the world's flattest continent.

On the basis of climate and relief, Australia has four major natural regions: the humid eastern highlands, the tropical savannas of northern Australia, the "mediterranean" lands of southwestern and southern Australia, and the dry interior (see Fig. 2.8 and Fig. 16.2).

The Humid Eastern Highlands

Australia's only major highlands extend along the east coast from just north of the Tropic of Capricorn to southern Tasmania in a belt 100 to 250 miles wide (161–410 km). Although complex in form and often rugged, these highlands seldom reach elevations of 3000 feet (914 m). Their highest summit and the highest point in Australia is Mount Kosciusko, which rises only 7310 feet (2228 m). The highlands and the narrow and fragmented coastal plains at their base constitute the only part of Australia that does not experience a marked period of drought each year. However, although onshore winds from the Pacific bring rain each month, the strong relief reduces the amount of agricultural land in this most favored of Australia's climatic areas. South of the city of Sydney and at higher elevations to the north, the climate is marine west coast, despite the location. North of Sydney, higher summer temperatures change the classification to humid subtropical, while still farther north, beyond approximately the parallel of 20°S, hotter temperatures and greater seasonality of rain cause subhumid conditions.

Tasmania is a rugged, beautiful island off Australia's southeast coast (Fig. 16.3). About 7 percent of the land is dedicated to national parks and other protected areas. The state's western region is heavily forested and dotted with lakes. Numerous rivers cut rapid, short courses to the sea, representing potential sources of hydroelectric power. At this latitude—roughly comparable to that of the U.S. state of Wyoming—snow sometimes falls on the high peaks even in the height of summer (November–January).

The Tropical Savannas of Northern Australia

Northern Australia, from near Broome on the Indian Ocean to the coast of the Coral Sea, receives heavy rainfall during the season locals call "the wet"—the Southern Hemisphere summer season. The coastal zone is subjected in summer to hurricanes, known as cyclones in Australia; on Christmas Eve in 1974, a powerful cyclone flattened the city of Darwin (which has since been rebuilt). After such deluges, northern Australia experiences almost complete drought during the winter. This highly seasonal distribution of rainfall is the result of monsoonal winds that blow onshore during the summer and offshore during the winter. The seasonality of the rainfall, combined with the tropical heat of the area, has produced a savanna vegetation of coarse grasses with scattered trees and patches of woodland (Figure 16.4). The long season of drought, the poverty of the soils, and the lack of highlands which would nourish large perennial streams all combine to reduce agricultural possibilities. The alluvial and volcanic soils that support large populations in some tropical areas are almost completely absent in northern Australia.

The "Mediterranean" Lands of the South and Southwest

The southwestern corner of Australia and the lands around Spencer Gulf have a mediterranean or dry-summer subtropical type of climate, with subtropical temperatures, winter rain, and summer drought. In winter, the Southern Hemisphere belt of the westerly winds shifts far enough north to bring precipitation to these districts, while in summer this belt lies offshore to the south and the land is dry. Crops introduced from the mediterranean-climate lands of Europe generally do well in these parts of Australia, but the shortage of highlands to catch moisture and supply irrigation water to the lowlands limits agricultural possibilities. Australian wines produced in this region, particularly around Adelaide, are becoming increasingly popular abroad. About 1000 wineries in Australia now produce about 2 percent of the world's total wine output.

The Dry Interior

The huge interior of Australia is desert, surrounded by a broad fringe of semiarid grassland (steppe) that is transitional to the more humid areas around the edges of the continent. This is the "outback" that has lent so much to Australian life and lore. Locals know it as "the bush," the "back of beyond," and the "never-never." Altogether, the interior desert and steppe cover more than one-half of the continent and extend to the coast in the northwest and along the Great Australian Bight in the south. This tremendous area of arid and semiarid land is too far south to get much rain from the summer monsoon, too far north to

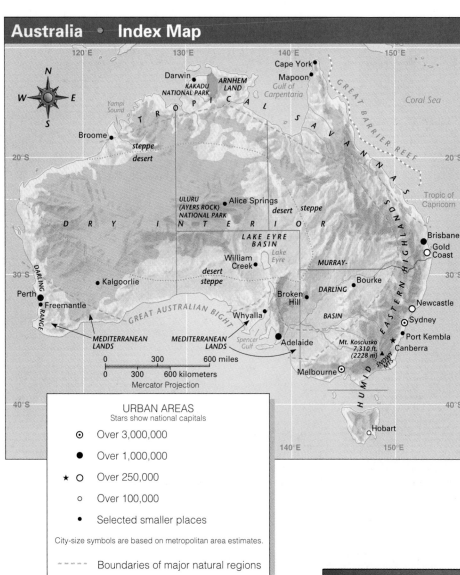

Figure 16.2 Reference map of natural features, political units, and cities in Australia.

Figure 16.3 Large parts of Tasmania are rugged and wild. This is Wineglass Bay in Freycinet National Park. *Tony Stone/Margaret Gowan*

Figure 16.4 Tropical savanna prevails in northern Australia. This is typical vegetation and terrain of the Kimberley region in the continent's extreme northwest. *Joseph J. Hobbs*

Figure 16.5 Aerial view of legendary Alice Springs, in the "red center" of Australia. A salt lake bed is in the foreground. *Joseph J. Hobbs*

benefit from rainfall brought by the westerlies in winter, and deprived of Pacific winds by the eastern highlands. The region is sparsely populated. The chief city is Alice Springs (Fig. 16.5; population: 28,000), a famous frontier settlement that is the jumping-off point for visitors to Uluru-Ayers Rock.

The western half of Australia is occupied by a vast plateau of ancient igneous, metamorphic, and hardened sedimentary rocks. Its general elevation is only 1000 to 1600 feet (*c.* 300–500 m) and its few isolated mountain ranges are too low to influence the climate or to supply perennial streams for irrigation. To the east, between the plateau and the eastern highlands, the land is even lower—in the great central lowland that stretches across the continent between the Gulf of Carpentaria and the Great Australian Bight. One effect of elevation on the climate is that the Lake Eyre Basin, the lowest part of this lowland—and of Australia—is the driest part of the continent. However, another part of the lowland, the Murray-Darling Basin, contains Australia's only major river system and has the most extensive development of irrigation works on the continent.

16.3 Australia's Natural-Resource-Based Economy

One important result of the widespread aridity and of the presence of highlands in the only humid area is that Australia has little good agricultural land (Fig. 16.6). The impact of this limiting factor is evident in the distribution of population, which generally follows that of arable land. Most of the country's small population is found in the humid eastern highlands and coastal plains, especially in the cooler south; in the areas of mediterranean climate; and in the more humid grasslands adjoining the southern part of the eastern highlands and the mediterranean areas.

Although as much as 15 percent of the total land area is potentially cultivable, only 6 percent of Australia is actually cultivated today. Fourteen percent of Australia's area is classified as forest and woodland, the greater part of which is of little economic value. More than one-half is classified as natural grazing land and about one-fifth as unproductive land. Despite its small proportion of arable area, however, Australia has one of the most favorable ratios of crop-producing land to population in the world.

The sectors of the Australian economy that regularly produce the country's leading exports are mining, ranching, and—

even with its small proportion of cultivable land—agriculture. Several factors support the country's emphasis on surplus production and export of primary commodities, and there are large and dependable markets for Australian primary products in Japan, the United States, Europe, and elsewhere. These products can be taken to market with inexpensive and efficient transportation by sea. The Australian people can supply and import the necessary skills and capital to exploit the continent extensively, utilizing a minimum of labor and a maximum of land and equipment.

Sheep and Cattle Ranching

In founding the first settlement around the excellent natural harbor of Sydney in 1788, the British government intended merely to establish a penal colony. There was to be enough agriculture on this "Fatal Shore" to make the colony self-sufficient in food and perhaps provide a small surplus of some products for export. Although self-sufficiency was elusive in the early days due to the poverty of the leached soils around Sydney, the colonists soon discovered that sheep did well, particularly the Merino

breed which was first imported from Spain in 1796. The market for wool expanded rapidly with the mechanization of Britain's woolen textile industry. Australia thus found its major export staple in the early years of the 19th century, and until recently sheep ranching has continued to be the most important of the country's rural industries (Fig. 16.7). Sheep graziers rapidly penetrated the interior of the continent, and by 1850 Australia was already the largest supplier of wool on the world market, a position it has never lost.

People have tried to raise sheep everywhere in Australia that seemed to offer any hope of success. Much of the interior has been proven too dry, parts of the eastern mountain belt too rugged and wet, and most of the north too hot and wet in summer. The sheep ranching industry has thus become localized, concentrated mainly in a crescent-shaped belt of territory that follows roughly the gentle western slope of the eastern highlands, from the northern border of New South Wales to the western border of Victoria (see Fig. 16.6). Beyond this concentration, the sheep industry spreads northward on poorer pastures into Queensland and westward into South Australia. Still another area of production rims the west coast, especially near

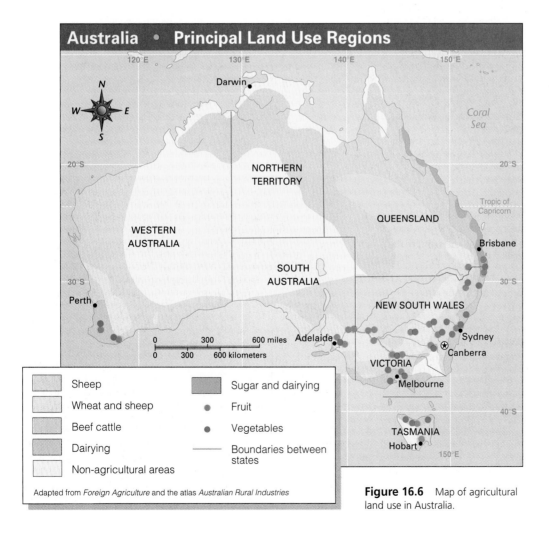

Figure 16.6 Map of agricultural land use in Australia.

Definitions & Insights

EXOTIC SPECIES ON THE ISLAND CONTINENT

Exotic species are nonnative plant and animal species introduced through natural or human-induced means into an ecosystem. Their impact tends to be pronounced and often catastrophic to native species, particularly on island ecosystems where resources and territories are limited. In Ecuador's Galapagos Island National Park, for example, cats, rats, pigs, goats, wasps, and other animals introduced by people have disrupted nests and food supplies and therefore reduced populations of endemic tortoises and other species.

The impact of exotic species suggests that Australia truly is an island. Both inadvertently and purposely, people have introduced to Australia foreign plants and animals that have multiplied and affected the land in biblical proportions. English settlers imported the rabbit for sport hunting, confident in the belief that, just as in England, there would be natural checks on the animal's population. The belief was mistaken. The rabbits multiplied indeed like rabbits, eating everything green they could find, with a ruinous effect on the vegetation. Observers describe the unbelievable concentrations of the lagomorphs as "seething carpets of brown fur." In the mid–20th century, the government mounted a huge eradication effort, employing fences, snares, dogs, guns, fires, and numerous other means, and exhorting combatants with the slogan "The Rabbit War Must Be Won!" An introduced virus called "white blindness" eventually helped reduce the vast numbers, but the animal is still prolific throughout large parts of the country. Other exotic remedies have been applied, with varying success, to correct numerous other problem species,

including mice, water buffalo, and prickly pear cactus. Many of the problem populations, notably cats and water buffalo, are **feral animals**—domesticated animals which have abandoned their dependence upon people to resume life in the wild. An estimated 12 million feral cats infest the country and are held responsible for causing the extinction or endangerment of 39 native animal species. One Australian politician has been promoting legislation that would kill all cats on the continent by 2020.

Exotic livestock such as sheep and cattle are also a problem in the Australian environment. These hard-hoofed imports cut into the soil, promoting erosion and desertification. Some range-management scientists and conservationists believe that the future of Australian ranching lies with kangaroos. These soft-hoofed marsupials are adapted to the Australian landscape and do not have such a damaging impact on the soil. Farmers and ranchers have traditionally eradicated them as pests; however, if consumers could learn to appreciate kangaroo meat, there would be a strong incentive for ranchers to reduce their cattle and sheep herds and allow kangaroos to proliferate—and be harvested.

Australia is an unlikely exporter of another quadruped, the dromedary camel. The more than 200,000 animals in Australia are feral descendants of those brought from South Asia and Iran early in the 20th century to provide transport across the arid continent. Australian camels are exported even to Saudi Arabia, where their wild temperaments suit them to racing.

Perth. The sheep ranches, or "stations," in Australia vary in size from a few hundred to many thousands of acres.

The main product of the sheep stations is wool, which has been of great importance in Australia's development. Through most of the country's history, wool has been the principal export, amounting to as much as a quarter of all exports as late as the 1960s. Recently, however, it has declined rapidly in importance, partly because of the rise of other commodities in the Australian economy and partly because of wool's slowly losing battle with synthetic fibers. Today, Australian sheep are raised increasingly for meat. Each year Australia exports large numbers of live animals to Saudi Arabia and other Muslim countries of the Middle East, where the devout slaughter them on the Feast of the Sacrifice commemorating the end of the Pilgrimage to Mecca.

Figure 16.7   Sheep are big business in both Australia and New Zealand. These motorized shepherds are mustering Merino sheep at the Narolah Station in southwestern Western Australia, on land that is productive for both sheep and wheat.   *Bill Bachman/Photo Researchers Inc.*

Lands in Australia that are unsuitable for sheep or for intensive agriculture are generally devoted to cattle ranching. Cattle stations are spread across the northern lands, with their hot, humid summers and coarse forage. The main belt of cattle ranching extends east-west across the northern part of the country, with the greatest concentration in Queensland (see Fig. 16.6). On many of the remote ranches, especially in the Northern Territory, Aborigines are the principal ranch hands, known as "drovers" and "jackaroos." The beef cattle industry has expanded rapidly in recent decades, spurred by the growing demand for meat that has accompanied increasing affluence in Japan, Europe, and the United States. Construction of roads for trucking Australian cattle to the coast has also aided this expansion. Australia is now one of the nations that dominate the world's beef exports.

Wheat Farming

Australia is one of the world's leading wheat exporters (Fig. 16.8). Wheat was the key crop in early attempts to make the settlement at Sydney self-sufficient. In the early years, the colony almost starved as a result of wheat failures, and the problem of grain supply was not solved until the development of the colony (now the state) of South Australia. There, settlers found fertile land and a favorable climate near the port of Adelaide, where wheat could be grown and shipped by sea to other coastal points.

Australian wheat production today is an extensive and highly mechanized form of agriculture, characterized by a small labor force, large acreages, and low yields per acre, but with very high yields per worker. The main belt of wheat production spread from the eastern coastal districts of South Australia into Victoria and New South Wales, generally following the semi-arid and subhumid lands that lie inland from the crests of the eastern highlands (see Fig. 16.6). The main wheat belt has thus come to occupy nearly the same position as the main belt of sheep production. The two types of production are in fact often combined in this area.

Dairy and Sugar Farming

Although far less important than grazing and wheat farming for export revenues, dairy farming is Australia's leading type of agriculture in numbers of people employed. It has developed mainly in the humid coastal plains of Queensland, New South Wales, and Victoria (see Fig. 16.6). Dairy farming has grown largely in response to the needs of the country's rapidly growing urban populations.

Australia is a sugar exporter with a unique history. The Queensland sugar-growing area is unusual in that it was a rare tropical area settled by Europeans doing hard manual work without the benefit of indigenous labor. Sugar production began on the coastal plain of Queensland in the middle of the 19th century to supply the Australian market. Laborers were initially imported from Melanesia to work as virtual slaves in the cane

Figure 16.8 Despite its limited arable land, Australia is a prodigious wheat producer. This is an aerial view of center pivot irrigation for wheat near Wyndham, in northwest Australia. *Joseph J. Hobbs*

fields. However, when the Commonwealth of Australia was formed in 1901 by union of the seven states (New South Wales, Victoria, Queensland, Northern Territory, South Australia, Western Australia, and Tasmania), Queensland was required to expel these imported "Kanaka" laborers in the interest of preserving a "white Australia." Anticipating that production costs would soar with the employment of exclusively white labor, the Commonwealth gave Queensland a high protective tariff on sugar. Operating behind this protective wall, the sugar industry has prospered there, though most of the original large estates have been divided into family farms worked by individual farmers.

Australia's Mineral Wealth

Unification of the separate Australian colonies into the independent federal Commonwealth of Australia encouraged the rise of industry, creating a unified internal market and tariff protection for manufacturing as well as for agricultural enterprises. Mineral wealth has also facilitated industrial development (Fig. 16.9). As in the case of agriculture, Australia's small population is able to draw upon the resources of an entire continent, and in mineral resources the continent is relatively rich. In addition to supplying most of its own needs for minerals, Australia has long been an important supplier to the rest of the world. In recent years, large new discoveries of iron ore, diamonds, and gold have been expanding this longtime role in international trade.

Mining has long had an impact on Australian life and landscapes. Gold played a large role in Australia's history, as it did in America's; some of the California "forty-niners" even continued their prospecting careers in Australia. Gold strikes in

Figure 16.9 Mineral map of Australia.

Victoria in the 1850s set off one of the world's major gold rushes and produced a threefold increase in the continent's population in just 10 years.

Ample supplies of coal also have been fundamental to Australian industrialization. Every Australian state contains coal, but the major reserves are near the coast in New South Wales and Queensland. Most of Australia's steel-making capacity is located near the coal mines north of Sydney at Newcastle and at Wollongong to the south. Iron ore is brought from distant deposits along and near other sections of the Australian coast (Fig. 16.10). Major new iron-ore discoveries and developments in Western Australia have made Australia a leading world supplier.

Australia is by far the world's leading bauxite producer. Major reserves of bauxite have been found near the north Queensland coast, in the Darling Range inland from Perth, and on the coast of Arnhem Land. New aluminum plants are supplying the Australian market, and large quantities of the partially processed material called alumina are being exported for final processing into aluminum metal at overseas plants.

Among Australia's other numerous important mineral products are copper, lead, titanium, zinc, tin, silver, manganese, nickel, tungsten, uranium, and diamonds. Australia now produces 40 percent of the world's diamonds, most of them industrial cutting diamonds rather than gem diamonds. A diamond "rush" is also under way, with prospecting and new finds in the country's northwest, both onshore and at sea. Gold production is also on the rise. In 1994 Australia replaced Russia as the world's third largest gold-producing area, after North America and South Africa. Australia's most famous mining centers are three old settlements in the dry interior—Broken Hill in New South Wales, Kalgoorlie in Western Australia, and Mount Isa in Queensland—all of which are still very active. The Broken Hill Proprietary Company, which grew with the mines of that town, later branched into iron and steel production and other businesses and has become Australia's largest corporation.

A shortage of petroleum has marred Australia's general picture of mineral abundance in the past. Exploration that located a number of relatively small oil fields in the 1960s and 1970s began to correct this shortfall. Most of these are on the mainland, but the main producing field is offshore on the continental shelf of Victoria. While local oil and natural gas supplies have aided the country's economy, Australia continues to import oil and to look for more domestic sources.

Figure 16.10 Australia is an important exporter of iron ore. This open-pit mine is in the Kimberley region of northwestern Australia. *Joseph J. Hobbs*

Urbanization and Industrialization

Only a small labor force is needed for the extensive forms of agriculture and grazing that are characteristic of Australia. Although mining, grazing, and agriculture produce most of Australia's exports, most Australians—about 85 percent—are city dwellers, and a high degree of urbanization is one of the country's notable characteristics. About ten million people, or 55 percent of the population, live in the five largest cities: Sydney (population: 3.7 million, city proper; 4 million, metropolitan area; Fig. 16.11); Melbourne (population: 2 million, city proper; 3.05 million, metropolitan area); Brisbane (population: 1.4 million, city proper; 1.6 million, metropolitan area); Perth (population: 1.2 million, city proper; 1.3 million, metropolitan area); and Adelaide (population: 1 million, city proper; 1.1 million, metropolitan area). The country's capital, Canberra (meaning "meeting place" in the Aboriginal language), is one of its smaller cities (population: 298,200, city proper; 305,000, metropolitan area). Located inland about 190 miles (310 km) southwest of Sydney, the site was empty grazing land when it was designated in 1908 to become the nation's capital. Australia's involvement in the two world wars delayed work on the city for decades. Now Canberra is widely regarded as one of the world's most beautiful modern planned cities.

All of the five largest cities are seaports, and each is the capital of one of the five mainland states. One reason for the striking degree of urban development and its concentration in the port cities is the heavily commercial nature of the Australian economy. A large proportion of the production from rural areas is destined for foreign export by sea. In addition, much of Australia's internal trade moves by coastal steamer, increasing the

Figure 16.11 One great attraction to international tourists is the beautiful port city of Sydney. The white structure is Sydney's famous Opera House. *Joseph J. Hobbs*

Regional Perspective

Australia and New Zealand on the Pacific Rim

Australia and New Zealand have been remote satellites of Great Britain throughout most of their short histories. Even with independence they pledged loyalty to the British Commonwealth of Nations. The United Kingdom was far more important than any other country in their trade relations, and as recently as the early 1970s it accounted for about 30 percent of New Zealand's trade and 15 percent of Australia's. Their orientation is beginning to change, however, as Australia and New Zealand seek stronger roles in the regional growth projected for the Pacific Basin and its strong poles, Japan and the United States. By 1990, Britain was relatively insignificant in Australia's trade and accounted for well under one-tenth of New Zealand's. Already, seven of Australia's ten largest markets are in the Asia-Pacific region, with Japan and the United States its largest and second-largest trading partners respectively. Australia, Japan, and the United States have become New Zealand's leading partners in trade.

Perhaps the best symbol of Australia's new Pacific perspective is the debate over whether or not Australians should

convert the country into a republic, ending more than 200 years of formal ties with Britain. This would mean that Britain's Queen Elizabeth would no longer be regarded as the chief of state, and perhaps that Australians could elect their own president. The monarch's portrait would be removed from Australian currency. The British Union Jack would be lifted from its position in the upper left corner of the Australian flag. In 1998, an Australian Constitutional Convention voted to adopt the republican model, and Australia's avowedly monarchist prime minister, John Howard, reluctantly agreed to put the issue to a referendum in 1999. Australians wanted the question to be resolved before hosting the Olympic Games of 2000 in Sydney and celebrating the centenary of the country's constitution in 2001. In essence, the referendum would ask Australians where they preferred their economic and political future to rest —either with the other nations of Asia and the Pacific, or with Great Britain and the Commonwealth.

The monarchy/republic debate has led to something of an identity crisis for Australia, and has tended to focus more

scrutiny on the sensitive issue of Asian immigration. Between 1945 and 1972, two million people, mainly British and continental Europeans, emigrated to Australia. They were allowed in by a "White Australia" policy that excluded Asian and black immigrants because of fears of invasion from Asia and a desire to increase the country's population with white, skilled, English-speaking immigrants. The rate of immigration has slowed since then, but it continues and has been opened considerably to skilled nonwhites. About 40 percent of the annual quota of 70,000 immigrants are Asian, who now make up about 5 percent of Australia's population. But Asians will form up to one-fourth of Australia's population by 2025, according to some projections. Recent years have seen a marked surge in racism targeted mainly at Asians. Pauline Hanson, a member of parliament and a candidate for prime minister in the 1998 election, said "I believe we are in danger of being swamped by Asians," and vowed to halt Asian immigration if her One Nation Party took power. "A truly multicultural country can never be strong or united," she declared.[6]

[6]From Mydans, Seth. "Sea Change Down Under: Drifting to the Orient." *New York Times,* February 7, 1997. P. A4.

need for port facilities. Growth of the respective capital-city ports was further stimulated by the fact that each state originally built its own individual rail system focusing on its particular port.

The Australian government has encouraged industrial development to support the growing population, to provide more adequate armaments for defense, and to increase stability through a more diversified and self-contained economy. A wealthy, urbanized, and resource-rich country such as Australia should, like most MDCs, be a major exporter of finished industrial products. However, approximately three-fourths of the country's exports come directly, or with only early-stage processing, from mining, agriculture, and grazing. Far from being an international industrial power of consequence, Australia

does not meet its own demands for manufactured goods and relies heavily on overseas production. Australia is a big importer of manufactured goods from Japan and the United States. The Labor Party government of Prime Minister Paul Keating was elected twice in part on its promise to correct this problem and to transform Australia from a protected agriculture- and mining-based economy to a lean, high-technology export economy. Its failure to achieve this goal contributed to the party's defeat in the 1996 and 1998 elections.

Tourism has recently become an important industry in Australia, particularly since the worldwide acclaim of the 1986 film *Crocodile Dundee.* (Crocodiles themselves occupy an important place in Australia's tourist business. In Kakadu National Park and other sections of northern and western Australia,

Ms. Hanson also insisted that Aborigines once practiced cannibalism, and that therefore they are unworthy of political sympathy today. Her party was defeated soundly in the election.

Ethnic tensions aside, Australia and New Zealand continue to forge important ties with their neighbors around the vast Pacific, and with each other. Both countries adopted a free-trade pact in 1990 called the Closer Economic Relations Agreement, which eliminated almost all barriers between them to trade in farm and industrial goods and services. Both countries belong to the 18-member Asia-Pacific Economic Cooperation (APEC) group, which in 1995 agreed to establish "free and open" trade and investment between member states by the year 2020. Australia and the United States are the most aggressive member states in arguing for the abolition of trade quotas and the reduction of tariffs imposed on imported goods.

Australia's Northern Territory city of Darwin is well-positioned geographically to take advantage of new trading relations with Asian countries of the Pacific Rim. While it is a five-hour flight from Darwin to Sydney, it is just two hours from Darwin to Jakarta, the capital of Indonesia. Thus, Darwin is promoting itself as Australia's "Gateway to Asia." The city has established a Trade Development Zone to provide manufacturing facilities and incentives for overseas companies and for Australian companies wanting to do business overseas. A new deep-water port is under construction, and the regional government is pushing for the completion of a transcontinental Darwin-Adelaide railway to help handle an anticipated boom in freight to be traded between Australia and its neighbors to the northeast.

Australia's Asia-oriented perspective is also changing the nature of one of the country's export staples: beef. Ranchers are shifting to breeds of cattle better suited for live shipment by sea to Indonesia, the Philippines, and Thailand. The country's live-cattle exports jumped from 125,000 in 1991 to 360,000 in 1995. In 1994, Japan surpassed the United States as Australia's largest recipient of processed beef. But the Asian economic crisis that began in 1997 severely curtailed exports of beef—and most other Australian products—to Asia. The resulting economic downturn helped keep the less Asian-oriented Liberal Party of John Howard in power for another term after the 1998 election.

Australia and New Zealand have Pacific-oriented defense and security agreements. For example, in 1951, they joined the United States to form the security alliance called Australia New Zealand United States (ANZUS). Australian troops supported American forces in the Vietnam War. Since 1987, however, when its liberal government declared that New Zealand would henceforth be a nuclear-free country, New Zealand's relations with the United States have been strained. Legislation banned ships carrying nuclear weapons or powered by nuclear energy from New Zealand's ports. The United States refused to confirm or deny whether its ships violated either restriction, so New Zealand denied port access to them. This action removed New Zealand from ANZUS.

many tours promise visitors a close-up look at "salties," the great saltwater crocodiles that are ancient and often fatal neighbors of people in the region.) Australia's attractions are sufficiently diverse to draw many different styles of tourists to various destinations on the continent. The 1200-mile-long Great Barrier Reef, located off the northeast coast, is one of the world's most prized scuba-diving areas. Hotels and other services on nearby Queensland beaches of the "Gold Coast" cater to those looking for an idyllic, tropical, restful holiday. The geological features of Uluru (Ayers Rock) and the nearby Olgas in the country's center are on the itinerary of most international tourists. For many visitors, Sydney is as lovely a port city as San Francisco and Venice. Adventurous tourists raft the whitewaters and hike the mountains of Tasmania.

16.4 New Zealand: Pastoral and Urban

Located more than 1000 miles (c. 1600 km) southeast of Australia, New Zealand consists of two large islands, North Island and South Island (separated by Cook Strait), and a number of smaller islands (Fig. 16.12). North Island is smaller than South Island but contains the majority of the population, which numbered about 3.8 million in 1998. With a total area of 104,000 square miles (269,000 sq km), the population density averages just 37 per square mile (15 per sq km). New Zealand is thus a sparsely populated country, although not so much so as Australia.

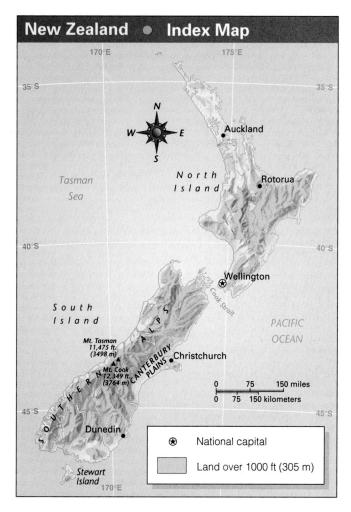

Figure 16.12 Index map of New Zealand.

Environment and Rural Livelihood

Much of New Zealand is rugged. The Southern Alps, often described as one of the world's most spectacular mountain ranges, dominate South Island (Fig. 16.13). These mountains rise above 5000 feet (*c.* 1500 m), with many glaciated summits over 10,000 feet, and in the southwest are deeply cut by coastal fjords. The mountains of North Island are less imposing and extensive, but many peaks exceed 5000 feet. New Zealand's highlands lie in the Southern Hemisphere belt of westerly winds and receive abundant precipitation. While lowlands generally receive more than 30 inches (76 cm), distributed fairly evenly throughout the year, highlands often receive more than 130 inches (330 cm). Precipitation drops to less than 20 inches 51 cm) in small areas of rain shadow east of South Island's Southern Alps. Temperatures typical of the middle latitudes are moderated by a pervasive maritime influence. The result is a marine west coast climate, with warm-month temperatures generally averaging 60° to 70°F (16° to 21°C) and cool-month temperatures of 40° to 50°F (4° to 10°C). Highland temperatures are more severe, and a few glaciers flow on both islands.

Rugged terrain and heavy precipitation in the Southern Alps and in the mountainous core of North Island explain an almost total absence of people from one-third of New Zealand. About one-fifth of the country is completely unproductive, except for the appeal of the mountains to an expanding tourist industry. New Zealand's attractions for visitors include golfing, fishing, hunting, hiking, and camping in spectacular mountainous national parks, and geyser watching around the Rotorua thermal area of North Island. Despite all this wilderness, New Zealand has suffered the impacts that typically occur as a result of human activities in island ecosystems. Nearly 85 percent of the country's lowland forests and wetlands have been destroyed

Figure 16.13 The Southern Alps tower over pastoral landscapes on New Zealand's South Island. *Paul Chesley/Tony Stone Images*

Figure 16.14 Dairy cattle on the lush pasture associated with New Zealand's cool and humid climate. The view is from a locality in the North Island not far from New Zealand's world-famous district of hot springs and geysers. *Thomas Brown*

since the Maori arrived here about 800 years ago. Of New Zealand's presettlement 93 bird species, 43 have become extinct and 37 are endangered (including the Kiwi, the national bird and the informal namesake of all New Zealanders). As in Australia, exotic plant and animal species are a major threat to the indigenous varieties.

Well-populated areas are restricted to fringing lowlands around the periphery of North Island and along the drier east and south coasts of South Island. The climate of New Zealand's lowlands is ideal for growing grass and raising livestock. More than one-half of the country's total area is in pastures and meadows that support a major sheep and cattle industry. Pastoral industries contribute greatly to the country's export trade. Meat, wool, dairy products, and hides together account for about one-half of what New Zealand sells abroad.

The earning power of these pastoral exports has given New Zealand a moderately high standard of living (per capita GNP in 1996 of $15,720, comparable to that of Ireland). The United Kingdom was the country's main market until recently. When the U.K. joined the European Common Market in the 1970s, however, a new situation emerged because the Common Market (now the European Union) included countries with surpluses competitive with New Zealand's. In response, New Zealand has tried to foster industrial growth and to expand trade

with other partners, particularly in Asia. These efforts have not been entirely successful. The economy suffered during the 1970s and 1980s, and the country slipped downward in the ranks of the world's affluent countries. A sizable emigration took place, mainly to Australia. These emigrants complained of not only the economic troubles, but also boredom in an isolated, placid society in which tax policies made it difficult for almost anyone to make or keep a high income. Today there is a reverse flow of migrants, particularly of Asians immigrating to New Zealand's cities. Like Australia, New Zealand is seeking to loosen or break its political and economic ties with the British Commonwealth of Nations and to reorient its economic orbit more toward Asia. However, like Australia, New Zealand's economy is feeling the negative impact of the recent Asian economic crisis as its exports to the region have declined.

Industrial and Urban Development

In New Zealand, as in Australia, a high degree of urbanization and active attempts to develop manufacturing supplement the basic dependence on pastoral industries (Fig. 16.14). The two countries are similar in their conditions and purposes of urban and industrial development. New Zealand's resources for manufacturing do not equal those of Australia, but coal, iron, and other minerals are present in modest quantities. There is much potential for hydropower development. New Zealand's magnificent natural forests enjoy excellent growing conditions, and production and export of forest products have become increasingly important as the country attempts to reduce its dependence on sheep and dairy exports.

New Zealand is not as urbanized as Australia. The country has only six urban areas with estimated metropolitan populations of more than 100,000: Auckland (population: 321,100, city proper; 910,200, metropolitan area) and the capital of Wellington (population: 150,800, city proper; 326,900, metropolitan area), at opposite ends of North Island; Napier-Hastings (population: 110,800, city proper), on the southeastern coast of North Island; Hamilton (population: 103,600, city proper), not far south of Auckland; and Christchurch (population: 297,600, city proper) and Dunedin (population: 114,600, city proper), on the drier east coast of South Island. Christchurch is the main urban center of the Canterbury Plains, which contain the largest

concentration of cultivated land in New Zealand. About one-half of the country's population lives in these six modest-sized metropolises (by comparison, the Sydney metropolitan area in Australia has approximately the same population as all of New Zealand). Note that all of these cities have English names; as evidence of New Zealand's struggle to reidentify itself, there is a strong movement afoot to replace many of the country's Anglified place names with their original Maori designations.

This concludes our survey of the Pacific World, with its far-flung islands and a single great island-continent. Although islands have an insular quality, this world region is, perhaps more than most, subject to influences from abroad because it depends so much on aid from and trade with greater continental powers. In the years to come, many islands will likely be forced to "go it alone" as foreign aid is withdrawn, and Australia will bear watching as it forges a new identity as a republic free of Britain. Rising sea levels (as a result of global warming) may also pose interesting dilemmas for the Region.

SUMMARY WITH SELECTED KEY TERMS

- **Australia and New Zealand** are products of **British colonization.** The British heritage is strongly reflected in the ethnic composition and culture of Australia and New Zealand's majority population.
- Both Australia and New Zealand have **minorities and indigenous inhabitants.** Native Australians are known as **"Aborigines,"** while the **Maori** are the dominant indigenous group in New Zealand.
- In Australia, one of the largest issues of contention between the Aborigines and the white majority is **land rights.** Aboriginal demands were declared in the **Eva Valley statement.** In 1993, the **Native Title Bill** addressed the Aborigines' major concerns and made several land rights concessions.
- On the basis of climate and relief, Australia has four major natural regions: **the humid eastern highlands, the tropical savannas of northern Australia, the "Mediterranean" lands of southwestern and southern Australia, and the dry interior.**
- **Exotic species** are nonnative plants and animals introduced through natural or human-induced means into an ecosystem. Exotic species

often have catastrophic effects on native species and the natural environment.
- The sectors of the Australian economy that regularly produce the country's leading exports are **mining, ranching, and agriculture.**
- Most Australians—about 85 percent—live in cities. Nearly ten million people, or 55 percent of the population, live in the five largest cities: **Sydney, Melbourne, Brisbane, Perth, and Adelaide.**
- New Zealand is located more than 1000 miles southeast of Australia. The country consists of two large islands, **North Island and South Island,** and a number of smaller islands.
- **Pastoral industries,** such as meat, wool, dairy products, and hides, contribute greatly to New Zealand's export trade.
- Australia and New Zealand are beginning to reorient their focus toward the **Pacific Rim** rather than the United Kingdom. The ultimate expression of this change in focus is the **referendum in 1999** to decide whether or not Australia will become a **republic,** thus ending more than 200 years of formal ties with Britain.

REVIEW QUESTIONS

1. What factors foster a basic kinship between Australia and New Zealand?
2. What was the original purpose of Australian settlement?
3. What is the significance of the "Dreamtime" to Australia's Aborigines?
4. Using maps and the text, identify the four natural regions of Australia.
5. Using maps and the text, identify the major agricultural regions of Australia.

6. Explain the factors that support Australia's emphasis on surplus production and export of primary commodities.
7. Describe wheat production in Australia today.
8. List the notable natural resource assets of Australia. What natural resources does Australia lack?
9. Using maps and the text, locate Australia's five largest cities. What characteristics do these cities have in common?
10. List the notable natural resource assets in New Zealand.

DISCUSSION QUESTIONS

1. Discuss the influence of Great Britain on the cultures of Australia and New Zealand.
2. Describe some of the similarities between Australia's Aborigines and Native Americans.
3. Describe some of the demands made by Aboriginal leaders in the Eva Valley statement. Were these demands met by the Native Title Bill of 1993?
4. Describe some of the factors that make Australia unique as a continent.
5. Provide examples of exotic animals that have been introduced to Australia. What have been the effects of exotic animals on Australia's natural environment?
6. What are some indications that Australia and New Zealand's orientation is beginning to focus on the Pacific Rim rather than the United Kingdom?

7 Africa South of the Sahara

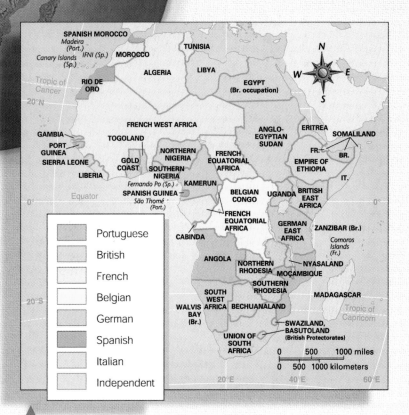

Legend:
- Portuguese
- British
- French
- Belgian
- German
- Spanish
- Italian
- Independent

European powers colonized almost all of Africa in the 19th and early 20th centuries. Postcolonial Africa continues, as it did during colonial times, to export mainly raw materials to the industrialized world.

Minerals are the chief economic asset of Africa south of the Sahara, but are distributed unevenly. South Africa enjoys the greatest mineral wealth and by far the highest standard of living on the continent. *Manaud / Elbaz / Matrix*

Low overall population density characterizes Africa south of the Sahara, but there are several clusters of dense population. Population growth rates rose rapidly after the mid-20th century, but have recently begun to decline sharply due to falling birth rates and the ravages of AIDS. *Sally Mayman / Tony Stone Images*

Regional Snapshot

Africa south of the Sahara is the poorest world region, with 18 of the world's 20 poorest countries. Its prospects for economic development trail behind those of other regions due to the legacy of colonialism, corrupt administration, civil and international conflict, and natural hazards. *Scott Daniel Peterson / Gamma Liaison*

Africa is the cradle of humankind where hominids originated and from where they diffused. The region has seen many indigenous civilizations and empires, and is ethnically and linguistically diverse. *Art Wolfe / Tony Stone Images*

Africa south of the Sahara is mostly tropical, has numerous large plateaus, large river systems, and deep rift valley lakes. Cycles of drought and flood are common, inflicting crop losses, famine, and loss of life, and evoking international relief efforts. *Herman Emmet / Photo Researchers, Inc.*

Chapter 17

A Geographic Profile of Africa South of the Sahara

▲ *This famous Apollo 17 view of the Earth from space features almost all of the continent of Africa. Note also that much of the ice-covered continent of Antarctica is visible.* NASA

CHAPTER OUTLINE

17.1 Area and Population

17.2 The African Environment

17.3 Cultures of the Continent

17.4 The Geographic Impact of Slavery

17.5 Colonialism

17.6 Shared Traits of African Countries Today

17.7 Agricultural and Mineral Wealth

*g*eographers typically recognize that the continent of Africa has two major divisions: North Africa—the predominantly Arab and Berber realm of the continent—and culturally complex Africa south of the Sahara, often called sub-Saharan or Black Africa. In this text, North Africa is regarded as part of the greater Middle East (see Chapters 8 and 9). This chapter sketches the broad outlines of the geography of Africa south of the Sahara and Chapter 18 examines the major African subregions (Fig. 17.1): West Africa,

the Sahel, the Horn of Africa, East Africa, Southern Africa, West Central Africa, and the Indian Ocean islands. Area and population data are in Table 17.1.

The region of Africa south of the Sahara is culturally complex, physically beautiful, and problem-ridden. In presenting an introductory perspective, we focus on the region's diverse environments, peoples and modes of life, major population concentrations, European colonial legacies, and major current problems.

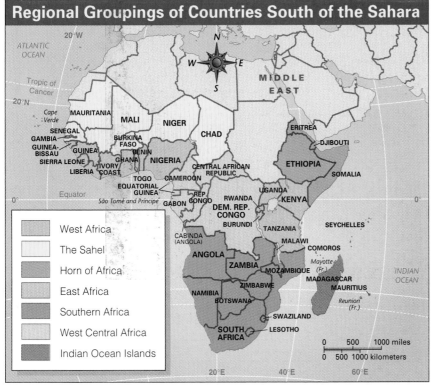

Figure 17.1 Informal regional groupings of countries south of the Sahara.

Source: Stock, Robert. *Africa South of the Sahara.*

419

Table 17.1 Africa: Basic Data

Political Unit	Area (thousand/ sq mi)	Area (thousand/ sq km)	Estimated Population (millions)	Annual Rate of Increase (%)	Estimated Population Density (sq mi)	Infant Mortality Rate (per thousand)	Urban Population (%)	Arable Land (%)	Per Capita GNP ($US)	
The Horn of Africa										
Ethiopia	386.1	999.9	58.4	2.5	151	128	16	12	100	
Eritrea	39	101	3.8	3	99	82	16	12	570	(GDP)
Somalia	242.2	627.3	10.7	3.2	44	122	24	2	500	(GDP)
Djibouti	9	23.3	0.7	2.3	73	115	81	0.0002	1200	(GDP)
Total	**676.3**	**1751.5**	**73.6**	**2.6**	**109**	**125**	**18**	**8.3**	**132**	
The Sahel										
Mauritania	395.8	1025.1	2.5	2.5	6	101	13	1	470	
Senegal	74.3	192.4	9	2.7	122	68	42	12	570	
Mali	471.1	1220.1	10.1	3.1	21	123	26	2	240	
Niger	489.1	1266.8	10.1	3.4	21	123	15	3	200	
Burkina Faso	105.6	273.5	11.3	2.9	107	94	15	10	230	
Gambia	3.9	10.1	1.2	2.4	309	90	37	18	1100	(GDP)
Cape Verde	1.6	4.1	0.4	2.9	257	52	44	11	1010	
Chad	486.2	1259.3	7.4	3.3	15	110	22	3	160	
Total	**2027.6**	**5251.4**	**52**	**3**	**26**	**103**	**23.4**	**3.1**	**313**	
West Africa										
Guinea	94.9	245.8	7.5	2.4	79	153	29	6	560	
Ivory Coast (Côte d'Ivoire)	122.8	318.1	15.6	2.6	127	89	46	8	660	
Togo	21	54.4	4.9	3.6	234	84	31	38	300	
Benin	42.7	110.6	6	3.2	139	94	36	13	350	
Sierra Leone	27.7	71.7	4.6	1.9	166	195	36	7	200	
Ghana	87.9	227.7	18.9	2.9	215	66	35	12	360	
Nigeria	351.6	910.6	121.8	3	346	84	16	33	240	
Guinea-Bissau	10.9	28.2	1.1	2.1	104	141	22	11	250	
Liberia	37.2	96.3	2.8	3.1	75	108	45	1	1100	(GDP)
Total	**796.7**	**2063.4**	**183.2**	**2.9**	**229.7**	**89.2**	**23.1**	**20**	**319**	
West Central Africa										
Cameroon	179.7	465.4	14.3	2.8	80	65	44	13	610	
Gabon	99.5	257.7	1.2	2	12	94	73	1	3950	
Republic of the Congo	131.9	341.6	2.7	2.3	20	107	58	0	640	
Central African Republic	240.5	622.9	3.4	2.1	14	97	39	3	310	
Democratic Republic of the Congo	875.3	2267	49	3.2	56	106	29	3	130	

Table 17.1 Africa: Basic Data *(continued)*

Political Unit	Area (thousand/ sq mi)	Area (thousand/ sq km)	Estimated Population (millions)	Annual Rate of Increase (%)	Estimated Population Density (sq mi)	Infant Mortality Rate (per thousand)	Urban Population (%)	Arable Land (%)	Per Capita GNP ($US)	
Equatorial Guinea	10.8	28	0.4	2.6	40	117	37	5	530	
Sao Tome and Principe	293	758.9	0.2	3.4	511	51	46	2	330	
Total	**1830.7**	**4741.5**	**71.2**	**3**	**120.6**	**97.1**	**34.4**	**3.5**	**322**	
East Africa										
Kenya	219.7	569	28.3	2	129	62	27	7	320	
Tanzania	341.2	883.7	30.6	2.5	90	100	13	3	170	
Uganda	77.1	200	21	2.7	273	81	14	25	300	
Rwanda	9.5	24.6	8	2.1	835	114	5	35	190	
Burundi	9.9	25.6	5.5	2.5	558	105	5	44	170	
Total	**657.4**	**1702.9**	**93.4**	**2.4**	**142.3**	**85.7**	**16.3**	**8**	**246**	
Southern Africa										
Zambia	287	743.3	9.5	1.9	33	109	39	7	360	
Malawi	36.3	94	9.8	1.7	269	140	20	18	180	
Zimbabwe	149.4	386.9	11	1.5	74	53	31	7	610	
Angola	481.4	1246.8	12	3.2	25	124	42	2	270	
Mozambique	302.7	784	18.6	2.2	62	134	28	4	80	
South Africa	471.4	1220.9	38.9	1.6	82	52	57	10	3520	
Namibia	317.9	823.4	1.6	1.7	5	68	27	1	2250	
Botswana	218.8	566.7	1.4	1.2	6	60	48	1	3100	(GDP)
Lesotho	11.7	30.3	2.1	2.1	178	80	16	11	660	
Swaziland	6.6	17.1	1	3.3	145	72	22	11	1210	
Total	**2283.2**	**5913.4**	**105.9**	**1.9**	**46.3**	**89**	**40.8**	**5**	**1549**	
Indian Ocean Islands										
Madagascar	224.5	581.5	14	3	62	96	22	4	250	
Mauritius	0.8	2.1	1.2	1	1483	21	43	49	3710	
Mayotte	0.1	0.3	0.1	4.7	1072	73	*	*	600	(GDP)
Comoros	0.9	2.3	0.5	2.7	861	77	29	35	450	
Reunion	1	2.59	0.7	1.6	721	9	73	17	4300	
Seychelles	0.2	0.5	0.1	1.4	432	7	59	2	6850	
Total	**227.5**	**589.3**	**16.6**	**2.8**	**74**	**85.7**	**26**	**4.3**	**719**	
Summary Total	**8499.4**	**22013.4**	**595.9**	**2.6**	**87.8**	**95.1**	**26**	**6.1**	**514**	

Sources: *World Population Data Sheet*, 1998. United Nations Statistics Division, 1998. *Almanac of Politics and Government*, 1998. *World Factbook*, 1997.

17.1 Area and Population

Africa south of the Sahara (including Madagascar and other Indian Ocean islands) is the largest in land area of all the major world regions discussed in this book. Its 8 million square miles (21.8 million sq km) make it more than twice the size of the United States. "People overpopulation" is apparent in many areas, and yet much of the region is sparsely populated. With a population of 596 million as of 1998, the region's average population density is substantially less than that of the United States. However, this density gap will close rapidly; the current rate of natural population increase in Africa south of the Sahara is 2.6 percent per year, or about four times that of the United States.

The relatively low population density of Africa south of the Sahara as a whole obscures the fact that a majority of this region's people live in a small number of densely populated areas that together occupy a small share of the region's total area. The principal areas of population are (1) the coastal belt bordering the Gulf of Guinea in West Africa, from the southern part of Africa's most populous country, Nigeria, westward to southern Ghana; (2) the savanna lands in the northern third of Nigeria; (3) the highlands of Ethiopia; (4) the highland region surrounding Lake Victoria in Kenya, Tanzania, Uganda, Rwanda, and Burundi; and (5) the eastern coast and parts of the high interior plateau (High Veld) of the Republic of South Africa (for population distribution, see Fig. 2.17, p. 51; for place locations, see Fig. 17.2). Each of these population concentrations has a strong impact on the political and economic geography of Africa. Lesser population concentrations are scattered irregularly through the sparsely inhabited deserts, steppes, and grassy and forested expanses of tropical "bush" that make up most of Africa.

17.2 The African Environment

Africa south of the Sahara is both rich in natural resources and beset with environmental challenges that make economic development difficult. It is home to some of the world's greatest concentrations of wildlife and to some of the most degraded habitats.

Large Plateaus and Major Rivers

Most of Africa is a vast plateau—actually a series of plateaus—rising to varying elevations (see inset, Fig. 17.2). The plateau surfaces are interrupted by prominent river systems such as the Nile, Niger, Congo, Zambezi, and Orange. Lowland plains form a narrow band around the coasts. Inland from the coast, escarpments mark the transition to the plateau surface, which typically lies at an elevation of more than 1000 feet (305 m). Near the Great Rift Valley in the horn of Africa and in southern and eastern Africa (see Regional Perspective, below), the general elevation rises to 2000 to 3000 feet, with considerable areas at 5000 feet and higher (see inset, Fig. 17.3). The highest peaks and largest lakes of the continent are located in this highland belt. The loftiest summits lie within a 250-mile radius (*c.* 400 km) of Lake Victoria. They include Mount Kilimanjaro (19,340 ft/5895 m) and Mount Kirinyaga (Mount Kenya; 17,058 ft/5200 m; Figure 17.4), which are volcanic cones, and the

Regional Perspective

The Great Rift Valley

One of the most spectacular features of Africa's physical geography is the **Great Rift Valley**, a broad, steep-walled trough extending from the Zambezi valley (on the border between Zimbabwe and Zambia) northward to the Red Sea and the valley of the Jordan River in southwestern Asia (see Fig. 17.3). Its relationship to the tectonic movement of crustal plates is still poorly understood. However, most Earth scientists believe it marks the boundary of two crustal plates which are rifting, or tearing apart, causing a central block between two parallel fault lines to be displaced downward, creating a linear valley. This movement will eventually cut much of southern and eastern Africa away from the rest of the continent and allow sea water to fill the valley.

The Great Rift Valley actually has several branches. Lakes, rivers, seas, and gulfs already occupy much of it. It contains most of the larger lakes of Africa, although Lake Victoria, located in a depression between two of its principal arms, is an exception. Some, like the 4823-foot-deep (1470-m) Lake Tanganyika (the world's second deepest lake after Russia's Lake Baikal), are extremely deep. Most have no outlet. Volcanic activity associated with the Great Rift Valley has created Mount Kilimanjaro, Mount Kirinyaga, and some of the other great African peaks, along with lava flows, hot springs, and other thermal features. Faulting along the Great Rift Valley in Ethiopia, Kenya, and Tanzania has also exposed remains of the earliest known ancestors of *Homo sapiens*.

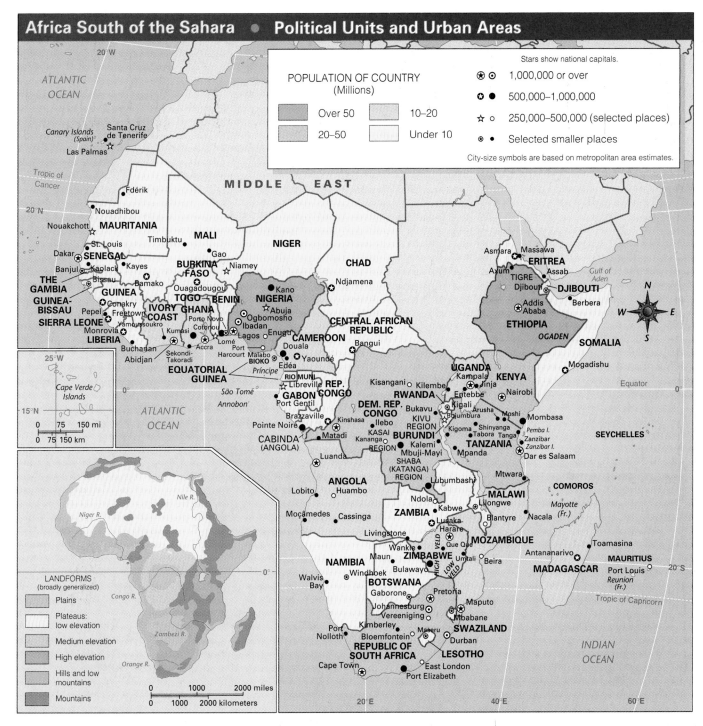

Figure 17.2 Reference map of countries, cities, and landforms of Africa south of the Sahara.

Ruwenzori Range (up to 16,763 ft/5109 m), a nonvolcanic massif produced by faulting. Lake Victoria, the largest lake in Africa, is surpassed in area among inland waters of the world only by the Caspian Sea and Lake Superior. There are several other large lakes in East Africa, including Lake Tanganyika and Lake Malawi.

The physical structure of Africa has influenced the character of African rivers. The main rivers rise in interior uplands and descend by stages to the sea. At various points they descend abruptly, particularly at plateau escarpments, so that their courses are interrupted by rapids and waterfalls. These often block navigation a short distance inland. Low water at certain

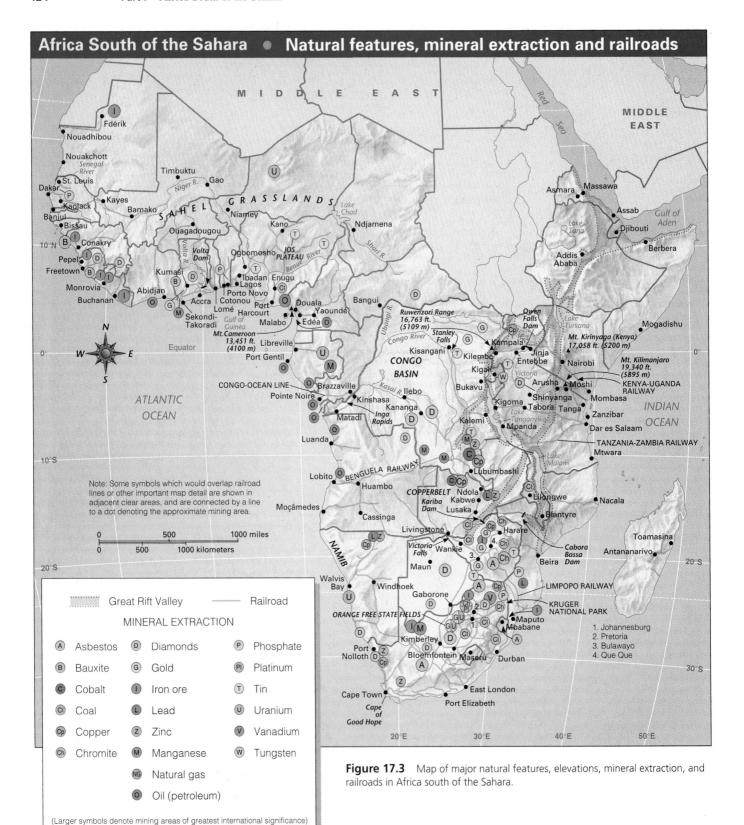

Figure 17.3 Map of major natural features, elevations, mineral extraction, and railroads in Africa south of the Sahara.

Figure 17.4 Kenya's Mount Kirinyaga (Mt. Kenya: 17,058 ft/5200 m), an extinct volcano, is Africa's second highest mountain. *Joseph J. Hobbs*

seasons and shallow and shifting delta channels also hinder navigation of many African rivers. Africa's discontinuous inland waterways are interconnected by rail and highways more than on any other continent.

Among the important rivers that have built deltas are the Niger, Zambezi, Limpopo, and Orange. In contrast, the Congo River has scoured a deep estuary 6 to 10 miles (10–16 km) wide which ocean vessels can navigate to the seaport of Matadi in the Democratic Republic of Congo, about 85 miles (*c.* 135 km) inland. The Congo is used more for transportation than is any other African river.

The frequent falls and rapids of rivers in Africa south of the Sahara have a positive side: They represent a great potential source of hydroelectric energy. Large power stations exist on the Zambezi River at the Cabora Bassa Dam in Mozambique and at the Kariba Dam (Fig. 17.5), which Zimbabwe and Zambia share; at the Kainji Dam on the Niger River in Nigeria; and at the Akosombo Dam on the Volta River in Ghana. Many other power stations of varying size are scattered over the continent. However, only about 5 percent of Africa's hydropower potential has been realized (compared to 59 percent in North America, for example). Many of the best sites are remote from large markets for power. The largest single potential source of hydroelectricity in Africa, and possibly in the world, is the stretch of rapids on the Congo River between Kinshasa and Matadi. The

Democratic Republic of Congo's Inga Project is developing this resource.

Climate, Vegetation, and Moisture

The equator bisects Africa, so most (about two-thirds) of the region lies within the low latitudes and has tropical climates; Africa is the most tropical of the world's continents. One of the most striking characteristics of Africa's climatic pattern is its symmetry or regularity. This is due mainly to the continent's position astride the equator, coupled with its generally level surface. (The broad pattern of climates is depicted in Fig. 2.7 on p. 30; see also the biomes map, Fig. 2.8, p. 32.) Areas of tropical rain forest climate center around the great rain forest of the Congo basin in central and western Africa. The forest merges gradually into a tropical savanna climate on the north, south, and east. This is the climatic and biotic zone that supports the famous large mammals of Africa. The savanna areas, in turn, trend into steppe and desert on the north and southwest. A broad belt of drought-prone tropical steppe and savanna bordering the Sahara Desert on the south is known as the Sahel (see Regional Perspective, Chapter 18). There is desert on the coasts of Eritrea, Djibouti, and Somalia in the Horn of Africa. In South Africa and Namibia, a coastal desert, the Namib, borders the Atlantic. The "Kalahari Desert," which lies inland from the Namib, is better described as steppe or semidesert than as true desert. Along the northwestern and southwestern fringes of the

Figure 17.5 The Kariba Dam straddles the Zambezi River between Zimbabwe and Zambia and provides hydroelectric power to both countries. *Joseph J. Hobbs*

continent are small but important areas of mediterranean climate, while eastern coastal sections and adjoining interior areas of South Africa have a humid subtropical climate. High elevations moderate the temperatures of extensive interior areas that lie within the realm of tropical savanna climate in the east and south of the continent.

Total precipitation on the African continent is very high but unevenly distributed. Some parts of Africa receive an overabundance of rain while other areas have scarcely any. Even in the rainier parts of the continent, large areas have a long dry season, and wide fluctuations occur from year to year in the total amount of precipitation. As a result, one of the major needs in Africa is better control over water. In the typical village household, women carry water by hand from a stream or lake or a shallow (often polluted) well. Use of more small dams would help provide water storage throughout the year, especially in the seasonally rainy areas.

Soils

Among Africa's most productive lands are alluvial soils on river plains. Other especially fertile soils are found in scattered areas where volcanic parent materials occur, particularly in parts of the East African and Ethiopian highlands. A third group of better than average soils are the grassland soils found in some areas of tropical steppe and tropical highland and in the midlatitude grasslands of the High Veld in South Africa. These soils are not entirely comparable to the highly fertile grassland soils of North America, as they are more difficult to cultivate and lose their fertility more quickly under continuous cropping.

Soils of the deserts and regions of mediterranean climate are often poor. True soils are absent over broad areas of desert. The same is true in the mediterranean areas, with the exception of some valleys that have fertile soils where materials transported from adjoining slopes have accumulated. In the tropical rain forests and savannas, reddish tropical soils generally dominate. These are characteristically infertile once the natural vegetation is removed and can support only shifting agriculture (see discussion on p. 436).

Animal Life

Africa has the planet's most spectacular and numerous populations of large mammals. The tropical grasslands and open forests of Africa are the habitats of large herbivorous animals—such as the elephant, buffalo, antelope, zebra, and giraffe—as well as of carnivorous and scavenging animals, such as the lion, leopard, and hyena. The tropical rain forests have fewer of these "game" animals (as Africans call them); the most abundant species here are insects, birds, and monkeys, with the hippopotamus, the crocodile, and a great variety of fish present in the streams and rivers draining the forests and wetter savannas.

While film documentaries promote a perception outside Africa that the continent is a vast animal Eden, the reality is less positive. Human population growth, urbanization, and agricultural expansion are taking place in Africa, as elsewhere in the world, at the expense of wildlife. Hunting and competition with domesticated livestock also take their toll. The numbers of many species have now diminished to the point that these animals are protected by law. Such laws are difficult to enforce, and poaching on a large scale has devastated some species (see Regional Perspective, p. 428). However, Africa is still home to some of the world's most extraordinary and successfully managed national parks, such as the Kruger National Park in South Africa and the Ngorogoro Crater National Park in Tanzania.

17.3 Cultures of the Continent

Many non-Africans are unaware of the achievements and contributions of the cultures of Africa south of the Sahara. The African continent was the original home of the human race. Recent DNA studies suggest that the first modern people (*Homo sapiens*) to inhabit Asia, Europe, and the Americas were descendants of a small group that left Africa via the isthmus of Suez about 100,000 years ago. After about 5000 B.C., indigenous people were responsible for agricultural innovations in four culture hearths: the Ethiopian Plateau, the West African savanna, the West African forest, and the forest-savanna boundary of West Central Africa. Africans in these areas domesticated such important crops as millet, sorghum, yams, cowpeas, okra, watermelons, coffee, and cotton. From Africa, these diffused to agricultures in other world regions.

Civilizations and empires emerged in Ethiopia, West Africa, West Central Africa, and Southern Africa. In the first century A.D., the Christian empire based in the northern Ethiopian city of Axum controlled the ivory trade from Africa to Arabia. Ethiopian tradition holds that a shrine in Axum still contains the Biblical Ark of the Covenant and the tablets of the Ten Commandments, which disappeared from the Temple in Jerusalem in 586 B.C. (see Chapter 9, p. 239). Several Islamic empires, including the Ghani, Mali, and Hausa states, emerged in West Africa between the 9th and 19th centuries. All of these agriculturally based civilizations controlled major trade routes across the Sahara. They profited from the exchange of slaves, gold, and ostrich feathers for weapons, coins, and cloth from North Africa. Three kingdoms arose between the 14th and 18th centuries in what is the present area of southern Democratic Republic of Congo and northern Angola. These included the Kongo kingdom, which had productive agriculture and was the hub of an interregional trade network for food, metals, and salt. In what is now Zimbabwe, the Karanga kingdom of the 13th to 15th centuries built its capital city at the site known as Great Zimbabwe. Its skilled metalworkers mined and crafted gold, copper, and iron, and merchants traded these metals with faraway India and China.

In the 16th century, European colonialism began to overshadow and inhibit the evolution of indigenous African civilization. However, the artistic, technical, and entrepreneurial skills of the region's peoples continued to flourish. These traits are today part of a greater African culture which is poorly understood and often stereotyped in the wider world. Surprisingly, in this diverse and often fragmented continent, there are many shared cultural traits which together comprise what geographer Robert Stock describes as "Africanity." He identifies the following eight constituent elements of the African identity:

1. A black skin color.
2. A unique conceptualization of the relationship between people and nature. Indigenous African religions emphasize that spiritual forces are manifested everywhere in the environment, in contrast with the introduced Christian and Muslim faiths, which tend to see nature as separate from God and people as apart from and superior to nature.
3. An identity tied closely to the land, with many people dependent on hunting, herding, and farming. This dependence on the land reinforces the sense of closeness to nature. Africans tend to treat the land as communal rather than individual property.
4. Emphasis on the arts, including sculpture, music, dance, and storytelling, as essential to the expression of African identity.
5. A view of Africans as individuals making up links in a continuing "chain of life," in which reverence of ancestors and nurturing of children are virtues. Parents prefer to have many children to keep the chain growing, and strive to educate them in the traditions of the ethnic group.
6. Extended rather than nuclear families, with parents and their children living and interacting with grandparents, cousins, nieces, nephews, and other relatives.
7. Respect for wise and fair authority, with village elders, "big men," and tribal chiefs endowed with powers they are expected to wield to benefit the group.
8. A shared history of colonial occupation that contributes to a unified sense that, in the past, Africans were humiliated and oppressed by outsiders.

Despite the common features of "Africanity," however, Africa south of the Sahara is culturally and ethnically diverse. African nations in this region vary greatly in their tribal or ethnic composition. (Politically and socially, Africans traditionally identified themselves by their tribe, recognizing members of the tribe as all those descended in kinship from a single tribal founder, or "eponym." However, because "tribe" and "tribalism" have acquired connotations of primitive feuding between hostile rivals, many Africanists now prefer to use the terms "ethnic group" and "ethnicity" in their place.) Some countries, like Somalia, Lesotho, Swaziland, and Botswana, are very homogeneous, while Tanzania, Cameroon, and Nigeria have hundreds of ethnic groups. Conflict between tribes or ethnic groups is now rather rare in Africa south of the Sahara, with such notable exceptions as the recent bloodshed between Hutu and Tutsi in Rwanda and Burundi.

There is also great linguistic diversity in Africa south of the Sahara. By one count, the peoples of this region speak more than 1000 languages. Most of the peoples of this region belong to one of four broad language groupings:

1. The Niger-Congo language family—the largest—which includes the many West African languages and the roughly 400 Bantu subfamily languages. Most of these are spoken south of the equator.
2. The Afro-Asiatic language family, including Semitic branch languages (such as Arabic and Amharic) and tongues of the Cushitic (such as Oromo and Somali) and Chadic (such as Hausa) branches. People living in the area adjoining the Sahara, from West Africa to the Horn of Africa, speak these languages. Even some of the Niger-Congo languages originating south of the Sahara, such as the Swahili (Kiswahili) tongue spoken widely in East Africa, have borrowed much from Arabic and other languages with roots elsewhere. The prominence of Arabic words in Swahili reflects a long history of Arab seafaring along the Indian Ocean coast of Africa; in fact, Swahili means "coastal" in Arabic.
3. The Nilo-Saharan language family of the central Sahel region, the northern region of West Central Africa, and parts of East Africa.
4. The Khoisan languages of the western portion of southern Africa, which the Bushmen (San) and related peoples speak.

17.4 The Geographic Impact of Slavery

Until about a thousand years ago, the cultures of Africa south of the Saharan desert barrier remained largely unknown to the peoples north of the desert. Egyptians, Romans, and Arabs developed contacts with the northern fringes of this region and some trade filtered across the Sahara, but to most outsiders the "Dark Continent" was a self-contained, tribalized land of mystery. Even at the opening of the 20th century, vast areas of interior tropical Africa were still little known to Westerners.

The tragic impetus for growing contact between Africa and the wider world was slavery (Fig. 17.6). Over a period of 12 centuries, as many as 25 million people from Africa south of the Sahara were forced to become slaves, exported as merchandise from their homelands. The trade began in the 7th century, with Arab merchants using trans-Saharan camel caravan routes to exchange guns, books, textiles, and beads from North Africa for slaves, gold, and ivory from Africa south of the Sahara. As many as two-thirds of the estimated nine-and-a-half million slaves exported between the years 650 and 1900 along this route were young women who became concubines and household servants in North Africa and Turkey. Male slaves usually became soldiers or court attendants. From the 8th to 19th centuries, about five

Regional Perspective

Managing the Great Herbivores

Elephants and rhinoceroses are Africa's largest and most endangered herbivores. Their plight has accelerated in recent years, and so have local and international efforts to maintain their populations in the wild. The problem has compelled countries with very different wildlife resources to work together toward solutions.

In 1970, there were about two-and-a-half million African elephants living on the continent. Poaching and habitat destruction reduced their numbers to 1.8 million by 1978. There are now an estimated 350,000. The main reason for the sharp decline is that elephants have something people prize: ivory. Poaching for ivory was reducing African elephants at a rate of 10 percent annually when delegates of the 112 signatory nations of the Convention on International Trade in Endangered Species (CITES) met in 1989. The organization succeeded in passing a worldwide ban on the ivory

trade, and since then the precipitous decline has halted.

CITES member states won the ivory ban over the strong objections of southern African states, led by Zimbabwe. While the East African nations of Uganda, Kenya, and Tanzania were suffering crashing populations, Zimbabwe was experiencing what it regarded as an elephant overpopulation problem (Fig. 17.A). Zimbabwe had 5000 elephants in 1900. Today there are an estimated 77,000, and they are increasing at a rate of 4 percent annually. There are also healthy and growing elephant populations in Botswana, Malawi, Namibia, and South Africa. Before the worldwide ivory ban, these countries profited from the sustainable harvest and sale of elephant ivory, hides, and meat. At the 1989 CITES meeting, these countries argued that they should not be punished for their success in protecting the great mammals. They

appealed for an exemption from the ivory ban so that they could earn foreign export revenue from a sustainable yield of their elephant populations. The majority of CITES members rejected this appeal, arguing that any loophole in a complete ban would subject elephants everywhere to illegal poaching, resulting in the loss of the African elephant. The elephant-rich countries reluctantly supported this position. In 1997, however, CITES reversed its policy and elected to allow Zimbabwe, Namibia, and Botswana to sell ivory as a reward for their positive wildlife policies. These countries are permitted to export ivory only to Japan, and only under the most tightly-monitored circumstances.

These elephant-rich nations continue to cull (kill) "excess" elephants. Zimbabwean officials argue that their country can support only 45,000 animals. Meat from the cull of about 5000 elephants

Figure 17.A   Zimbabwe is blessed with elephants. These animals in Hwange National Park are feeding from a box of treats provided to them by the country's preeminent pachyderm ecologist, Alan Elliot (standing, wearing cap). *Joseph J. Hobbs*

Figure 17.B Daggers are a nearly universal dress accessory for men in the Arabian peninsula nation of Yemen. The most prized dagger handles are of rhino horn, a custom that has had a devastating impact on rhinos thousands of miles distant, in Africa. *Joseph J. Hobbs*

yearly goes to needy villagers and crocodile farms. This resource helped the country weather the drought and near-famine of 1992.

If elephant ivory is like gold, rhinoceros horns are like diamonds. Men in the Arabian-peninsula nation of Yemen prize daggers with rhino horn handles (Fig. 17.B). Although Western scientists deny the medicinal efficacy of powdered rhino horn, traditional medicine in East Asia (particularly in Taiwan, China, and Malaysia) makes wide use of it, including as an aphrodisiac. These demands, and the current black-market value of about $25,000 per horn, have led to a precipitous decline in population of black rhinoceroses in Africa. There were an estimated 65,000 black rhinos in Africa in 1982; poaching has reduced that number to less than 2500.

Zimbabwe is on the front line in the war to protect the black rhino and appears to be losing. As recently as 1984 there were as many as 2000 of the animals in the country. Having reduced the

numbers in countries to the north, poachers turned to rhino-rich Zimbabwe. Since 1984 there has been a steady increase in the numbers of poachers crossing international boundaries to kill rhinos in Zimbabwe. Most of them come across the Zambezi River from Zambia. Since per capita gross national product in Zambia is only about $350, the prospect of making hundreds or thousands of dollars in a night's work is irresistible to many. Even the order to Zimbabwean wildlife rangers to shoot poachers on sight, in effect since 1985, has not slowed the slaughter. Armed with automatic weapons, poachers have killed more than 1500 rhinos since 1984. In the same period, poachers have killed four rangers, while rangers have killed more than 150 poachers.

With fewer than 300 rhinos surviving in Zimbabwe, wildlife officials have turned to more desperate measures. In 1991 they began dehorning rhinos to make them unattractive to poachers. To do this, a marksman tranquilizes the animal and two assistants use a chainsaw to

remove the two horns. It is not a permanent solution; the horn grows back at a rate of three inches yearly, so each animal must be regularly re-dehorned. Wildlife authorities have admitted that the program has failed. Poachers continue to kill the animals, perhaps out of spite, or because they cannot tell in the dark whether or not the prey has horns—or perhaps because they are after even a few inches of horn stump. Zimbabwean wildlife officials are now considering the possibility of opening a legal trade in rhino horns. They would raise rhino herds on state farms and regularly harvest their regrowing horns for sale. South Africa supports this idea of sustainable harvest. Like Zimbabwe, South Africa is sitting on a stockpile of tons of confiscated rhino horns, and would profit greatly from their legalized trade. In addition, with about 900 living animals, South Africa is the last stronghold of the black rhino and does not want to become the next frontline state in the rhinoceros war.

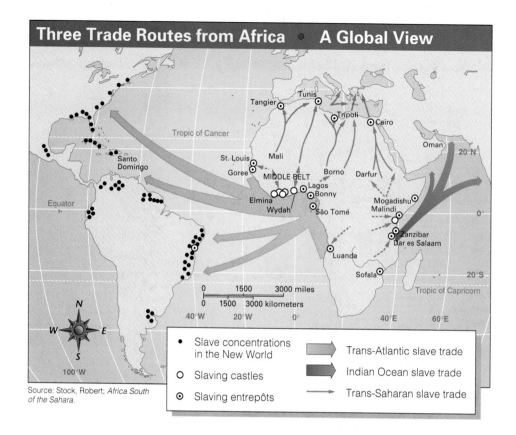

Figure 17.6 Slave export trade routes from Africa south of the Sahara.

Three Trade Routes from Africa • A Global View

Source: Stock, Robert; *Africa South of the Sahara.*

- Slave concentrations in the New World
- ○ Slaving castles
- ⊙ Slaving entrepôts
- Trans-Atlantic slave trade
- Indian Ocean slave trade
- Trans-Saharan slave trade

million more slaves were exported from East Africa to Arabia, Oman, Persia (modern Iran), India, and China. Again, most were women who became concubines and servants.

The notorious and lucrative traffic in slaves provided the main early motivation for European commerce along the African coasts, and it inaugurated the long era of European exploitation of Africa for profit and political advantage. The European-controlled slave trade was the largest by far. Between the 16th and 19th centuries, the capture, transport, and sale of slaves was the exclusive preoccupation of trade between the European world and West Africa. Portuguese and Spaniards began the trade in the 15th century, and a century later, English, Danish, Dutch, Swedish, and French slavers were active.

The business boomed with the development of plantations and mines in the New World. Populations of Native Americans in Anglo and Latin America were insufficient for these industries, so the Europeans turned to Africa as a source of labor (see Chapter 2, p. 44). The peak of the trans-Atlantic slave trade was between 1700 and 1870, when about 80 percent of an estimated 10 million slaves made the crossing. In escape attempts, in transit, and in the famines and epidemics that followed slave raids, probably more than 10 million others died.

Slaves were a prized commodity in the **triangular trade** linking West Africa with Europe and the Americas. European ships carried guns, alcohol, and manufactured goods to West Africa, exchanging them there for slaves. They then transported the slaves to the Americas, exchanging them for gold, silver, tobacco, sugar, and rum to be carried back to Europe. As "raw material" and as the labor working the mines and plantations of

Latin America, the West Indies, and Anglo America, slaves generated much of the wealth that made Europe prosperous and helped spark the Industrial Revolution.

While Europeans carried out the trade, their physical presence was limited to coastal shipping points. Africans were the intermediaries who actually raided inland communities to capture the slaves and assemble them at the coast for transit shipment. West African kingdoms initially acquired their own slaves in the course of waging local wars. As the demand for slaves grew, these kingdoms increasingly went to war for the sole purpose of capturing people for the trade. As the exports grew, so did the practice of Africans keeping African slaves. Even after Great Britain (in 1807) and the other European countries abolished slavery—finally bringing an end to the trans-Atlantic trade in 1870—slavery flourished within Africa. By the end of the century slaves made up half the populations of many African states. Slavery has not yet died out in the region; in Mauritania, some light-skinned Moors still enslave blacks, although the national government has outlawed this practice three times. Slavery also exists in the Sudan.

17.5 Colonialism

European penetration of the African interior began in 1850 with a series of journeys of exploration. Missionaries like David Livingstone, as well as traders, government officials, and now-famous adventurers and scientific explorers such as James

Bruce, Richard Burton, and John Speke, undertook these expeditions. By 1881, when Africans still ruled about 90 percent of the region, these exploits had revealed the main outlines of inner African geography, and the European powers began to scramble for colonial territory in the interior. By 1900, only Ethiopia and Liberia had not been colonized. Much of the carving up of Africa took place at the Conference of Berlin in 1884 and 1885, when the French, British, Germans, Belgians, Portuguese, Italians, and Spanish established their respective spheres of influence in the region. Africa south of the Sahara became a patchwork of European colonies—a status it retained for more than half a century (Fig. 17.7)—and Europeans in these possessions were a privileged social and economic class.

At the outbreak of World War II in 1939, only three countries—South Africa, Egypt, and Liberia—were independent. The United Kingdom, France, Belgium, Italy, Portugal, and Spain controlled the rest. But after the war, mainly in the 1960s and 1970s, a sustained drive for independence changed Africa from a colonial region to one comprising more than one-fourth of the world's independent states. This was a peaceful process in most instances, but bloodshed accompanied or followed independence in several countries, including what are now Angola and Democratic Republic of Congo.

European colonization of Africa south of the Sahara produced many of the negative attributes of underdevelopment described in Chapter 2 (see pp. 44–45). These included marginalization of subsistence farmers, notably those who colonial authorities—intent on cash crop production—displaced from quality soils to inferior land. In addition, European use of indigenous labor to build railways and roads often took a high toll in human lives and disrupted countless families. The colonizers also often corrupted traditional systems of political organization to suit their needs, sowing seeds of dissent and interethnic conflict.

The European colonial enterprise did have some positive impact. The colonies, and the independent nations that succeeded them, were the beneficiaries of new cities and the transport links built with forced or cheap African labor; new medical and educational facilities (often developed through Christian missions); new crops and better agricultural techniques; employment and income provided by new mines and modern industries; new governmental institutions; and government-made maps useful for administration and planning. Such innovations were very helpful, but they were distributed unequally from one colony to another, and were inadequate for the needs of modern societies when independence came.

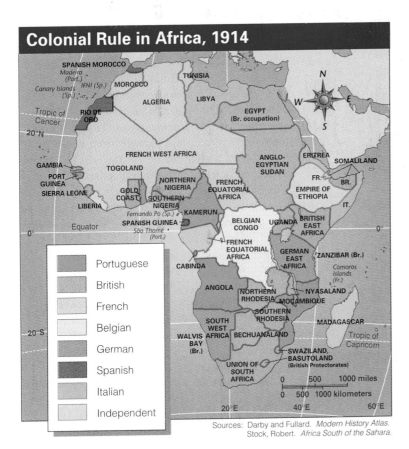

Sources: Darby and Fullard. *Modern History Atlas.* Stock, Robert. *Africa South of the Sahara.*

Figure 17.7 Colonial rule in 1914. Germany lost its colonies after World War I.

17.6 Shared Traits of African Countries Today

It is impossible in the scope of this book to examine in detail every nation in this complex region. The following generalizations provide an introductory overview, and subsequent sections in Chapter 18 deal with more specific regional and national issues. (The country of South Africa, which has a distinct historic and economic legacy, is an important exception to many of these generalizations, as discussed in Chapter 18.)

1. *Great poverty is characteristic.* Africa south of the Sahara is the most impoverished of the world's regions; 18 of the world's 20 poorest countries are there (see Table 17.1). The region's poverty-related problems include a high incidence of illiteracy, hunger, and disease; inadequate facilities for transportation and communication; and a lack of domestic capital to foster increased agricultural and industrial production. The average African eats 10 percent less than he or she did 20 years ago. Most African societies lack a middle class and the prospect of upward economic mobility. Instead, most are hierarchical, with only a very small (often tribally based) elite controlling the lion's share of the nation's wealth.

   Despite these grim features, in recent decades there have been some changes for the better. Improved standards of health and literacy in many areas have resulted from the work of national and international governmental and nongovernmental agencies. The extension of roads, airways, and other transportation facilities has promoted the marketing of farm products, including perishable items, from formerly inaccessible areas. Stores and markets in both rural and urban areas stock a variety of manufactured goods from overseas and from African sources. Modern factories have been established in many urban centers, and improved agricultural techniques have been introduced in many areas. Such changes have affected some peoples and areas much more than others, and the total impact of change represents only a beginning in lifting the region from its current state of underdevelopment.

2. *Most people live in rural areas.* The rural populations of individual countries range generally between an estimated 65 and 85 percent. The most rural include Rwanda (95 percent); Burundi (95 percent); Uganda (86 percent); Niger and Burkina Faso (85 percent); and Ethiopia, Nigeria, and Lesotho (84 percent). (See Table 17.1.) Life in villages is the rule, although dispersed homes on individual farms are common in some areas. The typical rural home is a small hut made of sticks and mud, with a dirt floor, a thatched roof, and no electricity or plumbing (Fig. 17.8).

3. *Subsistence agriculture is the main occupation in nearly all countries.* The great majority of farmers in Africa south of the Sahara operate on a subsistence basis, producing so lit-

Figure 17.8 The typical rural dwelling in Africa south of the Sahara is made of locally available materials and, although having an earthen floor, is kept meticulously clean. This home is in northern Zimbabwe. *Joseph J. Hobbs*

tle surplus for sale that they have little cash income and few savings. Women do a large share of the farmwork—they produce 80 to 90 percent of Africa's food—in addition to household chores and the bearing and nurturing of children (Fig. 17.9). Mechanization is rare, fertilizers are expensive, and therefore crop yields are low. Storage facilities are often so poor that pests and bad weather ruin much of the harvest. Subsistence agriculture is discussed in more detail on p. 436.

4. *Per capita food output in most countries has declined or has not increased since independence.* The world's most rapid rates of population increase, government policies unfavorable to agriculture, and damage from warfare and drought have been among the factors thwarting efforts to maintain self-sufficiency in food production. Substantial imports of food are now common in the majority of countries. A traditional emphasis on large families threatens to nullify any gains in food production that could be achieved. As in most LDCs, African parents want to have large families for several reasons: so that they will have extra hands to work the fields; so that they will be looked after when they are old or sick; and, in the case of girls, so that they will receive the "bride wealth" a groom pays in a marriage settlement (see Chapter 2, p. 48). Large families also convey status.

   The Malthusian scenario seems constantly to loom large over Africa (see Chapter 2, p. 48). Analysts fear the consequences of what they call the "one percent gap": Since the 1960s, the population of Africa south of the Sahara has grown at a rate of about 3 percent annually, while food production in the region has grown at only about 2 percent per year. Providing an adequate domestic food supply has be-

Figure 17.9 Mother and child in Zimbabwe. Women plant and harvest most of Africa's food, while also rearing its children.   *Joseph J. Hobbs*

come increasingly difficult because of prolonged droughts, rapid population increases, and unwise land-use practices brought on by population pressure. Inadequate transportation and maintenance and the high cost of imported oil, fertilizer, food, and other necessities reduce food output. Governments have contributed to food shortages by favoring industry over agriculture and export crops over food crops (see Chapter 2, p. 45). Many regimes have propelled food shortages into full-blown famines by investing in domestic or international warfare rather than in getting food to people.

5. *Drought is a persistent problem in most countries.* Several countries include deserts where rain falls infrequently, but annual dry seasons are the rule even in the more humid bulk of Africa south of the Sahara. Although all droughts create problems, the most devastating effects in this heavily agricultural part of the world occur when the normal rainy season fails to come for a year or more. This condition has been very widespread in the region since the late 1960s. One area of repeated drought and human suffering is the Sahel, the east-west belt of dry grass and shrub lands south of the Sahara Desert (see Chapter 18, p. 446, and Fig. 17.3, p. 424). In the early 1990s, severe drought afflicted Somalia, in the Horn of Africa, and South Africa, Zimbabwe, and several neighboring countries in southern Africa.

6. *Lack of education hinders development.* In only about one-third of the countries is more than one-half the population reported to be literate. In Mauritania, literacy in adults (those over age 15) was estimated at 38 percent in 1998; in Burkina Faso, that figure was 19 percent. In Burkina Faso, as in many countries of the region, the literacy rate for women is less than half that for men. Authorities on population and development concur that much higher literacy rates for women will be needed to begin to bring down the high population growth rates that underlie many problems. Lack of schooling for women in much of Africa south of the Sahara is a formidable barrier to improved family health and the creation of a more skilled labor force. All countries except South Africa are short of skilled workers, particularly workers with administrative and managerial skills. In most countries, college graduates were few at the time of independence, and more schools of higher education are needed.

7. *Poor transportation hinders development* (Fig. 17.10). Since independence, the new nations have been able to build relatively few surfaced roads or railroads. Those left

Figure 17.10 Many African river crossings have no bridges and it is necessary to use ferries, although some streams can be forded during the dry season. This photo shows vehicles and people boarding a ferry to cross an estuary north of Mombasa, on the Kenyan coast. Backup of traffic at these bottlenecks may take hours or even days to clear.   *Joseph J. Hobbs*

over from colonial days have often deteriorated. Overland transportation is inadequate in extent and is plagued by maintenance problems affecting both vehicles and routes. Some of the reasons include tropical heat, high humidity, dust, prevalence of unpaved roads, lack of lubricants and spare parts, low levels of technical skill, and governments lacking a sense of public service. Africa south of the Sahara critically needs a good international transportation network, coupled with a lowering of trade barriers, in order to create market opportunities on a vastly enlarged scale.

8. *There are serious public health problems.* A high incidence of disease and parasites affecting people, domesticated animals, and cultivated crops has been one of the main hindrances to African development. Insects carry many of the major diseases; mosquitoes, for instance, carry malaria, yellow fever, and dengue, or "breakbone fever." The tsetse fly carries sleeping sickness and nagana, a destructive disease affecting cattle and horses (see Definitions and Insights, p. 437). Large numbers of Africans are afflicted with digestive diseases and parasites, including dysentery, typhoid and paratyphoid fever, bilharziasis (schistosomiasis), hookworm, and other types of intestinal worms. Contaminated water and other unsanitary conditions are largely responsible for such afflictions. Other diseases common in Africa include tuberculosis, filariasis, nutritional deficiency diseases, pneumonia, yaws, leprosy, influenza, trachoma, venereal diseases, and many fungoid diseases of the skin. Limited and localized outbreaks of the fatal Ebola virus in West Africa and West Central Africa in 1995 caused great concern in the international medical community.

Recently, AIDS (acquired immunodeficiency syndrome), which apparently originated in Africa, has spread so extensively in this region that it is a new and major plague. In 1998, 86 percent of the world's estimated 30 million persons infected with HIV, the virus that causes AIDS, lived in Africa south of the Sahara. In some countries, including Botswana and Zimbabwe, as much as one quarter of the population is HIV-positive. Such pandemics have not been seen on Earth since the bubonic plague devastated 14th-century Europe and smallpox struck the Aztecs of 16th-century Mexico. This scourge is causing sharp reductions in life expectancy in Africa and, unless contained, will dramatically alter recent projections of the region's population growth. Life expectancy in Botswana was 61 years in 1993, but, because of the virus, was 41 in 1998. In neighboring Zimbabwe, the population growth rate in 1998 was 1.5 percent, rather than the projected 2.4 percent that would have been the rate were it not for AIDS-related deaths. It is impossible to calculate the economic costs of this mortality. While it may seem logical that a lower population would mean more prosperity, the incidence of HIV infection is particularly high among the region's most educated, skilled, and ambitious young urban professionals, including white-

collar workers and government employees; these are the ones who travel the most and are therefore the most likely to have encounters with prostitutes. The effects of the epidemic on corporate profits, education, research, health care, and other indices of progress are therefore probably great, and are certainly incalculable.

Despite the many health problems, there has been much progress recently in combating diseases such as river blindness in Africa. Enough is known about the control of tropical diseases and parasites to greatly reduce their incidence if means are available for the technical knowledge to be fully applied. But in vast, poverty-stricken Africa, the need for medical assistance far outruns the money and personnel available for such assistance.

9. *The national economies of all countries except South Africa are underindustrialized and overly dependent on the export of a few primary products, particularly minerals and cash crops.* Dependency theorists often point to Africa as a prime example of how colonialism created lasting disadvantages for the colonized. Africa's place in the commercial world is mainly that of a producer of foods and raw materials for sale outside the region. In most nations, one or two products supply more than two-fifths of all exports—for example, coffee and tea in Kenya (Fig. 17.11). Each country is therefore vulnerable to international oversupply of an export on which it is vitally dependent (see Chapter 2, p. 45). The value of imports generally exceeds that of exports in Africa south of the Sahara, with imports consisting overwhelmingly of manufactured goods, oil products, and/or food. Cash-crop and mineral exports are described in more detail on p. 437.

10. *Almost all countries are heavily in debt to foreign lenders.* Many billions of dollars in outside grants and loans during the postcolonial era have failed to eliminate poverty in the region. Instead, the debts which African governments owe to international lenders such as the International Monetary Fund (IMF), the World Bank, and private banks often compound economic woes and contribute to destructive resource uses (see Chapter 2, p. 45).

Much of the indebtedness dates to the late 1970s, when large amounts of loan capital became available to lending institutions in the form of "petrodollars" recycled from Middle Eastern oil states. Nations undertook many costly development projects with borrowed money. Western financiers, planners, and contractors gave optimistic assessments of the benefits to be expected from such projects, and African leaders were ready to accept loans as a way to reap quick benefits from newly won independence. Now, however, many countries are having great difficulty in meeting even the interest payments on their debts. The IMF and other lenders have resisted requests of African debtor nations that these institutions reschedule interest payments and advance new loans. As a condition of further support,

Figure 17.11 Kenya's economy is highly dependent on the export of coffee. Cash crops and other raw materials are typical exports of Africa south of the Sahara.   *Joseph J. Hobbs*

the lenders often demand that African debtors put their finances in better order by such measures as revaluating national currencies downward to make exports more attractive, increasing the prices paid to farmers for their products, and reducing corruption, especially among government officials and urban elites. Although many African debtor nations have attempted to conform to such demands, internal political factors slow progress. Loss of economic privileges angers urban elites on whom governments depend for support. Relaxation of price controls on food to stimulate production is often met by rioting among city-dwellers who have tight budgets already allocated for other costs of living.

11. *Economic and humanitarian assistance to the region has slowed since the end of the Cold War.* During the Cold War years, the Soviet Union and United States and their respective allies extended aid generously to many African nations to boost their competing interests. The ending of the Cold War has constricted these aid pipelines; the United States, for example, cut its economic assistance to Africa south of the Sahara by 30 percent between 1985 and 1992. The bitter experience of the United States in Somalia in the early 1990s also helped to suppress the West's appetite to extend humanitarian assistance. After 1984 and 1985, when television images of famine-wracked Ethiopia prompted American and Western European citizens to give generously,

popular sympathy and support during a series of subsequent African emergencies waned. This lack of apparent public and official interest in crises abroad is known as **donor fatigue**.

12. *Authoritarian governments have been the rule since independence, but progress toward democracy is now widespread.* True democracy has been slow to take root in the region. Military governments and one-party states (often dominated by a single ethnic group) have been common (Fig. 17.12). In the 1990s, military and other special interests seized power, overturned the results of popular elections, or stifled opposition in the Central African Republic, Nigeria, Kenya, Cameroon, Ivory Coast, and elsewhere. Some postcolonial regimes have been guilty of atrocities, and human rights violations by governments have been commonplace. In contrast, free and fair elections have brought new hope and better human rights records to Ghana, Mali, Benin, Tanzania, Zambia, Mozambique, Malawi, Nigeria (in 1999), and South Africa. Namibia, Botswana, and Zimbabwe have also dealt effectively with many of the challenges of democratization and development.

13. *There is serious political instability in many countries.* This is often based on tribal rivalries and antagonisms of long standing, although many of what appear to be tribally based differences are actually related to issues of economic class and political representation, as between the Tutsis and Hutus in Rwanda and Burundi (see Chapter 18). Coups, failed coup attempts, political murders, armed rebellions, and civil wars have punctuated the political histories of many countries since independence. There has been large-scale slaughter, often including massacres of unarmed civilians, in a long list of countries including Liberia, Sierra Leone, Ethiopia, Uganda, Mozambique, Angola, Rwanda, Burundi, Democratic Republic of Congo, and Zimbabwe. Many other countries have experienced serious civil disorder. Such conflicts and crises have drained national treasuries, discouraged foreign investment, and sidetracked progress toward nationhood.

14. *A diverse array of political, economic, and social ideas from the West, the Communist bloc, the Muslim world, and Africa south of the Sahara itself has influenced governments since independence.* Western democratic ideas have rubbed shoulders with African ideas of chieftainship, Communist ideas of one-party dictatorship, and Muslim ideas of theocratic government. Capitalism has coexisted with socialism, and individualism with collectivism. Different governments have adopted different combinations of such ideas. Various forms of socialism have been particularly widespread, resulting in centrally planned national economies, tight state control over wages and prices, state ownership of major economic enterprises, and attempts at collectivization of agriculture and rural life. Such measures have produced disappointing economic results in most

countries, although some socialistic governments have been able to expand their social services. The poor economic performance of socialism has sparked a widespread recent trend toward freer enterprise.

15. *Although formal political colonialism has vanished, most countries still have important links with the colonial powers that formerly controlled them,* and many foreign corporations that operated in colonial days still maintain an important presence. A good example is the Democratic Republic of Congo, where all mineral deposits and production have been nationalized, but where the same Belgian interests that monopolized the mining industry under the colonial regime still carry on mining for the Democratic Republic of Congo government. France, but not Britain, has a long history of postindependence intervention in the political and military affairs of its former African colonies. France is the only ex-colonial power to keep troops in Africa (with the highest numbers in Djibouti, Senegal, and the Central African Republic). In postcolonial Africa, France also has taken steps to ensure that most of its former colonies trade almost exclusively with France, and it in turn has supported national currencies with the French treasury. But France has found its paternalistic approach to be extremely expensive and, in the late 1990s, began reducing its military presence and other costly assistance to its African clients.

17.7 Agricultural and Mineral Wealth

Because most people of the region are subsistence farmers or livestock herders, and since the region's export economy is heavily dependent upon cash cropping and mineral production, these activities are described here in more detail.

Subsistence Agriculture

Over half of the people of Africa south of the Sahara depend directly on agriculture or pastoralism for a livelihood. In most areas, cultivators and their families grow a large share of the crops and livestock for their own use or for local sale; this is subsistence agriculture. In the steppe of the northern Sahel, both rainfall and cultivation are scarce and precarious. The savanna of the southern Sahel, with its greater and more dependable rainfall, is a major area of rainfed cropping despite its long dry season. Unirrigated millet, sorghum, corn (maize), and peanuts are major subsistence crops in the savanna. In the tropical savannas south of the equator, corn is a major subsistence crop in most areas (see Definitions and Insights, p. 439), with manioc and millet also widely grown. Corn, manioc, bananas, and yams are major food crops grown in the rain forests for subsistence and local sale. Oil palms provide household cooking oil in these areas.

Figure 17.12 Many countries in Africa have experienced military rule. These soldiers are on alert during a political crisis on the Indian Ocean nation of Comoros. *Joseph J. Hobbs*

Many African farmers practice land rotation between crop and fallow years. This practice is commonly known as **shifting cultivation**. Most tropical soils lose their natural fertility quickly when they are cropped, and after two or three years must be rested for several years (often 10 to 15 years and more) before they will again produce a crop. During the fallow period, the land reverts to wild vegetation. The need for fallowing can be lessened or eliminated by fertilization, but most African farmers cannot afford to buy chemical fertilizer, animal manure is generally not available, and there is no tradition of systematic fertilization with human wastes as in China and Japan. Some garden plots adjoining huts (known as the "women's land" because they are cultivated by the housewives) are kept more continuously productive by applications of ashes, house sweepings, and goat, chicken, and human manures. Such plots grow vegetables, melons, and bananas for household use, whereas the fields tilled in shifting cultivation generally grow staple crops such as corn, millet, and manioc.

People of the village generally work together to clear fresh land and harvest crops. Fire is a major tool in clearing, and the resultant ashes provide temporary fertilization. Farmers cultivate with hoes, and leave large stumps to decay in the fields.

The system of shifting agriculture is widespread in the world's tropics. Although it is not a very productive system, it traditionally has minimized soil erosion because most of the land is not in cultivation at any given time. It does provide a bare living for people too poor to afford fertilizer or farm machinery. Unfortunately, there has been a recent widespread trend in Africa for farmers to reduce the length of the fallow period in the cycle of shifting cultivation, or to abandon it altogether. This is a direct result of a surging growth in population: With more mouths to feed, in many places there is simply not enough land to provide the "luxury" of fallow. The tragic consequence is that short-term overuse of the land is eroding the soil and rendering it less able to support agriculture in the long run. When the best available lands are exhausted, farmers are often "marginalized" to places unsuitable for farming, such as semiarid lands or slopes that are too steep to till without the consequence of disastrous erosion (see Chapter 2, p. 45). These farmers often drive pastoralists from traditional grazing lands. Confined to smaller areas in which to browse and graze, the nomads' cattle, sheep, and goats often overgraze vegetation and compact the soil. Such dilemmas of land use are the root of the chronic proximity to famine in many countries of Africa south of the Sahara. Unwise political leadership is a common and dangerous addition to this volatile mix.

The Importance of Livestock

Many peoples in Africa south of the Sahara are pastoral, particularly in the enormous tropical grasslands both north and south of the equator. Herding of sheep and hardy breeds of cattle is particularly important in the Sahel. Although cattle raising is widespread through the savannas, cattle are largely ruled out over extensive sections both north and south of the equator by the disease called nagana, which is carried by the tsetse flies that also transmit sleeping sickness to humans (see Definitions and Insights, opposite). In Africa's tropical rain forests, tsetse flies are even more prevalent and few cattle are raised, but goats and poultry are common (as they are also in tsetse-frequented savanna areas).

Most Africans who live by tilling the soil also keep some animals, even if only goats and poultry. Among African peoples such as the Masai of Kenya and Tanzania and the Tutsi (Watusi) of Rwanda and Burundi, livestock not only contribute to daily diet but are an indispensable part of customary social, cultural, and economic arrangements. Cattle are particularly important, with sheep and goats playing a lesser role. The Masai are probably the most famous African example of close dependence on cattle. They milk and carefully bleed the animals for each day's food, and tend them with great care. The Masai give a name to each animal, and herds play a central role in the main Masai social and economic events through the year.

Zebu cattle represent status and wealth in Madagascar, and livestock owners tend to want higher numbers of the animals, rather than better quality stock, to enhance their standing (Fig. 17.13). The resulting population of zebu on the island is about 10 million. Their forage needs have grave consequences for Madagascar's rain forests and other wild habitats, however. People clear the forests and repeatedly set fire to the cleared lands to provide a flush of green pasture for their livestock, causing a rapid retreat of the island's natural vegetation.

Definitions & Insights

AFRICA'S GREATEST CONSERVATIONIST

Not much larger than the common housefly, the tsetse fly of sub-Saharan Africa packs a wallop. This insect carries two diseases, both known as trypanosomiasis, which are extremely debilitating to people and their domesticated animals. People contract "sleeping sickness" from the fly's bite, while cattle contract nagana. Since the 1950s there have been widespread efforts to eradicate tsetse flies so that people can grow crops and herd animals in fly-infested wilderness areas. Where the efforts have been successful, people have cleared, cultivated, and grazed the land. The results are mixed. While people have been able to feed growing populations in the process of opening up these lands, they have also eliminated important wild resources and in many cases caused erosion, desertification, and salinization of the land. Where the tsetse fly has been eliminated, so has the wilderness. The diminutive tsetse fly thus may be characterized as a **keystone species**—one which affects many other organisms in an ecosystem. The loss of a keystone species—in this case, a fly that keeps out humans and cattle—can have a series of destructive impacts throughout the ecosystem. For its role in maintaining wilderness in Africa south of the Sahara, some wildlife experts know the tsetse fly as "Africa's Greatest Conservationist."

As in Madagascar, cattle represent wealth in many societies of Africa south of the Sahara, and people tend to value them for their numbers rather than their quality. They have traditionally constituted the "bride wealth" that changes hands in the marital arrangements of African tribal societies. While men do not "buy" wives, it is a universal custom for the bridegroom to make a large gift to the bride's family in advance of the marriage; men customarily pay this "bride price" in cattle or other livestock, although they may also pay it in cash or merchandise. The gift signifies that the groom's intentions are serious and that he has financial prospects for supporting the bride. It also compensates the bride's father for the loss of her labor in the household and the fields. Should the marriage break up on account of the wife's misdeeds, the husband is entitled to the return of the bride wealth.

Cash Crops

The proportion of crops grown for export to overseas destinations and for sale in African urban centers has risen significantly in recent times. This has been partly a result of taxation and other governmental pressures. It has also been due to people's desire for cash with which to purchase food and manufactured goods such as hardware, utensils, bicycles, radios, and clothing.

Figure 17.13   Zebu cattle are extremely important in the cultures and household economies of rural Madagascar. There are about 10 million cattle in this country of 14 million people.   *Joseph J. Hobbs*

Governments also tend to favor such crops as a means of gaining foreign exchange with which to buy foreign technology, industrial equipment, arms, and consumption items for the elite.

Most export crops are grown on small farms, with some also grown on plantations and estates. Large plantations and estates have never become established to the degree they have in Latin America and Southeast Asia. Political and economic pressures forced many of them out of business during the period of transition from European colonialism to independence. Governments of newly independent states nationalized white-owned plantations, and many whites were obliged to sell their land to Africans (this process is still ongoing in Zimbabwe, which became independent in 1980). Now operated by Africans, many of these units provide a very important source of tax revenue and foreign exchange. One exception is South Africa, where the most important producers and exporters are the ethnic Europeans, who raise livestock, grains, fruit, and sugarcane. But throughout tropical Africa as a whole, there is a growing trend in export production from small, black-owned farms. The most valuable export crops are coffee, cacao, cotton, peanuts, and oil palm products. Crops of lesser significance, but very important in some regions, include sisal (grown for its fibers; Fig. 17.14), pyrethrum (used in insecticides), tea, tobacco, rubber, pineapples, bananas, cloves, vanilla, cane sugar, and cashew nuts.

Figure 17.14   A sisal plantation in eastern Kenya.   *Joseph J. Hobbs*

Definitions & Insights

CROP AND LIVESTOCK INTRODUCTIONS AS THE "CURSE OF AFRICA"

Many of the important food and export crops of Africa are not native to the continent. For example, corn (maize), manioc, peanuts, cacao, tobacco, and sweet potatoes were introduced from the New World. Cattle, sheep, chickens, and other domesticated animals now characteristic of Africa south of the Sahara also originated outside the region. Such nonnative species are known as **exotic species**, or introduced species.

Some exotic species such as manioc and chickens have adapted well to the environments and human needs in Africa south of the Sahara. Others are more problematic. Known in the human diet as "mealie-meal," corn has become the major staple food for southern Africans. However, the people of southern Africa came to regard corn as more of a scourge than a blessing during the severe regional drought of 1992. Introduced from relatively well-watered Central America, corn has flourished in African regions as long as rains or irrigation water have been sufficient. But the crop failed altogether during the 1992 drought. The situation was particularly critical in Zimbabwe, where the government had just sold its entire stock of corn reserves in an effort to repay international debts. Zimbabwe suddenly became a large importer of corn. Many in the country proclaimed corn to be "the curse of Africa," lamenting that, with its high water requirements, it should never have been allowed to become the principal crop of this drought-prone region. Agricultural specialists argued that native sorghum and cowpeas are much better adapted to African conditions and should be cultivated as the major foods of the future.

There are also growing questions about whether the cow is an appropriate livestock species for Africa. Cattle frequently overgraze and erode their own rangelands. The traditional approach to improving cattle pastoralism in Africa south of the Sahara, and to boosting livestock exports, has focused on the need to develop new strains of grasses that can feed cattle and other nonindigenous strains of livestock, since the native grasses of African savannas suit the wild herbivores but generally not the domesticated livestock.

A new approach turns this problem around by asking whether it might be more appropriate to commercially develop the indigenous animals already suited to the native fodder and soil conditions. Kenya, Zimbabwe, South Africa, and other countries are now maintaining "game ranches" where antelopes—such as eland, impala, sable, waterbuck, and wildebeest—along with zebras, warthogs, ostriches, crocodiles, and even pythons, are bred in semicaptive and captive conditions to be slaughtered for their meat, hides, and other useful products (Fig. 17.C). Safari hunting and tourism provide supplemen-

Figure 17.C Oryx antelopes maintained in a semidomesticated state on the Galana game ranch in Kenya. Husbandry of indigenous wildlife may prove to be a viable alternative to keeping cattle and goats, which have many harmful impacts on soil and vegetation in Africa south of the Sahara. *Sven-Olof Lindbald/Photo Researchers*

tal revenue in some of these operations. So far, these enterprises have demonstrated that indigenous large herbivores produce as much or more protein per area than domestic livestock under the same conditions. In commercial terms, the systems that have incorporated both wild and domestic stock have proven to be more profitable than systems employing only domestic animals. The World Wildlife Fund and other proponents of this "multispecies" system argue that it is an economically and ecologically sustainable option of land use in Africa. They emphasize that ranchers who keep both wild and domesticated animals will be better able economically to weather periods of drought, and will require a much lower capital input than they would expend on cattle or other domesticated livestock alone.

Having convinced many ranchers of the economic value of wildlife, the government of Zimbabwe has begun to enlist peasant farmers in protecting, and thus benefiting from, game animals. In an innovative and so far successful program known as CAMPFIRE (Communal Area Management Program for Indigenous Resources), the government has turned over management of wild animals to local councils, which are permitted to sell quotas of hunting licenses for elephant and other game. Profits from safari hunting return to the local communities to fund a wide range of development projects.

Minerals

The export of minerals has had a particularly strong impact on the physical and social geography of Africa south of the Sahara. The three primary source areas are: South Africa and Namibia, the Democratic Republic of Congo–Zambia–Zimbabwe region, and West Africa, especially the areas near the Atlantic

Ocean (see Fig. 17.3). Notable mineral exports from these regions include precious metals and precious stones, ferroalloys, copper, phosphate, uranium, petroleum, and high-grade iron ore, all destined principally for Europe and the United States.

Large multinational corporations, financed initially by investors in Europe or America, do most of the mining in Africa.

Figure 17.15　Deep beneath the surface of South Africa, miners are placing dynamite to blast loose the ore that contains diamonds. *Jason Lauré Photography, NYC*

Figure 17.16　There is no electricity in this village in Niger, but students receive some education and the entire community obtains news and entertainment from a television powered by photovoltaic cells. *John Chiasson/Gamma Liaison*

Mining has attracted far more investment capital to Africa than any other economic activity; money is invested directly in the mines, and a great many of the transportation lines, port facilities, power stations, housing and commercial areas, manufacturing plants, and other elements in the continent's infrastructure have been developed primarily to serve the needs of the mining industry.

Great numbers of workers in the mines are temporary migrants from rural areas, often hundreds of miles away (Fig. 17.15). The recruitment of migrant workers has had important cultural and public health effects. Millions of Africans who work for mining companies have come into contact with Western ideas as well as those of other ethnic groups, and have carried these back to their villages. The growing numbers of radios and televisions appearing in homes across the continent, even in remote villages, also convey an unprecedented wealth of information about the wider world (Fig. 17.16). Unfortunately, the phenomenon of migrant labor has also helped spread HIV, the virus that causes AIDS. Since most miners are single or married men who spend extended periods away from their wives or girlfriends, it is common that, in their loneliness, they contract the virus from prostitutes and then return home to their villages, where they spread it still further. The resolution of this dilemma is one of the key challenges facing the huge resource-rich, problem-ridden region of Africa south of the Sahara.

SUMMARY WITH SELECTED KEY TERMS

- According to many geographers, **Africa** has two major divisions: **North Africa and Africa south of the Sahara.** North Africa, the predominantly Arab and Berber realm of the continent, is regarded as part of the greater Middle East in this text (see Chapters 8 and 9). This chapter examines the **culturally complex** portion of Africa south of the Sahara, a region often called **sub-Saharan or Black Africa.**
- A number of shared cultural traits may be referred to as comprising **"Africanity."**
- Africa south of the Sahara can be divided into seven subregions: **West Africa, the Sahel, the Horn of Africa, East Africa, Southern Africa, West Central Africa, and the Indian Ocean islands.**
- The **relatively low population density** of Africa south of the Sahara obscures the fact that a majority of this region's people live in a **small number of densely populated areas** that together occupy a small share of the region's total area of 8 million square miles.
- Most of Africa consists of a **series of plateau surfaces** dissected by prominent river systems such as **the Nile, Niger, Congo, Zambezi, and Orange.**
- One of the most spectacular features of Africa's physical geography is the **Great Rift Valley**, a broad, steep-walled trough extending from the Zambezi valley northward to the Red Sea and the valley of the Jordan River in southwestern Asia. In terms of **plate tectonics**, the feature is believed to mark the boundary of two **crustal plates** which are rifting, or tearing apart.

- Although most (about two-thirds) of the region lies with the **low latitudes** and has **tropical climates,** Africa south of the Sahara contains a **great diversity of climate regimes and biomes.**
- Africa's **diverse wildlife** is often threatened by **human population growth, urbanization, and agricultural expansion within the continent.** Two of Africa's largest herbivores, **elephants and rhinoceroses,** are striking examples of Africa's endangered wildlife.
- Within the four **culture hearths** of Africa south of the Sahara—the Ethiopian Plateau, the West African savanna, the West African forest, and the forest-savanna boundary of West Central Africa—early indigenous people were responsible for several **agricultural innovations.** Within these culture hearths, Africans domesticated such important crops as millet, sorghum, yams, cowpeas, okra, watermelons, coffee, and cotton.
- The tragic impetus for growing contact between Africa and the wider world was **slavery.** Over a period of 12 centuries, as many as **25 million people** from Africa south of the Sahara were forced into slavery.
- Most of Africa south of the Sahara fell under **European colonialism** after the **Conference of Berlin** in 1884 and 1885. During this conference, Africa was carved up as the French, British, Germans, Belgians, Portuguese, Italians, and Spanish established their respective spheres of influence in the region.
- In general, the people of Africa south of the Sahara are **poor,** live in **rural areas, and practice subsistence agriculture** as their main occupation. In most countries, **per capita food output has declined or has not increased** since independence.

- Frequent **droughts, lack of education, poor transportation, and serious public health issues** have hindered development within Africa south of the Sahara. Most countries are **underindustrialized and overly dependent on the export of a few primary products.** In addition, while most countries are **heavily in debt** to foreign lenders, the amount of **economic and humanitarian assistance to the region has slowed considerably since the end of the Cold War.**
- Although many countries have been under **authoritarian governments** since independence, **progress toward democracy is now widespread.** Serious **political instability** is characteristic of many African countries south of the Sahara. A **diverse array of political, economic, and social ideas** from the West, the Communist bloc, the Muslim world, and Africa south of the Sahara itself has influenced governments since independence. In addition, **important links with the colonial powers that formerly controlled them** exist in many countries of the region.
- Over half of the people of Africa south of the Sahara depend directly on **agriculture or pastoralism** for a livelihood. The cultivation of **export crops** such as coffee, cacao, cotton, peanuts, and oil palm products is increasing throughout the region.
- The export of **minerals** has had a particularly strong impact on the physical and social geography of Africa south of the Sahara. Notable mineral exports include precious metals and stones, ferroalloys, copper, phosphate, uranium, petroleum, and high-grade iron ore.

REVIEW QUESTIONS

1. Using maps and the text, locate the seven major African subregions.
2. Using maps and the text, locate the five principal areas of population within Africa south of the Sahara.
3. Using maps and the text, locate the principal climatic zones of Africa south of the Sahara.
4. Using maps and the text, locate and describe the major slave trade routes from Africa south of the Sahara. Where in Africa did they originate? What were some of the major destinations of African slaves?
5. Using maps and the text, locate the major colonial possessions of the European powers in Africa south of the Sahara.

6. Explain why African parents typically want to have large families.
7. What is donor fatigue? How does it relate to Africa south of the Sahara?
8. What is the significance of bride wealth in Africa south of the Sahara? How is it often "paid"?
9. What is a keystone species? Why is the tsetse fly considered a keystone species in Africa south of the Sahara?
10. What is an exotic species? Provide some examples of exotic species that are important to agriculture in Africa south of the Sahara.

DISCUSSION QUESTIONS

1. Describe the tectonic processes that have led to the development of the Great Rift Valley. What physical features are associated with the Great Rift Valley? What will be the "final fate" of areas adjacent to the rift?
2. Discuss the main reasons behind the sharp decline in elephants and rhinoceroses in Africa south of the Sahara. What efforts have been taken to preserve these animals?
3. Describe the eight constituent elements of the "African identity" as identified by Robert Stock.
4. Describe the pattern of triangular trade. In addition to Africa south of the Sahara, what other regions of the world were involved in this pattern of trade? What goods were traditionally exchanged?
5. Describe the positive and negative effects of European colonization of Africa south of the Sahara.

6. Describe the major impacts AIDS has had on Africa south of the Sahara.
7. What is shifting cultivation? Why is this agricultural practice often implemented in tropical areas such as Africa south of the Sahara? What are the benefits and drawbacks of shifting agriculture?
8. Why has corn been called "the curse of Africa" in Zimbabwe? What are some crops that may be better adapted to African conditions?
9. Describe the positive and negative effects of mining on Africa south of the Sahara. What have been the cultural and public health effects of the recruitment of migrant mine workers?

Chapter 18

The Assets and Afflictions of Countries South of the Sahara

▲ *Boys in central Ethiopia. The future of Africa south of the Sahara lies with its children; 46 percent of the region's population is under age 15.* Joseph J. Hobbs

CHAPTER OUTLINE

18.1 The Horn of Africa

18.2 The Sahel

18.3 West Africa

18.4 West Central Africa

18.5 East Africa

18.6 Southern Africa

18.7 Indian Ocean Islands

*t*he countries of Africa south of the Sahara are diverse in size and geographic personality. In this chapter, sketches of the major subregions and their countries —and more in-depth portraits of Ethiopia, the Democratic Republic of Congo, and South Africa—illustrate some of the region's characteristic and unique challenges.

18.1 The Horn of Africa

East of the Sudan section of the Nile basin, a great volcanic plateau rises steeply from the desert. This highland and adjacent areas occupy the greater part of the "Horn of Africa"—named for its prominent protrusion from the continent into the Indian Ocean—and include the countries of Ethiopia, Eritrea, Somalia, and Djibouti (see Fig. 17.1, p. 419). Much of the area lies at elevations above 10,000 feet (3050 m), and one peak in Ethiopia reaches 15,158 feet (4620 m). The highland, where the Blue Nile, Atbara, and other Nile tributaries rise, receives its rainfall during the summer half of the year. Temperature conditions vary from tropical to temperate as elevation increases. Crops ranging from bananas, coffee, and dates to oranges, figs, and temperate fruits to cereals can be produced without irrigation. Large expanses of upland provide pasture for a variety of livestock, primarily cattle and sheep.

European Imperialism

European powers seized coastal strips of the Horn of Africa in the latter 19th century (see Fig. 17.7, p. 431). Britain was first, with British Somaliland in 1882; then France annexed French Somaliland in 1884; and, finally, Italy asserted dominance over Italian Somaliland and Eritrea in 1889. These areas along the Suez–Red Sea route have long had strategic importance, although their economic importance has been negligible.

Italy attempted to extend its domain from Eritrea over the much more attractive and potentially valuable land of Ethiopia

in 1896, but Ethiopian tribesmen annihilated the Italian forces at Aduwa. Forty years later, in 1936, a second attempt was successful. Hopes of developing the country's potential wealth and of using Ethiopia as a location in which to resettle the surplus Italian population were frustrated in World War II when British Commonwealth forces (primarily South Africans) defeated the Italian forces throughout the region. Ethiopia was restored to independence, and Eritrea was federated with it as an autonomous unit in 1952. Ethiopia incorporated Eritrea as a province in 1962.

Following World War II, Italian Somaliland was returned to Italian control to be administered as the Trust Territory of Somalia under the United Nations, pending independence in 1960 as a republic. The present Somali Democratic Republic, or Somalia, includes both the former Italian territory and the former British Somaliland, which chose to unite with Somalia. The people of much smaller French Somaliland remained separate and eventually became independent as the Republic of Djibouti, with its capital at the seaport of the same name.

The Peoples of Ethiopia

Ethiopia is inhabited by an estimated 58 million people of diverse ethnic and cultural origins. About 45 percent, including the often politically dominant Amhara peoples, adhere to the Coptic Christian faith, an ancient branch of Christianity that came to Ethiopia in the fourth century from Egypt, where there is still a sizable Coptic minority. The entire area has had important cultural and historical links with Egypt, the Fertile Crescent, and Arabia. The Ethiopian monarchy even based its origins and legitimacy on the union of the Biblical King Solomon and the Queen of Sheba, who, tradition holds, gave birth to the first Ethiopian emperor, Menelik. Until a Marxist coup brought an end to the emperorship in the 1970s, Ethiopia's rulers were always Christian. Ethiopia has many outstanding Christian artistic and architectural treasures, including the 11 churches of Lalibela, carved from solid rock in the 12th and 13th centuries (Fig. 18.1).

443

Figure 18.1 The churches of Lalibela, carved from volcanic rock in the 12th and 13th centuries, are among Ethiopia's many Christian cultural treasures. *Joseph J. Hobbs*

The rest of Ethiopia's people are divided among those of the Muslim faith (who make up about 40 percent of the population), those practicing a variety of indigenous religions, and some Protestant, Evangelical, and Roman Catholic Christians. Most Ethiopians live in the country's highlands. East of the mountain mass, lower plateaus and coastal plains descend to the Red Sea, the Gulf of Aden, and the Indian Ocean. Extreme heat and aridity prevail at these lower levels, and nomadic and seminomadic Muslim tribesmen make a living by herding camels, goats, and sheep. Dwellers of scattered oases carry on a precarious agriculture. The arid lowland sections have many characteristics more typical of the Middle East than other areas of Africa south of the Sahara.

Problems of Development in Ethiopia and Eritrea

In 1974, the aging emperor Haile Selassie, who had held Ethiopia's population together in a loose political union, was deposed by a civilian and military revolt against the country's feudal order. Causes of the revolt included widespread drought and famine during the early 1970s, separatist dissension in some areas, and general discontent with the country's social and economic backwardness. A Marxist dictatorship, strongly oriented to the former Soviet Union, emerged. Thousands of middle-class intellectuals were jailed or executed. Privately owned land and industries were nationalized, and collectivization of agriculture began. American influence, which had been strong under Haile Selassie, was eliminated. Munitions and advisors from the Soviet Union and other East-bloc nations poured in to equip and train one of Africa's largest armies, and

the Soviets gained access to base facilities in Ethiopian ports. Such foreign interventions contributed to the strife that has continued to hinder development in the region.

Ethiopia has substantial resources, including possibly significant mineral wealth. However, its resource base remains largely undeveloped, and the country is impoverished. With foreign assistance, the government has been able to develop a limited number of factories and a few dams to harness some of the country's large hydroelectric potential. A poor transportation system—whose best feature is the 3000 miles (*c.* 4800 km) of good roads built by the Italians during their occupation in the 1930s—is a major factor keeping the country isolated and underdeveloped. The French-built railroad from Djibouti to Ethiopia's capital Addis Ababa (population: 2.3 million, city proper) lacks feeder lines and has never been as successful as hoped. Many roads into the Ethiopian heartland are rough tracks, although better connections have gradually been developing. Air services—especially Ethiopia's remarkable fleet of DC-3s more than 50 years old but kept in top condition—help somewhat in overcoming the deficiencies in ground transportation. However, the high and rugged part of Ethiopia is badly handicapped by inaccessibility. A good share of its population lives in villages on uplands cut off from each other and from the outside world by precipitous chasms (Fig. 18.2). These formidable valleys were cut by streams that carried huge amounts of fertile volcanic silt to the Nile and thus provided material for the nourishing floods in Egypt (see Chapter 9, p. 253).

Ethiopia has been further troubled in recent times by political and military turmoil in the former Ethiopian province of Eritrea, the adjacent parts of northern Ethiopia, and in the borderland between Ethiopia and Somalia. In Eritrea, both Muslims and Christians (each roughly 50 percent of the population) resented their political subjugation to Ethiopia's Christian Amhara majority. Terrorism and guerrilla actions escalated into civil war in the early 1960s. The Ethiopian army made a major effort to stamp out the revolts of two Eritrean "liberation fronts" (one Marxist and the other not), but resistance continued. At the same time there was also a rebellion in Tigre Province adjoining Eritrea. In 1991, these combined rebel forces captured Addis Ababa and took over the Ethiopian government. This allowed the rebels in Eritrea to establish an autonomous regime, and Eritrea became independent in May 1993 following an overwhelmingly favorable vote in a referendum. Eritrea's main city and manufacturing center, Asmara (population: 358,100, city proper), is located in highlands at an elevation of more than 7000 feet (2130 m) and has a rail link with the Red Sea port of Massawa (see Fig. 17.3, p. 424).

Many geographers subsequently pointed to Eritrea as a model for sustainable development in Africa south of the Sahara; after independence, the country reduced its foreign debt, cut its population growth rate, and took other promising steps toward eliminating the burdens of underdevelopment. However, in 1998 Eritrea dimmed its bright prospects by launching a war with Ethiopia. The conflict emerged from economic

Figure 18.2 This highland village in north-central Ethiopia, like many in the country, is isolated by deep gorges (not visible in this view). The round structure to which all the village roads lead is the community's Christian church. *Joseph J. Hobbs*

rivalries and unresolved territorial issues between the nations. With Eritrea's independence, Ethiopia had agreed to become landlocked, so long as it could freely use Eritrea's Red Sea ports of Massawa and Assab. The concession was in large part Ethiopia's reward to Eritreans for helping with the overthrow of Ethiopia's Marxist regime. The countries' leaders in effect agreed upon an economic union in which Eritrea would be the industrial power and Ethiopia the agricultural one. They promised to practice free movement of people and goods between the two states and to share a single currency, the Ethiopian birr. But Eritrea introduced its own currency and announced steep duties and fees on goods shipped to and from Ethiopia through Eritrean ports. Ethiopia then turned to Djibouti and Kenya to find new outlets to the sea. Deprived of an anticipated major source of revenue, Eritrea attacked Ethiopia. The war continued into early 1999, when Ethiopia gained the upper hand by securing a border region disputed between the two countries. Even after a cease-fire went into effect, there was concern in Eritrea and in international diplomatic circles that Ethiopia might launch an offensive to establish a new foothold on Eritrea's Red Sea coast.

In addition to the devastation wrought by war, droughts centering in its war-torn north have stricken Ethiopia since the late 1960s. For years there has been little or no rain in areas inhabited by millions of people. Hunger has beset both people and livestock to the point that desperate farmers have fed their families with grain needed for seed and have sold their oxen needed for plowing. Ethiopian villagers are especially vulnerable to drought because of their heavy dependence on rainfed agriculture. The volcanic soils are fertile, but the streams are entrenched so far below the upland fields that little irrigation

is possible. In 1984 and 1985, drought in Ethiopia reached the point of widespread catastrophe. An estimated one million people perished. However, the massive and sustained relief effort helped reduce the impact of the country's worst drought.

Recent economic statistics for Ethiopia depict a country that is still desperately underdeveloped. Coffee, often picked from wild trees, made up 60 percent of the scarce exports in 1998. Manufacturing is largely confined to simple textiles and processed food. Addis Ababa and Eritrea account for the bulk of the area's manufacturing. Wood is in very short supply as a result of deforestation.

Homogeneous, War-Torn Somalia

Between Ethiopia and the Indian Ocean lies the poor, lowland country of Somalia. Some 99.8 percent of the people here are Sunni Muslims and 98.3 percent are ethnic Somalis. This is one of the most homogeneous populations in the world, but inter-clan rivalries have torn the country apart.

Moisture is so scarce that only 2 percent of the country is cultivated. Livestock raising on a nomadic and seminomadic basis is the main support for Somalia's simple economy; live animals (goats, sheep, camels, cattle), shipped mainly to Saudi Arabia, Yemen, and the United Arab Emirates, are the main exports, along with fish products and bananas. Food processing is the chief industry, and there is almost no mineral production. Somalia once received aid from the former East bloc but lost that resource when the bloc transferred its military aid to Ethiopia. Somalia then granted the United States access to base facilities at Mogadishu (population, 830,000), the capital, located in former Italian Somaliland, and to the much smaller city of Berbera in former British Somaliland.

Somalia was badly mauled in a rebellion that overthrew its government in early 1991. Fighting between the new rulers and resistance forces continued in 1992. The opposing forces in this destructive civil war were regional clan groups of the Somali people (the country has six major clans and numerous subclans). Contending forces inflicted great damage on Mogadishu, from which large numbers of people fled to the countryside as refugees. Massive famine threatened the devastated country in mid-1992. International efforts to provide relief were hindered by banditry in a country where law and order had disintegrated.

United States military forces intervened in late 1992 with "Operation Restore Hope," a mission to help protect relief workers and aid shipments. However, American policy makers decided to use military force to neutralize the power of one of the country's leading "warlords," Mohamed Farrah Aidid. In a gruesome sequence of events, 18 U.S. servicemen stranded in Aidid's territory in Mogadishu were killed in a single day in 1993. American public outcry at the tragedy brought an end to the American mission in 1994. The United States then handed over the main responsibility for aiding Somalia to United Nations forces.

Regional Perspective

Drought and Desertification in the Sahel

The Sahel region, like much of Africa south of the Sahara, experiences periodic drought. This is a naturally occurring climatic event in which rain fails to fall over an area for an extended period, often years. When rain does return, the arid and semiarid ecosystems of the Sahel come to life with a profusion of flowering plants, insects, and herbivores like gazelles, whose populations climb when foods are abundant. These ecosystems are resilient, meaning they are able to recover from the stress of drought and have mechanisms to cope with a natural cycle that includes periods of dryness and rain.

Desertification is the destruction of that resilience and the biological potential of arid and semiarid ecosystems. It is an unnatural, human-induced condition that has afflicted the Sahel severely since the late 1960s, and, as such, it is different from the natural phenomenon of drought (Fig. 18.A). However, the recent history of the Sahel suggests that drought can be the catalyst which initiates the process of desertification. During several years of good rains in the 1950s and early 1960s, the Fulani (Fulbe) and other Sahelian pastoralists allowed their herds of cattle and goats to grow very large, since the animals represent wealth and capital and

people naturally wanted their numbers to increase beyond the immediate subsistence needs of their families. However, due to declining death rates, the numbers and sizes of families keeping livestock were also very large. Thus, the unprecedented numbers of livestock built up during the rainy years made the Sahelian

ecosystem vulnerable to the impact of drought on an unprecedented scale.

Drought struck the region in 1968 and persisted through 1973. Annual plants failed to grow, so the large herds of cattle and goats turned to acacia trees and other perennial sources of fodder. They ate all of the palatable vegetation.

Figure 18.A Desertification in progress is shown graphically in this photo from Mali, in the Sahel. As growing human populations intensify the impact of naturally occurring drought, the Sahara creeps southward. *Steve McCurry/Magnum*

18.2 The Sahel

The Sahel region extends eastward from the Cape Verde Islands to the Atlantic shore nations of Mauritania, Senegal, and The Gambia and inland to Mali, Burkina Faso ("Land of the Upright Men"; formerly Upper Volta), Niger, and Chad (see Fig. 17.1). In colonial times, The Gambia was a British dependency; Mauritania, Senegal, Mali, Burkina Faso, Niger, and Chad were French; and Cape Verde was Portuguese. Cape Verde's dry volcanic islands lie well out in the Atlantic off Senegal, but in population, culture, economy, and historical relationships the islands are so akin to the adjacent mainland that they fit easily into a Sahelian context. Ancient Mali, with its legendary port of

Timbuktu on the Niger River, was an important trading intermediary between the Guinea Coast and Islamic North Africa before seafaring Europeans took over this trade.

Climatically, the region ranges from desert (in parts of the Sahara) in the north through belts of tropical steppe and dry savanna in the south. The name Sahel in Arabic means "coast" or "shore," referring to the region as a front on the great desert "sea" of the Sahara. Since the late 1960s the area has been subjected to severe droughts, which, in combination with increased human pressure on resources, has prompted a process of desertification (see Regional Perspective, above).

This area of Africa has seen many dramatic changes in climate and vegetation over its long history. There is abundant evidence, particularly in the form of prehistoric rock drawings,

This destruction was a critical problem, because plants play an important role in maintaining soil integrity, helping to intercept moisture and funnel it downward through the root system. Plant litter and decomposer organisms working around the plant contribute to soil fertility and help stabilize the soil. With hungry livestock in the region eating the plants, the landscape changed. No longer anchored and replenished, good soils eroded. "Junk" plants like Sodom apple (*Calotropis*) replaced palatable plants, and an almost impermeable surface formed on the land. This degraded ecosystem lost its resilience, so that even when rains returned, vegetation did not recover.

Meteorologists studying the Sahel drought discovered another important link between precipitation and the removal of vegetation. They noticed that when people and livestock reduced the vegetative cover, they increased the **albedo,** or the amount of the sun's energy reflected by the ground. Fewer plants meant more solar energy was deflected back into the atmosphere, which then reduced local precipitation. This connection is known as the **Charney Effect,** named for the scientist who documented

it. It was a tragic sequence: people responding to drought actually perpetuated further drought conditions.

The 1968–1973 drought devastated the great herds of the Sahel. An estimated three-and-one-half million cattle died. Two million pastoral nomads lost at least half of their herds, and many lost as much as 90 percent. Farmers dependent on unirrigated cropping were also affected. Harvests were less than half of the usual crop for 15 million farmers, and for many there was no harvest. The Niger and Senegal rivers dried up completely, depriving irrigated croplands in Niger, Mali, Burkina Faso, Senegal, and Mauritania. The cost in human lives was estimated at 100,000 to 250,000. Environmental refugees poured into towns like Nouakchatt, the capital of Mauritania, which were unprepared to deal with such a large and sudden influx.

This crisis resulted in an international effort to combat desertification using recommendations developed by the United Nations in 1977. The U.N. report, however, soon became a case study in how *not* to deal with an environmental crisis in the developing world. Most of the recommendations were universal, high-technology solutions that proved impos-

sible to implement in the villages and degraded pastures of the Sahel. For example, following the recommended guidelines, many international agencies attempted to relieve the suffering of Sahelian pastoralists using techniques such as digging deep wells to water livestock. But these agencies failed to anticipate that these wells would act like magnets for great numbers of people and animals; while there was plenty to drink, the animals starved to death when they decimated what vegetation remained around the new water supplies.

Since the disappointment of the 1977 U.N. plan, and with the lessons relief organizations learned as a result, there has been greater focus on sustainable development, with its local rather than universal solutions. An example is the technique of "rainwater harvesting," in which local villagers use local stones and tools to construct short rock barriers that follow elevation contours. Water striking these barriers backs up to saturate and conserve the soil, allowing more productive farming. The challenge to find such solutions is a continuous one; since 1968, drought has established an almost permanent presence in the Sahel.

that much of the now extremely arid northern Sahel and Sahara region was a grassy, well-watered savanna between approximately six and ten thousand years ago. Many of the large mammals now associated with East Africa frequented the region, along with bands of pastoralists who kept large cattle herds. Now, even though under much less favorable conditions, the raising of sheep, goats, camels, and cattle, combined where possible with subsistence farming, continues to be the major livelihood for peoples of the Sahel.

One of the most prominent features of the Sahel's landscape is Lake Chad, located north of landlocked Chad's capital, N'Djamena (population: 529,555, city proper). This huge lake is situated in a vast open plain. It has no outlets, and yet it is not very salty because groundwater carries most of the salts away.

Lake Chad is very shallow, with maximum depths between 7 and 23 feet (2–7 m). Characteristic of shallow lakes in arid lands, its surface area increases dramatically with rainy periods and shrinks in times of drought. Recent decades of drought have seen its area diminish substantially, but prolonged rains would quickly enlarge it. The waters of Lake Chad are very important economically for fish and irrigated agriculture.

Mining, manufacturing, and urbanization are generally insignificant in this subsistence-oriented region of Africa. Mauritania exports iron ore and Senegal exports phosphates. Niger exports uranium for France's nuclear power program. The largest metropolises are Senegal's main port and capital of Dakar (population: 1.64 million, city proper) and Bamako (population: 658,275, city proper), the capital and main city of

Figure 18.3   The faith of Islam unites most of the disparate peoples of the Sahel. This is the mosque and market at Djenne, near a tributary of the Niger River in southern Mali.   *Explorer/Photo Researchers*

Mali (Djenne, another Malian city, is shown in Fig. 18.3). Dakar was formerly the capital of the immense group of eight colonial territories known as French West Africa. It is the commercial outlet for much of the peanut belt in the Sahelian and West African savanna, and is a considerable industrial center by African standards. Dakar is also a tourist and resort center with fine beaches and an exotic blend of French and Senegalese culture.

The Sahel is also the border region between the predominantly Arab and Berber North Africa and the predominantly black Africa south of the Sahara. Relations between the ethnic groups are good in most countries, but there are problems in some. In Mauritania, for example, the government—led by the country's majority Arabic-speaking Moors—began expelling black Mauritanians across the border into Senegal in 1995. Analysts continue to fear that the Mauritanian government aims to "cleanse" the country of all of its 40 percent black population.

18.3 West Africa

Area, Population, and Physical Environment

West Africa, here defined as extending from Guinea-Bissau eastward to Nigeria (see Fig. 17.1), is larger than one might expect from looking at the map. The nine political units in the region make up almost 800,000 square miles (about 2 million sq km), or nearly one-fourth of the area of the United States.

The region also has impressive population totals. In 1998, the nine countries had an estimated total population of 183 million, or 87 million fewer than that of the United States. Four countries had under 5 million people, and only three—Ivory

Coast, with 16 million; Ghana, with 19 million; and Nigeria, with 122 million—exceeded 10 million. The most striking single aspect of West Africa's population distribution is that about two-thirds of the region's population resides within one country, Nigeria (which also has 20 percent of the population of the entire region of Africa south of the Sahara). This unusual concentration of people, along with advanced commercialization and urbanization, existed even before European contact. It may have been related both to the existence of powerful kingdoms that were able to impose a degree of internal order and security, and to the region's agricultural productivity and resource wealth.

Environmental contrasts within West Africa are extreme. Climatically, the region ranges from belts of tropical steppe, dry savanna, and wetter savanna, to some areas of tropical rain forest along the southern and southeastern coasts (see Figs. 2.7 and 2.8). Most of the countries' more populous areas are in the zone of tropical savanna climate.

Ethnic and Political Complexity

West Africa was an advanced part of Africa south of the Sahara when the European Age of Discovery began. A series of strong pre-European kingdoms and empires developed there, both in the forest belt of the south and in the grasslands of the north. In colonial times it became a French and British realm, except for a small Portuguese dependency and independent Liberia. Of the present countries, four—The Gambia, Sierra Leone, Ghana, and Nigeria—were British colonies; four—Guinea, Ivory Coast (officially Côte d'Ivoire), Togo, and Benin (formerly Dahomey)—were French; and Guinea-Bissau was Portuguese. Liberia is unique, becoming Africa's first republic after 1822, when 5000 freed American slaves sailed there and settled with support from the U.S. treasury. Problems between Americo-Liberians and the indigenous people have continued ever since.

A full-scale civil war broke out in 1989 and persisted for seven years. In 1997 the rebel leader who sparked the fighting, American-educated Charles Taylor, was popularly elected as Liberia's president.

In West Africa today, the English, French, and Portuguese languages continue in widespread use, are taught in the schools, and are official languages in the respective countries. European languages are a highly useful means of communication in this region, where hundreds of indigenous languages and dialects, often unrelated, are spoken. Economic, political, and cultural relationships between West African countries and the respective European powers that formerly controlled them continue to be close. France, for instance, supplies economic and military aid to its former colonies in this and other parts of "Francophone" (French-speaking) Africa. France further cultivates good relations with its former dependencies by inviting their leaders to summit conferences to discuss matters of common interest. Similarly, Britain maintains ties with "Anglophone" Africa through meetings of the Commonwealth of Nations, to which all its former colonies in Africa south of the Sahara belong.

As is often the case in Africa, the governments of the region must deal with serious problems growing out of the ethnic complexity of national populations. European governments contributed initially to some present troubles by drawing arbitrary boundary lines around colonial units without proper concern for ethnicity. As a result, the typical West African country today, like most countries elsewhere in Africa south of the Sahara, is a collection of ethnic groups that have little sense of identification with the national unit. The region is inhabited by peoples who speak a large variety of languages and dialects, and who in some cases have a history of suspicion and hostility toward one another. A variety of religious faiths also coexist here. Many people hold traditional animistic beliefs, in which they associate spiritualism with a variety of natural features and ancestral beings, rather than with a single deity. The Muslim faith is dominant in the north and over centuries has spread southward. Protestant and Roman Catholic missionaries converted many West Africans to Christianity.

Agriculture and Mining

The West African countries exhibit the customary economic pattern of the world's LDCs. They depend heavily on subsistence crop farming and livestock grazing, along with a few agricultural and/or mineral export specialties. In the wetter south, subsistence agriculture relies mainly on root crops such as manioc and yams and on maize, the oil palm, and, in some areas, irrigated rice. In the drier, seasonally rainy grasslands of the north (in northern Nigeria and northern Ghana, for example), nomadic and seminomadic herding of cattle and goats is important, along with the subsistence grain crops of millet and sorghum.

The major export specialties of West African agriculture are cacao (from which cocoa and chocolate are made), coffee, and oil-palm products in the wetter, forested south, and peanuts and cotton in the drier north. Ivory Coast leads the world in cacao exports. Nigeria was once the world leader in exports of palm oil and peanuts, but these exports largely vanished when Nigerian agriculture was allowed to deteriorate during the oil boom of recent decades. There is a marked correlation between the distribution of these resources and the distribution of Nigeria's major ethnic groups. The main oil-palm belt of Nigeria is in the denser rain forest of the southeast, inhabited by the Ibo people. The main cacao belt is in the lighter forests of the southwest, inhabited by the Yoruba people. The peanut belt is in the open grasslands of the north, inhabited by the Hausa and Fulani peoples.

The other main export industry of West Africa is mining (see Fig. 17.3, p. 424). The most spectacular and significant mining development has been Nigeria's emergence as a producer and major exporter of oil (see Problem Landscape, p. 450). Commercial oil development began here in the 1950s. By 1983, crude oil constituted 96 percent of Nigeria's exports by value, and, in 1998, when Nigeria stood as the world's sixth-largest oil exporter, it still made up 90 percent of the country's foreign export earnings. Extraction thus far has taken place mainly in the mangrove-forested Niger Delta and offshore in the adjacent Gulf of Guinea.

Other prominent mineral products of West Africa include iron ore, bauxite, diamonds (Fig. 18.4), phosphate, and uranium. Although the region supplies only a tiny fraction of the world's iron ore, its export is very important to Liberia. The

Figure 18.4 Small-scale diamond mining in Ivory Coast (Côte d'Ivoire), West Africa. The miners will screen for diamonds the small piles of dirt they have extracted by hand from shallow holes. The scrubby vegetation, a secondary growth that is typical of a deforested area, is common in West African tropical rain forest regions. *Manaud/Elbaz/Matrix*

PROBLEM LANDSCAPE

The Poor, Oil-Rich Delta of Nigeria

OIL IS THE ECONOMIC LIFEBLOOD OF POP-ulous Nigeria. Until the early 1990s it helped to make Nigeria one of the more prosperous and enviable countries of Africa south of the Sahara. But after 1993, when the military government of General Sani Abacha annulled democratic elections and imprisoned the apparent victor, corruption, human rights abuses, and mismanagement of the country's resource wealth made Nigeria an outcast in the world community.

The multinational Royal Dutch/Shell Group produces (as of 1999) about half of this OPEC member's output of two million barrels of oil per day. Most of the production is concentrated in the Niger River Delta, home to about seven million mostly Christian people. Some of that production comes from wells drilled in a 350-square-mile section of the Niger Delta region inhabited by 500,000 people of the Ogoni tribal group. Shell has generated billions of dollars in revenue from oil drilled since 1958 on Ogoni lands in this

region, known as the Rivers State. But, according to their spokesmen, the Ogoni have derived few benefits and have suffered much from oil development in their homeland. They complain that for more than 30 years, oil spills have tainted their croplands and water, destroying their crops and fisheries, while the flaring off of natural gas has polluted their air and caused acid rain. They note that despite the enormous revenue generated from oil drilled on their land, little money has returned to the area; most goes to the oil companies and to the Muslim-dominated government in the north. Meanwhile, most Ogoni live in palm-roofed mud huts; half of the delta region lacks adequate roads, water supplies, and electricity; schools have few books; and clinics have few medical supplies.

In an attempt to improve their plight, Ogoni activists founded the Movement for the Survival of the Ogoni People (MOSOP) in 1990. Thirty-two Ogoni chiefs and elders issued an "Ogoni Bill of Rights" in which they declared the right to a safe environment and more federal support of their people. One of the movement's leaders, a popular author, play-

wright, and television producer named Ken Saro-Wiwa, also called for self-determination for the Ogoni. However, after an antigovernment rally by hundreds of thousands of Ogoni people in January 1993, government police razed 27 villages, killing about 2000 Ogonis and displacing a further 80,000.

In another violent incident, Ogoni activists killed four founding members of MOSOP in May 1994. According to family members and supporters of the slain men, Saro-Wiwa incited their murders by stating that they "deserved to die" for not taking a more active position against the government and Shell. Government authorities arrested Saro-Wiwa and thirteen of his associates, charging them with responsibility for the murders. On October 31, 1995, the government court found Saro-Wiwa and eight other defendants guilty, and ordered their execution.

An international outcry ensued. Saro-Wiwa had already earned an international reputation as an environmentalist (he received the 1995 Goldman Environmental Prize) and as a human rights activist (he was a 1995 nominee for the Nobel Peace Prize). Environmental and human rights

---

shipments consist largely of high-grade ore. The main bauxite resources lie near the southwestern bulge of the West African coast; Guinea, by far the main producer and exporter, derives the bulk of its export revenue from bauxite and alumina (the second-stage product between bauxite and aluminum). Togo exports phosphate. Gold, bauxite, and manganese are important to Ghana, as are bauxite, titanium, and diamonds to Sierra Leone. However, much of Sierra Leone's potential wealth from diamonds is lost to smugglers, many reportedly from the entrepreneurial class of Syrians and Lebanese who settled in the country decades ago. The diamond purge and antigovernment guerrilla fighting since 1991 have contributed to a steady decline recently in Sierra Leone's standard of living. The country's prospects improved when its first democratically elected president, Ahmed Tejan Kabbah, took office in 1996. However, a military junta seized power a year later, forcing Kabbah into

exile. Asserting its self-appointed identity as West Africa's regional superpower—and perhaps with an eye on Sierra Leone's mineral wealth—Nigeria responded in 1998 with a military attack on Sierra Leone that in turn forced the junta to flee and restored President Kabbah to office.

Urbanization

The world's large cities are associated mostly with trade, centralized administration, and large-scale manufacturing. West Africa's principal cities are also national capitals and seaports, but, as the region has only minor manufacturing, its cities have developed primarily in its areas of major commercial agriculture. In nearly every country in the region, the capital is a primate city (see Chapter 4, p. 111), far larger than any other city.

groups mounted an international media and diplomatic campaign to win clemency for him. These defenders claimed that Saro-Wiwa was framed and did not receive a fair trial.

On November 10, 1995, the government hanged Ken Saro-Wiwa and his eight codefendants. Many analysts regarded the executions as General Abacha's effort to deter actions by other more serious rivals, especially within the army. As Nigeria is a thinly glued collection of 250 ethnic groups, Abacha may also have seen the Ogoni struggle for minority rights as a Pandora's box that would lead to the breakup of the nation. Redistribution of oil revenue also would have diminished his income and that of the country's other ruling senior military officers.

The international community weighed its options in the months following the executions. The United States, Britain, and other countries recalled their ambassadors to Nigeria. The World Bank announced that it would not extend a loan of $100 million for a project to develop liquefied natural gas (which, ironically, would have helped eliminate one of Saro-Wiwa's main concerns, the problem of pollution

caused by the flaring-off of natural gas). At the urging of South African president Nelson Mandela, the 52-member organization of the British Commonwealth suspended Nigeria from membership. Mandela pushed the United States and Britain for even stronger measures to punish the Nigerian government, including an embargo of Nigerian oil. That action would certainly hurt the country, which exports 45 percent of its oil to the United States and 40 percent to Europe. Fearing, however, the prospects of instability, civil war, and refugee migrations that such an action might touch off in an already unstable region, the United States and other major members of the international community declined to embargo Nigeria's oil or take any additional steps against its government.

Abacha died unexpectedly in 1998 and was replaced by a military general who returned civilian rule to the country in 1999, and made many important gestures of reconciliation with the peoples of the Delta. General Abubakar permitted demonstrations in memory of Saro-Wiwa, and agreed to return his remains and those of his executed colleagues for burial in their

homeland. Despite such gestures, however, dissent in the Delta has continued to grow. The Ogoni defiance and contempt of government and the oil companies has spread to the region's other major ethnic groups—the Ibo, Yoruba, and Ijaw. Reiterating the Ogoni protest that the region's oil generates $10 billion a year in revenue for the government, the Ijaw (who make up about 5 percent of Nigeria's population) in particular have pointed to the poverty and squalor of their settlements and demanded that some of the oil money be used to fund roads, electricity, running water, and medical clinics. Ijaw activists have spoken of secession and self-determination if their demands are not met. In 1998 they seized and held about 20 oil stations belonging to Royal Dutch/Shell and Chevron, temporarily cutting Nigeria's oil exports by one-third. Questions of equity in a potentially rich country thus continue to pose a threat to national, regional, and international stability. Nigerians hoped that after 16 years of self-appointed leadership, the 1999 popular election of President Olusegun Obasanjo would bring political stability and economic reform to their land of promise.

A total of only 23 percent of West Africa's inhabitants live in urban areas (as of 1998), but the urban population is increasing very rapidly. Nigeria has the most impressive urban development: Of the 11 largest metropolitan cities in West Africa, three are clustered in the Yoruba-dominated southwestern part of Nigeria, in and near the densely populated belt of commercial cacao production.

The largest of these three—and by far the largest in West Africa—is Lagos (population: 1.2 million, city proper; 3.8 million, metropolitan area). It had long been Nigeria's political capital, but by 1992 most governmental functions were moved to a new capital at Abuja in the interior. The development of Lagos surged in the early 20th century when the British colonial administration chose Lagos as the ocean terminus for a trans-Nigerian railway linking the Gulf of Guinea with the north. Harbor facilities able to handle large modern ships had to be

developed. Lagos is now West Africa's leading cargo port and industrial center. It is a sprawl of shanties on dirt streets surrounding a modest knot of modern high-rise buildings and older low-rise structures in the commercial, governmental, and residential core near the harbor. During the oil boom of recent decades, migrants from all parts of Nigeria (and many from outside the country) flooded into the city. With its severe traffic congestion, rampant crime, inflated prices, and inadequate public facilities, Lagos gives the impression of urbanism out of control. It is also indicative of a disorderly national economy and political system. Nigeria has ranked over many successive years as the world's most corrupt country. One observer cited some of the effects typical of corruption: "Sugar cane, for example, would thrive here, but its cultivation would end the highly lucrative trade in smuggled sugar controlled by a coterie of powerful business people. Similarly, repair of the refineries would halt senior military officers' profits from transactions in imported fuel, and the refurbishing of petrochemical plants would shave the big profits on imported fertilizers."[1]

Close to the border with Niger is Kano (population: 700,000, city proper), the most important commercial and administrative center of Nigeria's north and long a major West African focus of Islamic culture. Immediately west of Nigeria on the Guinea Coast lies Cotonou (population: 536,800, city proper), the main city and port of Benin. Further west, in southern Ghana—another agricultural area with a high degree of commercialization—are two more of the 11 largest West African metropolises. The larger of the two, Accra (population: 1 million, city proper; 1.39 million, metropolitan area), is Ghana's national capital. Until recently it was the main seaport, despite the lack of deep-water harbor facilities and the consequent need to transfer cargo from ship to shore through the surf. A modern deep-water port has been developed at Tema a short distance east of Accra, and that city is now Ghana's main port and industrial center. Well inland from Accra lies the country's second largest city, Kumasi (population: 385,200, city proper), in the heart of Ghana's cacao belt, which was once the world's greatest cacao-producing area.

Today, the world's leading cacao producer and exporter is Ivory Coast (Côte d'Ivoire), which contains yet another of West Africa's 11 largest cities, Abidjan (population: 2 million, city proper). Abidjan, which has a considerable Western veneer in its buildings and businesses, is Ivory Coast's main seaport (Fig. 18.5). Although the inland city of Yamoussoukro is the official capital, Abidjan is the de facto capital of a country which has vowed to achieve economic prosperity comparable to that of the "Asian tiger" countries. Also on the Guinea Coast are three more seaport capitals: Monrovia (population: 1 million, city proper) in Liberia; Freetown (population: 500,000, city proper) in Sierra Leone, on West Africa's finest natural harbor; and Conakry (population: 800,000, city proper) in Guinea.

Transportation

Inadequate transportation is a problem for West Africa. Good natural harbors are scarce due to the abundance of offshore sandbars and the tendency of river mouths to be choked with silt. It has often been necessary to transfer goods by tender between ship and shore, and these countries incur large expenses as they attempt to provide modern port conditions and facilities. Inland from the coast lies the formidable problem of adequate connection between the ports and the interior. Rapids, together with annual seasons of low water, limit the utility of rivers for transport. Some rivers do carry traffic, notably the Niger, its major tributary the Benue, the Senegal River, and the Gambia River. The Gambia River is the main transport artery for The Gambia—a remarkable country nearly surrounded by Senegal

[1]From Roger Cohen, "Nigeria's Painful Paradox of 'Self-Imposed Poverty.'" *New York Times,* August 23, 1998, p. 1.

Figure 18.5 Only about one-third of the people of Africa south of the Sahara live in cities, but, with considerable rural to urban migration, that number is growing. This is very modern Abidjan, the leading city of Ivory Coast (Côte d'Ivoire), with St. Paul's Cathedral in the foreground. *Charles O. Cecil/Visuals Unlimited*

and consisting merely of a 20-mile-wide strip of savanna grassland and woodland extending inland for 300 miles (c. 500 km) along either side of the river.

Railways are much more important than rivers in the transport structure of West Africa but are few and widely spaced. Instead of forming a network, the rail pattern comprises a series of individual fingers that extend inland from various ports but which usually do not connect with other rail lines (see Fig. 17.3, p. 424). The region's road network is dense in well-populated areas, but most roads are unpaved or unreliable in bad weather. Much of the region's area and many of its people are still remote from any good highway. Not surprisingly, a very large part of the limited long-distance passenger travel within the region moves by air.

18.4 West Central Africa

The subregion of West Central Africa is flanked by Cameroon and the Central African Republic to the north and the Democratic Republic of Congo (formerly Zaire, and widely known simply as "Congo") to the south (see Fig. 17.1, p. 419). In colonial days, France held three of the present units—Central African Republic, Republic of the Congo (widely known as "Congo Republic"), and Gabon—in the federation of colonies called French Equatorial Africa. Cameroon consisted of two trust territories administered by France and Britain. Belgium held what is now the Democratic Republic of Congo, Spain held Equatorial Guinea, and Portugal held the islands of São Tomé and Principe.

A narrow band along the immediate coast of West Central Africa is composed of plains below 500 feet (*c.* 150 m) in elevation. As in West Africa, the coast is fairly straight and has few natural harbors. Most of the region is composed of plateaus with surfaces that are undulating, rolling, or hilly. A conspicuous exception to the generally uneven terrain is a broad area of flat land in the inner Congo basin, once the bed of an immense lake. The greater part of West Central Africa lies within the drainage basin of the Congo River. The heart of the region has a tropical rain forest climate, grading into tropical savanna to the north and south (see Fig. 2.8, p. 33). Only scattered parts of West Central Africa, primarily along the eastern and western margins, are mountainous. The most prominent mountain ranges lie in the eastern part of the Democratic Republic of Congo near the Great Rift Valley and in the west of Cameroon along and near the border with Nigeria. Vulcanism has been important in mountain-building in both instances.

Hundreds of ethnic groups speaking many languages and dialects live in West Central Africa. The majority speak Bantu languages of the Niger-Congo family, while Nilo-Saharan language speakers predominate in the drier areas of grassland north of the rain forest. Small bands of Pygmies (Twa) live in the rain forest of the Congo basin. Within West Central Africa as a whole, population densities are low and populated areas are few and far between. Rough terrain, thick forests, wetlands, and areas with tsetse flies tend to isolate clusters of people from each other. The savanna lands and highlands are generally more densely populated than the rain forests. Most countries have an overall density well below that of Africa south of the Sahara as a whole (see Table 17.1, pp. 420–421).

Subsistence agriculture supports people in most areas. Production of export crops is small and takes place mainly in areas of rain forest, although some cotton is exported from savanna areas. Southern Cameroon is an important cash crop area, with coffee and cacao the main export crops, supplemented by bananas, oil-palm products, natural rubber, and tea. The farms occupy an area of volcanic soil at the foot of Mt. Cameroon (13,451 ft/4100 m), a volcano that is still intermittently active. The mountain has extraordinarily heavy rainfall, averaging around 400 inches (1000 cm) a year.

The island of Bioko is notable for its cacao exports. Bioko and the adjacent mainland territory of Río Muni, formerly held by the Spanish, received independence in 1968 as the Republic of Equatorial Guinea. Cacao production in Bioko developed on European-owned plantations, but the newly independent country's leaders expelled the plantation owners.

Exports of minerals, principally oil, aid the financing of some countries—most notably Gabon, Republic of the Congo, Cameroon, and most recently Equatorial Guinea, where oil revenues have created per capita GNPs that are somewhat high for Africa south of the Sahara (see Table 17.1). Cameroon has been engaged in an ongoing oil-related dispute with Nigeria over control of the Bakassi Peninsula, which juts into the Gulf of Guinea near the border between the two countries. Control of the penin-

sula will give the dispute's victor ownership of promising offshore oil reserves of the Gulf of Guinea.

Cities are few in these countries. Most of the larger ones are political capitals and/or seaports, including the Republic of the Congo's capital city of Brazzaville (population: 693,712, city proper) and its seaport of Pointe Noire (population: 351,000, city proper). Brazzaville, located directly across the Congo River from the much larger capital city of Kinshasa in the Democratic Republic of Congo, was once the administrative center of French Equatorial Africa. Despite the Republic of the Congo's huge offshore oil wealth, years of political mismanagement have perpetuated underdevelopment, and rival militias destroyed much of the country's infrastructure in fierce fighting in 1997. There is not yet a paved road between Brazzaville and Pointe Noire. The 320-mile (515-km) Congo-Océan Railway does connect the cities. Another pair of cities connected by rail is Cameroon's seaport of Douala (population: 1.03 million, city proper) and the country's inland capital, Yaoundé (population: 653,700, city proper). But, aside from a handful of rail fingers reaching inland from seaports, West Central Africa is largely devoid of railways, and roads only poorly serve most areas. Isolation and poor transport are major factors limiting economic development.

The Democratic Republic of Congo: A Case Study in Colonialism and Its Legacy

The Democratic Republic of Congo is Africa's second largest country in area (after Sudan), but its estimated population in 1998 was only 49 million, or well under one half that of the continent's most populous country, Nigeria. Nearly half of the Democratic Republic of Congo is an area of tropical rain forest in which the population is extremely sparse. The national population comprises some 250 ethno-linguistic groups speaking more than 75 languages. The Democratic Republic of Congo has a colonial history and faces modern challenges which are representative of the region of West Central Africa, and so merits in-depth discussion.

Colonial Background and Independence

Belgian penetration of the Congo basin began in the last quarter of the 19th century, following exploration by an American, Henry M. Stanley. During this early period, the colony was virtually a personal possession of Belgium's King Léopold II, whose agents ransacked it ruthlessly for wild rubber, ivory, and other tropical products gathered by Africans. This lawless era inspired Joseph Conrad's famous novel *Heart of Darkness* (1902), in which the trader Kurtz, at the point of death, evokes the ravaged Congo Region with the cry "The horror! The horror!" In the 1890s, colonial authorities built a railroad to connect the port of Matadi with the capital of Léopoldville (now Kinshasa), thus avoiding a stretch of rapids on the Congo River and providing access to the interior. In 1908, the Belgian government formally annexed the greater part of the Congo basin, creating the colony of Belgian Congo.

Pressure for independence built up rapidly in 1959, and Belgium yielded to the demands of Congolese leaders. In 1960, three months before adjacent French Equatorial Africa became the independent People's Republic of the Congo, the Congo colony became the independent Republic of the Congo. (In 1971 it took the name Zaire, meaning "River." Following the overthrow of Zaire's government in 1997, the country was renamed "Democratic Republic of the Congo," and, as mentioned earlier, is now generally referred to simply as "Congo." The neighboring People's Republic of the Congo is now generally referred to as "Congo Republic.") The new country quickly faced major crises that its government was unable to handle. These included a large-scale mutiny within the Belgian-trained Congolese army, regional separatist movements and rebellions—notably a declaration of independence by the mineral-rich Katanga (now Shaba) Province in the southeast—and ethnic warfare, especially in the diamond-mining area of Kasai Province.

With the Republic of the Congo disintegrating into chaos, Belgian troops intervened to protect and evacuate Belgian nationals. Shortly afterward, the United Nations dispatched an emergency force to restore order and assist the government in Katanga. After intense fighting between United Nations and Katanga forces, the secession movement in Katanga ended in 1963.

The country's economy then entered a difficult era that continues today. Major problems have included governmental corruption and mismanagement, squandering of funds on ill-advised projects, large withdrawals of foreign aid, a decline in the world price of copper (the country's chief export), the high cost of imported oil products after 1973, and civil war beginning in 1997 (see below).

Economic and Urban Geography Today

Like many other states in Africa south of the Sahara, the Democratic Republic of Congo has only a modest number of simple manufacturing plants. Typical commodities are cotton textiles, shoes, bricks, cement, wood products, processed foods, cigarettes, some chemicals, and metal products. In peaceful times, the country has been important in the world economy mainly as an exporter of minerals and tropical agricultural products. A large proportion of the minerals are smelted or concentrated in the Democratic Republic of Congo and then are further refined at plants in Belgium and in other overseas industrial countries. Among the leading mineral exports, copper, cobalt, and other metals come principally from the Shaba region and diamonds from the adjoining Kasai region (see Fig. 17.3). Copper is by far the leading export. The Shaba region has also been a major world producer of cobalt, an alloy with vital high-technology uses (for example, to resist heat and abrasion in jet aircraft engines). Diamonds, primarily industrial but including some gemstones, are mined in the Kasai region to the east and west of Kananga (population: 410,000, city proper).

Figure 18.6 Fishermen working shallows along the Congo River, the Democratic Republic of Congo's lifeline. *George Holton/Photo Researchers*

Most nonmineral exports of the Democratic Republic of Congo, including palm products, rubber, some coffee, and wood, come from areas of tropical rain forest near the Congo River (Fig. 18.6). Coffee is grown in many different parts of the Democratic Republic of Congo.

Matadi is the Democratic Republic of Congo's only important seaport. Rapids block navigation on the Congo just upstream from Matadi, and goods destined for Kinshasa, located on the Congo River at the lower end of another stretch of navigable water about 1000 miles (c. 1600 km) long, must be transshipped by rail or truck about 230 miles (370 km). Kinshasa (population: 3 million, city proper) is the Democratic Republic of Congo's capital, largest city, and principal manufacturing center. Another important city is Lubumbashi (population: 851,381, city proper), formerly Elisabethville, in the copper-mining region of the southeast. One of the Democratic Republic of Congo's major arteries of transportation connects these two cities. It is composed of a railway from Lubumbashi to Ilebo on the lower Kasai River (see Fig. 17.3) and a navigable water connection from Ilebo to Kinshasa via the Kasai River and the Congo River. Deterioration of roads handicaps travel within the country.

The Democratic Republic of Congo is so deeply mired in poverty that its per capita GNP is only about $130, a remarkably low figure for a country with such an abundance of resources. Only a handful of the world's countries are poorer. At present there is little optimism that the Democratic Republic of Congo's economy will soon improve.

The country's longstanding autocratic ruler, Mobutu Sese Seko, acquired a massive fortune during his reign of more than 30 years, from 1965 to 1997. Political opposition and public discontent with the 80 percent unemployment rate and staggering rates of inflation led to riots and unrest between 1991 and

1993. Production of copper and cobalt plummeted, and by 1997 the country's economy appeared to be on the brink of collapse.

Civil war rocked the country in 1997. Late in 1996 and early in 1997, Zairean army troops attempted to demilitarize and expel a perceived weak force of indigenous Tutsis living in the country's far eastern region. The assault backfired when these Tutsis, supported by neighboring Tutsi-dominated governments in Rwanda and Burundi and by other sympathetic countries in the region, took up yet more arms and initiated fierce attacks on government forces. The Tutsi counterattack quickly widened into an antigovernment revolt comprised of many ethnic factions and led by a non-Tutsi guerrilla named Laurent Kabila. In spring 1997, Kabila's forces pushed westward virtually unopposed through the huge country and in May occupied Kinshasa with little loss of life. President Mobutu fled Zaire on the eve of Kabila's triumph. Kabila then declared himself president of the country he now called the Democratic Republic of Congo and promised that political reforms would live up to the country's new name. Foreign companies rushed to sign mining contracts with the fledgling government, and throughout Congo there was hope that decades of neglect and poverty could be reversed in this resource-rich giant of West Central Africa. But it was not to be. A new rebellion emerged against the former rebels and developed into warfare, which quickly drew in numerous countries in the region (see Regional Perspective, p. 456).

18.5 East Africa

East Africa comprises three former British dependencies (Kenya, Tanzania, and Uganda) and two former Belgian dependencies (Rwanda and Burundi) that lie between the Congo basin and the Indian Ocean (see Fig. 17.1, p. 419). Kenya and Tanzania front the Indian Ocean, but Uganda, Rwanda, and Burundi are landlocked. Prior to their independence, the three British units varied in political status. Uganda was a protectorate administered with African interests paramount. Kenya was a colony and protectorate in which a white settler class had great political influence. Both countries are now republics. Tanganyika (formerly German East Africa, now Tanzania) was a United Nations trust territory administered by Britain. Now a part of Tanzania, the islands of Zanzibar and Pemba, which lie north of the city of Dar es Salaam and some 25 to 40 miles (40–64 km) off the coast, were once a British-protected Arab sultanate. The city of Zanzibar was the major political and slave-trading center and **entrepôt** of an Arab empire in East Africa. While Arabs formerly controlled political and economic life, the majority of the population was, and is, African. Africans have been in control since 1964, when a revolution overthrew the ruling sultan. That same year, the political union of Tanganyika and Zanzibar was formed, with the name "United Republic of Tanzania." However, Zanzibar and Pemba have a parliament and president separate from Tanzania's.

The small but densely populated highland countries of Rwanda and Burundi were also once a part of German East Africa. When Germany lost its colonies at the end of World War I, the League of Nations placed the two units under Belgian administration as the mandated territory of Ruanda-Urundi. This area later became a Belgian trust territory under the United Nations and then received independence in 1962 as two countries.

Area, Population, and Physical Environment

Although a coastal lowland fringed by coral reefs and mangrove swamps occupies the eastern margins of Kenya and mainland Tanzania, the terrain of the five East African countries is composed mainly of plateaus and mountains. The plateaus lie generally at elevations of 3000 to 6000 feet (c. 900–1800 m) but rise higher in some places, and in much of eastern and northern Kenya they descend to low plains. All five countries include sections of the region's Great Rift Valley. International frontiers follow the floor of the Western Rift Valley for long distances and divide lakes Tanganyika, Malawi, and Albert and smaller lakes among different countries. The more discontinuous Eastern Rift Valley crosses the heart of Tanzania and Kenya. Lake Victoria, lying in a shallow downwarp between the two major rifts, is divided among Uganda, Kenya, and Tanzania.

The two highest mountains in Africa, Mt. Kilimanjaro (19,340 ft/5895 m) and Mt. Kirinyaga (formerly Mt. Kenya, 17,058 ft/5200 m; see Fig. 17.4), are extinct volcanoes. Both are majestic peaks crowned by permanent snow and ice and visible for great distances across the surrounding plains. East Africa's most productive agricultural districts, including many that contribute to the export trade, tend to be localized in these volcanic areas of unusually fertile soils (Fig. 18.7). Vulcanism was also important in building the fertile highlands of Rwanda

Figure 18.7 East Africa is blessed with extensive areas of volcanic soil where commercial and subsistence crops thrive. This is a wheat field in the central highlands of Kenya. *Joseph J. Hobbs*

Regional Perspective

Countries Rush to Join Congo's War

There have been numerous civil wars in Africa south of the Sahara, but recent conflict in the Democratic Republic of Congo is something rather rare in the region, a broader fight that has involved numerous countries, some of which are not even Congo's neighbors (Fig. 18.B).

President Laurent Kabila's new regime proved to be a bitter disappointment to most of Congo's citizens and neighbors, and to the international community. Like his predecessor, former president Mobutu, he funneled much of the country's wealth and power into the hands of family and friends. He also obstructed international human rights investigations into the deaths and disappearances of thousands of Hutu refugees during the heady days of rebellion against Mobutu. And he failed to bring stability to the far east of his huge country, where an environment of chaos had given birth to his own bid for power.

The widening of the conflict began in 1998 as Kabila failed to crack down on Hutu forces, based in the eastern Congo, who launched successive devastating attacks on Tutsi interests in Rwanda (see discussion of Rwanda and Burundi on p. 460). Although the Tutsis of Rwanda had helped bring Kabila to power in Congo, they now turned against him, forming the backbone of a rebel alliance bent on bringing down Kabila's young regime.

Superficially, the conflict appeared to be a tribal one, with Nilotic ethnic groups

from Rwanda, Burundi, and Uganda fighting against a coalition of Bantu peoples of central and southern Africa and

Arabs from Chad and Sudan. Upon closer inspection, however, the conflict revealed that each player became in-

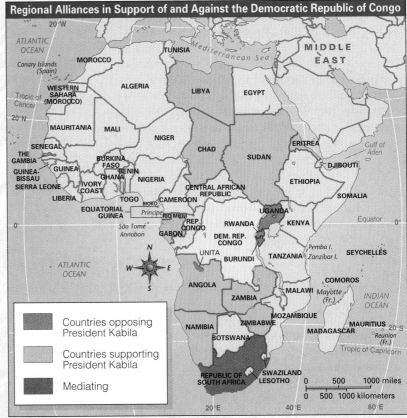

Source: Santoro and Dunn. *Africa's War of Nations, With Congo as the Pawn. The Christian Science Monitor.* August 27, 1998: p. 6.

Figure 18.B What began as a localized conflict in Congo has spread to neighboring countries, each having a special interest at stake.

and Burundi. The Virguna Mountains along the Rwanda-Uganda border include many active volcanos. One of the region's most active volcanos is the prominent Mt. Elgon (14,176 ft/4321 m) on the Uganda-Kenya border. On the Uganda–Democratic Republic of Congo border is Margherita Peak (16,762 ft/5109 m), part of a nonvolcanic range known as the Ruwenzori, which was commonly identified with the "Mountains of the Moon" depicted by early Greek geographers on their maps of interior Africa.

The climate of most of East Africa is classified broadly as tropical savanna, a category embracing a wide range of temperature and moisture conditions. Elevation has a strong effect on temperatures, and some agricultural districts, even on the equator, are so tempered by elevation that midlatitude crops like wheat, apples, and strawberries do well. Kenyan tourist brochures boast accurately that Nairobi (elevation: 5500 ft/ 1676 m) has a springlike climate year-round. There are great variations in the amount, effectiveness, and dependability of

volved to protect its own national, separatist, or economic interests—largely at the expense of anything that would resemble national cohesion and stability in the Democratic Republic of Congo.

These countries and entities—the Rebel Coalition—sent forces to oppose the government in Kinshasa:

Rwanda: To avenge Kabila's refusal to crack down on Hutus operating against this Tutsi-led country, and perhaps to establish a security buffer zone on its border with Congo, Rwanda supported rebels fighting Kabila. Rwanda also wanted protection for the Tutsi minority (known as Banyamulenge) living in extreme eastern Congo.

Burundi: Like Rwanda, Burundi's Tutsi-dominated government fought Congo to protect itself from Hutu attacks launched from inside Congo.

Uganda: This nation acted against Congo in part to establish a security buffer zone which would help protect it from attacks by Ugandan rebels operating from inside Congo.

Angola's rebel UNITA forces: The National Union for the Total Independence of Angola (UNITA) operates against the established government in Angola from bases inside Congo, and took on Kinshasa in a bid to retain those bases.

The following countries, representing the so-called "Congo Coalition," sent forces to support the government in Kinshasa:

Zimbabwe: Congo's President Kabila and Zimbabwe's President Robert Mugabe are old friends. But Mugabe's support of Kabila was widely viewed as an effort to turn attention away from Zimbabwe's internal problems, which included 45 percent unemployment, a large devaluation of its currency, high taxation, and widespread corruption. By sending forces to aid Kabila, Zimbabwe also may have wanted to establish itself as a regional superpower, akin to the position enjoyed by South Africa. And Zimbabwe acted to protect mining concessions and lucrative contracts that Congo awarded to Zimbabwean officials.

Angola: The Angolan government wanted to crack down on UNITA rebel bases inside Congo, and so joined the fray on Kabila's side.

Zambia: The government in Lusaka perceived greater regional stability if Congo remained a single entity governed from Kinshasa.

Namibia: Namibian firms have mining contracts with Congo, so the government in Windhoek supported Kinshasa to protect these interests.

Sudan: The national government in Khartoum supported Kabila's Congo mainly because of Sudan's bitter enmity with Uganda, which supported southern Sudan's rebellion against Khartoum.

Chad: Chadian troops were airlifted directly into the diamond-mining region of south-central Congo to protect Chad's investments there.

Libya, Gabon, Eritrea, and Equatorial Guinea: These countries indicated their willingness to join the Coalition supporting Kabila.

Only one country engaged in a non-aligned effort to mediate the conflict: South Africa. On the eve of his retirement from public life, South African president Nelson Mandela sought to solidify his legacy as an honest broker and peacemaker on the continent. A cease-fire between the combatants was signed in December 1998, but there were few reasons to be sanguine about the prospects for peace. The country in effect had already split into three entities: the rebel-held east, which imports and sells goods through Uganda and Kenya; a south that trades through Zambia and South Africa; and a west, operating from Kinshasa, that uses the country's outlet to the sea. These geographical divisions and alignments may become formalized should the country dissolve. Meanwhile the expatriate firms that would exploit Congo's mineral wealth and perhaps help bring this country out of dire poverty remain on the sidelines, along with numerous contenders to Kabila's seat of power.

---

precipitation and in the length and time of occurrence of the dry season (or seasons). Over much of East Africa, moisture is too scarce for a truly productive nonirrigated agriculture. Long droughts occur, even in what would normally be the rainy season. Relatively little of the present crop acreage is irrigated. Adequate moisture and good soil are linked only in the scattered areas that support the bulk of East Africa's population.

In Uganda, Kenya, and Tanzania, park savanna—composed of grasses and scattered flat-topped acacia trees—stretches over broad areas and is the habitat for some of Africa's largest populations of large mammals. Great herds of herbivores and their predators and scavengers migrate seasonally across these countries to follow the changing availability of food and water. The countries manage many parks and wildlife reserves, and in Kenya and Tanzania these are responsible for a huge share of the nations' hard currency revenues. Among the most famous are the Serengeti Plains and Ngorogoro Crater of Tanzania, and the Masai Mara and Amboseli National Parks of Kenya (Fig. 18.8).

Figure 18.8 Savanna grasslands in east and southern Africa host the world's greatest concentrations of large mammals. This is Kenya's Masai Mara National Park, in part of the greater Serengeti ecosystem that also extends into northern Tanzania. *Joseph J. Hobbs*

Most parts of East Africa are not thickly populated. The five countries make up some 657,000 square miles (1.7 million sq km), roughly the size of the U.S. state of Texas, with a total population in 1998 of 93 million (about one-third the population of the United States). Following sharp reductions as a result of falling birth rates and increasing AIDS-related deaths, the population growth rates of the five East African countries were moderate as of 1998. The population tends to be heavily concentrated in a few areas of dense settlement that support hundreds of people per square mile. The three largest and most important areas with a greater-than-average density are (1) in a belt along the northern, southern, and eastern shores of Lake Victoria, (2) in south central Kenya around and north of Nairobi, and (3) in most of Rwanda and Burundi.

Economic and Urban Geography

Most of East Africa's subsistence agriculturalists are cultivators, but some are pastoralists. Nearly the entire range of subsistence and export crops typical of Africa south of the Sahara is to be found in the region. For the area as a whole, the main subsistence crops are maize, millets, sorghums, sweet potatoes, plantains, beans, and manioc. Manioc (cassava) is a particularly valuable plant in areas prone to crop failure because the roots can be left in the ground for several years as a reserve food supply. Some cultivators depend exclusively on crops for subsistence, and others also keep livestock. Cattle are the most important livestock animal.

The most valuable export crop is coffee, but tea is also important in Kenya. Kenya and Tanzania also produce sisal for its valuable fiber (see Fig. 17.14). The islands of Zanzibar and Pemba rely mainly on millions of clove trees (located primarily

on Pemba) that provide the bulk of the world's supply of cloves. The islands' economies have been hurt drastically by a recent fall in world clove prices and by the economic turmoil in Indonesia, which had long imported most of the clove crop for use in its scented *kretek* cigarettes. Seeking a new future, Zanzibar has announced plans to transform itself into a free economic zone modeled after Singapore and Hong Kong.

Prior to the new era of political independence, East Africa was home to some 300,000 Asians (Indians, Pakistanis, and Goans from the former Portuguese colony of Goa on the west coast of India), 100,000 Europeans, and 60,000 Arabs (the figures are estimates for 1958). Most Asians had commercial, industrial, and clerical occupations, although some owned estates producing export crops. They dominated retail trade and owned factories. Some were wealthy, and practically all were more prosperous than most Africans. Arabs were primarily a mercantile class. Europeans were a professional, managerial, and administrative class. In the "White Highlands" of southwestern Kenya, about 4000 European families owned and operated estates that were worked by African laborers to produce export crops and cattle. Since independence, many Europeans have left, and Africans occupy much of the farmland formerly on European estates.

The most productive parts of East Africa are bound together by railways that form a connected system leading inland from the seaports of Mombasa in Kenya and Dar es Salaam and Tanga in Tanzania (see Fig. 17.3). Mombasa (population: 465,000, city proper) is the most important seaport in East Africa. Situated on an island connected with the mainland by causeways, the city has two harbors: a picturesque old harbor still frequented by Arab sailing vessels *(dhows)* and other small ships, and a deep-water harbor that has modern facilities for handling large vessels (Fig. 18.9).

From Mombasa, the main line of the Kenya-Uganda Railway leads inland to Kenya's capital, Nairobi (population: 1.16

Figure 18.9 Naval ships and grain silos in Kenya's Indian Ocean port of Mombasa. *Joseph J. Hobbs*

million, city proper), the largest city and most important industrial center in East Africa. It is also the busiest crossroads of international air traffic and the main outfitting and departure point for safaris into East Africa's world-renowned national parks. Nairobi was founded early in the 20th century as a construction camp on the railway, which continues westward to Kampala (population: 773,463, city proper), the capital and largest city of Uganda. Dar es Salaam ("House of Peace," population: 1.1 million, city proper) is the capital, main city, main port, and main industrial center of Tanzania, with rail connections to Lake Tanganyika, Lake Victoria, and Zambia. Rwanda's capital is in the highlands at Kigali (population: 237,782, city proper), and Burundi's capital is on the northernmost point of Lake Tanganyika at Bujumbura (population: 235,400, city proper). Only about 5 percent of the populations of Rwanda and Burundi is urban.

In comparison with many other African countries, the East African nations lack mineral reserves and mineral production. Manufacturing is in an early stage of development, limited mainly to agricultural processing, some textile milling, and the making of simple consumer items. Both Mombasa and Dar es Salaam have refineries that process imported crude oil, and there is a large cotton-textile mill at Jinja, Uganda, that operates with hydroelectricity from a station at the nearby Owen Stanley Dam on the Nile. The Tana River scheme in Kenya produces hydropower and supplies irrigation waters.

Social and Political Issues

The five East African countries face many social and geopolitical dilemmas. None has had much internal peace, order, or prosperity, and none has had easy relations with its neighbors. Ethnic rivalries and conflicts have beset all of them.

Aside from small minorities of Asians, Europeans, and Arabs, the population of East Africa is composed of a large number of African tribes; Tanzania alone, for example, has about 120 tribal groups. Among the large tribes are the Kikuyu and Luo of Kenya, the Baganda of Uganda, the Sukuma of Tanzania, the pastoral Masai of Kenya and Tanzania, and the Tutsi and Hutu of Rwanda and Burundi. Most East African peoples speak Bantu tongues of the Niger-Congo language family, but some in northern Uganda, southern and western Kenya, and northern Tanzania speak Nilo-Saharan languages. Swahili, a Bantu language drawing heavily on Arabic for vocabulary, is a widespread lingua franca (and is the national language of Tanzania), as is English. Religions in all five countries include animist, Muslim, and Christian elements. The Muslim religion reflects a long history of Arab commercial enterprise in the region, including slave trading.

Since independence, African governments have driven the Asian minority from its commercially predominant position. In Uganda, where Asians were the backbone of commerce and trade, the country's former leader Idi Amin abruptly expelled them from the country as refugees in the 1970s. Some have recently begun to return.

Episodes of bloodshed have punctuated East Africa's history. From 1905 to 1907, when mainland Tanzania was German-controlled, German authorities put down a great revolt against colonial control, costing 75,000 African lives. Immediately prior to its independence, Kenya experienced a bloody guerrilla rebellion led by a secret society within the large Kikuyu tribe called the Mau Mau. Antagonism directed against British colonialism and European ways in Kenya cost the lives of about 11,000 Africans and 95 white settlers over the period 1951–1956.

From the 1960s through the 1980s, Kenya was regarded as a relative economic success story for Africa south of the Sahara. However, corruption, ethnic strife, crime, terrorism (including the 1998 bombing of the U.S. Embassy in Nairobi), drought, and floods combined in the 1990s to reduce Kenya's prospects for economic development. Recent unrest has threatened the vital business of international tourism in Kenya, which normally contributes 50 percent of the country's hard currency revenue. In addition, since 1990, the international community has accused President Daniel Arap Moi's government of corruption and human rights abuses against political and tribal opponents. In 1992, Britain, the United States, and other donor countries threatened to cut off aid to Kenya unless Moi implemented democratic reforms. He appeared to do so, but Moi's victory in elections late that year was regarded widely as fraudulent. Limited international economic assistance resumed, however, in part to encourage greater democratization in advance of the 1997 elections that again resulted in Moi's reelection as president for another five-year term.

Discrepancies between the haves and have-nots are much greater in Kenya than in neighboring Tanzania, where former president Julius Nyerere instituted an order based on the principle of African socialism, or *ujamaa*. Intended mainly to create self-sufficiency and redistribute land, the system did improve social services but also led the country's economy to near collapse. The effort to eradicate rural poverty and increase harvests through the collectivization of scattered farming communities met with little success.

In postcolonial Uganda, a savage orgy of murder, rape, torture, and looting during the notorious regimes of Idi Amin and Milton Obote dwarfed the country's preindependence turmoil. The ravaged country finally had a respite in 1986 when Yoweri Museveni, the leader of a southern guerrilla force that had developed in opposition to Obote, took over the government. The formidable task of rebuilding began, beset by great difficulties because of the collapsed economy and the continuing cultural and political fragmentation of Uganda among some 40 tribes, mostly unable to communicate with each other in their native tongues and bearing heightened suspicions and animosities from the recent disastrous decades. Sporadic internal violence, border conflicts, and economic problems continued into the 1990s, but Museveni's Uganda came to be regarded widely as a model for democracy, social progress, and economic development in Africa south of the Sahara.

In the mid-1990s, Rwanda and Burundi engaged in disputes that had tragic consequences. About 85 percent of the population in Burundi and 90 percent in Rwanda is composed of the Hutu (Bahutu), and most of the remainder of the two populations is Tutsi (Watusi). These originally were not separate tribal or ethnic groups; the peoples speak the same language, share a common culture, and often intermarry. The distinction between Hutu and Tutsi rather was based on socioeconomic classification. Historically, the Tutsi were a ruling class who dominated the Hutu majority. Their power and influence were measured especially by the vast numbers of cattle the Tutsi owned. Tutsi who lost cattle and became poor came to be identified as Hutu, and Hutu who acquired cattle and wealth often became Tutsi.

Colonial rule in East Africa attempted to polarize these groups, thereby increasing antagonism between them. German and Belgian administrators differentiated between them exclusively on the basis of cattle ownership: Anyone with fewer than ten animals was Hutu and anyone with more was Tutsi. The colonists replaced all Hutu chiefs with Tutsi chiefs. These leaders carried out colonial policies that often imposed forced labor and heavy taxation on the Hutus. Education and other privileges were reserved almost exclusively for the Tutsi. Europeans mythologized the Tutsis as "black Caucasian" conquerors from Ethiopia who were a naturally superior, aristocratic race whose role was to rule the peasant Hutus.

Ferocious violence between these two peoples has marked Rwanda and Burundi intermittently since the states became independent in the early 1960s. After independence, the majority Hutus came to dominate the governments of both Rwanda and Burundi. In 1994, the death of Rwanda's Hutu president in a plane crash (in which Burundi's president also died) sparked civil war in Rwanda between the Hutu-dominated government and Tutsi rebels of the Rwandan Patriotic Front (the RPF, based in Uganda). Hutu government paramilitary troops systematically massacred Tutsi civilians with automatic weapons, grenades, machetes, and nail-studded clubs. Women, children, orphans, and hospital patients were not spared. An estimated 500,000 died. International media tracked these horrors, but outside nations did nothing to stop the fighting.

Well-organized and motivated Tutsi rebels seized control of most of the country and took power in the capital, Kigali, in 1994. That precipitated a flood of two million Hutu refugees mainly into neighboring Zaire, now the Democratic Republic of Congo (Fig. 18.10). Although the Tutsi-dominated government promised to work with the Hutus to build a new multiethnic democracy, Hutu insurgents, based mainly in eastern Zaire, continued to carry out attacks against Tutsi targets in Rwanda. These Hutu operations were suspended in 1997, when the mostly Tutsi forces of Laurent Kabila's rebel alliance established dominance in the region, but resumed in 1998 (see Regional Perspective, pp. 456–457).

A parallel situation existed in Burundi in 1996, when the Tutsi-dominated army overthrew the country's Hutu president. The new Tutsi president vowed to end the ethnic bloodshed that had cost 150,000 lives since October 1993, when Tutsis murdered Burundi's first elected Hutu president, sparking large-scale Hutu retaliatory slaughter of Tutsis. But the civil war instead dragged on. Prospects for peace brightened in 1998 when Hutus and Tutsis in Burundi called for a power-sharing transitional government with an equal number of ministerial seats drawn from each group and with the replacement of the prime minister by two vice presidents, one Hutu and one Tutsi.

18.6 Southern Africa

In the southern part of Africa south of the Sahara are five countries—Angola, Mozambique, Zimbabwe, Zambia, and Malawi—that share the basin of the Zambezi River and have long had important relations with each other. The first two were

Figure 18.10 Warfare and drought in postcolonial Africa have driven millions of Africans from their homes as refugees. This harrowing scene shows terrorized Rwandans fleeing into Tanzania in 1994 to escape the savage massacres of civilians in the country's civil war. *Scott Daniel Peterson/Gamma Liaison*

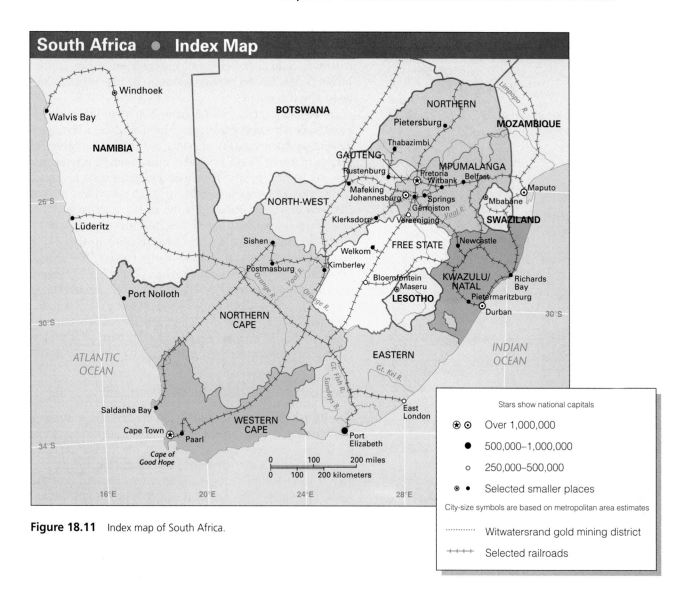

Figure 18.11   Index map of South Africa.

former Portuguese colonies and the latter three were formerly British. Still farther south are South Africa and four countries whose geographical problems are closely linked to South Africa: the three former British dependencies of Botswana, Swaziland, and Lesotho, and the former German colony of Namibia. These ten states are grouped here as the countries of southern Africa (see Fig. 17.1, p. 419). Other than their geographical location, there is so far little in the way of economic or political integration that binds these diverse states together. This may change as a result of goals established by the Southern African Development Community (SADC), a regional bloc composed of these ten countries plus Tanzania and Mauritius. SADC's member nations aim to establish by 2004 an economic free-trade zone in which goods can cross borders without tariffs and taxes. The dominant power in that group and in the region is South Africa, which is described here in the greatest detail; the other countries are examined in groups.

South Africa

South Africa (Fig. 18.11) was one of the world's most controversial countries until its new beginning in 1994. Organized from four British-controlled units in 1910 as the Union of South Africa, it was for five decades a self-governing constitutional monarchy under the British crown. In 1961, following a close majority vote by the ruling white population, it became a republic outside the British Commonwealth.

South Africa (1998 population: 39 million) accounts for about one-third of Africa's output of manufactured goods and one-third of the electricity generated on the entire continent. Its diversified industries include iron and steel, machinery, metal manufacturing, weapons, automobiles, tractors, textiles, chemicals, processed foods, and others. The United Kingdom and other Western countries have invested heavily in South Africa, and many thousands of jobs in those countries are directly linked to trade with it. South Africa's military strength, which

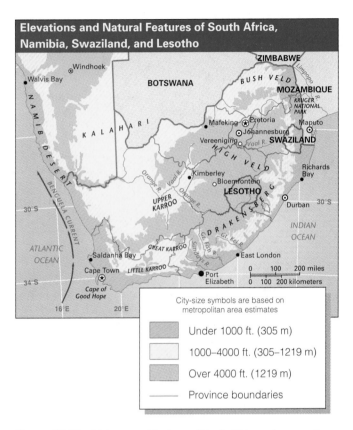

Elevations and Natural Features of South Africa, Namibia, Swaziland, and Lesotho

City-size symbols are based on metropolitan area estimates

- Under 1000 ft. (305 m)
- 1000–4000 ft. (305–1219 m)
- Over 4000 ft. (1219 m)
- —— Province boundaries

Figure 18.12   Major natural features of South Africa and surroundings.

included nuclear weapons until they were dismantled in 1990, contributes to its global significance. It is strategically important for its frontages on both the Atlantic and Indian oceans, its possession of Africa's finest transport network (vitally important to many African countries that trade with and through South Africa), and its diversified mineral wealth.

South Africa's physical environment has played a major role in its unusual course of development (Fig. 18.12). Lying almost entirely in the middle latitudes, it is a cooler land than most of Africa south of the Sahara, with annual temperatures generally between 60° and 70°F (16°–21°C). In combination with the country's natural resource wealth, this climate made the area very attractive for European settlement; it was one of the "neo-Europes" of the colonial era (see Chapter 2). Consequently, South Africa has been the main center of European settlement and advanced economic development in an underdeveloped continent that Europeans dominated until very recently.

Visitors from western Europe and North America will find in South Africa most of the institutions and facilities to which they are accustomed. However, the Europeanized cultural landscape does not reflect the majority culture of this unusual country. Whites represent only about 13 percent of the total population, and blacks outnumber them more than five to one. There are two other large racial groups: the so-called "coloreds," of mixed origin (9 percent), and the Asians (3 percent),

most of whom are Indians. Many peoples make up the black African majority of 75 percent. The Zulu and Xhosa tribes, both concentrated in hilly sections of eastern South Africa near the Indian Ocean, are the most populous of the nine officially recognized tribal groups.

A huge economic gulf separates South Africa's impoverished black Africans from about 5.3 million whites of European descent who dominate the economy and enjoy a lifestyle comparable to that of people in western Europe and North America. Because of the poverty of the non-European majority, South Africa's overall per capita GNP of $3520 is typical of such LDCs as Brazil and Turkey. For the unskilled and semiskilled labor needed to operate their mines, factories, farms, and services, the whites of South Africa have always relied mainly on low-paid black workers in most of the country and colored workers in the Western Cape Province. The white-dominated economy could not operate without them, and the Africans in turn are extremely dependent on white payrolls. While the races are interlocked economically, however, relations between them have historically been very poor and have been structured to benefit the whites.

How South Africa's Racial Situation Developed

The British-Afrikaner Division.   South Africa's white population is split between the British South Africans and the Afrikaners, or Boers (Dutch: "farmers"). The Afrikaners speak Afrikaans, a derivative of Dutch, as their preferred language. They outnumber the British approximately three to two. The Afrikaners are the descendants of Dutch, French Huguenot, and German settlers who began coming to South Africa more than three centuries ago. The Dutch East India Company established the earliest permanent settlement at Cape Town in 1652 as a way station to provide water, fresh vegetables, meat, and repairs for company vessels plying the Cape of Good Hope route to the Far East. Although the company did not intend to annex large areas of land, agricultural settlements slowly expanded in valleys near Cape Town, and a pastoral frontier society developed farther inland. Its impact on indigenous Africans was catastrophic. White settlers killed, drove out, and, through the unintentional introduction of smallpox, decimated most of the original Khoi and San (Bushman) inhabitants. They put the survivors to work as servants and slaves. When these proved to be insufficient in number, the colonists brought slaves from West Africa, Madagascar, East Africa, Malaya, India, and Ceylon (now Sri Lanka).

Great Britain occupied what it called the Cape Colony in 1806, during the Napoleonic Wars. Friction developed between many Boers and the British authorities, who imposed tighter administrative and legal controls than the Boers were accustomed to. The British abolished slavery throughout their empire in 1833, contributing to the anti-British sentiments of the slave-holding Boers. Boer discontent resulted in the Great Trek, a series of northward migrations by which groups of Boers, primarily from the eastern part of the Cape Colony, sought to find new interior grazing lands and establish new political units beyond British reach. After some earlier exploratory expeditions, the main trek by horse and ox-wagon began in 1836. It resulted in the founding of the Orange Free State, the Transvaal, and Natal, as Boer republics. Britain annexed Natal in 1845, but recognized Boer sovereignty in the Transvaal and the Orange Free State in 1852 and 1854 respectively.

While Boer disaffection with Britain was building, British settlers were coming to South Africa in increasing numbers. The area's natural resource wealth helped shape subsequent events. The British colonies and Boer republics might have developed peaceably side by side if diamonds had not been discovered in the Orange Free State in 1867 and gold in the Transvaal in 1886. These discoveries set off a rush to these republics of prospectors and other fortune hunters and entrepreneurs from outside the region—particularly from Britain. Ill feeling led to the Anglo-Boer War in 1899. British troops defeated the Boer forces decisively, ending the war in 1902 but leaving behind a reservoir of animosity that exists to this day.

The Union of South Africa was established in 1910 as a self-governing dominion under the British crown. Dutch became an official language on par with English (this was later altered to specify Afrikaans rather than Dutch). Pretoria in the Transvaal became the administrative capital as a concession to Boer sentiment. However, the national parliament now meets at Cape Town and the supreme court sits at Bloemfontein (population: 127,000, city proper) in the Orange Free State.

Since 1910, and particularly since the Nationalist (now National) Party took power in 1948, Afrikaners have dominated the political life of South Africa by virtue of their greater numbers and cohesion. Racial segregation characterized South African life from 1652 onward, but it was first systematized after 1948 under a comprehensive body of national laws supporting the official governmental policy of "separate development of the races" (subsequently called "multinational development" and commonly known as *apartheid;* see Definitions and Insights, opposite). The new laws imposed racially based restrictions and prohibitions on the entire population, but they weighed most heavily on black Africans. They ruled out significant sharing of power by black people in a unified South Africa. The apartheid system drew furious criticism from many other nations. Most of the world community ostracized South Africa, and many countries, including the United States, imposed economic sanctions against it.

The Homelands Scheme. The white government declared that racial peace and justice would exist only if the races were separated territorially and socially so that each race could develop within its allotted niche. There was to be no such thing as a unified South African nationality. Instead, the country was to be a collection of "nationalities" living side by side. The Nationalist Party policy of apartheid culminated in the scheme to establish tribal states or "homelands" (also known as "native re-

Definitions & Insights

APARTHEID

The official South African policy of "separate development of the races," known best by its Afrikaner name, **apartheid,** was the world's most anachronistic vestige of colonialism and formalized racism in the late 20th century. The apartheid laws mandated that all persons be classified and registered by race. Marriage and sexual relations between whites and nonwhites were prohibited. Segregated educational facilities at all levels existed for whites, coloreds (mixed-blood people), Asians, and black Africans. Public facilities were segregated by race, and political rights were denied to nonwhites. With limited exceptions, black Africans obtained citizenship only in tribal "homelands." They could not vote or hold office in "white" South Africa and had no representation in the white-controlled national parliament. Coloreds and Asians had no parliamentary representation until 1984, when a new, white-adopted constitution granted them some representation. A system of permits known as the "pass laws," applying mainly to the black population, restricted travel within South Africa. Law and custom imposed job discrimination in favor of whites. The broad powers of South Africa's minister of justice and police forces repressed dissent, and accused blacks had few rights. Authorities often broke up protest gatherings on the grounds that organizers did not have a proper permit or that the protest represented a threat to public order. The most publicized such incident happened in 1960 at Sharpeville near Johannesburg, where police killed 69 blacks protesting the "pass laws."

The races face a long period of reconciliation in the new South Africa. Some of the healing process has been formalized in a national "Truth and Reconciliation" Commission which allows those who were party to racial violence during the apartheid era to confess their misdeeds and, in effect, be absolved of them.

serves," "Bantustans," and "national states"), which were nominally intended to achieve "independence." These prescribed racial areas for homes, business, and private lands included numerous homelands and the "independent republics" of Transkei, Ciskei, Bophuthatswana, and Venda.

The ten South African homelands were territorial units reserved for Africans and run by elected African governments. Organized essentially on a tribal basis, they collectively incorporated hundreds of separate tracts of land that began to be set apart as African reserves soon after the Union of South Africa was formed. South Africa's laws permitted the government to classify a black South African as a legal resident of one of the homelands, even though he might never have seen it and had no sense of affiliation with the tribe on which the homeland was based. Whites ejected several million Africans not wanted in "white" South Africa from their homes and transferred them to

Figure 18.13 Johannesburg, the "City of Gold," is surrounded by huge dumps of waste material from former gold-mining operations. One of these dumps occupies the center of the view. Note the automobile expressway at the bottom of the view. In the distance lie residential areas occupied by whites. *Jesse H. Wheeler, Jr.*

homelands. They removed great numbers of blacks from urban areas and white farms and dumped them in poverty-stricken and crowded areas unequipped to handle new influxes of people. Millions of homeland residents continued to work in areas of South Africa outside their homelands. The government regarded them as temporary visitors, to be permitted in "white" South Africa only in such numbers as the white economy needed. Black breadwinners often migrated for long periods to jobs outside their homelands while the rest of the family stayed behind. The magnitude of destruction of black family life that this system caused in South Africa is beyond calculation.

The New South Africa

Black unrest became so widespread and violent in 1984–1986 that the government declared a state of emergency. Blacks killed many other blacks whom they perceived to be collaborating with the whites. An intensive government crackdown resulted in the wholesale banning of political activity by anti-apartheid groups, intensified restrictions on media, and the jailing of thousands of black Africans perceived as protest leaders. The government lifted the state of emergency in 1990, but widespread violence continued. Much of the fighting was between two large, tribally based rival factions, the Xhosa-dominated African National Congress (ANC), led by Nelson Mandela, and the Zulu-dominated Inkatha Freedom Party (IFP), led by Chief Mangosuthu Buthelezi. Tension and outright violence between these groups has persisted ever since.

A complete official turnabout on the issue of apartheid resolved the crisis. It began in 1989 after Frederik W. de Klerk, an insider in the Afrikaner political establishment, became president. Urged on by white South African business leaders, anti-apartheid activists, the powerful force of South African and international political opinion, the impact of international economic boycotts against South Africa, and his own stated convictions concerning the injustices and unworkability of apartheid, de Klerk launched a broad-scale program to repeal the apartheid laws and put South Africa on the road to revolu-

tionary governmental changes under a new constitution. Four years of intensive negotiation involving all interested parties resulted in an agreement to hold a countrywide, all-race election to create an interim government that would hold office for not more than five years. In April 1994, voters turned out in massive numbers to elect a new national parliament, which then chose Nelson Mandela as president. (Only 4 years earlier the white government had freed Mandela after a 27-year period of political imprisonment.) Mandela's African National Congress received 62.6 percent of the votes in the nation as a whole, followed by the Nationalist party (20.4 percent) and the Inkatha Freedom Party (10.5 percent).

This historic election ended white European political control in the last bastion of European colonialism on the African continent. The apartheid laws themselves became null and void. The bizarre homeland units were abolished and merged into South Africa's nine new provinces (see Fig. 18.11).

South African Regions

Northern, Free State, and Mpumalanga Provinces. The interior of South Africa is a plateau with a general elevation of 3000 to 6000 feet (*c.* 900–1800 m; see Fig. 18.12). Two river systems, the Orange (including its tributary, the Vaal) and the Limpopo, drain most of it. At the extreme east is the High Veld, originally a grassland similar to the midwestern prairies of the United States. The northeastern province of Northern Province is an area of woodlands and savanna grasses known as the "Bush Veld."

South Africa is the world's largest producer of gold (Fig. 18.13). The gold-mining industry, which supplies the country's most important export, is almost entirely confined to the High Veld. The greater part of the gold mined thus far has come from the Witwatersrand (formerly Rand), an area in the province of Gauteng (formerly Pretoria/Witwatersrand/Vaal [PWV]; see Fig. 18.11). Johannesburg, South Africa's largest city (population: 712,507, city proper; 2 million, metropolitan area), plus the adjacent impoverished and overcrowded black township of

464

Figure 18.14 A characteristic rural landscape in the mediterranean climate of South Africa's Western Cape Province. The middle distance is dominated by fields of ripe winter wheat at the base of one of the Cape Ranges. *Jesse H. Wheeler, Jr.*

Soweto, and nearby Pretoria (population: 525,583, city proper; 1.1 million, metropolitan area) are located in this province. Immense amounts of capital from British and other sources are invested in the mines (the world's deepest), which are operated by large corporations and employ African workers recruited from South Africa and other African countries.

Recovery of uranium from mine tailings and as a byproduct of current gold mining has made South Africa the largest uranium producer on the continent. And since 1867, the country has been an important source of diamonds from alluvial deposits along the Vaal and Orange River valleys and from kimberlite "pipes" (circular rock formations of volcanic origin). Kimberley (population: 155,000, city proper) in Northern Cape is the administrative center for the diamond industry.

The mining industry in South Africa requires large quantities of electricity. Most of this is supplied by generating plants powered by bituminous coal, which is present in vast quantities near the surface on the interior plateau. These coal deposits, which are the largest in Africa, are immensely important to South Africa, as the country has no petroleum or natural gas output and has only a modest hydroelectric potential. Coal is a major export, and it has facilitated development of Africa's leading iron and steel industry, concentrated at Pretoria and around Johannesburg. Iron ore comes by rail from large deposits in the northeast and central parts of the country.

The Transvaal provinces and adjacent areas also compose one of the world's most vital mineral-exporting regions, with exports including copper, asbestos, chromium, platinum, vanadium, antimony, and phosphate.

The High Veld is the country's leading area of crop and livestock production. The principal farm animals of the High Veld are beef cattle, dairy cattle, and sheep. Its main crops are corn (maize) and wheat. Much of the corn production becomes livestock feed, and surpluses become exports, although corn is also a major item of diet for the country's black population. Much of

South Africa is too dry for nonirrigated farming. In 1992, South Africa suffered from a regional drought and had to import large quantities of food and livestock feed.

Many dams have been constructed on South Africa's rivers to conserve water and provide hydroelectricity. In the early 1960s, the country inaugurated its most important water-control scheme, the Orange River Project. Water stored behind Orange River and Vaal River dams meets nearby urban and industrial needs, and a major share irrigates fields and orchards.

KwaZulu/Natal. KwaZulu/Natal occupies hilly terrain between the Indian Ocean and the Drakensberg, a mountainous escarpment at the edge of the interior plateau. The leading commercial crop of the area is sugarcane, grown mainly on large, white-owned farms worked by African laborers, including those drawn from the Zulu tribe—the region's most populous ethnic group. Historically, however, most workers in the cane fields were indentured laborers brought from India beginning in the 1860s. The majority chose to remain in South Africa when their terms of indenture ended, and some Indians came as free immigrants. Currently, about 85 percent of South Africa's Indian population is in KwaZulu/Natal, mainly in and near Durban (population: 715,669, city proper; 1.7 million, metropolitan area), KwaZulu/Natal's largest city and main industrial center and port. Most Indians are employed in commercial, industrial, and service occupations.

Eastern, Northern Cape, and Western Cape Provinces. Except for a mountainous fringe along the southern and southeastern coasts, nearly all of the Cape provinces area is semidesert and desert. Sections adjoining KwaZulu/Natal share the humid subtropical climate of that area. The vicinity of Western Cape Province has a mediterranean climate—the only African area south of the Sahara to have this climate type (Fig. 18.14). Average temperatures are similar to those of the same climate area in

465

southern California. The early settlers established vineyards within this region, and today the area's grapes and wines sell both in South Africa and abroad, where they are acquiring a growing share of the world market. The province also produces a variety of other fruits, including citrus fruits (mostly oranges), deciduous fruits, and pineapples. Sheep ranching is a major occupation in the semiarid interior. The arid western coast, swept by the cold, northward-moving Benguela Current, supports Africa's most important fisheries (see Fig. 18.12). Upwelling cool water brings up nutrients from the ocean floor and provides a fertile habitat for plankton, the basic food of fish.

Like KwaZulu/Natal, the Cape provinces have a distinctive racial group—South Africa's coloreds (widely known as the "Cape Coloreds"). Nearly nine-tenths live in the provinces, primarily in and near Cape Town (population: 854,616, city proper; 1.9 million, metropolitan area), the country's second largest city and one of its four main ports. This group originated in the early days of white settlement as a product of sexual relations between Europeans (Dutch East India Company employees, settlers, and sailors) and non-Europeans, including slaves and Khoi (Hottentot), Malagasy, West African, and various south Asian peoples. The resulting mixed-blood people vary in appearance from those with pronounced African features to others who are physically indistinguishable from Europeans. About nine-tenths speak Afrikaans as a customary language, and most of the remainder speak English.

Culturally, the coloreds are much closer to Europeans than to Africans, and many with light skins have successfully "passed" into the European community. Coloreds have always had a higher social standing and greater political and economic rights than black South Africans, although they have ranked below the Europeans in these respects. Most coloreds work as domestic servants, factory workers, and farm laborers, and perform other types of unskilled and semiskilled labor. Many coloreds are employed in fishing. There is a small but growing professional and white-collar class. White owners of grape vineyards are donating portions of their fields to an increasing number of the coloreds who once worked as indentured servants on the white lands, leading to a sharp increase in the region's output and exports of wines. Such progress is a sign of how truly different the new South Africa is from the old.

Namibia

Namibia (population: 1.6 million) has a long history of association with neighboring South Africa. South African forces overran this former German colony during World War I, and after the war the League of Nations mandated it to South Africa under the name of South West Africa. The terms of the mandate called for administration of the territory as an integral part of South Africa.

Independence in Namibia was linked to events in neighboring Angola. South Africa had long supported an Angolan guerrilla movement opposed to Angola's Cuban-backed Marxist government, and South African military forces actively intervened in the Angolan struggle in the mid-1970s. Eventually South African forces withdrew but continued to attack Angolan bases of the South West Africa People's Organization (SWAPO), a guerrilla group fighting for independence. In 1988, South Africa agreed to grant independence to Namibia, but only if the would-be Namibian government under SWAPO leadership could convince its main source of support, the government of Angola, to expel Cuban troops from Angola. The South African government granted independence to Namibia on March 21, 1990. SWAPO is now the major political party, but its Marxist leanings have not prevented the country from promoting capitalist investment.

The anomalous sliver of territory in Namibia's extreme northeast known as the Caprivi Strip, which extends fully 280 miles (450 km) eastward to the waters of the Zambezi River, was important to the German colonists who ruled Namibia early in the 20th century. They insisted on this access to Zambezi in the mistaken belief that it would allow navigation all the way to the Indian Ocean (the Victoria Falls stand in the way). Independent Namibia was permitted to retain the strip, despite protests from neighboring Botswana. The Caprivi is home to the Lozi ethnic group, whose tribal lands also extend into Botswana, Angola, and Zambia. Lozi efforts both in the Caprivi Strip and in Zambia to establish an autonomous or independent homeland are strongly opposed by both Namibia and Zambia.

Namibia is an extremely dry country. The Namib Desert extends the full length of Namibia in the coastal areas (Fig. 18.15). Inland is a broad belt of semiarid plateau country, merging at the east with the Kalahari semidesert of Botswana. Chronic shortages of fresh water have led the government to

Figure 18.15 A surprising abundance and diversity of wild animals and plants are adapted to the harsh conditions of the Namib Desert in Namibia. This is an oryx antelope known as the gemsbok. *Peter Lamberti/Tony Stone Images*

consider desalinating Atlantic Ocean waters (this option was rejected as too expensive); building the Epupa Dam on the Kunene River, which divides Namibia from Angola (this project is opposed by environmentalists and by human rights groups, who argue that the homeland of the Himba tribespeople would be inundated); and diverting water southward from the Okavango River (see following discussion of Botswana).

With its scattered livestock forage, the Kalahari section of Namibia is peopled mainly by Bantu-speaking herders and by remnants of the Khoi and San (Bushman) populations. Some of the San are among the world's last remaining hunters and gatherers, but the majority have given up the nomadic way of life in recent decades and have settled as farmers and wage laborers. The roughly 80,000 whites in Namibia live mostly on the interior plateau. Windhoek (population: 147,000, city proper), widely regarded as Africa's cleanest and safest capital, is located there. Walvis Bay, which South Africa ceded to Namibia in 1994, has a deep-water harbor and is the port for Namibia's seaborne trade.

Namibia's economy is tied closely to that of South Africa. South African, British, and U.S. mining companies export Namibian diamonds and various metals. Namibia has established an economic zone in the Atlantic stretching 200 miles offshore, in which fishing and fish processing based at Walvis Bay have become important industries. Large wildlife populations responding to dramatic fluctuations in wet and dry conditions make Etosha National Park an attraction for many international tourists who represent an important source of hard currency to Namibia. The young country is attempting to provide conditions favorable both to foreign investors and to Namibia's white community of business entrepreneurs and ranchers.

Botswana, Lesotho, and Swaziland

Three states bordering South Africa—Botswana (formerly Bechuanaland), Lesotho (formerly Basutoland), and Swaziland—were protectorates ("High Commission Territories") of Britain until the 1960s. Botswana and Lesotho gained independence in 1966, and Swaziland, the last absolute monarchy in southern Africa, became independent in 1968. Lesotho, a mountainous country containing the highest summits in the Drakensberg escarpment, is surrounded by South Africa, and Swaziland is nearly so, although it has a short frontier with Mozambique. Botswana, most of which is semidesert with wetlands in the north, is bordered on three sides by South Africa and Namibia and on the fourth side by Zimbabwe. With a wealth of wildlife attractions in the Okavango Delta and adjacent national parks, Botswana is a growing destination for international nature-loving tourists. Arguing that the World Heritage Site of the Okavango Delta would be damaged or destroyed, Botswana is strongly opposed to a Namibian plan to pipe water from the Namibian portion of the Okavango River southward to Windhoek.

Of the three countries, Swaziland has the most varied resources and the most diversified economy. Western Swaziland has mountains along the Drakensberg escarpment that receive heavy rainfall. Rivers that rise there and flow to the Indian Ocean supply irrigation projects in the lower and drier east. Mountain slopes planted with conifers form the basis for wood pulp and lumber industries. Mining is an important source of revenue, as is tourism—which features gambling casinos— and sugar production. Europeans using Swazi labor have developed most of the larger enterprises. Race relations have been better than in many African countries, and the economic benefits to Swaziland have been considerable.

Mining companies have been prospecting actively in Botswana and Lesotho; diamond yields in 1998 amounted to 71 percent of the total value of exports from Botswana. The black populations of these two countries raise livestock and a little grain, largely on a subsistence basis, and some Europeans own cattle ranches in eastern Botswana. Large numbers of African workers from all three countries migrate to jobs in South Africa, primarily in mining. All three countries are highly dependent on South Africa for trade and many kinds of services, and South Africa has established a customs union with the three and with Namibia.

South Africa occasionally exerts its power on behalf of its interests in these weaker countries. In 1998, for example, the prime minister of Lesotho perceived that a coup might be imminent in his nation, and called on South Africa for assistance. South African troops immediately invaded Lesotho, but rather than quelling the unrest, this intervention led to widespread condemnation of South Africa for meddling in the internal affairs of a sovereign nation.

Angola and Mozambique

Portugal played an active role as an imperial power in southern Africa from a very early time during the age of European colonial expansion. Portugal was, in fact, the earliest colonial power to build an African empire. The epic voyage of Vasco da Gama to India in 1497–1499 by the Cape route was the culmination of several decades of Portuguese exploration along Africa's western coasts. During the 16th century, Portugal controlled an extensive series of strong points and trading stations along both the Atlantic and Indian Ocean coasts of the continent. Later, as stronger powers outdistanced this earliest empire builder, Portugal managed to retain footholds along both coasts. By 1964, Portugal's principal African possessions, Angola (Portuguese West Africa) and Mozambique (Portuguese East Africa), were the most populous European colonies still remaining in the world. In both units, Portugal fielded a large military force to combat guerrilla opposition. In 1974, Portugal's resistance came to a breaking point, and an army coup in Portugal installed a regime in Lisbon that granted the two colonies independence.

Figure 18.16 Luanda, the capital and largest city of Portugal's former colony of Angola, exhibits the common pattern of cities in Africa south of the Sahara: a modest cluster of high-rise buildings in the city center surrounded by a sea of small one-story houses, often on dirt streets and generally roofed with metal sheeting. *Sarah Errington/Hutchison Library*

When the Portuguese withdrew from Angola in 1975, a Marxist-Leninist liberation movement supported by the Soviet Union and Cuba took over the central government in the seaport capital and largest city, Luanda (population: 1.46 million, city proper; Fig. 18.16). But this movement's authority was not recognized by other liberation movements, one of which—the National Union for the Total Independence of Angola (UNITA) —came to control large parts of the country. South Africa and the United States supplied aid to this anti-Marxist movement headed by Jonas Savimbi, a member of Angola's largest tribe. Large quantities of Soviet arms, thousands of Cuban troops, and many Soviet and East German military advisors supported the government in Luanda. Cuban contingents guarded vital oil fields, particularly in Angola's coastal exclave of Cabinda, separated from the rest of the country by the Congo River mouth (which is controlled by the Democratic Republic of Congo).

In 1988, representatives of Angola, Cuba, and South Africa signed an American-mediated agreement under which Cuba and South Africa would withdraw militarily from Angola, and South Africa would grant independence to Namibia. But even after the withdrawals, hostilities continued between UNITA and the government. In May 1991, Angola's warring factions signed a cease-fire accord, but it fell apart, and warfare continued until a peace agreement allowing UNITA's participation in all levels of government was reached in 1994. This agreement soon faltered, however, and while some UNITA members took up their legitimate government posts, others, including UNITA leader Savimbi, again took up arms, fighting especially hard to secure the oil fields of Cabinda, the country's main oil source.

Cabinda and other fields are scattered along the Angolan coast between the Congo mouth and Lobito. Crude oil and modest amounts of refined products make up nine-tenths of Angola's exports, and by 1996 Angolan oil comprised 7 percent of U.S. oil imports. In addition to oil and its products, Angola exports small amounts of diamonds, coffee, and other items. Diamonds are concentrated in the northeast where UNITA rebels have mined them for years to finance their antigovernment operations. A United Nations–sponsored embargo designed to cut off sales of UNITA-mined diamonds has failed, largely because of the ease of smuggling the gems and the difficulty of identifying their geographic origins once they reach the world market.

Portugal's other ex-colony, Mozambique, has the distinction of being the world's poorest country—but one with bright prospects. It consists of a coastal lowland rising to interior uplands and plateaus. The country has a tropical savanna climate, with a natural vegetation composed primarily of grasslands with scattered trees. The lower Zambezi River divides Mozambique into two fairly equal parts. There are two main port cities: Maputo (population: 1.07 million, city proper), the capital, in the extreme south, and Beira (population: 298,847, city proper), located about 125 miles (200 km) south of the Zambezi mouth. Under Portuguese rule, much of Mozambique's commercial economy was closely linked with the economies of nearby states, and the two ports had rail connections with adjoining landlocked countries (see Fig 17.3, p. 424)

This transportation net focusing on the Mozambican ports suffered severely from many years of guerrilla warfare, which first involved African insurgencies against the Portuguese and

468

then warfare by African dissidents against the Marxist government that took over from the Portuguese in Maputo. The highly destructive civil war—costing one million lives—between the Maputo government and the South African–backed rebel organization called RENAMO (Mozambique National Resistance) dragged on until October 1992 when a peace accord was signed and the United Nations sent a military peacekeeping force to Mozambique. Elections took place peacefully in October 1994, and a large share of the five million people displaced by the war began returning to their villages.

Despite its almost unimaginable poverty, Mozambique's future is promising. The government has been shifting away from Marxism-Leninism in favor of a more market-oriented economy and greater democracy. Transportation facilities in the "Beira Corridor" to Zimbabwe are being upgraded, and a new regional economic link with South Africa has been established. Known as the Maputo Development Corridor, this toll road joins Johannesburg's industrial and mining center with Maputo's deep-water port. Low-tax zones were created to attract new businesses to the corridor. The country's estimated 30 million land mines are being cleared, and war refugees are returning to farm this fertile land. By 1998 the country was self-sufficient in food production. Major exports are Mozambique's famous huge prawns, cashew nuts, cotton, tea, sugar, and timber. Multinational efforts are underway to process titanium for export and to develop the country's pristine beaches into world-class tourist destinations. Impressed with Mozambique's progress, foreign lenders have agreed to cancel about 80 percent of the country's debt.

Zimbabwe, Zambia, and Malawi

Prior to the post–World War II movement for African independence, the countries of Zimbabwe, Zambia, and Malawi (then known as Southern Rhodesia, Northern Rhodesia, and Nyasaland) were British colonies. In 1953, the three became linked politically in the Federation of Rhodesia and Nyasaland, or the Central African Federation. It had its own parliament and prime minister but did not achieve full independence. The racial policies and attitudes of Southern Rhodesia's white-controlled government created serious strains within the Federation helping to bring about its dissolution in 1963. Northern Rhodesia then achieved independence as the Republic of Zambia, and Nyasaland became the independent Republic of Malawi.

But the government of Southern Rhodesia, the main area of European settlement in the Federation, was unable to come to an agreement with Britain concerning Britain's demand for the political participation of the colony's African majority. In 1965, the government of Southern Rhodesia issued a Unilateral Declaration of Independence (UDI), and in 1970 it took a further step in separation from Britain by declaring itself to be Rhodesia, a republic that would no longer recognize the symbolic sovereignty of the British crown. Britain and the Commonwealth responded with a trade embargo. Negotiators for Britain

and the Rhodesian government unsuccessfully attempted to arrive at a formula to allow political rights and participation for the 95 percent of Rhodesia's population that was black and thus to achieve reconciliation and world recognition of Rhodesia's independence. Although the economic sanctions were ineffective, pressure on the white government increased in the mid-1970s with black guerrilla warfare against the government and the end of white (Portuguese) control in adjacent Mozambique. Independence for Rhodesia (renamed Zimbabwe) as a parliamentary state within the Commonwealth of Nations finally came in 1980.

This young country faces several ethnic dilemmas. First, the future course of relations between the country's two largest ethnic groups—the majority Shona and the Ndebele—remains uncertain; many Ndebele people want autonomy for the Ndebele stronghold in the south known as Matabeleland. Also, while most whites left Zimbabwe when it became independent, about 100,000 remained. These whites have been participating in Zimbabwe's political life, setting an early example to South Africa of an effective transition to majority black rule. But, while millions of rural blacks are crowded onto agriculturally inferior "communal lands," whites own about 70 percent of Zimbabwe's land—a situation that has periodically threatened to tear the country apart. Zimbabwe's president Robert Mugabe raised the prospect of strife in 1997 when he announced that the government would seize half of Zimbabwe's white-owned farms, "not pay a penny for the soil," and distribute the land to black peasants living in the overcrowded communal areas. Zimbabwe's stock market plummeted amidst fears of turmoil, and international lenders threatened to withdraw assistance to Zimbabwe. Mugabe responded both by insisting that there would be a carefully planned transfer of farmlands to black ownership, and by evicting black squatters from white-owned farms. But, under intense domestic pressure, Mugabe announced late in 1998 that more than 800 white-owned farms, representing one-quarter of the country's commercial agricultural productivity, would become government property. Insisting that the seizures violated its agreements with the government, the International Monetary Fund froze $55 million in aid to Zimbabwe. While spending an estimated $1 million daily to keep its military in the field in Congo (see Regional Perspective, p. 457), the country stood on the edge of crisis.

Zimbabwe, Zambia, and Malawi occupy highlands with a tropical savanna climate. Over nine-tenths of the annual rainfall comes in the six months from November to April. The natural vegetation is primarily open woodland, although tall-grass savanna predominates on the High Veld of Zimbabwe. The Zimbabwean High Veld, lying generally at 4000 feet (c. 1200 m) and higher, forms a broad divide between the drainage basins of the Zambezi and Limpopo rivers. Composed of a band of territory about 50 miles (80 km) broad by 400 miles (c. 650 km) long, oriented southwest-northeast across the center of Zimbabwe, the High Veld has been the main area of European settlement in the three countries. Most of Zimbabwe's whites live

Figure 18.17   The stripe running top to bottom across this photo is Zimbabwe's Great Dike, seen from space. About 300 miles (500 km) long and 6 miles (10 km) wide, the mineral-rich Great Dike was formed about 2.5 billion years ago when molten lava filled and widened a fracture in the Earth's crust. Note how faulting has offset the dike in two places. *NASA/Science Photo Library/Photo Researchers, Inc.*

Figure 18.18   On November 16, 1855, the Scottish missionary Dr. David Livingstone became the first non-African to view a waterfall known to Africans as Mosi Oa Tunya, "the Smoke That Thunders." "The falls are singularly formed," Livingstone wrote. "They are simply the whole mass of the Zambezi waters rushing into a fissure or rent made right across the bed of the river. The falls had never been seen before by European eyes, but scenes so lovely must have been gazed upon by angels in their flight." Victoria Falls—named by Livingstone for the Queen of England—were soon dubbed one of the "Seven Natural Wonders of the World." This is a view during the severe drought of 1992, when a trickle replaced what is normally a wall of water. The Zambezi River here forms the border between Zambia and Zimbabwe.   *Joseph J. Hobbs*

on it, including the 4500 farmers whose large holdings, worked by black laborers, produce most of the country's agricultural surplus to feed urban populations and to export. Tobacco grown on these white-owned farms is an important export. Other agricultural exports include cotton, cane sugar, corn, and roses.

The main axis of economic development in Zimbabwe lies along the railway connecting the largest city and capital, Harare (formerly Salisbury; population: 1.19 million, city proper), in the northeast with the second largest city, Bulawayo (population: 622,000, city proper), in the southwest. Along and near the axis are the principal areas of European farming and mines producing gold, nickel, asbestos, copper, chromium, and other minerals, primarily for export. This axis also corresponds with an extraordinary geological feature called the Great Dike, an intrusion of younger, mineral-rich material into older rock. Ranging in width from 2 to 7 miles (3 to 11 km), the Great Dike crosses the entire nation and is readily visible to astronauts in space as a giant stripe on the landscape (Fig. 18.17).

Iron ore mined at Que Que, about midway between Harare and Bulawayo, is used there to produce steel in southern Africa's only iron and steel plant. The Wankie coal field in western Zimbabwe supplies coal to smelt the ore.

Zimbabwe's manufacturing industries are more important, both in value of output and in diversity of production, than those of any other African nation south of the Sahara except Nigeria and South Africa. Despite this production, the country's exports are dominated overwhelmingly by unprocessed minerals, tobacco, cotton, and corn, and many manufactured goods must still be imported. Tourism is of growing importance today,

thanks to such attractions as the ruins of Great Zimbabwe, a wealth of animal life and other natural marvels, and good air and road links.

Below the spectacular Victoria Falls (regarded traditionally as one of the seven "Natural Wonders of the World," and another significant international tourist destination today; Fig. 18.18), the middle course of the Zambezi River separates Zambia from Zimbabwe. Hydropower from the Kariba Dam (completed in 1962; see Fig. 17.5, p. 425) on this stretch of the Zambezi flows to both countries. Zambia is much poorer than its southern neighbor, and much of it is wilderness with a sparse human population. As in Zimbabwe, the population lives mainly on a subsistence basis, but in Zambia, agriculture and cattle-keeping are less productive than in Zimbabwe.

Zambia's economy depends heavily on copper exported from several large mines developed with British and American capital. An urban area has developed at each mine, and the mining area, known as the Copperbelt, consists of a series of separate population nodes strung close to Zambia's frontier with the adjacent copper-mining Shaba region of southeastern Democratic Republic of Congo. The Copperbelt, which contains the majority of Zambia's manufacturing plants and copper mines, is largely an urban region with a high proportion of the country's small European population. Copper, most of which is shipped in refined form, composed 86 percent of Zambia's exports in 1998. Other mineral production includes cobalt, a byproduct of copper mining; zinc and lead mined along the railway between the Copperbelt and Zambia's capital and largest city, Lusaka (population: 982,362, city proper); and coal production from extreme southern Zambia.

Malawi is a long, narrow, densely populated country along the western and southern margins of Lake Malawi. Its population is almost entirely black African and is even more rural and agricultural than that of most African states. Tobacco is the main item in Malawi's export trade, with tea and cane sugar also of some importance. There is little mineral wealth, and manufacturing industries are meager. The country has long supplied many migrant laborers to mines in Zimbabwe and South Africa, and their earnings are an important adjunct to the internal economy of this poor country.

18.7 Indian Ocean Islands

The islands and island groups off the Indian Ocean coast of Africa are unique in their cultures and natural histories. Madagascar, the Comoro Islands, Réunion, Mauritius, and the Seychelles exhibit African, Asian, Arab, European, and even Polynesian ethnic and cultural influences (Fig. 18.19). As island ecosystems, they are home to many endemic plant and animal species.

Madagascar and the smaller Comoro Islands are former colonies of France, while Réunion remains an overseas possession of France. Mauritius became independent of Britain in 1968 as a constitutional monarchy under the British crown. The Seychelles, which gained independence in 1976, is a republic within the Commonwealth of Nations.

Madagascar (in French, "La Grande Ile") is the fourth largest island in the world, nearly 1000 miles (c. 1600 km) long and about 350 miles (c. 560 km) wide. It lies off the southeast coast of Africa and has geological formations similar to those of the African mainland. Its distinctive flora and fauna include most of the world's lemur and chameleon species (see Definitions and Insights, p. 472). Some of Madagascar's early human inhabitants migrated from the southwestern Pacific region, bringing with them the cultivation of irrigated rice and other culture traits. There was also a later influx of Africans from the

Figure 18.19 The 14 million people of Madagascar have a unique culture, with roots in both Africa and the Malayo-Polynesian realm. *Joseph J. Hobbs*

mainland. The official language is Malagasy, a Malayo-Polynesian tongue that has acquired many French, Arabic, and Swahili words.

The east coast of Madagascar rises steeply from the Indian Ocean to heights of over 6000 feet (1829 m). Because the island lies in the path of trade winds blowing across the Indian Ocean, the east side receives the heaviest rain and has a natural vegetation of tropical rain forest. The remainder of the island has a vegetation consisting primarily of savanna grasses with scattered woody growth. An eastern coastal strip—low, flat, and sandy—is backed by hills and then by escarpments of the central highlands. Paddy rice is the principal food crop of the coastal zone, and coffee, vanilla, cloves, and sugar are grown for export. The most densely populated part of the island is the central highlands, where the economy is a combination of rice

Definitions & Insights

MADAGASCAR AND THE THEORY OF ISLAND BIOGEOGRAPHY

In a recent issue of the journal *Scientific American*, Harvard University entomologist Edward Wilson and Stanford University biologist Paul Ehrlich asserted that if tropical rain forests continue to be cut down at the present rate, a quarter of all of the plant and animal species on Earth will become extinct within the next 50 years.[2] They based their estimate on a model which correlates habitat area with the number of species living in the habitat. This **theory of island biogeography** emerged from observations of island ecosystems in the West Indies and states that the number of species found on an individual island corresponds with the island's area, with a tenfold increase in area normally resulting in a doubling of the number of species. If island "A," for example, is 10 square miles in area and has 50 species, 100-square-mile island "B" may be expected to support 100 species.

What makes the theory useful in projecting species losses is the inverse of this equation: that a tenfold diminution in area will result in a halving of the number of species. Therefore, one-square-mile island "C" can be expected to hold just 25 species. In applying the model, ecologists treat habitat areas as if they were islands. So, if people cut down 90 percent of the tropical rain forest of the Amazon Basin, for example, the theory of island biogeography suggests they would eliminate half of the species of that ecosystem. Scientists caution that the theory is only a tool meant to help in making rough estimates; the actual number of species lost with habitat removal may be higher or lower.

As a rough guideline, the theory of island biogeography is useful in projecting and attempting to slow the rate of extinction in the world's biodiversity "hot spots," such as Madagascar. Over 90 percent of Madagascar's plant and animal species are endemic, occurring nowhere else on Earth. Extinction of species was well underway soon after people arrived on the island; the giant, flightless elephant bird *(Aepyornis)* was among the early casualties. But human activities, particularly the clearing of forests to grow rice and provide pasture for zebu cattle, are eliminating habitat areas on the island at a more rapid rate than ever before. Meanwhile, scientists are anxious to learn whether some of Madagascar's remaining plants might be useful in fighting diseases such as AIDS and cancer. Already Madagascar's rosy periwinkle has yielded compounds effective against Hodgkin's disease and lymphocytic leukemia. Other species like the rosy periwinkle could become extinct before their useful properties even become known.

How urgent is the task to study and attempt to protect plant and animal species in Madagascar? Scientists turn to the theory of island biogeography for an answer. Although people have lived on Madagascar for less than 2000 years, they have succeeded in removing 90 percent of the island's forest, setting the stage for some of the most ruinous erosion seen anywhere on Earth (Fig. 18.C). The theory of island biogeography suggests that in the process, they have caused the extinction of roughly half of the island's species. With Madagascar's human population expected to double in 23 years, and with pressure on the island's remaining wild habitats expected to increase accordingly, the task of conservation is extremely urgent.

Figure 18.C Although people have lived in Madagascar for less than 2000 years, it is now Earth's most eroded country. Since arriving on the island, people have cleared and burned its forest to cultivate crops and provide pastures for zebu cattle. As heavy rains strip the topsoil, people create new fields and pastures from the forest, perpetuating the destruction. In this aerial view of the north-central highlands, the rivers run red with the country's lost topsoil. "Madagascar is bleeding to death," one observer commented. *Joseph J. Hobbs*

[2]Edward O. Wilson, 1989. "Threats to Biodiversity." *Scientific American* 261(3): 60–66.

growing in valleys and cattle raising on higher lands. Antananarivo (formerly Tananarive; population: 1.25 million, city proper), the capital and largest city, is located in the central highlands. It is connected by rail with Toamasina (formerly Tamatave; 127,441, city proper), the main seaport, located on the east coast.

Madagascar has experienced a much smaller degree of economic development than such large tropical islands as Java, Taiwan, and Sri Lanka. Deep economic troubles and political dissension have marked its recent history under a socialist government. Unwise decision making made Madagascar fertile ground in the 1990s for international drug-money launderers,

Figure 18.20   Women of the Comoros.   *Joseph J. Hobbs*

loan sharks, and con artists who purloined much of this poor nation's financial assets.

The Comoro Islands, located about midway between northern Madagascar and northern Mozambique, are of volcanic origin. They exhibit a complex mixture of African, Arab, Malayan, and European influences (Fig. 18.20). Aside from the island of Mayotte, which has a Christian majority and which, by its own choice, remains a dependency of France, the Comoro group forms the Federal Islamic Republic of the Comoros (short form: Comoros), in which Islam is the official religion and both Arabic and French are official languages. The capital is Moroni (population: 20,000), on the island of Grand Comore. France has a defense treaty with the Comoros, which it invoked in 1995 to put down a coup led by a French mercenary. In 1997, two more Comoros islands (Mohéli and Anjouan) declared secession from the national government in Moroni and pleaded to become French dependencies—on the hope they could enjoy the somewhat higher standard of living prevailing on nearby Mayotte. The islands' futures remained uncertain as of early 1999.

Limited fishing, subsistence agriculture, and exports of vanilla provide livelihoods to the 600,000 inhabitants of the these islands. The coelacanth, a primitive fish thought to have become extinct soon after the end of the Mesozoic Era (the age of dinosaurs), has been discovered alive in Comoro Islands waters since 1938.

Réunion, east of Madagascar, is the tropical island home to 700,000 people, mainly of French and African origin. The island was uninhabited prior to the arrival of the French in the 17th century. Elevations reach 10,000 feet (3048 m). The volcanic soils are fertile, and tropical crops of great variety are grown. Cane sugar from plantations supplies 63 percent of all exports by value, with rum and molasses (derived from cane) supplying an additional 4 percent. Réunion also has a significant international tourist industry.

Sugar is the mainstay of the larger island of Mauritius, which lies nearby. Sugarcane, grown mainly on plantations, oc-

cupies most of the cultivated fields on this volcanic landscape. Mauritius has a larger proportion of lowland area than Réunion. Its population is composed mainly of descendants of Indians brought in by the British to work the cane fields, and also includes descendants of 18th-century French planters as well as some blacks, Chinese, and racial mixtures. The island has developed a clothing industry which supplied more export value than sugar in 1993 (year of latest available figures; clothing was 55 percent of all exports; sugar, 24 percent). International tourism also produces valuable revenues (about $300 million annually) for its 1.2 million people.

The unusual core of the Seychelles island group is composed of granites, which typically are found only on continental mainlands. The exceptional physical beauty of the islands, and the renowned friendliness of its people of French, Mauritian, and African descent, have made the Seychelles a famous tourist destination among those wealthy enough to afford travel to this remote area. The tourist boom began when an airport was built in 1971.

Many food requirements have to be imported to the islands. The country's largest exports consist of petroleum products, although the islands have no crude petroleum and no refining industry; the products are simply imported and then reexported. Fish products are the only other important exports. The Aldabra Atoll, far to the south of the main group of islands, is home to a population of giant tortoises and other endemic animals (Fig. 18.21).

As a group, the Seychelles are a fitting symbol of the plight and potential of the entire region of Africa south of the Sahara. They are culturally diverse, have an abundance of some resources, and utterly lack other vital assets. Outsiders are drawn to the often imaginary Eden evoked by both the Seychelles and Africa, but would do well to make a stronger effort to understand the even more marvelous facts of land and life in the region.

Figure 18.21   A giant Aldabran tortoise on the beach of Aldabra Atoll, in the Seychelles. As in the Galapagos Islands, unique life-forms have evolved through long periods of isolation from other animal populations. *Joseph J. Hobbs*

SUMMARY WITH SELECTED KEY TERMS

- Comprised of a great **volcanic plateau** and adjacent areas, **the Horn of Africa** includes the countries of **Ethiopia, Eritrea, Somalia, and Djibouti.**
- After nearly thirty years of **terrorism, guerilla actions, and civil war,** the former Ethiopian province of Eritrea became independent in May 1993.
- In addition to the devastation wrought by war with Eritrea, **droughts** centering in its war-torn north have stricken Ethiopia since the late 1960s. In 1984 and 1985 an estimated **one million people perished** when the drought reached the point of widespread catastrophe.
- The **Sahel** region extends eastward from the **Cape Verde Islands** to the Atlantic shore nations of **Mauritania, Senegal and The Gambia and inland to Mali, Burkina Faso, Niger, and Chad.** The name Sahel in Arabic means **"coast" or "shore,"** referring to the region as a front on the great desert "sea" of the **Sahara.**
- Since the late 1960s, the Sahel has been subjected to severe **droughts,** which, in combination with increased **human pressure on resources,** has prompted a process of **desertification.**
- **West Africa** is defined as the region extending from **Guinea-Bissau eastward to Nigeria.** Most West African countries maintain **strong economic, political, and cultural relationships** with the European powers that formerly controlled them.
- The West African countries' heavy dependence on **subsistence crop farming, livestock grazing, and a few agricultural and/or export specialties** is the customary economic pattern of the world's **LDCs.**
- Although a number of prominent **mineral products** are mined in West Africa, the most spectacular and significant mining development has been **Nigeria's** emergence as a **producer and major exporter of oil.**
- West Africa's principal cities are also **national capitals and seaports,** but as the region has only minor manufacturing, its cities have developed primarily in its **areas of major commercial agriculture.**
- The subregion of **West Central Africa** is flanked by **Cameroon and the Central African Republic to the north and the Democratic Republic of Congo to the south.**
- **Subsistence agriculture** supports people in most areas of West Central Africa. In addition, exports of **cash crops and minerals** aid the financing of some countries.

- Originally a colony of Belgium, the **Democratic Republic of Congo** has faced a number of challenges since its independence in 1960. In addition to **civil war, the unstable economy** of the country has cast the Democratic Republic of Congo into a state of **extreme poverty** despite its abundant natural resources.
- **East Africa** includes the countries of **Kenya, Tanzania, Uganda, Rwanda, and Burundi.**
- A wide array of crops are grown by the **subsistence agriculturalists** of East Africa. In addition, several **export crops,** such as coffee and tea, provide a source of income for the region's inhabitants. In comparison with many other African countries, however, the East African nations **lack mineral reserves and mineral production.**
- **Ethnic rivalries and conflicts** have beset all of the East African countries. Warfare between the **Hutus and Tutsis** of Rwanda and Burundi serves as a recent example of the tension and violence that can occur between the many ethnic groups of the region.
- **Southern Africa** includes the countries of **Angola, Mozambique, Zimbabwe, Zambia, Malawi, South Africa, Botswana, Swaziland, Lesotho, and Namibia.**
- The dominant power in Southern Africa is the country of **South Africa.** Its economic power, which is based on **diversified industries and its mineral wealth,** is now comparable to that of the rest of Africa combined.
- **Racial segregation** characterized South African life from 1652 onward, but it was first systematized after 1948 under a comprehensive body of national laws supporting the official governmental policy commonly known as **apartheid.**
- In April 1994, **Nelson Mandela** was elected president of South Africa after an all-race election was held in the country. After years of conflict, this historic election **ended white European political control** in the last bastion of European colonialism on the African continent. The **apartheid laws themselves became null and void.**
- The **Indian Ocean islands** of **Madagascar, the Comoro Islands, Réunion, Mauritius, and the Seychelles** exhibit African, Asian, Arab, European, and even Polynesian ethnic and cultural influences.

REVIEW QUESTIONS

1. Using maps and the text, identify the four countries of the Horn of Africa.
2. Describe the ethnic and cultural diversity of Ethiopia's population.
3. Using maps and the text, identify the eight countries of the Sahel.
4. In what ways is the Sahel both a physical and cultural boundary in Africa?
5. Using maps and the text, identify the nine countries of West Africa.
6. Describe the inadequacies of West African transportation.
7. Using maps and the text, identify the seven countries of West Central Africa.

8. List the notable natural resource assets of the Democratic Republic of Congo.
9. Using maps and the text, identify the five countries of East Africa.
10. Describe the ethnic diversity of East Africa.
11. What was the original distinction between the Hutus and the Tutsis?
12. Using maps and the text, identify the ten countries of Southern Africa.
13. What were the main causes of the Anglo-Boer War of 1899?
14. List the notable natural resource assets of South Africa.

DISCUSSION QUESTIONS

1. What motivated Eritrea to attack Ethiopia in 1998?
2. What was the purpose of "Operation Restore Hope"? What was the outcome of the United States' intervention in Somalia?
3. What is desertification? What are the major causes of desertification?
4. Discuss the major complaints raised by the Ogoni and Ijaw tribal groups concerning oil development in Nigeria.
5. Discuss the major reasons for the Democratic Republic of Congo's present economic problems.
6. What African countries became involved in the Democratic Republic of Congo's recent civil war? What reasons motivated each of these countries to enter into the conflict?
7. How has the Great Rift Valley affected the physical environment of East Africa?
8. What factors led to South Africa's advanced economic development as compared to the rest of Africa?
9. Discuss the major policies of the former apartheid laws of South Africa. What was the purpose of homelands?
10. What is the theory of island biogeography? Why might the theory be useful in understanding the relationship between habitat loss and species loss?

8 Latin America

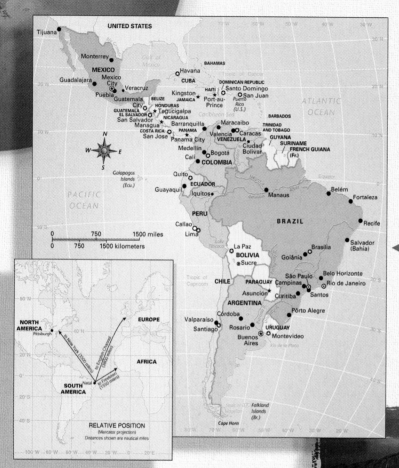

Settlement patterns in Latin America have long circled around the rim of Middle and South America and thrived in major upland centers. This so-called "rimland" and "mainland" demographic pattern has provided a distinctive geography to all of Latin America.

Latin America, in both Middle America and South America, has grand ceremonial landscapes that bespeak powerful cultural development prior to the arrival of Columbus. Tradition suggests that the peoples of Latin America had considerable sophistication in their architecture, cities, economic networks, and military control systems. *Robert Fried / Tom Stack & Associates*

Spectacular natural landscape features blended with modern city landscapes and urban amenities characterize many tourist destinations in Latin America. These locales are not only economically significant, but they lead to powerful images both of such places and the people who visit them. *Brissand–Figaro / Gamma Liaison*

Regional Snapshot

Illegal migration to the United States plays a role in the steady migration from Latin America to the U.S. and Canada. The migrants may come in illegally through the 2000 mile Mexican border with the U.S. or they may utilize formal quota systems that attempt to regulate such migration. *Alon Reininger / Contact Press Images*

In major sections of Middle America and the tropics of South America the burning of forest resources for short-term farming and cattle grazing is having a considerable impact on the envronment of Latin America. Forest lands once burned never again return to their natural richness or biotic diversity.
Paul Edmonson

The maquiladora movement has promoted Mexican development of manufacturing firms that capitalize on proximity to the U.S. border and major markets of the U.S. and Canada, and the eagerness of Latin American youth to make themselves more marketable to American firms. This influence spreads far south of Mexico. *Gabriel Corian / The Image Bank*

Chapter 19

A Geographic Profile of Latin America

▲ One of the most dramatic landscapes of South America is the remains of Machu Picchu—known also as the "Lost City of the Incas." This highland (8000 ft [c. 2400 m]) city and ceremonial center of the Incas is atop the Urubamba Valley in Peru and serves both as a major tourist destination and an archaeologic treasure of major consequence. It became known to the West in 1911.   Robert Fried/Tom Stack & Associates

CHAPTER OUTLINE

19.1 Definition and Basic Magnitudes

19.2 Latin America: A Rich Diversity of Geographic Habitats

19.3 Population Geography: Rimland-Mainland Pattern, Diverse Stocks, and Expanding Cities

19.4 "Emerging" or "Developing" Nations: Characteristics and Frustrations

*t*he land portion of the Western Hemisphere south and southeast of the United States is commonly known as Latin America. This region has a close and long-term relationship with the United States. The name reflects the importance of culture traits inherited primarily from the Latin European colonizing nations of Spain and Portugal (Fig. 19.1). Spanish is the major language, while Portuguese is the language only of the region's largest country, Brazil. Dialects of French are spoken in Haiti and in France's dependencies of French Guiana, Guadeloupe, and Martinique. Some small units in the Caribbean Sea or on the nearby mainland do not exhibit Latin culture except in minor ways. They speak mainly English or Dutch, and some are still dependencies of the United Kingdom, the United States, or the Netherlands (Table 19.1). Their presence in an area of primarily early Spanish control reflects early colonial struggles among European nations for territories suitable for sugarcane plantations. Important precolonial influences in some Latin American countries are manifest in native Indian languages, customs, and costumes. But Latin influences are predominant today in the region as a whole and can be seen not only in Latin-derived (Romance) languages but also in the dominance of the Roman Catholicism that arrived with the first ships from Latin Europe. This religion has been one of the most powerful influences diffused to the so-called New World in the centuries that followed the 1492 arrival of Columbus and the subsequent drive for "Glory, God, and Gold" that led to literally centuries of exploration, colonization, and settlement. However, language and religion are merely the most obvious cultural importations, which also included such major elements as land tenure arrangements, governmental practices, legal systems, social structures, and economic systems.[1]

European cultures have been transplanted to Latin America with vital modifications. Underneath a veneer of sameness—

promoted by the widespread use of the Spanish language—important distinctions exist between regions, between countries, and within countries. Among the many contributing factors are diverse pre-European and colonial experiences, different resource bases, divergent political systems, and differential rates of development. The result is that Latin America presents many faces to the world, including:

1. extraordinary variety in environmental settings;
2. population groups of considerable diversity;
3. local economies that run the scale from technologically advanced to extremely underdeveloped; and that range from being very closely linked to North America to being largely self-sufficient at the local level;
4. diverse governmental philosophies in political control and structures, characterized by relatively frequent shifts;
5. countries that have distinctive personalities despite enough overall similarity and cohesion to fit together reasonably well in an identifiable major world region; and

Figure 19.1 Roman Catholicism is the leading religion in all Latin American countries except some relatively small units colonized by countries other than Spain or Portugal. The old Cathedral of Lima, Peru, is one of the more majestic among great numbers of Catholic Church buildings spread through Latin America from Mexico and Cuba to Argentina and Chile. *Guilio J. Barbero*

[1] The authors wish to thank Brian K. Long for his valuable assistance in the initial writing of Chapters 19 and 20 in the 1995 edition. These chapters also contain some material written originally by Dr. Richard S. Thoman but updated here and presented in a revised format.

Table 19.1 Latin America: Basic Data

Political Unit	Area (thousand/ sq mi)	Area (thousand/ sq km)	Estimated Population (millions)	Estimated Annual Rate of Increase (%)	Estimated Population Density (sq mi)	Infant Mortality Rate	Urban Population (%)	Arable Land (% of total area)	Per Capita GNP ($US)
Middle America									
Mexico	736.9	1908.6	97.5	2.2	132	28	74	12	3670
Guatemala	42	108.9	11.6	3.1	277	51	38	12	1470
Belize	8.9	23	0.2	2.7	27	34	51	2	2700
El Salvador	8.1	21	5.8	2.5	724	41	50	27	1700
Honduras	43.2	111.9	5.9	2.8	137	42	44	14	660
Nicaragua	46.9	121.5	4.8	3.2	102	46	63	9	380
Costa Rica	19.7	51.1	3.5	1.9	179	11.8	44	6	2640
Panama	28.7	74.3	2.8	1.8	96	22	55	6	3080
Cuba	42.4	109.8	11.1	0.6	262	7.2	74	23	1370
Dominican Republic	18.7	48.4	8.3	2.1	446	47	62	23	1600
Haiti	10.7	27.8	7.5	2.1	708	74	33	20	310
Jamaica	4.2	11	2.6	1.8	613	16	50	19	1600
Trinidad and Tobago	2	5.1	1.3	0.8	646	17.1	65	14	3870
Bahamas	3.9	10.1	0.3	1.7	76	19	86	1	3865
Barbados	0.2	0.4	0.3	0.5	1596	14.2	38	77	7000
Antigua and Barbuda	0.2	0.4	0.1	1.2	394	18	36	18	7330
Dominica	0.3	0.8	0.1	1.1	256	16.2	*	9	2100
Grenada	0.1	0.3	0.1	2.3	731	12	32	15	2880
St. Christopher and Nevis	0.1	0.3	0.04	1	302	25	43	22	5870
St. Lucia	0.2	0.6	0.1	1.9	628	18	48	8	3500
St. Vincent and the Grenadines	0.1	0.3	0.1	1.6	797	19	25	38	2370
Puerto Rico	3.5	9.1	3.9	1	1129	11.5	71	8	6200
Guadeloupe	0.7	1.8	0.4	1.2	673	7.9	*	18	4700
Martinique	0.4	1.1	0.4	0.9	982	6	*	10	6000
Netherlands Antilles	0.3	0.8	0.2	1.2	690	11.5	*	8	8700
Total	**1022.4**	**2648.4**	**168.9**	**2.1**	**165.2**	**24.6**	**65**	**12.4**	**2867**

6. a mainland region that extends from Mexico south to Argentina, and an island world that is also called Caribbean Middle America.

19.1 Definition and Basic Magnitudes

With a land area of slightly more than 7.9 million square miles (over 20.5 million sq km), Latin America—the combination of Middle and South America—is surpassed in size by Africa,

Russia and the Near Abroad, North America (including Greenland), and Asia (see Table 1.2). However, its maximum latitudinal extent of more than 85 degrees, or nearly 5900 miles (c. 9500 km), is greater than that of any other major world region, and its maximum east-west measurement, amounting to more than 82 degrees of longitude, is also impressive. Yet Latin America is not so large as these figures might suggest, for its two main parts are offset from each other (Fig. 19.2). The northern part, known as Middle America (see Chapter 20)—which includes Mexico, Central America, and the islands of the Caribbean—trends sharply northwest from the north-south ori-

Table 19.1　Latin America: Basic Data *(continued)*

Political Unit	Area (thousand/ sq mi)	Area (thousand/ sq km)	Estimated Population (millions)	Estimated Annual Rate of Increase (%)	Estimated Population Density (sq mi)	Infant Mortality Rate	Urban Population (%)	Arable Land (% of total area)	Per Capita GNP ($US)
South America									
Brazil	3265.1	8456.6	162.1	1.4	50	43	76	7	4400
Colombia	401	1038.6	38.6	2.1	96	28	71	4	2140
Venezuela	340.6	882.2	23.3	2.1	68	20.9	86	3	3020
Ecuador	106.9	2768.7	12.2	2.2	114	40	61	6	1500
Peru	494.2	1280	26.1	2.2	53	43	71	3	2420
Bolivia	418.7	1084.4	8	2.6	19	75	58	3	830
Argentina	1056.6	2736.6	36.1	1.1	34	22.2	87	9	8380
Chile	289.1	748.8	14.8	1.4	51	11.1	85	7	4860
Uruguay	68	176.2	3.2	0.8	47	19.6	90	8	5760
Paraguay	153.4	397.3	5.2	2.7	34	27	52	20	1850
Guyana	76	196.8	0.7	1.7	9	63	36	3	690
French Guiana	34	88.1	0.2	2.5	5	14	*	0	4390
Suriname	60.2	155.9	0.4	1.8	7	29	70	0	1000
Total	**6763.8**	**10,377.4**	**330.9**	**1.6**	**48.9**	**33.5**	**76**	**6.5**	**4106**
Summary Total	**7786.2**	**13,025.8**	**499.8**	**1.8**	**64.2**	**29.1**	**72.3**	**7.3**	**3687**

Sources: *World Population Data Sheet,* 1998. United Nations Statistics Division, 1998. *Almanac of Politics and Government,* 1998. *World Factbook,* 1998.

entation of the continent of South America. The latter is thrust much farther into the Atlantic Ocean than is the Caribbean realm or Latin America's northern neighbor, North America. In fact, the meridian of 80°W, which intersects the west coast of South America in Ecuador and Peru, passes through Pittsburgh, Pennsylvania. Brazil lies less than 2000 miles (under 3200 km) west of Africa (see inset, Fig. 19.2).

19.2 Latin America: A Rich Diversity of Geographic Habitats

Topographic and Climatic Variations

Latin America is characterized by pronounced differences in elevation and topography from one area to another (Fig. 19.3). Low-lying plains drained by the Orinoco, Amazon, and Paraná-Paraguay river systems dominate the north and central part of South America and separate older, lower highlands in the east from the rugged Andes of the west. In Mexico, a high interior plateau broken into many basins lies between north-south trending arms of the mountains called the Sierra Madre. High mountains within Latin America—largely contained within the

Sierra Madre and the Andes—form a nearly continuous line from northern Mexico and the southern United States to Tierra del Fuego at the southern tip of South America. Most of the smaller islands of the West Indies are volcanic mountains, although some islands made of limestone or coral are lower and flatter. The largest islands have a more diverse topography, including low mountains.

Humid Lowland Climates

Most Latin American lowlands are associated with a dominant form of natural vegetation or ecosystem. The tropical rain forest climate, with its heavy year-round rainfall, monotonous heat and humidity, and associated superabundant vegetation dominated by large broadleaf evergreen trees, lies primarily along or very near the equator, although segments extend to the margin of the tropics in both the Northern and Southern hemispheres. The largest segment lies in the basin of the Amazon River system. Additional areas are found in southeastern Brazil, eastern Panama, the western coastal plain of Colombia, on the Caribbean side of Central America and southern Mexico, and along the eastern (windward) shores of some Caribbean islands, particularly Hispaniola and Puerto Rico (Fig. 19.4).

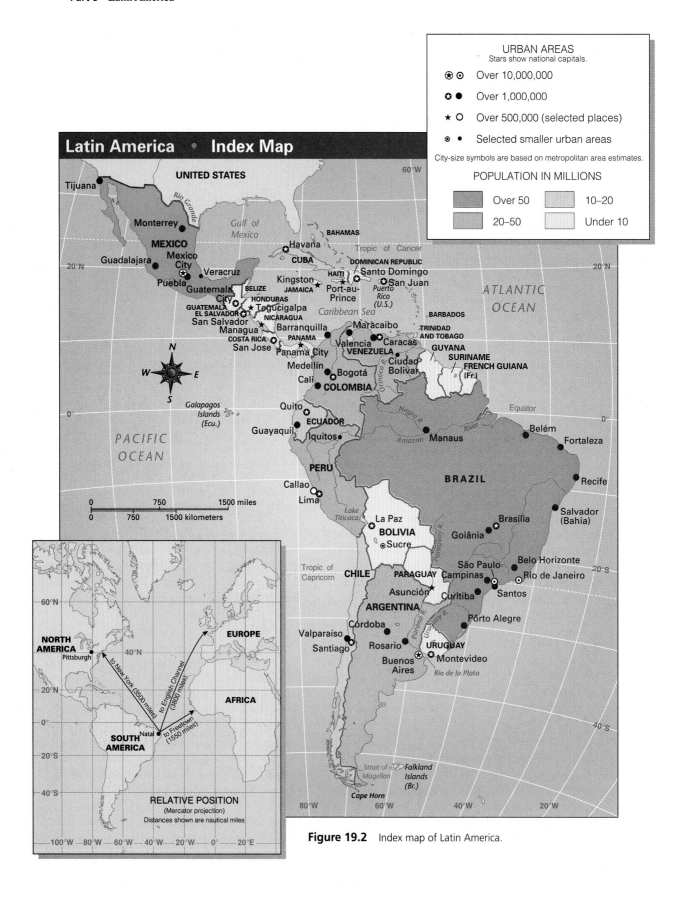

Figure 19.2 Index map of Latin America.

Figure 19.3 A map of physical features in Latin America.

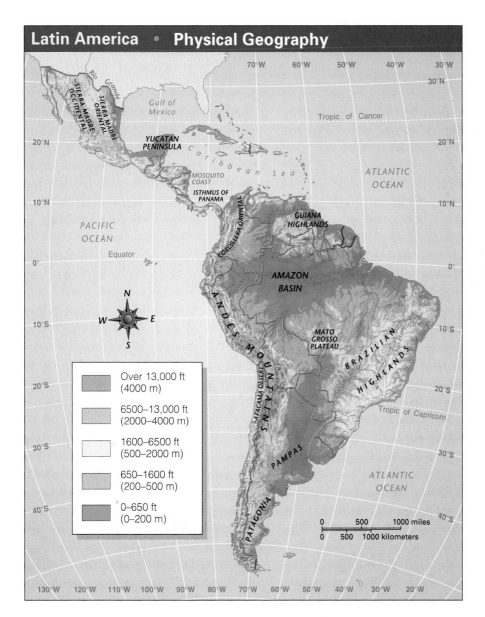

On either side of the principal region of tropical rain forest climate, the tropical savanna climate extends to the vicinity of the Tropic of Capricorn in the Southern Hemisphere and, more discontinuously, to the Tropic of Cancer in the Northern Hemisphere. In this climate zone, the average annual precipitation decreases and becomes more seasonal as one approaches the poles. The mean temperature decreases and the broadleaf evergreen trees of the rain forest grade into tall savanna grasses, woodlands, or deciduous forests that lose their leaves in the dry season.

Still farther poleward in the eastern portion of South America lies a sizable area of humid subtropical climate, which has cool winters unknown to the tropical zones. Its Northern Hemisphere counterpart is north of the Mexican border in the southeastern United States. In South America, this climate is associated principally with a natural vegetation of prairie grasses in the humid pampas of Argentina, Uruguay, and extreme southern Brazil. On the Pacific side of South America, a small strip of mediterranean or dry-summer subtropical climate in central Chile is similar to that in southern California.

Dry Climates and the Factors That Produce Them

The humid climates of Latin America have a more or less orderly and repetitious spatial arrangement. One may expect to find similar climates in generally similar positions on all major land masses of the world. But the region's dry climates—desert and steppe—are often due, at least in part, to local circumstances such as the presence of high landforms that block moisture-bearing

Figure 19.4   A map of climate patterns in Latin America.

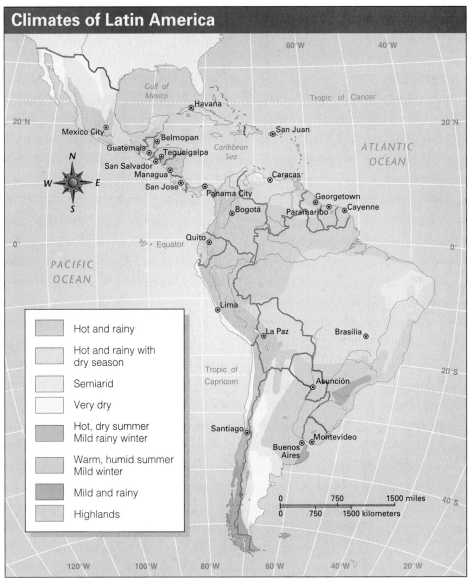

Source: Rand McNally. *Atlas of World Geography.*

air masses or the presence of offshore cool ocean water. In northern Mexico, such climates are partly associated with the global pattern of semipermanent zones of high pressure which create arid conditions in many areas of the world along the Tropics of Cancer and Capricorn. However, the aridity is also due in part to the rain-shadow effect caused by the presence of high mountain ranges on either side of the Mexican plateau. In Argentina, the extreme height and continuity of the Andean mountain wall are largely responsible for the aridity of large areas, particularly the southern region called Patagonia. Here, the Andes block the path of the prevailing westerly winds, creating heavy orographic precipitation on the Chilean side of the border but leaving Patagonia in rain shadow. But in the west coast tropics and subtropics of South America, the Atacama and Pe-

ruvian deserts cannot be explained so simply. Here, shifting winds that parallel the coast, cold offshore currents, and other complexities, as well as the Andes Mountains, are important conditions that combine to create the world's driest area. However, the mountains serve to restrict this area of desert to the coastal strip. Arid and semiarid conditions also are experienced along the northernmost coastal regions of Colombia and Venezuela, and in the Sertão of northeastern Brazil.

Mountains play such a major role in the development of climate patterns and overall demographic patterns of Latin America that a whole pattern of regionally unique zonation has been developed for the region. It is built around **altitudinal zonation** and is characterized by four dominant regional zones: the *tierra caliente, tierra templada, tierra fría,* and *tierra helada.*

19.3 Population Geography: Rimland-Mainland Pattern, Diverse Stocks, and Expanding Cities

In mid-1999, the world was approaching nearly 6 billion people, of whom Latin Americans totaled approximately 500 million, or almost 8.4 percent of the world's population. Because Latin America is a large and diverse region with well over twice the land area of the United States, it is not surprising that its population is unevenly distributed and exhibits enormous variations in density. There is, however, some overall order in the fact that most of Latin America's people are packed into two major geographic alignments. The larger of these two areas—called the rim—is a discontinuous ring around the margins of South America. The second—a relative highland—extends along a volcanic belt from central Mexico southward into Central America. The two alignments are often defined as the **Rimland** and **Mainland**—meaning the coastal regions in contrast to the interior and upland regions—pattern of Latin American population distribution and are outlined reasonably well by the pattern of population shown in Figure 2.15.

Major Populated Areas

The South American "rim" of clustered population contains roughly two-thirds of Latin America's people. Two major segments of the rim can be discerned. One—much the larger in population and area—extends along the eastern margin of the continent from the mouth of the Amazon River in Brazil southward to the humid pampa (subtropical grassland) around Buenos Aires, Argentina. The second segment, located partly on the coast and partly in the high valleys and plateaus of the adjacent Andes Mountains, stretches around the north end and down the west side of South America. This crescent begins in the vicinity of Caracas, Venezuela, on the Atlantic coast and arches around to the vicinity of Santiago, Chile, on the Pacific coast. This second segment is more fragmented than the first, as it is broken in many places by steep Andean slopes or coastal desert. For example, a strip of hot, rainy coast lies between the Amazon mouth and Caracas in which the thinly populated interior of South America extends to the ocean. In the far south, the population rim is again broken in the rugged, rainswept southern Andes and the dry lands of Argentina's Patagonia that lie in the Andean rain shadow on the Atlantic side of the continent. These latter territories have only a scanty human population, although millions of sheep graze in Patagonia and adjacent Tierra del Fuego (Fig. 19.5).

It is thought that few native Indians lived in the Rimland along the Latin American coasts before the European conquest. Major population centers were located in the Andes in South America, and the coasts were only lightly inhabited by migratory native peoples. The high intermontane areas within the Andes already exhibited a pattern of settled communities, ruled by

Figure 19.5 This map shows clearly the power of the rimland in the distribution of Latin American population. It highlights the importance of upland locations as well, particularly in the northwest sector of South America. The many open, sparsely settled regions continue to be looked at by government planners as possible frontiers for new agricultural development and human settlement.

the Inca Empire, approximately parallel to contemporary demographic patterns (Fig. 19.6). But when the Europeans began their quest for and settlement of South America in the 1490s, the coasts largely lay open to whatever settlement pattern the newcomers might devise. Because the Europeans approached from the sea at many separate points, the development of ports as bases for subsequent penetration of the interior was the first major settlement task. Many early ports eventually grew into sizable cities, and a few became major metropolitan centers. In some instances—notably Lima, Caracas, and Santiago—the main city developed a bit inland but retained a close connection with a smaller coastal city that was the actual port. Around the ports, agricultural districts developed, spread, and shipped an increasing volume of trade products overseas. The ships that took these goods away returned with more from Europe. They also carried passengers, including government officials, in both directions. Slave ships brought the Africans who provided the principal labor on European-owned plantations, primarily in the

Definitions & Insights

EL NIÑO AND NATURAL DISASTER

The years 1997 and 1998 were classic El Niño years in Latin America. Across the Pacific came an unusually warm current, washing up against the western shores of tropical South and North America, displacing colder normal currents. Not only did this temperature change have a major negative effect on traditional fishing patterns on the west side of Latin America, the warmer surface water in the Pacific played havoc with normal air movement and weather patterns. As can be seen in Figure 19.A, storms lashed a wide band of Middle and northeastern South America, while drought struck broad sections of Brazil, Venezuela, Colombia, and Central America. Fires came after the unseasonably dry periods caused by this cyclic hazard called El Niño, adding considerable forest loss to the fishing loss.

The impact of this weather phenomenon is felt through all bands of society in Latin America and across a great variety of landscapes. In 1997–1998, for example, Bolivia experienced six months of highland droughts and lowland floods—caused in part by the fact that westerly winds that ordinarily would have delivered snow to the Andes came as unusually warm rains; this precipitation washed directly down into the lowlands as floods. Thus moisture was lost for spring farming and devastation was created by the floods that brought their own horror to the lowlands. The Panama Canal had to restrict freighter travel through its locks because the unseasonal drought brought water levels to nearly unprecedented lows, leaving larger ships with deeper droughts unable to deal with the artificial waterways of canal passage. The approximate costs of El Niño in these two years was put at $20 billion, but in fact the costs will go much higher, for not only must there be enormous repair projects organized and paid for, but the loss of crops and cropland will have a long-term impact on Latin American productivity.

In the same family of natural hazards that plague Latin America was the hurricane called Mitch that wandered leisurely across the Caribbean Sea, the Gulf of Mexico, and the northern sector of Central America in the fall of 1998. It was the rain, not the more common hurricane winds, that made Mitch the worst storm Middle America has known in modern times. In the four days that the storm system hovered over the western Gulf of Mexico and the countries of Guatemala, Honduras, Belize, Nicaragua, and El Salvador, it dumped more than two feet of rain not only on the coastal regions (the most common focus of hurricane devastation) but also on inland and upland regions. Even the older residents of this Central American region, who had endured the strong Hurricane Fifi in 1974 and Gilbert in 1988, were caught unready as Mitch came inland. More than 15,000 people died as a result of Mitch, and the disease and weakness caused by the loss of potable water, food crops, and shelter were compounded by the disaster worked upon transportation infrastructure, making outside help virtually unable to reach the region except by helicopter. No storm like Mitch has been seen before by anyone alive in Middle America now, and, in combination with the hardships worked upon Latin America by El Niño, this has been a period of extraordinary demand and difficulty for Latin Americans attempting to attain and maintain landscapes and lifestyles of aggressively developing countries. The web of influences worked on the people by these natural disasters spills beyond the actual locales visited by these physical systems, for migration is often seen as the only real solution to surviving the disasters associated with such storms and cyclic weather changes—and Anglo America and major cities in Latin America are the general target destinations of these destitute migrants.

tropical lowlands with the generally fertile alluvial soils of the coastal plains. Populations along or near the coasts multiplied and are still doing so today.

Meanwhile, the ever powerful lure of gold and silver stimulated the penetration of the Andes and the Brazilian Highlands. After looting the stores of precious metals that had been accumulated by the indigenous peoples for ceremonial use, tribute,

Figure 19.6 The old city of Cuzco in the Peruvian Andes, shown here, was the capital of the Inca Empire when the Spanish conquerors came to Latin America. The city, called "The City of the Sun" by the Incas, is located in a basin at an altitude of about 11,000 feet (c. 3400 m). *Guilio J. Barbero*

486

Latin America • **Impacts of El Niño in 1997–1998**

Figure 19.A The impact of El Niño in 1997–1998 was both far reaching and geographically varied, further supporting the lore that this weather phenomenon is becoming increasingly significant to the region. Consequences of this shift in ocean current patterns and associated air mass movement were felt far beyond Latin America as well.

Storm
Fire
Drought
Heavy rain and flood

Source: *The Economist*, May 9, 1998. p.36.

and trade, the newcomers opened mines, or in many places took over old mines, and urban centers arose to service the mines. Some highland cities gained a wider importance as centers for new ranching or plantation areas. Such functions still endure, although great numbers of older mines have closed, and mining today is far more diversified and less dependent on gold and silver than was true in the early colonial age. A few highland cities—of which the largest is Bogotá in Colombia—have grown into sizable metropolises. These places are separated from the seaports by difficult terrain that required feats of considerable engineering skill before satisfactory transportation links were developed. Some of the main seaports and highland cities became important centers of colonial government as well as economic nodes, and several are national capitals today, such as La Paz in Bolivia and Quito in Ecuador. Rapid development

of manufacturing industries and explosive population growth have characterized such cities in recent times.

The second major alignment of populated areas in Latin America lies on the mainland of Middle America to the northwest of the South American "rim." Composed of the majority of Middle America's people, it extends along an axis of volcanic land that dominates central Mexico and from there reaches southeastward through southern Mexico and along the Pacific side of Central America to Costa Rica (see Fig. 19.5). This belt is characterized by good soils, rainfall adequate for crops but not excessive, and enough elevation in most places to moderate the tropical heat but still permit the raising of many tropical crops such as coffee. The belt of high population density that extends across central Mexico from the Gulf of Mexico to the Pacific Ocean was already Mexico's center of population when

487

Regional Perspective

Altitudinal Zonation in Latin America

One of the most significant features of Latin America's climatic pattern with respect to the distribution of economic activity and population is a series of highland climates arranged into zones by elevation. This zonation results from the fact that air temperature decreases with elevation, at a normal rate of approximately 3.6°F (1.7°C) per 1000 feet (304.8 m) of elevation. At least four major zones (Fig. 19.B) are commonly recognized in Latin America: the *tierra caliente* (hot country), the *tierra templada* (cool country), the *tierra fría* (cold country), and the *tierra helada* (frost country). At the foot of the highlands, the *tierra caliente* is a zone embracing the tropical rain forest and tropical savanna climates discussed earlier. The zone reaches upward to approximately 2500–3000 feet above sea level at or near the equator and to slightly lower elevations in parts of Mexico and other areas near the margins of the tropics. In this hot, wet environment are grown such crops as rice, sugarcane, bananas, and cacao, often on a plantation basis. Populations of black Africans, brought to the New World as slaves, are concentrated in many of the *tierra caliente* zones.

The *tierra caliente* merges almost imperceptibly into the *tierra templada*. Although sugarcane, cacao, bananas, oranges, and other lowland products reach their respective uppermost limits at some point in this higher level, the *tierra templada* is most notably the zone of the coffee tree. In the *tierra templada,* coffee can be grown with relative ease (although by no means are all the soils suitable); at lower elevations the crop encounters difficulties because of excessive heat and/or moisture. The upper limits of this zone—approximately 6000 feet (1800 m) above sea level—tend also to be the upper limit of European-induced plantation agriculture in Latin America. In its distribution, the *tierra templada* flanks the rugged western mountain ranges and, in addition, is the uppermost climate in the lower uplands and highlands to the east. Thickly inhabited sections occupy large areas in southeastern Brazil, Colombia, Central America, and Mexico. Although broadleaf evergreen trees characterize the moister, hotter parts of this zone, coniferous evergreens replace them to some degree toward the zone's poleward margins. In such places as the highlands of Brazil or Venezuela where there is less moisture, scrub forest or savanna grasses appear—the latter generally requiring more water.

In brief, the *tierra templada* is a prominent zone of European-influenced settlement and of commercial agriculture. Six metropolises exceeding 2 million in population—São Paulo, Belo Horizonte, Caracas, Medellín, Cali, and Guadalajara—are in this zone, while two others—Mexico City and Bogotá—lie just above it. Four smaller cities that are national capitals and the largest cities in their respective countries are located in the *tierra templada:* Guatemala City, San Salvador, Tegucigalpa, and San Jose. Others, like Rio de Janeiro, which are situated at lower elevations, have close ties with predominantly residential or resort towns in these cooler temperatures.

The *tierra fría* (at 6,000–10,000 ft) can be distinguished from the other zones by the related facts that it often experiences frost and is often the habitat of a native Indian economy with a strong subsistence component. European colonization in Latin America has driven some native Indian settlement upslope and into the *tierra fría* zone, although some major populations—the Inca of Peru for example—had already selected upland locations for their settlement before the arrival of Columbus. These upland settlements are most extensive in Ecuador, Peru, and Bolivia, and this type of economy is also very evident in Colombia, Guatemala, and southern Mexico. The *tierra fría* is comprised of high plateaus, basins, valleys, and mountain slopes within the great mountain chain that extends from northern Mexico to Cape Horn. By far the largest areas are in the Andes, although areas in Mexico are siz-

---

the Spanish conquerors came. This was the principal domain of the large and technologically advanced Aztec Empire, which, together with its vassal states, composed much of Middle America's population at that time. Prior to the Aztecs, the Maya (to the east) had been another dominant and sophisticated Amerindian population. Then, as today, the soils of this area's high volcanic plateaus and flat-floored basins provided the foundation for a productive agriculture based primarily on maize. Spanish occupation did not alter the dominance of the area within Mexico, and roughly half of Mexico's 94 million people (1998) now live in, around, or between the region's two principal cities of Mexico City and Guadalajara.

The earliest major Mexican urban center was in full growth long before any European colonization brought either settlement patterns or urban customs to Latin America. The vast city of Teotihuacan, located in the valley of Mexico, was the first true urban center in the Western Hemisphere. Its construction is thought to have begun approximately 2000 years ago, and by 500 A.D. it was as large as London was in 1500 A.D. Located some 30 miles from present day Mexico City, it covered about eight square miles. The city was laid out on a grid that was aligned to mesh with stars and constellations that were central to the time keeping of the Teotihuacanos. As of yet, the origins, language, and culture of these people are not fully understood.

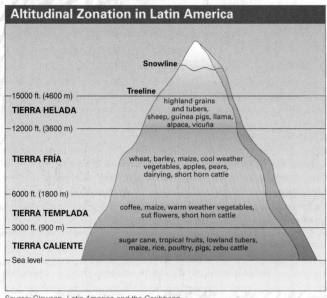

Altitudinal Zonation in Latin America

Snowline

Treeline

15000 ft. (4600 m)

TIERRA HELADA — highland grains and tubers, sheep, guinea pigs, llama, alpaca, vicuña

12000 ft. (3600 m)

TIERRA FRÍA — wheat, barley, maize, cool weather vegetables, apples, pears, dairying, short horn cattle

6000 ft. (1800 m)

TIERRA TEMPLADA — coffee, maize, warm weather vegetables, cut flowers, short horn cattle

3000 ft. (900 m)

TIERRA CALIENTE — sugar cane, tropical fruits, lowland tubers, maize, rice, poultry, pigs, zebu cattle

Sea level

Source: Clawson, *Latin America and the Caribbean*.

Figure 19.B A map of altitudinal zonation in Latin America. Because of the profound differences in relief in the landscapes of Latin America, specific patterns of land use have evolved in relation to differing altitudinal zones. This map shows their elevations and general land use patterns.

able. The upper limit of the zone is generally placed at about 10,000 feet for locations near the equator and at lower elevations toward either pole. This line is usually drawn on the basis of two generalized phenomena: (1) the upper limit of agriculture, as represented by such hardy crops as potatoes and barley, and (2) the upper limit of natural tree growth.

Another zone lies above the other three and consists of the alpine meadows, sometimes called paramos, along with still higher barren rocks and permanent fields of snow and ice. This zone is known as the *tierra helada*. It has some grains and animals (llama, alpaca, sheep) but is largely above the mountain flanks that are central to upland Indian settlement and agriculture. In the highlands of Latin America, the *tierra fría* tends to be a last retreat and a major home of the indigenous peoples, except in Guatemala and Mexico (see Fig. 19.B), and is characterized by small settlements and what Europeans and Americans might consider rather primitive ways of life. However, certain valuable minerals like tin and copper are located here, attracting modern types of large-scale mining enterprise into the *tierra fría* and the *tierra helada* of Bolivia and Peru.

The city may have had 200,000 residents at the time of its zenith as a ceremonial center in 500 A.D. There were some 2000 apartment compounds, a twenty-story Pyramid of the Sun, a Pyramid of the Moon, and a 39-acre civic and religious complex called the Great Compound or the Citadel (Ciudadela). As described by National Geographic Society anthropologist George Stuart, "The whole is a masterpiece of architectural and natural harmony."[2] It still stands as a monument to native American engineering skill and cultural intensity (Fig. 19.7).

[2]George E. Stuart, "The Timeless Vision of Teotihuacan," *National Geographic*, Vol. 188, No. 6 (December 1995), 2–35.

In Pacific Central America also, the basic pattern of population distribution already was established when the Spanish took over. Today, the majority of people live in highland environments, but considerable numbers live in lowlands as well. The ongoing vulcanism of the highlands periodically deposits ash into the highland basins, and this weathers into fertile soils that grow maize, beans, and squash for local food and the coffee that is Central America's largest export. Between the highlands and the Pacific Ocean lies a coastal strip of seasonally dry lowlands where cattle and cotton are grown largely for an export market. By contrast, the Atlantic side of Central America is lower, hotter, rainier, and far less populous than the Pacific side. Here, on

Figure 19.7 The Pyramid of the Sun in the pre-Aztec center of Teotihuacan in the Valley of Mexico is a bold reminder of the architectural skill local Middle American cultures had prior to the arrival of Cortez and the West. This pyramid covered 10 acres (4 ha) at its base and was the dominant structure in a ceremonial and commercial city complex that was about 7 sq mi (18 sq km) in extent and reached its zenith in approximately A.D. 300 to 900. It was virtually abandoned at the time of the Spanish conquest of Mexico in the early 16th century. *Cameraman Int'l Ltd.*

large corporate-owned plantations, are grown most of the bananas that are Central America's agricultural trademark and second most valuable export after coffee.

Variations in Population Density

Average population densities for entire countries conceal the fact that in nearly every Latin American mainland country there is a basic spatial configuration consisting of a well-defined population core (or cores) with an outlying sparsely populated hinterland. Densities within the cores are often greater than anything to be found within areas of comparable size in the United States. On the other hand, most of the mainland countries have a larger proportion of sparsely populated and relatively unorganized terrain than is true of the United

States. Figures on population density for the greater part of Latin America are extraordinarily low, averaging under two persons per square mile over approximately half of the entire region. This pulls down Latin America's average density—and that of most major Latin American countries—to levels somewhat below the average density for the United States.

The pattern of core versus "outback" (to use an Australian term) can be seen in Brazil and Argentina. Most Brazilians, for example, live along or very near the eastern seaboard south of Recife, a port city in northeastern Brazil, whereas large areas in the interior are still thinly populated. In Argentina, approximately three-fourths of the population is clustered in Buenos Aires or the adjacent humid pampa—an area containing a little over one-fifth of Argentina's total land. The less densely settled regions in Latin America are generally quite distinctive from the core. While there is a steady migration from the outback to the Rimland, there is very little reverse migration of the core population going to the outback, with its relatively more demanding and less convenient setting for people who have grown accustomed to the environmental and urban amenities of the Rimland region. In Latin America, population patterns in Brazil, Argentina, and Mexico as well as in the Andean countries reflect this demographic imbalance (see Fig. 19.5).

Not every country displays the pattern of core versus outback. A very different picture of population density and distribution is evident in El Salvador and Costa Rica—both of which are composed of thickly settled volcanic land—and in most Caribbean islands (see Table 19.1). Because the combined population of the islands in the Caribbean is considerably less than that of central Mexico alone, these islands do not contain a concentration of people comparable in scale to either of the two major Latin American concentrations—the Rimland and Mainland—discussed earlier. However, they do stand out for exceptionally heavy densities, which have been created over a long period by the accumulation of expanding population on territories of very restricted size. Most often the crowding is aggravated by high rates of natural population increase and/or the prevalence of steep slopes that restrict possibilities for further farming or settlement; this is also true of parts of El Salvador and Costa Rica.

Ethnic Diversity

Despite the overall dominance in Latin America of culture traits derived from Europe, only three of this region's 35 nations—Argentina, Uruguay, and Costa Rica—have preserved white European racial strains on a large scale with little admixture by Indians or blacks. This is due in part to the fact that these nations received relatively large numbers of European immigrants during the late 19th and early 20th centuries. Scattered districts in other countries also are predominantly European or profoundly Amerindian (native Indian).

Various sources estimate that 40 to 82 percent of the population in highland nations of South America are composed of

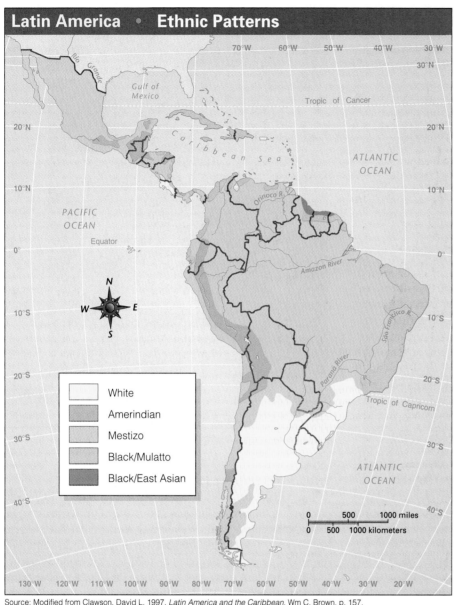

Figure 19.8 While this map illustrates the broad patterns of ethnic distribution in Latin America, there is always more variety in the local scene—especially in the cities—than is shown on such a regional map.

Source: Modified from Clawson, David L. 1997. *Latin America and the Caribbean.* Wm C. Brown. p. 157.

native Indians or Amerindians. Ecuador at 60 percent and Bolivia at 82 percent are the highest. Native Indians are a significant minority overall in Mexico at 25 to 30 percent, but they comprise over 50 percent in some southern states of the country. Guatemala is more than 40 percent Amerindian (Fig. 19.8).

Native Indians are also a major population element in the basin of the Amazon River and in Panama, where scattered lowland Indians maintain their aboriginal cultures. However, the outer world is pushing in on them despite some efforts on the part of governments to protect their ways of life. Over the centuries after Columbus, the encounters of lowland Indians with Europeans generally had catastrophic results for the aborigines, who often resisted exploitation fiercely but were no match for the Europeans' diseases (especially smallpox), guns, liquor, and

relentless hunger for more land and minerals. The more numerous highland Indians, who were more advanced in technology, more distant, and better organized than the lowlanders, were somewhat more successful in dealing with European encroachment and exploitation but still suffered enormous population losses at the hands of their discoverers. Disease was the most active killer.

Black Latin Americans of relatively unmixed African descent are found in the greatest numbers on the Caribbean islands and along hot, wet Atlantic coastal lowlands in Middle and South America. These are the areas to which African slaves were brought during the colonial period, primarily as a source of labor for sugar plantations. Slavery was gradually abolished during the 19th century, although not until the 1880s in Brazil and Cuba. By that time, slavery had generated large fortunes for many owners of plantations and slave ships and had introduced African peoples to the region who have made cultural contributions that today are a varied, colorful, and important part of Latin American civilization.

Latin America has escaped many of the racial tensions that grip much of the world. This results, in part, from the fact that the majority of Latin Americans have a mixed racial composition. Much of the region exhibits a primary mixture of Spanish and native Indians, resulting in a heterogeneous group known as *mestizos.*

Population Growth and Expansive Urbanization

Irrespective of ethnic composition and type of culture, most Latin American countries today find themselves in the second stage of what has been called the **demographic transition** (see Chapter 2). Startling rates of urban and metropolitan growth have been a major Latin American characteristic during recent times. Recent estimates (1998) give a figure of 72 percent urban for the population of the region as a whole, compared to a world average of 43 percent. A major element in the rapid growth of Latin American cities is a huge and continuing migration from countrysides that often are so overpopulated and poor as to be little better than rural slums.

In any consideration of human migration, there is a discussion of the environmental and social factors that cause someone to leave an area (the **push factor**) and the oftentimes corresponding factor that attracts (the **pull factor**) someone to a specific locale. Generally economic factors are given the most importance in assessing migration flows, particularly those from the countryside to the city. However, very often a young man or woman will leave a position of steady underemployment in the rural sector and go to the city, even if there is no certainty about employment there at all. Images of the city are often filled with "the good life" and mythic opportunities for economic gain. As a result, these urban centers become overcrowded with unskilled, generally uneducated rural youth who have left their very simple homes in the countryside hoping to gain money with which they can help support rural family members and improve their own livelihood.

Very often this critical migration ends with economic failure, but the pressure to hide such lack of success is so strong that there is very little reverse migration, or return to the countryside. Even where there is active industrial growth and employment opportunity—such as in Mexico City and São Paulo, Brazil (two of the four largest cities in the world)—there are many more migrants and applicants than there are factory or service-sector positions. As a result, Latin American slums grow with a steadiness that challenges every urban planner and municipal administrator.

Even in rural areas that are better off, cities are viewed by the ambitious young as places to get ahead rapidly and enjoy

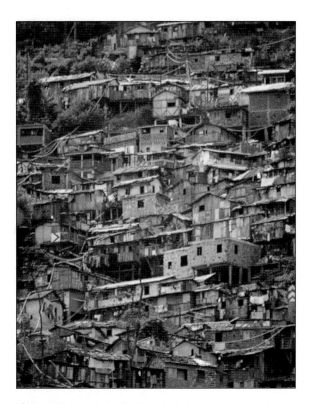

Figure 19.9   Many of Latin America's poorest people live in ramshackle housing that clings to nearly vertical slopes. The shantytown in this photo overlooks scenic Rio de Janeiro; the locale is the suburb of Santa Marta. *Stephanie Maze/Woodfin Camp & Associates*

exciting amenities while doing so. Deeply rural and tradition-bound villages offer little competition to the bright lights and swirling action of the images and realities of the metropolis. But a great many rural dwellers flee to the cities out of sheer desperation induced by drought, environmental deterioration, depressed farm prices, runaway inflation, chronic unemployment, guerrilla warfare, or other ills that beset the countrysides of many developing parts of the world in our time. In Latin America, the urban explosion is especially problematic in the major urban centers in the Rimland on the margins of the agricultural lowlands in both South and Middle America. Caring for the huge influxes of people has strained the services of Latin American cities beyond their limits. As a result, every large city has big slums, which often take the form of ramshackle shantytowns on the urban periphery. Sometimes they are built on hillsides overlooking the city, as in Rio de Janeiro, where some of the world's more unsightly housing (Fig. 19.9)—called *favelas* in Brazil and *barrios* in Spanish-speaking countries—commands one of the world's more imposing views. Such shantytowns, which are practically universal in the world's less developed countries, are full of underemployed and ill-fed people who still may prefer their present plight to the conditions of the depressed rural landscapes from which they came. They do, however, keep people eager to work in the hub of economic activity. As such, they are sometimes called "slums of hope."

19.4 "Emerging" or "Developing" Nations: Characteristics and Frustrations

Latin America is an "emerging" or "developing" region that does not yet provide a good living for most of its people. It is not the worst off of the major world regions; in fact, its overall per capita GNP is almost four times that of Africa (see Table 1.2). But compared with Anglo America or Europe, the Latin American region as a whole is quite poor. This tends to be masked by the glitter of the great metropolises, with their forests of new skyscrapers. In an attempt to overcome their economic deficiencies and constrain popular discontent, many Latin American governments have borrowed heavily from the international banking community, including leading banks in the United States, in order to fund domestic development schemes. By the late 1990s, unpaid loans had reached staggering proportions, and the Latin American debtor nations were having great difficulty in mustering even the annual interest payments, let alone generating surplus capital to pay toward the principal of these troubling loans. This debtor situation is very often associated with the process of "emerging" or "developing."

Another way countries deal with the demands and dangers of being a developing or emerging nation is through membership in international organizations. In Latin America there are a number of such linkages that focus on tariff barriers and economic unity (Fig. 19.10). The best known of these is the North American Free Trade Association (NAFTA), of which Canada, the United States, and Mexico are currently members, but of which Chile is close to becoming a member as well. In addition to NAFTA is the organization comprised of a cluster of nations in South America—Brazil, Argentina, Uruguay, and Paraguay—called the Southern Cone Common Market. This group is better known under the name Mercosur and is expected to have negotiated a free-trade zone with the European Community by the year 2003. There is also the Andean Group, with Venezuela, Colombia, Ecuador, Peru, and Bolivia as current members. Additional groups are shown in Figure 19.10.

Two Systems of Agriculture: Latifundia and Minifundia

Although the total number of Latin Americans employed in agriculture has not declined very much in recent decades, there has been a sizable percentage decline in agriculture as a component of both employment and value of product in national economies. The actual percentages vary a great deal from country to country. Dependence on agricultural exports has also dropped as national economies have become more diversified, but in many countries more than half of all export revenue is still derived from products of agricultural origin (Table 19.2).

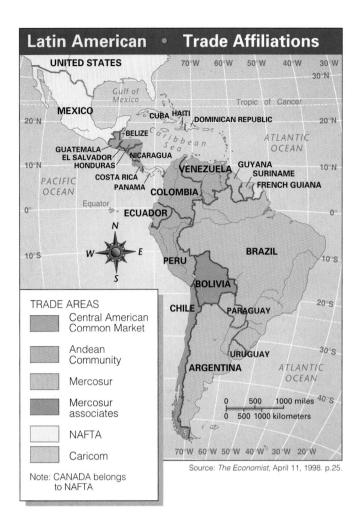

Source: *The Economist*, April 11, 1998. p.25.

Figure 19.10 Latin America is actively experimenting with a variety of alliances in an effort to enhance and expand the economic impact of the region overall, and of internal units in different combinations.

Farms in Latin America are often divided into two major classes by size and system of production. Large estates with a strong commercial orientation are **latifundia** (singular: latifundio). These estates, whether called *haciendas,* plantations, or some other name, are owned by families or corporations. Some have been in the hands of the same family literally for centuries. The desire to own land as a form of wealth and a symbol of prestige and power has always been a strong characteristic of Latin American societies. Huge tracts were granted by Spanish and Portuguese sovereigns to members of the military nobility who led the way in exploration and conquest. Some of this land has been reallocated to small farmers by government action from time to time; the *ejido*—an innovative communal land tenure program in Mexico that gave more land to the farmers and middle class—is a good example, but in many countries a very large share of the land is still in the hands of a small, wealthy, landowning class.

493

Table 19.2 Leading Exports of Selected Latin American Countries in 1999

Country	Commodity	Percent of Total Exports
Venezuela	Crude petroleum and petroleum products	77
Cuba	Sugar	63
Mexico	Metallic products, machinery, equipment	58
Chile	Industrial products	45
Ecuador	Food and live animals	48
Panama	Bananas	33
Colombia	Petroleum products	27
Belize	Sugar	30
El Salvador	Coffee	33
Costa Rica	Bananas	24
Dominican Republic	Ferronickel	32
Honduras	Coffee	32
Paraguay	Soybean products	31
Guatemala	Coffee	23
United States	Machinery and transport (for comparison)	47

Britannica Book of the Year 1999. Encyclopaedia Britannica, Inc., 1999.

Minifundia are smaller holdings with a strong subsistence component. The people who farm them generally lack the capital to purchase large and fertile properties, and hence they are relegated to marginal plots, often farmed on a sharecropping basis. Individuals who do own land are frequently burdened by indebtedness, the continued threat of foreclosure of farm loans, and the fragmented nature of farms, which are becoming smaller and smaller in size as they are subdivided through inheritance and the reversal of promised governmental programs of land distribution. Such farms produce food for family use but also for the local market. The crops most commonly raised—especially in Middle America—are maize (Fig. 19.11), beans, and squash, although many other crops are locally important, especially in the different climatic zones of the highland regions.

Although these small farmers make up the bulk of Latin America's agricultural labor force, food has to be imported into many areas, and many of the people are poorly nourished. Productivity is low because of the marginal quality of the land and the generally rudimentary agricultural techniques. There is little capital with which to buy machinery, fertilizers, and improved strains of seeds. Soil erosion and soil depletion are making serious inroads in many areas.

Land Reform

Land reform has been a topic of both political and economic significance in Latin America. The region's history is filled with attempts—sometimes very serious and bloody, sometimes less painful governmental innovations—to achieve a better balance between those who work the land and those who own the land. The main attention has been given to breaking up existing properties or bringing vacant land (whether owned by the public, by private individuals, or by the Catholic Church) into cultivation, usually by small farmers. Some new farms have been structured as communal holdings, reflecting indigenous traditions or 20th-century revolutionary plans of agrarian reform. In Cuba, for example, the Communist government that took control in 1959 placed the land from expropriated estates in large farms owned and operated by the state. Workers on these farms are paid wages. The Nicaraguan Sandinistas implemented land reform as part of their revolutionary efforts in the 1980s.

The details and success of land reform schemes have varied sharply from one Latin American country to another. Such schemes can help relieve poverty in the countryside, but they do little for the majority of Latin America's poor, who live in cities. In Mexico, there was great fanfare at the introduction of the *ejido* (communally farmed land in Mexican Indian villages) early in the 20th century. It has been so effective that *ejidos* currently account for 50 percent of the cultivated land in Mexico and produce nearly 70 percent of the beans, rice, and corn of this nation. The *ejido*, by making land available to farming communites, has been the most successful land reform program attempted in Latin America, including Cuba. Puerto Rico has had public battles over *"pan, tierra, y libertad"* (bread, land, and liberty) because of the difficulty in effecting real distribution of farmland to the farming people, and the January 1994 uprising in Chiapas, Mexico, had land distribution at its heart. Virtually every nation in Latin America can point to events, martyrs, and programs birthed in protest or blood in an effort to achieve more equitable land ownership. Nevertheless, the great majority of the arable land in Latin America continues to be owned by wealthy farming families, agribusiness operations (both domestically and foreign-owned), and the church. There is a world of small farmers in Latin America, but they seldom own as much as 20 percent of a nation's arable land.

Minerals and Mining: Spotty Distribution and Unequal Benefits

Latin America is a large-scale producer of a small number of key minerals that are very significant to the outside nations where they are re-marketed. Only a handful of Latin American nations gain large revenues from such exports; in most countries, mineral production is relatively minor. Even in the countries that do have a large value of mineral output—notably Mexico, Venezuela, Chile, Ecuador, and Brazil—much of the profit appears to be dissipated in the form of showy buildings, corruption, ill-advised development schemes, and enrichment of the upper classes and foreign investors. Benefits to the broader population have tended to be rather minimal. Nonetheless, significant infrastructure development, including many new highways, power stations, water systems, schools, hospitals, and employment opportunities, has been made possible by mineral revenues. Among Latin America's most important known mineral resources are petroleum, iron ore, bauxite, copper, tin, silver, lead, zinc, and sulfur. However, local use of these minerals for industrialization has been retarded by a lack of good coal, especially the coking coal that is very important in steel production, although Brazil has realized a major improvement in its production and now is the top automobile producer in Latin America.

All but a small proportion of Latin America's petroleum is extracted in the Caribbean Sea–Gulf of Mexico area, particularly the central and southern Gulf coast of Mexico and northern Venezuela (see Table 19.2). Other oil fields in Latin America are widely scattered, with the principal ones located along or near the Atlantic coast in Brazil, Argentina (in Patagonia), and Colombia, or in sedimentary lowlands along the eastern flanks of the Andes in every country from Trinidad and Tobago to Chile. Natural gas is extracted in many areas that produce oil, but Latin American production is not yet of major world consequence. Mexico, Venezuela, and Argentina are the largest producers.

Latin America is a major world area in the production of metal-bearing ores. Most of the extracted ore is shipped to overseas consumers in raw or concentrated form, although the region has scattered iron and steel plants, as well as smelters of nonferrous ores, which process metals for use in Latin American industries or for export. Deposits of high-grade iron ore in the eastern highlands of Brazil and Venezuela are the largest known in the Western Hemisphere and are among the largest in the world; the two countries are Latin America's main producers and exporters of ore. Brazil is the main producer of iron and steel by far, with Mexico second. Most of the region's production of bauxite—the major source for aluminum—comes from Jamaica, Brazil, Suriname, and Guyana. The deposits in these countries are located relatively near the sea and are of critical importance to the industrial economies of the United States and Canada. Huge unexploited bauxite deposits in Venezuela are a major resource for the future.

Chile is overwhelmingly the largest copper producer in Latin America and ranks number one in the world. Most of Latin America's known reserves of tin are in Bolivia or Brazil, and those countries produce most of the region's output. The silver of Mexico, Peru, and Bolivia—sought from the very beginning of Spanish colonization—is found principally in mountains or rough plateau country. Mexico and Peru are by far the largest Latin American producers of silver and of lead and zinc as well. Peru's deposits of all three minerals are in the Andes, and Mexico's are mainly in the dry northern and north-central sections of the country.

The Increasing Importance of Manufacturing and Services

Although they are still very spotty in distribution and often lacking in complexity and sophistication, manufacturing industries in Latin America are making an increasingly important contribution to national economies. Most parts of the region are

Figure 19.11 Maize (corn) is a major crop all over Latin America. In this photo, a government drug inspector in Mexico is about to burn some harvested corn, which is known to contain concealed heroin. Many ingenious means are used to smuggle illegal drugs into the United States. *Andy King/Sygma*

still a long way from full industrialization, but certain districts have an impressive array of factories, including some of the most modern types. Because the industrial revolution has come to Latin America very late, the region does not face large problems of industrial obsolescence. There is a regionwide push to attract industry from abroad with offers of tax exemptions, cheap labor, and other inducements. The development of the *maquiladora* (assembly operations in Latin America that combine components manufactured from other countries with local labor) movement and NAFTA has done a great deal to keep Latin American labor occupied in locally run factories.

By far the most numerous manufacturing establishments in Latin America are household enterprises or small factories that employ fewer than a dozen workers and sell their products mainly in home markets. Larger operations are almost always located in the larger cities and often are branch plants of overseas companies. The Latin American countries that have the largest output of factory goods are Brazil, Mexico, and Argentina. The industrial scene is dominated by a handful of large metropolises, mostly notably São Paulo, Mexico City, and Buenos Aires. Brazil actually ranks in the top ten of the world's nations in production of manufactured goods, although its per capita ranking is much lower than that.

The proportion of Latin Americans in service employment has risen sharply during the past quarter century as people from the countryside have flooded the cities. Expanding government and business bureaucracies, the increasing affluence of urban upper and middle classes, an increase in tourism, and a general increase in the facilities required for even a minimal servicing of hordes of new urban dwellers have created many millions of new service jobs. Even at general pay scales far below those of western Europe or Anglo America, such jobs have given new opportunity to desperately poor migrants from overcrowded rural areas, and they have helped provide employment to accommodate the explosive natural increase of population within the Latin American cities themselves.

The Increasing Importance of Tourism

In many Caribbean islands and scattered places on the mainland, tourism particularly has become a major economic asset (Table 19.3 and Fig. 19.12). The recent development of major resort complexes in Mexico, the Caribbean islands, Costa Rica, parts of Brazil, and other Latin American countries has expanded tourism as a major player in the generation of foreign exchange. Cancún, Mexico, for example, was recently described in a tourist brochure as "Cancún—Mexico's American Resort . . . Cancún may seem more American than Mexican." The region's proximity to the relatively wealthy and mobile U.S. and Canadian populations, the expansion and improvement of regional air transport, and the increasing inclination of the Anglo Americans to spend money for tourism have all brought considerable disposable income to the countries and coastal areas that have made the appropriate landscape in-

vestments. Passenger cruise ships now stop at ports of call in the Antilles, on the Pacific coast of Mexico as well as at Cancún, and at Santa Marta and Cartagena in Colombia. The transiting of the Panama Canal continues to be important as a tourist activity. The canal, however, continues to be geographically significant for all the ships that can gain passage in it, for it cuts some 8000 miles off the voyage from New York to San Francisco.

Cruise companies are also increasing the number of sailing routes along the east coast of South America with stops in the Amazon Basin and in coastal ports of Brazil, Uruguay, and Argentina. San Juan, Puerto Rico, is now a major port of embarkation for tourist cruises destined for the eastern Antilles, with passengers regularly being flown from U.S. and Canadian cities to San Juan for boarding and embarkation. There is also a growing demand for ecotourism destinations that present visitors with a chance to view relatively pristine landscapes with unusual fauna and flora. Countries in which tourist dollars amount to more than 20 percent of the nation's foreign exchange include Argentina, Dominican Republic, and the Bahamas.

Table 19.3 International Tourism to Selected Latin American Destinations

Country	Total No. Tourists (000s)	Percentage of Tourists by origin (%)		Total Tourist Receipts ($U.S.) (000,000s)
		Americas	Europe	
Argentina	3030.9	83.0	12.1	3090
Aruba	541.7	89.6	9.8	443
Bahamas	1398.9	89.1	8.7	1244
Barbados	385.4	59.0	40.0	463
Brazil	1474.9	72.4	23.2	1307
Chile	1283.3	89.8	8.0	706
Colombia	1075.9	92.3	7.1	705
Costa Rica	610.6	83.5	14.8	431
Cuba	460.6	49.1	47.5	382
Dominican Republic	1523.8	n.a.	n.a.	1054
Ecuador	403.2	80.2	16.9	192
El Salvador	314.5	92.4	6.6	49
Guatemala	541.0	77.5	19.6	243
Jamaica	909.0	77.1	20.1	858
Martinique	320.7	18.5	80.8	282
Mexico	17,271.0	97.0	2.1	5997
Panama	311.4	90.2	6.3	207
Puerto Rico	2639.8	70.2	n.a.	1511
Uruguay	1801.7	83.2	2.7	381
U.S. Virgin Islands	390.0	94.9	3.3	792
Venezuela	433.5	54.9	43.2	432

Adapted from David L. Clawson, *Latin America and the Caribbean: Lands and Peoples.* Dubuque, Iowa: Wm. C. Brown, Publishers, 1997: 281.

Figure 19.12 Almost nothing is as important as landscape elements in the selection of tourist destinations. This major hotel in Nassau in the Bahamas features natural-looking water displays, lush vegetation and flowering trees, and comfortable accommodations—all with English as the spoken language. The economic importance of becoming a major tourist destination means a great deal to the economies of many Latin American countries. *Timothy O'Keefe/Tom Stack & Associates*

Staging for the Future

All of the elements for steady development and democratic reform are present in the diversity of Latin America. In many of the 35 countries in this region, there are historical periods of success in both of these areas, but, at the same time, the region has known great frustration as single-crop or single-mineral economies have been depressed by rapid and deep drops in prices for these goods, leaving the Latin American export country in major trouble because of a too-narrow base for economic stability. Populations are varied in stock, but not so dense as to be an overall burden on the country—except in cases of overcrowded urban centers that have become migration destinations for many rural youth seeking new, urban opportunity. At the same time, most countries in Latin America have isolated farming populations who support themselves by traditional slash-and-burn farming practices—especially in the tropics—that show almost no evidence of the region's aggressive efforts to modernize (Fig. 19.13). However, the recent inflow of tourist dollars and the steady efforts to slow population growth, even within the context of a Roman Catholic region, are taken as signs of optimism by planners and citizens alike. It is with keen anticipation that the peoples and governments of Latin America will observe the continuing impact of NAFTA during these coming years, trying to assess both the economic and environmental effects of these significant attempts at regional cooperation.

Figure 19.13 Slash-and-burn agriculture is one of the most powerful agents of landscape change in Latin America. Major forest regions are often a long distance from governmental protection. A family can turn forest richness into a year or two of varied crops and then the infertility of the soils drives the family on to other forest reserves. In this process small populations are given traditional returns on family labor, but primary forest reserves are often lost to unproductive volunteer grasses and failed land use. The tension over this land use is further intensified if ranchers with machinery come in and clear-cut the forests in an effort to create rangeland for beef exports to North America. *Paul Edmonson*

SUMMARY WITH SELECTED KEY TERMS

- **Latin America** lies south of the United States and includes **Mexico, the Central American countries, the island nations of the Caribbean Sea,** and **all the countries of South America.** It is called Latin America because of the nearly universal use of Latin-based languages and the dominance of the **Roman Catholic influences** that are associated with the colonial period from the early 16th century through the middle of the 19th century. We use the term Middle America for Mexico, Central America, and the Caribbean.

- Major themes in Latin America include extraordinary **variety in environmental settings; population groups of great diversity; a wide variety of levels of economic development; countries that have highly diverse cultural and demographic profiles;** and **diverse governmental philosophies.**

- A major tool for analysis and description of Latin America is **altitudinal zonation,** which includes *tierra caliente, tierra templada, tierra fría,* and *tierra helada.* These zones go from sea level up through various altitudes, each with distinctive climate, soil, agriculture, and settlement patterns.

- Major **humid zones** are found in the lowlands of Latin America, with **tropical rain forests** occupying major river basins, especially in Brazil and Venezuela. **Grasslands** are found poleward of the tropics, and there is a zone of mediterranean climate on the western coast of Chile.

- Dry climates—desert and steppe—are often located where they are because of **orographic features** that influence movement of air masses. There are **major desert regions** on the west coast of both South America and in northern Mexico.

- Latin America's population amounts to nearly **9 percent of the world's total population,** and its growth rate tends to be above the world average, especially in **Central American** countries where there continues to be a relatively **large native American population.** The general demographic pattern in this region shows a majority of the people living on the rim of the land masses—an area often called the **Rimland.** There is also a major population band in the mountain valleys, and, in some cases, on high plateaus or on the flanks of the Andes Mountains in western South America.

- Before the arrival of the European colonists and settlers in the late 1400s, Latin America had been home to some productive and highly advanced native cultures, with the **Aztecs,** the **Maya,** and the **Incas** the best known and most widely studied. All of these had either declined before European colonization or declined rapidly after 1500. Although their demise is often associated with mining and other land-use demands, the real killer was **European disease** visited upon a population that had no immunity against Western diseases—especially **smallpox.**

- Latin America has become one of the **most urbanized world regions,** with now more than two-thirds of its population living in cities, compared to the world average of 43 percent. There continues to be steady **rural-urban migration** as well as a steady stream of both legal and illegal immigrants north into the United States. The **North American Free Trade Agreement (NAFTA)** has, as one of its goals, the promotion of industrial and commercial growth in Latin America so that migration to the United States becomes less attractive to the large underemployed Latin American populations.

- **Latifundia and minifundia** are two major agricultural systems of historical importance that continue to characterize major blocks of regional land use. There is also a strong **plantation economy,** especially in Central America, the Caribbean, and coastal Brazil and sugar and fruit all play dominant agricultural roles. Farm landscapes range from peasant households that are largely **subsistence farming,** to **haciendas,** to **ejidos,** to foreign-owned fruit and coffee plantations—all of which have their own distinctive looks and functions.

- **Minerals** have played a major role in the history, economic development, and trade networks of Latin America, and include **iron ore, bauxite, copper, tin, silver, lead, zinc,** and **sulfur. Petroleum** has been of dominant importance—especially in Mexico and Venezuela—but has lost some importance recently because of flat or declining world oil prices. Gold has had local importance, and silver is at the center of centuries of European interest in the region.

- In recent years **economic growth** has depended increasingly on **tourist** activity and on manufacturing goods for both domestic and foreign markets. **Foreign investment** has been influenced by **political instability** in various parts of Latin America, and the United States military has played a role in promoting and eliminating governments in a number of Middle America nations. The construction of the **Panama Canal** was enabled, largely, by the quick United States recognition of a coup that broke the Panama Isthmus away from Colombia in 1903. Construction of the canal began within a year and was finished a decade later, eliminating some 8000 miles in the voyage from New York City to San Francisco.

REVIEW QUESTIONS

1. What countries make up the Mainland of Middle America?
2. What are the major island nations of the Caribbean Sea?
3. Besides the Spanish spoken in most of Latin America, what other languages are spoken in this region, and where are those languages found?
4. Compare the area and the extent of latitude and longitude of Latin America with the same features in Anglo America.
5. What are the major climate patterns of Latin America? What are the physical features that play major roles in such patterns?
6. Explain the causes and the significance of hurricanes in Latin America. What other natural hazards are part of the geography of Latin America?
7. Explain the four zones discussed in altitudinal zonation in Latin America. What economic and demographic importance do those zones have in the region?
8. What are the major features of the population distribution in Latin America? What features characterize the patterns of population growth in the region? What role does migration play in these patterns?
9. Outline the major agricultural patterns and dominant land use in

the region. What role does export play in the overall agricultural pattern of Latin America? What goods and what destinations characterize such activity?

10. Tourism and mineral extraction both relate very closely to geographical conditions and characteristics. What are the traits important in each and how are they expressed in Latin America?

11. List the countries included in, and explain the regional breakdown of, Latin America, Middle America, Central America, Caribbean Sea, and South America.

DISCUSSION QUESTIONS

1. Discuss the possible patterns of development in Latin America had Columbus not come westward in his effort to find East and South Asia. What if no European explorers had found Latin America by 1600? By 1800?

2. Latin America is sometimes described as "a land of contrasts." How would you defend such a broad description for the region?

3. Where are the subregions in Latin America that were most densely settled before the Europeans came? How did overall regional demographic patterns change after 1492? What has been the impact of such changes?

4. Talk about the positive and the negative impact of NAFTA. Discuss it from the point of view of Latin Americans; from other points of view. Has it been largely beneficial or detrimental for Latin America? For Anglo America?

5. What major crops were diffused from Latin America to the broader world? What major crops in today's Latin American agriculture came from the Old World when the Europeans began to settle Latin America? What economic and cultural impact has this diffusion had?

6. What are the cultural forces working to reduce population growth rates in Latin America? What forces tend to keep growth rates high? What impact does population growth have on political stability and economic development?

7. Discuss the geography of isolation caused by mountains, vast river systems, islands, and other aspects of the region. How do these features influence patterns of political stability and human migration?

8. What were the causes and the impact of 1998's Hurricane Mitch in Middle America? Of El Niño?

9. Consider places as geographically diverse as the Andean Highlands and the Amazon Basin. What are the political and cultural forces that enable such distinct locales to both be part of the same world region?

10. Imagine that you are a young Latin American living in a small village or town in Middle America or South America with little business activity. What are the images, history, and knowledge that will cause you to think about migrating? What destinations might you consider and why?

Chapter 20

The Varied Worlds of Middle America

▲ In the mid-1990s, a pattern began of makeshift rafts setting sail from Cuba, and later Haiti, for the Florida peninsula. Often overcrowded and never prepared for bad weather or rough water, these flotillas were more interested in being intercepted by the U.S. Coast Guard than actually crossing the 90 miles of open sea that lies between Cuba and Florida. Once the United States changed its initial policy of bringing the so-called "boat people" to its shores, the flow of rafters decreased. These men have come ashore on the Grand Caymans, having launched themselves from the southern shore of Cuba. Their raft lies just offshore.   Susan Blanchet /Blanchet Photo/Dembinsky Photo Associates

20.1 Mexico: An Unrecognized Giant

20.2 Central America: Fragmented Development in an Isthmian Belt of States

20.3 The Caribbean Islands

*M*iddle America includes (1) Mexico, by far the largest country in area and population, (2) the small Central American states of Guatemala, El Salvador, Belize, Honduras, Nicaragua, Costa Rica, and Panama, and (3) the numerous islands in the Caribbean Sea or near it. Of the many island political units, the largest are Cuba (which is only 90 miles away from the United States), Haiti and the Dominican Republic (which share the second largest, Hispaniola), and Jamaica. The entire region presents a patchwork of physical features, races, cultures, political systems, population densities, and pursuits. Most of the political units are independent, but a few are still dependencies of outside nations (Fig. 20.1). We begin with the most prominent regional unit, Mexico.

20.1 Mexico: An Unrecognized Giant

The federal republic of Mexico—officially, the Estados Unidos Mexicanos (United Mexican States)—is the largest, most complex, and most influential country in Middle America. Compared to most of the world's countries, Mexico is a spatial giant, but its large size tends to go unrecognized, largely because it lies in the shadow of the United States. Its capital, Mexico City, stands now as the world's largest city in population. Triangular in shape and situated next to the compact bulk of the conterminous United States, Mexico looks comparatively small on a world map, but its area of 762,000 square miles (c. 2 million sq km) is nearly eight times that of the United Kingdom, and its elongated territory would stretch from the state of Washington to Florida. The country's huge population (an estimated 97.5 million in 1998, expected to reach 120 million by the year 2010) makes Mexico by the far the largest nation in which Spanish is the main language. Mexico is also a big country economically, ranking quite high among the world's nations in annual output of many different commodities, such as metals, oil, gas, sulfur, sugar, coffee, corn, citrus fruit, cacao, and cattle. It has also made substantial gains in its development of manufacturing and component assembly industries.

Mexico's remarkable diversity of mineral resources and agricultural products is made possible by great environmental variety. Metal-bearing ores exist in many places, mined particularly in mountainous terrain of northern and central Mexico. Iron ore from scattered locations supports a fairly sizable iron and steel industry, which also can draw on substantial coal deposits, some suitable for coking. The coastal lowlands along the Gulf of Mexico share the oil, natural gas, and sulfur deposits that are a marked feature of adjacent coastal Texas and Louisiana. In agriculture, great variety is achieved despite a mountainous environment lacking in tillable land. Only one-eighth of the country is cultivated; good cropland exists mainly (1) on the floors and lower slopes of mountain basins with volcanic soils, and (2) on scattered and generally small alluvial plains. Agricultural habitats vary according to elevation, rock types, physiographic history, and atmospheric influences. Mountains too steep to cultivate without excessive erosion dominate the terrain in most areas, but they are flanked by plains, plateaus, basins, and foothills that can be tilled productively if enough water is available.

A little over one-half of Mexico lies north of the Tropic of Cancer and is an area dominated by desert or steppe climates with hot summers and winters that are warm to cool depending on elevation. South of the Tropic of Cancer, where about four-fifths of all Mexicans live, seasonal temperatures vary less, but elevation differences create conditions ranging from high temperatures in the *tierra caliente* of the lowlands along the Gulf of Mexico, through moderate heat in the *tierra templada* of the low highlands, to cool temperatures in the *tierra fría* of the highlands (see Regional Perspective, Chapter 19, p. 488). Nearly all areas inside the tropics are humid enough to support rainfed crops, although irrigation is often used to increase the diversity of crops and/or the intensity and productivity of agriculture. Large areas have a dry season in the low-sun period when irrigation becomes a necessity for cropping.

On the hemispheric and world stage, Mexico speaks with a respected voice, although the country's relative poverty and its great economic dependence on the United States make its influence

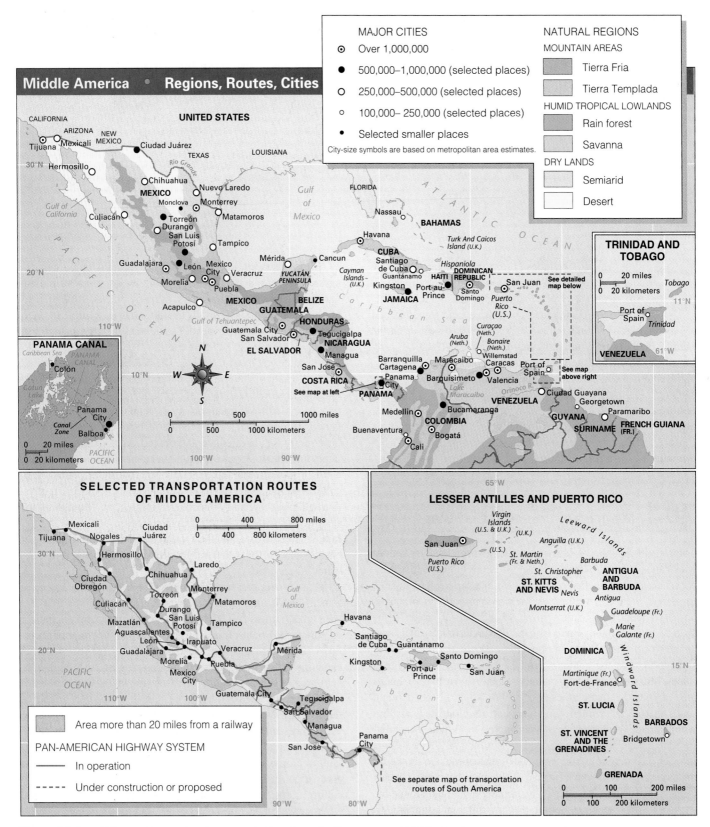

Figure 20.1   Middle America—regions, routes, and cities. General reference maps of Middle America. Curaçao and Bonaire comprise the Netherlands Antilles; Aruba is a separate country. Antigua and Barbuda form one political unit, as do St. Kitts and Nevis, and St. Vincent and the Grenadines. Major highways in the transportation inset are supplemented and interconnected by other surfaced roads not shown.

less potent than might otherwise be true of so large a nation. Mexico has over one-third as many people as the United States but in 1996 had a gross national product only 3 percent as great. Economic relations with the United States are crucial to Mexico because approximately two-thirds of the country's foreign trade is with the United States, and the latter nation is overwhelmingly the main source of outside capital. The 1994 North American Free Trade Agreement (NAFTA) between the United States, Canada, and Mexico has stimulated considerable economic activity on both sides of the Mexico–U.S. border.

There is inevitably a touchiness in relations between Mexico and the United States due to (1) the disparity between them in wealth, power, and cultural influence; (2) the historic fact that Mexico lost more than one-half its national territory to the United States in the 19th century; and (3) Mexicans' national pride, derived from great cultural longevity. Mexico's aboriginal Indian civilizations were more highly developed than those within the territory of the present United States, and in the early 17th century, when the first tiny English settlements of the future United States were struggling for survival at the edge of a vast wilderness, the court of the Spanish Viceroy in Mexico City was an opulent and far-reaching center of imperial power.

Geographic Signatures from Indian, Spanish, and Mexican Eras

Mexico's human geography exhibits many elements from three major eras: the Indian era prior to the Spanish conquest of the early 16th century, the Spanish colonial era of the 16th to the early 19th centuries, and the Mexican era following Mexican independence in 1821. In the Indian era, the country was the home of Indians speaking hundreds of languages or distinct dialects and possessing cultures more advanced than any others in the Americas except those within the Inca Empire in the Andes of South America. Great builders in stone, the aboriginal peoples constructed the mighty Pyramids near Mexico City (Fig. 20.2) and the huge temples and palaces of the Mayan culture in Yucatán. Today, richly varied Indian cultures persist and the Indian racial component in the population is pronounced; the greater part of the people of Mexico are *mestizos* with mixed Spanish and Indian blood. The present settlement pattern was powerfully influenced by the distribution of Indians at the time the Spanish came. The conquerors sought out concentrations of Indians as laborers and potential Christian converts; hence the locations of Spanish-inspired settlement tended to conform to the population pattern already established.

The Spanish colonial era lasted nearly four centuries. Following the overthrow of the Aztec Empire by Hernando Cortez (Hernán Cortés) in 1521, the Spanish Crown established the Viceroyalty of New Spain, which was ruled from Mexico City and eventually encompassed not only the area now included in Mexico but also most of Central America and about one-quarter of the territory now occupied by the 48 conterminous American states. A Hispanic pattern of life, still highly evident today,

Figure 20.2 The Pyramid of the Sun (shown in photo) and the Pyramid of the Moon are prominent pre-Aztec structures rising in the midst of maize and maguey fields near Mexico City. They are sited on broad avenues floored by stone and lined with ruined stone temples and priests' quarters. The entire assemblage, known as Teotihuacán, is what remains of a city that may have housed 200,000 people prior to the city's destruction by fires some 14 centuries ago. *Jesse H. Wheeler, Jr.*

was established by Spanish administrators, fortune hunters, settlers, and Catholic priests. Racial, ethnic, and cultural residues of their activities are apparent in recent estimates that *mestizos* comprise 60 percent of Mexico's population, while Caucasians (nearly all of Spanish ancestry) comprise 9 percent. Spanish is spoken as a first language by 93 percent of the populace, and religious affiliation in Mexico is 90 percent Roman Catholic. Other strong Spanish influences are found in architectural styles and urban layouts; towns and cities characteristically have a Spanish-inspired rectangular grid of streets surrounding a plaza at the town center. Cattle ranching begun by the Spanish also endures as a way of life in large sections of Mexico; the Spanish introduced not only cattle, but also wheat, sugarcane, sheep, horses, donkeys, and mules. The Spanish hunger for precious metals greatly expanded mining, particularly of silver. Many of Mexico's larger cities were founded as silver-mining camps or regional service centers for such mining areas. Innumerable towns and cities of today originated essentially as Catholic missions in the midst of clustered Indian populations.

Thus, when the Mexican era began in the early 19th century, Mexico was overwhelmingly rural, illiterate, village-centered, Church-oriented, and poor. In the capital, an educated elite governed and exploited the country with little benefit to the Indian and *mestizo* farmers in the countryside. The bloody and chaotic Mexican Revolution of 1810–1821 ended Spanish control and instituted a short-lived Mexican Empire that expired in 1823 and was followed by a federal republic in 1824. Then came a long period of political instability, with personalities of many

Definitions & Insights

THE NORTH AMERICAN FREE TRADE AGREEMENT (NAFTA)

The origins of NAFTA are varied, but three motivations have been central to this multinational agreement among the United States, Canada, and Mexico: (1) a wish by all three to promote Mexican economic growth and development; (2) hope for expanded export markets for the United States and Canada; and (3) a desire to decrease the flow of illegal migrants from Mexico to the United States through the expansion of Mexican manufacturing activity. The earlier success among the European nations in their creation of the European Economic Community (EEC) led NAFTA countries to assume that lower tariff barriers in Latin America and Anglo America would enable the affected economies to expand into markets for which they were best suited. In 1994, the first year that NAFTA was in effect, trade between Mexico and the United States alone went up 23 percent, totaling more than $100 billion in value. Even with the 1994 Mexican debt crisis and steep economic decline, the total U.S. export flow to Mexico in 1995 was 11 percent higher than the total for 1993, the last pre-NAFTA year. In 1995, the Mexican production of passenger cars rose 23 percent, and that of trucks rose 132 percent. The migration of the automotive industry southward from the United States has had a strong impact on the major industrial centers in Mexico because of the tariff reduction engineered by NAFTA.

In the United States there continues to be some populations who are discontented with NAFTA because of the steady relocation of textile, shoe, and component assembly industries. Opponents of this agreement say that the flight is simply toward cheaper labor costs, an absence of union activity, and much relaxed environmental constraints on factories. The final impact of this agreement will not be easily determined. The 1994 treaty allows fifteen years to pass before all of the targeted tariffs can be much reduced or eliminated. In the meantime, there continues to be regional economic competition at all levels as Mexico's economy climbs out of its 1994 near-collapse and American industries try to compete with labor costs that approximate $0.80 an hour compared to U.S. scales 10 to 15 times that amount.

persuasions contending for power. From the 1830s through the 1840s huge blocks of the nation's territory were lost, including the Central American part of former New Spain, as well as Texas, California, and areas now comprising New Mexico, Arizona, and parts of other states.

During the 1850s and 1860s, a dynamic Indian reformist, Benito Juárez, attained power for a time, but the country then settled into a long period of despotic personal rule by President Porfirio Díaz which began in 1876 and did not end until 1910. Democratic processes were largely suspended and the rural

population sank deeper into poverty even as heavy foreign investments were made in railroads, the oil industry, metal mining, coal mining, and manufacturing. Political discontent over national disunity and weakness, internal oppression and corruption, foreign exploitation, poverty, and a dynamic need for land reform then led to the Mexican Revolution of 1910–1920, which ousted Díaz and subsequently turned into a destructive struggle among personal armies of regional leaders. In 1917, a new constitution was enacted under which land reform was an urgent priority. Since then, huge acreages have been expropriated from *haciendas* by the government and redistributed or restored to landless farmers and farm workers, most particularly in a program that provided *ejidos*, or small farming units, for village farmers. Large *haciendas* still exist, especially in the dry north, but they contain only a very small proportion of Mexico's cultivated land.

Agriculture is still a highly important source of livelihood, although recent estimates report Mexico's population to be 71 percent urban. Corn is by far the most important crop. Mexico is almost certainly the area where corn was first domesticated and from which it spread to become the premier crop of the New World and many overseas areas. Although not an export of major importance, it is central to the Mexican diet. Beans and squash, also an inheritance from the Indian past, are other universal Mexican crops and foods.

Today, the total value of output from factories and mines is several times that of output from farms. All of Mexico's economic production needs to be seen in a regional context, however, as the output of particular commodities tends to be rather sharply localized within particular areas.

Regional Geography of Mexico

Mexico is strongly regionalized. Although innumerable areas have distinctive personalities, space restricts us here to regions of the broadest scale. We focus on (1) Central Mexico—the country's core region; (2) Northern Mexico—dry, mountainous, and increasingly intertwined with the United States; (3) the Gulf Tropics—composed of hot, wet coastal lowlands along the Gulf of Mexico, together with an inland fringe of *tierra templada* in low highlands; and (4) the Pacific Tropics—a poorly developed southern outlier of humid mountains and narrow coastal plains (Fig. 20.3).

Central Mexico: Volcanic Highland Core Region

It is very common for a Latin American country to have a sharply defined core region where the national life is centered. Mexico is a classic example of such a cultural and demographic pattern. About one-half of the population inhabits contiguous highland basins clustered along an axis from Guadalajara (population: 2.4 million) at the northwest to Puebla (population: 1.3 million) at the southeast. Toward the eastern end of the axis is Mexico City (population: 15 million; other estimates for a

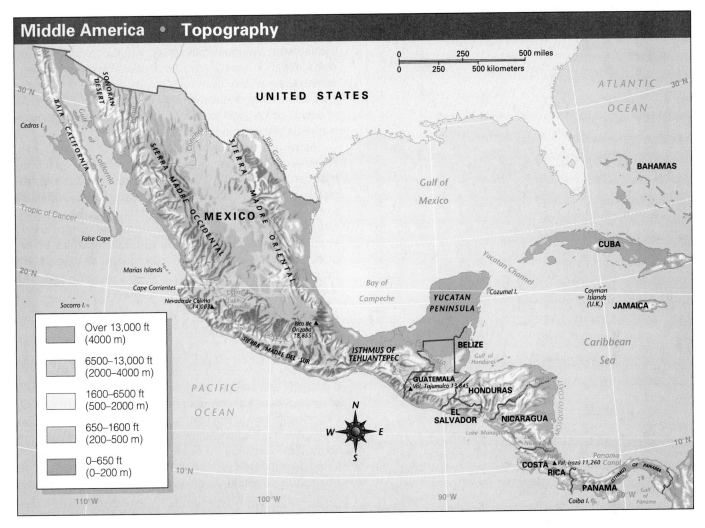

Figure 20.3 The variety of topographic features in Middle America as shown in this map help illustrate the importance of altitudinal zonation in the region. The significance of such topographic patterns has played a role in everything from mining and economic development to orographic precipitation and associated rain shadows.

broader metropolitan territory range upward to 20 million or more). The basins comprise Mexico's core region, which is often referred to as Central Mexico or the Central Plateau. The basin floors vary in elevation from about 5000 feet (1524 m) to 9000 feet (2727 m), with each basin separated from its neighbors by a hilly or mountainous rim. The centers of the basins contain flat land, which may be swampy. The swampiness has led populations to cluster on higher sloping land, and population pressure has induced cultivation of steep slopes, causing much soil erosion. Most areas are too high and cool for coffee or the other crops of the *tierra templada*. The prevailing *tierra fría* environment dictates growth of crops that are less subject to frost damage, such as corn, sorghums, wheat, and potatoes. In addition, cattle (both dairy and beef) and poultry are raised. A productive agriculture that still relies heavily on human labor

and animal draft power has long been maintained here for millennia, based on fertile and durable soils derived from lava and ash ejected by the many volcanoes.

As mentioned earlier, Mexico City was a major urban and political focus long before its conquest by Cortéz in 1521. Near the city are impressive pyramids and other monuments built by Indian precursors of the Aztecs. But Indian political power reached its apex under the Aztec Empire in the centuries immediately before the Spanish came. The Aztecs, notorious for their practice of ritual human sacrifice, came to the Basin of Mexico from the north and built their capital on an island in a shallow salt lake, with causeways to the shore (Fig. 20.4). This decision to build their most significant ritual and demographic center—Tenochtitlán—in the middle of what was then Lake Texcoco is a good example of simple origins confounding complex later

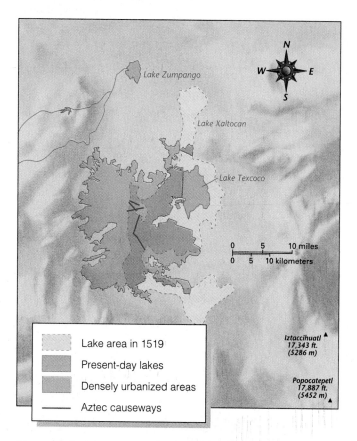

Lake Zumpango

Lake Xaltocan

Lake Texcoco

0   5   10 miles
0   5   10 kilometers

Iztaccihuatl ▲
17,343 ft.
(5286 m)

Popocatepetl
17,887 ft.
(5452 m) ▲

Lake area in 1519

Present-day lakes

Densely urbanized areas

—— Aztec causeways

Figure 20.4 Map of subsidence in Mexico City. This map of ongoing land subsidence in Mexico City highlights an environmental problem of major proportions. Having selected a lake bed for the area of initial urban settlement here, city dwellers have been plagued with periodic subsidence for centuries. The enormous growth of the city and associated increased withdrawal of water from below its surface, and the ever heavier urban burden placed upon this landscape, have led to the continuing and perplexing problem.

histories. Built in the mid–14th century, the city was lined with canals and dike constructions to control flooding and protect water resources for the city. It was here that Cortéz finally defeated the Aztecs. Mexico City was later founded in this same general locale. Most of the lake subsequently was drained, and for the past four and a half centuries, city structures have been built on the drained land. The result has been a gradual subsidence, or settling, of buildings (still in progress today) as the lake sediments contract. In recent times, this problem has been aggravated by overpumping of ground water to support a metropolitan region of more than 20 million people. But despite the unstable subsurface, including the fact that the city is underlain by a geologic fault, Mexico City now has imposing modern office towers with foundations designed to withstand both subsidence and earthquake shocks. This combination of two such significant environmental hazards underlying what is generally considered to be the largest city in the world reminds us that initial settlement decisions seldom anticipate the demographic, political, or cultural magnitude a ritual center can achieve as time passes.

Explosive urbanism has outrun the efforts of Mexico City's planners to cope with it. One mayor of the city commented that administering this monstrous sprawl was like "repairing an airplane in flight." Nonetheless, the city does have some spectacular planning achievements to its credit—not least of which are a modern subway system over 60 miles (100 km) long and a system to pump water to the city through the encircling mountain walls of the Basin of Mexico.

Amid the many memorials to Mexico City's dramatic past, the modern city carries on its functions as the nation's leading business and industrial center. It is estimated that approximately one-third of all manufacturing employees in Mexico work in thousands of factories in metropolitan Mexico City. Most factories are small, and manufacturing is devoted largely to miscellaneous consumer items, although some plants carry on heavier manufacturing such as metallurgy, oil refining, and the making of heavy chemicals.

Northern Mexico: Localized Development in a Dry and Rugged Outback

North of the Tropic of Cancer, Mexico is characterized by ruggedness, aridity, an extensive ranching economy supporting a generally sparse population, and some widely separated spots of more intense activity based principally on irrigated agriculture, mining, heavy industry, and diversified border industries adjacent to the United States. Most of Northern Mexico lies above 2000 feet (610 m) in elevation, but there are lowland strips along the Gulf of Mexico, the Gulf of California, and the Pacific. The climate is desert or steppe, with scanty xerophytic vegetation except where elevations are high enough to yield more rainfall. In the latter areas, some uplands are carpeted with steppe grasses, and mountains rising still higher have productive coniferous forests. Two major north-south ranges of high mountains—the Sierra Madre Occidental in the west and the Sierra Madre Oriental in the east—tower above the surrounding areas. Surface transport across them is poorly developed. Most forest land is in the Sierra Madre Occidental. This is Mexico's main source of timber, and lumbering is important in some places. Within Northern Mexico as a whole, ranching is the most widespread source of livelihood, although the total number of people supported by it is small (see Fig. 20.3).

In addition to being Mexico's main area of livestock ranching, Northern Mexico also provides the principal output of metals, coal, and products of irrigated agriculture. Iron ore, zinc, lead, copper, manganese, silver, and gold are the main metals produced. The ores, which often contain more than one metal, are smelted with coal mined in Northern Mexico. Some coal is made into coke and used for the manufacture of iron and steel at a number of places, particularly around the city of Monterrey (population: 2.1 million), which is Mexico's leading center of heavy industry and Northern Mexico's most important business and transportation hub.

Irrigated agriculture in the dry north has expanded greatly since World War II with the building of many multipurpose dams to store the water of rivers originating in the Sierra Madre and emptying into the Gulf of California or the Rio Grande. New or expanded irrigated districts now are widespread, with the main clusters located in the northwest and along or near the Rio Grande. Cotton is a major product of Northern Mexico's irrigated agriculture, and varied fruits and vegetables for shipment to U.S. markets also are grown. In some irrigated areas, the Mexican government has helped develop agricultural colonies of small landowning farmers who have migrated from more crowded parts of the country.

Since World War II, there has been a large upsurge of manufacturing in widely separated Mexican cities immediately adjacent to U.S. cities along the international boundary. The largest are Tijuana (population: 750,000, city proper; probably well over 1 million, metropolitan area) adjacent to San Diego, California, and Ciudad Juárez (population: 800,000, city proper) across the Rio Grande from El Paso, Texas.

These Mexican border cities are collecting points for Mexican and Central American migrants who attempt to enter the United States illegally and often succeed despite the efforts of the U.S. Border Patrol to apprehend them (Fig. 20.5). The tension in this zone has also grown because of the smuggling of drugs into the United States through the long border. A U.S. immigration law in 1986 gave amnesty to illegal migrants who could prove residence in the United States for a stated length of time, but the law prescribed tougher measures to try to halt the continuing illegal flow. It remains to be seen how well such measures can succeed along this bicultural border when great poverty lies closely adjacent to affluence, and strong demand for low-wage Mexican labor exists on the U.S. side. This tension came to a head in California during the election in 1994, when Proposition 187—prohibiting the state from providing any welfare or medical support to undocumented aliens (the great majority of whom in California come from Latin America)—brought national attention to the steadily increasing flow of illegal migrants to the United States.

The Gulf Tropics: Oil and Gas, Plantations, the Mayas, Tourism

On the Gulf of Mexico side, Mexico has tropical lowlands that have a much lower population density than Central Mexico but which are very important to the national economy. Their most valuable product is oil, discovered in huge quantities in the mid-1970s. Just as in coastal Texas and Louisiana, some fields are onshore and others lie underneath the Gulf. The Mexican oil industry is a government monopoly (carried on by a government corporation, Petróleos Mexicanos, or PEMEX), and the United States is its largest foreign customer. The bonanza oil strikes in the 1970s seemed so promising that the government borrowed heavily from foreign banks to finance both oil development and a wave of other economic and welfare schemes. Dependence on oil income grew to the point that crude oil represented 67 percent of all Mexican exports in value in 1983. Subsequently, however, a worldwide oil glut caused prices to

Figure 20.5 Due to the disparity in income and employment opportunities between the United States and Mexico, the 2000-mile-long border between the two countries sees endless attempts by Mexicans and Central Americans to enter the United States illegally. In this photo, the point of entry is a break in the fence along the Rio Grande River at El Paso, Texas. *Alon Reininger/Contact Press Images*

skid, and by 1990, crude oil accounted for only 33 percent of exports. Economic reforms in Mexico during the 1980s reduced inflation and spurred growth, but by 1994 the country's external debt was nearly $128 billion, and a sizable share of export revenue was being spent on debt service. The 1994 collapse of the Mexican peso came in large part because of the combination of massive foreign debt and diminishing oil revenues—a pattern that continued through the late 1990s. Nonetheless, recent restructuring of the economy has created a more hopeful outlook for the future.

The Gulf Tropics are Mexico's main producer of tropical plantation crops, which contribute in a relatively minor way to the export trade. Cacao, sugarcane, and rubber are leading export crops in the lowlands, and coffee exports come from a strip of *tierra templada* along the border between the lowlands and Mexico's Central Plateau.

The greater part of Mexico's Gulf Tropics lies in the large Yucatán Peninsula. In pre-Hispanic times, the Mayan people developed a notable civilization there and in adjacent parts of Mexico, Guatemala, and Belize. Its existence was unknown until ruined cities overgrown by jungle were found in the 19th century. Recent translations of Mayan writing and ethnographic studies of their descendants reveal an enduring culture with an elaborate past. Why they abandoned their cities is still a mystery.

The Pacific Tropics: Mountainous Southwestern Outlier

South of the Tropic of Cancer, rugged mountains with a humid climate lie between the densely populated highland basins of Central Mexico and the Pacific Ocean. This difficult terrain is far more thinly populated than Central Mexico, has a high incidence of relatively unmixed Indians living in traditional ways, and has

507

Regional Perspective

Maquiladoras

One of the most significant economic innovations that links Mexico and the United States are the fabrication centers that stretch along the northern border of Mexico. These are assembly and light-manufacturing businesses that either produce components of products to be exported abroad or assemble products from parts produced abroad and in Mexico. They then ship the finished product abroad—most often to the United States. These plants are called *maquiladoras.* They were begun in 1965 and expanded in their economic significance by 70 percent from 1970 to 1980. The number of these plants has doubled in the 1990s. They are generally located just south of the U.S.–Mexican border, where there is relatively inexpensive Mexican labor and easy transport of partially assembled goods from north of the border to these

free port facilities. The economic benefits of the *maquiladoras* to Mexico include the abundance of jobs linked with these factories and their ability to employ Middle American labor that would otherwise try to migrate to the United States. Since much of the capital supporting the *maquiladoras* is primarily from the United States and Japan, there is not as much trickle-down economic benefit from these activities as was anticipated, but the importance of the innovation has led it to generate more foreign exchange than tourism in the last decade.

Since 1994–1995, *maquiladoras* in Mexico have produced more than 14 percent worth of Mexico's export trade. In the region around Nuevo Laredo and Matamoros—just across the border from southern Texas—there are more than 2000 *maquiladora* operations. The oper-

ations have also been effective in providing economic opportunities for the Mexican laborers who settled at the border zone in order to participate in various legal seasonal labor programs in the American Southwest.

This economic innovation will continue to have a strong impact not only on Mexico's economic development, but on the economies of other nations as they consider this free port arrangement to encourage offshore factory growth. In addition, the manufacturing skills learned by maquiladora workers can be transferred to jobs in the industrial center around Mexico City and Monterrey, where there are solid job openings for skilled labor. While the *maquiladoras* seem a highly localized, borderlands economic phenomenon, their impact is in fact much more broadly regional.

relatively little economic production. There are two exceptions to the comparative underdevelopment. One is the coastal resort of Acapulco, located on a spectacular bay about 190 miles (*c.* 300 km) south of Mexico City (Fig. 20.6). This port city of the cliff-diver images continues to play an important role in attracting tourist dollars. Also, there is a major new iron and steel complex about 150 miles up the coast from Acapulco. Its operation is based on local deposits of iron ore and is intended to stimulate a more general economic growth within this region.

Figure 20.6   The resort hotels that have grown up around the early resort city of Acapulco play a strong role in the expanding importance of tourist dollars in the Mexican economy. The area's mountains, beaches, warm climate, and proximity to reasonably convenient airports have all helped this Mexican city and area grow rapidly.
Mark Lewis/Gamma Liaison

20.2 Central America: Fragmented Development in an Isthmian Belt of States

Between Mexico and South America, the North American continent tapers southward through an isthmian belt of small and poor countries known collectively as Central America. Five countries—Guatemala, Honduras, Nicaragua, Costa Rica, and Panama—were the initial members of the Central American Federation. They all have seacoasts on both the Pacific and the Atlantic oceans (Caribbean Sea; see map, Fig. 20.1). The other two Central American countries have coasts on one ocean only: El Salvador on the Pacific and Belize on the Gulf of Mexico, or the Atlantic. If the seven countries formed one political unit, they would have an area only about one-fourth that of Mexico (see Table 19.1) and a population only one-ninth that of the

United States. Geographical fragmentation is a major characteristic of Central America, manifesting itself in the fragmented pattern of its political units. Five of the present countries—Guatemala, El Salvador, Honduras, Nicaragua, and Costa Rica—originally were governed together as a part of New Spain, in a unit called the Captaincy-General of Guatemala. Composed of many small settlement nodes isolated from each other by mountains, empty backlands, and poor transportation, the unit never established strong geopolitical cohesion. Separate feelings of nationality became strong enough to fracture overall unity after the end of Spanish imperial rule in 1821. The area gained independence from Mexico in 1825 as the Central American Federation, but in 1838–1839, this loose association segmented into the five republics of today. In recent decades, there have been serious international tensions affecting the five-country area, associated in part with antigovernment guerrilla warfare in various countries. However, by the mid-1990s, tensions had lessened, and the area appeared to be entering a new era of greater political stability.

In Guatemala, El Salvador, Honduras, and Nicaragua, society is composed of groups that differ sharply in wealth, social standing, ethnicity, culture, and opinion. Wealthy owners of large estates have formed a social aristocracy since the first land grants were made in early colonial times by the Spanish Crown. Members of this group tend to be of relatively unmixed Spanish descent, although the great majority have some Indian blood. They have, in general, monopolized wealth, defended the status quo, and exercised great political power. Occupying a middle position in society are small land-holding farmers, largely *mestizos,* who own and work their own land. At the bottom of the scale are millions of landless *mestizo* tenant farmers and hired workers, and the many native peoples (largely in Guatemala) who live essentially outside of white and *mestizo* society. In the early 1990s, urban dwellers formed a majority in Nicaragua (62 percent of its population) but were in the minority in the other three nations. In each country, the national capital is overwhelmingly the largest city and serves not only as the demographic hub but also as the economic and political center of the nation.

Costa Rica has had a rather different historical development from the other four countries of the old Central American Federation. It was distant from the center of government in Guatemala and was left largely to its own devices in developing its economy and polity. Spanish settlers early on eliminated the smaller numbers of indigenous people there, and the ethnic pattern that developed was dominated overwhelmingly by whites of Spanish descent. By one estimate, more than 85 percent of the present Costa Rican population is white, whereas the same source shows an overwhelming predominance of *mestizos* and/or Indians in El Salvador, Honduras, Nicaragua, and Guatemala. No precious metals of significance were found and no large landed aristocracy developed. The basic population of small land-owning European immigrant farmers or their descendants managed to create one of Latin America's most

democratic societies despite some episodes of dictatorial rule. It has also evolved steadily as a nation undergoing productive economic development.

Panama and Belize (once British Honduras) were not part of the Central American Federation, and they achieved independence much later. Panama was governed from Bogotá as a part of Colombia until 1903, when it broke away as an independent republic. The United States was deeply involved in the political maneuvering that led to Panamanian independence, and in 1904 the United States was granted control over the Panama Canal Zone, site of the famous transoceanic Panama Canal, which was opened in 1914. In 1978, after long agitation by Panama for return of control of the Canal Zone, the United States ratified a treaty under which Panama will receive full control in stages, ending on December 31, 1999. Meanwhile, in 1989, a U.S. military invasion of Panama ousted its president, Manuel Noriega, who was captured, brought to the United States, and subsequently convicted of drug-smuggling charges by a U.S. court and jailed.

Belize, on the Caribbean side of Central America, was held by Great Britain during the colonial age under the name of British Honduras. Remote, poor, underdeveloped, and nearly uninhabited, the colony was used largely as a source of timber, with African slaves brought in to do the woodcutting. It was also claimed by Guatemala as part of its national territory. Not until 1981 did Belize become independent. It is now a parliamentary state governed by a prime minister and bicameral legislature; the British monarch, represented by a governor-general, is the chief of state. English is the official language and **ecotourism** (carefully planned tours to environmentally sensitive areas at relatively expensive rates) is steadily more important in the small country as the continuing prohibition of United States tourism to Cuba promotes the development of new tourist destinations in and on the margins of Caribbean Latin America.

The Central American states are among the poorest in Latin America. Mineral wealth is generally lacking, although Guatemala has some oil production and there is scattered mining of metals in all countries—including silver mining at a few sites that have been operational since early colonial times. Commercial agriculture producing coffee, bananas, cotton, sugarcane, beef, and a miscellany of other export commodities is very patchy in distribution and not very impressive in overall output, although it does provide the greater part of Central American exports. All the countries except overcrowded El Salvador still have pioneer zones where new agricultural settlement is taking place. There is some manufacturing in the few large cities, but the products are generally simple consumer goods for domestic markets. There are a few factories assembling foreign-made cars, but such bits and pieces of productive enterprise yield scanty returns in relation to the needs of the impoverished population, and the situation is worsened by the fact that a large share of the profit from such enterprises makes its way into very few pockets.

Physical Belts and Their Activities

The distribution of settlement—and the accompanying economic life—in Central America strongly relates to three parallel physical belts: the volcanic highlands, the Caribbean lowlands, and the Pacific lowlands. All Central American countries except Belize share the volcanic highlands that reach south from Mexico. Here lie the core regions, including the political capitals, of Guatemala, El Salvador, Honduras, Nicaragua, and Costa Rica. The only cities with over a million people in their metropolitan areas are San Salvador (1.5 million), Guatemala City (2 million), and San José (1.4 million)—all national capitals. Tegucigalpa (Honduras) and Managua (Nicaragua) are between 500,000 and 1 million in population. Dotted with majestic volcanic cones and scenic lakes, the highlands are the most densely settled large section of Central America. Climatically, they are classed largely as *tierra templada,* and most of Central America's coffee is grown there. Only a minor fraction of the land is still forested. The forces that drive such deforestation are, in part, the human migration into forests that traditionally have been landscapes of opportunity for landless peasants willing to clear forest and farm the ragged lands, generally with slash-and-burn methods.

There is popular sentiment in this region that the government has too much concern for governmental cabinets and high offices and too little for the needs of the common people (see Landscape in Literature, p. 511). Central America historically has had revolutionary movements that embody these fears, and the recent histories of Nicaragua and El Salvador have demonstrated this potential for popular uprising, and for U.S. involvement at one level or another. In Nicaragua in 1979, for example, the governing Somoza family—with a long history of U.S. support—was thrown out by the country's Sandinista revolutionaries. After a very contentious eleven-year rule characterized by continual battles with the military-backed Contras (also supported by the United States), the Sandinistas were replaced by an elected president, and by the early 1990s, a reasonably stable government was established with some of the earlier Sandinista reforms institutionalized. Nearly one-third of the countries in the broader Latin America have similar histories of U.S. military, economic, or political intervention.

By the 17th century, black African slaves were being brought into this region of Central America, providing the initial basis for Central America's relatively small black and **mulatto** or **Creole** populations. In the late 19th century, some black workers migrated there from Jamaica to work on the banana plantations being developed by U.S. fruit companies to serve the American market. At the turn of the century, several companies were consolidated into the United Fruit Company (now part of United Brands). An up-and-down history of banana growing saw many plantations eventually shift to the Pacific side of Central America because of devastating plant diseases on the Caribbean side. Today, exports of bananas,

grown in part on the Caribbean lowlands and in part on the Pacific lowlands, continue to be very important in the economies of Honduras, Panama, and Costa Rica. The 1999 near trade war between the United States and the European Union related to banana production from this area. The Caribbean lowlands represent the principal area in Central America where agricultural settlement is expanding on a pioneer basis. Settlers generally come from overcrowded highlands and are often aided by government colonization schemes featuring land grants and the building of roads to get products to market and bring in supplies. Even with this varied activity, however, there is a continuing problem of sluggish economic development.

The Panama Canal, providing an interocean passage through the narrow isthmus of Panama, is of great importance to both Panama and the world. Some 80 shipping routes, or about 5 percent of the world's cargo volume, use the 50-mile (80 km) shortcut from the Atlantic to the Pacific. On an ocean voyage from New York City to San Francisco, nearly 8000 miles are saved by using the Canal (Fig. 20.7). Approximately one-seventh of all United States trade is carried through it. The political and nearly military involvement of the United States in the early 20th century enabled Panama to become independent of Colombia and, with rapid U.S. diplomatic recognition, Panama opened the isthmus to a decade-long American construction effort. More than 25,000 lives were lost in the process, mostly because of disease in the hot and swampy lands that lay between the Atlantic and the Pacific. It opened for business in 1914, and

Figure 20.7 When the Panama Canal was completed in 1914, it was thought to have been much too large for the shipping that might move through the Isthmus of Panama. This recent photo of the Miraflores Locks in the canal shows how close these container ships come to the edge of the concrete lock channels. The truly massive contemporary oil tankers are too wide and too long for the locks of the canal system now. Even with such inadequacies, the Panama Canal continues to play a major role in shipping patterns from East Asia to the Atlantic seaboard and Europe. Panama City lies in the background.   *Will & Deni McIntyre/Photo Researchers*

Landscape in Literature

And We Sold the Rain

Targeting the predicament in which many of these countries find themselves, author Carmen Naranjo sets the stage for a satire entitled "And We Sold the Rain" by describing a very poor Central American country. It is plagued by the classic problems of too little food, too many people, and a crippling gap between poor and rich and city and countryside. In an effort to take the population's mind off of hunger and these problems, the country's president decides to sell a natural resource—the rain that pours down on this impoverished place continually, causing everything to flood and people to grow weary of the wetness. He sells the rain to a Middle Eastern king of a country called the Emirate of the Emirs. The following excerpt begins just after the president has explained to his people how the money gained from the sale of the rain will let his small Central American country be stronger, independent, and more prosperous.

The people smiled. A little less rain would be agreeable to everyone, and the best part was not having to deal with the six fat cows [the International Monetary Fund, the World Bank, the Agency for International Development, the Embassy, the International Development Bank, and perhaps the European Economic Commission], who were more than a little oppressive. Moreover, one couldn't count on those cows really being fat, since accepting them meant increasing all kinds of taxes, especially those on consumer goods,

lifting import restrictions, . . . paying the interest, which was now a little higher, and amortizing the debt that was increasingly at a rate only comparable to the spread of an epidemic. And as if this were not enough, it would be necessary to structure the cabinet in a certain way. . . .

The president added with demented glee, his face garlanded in sappy smiles, that French technicians, those guardians of European meritocracy, would build the rain funnels and the aqueduct, [and provide a] guarantee of honesty, efficiency, and effective transfer of technology.

By then we had already sold, to our great disadvantage, the tuna, the dolphins, and the thermal dome, along with the forests and all Indian artifacts. Also our talent, dignity, sovereignty, and the right to traffic in anything and everything illicit.

The first funnel was located on the Atlantic coast, which in a few months looked worse than the dry Pacific. The first payment from the emir arrived—in dollars!—and the country celebrated with a week's vacation. A little more effort was needed. Another funnel was added in the north and one more in the south. Both zones immediately dried up like raisins. The checks did not arrive. What happened? The IMF garnisheed them for interest payments. Another effort: a funnel was installed in the center of the country, where formerly it had rained and rained. It now stopped raining forever, which paralyzed brains, altered behavior, changed the climate, defoliated the corn, destroyed the coffee, poisoned aromas, devastated canefields, desiccated palm trees, ruined orchards, razed truck gardens, and narrowed faces, making people look and act like rats, ants,

and cockroaches, the only animals left alive in large numbers.

To remember what we once had been, people circulated photographs of an enormous oasis with great plantations, parks, and animal sanctuaries full of butterflies and flocks of birds, at the bottom of which was printed, "Come and visit us. The Emirate of Emirs is a paradise."

The first one to attempt it was a good swimmer who took the precaution of carrying food and medicine. Then a whole family left, then whole villages, large and small. The population dropped considerably. One fine day there was nobody left, with the exception of the president and his cabinet. Everyone else, even the deputies, followed the rest by opening the cover of the aqueduct and floating all the way to the cover at the other end, doorway to the Emirate of the Emirs.

In that country we were second-class citizens, something we were already accustomed to. We lived in a ghetto. We got work because we knew about coffee, sugar cane, cotton, fruit trees, and truck gardens. In a short time we were happy and felt as if these things too were ours, or at the very least, that the rain still belonged to us.

A few years passed; the price of oil began to plunge and plunge. The emir asked for a loan, then another, then many; eventually he had to beg and beg for money to service the loans. The story sounds all too familiar. Now the IMF has taken possession of the aqueducts. They have cut off the water because of a default in payments and because the sultan had the bright idea of receiving as a guest of honor a representative of that country that is a neighbor of ours.[1]

[1]From Carmen Naranjo, "And We Sold the Rain," trans. by Jo Anne Engelbert, *Worlds of Fiction*, eds. Roberta Rubenstein and Charles R. Lawson (New York: Macmillan Publishing Co., 1993, 948–952).

PROBLEM LANDSCAPE

The Burning of Central America

"THE SKY IS THICKLY OVERCAST; THE smoke makes eyes water and lungs gasp for breath; the muggy heat is unbearable. This is the Nicaraguan capital, Managua. Life was much the same this week in Mexico City, which was on full smog alert. Honduras reported a continuing sharp rise in respiratory disorders, and until Monday the country's four main airports had all been closed for some days for lack of visibility. At Tikal, in Guatemala, 700 fire fighters were struggling to protect historic Mayan temples. For weeks now, fires have raged in scrub and forests from Panama to southern Mexico. Their effects are devastating, and will long outlast the human discomfort."[2]

The records of smoke, haze, and widespread burning associated with Central America in 1998 are thought by some to be just another reflection of the impact of El Niño and associated drought (see Problem Landscape in Chapter 21). However, the real issue is not the cyclic uprising of warm ocean waters linked with El Niño, but the more pervasive human trait of attempting to harvest regional forests for economic gain.

There has been a traditional view that the forest lands are the resources of those who come to them, cut down the vegetation, burn out the trees to let the light shine in, and plant crops. This *milpa* method (also called **slash-and-burn** or **swidden**) of land use goes back to the earliest settlements in Middle America (Fig. 20.A). It is more surprising now because all governments have placed legal restrictions on the burning of forest lands. However, as this article shows, the force of law has not generally been as strong as the drive for peasant farmholds or forest products.

In Central America, the article goes on to say, there is a steady incursion of big ranchers clearing land for cattle and of timber companies "getting permission to cut 50 trees and taking 500 . . ." There is, in addition, the ongoing impact of the former Nicaraguan Contras guerrillas staking claim to forest lands in the face of few other economic opportunities. These forces, when totaled, amount to a major dynamic of landscape change. A Nicaraguan ecologist has proclaimed "Nicaragua will be a desert in 50 years' time." Even making such an assessment has a danger to it, for two Central American ecologists lobbying against the burning of the forests have been murdered in the 1990s while others have received death threats.

A section of the Nicaraguan coast in the northeast was selected as a United Nations (UNESCO) "biosphere" reserve in February 1998. UN scientists in charge of evaluating such landscapes for this UN designation have declared that this pristine part of the Caribbean coast lands and forest lands of Nicaragua will have to be given military protection by the government if Nicaragua is serious about preserving this area in its current state. The gloom that these widespread Central America fires evokes is captured by this quote from the newspaper *La Prensa,* printed in Managua: "Desolation and destruction are everywhere; if nothing is done Nicaraguans will be without flora, fauna, or food."

[2]"The Burning of Central America." *The Economist*, May 30, 1998, 33–34.

Figure 20.A   Slash-and-burn agriculture and unauthorized forest burning are very powerful agents of landscape change in all Latin America. Major forest regions are often a long distance from direct governmental protection. A family can turn forest richness into a family resource by cutting, burning, planting, and harvesting an acre or two of forest land. In this process small populations are given traditional returns on family labor, but primary forest reserves are often lost to unproductive volunteer grasses. This same burning process is also undertaken in an effort to raise cattle for the Anglo America beef market. For whatever reason this burning of forestland is undertaken, it reduces the biotic diversity of the region, and the forest that may ultimately grow back if human intervention is denied will never be as rich a resource as it was as primary forestland.   *Paul Edmonson*

The Panama Canal • A Global View

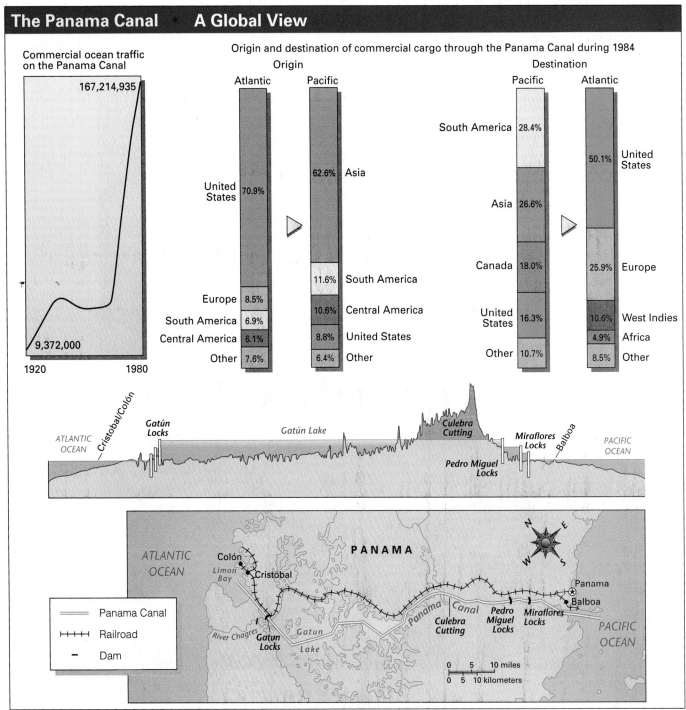

Source: *Atlas of Central America and the Caribbean.* Macmillan Publishing Co., New York, NY (1985).

Figure 20.8 In nine decades of using the Panama Canal, it has had a powerful influence on patterns of transoceanic shipping. In a voyage from New York harbor to the San Francisco harbor, using the canal cuts nearly 8000 miles from the transit. It pulls trade from all over the world not only because of the shipping time it saves but also because the rounding of the southern end of South America through the Straits of Magellan is a particularly difficult passage. This map with graphic shows the broad range of trade points that mark the origin and destination of cargo carried through the canal, and it shows the vertical profile of the actual canal route through the Isthmus of Panama. Because of the size of the locks, planners have begun, yet again, to think about alternative routes through the Isthmus. At the very end of 1999, the Panama Canal will be turned over to Panama authorities by the United States.

in 1996 it generated $486,000,000 in traffic income. The Panamanian government received $105,000,000 of that total.

More than one-half of Panama's population lives close to the canal, and, until 1997, there were some 10,000 U.S. troops stationed in and around it. The United States will pull out completely by December 31, 1999, and hand it over to the Panamanian government. This plan has generated some anxiety among the shipping nations for whom the canal is central to their routes. It is also estimated that it will take some $7 to $10 billion to modernize the locks and the channels so that the world's largest tankers can use the shortcut. Regardless of what the future holds for the canal, this geographic shortcut has been an innovation of enormous significance in global shipping patterns for nearly all of the past century (Fig. 20.8).

The eastern section of Panama adjoining Colombia contains a roadless stretch where the continuity of the Pan-American Highway System is broken, in what is called the Darien Gap (see Fig. 20.1). Thick rain forests, mountains, and swamps have presented serious obstacles to the bridging of this last remaining gap in a surfaced highway linkage from northern Canada and the United States to southern Argentina and Chile.

20.3 The Caribbean Islands

The islands of Caribbean America—commonly known as the West Indies—are very diverse. They encompass a wide range of sizes and physical types, great racial and cultural variations, and a notable variety of political arrangements and economic mainstays. Cuba, the largest island, is primarily lowland, although low mountains exist at the eastern and western ends of the island. The next largest islands—Hispaniola, Jamaica, and Puerto Rico—are steeply mountainous or hilly. Trinidad, just off South America, has a low mountain range in the north (a continuation of the Andes) and level to hilly areas elsewhere. Most of the remaining islands, all comparatively small, fall into two broad physical types: (1) low, flat limestone islands rimmed by coral reefs, including the Bahamas (northeast of Cuba) and a few others; and (2) volcanic islands, such as the Virgin Islands, Leeward Islands, and Windward Islands, stretching along the eastern margin of the Caribbean from Puerto Rico toward Trinidad (see Fig. 20.1). Each of the volcanic islands consists of one or more volcanic cones (most of which are extinct or inactive), with limited amounts of cultivable land on lower slopes and small plains. Barbados, east of the Windward Islands, is a limestone island with moderately elevated, rolling surfaces.

The islands are warm all year, although extremely hot weather is uncommon. Precipitation, however, varies notably, with windward slopes of mountains often receiving very heavy precipitation, and leeward slopes and very low islands having rainfall so scanty as to create semiaridity. The natural vegetation varies from luxuriant forests in wet areas to sparse woodland in dry areas. Most areas have two rainy seasons and two dry seasons a year. Hurricanes, approaching from the east and then curving northward, are a scourge of the northern islands in late summer and autumn.

Plantation agriculture and tourism are the economic activities for which these islands are best known. Sugarcane, the crop around which the island economies were originally built, is still the main export crop in Cuba, Barbados, and a few other places. Other commercial crops include coffee, bananas, tobacco, cacao, spices, citrus fruits, and coconuts. Production comes from large plantations or estates worked by tenant farmers or hired laborers, from small farms worked by their owners, or, in Cuba, from state-owned farms. Mineral production is absent or unimportant on most islands, the most conspicuous exceptions being Jamaica (bauxite), Trinidad (petroleum), and Cuba (metals). Manufacturing has made sizable gains during recent times, and it now provides the largest exports from some islands. Tourism is a major source of revenue in many islands, most notably the Bahamas, Puerto Rico, and the Dominican Republic. Most tourists come from the United States (see Table 19.3).

Emigration has long been an active option for this region's inhabitants, most of whom are economically underprivileged. Millions of West Indians now live in the United States, the United Kingdom, Canada, France, or the Netherlands—mainly in cities. The money they send back to relatives is a major source of island income.

The Greater Antilles: Major Islands of the Caribbean Sea

The four largest Caribbean islands are together called the Greater Antilles; the smaller islands are known as the Lesser Antilles. Of the five present political units in the Greater Antilles, Cuba, the Dominican Republic, and Puerto Rico were Spanish colonies, Haiti was French, and Jamaica was British. This diversity of colonial origins is only one of many ways in which these units have distinct geographic personalities.

By any standard, Cuba is the most important Caribbean island unit. Centrality of location, its large size, a favorable environment for growing sugarcane, some useful minerals, and proximity to the United States are important factors contributing to Cuba's leading role in the Caribbean. Cuba lies only 90 miles (145 km) across the Strait of Florida from the United States. At the western end of the island, Havana developed on a fine harbor as a major colonial Spanish stronghold. The island would stretch from New York to Chicago if superimposed on the United States and is nearly as large in area as the other Caribbean islands combined. Gentle relief, adequate rainfall, warm temperatures, and fertile soils create very favorable agricultural conditions. Agriculture has always been the core of the Cuban economy, and sugarcane has been the dominant crop. Cane sugar still accounted for the largest share of Cuban exports by value in 1995. The 1959 revolution, led by Fidel Castro, shifted the ownership of most Cuban land to the state. About 90 percent of all farmland is now under state control.

However, as in most Communist regimes, state-controlled agriculture has led to significant decreases in crop production in Cuba, and a more than 38-year U.S. trade embargo with Cuba has made it difficult to find effective replacement exports for the island.

In 1962, a great international crisis erupted when the Soviet Union began to build facilities in Cuba that would have accommodated nuclear missiles capable of striking Washington, D.C., New York, and other cities of the eastern United States. The presence of the Soviet missile silos and launchers was discovered through the use of aerial photo interpretation and remote sensing, geographic research tools of increasing military and commercial utility. Although the missiles were ultimately withdrawn under intense pressure from the U.S. government—led at the time by President John F. Kennedy—Cuba has remained a storm center of political life in the Western Hemisphere. One of the curious geographic outcomes of the tense geopolitics in

Definitions & Insights

TOURISM

Tourism is one of the classic two-sided coins of economic development. Consider, for the moment, the impact of the international tourist who arrives in a Middle American nation for a short pleasure or business trip. On the one side is the impact of the foreign exchange brought in by the traveler. The anticipated arrival of such people leads to the construction of hotel and recreational accommodations and to major investments in the infrastructure of ports, highway and rail networks, airports, and restaurants and entertainment facilities. At the same time, these dollars tend to promote the preservation of native crafts and historic landscapes to continue to attract visitors.

On the other side of the coin is the argument that much of the capital generated by tourism goes into transnational coffers, and that the wages paid to the local people who do everything from hotel work to playing tour guide to entertaining the tourists are at best modest. The sometimes powerful cultural tensions that arise when tour boats or buses arrive are very real (giving rise to the caustic phrase "If there is a tourist season, why can't we shoot 'em?"). The continued pressure on public space, transportation facilities, and natural resources such as beaches and national parks generated by tourist traffic adds to the tensions. However, looking at the magnitude of the foreign exchange generated annually (Mexico: $6 billion; Puerto Rico: $1.8 billion; Dominican Republic: $1.6 billion; Bahamas: $1.5 billion; Jamaica: $1 billion), it can be seen that there is a fiscal consequence to being a popular tourist destination. As a geographic theme, the interweaving of the landscapes, the physical settings, the international images of a place, and the network of travel systems that bring people to a popular tourist location presents interesting opportunities for study.

the Cold War era of the 1960s had Cuba shipping its sugar to mainland China, while Taiwan (then a close ally of the United States) shipped its sugar to the United States.

Aid to Cuba from the Soviet Union fell sharply when the USSR ended its support for the Soviet bloc in 1989 and the USSR and the Communist Party collapsed in 1991. This profound change has placed new strains on the Cuban economy. Castro's efforts in the mid-1990s to have the 1960 U.S. economic sanctions removed were brought to American attention in 1994 when tens of thousands of people—men, women, and children—from Cuba (and thousands more from Haiti at the same time) launched themselves from the northern and western shores of their island, hoping either to reach the Florida peninsula or to be picked up by the U.S. Coast Guard and ultimately to be granted asylum in the United States. Castro argued that if the United States relaxed its sanctions, the Cuban economy would be able to gain enough strength to make flight to the United States relatively less attractive for Cubans (see chapter opening photograph). With the mainland of the United States barely 90 miles distant from Cuba, the political and economic significance of this island continues to be a major thorn in U.S.–Latin American relations.

Havana (Habana; population: 2.2 million) is Cuba's main industrial center, as well as its political capital, largest city, and main seaport. Even with its economic troubles, Cuba's government, aided in a massive way by the former Soviet Union, has had some success in raising the adequacy of medical care, housing, and education over the years, particularly in comparison to the levels of such services during the 25 years of the Batista regime that Castro replaced in 1959.

East of Cuba, the mountainous island of Hispaniola is shared uneasily by the French-speaking Republic of Haiti and the Spanish-speaking Dominican Republic. In Haiti, the black population forms an overwhelming majority, although a small mulatto elite (perhaps 1 percent of the population) controls a major share of the wealth. Population density is extreme, urbanization is scanty, and poverty is rampant. The agricultural economy, largely of a subsistence character, reflects little influence from the former French colonial regime, which was driven out at the beginning of the 19th century. The republic, independent since 1804, has a West African flavor in its landscape, crops, and other cultural expressions. Deforestation, erosion, and poor transportation are major problems. Aside from subsistence farming, Haiti's small economy concentrates on light manufacturing (aided by low-wage labor), coffee growing, and tourism.

The Dominican Republic bears chiefly a Spanish cultural heritage. Santo Domingo, its capital, is the oldest European city established in the New World (c. 1496), possessing the first European church, school, and hospital as well. An independent country since 1844, the republic has a history of internal war, foreign intervention (including periods of occupation by U.S. military forces), and misrule, although recent peaceful transfers of power point toward greater governmental stability. The increasingly diversified economy relies on tourism, mining, oil

refining, and agricultural products such as sugar, meat, coffee, tobacco, and cacao. *Maquiladora* export production is playing an increased role in the republic's economy—a pattern developed in the NAFTA world of the Mexico–U.S. borderlands. The Dominican Republic also produces baseballs for the U.S. market, and has sent some of the best major league baseball players to the United States as well. Urbanization is more pronounced than in Haiti, and the capital of Santo Domingo (population: 2.5 million, city proper) is considerably larger than Haiti's capital of Port-au-Prince (metropolitan population: 1.3 million).

Puerto Rico, smallest and easternmost of the Greater Antilles, is a mountainous island with coastal strips of lowland. The island was ceded to the United States by Spain after the latter's defeat in the Spanish-American War of 1898, and was administered by Congress as a federal territory until 1952. In that year, it became a self-governing commonwealth voluntarily associated with the United States. Puerto Rico's pre–World War II economy was agriculturally based, with a strong emphasis on sugar. There were accompanying wide disparities within the populace in income and living conditions. After the war, a determined effort by Puerto Ricans to raise their living standard brought great changes. Hydroelectric resources were harnessed, a land-classification program provided a basis for sound agricultural planning, and a thriving development of manufacturing industries and tourism began, financed in large measure by capital from the United States. A complex package of tax incentives was organized to attract American businesses and industries, and a steady migration stream sent Puerto Ricans to the United States—primarily to the metropolitan New York area—also providing capital for the island in the form of family remittances.

Today, agriculture has been far outstripped in product value by manufacturing. Petrochemicals, pharmaceuticals, electrical equipment, food products, clothing, and textiles are among the major lines. Sizable copper and nickel deposits have recently been found. Although Puerto Ricans have chosen thus far to remain a commonwealth, there is strong pressure within the is-

land for statehood, and a small minority has expressed a desire for total independence (Fig. 20.9). San Juan (population: 1.8 million) is the capital and main city and port.

Jamaica, independent from Great Britain since 1962, is located in the Caribbean Sea about 100 miles (*c.* 160 km) west of Haiti and 90 miles (*c.* 145 km) south of Cuba. Its population of 2.6 million (as of 1998) is largely descended from black African slaves transported there by the Spanish and (after 1655) the British. When slavery was abolished in 1838, most blacks left the sugar plantations on the coastal lowlands and migrated to the mountainous interior. Tourism, bauxite, and agriculture have constituted Jamaica's economic base for decades.

The Lesser Antilles: Common Threads But Differing Circumstances

Over 3 million people live on the islands of the Lesser Antilles that are scattered in a 1400-mile (2250-km) arc from the United States and British Virgin Islands east of Puerto Rico to the vicinity of Venezuela. This screen of small islands defining the eastern edge of the Caribbean Sea contains a variety of political entities. The Netherlands Antilles and Aruba are Dutch dependencies evolving toward greater self-government. Guadeloupe and Martinique are French overseas departments, and Anguilla and Montserrat are still British dependencies. The British, however, have withdrawn from these islands in a massive way, starting with independence for Jamaica and Trinidad and Tobago in 1962. Since then, independence from British rule has been gained by Barbados (1966) and six smaller units. The British also granted independence to Bermuda in 1968, to the Bahamas in 1973, and to the mainland unit of Belize in 1981, but still retain their colonies of the British Virgin Islands, Cayman Islands, and Turks and Caicos Islands. All their former colonies have retained membership in the Commonwealth of Nations.

The economies of the islands suffer from many serious problems, including unstable agricultural exports, limited natural resources, a constantly unfavorable trade balance,

517

embryonic industries lacking economies of scale, and chronic unemployment. A shared trait of the islands is their history of European colonialism and plantation slavery. Such generalizations contribute to some understanding of the islands but give little hint of the individuality these small units present when viewed in detail. One example—Trinidad and Tobago—is discussed here.

Trinidad and Tobago is the largest, richest in natural resources, and most ethnically varied of the island units in the eastern Caribbean. Trinidad, by far the larger of the two islands, has offshore oil deposits in the south. Yet Trinidad refines more oil than it pumps. Imported crude oil from various sources is re-

fined and then shipped out, mainly to the United States. Some 43 percent of the population is black, most of whom live mainly in the urban areas and are often employed in the oil industry. Another 36 percent are the descendants of immigrants from the Indian subcontinent—a population stream that developed for indentured workers after the abolition of slavery. This element lives primarily in the intensively cultivated countryside. About 16 percent are classified as "mixed," and there are small numbers of Chinese, Madeirans, Syrians, European Jews, Venezuelans, and others. This ethnic complexity of Trinidad is reflected in religion, architecture, language, diet, social class, dress, and politics.

SUMMARY WITH SELECTED KEY TERMS

- **Middle America** is made up of **Mexico,** the **Central America** countries of **Guatemala, El Salvador, Belize, Honduras, Nicaragua, Costa Rica, and Panama,** and countries that are in the **Greater and the Lesser Antilles** in the Caribbean Sea. **Cuba, Haiti, the Dominican Republic, Puerto Rico, Jamaica, and Trinidad and Tobago** are the largest of the Caribbean Islands.

- **Mexico** is the **largest, most populated,** and **economically most developed** of the Middle America nations. Its area of **762,000 square miles** and its population of more than **100 million** gives Mexico a giant role in Middle America. It is also the largest nation in the world to have Spanish as its dominant language.

- The **North American Free Trade Agreement (NAFTA)** has played a major role in the pace of economic change in Mexico, particularly in the borderlands between Mexico and the United States. There has been steady demographic and economic growth in the cities lying just south of the U.S. border *(maquiladoras)*, and much of Mexico's light industry occurs in this region. The **major industrial manufacturing takes place further south.** NAFTA is seen as a source of both good news and bad news by Mexicans because of the activity that it stimulates and the migrants it collects along the northern edge of the country. **Mexico's GNP is 3 percent of that of the United States.**

- **Mexico's history is composed of three distinct eras: the Indian, the Spanish, and the Mexican. The Aztecs and Mayas** were the most important Indian populations, but there were others as well. There is still a strong Spanish presence evident, and the large population of *mestizos* represent the people who are a racial mix of the Spanish and the indigenous groups in Mexico. The Spanish were a major force in establishing land-use patterns that continue today. The Mexican era began with the 1810–1821 period of the **Mexican Revolution**, and the country suffered through many governments and the loss of much land to the United States in the following decades. The second period of revolution in Mexico was 1910–1920.

- Central Mexico is a **volcanic highland** area in which **Mexico City, the world's second largest city (15,000,000 or more)** is located. There is a steady problem of **urban subsidence** there, and **smog** is a growing problem also. It is Mexico's **leading industrial center.**

- The **Gulf Tropics of Mexico have gas, oil, plantations, Mayan ruins, and tourists** as major features. **Mexican oil** has been important in economic development as a nationalized resource, but it has also caused overspending and excessive debt as world oil prices

dip. **Tourism** has been a growing influence on settlement and development in this region as well as on the Pacific Coast and in northern Mexico.

- **Central America** is made up of **Guatemala, Belize, El Salvador, Honduras, Nicaragua, Costa Rica, and Panama.** In Guatemala, Honduras, and El Salvador there are sizable blocks of **native Indian peoples,** and there is a strong European presence in Costa Rica. Panama's history over the past century has been shaped powerfully by the construction of the **Panama Canal in 1904–1914,** and by the **American presence in support of the canal.** In all of the Central American countries—although less in Costa Rica—there is a **wide gap in economic prosperity between the very wealthy and the very poor.**

- **Plantation agriculture** continues to be important in the *tierra caliente* on both coasts of Middle America. The system of altitudinal zonation encompasses the *tierra caliente, tierra templada, tierra fría,* and *tierra helada.* **Bananas are the major plantation crop,** but **sugar** is also raised in Central and Caribbean America. There is new interest in **beef raising** for the Anglo American market, and that is causing some shift in land use—both legal and illegal. The illegal land use is associated primarily with unauthorized **deforestation** in both the *tierra caliente* and *tierra templada.*

- There are many **newly independent countries** in Caribbean America that have emerged from colonial connections to Europe in the past four decades. Some islands continue to have **colonial relationships** with Europe and maintain very strong economic and political affiliation with the countries that colonized them earlier.

- **Cuba is the most important Caribbean island because of its size, population, and very unique recent history. Sugar and tourism** have long been central to its economy, but since **Castro's clash with the United States** in the mid–20th century, both tourism and sugar trade have been powerfully constrained. Tourism is currently gaining economic significance again, but the United States continues to refuse to buy any sugar from the island. The **1962 Cuban Missile Crisis** continues to shape the U.S. disinclination to have any open political interaction with Cuba.

- **Puerto Rico** is the smallest and easternmost of the Greater Antilles and is a **self-governing commonwealth of the United States.** There continue to be groups on the island seeking either U.S. statehood or complete independence. There continues to be a **steady migration stream of Puerto Ricans** from the island to the American Northeast.

New York City is a **traditional migration destination**. Agriculture has been steadily losing ground to industry as Puerto Rico has actively been pursuing stronger economic development.

- Other major islands of settlement and of political and economic significance in the Caribbean include **Jamaica and, in the Lesser Antilles, the Netherlands Antilles, Aruba, Guadeloupe, Martinique, Barbados, the Bahamas, and the Virgin Islands.** These represent a variety of political affiliations. Banking, tourism, and varied resources are important in the economic profiles of these locales.

REVIEW QUESTIONS

1. Define Middle America, and list the countries in Central America and the major islands in Caribbean America.
2. What is the largest country in Middle America, and what is the population of its capital city?
3. What does NAFTA stand for? What countries have created this pact; and what is its primary significance for Middle America?
4. What are the zones defined in altitudinal zonation in this chapter, and what are the most significant settlement and agricultural characteristics of each zone?
5. What are the primary components of the regional geography of Mexico?
6. What is the name of the cooperative manufacturing innovation that brings together Mexican or Middle American labor and industrial components from all over the world? What is the general market for such activity?
7. Name the countries of Central America and describe the problems with forest burning that goes on in this region.
8. Describe the genesis, development, and significance of the Panama Canal.
9. What is the primary agricultural pattern of the Caribbean Islands? What is the role played by tourism in the local economies?
10. Compare the regional histories of Cuba and Puerto Rico.
11. What role have natural resources played in the development of the Caribbean Islands, and which have been most important where?

DISCUSSION QUESTIONS

1. Discuss the ways in which proximity to the United States has been a factor in the development of Middle America.
2. What have been the factors that have caused Mexico City to grow so rapidly? Discuss the ways in which such growth has been both positive and negative for Mexico.
3. Create a scenario that explains the role that Indian cultures have played in Middle America, bringing their influence into the present as well.
4. Discuss the role of migration in Middle America, with particular reference to the U.S.–Mexican border, the *maquiladoras,* and Mexican economic development.
5. Develop profiles of Guatemala, Costa Rica, and Panama as three distinct Central American nations, assessing how each of them has been influenced by the United States.
6. Discuss the role of the Panama Canal in the 20th century, and speculate on its role in the next century as well.
7. Lobby for and against the role of tourism in the Caribbean Islands, and in Middle America overall. What is the impact of such economic activity?
8. Discuss the role of burning forests in Middle America from the perspective of the government, the peasantry, environmentalists, and ranchers looking toward Anglo American markets.
9. Discuss the strongly distinct economic and geographic development of Cuba, Haiti, the Dominican Republic, and Puerto Rico. What has caused such differences?
10. What is the message of the literature excerpt, "And We Sold the Rain"?

Chapter 21

South America: A Giant Coming Awake

▲ In all of Latin America the question of children, family size, family planning, and the future of all of these elements is a feature both of the landscape and social consciousness. These two children on the Amazon River near Leticia, Colombia, carry much of the future in their youth. In their curiosity about the presence of a geographer with camera, they also represent all of Latin America in attempting to understand how to use the proximity and economic and cultural power of the United States in a productive manner.   Joseph J. Hobbs

21.1 The Andean Countries: Continuing Highland and Lowland Contrasts in Latin America

21.2 Brazil: Development in an Emerging Tropical Power

21.3 Countries of the Southern Midlatitudes

21.4 Antarctica: The Growing Fascination with the Last Continent

 outh America comprises the lion's share of Latin America's total land area. It is the fourth largest continent on Earth, amounting to 6,880,706 square miles (17,821,028 sq km). It is approximately 4500 miles from north to south, and a maximum of 3200 miles wide. It includes the world's largest Portuguese-speaking nation—Brazil —which is also home to the largest remaining tropical rain forest in the drainage of the broad network of the Amazon River. South America has a history of significant native American development and settlement on the west side of the Andes Mountains, which course south from Colombia in the north all the way to Tierra del Fuego at the very tip of the continent (Fig. 21.1). The continent also has a wide range of climates, landscapes, and ecological zones, ranging all the way from the parched dry desert of the Atacama in Chile to the lush tropical rain forests of central Brazil and eastern Peru and Ecuador. It sees itself as a region ready to be fully involved in the economic development of the Western Hemisphere and in the Pacific trade flows that will play a dominant role in the unfolding of the 21st century. Even though there are enormous areas still distant from transportation hubs and economic centers, and though there are ongoing struggles within its confines, South America sees itself as a giant coming awake (Fig. 21.2).

21.1 The Andean Countries: Continuing Highland and Lowland Contrasts in Latin America

The South American countries of Colombia, Venezuela, Ecuador, Peru, and Bolivia—all within the tropics and traversed by the Andes—are grouped in this text as the Andean countries (see Fig. 21.1). Common interests among them were recognized in 1969, when five countries signed the Cartagena (Colombia) Agreement and agreed to create a new free-trade zone called the Andean Group. Initially Chile was part of the group,

but it later withdrew. Venezuela came late to the group, which is now made up of Bolivia, Peru, Ecuador, Colombia, and Venezuela. A few outstanding traits shared by these countries are summarized here.

1. *Environmental zonation.* This region's zonation, both vertical and horizontal, is pronounced. The natural setting of each country features local environments that offer extreme contrasts and that often are very isolated by natural barriers. There are, in addition, demographic patterns that relate to these zones as well.

2. *Fragmented settlement patterns.* There is widespread fragmentation in patterns of settlement and economic development because of the broken topography that characterizes the Andean countries. This is particularly notable in Peru.

3. *Populous highlands and sparsely settled frontiers.* Every country contains both well-populated highlands and an extensive pioneer fringe of lightly populated interior lowlands or low uplands awaiting development. Some interior areas already yield minerals (notably oil) on an export basis.

4. *Dynamic coastal lowlands.* Every country except Bolivia has an economically productive coastal lowland region. Bolivia originally had a Pacific coastal zone but lost it to Chile during a 19th-century war, thus becoming a landlocked state—as is adjacent Paraguay.

5. *Significant indigenous populations.* All countries had relatively large Indian populations in the mountains, valleys, and basins at the time of the 16th-century Spanish conquest, and these old settlement areas are still major zones of dense population and strong political influence today, especially in the highlands.

6. *Significant native Indian landscape signatures.* The Incas of Peru were the most creative and productive pre-Columbian (pre-1492) culture in Latin America. At the time of the Spanish conquest, the Incas had authority over South America from the current border of Ecuador and Colombia, south for nearly 2000 miles to the Rio Maule in Chile. From their governing center in Cuzco, the Incas engineered a system of roads, bridges, and settlements. In building these structures,

521

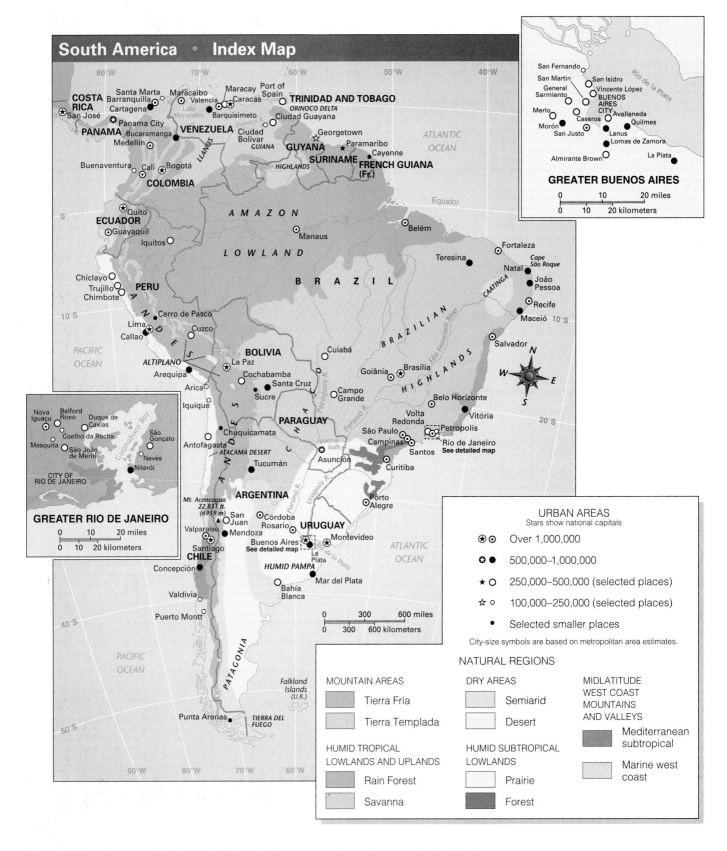

Figure 21.1 Index map of South America. General reference maps of natural areas and major cities in South America. Note the clearly defined arrangement of cities around the rim of the continent and the enormous and relatively empty expanses in the interior. See also the world landforms map in Chapter 2 (p. 36) on which the Brazilian Highlands are shown as "plateaus."

522

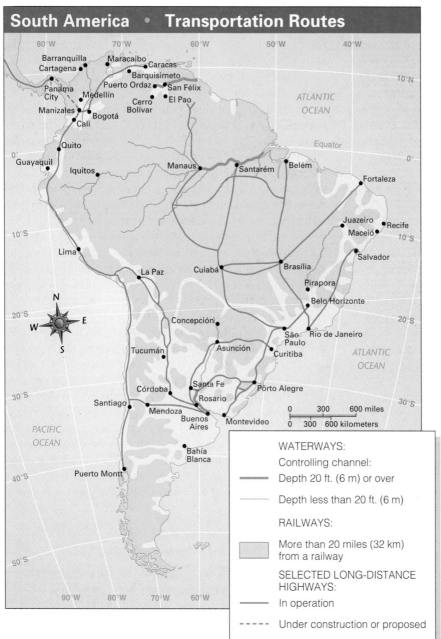

Figure 21. 2 Transportation routes in South America. The basic pattern of major highways in the Pan American Highway System is shown, together with new long-distance roads reaching deep into the interior of Brazil.

they demonstrated extraordinary skill, even though they had neither paper nor a writing system. (Records were kept by knotted ropes called *quipus*.) Their power was centered in two cities—Cuzco in the south and Quito in the north. They used geographic and topographic surveys and extended military ontrol throughout this empire from 1200 until the arrival of Spain's Francisco Pizarro in 1532. When Pizarro landed, he was warmly greeted by an Inca leader, Atahualpa. The Spanish conquistador welcomed the Incan dignitary into his camp and then killed him, launching the destruction of the empire that had ruled a great deal of the eastern side of the Andes for more than three centuries. Of all the riches Pizarro gained from this conquest, perhaps the most important is the potato, a tuber domesticated by predecessors of the Inca and diffused widely to the Old World during the Age of Discovery.

7. *Dominant Hispanic traditions.* Each one of the Andean countries maintains strong Hispanic traditions but at the same time has an important heritage from Andean Indians. The Spanish language is official in all five countries (although Indian languages also are official in Bolivia and Peru), and over nine-tenths of all religious adherents in every country embrace

Roman Catholicism. These badges of Iberian culture were imposed by the Spanish on the advanced Indian cultures within the Inca Empire. Today, Bolivia, Peru, and Ecuador have Indian majorities, and cultural survival from pre-Spanish times contributes to a keen sense of indigenous identity in each. In all the rest of Latin America, only Guatemala has an indigenous majority, although Indian strains are very pronounced in many countries with important mestizo elements. The latter is also true of Colombia and Venezuela, where native Indians amount to only 1 or 2 percent of the total but mestizos form the largest ethnic element.

8. *Endemic poverty.* All the countries in this region are relatively poor, although oil wealth makes Venezuela considerably better off than the others. However, considerable capital flows into many of these countries through the cultivation, processing, and marketing of illegal drugs—especially cocaine. It is difficult to track this money and accurately discuss its broader significance (see Problem Landscape, p. 526).

9. *Primate cities.* In most countries except Ecuador and Colombia, the national capital towers above all other cities in size and importance, reflecting the characteristics of primate cities.

Colombia: Large Resources and Problems in a Difficult Terrain

Within its wide expanse of the Andes, Colombia includes many populous valleys and basins, some in the *tierra fría* and others in the *tierra templada.* The great majority of Colombia's mestizos and whites are scattered among these upland settlement clusters, which tend to be separated by sparsely settled mountain country with extremely difficult terrain. Lower valleys and basins, along with some coastal districts, developed plantation agriculture employing slave labor. Consequently, 14 percent of the country's present population is mulatto and 4 percent is black.

The Highland Metropolises: Bogotá, Medellín, Cali

Colombia's inland towns and cities have tended to grow as local centers of productive mountain basins. The three largest —Bogotá, Medellín, and Cali—are diverse in geographic character. The largest Andean center in all of South America is Bogotá (metropolitan population: 4.0 million), Colombia's capital. It lies in the *tierra fría* at an elevation of about 8700 feet (*c.* 2650 m—much higher than Denver, Colorado). The Spanish seized the land in this region, put the Indians to work on estates, and improved agricultural productivity by introducing new crops and animals—notably wheat, barley, cattle, sheep, and horses. However, exploitation and disease reduced Indian numbers so drastically that the population took a long time to recover. The Spaniards were not able to develop an export agriculture, as night coolness and frosts precluded such crops as sugarcane and coffee, and early transport to the coast via the Magdalena River and its valley was extremely difficult. Slopes were so formidable

that a railroad connection with the Caribbean was not achieved until 1961. In recent decades, modern governmental functions and the provision of air, rail, and highway connections have led to explosive urban growth. The Bogotá area has even found a large agricultural export market in flowers grown for air shipment to the United States.

Medellín (metropolitan population: 1.5 million) lies in the *tierra templada* at about 5000 feet (1524 m) in a tributary valley above the Cauca River. The area lacked a dense Indian population to provide estate labor, was difficult to reach, and consequently remained for centuries a region of shifting subsistence cultivation emphasizing corn, beans, sugarcane, and bananas. The introduction of coffee and the provision of railroad connections changed this situation in the early 20th century. Excellent coffee could be produced under the cover of tall trees on the steep slopes, and the Medellín region became a major producer and exporter. Local enterprise and capital developed the city itself into a considerable textile-manufacturing center, and it grew from about 100,000 in population in the 1920s to its present size approaching 1.5 million.

Cali (metropolitan population: 1.4 million) is located in the *tierra templada* some 3000 feet (915 m) above sea level on a terrace overlooking the floodplain of the Cauca River. Cali had easier connections than did Medellín with the outside world via the Pacific port of Buenaventura. The Spanish developed sugarcane plantations in the Cali region and also produced tobacco, cacao, and beef cattle. Initial Indian labor was followed by that of African slaves. From the late 19th century onward, dams on mountain streams provided hydropower for industrial development. This asset, along with cheap labor, attracted foreign manufacturing firms, and Cali grew at a rate comparable to that of Medellín.

Colombian Coasts and Interior Lowlands

The Caribbean and Pacific coasts of Colombia contrast sharply with each other in environment and development. The Pacific coastal strip has tropical rain forest climate and vegetation. In the north, mountains rise steeply from the sea, but there is a marked coastal plain farther south. The area is thinly populated, with a large proportion of the population of African ancestry. The main settlement is Buenaventura (population: 165,829), an important port but not a large city. Its principal advantage is that it can be reached from Cali via a low Andean pass. The Caribbean coastal area is much more populous. Three ports have long competed for the trade of the upland areas lying to the south on either side of the Magdalena and Cauca rivers. Barranquilla (metropolitan population: 917,486) and Cartagena (population: 725,000) are the larger and more important cities in terms of legitimate trade. But the smaller city of Santa Marta (metropolitan population: *c.* 175,687), known as a beach resort as well as a port, may well be more important in trade today. It is reputed to be the control center for Colombia's huge illegal export trade in cocaine and marijuana.

Most of the Caribbean lowland section of Colombia is a large alluvial plain in which agricultural development has been handicapped by floods. Nevertheless, the region has a history of ranching and banana and sugarcane production. The Colombian government is now fostering settlement by small farmers who grow corn and rice.

Colombia's interior lowlands east of the Andes form a resource-rich but sparsely populated region that makes little contribution to the national economy as yet. It is a mixture of plains and hill country, with savanna grasslands in the north and rain forest in the south. The grasslands are used for ranching and the forest for shifting cultivation. Considerable mineral wealth awaits exploitation.

The Colombian Potential

Colombia has the resources for notable economic development in the future. Energy resources include: a large hydropower capacity; coal fields in the Andes, the Caribbean coastal area, and the interior lowlands; oil in the Magdalena Valley as well as in newly discovered and developing fields in the interior; and large supplies of natural gas from the area near Lake Maracaibo. Iron ore from the Andes already supplies a steel mill near Bogotá, and large Andean reserves of nickel are under development. At least one-half of the country is forested, with most of the forest made up of mixed hardwood species in the tropical rain forest. Because of the intermixture of many different kinds of trees, this type of growth has relatively low utility for forest industries other than the making of plywood. Of the foregoing resources, only oil yet plays much of a part in the country's export economy. In 1998, petroleum and its products represented approximately 27 percent of all legal Colombian exports by value, with coffee representing 15 percent. However, illegal cocaine and marijuana may well have been the largest category of exports.

Venezuela: The Emerging Dominance of Caracas

Venezuela's population is concentrated mainly in Andean valleys and basins not far from the Caribbean coast, as it was when the Spanish arrived in the 1500s. The highlands contain the main city and capital, Caracas (city population: 2 million). At an elevation of 3300 feet, Caracas has a *tierra templada* climate. Around Caracas, the Spanish found a relatively dense Indian population to draw on for forced labor, as well as some gold to mine and climatic conditions amenable to commercial production of sugarcane and, later, coffee. Labor needs associated with sugarcane led to the introduction of African slaves, but Venezuelan plantation development remained relatively small. Approximately one-tenth of the present population is classified as black. Until the middle of the 20th century, Caracas was a fairly small city and its valley was still an important and highly productive area of farming. Now, the city fills and overflows the valley, and Venezuelan agriculture is now more generally associated with other highland valleys and basins.

Discovery of oil resources in the coastal area around and under the large Caribbean inlet called Lake Maracaibo have made Venezuela a relatively fortunate Latin American country in recent decades. It was one of the most influential charter nations in the 1961 founding of the Organization of Petroleum Exporting Countries (OPEC). Oil revenues have provided a per capita income that is fairly high for Latin America (see Table 19.1), although Venezuela is far from being a rich nation. The oil-based income has fueled rapid urbanization, attracted foreign immigration, and financed industrialization. However, in recent years the country's income has suffered because of a decline in world oil prices.

While the highlands have supported Venezuela in the past, and the coastal lowland is crucial to the present, the much larger share of the country that lies inland from the Andes is a major hope for the future. This sparsely inhabited section is mostly tropical savanna in the Guiana highlands and in lowlands (often marshy) along the Orinoco River and its tributaries. For centuries, remote interior Venezuela has been Indian country (in the highlands) or poor and very sparsely settled ranching country (in the lowlands), but development is now quickening. Government-aided agricultural settlement projects are occupying parts of the lowlands, and major oil and gas reserves are beginning to be exploited. In the Guiana highlands after World War II, enormous iron-ore reserves south of Ciudad Bolívar and Ciudad Guayana began to be mined for export by American steel companies, and huge bauxite reserves in the highlands continue to await exploitation. Venezuela has pushed the development of a major industrial center at Ciudad Guayana (city population: over 523,500, city proper). Ocean freighters on the Orinoco carry export products away and bring in foreign coal for industrial fuel. Major steel, aluminum, oil-refining, and petrochemical industries are in operation, with further diversification planned.

In terms of the physical setting, Venezuela is also home to Angel Falls (3212 ft), the world's highest waterfall. Its existence is part of the reason that Venezuela has dedicated nearly one-third of its area to protected parks and reserves. This governmental decision regarding land use also reflects the increasingly important role that tourist dollars play in Venezuela's economic picture.

Ecuador: Andean Tenacity, Dynamic Coast, Oil-Rich Amazon

In Ecuador, the Andes form two roughly parallel north-south ranges. Between them lie basins floored with volcanic ash. These basins are mostly in the *tierra fría*, at elevations between 7000 and 9500 feet (c. 2100–2900 m). Quito (city population: 1.1 million), Ecuador's capital and second largest city, is located almost on the equator, high up on the rim of one of the northern basins. Its elevation of 9200 feet (2800 m) yields daytime

PROBLEM LANDSCAPE

The Geography of *Coca* in Latin America

ONE OF THE CLASSIC RESPONSES OF UP-land farmers to the difficulty and cost of transporting agricultural commodities to the markets in the adjacent lowlands has been to turn grain into liquor. The "white lightning" that is such a part of the mountain lore of the Appalachian region in the United States was a farmer's response to how little money he made carrying bags of grain to the markets compared to the money he could earn selling illegal or untaxed liquor. In Latin America, however, illegal liquor is not the farmer's response to this geographic isolation from grain markets; it is coca, the common name for the shrub *Erythroxylon* or *E. coca*.

This shrub is found in the upland regions of the Andes of South America, as well as in the mountain regions of Caribbean Middle America. Before the Spanish came to Latin America, coca was limited in its use. It was chewed as a stimulant only by the upper classes of the Inca. The Spanish, however, saw that when it was chewed by the "indios" who were forced into heavy mining labor in the extraction

of mountain gold and silver ores, they could endure longer hours and carry heavier loads. The chewing of coca also reduced hunger pangs, diminishing the realization of the inadequate diet so characteristic of the upland Indians during the Spanish demands for mine labor.

Today approximately 90 percent of the world's cocaine comes from Colombia, Peru, and Bolivia. The mountain uplands of these three countries provide good cover, an accommodating environment, and considerable independence from the control of the central government. This native plant can be brought to harvest in under a year and a half, whereas upland fruit trees will take four years to produce a harvest. An acre of coca shrubs will generate between $1000 and $1300 in cash yield—approximately four to six times the return on any other mountain agricultural crop. In Bolivia from 1960 to 1985, the return on raising coca increased an average of 11 percent annually. Given the modest return on upland farming and the associated costs of moving bulk commodities to lowland market centers, it is clear why coca has become the upland crop of choice for many subsistence farmers—and many export-oriented mountain farmers as well.

The cultural pattern that has developed around the raising of coca includes the involvement of lowland drug-traffickers, who contract to receive, process, transport, and sell cocaine from the coca raised by the upland farmers. This trafficking has in turn led to whole networks of transporters, judges, police, customs agents, pilots, and "mules" (the people used to carry the most valuable drugs). The scale of this coca business is such that in 1996 a major drug kingpin, Juan Garcia Abrego, was arrested by Mexican authorities and extradited to Houston for trial. Journalist Howard LaFranchi points out that it is alleged that Abrego was masterminding a trade that brought "100 tons of cocaine annually to United States markets over the past ten years, earning him annual revenues of more than $20 billion."[1]

One researcher on the role of coca in the economy of Latin America pointed out that "the coca industry . . . in less-developed countries . . . is labor intensive, decentralized, growth-pole oriented, cottage-industry promoting, and foreign-exchange earning. If the coca industry were completely licit . . . it could be the final answer to rural development. . . ."[2] Another writer pointed out, " . . . the

[1]Howard LaFranchi, "Mexico Polishes Images with US," *Christian Science Monitor.* January 17, 1996: 7.
[2]Ibid.

temperatures averaging 55°F (*c.* 13°C) every month of the year, so that daily fluctuations between warm daytimes and cold nights are the principal temperature variations.

The first Spanish invaders came to Ecuador from Peru in the 1530s. They found dense Indian populations in Ecuador's Andean basins. The conquerors appropriated the land and labor of the Indians and built Spanish colonial towns, often on the sites of previous Indian towns. But they found no great mineral wealth here, and both climate and isolation worked against the development of an export-oriented commercial agriculture. It was from this hearth, however, that the potato (*Solanum tubero-*

sum) was diffused into Europe and into world trade. Even today, these Andean basins have an agriculture with a strong subsistence component and sales mainly in local markets (Figure 21.3). Potatoes, grains, dairy cattle, and sheep are the main crop and livestock emphases. The isolated Ecuadorian Andes attracted relatively few Spaniards, except to Quito, and did not require African slave labor. Consequently, the mountain population has remained predominantly pure Indian, although with some mestizo and white admixture. The isolation has recently decreased somewhat with the growth of a certain amount of industry in Quito and lesser Andean cities.

cocaine industry is expanding swiftly to other Latin American republics—Ecuador, Panama, Venezuela, Brazil, Argentina, and Guatemala—at a speed linked, ironically, to the intensity of eradication in the Andean countries."[3]

Coca represents a truly dramatic and regional problem. Just as the engine that drives the economic development of much of Latin America is the United States market, so is it also for coca. But coca is also the crop and production activity that the United States most reviles. Even while informal market mechanisms work to promote expansion of *E. coca* in the United States, the government, the military, and the media all make major efforts to force the upland farmers and their associated traffickers to replace this crop with cocoa, cotton, sugarcane, or any other upland crop that could be accommodated by the microgeographies of the *tierra templada* or the *tierra fría* where the coca plants are now raised. Environmentalists claim that besides the more than 171,000 hectares of coca cultivated in just Bolivia and Peru, "producers of illicit drugs have deforested as much as 1 million hectares of fragile jungle forest lands in the Amazon region."[4]

Like so many agricultural patterns that hold sway over the peasant farmer, the success of this cropping pattern is dependent upon a world beyond the borders and beyond the uplands of the farmer in this problem landscape. The farmer's well-being is not only threatened by a drop in the world price (street price) of his crop, but it is also put under stress by local efforts to eradicate the raising of the crop—through everything from military encroachment to governmental efforts to interest the farmers in alternative crops. Such programs are almost always funded by external (most often United States) sources, so the peasant farmers never know when support for alternative farming and processing will collapse.

This confusion is confounded by the reality that coca leaves play a very regular role in the lives of those in the Andean countries. In Bolivia, for example, the 13,000-foot (4,000-m) upland setting of much of the population has made chewing the coca leaves a normal thing to alleviate the influences of elevation. To organize a program to begin to spray coca plants or rigorously enforce prohibitions on growing and processing coca could cause a major disturbance of cultural norms.

The problem of cocaine is, in fact, a global problem that links the Andean countries and its upland farmers to major blocks of the urban world. To see the global linkage that ties two very disparate parts of the world together, consider what happens when the New York and Los Angeles police departments effect a massive sweep to jail drug sellers in their cities and change the flow of cocaine on the streets of the United States for even a few months. The market forces shift back up the connecting linkages from urban drug dealer to smuggler to Peruvian or other Latin American trafficker to the processor in the mountain highland to, finally, the peasant farmer who raises the actual crop. In a sense, like all farmers, his family's success depends on forces over which he has absolutely no influence or control. While we think of this problem landscape as a major factor in our own urban and suburban instability because of drug abuse, the Andean or Central American or South American upland farmer sees it as a problem that is just as real and just as disturbing to his own patterns of life.

[3]Mario de Franco and Richard Godoy, "The Economic Consequences of Cocaine Production in Bolivia: Historical, Local, and Macroeconomic Perspectives." *Journal of Latin American Studies.* Vol. 24, no. 2 (May 1992): 375–406.

[4]U.S. State Department Dispatch, "Fact Sheet: Coca Production and the Environment." Vol. 3, no. 9 (1992): 165.

A drastic shift of Ecuador's economy and population toward the coast took place in the 20th century. Through the 19th century, approximately nine-tenths of the people still lived in the highlands, but that proportion now has been reduced to slightly under one-half. The growth zone near the coast is a mixture of plains, hills, and low mountains, with a climate varying from rain forest in the north, through savanna along most of the lowland, to steppe in the south. The main focus of development has been the port city of Guayaquil (population: 1.5 million), located at the mouth of the Guayas River, which drains a large alluvial plain where bananas for export (Ecuador is the world's leading banana exporter) and rice for domestic consumption are grown. On adjoining slopes, coffee is planted for export, and along the Pacific there is a considerable fishing industry. Anchovy harvests have long been a feature of the coastal fishing industry. The Galapagos Islands, located in the Pacific Ocean off the western coast of Ecuador, have been claimed by Ecuador since 1832 and currently are significant in the Latin American promotion of ecotourism—the careful touring of concerned (but well-paying) groups through delicate and often endangered environments. The population of the coastal region is mainly mestizo and white.

Figure 21.3 Hillside farming has long been a landscape feature of South American agriculture. In this scene from mountainside farming in south-central Ecuador, the row crops have been planted in such a way that erosion is diminished. In addition to this contour planting, there have also been eucalyptus trees planted around the margins of the field to grow into a windbreak function. *Joseph J. Hobbs*

In the 1960s, major oil finds in the little-populated trans-Andean interior began to be exploited. The area lies under a tropical rain forest climate in the upper reaches of Amazon drainage. A several-hundred-mile pipeline across the Andes was constructed from the oil fields to a small port on the Pacific, and Ecuador became an oil-exporting country. In 1997, oil and its products provided more than 40 percent of the country's foreign-exchange earnings. Other major mineral resources are not known to exist. Possible future mineral discoveries in the interior rain forest add urgency to a border dispute with Peru, which controls considerable territory of this type claimed by Ecuador. In 1981, and again in 1994–1995, a bout of armed conflict between the two countries occurred along their undefined boundary in the mountains of southeastern Ecuador. In late 1998 the two nations signed a treaty resolving this issue.

Peru: The Ongoing Struggle for Development in an Environment of Extremes

This region's Indian populations have been famous for their extensive road building—the Inca had a network consisting of more than 2000 miles. Accompanying this road development was the domestication of the potato *(Solanum tuberosum)*—one of the most significant crops to diffuse from the Andean uplands to Europe and Asia. It was initially raised by the Indians in the Andes and the Incas in Peru and Ecuador and taken to the Old World in around the 1570s. For two centuries the potato slowly took hold in Britain, and by the middle of the 19th century, it was particularly dominant in Ireland. (A potato blight in Ireland in the 1840s led to a decline in productivity and to famine, and millions of Irish fled their country for the east coast of the United States.) The tuber should truly be called the Peruvian or Andean potato rather than the Irish potato, but it got its name to differentiate it from the sweet potato that followed approximately the same diffusion path. The Irish potato became important because it could be grown in sandy soils, in a wide variety of environmental conditions, in high or low moisture, and did reasonably well when kept in the ground until needed. In terms of nutrition, the lowly potato gives a higher caloric yield per unit of land than wheat, rice, or corn.

Peru is a poor nation with an extraordinarily fragmented distribution of population, induced by an environment that exhibits extremes of ruggedness (in the Andes), dryness (along the coast), or wetness (in the trans-Andean Amazon lowland). More than one-half of the population, predominantly pure Indian, lives in the Andean altiplano. Like the Andean Indians in other countries, these highlanders are descendants of an ancient civilization ruled in the three centuries before Spanish times by the Inca Empire, which originated around Cuzco (city population: 302,700) in Peru's southern Andes. The highlands provide a notably poor environment for the development of surplus-producing agriculture. The mountain ranges are considerably higher on the average than those of Ecuador or Colombia, and cultivable valleys and basins tend to be small, densely populated, and in the *tierra fría*, where cool temperatures preclude such export crops as coffee, sugarcane, or cacao and have largely limited the options of farmers to potatoes, grains, and livestock (llamas and alpacas) raised for home consumption and sale in local markets. Farmed areas in central and southern sections of the Peruvian Andes tend to be so dry despite their elevation that high productivity is limited to irrigated areas. (Long before the Spanish invasion, Indian engineers were adept at constructing irrigation systems.) Today, considerable cash is earned as tourists travel to varied Inca sites, especially to the monumental city of Machu Picchu in the Andes Mountains (Fig. 21.4). At the same time, many farmers, as in other Andean countries, have turned to the illegal but profitable growing and sale of the coca leaves from which cocaine is extracted. Peru is reputed to be the world's largest supplier.

A narrow, discontinuous coastal lowland lies between the Peruvian Andes and the sea. Climatically, it is a desert of remarkable aridity associated with cold ocean waters along the shore. Air moving landward becomes chilled over these waters. Onshore, a relatively cool and heavy layer of surface air underlies warmer air. For this reason, the turbulence needed to generate precipitation is absent, although the surface air does produce heavy fogs and mists during the cooler part of the year, moderating temperatures somewhat. The great majority of rivers from the Andes carry so little water that they die out before reaching the sea. But irrigated agriculture has been practiced for millennia along these river valleys as they cross the desert coastal plains.

In 1535, the Spanish founded Lima (city population: 6.4 million) in an area of this type a few miles inland from a usable natural harbor. The city proved relatively well located for reaching the Indian communities and mineral deposits of the Andes, especially the Cerro de Pasco silver deposits which, for a time, led the world in silver production. Lima was designated the colonial capital of the Viceroyalty of Peru and became one of the main cities of Spanish America. In the late 19th century, foreign companies became interested in the commercial agricultural possibilities of the Peruvian coastal oases. As a result, commercially oriented oases now spot the coast, producing varying combinations of irrigated cotton, rice, sugarcane, grapes, and olives. Small ports associated with the oases are generally fishing ports and also fish-processing centers, as the cold ocean along the coast is exceptionally rich in marine life. But this resource provides an uncertain livelihood. For a few years in the 1960s and early 1970s, Peru led the world in volume of fish caught (largely anchovies for processing into fishmeal and oil), and the port of Chimbote (city population: over

Figure 21.4 One of the most dramatic landscapes of South America is the remains of Machu Picchu—known also as the "Lost City of the Incas." This highland (8000 ft [2400 m]) city and ceremonial center of the Incas overlooks the Urubamba Valley in Peru and serves as both a major tourist destination and an archaeologic treasure of major consequence. It became known to the West in 1911. *Robert Fried/Tom Stack & Associates*

Definitions & Insights

EL NIÑO

"El Niño" is the popular name given to a weather condition that has made major news for decades. The name means "The [Christ] Child," referring to a shift in the temperature of ocean currents off the west coast of South America that occurs most often in December with the arrival of warm tropical waters spreading from west to east across the Pacific Ocean. Normal ocean currents give the west coast of South America a resident cold current (historically called the Humboldt Current, but now known as the Peru Current). This cold current is vital not only to the fishing industry, but also to temperature patterns on the adjacent lands of Latin America. However, once every four to six years, El Niño comes and changes everything. This thick layer of warmer water comes across the Pacific from west to east in the late summer and fall, often arriving off the coast of South America around Christmas. It caps the usually upwelling cold water associated with the Peru Current. Fishing is changed completely because of the shift in water temperatures. Regional air movement becomes unstable because the warm ocean surface of El Niño sets in motion a whole different set of storm and precipitation patterns than those associated with the normal cold Peru Current. The somewhat mythic power of El Niño is given credit for stimulating weather changes, droughts, hurricanes, tornados, and even modifying human behavior.

314,700, city proper) was the world's leading fishing port. But during the 1970s, a change in water temperature associated with El Niño led to a drastic decline and widespread unemployment in the industry. However, by 1991, the fisheries had recovered to the point that Peru was the world's fourth nation in total fish catch (after China, the former Soviet Union, and Japan).

Overall, the resources and performance of Peru's agriculture are so poor that the country must import a large part of its food. Consequently, it remains dependent on minerals to supply the exports it needs to gain foreign exchange. Oil and metals are the main ones. Most of the oil comes from a field developed during the 1970s in the remote Amazonian rain forest of northeastern Peru. Some is transported via pipeline across the Andes and some by river across Brazil from Peru's Amazon port of Iquitos (population: 293,000, city proper). However, the aggregate export value of metallic ores—mainly copper, zinc, lead, and silver—far exceeds that of oil. Metal-mining sites are divided between the Andes and the coast and Peru ranks third in production among silver-mining countries.

More than one-half of Peru occupies interior Amazonian plains with a tropical rain forest climate. Fewer than 2 million people inhabit this large area. The main settlement, Iquitos, originated in the natural-rubber boom of the Amazon Basin during the 19th century and has tended to be more oriented to outside markets via the Amazon than to Andean and Pacific Peru. The Amazon River is now open to oceangoing vessels for 1900 miles (*c.* 3060 km) into the interior.

Economic development in Peru has been hampered by governmental instability. The country has gyrated wildly between civilian and military rule. Even the military governments have followed very divergent policies—sometimes leftist and sometimes rightist—with one government undoing the "reforms" of another. The 1980s saw widespread guerrilla activity by such movements as the "Shining Path"—which purportedly took its inspiration from Communist China's Mao Zedong (Mao Tsetung). By 1992, this challenge to national stability had resulted in more than 5000 deaths. Despite the unrest, Peru held a democratic election in 1985, which brought about the first peaceful and democratic change of government in four decades. Then, in April 1992, the elected president, Alberto Fujimori—the son of Japanese immigrants to Peru—and the armed forces dissolved the elected Congress in favor of rule by decree. The president claimed that the new powers would allow him to deal more effectively with economic problems, the illegal drug trade, and guerrilla warfare. By the mid-1990s, Peru had regained reasonable stability and was devoting itself to the timely payment of the interest costs on very large foreign debts. Fujimori was reelected president in 1995, even with his history of suspending certain civil rights to fight the rebels of the Shining Path movement. The 1996–1997 siege of Peru's Japanese Embassy by the Tupac Amaru rebel group is a further example of ongoing tensions between classes in Peru.

Landlocked Bolivia: Isolated Andean Core and Distant Lowland Fringe

Bolivia is a land of extreme Andean conditions and is by far the poorest Andean country. Like the other countries, it has a sparsely populated tropical lowland east of the mountains. But, unlike the others, it has no coastal lowland to supplement or surpass the population and production of its mountain basins and valleys. Agriculture still employs two-fifths of Bolivia's people and is limited by extreme *tierra fría* conditions combined with aridity. La Paz (city population: 784,976, city proper) is the main city and the de facto capital (Sucre is the legal capital, but the supreme court is the only branch of the government actually located there). Lying at an elevation of about 12,000 feet (*c.* 3700 m), La Paz averages 45°F (8°C) in the daytime in its coolest month and only 53°F (12°C) in its warmest month. Aridity is a further problem; the city averages only 22 inches (56 cm) of precipitation per year, with 5 months essentially dry. A more rural cluster of people lives at a comparable elevation around Lake Titicaca on the Peruvian border. At 12,507 feet, this lake is the highest navigable lake in the world and boasts regular steamer service among lakeside ports high in the Andes. Smaller settled clusters are scattered through Andean Bolivia, mainly along the easternmost of the two main An-

Figure 21.5 The central market plaza is a landscape feature of both South and Middle America. The fruits, fresh meat, handicraft goods, and the presence of *indio* and *mestizo* populations are all markers of village scenes that are repeated from Chile to Mexico. *Joseph J. Hobbs*

dean ranges and the Altiplano—a very high basin between the two towering ranges. At such high altitudes, export-oriented commercial agriculture has not been possible. Not only is the climate unfavorable, but products would have to reach the sea via 13,000-foot (*c.* 4000 m) passes to the Pacific coast or perhaps by a long route to the Rio de la Plata to the east. Accordingly, the infusion of Spanish blood was relatively light, and the country has remained predominantly (over 80 percent) Indian. There continues to be a strong Indian presence in the upland villages all through Andean South America (Fig. 21.5).

Known as "Upper Peru" by the Spanish, Bolivia was settled by Europeans partly for the land and labor of its Indians but in good part for its minerals. In the late 16th century, the Bolivian mine of Potosi was producing about one-half of the world's silver output. Since the late 19th century, tin has been the mineral most sought in the Bolivian Andes, although exports of tin have now become far less valuable than exports of natural gas from Bolivia's eastern savanna lowlands. Since the 1920s, oil and gas discoveries have been made around the old lowland frontier town of Santa Cruz (population: 767,260, city proper). Andean Indians, Europeans, and Japanese have settled around Santa

Cruz in growing agricultural areas. Farther north, where the lowlands have a tropical rain forest climate and where intermediate Andean slopes lie in the *tierra templada*, other pioneer agricultural settlements have been attracting immigrants from the highlands who come to pan for gold, raise cattle (meat is flown to La Paz), cut wood, and grow coffee, sugar, and, illegally, the coca plant. Nevertheless, even with the historic richness of the Potosi mine and the newer lowland settlements, Bolivia today is among the poorest nations in all of Latin America.

21.2 Brazil: Development in an Emerging Tropical Power

Brazil is a major subdivision of Latin America and South America. It has an area of about 3.3 million square miles (8.5 million sq km) and a population estimated at 162 million as of 1998. Although smaller in area than the entire United States, Brazil is larger than the 48 conterminous states (Fig. 21.6). This giant country had only about 61 percent as many people as the United States in 1998, but its annual rate of natural increase is so much higher (1.4 percent as opposed to 0.6 percent in the United States) that the gap between the two countries is expected to narrow rapidly as we move into the 21st century.

Rapid population growth is an important facet of the recent dynamism that is giving Brazil an increasingly important role in hemispheric and world affairs. The dynamism is also dramatically expressed in such related phenomena as (1) explosive urbanism that has made metropolitan São Paulo (metropolitan population: 17 million) and Rio de Janeiro (metropolitan population: 11 million) two of the world's largest cities; (2) the rise of manufacturing to the point that manufactured exports now exceed agricultural exports in value; and (3) the diversification of export agriculture to such a degree that the traditional overdependence on coffee has disappeared (see Table 19.2).

By providing an expanding domestic market and labor force, rapid population increase can make possible economies of scale and an accelerating economic development in Brazil. But today's burgeoning Brazilian population also poses serious problems for the nation. Brazil is hard-pressed to keep economic output ahead of population increase and to keep education and other social services at adequate levels. The country now has a sizable complement of millionaires, a substantial and growing middle class, and large numbers of technically skilled workers, but the benefits from development have been distributed so unevenly that many millions still live in gross poverty. And Brazil remains deeply mired in a huge foreign debt, owed by a government whose ambitions have outrun its finances in recent times. However, some portents are favorable: Brazil does have abundant natural resources of many kinds, many friends among the world's advanced economic powers, and a remarkable degree of internal harmony among the diverse racial, ethnic, and cultural elements that make up the Brazilian population.

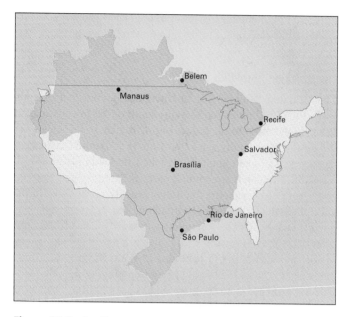

Figure 21.6 Brazil compared in area with the conterminous United States.

Environmental Regions and Their Utilization

Brazilian development must contend with many problems associated with humid tropical environments. The equator crosses Brazil near the country's northern border, and the Tropic of Capricorn lies just south of Rio de Janeiro and São Paulo. Only a relatively small southern projection extends into midlatitude subtropics comparable to those of adjoining sections of Uruguay and Argentina. Within the Brazilian tropics, the climate is classed as tropical rain forest or tropical savanna nearly everywhere, although relatively small areas of *tierra templada* exist in some highlands, and there is one section of tropical steppe climate near the Atlantic in the northeast. The overwhelming predominance of tropical rain forest and tropical savanna makes it hard to keep the soil fertile and, in places, poses severe maintenance and disease problems, creates general health and sanitation problems, and makes water supply difficult in the huge areas that have a dry season.

On the opposite side of the coin, Brazil is not handicapped to any marked degree by excessively high or rugged terrain. Nowhere does it reach the Andes. Mountains exist, but they are limited in extent and elevation. This leaves a country made up largely of relatively low uplands or extensive lowlands, the latter found primarily in the drainage basin of the Amazon River or in strips along the Atlantic coast.

The physiographic diversity of Brazil is very great when viewed in detail, but its major components can be subsumed under three landform divisions: the Brazilian Highlands, the Atlantic Coastal Lowlands, and the Amazon River Lowland. The country also includes a part of the Guiana Highlands that extends across the border from Venezuela, but this region of hills and low mountains is sparsely populated and has comparatively little importance in Brazil's present development.

The Brazilian Highlands

The Brazilian Highlands is a triangular upland that extends from about 200 miles (c. 325 km) south of the lower Amazon River in the north to the border with Uruguay in the south. On the east, it is generally separated from the Atlantic Ocean by a narrow coastal plain, though it reaches the sea in places. Westward, the Highlands stretch into parts of Paraguay and Bolivia. From here, the northern edge runs raggedly northeastward toward the Atlantic, steadily approaching the Amazon River but never reaching it. The topography is mostly one of hills and river valleys, interspersed with tablelands and scattered ranges of low mountains. From Salvador to Pôrto Alegre, the Highlands descend very steeply to the coastal plain or the sea. One-half of Brazil's metropolitan cities of a million or more population are seaports spotted along this coastal lowland: in the north are Fortaleza (city population: 2 million), Salvador (city population: 2.174 million), and Recife (city population: 1.315 million); in the south lie Rio de Janeiro, Santos (city population: 419,500), and Pôrto Alegre (city population: 1.28 million). Two of Brazil's three historic capitals are among them: Salvador (formerly Bahia), the colonial capital from the 16th century to 1763, and Rio de Janeiro (Fig. 21.7), which was made the colonial capital in 1763 and then, with independence in 1822, the national capital, until 1960. In that year, the capital

Figure 21.7   A dramatic view of one of the most frequently visited cities in Latin America, Rio de Janeiro, Brazil. The decision to create a new capital city in Brasília in the 1950s was an effort, in part, to pull some population away from this setting and into the interior of Brazil. *Brissaud-Figaro/Gamma Liaison*

was moved to the new inland city of Brasília (city population: 1.6 million). These locational shifts were motivated by changing economic and political alignments within the country, although the shift to Brasília is more symbolic of the country's hopes for future inland development than it is of the locational realities of present development (Fig. 21.8).

Another exceptional section of the Brazilian Highlands occurs in the northeast, where climatic maps show an interior area of semiarid steppe climate near Brazil's eastern tip. Actually, the steppe area is only a shade drier than a larger surrounding area in which the tropical savanna climate is so dry as to be almost steppe. The whole area, both climatic steppe and the marginal tropical savanna around it, is sometimes called the Caatinga, a word referring to its natural vegetation of sparse and stunted xerophytic forest. Wet enough to attract settlement at an early date, it has been a zone of recurrent disaster because of wide annual fluctuations in rainfall. Years of drought periodically witness agricultural failure, hunger, and mass emigration. The whole region called the Northeast is considered the poorest region of Brazil.

The Atlantic Coastal Lowlands

The narrow Atlantic Coastal Lowlands lie between the edge of the Brazilian Highlands and the Atlantic in most places, with the continuity broken occasionally by seaward extensions of the Highlands. Climatically, the northern part of the discontinuous ribbon of plain is tropical savanna, the central part is tropical rain forest, and the southern part is humid subtropical. This was the first part of Brazil to be settled by Europeans, and it has been in agricultural use for a changing variety of crops since the 1500s.

The Amazon River Lowland

The other major lowland area of Brazil is radically different from the coastal lowlands in development and size. This Amazon Lowland reaches the Atlantic along a coast that extends some distance on either side of the Amazon's mouth. From this coastal foothold, it extends inland between the Guiana Highlands and the northern edge of the Brazilian Highlands, gradually widening toward the west. Far in the interior, beyond Manaus (city population: 1.1 million, city proper), the Lowland widens abruptly and extends across the frontiers of Bolivia, Peru, and Colombia to the Andes.

The Amazon Lowland is composed generally of rolling or undulating plains except for the broad floodplains of the Amazon and its many large tributaries. The floodplains are often many miles wide and lie below the general plains level. They are nearly flat, but most of the lowland lies slightly higher, is more uneven, and is not subject to annual flooding. Tropical rain forest climate and vegetation prevail nearly everywhere, although southern fringes are classed as tropical savanna. Manaus and the ocean port of Belém (city population: 1.3 million) are great exceptions to the level of development in the rest of

Figure 21.8   It is alleged that Oscar Neimeyer, the primary architect of Brasília, drew his first designs for this new capital city on the back of an envelope. Created in thinly settled savanna in 1956 and assigned the role of national capital in 1960, Brasília has been only moderately successful in shifting the seat of national government and the country's demographic center from Rio de Janeiro into the country's interior. This is the Congress Building, designed by Neimeyer.   *George Holton/Photo Researchers*

the lowland; most of the Amazon region is still very undeveloped and sparsely populated. However, in recent decades the Brazilians have opened large areas by building trans-Amazon highways to augment river and air transportation. The result has been a considerable influx of people into Amazonia. These new settlers live, generally, on a semisubsistence basis, employing various forms and combinations of ranching, agriculture, mining, lumbering, and fishing. Many ecologists fear that the push into the Amazon will eventually destroy the trees and the irreplaceable ecology of the rain forest. The number of living species of plants and animals reaches its maximum here (for example, there are 600 types of palms, 80,000 plant species, 30,000,000 insect species, and 500 species of fish), and destruction of the forest would cause staggering losses to the world's biodiversity. It is for this reason that international government agencies have taken an interest in attempting to preserve aspects of this, and other, Latin American environments.

Eras of Brazilian Development

In the 1494 Treaty of Tordesillas, Spain and Portugal agreed on a line of demarcation in the so-called New World along a meridian which, as it turned out, intersected the South American coast just south of the mouth of the Amazon (Fig. 21.9). Because of this, the eastward bulge of the continent was allocated to Portugal. The Portuguese subsequently expanded beyond this line into the rest of what is now Brazil, with international boundaries drawn in remote and little-populated regions east of the Andes where Portuguese and Spanish penetration met.

The Sugarcane Era

The colonial foundations of Brazil were laid early in the 16th century. A Portuguese fleet, well off-course on a voyage around Africa, discovered the Brazilian coast in 1500, and a sporadic trade with local Indians for "Brazilwood" (a South American tree that produced a much-desired red dye in early trade) began. Settlement beyond the first small trading posts was promoted by Portugal in the 1530s in order to combat French and Dutch incursions. Meanwhile, the coastal settlements became increasingly valuable as they rapidly developed exports of cane sugar to Europe and brought shiploads of slaves from Africa to labor in the cane fields. Slave traders in search of Indian slaves penetrated the interior to some degree, and Catholic missions were founded among the Indians. Unfortunately, the missions often brought uncontrollable disease epidemics and slaving expeditions in their wake.

Expansion of coastal sugar plantations and increasing penetration of the interior dominated development in 17th-century Brazil. The colony became the world's leading source of sugar. Although some sugarcane was grown as far south as the vicinity of the Tropic of Capricorn, the main plantation areas developed farther north in more thoroughly tropical areas that also lay closer to Europe. Meanwhile, missionaries penetrated well up the Amazon, and a series of wide-ranging expeditions, originating largely from São Paulo, took place in search of either Indians to enslave or valuable new mineral deposits. In the late 1600s, sizable gold finds were made in the Brazilian Highlands some hundreds of miles to the north and northwest of São Paulo in what is now Minas Gerais state.

The Gold and Diamond Era

During the 18th century, Brazil's center of economic gravity shifted southward in response to the expansion of gold and diamond mining in the highlands north and northwest of São Paulo and Rio de Janeiro. By the mid-1700s, Brazil was producing nearly one-half the world's gold. In the latter part of the century this output declined, but by that time much of the highland around the present city of Belo Horizonte (founded in 1896 as a state capital) had been settled, along with areas near São Paulo and spots in the far interior. Despite the difficulty of traversing the Great Escarpment between Rio de Janeiro's magnificent harbor and the developing interior, the city became the port and economic focus for the newly settled territories. Gold and diamonds were ideal high-value commodities to move by pack animals over poor roads and trails to the port.

Besides the attraction of mineral wealth, a factor contributing to the rise of the new areas was the economic distress of the sugar coast to the north. This was a result of increasing development of competing sugar production in other New World colonies (notably in the West Indies). One solution was relocation of plantation owners with their slaves to the newly developing areas in Brazil. Another was movement into the Caatinga (Sertão) area inland from the sugar coast. There, a ranching economy began, augmented by cotton growing in the next century.

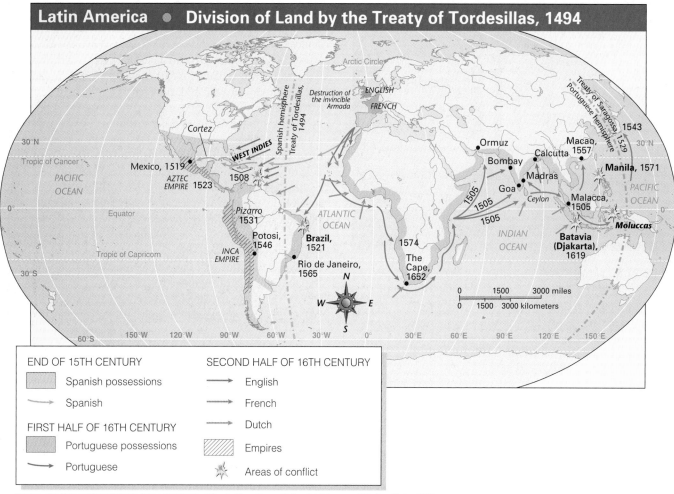

Latin America ● Division of Land by the Treaty of Tordesillas, 1494

END OF 15TH CENTURY

- Spanish possessions
- Spanish

FIRST HALF OF 16TH CENTURY

- Portuguese possessions
- Portuguese

SECOND HALF OF 16TH CENTURY

- English
- French
- Dutch
- Empires
- Areas of conflict

Source: *The Harper Atlas of World History*. The Treaty of Tordesillas. Harper Collins Publishers. p. 139. New York. 1992.

Figure 21.9   In 1494, Spain and Portugal met at Tordesillas, Spain, and made decisions about dividing up the non-Christian world. This map shows the Spanish domain lying to the west of the yellow line. The so-called "Spanish Hemisphere" gave Spain virtual control of all the lands of the New World except for the eastern wedge of Brazil. Portugal had its hemisphere lying east of those territories, and that gave the realm of South and East Asia to its sailors and settlers—except for the Philippines, which came under Spanish control in the late 16th century. This 15th-century cartographic division of the surface of the Earth played a dominant role in the patterns of colonization and economic control that were operative during the Age of Discovery.

The following 18th-century events, however, set a pattern of economic hardship for this northeastern part of Brazil that has never been broken: (1) continued dependence on hard-pressed coastal sugar and inland cotton production, (2) poverty of unusual severity even for Latin America, (3) recurrent disasters brought by drought, and sometimes by flood, in the climatically unreliable interior, and (4) flows of population to other parts of Brazil. Such flows turn into strong currents in the worst times.

Rubber and Coffee Booms

Nineteenth-century Brazil was changed by rubber and coffee booms, by much-intensified settlement of the southeastern section, by the beginnings of modern industry, and by extension of the ranching frontier far into the Brazilian Highlands. The country underwent these changes as a colony until 1822, as an independent empire with a strongly federal structure until 1889, as a military dictatorship until 1894, and finally as a civilian-ruled republic. Its political history at this time was turbulent, and not until 1888 did the country emancipate its large slave population.

Of the major economic developments of the 19th century in Brazil, the boom in wild-rubber gathering in the Amazon Basin had the most ephemeral impact. This followed the discovery of vulcanization—making rubber a valuable material—by the American Charles Goodyear in 1839. At that time, Amazonia was the only home of the rubber tree (*Hevea brasiliensis*). A "rush" up the rivers followed, and thousands of widely scattered settlers began to tap the rubber trees and send latex downstream. Manaus grew as the interior hub of this traffic and Belém as its port. But then some rubber tree seeds were smuggled to England, and seedlings transplanted to Southeast Asia became the basis for rubber plantations. By 1920 the Amazon production, which depended on tapping scattered wild trees, was practically dead.

Coffee was the preeminent boom product of Brazil during the 19th century. It had long been produced in Brazil's northeast in small quantities, and by the late 1700s it was an important crop around and inland from Rio de Janeiro. During the 1800s, production spread in the area around São Paulo and then increased explosively in output during the last decades of the century. This growth was related to the broadening world market for coffee, the emergence of expanded and faster ocean transport, the nearly ideal climatic and soil conditions in this part of the Brazilian Highlands, and the arrival of railway transport to move the product over land. A spectacular and critical transport development was the completion by British interests in 1867 of a railway linking the coffee-collecting city of São Paulo with the ocean port of Santos at the foot of the Great Escarpment. The coffee frontier crossed the Paraná River to the west and spread widely to the north and south of São Paulo.

Production doubled and redoubled. In the late 1800s, over a million immigrants a year from overseas poured into Brazil, mostly into the region near São Paulo, to work in the coffee industry (Fig. 21.10). São Paulo began its rapid growth as the business center and the collecting and forwarding point for the coffee industry; it also began a dynamic career as an industrial center. These developments were facilitated by very early hydroelectric production in the vicinity.

Twentieth-Century Development to the End of World War II

Brazil's development during the 20th century was not spectacular until after World War II. The rubber and coffee booms of the 19th century both collapsed in the early years of the 20th century. Brazil became a minor factor in world rubber production. It continued to lead the world in coffee production and export, as it does today, but increasing international competition depressed coffee prices, and the industry struggled economically.

The most positive developments of this period were:

1. The continued growth of early-stage industrialization in the São Paulo area;
2. Some expansion of the railway network;
3. The arrival, beginning in 1925, of Japanese immigrants who began to make important contributions to agriculture, especially truck farming;
4. Successful government attacks on two widespread debilitating diseases—malaria and yellow fever—during the 1930s; and
5. A major expansion of irrigated rice production in the Paraíba Valley between Rio de Janeiro and São Paulo. (Rice is a staple of the Brazilian diet, as are corn, beans, and manioc.)

Figure 21.10 Coffee continues to hold a major economic and agricultural role in both Middle and South America. The varied microclimates of the flanks of the Andes and other parts of the *tierra templada* allow for the production of many distinctive flavors and images. This mother and daughter are harvesting Robusta coffee in Rondonia, Brazil. *Nigel J. H. Smith/ Animals Animals/Earth Scenes*

Overall, however, the country struggled through this period without any major export boom. In 1930, at the beginning of the worldwide Great Depression, Brazil's economy practically collapsed, and in that year democratic government was replaced by a rightist dictatorship that lasted until 1945.

Quickened Development Since World War II

Since 1945, the pace of development in Brazil has generally been very rapid. However, fluctuations have occurred from time to time, including a downturn in the 1980s. Growth took place under elected governments between 1945 and 1964 but occurred most rapidly between 1964 and 1985 under a military dictatorship with technocratic leanings and great ambitions. In 1985, the dictatorship relinquished power to a new elected government, which then had to face the accumulated problems of a stalled "take-off." Among these were a crushing national debt that the military regime had contracted to finance expansion, and widespread economic misery caused by governmental attempts to keep up with payments on the debt. This pair of constraints (the Latin American plight alluded to directly in the literature selection in Chapter 20) on governmental stability has played a role all through the region. The International Monetary Fund has brokered a major proportion of the loans that have enabled some Latin American governments to stay in authority. Conditions for such loans generally include massive efforts to reduce government spending and control inflation, and monetary policies that support the local currency. Such policies are never popular, although the cash inflows that come from these IMF loans are always welcome.

The provision of a far more adequate internal transportation system has been an essential foundation of Brazil's growth in the last few decades. This has involved primarily the building of a network of long-distance paved highways and emphasis on trucks as movers of goods. Major urban centers have been connected, and additional highways have been pushed into sparsely populated and undeveloped areas. The new roads have attracted ribbons of settlement, have facilitated the flow of pioneer settlers to remote areas, and have furthered the economic development of such areas by connecting them to markets (for rubber, minerals, timber, fish, tourism, and some agricultural products) and sources of supply. But the single most spectacular instance of interior development has been the building of a new capital in a more interior and central location.

Brasília was a daring effort by the central government to lure people away from the coasts into the largely unsettled Brazilian interior. In 1956, a decision was made to treat the new capital as a growth pole—a term used to describe something that promotes growth of an industry or of a city through investment of new capital in a specific situation. The whole culture of the government in Rio de Janeiro—the traditional capital—was geared to the environmental amenities of that coastal location. In selecting central Brazil for the location of the new federal capital, the government was trying to find the most

effective way to prompt a significant demographic shift in Brazil, impelling migration toward the interior of the country. Built as a totally planned city on empty savanna in the Brazilian Highlands, Brasília was occupied by the government in 1960. Its growth, accompanied by unwanted adjacent shantytowns, has been such that the new federal district of about 2300 square miles (6000 sq km) had more than 1.85 million people by 1998 (see Fig. 21.8). However, Brasília remains only modestly settled, and even today its metropolitan population pales next to Rio de Janeiro, which continues to grow with nearly 11 million; São Paulo has about 17 million.

Agricultural Expansion: Sugar, Citrus, and Now, Soybeans

Three major agricultural boosts in older-settled territories have contributed to the country's recent economic growth. A program of the government in recent years to modernize and reinvigorate the old sugar industry had considerable success, and Brazil is again a major sugar exporter. In 1962, a severe freeze in Florida gave Brazil an entering wedge into the world market for orange concentrate, and by the 1980s Brazil was overwhelmingly the world's largest exporter. Also in the 1960s, Brazil became a major soybean producer and exporter, with production centered in the subtropical southeast and in a new district in the tropical Brazilian Highlands. By the mid-1990s, soy products had become Brazil's biggest agricultural export commodity.

Definitions & Insights

HUMAN MIGRATION: THE STEADY DEPARTURE FROM COUNTRYSIDE TO CITY

In 1960, São Paulo, the largest urban concentration in Brazil, had 3.8 million residents. In 1997, it had 16.5 million. In 1970, Rio de Janeiro had 3.4 million people; in 1997, it had more than 10 million. In the quarter century between 1970 and 1995, more than 30 million Brazilians abandoned traditional roots in the countryside and moved to urban centers. On the one hand, these migrants have been pushed out of their rural setting by the steady introduction of new farm machinery that has rendered the traditional roles of the rural underemployed even more marginal. On the other hand, however, the real catalyst in the rural-to-urban migration pattern in Latin America is the image of new opportunities, new landscapes, and wholly new lifestyles associated with the urban scene.[5]

The mayor of Caracas, Venezuela, who saw his city expand from just over a half million in 1975 to nearly 3 million in 1995, commented that "We have to be both firemen and philosophers" to deal with the demands of this unprecedented flow of urban immigrants. Tijuana, Mexico, and Medellín, Colombia, are both growing at rates

of approximately 10 percent a year. To deal with this growth and its negative as well as positive implications, South American cities are trying various innovations. In São Paulo—the largest city in South America—the Cingapura [Singapore] Project is attempting to replace hundreds of barrios (*favelas* in Portuguese) with low-rise blocks of flats. Generally, the *favelas* in all major Brazilian cities were bootstrapped (built with local scrap material and no governmental help) by self-reliant inhabitants. In the Cingapura Project, residents are moved to temporary barracks while the new flats are put up. Once the flats are finished, these same displaced, often recent rural-to-urban migrants move back in, paying for them with twenty-year, low-interest loans. The general response—particularly of those who are granted loans—has been positive. This project has been earmarked for 92,000 families, or half a million people. The scale of city-bound migration flows in South and all of Latin America is such that governments and agencies are looking to see how well the São Paulo effort works.

[5]"Latin America's Drift to the Cities," *Economist*, October 14, 1995.

The Energy in Brazil's New Industrial Age

Even more significant has been the extremely rapid expansion of Brazilian manufacturing. In the 1980s, the value of manufactured goods exported from Brazil began for the first time to exceed the combined value of all other exports. The products involved today are not just the simple ones of early-stage industrialization, although textiles and shoes are still major items. Steel, machinery, automobiles and trucks, ships, chemicals, and plastics are all important exports, as are weapons—especially types developed by the Brazilian military during its armed suppression of urban guerrilla opposition during the 1960s and 1970s. Aircraft are also manufactured, although mainly for the extensive domestic market that one would expect to find in a country so large. An industrial core area in Brazil has developed with São Paulo as its core, now producing nearly 60 percent of Brazil's manufactured goods.

Metals represent the other main contribution of Brazil's resource wealth to Brazilian industry. The Brazilian Highlands are highly mineralized with ores yielding numerous metals, outstanding among which are iron, bauxite, manganese, tin, and tungsten. Known reserves give Brazil about one-eighth of the world's iron ore, and the country is a leading producer and exporter. The first modern steel mill began production in 1946 at Volta Redonda, located in the Paraíba Valley between Rio de Janeiro and São Paulo. Others have been built in the industrial core zone since that time, and Brazil has become a considerable exporter of steel.

Brazil's greatest resource deficiency is in fossil fuels. Known supplies of coal are restricted to the extreme southeast and are very inadequate for the country's needs, although they do provide some coking coal. Domestic petroleum supplies are known thus far only in the northeast. Located offshore for the most part, they are the result of intense exploration after the drastic rise in world oil prices in 1973. But they supply only a little over half of Brazil's oil consumption (as of 1994), and crude oil is a major Brazilian import. This is true despite a relatively successful program to develop and use gasohol, whose 20 to 25 percent alcohol content is currently supplied by sugarcane grown in the northeast and in the area around São Paulo. Unless the problem of oil supply can be solved, the country's continuing industrial development drive may be seriously hampered.

21.3 Countries of the Southern Midlatitudes

Four countries of southern South America—Argentina, Chile, Uruguay, and Paraguay—differ from all other Latin American countries in that they are essentially midlatitude rather than low-latitude in location and environment. Northern parts of all the countries except Uruguay do extend into tropical latitudes, but the core areas lie south of the Tropic of Capricorn and are climatically subtropical. Argentina barely extends into the tropics in the extreme north, centers in subtropical plains around Buenos Aires, and reaches far southward into latitudes equivalent to those of northern Canada. Chile also extends south to these latitudes. Uruguay's comparatively small territory is entirely midlatitude and subtropical, and most Paraguayans live in the half of their country that lies south of the Tropic of Capricorn. The core areas of the three more eastern countries—Argentina, Paraguay, and Uruguay—have the same humid subtropical climatic classification as the southeastern United States. Chile's midsection or core area has a mediterranean (dry-summer subtropical) climate like that of California.

Besides their atypical locations and climates, these four Latin American countries share certain other regional characteristics. Agriculturally, for instance, they compete more with other midlatitude countries (in both the Northern and Southern Hemispheres) than with most other Latin American or other tropical countries. Argentina is an important export producer of such crops as wheat, corn, and soybeans, and thus competes with North American farmers. Both Uruguay and Argentina are sizable exporters of meat and other animal products. Paraguay exports soybeans and cotton. Chile imports more food than it exports, but it does export fruits, vegetables, and wine, as does California.

Because these countries never developed labor-intensive and export-oriented plantation agricultures in colonial times, they have a relatively small black population. Argentina and Uruguay have been major magnets for European immigration and have populations estimated to be 85 percent "pure" European in descent. Paraguay and Chile, in contrast, attracted fewer Europeans and are about 90 percent mestizo. Uruguay, Argentina, and Chile are among the most developed countries in Latin America. They have relatively low dependence on agricultural employment, with only 10 to 15 percent of their employed workers still in agriculture. Paraguay, with more than 45 percent of its labor force still on farms in the 1990s, is less developed and poorer than the other countries, partly because of its history of landlocked isolation from the outside world.

Argentina: The Forces of Nature, Immigration, Markets, and Politics in the Cultural Landscape

Argentina is a major Latin American country whose area of nearly 1.1 million square miles (2.8 million sq km), second in Latin America only to that of Brazil, is about one-third the area of the 48 conterminous American states. Argentina's estimated population of 36.1 million in mid-1998 ranked it far behind Brazil and Mexico within Latin America, but placed it in close proximity to Colombia (38.6 million) for fourth place.

Argentinian Agriculture

Agricultural products that Argentina exports through its capital city of Buenos Aires (population: 12 million metropolitan) and lesser ports are sufficient to make the country one of the world's principal sources of surplus food. The leading agricultural exports are wheat, corn, soybeans, and beef. Agriculture is concentrated heavily in the country's core region, known as the humid pampa. ("Pampa" comes from an Indian word for plains.) This is the Argentine part of the area identified on the map of natural regions as humid subtropical lowland prairie (see Fig. 21.1). The environment is outstanding for agriculture. Level to gently rolling plains lie under a climate similar to that of the southeastern United States, featuring hot summers, cool to warm winters, and an annual precipitation averaging 20 to 50 inches (c. 50–130 cm). But the area is specially advantaged by soils that are more fertile than might be expected from the climatic conditions. The grasses of the original prairie provided an unusually high content of organic material (humus), just as they did in the prairie areas of North America. The soils have a loose structure of fine particles, affording both a high concentration of plant nutrients and the maximum opportunity for plant roots to be nourished by feeding on individual soil particles. The local beef industry is fed from the grasses and grain that flourish in this setting, leading directly to the fact that the Argentine per capita beef consumption is the highest in the world.

From the 16th century to the late 1800s, the humid pampa was sparsely settled ranching country, and parts of it remained the domain of Indians. Then, in the late 19th and early 20th centuries, Argentina experienced a rapid and dramatic shift to commercial agriculture and urbanization. There was heavy

Figure 21.11 This scene, reminiscent of beef cattle and cowboys in the American West, shows a ranch in Argentina's humid subtropical climate. Argentine exports of meat and grain compete in world markets with comparable midlatitude products from farms and ranches in the United States. *Juan-Pablo Lira/The Image Bank*

European immigration, and population grew rapidly. Commercial contact with Great Britain contributed greatly to Argentina's new dynamism. At the time, the British market for imported food was increasing rapidly; refrigerated shipping came into use, allowing meat from distant sources such as Argentina to be imported; and a British-financed rail network (the most dense in Latin America) was built in Argentina to tap the countryside for trade. The network focused on Buenos Aires, which rapidly became a major port and a large city. British cattle breeds, favored by the tastes of British consumers, were introduced to Argentine ranches, or *estancias* (Fig. 21.11), along with alfalfa as a feed crop. Immigrant farmers assumed a major role in agriculture, generally as tenants on the large ranches that continued to dominate the pampa economically and socially. The immigrants came primarily from Italy and Spain but also from a number of other countries, giving present-day Argentina a complex mixture of Spanish-speaking citizens from various European backgrounds. Soon an export agriculture developed in which wheat, corn, and other crops supplemented and then surpassed the original beef exports.

Argentine areas outside the humid pampa are less environmentally favored, less productive, and less populous. The northwestern lowland just inside the tropics—also called El Gran Chaco—has a tropical savanna climate and supports cattle ranching and some cotton farming. Westward from the lower Paraná River and the southern part of the pampa, the humid plains give way to steppe (see Fig. 21.1) and then desert, with production and population declining accordingly. Population in these western areas is concentrated in oases along the eastern foot of the Andes, where mountain streams emerging onto the eastward-sloping plains have been impounded for irrigation. Vineyards and, in the hotter north, sugarcane are special emphases in a varied irrigated agriculture. Córdoba (population: 1.2 million) is the largest oasis service center and the second largest city in Argentina. To the south of the pampa lie thinly populated deserts and steppes in the Patagonian plateau, a bleak region of cold winters, cool summers, and incessant strong winds. Indian country until the early 20th century, Patagonia now has millions of sheep on large ranches, but little other agricultural production, although oil production has been gaining importance on the south coast.

Urban and Industrial Argentina

Only about one-tenth of Argentina's labor force is directly employed in agriculture and ranching. The great majority of Argentinians are city dwellers employed in an economy where service occupations are very important and manufacturing employs about twice as many people as does agriculture. Metropolitan Buenos Aires (and the surrounding province) has about one-third of the national population. It was settled in the late 16th century as a port on the broad Paraná estuary called the Rio de la Plata. But until Argentina became reoriented toward British and then world markets, the city remained a rather isolated outpost of a country focused on the north and the western oases, with relatively little trade by sea. In the 19th century, Buenos Aires first became the national capital and then underwent explosive growth as a seaport, rail center, manufacturing city, and receiver of immigrants.

The growth of industry in Argentina has conformed to a sequence often found in nations emerging from an underindustrialized state:

1. *Local markets.* Industrialization began when the government fixed tariffs to help small-scale industries supply consumer items for sale within the country and to support plants processing agricultural exports (for example, meat-packing plants). Called a policy of import substitution (encouragement of local firms to produce goods most frequently imported), it restricts consumption of popular foreign goods through the imposition of high tariffs on goods from overseas.

2. *Wartime isolation.* Some consumer industries such as textile manufacturing and shoemaking were expanded with government support when the country was cut off from imported supplies during World War I and again during World War II.

3. *State-led industrialization.* After World War II, a deliberate state-led policy of industrialization was instituted. There have been lapses in this policy and fluctuations in the mechanisms of support, but considerable success has been achieved in the quantitative growth of manufacturing and diversification of industrial output. Imported coal to supply power has been replaced by oil produced within Argentina (mainly from fields on the Atlantic coast of Patagonia), together with natural gas from the archipelago of Tierra del Fuego and the Andean piedmont. In addition, there is considerable hydroelectric output, and some production of atomic power. Steel plants using imported ore and coal have been built near Rosario on the navigable Paraná River, and diversification has extended into the manufacture of chemicals, machinery, and automobiles. The Iguassu Falls provide considerable additional engineering potential for hydroelectric power, but any government would face enormous resistance if there was an effort to modify the falls for power generation (Fig. 21.12).

However, the economic health of a relatively large urbanized service and industrial economy is open to considerable question in Argentina's case. Manufacturing has never become competitive on the world market, and the country's trade pattern is still dominated by agricultural exports and industrial imports. Manufacturing contributes considerably less to national production than might be expected in a country of Argentina's size and economic status, and recent rapid growth of the small-enterprise service sector of the economy reflects in part a decline in some industries, with resulting unemployment. And despite sizable industrial gains, Argentina has not become a prosperous country. In 1998, its per capita GNP of $8380 was lower than that of poor European countries such as Portugal ($10,160) or Greece ($11,460), and was a little more than half that of relatively poor Spain ($14,350).

Figure 21.12 Iguassu Falls, which Argentina shares with Brazil near the Paraguay border, are seen in this shot from the Argentine side. These 200-foot falls symbolize the great energy potential available to many South American countries in the form of falling water. Such resources are vital to "emerging" nations that often lack adequate deposits of mineral fuels. They are also valuable as tourist destinations in many parts of South and Middle America. *Robert Fried/Tom Stack & Associates*

The Disastrous Impact of Argentine Politics

Political turmoil and government mismanagement of the Argentine economy have taken a heavy toll on the country. After initial disturbances following independence, Argentina became a constitutional democracy that was actually controlled by wealthy landowners. This situation continued for many decades until 1930, when a military coup toppled the constitutional government. Between 1930 and 1983, the country was controlled either by military juntas or by Juan Perón or his party, with brief intervals of semidemocratic government. Although there were variations, both the military generals and the Perónists were ultranationalist and semifascist. They distrusted democracy, communism, and free-enterprise capitalism, and believed in authoritarianism, violent repression of opposition (which itself was often violent), and state ownership and/or direction of major parts of the economy. They courted the support of a growing urban working class by providing it with jobs, pay, and benefits beyond what was justified by its productivity and output, and they paid for these things by inflating the currency and borrowing abroad. Inflation rates eventually reached hundreds of percentage points a year, lowering savings and leading to the flight of investment capital and an economy heavily focused on currency manipulation and speculation rather than on productive investment.

Regional Perspective

The Southern Cone Common Market: Mercosur

In January 1995, the Southern Cone Common Market—known as Mercosur—became fully operative. It was built around the four nations of Brazil, Argentina, Uruguay, and Paraguay, and later added Bolivia and Chile as associate members. This union was created in part in response to the success that was being realized by the European Common Market (now the European Union). The benefits gained through a common market and its easier trade flows were appreciated by the South American nations, who were all trying aggressively to achieve greater economic development and international stature. The success of the North American Free Trade Agreement

(NAFTA) in northern Latin America after 1995 and 1996 gave additional momentum to the formation of Mercosur.

By the end of 1997, Mercosur was seen as an effective economic union of nations with a growing middle class, an expanding base of disposable income, and a particular interest in having Mercosur serve as a southern NAFTA. Relations between Argentina and Chile have improved with the construction of an oil pipeline allowing Argentine gas to flow to Chile. Two additional—and heretofore unprecedented—pipelines are slated for construction in the near future. Electricity is being sold back and forth across the southern Andes, and there is even talk

about expansion of the highway network to enable Argentine and Brazilian producers to transport soybeans, mineral resources, and manufactured goods to Chile in order to get into active Pacific Rim trade. Although the 1997 slowdown in almost all East Asian economies has slackened the pace of some of these innovations, Mercosur continues to set the stage for a new level of regional economic cooperation in the southern tip of South America. These nations are keenly interested in being more fully integrated into the economic dynamics of the northern end of the hemisphere, especially North America, and of East Asia.

Chile: Distinct Landscapes in Diverse Regions

Chile's peculiar elongated shape tends to conceal the country's sizable area, which is more than double that of Germany. From its border with Peru to its southern tip at Cape Horn, Chile stretches approximately 2600 miles (c. 4200 km), but in most places its width is only about 100–130 miles (c. 160–210 km). Given a favorable natural environment, Chile might not find its curious shape a serious obstacle to development, but the country has actually had to struggle with rugged terrain, extreme aridity, and cool wetness to such a degree that only one relatively small section in the center has any considerable population. The long interior boundary lies mostly near the high Andean crest, with the populous core areas of neighboring countries lying some distance away on the other side of the mountains. Although the Andean barrier has by no means been impassable during Chile's four and a half centuries of colonial and independent existence, traffic across the mountains has been difficult and is still light.

The Middle Chilean Core Region

Chile's populous central region of mediterranean (dry-summer subtropical) climate occupies lowlands between the Andes and the Pacific from about 31 to 37 South latitude. Mild wet winters and hot dry summers characterize this area, as they do southern California at similar latitudes on the west coast of North Amer-

ica (see Fig. 21.1). This strip of territory has always been the heart of Chile's core area. Topographically, it consists of lower slopes in the Andes, a hilly central valley, and low coastal mountains. The area was the southern outpost of the Inca Empire and was occupied in the mid-1500s by a small Spanish army that approached from the north. Members of the army made use of land grants from the king of Spain to institute a rather isolated ranching economy. Grants varied in size by the military rank of the grantees, and this fact, together with the conquered status of the Indians, created a society marked by great social and economic inequality. At one end of the scale was the aristocracy, composed of those holding enormous ranches. At the other end was a mass of landless cowboys and subsistence farmers. Men from all ranks of society married Indian women quite freely, sometimes several at a time. This has created the Chilean nation of today, over 90 percent of which is classified as mestizo.

The owners of great haciendas dominated the Chilean core area and the country itself well into the 20th century, and ranching remained the primary pursuit in middle Chile. Agriculture, both irrigated and nonirrigated, was carried on but was secondary in importance and generally supplemental to ranching. Wheat, vineyards, and feed crops became agricultural specialties (Fig. 21.13). During the 1960s and early 1970s, population pressure and resulting political demands led to large-scale land reform that somewhat reduced the role of many landowners. Growing population pressure in the core region has contributed to rapid urbanization in recent decades. The national capital of

Figure 21.13 The grape harvest in the central valley of Colchagua, Chile, is evocative of the images of California. Both locales share a mediterranean climate, and vineyards and wine production play important economic roles in both regions. *Carl Frank/Photo Researchers*

Santiago, founded at the foot of the Andes in the 16th century, now has a metropolitan population of more than 4 million. Metropolitan Valparaíso, the core region's main port and coastal resort, has another million, and the port of Concepción has about 700,000. Altogether, the area of mediterranean climate is the home of about 75 percent of Chile's 14.2 million people. The population of the country was 85 percent urban by a recent estimate—exceeded among South American republics only by that of Uruguay (90 percent), Argentina (87 percent), and Venezuela (86 percent).

To both north and south, the core area contains transition zones where population density lessens until areas of very scanty population are reached. On the north, the population declines irregularly across a narrow band of steppe to the arid and almost empty wastes of the Atacama desert. In the south, the transition zone is much larger and more important economically. South of Concepción, rainfall increases, average temperatures decline, and the mediterranean type of climate gives way to a notably wet version of the marine west coast climate, whose natural vegetation is impressively dense forests. The southernmost part of the mediterranean climatic area and the adjacent northern fringe of the marine west coast area as far south as the small seaport of Puerto Montt have become part of Chile's core in the past century, forming a less populous transitional end of the core. The Araucanian Indians native to these forests fought off Inca expansion, and they were also able to check most Spanish expansion for three centuries. They were eventually assimilated into Chilean society rather than truly conquered. Their full-blooded Indian descendants are still numerous in this part of Chile. Also distinctive is a less numerous German element descended from a few thousand immigrants who settled on this wild frontier in the mid-19th century. Today,

the German community is economically and culturally very important, even putting its stamp on architectural styles. Most people in the southern transition zone, however, are wholly or partly descendants of mestizo settlers who came from the core of Chile in the past century as population pressure increased. Agriculture in this green landscape of woods and pastures is devoted largely to the raising of beef cattle, dairy cattle, wheat, hay, deciduous fruits, and root crops.

Northern and Southern Chile

North of the core, Chile is desert, except for areas of greater precipitation high in the Andes. This part of Chile, the Atacama Desert, is one of the world's few nearly rainless areas. Mild temperatures, high relative humidity, and numerous fogs are the result of almost constant winds from the cool ocean current offshore. Chile gained most of the Atacama by defeating Bolivia and Peru in the War of the Pacific (1879–1883). Bolivia had previously included a corridor across the Atacama to the sea at the port of Antofogasta, and Peru owned the part of the desert farther north. The war was fought over control of mineral resources, primarily the Atacama's world monopoly on sodium nitrate, a material much in demand at that time for use in fertilizers and the manufacture of smokeless powder. Since then, other mineral resources of the Atacama and the adjacent Andes, principally copper, have eclipsed sodium nitrate in importance.

South of Puerto Montt, the marine west coast section of Chile continues to the country's southern tip in Tierra del Fuego, the cool island region of windswept sheep pastures and wooded mountains divided in ownership between Chile and Argentina. In this little-populated strip, the central valley and coastal mountains of areas to the north continue southward, but here the valley is submerged and the Pacific reaches the foot of the Andes along a rugged fjorded coastline. The coastal mountains project above sea level only in higher parts that form offshore islands. Extreme wetness, cool temperatures, violent storms, and dense mountain forests are characteristic. Little agriculture beyond some grazing of sheep has been possible, and population is very sparse. Some oil and gas are produced along or near the Strait of Magellan, which separates the South American mainland from Tierra del Fuego.

Like most Latin American countries, Chile must solve serious political problems if it is to achieve the development and prosperity its resources warrant. A period of democratic government led to the election of a Marxist administration in 1970 by a minority vote against badly divided opposition. This government's drastic changes in the political and economic structure of the country threatened Chile's democracy and created chaos in its economy. A military coup in 1973 then fastened the repressive rightist dictatorship of Augusto Pinochet on Chile. It undid many of the Marxist measures that had been instituted, and the country experienced considerable economic success in the new military regime's early years. This dictatorship achieved a degree of popularity with much of Chile's political right and center at the same time that it was violently repressing much of the left. Then the economy was buffeted by world recession and depressed copper prices in the late 1970s and early 1980s, and opposition to the regime expanded and intensified. In 1988, a referendum ousted Pinochet, and military rule was replaced by a new civilian government in 1990, but friction still exists between government and nongovernment factions. An energetic effort to bring Pinochet to court in Spain on charges of the murder and torture associated with his regime reminds all students of Latin America that the impact of the political regimes there may last lifetimes longer than the actual regime itself.

Uruguay: Costly Dependence on Agriculture in an Urbanized Welfare State

Uruguay, with an area the size of Missouri and bordered only by Argentina and Brazil, is a part of the humid pampa. Its independence from Argentina and Brazil resulted from its historic buffer position between them. In colonial times, there were repeated struggles between the Portuguese in Brazil and the Spanish to their south and west over possession of this territory flanking the mouth of the main Rio de la Plata river system in southern South America. After Argentine independence from Spain, these struggles continued at the same time that a movement for political separation grew in the territory north of the Rio de la Plata. Great Britain eventually intervened, fearing Portuguese expansion south of the river, and the two contenders signed a treaty in 1828 that recognized Uruguay as an independent buffer state between the two larger countries.

Montevideo, Uruguay's national capital, is the only large city in the country, and its city population of 1.36 million includes about one-half of all Uruguayans. The city has dominated the country almost from Montevideo's inception in the 17th century, as first a Brazilian (Portuguese) and then an Argentine (Spanish) fort. Its port facilities handle Uruguay's important waterborne trade. Montevideo directs an economy that developed such a high degree of government ownership that the majority of workers came to be employed directly or indirectly by the government. The city is the main center of Uruguay's manufacturing industries, which employed about one-fifth of the country's labor force in the mid-1980s but which export little except textiles. Meanwhile, agriculture and ranching, with 11 percent of the labor force, supply the country's meat-rich diet and the greater part of all Uruguayan exports. This structure began to fail in the 1960s as costs outran production, and the structure was devastated by the high oil prices imposed by oil exporters in the 1970s. Aside from hydropower, which supplies nearly all electricity, Uruguay has almost no natural resources to support industrialization and depends heavily on imports of fuels, raw materials, and manufactured goods, which must be purchased with the proceeds from agricultural and textile exports. In the early 1990s, Uruguay was attempting to cope with these problems and to preserve its recently restored democracy. Lower world oil prices were a factor easing the

burdens of this conflict-racked country. An extensive program is underway to privatize government-owned industries.

Development Problems and Prospects in Landlocked Paraguay

The landlocked and underdeveloped country of Paraguay has long been isolated from the main currents of world affairs. When a nation is landlocked, a wide range of geography-based conditions are set in motion. All transit into the landlocked nation depends on the capacity and inclination of the nation's neighbors to build, maintain, and keep open cross-country road networks, mountain passes, bridges, and other elements of a surface transport system. Beyond the transportation difficulties and the political and economic costs that have to be paid to move goods, and even people, from the landlocked nation to an ocean coast—even in this time of expanding air linkages—there is the psychological cost of such a geographic condition. The limits of the map have been the spark of conflict among countries for centuries. To be on the ocean or on major international rivers creates a view of one's place in the world quite distinct from that of the land-locked country.

The Spanish founded Paraguay's capital city of Asunción (city population: 546,637) in the 1530s at a site far enough north on the Paraguay River to promise security against the warlike Indians of the pampas to the south. In addition, high ground at the riverbank gave security from floods. Asunción became the base from which surrounding areas were occupied. But when independence came, Paraguay had no way to reach the sea except through Argentina or Brazil. The main route that developed reached down the Paraguay and Paraná Rivers to Buenos Aires. But this route is much longer than it appears to be on maps, owing to meandering of the rivers, and navigation was hindered by such difficulties as shallow and shifting channels, fluctuations in water depth between seasons, and sandbars. Transport costs were so high that Paraguay could not develop an export-oriented commercial economy, and the country remained locked into a predominantly self-contained economy based on the resources of its own territory.

West of the Paraguay River is the region known as the Chaco ("hunting ground"), which extends into Bolivia, Argentina, and Brazil. This is a very dry area, increasingly so toward the west. Precipitation averages 20 to 40 inches ($c.$ 50–100 cm) annually, but high temperatures cause rapid evapo-transpiration, and porous sandy soils absorb surface moisture. These conditions produce an open xerophytic forest in the east, which thins westward and then gives place to coarse grasses on a vast plain of flat alluvium where wide areas are flooded during the wet season. Paraguayans in the entire western region number only in the neighborhood of 100,000. A striking element in this sparse population are the scattered Mennonite colonies that aggregate approximately 20,000 people.

Paraguay's comparatively low level of development manifests itself in a number of ways, one of which is its total population. Only an estimated 5.2 million people occupied its sizable territory in 1998. This small population results partly from the fact that Paraguay has been a country of emigration for most of its history and has never attracted any large immigration. The latter circumstance has given the country a racial mix containing few whites and a high proportion of Guaraní Indian blood in the 95 percent of the population classed officially as mestizo. Urbanization and industrialization have not gone far. Asunción is the only urban place with more than 150,000 people, and about 45 percent of the country's labor force is still employed directly in agriculture. Per capita income is far below that of Argentina or Uruguay.

Recently, the pace of Paraguay's development has quickened. Modern road and air links with the outside world and between parts of Paraguay itself have been forged. With this access to outside markets, exports of soybeans and cotton have grown rapidly and have come to dominate the country's trade. The future may well see larger and more dramatic changes due to three huge new hydroelectric installations along the Paraná River on the southeastern border of the country (see Fig. 21.12). These were partly operative in 1996, with construction work continuing. Paraguay is exporting electricity to pay its share of the costs of these projects, which it undertook jointly with Argentina and Brazil. The new hydroelectric output will give Paraguay a far more favorable energy base in the future than this fuel-deficient country has had in the past.

21.4 Antarctica: The Growing Fascination with the Last Continent

Antarctica is the world's fifth largest continent, with an area of 5,500,000 square miles (14,245,000 sq km) lying south of the tip of South America—virtually filling the Antarctic Circle. It is the setting of enormous human dramas in exploration, bravery, and foolishness as people have crossed the continent with dog sleds, on skis, on foot, in airplanes, and now in tour helicopters. The mystery of the place comes from its winter of darkness, with continual problems of "whiteouts" caused by light refraction on an extensive snow and ice surface covering some 95 percent of the continent. Mirages are common, and average temperatures during the "summer" months barely reach $0°F$. Winter averages are the coldest in the world, with winter mean temperatures averaging $-70°F$ ($-57°C$).

The whole continent is alive with ice formations created by the 2–10 inches (5–25 cm) of annual precipitation that then move toward the open seas of the South Atlantic. Glacial action of all sorts provides a continual breaking away of valley glaciers, ice streams, and tongues all crashing into the sea at the margins of the continent, forming icebergs. There is virtually no human settlement beyond the research teams that have constructed an array of

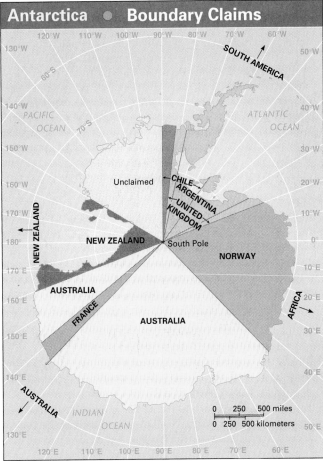

Source: Rubenstein, James M. *An Introduction to Human Geography*, 6th ed., Upper Saddle River, NJ.: Prentice Hall, 1999. p.271.

Figure 21.14 International land claims to Antarctica. Antarctica, the massive continent that surrounds the South Pole and lies at the southern extreme of South America and Tierra del Fuego, has become the focus of considerable global interest in the past three decades. The 1957–1958 International Geophysical Year (IGY) opened up significant international interest in exploration and the settlement of research teams. This map represents the declaration of territorial rights expressed by nations relatively close to Antarctica and by nations wishing to have control of at least some of the resources that are thought to be under the nearly total ice cover.

semipermanent structures on the small areas of exposed land that lie near the outer edges of the continent and on the island of Little America in the Ross Sea (Fig. 21.14).

There has long been speculation about the potential for considerable resource wealth lying beneath the sometimes-thousands-of-feet-thick cap of ice. Although many nations now claim some rights to the continent and support periodic settlements of research scholars, neither the resources nor the fauna are likely to lead to a significant exploitation of this enormous continental mass. It will continue to be, however, another frontier in the geography of the future.

SUMMARY WITH SELECTED KEY TERMS

- South America is made up of a number of **geographic subregions,** one of which consists of the **Andean countries,** including **Colombia, Venezuela, Ecuador, Peru,** and **Bolivia.** These countries have marked patterns of settlement, both along the coasts in the *tierra caliente* and in the uplands in the *tierra templada* or *tierra fría.*

- Major characteristics of the Andean countries include a varied **environmental zonation,** fragmented topographic patterns, populous highlands and sparsely settled frontiers, significant indigenous populations, significant **native landscape signatures,** dominant Hispanic traditions, endemic poverty, and **primate cities.**

- Colombia has three major highland metropolises: Bogotá, Medellín, and Cali. Each has distinct characteristics. Coastal Colombia has less settlement than the uplands, with the Caribbean shores having the larger cities. **Bananas** and **sugarcane** have been the traditional crops, with **corn** and **rice** gaining new importance.

- In Venezuela, major settlement is also concentrated in the uplands, with population in Caracas at more than 4 million. **Discovery of oil** in the coastal regions a century ago around **Lake Maracaibo** prompted rapid industrial growth. **Iron resources** in the Guiana Highlands support an active export flow, utilizing the Orinoco River in part.

- Ecuador also has major settlements in the Andean basins. There has been a steady increase in the proportion of Ecuadoran population locating in the lower slopes and the coastal areas. **Oil** has begun to be extracted from the eastern side of Ecuador in the **tropical lowlands** that drain into the **Amazon River.**

- Peru was the home of the Incas, whose empire extended over 2000 miles through Andean South America. The **Lost City of the Incas**—known now as **Machu Picchu**—represents a bold and well-constructed ceremonial center and city that had reached its zenith well before Columbus began his first voyage. The **potato was domesticated in Peru** and moved into active Old World trade after the arrival of the Europeans.

- **El Niño** is a **periodic weather phenomenon** that brings warm ocean waters from the western Pacific to the west coast of South and Middle America. The normally **cold Peru Current** is modified by the El Niño waters, influencing not only fishing along the coast but also weather both on the lands lying east of the current and globally.

- Bolivia is South America's poorest country, **landlocked** high in the Andes with its capital, La Paz, located at 12,000 ft (3700 m). The Bolivian mine at **Potosi** produced nearly **one-half of the world's silver in the 16th century** as a result of the enormous investment of **Indian labor** and, later, **African slaves** in **mining and refining precious metals** for world trade.

- **Brazil** is a clearly **emerging industrial giant** in South America. It is larger in area than the 48 conterminous United States, but has more than 100 million fewer people. Settlement tends to cluster more along the coast than in the thinly settled interior. In the late 1950s and early 1960s Brazil undertook a campaign to promote greater settlement in the interior savanna of the country. A new capital, **Brasília,** was constructed and, in 1960, the seat of federal government was formally moved from Rio de Janeiro on the eastern coast to the new capital. This move has caused some **demographic shift,** but it is still along the coast that most of Brazil's population is concentrated.

- Brazil has a large realm of **tropical rain forest** and **tropical savanna** climate patterns, and does not have the **highlands** so characteristic of South American Andean countries. There is a **broad coastal plain,** and the **Amazon River** drains more than 2 million square miles. Periods of economic change in Brazil can be identified as the **sugarcane era,** the **gold and diamond era,** the **rubber and coffee booms,** and now industrialization, **tourism,** and export agriculture, with increasing importance on soybeans.

- Countries of the **southern midlatitudes** include **Argentina, Uruguay, Paraguay,** and **Chile.** The first three all possess a humid subtropical climate like that in the southeastern United States. Chile, however, is more similar to southern California since both regions are mediterranean in their climate.

- Argentina is the second largest nation in South America, with an area of more than 1 million square miles (2,776,884 sq km). It has major urban growth along the coast and in the alluvial plain of the Rio de la Plata, but it is the agricultural productivity of the **pampas**—the dominant grasslands—of the interior that largely defines Argentina. **Beef** and **wheat** are major Argentine **export commodities.**

- Argentina made use of a policy of **import-substitution** in the meat and **food processing** industries in World War II. The isolation from normal trade flow allowed other industries to expand as well. This led to state promotion of many local industries, but in recent decades the government has shifted virtually all such manufacturing toward privatization.

- **Mercosur** is the name of the **Southern Cone Common Market.** Southern Cone is a term used to represent the cluster of countries that taper like a cone toward **Tierra del Fuego.** Mercosur includes Brazil, Argentina, Uruguay and Paraguay, with Bolivia and Chile as associate members.

- Chile has worked to diminish its dependence upon the export of **copper** for its foreign exchange, although it is still very important. Increased trade with East Asia and countries of the **Pacific Rim** was becoming increasingly important until the mid-1997 collapse of the Thai currency and consequent collapse of export markets to East Asia.

- Uruguay and Paraguay both represent further distinctive national scenes in South America. Uruguay embarked on an unusual experiment as a **total welfare state,** but in the last two decades the enormous expense of such a social structure has led the government away from that program. While a number of Latin American nations have been hurt by the **low oil prices in the late 1990s,** Uruguay has benefitted from that decline because it depends on imported oil for its petroleum needs.

- **Landlocked** Paraguay has been a traditional source of **out-migrating youth.** The country not only lacks an outlet to the sea that it controls, but it has **few mineral resources** and lost what proximate land that did have resources in a 19th-century war with Bolivia. Recent efforts to create major **hydroelectric installations** in cooperation with Argentina and Brazil have changed the outlook for its economic development.

- **Antarctica** is often perceived as a potential **resource outpost** of sorts for South America, although current interest in the continent lies primarily in **research.**

REVIEW QUESTIONS

1. What are the countries included among the Andean countries of South America?
2. Outline the most distinctive demographic patterns of South America, making special reference to lowland and highland differentiation.
3. Name some of the most significant pre-Columbian peoples in South America, and outline their contributions to culture and economy both in the past and now.
4. What are the most significant Hispanic landscape signatures that exist in South America?
5. Explain how topography has been a major influence in demographic patterns in the Andean countries.
6. What is the resource that has been so important for both Venezuela and Ecuador?
7. Discuss the various dimensions of the Colombian economy, and relate these to geography.

8. Discuss the impact of the discovery of petroleum on the eastern flank of the Andes.
9. What are the characteristics of El Niño, and where has it had the most impact?
10. List four of the eras that Brazil has experienced in its economic development. What role has geography played in each of these eras?
11. Define "growth pole" and explain how Brasília can be related to that theory.
12. Locate the major area of tropical rain forest in South America and describe its demography, its biological and environmental significance, and the reasons for its economic instability.
13. Outline the ways in which Antarctica might be a political hot spot in the next century.

DISCUSSION QUESTIONS

1. What role have the Andes Mountains had in the development of contemporary South America?
2. What are the geographic factors that make coca and marijuana such popular upland crops in South America?
3. What has caused the Rimland-Upland or Rimland-Mainland demographic patterns in South America?
4. How do you explain the locational and economic patterns of the indigenous people in this region? How and why would you change such patterns it if you were a government planner?
5. Discuss the theory of a growth pole and relate it to a capital city, and also to some new industry, explaining where you would locate a new venture and what impact it might have.
6. Discuss the ways in which Mercosur works, and discuss the

impact it might have if it were to unfold in ways similar to NAFTA or the European Union.
7. Discuss the ways in which El Niño has changed the face of both economics and short-term weather patterns in South America. Has it had influence elsewhere?
8. Comment on the images of Brazil's tropical rain forest and Antarctica in terms of resources, population densities, and governmental concern.
9. Discuss the significance of being a landlocked nation and suggest ways to modify the locational difficulties associated with such a position. Name examples.
10. What are the reasons Chile or any other South American country might like to be part of NAFTA and not Mercosur?

9 Anglo America

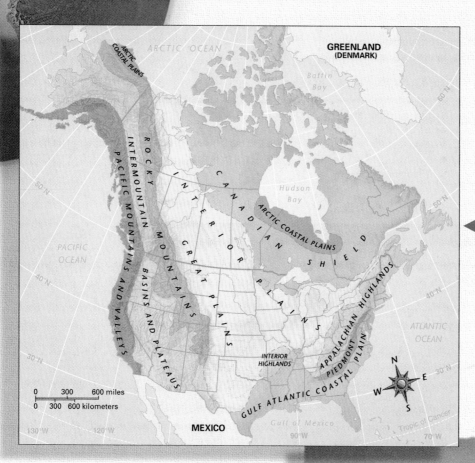

North America is a continent with a north-south trend in the major mountain systems that stretch from high latitudes to the southern borders. Human settlement has historically moved east to west across the grain of these mountain systems, sparking continuing episodes of historical dramas as people have undertaken continuing migrations and settlement.

Urban sprawl has oozed out of settlements all across Anglo America. Early intersections became village centers, then town Main Street intersections, and then nodes around which major financial and professional and commercial development grew. This pattern continues, but with some counter-patterns as population returns to smaller urban centers as well. *John Troha / Black Star*

Water is central to the whole development of Anglo America. From arid areas in continuing water deficit to lush areas with ongoing problems of flood control and water management, the challenge of bringing water to the service of human settlement has been a critical theme of Anglo America. *William Campbell / Time Magazine*

Regional Snapshot

Traffic movement is the lifeblood for the settled landscapes of Anglo America. Major highway systems have become the very heart of Anglo American lifestyles. The networks of limited access roadways mark the areas of most dynamic settlement and cultural and financial activity. *Scott Robinson / Tony Stone Images*

The populations of Anglo America are ever more diverse. Arrangements of ethnic and racial intermixing are being shaped anew every year in both countries and in almost all regions within Anglo America. This pattern is likely to grow more important as time goes on and as regional immigration continues. *Kit Salter*

Agricultural productivity is one of the major hallmarks of Anglo America. From the coasts to the interiors, from river flood plains to broad open prairies, agriculture has marked initial settlement, historic economic development, and now bespeaks current export activity. Agriculture continues to be at the heart of Anglo America. *Thomas Kitchin / Tom Stack & Associates*

Chapter 22

A Geographic Profile of Anglo America

▲ North America is not the largest landmass in Earth's collection of continents, but it is fortunate to have one of the most productive agricultural regions of any landmass in the world. In this photograph, the broad, extensive plains regions spread from the central United States up into Canada, serving as a resource center for the food needs of the entire world. This satellite image also gives clear evidence of the various north-south mountain systems as well as the arctic reaches and major river systems of the continent.   World Sat International/Science Source/Photo Researchers

22.1 Definition and Basic Magnitudes

22.2 A Region of Broadly Diverse Environmental Settings

22.3 The Geographic Abundance of Anglo America

22.4 Population Characteristics

22.5 Mechanization, Productivity, Services, and Information

22.6 The Dynamics of Neighborly Tensions in Anglo America

*i*n this final region of the text, we turn to the countries of the United States and Canada. They make up the continent of North America, but as a culture region they are called Anglo America (see Definitions & Insights, p. 553). As part of the so-called New World, they received active early settlement by numerous Amerindian groups who scattered all across the landscapes of the two countries before the arrival of Columbus in 1492. Since that date, this region has been a center of continual settlement and of economic and cultural development and diffusion, and has served as a target destination for migrating peoples for the past five centuries. The region is a complex mosaic of greatly varied geographic characteristics.

American countries themselves. Canada is slightly the larger in area (3.85 million sq mi/9.98 million sq km, compared with 3.62 million sq mi/9.37 million sq km for the United States), but is less wealthy and much less powerful on the global stage than is its neighbor, the United States. In population, Canada had nearly 30.6 million in 1998, while the United States total was 270 million that same year. Thus the geographical impact of size and population on either wealth or political power has yet to be fully determined (Fig. 22.1).

The large Arctic island of Greenland (population: about 57,000) is included in the regional definition of Anglo America because of its spatial proximity to Canada (Fig. 22.2), even though it has mainly an Inuit and Danish culture and is a self-governing part of Denmark. This range of international influences on both the landscape and the historical past of this region is continued today by an active pattern of international trade.

22.1 Definition and Basic Magnitudes

Possession of a large area assures a country of neither wealth nor power. But it does afford at least the possibility of finding and developing a wider variety of resources and, other things being equal, of supporting a larger population than might be expected in a small country (Table 22.1). The fact that there is no direct relationship between area on the one hand and wealth and power on the other may be seen by comparing the Anglo

22.2 A Region of Broadly Diverse Environmental Settings

Even in a broad-scale view that suppresses much detail, the natural environments of Anglo America are remarkably diversified. This diversity reflects a number of major circumstances

Table 22.1 Anglo America: Basic Data

Political Unit	Area (thousand/ sq mi)	Area (thousand/ sq km)	Estimated Population (millions)	Estimated Annual Rate of Increase (%) (actual)	Estimated Population Density (sq mi)	Estimated Population Density (sq km)	Infant Mortality Rate	Urban Population (%)	Arable Land (% of total area)	Per Capita GNP ($US)
Canada	3558.1	9215.5	30.6	0.5	9	3	6.3	77	5	19,020
United States	3536.3	9159	270.2	0.6	76	29	7	75	20	28,020
Total	7094.4	18,374.5	300.8	0.6	42	16	6.7	75	13	27,104

Sources: *World Population Data Sheet,* 1998. United Nations Statistics Division, 1998. *Almanac of Politics and Government,* 1998. *World Factbook,* 1998.

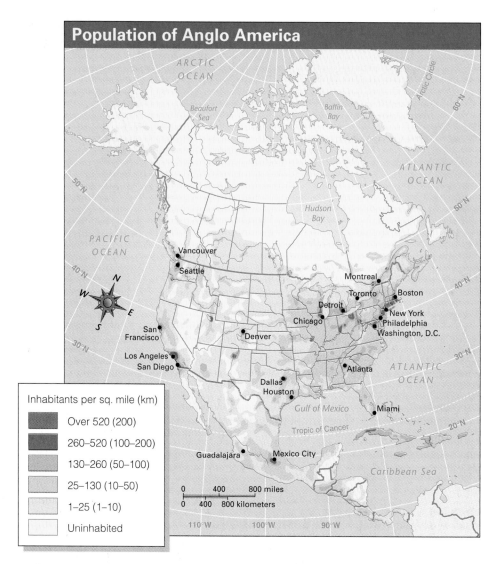

Population of Anglo America

Inhabitants per sq. mile (km)

Over 520 (200)

260–520 (100–200)

130–260 (50–100)

25–130 (10–50)

1–25 (1–10)

Uninhabited

Figure 22.1 Anglo America: population patterns.

and leads to numerous "microgeographies," or distinctive landscapes that have their own identities and use patterns.

Major Landform Divisions

The major landform divisions are shown in Figure 22.3. The glacially scoured and thinly populated Canadian Shield is formed of ancient rocks that are rich in metal-bearing ores. The surface is generally rolling or hilly. Agriculture, handicapped by poor soils and a harsh climate, is limited and marginal. Water power, wood, iron, nickel, and uranium are major resources. The United States section bordering Lake Superior is called the Superior Upland.

The Arctic Coastal Plains occur in two widely separated sections, each of which is largely unpopulated wilderness: The

Alaska section along the Arctic Ocean contains rich oil deposits, exploited on a major scale, but the section along Hudson Bay is economically insignificant. Mountainous Greenland is buried under permanent ice except for thinly inhabited coastal fringes of tundra. Fishing is the main economic activity there.

The Gulf-Atlantic Coastal Plain is low and relatively level, with soils that are generally sandy and relatively infertile. Agriculture involves many different specialities in scattered areas of intensive development. Pine or mixed oak-pine forests cover large areas. Swamps and marshes are abundant, and the coast is indented by river mouths and bays on which many prominent seaports are located. Stands of swamp hardwoods occur in poorly drained areas, and prairie grasslands were the original (but probably fire-modified) vegetation in certain coastal areas in Texas and southwestern Louisiana and in some inland areas

Figure 22.2 Anglo America: political divisions and main cities. City-size symbols conform to metropolitan area estimates given in the text.

Fig. 22.3   Anglo America: major landform divisions.

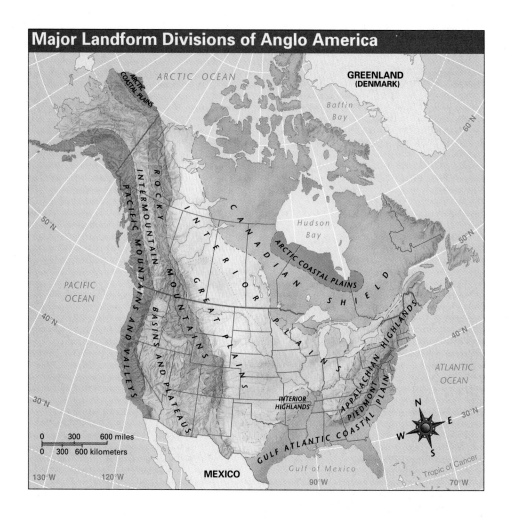

Figure 22.4   One of the most distinctive landscapes in Anglo America is the bayou swamp land in the American southeast. The great waters of the Mississippi and its river system and the natural rainfall of the region combine in broad level lands, often just barely above sea level, to create landscapes like this one in Atchafalaya swamp with its giant cypresses.   *Eastcott/Momatiuk/Photo Researchers, Inc.*

underlain by limestone (Fig. 22.4). The Lower Mississippi Valley is a subregion of good and productive soils. Oil, gas, sulfur, and salt are major resources in the Gulf Coast section, where they provide the basis for major oil-refining and chemical industries.

The Piedmont, inland from the Gulf-Atlantic Coastal Plain, is formed of harder and more ancient rocks, is somewhat higher and more hilly, and has somewhat better soils—though now much depleted and eroded in the center and south because of a past emphasis on the growing of clean-tilled cotton, tobacco, and corn in hilly areas with heavy rains. Many cities, such as Richmond, Virginia, Washington, D.C., and Baltimore, are spotted along the fall line between the two physiographic regions, where falls and rapids mark the head of navigation on many rivers and have long supplied water power.

The Appalachian Highlands form a complex system in which mountain ranges, ridges, isolated peaks, and rugged dissected plateau areas are interspersed with narrow valleys, lowland pockets, and rolling uplands. Elevations are relatively low, and most land is occupied by coniferous, deciduous, or mixed forests. Coal, water power, and wood are notable resources. The Appalachian Highlands have a surprisingly large population for a highland area, and their importance in the Anglo American economy is significant.

The Interior Highlands have many physical similarities to the Appalachian Highlands, but they are lower and have only minor coal deposits. The Interior Highlands are divided by the Arkansas River valley into the dissected Ozark Plateau on the north and the east-west oriented ridges and valleys of the low Ouachita Mountains on the south.

The Interior Plains lie mostly within the drainage basins of the Mississippi, St. Lawrence, Mackenzie, and Saskatchewan Rivers. The western third of this region is called the Great Plains and is generally delineated from the eastern Interior Plains by the 20-inch isohyet. Huge expanses of gently rolling or flat terrain exhibit some of the world's finest soils and most productive farms. The best soils have developed under prairie or steppe grasslands, but some very good soils in the east have developed under deciduous forest. Large quantities of beef, pork, corn, soybeans, and wheat are produced. Coal, oil, gas, and potash are present in major quantities.

The Rocky Mountains, generally high and rugged, consist of many linear ranges enclosing valleys and basins. Easy passes are almost nonexistent. Ranching, mining (of both metals and fuels), and recreation are major activities. Many major rivers rise in these mountains, including the Columbia, Colorado, Missouri, and Rio Grande.

The Intermountain Basins and Plateaus consist generally of high plateaus trenched by canyons or—in the extensive Basin and Range Country of Nevada, Utah, and several adjacent states—by hundreds of linear ranges separated by basins of varying size. Shielded from moisture-bearing winds by high mountains on both the west and the east, the Intermountain area is this continent's largest expanse of dry land. Mining of cop-

Definitions & Insights

GEOGRAPHIC NOMENCLATURE

Nomenclature poses a continual challenge for geographers. Names that make good sense at one time often make much less sense at another time (did you ever wonder why the university in Evanston, Illinois, is called Northwestern?). The world region we explore in this chapter consists of two major countries—the United States of America and Canada. The term Anglo America has been used as a regional title for this pair because of the pervasive and significant influences of Britain (the home of the Anglo-Saxons) on the development of the region. While the Anglo influences continue to be considerable in the United States and Canada, other powerful voices and presences are helping to shape current landscapes and culture. There is no term in current usage that embraces all of the ethnic and cultural diversity expressed in these two large countries and their distinct ethnic worlds within other worlds. Some would suggest the simple use of North America as a generalized solution to the growing cultural diversity of Canada and the United States.

North America, however, is also flawed as a collective term for these two countries. Often, Mexico is also included in the concept of North America—recall the North American Free Trade Agreement (NAFTA). Nevertheless, in this text we deal with Mexico as part of the world of Latin America because of similarities of language, culture, ethnic mixtures, and economic and cultural patterns among the nations that lie south of the southern border of the United States. In this text the term North America continues to be used for the continental landmass.

Another argument could be raised claiming that the names of the region ought to be derived from Native American groups (as so many U.S. state names are) because of their profound influence on the initial settlement of the region. However, in considering the 20th and 21st centuries, that argument loses power. Thus, we mark the region north of Mexico as Anglo America.

The enduring significance of the Anglo roots of not only majority populations but also many political, economic, and cultural institutions causes us to use the term Anglo America in this text for the two large countries north of Mexico. There will always be arguments about other ways to see and signify the world's regions. Discussion of the rationale for such nomenclature is fair game for provocative geographic discussions.

per, oil, gas, coal, uranium, and other minerals is prominent in scattered spots, and numerous major dams impound irrigation water and generate much hydroelectricity along the Columbia and Colorado rivers.

The Pacific Mountains and Valleys lie parallel to the Pacific shore and contain North America's highest mountains. In California, the Central Valley between the Sierra Nevada mountain system on the east and the Coast Ranges on the west is a huge exhibit of productive irrigated agriculture on flat alluvial land.

Figure 22.5   Anglo America: major climatic regions.

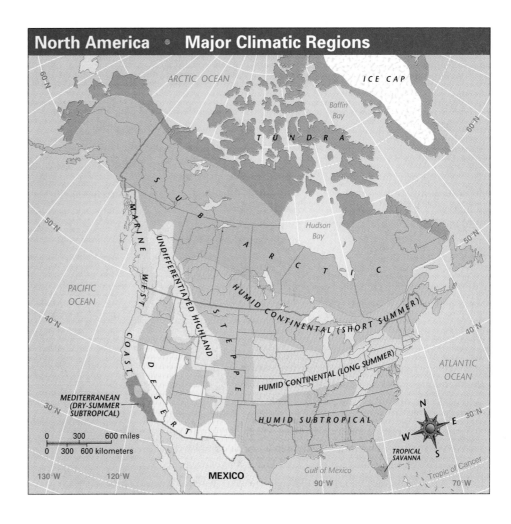

Farther north, the Willamette–Puget Sound Lowland forms another major valley between the Coast Ranges and the Cascade Range of Oregon and Washington. Extreme western Canada and southern Alaska are dominated by high mountains, with many spectacular glaciers. The economy of the Pacific Mountains and Valleys has grown rapidly on the basis of a major collection of natural resources: abundant forests, fisheries, rivers to produce hydroelectricity and supply irrigation water, oil and gas in California, gold that once drew settlers to California and British Columbia, and both climatic and scenic resources undergirding development in the California subtropics. This wide range of environmental settings has helped stimulate the mobility of the Anglo American population (see Regional Perspective, p. 556).

Major Climatic Regions

The continental climatic pattern in this region exhibits both regional variance and largeness of scale. The United States includes a greater number of major climatic types than does any other country in the world, and even Canada is more varied than is commonly assumed. The variety of economic opportunities and possibilities afforded by this wide range of climates is one of the basic factors underlying the economic strength of this world region.

Each climate region is shown by name and color on Figure 22.5. The tundra climate, characterized by long, cold winters and brief, cool summers, has an associated vegetation of mosses, lichens, sedges, hardy grasses, and low bushes. The subarctic climate has long, cold winters and short, mild summers, with a natural vegetation of coniferous snow forest resembling the Russian taiga. Population is extremely sparse in the tundra and subarctic climates and is practically nonexistent in Greenland's ice-cap climate. Scattered groups of people engage in trapping, hunting, fishing, mining, logging, and military activities; many live largely on welfare. The humid continental climate with short summers is characterized by long and quite cold winters and short warm summers. Agriculturally, there is a heavy emphasis on dairy farming except in the extreme west, where spring wheat production is dominant. The humid continental climate with long summers has cold winters and hot summers; agriculturally, this belt, which includes the agricultural richness of the Midwest, is marked by an emphasis on dairy farming in the east and a corn-soybeans-cattle-hogs combination in the midwestern (interior) portion. The diverse geographic characteristics that provide this range of environmental settings, however, also give rise to a wide variety of natural hazards. Figure 22.6 shows the hazards that sweep across not only North America, but also Middle America as part of a hemispheric characteristic.

In the humid subtropical climate, winters are short and cool, although with cold snaps, and summers are quite long and hot.

Agriculture is rather diverse from place to place, with some major specialized emphases on cattle, poultry, soybeans, tobacco, cotton, rice, peanuts, and a wide range of fruits and vegetables. Within each of the three humid climatic regions just described, the pattern of natural vegetation is complex, and each region has areas of coniferous evergreen softwoods, broadleaf deciduous hardwoods, mixed hardwoods and softwoods, and prairie grasses. Extreme southern Florida has a small area of tropical sa-

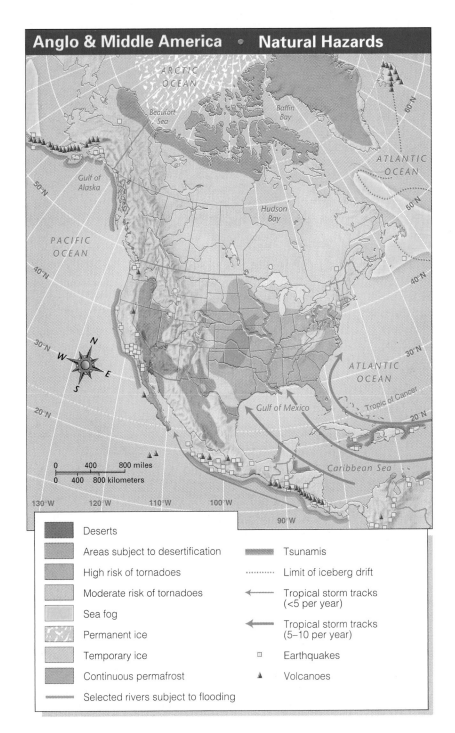

Anglo & Middle America ● Natural Hazards

Legend:
- Deserts
- Areas subject to desertification
- High risk of tornadoes
- Moderate risk of tornadoes
- Sea fog
- Permanent ice
- Temporary ice
- Continuous permafrost
- Selected rivers subject to flooding
- Tsunamis
- Limit of iceberg drift
- Tropical storm tracks (<5 per year)
- Tropical storm tracks (5–10 per year)
- Earthquakes
- Volcanoes

Figure 22.6 This map shows the wide variety of environmental hazards facing Middle and North America. Each of these hazards is caused by natural physical forces, but the impact of these phenomena varies in close relationship to the sorts of human settlement patterns in the location visited by a specific hazard.

Regional Perspective

Mobility and Recreation in Anglo America

One of the most dynamic regional images of Anglo American culture is that of vast highway networks, motels and roadside commercial strips, and an abundance of motor homes, recreational vehicles, and cars loaded with children and gear—especially in the summer months. This pattern of seemingly continual motion has led to the emergence of abundant car rental facilities, expanding networks of interstate and limited-access highways, hurried weekend trips to not-so-nearby parks or second homes and cabins, and increasing expansion of government-maintained parks and recreational facilities. Large-scale Disney World kinds of recreation destinations are especially popular, just as is the network in both Canada and the United States of magnificent national parks created around landscape features of stunning beauty. But there is also a growing world of smaller regional, even municipal, green space that holds retail development at bay and provides environmental and recreational amenities for generally modest entrance fees and a generally accepted tax burden. These landscapes of leisure and the pattern of steadily expanding mobility that increases their accessibility are major components of the important culture changes of the past three decades. The construction of new stadiums, aquariums, convention centers, riverboat casinos, auto racing tracks, golf courses, spas, artificial multiuse lakes, and theater complexes of all sizes, combined with a general governmental inclination to support such projects with tax breaks, demonstrates the economic benefits of such mobility and recreation. Add to those features the plethora of fast-food restaurants that explode upon the scene as soon as a new highway interchange is opened, and you can see the impact of our enormous fascination with staying in motion and spending considerable money on recreation, travel, and convenience.

vanna climate, which is a major climate in adjacent Latin America. The state of Hawaii has a tropical rain forest climate. Along the Pacific shore of the United States and Canada, the narrow strip of marine west coast climate is associated with the barrier effect of high mountains near the sea, ocean waters offshore which are warm in winter and cool in summer relative to the land, and winds prevailingly from the west throughout the year. The mild, moist conditions have produced a magnificent growth of giant conifers (most notably the redwood and the Douglas fir) that provides the basis for the large-scale development of lumbering. Lush pastures support dairy farming, as they do in the corresponding climatic region in northwestern Europe.

The mediterranean or dry-summer subtropical climate of central and southern California, in which nearly all precipitation comes in the winter season, is associated with irrigated production of cattle feeds, vineyards, vegetables, fruits, cotton, and a great range of other crops. These crops, together with associated livestock, dairy, and poultry production, make California the leading U.S. state in total agricultural output. In the semiarid steppe climate, occupying an immense area between the Pacific littoral of the United States and the landward margins of the humid East and extending north into Canada, temperatures range from continental in the north to subtropical in the south. The natural vegetation of short grass, bunch grass, shrubs, and stunted trees supplies forage for cattle ranching, which is the predominant form of agriculture. Areas that are less dry are often used for wheat, and both wheat and other crops are grown in scattered irrigated areas, often associated with major rivers such as the Columbia, Snake, Arkansas, or Rio Grande. The desert climate of the U.S. Southwest is associated with scattered irrigated districts and settlements emphasizing mining, recreation, and retirement. High, rugged mountains, such as the Rockies and the Sierra Nevada, have undifferentiated highland climates varying with latitude, elevation, and exposure to moisture-bearing winds and to sun. More detailed discussion of climatic regions and major landform divisions and associated landscapes, with many relevant maps and photographs, can be found in Chapters 23 and 24.

22.3 The Geographic Abundance of Anglo America

Both the United States and Canada are wealthy nations, with Canada much the lesser in this respect. Canada's gross national product (GNP) is a little less than one-tenth as great as that of the United States. On a per capita 1998 basis, Canada's is $19,020; in the United States, it is $28,020. The output of goods and services in the United States is more than 80 percent of that of all Europe. In producing these goods and services, the United States consumes more than 35 percent of the world's annual energy consumption. The ranking of the United States among the fifteen nations with the highest per capita GNPs can be seen in Table 22.2. The relative standing of the North American countries compared with other nations has been declining since World War II because of more rapid economic growth in some other countries.

Table 22.2 World's Top Fifteen Countries in Per Capita GNP

Political Unit	$US
1. Luxembourg	45,360
2. Switzerland	44,350
3. Japan	40,940
4. Norway	34,510
5. Denmark	32,100
6. Singapore	30,550
7. Germany	28,870
8. Austria	28,110
9. United States	28,020
10. Iceland	26,580
11. Belgium	26,440
12. France	26,270
13. Netherlands	25,940
14. Sweden	25,710
15. Finland	23,240

Sources: *World Population Data Sheet,* 1998. United Nations Statistics Division, 1998. *Almanac of Politics and Government,* 1998. *World Factbook,* 1998.

The development of such a high degree of wealth and power in Anglo America is a matter too complicated to be fully explained in the context of an introductory text and is a phenomenon subject to various interpretations. But it is clear that the United States and Canada possess, or have possessed, a number of specific geographic assets on which they have been able to capitalize. Some of the more important ones include:

1. *Size.* Both countries are large in area and are among the five largest in the world.
2. *Internal unity.* Both possess a comparatively effective internal unity, although Canada has had some difficult problems in this respect in recent decades as Quebec continues to attempt to withdraw from Canada and gain independence. There have also been major negotiations in both countries over Native American demands for expanded land rights and the right to their own self-governing territories.
3. *Resource wealth.* Both are outstandingly rich in natural resources and have well-developed capacities for making use of all levels of resource wealth.
4. *Size of populations.* The combined population of both countries is large, yet neither country is overpopulated—although there are images of urban densities and poverty, especially in the United States, that demonstrate clear problems.
5. *Role of technology.* Both countries developed mechanized economies rather early and under very favorable conditions. Recent innovations in robot technology have led to a contin-

ued replacement of human labor with machine labor in, for example, automobile manufacturing in both Canada and the United States. It is likely that such patterns of labor shifting will continue, even though labor unions do what they can to slow down the pace of such replacement because of the associated loss of employment.

6. *Neighborly relations.* Despite occasional irritations and quarrels, the relations between these Anglo American nations are generally friendly and cooperative.
7. *Geographical diversity.* A remarkable range of environmental settings has both allowed and stimulated full use of human ingenuity in economic development.
8. *Institutional stability.* In both the United States and Canada there has been a history of reasonable continuity of political, economic, and cultural institutions. Even in the face of some tensions and institutional evolution, the regional pattern has been one of stability.

The combination of these influences creates a regional base that has a rich history of economic development and relative abundance for its populations. This is not to say that North America is not without significant problems of poverty and imbalance in the distribution of the products of this abundance, but taken as a whole, North America is a region of continued strong potential for "the good life."

Agricultural and Forest Resources

The United States probably ranks first among the world's nations in arable land of high quality. A much smaller proportion of Canada is arable, but it still has a total of more arable acreage than all but a handful of countries. Such abundance has helped North America become the main food-exporting region in the world. Canada and the United States also have abundant forest resources. Canadian forests are larger in extent than those of the United States but are not so varied. For example, Canada has no forests comparable to the redwood stands of California or the fast-growing pine forests of the southeastern United States. Wood has been an abundant material in the development of both countries. Even now, the United States cuts more wood each year than any other country in the world. It both exports and imports large quantities, with the exports from private lands going largely to Japan (Fig. 22.7) and the imports coming overwhelmingly from Canada. Canada's timber industry is considerably smaller, although it is still one of the greatest in the world; the country's small population compared to the size of its forest resources allows it to be the world's greatest exporter of wood.

Mineral Resources

North America is also outstandingly rich in mineral resources. Such resources played a major role in the development of the region's economy, and mineral output is still huge. Canada is a major exporter of mineral products, largely to the United States.

Definitions & Insights

RESOURCES

One of the most important definitions you will learn in this text is: "Resources are cultural appraisals." This means that what we call "natural resources" are—with the exception of air, water, and space—actually Earth products we define as having value for us. For example, before it was seen to be an energy source, coal was a troublesome rock because it broke up too easily to be useful for building. Brea—the term used in Southern California for the tar-like substance that used to seep to the surface in oil-rich lands in the Los Angeles basin—was the reason some land was relatively cheap in the first Los Angeles land boom of the 1880s. Brea lands were the cheapest land you could buy then. Such lands now, of course, are among the most expensive because of petroleum's resource value to us now. Brea has not changed, but society has now appraised that sticky, black product—petroleum—to be a resource of considerable value.

Cultural appraisals change because of shifts in technology, in societal desires, and in availability. Coal, for example, is diminishing in its value as a resource not because its characteristics are changing but because—as a society—we currently place a higher value on clean air. Since burning natural gas is less harmful to our atmosphere, the resource value of natural gas is increasing while the value of coal is diminishing. The cultural forces that lead to changing appraisals are difficult to predict but are continually at work. Every time you go to a flea market or a garage sale, for example, you are likely to see things that you may consider to be outdated junk; yet someone else has appraised those old 33-rpm records or that tube radio as having value. Resources are cultural appraisals and therein lies much of the dynamic of the economic and geographic tensions that define our current world.

But the United States, although still a major producer, has also become a major importer. This situation reflects the partial depletion of some American resources, combined with growing foreign production. It also reflects the enormous demands and buying capacity of the U.S. economy.

Abundant energy resources have been basic to the rise of the Anglo American economy and remain crucial to its operation and expansion. Coal was the largest source of energy in the 19th-century industrialization of the United States. Today, the country ranks with China and Russia as one of the three world leaders in magnitude of estimated coal reserves. In coal production, China and the United States are the world leaders, each producing about one-fourth of the world output. The United States is the world's largest coal exporter by a wide margin. Canada is also an important coal exporter, with reserves and production that are small on a world scale but large in terms of Canada's needs.

Petroleum fueled much of the development of the United States in the 20th century, and for decades the country was the world's greatest producer, with reserves so large that the question of long-term adequacy received little thought. In 1994, the United States was still the largest producer of oil in the world (followed by Saudi Arabia and Russia), supplying approximately one-eighth of world output. Even with that production level, the United States still imports just under half of its own domestic needs. Canada's reserves and production were and are small in comparison but are sufficient to be the basis of a continuing energy boom in its western province of Alberta.

Natural gas became a fuel of rapidly growing significance in the later 20th century. In 1994, the United States was the world's largest consumer of gas (ahead of Russia by a small margin) and was a major importer (from Mexico and Canada). It ranked second in the world in output (behind Russia, but far ahead of any other country), producing one-quarter of the world's gas from 4

Figure 22.7 The economic importance of logging operations at all scales is central to British Columbia. From mountain flanks to forested valleys to rivers punctuated with great floating log rafts, the presence of this economic activity is common in the province. This log carrier is one of the many crafts moving logs into readiness for shipping lanes that go across the Pacific Ocean, destined most often for the Japanese market and other East Asia locales.
Richard Kolar/Animals Animals/Earth Scenes

percent of known world reserves. Canada, with smaller but rapidly increasing proven reserves, produced about 5 percent of world output in 1994 and was a major exporter.

Potential water power in the United States and Canada is small on a world scale (Table 22.3), but development has been so intense that the two countries rank with Russia as the world leaders in hydropower capacity and output (Fig. 22.8). Canada generates approximately 62 percent of its electricity from hydropower; the United States generates 9 percent. At the same time, nuclear power can draw on the large uranium resources in each country and is an important secondary source of electricity in both.

Nonfuel mineral resources in North America present a similar picture of abundance. Long exploitation on massive scales has had an impact on American reserves, especially of the ores easiest to mine and richest in content, and even large mineral outputs within the United States are insufficient to supply the

huge demands of the domestic economy. In many such cases, Canada has been able to assume the role of principal foreign supplier. A major example is iron ore. The United States originally had huge deposits of high-grade ore, principally in Minnesota's Mesabi Range and in lesser ore formations near western Lake Superior, but by the end of World War II this ore had been seriously depleted. New sources of high-grade ore then were developed in Quebec and Labrador, principally for the American market. Meanwhile, the Lake Superior region somewhat revived its mining industry by extracting lower-grade ore, which is present in vast quantities but must be processed in concentrating plants before it can be shipped at a profit to iron and steel plants.

As Table 22.3 shows, there is major production of many other nonfuel minerals by both the United States and Canada. In fact, the mineral wealth of North America makes it easier to summarize gaps in the resource array than to describe the

Table 22.3 Anglo America's Share of the World's Known Wealth and Production			
	Aproximate Percent of World Total		
Item	U.S.	Canada	Anglo America
Area	7	7	14
Population	4.6	0.5	5.1
Agricultural and Fishery Resources and Production			
Arable land	13	3	16
Meadow and permanent pasture	8	1	9
Production of:			
Wheat	12	5	17
Corn (grain)	46	1	47
Rice	1	0	1
Soybeans	52	1	53
Cotton	26	0	26
Milk	14	1	15
Number of:			
Cattle	11	1	12
Hogs	8	1	9
Fish (commercial catch)	6	1	7
Forest Resources and Production			
Area of forest/woodland	6	8	14
Wood cut	14	5	19
Sawn wood production	21	12	33
Wood pulp production	46	19	65
Minerals, Mining, and Manufacturing			
Coal reserves	24	1	25
Coal production	17	1	18
Coal exports	25	8	33

(continued)

Table 22.3 Anglo America's Share of the World's Known Wealth and Production *(continued)*

Item	Aproximate Percent of World Total		
	U.S.	Canada	Anglo America
Published proved oil reserves (tar sands and oil shale)	2	1	3
Crude oil production	12	2	14
Crude oil refinery capacity	21	4	25
Natural gas production	24	5	29
Natural gas reserves	4	3	7
Potential iron ore reserves	5	6	11
Iron ore mined	5	4	9
Pig iron produced	7	2	9
Steel produced	13	2	15
Aluminum produced	19	7	26
Copper reserves	17	5	22
Copper mined	19	8	27
Lead reserves	22	13	35
Lead mined	10	10	20
Zinc reserves	13	15	28
Zinc mined	11	13	24
Nickel mined	1	21	22
Gold mined	14	8	22
Phosphate rock mined	27	0	27
Potash produced	6	23	29
Native sulphur produced	21	0	21
Asbestos mined	1	18	19
Uranium reserves	9	12	21
Uranium produced	16	29	45
Electricity produced	26	7	33
Potential hydropower	4	3	7
Production of hydropower	4	4	8
Nuclear electricity produced	31	4	35
Sulfuric acid produced	26	3	29

Sources: United Nations Statistics Division, 1998. *Almanac of Politics and Government,* 1998. *World Population Data Sheet,* 1998.

resources that are present. Major minerals in which both countries appear to be truly deficient include chromite, tin, diamonds, high-grade bauxite, and high-grade manganese ore.

22.4 Population Characteristics

From the beginning of settlement until the 1960s and 1970s, Anglo America was a region of rapidly growing population. By 1800, approximately two centuries after the first European settlements, there were more than five million people in the United States and several hundred thousand in Canada. Growth since that time is summarized in Table 22.4. The population of this region long increased at a rate much more rapid than that of the world as a whole. As a result, one element in the mounting importance and power of the United States, where most of this growth occurred, was its possession of an increasing share of the world's population. By the 1980s, however, the population growth rates of both the United States and Canada were well below those of most of the world, except for Europe, Japan, and the former Soviet Union. Together, the United States and Canada now account for about 5 percent of the estimated world population and about 14 percent of the world's land area (excluding Antarctica). Most of Anglo America's 301 million people (as of 1998) live in the eastern half of the region, from the St. Lawrence Lowlands and Great Lakes south and east. In Canada, approximately 90 percent of the population live within 100 miles of the border between Canada and the United States. The estimated average population density of the two countries in 1998 was only 76 per square mile (28 per sq km) for the United States and 9 per square mile (3 per sq km) for Canada, with its tremendous expanse of sparsely settled northern lands. And by world standards, even the United States is far from overpopulated (see Fig. 22.1).

The original American Indian population was overwhelmed by European settlement, which began four centuries ago, and disease, and thus was reduced to a politically weak and economically underprivileged minority segregated mainly in western and northern areas. The role of the original peoples of the North American continent continues to be one of difficulty and governmental attention. Both Canada and the United States continue today to each have one exceptionally large ethnic minority which has been the focus of serious problems of national unity. For example, about 26 percent of Canada's people are French in language and culture. These French Canadians are concentrated mostly in the lowlands along the St. Lawrence River, where they form the majority in the province of Quebec. Their ancestors in Quebec, some 60,000 in number, were left in British hands when France was expelled from Canada in 1763. They did not join the English-speaking colonists of the Atlantic seaboard in the American Revolution, and by their refusal to do so they laid the basis for the division of the continent of North America into two separate countries. Until the late 20th century, they increased rapidly in numbers, and they have clung tenaciously to their distinctive language and culture.

At times, major controversies arise between this group and other Canadians, and a provincial government with separatist aims was elected by Quebec's voters in the 1970s. However, the

Figure 22.8 Anglo America's large-scale development of hydropower is symbolized by Glen Canyon Dam and Lake Powell, situated on the Colorado River between the states of Arizona (right) and Utah (left). Note the power station at the foot of the dam, the extreme desert environment, the escarpment in the background—separating two levels of the Colorado Plateau—and the mountains rising above the general surface of the plateau in the distance. *William Campbell/Time Magazine*

Quebec electorate then refused to approve its government's movement toward secession from Canada, and this possibility temporarily receded, only to flare up with renewed heat in the late 1980s. The Canadian government has attempted to preserve national unity through the recognition of both English and French as official languages and the provision of legal protection for French-Canadian institutions. In October 1995, the Québécois, by popular referendum, voted to remain a province within Canada. The vote, however, was 51 to 49 percent, so that

Table 22.4 Population Growth in the United States and Canada, 1850–1998					
Year	U.S. Population (millions)	Canadian Population (millions)	Total (millions)	Total Increment	
				(millions)	(%)
1850	23.3	2.4	25.7	—	—
1900*	76.1	5.4	81.5	55.8	217
1950*	151.1	14	165.1	83.6	103
1960*	179.3	18.2	197.5	32.4	20
1970*	203.2	21.6	224.8	27.3	14
1980*	226.5	24.3	250.8	26	12
1990*	248.7	28.1	276.8	26	10
1998*	270.2	30.6	300.8	24	9

*Actual dates for Canada are 1901, 1951, 1961, 1971, 1981, and 1991. The decennial census is taken one year later in Canada than in the United States. This table uses census data except for the 1998 midyear estimates.

Figure 22.9 In San Francisco—as in many of the major urban centers in Anglo America—a racial sharing of public space has occurred in recreation and in work. This scene from a city park is instructive of such casual interaction among differing population groups. *Kit Salter*

the issue of Quebec separatism is still very much alive in Canada and will likely stay on the national agenda for some time to come.

Black people of African descent, one-eighth of the United States population, are the principal minority in that country. Unlike the French Canadians, they are not largely confined spatially to only one part of the country, although they do predominate in a number of large cities, and particularly in the centers of those cities. Historically, African Americans as a group have not been an active force toward sectionalism in the United States. Nevertheless, a conflict of attitudes toward black slavery was among the important factors that nearly split the United States into two nations in the American Civil War, and race relations are still a major source of friction in American politics. The primary setting for most of the race relations difficulties continues to be urban America. There is a segment of the African-American population that has moved into the middle class and above, but a larger segment resides in the largely impoverished and troubled central cities, where problems in housing, jobs, crime, and racial isolation continue to tear at a traditional American sense of unity, or at least the hopes for a community of peoples within the diversity of the country.

Spanish-speaking people are a smaller ethnic minority in the United States than are African Americans, but their growth rate is more rapid. A little more than 9 percent of the United States population is Hispanic. The majority are of Mexican origin and are strongly concentrated in California, Texas, New Mexico, Arizona, and Colorado. Other important Hispanic groups include people of Puerto Rican origin—most heavily represented in New York City and in the northeast—and people of Cuban origin, who live largely in southern Florida. Estimates vary as to how many million other Hispanic people, mainly Mexicans and Central Americans, are actually resident in the United States as illegal immigrants. Like African Americans, American Hispanics are not numerically or politically dominant in any state. They are, however, gaining in relative proportion in many of the large urban centers, especially in the

Southwest. Nationally, Asian Americans equal nearly 3 percent of the U.S. population. While initial settlement was concentrated in the distinctive "Chinatown" landscapes that so boldly gave this population (whether Chinese or not) its image in the 19th and early 20th century, Asian Americans have been steadily moving out of the central city and into white and mixed neighborhoods. The closest approach to minority dominance of a state is in multiethnic Hawaii, where Asian Americans have become very powerful politically (although without any hint of separatism). Among the many Asian-American groups in Hawaii, Japanese Americans are by far the most numerous. In the United States as a whole in the recent decades, there have been large increases in the Filipino, Korean, Vietnamese, and Cambodian populations, thereby further expanding the characteristics of Asian Americans in the Anglo American landscape. Cities throughout the region tend to be the areas where there is the greatest amount of ethnic mixing, particularly in the workplace and in recreation (Fig. 22.9).

The governments of both the United States and Canada have assiduously fostered national unity through the provision of adequate internal transportation. In both countries, transportation networks have had to be built not only over long distances but primarily against the "grain" of the land. Effective transport links between east and west have been the most imperative, but most of the mountains and valleys have a north-south trend, which has presented a series of obstacles to cross-country transport. The coasts of both countries were first tied together effectively by heavily subsidized transcontinental railroads (Fig. 22.10) completed during the latter half of the 19th century. Subsequently, national highway and air networks further enhanced national unity. The populations of both countries have long held mobility to be an important right and the

Figure 22.10 Transcontinental railroads have played an enormously important role in the development of the United States and Canada, and they continue to do so today. The photo shows containers being double-stacked on specially designed low-slung rail cars. They are often shipped for thousands of miles to and from seaports. *Michael L. Abramson*

Figure 22.11 One of the most prominent landscape signatures of Anglo America is a busy freeway. This photo of a Los Angeles freeway at dusk illustrates the fascination "Angelenos" have with auto-mobility. While this image perhaps makes sense at the rush hour, it is instructive to know that such traffic density in Southern California is common for nearly 15 hours a day. This dependence on the automobile and the independence it allows—even though it means very crowded roadways—is a major characteristic of urban Anglo America. *Scott Robinson/ Tony Stone Images*

development of affordable automobile transportation in the early decades of the 20th century began a century-long fascination with the car that has not abated on either side of the border. The love affair with the car is not only a part of the pattern of regional mobility, but it has also served as a significant economic feature of industrial growth near the border as well (Fig. 22.11).

22.5 Mechanization, Productivity, Services, and Information

The high productivity and national incomes of the two Anglo American countries have come about essentially through the use of machines and mechanical energy on a lavish and ever-increasing scale during the past century and a half. From an early reliance on waterwheels to power simple machines, the United States increasingly exploited the power generated from coal by steam engines. Later, the total power used to drive machines increased enormously through use of oil, internal-combustion engines, and electricity. In achieving this unusually energy-abundant and mechanized economy, the United States was able to take advantage of a unique set of circumstances. For one thing, it had resources so abundant and varied that they attracted foreign capital and also made possible domestic accumulation of capital through large-scale and often wasteful exploitation. There was also a labor shortage, which attracted millions of immigrants as temporary low-wage workers but also promoted higher wages and labor-saving mechanization in the long run. This dynamic continues today, especially at the southeastern border of the United States and through the port of Los Angeles—often spoken of as the current Ellis Island (the port of immigration on the east coast during the 19th and first

half of the 20th century). A relatively free and fluid society encouraged striving for advancement; large segments of the population subscribed with an almost religious fervor to ideals of hard work and economic success. Seldom did national energies have to be diverted excessively into defense and war.

Most of the foregoing American conditions were duplicated in Canada, and some of them in a number of other countries, but one American advantage was not: the existence of a large, unified, and growing internal market that allowed producing organizations to specialize in the mechanized mass production of a few items, thus lowering the unit costs of production and further expanding the size both of the producing firms and of the market. In recent decades, however, this enormous American asset has been significantly eroded, along with some others, by the enormous expansion of the global market. Transportation and communication have become so cheap and rapid and so many artificial trade barriers have fallen that the world market rather than the domestic market has become the essential one for major industries.

These forces of mechanization and associated productivity have been increasingly replaced in the latter decades of the 20th century with a new labor sector: services and information management. The service sector has become a more significant employer than the manufacturing sector in North America as the manufacture of shoes, clothes, toys, and cars has migrated offshore, a widespread and economically dynamic sector has grown up around the acquisition, manipulation, analysis, and distribution of information. As telecommunication technologies have made all of urban North America become more global in both outlook and operation, the recent economic landscape of North America has seen blast furnaces and conveyor belts steadily replaced by keyboards, monitors, head-sets, and personal service occupations.

Figure 22.12   The North American Free Trade Agreement (NAFTA) has never had complete support. This demonstration in Austin, Texas, was organized by union leaders fearful of the loss of jobs that would occur as industries left the United States and went south into Mexico and Central America.   *Robert E. Daemmrich/Tony Stone Images, Inc.*

22.6 The Dynamics of Neighborly Tensions in Anglo America

For many years after the American Revolution, the political division of Anglo America between a group of British colonial possessions to the north and the independent United States to the south was accompanied by serious friction between the peoples and governments on either side of the boundary. This heritage of antagonism was a result of (1) the failure of the northern colonies to join the Revolution; (2) the use of those colonies as British bases during that war; (3) a large proportion of the northern population having come from Tory stock driven from U.S. homes during the Revolution; and (4) uncertainty and rivalry concerning who had ultimate control of the central and western reaches of the continent. The War of 1812 was fought largely as a United States effort to conquer Canada, though this intent was not specified at the time. Even after its failure, a series of border disputes occurred, and U.S. ambitions to possess this remaining British territory in North America were openly expressed; suggestions and threats of annexation

were made in official quarters throughout the 19th century and even into the 20th century.

In fact, Canada's emergence as a unified nation is in good part a result of American pressure. After the American Civil War, the military power of the United States took on a threatening aspect in Canadian and British eyes. Suggestions were made in some American quarters that Canadian territory would be a just recompense to the United States for British hostility to the Union during the Civil War. However, the British–North America Act, passed by the British Parliament in 1867, brought an independent Canada into existence by combining Nova Scotia, New Brunswick, and Canada into one Dominion under the name of Canada. With the Act, Great Britain sought to establish Canada as an independent nation capable of morally deterring U.S. conquest of the whole of North America. Although hostility between the United States and its northern neighbor did not immediately cease with the establishment of an independent Canada, relations have improved gradually, and the frontier between the two nations has ceased to be a source of insecurity. This frontier, stretching completely unfortified across a continent, has become more a symbol of friendship than of enmity.

The bases of Canadian-American friendship are cultural similarities, the material wealth of both nations, and the mutual need for and advantages of cooperation. A large volume of trade moves across the frontier and strengthens both countries economically and militarily. Each is a vital trading partner of the other, although Canada is much more dependent on the United States in this respect than is the United States on Canada. In 1996, for example, Canada supplied 19 percent of all United States imports by value and took 22 percent of all United States exports, making Canada the leading country in total trade with the United States. On the other hand, in 1996 the United States supplied 65 percent of Canada's imports and took 81 percent of its exports. The United States also supplies large quantities of capital to Canada, which has been an important factor in Canada's rapid economic development during recent decades but which also has been a frequent source of Canadian disquiet because of the amount of U.S. influence on the Canadian economy that the investment represents. However, Canadians in turn invest heavily in the United States. Except for Canadian exports of automobiles and auto parts to the United States, the main pattern of trade between the two countries is the exchange of Canadian raw and intermediate-state materials—primarily ores and metals, timber and newsprint, oil and natural gas—for American manufactured goods. A free-trade pact to unite the two in one market was signed in 1988. In 1992, Canada joined the United States and Mexico in signing the North American Free Trade Agreement (NAFTA) that was enacted in 1994 (Fig. 22.12).

SUMMARY WITH SELECTED KEY TERMS

- The countries of the **United States** and **Canada** comprise **Anglo America. Mexico** is defined as part of **Middle America**, even though the **North American Free Trade Agreement (NAFTA)** has linked the three countries. **Greenland** is also part of this region

because of its **proximity** to Canada, even though it is politically a part of **Denmark**.
- Both the United States and Canada have a sizable area, although the **population statistics** are quite dissimilar for the two countries.

The combined population for the two countries amounts to just over 300 million. Greenland, also large, has a population of fewer than 60,000.

- Anglo America has a broad range of **diverse environmental settings,** including the **Arctic Coastal Plains, Gulf-Atlantic Coastal Plain, Piedmont, Appalachian Highlands, Interior Highlands, Interior Plains, Rocky Mountains, Intermountain Basins and Plateaus,** and **Pacific Mountains** and **Valleys.**
- **Climatic patterns** for the region are also varied. Climate types include **tundra, subarctic, humid continental with short summers, humid continental with long summers, tropical savanna,** and **tropical rain forest.** There are also **undifferentiated highland and mountain zones.** Patterns of **vegetation** and settlement are associated with these climate patterns.
- The text suggests that a number of factors are central to the **geographic richness** that characterizes Anglo America. These factors include **size, internal unity, resource wealth, population size, technology, intraregional relations,** and **geographic diversity.** Many regions possess some of these features, but Anglo America is particularly fortunate in the presence of all these elements.
- Particularly important resources for Anglo America include **agricultural land and forests.** Both of these play a central role in exports of the two countries. Additional major resources include **coal, petroleum and natural gas, water power, and minerals such as iron ore.** The global importance of these sources in this region is shown clearly in **Table 22.3.**

- Demographic characteristics of the two countries show similar patterns of early settlement **displacing native populations,** and then subsequent **waves of immigration** continually changing the demographic, ethnic, and racial characteristics of the resident populations. In Canada, the **French** are the most significant minority. In the United States, **African Americans** fill that role. Both countries have an increasingly **diverse ethnic blend** created by active east, west, and south border **immigration flows. Hispanic and Asian-American** populations are increasingly gaining importance in the United States.
- Both countries have a history of active and significant **industrialization in the 19th and 20th centuries.** In the past three or four decades, the economic profiles of the countries have changed as manufacturing has become less important and employment in **service industries and information management** has grown more significant. **Small towns** have given way to **middle-sized and large urban centers** in both countries.
- One of the most critical facets of regional success in Anglo America has been the **fundamentally amicable relationship** that has existed between the two countries. The **United States is the major customer for Canadian exports,** and **Canada plays a similar role for the United States. Ninety percent of the Canadian population lives within 100 miles of the international border.** This proximity also causes some frustrations as cultural influences from the United States sometimes seem overpowering to the Canadians.

REVIEW QUESTIONS

1. What are the three political units that make up Anglo America?
2. Explain the reasons for the various traditional names given to this region. How does the name NAFTA fit into this pattern?
3. What are the basic area and population figures for Anglo America? How do they relate to the global picture?
4. List the major landform characteristics of the region, taking three of those characteristics and giving some detail on the landscapes and land-use patterns associated with them.
5. Do the same thing for the climatic characteristics in North America. Relate three of the climatic zones to similar zones you have read about in other regions of the world.
6. What are the most important characteristics of the geographic richness of Anglo America? List four of the eight traits given in the text and describe your perceptions of them if you grew up in Anglo America.

7. What aspects of the mineral wealth of Anglo America seem most significant to you? Why?
8. Which population characteristics are most significant globally? Which ones seem to be the most important locally? Does this relationship change with time?
9. What role has immigration had in shaping the current demographic scene in Anglo America?
10. Give examples from your own knowledge that represent the issues raised in the section on mechanization, productivity, services, and information. How are these likely to change over time?
11. Describe the relations between Canada and the United States, pointing out both the positive and the negative features of that relationship. How is it likely to change in the next decades?

DISCUSSION QUESTIONS

1. Discuss the geographic factors that might explain the quite distinct history and geographic development of Canada and the United States.
2. Take the topics of area and population size and talk about which patterns of settlement, economics, politics, and culture seem to be dominant in visibility and global consequence. Find exceptions to explain.
3. Utilizing Table 22.3, talk about the regional significance of Anglo America during the past fifty years and speculate on what might happen in the next fifty years.

4. Discuss the ways in which the American Revolution and the American Civil War had consequences for Canadian development.
5. Discuss the fact that approximately 90 percent of the Canadian population lives within 100 miles of the U.S. border.
6. What are the historic and geographic features that have made the United States and Canada exist fundamentally as good neighbors?
7. What factors might change that historical pattern significantly?

Chapter 23

The Regional Mosaic of Canada

▲ *In Sparwood, British Columbia, coal mining is big business. This enormous truck has been designed as the most efficient means of moving vast amounts of coal from the pit to the loading area for both domestic shipping and export. In Canada, major investments in such technology have been made because of the importance of resources in the development of local regions and the overall national importance of such economic activities.* Linda Brown

CHAPTER OUTLINE

23.1 Canada: Highly Developed and Highly Varied

23.2 Canada's Regional Landscape Mosaics

23.3 Atlantic Canada: The Causes and Frustration of Steady Decline

23.4 The Two Dissimilar Profiles of Canada's Core Region: Ontario and Quebec

23.5 The Prairie Region: Wheat, Resources, and Isolation

23.6 The Vancouver Region: Core of British Columbia

23.7 The Canadian North: Reshaping a Wilderness and Resource Frontier

 atterns of active migration create regional mosaics that can be quite distinctive. Both of the major nations of Anglo America have been peopled by migrants moving across the oceans and land in search of a new set of opportunities. The patterns of human development that have emerged from these migration histories are distinctive in that they have provided both urban and rural pockets of ethnic continuity. At the same time, the immigrants have endured travails of enormous similarity no matter the place of origin, the ethnic makeup of the migrating population, or the world into which these people came. Migration is truly a pattern profoundly associated with world regional geography. In this chapter on Canada, it can be seen that one nation can be home to quite distinctive mosaics of regional differentiation. Migration is one of the cultural forces that creates such mosaics.

As you learn of Canada and Anglo America, keep in mind the demands of human migration, immigration, and being an outsider trying to find a new home and future in a world very different from one's own origins. Not only has this dynamic been important for the cities of Canada, but it also has shaped much of the urban, town, and even rural growth of Anglo America.

Canada is a highly developed nation—affluent, industrialized, technologically advanced, and urbanized. But in some ways it is curiously unlike most developed countries. These differences are closely related to Canada's internal geography and location adjacent to the United States.

23.1 Canada: Highly Developed and Highly Varied

Canada's prosperity is derived in considerable part from an industrial output sufficiently great to place the country among the world's top 10 or 11 manufacturing nations. Products in which it ranks exceptionally high include newsprint, hydroelectricity, commercial motor vehicles, and aluminum. The country is very

urbanized (Fig. 23.1), with 77 percent of its population classed as urban in 1996 (compared to 75 percent in the United States) and with about one-third of its people living in its three largest metropolitan areas: Toronto (population: 3.9 million); Montreal (population: 3.1 million); and Vancouver (population: 1.6 million). These dynamic metropolises convey a positive image of an advanced and prosperous country. Impressively good national statistics on health, housing, and education provide further evidence of the same. However, the enormous clustering of the Canadian population up against its southern border further demonstrates the country's bittersweet relationship with its neighbor, the United States.

Along with its prosperity, Canada exhibits some striking departures from the expected patterns of developed countries. One of these is an unusual pattern of trade. Despite a fair-sized export market for automobiles and parts and some other fabricated items, Canada is mainly an exporter of raw or semifinished materials, energy, and agricultural products, and is a major importer of manufactured goods and specialty food items. By contrast, most developed nations export mainly manufactured goods and import a mixture of manufactured and primary products. Canada, because of its proximity to the complex economy of the United States, has developed this somewhat unusual pattern.

A second departure from other developed nations is Canada's overwhelming dependence on trade with a single partner, the United States. In 1997, the United States took 82 percent of Canada's exports and supplied 67 percent of Canada's imports. No other developed nation comes near these percentages of trade with just one other country.

Still another major divergence from other developed countries lies in the degree to which the Canadian economy is financed and controlled from outside the country, and in this Canada again has an overwhelming dependence on the United States. Somewhat more than three quarters of all foreign investment in Canada is American. The extent of United States economic dominance in Canada has long been a sore point with

Landscape in Literature

In the Skin of a Lion

Writer Michael Ondaatje's novel about European migrants coming to Canada to find a new life in the early decades of the 20th century provides a window on the difficulties—as well as some of the humor—of this particular migration and transition. The struggles of adjusting to a new home, learning a new language, and finding an adequate job are all illustrated in these images of Nicholas on his move from the Balkans to Toronto.

Nicholas was twenty-five years old when war in the Balkans began. After his village was burned he left with three friends on horseback. They rode one day and a whole night and another day down to Trikala, carrying food and a sack of clothes. Then they jumped on a train that was bound for Athens. Nicholas had a fever, he was delirious, needing air in the thick smoky compartments, waiting to climb up on the roof. In Greece they bribed the captain of a boat a napoleon each to carry them over to Trieste. By now they all had fevers. They slept in the basement of a deserted factory, doing nothing, just trying to keep warm. . . .

Two of Nicholas' friends died on the trip. An Italian showed him how to drink blood in the animal pens to keep strong. It was a French boat called La Siciliana. He still remembered the name, remembered landing in Saint John and everyone thinking how primitive it looked. How primitive Canada was. They had to walk half a mile to the station where they were to be examined. They took whatever they needed

from the sacks of the two who had died and walked towards Canada.

Their boat had been so filthy they were covered with lice. The steerage passengers put down their baggage by the outdoor taps near the toilets. They stripped naked and stood in front of their partners as if looking into a mirror. They began to remove the lice from each other and washed the dirt off with cold water and a cloth, working down the body. It was late November. They put on their clothes and went into the Customs sheds.

Nicholas had no passport, he could not speak a word of English. He had ten napoleons, which he showed them to explain he wouldn't be dependent. They let him through. He was in Upper America.

He took a train for Toronto, where there were many from his village; he would not be among strangers. But there was no work. So he took a train north to Copper Cliff, near Sudbury, and worked there in a Macedonian bakery. He was paid seven dollars a month with food and sleeping quarters. After six months he went to Sault Ste. Marie. He still could hardly speak English and decided to go to school, working nights in another Macedonian bakery. If he did not learn the language he would be lost.

The school was free. The children in the class were ten years old and he was twenty-six. He used to get up at two in the morning and make dough and bake until 8:30. At nine he would go to school. The teachers were all young ladies and were very good people. During this time in Sault he had translation dreams—because of his fast and obsessive studying of En-

glish. In the dreams trees changed not just their names but their looks and characters. Men started answering in falsettos. Dogs spoke out fast to him as they passed him on the street.

When he returned to Toronto all he needed was a voice for all this language. Most immigrants learned their English from recorded songs or, until the talkies came, through mimicking actors on stage. It was a common habit to select one actor and follow him throughout his career, annoyed when he was given a small part, and seeing each of his plays as often as possible—sometimes as often as ten times during a run. Usually by the end of an east-end production at the Fox or Parrot Theater the actors' speeches would be followed by growing echoes as Macedonians, Finns, and Greeks repeated the phrases after a half-second pause, trying to get the pronunciation right.

This infuriated the actors, especially when a line such as "Who put the stove in the living room, Kristin?"—which had originally brought the house down—was spoken simultaneously by at least seventy people and so tended to lose its spontaneity. When the matinee idol Wayne Burnett dropped dead during a performance, a Sicilian butcher took over, knowing his lines and his blocking meticulously, and money did not have to be refunded.

Certain actors were popular because they spoke slowly. Lethargic ballads, and a kind of blues where the first line of a verse is repeated three times, were in great demand. Sojourners walked out of their accent into regional American voices.

From Michael Ondaatje. *In the Skin of a Lion* (New York: Penguin Books, 1987): 45–47.

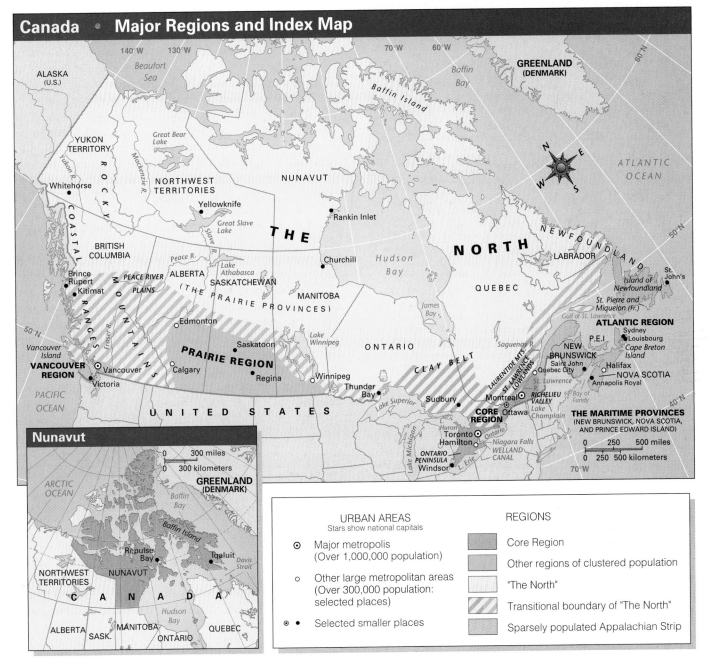

Figure 23.1 Major regions of Canada. In 1991 the Canadian government undertook to create a new Inuit (Eskimo)-controlled Nunavut Territory taken from portions of the Northwest Territories. In May 1993, Canada's prime minister signed the Nunavut Agreement, setting in motion the final steps to create Nunavut (Inuit: "our land") as a separate territory. The final transfer took place on April 1, 1999.

many Canadians, who resent imputations that their country is an American economic "colony" (Fig. 23.2). Such irritations tend to be exacerbated by the degree to which American mass-produced culture has permeated Canada and the degree to which Canada is taken for granted in the United States. Few U.S. schools offer courses on Canada, and a high percentage of American students cannot even name Canada's capital. American news media give relatively scanty coverage to Canada.

There has been for some time a movement in Canada to achieve a more predominant position for Canadians within their own economy, to strengthen economic relations with industrial countries other than the United States, and to heighten cultural self-determination. But there seems to be no way for Canada to pull away from the United States without risking its own prosperity. The country's economy is geared to a close interchange with the gigantic economy next door. In 1988, the two countries

Figure 23.2 In Canada, there is a general unease with the presence of American cultural icons all through the Canadian culture. This evidence of cultural diffusion (more often going south to north than north to south) is particularly frustrating to the French Canadians in Montreal and Quebec City. Yet, as is so often the case with the diffusion of popular culture, once one part of a culture complex finds its way to a new destination, everything from blue jeans to fast food finds its way there sooner or later. This is a Wendy's in Montreal. *Jeff Greenberg Photo Researchers, Inc.*

moved even closer when they signed a free-trade pact to abolish all tariffs on trade between them and to remove or lower many other restraints on trade. In 1994 this arrangement was expanded when the United States, Canada, and Mexico initiated the North American Free Trade Agreement (NAFTA), but in a process that was not without dissent.

23.2 Canada's Regional Landscape Mosaics

In addition to the problem of maintaining satisfactory relationships with the United States, Canada also has problems posed by its own regional structure. The country is divided between a thinly settled northern wilderness that occupies most of Canada's area and a narrow and discontinuous band of more populous regions stretching from ocean to ocean in the extreme south (see Fig. 23.1). These southern regions are sharply different from each other, and are often in political conflict with each other and with the federal government at Ottawa.

About four-fifths of Canada has cold tundra and subarctic climates; more than one-half is in the glacially scoured and agriculturally sterile Canadian Shield (Fig. 23.3), and the part near the Pacific is dominated by rugged mountains. Such conditions create the huge expanse of nearly empty land that Canadians call "the North." More than one-half of it lies in territories administered from the country's capital in Ottawa, Ontario, but

large parts extend into seven of the ten provinces that stretch across southern Canada and make up the major political subdivisions of Canada's federal system. Even within these provinces, most of the land is lightly inhabited, and nearly all the people are concentrated in limited areas of relatively close and continuous settlement, separated by wedges of thinly populated terrain. These population clusters lie just north of the country's boundary with the 48 contiguous United States. Nearly 90 percent of Canada's people live within 100 miles (161 km) of the border. This has facilitated the development of strong ties between Canadian regions and adjoining American regions.

Canada's population clusters can be grouped into four main regions: (1) an Atlantic region of peninsulas and islands at Canada's eastern edge; (2) a culturally divided core region of maximum population and development along the lower Great Lakes and St. Lawrence River in Ontario and Quebec provinces; (3) a Prairie region in the interior plains between the Canadian Shield on the east and the Rocky Mountains on the west; and (4) a Vancouver region on or near the Pacific coast at Canada's southwestern corner (Fig. 23.1). Each region is marked by a distinctive physical environment and cultural landscape; each has a different ethnic mix and distinctive economic emphases. Each plays a different role within Canada and has its own set of relationships with the United States and other countries. All the regions have important links with the Canadian North. (For statistical data on Canadian areas, see Table 22.1.)

23.3 Atlantic Canada: The Causes and Frustration of Steady Decline

The easternmost of Canada's four main populated areas is characterized by a series of strips or nodes of population, largely along or very near the seacoast, in the provinces of Newfoundland, New Brunswick, Nova Scotia, and Prince Edward Island. The four provinces are commonly called the Atlantic Provinces, and three, excluding Newfoundland, have long been known as the Maritime Provinces. Their main populated areas are separated from the far more populous Canadian core region in Quebec and Ontario by areas of mountainous wilderness. Prince Edward Island, which is Canada's smallest but most densely populated province (Table 23.1), is a lowland occupied largely by a continuous mat of farms and small settlements, but the main populated areas of New Brunswick, Nova Scotia, and Newfoundland occupy valleys and coastal strips within a frame of lightly settled upland. As a province, Newfoundland includes the dependent territory of Labrador on the mainland, but in common usage the name Newfoundland is restricted to the large island opposite the mouth of the St. Lawrence River. Formerly a self-governing dominion, Newfoundland was the last province to join Canada, which it did in 1949. Bleak and nearly uninhabited Labrador, situated on the Canadian Shield, is very

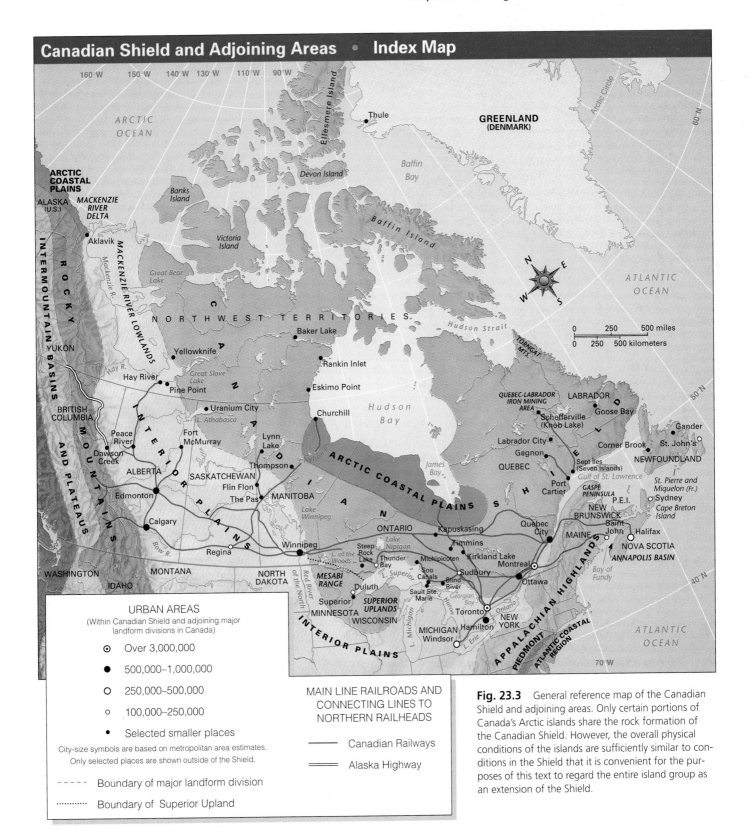

Canadian Shield and Adjoining Areas • Index Map

URBAN AREAS
(Within Canadian Shield and adjoining major landform divisions in Canada)

- ⊙ Over 3,000,000
- ● 500,000–1,000,000
- ○ 250,000–500,000
- ∘ 100,000–250,000
- • Selected smaller places

City-size symbols are based on metropolitan area estimates. Only selected places are shown outside of the Shield.

----- Boundary of major landform division

············ Boundary of Superior Upland

MAIN LINE RAILROADS AND CONNECTING LINES TO NORTHERN RAILHEADS

—— Canadian Railways

═══ Alaska Highway

Fig. 23.3 General reference map of the Canadian Shield and adjoining areas. Only certain portions of Canada's Arctic islands share the rock formation of the Canadian Shield. However, the overall physical conditions of the islands are sufficiently similar to conditions in the Shield that it is convenient for the purposes of this text to regard the entire island group as an extension of the Shield.

PROBLEM LANDSCAPE

Border Wars

"THE CANADIAN-AMERICAN FRONTIER IS the longest and most peaceful in the world, right? Right—but that hasn't stopped economic wars smouldering, even blazing along it this summer. The potato war . . ."

This article in The *Economist*[1] goes on to chronicle a number of contentious disagreements between Canada and the United States. One of the most frequently cited facts about Canada and the United States is the fact that they are such good neighbors. They share the longest border in the world between two nations—5527 miles (8895 km)—and they have a history that is more generally amicable than confrontational. There is a persistent frustration among Canadians concerning being in the cultural and material shadow of their southern neighbor, but nonetheless, nearly 90 percent of the Canadian population lives within 100 miles of the U.S. border—so the irritation must be a bittersweet one (90 per cent of the population lives on 12 percent of the land in these southern reaches).

However, there are some specific items that have bred an ongoing irritation and that reflect themes that comprise the elements of a problem landscape. Here are two:

The wheat war. Even though less than 5 percent of Canada's area is suitable for farming, it is a major world wheat exporter because of the extent of the spring-wheat

growing area in the Prairie Region, west of Lake Superior. It is in this area that farmers have invested heavily in large-scale mechanization. Although only about 3 percent of the Canadian labor force is in agriculture, Canada is a major provider of wheat to markets all over the world. The Great Plains region of the United States is a direct topographic continuation of the immense open plain that courses south-southeast from the Prairie provinces in Canada. Spring wheat is the major crop in that region, and it is also vital to the Dakotas and adjacent states; it is replaced by winter wheat further south where there is a longer frost-free growing season.

The wheat war between the two countries was particularly difficult during the late 1990s. With agricultural prices for grains and pork dropping to the lowest levels in more than 50 years, farmers both south and north of the U.S.-Canada border went bankrupt. They left farming, going into local towns and cities or migrating to wholly new lives in new cities altogether. In late 1997 this tension was made more intense by American claims that Canada was "dumping" its durham wheat on the United States market.

One of the ironies of this tension is the fact that there is not adequate global food production in a world just at six billion in total population. Yet here is a resource problem that relates much more to marketing and transportation than resource availability itself.

The salmon war. Salmon migrate from the waters at the edge of the Pacific Ocean up the coast seeking the rivers of British Columbia (and points north and south) to spawn. With the expiration of a 1985 treaty that had regulated the harvest of such fish, Alaska fishermen moved in 1997 and 1998 into the shallows and harvested enormous numbers of river-seeking salmon. Canada retaliated by blocking the route of an Alaskan ferry for three days, keeping it in Prince Rupert Harbor (on the western edge of British Columbia by the mouth of the Skeena River) as a hostage to negotiations to limit salmon harvest. This act has been followed by other economic boycotts and by Canadian efforts to fully harvest the waters around Vancouver Island so as to deny American fishermen access to these fish. Volleys in this border "skirmish" are still being fired on both sides of the border, reminding us again of the continuing need to carefully chart the use of resources in a border zone.

Being good neighbors is never easy. Virtually every aspect of sharing borders, especially sharing patterns of agricultural or industrial production, means that there are continual areas of friction that crop up, even in a context that is fundamentally agreeable. These two nations, after all, do exchange approximately $1 billion a day in goods and services.

[1]"Canada–USA Border Wars." The *Economist*. August 16, 1997: 29.

---

much a part of Canada's North. It is important to Newfoundland for revenues from iron mining and hydroelectric production.

Isolation from markets afflicts almost all major enterprises in the Atlantic region. The market within the region is small and divided into many small clusters of people separated by sparsely populated land. The largest city in the Atlantic provinces—

Halifax, Nova Scotia—has a metropolitan population of only about 320,000. Only two other cities—St. John's, Newfoundland, and Saint John, New Brunswick—surpass 125,000. The entire area has only 2.4 million people, or fewer than metropolitan Toronto or Montreal. It is a region of out-migration and is the poorest of the four populated regions of Canada.

Table 23.1   Canadian Provinces and Territories: Basic Data

Political Unit	Land Area (thousand/ sq mi)	Land Area (thousand/ sq km)	Estimated Population (thousands)	Estimated Population Density (sq mi)	Estimated Population Density (sq km)	Economic Production (% of Value) (farm receipts)	Economic Production (% of Value) (mineral output)	Economic Production (% of Value) (manufacturing)
Atlantic Provinces								
New Brunswick	27.8	72.1	753.6	27	10	1.3	2.5	2
Nova Scotia	20.4	52.8	934.6	46	18	1.5	1.5	1.7
Prince Edward Island	2.2	5.7	136.4	62	24	1.1	0.01	0.1
Newfoundland-Labrador	143.5	371.7	544.4	4	2	0.3	2.1	0.5
Total	**193.9**	**502.3**	**2369**	**12**	**5**	**4.2**	**6.1**	**4.3**
Core Provinces								
Quebec	523.9	1356.8	7333.3	14	5	16.2	7.4	25.7
Ontario	344.1	891.2	11,411.5	33	13	21.5	13.5	47.5
Total	**867.9**	**2248.0**	**18,744.8**	**22**	**9**	**37.7**	**20.9**	**73.2**
Prairie Provinces								
Manitoba	211.7	548.4	1138.9	5	2	10.3	3.2	2.2
Saskatchewan	220.3	570.7	1024.4	5	2	20.3	8.6	4.8
Alberta	248.8	644.4	2914.9	12	5	19.8	48.2	6.5
Total	**680.9**	**1763.5**	**5078.2**	**8**	**3**	**50.4**	**60**	**13.5**
Pacific Province								
British Columbia	359.0	929.7	4009.9	11	4	7.1	11	9.1
Territories								
Yukon	184.9	479.0	31.7	0.17	0.07	0.01	1	—
Northwest Territories	1271.4	3293.0	67.5	0.05	0.02	—	2	—
Total	**1456.4**	**3772.0**	**99.2**	**0.07**	**0.03**	**0.01**	**3**	**—**
Summary Total	**3558.1**	**9215.5**	**30,600**	**9**	**3**	**100**	**100**	**100**

Source: *Statistics Canada* 1998.

A Centuries-Old Tradition of Fishing

The fishing industry has always been of outstanding importance in Canada's Atlantic region. Before the beginning of settlement in the early 17th century, and probably before the discovery of the Americas by Columbus in 1492, European fishing fleets began operating in the waters—primarily in the "banks" —along and near these shores. The area lies relatively close to Europe, and its waters have always been exceptionally rich in fish. However, it has suffered from steady overfishing in recent years and this has brought about the near disappearance of the cod that have been the mainstay of this region's productivity.

Agricultural Hardships

In general, Atlantic Canada provides hard environments for farming. The four provinces lie at the northern end of the Ap-

palachian Highlands and have topographies dominated by hills and low mountains. Soils that are predominantly mediocre to poor exist under a cover of mixed forest. Many soils that once were farmed wore out quickly and were abandoned. In the 20th century, agriculture has been concentrated on relatively small patches of the best land.

Climate also handicaps agriculture. The Maritime Provinces have a humid continental short-summer climate, while Newfoundland has a subarctic climate. The entire region is humid and windy, with much cloud cover and fog. Strong gales are frequent in the winter. Summers are cool, and colder conditions occur as elevation increases; a good part of upland Newfoundland is tundra. Thus, most areas in the Maritimes and Newfoundland have always been agriculturally marginal, although some more favored lowlands with better soils—notably Prince Edward Island and the Annapolis-Cornwallis Valley of Nova

Definitions & Insights

FISHING BANKS: GENESIS AND DECLINE

Elevated portions of the sea bottom known as "banks" are located off the Atlantic coast from near Cape Cod to the Grand Banks, which lie just off southeastern Newfoundland. The Grand Banks have been at the heart of the fishing economy in Canada's northeastern region for nearly five centuries. The shallowness of the ocean and the mixing of waters from the cold Labrador Current and the warm Gulf Stream foster a rich development of the tiny organisms called plankton, on which fish feed. Inshore fisheries supplement the catch from the banks. Although many kinds of fish and shellfish are caught, the early fleets fished mainly for cod, which were cured on land before making the trip to European markets. Both the British and the French established early fishing settlements in Newfoundland, but the French—who had 150 vessels fishing these banks in 1577—were driven out by the British in 1763. The island has remained strongly British in population, as its limited development has attracted very few other immigrants.

In 1977, worried about the effects of overfishing in the Grand Banks, the Canadian government began to enforce a 200-mile offshore jurisdiction which prohibited overt foreign competition in the fishing of the Grand Banks. This held off the decline of available fish somewhat, but the greater efficiency of fishing technology propelled this industry toward difficult times. In the 1990s, the overfishing of cod in the banks led to a serious decline in fishing productivity. In Newfoundland this decrease in yield has led to unemployment that runs higher than 20 percent, more than three times the overall Canadian level.

Scotia—are exceptions. Potatoes, dairy products, and apples are major agricultural specialties in the Maritimes.

Geographic Profile of a Declining Region

During the first two-thirds of the 19th century, the Maritime Provinces became a relatively prosperous area with a preindustrial economy. Local resources supported this development, including many harbors along indented coastlines, abundant fish for local consumption and export, timber for export and for use in building wooden sailing ships, and land—albeit not very good—for the expansion of an agriculture partly subsistence in character and partly serving local markets. The region carried on an extensive commerce with Great Britain and the West Indies. Then followed an era (still continuing) of relative economic decline within an industrializing Canada. The ports were so far from the developing interior of the continent that most shipping bypassed them in favor of the St. Lawrence ports or the Atlantic ports of the United States. Halifax, Nova Scotia, and Saint John, New Brunswick, are the main ports of the region today, and—while each handles considerable traffic, with an emphasis on containers—neither is in a class with Montreal or the main U.S. ports. St. John's and Gander, in Newfoundland, continues to play a significant role in serving as a refueling stop for many trans-Atlantic air flights.

In the meantime, fishing, forestry, and agriculture suffered various disabilities. Fishing did not prove to be an adequate basis for a prosperous modern economy (Fig. 23.4), primarily for the following reasons:

1. *Tariff barriers.* Tariff barriers were imposed by the United States against Canadian fish during the competitive efforts of Canada and the United States to preserve their own domestic markets.
2. *Global competition.* Competition from major world meat exporters and newly developed fishing areas in other parts of the world, particularly after the end of World War II, had grown to such size that it began to cripple the Canadian fishing economy.
3. *Foreign competition on the local scene.* The fishing conditions of the Grand Banks and other nearby waters of the Maritime Provinces were so productive that fishing fleets from other nations and regions—many possessing high-tech fishing operations—began to drive Canadian fishermen from their own waters.
4. *Disappearance of the resource.* The combination of expanding numbers of fishermen, more sophisticated and effective technology, and steady overfishing led to a major and continuing decline in the number of fish in this traditionally rich area.

Wood industries also suffered when forests became depleted by overcutting, ships began to be made of iron and steel, and competition set in from new areas of forest exploitation farther west. The position of agriculture in the Maritimes was undermined by the building of railroads, which facilitated settlement of Canada's interior and thus brought cheaper farm products into the Atlantic region. It also provided easier access for local farmers considering moving west. Today, quite a few people in the Maritime Provinces still farm, but they do so largely on a small-scale, part-time basis while earning their living mainly from other employment.

As these economic challenges have arisen, the Atlantic region has attempted to meet them by developing industry. For example, many small cotton-textile factories were built during the 19th century. But these early industrial ventures largely failed when they were undercut by competition from mills in New England, central Canada, and Europe that were located closer to major markets. Also, sizable coal reserves existed near the town of Sydney on northern Cape Breton Island in Nova Scotia, and iron ore was present on Newfoundland's Bell Island. On the basis of these resources, a coal-mining industry and an iron-and-steel plant were developed at Sydney which flourished for some time, largely by supplying steel for Canadian railway construction. But eventually both the plant and the coal mines declined, with consequent economic depression in the Sydney area. The plant at Sydney was bought by the provincial government after World War II to forestall its closure; it continues to operate at a reduced scale and on a subsidized basis in order to provide employment. The local coal industry has been handicapped by both the increasing cost of extracting the coal, which is deep-lying, and the shrinking market for coal due to its widespread replacement by oil as a fuel. Many coal mines have closed.

Currently, the basic economic activities that support Atlantic Canada's struggling economy are numerous but relatively small:

1. *Specialized agriculture.* Specialized commercial potato farming is somewhat successful in Prince Edward Island and the Saint John River Valley of eastern New Brunswick.
2. *Continued fishing.* Protected from foreign vessels since 1977 by the extension of Canadian control to waters 200 miles (322 km) offshore, commercial fishing continues on a small scale.
3. *Tourism.* Based on the region's scenic beauty, cool summers, and sense of splendid isolation, tourism is also handicapped by some of those same qualities because of the region's distance from population centers with their popular urban amenities.
4. *Forestry.* Pulp and paper manufacturing are making a modest comeback and undergoing steady growth because of the expansion of the international market for paper and wood products.
5. *Energy.* The export of electricity to adjacent provinces and the United States, especially from Labrador's large Churchill Falls hydropower installation and from New Brunswick, has been a steady factor in the modest expansion of economic health in the region.
6. *Minerals.* Atlantic Canada does possess a few large metal mines. Both coal and iron ore have a traditional importance in the region's mineral resources, but they pale compared to current anticipation of petroleum wealth. Offshore oil and gas exploration in the 1970s and 1980s found the Hibernia Field off southeastern Newfoundland. While it is thought that these fields might possess enormous reserves, the extraction is particularly difficult because of local weather conditions and ocean floor characteristics. Production has only moved ahead since the mid-1980s. This potential source of wealth for a region that has been traditionally the poorest in Canada has led to considerable intranational bickering over the ownership of the petroleum reserves.

Extremely important to the economy of the Atlantic region is a high level of subsidies from the federal government in the form of (1) pension and welfare payments, (2) the stationing of military forces in the region (Halifax is a major naval base), (3) the presence of many federal administrative offices in the region, (4) direct payments to provincial treasuries, and (5) funding for economic development projects and efforts to attract industry.

Figure 23.4 The fishing villages of Nova Scotia suggest a link with the past that is more a mark of marginal fishing than tourist delight in the contemporary context of Canada's Maritime Provinces. The small craft and family operations of this photo have a very difficult time competing in a world fishing economy that is increasingly high-tech in its design and function. *Thomas Kitchin/Tom Stack & Associates*

23.4 The Two Dissimilar Profiles of Canada's Core Region: Ontario and Quebec

The core region of Canada, with about two-thirds of the nation's population, has developed along lakes Erie and Ontario and thence seaward along the St. Lawrence River. In this area, the Interior Plains of North America extend northeastward to the Atlantic. They are increasingly constricted seaward—on the north by the edge of the Canadian Shield and on the south by the Appalachian Highlands. The two provinces that divide the core region—Ontario and Quebec—both incorporate large and little-populated expanses of the Shield; Quebec has a strip of Appalachian country along its border with the United States. But most of the core region lies in lowlands bordering the Great Lakes or the St. Lawrence River from the vicinity of Windsor, Ontario, to Quebec City, Quebec. The entire lowland area is often loosely termed the St. Lawrence Lowlands, although the Ontario peninsula between lakes Huron, Erie, and Ontario is frequently recognized as a separate section.

Contrasts between the Ontario and Quebec Lowlands

The Ontario and Quebec sections of the core region show marked differences in types of agriculture and cultural features. The Ontario Peninsula—strongly British in its roots—echoes the U.S. Midwest's cropbelts on a small scale, with corn and livestock production, dairy farming, and growth of various specialty crops such as tobacco. Along the St. Lawrence in Quebec,

Figure 23.5 Broad open plains with checkerboard field patterns—so common in the Great Plains of the United States—are the farming images that are evoked when one thinks of Canadian agriculture. But, in the rolling and even mountainous lands of the Maritime Provinces and the Atlantic Region in Canada, a wholly distinct landscape pattern emerges. This agricultural region on the Gaspe Peninsula in Quebec reminds the reader that settlement and farming began along the Atlantic Coast and only slowly worked west to the more productive Prairie Region of Canada. *Francis Lepine/Animals Animals/Earth Scenes*

the French heritage is clearly evident in the landscape signatures of the French "long-lot" pattern of strip-shaped agricultural landholdings (Fig. 23.5).

In the Quebec landscape along the St. Lawrence, large Catholic churches are prominent landscape features. The Quebec lowlands lie farther north than the Ontario Peninsula and have a harsher winter climate. Dairy farming is the predominant form of agriculture. Together, Ontario and Quebec account for about two-fifths of the value of products marketed from Canadian farms.

The core area contains more than 18 million people. Ontario's population is basically British in origin but has many minority ethnic groups (see Landscape in Literature, p. 568). By contrast, about four-fifths of the people in Quebec are of French origin (Fig. 23.6). This represents the biggest exception to the dominance of English culture in Anglo America and is the only instance in which a national minority controls a provincial or state government. Nunavut Territory on the northwest edge of Hudson Bay and in the Northwest Territories becomes a major politcal feature of Canada peopled almost entirely by a national minority.

Within the core, life focuses on Canada's two main cities: Toronto in Ontario (population: 4,338,400) and Montreal, the second largest French city in the world, in Quebec (population: 3,328,000). Other large cities of the core include Ottawa (population: 925,000), Quebec City (population: 695,000), and Hamilton (population: 641,000). Ottawa, the federal capital, is located just inside Ontario on the Ottawa River boundary between the two provinces. Quebec City, the original center of French administration in Canada, is the capital of Quebec Province. Hamilton, a Lake Ontario port, is the main center of Canada's steel industry, supplying steel for the automobile and other metal-fabricating industries of Ontario. A good part of the urban development in the core is associated with manufacturing. Three-fourths of Canada's total manufacturing development is here, with somewhat more in Ontario than in Quebec. But almost three times as many people are employed in the service industry, reflecting the region's status as the business and political center of the country.

Historical Evolution

The coreland began as an entry to the interior of North America, and this gateway function continues today. The French founded Quebec City in 1608 at the point where the St. Lawrence estuary leading to the Atlantic narrows sharply. Fortifications on a bold eminence allowed control of the river (Fig. 23.7). From Quebec, fur traders, missionaries, and soldier-explorers soon discovered an extensive network of river and lake routes—with connecting portages—reaching as far as the Great Plains and the Gulf of Mexico. Montreal, founded later on an island in the St. Lawrence River, became the fur trade's forward post toward the interior Indian-dominated wilderness. Between Quebec City and Montreal, a thin line of settlement evolved along the St. Lawrence and formed the agricultural base for the colony. Population grew slowly in this northerly outpost where the winter was

Figure 23.6 This photo shows Quebec City, once the capital of France's North American empire and now the capital of Quebec Province. At the right is the Lower Town along the St. Lawrence River; at the left, the Upper Town. Grain elevators symbolize the city's seaport function. The hotel called the Chateau Frontenac is a reminder of the city's heritage from Imperial France with its distinctive copper sheathing. *Jesse H. Wheeler, Jr.*

576

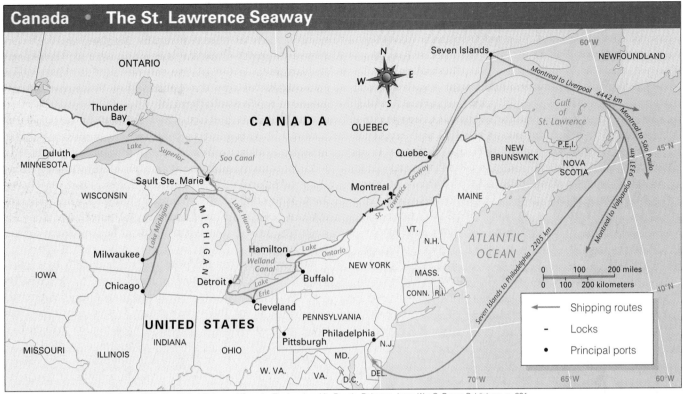

Canada • The St. Lawrence Seaway

Source: Getis, Arthur and Judith Getis. 1995. *The United States and Canada: The Land and Its People.* Dubuque, Iowa: Wm C. Brown Publishers. p. 231.

Figure 23.7 This waterway, more than 2300 miles (c. 3700 km) long, connects the Atlantic Ocean with the Great Lakes in the heart of the continent. This means that cities like Duluth, Minnesota, and Thunder Bay, Ontario, are both international ports and serve international vessels. The Seaway was opened in 1959.

harsh and only French Catholics were welcome. When finally conquered by Britain in 1759, French Canadians numbered only about 60,000.

A few British settlers came to Quebec after the conquest, and this immigration increased rapidly during and after the American Revolution, thus laying the foundations for the British segment of Canada's core. Many of the early English-speaking immigrants were Loyalist refugees from the newly independent United States, whose rebellion French Canada had refused to join. They were soon joined by more English settlers immigrating directly from Europe. These newcomers formed a sizable minority in French Quebec (Fig. 23.8), including a British commercial elite in Montreal that persists to this day, but they also settled west of the French on the Ontario peninsula. By 1791, the British government found it expedient to separate the two cultural areas into different political

divisions—known at that time as Lower Canada (Quebec) and Upper Canada (Ontario), taking their directional terms from their positions on the St. Lawrence River.

The St. Lawrence Seaway

Very important among the core region's advantages were its lowland and water connections between the Atlantic and the interior of North America. In eastern North America, the only other connection of this sort through the Appalachian barrier is

Figure 23.8 An October 1995 referendum held in Quebec Province in Canada pitted the French-speaking citizens who wanted to achieve an independent Quebec against people of the same province who wanted to maintain the bilingual lifestyle that had come to characterize this early 18th-century home to French settlers. The move to secede failed by the narrowest of margins and the issue is likely to be revived again in the near future. *A. McInnis/Gamma Liaison*

the lowland route connecting New York City with Lake Erie via the Hudson River and Mohawk Valley. Rapids on the St. Lawrence River long barred ship traffic upstream from Montreal, but the lowland along the river facilitated railway construction, and the building of some 19th-century canals bypassed the rapids and allowed small ocean vessels to enter the Great Lakes. Finally, in the 1950s, the river itself was tamed by a series of dams and locks in the St. Lawrence Seaway project (see Fig. 23.7). Since then, good-sized ships have been able to reach the Great Lakes via the river. The Welland Canal, which bypasses Niagara Falls between lakes Ontario and Erie by means of a series of stairstepped locks, predated the Seaway. The canal admits shipping vessels to the four Great Lakes above the falls. Farther up the lakes, the "Soo" Locks and Canals enable ships to pass between Lake Huron and Lake Superior, and natural channels interconnect lakes Erie, Huron, and Michigan. The total length of the St. Lawrence Seaway project is 2342 miles (3769 km), and cities on the margins of the Great Lakes—both in Canada and in the United States—now serve as international ports. The Seaway provides an eight-meter-deep waterway in the heart of North America, and it moves wheat, iron ore, petroleum, and durable goods from mid-April until approximately mid-December, with the assistance of ice breakers. Construction was begun in the 1950s, although its full complement of canals, locks, and dams include transportation components that had been built decades earlier. The project has changed commercial geography in the heartland of North America in many ways. It also, through unwanted events, has changed the ecology of the region as well, particularly in the case of the arrival of zebra mussels that have been diffused into these interior waters from Europe and that are now a major blight on shipping and machinery efficiency because of their skyrocketing numbers and persistence.

The Rise of Commerce, Industry, and Urbanism

Until Canada was formed as an independent country in 1867—an act fomented in good part because of the success of the Union army in the American Civil War—Quebec and Ontario developed largely in a common preindustrial fashion, with economies based primarily on agriculture and forest exploitation. With its larger expanses of good land and a somewhat milder climate, the Ontario peninsula became more agriculturally productive and prosperous than Quebec. It specialized in surplus production and export of wheat. On a superior natural harbor, the Lake Ontario port of Toronto developed and became the colonial (now the provincial) capital and Ontario's largest city. But the bulk of Canada's exports at this time moved eastward to meet ocean shipping, and this movement favored Montreal. Until about 1850, most shipping from the Atlantic went no farther upstream than Quebec City. But subsequent improvements along turbulent sections of the river enabled ocean ships to navigate safely as far inland as the major rapids (La-

chine Rapids) at Montreal. Quebec City was bypassed and its growth began to slow, although it has continued to be an active ocean port. In the later 19th century, Montreal decisively surpassed its French-Canadian rival and became the leading port and largest city in Canada. But only recently has Montreal been exceeded in trade by the Pacific port of Vancouver and surpassed in population by Toronto.

As Montreal's port traffic grew, the city's business interests facilitated it by building a pioneering railway network based in Montreal. This network made use of lowland routes that radiated from the city: along the St. Lawrence Valley leading east to the Atlantic and west to the Great Lakes; along the Ottawa River Valley through which Canada's transcontinental railroads found a passage west and north into the Shield; and along the Richelieu Valley–Champlain Lowland leading south toward New York City.

Britain created an independent Canada in 1867 by confederating Quebec, Ontario, New Brunswick, and Nova Scotia. By this time, the core area was already dominant in agriculture and population. In the following century, it was transformed into today's urban-industrial region. In its evolution to industrial dominance, the coreland profited from a number of clear advantages over the rest of Canada: superior position with respect to transport and trade, the best access to significant resources, accessibility to the largest markets, and the most advantageous labor conditions.

Resource Availability

In addition to the water routes that connect it to the world and to all the shores of the Great Lakes, the Canadian core area has had a number of other major resources to support industrialization:

1. *Agricultural resources.* Farmland good enough for a large commercial agriculture, especially in Ontario, has supplied materials for food processing industries and a market for farm equipment industries.
2. *Forest resources.* Forests, mainly in the adjacent southern edge of the Canadian Shield, once supplied lumber for export and now supply wood for a very large pulp and paper industry, mainly in Quebec along the St. Lawrence and its north bank tributaries.
3. *Energy generation.* Especially at the descent from the Shield to the plains, many high-volume, steeply falling rivers drive one of the world's larger concentrations of hydroelectric plants. Major hydroelectric plants are also present at Niagara Falls, along the St. Lawrence as part of the Seaway project, and, more recently, at far northern sites in the Shield. Electricity is cheap in the coreland and is exported to the United States. Coal requirements are met from nearby Appalachian fields in the United States, while oil comes from Atlantic or Middle Eastern ports by tanker up the St. Lawrence, and both oil and natural gas come by pipeline from western Canada.
4. *Bulk commodity transport.* Metallic minerals are varied and

abundant in the Shield to the north and west of the core area. Among them are very large deposits of iron ore located near the western end of Lake Superior on both sides of the international border, while others, more recently developed, are located in northeastern Quebec and adjacent Labrador. A large development of metal processing and fabricating industries, especially in Ontario, uses these Shield minerals, which also are exported.

Through the combination of these resources, their processing, and their distribution to both domestic and foreign markets, the Canadian core region has established a solid base for continuing economic development and relative prosperity.

Market and Labor Advantages

Ever since industrialization began, the population of the core area has represented the main cluster of Canadian consumers and labor. A high-tariff policy to forestall American competition was adopted in Canada shortly after its confederation. While it fell short of shutting out foreign industrial products, which Canada imports in large quantities, it did favor the growth of a large variety of Canadian industries, mainly in the core region and often including branch plants of U.S. companies "jumping" the tariff wall. However, the policy has long aroused protest from other parts of Canada in which consumers preferred to buy cheaper American goods. It also has fostered many high-cost Canadian industries that lack international competitiveness. The exceptions to this latter phenomenon tend to be mainly industries making products desired in the United States.

Thus, the Canadian pulp-and-paper and mining industries, with the U.S. market available for their products, tend to be large and efficient. In 1965, the two countries negotiated a free-trade agreement with respect to the production and import and export of automobiles. This sparked rapid expansion of the auto industry into southern Ontario. General Motors, Ford, and Chrysler all have plants there—some of which are in Detroit's Canadian satellite of Windsor (population: 260,000). However, metropolitan Toronto is the main center of the industry. More recently, the inauguration of NAFTA has begun to cause major industrial readjustments—generally characterized by the more labor-intensive firms (clothing especially) moving to Mexican locales—in Canada's core region, just as it has in other parts of North America.

Industrial Differences Between Ontario and Quebec

Canada's core region produces a wide range of manufactured goods, and for most of these products it accounts for more than three-fourths of Canada's national output. Within the core, there are marked differences between the industrial emphases of Ontario and those of Quebec. During the formative decades of industrialization, Ontario had cheaper access to United States

coal. This favored the development of iron-and-steel capacity and metal products industries there, and these types of industries are still more prominent in Ontario than in Quebec. Montreal also has industries focused on making transport equipment and other metal goods, but these are less typical of Quebec than are more labor-intensive industries such as apparel manufacturing. Such industries were established primarily to deal with a traditional surplus of labor in Quebec created by an unusually rapid natural increase of Quebec's Roman Catholic population. This labor surplus led to migration into other parts of Canada and the United States and also to the establishment of manufacturing industries that were in need of notably cheap labor. Textiles, shoes, and clothing are typical products. The differences between the two provinces still persist, and they have resulted in lower average wages and incomes in Quebec, which has also experienced a steady decline in its population growth.

Cultural Divergence and Its Consequences

The economic differences between Quebec and Ontario are a serious matter but probably less so than the cultural differences. In Quebec, 82 percent of the people speak French as a preferred language and 11 percent speak English. This is the only province in which French speakers predominate and control the provincial government. On the other hand, Ontario, where 77 percent speak English as a first language and only 5 percent speak French, is the largest in population of the nine predominantly English-speaking provinces. Major political consequences have arisen from this situation (see Fig. 23.8 and Definitions and Insights, p. 580).

23.5 The Prairie Region: Wheat, Resources, and Isolation

Manitoba, Saskatchewan, and Alberta are identified as a group under the name "the Prairie Provinces." Isolation is one factor that has led to this grouping. Their populations are separated from other populous areas by hundreds of miles of very thinly inhabited territory. To the east, the Canadian Shield separates them from the populous parts of Ontario, and on the west, the Rocky Mountains and other highlands separate them from the Vancouver region. To the north lies subarctic wilderness, and to the south are lightly populated areas in Minnesota, North Dakota, and Montana.

The prairie environment that gives the provinces their regional name exists only in some southern sections of the three, and it is to these sections that we ascribe the name "Prairie region" in this text. The region is a triangle of natural grasslands extending to an apex about 300 miles (c. 500 km) north of the United States border. It lies within the Interior Plains between the Shield and the Rockies, with the southern base of the triangle resting on the United States boundary. On all sides except

Definitions & Insights

THE QUEBEC SEPARATIST MOVEMENT

Increasingly common on the global stage are efforts of specific ethnic, linguistic, or religious population segments to withdraw from a larger body politic and gain partial or full independence. The history of such a movement in Canada is particularly instructive of the types of demographics and political geography that fuel such campaigns. The roots of the tension in Canada can be seen in early Canadian history: The French were early settlers in these lands, but after the 1763 conquest of the British over the French, the broad sweep of language, government, and authority became clearly British in cast. A cultural dichotomy developed that gave the French dominant influence in Quebec, but established only modest influence for them beyond the St. Lawrence lowlands region.

After the end of World War II, more widespread French Canadian dissatisfaction developed, possibly because of increasing attention given worldwide to the plight of minorities. This led, in Quebec, to the formation of a separatist political party—the Parti Québécois—dedicated to the ultimate full independence of Quebec from Canada. In 1976, the party achieved control of the Quebec Provincial Government and made French the language of commerce, requiring that all immigrants settling in the province be educated in French. In 1980 and again in 1985, the separatists forced referenda on independence which lost both times. In 1995, the issue was taken to provincial voters again, and the number of voters who favored staying a part of Canada won by a very small margin.

The effects of 40 years of controversy have been considerable. The Ottawa government (the seat of the Canadian federal government) has made French an official language of Canada—along with English—all across the nation. Laws have now stipulated that the children of French Canadians born outside of Quebec may have their children educated in French. And there continues to be political effort in the corridors of power in Quebec as separatist groups try to time another effort for independence. However, a steady influx of non-French-speaking immigrants who seek jobs and settlement in Quebec City and other cities and towns along the St. Lawrence Seaway tend to vote with the pro-English bloc, which continues to thwart the Quebec separatists.

The other implication of all this has been that other Canadian provinces have utilized the separatist strategies to gain more control of their own provincial governments. As Ottawa has given the French Canadians more authority in Quebec in the hope that they might find the idea of full independence less attractive, other provinces—especially in the Prairie Region—have come forward with ever more demanding requests regarding governmental autonomy. Canadian law supposedly requires that there be no more than one separatist referendum every five years, but even that regulation may be overlooked as this process continues. It has the potential for leaving Canada in political turmoil for a long time to come, and the world is watching to see how this pursuit of cultural independence plays itself out.

---

the south, the grasslands are bordered by vast reaches of forest. Most of the southern part of the triangle is a northward continuation of steppe grasslands from the Great Plains of the United States. However, toward the forest edges the soils are moister, and the original settlers found true prairies: taller grasses with scattered clumps and riverine strips of trees. It was to the prairies that settlers were most attracted, and these moister grasslands still form a more densely populated, arc-shaped band near the forest edges.

The Prairie region was settled late, but quite rapidly, in a few decades after 1890. Settlers found reasonably good land, with a climate of long and harsh winters, short cool summers, and marginal precipitation. Only hardy crops could be grown. Early subsistence farming was succeeded by specialization in spring wheat for export, and the Prairie region became the Canadian part of the North American Spring Wheat Belt (Fig. 23.9). The settlers were predominantly English-speaking people from eastern Canada and the United States, but they included notable minorities of French Canadians, Germans, and Ukrainians. The ethnic composition of the region still reflects these elements.

Although the economy has diversified, agriculture is still basic. The region produces about one-half the value of farm

products in Canada. Agriculture has been diversified to include other hardy crops such as barley and rapeseed (for vegetable oil and fodder), together with raising of livestock.

Urbanization and nonagricultural industries have become increasingly important, especially in Alberta and Manitoba. The three main metropolises are Edmonton (population: 840,000) and Calgary (population: 755,000) in Alberta, and Winnipeg (population: 650,000) in Manitoba. These cities have all grown near the corners of the Prairie triangle where connections to the outside world are focused. Winnipeg developed where the rail lines through the Shield crossed the Red River and entered the Prairie region; Calgary settled near a pass over the Rocky Mountains that gave a route toward Vancouver; and Edmonton was established near another Rocky Mountain pass leading to the Pacific. Both Winnipeg and Edmonton have developed additional functions as metropolitan bases for huge sections of Canada's North. None of the three Prairie provinces has yet become a notable manufacturing center, but the richness of the environmental setting has turned this region into an important tourist destination for Canadians and Americans alike.

Minerals are the foundation of recent advances in the region's economy, with this region now producing more than 60 percent of Canadian mineral production. Coal deposits underlie

Figure 23. 9 The landscapes of farming and particularly wheat cropping are of major importance in the heart of Canada. This scene from Thunder Bay on the northwest shore of Lake Superior shows the grain elevators that are so common a signature along the rail lines and in large shipping facilities. Thunder Bay is an international port because of the connection to the Atlantic provided by the St. Lawrence Seaway and it is from here that a major portion of Canada's large annual grain export is launched toward international destinations. *Thomas Kitchin/Tom Stack & Associates*

both the Rocky Mountain foothills and the plains in Alberta, as well as western Saskatchewan. Coal has recently begun to provide a major export market. Nickel, copper, zinc, and other metals come from the Canadian Shield of Manitoba; the Saskatchewan Shield produces uranium and southern Saskatchewan produces major quantities of potash. But it is petroleum and associated natural gas that have had the most economic impact on the Prairie region.

The region's oil industry began near Calgary, Alberta, in the 1940s. As production has multiplied and new fields have been brought in, the industry has remained predominantly in Alberta, with a minor portion in Saskatchewan and a tiny share in southern Manitoba. An important product extracted from natural gas in Alberta is sulfur, of which Canada is a dominant world exporter.

23.6 The Vancouver Region: Core of British Columbia

Canada's transcontinental belt of population clusters is anchored at the Pacific end by a small region centering on the seaport of Vancouver at the southwestern corner of British Columbia. The region contains no more than 5 or 10 percent of British Columbia's area but has over one-half of the province's population. Metropolitan Vancouver has 1.8 million people, and another 300,000 reside in the metropolitan area of the province's capital city, Victoria (population: 311,000). Vancouver is located on a superb natural harbor at the mouth of the Fraser River. This valley provides a natural route for railways and highways across the grain of rugged mountains and plateaus in Alberta and British Columbia. Victoria is located at the southern end of Vancouver Island—the southernmost and largest of a chain of islands along Canada's Pacific coast.

Victoria was the leading commercial and industrial center in British Columbia before the completion of Canada's first transcontinental railroad, the Canadian Pacific, in 1886. Vancouver then became the country's main Pacific port, and it has recently become the largest seaport of the entire country (Fig. 23.10). It has profited from continuing development and a buildup of foreign trade in British Columbia and the Prairie provinces, particularly Canada's increasing exports of resource

Figure 23.10 A portion of the harbor at Vancouver, British Columbia, Canada's busiest seaport. The container port is in the center of the photo, which was taken from the city's central business district. The growth and expansion of this city in recent years reflects Canada's shifting and expanding trade connections with the nations of the Pacific Rim. *Porterfield Chickering/Photo Researchers*

commodities to East Asia, primarily Japan. This city served as a major migration destination for Hong Kong Chinese in the years preceding the British handover of Hong Kong to the People's Republic of China. Vancouver now is home to Anglo America's second largest Chinatown, second only to that in San Francisco.

The recent years of the so-called Asian Economic Flu (recession) have reduced the Asian tourist flow to British Columbia by about 20 percent. Japanese tourists have traditionally had the biggest impact on local economies, and British Columbia has, at the provincial level, seen a drop of 17 percent of Asian tourist flow into the region. Korean numbers dropped by 60 percent in 1998, and those of mainland Chinese dropped by 25 percent. The only good news in this situation is that the number of American tourists has increased during this same period of Asian decline.

Vancouver also exports huge quantities of such bulk commodities as grain, wood in various forms, coal, sulfur, metals, potash, and asbestos. Sea trade is fundamental to Vancouver's economy, but the city is far more than just a seaport. Despite its peripheral location within British Columbia, it is the province's main industrial, financial, and corporate administrative center. As such, it has numerous links with production nodes scattered through a sparsely populated and mountainous province larger in area than Texas and California combined.

The Province of British Columbia

British Columbia, along with the Yukon Territory to its north and a narrow fringe of Alberta, is the high part of Canada. The fjorded Coast Ranges along the Pacific and the Rockies of the interior are high and spectacular, and the Intermountain Plateaus between them are largely rugged country. Only the northeastern corner of the province is plains country. This is "Peace River Country," the extreme northern outpost of Prairie-region grain and livestock agriculture. The province is also distinctive climatically within Canada by virtue of its coastal strip of marine west coast climate. Here, moderate temperatures and heavy precipitation are characteristic. The moisture supports the dense coniferous forests that are one of the province's principal resources, and the mild temperatures, combined with spectacular scenery, attract tourism and retirees. The climate of the interior is subarctic in the north and varies according to elevation and exposure in the south. Like the Coast Ranges, the interior is largely forested, although not so luxuriantly.

Scattered small communities exploit abundant natural resources at increasing scales from an increasing number of locations. Forest products come from both the interior and the coast (Fig. 23.11). Salmon fishing continues along the coast, as it has for a century. More important than forestry and fishing today is a boom in the mining of metals (copper, molybdenum, silver, lead, and zinc) and coal. Older metal production sites in the south are being supplemented by operations farther and farther north. The driving factor is Japanese and United States demand,

Figure 23.11 Timber continues to play a major role in the economy of Canada, especially in the province of British Columbia. This resource has found a vast market in the Pacific Rim countries, and this photo shows logs on Slocan Lake in British Columbia, working their way toward a local sawmill. *Gordon Fisher/Tony Stone Images*

with development being supported by foreign capital, largely Japanese. The province is a storehouse of energy resources in the form of both coal and hydropower, with coal shipped to Japan and electricity exported to the United States. Near the coastal town of Kitimat, hydroelectricity is used to produce aluminum from imported alumina (the second-stage material from bauxite ore). Wages tend to be high in the small and isolated production nodes scattered over British Columbia, but so is the cost of living, and cultural amenities are few. Most new residents of British Columbia settle in cosmopolitan Vancouver or in Victoria, known as the most "English" of Canadian cities.

23.7 The Canadian North: Reshaping a Wilderness and Resource Frontier

The North comprises most of Canada, but where its southern edge lies is arguable. It certainly includes that major section of Canada that is very sparsely populated because of harsh physical environments, but population density fades gradually northward. In this discussion, we use the southern edge of the subarctic climate as, for the most part, the southern edge of the North, but do not include the island of Newfoundland—with its

Regional Perspective

The Downside of Resource Wealth

There is a good deal of luck involved in the discovery of riches in the soil of a specific piece of real estate. This fact applies both at the level of a single miner who strikes it rich on a random piece of land and at the more broad-based and powerful level of settlement across a wide-open expanse, as in the westward movement of pioneers across Canada beyond Lake Superior in the 19th century. The Prairie provinces that were settled in this inexorable movement toward British Columbia's Pacific coast have come to realize that there is considerable wealth in the lands they received in the allocation of provincial territories.

However, there is a frustration that has built up—particularly in western Canada—because of the very wealth these mineral and soil resources represent. It has led to a political posture called **resource separatism.** Resource separatism occurs when, within the boundaries of a single nation, one region has a particularly rich resource base but sees the cash rewards for such good fortune, or regional productivity, go to the federal government and other parts of the country in what the source area considers to be an unbalanced manner.

In the province of Alberta, the presence of fossil fuels has led to the potential for resource separatism. On a world scale, Canada's oil output is still small, but it has nevertheless led to explosive growth for Calgary and Edmonton, which are the Canadian oil industry's main business centers. Another result of the oil boom has been a running feud between Alberta and the federal government concerning resource control and, especially, price control of oil piped from Alberta to eastern Canada. Alberta's economic dynamism and aggressive political stance seem likely to continue, as the province is much richer in potential energy than in present energy production. It is estimated that "tar sands" in northern Alberta may contain two to three times the energy equivalent of all the oil now thought to exist in the Middle East. Major extraction must await improved technology and/or a time when greater scarcity of oil yields higher prices.

The oil fields that enrich Alberta and Saskatchewan have their origins south of the Gulf of Mexico, course up northward through Texas (where they produce one sixth of the United States total oil output) and extend further north through the Prairie provinces. Alberta, particularly, is interested in gaining a greater percentage of the foreign exchange generated by the gas, tar sands, and crude oil of these beds. Resource-based domestic tension such as this reminds us that one can never tell exactly what issues will crop up from the discovery of resource wealth.

Atlantic associations and somewhat denser population. This definition places large sections of all the provinces except the Maritimes in the North. In addition, we may justify inclusion in the North of parts of the Canadian Shield which project south of the subarctic climate into southeastern Manitoba, Ontario, and Quebec. These areas are so handicapped by poor soils as to be sparsely populated, possessing the demographic patterns that are generally associated with the North (Fig. 23.12).

The southern fringes of the North, often called the "Near North," are spotted with widely scattered islands of development of considerable importance to Canada's economy and, through trade connections, to the economy of the United States. Many of these are mining settlements. The Shield here has a wide variety of metallic ores, although geological processes have left it lacking in fossil fuels. The mining settlements range in size from small hamlets to small metropolitan areas. The largest is Sudbury (metropolitan population: 158,000), located in Ontario north of Lake Huron. Particularly noteworthy among many metals produced in the North are nickel and copper around Sudbury, iron ore from Quebec and Labrador, and uranium mined just north of Lake Huron and in northern Saskatchewan.

Some Northern settlements manufacture pulp and paper or smelt metals. A scattering of towns producing wood pulp and paper is spread across the southern fringes of the North. Most are small, but in two places clusters of mills combine with other functions to produce fairly sizable agglomerations. A string of towns along the Saguenay River tributary of the St. Lawrence in Quebec has a combined population of more than 180,000 people, supported by pulp-and-paper plants and by the manufacture of aluminum from imported raw materials brought in by ship. The other agglomeration is at Thunder Bay, Ontario (population: 140,000), on the northwestern shore of Lake Superior. In addition to pulp and paper, Thunder Bay derives support from being at the Canadian head of Great Lakes navigation. Its port links the Prairie provinces with the Ontario-Quebec core area, the United States, and overseas points.

Water resources are of major importance in the southern fringe of the North. Hydroelectricity supports local mines and mills and is sent to southern Canada and the United States. Rivers and lakes are so characteristic of the Shield that much of the latter appears from the air to be an amphibious landscape. There is a particular cluster of hydroelectric plants where

Figure 23.12 Canada has been negotiating with First Peoples and the Inuit peoples for years in an effort to gain resolution of seemingly endless treaty and boundary litigation. In 1999, a new realm of nearly 800,000 square miles (2,072,000 sq km) will be deeded to the Inuit peoples. This is one-fifth of Canada's area, and the population resident in these northern lands will amount to fewer than 28,000. The Nunavut population shares ancestry with peoples that range all the way from Siberia to Greenland.

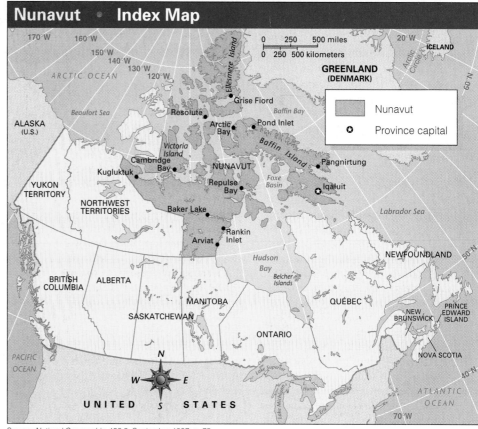

Source: *National Geographic*, 192:3, September 1997. p. 75.

St. Lawrence tributaries from the north rush down through the Laurentide "Mountains"—the somewhat raised southern edge of the Shield in Quebec. But increasing demand, plus increasing ability to transmit electricity for long distances, has led to major developments much farther north. These are located near James Bay, in Labrador, and in Manitoba near Hudson Bay. A remarkably small number of people live in this vast area. With agricultural possibilities so limited, a large increase in population seems unlikely in the foreseeable future. Agricultural settlements are rare, and most of them are declining.

"Territorial" Canada: Bleak and Empty Federal Domain

North of 60 degrees latitude lies the part of Canada not yet considered sufficiently developed and populous for provincial status. This bleak area—known as "Territorial Canada"—is under the direct administration of the Canadian federal government. Two administrative territories have been established, and a third one was formed in 1999 (see Fig. 23.1). In the northwest, the Yukon Territory is cut off from the Pacific by the panhandle of Alaska, once claimed by Canada but acquired by the United States. It is mostly rugged plateau and mountain country. The

much larger Northwest Territories actually lie east of Yukon Territory. They include a fringe of the Rockies, a section of the Interior Plains along which the Mackenzie River flows northward to the Arctic Ocean, a huge area of plains and hills in the Canadian Shield, and the many Arctic islands. Two huge lakes, Great Bear Lake and Great Slave Lake, both drained by the Mackenzie River, mark the western edge of the Shield. The Yukon Territory and the western part of the Northwest Territories have mainly a subarctic climate and taiga vegetation, with tundra near the Arctic Ocean, but in the eastern part of Canada tundra extends far to the south (see Fig. 23.1). The Nunavut Territory occupies eastern and northern areas withdrawn from the Northwest Territories.

The population of Territorial Canada is only about 97,000 (as of 1997). About one-fifth of the inhabitants in each of the present territories are the native "First Nation" peoples, and in the Northwest Territories about one-third—some 15,000 to 20,000—are Inuit (Eskimos). Most whites are government, corporate, or church employees, often relatively transient. First Nation peoples and Inuit still carry on some hunting, gathering, and fishing from homes in small fixed settlements; practically all Inuit inhabit prefabricated government-built housing in widely separated villages along the Arctic Ocean or Hudson

Bay (Fig. 23.13). Unemployment among both First Nations folk and Inuit is very high, and a large proportion live essentially on welfare.

Europeans and white Anglo Americans have been seeking resources from these far northern reaches since the 17th century, when England's Hudson's Bay Company first established fur-trading posts on the shore of Hudson Bay. The chief 20th-century quest has been for minerals. Mines produce metals or, in one instance, asbestos. Exploration for minerals is active, focusing on oil and gas known to exist in quantity in the Mackenzie Valley and in the Beaufort Sea section of the Arctic Ocean near the Mackenzie delta. In addition to income from the few mining enterprises, the territories also receive funding from federal government payrolls in the territorial capitals of Whitehorse (Yukon) and Yellowknife (Northwest Territories). The nature of the lives of the First National peoples has changed immeasurably, however, with the diffusion of things both good and bad from Canada and the United States (see Fig. 23.13).

In April 1999 there was a major shift in Canada's geopolitical map (see Fig. 23.12). An area of approximately 800,000 square miles (twice the size of the province of Alberta) now divides the Northwest Territories into two realms. The eastern part is known as "Nunavut" ("Our Land" in Inuktitut) and the western portion still has not been given an official name. This is the first major change in the map of Canada since the 1949 inclusion of Newfoundland as a province. In addition to this land for the Inuit First Nations peoples, the Canadian government has allocated $1.2 billion to be distributed over the next 14 years to the Inuit population. This new land, with its population of approximately 24,000, represents another effort to achieve some more equitable alignment between the early treaties signed by the ever-westering settlers and the indigenous peoples resident both in Lower and Upper Canada and in the North and Northwest Territories. A part of this resolution has been the creation of a number of governmental associations to help guide the use of the land and the capital connected with this agreement in ways that are as positive as possible for the indigenous peoples of this region.

This allocation of land and financial resources is part of the Ottawa government's efforts to cope with the problems of the territorial North. Isolation has been mitigated by the airplane

Figure 23. 13 Subsistence hunting and trapping are relatively rare occupations in the present world, although small numbers of people still gain some income from such activities. This photo shows hides of arctic wolves curing in the sun at the (newly renamed) village of Arviat on Hudson Bay in Canada's Northwest Territories. The month is August, but the Inuit woman is warmly dressed for this tundra climate. Today, Canada's Inuit live in prefabricated houses like the one seen here. Fuel oil for household heating and cooking comes in oil drums brought by ship. The hides came from a hunt by the woman's husband during the preceding winter; pursuing a pack of wolves on his snowmobile, he shot them when they could run no more. *Jesse H. Wheeler, Jr.*

and by telecommunications that now bring telephone service, radio, and even television to remote settlements. A network of schools and medical clinics blankets the area. Air transportation enables the seriously ill to be flown to the few hospitals in the region or to hospitals in cities such as Edmonton or Winnipeg that serve as metropolitan bases for both the provincial and the territorial North. Meanwhile, the increasingly assertive Indian and Inuit populations are pressing land claims to mineralized areas and pipeline routes and are demanding regulations to prevent further environmental disruptions by mining companies.

SUMMARY WITH SELECTED KEY TERMS

- **Canada** has a vital history of active immigration of people into its vast land. **Michael Ondaatje's novel *In the Skin of a Lion*** gives a good chronicle of the process of **European** migrants trying to find work, learn English, and adjust to a very different world. One of the results of this process is the continuing presence of pockets—both in cities and in the countryside—of ethnic niches that reflect such migration.

- Canada is **highly urbanized, with 77 percent** of its population living in cities. It has a major **global role** in the production of **newsprint, hydroelectricity, commercial motor vehicles, and aluminum.** Nearly **one-third of its total population** lives in **Toronto, Montreal, and Vancouver.**

- **Ten provinces** make up the political administration of Canada, but **more than one-half** of the area of Canada **is contained within the**

North—with its very **low population densities.** Nearly **90 percent** of the Canadian population **is settled within 100 miles (161 km) of the U.S. border**, with the largest bloc of population settled in the southeast near the valley of the St. Lawrence River and Seaway. The more than **5000 miles (8050 km) of border** between the two countries is generally open and allows **easy cultural and economic connections** between the populations on both sides of the boundary.

- **Atlantic Canada** consists of the provinces of Newfoundland, New Brunswick, Nova Scotia, and Prince Edward Island. **The region is also called the Maritime Provinces, or Atlantic Provinces. Cod fishing** has been the mainstay of its economy ever since earliest settlement. **The Grand Banks** on the east of this region have long been a rich source of fish harvest because of the **mixing of warm waters from the south and cold from the north.** Recently, the Banks have become so **overfished** that they have lost their capacity to support the region's fishing population.

- **The Atlantic Provinces are now promoting specialized agriculture, tourism, sophisticated fishing gear, forestry, energy, and minerals** as the economic activities that will help diminish the importance of federal government support and increase local economic strength.

- **Ontario and Quebec** are the two provinces that make up the core of Canadian settlement and political influence. **Two-thirds** of Canada's **population lives in this region.** The **margins** of the **St. Lawrence River and the Great Lakes** make up the areas of primary settlement and industrial activity. The **Ontario peninsula** is strongly **British** while **Quebec** is just as powerfully **French** in its cultural flavor, with 80 percent of the Quebec population of French origin.

- **Industrial development and trade** moved upstream on the St. Lawrence steadily until the completion of the St. Lawrence Seaway in the 1950s. **Quebec City**, the earliest city settlement in Canada, was outgrown by **Montreal**, and now Montreal has been surpassed by **Toronto** in size and economic importance, although **Vancouver** on the west coast has grown to be the biggest port center in Canada.

- **The resources** of greatest importance **in Quebec and Ontario** are **agriculture and forestry**, while **energy generation and bulk commodity transport** have also been central to the region's economy.

- **Hamilton, Ontario, is the center of the steel industry,** benefitting from good access to American coal and proximity to automobile manufacturing; **apparel manufacturing** is centered in **Quebec**, and **Montreal** is a node for a variety of **metal fabrication industries**.

- **Quebec separatists** have been a political presence for the **past 40 years,** and in **1995, the push for an independent Quebec was defeated by less than 50,000 votes** (or less than 1 percent). In its battle to gain additional accommodations from Ottawa—the head of Canada's federal government—Quebec has gained additional provincial autonomy. **Other regions, especially the Prairie Region, have sought the same benefits by threatening to change their relationship with Ottawa.**

- **The Prairie Region** is made up of **Manitoba, Saskatchewan, and Alberta. Wheat, petroleum, and coal** are the **major economic resources. Major urban centers include Edmonton, Calgary, and Winnipeg.** These centers provide linkage east and west with other major Canadian regions, but also to the North. This region has made continuing efforts to gain more provincial fiscal benefits as a source area for major resources—particularly oil—that flow into export trade. This is sometimes called **"resource separatism."**

- **The Vancouver Region is centered on Vancouver, British Columbia, at the mouth of the Fraser River.** More than one-half of the province's population lives in the Vancouver area, which is the region's main industrial, administrative, financial, and cultural center. It is also home to the second largest Chinatown in North America. **British Columbia** has become a major tourist and retirement destination. **Trade is increasingly focused on East Asia,** although the economic slowdown in that region has changed trading patterns for this western part of Canada since 1997.

- **The Canadian North** has an ambiguous southern border. The southern limit of the region of subarctic climate—excluding the island of Newfoundland—can serve to define the southern edge of the North, and population densities tend to decline the farther north one goes into the North. Nickel, copper, and uranium are the major resource metals mined and exported from the North. **Forestry, pulp manufacture, and hydroelectricity** are three additional economic resources.

REVIEW QUESTIONS

1. Name the provinces in Canada, from east to west.
2. Where is the migrant from in the Landscape in Literature selection? Name three different places he tried to find work and learn language in Canada.
3. Name the major import-export patterns for Canada, and cite the most important goods and trading partners in the trade.
4. List the political units that make up the Atlantic Provinces, give another name for that region, and outline the reasons for steady economic decline there.
5. What two political units make up the core region of Canada? Outline some of the major similarities and differences in their geographic characteristics.
6. Explain the origins and evolution of the Quebec separatist movement.
7. What role has the St. Lawrence River had in the development of Canada? Describe what has been done in the way of landscape transformation to enhance its utility.
8. What impact has the North American Free Trade Agreement (NAFTA) had on Canadian economic activity?
9. What is the role of automobile manufacturing and wheat production in the economic geography of Canada?
10. What resources have given rise to resource separatism in the Prairie Region?
11. What is the major urban center of British Columbia, and what trade patterns are of particular importance to that city and that region?
12. What is the name of the new territory turned over to the Inuit in the Northwest Territories in 1999, and what are the reasons for this decision?

DISCUSSION QUESTIONS

1. What are the geographic characteristics of North America that allow Canada and the United States to share a border with basic cooperation? What issues could cause growing friction?
2. What are the historical events in Michael Ondaatje's literature excerpt that have relevance for other parts of the world that you have studied thus far?
3. Discuss the reasons for the Canadian frustration at being so close to the United States, and suggest ways in which such feelings might be reversed.
4. Role-play a discussion of Quebec's possible independence, with part of a group representing the French in Quebec and another part representing the English in Ontario. Discuss the reasons for, and the dangers of, independence for Quebec.
5. Discuss the ways in which regions of the United States have experienced similar cultural and economic histories as the Atlantic Provinces, the Prairie Region, and the Core Region.

6. What are the geographic factors that led to Canada's particular demographic patterns?
7. Speculate on how Canada's development and historical geography would have been different had there been no St. Lawrence River.
8. Discuss the geographic and cultural factors that led to the Nunavut Agreement in 1993 and the subsequent 1999 handing over of approximately 500,000 square miles of land by the Canadian government.
9. Discuss the ways in which the United States has both benefitted from and been frustrated by its proximity to Canada. What role has this proximity had in the development of the United States?
10. Speculate on how the map of Canadian settlement will change over the next decade; quarter century; half century. What will be the role of immigrants? Where are they likely to come from?

Chapter 24

The United States: A Landscape Transformed

▲ Rivers, barges, bridges, ceremonial architecture, urban parkland, skyscrapers, and low-lying sprawl—all of these features are part of the fabric of urban America. This aerial photograph of St. Louis, Missouri, Gateway Arch possesses them all, stretching out from the banks of the Mississippi River. The Arch is a 63-story structure intended to reveal this city's role as a jumping-off place in the movement west—beyond the Mississippi, across the state, and eventually across the continent—for the country's ever westward-moving population.   Charlie Palek/Tom Stack & Associates.

CHAPTER OUTLINE

24.1 The Northeast: The Traditional Core Region

24.2 The South: Old Images, New Images

24.3 The Midwest: Agricultural and Industrial Heartland

24.4 The West: Oasislike Development in a Dry and Dramatic Setting

24.5 Alaska and Hawaii: Exotic Outposts

*t*he historical geography of the United States reflects a continuing series of migrations. The country was initially peopled by the earliest migrants coming from northeast Asia across the Bering Strait land bridge and down into the vastness of North America and beyond. Then, thousands of years later, came immigrants from Europe and slaves from Africa, and then from all over the world. Those who came of their own volition were seeking some new environment, some new setting that would be safer, more productive, more satisfying. American history and geography have been made up of the products of this restlessness since this continent was first found by migrating peoples.

Such steady movement means a continuing change in the country's cultural landscape. To the places peopled by new arrivals, there come new languages, new customs and costumes, new architecture, and new mixes of the cultural personalities created by these demographic blendings. To the landscapes left behind when migrants leave, there is also profound change. The nature and magnitude of such change depends in part on the factors that set a particular migration in motion. However, while different migrations have distinct sets of catalysts and migration paths, there are elements of similarity in all the migrations that the United States has known. For example, people move because industrial advances replace human labor, and as a result, people who worked suddenly have no work. Leaving behind their past and its worn landscape, they head west—most often the case in American history—and try to find a place to begin again, and anew. These dynamics have traveled with migrants coming to the United States—indeed, to all migration target destinations—since humans began to explore and "hit the road" for new havens, new settings, and new beginnings. Our cultural landscape is rich in the evidence of this human search (see photo, p. 588).

24.1 The Northeast: The Traditional Core Region

Of the four major regions of the United States, the Northeast is the most intensively developed, densely populated, ethnically diverse, and culturally intricate. It incorporates the nation's main centers of political and financial activity, and it is the area in which relationships with foreign nations are the most elaborate (for example, the United Nations and the nation's capital at Washington, D.C., are both located in this region). Strong traditions of intellectual, cultural, scientific, technological, political, and business leadership persist there. The country's political and economic systems had their main beginnings in the Northeast, and it traditionally has been the chief reception center for overseas immigrants.

The Northeast consists of six New England states (Maine, New Hampshire, Vermont, Massachusetts, Connecticut, and Rhode Island), plus five Middle Atlantic states (New York, New Jersey, Pennsylvania, Delaware, and Maryland), and the District of Columbia. Delaware and Maryland have important historical links with the states of the U.S. South, but their present character classifies them more appropriately as Middle Atlantic states; hence they are included in this region (Fig. 24.1).

As of 1998, about 54 million people, or 19 percent of the national population, lived within the Northeast, on 5 percent of the nation's area (Table 24.1). Hence, the regional population density is several times that of the South, Midwest, or West. But the density is much more extreme within a narrow and highly urbanized belt stretching about 500 miles (c. 800 km) along the Atlantic coast from metropolitan Boston (in Massachusetts) through metropolitan Washington, D.C. The belt is often called the Northeastern Seaboard, Northeast Corridor,

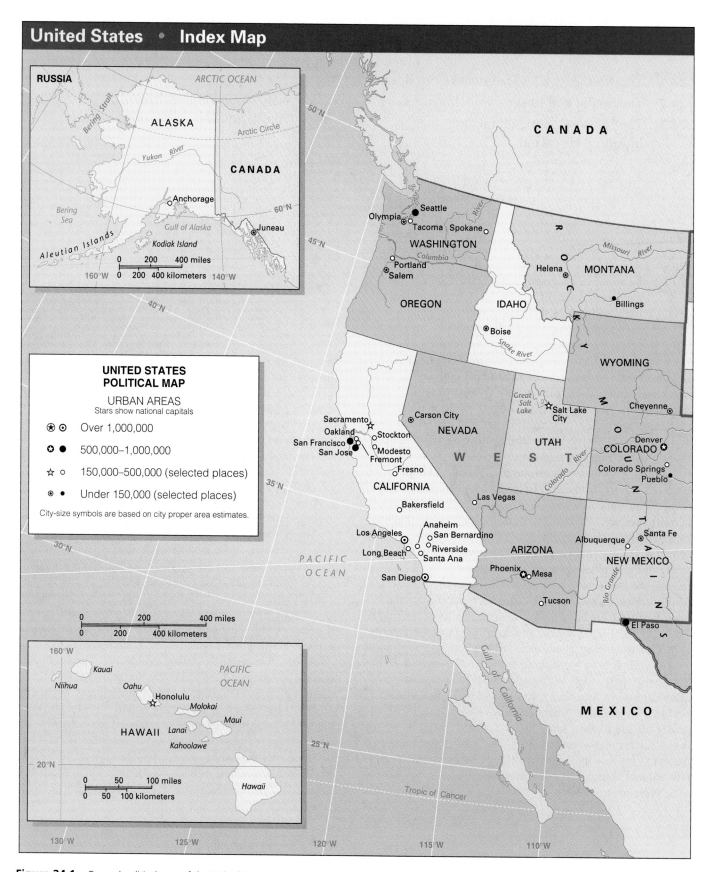

Figure 24.1　General political map of the United States.

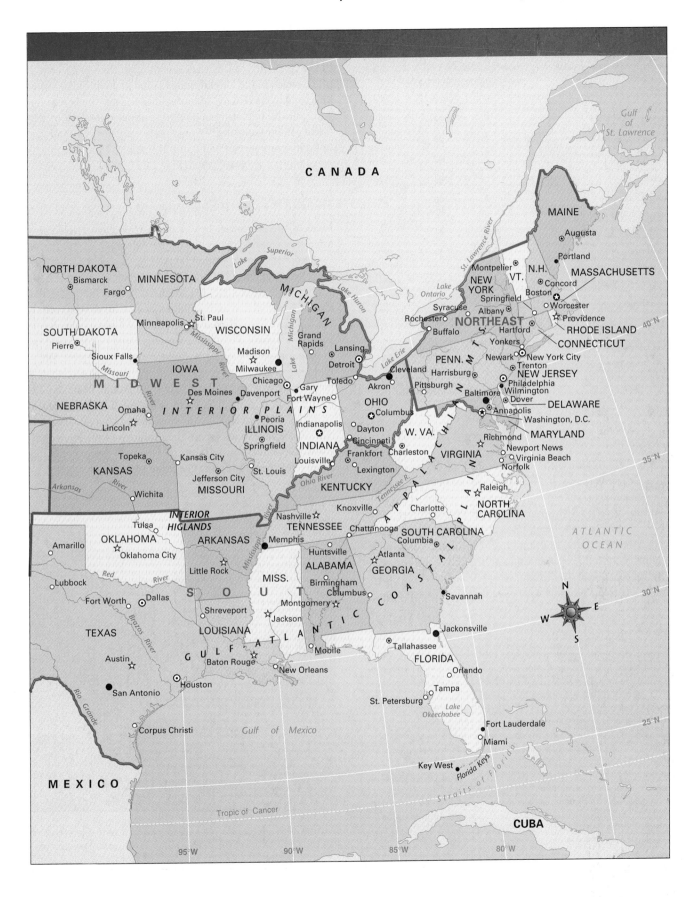

Table 24.1 **United States Regions: Comparative Data**

	Percent of National Totals			
	Northeast	South	Midwest	West
Total area	5	24	22	49
Total population	19	35	23	23
Land in farms	2	30	37	31
Cropland	3	19	66	12
Value of farm products sold	5	30	40	25
Value of livestock and products sold	8	33	41	18
Value of crops sold	5	26	39	30
Value added by manufacture shipments	19	33	31	17
Value of mineral output	3	62	13	22
Value of retail sales	22	32	25	21
Total personal income	25	30	23	22
Total metropolitan population	25	30	22	23

Source: U.S. Bureau of the Census, *Statistical Abstract of the United States*, 1998.

Boston-to-Washington Axis, "Boswash," or "Megalopolis" (see Fig. 24.1 and photo, p. 592). Here, seven main metropolitan areas contain about 40 million people: Boston (5.4 million), Providence (1.1 million), Hartford (1.2 million), New York (19.5 million), Philadelphia (5.9 million), Washington, D.C. (4.5 million), and Baltimore (2.4 million).[1] Additional population within this belt amounts to roughly 6 million, giving it a total of about 46 million, or approximately 85 percent of the people in the Northeast. The leading cities of the belt are

[1]City populations in this chapter are rounded 1998 estimates for Metropolitan Statistical Areas (MSAs), Primary Metropolitan Statistical Areas (PMSAs), or Consolidated Metropolitan Statistical Areas (CMSAs), with the most inclusive figure being used (extrapolated from U.S. Bureau of the Census's *Statistical Abstract of the United States*, 1998). In Census Bureau terminology, MSA refers to an urban aggregate composed of a county, or two or more contiguous counties, meeting specified requirements as to metropolitan status. In New England, MSAs are defined by cities and towns rather than by counties. Designation of MSAs represents an attempt to give a realistic picture of functional urban aggregates, each of which often includes more than one incorporated urban place ("political city"), plus unincorporated urbanized areas and stretches of countryside where agriculture may be important. But even the countryside is closely bound to the services, markets, and employment afforded by the urban places. A PMSA is, in general, a larger aggregate, and a CMSA is the most inclusive unit of all. Figures cited in this chapter for the very largest cities, such as New York, Los Angeles, and a considerable list of others, are for CMSAs. Any of the three types of metropolitan units may include counties in more than one state.

major centers of political decision making, corporate headquarters, finance, trade (retailing), and services. An indicator of the centrality of this region is the fact that 40 percent of all office space in the United States is located within 50 miles (80 km) of New York City. Manufacturing is relatively less important, but the belt still accounts for about one-sixth of the nation's manufacturing. Congested trafficways and active intercity commercial and cultural linkages bind this mosaic of metropolitan areas together (Fig. 24.2).

The Northeastern Environment

Most of the Northeast lies in the Appalachian Highlands, but relatively small sections lie in the Atlantic Coastal Plain, the Piedmont, or the Interior Plains (Fig. 24.3). Atlantic Coastal Plain areas include (1) Cape Cod in Massachusetts, (2) Long Island in New York state, (3) southern New Jersey, (4) the Delmarva Peninsula, which includes nearly all of Delaware plus the eastern shore of Maryland (east of Chesapeake Bay) and parts of Virginia, and (5) the western shore of Maryland inland to the boundary between Baltimore and Washington, D.C. The Plain is low and flat to gently rolling, with many sand dunes and marshes or swamps. Soils are sandy and range from very low to mediocre in fertility. The Plain is indented by broad and deep estuaries ("drowned" lower portions of rivers). The largest Northeastern seaports are on the estuaries of the Hudson River (New York); the lower Delaware River (Philadelphia); and the Patapsco River (Baltimore) near the head of Chesapeake Bay.

Figure 24.2 Rush-hour traffic in downtown Boston, Massachusetts, is representative of the congested Boston-to-Washington "megalopolis" along the Atlantic shore of the Northeast. Downtown Boston is notorious for its winding, cluttered maze of streets, some of which originated as cow paths in colonial times. *Neal J. Manschel/Christian Science Monitor*

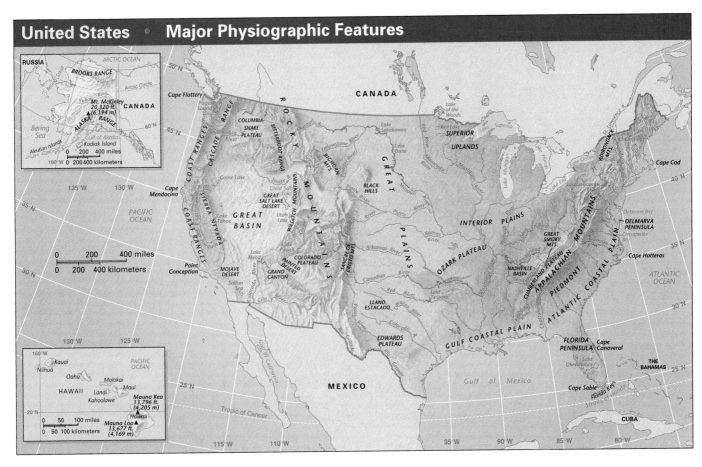

Figure 24.3 Major physiographic features of the United States.

Between the Coastal Plain and the Appalachians, the northern part of the Piedmont extends across Maryland, Pennsylvania, and New Jersey to New York. The Piedmont is higher than the Coastal Plain but lower than the Appalachians. Where the old, erosion-resistant igneous and metamorphic rocks of the Piedmont meet the younger, softer sedimentary rocks of the Coastal Plain, many streams descend abruptly in rapids or waterfalls, and the boundary between the two physical regions is known as the fall line. It is marked by a line of cities that have developed along it: Washington, D.C., Baltimore, Wilmington (Delaware), Philadelphia, Trenton (New Jersey), and New Jersey suburbs of New York City (see Fig. 24.1). Most of the Piedmont is rolling, with soils generally better than those of the Coastal Plain or Appalachians.

Aside from a narrow strip of the Interior Plains along the Great Lakes and St. Lawrence River (see Figs. 24.1 and 24.3), the rest of the Northeast lies in the Appalachian Highlands. But in New York, a narrow lowland corridor—the Hudson-Mohawk Trough—breaks the Appalachians into two quite different subdivisions. This Trough is composed of the Hudson Valley from New York City northward to Albany, and the Mohawk Valley westward from Albany to the Lake Ontario Plain.

Northeast and north of the Trough, the old hard-rock mountains of New England and northern New York form several ranges, characterized by rough terrain, cool summers, snowy winters, and poor soils. The Adirondack Mountains of northern New York rise as a roughly circular mass, surrounded by lowlands. Heavily forested and pocked by numerous lakes, these glaciated mountains reach over 5000 feet (1524 m) in elevation. They are bounded on the east by the [Lake] Champlain Lowland, which reaches northward into Canada as a continuation of the Hudson Valley. East of it, the relatively low Green Mountains occupy most of Vermont and extend southward to become the Berkshire Hills of western Massachusetts and Connecticut. Farther east across the narrow valley of the upper Connecticut River, the White Mountains occupy northern New Hampshire and extend into Maine. Here, Mount Washington in New Hampshire rises to 6288 feet (1917 m), the highest elevation in the Northeast. The Adirondacks and the mountains of New England are major recreation areas for the Northeast's urban populations.

In New England, the hilly areas between the mountains and the sea, sometimes termed the New England Upland, are considered to be part of the Appalachians, but they are relatively

Figure 24.4   Originally, the farmland in New England had to be cleared of boulders left behind by continental glaciers. With the opening of the Erie Canal in 1825, a steady abandonment of New England farmland began. So much farmland has been abandoned that one often finds stone walls winding through patches of woodland where crops once were grown. Trying to interpret symbols of landscape change like this makes geography a kind of detective work, leading to a clearer understanding of what humans have done in their creation of cultural landscapes.   *Jesse H. Wheeler, Jr.*

low rather than mountainous. The original soils were persistently stony (caused by the deposition of stones by glaciation), but the early settlers were able to use this area for subsistence agriculture once the stones and trees were laboriously cleared. Boulders from the fields were used to build New England's famous stone fences. Today, many of these have been torn down to facilitate modern farming or urban development, but some derelict fences may still be seen keeping quiet guard in the woodlands where farming has been abandoned (Fig. 24.4). The poor soils of the Upland proved increasingly unable to support a stable long-term commercial agriculture, and the late 18th and the 19th centuries saw a large movement of New Englanders from unproductive farms to Northeastern cities or—especially with the completion in 1825 of the Erie Canal—to more fertile agricultural areas to the west. This migration gave a strong New England cultural flavor to localities scattered all the way to the Pacific coast. Figure 24.5 sets the stage for the settlement patterns created in large part by migrants from the Northeast all across the rest of the country.

South of the Hudson-Mohawk Trough, the Northeastern Appalachians include three distinct sections: the Blue Ridge in the east, the Ridge and Valley Section in the center, and the Appalachian Plateau in the north and west. They lie roughly parallel, with each trending northeast-southwest. The Blue Ridge—long, narrow, and characterized by old igneous and metamorphic rocks—has various local names. Often it forms the single ridge its name implies, but it broadens into many ridges in the South. West of it, the Ridge and Valley Section, characterized by folded sedimentary rocks, consists of long, narrow, and roughly parallel ridges trending generally north and south and separated by narrow valleys. This valley-dominated eastern strip is essentially a single large valley known as the Great Appalachian Valley. Its limestone floor has decomposed into some of the better soils of the Appalachians. The Appalachian Plateau lies west and north of the Ridge and Valley Section. The northern part is often called the Allegheny Plateau, whereas, in parts of eastern Kentucky and farther south, the plateau becomes known as the Cumberland Plateau. Notable east-facing escarpments—the Allegheny Front in the north and the Cumberland Front in the south—mark the eastern edge, from which elevations gradually decline toward the west. Except in New York, the Plateau is underlain by enormous and easily worked deposits of high-quality bituminous coal which have been extremely important in the economic development of the United States.

Early Development in the Seaboard Cities: The Crucial Role of Commerce

The roots of the Northeast's urban development go far back in time. Boston was founded by English settlers and New Amsterdam (presently New York City) by the Dutch in the early 1600s, although the latter was subsequently annexed and renamed by England in 1664. Philadelphia was founded by the English in the later 1600s and Baltimore in the early 18th century. All four cities were major urban centers by the time of the American Revolution (1775–1783), although by today's standards they were quite small.

Except for Washington, D.C., the largest metropolises of the Northeastern Seaboard gained their initial impetus as seaports, frequently located at river mouths. The federal capital was given the site it has as part of a political compromise. It was on the border between the agrarian southern states—with their large slave populations—and the more diversified northern states. The site also placed the capital on the fall line between seaboard and upcountry sectional interests. Many state capitals in the United States have been placed in response to similar locational compromises.

New York City was founded at the southern tip of Manhattan Island, on the great natural harbor of the Upper Bay (Fig. 24.6). It shipped farm produce from lands now occupied by built-up areas of metropolitan New York City and from estates farther up the Hudson River. Philadelphia was sited at the point where the Delaware River is joined by a western tributary, the Schuylkill. Philadelphia's local hinterland—the Pennsylvania Piedmont and Great Valley, plus parts of southern New Jersey—was productive enough to make the city a major exporter of wheat and helped it become the largest city (population: about 40,000 in 1776) in the thirteen colonies. Baltimore

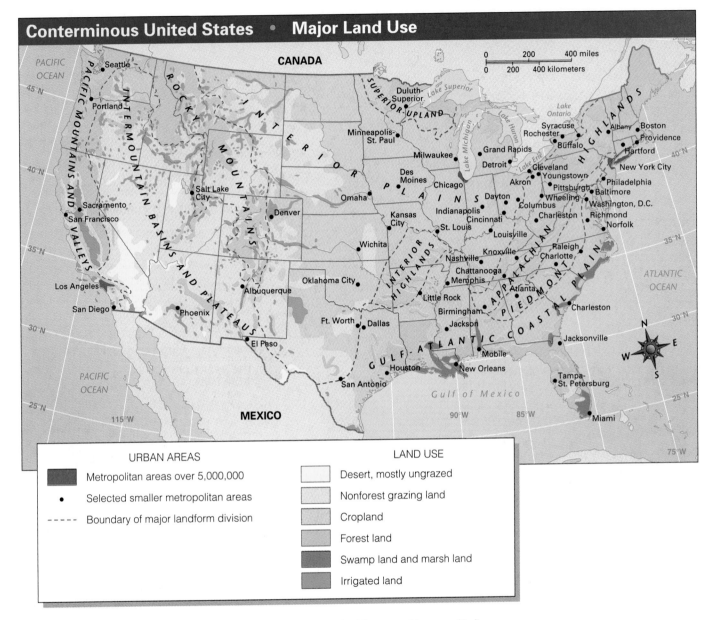

Figure 24.5 This map of generalized land use in the conterminous United States provides a good indication of the variety of land use, and also of the importance of moisture and irrigation water.

was founded on a Chesapeake Bay harbor at the mouth of the Patapsco River and shipped tobacco and other farm products from its local hinterland in the Piedmont and the Coastal Plain. Boston had a somewhat different early development, since its New England hinterland produced little agricultural surplus for export. Instead, colonial New England placed its main emphasis on activities connected with the sea. Its forests yielded superior timber with which to build ships, and both timber and ships were exported, as were cod and other fish that were initially abundant along this coast. Whaling was important in the 18th and 19th centuries. New Englanders also developed a wide-ranging merchant fleet and trading firms with far-flung interests. Such activities characterized many ports, of which Boston was the largest.

The Race for Midwestern Trade

As the present Midwest was settled, primarily between 1800 and 1860, its agricultural surpluses and need for manufactured goods greatly expanded the trade of the cities of New York, Philadelphia, and Baltimore. The ports engaged in a race to establish transport connections into the interior. New York City

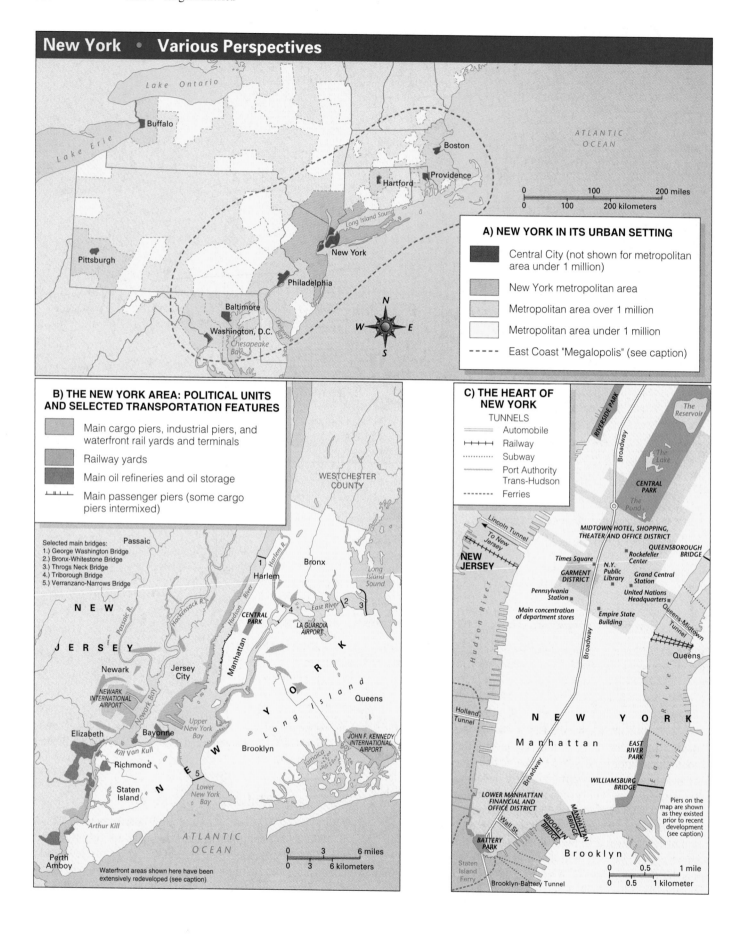

New York • Various Perspectives

A) NEW YORK IN ITS URBAN SETTING

- Central City (not shown for metropolitan area under 1 million)
- New York metropolitan area
- Metropolitan area over 1 million
- Metropolitan area under 1 million
- ----- East Coast "Megalopolis" (see caption)

B) THE NEW YORK AREA: POLITICAL UNITS AND SELECTED TRANSPORTATION FEATURES

- Main cargo piers, industrial piers, and waterfront rail yards and terminals
- Railway yards
- Main oil refineries and oil storage
- Main passenger piers (some cargo piers intermixed)

Selected main bridges:
1.) George Washington Bridge
2.) Bronx-Whitestone Bridge
3.) Throgs Neck Bridge
4.) Triborough Bridge
5.) Verranzano-Narrows Bridge

Waterfront areas shown here have been extensively redeveloped (see caption)

C) THE HEART OF NEW YORK

TUNNELS
- Automobile
- Railway
- Subway
- Port Authority Trans-Hudson
- Ferries

Piers on the map are shown as they existed prior to recent development (see caption)

Figure 24.6 (at left) New York stands at the center of a strip of highly urbanized land that was christened "megalopolis" by the French geographer Jean Gottmann in 1961. The dashed line is not intended to indicate a precise boundary but shows broadly the area included by Gottmann in his megalopolis concept. Many waterfronts and other areas shown on the two maps of New York are being redeveloped for new commercial, residential, and recreational uses. This is especially true of the piers along both sides of the Hudson River. In the map "The Heart of New York," the shading for the "Midtown hotel, shopping, theater, and office district" covers the areas in which most of Manhattan's well-known hotels, restaurants, department stores, specialty shops, theaters, concert halls, and museums are found. The Midtown area also includes numerous office buildings, most notably the cluster of skyscrapers in Rockefeller Center. At the southern end of the area near the main department stores is New York's garment district, with its large workforce and numerous workrooms, sweatshops, and showrooms crowded into a remarkably small segment of this densely built-up city. The lower Manhattan financial and office district includes the imposing cluster of office skyscrapers in the Wall Street area and the World Trade Towers. Farther north, City Hall and a large assemblage of other government buildings occupied by city, county, state, and federal offices dominate the cityscape. Note the complex of bridges, tunnels, and ferries that ties Manhattan Island to the rest of metropolitan New York.

surpassed its rivals, largely because of its access to the Hudson-Mohawk Trough—the only continuous lowland passageway in the United States through the Appalachians (see Figs. 24.1 and 24.3). The Erie Canal was completed along this corridor in 1825. Connecting Lake Erie at Buffalo to the navigable Hudson River near Albany, the canal was a key link in an all-water route from the Great Lakes to New York City and it helped reduce transport costs between New York and the Midwest to a small fraction of their previous level. Settlement of the Midwest was stimulated, and a flood of trade was directed through the port of New York. In 1853, using the same route, New York interests became the first to connect a growing Chicago with an east coast port by rail. By the 1850s, New York had become the leading United States port and city by a wide margin. The competing ports of Baltimore and Philadelphia also shared in Midwestern trade. Major railways were eventually pushed through the Appalachians into the Midwest from both ports. Boston was never a real competitor in this race, although it has remained the largest seaport in New England.

Early Industrial Emphases on the Northeastern Seaboard

Industrialization got its first big boost during the same period that the Northeast established connections to the interior. Favorable factors included accumulated capital from commercial profits; access to industrial techniques pioneered in Great Britain; skills from previous commercial operations and handicraft manufacturing; cheap labor supplied by immigrants (see Chapter 23, Landscape in Literature, p. 568); and the presence of energy resources and raw materials that were more important in early industrialization than they are today. Also favorable was a United States high-tariff policy that restricted imports and secured much of the nation's market for its own emerging manufacturers. The first American factory, a water-powered textile plant using pirated British technology, was established by Samuel Slater on the Blackstone River at Pawtucket (near Providence), Rhode Island, in 1790, and the subsequent growth of factory industry in the Northeast was rapid.

The circumstances and results of industrialization were different in different parts of the Northeast. New England had suitable water power near its ports, with many small and swift streams to turn early waterwheels, but this region lacked coal for metalworking. It specialized in textiles, drawing wool from New England farms, cotton from the South, and abundant labor from recent European immigrants, especially the Irish. However, one New England state, Connecticut, took a different path. It expanded a colonial specialty in metal goods—based on small local ore deposits—by becoming a major manufacturer of machinery, guns, and hardware (which it still is). Mill towns sprang up across southern New England and up the coast into southern Maine. In the Middle Atlantic states, early industrialization also included textiles, but there was more emphasis on metals, clothing, and chemicals. The emphasis on metals reflected small local deposits of iron ore, especially in southeastern Pennsylvania. Then, in the second quarter of the 19th century, the largest deposits of anthracite coal in the United States became a major source of power. This coal lay deep underneath the Pennsylvania Ridge and Valley Section between Harrisburg (population: 621,000) and Scranton–Wilkes-Barre (combined population: 637,000), but was made accessible despite the terrain by the determined building of canals and railroads and by large capital investment in the mines themselves. For many decades, anthracite was very significant industrially. Usable in blast furnaces without coking, it fueled large-scale iron and steelmaking in various eastern Pennsylvania cities. Iron and steel from the early plants contributed greatly to the first industrial surge in the Northeast by furnishing material for the manufacture of machinery, steam engines, railway equipment, and other metal goods.

The Primacy of New York City

The United States' largest city is centrally located within the Northeastern Seaboard's urban strip, midway between Boston and Washington, D.C. An enormous harbor, an active business enterprise, and a central location within the economy of the colonies and the young United States had already made New

York City the country's leading seaport before access to the Hudson-Mohawk route gave it an even more decisive advantage. Superior access to the developing Midwest then moved New York rapidly to unquestioned leadership among American cities in size, commerce, and economic impact.

New York City is centered on the less than 24 square miles of Manhattan Island (see Fig. 24.6). The island is narrowly separated from the southern mainland of New York state on the north by the Harlem River, from Long Island on the east by the East River, and from New Jersey on the west by the Hudson River. Manhattan is one of five "boroughs" of the city proper. The others are the Bronx on the mainland to the north, Brooklyn and Queens on the western end of Long Island, and Staten Island to the south across Upper New York Bay. In addition, the broader New York metropolitan area includes much of the population and industry of northern and central New Jersey, communities on Long Island east of Brooklyn and Queens, southernmost New York state just north and west of the Bronx, and southwestern Connecticut. Expressways and mass-transit lines converge on bridges, tunnels, and ferries that link the boroughs to each other and to northern New Jersey (see Fig. 24.6). Port activity is heavily concentrated in New Jersey along the Upper Bay and Newark Bay, although some traffic continues to be handled in various New York sections of the waterfront. The whole area is held together in part by an amazing 722 miles of subway system, parts of which are more than a century old.

Industrial Emphases of the Northeast Today

The economic dominance of the Northeast to the whole of the U.S. economy is apparent in a brief look at some of its major specialties:

1. *Clothing design and manufacturing.* Clothing manufacturing is carried on in many locations but is concentrated in metropolitan New York City. The clothing industry in Manhattan was originally an outgrowth of New York's role as an importer of European-made cloth and clothing. It was favored by a large number of skilled immigrant workers in the later 19th century, among them Jewish tailors fleeing from persecution in Tsarist Russia.

2. *Iron and steel manufacturing.* Iron and steel production is still found in the Northeast, although employment and output have been drastically reduced in recent years because of competition from cheaper imported steel, large imports of products made from foreign rather than American steel, and the substitution of aluminum or plastics for steel in industrial processes, especially in automobile manufacturing.

3. *Chemical industries.* Chemical manufacturing is very diversified and widespread, and includes production of both industrial chemicals in bulk and a multiplicity of consumer products. The outlying areas of metropolitan New York City form the largest center, although the DuPont Company, which is the world's largest chemical manufacturing concern, has both its headquarters and a major share of its manufacturing in Wilmington, Delaware.

4. *Photographic film and equipment.* Photographic equipment manufacturing is dominated by the world's largest photographic company, Eastman Kodak, which has both its headquarters and main plant in Rochester, New York. The Polaroid Corporation, based principally in the Boston area, is another major name in this industry.

5. *Electronics manufacturing.* Advanced electronics manufacturing, centering on computers, is widespread in the Northeast. This industry is a major specialty in eastern Massachusetts, being especially concentrated in industrial parks along Route 128 and other beltways that ring the Boston area. Some old textile towns such as Lowell and Lawrence on the Merrimack River have largely converted to electronics or other high-technology industries. They are examples of the widespread adaptive reuse of old mill towns in the Northeast. Eastern Massachusetts was once the nation's leading center of textile milling, but most of this industry has long since moved to the South—and recently, offshore—leaving behind substantial turn-of-the-century building stock that is now being adapted to high-tech industrial use or, increasingly, factory outlet malls.

6. *Electrical equipment manufacturing.* Concentrated in southern New England, New York, and Pennsylvania, electrical equipment manufacturing is dominated by the General Electric Company, headquartered at Stamford, Connecticut, with major production facilities in Schenectady, New York. Westinghouse Electric, headquartered in Pittsburgh, is another major company with plants in the Northeast and other areas.

7. *Aircraft engines.* The manufacture of aircraft engines and helicopters is heavily concentrated near Hartford, Connecticut, where it is carried on by United Technologies Corporation, maker of the famous Pratt and Whitney engines and Sikorsky helicopters.

8. *Nuclear submarines.* The manufacture of Trident nuclear submarines in the New London, Connecticut, area continues New England's long shipbuilding tradition as well as Connecticut's long tradition of armaments manufacturing. It was at a factory built near New Haven, Connecticut, in 1798 to supply muskets under a federal contract that Eli Whitney originated the use of interchangeable product parts in the manufacture of firearms. This practice is central to modern mass production of innumerable types of goods.

9. *Publishing and printing.* Publishing and printing are industries in which the Northeast decisively overshadows the rest of the country. Most of the publishing industry is in the New York metropolis, although the Government Printing Office in Washington, D.C., is the world's largest publishing enterprise and printing plant. Many of the book and magazine printing phases of the industry have been steadily decentralized to other parts of the country, but editorial and management functions remain concentrated in the Northeast.

This assemblage of manufacturing, planning, design, decision-making, and management capabilities has also prompted the establishment of tens of thousands of smaller, complementary, and ancillary industries that make their living by supplying materials and information to this concentration of these varied dominant businesses. In the geography of commerce, there is great importance placed upon proximity, interaction, timely supply of goods and information, and complex networks of spatial interaction. New York City and the Northeast have historically been very effective in this union.

The "Rust Belt" versus the "Sun Belt": Industrial Change and Demographic Shifts in the Core Region

The vernacular term "Rust Belt" has come to describe this region of early urban growth and manufacturing. The region contains the great majority of cities that were already prominent by the early 20th century. Most of the country's old "smokestack industries"—a term applied particularly to the steel and other heavy metallurgy industries—are in the Middle Atlantic states and eastern Midwest. Today, these industries are often characterized by outmoded buildings and equipment, depressed sales and profits, high unemployment, abandoned industrial buildings, and landscapes of decline and abandonment.

By contrast, the Sun Belt in the South and West has been enjoying faster growth in population and jobs, including those in manufacturing. ("Sun Belt" is a very elastic term, but common usage and geographic logic both suggest an area encompassing most of the South, plus the West [in this text's regional system for the United States] at least as far north as Denver, Salt Lake City, and San Francisco.) The regional loss of manufacturing employment reflects such factors as (1) closing of obsolete plants, (2) shifting of production to more profitable locations in other regions or overseas, (3) reduced employment in surviving plants reequipped with new labor-saving machinery, and (4) failure of new plants to locate in the region in sufficient numbers to offset the job losses in existing industries.

Employment in service industries has been expanding quite rapidly in the Northeast, resulting in an expansion of total employment. But despite this, the region has been growing in population relatively slowly, with heavy outmigration from the Middle Atlantic region. The main exceptions to this slow growth are the peripheries of some metropolitan areas into which suburban and exurban development are expanding. Notable examples include the expansion into New Hampshire from Boston; into Connecticut and New Jersey from New York City; and into Maryland from Washington, D.C. The state of Vermont, whose population growth has recently been high, is not suburban, but much of this quiet and scenic state is exurban, attracting "refugees" from metropolitan areas, owners of second homes and vacation homes, and some diligent commuters, including those traveling by air and, increasingly, professionals

Definitions & Insights

ADAPTIVE REUSE

One of the key factors in any effort to keep vitality in any urban place is the adaptive reuse of earlier building stock. This means taking abandoned commercial or factory buildings and converting them to some economic function that brings jobs, commercial activity, and people back to a once relatively productive area. The process is also associated with gentrification—the refurbishment of old building stock by urban professionals for residential or office space. This has been done with particular success in New England. In any downtown that has been losing customers and businesses for some years, the transformation of empty retail stores into specialty shops or trendy restaurants has become a commonplace example of adaptive reuse of older buildings. This process is important because (a) it gives new vitality to places and often well-designed buildings that once had economic and social centrality but subsequently lost it; (b) it helps create a sense of the value of the past as old landscapes are visited and utilized again; and (c) it maintains the structures in their historic context as landscape reference points or pivotal buildings. While adaptive reuse is most often considered in the context of old central business districts, the process has become increasingly appropriate in the resettlement of old residential neighborhoods as well. It serves as a bellwether for the revitalization of neighborhoods and helps in the fuller use of at least some sections of a graying, declining, and sometimes abandoned urban landscape.

who are linked to urban centers through telecommunication networks. This expanding population can thus live in the country and still derive their income from offices set right in the heart of the eastern cities. And despite the weight of economic and urban problems, the Northeast has recently been the scene of spectacular new development, such as the Baltimore Inner Harbor area, in the great metropolises, both in old decaying cores and around the peripheries (Fig. 24.7).

24.2 The South: Old Images, New Images

Various definitions of the South are possible, and none is entirely satisfactory (see Fig. 24.1). The region is discussed here as a block of 14 states: 5 along the Atlantic (Virginia, North Carolina, South Carolina, Georgia, and Florida); 4 along the Gulf of Mexico (Alabama, Mississippi, Louisiana, and Texas); and the 5 interior states of West Virginia, Kentucky, Tennessee, Arkansas, and Oklahoma. Arguments could be made on various grounds for excluding some of these states. For example, West

Figure 24.7 The Baltimore Inner Harbor Development—called Harborplace—of James Rouse is one of the most successful reclamations of an old warehouse and wharf area in the country. The current uses of this initial 18th-century harbor facility feature a two-story shopping mall with abundant outdoor restaurants and cafes, a new aquarium, relic brick factory power plant buildings converted into elegant townhouses and condominiums, and easy access to the recently constructed Camden Yards, home of the Baltimore Orioles. Rouse called such (re)developments "festival marketplaces." Berthing for local water-craft and tour ships that cruise the Chesapeake Bay are all part of this urban area that, thirty years ago, looked as though it would never be presentable to the public eye again, or economically useful. It now serves as one of Baltimore's most popular tourist destinations. *Jerry Watcher/Photo Researchers*

Virginia and Kentucky did not secede from the Union during the Civil War, and the western parts of Oklahoma and Texas lie in dry environments more typical of the West than the South. Although Missouri was a slave state and thus part of the South before the Civil War, it today seems more typically Midwestern and thus is not included here.

Physical Environments and Their Utilization

Most of the South has a humid subtropical climate, with summers that are long, hot, and wet. January average temperatures range from the 30s (degrees Fahrenheit) along the northern fringe to the 50s near the Gulf Coast, and the 60s in southern Florida and the southern tip of Texas. More than 40 inches (*c.* 100 cm) average annual precipitation is characteristic, rising to over 50 inches in many Gulf Coast and Florida areas and more than 80 inches in parts of the Great Smoky Mountains. High humidity is characteristic, intensifying heat discomfort in summer and chilliness in winter. There are both advantages and disadvantages to these climatic conditions. They foster rapid and abundant forest growth, and forest-based industries are of major importance. Agriculturally, they provide long growing seasons and heat and moisture for a wide variety of crops. But insects and pests also flourish, and the heavy rainfall leaches out soil nutrients and causes severe erosion and flooding.

The Predominance of Plains

Most of the South is classified topographically as plains, but the plains form sections of three major landform divisions—the Gulf-Atlantic Coastal Plain, the Piedmont, and the Interior Plains (see Fig. 24.3). The Coastal Plain occupies the seaward margin from Virginia to the southern tip of Texas, including all of Florida, Mississippi, and Louisiana and portions of all the other Southern states except West Virginia. It is low in elevation, and large areas near the sea and along the Mississippi River are flat, but most inland sections have an irregular surface.

The Piedmont extends across central Virginia, the western Carolinas, northern Georgia, and into east-central Alabama. Its surface is mainly a rolling plain, with some hilly areas. The natural cover is mixed forest in which broadleaved hardwoods, especially oak, tend to predominate over pines. Large areas have reverted to forest from former agricultural use. Soils tend to be poor, and many have suffered serious damage through past cultivation of clean-tilled row crops (cotton, corn, and tobacco) on easily eroded slopes subject to frequent downpours. The region's fall line (often where early industry began) lies between the Piedmont and the Coastal Plain.

Hills, Mountains, and Small Plains of the Upland South

Large parts of the South are hilly or mountainous. In the central and eastern South, two large embayments of rough country project into the South from the Northeast and Midwest. The larger of the two lies mainly in the Appalachian Highlands but includes some rough land west of the Appalachians known as the Unglaciated Southeastern Interior Plain (see Fig. 24.3). Farther west, the second large embayment is the southern part of the Interior Highlands. The three physical areas—Appalachians, Unglaciated Southeastern Interior Plain, and Interior Highlands—are often termed the Upland South.

Southern areas within the Appalachian Highlands extend from western Virginia to eastern Kentucky and southward to northern Alabama. The Highlands include all the major physical subdivisions already distinguished in the Middle Atlantic states, with the total complex narrowing southward (see Figs. 24.1 and 24.3). The Blue Ridge extends from the Potomac River to northern Georgia. Its highest section is called the Great

600

A

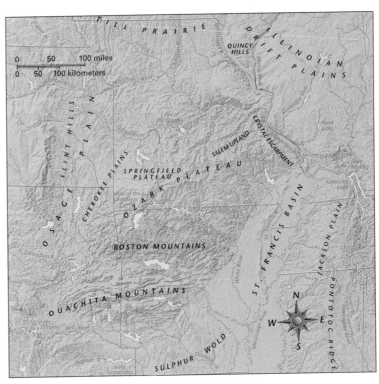

B

Figure 24.8 A scene in the southwestern edge of the Missouri Ozarks. (a) The water body lined with boathouses is an arm of one of the many artificial lakes in the interior highlands. (b) The subdued hilly topography evident on the map is characteristic of this physiographic area. *Jesse H. Wheeler, Jr.*

Smoky Mountains. Here, Mount Mitchell in western North Carolina is the highest point in the eastern United States, at 6684 feet (2037 m). The Ridge and Valley Section, including the Great Appalachian Valley, runs parallel to the Blue Ridge and inland from it, from western Virginia into Alabama. The southern section of the Appalachian Plateau, known as the Cumberland Plateau in parts of Kentucky and areas to the south, is quite wide in the north, where it occupies eastern Kentucky, most of West Virginia, and a small section of Virginia. It narrows rapidly across eastern Tennessee and then widens to occupy much of northern Alabama. As in Pennsylvania, it is edged on its east by an abrupt escarpment, is mostly dissected into hills and low mountains, and is underlain by very large deposits of high-quality coal.

Central Kentucky, central Tennessee, and part of northern Alabama are in the Unglaciated Southeastern Interior Plain. This part of the Interior Plain, bounded by the Tennessee River on the west and south, was not subject to the glacial smoothing that occurred on the Plains farther north. Hence, a good part of this "plains" area is actually hill country, the result of dissection by streams over a long period of time. True plains are present but are not extensive. Two of them—the Kentucky Bluegrass Region around Lexington and the hill-studded Nashville Basin around and south from Nashville, Tennessee—are underlain by limestone that has weathered into good soils. Both areas are famous for their agricultural quality in a region of poor to mediocre soils and difficult slopes.

In the South, the Interior Highlands occupy northwestern Arkansas and most of the eastern margin of Oklahoma. From Arkansas, they extend northward into the southern part of the Midwestern state of Missouri. Composed of hill country, low mountains, and some areas of plains, these highlands are split by the broad east-west Arkansas Valley followed by the Arkansas River. North of the valley, the Boston Mountains form the southern edge of the extensive Ozark Plateau (Fig. 24.8). South of the valley lie the Ouachita Mountains of west central Arkansas and adjacent Oklahoma. Most of the Ozark Plateau has been dissected into somewhat subdued hill country, but some parts of the Plateau, especially the Boston Mountains, attain mountainous character. The Ouachita Mountains consist of roughly parallel ridges and valleys running east and west.

The Southern Interior Highlands have not been rich in resources. Most soils are quite poor, with many very stony. Most of the area is still forested, but it has been heavily cut over, and the present timber is generally rather poor. Mineral deposits are varied but often not of sufficient value to justify present exploitation. The main mineral wealth extracted today from the Interior Highlands, primarily lead, comes from the section outside the South in Missouri. A number of large artificial lakes used for recreation are major resources in both Missouri and Arkansas.

Regional patterns of the South in population density, urbanism, ethnicity, and income are distinct from those of the other major regions. Until recently, it was a region of farms, villages, and small cities, with low population density compared to that of

Figure 24.9 This scene of the Dallas skyline at dusk does a good job of illustrating the drama of this Texas financial and commercial center. Dallas, in combination with Ft. Worth, exists as the major commercial and financial center for Central Texas. Architects and engineers have given much attention to the clear light of this setting, draping glass over the surfaces of the tall buildings that often serve as icons or billboards for specific Texas firms. *James Blank/FPG International*

the Northeast and Midwest. But for the past several decades, the South has been a region of rapid population growth, second only to the West, while its farm population has decreased drastically. It has now surpassed the Midwest in overall population density, although both are far behind the Northeast. Because recent Southern growth has been overwhelmingly urban, previous contrasts in this respect also are rapidly fading. The South has more small and medium-sized metropolitan areas than other regions of the country (because of recent rapid growth of many small cities). However, it has fewer centers of truly large size than might be expected from its population. The largest urban complex, Dallas–Fort Worth, has only 4.3 million people, and only it and Houston (population: 4.1 million), Miami (3.4 million), Atlanta (3.3 million), and Tampa–St. Petersburg (2.2 million) are among the more than 20 American metropolitan areas with over 2 million people (Fig. 24.9). But these Southern urban complexes continue to grow very rapidly, and some of them have imagery that surpasses their demographic size (Fig. 24.10).

The population of the South has less ethnic and racial variety than that of the other regions. The overwhelming majority is comprised of white Protestants of British ancestry, African-American Protestants, or Mexican-American Catholics. Most striking is the heavy preponderance of white Protestants with Anglo-Saxon origins. The original European settlers along the Atlantic were British. While small groups of Spanish Catholics settled in Florida and Texas, and some French Catholics settled in Louisiana and elsewhere along the Gulf Coast, their numbers were insignificant compared with the tide from Great Britain.

Another way in which the South stands out ethnically from the other major regions is in its high proportion of African-American population. This is largely a result of the importation of African slaves for plantation labor in the pre–Civil War South. The five American states that rank highest in percentage of African Americans—from about one-fourth to somewhat over one-third—are Mississippi, Louisiana, South Carolina, Georgia, and Alabama, which together form a Coastal Plain belt. Migration out of the South, mostly in the past 50 years, has given some northern states and California large African-American minorities. New York and California both have larger African-American populations than any Southern state, but the percentage is only 16 percent in New York and 7 percent in California. African-American populations outside the South are very concentrated in large metropolitan areas. Recently, a sizable reverse migration of African Americans from North to South has occurred as economic and social conditions in the South have improved at the same time economic conditions of many Northeastern cities have deteriorated.

Other distinctive racial and ethnic elements in the South's population are less widely distributed. Especially notable are Mexican Americans, Cuban Americans, and Cajuns (of French descent). The 1990 census reported that the four states leading in percentage of population "of Hispanic origin" were New Mexico (38 percent), Texas (26 percent), California (26 percent), and Arizona (19 percent). In Texas, Hispanic people primarily of Mexican descent live mainly in a broad band of territory along the Gulf of Mexico and the Rio Grande, where

Figure 24.10 New Orleans has done more to capitalize on its riverfront location than perhaps any other city in the United States. Early settlement by the French led to the establishment of the French Quarter, and today that landscape is at the heart of the tourist role in the New Orleans economy. This night time shot of Bourbon Street in the French Quarter shows the gaudy way entertainment can have an influence on maintaining architecture and neighborhood flavor and image. *Jeff Greenberg Photo Researchers, Inc.*

they form a much higher proportion than the statewide 26 percent. Natural increase and immigration have rapidly expanded their numbers (in Texas and elsewhere) during recent decades.

South Florida has received intermittent immigration of another "Hispanic origin" group—Cuban Americans. Almost a million refugees have fled from Communist Cuba since the revolution of 1959, and the majority of them or their descendants now live in metropolitan Miami. Cuban Americans, drawn largely from the middle and upper classes of pre-Communist Cuba, have shown rather rapid upward mobility economically. They also have been assimilating into the general population through marriage and rapid adoption of U.S. culture traits. Meanwhile, they have imprinted aspects of their own culture on south Florida. The country's Mexican-American population, on the other hand, largely continues to be a group whose Hispanic identity remains strong. The largest non-Hispanic minority language group in the South are the French-descended Cajuns of southern Louisiana.

Agricultural Geography of the South Today

What are the main features of the South's agricultural geography after a half-century of revolutionary departure from regional dependence on cotton, changes in crop and livestock emphases, farm mechanization, farm markets, and the number of farmers? In detail the picture is intricate, but some generalizations are possible:

1. In appearance and land use, the greater part of the South today is not agricultural. Over large areas, less than half the land is in farms, and even where land is farmed, the amount of cropland is apt to be greatly exceeded by woodland and/or pasture.

2. The present South is not outstandingly agricultural compared with the country as a whole. Only about 2 percent of the national population still lives on farms, and the proportion in the South is very close to this—higher than in the Northeast or West, but much lower than in the Midwest.

3. Southern farms tend to be small, low in production, and often part-time. The prevalence of part-time operations and of small full-time farms means that output per farm over most of the South is far below the national average.

4. Within the South as a whole, animal products decisively surpass crops in total farm sales. Beef cattle, broiler chickens, and dairy cattle are the main sources of animal products sold. Rising cattle production has been fostered by (1) the presence of much land that is too poor for crops but amenable to pasture, (2) improvements in pasture grasses and cattle breeds, (3) increased availability of chemicals for pest and disease control, and (4) greater emphasis on soybeans, sorghums, and citrus by-products for cattle feed. Soybeans have become the most widespread and important Southern crop. The leguminous soybean plant adds nitrogen to the soil and can be used for hay. But its greatest value lies in the beans that are pressed for soybean oil (for use in many chemical processes and, now, as a fuel additive as well), with the residue (oilcake) being fed to cattle and poultry. Grain sorghums, which have low moisture requirements, have become a mainstay for cattle feeding in drier parts of Texas and Oklahoma. The greater part of Florida citrus fruits are now processed for frozen concentrate, and the residue is fed to cattle; the main market is the large ranch industry in southern Florida. Broiler chicken production has become a specialty of poorer agricultural areas, with major concentrations in the Interior Highlands of Arkansas and the Appalachians of Georgia and Alabama. Dairy farming is growing in importance as Southern urban markets grow.

5. The historic staples of tobacco and cotton are still major elements in the agriculture of some parts of the South. Most of the nation's tobacco is grown in the eastern and northern South. In 1996, it was the leading source of income in the agriculture of North and South Carolina, Kentucky, and Tennessee. Nearly all the South's remaining cotton is grown on large mechanized farms in the Mississippi Alluvial Plain or in west Texas.

6. Many islandlike districts of intensive agriculture produce a variety of other specialties. Often such districts represent the main producing areas of their kind in the United States. Some major specialties and areas include citrus fruits in the central part of the Florida peninsula; truck crops in the Florida peninsula; sugarcane in the Louisiana part of the Mississippi Lowland and near Lake Okeechobee in Florida (on a drained section of the northern Everglades); rice in the Mississippi Lowland and the prairies along the coasts of Louisiana and Texas; peanuts in southern Georgia and adjoining parts of Alabama and Florida; and race horses, bred particularly in the Kentucky Bluegrass region around Lexington, where the horse farms are showpieces of productive and conspicuous wealth.

Industrial Change in the South and Its Geographic Outcome

After the Civil War, Southern leaders began a drive to industrialize the region, and this effort is still in progress. The earliest large-scale industries to develop were cotton-textile and apparel factories using cheap labor from the economically depressed countryside. In the late 19th and early 20th centuries, this development gained regional importance, and from the 1930s to the 1950s it sapped the cotton-textile industry of New England and eventually gave national preeminence in the industry to the Southern Piedmont from Virginia to Alabama. The new factories were built in small towns and cities—many of which offered tax advantages and free land to attract mills. Southern mills in the Piedmont sold cloth first to apparel factories in the Northeast, then to new Southern factories built at many locations in the Piedmont and Coastal Plain. In the 1940s and 1950s, Southern industrialization accelerated and diversified. Low wages remained a powerful factor, but many other advantages developed, including new hydroelectric facilities built by the federal government in the Tennessee River Basin and elsewhere; federal improvements in river transportation and harbors; and joint federal and state construction of the Interstate Highway System, which gave the South cheap and fast new connections with markets elsewhere. Air-conditioning made Southern workplaces more comfortable and productive. When synthetic fabrics arose to challenge cottons and woolens, the South had abundant raw materials (wood and hydrocarbons) from which to make synthetic fibers. Unfortunately, the offshore movement of textile plants in the past decade has been most significant in the recent decline of the Southern textile industry.

An extraordinarily valuable regional store of energy resources—oil, natural gas, coal, and falling water—powered new Southern industries and provided large energy exports to markets outside the region. Increasing technical skills became available through a combination of on-the-job experience, improved research and teaching by Southern universities, and import of technology to such places as the Oak Ridge atomic research facility near Knoxville, Tennessee, and the space flight facilities of the National Aeronautics and Space Administration (NASA) at Huntsville, Alabama; Houston, Texas; and Cape Canaveral, Florida. As population increased and incomes grew, Southern regional markets became large enough to attract

plants that could profit from proximity to those markets. By the early 1990s, the South was reaching a degree of industrialization generally commensurate with that of the entire nation; at least two Southern states—Texas and North Carolina—are now major industrial states.

The Major Southern Industrial Belts

The largest cluster of industrial development in the South lies along the Piedmont, with extensions into the adjacent Coastal Plain and Appalachians. Within this concentration, which may be termed the Eastern South Industrial Belt, textiles are the most prominent branch of manufacturing, but clothing, chemical products (including synthetic fibers), furniture, tobacco products, and machinery are also important. Much of this region's development, as well as that of adjacent areas in the South, has been fostered by the Tennessee Valley Authority (TVA). The TVA was founded in 1933 by the federal government during Franklin Roosevelt's New Deal era in order to promote economic rehabilitation of the deeply depressed Tennessee River Basin. It is a public corporation that built many dams, locks, and hydrostations on both the Tennessee and its tributaries. These installations controlled floods; allowed barge navigation as far upstream as Knoxville, Tennessee; generated hydroelectricity; and provided recreation at numerous reservoirs. The demand for electricity soon outran the capacity of the hydrostations, and many thermal plants were then built, fired by coal or nuclear energy. The success of the TVA in helping industrialize several Southern states is a well-publicized facet of the industrial surge that has spread ever more widely across the South in recent decades. While in the North industrial development focused on the huge metropolises that arose in the railway age, in the South most development came later and took advantage of long-distance power transmission and high-speed truck transportation to spread widely into small cities and towns.

The second main belt of industrial development in the South lies along and near the Gulf Coast from Corpus Christi, Texas, to New Orleans and Baton Rouge, Louisiana. The chemical industry, including oil refining, is the dominant industrial sector within this West Gulf Coast Industrial Belt. It is based on deposits of oil, natural gas, sulfur, and salt. Texas and Louisiana produce nearly one-third of the oil, over three-fifths of the natural gas, and practically all the native sulfur output of the United States, with the Gulf Coast strip the leading area of production for all three. Machinery and equipment for oil and gas extraction, oil refining, and chemical manufacturing are also important products. Major refineries and petrochemical plants are clustered at several locations, most notably metropolitan Houston and Beaumont, Texas, and Baton Rouge, Louisiana.

Many other industries are present in the South, some of which are widespread while others are localized in a few places. Food processing, machinery, furniture, and pulp-and-paper industries are widely distributed, with food industries especially important in Florida and pulp-and-paper plants particularly

Figure 24.11 The Houston Ship Channel (also called Canal) has linked this manufacturing and food processing city to the Bay of Galveston and the Gulf of Mexico. The Channel was cut through the former Buffalo Bayou in 1914 and has a maintained depth of 34 ft and a width of 200 ft and is 57.3 miles long. Houston has become the country's number one port. In this photo oil tankers are loading a major Texas product for transport. Just upstream freighters are being filled with wheat, another major Texas export product. *David R. Frazier Photo Researchers, Inc.*

prominent near the Gulf Coast from eastern Texas to Florida and along the Atlantic Coast in Georgia. (For raw material, the latter industry depends largely on the pine trees that grow rapidly in the wet subtropical climate.) Aircraft and aerospace industries are prominent in the Atlanta and Dallas–Fort Worth areas, and both auto assembly and the manufacture of auto components are expanding, especially in Tennessee and Kentucky.

Texas-Oklahoma Metropolitan Zone

The Texas-Oklahoma Metropolitan Zone contains the two largest urban clusters in the South—Dallas–Fort Worth and Houston—plus four other metropolises with populations of roughly 750,000 or over: San Antonio (population: 1.4 million) and Austin (population: 964,000), Texas, and Oklahoma City (population: 1.07 million) and Tulsa (population: 745,000), Oklahoma. Houston is a seaport by virtue of the Houston Ship Channel, constructed inland to the city in the late 19th and early 20th centuries (Fig. 24.11). The city has become a major port serving Texas and the southern Great Plains, in addition to being the principal control, supply, and processing center for the oil and gas fields and the chemical industry of the Gulf Coast. Dallas–Fort Worth, San Antonio, and Austin developed at or near the western edge of the Coastal Plain in a north-south strip of territory known as the Black Land Prairie, where unusually fertile limestone-derived soils supported a productive agriculture and towns and cities to service it (Fig. 24.12). Dallas–Fort Worth and San Antonio also serve large trading hinterlands stretching far to the west, within which there are no comparable

competing centers. In the case of Dallas–Fort Worth, the same is true in other directions except for the competition of Houston to the south. Dallas–Fort Worth is the dominant business center for this huge region. San Antonio is a major military base and retirement center, and Austin is the site of the strongly funded University of Texas, a center for high-technology industries, and the seat of Texas state government. All these cities are involved in the oil and gas industry.

Florida Metropolitan Zone

A second area of exceptional metropolitan development is the Florida peninsula. Two Florida metropolitan areas here have estimated populations of more than 2 million each—Miami (3.4 million) and Tampa (2.2 million)—and three more have over 900,000 each: Orlando (1.4 million), Jacksonville (975,000), and West Palm Beach (955,000). The distribution of Florida's main metropolitan areas reflects the importance of amenities attractive to tourists and retirees (Fig. 24.13). Proximity to the ocean is important, but winter warmth is even more crucial. Miami's marginally tropical winters have undoubtedly helped it to grow faster than Jacksonville in the extreme northeast. But Florida's cities are by no means totally devoted to vacation and

Figure 24.13 Florida, especially southern Florida, has become a major destination for retirement populations. They come from all over the country, but especially from the Northeast. The impact of this migration pattern has had an influence on the whole urban and residential scene. The region is also a destination for "snowbirds," the senior citizens who maintain homes in the North and the Midwest but who, in the cooler winds of fall, get in all sorts of vehicles and work their ways southwest and southeast toward warmer winter months. This Miami scene is one repeated again and again in the cities of the South. *Leo de Wys Inc. / J.B. Grant*

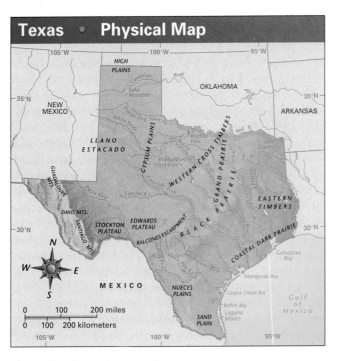

Figure 24.12 Physical map of Texas.

retirement functions. Miami has become a major point of contact between the United States and Latin America in regard to air traffic, tourist cruises, and banking, as well as illegal drug traffic. Tampa Bay affords a major harbor well placed for the trade of central Florida's citrus belt and phosphate mines. High-technology industries and regional corporate offices are increasingly attracted to Florida as the state's population grows and improved transportation and communication reduce the disadvantages of location in places remote from the Old Metropolitan Belt of the Northeast. For example, Orlando's extraordinarily rapid recent growth has been spurred by corporate office development as well as by the presence of Disney World—the world's largest tourist center and tourist attraction.

Southern Piedmont and Coastal Virginia Metropolitan Zone

The third major Southern concentration of large metropolises lies along the Piedmont, the fall-line, and the nearby Virginia Coastal Plain. Piedmont and fall-line metropolises include Richmond (population: 917,000), Virginia; Charlotte (population: 1.3 million), Greensboro (population: 1.1 million), and Raleigh (population: 965,000), North Carolina; and Greenville (population: 873,000), South Carolina. These metropolises have tended to form by the coalescing and integrating of cities located near each other, resulting in multiple-centered city clusters, although each is identified here by the name of its leading central city.

However, the two largest metropolises of this metropolitan zone differ in function from those just named. In the south, Atlanta (population: 3.3 million), Georgia, ranks fourth in metropolitan population within the entire South. It has diverse industries but stands out as a major inland transport center. Railroads and highways have generally avoided the Great Smoky Mountains, which lie just north of Atlanta. Instead, they skirt the southern end and converge on the city. Its ascendancy as a regional transport center today is seen in its airport, which is the second busiest in the United States as measured by volume of passenger traffic (54 million). Chicago-O'Hare has 66 million; Dallas–Fort Worth, 52 million; and Los Angeles International, 51 million. These transport facilities and its superior location have been exploited to make Atlanta the largest Southeastern business and governmental center.

In the north on the Coastal Plain is the seaport of Norfolk (population: 1.5 million), Virginia. Located on the spacious Hampton Roads natural harbor at the mouth of the James River, Norfolk ships Appalachian coal by sea and is the main Atlantic base for the United States Navy.

Metropolises of the Upland South

The entire Upland South (Appalachians, Unglaciated Southeastern Interior Plain, and Southern Interior Highlands) contains only five metropolitan areas of over 500,000 people.

Louisville (population: 981,000), Kentucky, began as a river port at the "Falls of the Ohio," where goods had to be trans-shipped around this break in transportation. It was also the closest point of contact on the Ohio River for the earliest settled and most productive agricultural area in the state—the Bluegrass Region. By the time a canal had been built to bypass the falls (in 1830), Louisville was a leading presence on the river, and it has continued to grow as a diversified regional center.

Two comparable centers, Nashville (population: 1.1 million) and Knoxville (population: 631,000), Tennessee, also began in locations associated with river transport and limestone soils. Nashville is located where the Cumberland River, a tributary of the Ohio, extends farthest south into the hilly but fertile Nashville Basin of central Tennessee. Knoxville is situated far up the Tennessee River from the Ohio in an area of limestone valleys within the Ridge and Valley Section of the Appalachians.

The same themes of transport advantages, relatively good soils, and growth as the major center for a considerable region are repeated in the case of Little Rock (population: 538,000), Arkansas. In this case, the exceptional land lies mainly in the Mississippi Alluvial Plain to the immediate east and secondarily in the Arkansas Valley to the west. The valley also provides a lowland route channeling traffic through the Interior Highlands, and the Arkansas River flows into the Mississippi.

The major exception, however, to the general pattern of urban location and development in the Upland South just described is Birmingham (population: 872,000), Alabama, which is the only large center of iron and steel industry in the South. Its development in the Ridge and Valley Section of the Appalachians was based on large deposits of both coal (from the nearby Cumberland Plateau) and iron ore.

Other Southern Metropolises

Other Southern cities with metropolitan populations of over 500,000 are New Orleans (population: 1.3 million) and Baton Rouge (population: 558,000), Louisiana; Memphis (population: 1.1 million), Tennessee; Charleston (population: 522,000), South Carolina; and El Paso (population: 665,000), Texas. The first three are Mississippi River ports. New Orleans was founded by the French just upstream from the head of the "bird's foot" delta where the Mississippi divides into separate channels flowing to the Gulf. In the riverboat era before the Civil War, New Orleans was New York's main competitor as a port. It then underwent relative decline as a port in the railway era but it remains a major seaport, especially for shipping grains and soybeans from the Midwest and the Mississippi Alluvial Plain—as well as a major tourist center. Somewhat upriver, Baton Rouge is accessible to medium-sized ocean tankers and is a major center of the Gulf Coast oil refining and petrochemical industries. Farther upriver, Memphis developed as a river port where the Mississippi flows adjacent to higher ground at the east side of its broad flood plain. The city's function as the main business center for the greater part of the Mississippi Alluvial

Plain (and other nearby areas) dates back to antebellum days when a major sector of the "Cotton Kingdom" developed in "The Delta"—a regional name then in use for the large part of the Mississippi Alluvial Plain adjacent to Memphis in northwestern Mississippi. The remaining metropolises lie at opposite ends of the South. Charleston was the port and commercial capital of the first major plantation area outside of Virginia. It was the leading city of the South at the time of the American Revolution, then languished as the plantation economy moved westward, and has resumed rapid growth with the rise of the modern Southern economy.

24.3 The Midwest: Agricultural and Industrial Heartland

The existence of a Midwest (Middle West) in the United States is widely recognized, but perceptions differ as to its extent. In this chapter, the term Midwest is applied to 12 states called the North Central states in federal publications. They are commonly subgrouped into the East North Central states of Ohio, Indiana, Illinois, Michigan, and Wisconsin and the West North Central states of Minnesota, Iowa, North Dakota, South Dakota, Nebraska, Kansas, and Missouri. The two subgroups are referred to here as the eastern Midwest and the western Midwest. The Mississippi River separates them except for part of the Wisconsin-Minnesota boundary (see Figs. 24.1 and 24.3). All states in the eastern Midwest border one or more of the Great Lakes, but only Minnesota does so in the western Midwest. In the eastern Midwest, Ohio, Indiana, and Illinois are bounded on the south by the Ohio River; in the western Midwest all states except Minnesota front on the Missouri River.

The use of the term "Midwest" for these 12 states is arbitrary to some degree but can be justified by spatial (mappable) characteristics distinguishing them as a group from the Northeast, South, and West. Among such characteristics are the Midwest's (1) interior "heartland" location; (2) distinctive plains environment; (3) important associations with major regional lakes and rivers; (4) ethnic patterns; (5) outstandingly productive combination of large-scale agriculture, industry, and transportation; and (6) rural landscapes famed for their rectangularity, symmetry, and prosperous appearance.

Midwestern Agriculture and Its Environmental Setting

These vast Midwestern plains produce a greater output of foods and feeds than any other area of comparable size in the world. The Midwest commonly accounts for more than two-fifths of all United States agricultural production by value. This output is unevenly spread among the Midwestern states, with Iowa leading. Within the nation, Iowa is the third-ranking state in value of farm products, exceeded only by much larger Califor-

nia with its irrigation-based agriculture, and by Texas. Ordinarily, five Midwestern states—Iowa, Nebraska, Illinois, Kansas, and Minnesota—are among the nation's top seven in total agricultural output, and nine Midwestern states are among the top fifteen. The lowest-ranking Midwestern states are Michigan, handicapped by large areas of infertile sandy soil, and the Dakotas, handicapped by dryness, rough land toward the west, and a shorter frost-free season.

Within most of the Midwest, soils are exceptionally good. The best soils developed under grasslands, which supplied abundant humus. Such outstandingly fertile soils normally accompany steppe climates in the middle latitudes, where the fertility is often counterbalanced by precipitation so low as to be marginal for agriculture. However, the remarkable feature of the Midwest is that its soils—located in a more humid continental climatic area—still developed under a tall-grass prairie. Such climatic conditions normally produce a natural vegetation of forest, and trees grow well if they are planted. But in the early 19th century, the first settlers found large expanses of prairie, or a prairie-forest mixture, as far east as northwestern Indiana. From there, the prairie region extended west in a broadening triangle to the eastern parts of the Dakotas, Nebraska, and Kansas, and on into the South in Oklahoma and Texas. Of the Midwestern states, only Michigan and Ohio lacked considerable expanses of prairie. Thus, a very large part of the region had soils of a richness generally found only in semiarid areas but which in the Midwest were combined with humid conditions more favorable to most crops. The existence of the anomalous triangle of prairie bordered by forests both north and south can be accounted for in two main ways. One explanation holds that the prairie resulted from periodic drought conditions, discouraging tree growth in this normally humid area. The other explanation suggests that Native Americans may have burned the forest to improve hunting conditions and create grazing range for game. Possibly both factors were important, and prairie fires set by lightning also may have helped perpetuate the prairie once it was established (grass is less susceptible than tree seedlings to lasting damage by fire). Whatever the causes, the occurrence of grassland soils in a humid area where forests might normally be expected provided an extraordinarily favorable agricultural environment.

Another natural condition contributing to the unusual excellence of many Midwestern soils is the widespread presence of loess soil deposits made of dust particles. These wind-borne soil materials are thought to have accumulated in glacial times when strong winds picked up the particles from dry surfaces where the ice had melted but where no protective mantle of vegetation had yet become established. Because most of the nutrients that nourish plants are contained in these finer soil particles, loess is associated with soils of exceptional fertility in various parts of the world.

In the eastern Midwest, the soils formed under natural broadleaf forests are not exceptionally fertile by nature, but neither are they poor (see Fig. 24.5). The principal areas of poor

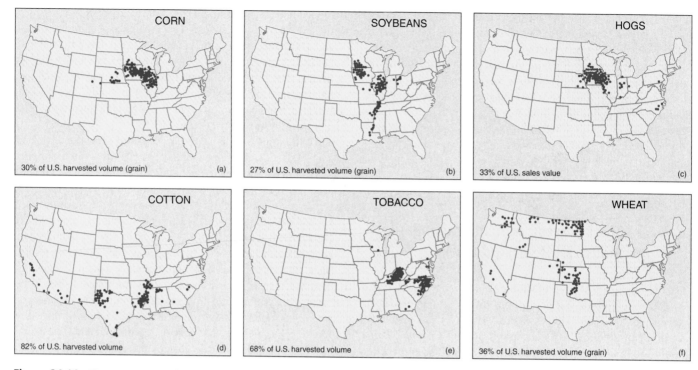

Figure 24.14 These two panels of maps show the 100 leading U.S. counties for selected crops and farm products. Each map also tells the reader what percent of the U.S. total production of this commodity comes from the aggregate of these 100 counties (except for rice, where only the top 50 counties are shown). *Cartography by Joseph H. Astroth, Jr., and Jennifer Nichols*

soil in the Midwest lie in (1) the uplands of the Ozarks and the Appalachian Plateau; (2) sandy areas of needleleaf forest in northern Michigan, Minnesota, and Wisconsin; and (3) rougher western parts of the four westernmost states. After long use, the present quality of Midwestern soils depends heavily on how they have been handled. Where properly fertilized and protected against depletion and erosion, they have proved remarkably durable. Although unwise practices have caused much soil damage, huge expanses remain highly productive. Aided by science and technology, the soils of the Midwest as a whole have higher yields today than ever before.

The Legendary American Corn Belt

Corn is the single leading product of Midwestern farms, and the Corn Belt, where this crop dominates the landscape, has become an agricultural legend for its productivity. Sales of beef cattle in the Midwest are actually greater than receipts from corn, but the cattle are raised primarily on corn. Over four-fifths of American corn production and around two-fifths of world production come from the Midwest. Highly favorable conditions for large-scale corn growing are provided by fertile soils, hot and wet summers, and huge areas of land level enough to permit mechanized operations. The corn plant itself and the methods of production have been greatly improved by extensive research.

Soybeans are another major element in Corn Belt agriculture (Fig. 24.14). Long cultivated in China, they have had a rapidly increasing impact on American and world agriculture during recent decades. In World War II, it was learned that oil of the bean could be used in the manufacture of paints, soap, glycerin, printing ink, and many other chemical-based products. The growth of the bean also adds nitrogen to the soil, provides livestock feed, and the oil is now also being experimented with as a fuel additive (similar to the one derived from corn) for Midwestern gas stations. Per capita annual consumption in the United States has grown to 250 pounds, a clear indication of the growing importance of this unusual resource. Soybeans have recently turned the Corn Belt into a region that could be more realistically named the Corn and Soybean Belt.

The Rise of Midwestern Agriculture

Despite its impressive resources and accomplishments, Midwestern agriculture is a risky enterprise. Natural hazards such as drought and plant diseases increase the normal business risk, but the main hazards are sharp fluctuations in world supply and demand and therefore in market prices. These forces have driven waves of farmers out of business at various times in the past. Between 1930 and 1992, for example, the number of farms in the Midwest decreased from slightly over 2 million to approximately 777,000. Most of the land of defunct farms was

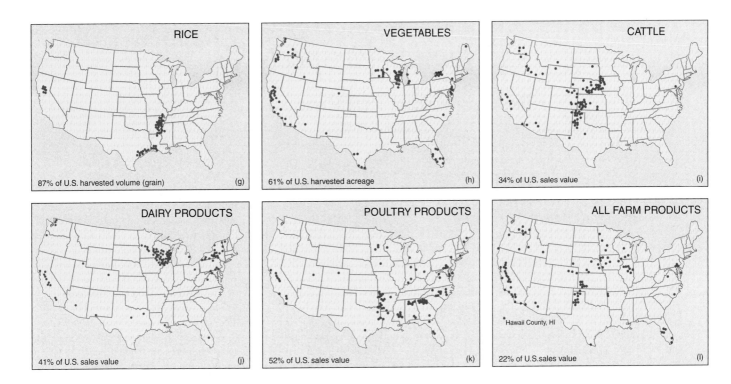

RICE
87% of U.S. harvested volume (grain) (g)

VEGETABLES
61% of U.S. harvested acreage (h)

CATTLE
34% of U.S. sales value (i)

DAIRY PRODUCTS
41% of U.S. sales value (j)

POULTRY PRODUCTS
52% of U.S. sales value (k)

ALL FARM PRODUCTS
Hawaii County, HI
22% of U.S.sales value (l)

bought or rented by surviving farmers who expanded their acreage. The amount of land in farming decreased only 6 percent, but the mean size of farms increased from about 180 acres (73 ha) to about 440 acres (166 ha). Increased mechanization allowed farmers to work larger acreages but also increased their capital costs and hence increased their borrowing and their household vulnerability to low market prices. Many of the farmers who left farming in the past two decades were victims of excessive indebtedness due to enormous machinery costs. And, as U.S. farms become fewer, larger, and more highly capitalized (in the Midwest and elsewhere), corporate forms of farm ownership and enterprise are increasingly evident.

Agricultural Foundations of Midwestern Industry

Midwestern agriculture has been, and remains, one of the main foundations of regional industry. Meatpacking, grain milling, and other food processing industries have always been of major importance. Cincinnati first developed as an early hog-packing "Porkopolis" before the Civil War, and Carl Sandburg's poem "Chicago" celebrated that city's role as the "hog butcher for the world." The center of the Midwestern packing industry has moved westward with time to a present location in and near east central Nebraska and western Iowa. The same westward migration also has been true of wheat milling, in which Minneapolis– St. Paul is now a leader. In addition to processing agricultural

products, Midwestern industry supplies many needs of the region's farms. Many technical advances in farm equipment— such as Cyrus Hall McCormick's 1840s mechanical reaper for harvesting wheat—originated in the Midwest. The world's largest farm-machinery corporations (John Deere, Caterpillar, Navistar) are headquartered in the region, and many Midwestern cities are heavily dependent on the industry. Such activity has helped foster the development of a broad range of other types of machine manufacture. By the mid-1990s, St. Louis had become second only to Detroit in the manufacture of automobiles.

The Midwestern Steel Industry

The emergence of over half of the United States steel industry in the eastern Midwest was also based in part on Midwestern agriculture. In the late 19th century and early 20th century, huge steel plants were built on the Great Lakes in Ohio, Illinois, Indiana, and Michigan. Much of their output was used to build and equip the transport network employed in hauling agricultural products. Additional steel was used by machinery industries serving agriculture and to construct the region's burgeoning cities, which were themselves heavily dependent on agricultural trade. Then the growth of the automobile industry in the Midwest created a major new demand for steel.

Besides a favorable market situation, the Midwestern steel industry had an excellent resource base in and near the region. High-grade coal was available from the Appalachian Plateau.

Regional Perspective

The Abandonment of Small Towns . . . or Is It?

A statistic of real magnitude is this simple dimension of change: Between 1900 and 1990, the American population more than tripled. That is, from a national population of 76.2 million in 1900, we grew to a total population of more than 250 million nine decades later. In that same nine decades, however, 8 out of 10 counties all across America had lost population. Or, 2400 counties had fewer people in 1990 than they had had at the turn of the century in 1900. Thus: National population total = threefold increase; 80 percent of our counties in the same period lost population. What can such a statistic mean? Where did all of the people who made up the threefold national growth go? They had gone to the major metropolitan centers that we have been talking about in the earlier pages in this chapter, and they went to cities yet undiscussed in the Midwest and West. What they left was the American small town.

Geography is at the heart of this extraordinary demographic shift. Small-town settlements had emerged as vital clusters of merchants and townfolk located on some avenue of transportation —the highway, railway, river, or a combination. These clusters were surrounded by the broad, open farmlands being cleared and farmed in the late 18th and 19th centuries. Dirt roads made moving farm produce to local markets, to train stations, and to riverside wharfs difficult, so people quickly learned that there was a geographic wisdom for businesspeople and farmers to locate close together. The small towns that became county seats had a particular advantage because they were the seat of local government as well. Particularly fortunate towns became home to

food processing firms, shoe or textile factories, or any number of small-scale industries that added a factory payroll to these centers.

Then, in the steady flow of landscape transformation so central to the expanding settlement of the United States, roads began to be surfaced so that they could function in all seasons. Some gravel and blacktop roads were replaced by concrete and were made into four lanes, while many were left as two lanes. The larger roads with better surfaces led into hubs of nodality in ever larger settlements. Smaller towns not only lost economic centrality, they lost residential utility also, as more and more people moved to the larger centers where "things seemed to happen." Now, as farm families get into their sedans and drive through the increasingly empty streets of the small towns that lie between their farmsteads and the regional shopping centers or massive discount stores that are their destination for shopping or entertainment, they become witnesses to a really monumental demographic and cultural change that has left Main Street unused and small towns tired and increasingly empty.

This shift is broadly regional. The distribution of the 80 percent of the American counties that have lost population in the past century is scattered through all of the states. The 20 percent that have gained—and many have gained enormously—are generally nested around the major urban centers that have been at the heart of decades of urban growth and sprawl.

However, there is another side to this geographic equation. Two things (at least) are happening to change some of

this story. In the past decade, there has been a significant out-migration of urban professionals from large urban centers. The numbers have not been major yet, but these reverse pioneers are trying to see if they can take up residence in small-town America and still maintain their professional and career links with the major cities in which their offices exist. Can networks of phone lines, Fax machines, and the Internet, and occasional trips back to the big city, substitute for full-time residence in the metropolitan area? And, if so, can residence in small towns—with their lower prices for shelter, their more homey atmosphere, and their continued link with the American past, a golden age of simpler lifestyles —satisfy the needs of these ex-urbanites?

Like so many things about the human environment, this development is not quite one thing or another. This new pattern may, in fact, help return vitality to the small towns that were so much at the heart of our culture a century ago. But the vitality will be of a different nature than that which provides our images of the past. The beauty of this speculation is that you—the very student reading this text—will be able to explore the merits of both the big city and the small town yourselves as you decide where it is you settle after graduation from college and see which built environment works the strongest pull on you. After all, you have been born into a generation that—because of fiber-optic networks—will have a greater range of possible living situations than any prior American generation. Make use of those options!

Supplements of lower-grade coal came from mines in Illinois, southwestern Indiana, and western Kentucky. Huge quantities of high-grade iron ore came cheaply by lake boat from the Mesabi Range inland from Duluth, Minnesota. By the end of World War II, this high-grade ore was largely exhausted, but

completion of the St. Lawrence Seaway in the 1950s made it feasible to bring ore from newly opened deposits in Labrador and Quebec. In addition, changes in smelting technologies increased the utility of the remaining lower-grade ores in the Mesabi and other fields near Lake Superior.

Figure 24.15 The combination of land near the southern edge of Lake Erie, placement of one of the world's most complete rail networks, and proximity to a labor pool that had experience with steel, rubber, and machinery made Youngstown a productive steelmaking center up into the decades following World War II. In the past several decades, however, foreign competition and major domestic shifts in U.S. steel production have left places like this Youngstown steel mill abandoned. *Mary Altier, 1999*

The steel-making districts of the eastern Midwest have been crucial for American economic development. The leading districts are located (1) within metropolitan Chicago along and near Lake Michigan in northwestern Indiana and south Chicago (Fig. 24.15); (2) along and near Lake Erie in Cleveland and other Ohio cities; and (3) around Detroit, Michigan. The areas of Buffalo, New York, and Pittsburgh, Pennsylvania, also belong to this complex of Great Lakes steel districts. But by the 1970s, these smokestack cities had also become major economic problem areas. The main plants in these areas are enormous, and many are old and now abandoned. Consequently, old steel districts have been saddled with major unemployment problems due to plant closings, production cutbacks, and automation. The high wage rates associated with the unionized steel industry tend to discourage the entry of new types of employment. An additional deterrent lies in the blighted landscapes and lack of amenities left over from an earlier age of industrialization.

The Enormous Automobile Industry

The availability of steel and a central location vis-à-vis markets were factors contributing to a heavy concentration of the United States automobile industry in Detroit and nearby Midwestern cities. When the industry began to develop during the early 20th century, production on a very large scale was necessary to achieve costs and prices low enough for mass sales. This required that production be centered in a few large plants and the vehicles marketed throughout the country. Steel was needed in huge quantities, as was a location from which distribution could reach the whole national market relatively cheaply. Detroit was

nearly optimum for the latter, as the national center of population at that time was in southern Indiana (it has now moved west to central Missouri), and buying power was higher toward the north and east. Detroit labor and management had a background in building lake boats and horse carriages, and experiments with internal combustion motors had been carried on by various inventors, including Henry Ford—a Detroit native.

From southeastern Michigan, the auto industry spread over much of the eastern Midwest: north to Flint and Lansing, east to Cleveland, south to Dayton and Indianapolis, and west to the Milwaukee area. Many cities within this circumference became very dependent on supplying auto parts. Assembly itself expanded to St. Louis and Kansas City and eventually to other major regions of the country. The American corporate giants of the industry—General Motors, Ford, and Chrysler (now Daimler-Chrysler)—still have headquarters and major plants in metropolitan Detroit. As in the case of steelmaking, the motor vehicle industry has provided much of the Midwest with a basic industry of great size, vital to employment, and paying high wages. However, like the steel industry, it became a focus of employment problems in the 1980s because of decreasing plant productivity and growing sales of competing operations overseas, especially from Japanese plants. This industry has a history of boom and bust cycles, and it is now having a vigorous resurgence, but its operations have diffused south and west to many additional cities, both in the Midwest and in other regions.

Regional Unity of Midwestern Metropolises

A large majority of the Midwest's population is metropolitan. As in other parts of the United States, an increasing amount of commuting has been integrating more and more counties with the cities at or near the centers of metropolitan areas. Hence, most large metropolitan areas are now sprawling multicounty districts. These are essentially urban in employment and function, even though outlying sections may not be highly urban in appearance. Even including the world of small towns and farm communities in the outlying rural landscapes, there is a regional unity of these Midwestern urban centers that is both economic and cultural. Farm families go to the regional malls and city folk go out to the farmers' markets and seasonal fairs.

The larger Midwestern cities around which the main metropolitan areas have clustered had their beginnings in the early 19th century or occasionally earlier. Water transport was relatively more important then than it is now, and city locations tend to be related to lake and/or river features that conferred early importance in a developing pattern of water transportation and human settlement. Even today, most major centers still benefit from advantageous locations for freight movement by water. When railroads came, the cities that were already established generally became rail centers as well. Then, in the mid–20th century, highways and airways also focused on the major existing cities.

Great Lakes Metropolises: Chicago, Detroit, Cleveland

The three largest Midwestern metropolitan concentrations are around Chicago, Illinois; Detroit, Michigan; and Cleveland, Ohio (see Fig. 24.1). By far the largest is Chicago—the country's third largest city (population: 8.5 million). The city began at the mouth of the small Chicago River, which afforded a harbor near the head of Great Lakes navigation from which the river projected farther into the prime farmlands of the Midwest. A few miles up this river, a portage to tributaries of the Illinois River provided a southward flowing connection via the Illinois to the Mississippi River near St. Louis. A canal built in 1848 along the portage route connected Lake Michigan with the Illinois River and helped Chicago become the major shipping point for farm products moving eastward by lake transport from the rich farming areas in the prairies to the city's south and west. Then, in the 1850s, railways fanned out from Chicago into an ever broader hinterland and also connected the city with Atlantic ports. Chicago became the country's leading center of railway routes and traffic and has remained so. In the 20th century, new harbors were built south of the city center, and new large-scale canal connections facilitated an increasing barge traffic using the Illinois River. Hence, at this major focus of transport and population growth, Chicago's huge industrial structure rose. Several major industries developed very early: food processing, especially meatpacking (which has now almost disappeared); wood products made from white pine and other timber cut in the forests to the north; clothing; agricultural equipment; steel; and a broad range of machinery manufacturing. Today, the most important lines of manufacturing are machinery and steel, for both of which Chicago is the leading center in the country. The city's banks, corporate offices, commodity exchanges, and other economic institutions make it the overall economic capital of the Midwest. To the east its metropolitan area reaches into northwestern Indiana, and to the north its area coalesces with that of Milwaukee (population: 1.6 million), Wisconsin, which also has grown around a natural harbor on the lake.

The second greatest concentration of people and production in the Midwest is in metropolitan Detroit, together with the smaller adjoining area of Toledo, Ohio. The Detroit area extends along the water passages separating Michigan from the Ontario peninsula in Canada and connecting Lake Huron with Lake Erie. From north to south, these passages include the St. Clair River, Lake St. Clair, and the Detroit River (see Fig. 24.1). Detroit's metropolitan population is 5.3 million, and the adjacent Toledo metropolis contains 613,945 more. A closely connected Canadian population in metropolitan Windsor, Ontario, just across the Detroit River from Detroit, adds another 262,000 to the metropolitan complex.

Detroit was founded as a French fort and settlement along the Detroit River in 1701. It grew as a service center for a Michigan hinterland that was settled rapidly following completion of the Erie Canal in 1825. It remained modest in size until explosive growth took place during the automobile boom of the 20th century. Toledo was founded on a natural harbor at the western end of Lake Erie. In the 19th century, natural gas and oil from early fields in western Ohio furnished fuel for a glass industry there, and the metropolitan area now produces much of the glass for Detroit automobile production.

The third-ranking metropolitan concentration in the Midwest is in northeastern Ohio, where the Lake Erie port of Cleveland has a metropolitan population (including nearby Akron) of 2.9 million. Cleveland developed in the 19th century as a center of heavy industry on a harbor at the mouth of the Cuyahoga River. Ore from the Lake Superior fields was offloaded there for shipment to Pittsburgh, and there was a return flow of Appalachian coal for Great Lakes destinations. Smaller Lake Erie ports also participated in this traffic. Movement of ore one way and coal the other made it logical to develop the iron and steel industry at Cleveland, at other Lake Erie ports, and at Youngstown on the route between Pittsburgh and the lake. Machinery and chemical plants developed, and eventually the auto industry spread into the district. Part of the latter development was the rise of Akron as the leading center of tire manufacturing in the United States. Today, northeastern Ohio is afflicted by the slow growth, contraction, or closure of many older industrial enterprises.

River Cities: St. Louis, Minneapolis–St. Paul, Cincinnati, Kansas City, Omaha

The remaining major metropolises of the Midwest originated at strategic places on the major rivers that were early arteries of Midwestern traffic. These cities still benefit from riverside locations, although other forms of transportation have now relegated river traffic to a subordinate position.

St. Louis, Missouri, is the center of a metropolitan population of about 2.5 million in Missouri and Illinois. It was founded by the French in the 18th century and remained the largest city in the Midwest until the later 19th century. It is located at a natural crossroads of river traffic using the Mississippi to the north and south, the Missouri to the west, the Illinois toward Lake Michigan, and (via a southward connection on the Mississippi) the Ohio toward the east. Early traffic had to adjust to the fact that the Mississippi downstream from the Missouri mouth was notably deeper than the upstream channel or the Missouri. Hence, many riverboats plying the Mississippi below St. Louis drew too much water to use the channels above the city, and St. Louis became a break-of-bulk point where goods were transferred between different kinds of craft. Coal from southern Illinois was available to support industrialization, and the metropolitan area eventually developed an industrial structure noted for its variety (see photo, p. 588).

Minneapolis–St. Paul (population: 2.7 million), Minnesota, developed at the head of navigation on the Mississippi River.

Definitions & Insights

THE MIDDLE-SIZED CITY

The middle-sized city in the United States (750,000–2,000,000 in population) is taking on a new importance in American migration and development patterns. Cities of this general size—and Cincinnati, Ohio, is a prime example—have emerged in the past decade as target destinations for both industry and finance. The blend of established urban infrastructure and generally well-educated labor pools that are not as expensive as those resident in the really large metropolitan areas has stimulated steady growth. As in the Regional Perspective in this chapter, the ever-expanding capacity of telecommunications to link virtually any node with other nodes has reduced the traditional importance of, for example, financial firms locating in the central city. The network afforded by computers, Fax machines, the Internet, and expanding local airline links between middle-sized cities and large urban centers has changed the geography of business location. There are often two critical population blocks resident in the middle-sized cities—or willing to move to them for solid employment. First are the semiskilled high school and college graduates or near-graduates who are willing to take beginning work in firms at annual salaries of $20,000–40,000. As is the case so often in migration to the city, they will take such jobs on the hope that they will either advance in a Proctor & Gamble or CitiCorp or put them-

selves in a good position to take advantage of another opportunity in that city. Smaller cities often lack the "pull" to bring in growth firms because they do not have the urban amenities—in music, museums, shopping, and schools—that tend toward the larger urban places. At the same time, these middle-sized cities do not generate the images of crowding, crime, environmental decline, and gridlock that really large cities often do.

The other population block in middle-sized cities is the well-educated young professionals who find strong enough amenities in the middle-sized cities that they will accept a somewhat lower salary as a tradeoff for the urban characteristics of these cities. Professional sports complexes, energetic renewal of traditional downtown landscapes with trendy cafes and restaurants, and expanding airport facilities are often major aspects of this growth. These are often complemented by the willingness of city administrators to give incoming firms a sweet deal on land, taxes, and transportation so that the move to the smaller urban center looks more attractive to the major firms that have grown up generally in much larger cities. This demographic shift is yet another aspect of the changing geography of our settlement patterns in the United States.

The Falls of St. Anthony, marking this point, supplied water power for early industries, especially sawmilling. Then transcontinental rail lines were built west from these "Twin Cities" to the Pacific. The spread of business westward along these lines made Minneapolis–St. Paul the economic capital of much of the northern interior United States. No competing center of comparable size emerged between Minneapolis–St. Paul and Seattle. The business hinterland of the Twin Cities thus stretches across the Spring Wheat Belt to the foot of the Montana Rockies. Minneapolis–St. Paul has long been a major focus of the grain trade and flour milling and now has a diversified industrial structure. It is also the home to the nation's largest retail mall—The Mall of America—which has a regular flow of international tourists who come to the Twin Cities largely for shopping there.

A much older river city is Cincinnati (population: 1.9 million) on the Ohio River. Its metropolitan area includes the southwest corner of Ohio and adjacent parts of Indiana and Kentucky. For a time, its situation resembled Minneapolis–St. Paul in that it dominated a large hinterland to the west and north. The Ohio River turns sharply southward from the city, and early 19th-century settlers floating down the river often disembarked at Cincinnati to move overland into Indiana. The city quickly developed a major specialty in slaughtering hogs from Ohio and Indiana farms and shipping salt pork downriver to markets in the South. This early meatpacking function eventu-

ally was lost, but coal mining developed in the nearby Appalachian Plateau, and Cincinnati became the economic capital for much of the mining region. It also developed a diversified and relatively stable industrial base, built in part around tool and die machine-working for the region's industrial centers.

Kansas City, Missouri (population: 1.7 million in Missouri and Kansas), strongly resembles Minneapolis–St. Paul in its situation and development. It began at what was essentially the head of navigation westward on the Missouri River. Although river transport was possible upstream, it was subject to navigational hazards and led sharply northwest instead of directly westward. Consequently, two of the major overland trails to the West—the Santa Fe Trail and the Oregon Trail—had their main eastern termini and connection with river transport in the Kansas City area. Later on, cattle drives came to Kansas City from the Great Plains, and railways reached westward from the city. The Pony Express played out its one year of history in the early 1860s from St. Joseph, just north of Kansas City. Today, despite a major competitor to the west in Denver, Colorado, Kansas City has a hinterland extending far to the west and is the economic capital of the Winter Wheat Belt. Omaha, Nebraska (metropolitan population: 663,000 in Nebraska and Iowa), is a smaller riverside city resembling Kansas City and Minneapolis–St. Paul in its functions and westward reach. Like them, Omaha is in the western edge of the intensively farmed part of the Midwest, with drier, less productive wheat and cattle country to the west.

24.4 The West: Oasislike Development in a Dry and Dramatic Setting

Eleven states make up the West (see Figs. 24.1, 24.3, and 24.16). The eight interior states (New Mexico, Arizona, Colorado, Utah, Nevada, Wyoming, Montana, and Idaho) are often termed the Mountain states or Mountain West. Farther west are the three Pacific coast states of California, Oregon, and Washington. The latter two states are often termed the Pacific Northwest. Federal publications include Alaska and Hawaii in statistics for the Western states, but these two states are so different that they are discussed here in a separate section and are not included in generalizations about the West.

Thus defined, the West comprises 49 percent of the United States by area, or 38 percent of the conterminous 48 states. It contains large percentages of land owned by the federal government and accommodates major military facilities all across its predominantly arid landscapes. The region is still lightly settled except for a huge cluster of people and development in central and southern California and a handful of smaller clusters in other widely dispersed locations. In 1998, the West had 58.5 million people (about one-fifth of the nation), 32.7 million of whom lived in California. In 1998, it accounted for 22 percent of the nation's agricultural output and about one-sixth of its value added in manufacturing. California is the West's economic colossus, producing almost two-thirds of the region's manufactured goods and one-half of its farm products. The eight Mountain states contrast sharply with California, with 16.8 million people in 1998 (just over 6 percent of the national population) in eight states with large areas. In 1998, these states produced 8 percent of the country's agricultural output and 3 percent of its value added by manufacturing.

Challenges to Development in the West

The generally low intensity of development in the West results principally from dry climates, rough topography, and absence of water transportation except along the Pacific coast. Because settlement is clustered in places where water is adequate, the population pattern is oasislike—with oases separated from each other by little-populated steppe, desert, and/or mountains. The climates that so affect Western settlement are strongly related to topography (see Fig. 24.16). Precipitation in this region is largely orographic, and dryness is largely due to rain shadow. That dryness is caused by being on the wrong side (the leeward side) of north-south trending mountains that block the moisture-bearing air masses that come from the west. This is not only a strong regional characteristic but also a source of continuing threats from wildfires, floods, landslides, and droughts. Thus, an understanding of topographic layout is needed to understand the climatic conditions that either handicap or facilitate settlement. If the mountains of the West ran east and west, everything would be different.

Western Topography and Its Effect on Climate

The Pacific shore is bordered by mountains. The relatively low Coast Ranges provide an often spectacular coastline from the Olympic Mountains in Washington to just north of Los Angeles, where the mountains swing inland and then continue south to Mexico. Inland, a line of truly high mountains parallels the Coast Ranges. The northernmost of these inland ranges are the Cascade Mountains, extending southward from British Columbia across Washington and Oregon to northern California. The Cascades are surmounted by many volcanic cones, among which Mt. Hood (near Portland) reaches over 11,000 feet and Mt. Rainier (near Seattle) over 14,000 feet. At its southern end, the Cascade range joins the rugged Klamath Mountains, a deeply dissected plateau where the Cascades, Coast Ranges, and the Sierra Nevada range almost meet. The Sierra Nevada range increases in elevation southward to the "High Sierra," where Mt. Whitney, at 14,494 feet (4418 m), is the highest peak in the 48 conterminous states.

Two large lowlands lie between the Coast Ranges and the inner line of Pacific mountains, and a smaller but highly important lowland lies outside the Coast Ranges in the extreme south. In Washington and Oregon, the Willamette–Puget Sound Lowland separates the Cascades from the Coast Ranges. The Oregon sector, extending from the lower Columbia River to the Klamath Mountains, is called the Willamette Valley. The Washington sector, north of the Columbia, is the Puget Sound Lowland, which is deeply penetrated by the deepwater inlet from the Pacific called the Puget Sound. The lowland surface in both states is generally hilly. The second large lowland is the Central Valley, or Great Valley, of California, enclosed by the Klamath, Sierra Nevada, and Coast ranges or by outliers that seem to weave the systems together. The Central Valley is a practically flat alluvial plain, some 500 miles (c. 800 km) long by 50 miles wide. In California's far southwest, a smaller lowland lies between the coastal mountains and the shore. The wider northern part of this lowland is the Los Angeles Basin, and a narrower coastal strip extends southward to San Diego and Mexico. The Los Angeles Basin is a plain dotted by isolated high hills and overlooked by mountains.

Precipitation is brought to this terrain by air masses moving eastward from the Pacific Ocean. This air movement occurs more frequently in the winter than in the summer, so summer is much the drier season. It also occurs more frequently and generates more precipitation in the north than in the south. Consequently, summer conditions are desertlike in the south, although winter brings some rain and mountain snow. The mountains of the north receive the most precipitation, including the deepest snows in the United States in the higher areas. South of San Francisco, California, the mountains of most areas receive only modest amounts of precipitation at best, and the magnificent Douglas fir and redwood forests of the north give way to the sparse forests and chaparral (drought-tolerant grasses and shrubs) that characterize the southern half of California.

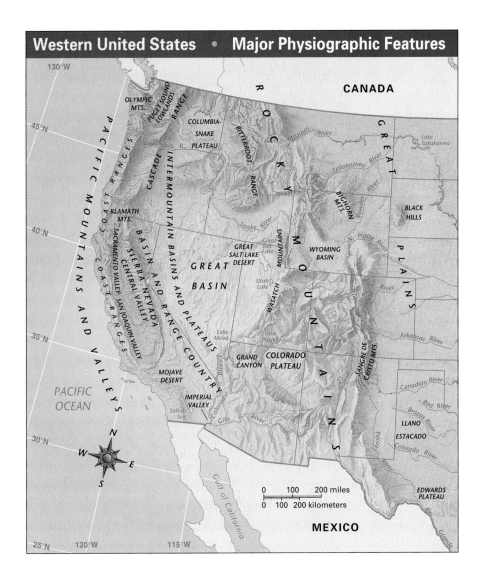

Figure 24.16 Major physiographic features of the western United States.

The lowlands of this Pacific strip vary from quite wet to arid. The Coast Ranges of Washington and Oregon are mostly low enough, and the incoming moisture is great enough, that the Willamette–Puget Sound Lowland is not severely affected by its position on the lee side of the mountains. Hence, this lowland is wet and naturally forested. Because its temperatures are moderated in both winter and summer by incoming marine air, this area has a marine west coast climate similar to that of western Europe. The Central Valley of California is affected more by rain shadow cast by the Coast Ranges. Although generally classed as having a mediterranean climate—with subtropical temperatures, wet winters, and dry summers—this valley is dry enough that its natural vegetation was steppe grass for the most part, grading into desert shrub in the extreme south. The Los Angeles Basin also has a climate usually classified as mediterranean subtropical, but the precipitation is so low that a steppe or even a desert classification could be argued.

But it is east of the high mountains that rain shadow becomes extreme. A vast area reaching east to the Rockies and around their southern end in New Mexico is predominantly desert or steppe. Little moisture is received from air masses that already have provided precipitation to the mountains near the Pacific. This region is known physiographically as the Intermountain Basins and Plateaus. It includes large parts or all (in the case of Arizona) of nine Western states (see Figs. 24.1 and 24.3).

Three major physical subdivisions of the Intermountain Basins and Plateaus are commonly recognized. The Columbia–Snake River Plateau occupies most of eastern Washington and Oregon and southern Idaho. It is characterized by areas of relatively level terrain formed by past massive lava flows, together with scattered hills and mountains and deep-cut river canyons. The Colorado Plateau occupies northern Arizona, northwestern New Mexico, most of eastern Utah, and western Colorado. Here, rolling uplands lie at varying levels, cut into isolated sections by the deep river canyons of the Colorado and its tributaries. The rest of the Intermountain area is the Basin and Range Country, which continues beyond New Mexico into western

Texas. The part between the Wasatch Range of the Rockies and the Sierra Nevada in California has no river reaching the sea and is called the Great Basin. But the whole Basin and Range Country consists of small basins separated from each other by blocklike mountain ranges. This is the driest part of the West.

East of the Intermountain Basins and Plateaus, the high and rugged Rocky Mountains extend southward out of Canada through parts of Idaho, Montana, Wyoming, Utah, and Colorado, where a number of peaks rise above 14,000 feet (4267 m). The Rockies are comprised of many separate ranges with individual names. Valleys and basins separate these ranges. The continuity of the system is almost broken in the Wyoming Basin, which provides the easiest natural passageway through the Rockies and was the route through which the first transcontinental railroad—the Union Pacific—passed. The Rockies also are a formidable barrier to moisture-bearing air masses from either west or east. Hence, they create rain shadows for both the Great Plains at their eastern foot and the Intermountain area at their western foot. The mountains themselves receive much orographic precipitation and are forested. The highest elevations extend above the tree line to mountain tundra and snow fields.

The higher western section of the Great Plains (part of the Interior Plains) borders the Rockies in Montana, Wyoming, Colorado, and New Mexico. So elevated are these plains that Denver, at their western edge, is famed as the Mile High City. The most common landscape is that of an irregular surface interrupted at intervals by conspicuous valleys. Variations include hilly areas, smooth plains, and isolated mountains, as in the Black Hills of Wyoming and South Dakota. The plains are covered by steppe grassland, marked by semiaridity and temperatures that are continental in the north and subtropical in the south.

Water, Agriculture, and Population

Most land east of the coastal states in the West is used for grazing, if at all, and the carrying capacity is so small that only extremely low densities of animals and ranch population are possible. Unirrigated farming of some steppe areas supports only slightly higher population densities. The most important of the latter areas are the wheat-growing areas of the Great Plains and the Columbia Plateau (known also as the Palouse) of Washington. There, yields are relatively low, farms are large and highly mechanized, and population is sparse. Much the greater part of Western agricultural output comes from a few large irrigated oases that grow both food crops and feeds—particularly alfalfa—for dairy and beef cattle kept on feedlots in the oases, or for sale as supplementary feed to ranchers outside the oases. Some oases in California and Arizona grow cotton, fruits, vegetables, and sugar beets, and they are highly dependent on irrigation (Fig. 24.17).

The West is strongly metropolitan in population. Most of these urban Westerners are crowded into big clusters located in

Figure 24.17 Sprinkler irrigation in the flat, rich Imperial Valley. Lettuce and other vegetables, cotton, alfalfa, hay, and feeder cattle are important products of the valley. *Steve Liss / Time Magazine*

or near major oases. The Pacific Northwest cluster centering on Seattle and Portland might be considered an exception, but even this rainfed area is a kind of oasis within the larger area of Western drylands. These major urban clusters have very diversified functions, among which the servicing of agriculture is generally the oldest and continues to be important.

Southern California: The Evolution of a Pacific Megalopolis

The Southern California population cluster is by far the West's largest. The name Southern California is commonly used for the Los Angeles Basin and the narrow lowlands north to Santa Barbara and south to San Diego, along with their interior frame of mountains (Fig. 24.18). The Indian villagers who originally inhabited this area practiced no agriculture but subsisted on grass seeds, acorns, other wild foods, and some coastal fishing. In the 18th century, small-scale irrigated agriculture and cattle ranching were introduced by Spanish settlers from Mexico. By the 20th century the Los Angeles area had become a large and important agricultural district using Colorado River water (see Figs. 24.19 and Problem Landscape, p. 621). Today, agriculture has been largely crowded out of the area by a sprawling urbanization that in 1994 comprised an estimated 18 million people in the combined metropolitan areas of Los Angeles (population: 15.4 million) and San Diego (population: 2.6 million). Packed into this small corner of the West are about three-fifths of California's people, one-third of the population of the entire West, and one of every 15 people in the nation.

San Diego and Los Angeles were founded as remote footholds on the northern frontier of Spain's empire in the Americas. Catholic priests founded missions in coastal loca-

Figure 24.18   A relatively small but representative sector of the immense Los Angeles metropolis. Homes of the wealthy movie community perch on the chaparral-clad slopes below the "Hollywood" sign. Pushing residential pads up into the heights of the Santa Monica Mountains continues to present the city with problems in fire control, water supply, and even flood control. However, the combined status of living high in the hills and the views of such locations (especially at night) makes this a very distinctive and popular landscape in Los Angeles.   *Will & Deni McIntyre / Tony Stone Images*

Figure 24.19   A view of desert and a major irrigation canal in Arizona's Basin and Range country near Phoenix. The canal brings water from the Colorado River to central and southern Arizona; it is one link in the system of canals, pipelines, tunnels, and pumping stations called the Central Arizona Project. A segment of this system delivers Colorado River water to Southern California as well. Note the natural spacing of the xerophytic vegetation to take advantage of all available moisture.   *U.S. Bureau of Reclamation*

tions as far north as San Francisco, including those in San Diego (founded in 1769) and San Gabriel (founded in 1781) in the Los Angeles area. The settlement at San Diego was begun on a magnificent natural harbor that is now the main continental base for the United States Navy's Pacific Fleet. The San Gabriel Mission and the Pueblo (civil settlement) of Los Angeles were sited near the small Los Angeles River. When Mexico gained independence from Spain in the early 19th century, the California settlements became part of the new country. They subsequently came into the possession of the United States in 1848 at the end of the Mexican-American War. Until the late 1800s, San Diego and Los Angeles were little more than hamlets in one of the world's more isolated corners.

A nearly continuous Southern California boom that has lasted to the present began with the arrival of railroad connections to the eastern United States in the 1870s and 1880s. The main lines (the Southern Pacific and the Santa Fe) crossed the area's encircling mountains via the Cajon Pass near San Bernardino. To promote settlement that would create a demand for rail traffic, the railroads offered extremely low fares westward. Oranges shipped in refrigerated cars were the main local product that could stand the transportation cost to far-off eastern markets, and irrigated orchards spread rapidly. Los Angeles was the main transport and business center. In the 1890s, a freeze in Florida put California in the lead in U.S. citrus production, and the state maintained this position until urban sprawl replaced many of the Southern California groves in the mid–20th century. (The Disneyland amusement park was constructed in a

huge orange grove in 1955.) The climate that favored citrus also attracted a new population seeking health and comfort.

In fact, the Southern California climate, water from streams descending the mountains, and other natural resources have continued to support development and attract population. Oil was discovered in the Los Angeles Basin in the 1890s, and the region is still an important producer and refiner. Fishing fleets in the Pacific (primarily tuna boats) soon became large and wide-ranging. In the early 1900s, pioneering moviemakers moved from the Northeast to Los Angeles to take advantage of bright, warm weather and varied outdoor locales. The success of the movie industry later led to production of programs for national television when that medium burgeoned after 1950.

In the 1920s and 1930s, aircraft manufacturing was attracted by the favorable climate, the long hours of sunlight year-round, the amount of space still available for factories and runways, and the presence of a considerable labor force. This industry boomed during World War II and afterwards joined with the electronics industry in aerospace development. Prior to the 1990s the aerospace industry had been Southern California's largest manufacturer, heavily supported by federal military expenditures. Thus, beneath the glittery film-industry reputation, Los Angeles had actually become a major industrial city—an aerospace Detroit with palm trees. In the early 1990s, however, its economy (and that of California generally) slumped, partly as a result of declining federal defense outlays, and—for the first time in decades—California saw more people leave the state than enter it. But, by 1995, the patterns had reversed, and, with the economy growing again—in part from the enormous construction required in recovery from the massive Northridge earthquake in January 1994, and in part from a business shift from defense to industrial operations—population also began to grow through migration. Tourism and a growing regional financial sector supported by the booming Pacific trade have also contributed significantly to the resurging economic dominance of Southern California in this region.

All this would have been impossible without the supply of vast and increasing amounts of water to this dry area. Streams from the surrounding mountains and wells tapping local groundwater resources became inadequate by late in the 19th century, and water is now brought by aqueduct from the Sierra Nevada and the Colorado River (see Fig. 24.19), both of which are hundreds of miles away. Schemes have been proposed to acquire water from still more distant areas as far north as Canada. Economical desalinization of sea water would be one solution if technology could provide a low-cost process.

The San Francisco Metropolis

In 1995, 6.5 million Californians lived in the San Francisco metropolis, which is the fifth largest in the United States. San Francisco is located on a hilly peninsula between the Pacific Ocean and the San Francisco Bay, but the metropolitan area includes counties on all sides of the Bay. The climate is unusual. Marine conditions are evidenced in San Francisco's very moderate temperatures, which range from an average of 48°F (9°C) in December to only 64°F (18°C) in September. However, the area is in the border zone between the marine west coast and mediterranean climates, with the mediterranean influence evident in the fact that about 80 percent of the area's rainfall comes in five wet months from November through March. Mean annual precipitation is only 20 inches in San Francisco, but the moderate temperatures and low rates of evaporation have enabled this amount of moisture to produce a naturally forested landscape. There are even groves of redwoods within a short distance both north and south of the Golden Gate strait (Fig. 24.20). East of San Francisco Bay, somewhat warmer temperatures and lower precipitation have produced a scrubby forest (chaparral or dispersed California oaks) or grassland.

San Francisco began as a Spanish fort (presidio) and a mission on the south side of the Golden Gate strait and subsequently was developed, first as a port for the gold-mining industry and then for the expanding agricultural economy of central California. The Gold Rush of 1849 and subsequent

Figure 24.20 The California Redwoods *(Sequoia sempervirens)* are one of the most evocative images of forest in the West. The Avenue of the Giants just north of the San Francisco Bay area in California helps provide a visual cue to the size of these trees. Because the wood is so valuable as a building material—valued because of both color and its durability for deck construction—this tree has been a focus of tensions between environmentalists and logging interests for decades.
S. Bisceglie/Animals Animals/Earth Scenes

years centered on the gold-bearing alluvial gravels of Sierra Nevada streams entering the Central Valley north and south of the current city of Sacramento. San Francisco, with its spacious natural harbor, was the nearest port to receive immigrants and supplies for the gold camps. Today San Francisco is a major domestic and international tourist destination for its scenery, wines, and arts and other cultural attractions.

San Francisco has added a sizable industrial sector to its original functions as a port and regional economic center. Manufacturing in the metropolitan area is highly varied, but in recent years computer-related and other electronics businesses in a strip of land near the southwestern shore of San Francisco Bay have become the outstanding industry. This area is now commonly referred to as Silicon Valley, after the chips that carry microcircuits. The rise of Silicon Valley as a center of high technology is due in part to the attractiveness of the San Francisco area as a place to live and the nearby presence of two outstanding universities—the University of California at Berkeley near the industrial center of Oakland on the east side of the Bay, and Stanford University in nearby Palo Alto. Software firms have moved to a number of other regions for development, but Silicon Valley continues to be the dominant center for such innovation, development, and venture capital.

Seattle and Portland and Their Pacific Northwest Setting

The Puget Sound Lowland of Washington and the Willamette Valley of Oregon to its south are not irrigated oases. They have abundant precipitation and luxuriant natural forests, as do the adjoining Coast Ranges and Cascades. But these humid conditions end abruptly near the crest of the Cascades. Hence, the area is oasislike—a sharply limited area of abundant water and clustered population within the generally dry West. This population is centered in two port cities, each located on a navigable channel extending inland through the Coast Ranges. Seattle (population: 3.2 million), Washington, located on the eastern shore of the Puget Sound, has access to the interior via passes over the Cascades. Even with its pattern of rainfall, the Seattle setting is home to the Boeing Aircraft Company—the world's largest aircraft producer. The local early abundance of the Sitka spruce (used for early wooden plane construction) and the availability of cheap electricity led to the 1917 founding of the firm by William Boeing. Portland (population: 2 million), Oregon, is on the navigable lower portion of the Columbia River where the Willamette tributary enters it, and has a natural route to the interior via the Columbia River Gorge through the Cascades. Outside these metropolises, the Lowland contains somewhat over a million other people.

Fishing, logging, and trade built the cities of the Lowland in the latter 19th and early 20th centuries. But it was agricultural land in the Willamette Valley that drew the first large contingents of settlers to "Oregon Country" in the 1840s. Agriculture is still of some importance, although it is handicapped by hilly terrain and leached soils. The leading products are milk and truck crops, and a local Oregon wine industry is expanding in importance. Fishing, primarily for salmon, is still carried on, but the Pacific Northwest fish catch is now much smaller than that of any one of the three leading states in this industry—Alaska, Massachusetts, and Louisiana (as of 1993). However, Oregon and Washington are still the leading states in production of lumber, and they still have one quarter of the sawtimber resources of the United States. The trees are mostly coniferous softwoods, and many are enormous in size, especially the famed Douglas fir. The timber industry, now under heavy attack by environmentalists, is still basic to the economies of both states.

The trading hinterlands of Seattle and Portland extend across the Cascades and the Columbia–Snake River Plateau to the Rocky Mountains. Along or near the eastern slope of the Cascades are scattered irrigated areas, including the Wenatchee and Yakima valleys, that make Washington the leading state in apple production. Near the eastern foot of the Cascades in Washington, the mighty Columbia River flows southward. Beginning in the 1930s, it was harnessed by a series of federal dams (Fig. 24.21), of which the largest is Grand Coulee Dam. These have supplied power to an area lacking in coal and oil, have improved navigation on the lower Columbia, and have attracted a major concentration of power-hungry aluminum

Figure 24.21   The Dalles Dam on the Columbia River permits salmon to migrate via the fish ladder in the foreground. Note the power-transmission lines in the distance. The dam, located on the border between Oregon and Washington, is one of 56 major dams in the Columbia River watershed.   *Robert Harbison / Christian Science Monitor*

plants to the Pacific Northwest. The dams also have supplied the water for a huge expansion of irrigation in the Columbia Plateau of both Washington and Oregon. However, they have also helped kill the area's significant salmon fishing industry.

In southeastern Washington, the Columbia Plateau rises high enough to have a steppe rather than a desert climate and is mantled by a deep hilly covering of fertile and water-retentive loess. This area, called the Palouse, grows unirrigated wheat, despite an annual precipitation averaging as low as 9 inches (23 cm) in some places. At the northeastern edge of the Palouse, Spokane (population: 386,000) serves as the local capital of the "Inland Empire" in the northern Columbia Plateau and adjacent Rockies.

Another major industry that has been sited in the Pacific Northwest because of owner environmental preferences is the Microsoft Corporation of Redmond, Washington. The firm was initially founded in Albuquerque, New Mexico, in 1975, but by 1979 the two founders—Bill Gates and Paul Allen—moved the headquarters to the Seattle area. It has since grown to be the largest computer software company in the world, and is now a source of major payroll and export revenues for the Northwest and the United States.

Population Clusters in the Mountain States

The eight inland states in the Mountain West are drier than those in the Pacific Mountains and Valleys, lack the abundant irrigation water of California, and lack the advantage of ocean transportation. These handicaps have been so constricting that the eight states had only 17 million people in 1998. About half of these are in a handful of metropolitan areas in irrigated districts. Outside of them, the common pattern in a Mountain West state is that of a few hundred thousand people scattered across a huge area—on ranches and farms, on Indian reservations, or in small service centers and mining towns.

Phoenix and Tucson. The largest population cluster in the Mountain West lies in the Basin and Range Country of southern Arizona. Here live about three quarters of Arizona's people, in the adjacent metropolitan areas of Phoenix (population: 2.5 million) and Tucson (population: 731,523). Water from the Salt and Colorado rivers (see Fig. 24.19), plus groundwater, makes it possible to support this development in a subtropical desert.

Phoenix was originally the business center for an irrigated district along the Salt River. The district expanded after 1912 when the first federal irrigation dam in the West (Roosevelt Dam) was completed on the river. Today, cotton is the area's main crop, although hay to support dairy and beef cattle is important. Growth in both Phoenix and Tucson accelerated after the introduction of air-conditioning made them attractive retirement and business centers. Each is the site of a major state university, and Phoenix is the state capital. Tourism to both urban and nonurban destinations, the increasing income level of Sunbelt retirees, and copper mining are also basic to Arizona's economy.

The Colorado Piedmont. Another important population cluster stretches along the Great Plains at the eastern foot of the Rocky Mountains. Within this area, called the Colorado Piedmont, metropolitan Denver contained 2.2 million people in 1998, and more than 800,000 lived in metropolitan Colorado Springs (452,000, city proper) or smaller cities. These cities developed on streams emerging from the Rockies. The stream valleys provided routes into the mountains, where a series of mining booms began with a gold rush in 1859. One mineral rush followed another through the later 19th century, with various minerals being exploited at scattered sites. The Colorado Piedmont towns became supply and service bases for the mining communities, which often were short-lived. Irrigable land along Piedmont rivers afforded a farming base to provide food for the mines, and agricultural servicing became an important function of the towns. In time, agriculture grew into a prosperous business with more stability than mining.

During the 20th century, the Piedmont centers have grown rapidly. Irrigated acreage has been expanded greatly, fostering a boom in irrigated feed to support some of the nation's largest cattle feedlots. The Colorado Rockies have also become an important producer of molybdenum, vanadium, and tungsten—comparatively rare alloys that are much in demand for high-technology uses. High energy prices in the 1970s led to expanded extraction of fuels from the state's huge coal reserves and scattered oil and gas deposits. And tourism in the Rockies has profited greatly from increasing American affluence, better transportation, and the rise in the popularity of skiing. Finally, Denver also has become a major center for the western regional offices of the federal government, with Colorado Springs the site of the United States Air Force Academy and the nearest large city to the headquarters of the North American Air Defense Command, located in a redoubt inside a mountain.

The Wasatch Front (Salt Lake) Oasis. Third in size among the population clusters of the interior West is the Wasatch Front, or Salt Lake, Oasis of Utah, composed of metropolitan Salt Lake City–Ogden (population: 1.2 million) and Provo (population: 291,000) to its south. It contains about three-fourths of Utah's people, in a north-south strip between the Wasatch Range of the Rockies and the Great Salt Lake, with a southward extension along the foot of the Wasatch. In the rest of Utah, there are only about 400,000 people. The oasis was developed as a haven for people of the Mormon faith when they were expelled from the eastern United States in the 1840s. After a difficult trek westward, they began irrigation development based on water from the Wasatch Range, and they built at Salt Lake City the temple that is the symbolic center of the Mormon religion and culture. From this center, Mormon settlers have spread widely into all the surrounding states and California. Thus, Salt Lake City is a regional religious and cultural capital; in addition, with no large competing metropolis within hundreds of miles, the city is the business capital for a huge although sparsely populated area outside the Mormon core region at the foot of the Wasatch Mountains.

The Less Populous States of the Interior West. The other five states of the interior West are even less populous than Utah. Their aggregate population of about 6 million, in states that are very large in area, is less than that of the San Francisco metropolis. The people of New Mexico are largely strung along the north-south valley of the middle and upper Rio Grande, where Albuquerque (population: 646,000) is the main city, although it competes economically with El Paso, Texas, which is located on the Rio Grande just south of New Mexico. New Mexico's economy is a typical Western mix of irrigated agriculture, ranching, mining, tourism, some forestry in the mountains, and federally funded defense contracting. Similarly, most of the population of Idaho is in oases strung along a river—in this case, the Snake River, as it crosses the volcanic Snake River Plains in the southern part of the state. Agricultural output is more substantial in Idaho, nationally known for its potatoes. But tourism and federal money yield less income than they do in New Mexico, and the largest city of the state, Boise, has only about 150,000 people in the city proper.

Except for a small section that lies in the Sierra Nevada, the state of Nevada is desert, with very little irrigation. Metropolitan Las Vegas (population: 1.1 million) and Reno (population: 283,000) have more than four-fifths of the state's population, and the state's laws allow gambling and related

PROBLEM   LANDSCAPE

Buying a Gulp of the Colorado

"THE BASIN OF THE COLORADO RIVER cuts a huge, wriggling swathe across the western United States. It covers 244,000 square miles [632,000 sq km] and provides water to 25 million people, 16 million in southern California alone. Who gets this water, and why, obsesses the West. So much importance is attached to the distribution of the river's annual surplus of water that the decision is made by the secretary of the interior, no less."[2]

Water for the western United States is a most critical issue. From the crests of the various coastal mountain ranges that course from Southern California through the state of Washington to the lands east of the Rocky Mountains, there is general water deficiency. For millennia after the initial arrival of the first human migrants from northeast Asia, water needs were met by keeping settlement small. For centuries after the arrival of Columbus in North America, water was a problem in the West, but it was generally seen as a local concern and, again, settlement size in the most arid areas seldom grew beyond the capacity for landscapes to support local hamlets and towns. But, with the arrival of the 20th century, the completion of the transcontinental railway, and the active promotion of real estate development in Southern California especially, water needs grew dominant in the whole equation of American demographic shifts to the southwest of the country. What has been a nagging problem for decades may now be growing to a problem of much larger dimension.

There are two significant "exotic" water systems in the American West. The largest is the flow of the Colorado River primarily from the mountain peaks in north central Colorado. The river flows through the Grand Canyon, piles up behind Hoover Dam, and—from that dam—supplies a major portion of the water needs of three lower-basin states: California, Arizona, and Nevada. Lake Mead, which is the largest reservoir in the United States, also generates a major portion of the electricity of the region. This water has been a source of litigation and federal and state argument virtually since the dam was completed in the mid-1930s. Because of the fast growth of Arizona in the past decade, that state is pursuing its share of the Colorado River water with more energy than it has before. The lion's share of the Colorado River water has gone to Southern California, where it is critical to both Los Angeles and San Diego and the urban sprawl between the two. The second "exotic" system is the 223-mile gravity flow canal that brings water from the Sierra Nevada snowmelt through the Owens Valley east and north of Los Angeles down to the enormous metropolitan region.

Both of these systems are now in conflict. The Owens Valley has developed such dust storms for the population resident in the high valley (which sold away its water rights piece by piece a century ago) that the California Air Resources Board may require Los Angeles to diminish the amount of water it removes from the area annually. This would enable the valley to expand the area of the water on the Owens lake bed (Mono Lake) so that there would be less potential for dust disturbance. Concerning the Colorado, there is increasing tension for Southern California because both Arizona and Nevada—which have historically drawn off less water from the Colorado flow than they were allowed—are both seeking their full share because of urban growth. Las Vegas is the fastest growing large city in the United States, and Arizona continues to grow beyond its water resources as well. There is a complex "water banking" system being devised that will enable Nevada and Arizona to come closer to the goals they have, but each extra amount they are able to siphon off heightens Southern California's anxiety about its own water budget (Fig. 24.A).

There are few geographic features that have stimulated tension, fights, and even warfare the way water rights have. The problem landscape that spreads across a half a dozen states in the American West now has parallels—at a smaller scale—that run north and south, east and west across the country. Water is at the heart of settlement in most regions, and its management and availability are often at the very pinnacle of success—or disaster—in human settlement.

Figure 24.A   The Colorado River Basin.

[2]"Buying a Gulp of the Colorado," *Economist*, January 24, 1998: 28.

Figure 24.22   The fastest growing city in America for the past decade has been Las Vegas, Nevada. A settlement literally extracted from the Nevada desert through the use of exotic water and the economic decision to offer uncommon entertainments, the city has grown as a bold and sometimes overwhelming testimony to the human capacity to remake the surface of the earth.   *Breck P. Kent/Animals Animals/Earth Scenes*

tourism to overshadow the usual Western mix of economic activities. Casinos and associated entertainment and convention business play a dominant role in the state's wealth. The current pace of population growth in the state is also a testimony, again, to the importance of air-conditioning (Fig. 24.22).

In Montana and Wyoming, there is little irrigation and no warm-winter climate to encourage year-round tourism or to attract retirees. Economies based on farming, ranching, mining, and summer tourism support extremely sparse populations.

24.5 Alaska and Hawaii: Exotic Outposts

In the middle of the 19th century, the United States began to acquire Pacific dependencies. Two of these, Alaska and Hawaii, were admitted to the Union as states in 1959. The United States acquired Alaska by purchase from Russia in 1867, and Hawaii was annexed in 1898. Acquisition of a Pacific empire was motivated partly by defense considerations, and these have continued to play an important role in both Alaska and Hawaii. Large military installations exist in both states, and defense expenditures are a major element in the economies of these states. In terms of geography, these two most recent states are about as different from each other as is possible, and in many ways they stand apart from the other 48 as well.

Alaska's Difficult Environment

Alaska (from Alyeska, an Aleut word meaning "the great land") makes up about one-sixth of the United States by area, but is so rugged, cold, and remote that its total population was only 614,000 in 1998. The state's difficult environment can be conveniently assessed in four major physical areas (Fig. 24.23):

1. Alaska's Arctic Coastal Plain is an area of tundra along the Arctic Ocean north of the Brooks Range. Known as "the North Slope," it contains very large deposits of oil and natural gas.
2. The barren Brooks Range forms the northwestern end of the Rocky Mountains. Summit elevations range from about 4000 to 9000 feet (*c.* 1200–2800 m).
3. The Yukon River Basin lies between the Brooks Range on the north and the Alaska Range on the south. This is the Alaskan portion of the Intermountain Basins and Plateaus. The Yukon is a major river, navigable by riverboats for 1700 miles (*c.* 2700 km) from the Bering Sea to Whitehorse in Canada's Yukon Territory. In Alaska, the Basin is generally rolling or hilly, with some sizable areas of flat alluvium that often are swampy. Its subarctic climate is associated with coniferous forest that generally is thin and composed of relatively small trees. Fairbanks (population: 31,000, city proper) is interior Alaska's main town. The Alaska Highway connects it with the "Lower 48" (states) via western Canada, and both paved highways and the Alaska Railroad lead southward to Anchorage and other ports on the Gulf of Alaska. Fairbanks is a major service center for the Trans-Alaska Pipeline from the North Slope oil area to the port of Valdez in the south. The city is located on the Tanana River tributary of the Yukon and is associated with a small agricultural area in the Tanana Valley.
4. The Pacific Mountains and Valleys occupy southern Alaska, including the southeastern panhandle. The mainland ranges of the panhandle are an extension of the Cascade Range and the British Columbia Coastal Ranges, and the mountainous offshore islands are an extension of the Coast Ranges of the Pacific Northwest and the islands of British Columbia. The Inside Passage from Vancouver and Seattle lies between the islands and the mainland. On the mainland, some mountain peaks exceed 15,000 feet (4572 m). Fjorded valleys extend long arms of the sea inland. A very cool and wet marine west coast climate prevails, fostering coniferous forests that are a major resource, exploited primarily for the Japanese market. From Juneau (population: 27,000, city proper) northwestward, a majestic display of glaciers defines a long stretch of coast (Fig. 24.24). However, the highest mountains of Alaska do not lie along the coast but are well inland in the Alaska Range, where Mt. McKinley (Denali) rises to 20,320 feet (6194 m) about 100 miles north of Anchorage. This is the highest peak in North America. Anchorage is the state's largest city (population: 258,000) and has more than two-fifths of the state's total population. At the southwest, the

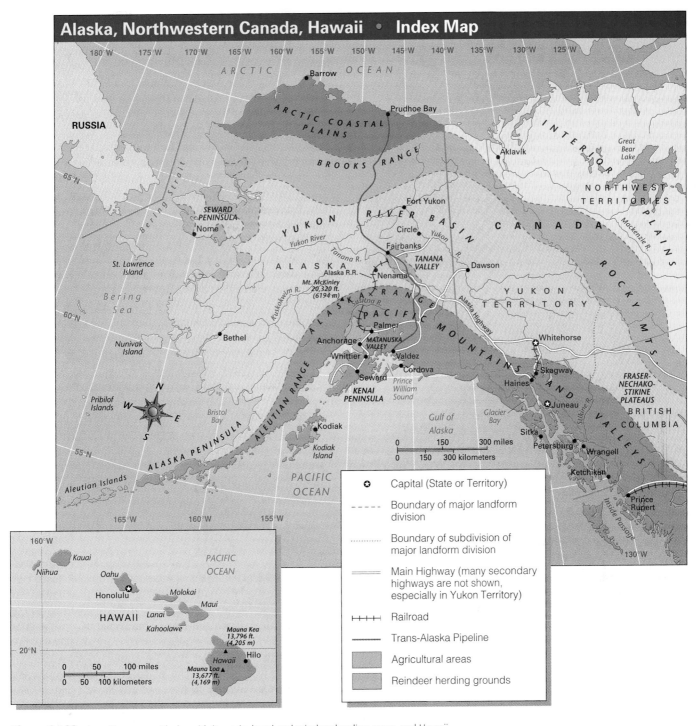

Figure 24.23   Location maps: Alaska with its agricultural and reindeer herding areas, and Hawaii.

Alaska Range merges into the lower mountains occupying the long and narrow Alaska Peninsula and the rough, almost treeless Aleutian Islands.

Immediately south of Anchorage, the Kenai Peninsula lies between Cook Inlet and the Gulf of Alaska, and still farther to the southwest is Alaska's largest island, Kodiak. These features mark the western end of the horseshoe of coastal mountains that curves across the north of the Gulf of Alaska from the southeastern panhandle. The climate of lower areas in this western coastal section is harsher than in the panhandle and is classified as relatively mild subarctic.

Figure 24.24 The spectacular river of ice is the Margerie Glacier at Glacier Bay, Alaska. In the foreground is the cruise ship Noordam of the Holland America Line. Such dramatic landscapes as this one have made Alaska grow in popularity as a tourist destination, bringing increasingly significant tourist dollars to the state. *Holland America Line*

Part of the geographic drama of Alaska comes from the fact that it lies within the chain of mountains that comes up from Tierra del Fuego and the Andean Cordillera in South America, through varied mountain chains in Middle and North America, and finally arcs to the west southwest through the Aleutian Islands of Alaska. This chain is called the **Ring of Fire** and it continues to the south and west as it courses through Japan, Taiwan, the Philippines, additional islands in the southwest Pacific and finally terminates in the North Island of New Zealand. This long arc got its name from the volcanoes which are caused by the intersection of massive plates that sit atop the mantle of the Earth's core. Volcanic eruptions at Mount St. Helens in Washington in 1986 and Mt. Pinatubo in the Philippines in 1990, and the massive earthquakes in 1964 in Anchorage and in 1995 in Kobe (Japan), all represent recent expressions of the hazards of draping human settlements across the landscape of the Ring of Fire (Fig. 24.25).

From Fur Trade to Oil Age

After approximately two centuries of occupation by Russia and the United States, the population of Alaska was still only 73,000 in 1940. Eighteenth-century Russian fur traders were spread thinly along the southern and southeastern coasts. Evidence of this era persists in the many Russian place names. In 1812, the Russians established an agricultural colony on the California coast at Fort Ross north of San Francisco Bay. This effort to more adequately provision the Alaskan posts was abandoned in 1841, and then, in 1867, remote and unprofitable Alaska was sold to the United States. By that time, hunting had

nearly wiped out the sea otters, whose expensive pelts had provided the main incentive for Russia's Alaskan fur trade. Only recently under American and Canadian governmental protection has a substantial beginning been made at replenishing these mammals. Under the Americans, there was a burst of settlement connected with gold rushes in the 1890s and for a few years thereafter, followed by a very slow increase connected largely with fishing and military installations.

Beginning with World War II, during which Japanese troops invaded the Aleutian Islands, growth became more rapid. Today, metropolitan Anchorage at the head of Cook Inlet is the main population cluster. The rest of Alaska's people are erratically distributed over the state but living mainly in or near Fairbanks, in the state capital of Juneau in the panhandle, or in other small and widely spaced towns and villages. Of the total population, about one-sixth is composed of "Alaska Natives" (Inuit, or Eskimos; American Indians; or Aleuts). Long the victims of exploitation and neglect by whites, these peoples have recently received some redress in large land grants and cash payments under the Alaska Native Claims Settlement Act passed by Congress in 1971.

The burst of growth since 1940 has been principally due to (1) increased military and governmental employment, (2) the rise of the fishing industry to major importance, with salmon by far the main product, (3) revenues accruing from the development of Arctic oil (discovered in 1965), (4) the rise of air transport as a major carrier of both people and freight, and (5) visits by increasing numbers of tourists, both domestic and foreign. Agriculture consists largely of high-cost dairy farming in a few restricted spots. Both food and other consumer goods come overwhelmingly from the "Lower 48" states.

Hawaii: Americanization in a South Seas Setting

The state of Hawaii consists of eight main tropical islands (see Fig. 24.23) lying just south of the Tropic of Cancer and somewhat over 2000 miles (c. 3200 km) from California. These islands are all volcanic and largely mountainous, with a combined land area of about 6400 square miles (c. 17,000 sq km), which is somewhat greater than the combined areas of Connecticut and Rhode Island. The island of Hawaii is by far the largest and the only one to have active volcanic craters. The prevalent trade winds and mountains produce extreme precipitation contrasts within short distances. Windward slopes receive heavy precipitation throughout the year (rising to an annual average of 460 inches [1168 cm] on Mt. Waialeale on the island of Kauai—the wettest spot on the globe), while nearby leeward and low-lying areas may receive as little as 10 to 20 inches (25–50 cm) annually. Some areas are so dry that growing crops requires irrigation. Temperatures are tropical except for colder spots on some mountaintops. Honolulu averages 81°F (27°C) in August and 73°F (23°C) in January; the average annual precipitation in the city is 23 inches (58 cm), with a rel-

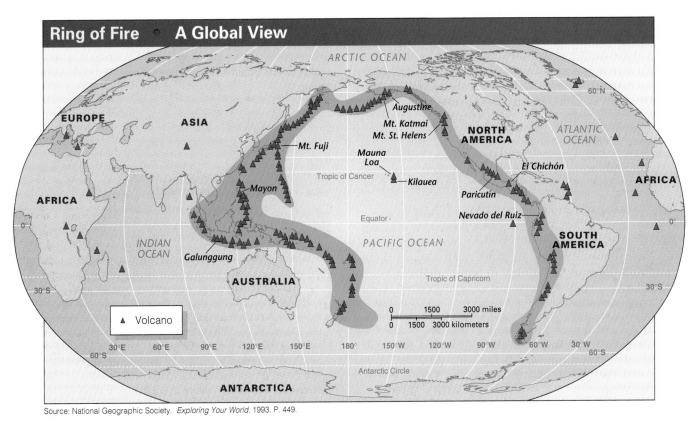

Source: National Geographic Society. *Exploring Your World*. 1993. P. 449.

Figure 25.25 More than half of Earth's active volcanoes lie along this arc of mountains from Tierra del Fuego in South America to the islands of New Zealand. The arc is known as the Ring of Fire.

atively dry high-sun period ("summer") and a wetter low-sun period ("winter").

Hawaii was more populous than Alaska and was the site of a Polynesian monarchy when it was acquired by the United States. By 1997 its population was 1.2 million, of whom about 875,000 were concentrated in metropolitan Honolulu. The metropolis occupies the island of Oahu, with the city proper located near the southeastern corner of the island. Immigration from the Pacific Basin has given the state's population a unique mix by ethnic origin: one-third white Caucasian, one quarter Japanese, somewhat under one-fifth native Polynesian, 14 percent Filipino, 6 percent Chinese, and a small contingent of Koreans and Samoans. Although there has been ongoing tension between native islanders and the immigrants who come from the U.S. mainland and East Asia, much racial and ethnic mixing has taken place. In general, the population is thoroughly Americanized.

Hawaii's economy is based overwhelmingly on military expenditures, tourism, and commercial agriculture. The main defense installations, including the Pearl Harbor naval base, are on Oahu. The main agricultural products and exports—cane sugar and pineapples—have long been grown on large estates owned and administered by Caucasian families and corporate interests but worked mainly by non-Caucasians. Work on such estates has often been the first employment of new immigrants. However, changing demographic patterns, land values, and the rapidly declining role of agriculture in Hawaii have all led to steadily changing landscapes in the islands. Agriculture now amounts to only about 1 percent of the annual gross state product. Sugar has lost importance, while tourism has been the great growth industry of recent decades, propelled by the development of air transportation and the growing affluence of mainland America and East Asia. Visitors from Japan and other Pacific rim countries have become more common with the rising level of economic development in the past 15 years. The main attractions for tourists are the tropical warmth, beaches, spectacular scenery, exotic cultural mix, and highly developed resorts. However, some of the "South Seas" glamour of the islands has been eroded by the remorseless spread of commercial and residential development.

A factor that has changed the role of tourism in the state economy is the major slowdown in the economic growth of East Asia since 1997. (Japan, especially, has been a traditional source of a wealthy tourist population, for which Hawaii was a major tourist destination.) A broad-scale decline in tourist numbers and subsequent business activity took place in 1998, when Hawaii's unemployment rate also climbed to 33 percent higher than the national average. Major hotel chains and retailers closed branches or downsized staff and services. It has become apparent that Hawaii's economic vitality is more closely tied to business health in Japan and East Asia than was beneficial. It is assumed that the state will not realize a turnaround in its own recent declines until East Asia gets over its own Asian economic "flu."

SUMMARY WITH SELECTED KEY TERMS

- This chapter divides the **United States into four regions.** The first is the **Northeast,** comprised of **Maine, New Hampshire, Vermont, Massachusetts, Connecticut, Rhode Island, New York, New Jersey, Pennsylvania, Delaware, Maryland, and the District of Columbia.** In this region live about **20 percent of the nation's** population, or some 55 million people. The primary center of this population is a region called **"Boswash,"** which is a belt of **highly urbanized landscapes** running from Boston at the north to Washington, D.C., at the south. There are seven major urban centers in this "Boswash" region: **Boston, Providence, Hartford, New York, Philadelphia, Washington, D.C., and Baltimore.**
- The **Appalachian Highlands** is the overall physiographic term for the Northeast, and the three most significant component parts are: the **Coastal Plain, the Piedmont, and the Interior Plains.** The **fall line** is the term that represents the break in topography between the low-lying Coastal Plain and the inland rise in elevation in the Piedmont. The fall line has been significant as a **break in bulk point for ships and small freighters** coming inland that cannot negotiate the falls that occur. Early manufacturing centers were also built around these drops in elevation. **Baltimore, Washington, D.C., Philadelphia, and Trenton, New Jersey, are examples of fall-line cities.**
- Roots of early regional development go back to **English settlers in Boston, Dutch settlers in New Amsterdam (New York), and later the English in Baltimore and Philadelphia.** The early locations that were favored were on fall lines (Philadelphia) or good harbors (New York).
- The **Erie Canal**—in conjunction with the **Hudson-Mohawk Trough**—stimulated a major migration flow to the northwestern territories, now the Midwest. By the 1850s, so much trade had flowed through that corridor that **New York harbor became the nation's leading harbor,** and it still is of major importance, although Houston has become the country's largest by volume. This passageway also led to significant **migration** from New England farming as whole farm households traveled to the states at the southern margin of Lake Erie in quest of better soils and farming situations.
- **Early industrialization** took place along the fall-line cities in the region and then expanded to western Pennsylvania as steel began to be produced along the southern **margins of Lake Erie, and especially in Pittsburgh.** Additional development in **flour milling, electrical goods, chemical and textile production, and photographic equipment** all took on regional and national importance. However, the real **center of major industry, finance, textile manufacturing, and varied manufacturing was in New York City.** This city's primacy has been maintained to the present in many of those industries.

- The **South** includes: **Virginia, North Carolina, South Carolina, Georgia, Florida, Alabama, Mississippi, Louisiana, Texas, West Virginia, Kentucky, Tennessee, Arkansas, and Oklahoma.**
- A **humid subtropical climate** with long hot summers is common in the South, but rainfall diminishes as you go west in the region. There is considerable regional variety in physiographic units, from the **Gulf-Atlantic Coastal Plain; the Piedmont; and the Interior Plains, with, again, the fall line lying between the Piedmont and the Coastal Plain. The Upland South** has hills and low mountains, and the subunits of that area include the **Appalachians, Unglaciated Southeastern Interior Plain, and the Interior Highlands.**
- The South's main agricultural characteristics today include: a **diminishing role of agriculture** overall in the region (only 2 percent of the population still lives on farms); **the reduction of farming to a more part-time occupation,** with wages being earned in nearby part-time city jobs; the rise in **beef cattle, broiler chickens, and dairy cattle, which have all become more important than crops. Tobacco** has been the most recent crop to **begin a strong decline. Citrus, peanuts, truck crops, and race horses** are some of the special farming activities that now characterize the region.
- New **industrial development** has included **textiles** (although this industry is moving on toward Mexico and Middle America and other offshore sites), and the **development of synthetic fibers** from local resources have been important. Also, **energy from hydroelectric development, oil and natural gas, and, to a lesser degree, coal** has been developed. Texas and North Carolina now stand as two major industrial states.
- The **industrial belts of the region** lie along the **Piedmont** and in part are connected to the **Tennessee Valley Authority** of the 1930s. This regional industrial development was unusual because it was focused as much on **small towns and middle-sized cities** rather than the major metropolitan centers of the Northeast. The second major belt was along and near the **Gulf Coast. The Texas-Oklahoma Metropolitan Zone** has the two largest urban centers in the South: Dallas–Ft. Worth and Houston. The **Houston Ship Channel** has made that city a major shipping center. Florida is another major metropolitan zone, but with a focus more on service and tourism than on manufacturing. There are other metropolitan zones in the South, such as cities of the **Upland South and the Mississippi Delta,** and others.
- The Midwest as a region encompasses **twelve states: Ohio, Indiana, Illinois, Michigan, Wisconsin, Minnesota, Iowa, North Dakota, South Dakota, Nebraska, Kansas, and Missouri.** As a **region** it would rank **13th or 14th in the world in area and popu-**

lation. In the production of **pork, farm machinery, and soybeans,** no nation in the world other than the United States would surpass the Midwest.

- **Plains** are the most distinctive topographic feature of this region, with the areas of major relief lying at the margins of the region. **Soils are generally very good,** and toward the east rainfall is generally adequate for productive farming. **Rainfall diminishes as you go west**, making irrigation more an agricultural feature there. Soils also benefit from the presence of **loess—windborne dust particles** from the west and north, probably glacial in its origin.

- **The Corn Belt** is a major part of the Midwest, and corn continues to be the **single most important crop** in Midwest agriculture, although **soybeans** are gaining in importance each year. Approximately **40 percent of the world's corn comes from the Midwest.**

- **Wheat** is very important in the Dakotas and south, with **spring wheat** major in the north and **winter wheat** most important in Kansas and south. Associated with wheat are sorghums, barley, and livestock production. In the region overall, **sales of animal products exceed those of crops. Cattle and hogs** are the two primary animals raised in the Midwest. There are also specialty crops that include **fruit, sunflowers, sugar beets, and potatoes.**

- **Detroit** has long been the **center of automobile manufacturing** in the country, related in part to the development of a major steel concentration along the southern shore of Lake Erie, **utilizing iron ore from Minnesota, coal from Pennsylvania, and the Great Lakes as the avenue of major movement** into the St. Lawrence Seaway and world trade. Slowly the steel industry has been moving away from its early-20th-century locales, going further south and west. **St. Louis** has now become second only to Detroit in the **manufacture of automobiles** because of shifting industrial locations.

- The region of the **West**, in this text, includes **New Mexico, Arizona, Colorado, Utah, Nevada, Wyoming, Montana, Idaho, California, Oregon, and Washington.** The region includes **Alaska and Hawaii,** as two very distinct variants on the themes of the West. The region's landscape is predominately arid and amounts to approximately 38 percent of the conterminous United States. **One-fifth of the nation's population lives in the West.**

- **Aridity** is the most pervasive geographic characteristic of the West, and **settlement is described as oasislike.** Topography plays a strong role in the climate patterns of the region, with **windward and leeward locations** very important. Major ranges along the west coast (including the **Cascades, Klamath Mountains, Coast Ranges, and Sierra Nevadas**) have the largest population centers on their windward sides, while population densities generally diminish quickly as you go east from those ranges. The **Willamette–Puget Sound Lowland in the north and the Central Valley in California** in the south are two significant lowlands in terms of population size.

- **Grazing is the most common use of the land** in the West, if it is brought into agriculture at all—unless irrigation is possible. **Irrigated farmlands raise alfalfa for beef cattle and dairy** herds, or they may be used for **cotton or specialty crops,** particularly in California. In the regions that have access to **exotic water,** there has been a continual tension between the use of that resource for irrigated farmland or urban settlement and expansion.

- **Southern California is the West's largest population cluster.** It did not begin to grow significantly until water was brought from the distant **Owens Valley to Los Angeles early in the 20th century. As** water from the Colorado River Project was also made available to cities in the southwest and in Southern California, growth accelerated. Presently, about **one-seventh of the United States population lives in Southern California,** in part because of the networks of water movement and late-19th-century railroad developments. The continual presence of **natural hazards including fire, flood, earthquakes, and intense drought** have done little to retard the growth of this area.

- **San Francisco** is also a major urban center, sharing many of the same hazards as Southern California. It has grown from a different set of origins and has become well-known as the home area of **Silicon Valley,** the dominant **software development center** in the West, and in the United States. Even with the continuing dispersal of software development (especially to the Seattle area in association with the location of Microsoft), the San Francisco area is **dominant in that industrial activity.**

- **California's Central Valley** is a very productive agricultural area, with **grapes and cotton** as the two lead crops. **Irrigation**, primarily by waters from the Sierra Nevada range, has long been central to such productivity. **Milk and cattle** are also major products from this region and many cities have developed in support of this productivity, **Sacramento, Fresno, and Stockton** among them.

- **Seattle, Washington,** its surrounding **Puget Sound Lowland,** and the **Willamette Valley of Oregon** have a **marine west coast** climate that provides abundant natural rainfall. **Fishing, logging, and trade built these cities in the late 19th century, and aircraft manufacture (Boeing) and software industries of all sizes (Microsoft as the largest) continue to promote development of this region.**

- In the **Mountain West—the states that lie east of the Pacific coast states**—water is less available than on the coast, and **population is clustered** around centers that have a reliable **water source and transportation lines** running at least both east and west. **Phoenix and Tucson, Arizona; Denver, Colorado; and Salt Lake City, Utah,** are the four largest cities in this region, and they all have distinct development histories and identities.

- **Alaska and Hawaii** are the two most recent additions to the United States (they both became states in 1959), and they represent very distinctive landscapes and environmental locales. Large **military installations** exist in both states. Alaska stretches north of the **Arctic Circle** and has a total population of approximately 600,000. Although there is a variety of landscapes in Alaska, all are characterized by intense winter seasons and a demanding seasonal range of weather. **Anchorage** is the largest city and has **nearly one-half of the state's total population.**

- Alaskan history chronicles **fur trapping, gold rushes, oil exploration, and military installations.** Sports fishing and hunting and general tourism play an increasingly significant role in the state's economy.

- **Hawaii** lies some 2000 miles west of California and is comprised of eight major islands. It has a profound range of rainfall patterns, with **agriculture playing a very important role in its early development.** Today agriculture amounts to only 1 percent of its annual gross state product, even though it continues to be a source of sugar and pineapple for mainland markets. **Military expenditures** have also been important, and, during the past two to three decades, **tourism** has brought major capital to the islands. The **Asian economic downturn in 1997** has caused a significant decrease in tourist arrivals.

REVIEW QUESTIONS

1. Name the states that make up the four regions that this text uses to regionalize the United States.
2. Approximately what percent of the United States population lives in the Northeast? In the West?
3. What are the major topographic features that characterize the Northeast?
4. Explain the "fall-line cities" and discuss their role in early industrialization and settlement.
5. When did the Erie Canal open, what route did it follow, and what demographic impact did it have?
6. What are the reasons for the primacy of New York City?
7. What are some of the old images and the new images of the South?
8. Outline the topographic characteristics of the South, and explain how they relate to patterns of urban settlement and agriculture of the region.
9. Explain the significance—and the process—of mechanization of cotton farming in the South.
10. What are the major aspects of patterns of urbanism in the South, and what are the most important cities?
11. Why are the soils so good in the Midwest? How do these characteristics relate to topography?
12. Explain the pattern of small-town growth, decline, and potential resurgence in the Midwest.
13. What are the major characteristics of industrial growth in the Midwest, and what cities have been most influenced by these changes?
14. Why is the term "oasislike" used for urban settlement in the West?
15. List and give the importance of the major topographic features in the West. What role do such features have in patterns of agriculture and urban settlement in the region?
16. What is the meaning of "exotic water" and explain its importance to the West.
17. Compare Los Angeles, San Francisco, and Seattle in terms of their natural settings, their histories, and their contemporary economic and cultural characteristics.
18. How much of the United States agricultural production comes from California? Outline what makes this possible, and what products it entails.
19. Outline the geographic characteristics of Alaska and Hawaii, and relate such features to their individual histories, their contemporary settlement patterns, and their future development.
20. What is the Ring of Fire? Relate it to a global pattern.

DISCUSSION QUESTIONS

1. Discuss criteria that you might use to regionalize a country as diverse and as large as the United States. How successful do you feel this text is in dividing the country into four regions?
2. What has been role of "fall-line" locations in the urban and economic history of the country?
3. Discuss the impact changes in transport technology have had on settlement and demographic patterns in the United States.
4. Discuss the distinctive roles of cotton, corn, soybeans, and wheat on the landscapes of the regions of the country.
5. What roles have rivers played in the patterns of settlement, economic activity, and tourism in the United States?
6. Water has been a key environmental feature of which regions, and how has it been utilized in their economic development and environmental management?
7. Discuss the shifting patterns of American steel production, explaining what geographic, economic, and demographic features have been most influential in today's patterns.
8. What do you think will become of the small town in America? Why?
9. Name the three most highly visible cities in the United States and explain why they have such visibility. What is the significance of such characteristics?

References and Readings

Current Sources

Chapter 1

Birdsall, Stephen S. "Regard, Respect, and Responsibility: Sketches for a Moral Geography of the Everyday." *Annals of the Association of American Geographers,* 86(4): 619–629, 1996.

Demko, George J., et al. *Why in the World: Adventures in Geography.* New York: Anchorbooks, published by Bantam Doubleday, 1992.

Gaile, Gary L., and Cort J. Wilmott, eds. *Geography in America.* Columbus, Ohio: Merrill Publishing, 1989.

Geography Education Standards Project. *Geography for Life: National Geography Standards 1994.* Washington, D.C.: National Geographic Research and Exploration, 1994.

Lineback, Neal. *Geography in the News.* Southern Pines, NC: Karo Hollow Press, 1995.

Marshall, Bruce, ed. *The Real World.* Boston: Houghton Mifflin Company, 1991.

Rediscovering Geography: New Relevance for Science and Society. Washington, D.C.: National Academy Press, 1997.

Chapter 2

Bennett, Charles F. *Man and Earth's Ecosystems: An Introduction to the Geography of Human Modification of the Earth.* New York: John Wiley & Sons, 1975.

Bohlen, Celestine. "Markets Jolted, Dow off 4.1% as the Russian Economic Slide Adds to Pressures on Yeltsin. His Opponents Want Changes, Including Him." *New York Times,* August 28, 1998, 1.

Botkin, Daniel B., Margriet F. Caswell, John E. Estes, and Angelo A. Ario, eds. *Changing the Global Environment.* Boston: Academic Press, 1989.

Bronner, Ethan. "Moscow Dashes American Illusions." *New York Times,* August 30, 1998, 4–1.

———. *State of the World 1990.* New York: W.W. Norton & Co., 1990.

Brown Lester, ed. *State of the World 1994.* New York: W.W. Norton & Co., 1994.

Callaghy, Thomas M. "Globalization and Marginalization: Debt and the International Underclass." *Current History,* 96(613):392–396, 1997.

Clark, William C. "Managing Planet Earth." *Scientific American,* 261(3): 19–26, 1989.

Crossette, Barbara. "How to Fix a Crowded World: Add People." *New York Times,* November 2, 1997, 4–1.

Dasmann, Raymond F. *Environmental Conservation.* New York: John Wiley & Sons, 1984.

———. *Ecodevelopment—An Ecological Perspective: Tropical Ecology and Development, 1331–1335.* Kuala Lumpur: International Society of Tropical Ecology, 1980.

De Souza, Anthony R., and Frederick P. Stutz. *The World Economy: Resources, Location, Trade and Development.* New York: Macmillan, 1994.

Detwyler, Thomas R., ed. *Man's Impact on the Environment.* New York: McGraw-Hill, 1988.

Dubos, Réne. *A God Within.* New York: Charles Scribner's Sons, 1972.

English, Paul, and James A. Miller. *World Regional Geography: A Question of Place.* New York: John Wiley & Sons, 1989.

Francis, David R. "Why Markets Tumble Together." *Christian Science Monitor,* September 1, 1998, 1.

———. "Global Crowd Control Starts to Take Effect." *Christian Science Monitor,* October 22, 1997, 1.

———. "Welcome Mat Now Offered Foreign Firms." *Christian Science Monitor,* February 28, 1996, 1.

Fuerbringer, Jonathan. "1998: A Year Emerging Markets Seem Synonymous With Anguish." *New York Times,* August 27, 1998, C1.

Gabler, Robert E., Robert J. Sager, and Daniel L. Wise. *Essentials of Physical Geography,* 4th ed. Philadelphia: Saunders College Publishing, 1994.

Glacken, C.J. "Changing Ideas of the Habitable World." In *Man's Role in Changing the Face of the Earth,* edited by W. L. Thomas, 70–92. Chicago: University of Chicago Press, 1956.

Glaeser, Bernhard, ed. *Ecodevelopment: Concepts, Projects, Strategies.* Oxford: Pergamon Press, 1984.

Gordon, Michael R. "The Fellow Travelers." *New York Times,* August 28, 1998, 1.

Goudie, Andrew. *The Human Impact on the Natural Environment.* Oxford: Basil Blackwell, 1986.

Greenhouse, Steven. "The Greening of American Diplomacy." *New York Times,* October 9, 1995, A4.

Grossman, Larry. "Man-Environment Relations in Anthropology and Geography." *Annals of the Association of American Geographers,* 67: 126–144 (1977).

Gupta, Ajavit. *Ecology and Development in the Third World.* New York: Routledge, Chapman & Hall, 1988.

Hanson, Herbert C. *Dictionary of Ecology.* New York: Philosophical Library, 1962.

Hardesty, Donald L. *Ecological Anthropology.* New York: John Wiley & Sons, 1977.

Hardin, Garrett. "Living on a Lifeboat." In *Readings in Ecology, Energy and Human Society: Contemporary Perspectives,* edited by William R. Burch, Jr. New York: HarperCollins, 1977.

Holmes, Steven A. "Global Crisis in Population Still Serious, Group Warns." *New York Times,* December 31, 1997, A7.

International Union for the Conservation of Nature, United Nations Environmental Program, World Wildlife Fund. *World Conservation Strategy: Living Resource Conservation for Sustainable Development:* IUCN, Gland, 1980.

Jackson, Robert M. "Political Economy." In *Global Issues 98/99,* edited by Robert M. Jackson, 120–121. Guilford, Conn.: Dushkin/McGraw-Hill, 1998.

Kahn, Joseph. "The Bear Draws Blood." *New York Times,* August 30, 1998, 3–1.

Kates, Robert W., and Ian Burton, eds. *Geography, Resources, and Environment, Vol. II.* Chicago: University of Chicago Press, 1986.

Kaufman, Les, and Kenneth Mallory. *The Last Extinction.* Cambridge, Mass.: MIT Press, 1986.

Kiefer, Francine. "Steamy White House Fight Over Global Warming." *Christian Science Monitor,* July 16, 1998, 3.

Kopp, Raymond J., Richard D. Morgenstern, and Michael A. Toman. "Climate Change Policy After Kyoto." *Resources,* 130: 4–6, 1998.

Kristof, Nicholas D. "Asians Worry That U.S. Aid Is a New Colonialism." *New York Times,* February 17, 1998, A4.

Lovelock, J.E. *Gaia.* Oxford: Oxford University Press, 1979.

MacNeill, Jim. "Strategies for Sustainable Economic Development." *Scientific American,* 261: 104–113 (1989).

Martin, Geoffrey J., and Preston E. James. *All Possible Worlds: A History of Geographical Ideas.* New York: John Wiley & Sons, 1993.

McLaren, Digby J., and Brian J. Skinner, eds. *Resources and World Development.* New York: John Wiley & Sons, 1987.

Miller, G. Tyler. *Living in the Environment.* Belmont, California: Wadsworth, 1996.

Moran, Emilio F. *Human Adaptability.* Boulder, Colo.: Westview Press, 1982.

Morgenson, Gretchen. "Markets Jolted, Dow Off 4.1% As the Russian Economic Slide Adds to Pressures on Yeltsin. 357.36 Point Fall. Wall Street Takes a Hit, but Other Countries Suffer Even More." *New York Times,* August 28, 1998, 1.

Myers, Norman. "Population, Environment and Conflict." *Environmental Conservation,* 14: 15–22 (?).

———. *A Wealth of Wild Species: Storehouse for Human Welfare.* Boulder, Colo.: Westview Press, 1983.

Nasar, Sylvia. "Why International Statistical Comparisons Don't Work." *New York Times,* March 8, 1992, E5.

O'Riordan, T. *Environmentalism.* London: Pion, 1976.

Pearce, David W., and R. Kerry Turner, eds. *Economics of Natural Resources and the Environment.* Baltimore: Johns Hopkins University Press, 1990.

Population Reference Bureau. *World Population Data Sheet.* Washington, D.C.: Population Reference Bureau, 1996.

Postel, Sandra. "Carrying Capacity: Earth's Bottom Line." In *State of the World,* edited by Lester Brown, 3–21. New York: W.W. Norton & Co., 1994.

Potter, Thomas Michael. "Social Organization and Environmental Destruction: Capitalism, Socialism and the Environment." In *International Dimensions of the Environmental Crisis,* edited by Richard N. Barrett, 21–29. Boulder, Colo.: Westview Press, 1982.

Prince, Cathryn. "As World Heats Up, U.S. Plan Would Make Firms Pay to Pollute." *Christian Science Monitor,* July 19, 1996, 6.

Repetto, R., ed. *The Global Possible.* New Haven: Yale University Press, 1985.

Richbury, Keith B. "Spreading the Wealth: How 'Globalization' Is Helping Shift Cash from Rich Nations to Poor Ones." pp. 127–130 in *Global Issues 98/99,* edited by Robert M. Jackson, 127–130. Guilford, Conn.: Dushkin/McGraw Hill, 1998.

Roberts, Neil. *The Holocene: An Environmental History.* Oxford: Basil Blackwell, 1989.

Rosenau, James N. "Complexities and Contradictions of Globalization." In *Global Issues 98/99,* edited by Robert M. Jackson, 122–130. Guilford, Conn.: Dushkin/McGraw Hill, 1998

Rubenstein, James M. *The Cultural Landscape: An Introduction to Human Geography.* New York: Macmillan, 1994.

Ruckelshaus, William D. "Toward a Sustainable World." *Scientific American,* 261: 114–120, 1989.

Sahlins, Marshall. *Stone Age Economics.* Walter DeGruyter Press, 1972.

Simmons, I.G. *Changing the Face of the Earth.* Oxford: Basil Blackwell, 1989.

Soule, Michael E. *Conservation Biology: The Science of Scarcity and Diversity.* Sunderland, Mass.: Sinaeuer Press, 1986.

Stevens, William K. "New Evidence Finds This Is Warmest Century in 600 Years." *New York Times,* April 28, 1998, B13.

———. " '95 the Hottest Year on Record as the Global Trend Keeps Up." *New York Times,* January 4, 1996, A1.

———. "Experts Confirm Human Role in Global Warming." *New York Times,* September 10, 1995, A1.

Tivy, Joy, and Greg O'Hare. *Human Impact on the Ecosystem.* Edinburgh: Oliver and Boyd, 1981.

Uchitelle, Louis. "I.M.F., Pressing Russia on Earlier Loan Pact, Takes Time to Rule on New Cash." *New York Times,* August 28, 1998, A10.

United Nations Development Program. *HDR 1997 Country Rankings.* 1998. World Wide Web document http://www.undp.org/undp/hdro/table2.htm

Wattenberg, Ben J. "The Population Explosion Is Over." *New York Times,* November 23, 1997, 60–63.

Wilson, Edward O. "Threats to Biodiversity." *Scientific American* 261 (3): 60–66, 1989.

———. *Biodiversity.* Washington, D.C.: Smithsonian Institution Press, 1987.

Wines, Michael. "As Ruble Falls, Moscow Unravels Faster and Faster." *New York Times,* September 4, 1998, A4.

World Commission on Environment and Development. *Our Common Future.* Oxford: Oxford University Press, 1987.

World Resources Institute. *World Resources, 1994–95.* New York: Oxford University Press, 1994.

———. *The Global Possible: Resources, Development and the New Century.* Washington, D.C.: World Resources Institute, 1984.

WuDunn, Sheryl. "Asia Fears Crisis It Exported May Be Headed Back Again." *New York Times,* September 2, 1998, C8.

Chapter 3

"An Awfully Big Adventure." *Economist,* April 11, 1998, 3–5.

Blouet, Brian W. "The Political Geography of Europe: 1900–2000 A.D." *Journal of Geography,* 95(1): 5–14, 1996.

"European Business." *Economist,* 340(7974): 22–23, 1996.

"For NATO, Eastward Ho!" *Economist,* March 1, 1997, 49–51.

"How to Say No." *Economist,* November 1, 1997, 51–52.

Jordan, Terry G. *The European Culture Area: A Systematic Geography.* New York: HarperCollins, 1996.

"Millions Want to Come." *Economist,* April 4, 1998, 55–56.

Renfrew, Colin. "The Origins of Indo-European Languages." *Scientific American,* 261(4): 106–114, 1989.

Chapter 4

"Beyond Omagh." *Economist,* August 22, 1998, 15.

Cocleman, D.A. "International Migration: Demographic and Socioeconomic Consequences in the United Kingdom and Europe." *International Migration Review,* 29(1): 155(52), 1995.

"Crossed Fingers in France." *Economist* 343(8014): 45–47, 1997.

"The Disturbing Niceness of George Mitchell." *Economist,* April 11, 1998, 45.

"From Workshop to Melting Pot." *Economist,* August 8, 1998, 48–49.

Geddes, Andrew. "Immigrant and Ethnic Minorities and the EU's 'Democratic Deficit'." *Journal of Common Market Studies,* 33(2): 197, 1995.

Irving, Clive. "The New Metropolis." *Conde Nast Traveler,* February:96–108, 1998.

Katzenstein, Peter J. "United Germany in an Integrated Europe." *Current History,* 96(608): 116–123, 1997

"The Leaving of Liverpool." *Economist,* August 15, 1998, 48–50.

Lueschen, Leila S. "French Agriculture: Trends and Policies." *Agribusiness* 11(5): 447–462, 1995.

"Omagh's Terrible Sacrifice." *Economist,* August 22, 1998, 46.

"Refitting on the Clyde." *Economist,* August 22, 1998, 47–48.

Stanglin, Douglas. "Ready and Waiting." *U.S. News and World Report,* 121(25): 23(2), 1996.

"Who Should Be German Then?" *Economist,* July 4, 1998, 45.

Wild, Trevor, and Philip N. Jones. "Spatial Impacts of German Unification." *The Geographical Journal,* 160 (1):1–16, 1994.

Chapter 5

Clark, Arthur L. *Bosnia: What Every American Should Know.* New York: Berkley Books, 1996.

"Dual Market." *Economist*, May 31, 1997, 69–70.

"Europe Grows Apart." *Economist*, March 7, 1998, 75–76.

Hoge, Warren. "Norway's Awesome Nature, Awesomely Overcome." *New York Times*, August 28, 1998.

————. "Sweden's Frozen North Can Be Balmy for Business." *New York Times*, August 21, 1998.

Murphy, Alexander B. and Anne Hunderi-Ely. "The Geography of the 1994 Nordic Vote on European Union Membership." *Professional Geographer* 48(3): 2184–2197, 1996.

O'Connor, Mike. "Kosovo Refugees: Pawns in a NATO-Serb Clash?" *New York Times*, August 24, 1998.

————. "Serbian Cannon Fire Kills 3 Mother Teresa Aid Workers." *New York Times*, August 26, 1998.

Osborn, Alan. "The European Parliament Supports the Kyoto Accord." *Europe*, 373:30(1), 1998.

Silber, Laura, and Allan Little. *Yugoslavia: Death of a Nation.* Harmondsworth, U.K.: Penguin, 1995.

"Will Yugoslavia Break Again?" *Economist*, April 18, 1998, 45–46.

Woodward, Susan L. "Bosnia after Dayton: Year Two." *Current History*, 96(608): 97–103, 1997.

Chapter 6

Bohlen, Celestine. "Markets Jolted, Dow off 4.1% As the Russian Economic Slide Adds to Pressures on Yeltsin. His Opponents Want Changes, Including Him." *New York Times*, August 28, 1998, 1.

Bronner, Ethan. "Moscow Dashes American Illusions." *New York Times*, August 30, 1998, 4-1.

Burke, Justin. "Volgograd: The Living World War II Memorial." *Christian Science Monitor*, September 10, 1993, 9.

DeBlij, Harm J., and Peter O. Muller. *Geography: Realms, Regions and Concepts.* New York: John Wiley & Sons, 1994.

Fuerbringer, Jonathan. "1998: A Year Emerging Markets Seem Synonymous With Anguish." *New York Times*, August 27, 1998, C1.

Gordon, Michael R. "Ruble Withering? So What—Russians Survive on Barter." *New York Times*, September 6, 1998, 1.

————. "The Fellow Travelers." *New York Times*, August 28, 1998, 1.

Ingwerson, Marshall. "Russia's Bear Economy Goes Bullish as Reforms Kick In." *Christian Science Monitor*, October 10, 1995, 6.

Kahn, Joseph. "The Bear Draws Blood." *New York Times*, August 30, 1998, 3–1.

McElroy, Claudia. "An Islamic Enigma Rises in Central Asia." *Christian Science Monitor*, April 14, 1998, 5.

Moffett, George. "Assessing a Russia Without Yeltsin at the Helm." *Christian Science Monitor*, November 14, 1995, 1.

Morgenson, Gretchen. "Markets Jolted, Dow off 4.1% as the Russian Economic Slide Adds to Pressures on Yeltsin. 357.36 Point Fall. Wall Street Takes a Hit, but Other Countries Suffer Even More." *New York Times*, August 28, 1998, 1.

Poletz, Lida. "Ukraine and Belarus Elect New Pro-Russian Leaders." *Christian Science Monitor*, July 12, 1994, 3.

"The Selling of Russia." *Economist*, November 18, 1995, 88.

Shepherd, Leslie. "Russians Turn to Bartering." Columbia Missourian, September 7, 1998, 3A.

Sneider, Daniel. "Russia's Future Shadowed by Centuries of Conquest." *Christian Science Monitor*, August 10, 1994, 7.

Specter, Michael. "After Lull, War Revives in Chechnya's Ruins." *New York Times*, November 26, 1995, A1.

————. "Chechen Insurgents Take Their Struggle to a Moscow Park." *New York Times*, November 24, 1995, A1.

————. "Pro-Russian Chechen Leader Survives Bombing in Capital." *New York Times*, November 21, 1995, A4.

————. "Russia's Fall Grain Harvest Seen as Worst in 30 Years." *New York Times*, October 10, 1995, A6.

————. "Climb in Russia's Death Rate Sets Off Population Implosion." *New York Times*, A1.

Stanley, Alessandra. "From Repression to Respect, Russian Church in Comeback." *New York Times*, October 3, 1994, A1.

Uchitelle, Louis. "I.M.F., Pressing Russia on Earlier Loan Pact, Takes Time to Rule on New Cash." *New York Times*, August 28, 1998, A10.

Wines, Michael. "As Ruble Falls, Moscow Unravels Faster and Faster." *New York Times*, September 4, 1998, A4.

Chapter 7

Andrews, Edmund L. "Vodka Bottleneck Contains a Volatile Political Mix." *New York Times*, A4.

Barraclough, Colin. "Azerbaijan Keen to Seal Oil Deal with West." *Christian Science Monitor*, August 9, 1994, 9.

————. "Turkmenistan Ripe for Capitalism." *Christian Science Monitor*, June 24, 1992, 7.

Bohlen, Celestine. "Yeltsin and Ukraine's Chief in Pact on Black Sea Fleet." *New York Times*, April 16, 1994, 5.

————. "Poor Region in Russia Lays Claim to Its Diamonds." *New York Times*, November 1, 1992, 3.

————. "Russia Permits Just a Peek at Nature in the Raw." *New York Times*, October 28, 1992, A4.

Broad, William J. "Russians Describe Extensive Dumping of Nuclear Waste." *New York Times*, April 27, 1993, A1.

Burke, Justin. "In Caviar Capital, Cultures Clash." *Christian Science Monitor*, September 14, 1993, 10.

————. "Volga Germans Seek Lost Homeland." *Christian Science Monitor*, September 10, 1993, 8.

————. "Samara's Defense Industry Moves Reluctantly Toward the Free Market." *Christian Science Monitor*, September 9, 1993, 8.

————. "A Part of Russia That Wants Out." *Christian Science Monitor*, September 3, 1993, 6.

————. "The Volga Basin's Merchant Capital Is New Russia's Economic Showcase." *Christian Science Monitor*, September 2, 1993, 8.

————. "A Russian Republic Looks Away." *Christian Science Monitor*, March 10, 1993, 6.

————. "Uzbek Leaders Clamp Down to Keep Peace." *Christian Science Monitor*, January 5, 1993, 8.

————. "Uzbekistan Moves Slowly Toward Economic Reform." *Christian Science Monitor*, January 5, 1993, 8.

————. "Uzbek Leaders Pick Stability Over Reform." *Christian Science Monitor*, December 11, 1992, 10.

————. "Tajiks Struggle for National Identity." *Christian Science Monitor*, September 30, 1992, 10.

————. "Siberia's Buryats Hope for Recovery." *Christian Science Monitor*, September 3, 1992, 7.

"Caspian Carve-up." *Economist*, March 7, 1998, 66.

"Chechnya: Taiwan on the Caucasus." *Economist*, February 22, 1997, 59.

Colarusso, John. "Chechnya: The War Without Winners." *Current History*, 94(594): 329–336, 1995.

"Communists Lead as Belarus Votes." *Christian Science Monitor*, November 30, 1995, 9.

Dudaev, Dzhohar. "Chechnya is World's Concern." *Christian Science Monitor,* November 1, 1995, 19.

"End of a Road for Russia?" *Economist,* September 5, 1998, 57–58.

Erlanger, Steven. "U.S. Tries to Defuse Russia-Latvia Dispute." *New York Times,* April 16, 1998, A9.

―――. "U.S. Aid for Huge Russian Lake Is in Jeopardy." *New York Times,* September 3, 1995, A3.

―――. "In Russia, Turning Oil into Money Is Actually Hard." *New York Times,* March 20, 1994, E5.

―――. "A Cry of Pain Rises From the Cradle of Soviet Science." *New York Times,* November 21, 1993, E18.

―――. "Heirs of the Golden Horde Reclaim a Tatar Culture." *New York Times,* August 13, 1993, A3.

―――. "To Kazakhstan, With Gold in Mind." *New York Times,* March 10, 1993, A4.

―――. "Russia and Ukraine: Condemned to Get Along." *New York Times,* June 21, 1992, E3.

―――. "Uzbeks, Free of Soviets, Dethrone Czar Cotton." *New York Times,* June 20, 1992, A4.

"Fishing for Gas, Sturgeon and Influence." *Economist,* January 24, 1998, 43.

Ford, Peter. "A Game of Musical Chairs That No One Wins." *Christian Science Monitor,* December 17, 1997, 10–11.

―――. "Rebels Keep Guns While Russia Tries to Call Shots in Chechnya." *Christian Science Monitor,* September 21, 1995, 8.

―――. "Why Georgia (the Country) Tries to Act Big for Its Size." *Christian Science Monitor,* September 6, 1995, 6.

―――. "A Remote Republic Rattles Moscow." *Christian Science Monitor,* September 16, 1994, 7.

Gordon, Michael R. "Facing Severe Shortage of Food, Russia Seeks Foreign Relief Aid." *New York Times,* October 10, 1998, 1.

―――. "Georgia Plots a Future Free of Russia's Sway." *New York Times,* January 12, 1997, 4.

―――. "Chechnya Toll Is Far Higher; 80,000 Dead, Lebed Asserts." *New York Times,* September 4, 1996, A3.

―――. "Russia Agrees to Closer Links With Three Ex-Soviet Lands." *New York Times,* March 30, 1996, A4.

Greenhouse, Steven. "Ukraine Votes to Become a Nuclear-Free Country." *New York Times,* November 17, 1994, A6.

Greenwald, Igor. "Four Years Old, Ukraine Improvises Its Identity." *Christian Science Monitor,* August 24, 1995, 6.

Hottelet, Richard C. "Now, the New 'Great Game' for Control in Central Asia." *Christian Science Monitor,* January 13, 1998, 18.

Ingwerson, Marshall. "Peace with No Honor: Chechnya Pact Leaves Russian Troops Bitter." *Christian Science Monitor,* September 3, 1996, 1.

―――. "Marketing Nuclear Plants for an Energy-Hungry World." *Christian Science Monitor,* November 8, 1995, 10.

―――. "Foreign Investors in Kazakstan Face Obstacles." *Christian Science Monitor,* October 12, 1995, 6.

―――. "Kyrgyz Brokers Dream of a Trading Frenzy." *Christian Science Monitor,* September 20, 1995, 7.

―――. "Empire Lost, Russian People Stream Out of Central Asia." *Christian Science Monitor,* September 19, 1995, 6.

―――. "New Nation Digs Deep for Its Turkic Roots." *Christian Science Monitor,* September 12, 1995, 1.

―――. "'The Turks Are Coming,' Cry Russians in Their Ex-Empire." *Christian Science Monitor,* September 26, 1995, 1.

Ingwerson, Marshall, and Sami Kohen. "At Your Local Gas Pump Soon: Caspian Sea Oil." *Christian Science Monitor,* October 11, 1995, 6.

Kamm, Henry. "With 'No Enemies,' Russia's Baltic Fleet Rusts." *New York Times,* November 27, 1995, A6.

―――. "Turks Fear Role in Asia of Russians." *New York Times,* June 18, 1994, A6.

"Kazakhstan: High Stepping." *Economist,* March 16, 1996, 41.

Kim, Lucian. "Taliban Jars Central Asia." *Christian Science Monitor,* August 14, 1998, 1.

Kinzer, Stephen. "A Defiant Satellite Nation Finds Russia's Orbit Inescapable." *New York Times,* May 3, 1998, 12.

―――. "Turks Fearful of an Oil Disaster As the Bosporus Gets Busier." *New York Times,* January 11, 1998, 4.

Lee, Rensselaer W. "Post-Soviet Nuclear Trafficking: Myths, Half-Truths, and the Reality." *Current History,* 94 (594): 343–348, 1995.

LeVine, Steve. "Iran Opens Big Gas Pipeline to Neighbor, Defying U.S." *New York Times,* December 30, 1997, A5.

―――. "Free of the Russians, But Imprisoned in Cotton." *New York Times,* November 20, 1997, A4.

―――. "Daubs of Oil and Exultation As Caspian Pipeline Opens." *New York Times,* November 13, 1997, C1.

―――. "Caspian's Sun on Rise, Cashing Russian Shadow." *New York Times,* October 27, 1995, A6.

―――. "Oil Consortium to Skirt Russia in Its Shipments." *New York Times,* October 7, 1995, A22.

―――. "U.S. and Russia at Odds Over Caspian Oil." *New York Times,* October 4, 1995, C2.

―――. "After Karl Marx, a 1,000-Year-Old Superman." *New York Times,* August 31, 1995, A4.

Lewison, Arlene. "A Century in Mourning." *Columbia Missourian,* September 17, 1995, G1.

Linden, Eugene. "The Rape of Siberia." *Time,* September 4, 1995, 42–53.

Matloff, Judith. "Desperate to Collect Taxes, Russia Can't." *Christian Science Monitor,* September 4, 1998, 1.

―――. "Russians Ask: What Reform?" *Christian Science Monitor,* September 1, 1998, 1.

―――. "Union With Russia Faces Dim Prospects for Belarus Leader." *Christian Science Monitor,* July 9, 1998, 7.

―――. "Old Roots Trip Russia's Farms." *Christian Science Monitor,* June 26, 1998, 1.

―――. "Russians Find Becoming a Latvian Isn't Easy." *Christian Science Monitor,* April 21, 1998, 6.

―――. "Heart of Europe Waits to Join Europe's Alliance." *Christian Science Monitor,* March 6, 1998, 7.

McElroy, Claudia. "An Islamic Enigma Rises in Central Asia." *Christian Science Monitor,* April 14, 1998, 5.

Meyer, Stephen M. "The Devolution of Russian Military Power." *Current History,* 94 (594): 322–328, 1995.

Misiunas, Romauld J. "This Tiny Russian Enclave in Europe Could Explode." *Christian Science Monitor,* April 6, 1994, 23.

"Moldova Holds Its Breath, Too." *Economist,* March 9, 1996, 50.

Moffett, George. "Rail Line Could Put Iran in the Middle of a New 'Silk Road' Between Asia and West." *Christian Science Monitor,* April 25, 1996, 1.

Nelan, Bruce W. "The Rush for Caspian Oil." *Time,* May 4, 1998, 38–42.

Niebuhr, Gustav. "The Armenian Pontiff and His Enduring Flock." *New York Times,* May 9, 1998, A9.

"The Northern Sea Route: Plain Sailing or Environmental Disaster?" *WWF Arctic Bulletin,* 3.94: 10–12, 1994.

Perlez, Jane. "Despite U.S. Help and Hope, Ukraine's Star Fades." *New York Times,* June 27, 1996, A3.

―――. "Ukraine Sells Its Companies, But Buyers Are Few." *New York Times,* November 2, 1995, A1.

Poletz, Lida. "Ukraine's Religious Standoff Makes Unlikely Political Allies." *Christian Science Monitor,* July 21, 1995, 6.

―――. "Ukraine Between East and West." *Christian Science Monitor,* June 16, 1994, 6.

Rasputin, Valentine. *Siberia on Fire: Stories and Essays by Valentin Rasputin.* De Kalb: Northern Illinois University Press, 1989.

Rosen, Yereth. "USSR Leaves Radioactive Legacy." *Christian Science Monitor,* August 26, 1992, 8.

"Russia and Dagestan: Losing Control?" *Economist,* August 1, 1998, 45.

"Russian Energy: A Dangerous Bear Dance." *Economist,* August 29, 1998, 45–47

"Russian Oil: Not a Gusher." *Economist,* October 14, 1995, 78.

"Russia's Other Governments." *Economist,* January 3, 1998, 47–48.

"Russia's Regions: Fiefs and Chiefs." *Economist,* January 25, 1997, 46–47.

"Russia's Riddle of the Regions." *Economist,* March 23, 1996, 47–48.

Salpukas, Agis. "Kazakhstan and Chevron Try to End Pipeline Impasse." *New York Times,* October 28, 1995, A18.

———. "Siberian Oil Venture by Four Companies." *New York Times,* April 12, 1994, C4.

Schmemann, Serge. "War Bleeds Ex-Soviet Land at Central Asia's Heart." *New York Times,* February 21, 1993, A1.

Shaw, Denis J.B. *The Post Soviet Republics: A Systematic Geography.* Harlow, England: Longman Press, 1995.

Sloane, Wendy. "Mills Quiet in Russia's Manchester." *Christian Science Monitor,* August 5, 1994, 8.

———. "Russia Backs Overthrow of Rebel Leader in Caucasus." *Christian Science Monitor,* August 4, 1994, 6.

———. "Russia-Estonia Summit Seeks Troops Solution." *Christian Science Monitor,* July 27, 1994, 4.

———. "Baltics Accuse Russia of Reneging on Troop Pullout." *Christian Science Monitor,* April 8, 1994, 5.

———. "Russia's Bashkir Republic Pushes for Its Autonomy." *Christian Science Monitor,* October 26, 1993, 6.

Sneider, Daniel. "Russian Leaders Watch Closely As Ukraine and Belarus Vote." *Christian Science Monitor,* June 22, 1994, 3.

———. "Tajikistan, A Tangle of Diverse Identities." *Christian Science Monitor,* May 13, 1994, 6.

———. "Russian Bear Roams in Battered Tajikistan." *Christian Science Monitor,* May 12, 1994, 7.

———. "Ukraine's Economy Stumbles." *Christian Science Monitor,* March 30, 1994, 7.

———. "Turkmenistan, Out from Under Soviet Masters, Reaches to Iran." *Christian Science Monitor,* April 13, 1993, 7.

———. "Turkmenistan: Slow Reforms, No Dissent." *Christian Science Monitor,* March 25, 1993, 7.

———. "Estonia Leads Baltic States into New Era." *Christian Science Monitor,* January 25, 1993, 8.

———. "Oil Fuels Azeris' Hopes for Future." *Christian Science Monitor,* January 13, 1993, 10.

———. "Central Asians to Build Own Common Market." *Christian Science Monitor,* January 7, 1993, 2.

Specter, Michael. "Yogurt? Caucasus Centenarians Never Eat It." *New York Times,* March 14, 1998, 1.

———. "Russia and Ukraine Sign a Friendship Treaty." *New York Times,* June 1, 1997, 8.

———. "Decisive Battle, Vague Promises." *New York Times,* September 1, 1996, A7.

———. "Belarus and Russia Form Union, Reuniting Two Former Soviet Lands." *New York Times,* March 24, 1996, A1.

———. "Pro-Russian Chechen Leader Survives Bombing in Capital." *New York Times,* November 21, 1995, A4.

———. "Yeltsin Threatens Action on Warring Secessionist Area." *New York Times,* November 30, 1994, A3.

Stanley, Alessandra. "Yeltsin Signs Peace Treaty With Chechnya." *New York Times,* May 13, 1997, A6.

———. "Yeltsin, Echoing Cold War, Rails Against West." *New York Times,* October 20, 1995, A3.

Sullivan, Walter. "Soviet Nuclear Dumps Disclosed." *New York Times,* November 24, 1992, B9.

Tyler, Patrick F. "Soviets' Secret Nuclear Dumping Raises Fears for Arctic Waters." *New York Times,* May 4, 1992, A1.

Walker, Sam. "No Time Like the Present to Invest in Eastern Siberia's Sakha Republic." *Christian Science Monitor,* March 9, 1994, 9.

Wines, Michael. "The 'Bunny' Breeds Rapidly, Sinking Belarus' Economy." *New York Times,* August 2, 1998, 1.

Chapter 8

Bates, Daniel G., and Amal Rassam. *Peoples and Cultures of the Middle East.* Englewood Cliffs, N. J.: Prentice-Hall, 1983.

Coon, Carleton S. "Point Four and the Middle East." *Annals of the American Academy of Political and Social Science,* 270: 88–92, 1950.

English, Paul Ward. "Geographical Perspectives on the Middle East: The Passing of the Ecological Trilogy." In *Geographers Abroad: Essays on the Problems and Prospects of Research in Foreign Areas,* University of Chicago Development of Geography Research Paper No. 152, edited by Marvin W. Mikesell, 134–164, 1973.

———. "Urbanities, Peasants, and Nomads: The Middle Eastern Ecological Trilogy." *Journal of Geography,* 61: 54–59, 1967.

Held, Colbert. *Middle East Patterns: Places, People and Politics.* Boulder, Colo.: Westview Press, 1994.

Chapter 9

"Algeria: Villages into Killing Fields." *Economist,* January 10, 1998, 36.

Beaumont, Peter, Gerald H. Blake, and J. Malcolm Wagstaff. *The Middle East: A Geographical Study.* New York: Halstead Press, 1988.

Bennet, James. "U.S. Cruise Missiles Strike Sudan and Afghan Targets Tied to Terrorist Network." *New York Times,* August 21, 1998, 1.

Britannica Book of the Year. 1994.

Bulloch, John, and Adel Darwish. *Water Wars: Coming Conflicts in the Middle East.* London: Victor Gollancz, 1993.

Burns, John F. "For Afghans, Full Circle." *New York Times,* August 13, 1998, 1.

———. "The West in Afghanistan, Before and After." *New York Times,* February 18, 1996, E5.

———. "From Cold War, Afghans Inherit Brutal New Age." *New York Times,* February 14, 1996, A1.

———. "Afghan Capital Grim As War Follows War." *New York Times,* February 5, 1996, A1.

Crossette, Barbara. "Forgotten Moroccan P.O.W.s Freed From Sandy Squalor." *New York Times,* December 8, 1995, A1.

Drysdale, Alasdair, and Gerald H. Blake. *The Middle East and North Africa: A Political Geography.* Oxford: Oxford University Press, 1985.

"Egypt: Back from the Desert." *Economist,* April 12, 1997, 36.

Frank, Harry Thomas. *Discovering the Biblical World.* Maplewood, N. J.: Hammond Inc., 1988.

George, Allen. "Dry Them Out." *The Middle East,* May 1993, 17–18.

Gibbs, Walter. "World's Small Powers Bypass U.S. and UN to Ban Land Mines." *Christian Science Monitor,* September 19, 1997, 1.

Girardet, Edward. "Land Mines: Soviets Leave Dangerous Legacy Behind in Afghanistan." *Christian Science Monitor,* June 22, 1988, 7.

Goldberg, Cary. "Peace Prize Goes to Land-Mine Opponents." *New York Times,* October 11, 1997, 1.

Goode's World Atlas. (For Metropolitan Area Population Figures.) New York: Rand McNally, 1995.

Hawkes, Jacquetta, ed. *Atlas of Ancient Archaeology.* New York: McGraw-Hill, 1974.

Hedges, Chris. "In a Remote Southern Marsh, Iraq Is Strangling the Shiites." *New York Times,* November 16, 1993, A1.

Held, Colbert. *Middle East Patterns.* Boulder, Colo.: Westview Press, 1994.

Ibrahim, Youssef M. "As Algerian Civil War Drags On, Atrocities Grow." *New York Times,* December 28, 1997, 1.

————. "Algeria Is Edging Toward Breakup." *New York Times,* April 4, 1994, A1.

"Iran and Afghanistan: Techy Neighbours." *Economist,* August 15, 1998, 35.

"Iran's War on Fundamentalism." *Economist,* September 12, 1998, 47.

"Iraqis Are Said to Wage War on Marsh Arabs." *New York Times,* October 19, 1993, A6.

Jehl, Douglas. "Holier Than Thou: Behind the Iranian-Afghan Rift." *New York Times,* September 7, 1998, A3.

Kinzer, Stephen. "A Delicate Dance With Islam." *New York Times,* March 29, 1998, 8.

Landay, Jonathan S. "Bill to Restrict Land Mines May Be Defused on Hill." *Christian Science Monitor,* September 21, 1995, 3.

"Landmines: Diana's Legacy." *Economist,* September 6, 1997, 47.

Leary, Warren E. "Better Weapons Emerge for War Against Mines." *New York Times,* December 16, 1997, B11.

Lev, Martin. *Traveler's Key to Jerusalem.* New York: Alfred A. Knopf, 1989.

Nelan, Bruce. "No Clean Sweep for Mines." *Time,* September 29, 1997, 38.

North, Andrew. "Flight of the Marsh Arabs." *The Middle East,* February 1994, 37–39.

————. "New Evidence Shows Marshlands Draining Away." *The Middle East,* October 1993, 22–23.

Peterson, Scott. "Struck Down in Turkey, Islam Coils." *Christian Science Monitor,* January 20, 1998, 6.

————. "James Baker Steps Into a N. African Sandstorm." *Christian Science Monitor,* August 19, 1997, 7.

Prusher, Ilene R. "Water Lies at Heart of Middle East Fight." *Christian Science Monitor,* April 17, 1998, 1.

"Report on Land Mines Shows Number Declining." *New York Times,* September 4, 1998, A7.

Schmemann, Serge. "Israel and the Ethiopians: Welcoming Newcomers Isn't Always So Easy." *New York Times,* February 4, 1996, E3.

"Southern Sudan's Starvation." *Economist,* July 18, 1998, 40.

Stauffer, T.R. "Water Flows in Qaddafi's Pharaonic Project." *Christian Science Monitor,* January 8, 1996, 1.

Theodoulou, Michael. "Of Missiles and Mediators: On Cyprus, Tensions Rise." *Christian Science Monitor,* July 21, 1998, 6.

"To Be, or Not to Be, a Saharawi?" *Economist,* March 21, 1998, 52.

Tregenza, Leon Arthur. *Egyptian Years.* London: Oxford University Press, 1958.

Velin, Jo-Anne. "Efforts to Ban Mines Persist After Setback." *Christian Science Monitor,* October 18, 1995, 6.

"Water in the Middle East: As Thick As Blood." *Economist,* December 23, 1995, 53–55.

Weiner, Tim, and James Risen. "Decision to Strike Factory in Sudan Based on Surmise." *New York Times,* September 21, 1998, 1.

World Book Encyclopedia Yearbook. New York: World Book Encyclopedia, 1994.

Wren, Christopher S. "Everywhere, Weapons That Keep on Killing." *New York Times,* October 8, 1995, E3.

Chapter 10

"Asia's Precarious Miracle." *Economist,* March 1, 1998, 18.

Chapman, Graham P., and Kathleen M. Baker. *The Changing Geography of Asia.* London: Routledge, 1992.

"Chronology of the Asian Currency Crisis." *Washington Post,* 1997.

Drake, Christine. "National Integration in China and Indonesia." *Geographical Review,* 82: 209–220, 1992.

"East Asia's Delicate Balance." *Economist,* July 25, 1998, 10.

"The El Niño and the North Korean Famine." Online at *http://www.kimsoft.com/korea/nk-nino.htm*

Gellert, Paul. "A Brief History and Analysis of Indonesia's Forest Fire Crisis." *Indonesia,* 65: 63– 85, 1998.

Gidwani, Vinay. "India's Nuclear Test." *Economic and Political Weekly,* 32(22): 1312–1315, 1998.

"India's Next 50 Years." *Economist,* August 16, 1997, 11–12.

"Indonesia Fire News on Line." Online at *http://www.suite101.com/articles/article*

Jones, Gavin. "Population and the Family in Southeast Asia." *Journal of Southeast Asian Studies,* 26(10): 184(12), 1995.

Lee, Manwoo. "North Korea: The Cold War Continues." *Current History,* 95(605): 438–442, 1996.

Malik, Iftikhar. "A Nuclearised South Asia: View From the Other Side." *Economic and Political Weekly,* 32(22): 1304–1305, 1998.

Shinn, James. "Japan as an 'Ordinary Country.'" *Current History,* 96(605): 401–407, 1996.

Stiglitz, Joseph. "Bad Private Sector Decisions." *Wall Street Journal,* February 4, 1998.

Takashi, Inoguchi. "The Coming Pacific Century." *Current History,* 93(597): 25–30, 1994.

Chapter 11

Barr, Cameron. "India's Poor Assess Damage from Rioting." *Christian Science Monitor,* December 21, 1992, 2.

Basham, A.L. *The Wonder That Was India.* New York: Grove Press, 1959.

Bearak, Barry. "Get-Tough Rule Puts Lid on Chaotic Karachi." *New York Times,* February 10, 1999, 1.

————. "Poor Face Grimmest Choices As Pakistan's Economy, and Regime, Unravel." *New York Times,* August 30, 1998, 12.

Bhardwaj, Surinder Mohan. *Hindu Places of Pilgrimage in India.* Berkeley: University of California Press, 1983.

"Blast Kills 53 in Sri Lanka; 1,400 Injured." *New York Times,* February 1, 1996, A1.

Bokhari, Farhan. "Cotton Farmers Face Big Setback." *Christian Science Monitor,* December 29, 1993, 8.

Burns, John F. "India Sets 3 Nuclear Blasts, Defying a Worldwide Ban; Tests Bring a Sharp Outcry." *New York Times,* May 12, 1998, 1.

————. "Hindu Party Is Candid on Arms But Not Much Else." *New York Times,* March 19, 1998, 1.

————. "Hindu Bloc Gains Weak Grip on Helm." *New York Times,* March 16, 1998, 1.

————. "India Counts on Vote to Blunt Kashmir Insurgency." *New York Times,* September 7, 1996, A3.

"Casteing Stones." *Economist,* July 19, 1997, 38.

Cressey, George B. *Asia's Lands and Peoples.* New York: McGraw-Hill, 1963.

Crossette, Barbara. "New Delhi Pledges to Sign World Ban on Nuclear Tests." *New York Times,* September 25, 1998, 1.

————. "Pakistan Pledges A-Bomb Test Ban If Sanctions End." *New York Times,* September 24, 1998, 1.

Dugger, Celia W. "A Sacred Indian Site But, Still, the Font of Strife." *New York Times,* August 30, 1998, 3.

"11 Killed in Truck Bombing at Sri Lanka Buddhist Site." *New York Times,* January 26, 1998, A3.

Elhance, Arun P. "From War to Water Pacts in Turbulent South Asia." *Christian Science Monitor,* January 15, 1997, 19.

Gargan, Edward A. "Shackled by Past, Racked by Unrest, India Lurches Toward Uncertain Future." *New York Times,* February 18, 1994, A4.

————. "For Many Brides in India, a Dowry Buys Death." *New York Times,* December 30, 1993, A5.

———. "Though Sikh Rebellion Is Quelled, India's Punjab State Still Seethes." *New York Times,* October 26, 1993, A1, A4.

Ginsburg, Norton, ed. *The Pattern of Asia.* Englewood Cliffs, N.J.: Prentice-Hall, 1958.

Girardet, Edward. "Helping Farmers Shake Poppy Habit." *Christian Science Monitor,* January 12, 1989, 6.

Hazarika, Sanjoy. "Plan to Clean Holy River Fails to Stem Tide of Filth." *New York Times,* October 18, 1994, B7, B10.

———. "India Dam Project Brings a Quandry." *New York Times,* June 2, 1992, A5.

"The Indian Election: Caste and Votes." *Economist,* February 28, 1998, 47.

Ijaz, Mansoor. "Dispel South Asia's Nuclear Cloud." *Christian Science Monitor,* February 20, 1998, 14.

Kinzer, Stephen. "Pakistan Is Under Growing Pressure Not to Respond to India with Atom Test." *New York Times,* May 15, 1998, A5.

Kour, Samsar Chand. *Beautiful Valleys of Kashmir and Ladakh.* Mysore, India: Wesley Press, 1942.

Landay, Jonathan S. "The New Nuclear World Order." *Christian Science Monitor,* 1.

Lodrick, Deryck O. *Sacred Cows, Sacred Places: Origins and Survivals of Animal Homes in India.* Berkeley: University of California Press, 1981.

Malik, Rajeev. "The Threat to India's Economic Reforms." *Christian Science Monitor,* December 21, 1992, 19.

Marquand, Robert. "After Nuclear Tests, Tests of Will in Kashmir." *Christian Science Monitor,* August 10, 1998, 6.

McGeary, Johanna. "Nukes . . . They're Back." *Time,* May 25, 1998, 34–42.

McGirk, Tim. "The Sword of Islam." *Time,* September 28, 1998, 56–57.

Miller, G. Tyler. *Living in the Environment,* 9th ed. Belmont, California: Wadsworth Press, 1996.

Moffett, George. "Pakistan's Population Growth Saps Economic Prosperity." *Christian Science Monitor,* September 6, 1994, 7.

"Pakistan Takes a Beating." *Economist,* May 16, 1998, 37.

Schwartzberg, Joseph E. "South Asia." In *World Geography,* 3rd ed., edited by John W. Morris, 513–552. New York: McGraw-Hill, 1972.

Spencer, J.E., and William L. Thomas. *Asia, East by South: A Cultural Geography.* New York: John Wiley & Sons, 1971.

"Sri Lanka: The Victory Still to Come." *Economist,* December 9, 1995, 39.

"Sri Lanka's Unhappy Birthday." *Economist,* February 7, 1998, 41.

Tefft, Sheila. "Caste Dispute Deepens India's Political Crisis." *Christian Science Monitor,* December 28, 1990, 8.

———. "Political and Ethnic Hot Spots in South Asia." *Christian Science Monitor,* May 8, 1990, 8.

"Violence Comes to Shangri-La." *Economist,* October 6, 1990, 40.

"Why India Loves the Bomb." *Economist,* August 22, 1998, 37.

Zubrzycki, John. "War in Sri Lanka Feeds on Itself." *Christian Science Monitor,* August 12, 1998, 5.

———. "The Bombs Work—Now What?" *Christian Science Monitor,* June 2, 1998, 1.

———. "Defiant India May Ignore Economic Sanctions." *Christian Science Monitor,* May 15, 1998, 6.

———. "Lower Castes Still Stuck on India's Bottom Rung" *Christian Science Monitor,* August 29, 1997, 1.

Chapter 12

"After the Smoke." *Economist,* September 30, 1995, 40.

"Aung San at the Bridge." *Economist,* August 1, 1998, 34.

Aung-Thwin, Maureen. "Suu Kyi's Release: Act I in Burma's Drama." *Christian Science Monitor,* August 2, 1995, 19.

Barr, Cameron. "Asian Tiger's New Path: Blocking the Flow of Capital." *Christian Science Monitor,* September 24, 1998, 1.

———. "More Nations Whip Up Squalls Over Tiny Isles." *Christian Science Monitor,* March 27, 1996, 1.

———. "A Democracy Fighter's Plight." *Christian Science Monitor,* September 1, 1995, 1.

Browne, Malcolm W. "Crowding and Managerial Gaps Imperil Vietnam." *New York Times,* May 8, 1994, A1.

———. "More Species May Be Discovered in Vietnam." *New York Times,* May 3, 1994, B9.

"Burma and Bangladesh at Odds Over Muslim Refugees." *Christian Science Monitor,* January 28, 1992, 4.

"Cambodia: Unhappy Returns." *Economist,* October 3, 1998, 44.

Chenault, Kathy. "Cambodia: Victory at the Ballot Box, But Threat of War Lingers." *Christian Science Monitor,* October 6, 1993, 11.

Cohen, Yvan. "U.S. Sanctions Fail to Bring Democracy to Burma." *Christian Science Monitor,* January 28, 1998, 6.

Cox, C. Berry, and Peter D. Moore. *Biogeography: An Ecological and Evolutionary Approach.* Oxford: Basil Blackwell, 1993.

Crossette, Barbara. "Indonesia Agrees to an Autonomy Plan for East Timor." *Christian Science Monitor,* August 6, 1998, A3.

Crowell, Todd. "Competing to Be Asia's No. 2." *Christian Science Monitor,* April 22, 1998, 1.

Dixon, Lily. "U.S. Ends Era of Welcome for Vietnam's Refugees." *Christian Science Monitor,* October 14, 1997, 1.

Erlanger, Steven. "America Opens the Door to a Vietnam It Never Knew." *New York Times,* February 6, 1994, 4A–4.

Erlich, Reese. "Low Wages in Thailand Despite Rapid Growth." *Christian Science Monitor,* August 30, 1995, 9.

———. "Philippines Population Efforts Raise Hackles." *Christian Science Monitor,* January 11, 1994, 10.

Esper, George. "The Rebuilding of Vietnam." *Columbia Missourian,* January 9, 1994, 4.

Gargan, Edward A. "Last Laugh for the Philippines: Onetime Joke Economy Avoids Much of Asia's Turmoil." *New York Times,* December 11, 1997, C1.

———. "A Boom in Malaysia Reaches for the Sky." *New York Times,* February 2, 1996, C1.

Gillotte, Tony. "Vietnamese Ecologist Assists Village Farmers." *Christian Science Monitor,* August 2, 1995, 14.

"The Grapes of Wrath." *Economist,* May 23, 1998, 36.

"The Hopes for Recovery Fade." *Economist,* June 6, 1998, 38.

Hottelet, Richard C. "Dangerous Isles: Even Barren Rocks Attract Conflict." *Christian Science Monitor,* March 7, 1996, 18.

Jones, Clayton. "Economic Cooperation Zones Create New Asian Geometry." *Christian Science Monitor,* December 1, 1993, 12.

———. "Paradise Islands or an Asian Powder Keg?" *Christian Science Monitor,* December 1, 1993, 14.

———. "Search for Security in the Pacific." *Christian Science Monitor,* November 17, 1993, 11.

———. "From Carpet Bombing to Capitalism in Laos." *Christian Science Monitor,* November 10, 1993, 10.

Kamm, Henry. "Decades-Old U.S. Bombs Still Killing and Maiming Laotians." *New York Times,* August 10, 1995, A5.

———. "Laos Capital Defies Time and Change." *New York Times,* August 6, 1995, A5.

———. "Communism in Laos: Poverty and a Thriving Elite." *New York Times,* July 30, 1995, A6.

Kemf, Elizabeth. *The Month of Pure Light: The Regreening of Vietnam.* London: The Women's Press, 1990.

"Laos: Dammed If They Do." *Economist,* August 30, 1997, 27.

"Malaysia Output Down 6.8%; Nation Formally in Recession." *New York Times,* August 28, 1998, C2.

"Myanmar's Secret Plague." *Economist,* August 23, 1997, 31.

"Myanmar: Trouble in the Pipeline." *Economist,* January 18, 1997, 39.

Mydans, Seth. "Cambodia: A New Try." *New York Times,* September 20, 1998, 7.

———. "Cambodia Result: Maybe Stability." *New York Times,* August 7, 1998, A6.

———. "As Boom Fails, Malaysia Sends Migrants Home." *New York Times,* April 9, 1998, 1.

———. "Asia's Wealth Ebbs, But Laos Is Too Poor to Care." *New York Times,* January 1, 1998, A4.

———. "Malaysia Taking Pledge of Austerity." *New York Times,* December 16, 1997, A9.

———. "Thailand Economic Crash Crushes the Working Poor." *New York Times,* December 15, 1997, A9.

———. "Its Mood Dark As the Haze, Southeast Asia Aches." *New York Times,* October 26, 1997, 3.

———. "Southeast Asia Chokes and Indonesian Forests Burn." *New York Times,* September 25, 1997, 1.

———. "Gas Pipeline Project Angers Critics of Burmese Rights Abuses." *New York Times,* December 8, 1996, 12.

———. "New Boat People Exodus: Back to Vietnam." *New York Times,* April 17, 1996, A1.

Olson, Elizabeth. "U.N. Urges Fiscal Accounting Including Sex Trade." *New York Times,* August 20, 1998, A11.

Peck, Grant. "Asia's Urban Morass: Exhausted Engines of the World Economy." *Columbia Missourian,* March 6, 1994, 4.

Porter, Gareth. "The Environmental Hazards of Asia-Pacific Development: The Southeast Asian Rain Forests." *Current History,* 93(587): 430–434, 1994.

Scott, David Clark. "Surging Sales for the Golden Triangle." *Christian Science Monitor,* January 6, 1989, 4.

Seper, Chris. "Unlike in Past, World Yawns at Cambodia's Latest Crisis." *Christian Science Monitor,* August 28, 1998, 6.

Shari, Michael. "Indonesia's Brass Polishes Itself." *Christian Science Monitor,* September 18, 1995, 1.

Shenon, Philip. "AIDS Epidemic, Late to Arrive, Now Explodes in Populous Asia." *New York Times,* January 21, 1996, A1.

———. "Filipino Victims May Share in Marcos's Loot." *New York Times,* October 28, 1995, A1.

———. "Burmese Cry Intrusion." *New York Times,* March 29, 1994, A5.

———. "AIDS Onslaught Breaches the Burmese Citadel." *New York Times,* March 11, 1994, A7.

———. "Pol Pot, the Mass Murderer Who Is Still Alive and Well." *New York Times,* February 6, 1994, A1.

Simpson, Beryl Brintnall, and Molly Conner-Ogorzaly. *Economic Botany: Plants in Our World.* New York: McGraw-Hill, 1986.

"Singapore Celebrates Economic Success." *Dodge City* (Kan.) *Daily Globe,* August 12, 1995, 5.

Stanley, Bruce. "Ho's Heirs in Hanoi Divide Power." *Christian Science Monitor,* August 4, 1995, 1.

———. "Vietnam Revels As the World Beats a Path to Its Open Door." *Christian Science Monitor,* July 18, 1995, 7.

Switow, Michael. "Asia Sputters Into Low Gear." *Christian Science Monitor,* December 1, 1997, B6.

"Tamed Tigers." *American Century Investor Perspective,* First Quarter: 6–7, 1998.

Tan, Abby. "Burma's Star Dissident Emerges." *Christian Science Monitor,* July 13, 1995, 6.

Tefft, Sheila. "Boat People Languish: Asia's Mired Huddled Masses." *Christian Science Monitor,* December 7, 1995, 1.

———. "Indonesian Regime Retains Grip but Faces, and Allows, More Dissent." *Christian Science Monitor,* February 25, 1994, 8.

———. "A New Kind of Battle Rages at Angkor Wat." *Christian Science Monitor,* September 18, 1992, 10.

———. "Cambodia's Long, Tough Road Home." *Christian Science Monitor,* May 20, 1992, 9.

———. "Thai Spirit Houses." *Christian Science Monitor,* August 30, 1991, 13.

"Thailand's Tourist Industry: Beached." *Economist,* July 6, 1991, 72.

Thong, Huynh Sanh. *The Heritage of Vietnamese Poetry,* 7, 69, 127, 209. New Haven: Yale University Press, 1979.

Tyler, Patrick E. "China Battles a Spreading Scourge of Illicit Drugs." *New York Times,* November 15, 1995, A1.

United Nations International Drug Control Programme. *World Drug Report.* New York: Oxford University Press, 1997.

"Vietnam: Re-mob Happy." *Economist,* April 26, 1997, 30.

Wallerstein, Claire. "Asia Enterprise: Subic Bay Thrives in Post–U.S. Era." *Christian Science Monitor,* November 10, 1997, 1.

"With the Rebels." *Economist,* October 14, 1995, 39.

WuDunn, Sheryl. "Asia's Tigers Lick Wounds and Worry." *New York Times,* September 16, 1998, C1, C6.

Chapter 13

"An Engine for Growth." *China Economic Review* 8(2): 16, 1998.

Bian, Wu. "Three Gorges Resettlement Project on a Sound Track." *Beijing Review,* 40(50): 13, 3, 1997.

"Can a Bear Love a Dragon?" *Economist,* 343(8014): 19–21, 1997.

Cannon, Terry and Alan Jenkins. *The Geography of Contemporary China: The Impact of Deng Xiaoping's Decade.* London: Routledge, 1992.

Fan, C. Cindy. "Economic Opportunities and Internal Migration: A Case Study of Guangdong Province." *Professional Geographer,* 48(1): 28–45, 1996.

Gilbert, Elizabeth. "Valley of the Dammed." *Utne Reader,* July-August 1996, 84–94.

Guldin, Gregory Eliyu. "'Desakotas' and Beyond: Urbanization in Southern China." *Ethnology,* 35(4): 265(19), 1996.

Hoh, Erling. "China's Three Gorges Dam Will Soon Transform the Yangtze." *Discovery,* July 1996, 29–42.

Jisen, Ma. "1.2 billion—Retrospect and Prospect of Population in China." *International Social Science Journal,* 48(2): 261(8), 1996.

Kaye, Lincoln. "The Grip Slips." *Far Eastern Economic Review,* 158(19): 18–20, 1995.

McCarthy, Terry. "The Pulse of China." *Time,* June 29, 1998, 31–35.

Minqi, Le. "China: Six Years After Tiananmen." *Monthly Review,* (20): 1–13, 1996.

"No Truck with Trains: China's Joint Ventures Take to the Highway." *Far Eastern Economic Review,* 159(3): 46, 1996.

Pannell, Clifton W., and Laurence J.C. Ma. *China: The Geography of Development and Modernization.* New York: John Wiley & Sons, 1983.

"The People's Dictatorship." *Economist,* 334(7906): 4–22, 1996.

Rongxia, Li. "Water, Land and Air Transportation: Competition Sharpens." *Beijing Review,* 40(3): 16–20.

Rosenthal, Elizabeth. "In China's Countryside, It Is Time to Grow Rich." *New York Times,* May 30, 1998, A4.

———. "China's Middle Class Savors Its New Wealth." *New York Times,* June 19, 1998, A1–A8.

"Scraply Islands." *Economist,* May 24, 1997, 39.

"South-East Asia's Learning Difficulties." *Economist,* August 16, 1997, 30.

"Stay Back, China." *Economist,* 338(7957): 39–41, 1996.

"Storm on the Yangtze." *Economist,* 346(8960): 45, 1998.

Toops, Stanley. "Recent Uyger Leaders in Xinjiang." *Central Asian Survey,* 11(2): 77–92, 1992.

Topping, Audrey R. "Ecological Roulette: Damming the Yangtze." *Foreign Affairs,* 74(5): 132–147, 1995.

Tyler, Patrick. "For Taiwan's Frontier Islands, the War Is Over." *New York Times,* October 4, 1995, A3.

Chapter 14

Baker, Michael. "S. Korea Keeps Smile Aimed at North." *Christian Science Monitor,* September 29, 1998, 6.

Barr, Cameron W. "More Freedom, Less Security for Japan." *Christian Science Monitor,* August 21, 1998, 8.

———. "Tiny Islands Emerge As Big Dispute." *Christian Science Monitor,* February 16, 1996, 6.

———. "Kobe's Political Aftershock: Defiant Japanese." *Christian Science Monitor,* July 20, 1995, 1.

Basic Facts on the Nanjing Massacre and the Tokyo War Crimes Trial. Bound Brook, N.J.: New Jersey–Hong Kong Network, 1996 (pamphlet and WWW document).

Blustein, Paul. "Women Need Not Apply." *Washington Post,* National Weekly Edition, August 28–September 3, 1995, 18.

Brauchli, Marcus W., and David P. Hamilton. "Watch Out, Investors: Asia's Hot Spots Are Flaring." *Wall Street Journal,* August 18, 1995, A6.

Broad, William J. "Japan Plans to Conquer Sea's Depths." *New York Times,* October 18, 1994, B7.

———. "New Study Questions Hiroshima Radiation." *New York Times,* October 13, 1992, A6.

Clark, Gregory. "The Confidence Gap Widens in a Shaken Japan." *International Herald Tribune,* February 1, 1995, 8.

Cutler, B.J. "Four Small Islands Prevent Treaty." *Columbia Missourian,* October 21, 1990, 5F.

De Souza, Anthony R., and Frederick P. Stutz. *The World Economy: Resources, Location, Trade and Development.* New York: Macmillan, 1994.

Ding, Ignatius. "Opposition to Japan's Bid for Permanent Seat in UN Security Duffy, Michael. "The Rubin Rescue" *Time,* January 12, 1998, 46–57. Council." *New York Times,* December 15, 1996, 14.

Erlanger, Steven. "As Clinton Visits Changing Asia, Military Concerns Gain Urgency." *New York Times,* April 15, 1996, A1.

Gaouette, Nicole. "Salting Away $116,000 Just Isn't Enough for Rainy Days." *Christian Science Monitor,* August 17, 1998, 7.

Gurdon, Meghan Cox. "Calm in Korea Belies Tension in Nuclear Dispute." *Christian Science Monitor,* February 1, 1994, 1.

Holstein, William J., and Kaxmi Nakarmi. "Korea." *Business Week,* July 31, 1995, 32–38.

"Japan: Political Meltdown." *Economist,* April 19, 1997, 1.

"Japan's Capital: Heading for the Hills?" *Economist,* February 1, 1997, 38.

"Japan's Unspoken Fears." *Economist,* October 7, 1995, 35–36.

Kambara, Keiko. "Japan's Birthrate Hits Low." *Christian Science Monitor,* March 20, 1990, 4.

Kristof, Nicholas D. "Japan Apologizes Forcefully for Its Occupation of Korea." *New York Times,* October 9, 1998, A3.

———. "Japan Is Torn Between Efficiency and Egalitarian Values." *New York Times,* October 26, 1998, 1.

———. "North Korea Wouldn't Invade the South, Would It?" *New York Times,* April 14, 1996, D1.

———. "South Korea's President, in a Rift With North, Delays Talks." *New York Times,* October 15, 1995, A4.

———. "In Japan, Chicken Little Lays the Golden Egg." *New York Times,* July 10, 1995, E5.

MacDonald, Donald. *A Geography of Modern Japan.* Woodchurch, Ashford, Kent (England): Paul Norbury Publications, 1985.

Matthews, Rupert O. *The Atlas of Natural Wonders.* New York: Facts on File, 1988.

McGill, Douglas C. "Scour Technology's Stain With Technology." *New York Times Magazine,* October 4, 1992, 32–60.

Moffett, George. "North Korea Curtails Its Nuclear Program." *New York Times,* September 21, 1995, A1.

"Nature Rarely Repeats Itself." *Economist,* August 2, 1997, 63.

"North Korea: Curiouser and Curiouser." *Economist,* August 22, 1998, 32.

Newcomb, Amelia A. "In Japan, Single Life Looks Good." *Christian Science Monitor,* January 2, 1998, 1.

Ozawa, Ichiro. *Blueprint for a New Japan: The Rethinking of a Nation.* New York: Kodansha International, 1995.

Palmer, R.R., and Joel Colton. *A History of the Modern World.* New York: Alfred A. Knopf, 1971.

Pollack, Andrew. "Japan Questions Its Costly Program to Predict Earthquakes." *New York Times,* January 13, 1998, B14.

———. "After Accident, Japan Rethinks Its Nuclear Hopes." *New York Times,* March 25, 1997, A6.

———. "Behind North Korea's Barbed Wire: Capitalism." *New York Times,* September 15, 1996, A3.

———. "Okinawans Send Message to Tokyo and U.S. to Cut Bases." *New York Times,* September 9, 1996, A3.

———. "Armed North Korea Troops Again Violate the DMZ." *New York Times,* April 8, 1996, A7.

———. "The Creed on the Farm: Rice Land is Sacred." *New York Times,* February 18, 1993, A4.

———. "Japan's Role in Ecology: Leadership That Has Had a Slow Start at Home." *New York Times,* July 31, 1992, A7.

Powell, Bill. "End of the Age of Hubris." *Newsweek,* January 30, 1995, 33.

Reid, T.R. "Kobe Wakes to a Nightmare." *National Geographic* 188(1): 112–136 (July 1995).

Reid, T.R., and Paul Blustein. "What a Difference a Half Century Makes." *Washington Post,* National Weekly Edition, August 21–27, 1995, 21.

Rosenthal, Elisabeth. "North Korea Says It Will Unseal Reactor." *New York Times,* May 13, 1998, A10.

Sanger, David E. "U.S. to Send North Korea Food Despite Missile Launchings." *New York Times,* September 10, 1998, A3.

———. "South Korea's Crisis Hinders Nuclear Deal With the North." *New York Times,* February 5, 1998, 1.

———. "Japan, Bowing to Pressure, Defers Plutonium Projects." *New York Times,* February 23, 1994, A2.

———. "Japan Denies Any Plans to Build Nuclear Bombs." *New York Times,* February 2, 1994, A5.

Shinohara, Makiko. "Scientists Clash Over Whaling." *Christian Science Monitor,* February 27, 1992, 10.

Spaeth, Anthony. "Engineer of Doom." *Time,* June 12, 1995, 57.

Spencer, E.W. "Japan: Stimulus or Scapegoat?" *Foreign Affairs,* 62(1): 123–137, 1983.

Sterngold, James. "Life in a Box: Japanese Question Fruits of Success." *New York Times,* January 2, 1994, A1.

"A Striptease in Japan." *Economist,* October 14, 1995, 20.

Weinstein, Michael M. "Newfangled Econ 101: Throw Caution to the Winds." *New York Times,* September 20, 1998.

Wood, Daniel B. "California Rice Growers Anticipate Upturn in Production, Demand." *Christian Science Monitor,* December 13, 1993, 7.

Chapter 15

Associated Press. "Cease-Fire Is Signed on Bougainville." *New York Times,* May 1, 1998, A13.

Chaddock, Gail Russell. "How French Islanders Live Above Ground Zero." *Christian Science Monitor,* August 31, 1995, 6.

———. "Avast, Ye Nuclear Protestors! Prepare to Be Boarded!" *Christian Science Monitor,* August 25, 1995, 6.

———. "French Wallets Take Hit in Protests Over Nuclear Tests." *Christian Science Monitor,* August 25, 1995, 1

———. "France Drops an Economic Bomb on Tahiti." *Christian Science Monitor,* August 14, 1995, 8.

Chaddock, Gail Russell, and David Rohde. "France, Colonies Stay Mum on Nukes." *Christian Science Monitor,* July 12, 1995, 7.

Cox, C. Barry, and Peter D. Moore. *Biogeography: An Ecological and Evolutionary Approach.* Oxford: Basil Blackwell, 1993.

DeBlij, H.J. *Human Geography.* New York: John Wiley & Sons, 1993.

Detwyler, Thomas R., ed. *Man's Impact on the Environment.* New York: McGraw-Hill, 1971.

Diamond, Jared. "Easter's End." *Discover,* 16(8): 62–69, 1995.

Foster, Catherine. "War in the Pacific: Legacy of a Copper Mine." *Christian Science Monitor,* July 20, 1994, 11.

Freeman, Otis W. "The Pacific Island World." *Journal of Geography,* 44: 25, 1945.

Gabler, Robert E., et al. *Essentials of Physical Geography.* Philadelphia: Saunders College Publishing, 1987.

Jordan, Terry G., and Lester Rowntree. *The Human Mosaic.* New York: Harper and Row, 1986.

Kristof, Nicholas D. "An Atomic Age Eden (But Don't Eat the Coconuts)." *New York Times,* March 5, 1997, A4.

———. "In Pacific, Growing Fear of Paradise Engulfed." *New York Times,* March 2, 1997, 1.

———. "U.S., Hands Off Guam! (Please?)." *New York Times,* November 17, 1996, E14.

Miller, G. Tyler. *Living in the Environment.* Belmont, California: Wadsworth, 1994.

Mitchell, Andrew. *The Fragile South Pacific: An Ecological Odyssey.* Austin: University of Texas Press, 1989.

Oliver, Douglas L. *The Pacific Islands.* New York: Doubleday Anchor Books, 1962.

Scott, David Clark. "Landowners Claim Bigger Slice of Mineral Pie in Papua New Guinea." *Christian Science Monitor,* December 27, 1988, 9.

"Settled in the Pacific." *Economist,* May 2, 1998, 40.

Shenon, Philip. "A Pacific Island Nation Is Stripped of Everything." *New York Times,* December 10, 1995, A3.

———. "Tahiti's Antinuclear Protests Turn Violent." *New York Times,* September 8, 1995, A4.

Silber, Judy. "New Find May Provide Insight Into Rising Seas." *Christian Science Monitor,* July 3, 1998, 3.

Whitney, Craig R. "Under Pressure, France Is Ending Its Nuclear Tests." *New York Times,* January 30, 1996, A1.

Woodward, Colin. "Past Atomic Tests Make Islanders Wary of Returning." *Christian Science Monitor,* August 11, 1998, 6.

———. "Uncle Sam Goes Home: Trouble in Paradise." *Christian Science Monitor,* August 11, 1998, 6.

Chapter 16

Arkley, Lindsey. "A Continental Divide: Who Owns Aboriginal Lands?" *Christian Science Monitor,* December 8, 1997, 1.

"Australia." *Nature* (film series). Narrated by George Page. Washington, D.C.: Public Broadcasting System, 1988.

"Australia: Neither Black Nor White." *Economist,* May 3, 1997, 28.

"Australia's Election: The Cry from the Outback." *Economist,* September 5, 1998, 37.

Blainey, Geoffrey. *Triumph of the Nomads: A History of Ancient Australia.* Melbourne: Macmillan, 1982.

Chaddock, Gail Russell. "Australian Farmers Consider Commercializing Kangaroo." *Christian Science Monitor,* November 29, 1995, 15.

———. "Australia's Hidden Strength in Asia: Chinese Immigrants." *Christian Science Monitor,* September 21, 1995, 7.

Coatney, Caryn. "The Queen's Defenders Down Under." *Christian Science Monitor,* February 11, 1998, 1.

Cohen, David. "Driving a Canoe? Defining Rights of Native Kiwis Can Be a Stretch." *Christian Science Monitor,* April 15, 1997, 1.

———. "New Zealand No Longer Tries to Be 'More English Than England," *Christian Science Monitor,* November 2, 1998, 6.

———. "N. Zealand's Identity Crisis." *Christian Science Monitor,* February 6, 1998, 6.

———. "New Zealand's Eco-Myth." *Christian Science Monitor,* November 7, 1997, 7.

———. "No Nukes Is Good Nukes? Ten Years After, Kiwis Soften Stance." *Christian Science Monitor,* June 17, 1997, 7.

Evans, Howard Ensign, and Mary Alice Evans. *Australia: A Natural History.* Washington, D.C.: Smithsonian Institution Press, 1983.

Farnsworth, Clyde H. "Land Rights of Aborigines Set Off Australian Battle." *New York Times,* December 17, 1997, A5.

———. "Australians Enjoy a Bullish Market in Wild Camels." *New York Times,* August 12,1997, A4.

———. "Rich Nation, But Illness Still Plagues Aborigines." *New York Times,* June 1, 1997, 5.

———. "Australia's Other Drink." *New York Times,* May 24, 1997, 21.

———. "The Betrayed Maori Are Calling for a Reckoning." *New York Times,* March 20,1997, A4.

Foster, Catherine. "Darwin, Aussie Pioneer Town, Seeks Survival in Asian Markets." *Christian Science Monitor,* June 22, 1994, 10.

———. "U.S., New Zealand Mend Ties Despite Differences." *Christian Science Monitor,* February 22, 1994, 6.

———. "Australia's Diamonds Lead Market Uptick." *Christian Science Monitor,* February 2, 1994, 7.

———. "New Gold Mining Activity Surges Ahead in Australia." *Christian Science Monitor,* January 12, 1994, 8.

———. "Australia Grants Aborigines Right to Claim Native Title." *Christian Science Monitor,* December 23, 1993, 3.

———. "Family Setting Fights Addiction." *Christian Science Monitor,* November 17, 1993, 18.

———. "Aborigines Unite to Fight for Australian Land Claims." *Christian Science Monitor,* August 9, 1993, 4.

———. "Aboriginal Families Get Their Land." *Christian Science Monitor,* January 6, 1993, 7.

"Gold Diggers." *Economist,* September 23, 1995, 5.

Grant, Bruce. "Human Rights in Asia: Australia Confronts an Identity Crisis." *New York Times,* March 20, 1994, E5.

MacLeod, Alexander. "New Zealand Shepherds in a Contentious Revolution." *Christian Science Monitor,* November 22, 1995, 6.

McBride, Eileen. "As Old Certainties Fade in Australia, a New Self-Image Evolves." *Christian Science Monitor,* February 20, 1997, 7.

———. "Australia Takes Barriers Off World's Greatest Reef." *Christian Science Monitor,* January 8, 1997, 10–11.

Moorehead, Allen. *Cooper's Creek.* New York: Macmillan, 1963.

Mydans, Seth. "Sea Change Down Under: Drifting to the Orient." *New York Times,* February 7, 1997, A4.

———. "The Stray Cats of Australia: 9 Lives Seen as 9 Too Many." *New York Times,* January 28, 1997, 1.

Pool, Gail. "Traveling in the Dreaming Tracks of Aboriginal Australia." *Christian Science Monitor,* September 2, 1987, 20.

"The Republican Rumble Down Under." *Economist,* February 14, 1998, 41.

Rohde, David. "Outback and New Market Needs Challenge Australian Cattleman." *Christian Science Monitor,* July 26, 1995, 9.

———. "Aussie Voters, Wary of Rapid Change, Put Heat on Leaders." *Christian Science Monitor,* July 21, 1995, 7.

Scherer, Ron. "Looking Up Down Under: An Aussie Bounce." *Christian Science Monitor,* May 15, 1997, 9.

_____. "How Native Land Disputes Can Be Win-Win for All." *Christian Science Monitor,* May 13, 1997, 11.

_____. "Living Off the Sheep's Back." *Christian Science Monitor,* April 16, 1992, 10.

_____. "Australia to Improve Native-Rights System." *Christian Science Monitor,* April 1, 1992, 6.

_____. "Australian Aborigines Turn to Antidrinking Programs." *Christian Science Monitor,* March 30, 1992, 15.

_____. "Australians Confront Poor Treatment of Aborigines." *Christian Science Monitor,* May 22, 1991, 6.

_____. "What's Hot, Dry and Big? Australia's Outback." *Christian Science Monitor,* December 19, 1990, 10.

Scott, David Clark. "Knocking Down Trade Walls." *Christian Science Monitor,* August 19, 1988, 7.

Shenon, Philip. "Australian Labor Party Retains Power in Election." *New York Times,* March 14, 1993, A3.

Sullivan, Jack. *Banggaiyerri: The Story of Jack Sullivan As Told to Bruce Shaw.* Canberra: Australian Institute of Aboriginal Studies, 1983.

Terrill, Ross. "The Aborigines' Search for Justice." *World Monitor,* May 1989, 48–55.

_____. "Australia's Next Frontier." *World Monitor,* October 1988, 36–46.

"Trade in the Pacific: No Action, No Agenda." *Economist,* November 25, 1995, 75.

Twain, Mark. *Following the Equator.* Reprint. New York: Dover, 1989.

Walker, Ruth. "Redefining Australia Takes a Backseat for This Election." *Christian Science Monitor,* October 1, 1998, 6.

_____. "Race, Asia Crisis Will Underlie Australia's Vote." *Christian Science Monitor,* August 31, 1998, 6.

_____. "'Whose Land?' Divides Ranchers and Aborigines." *Christian Science Monitor,* May 13, 1998, 7.

Chapter 17

"Africa's Colonial Favourites." *Economist,* October 28, 1995, 20.

Anderson, Terry L. "Zimbabwe Makes Living With Wildlife Pay." *Wall Street Journal,* October 25, 1991.

Associated Press. "New Study Bolsters a Theory on Migration of Early Humans." *New York Times,* March 8, 1996, A11.

Battersby, John. "Severe Drought Threatens Reform in Southern Africa." *Christian Science Monitor,* March 23, 1992, 1.

Butler, Victoria. "Is This the Way to Save Africa's Wildlife?" *International Wildlife,* March/April 1995, 38–43.

Camerapix Publishers International. *Spectrum Guide to Zimbabwe.* Edison, N.J.: Hunter Publishing, Inc., 1991.

French, Howard W. "African Democracies Worry Aid Will Dry Up." *New York Times,* March 19, 1995, A1.

_____. "An Ignorance of Africa As Vast As the Continent." *New York Times,* November 20, 1994, E3.

Hanley, Charles J. "Starting From Zero: Self-Sufficiency is a Fantasy for Nations of Sub-Saharan Africa." *Columbia Missourian,* February 28, 1993, 6.

Ibrahim, Youssef M. "AIDS Is Slashing Africa's Population, U.N. Survey Finds." *New York Times,* October 28, 1998, A3.

International Union for the Conservation of Nature and Natural Resources. *The Nature of Zimbabwe: A Guide to Conservation and Development.* Gland, Switzerland: IUCN, 1988.

Keller, Bill. "Southern Africa's Old Front Line Ponders Its Future in Mainstream." *New York Times,* November 20, 1994, A1.

_____. "Even Short of Horns, Rhinos of Zimbabwe Face Poacher Calamity." *New York Times,* October 11, 1994, B7.

Lederer, Edith. "The Next Famine." *Columbia Missourian,* August 21, 1994, 4.

Matloff, Judith. "Ivory Trade Made Legal—For Some." *Christian Science Monitor,* June 20, 1997, 5.

McNeil, Donald G. "AIDS Is the Silent Killer in Africa's Economies." *New York Times,* November 15, 1998, 1.

McPherson, James M. "Involuntary Immigrants." *New York Times Book Review,* 24.

Miller, G. Tyler. *Living in the Environment.* Belmont, California: Wadsworth Press, 1994.

Moffett, George D. "African Elephants to Remain Protected by International Treaty." *Christian Science Monitor,* March 13, 1992, 7.

Nasar, Sylvia. "Political Causes of Famine: It's Never Fair Just to Blame the Weather." *New York Times,* January 17, 1993, A1.

Nelson, Joan M. "Democracy in Africa." *Christian Science Monitor,* March 30, 1993, 18.

Noble, Kenneth B. "Political Chaos in Zaire Disrupts Efforts to Control AIDS Epidemic." *New York Times,* A1.

Perlez, Jane. "Rhino Near Last Stand, Animal Experts Warn." *New York Times,* July 7, 1992, A5.

_____. "Zimbabwe Kills Elephants to Help Save Lives." *New York Times,* July 5, 1992, A1.

Press, Robert M. "Mali Elections Break New Ground." *Christian Science Monitor,* February 13, 1992, 4.

Shiner, Cindy. "Wests Finds a Darling in Africa." *Christian Science Monitor,* October 5, 1995, 6.

Specter, Michael. "Doctors Powerless As AIDS Rakes Africa." *New York Times,* August 6, 1998, 1.

Stock, Robert. *Africa South of the Sahara: A Geographical Interpretation.* New York: Guilford, 1995.

Varisco, Daniel Martin. "Rhinoceros Horn Is Also the Animal's Achilles' Heel." *Christian Science Monitor,* June 28, 1988, 19.

Weisman, Steven R. "Bluefin Tuna and African Elephants Win Some Help at a Global Meeting." *New York Times,* March 11, 1992, A7.

Whitney, Craig. "Paris Snips Ties Binding It to Africa." *New York Times,* July 25, 1997, A5.

Zweifel, Thomas D. "New, Genuine Leaders in Africa." *Christian Science Monitor,* September 6, 1995, 19.

Chapter 18

Adams, R. and M., and A. Willens. *Dry Lands: Man and Plants.* London: Architectural Press, 1978.

"After the Hangings." *Economist,* November 18, 1995, 41.

"Africa's Pointing Finger." *Economist,* November 21, 1998, 48.

"Angola: Rebel Without a Cause." *Economist,* June 27, 1998, 48.

Battersby, John. "Going Home in Southern Africa." *Christian Science Monitor,* March 11, 1994, 6.

_____. "Namibian Sovereignty Complete As South Africa Withdraws from Port." *Christian Science Monitor,* March 1, 1994, 7.

_____. "Zimbabwe: Unity Pact Papers Over Persistent Tensions." *Christian Science Monitor,* December 28, 1992, 11.

Beran, Pau. "The Only Way to Dislodge Nigeria's Dictator." *Christian Science Monitor,* November 24, 1995, 19.

Biswas, M.R., and A.K. Biswas. *Desertification: Environmental Science and Application,* vol. 12. New York: Pergamon Press, 1980.

Botkin, Daniel B., Margriet F. Caswell, John E. Estes, and Angelo A. Ario, eds. *Changing the Global Environment.* Boston: Academic Press, 1989.

"Botswana and Namibia: Thirst." *Economist,* July 5, 1997, 48.

Browne, Malcolm W. "Fish That Dates Back to Age of Dinosaurs Is Verging on Extinction." *New York Times,* April 18, 1995.

"Burundi—the Next Bloodbath?" *Economist,* October 14, 1995, 49.

"Burundi on the Brink of Peace?" *Economist,* June 20, 1998, 49.

"Cash, Please." *Economist,* September 5, 1998, 44.

Cohen, Roger. "Mobil Spill Bares Ebb Tide of Nigerian Life." *New York Times,* September 20, 1998, 1.

"Congo: War Turns Commercial." *Economist,* October 24, 1998, 44.

Crown, Sarah. "In Adopting Black Babies, Whites 'Become African.'" *Christian Science Monitor,* December 26, 1995, 1.

————. "Edgy Whites Still Fleeing South Africa." *Christian Science Monitor,* September 20, 1995, 1.

Daley, Suzanne. "Angolan Rebels Still Break Diamond Embargo, Rights Group Says." *New York Times,* December 15, 1998, A7.

————. "How Did Pretoria Err? Lesotho Counts the Ways." *New York Times,* September 27, 1998, 6.

————. "Indian Ocean Island Yearns to Retie Colonial Bond." *New York Times,* September 29, 1997, A4.

————. "South African Democracy Stumbles in Old Rivalry." *New York Times,* January 7, 1996, A1.

————. "Tradition-Bound Swazis Chafing Under Old Ties." *New York Times,* December 16, 1995, A4.

————. "As Crime Soars, South African Whites Leave." *New York Times,* December 12, 1995, A1.

————. "Foes in Angola Still at Odds Over Diamonds." *New York Times,* September 15, 1995, A1.

Darnton, John. "Intervening with Élan and No Regrets." *New York Times,* June 26, 1994, E3.

De Waal, Alex, and Rakiya Omaar. "The Genocide in Rwanda and the International Response." *Current History,* 94(591): 156–161, 1995.

Dunn, Kate. "Zimbabwe Land Grab Rattles Region." *Christian Science Monitor,* November 25, 1998, 5.

————. "Why Zimbabwe Entered the Fray." *Christian Science Monitor,* August 27, 1998, 6.

————. "Power of Forgiveness in South Africa." *Christian Science Monitor* July 3, 1998, 1.

————. "Hoes and Hope Turn Around a War-Wracked African Nation." *Christian Science Monitor,* January 15, 1998, 7.

"Eritrea and Ethiopia: Spit and Slug." *Economist,* September 19, 1998, 60.

French, Howard W. "Congo Replay." *New York Times,* August 5, 1998, 1.

————. "West Africa's New Oil Barons." *New York Times,* March 7, 1998, B1.

————. "Nigerians Take Capital of Sierra Leone As Junta Flees." *New York Times,* February 14, 1998, A3.

————. "A West African Border With Back-to-Back Wars." *New York Times,* January 25, 1998, 3.

————. "Liberia Waits: Which Charles Taylor Won?" *New York Times,* January 17, 1998, A3.

————. "Outcome of Cameroon Vote: Fear of the Future." *New York Times,* October 14, 1997, A3.

————. "Where Proud Moors Rule, Blacks Are Outcasts." *New York Times,* January 11, 1996, A5.

————. "Nigeria Comes on Too Strong." *New York Times,* November 19, 1995, E3.

————. "Repression in Nigeria." *New York Times,* November 12, 1995, 9.

————. "Nigeria Executes Critic of Regime; Nations Protest." *New York Times,* November 11, 1995, A1.

————. "Africa's Ballot Box: Look Out for Sleight of Hand." *New York Times,* October 24, 1995, A3.

————. "Africa's Nations Start to Be Their Brothers' Keepers." *New York Times,* October 15, 1995, E6.

————. "Freetown Journal: Darkness, Not the Diamond's Dazzle." *New York Times,* October 9, 1995, A4.

————. "Nigeria Chief Offers Foes Concessions." *New York Times,* October 2, 1995, A5.

————. "Sleepy Congo, a Poor Land Once Very Rich." *New York Times,* June 18, 1995, 3.

Goldman Environmental Foundation. "Save One of the World's Brave and Honorable Environmental Heroes." *New York Times,* November 3, 1995, A7.

Gordimer, Nadine. "In Nigeria, the Price for Oil Is Blood." *New York Times,* May 25, 1997, E11.

Gritzner, Jeffrey Allman. *West African Sahel: Human Agency and Environmental Change.* University of Chicago Geography Research Paper No. 226, 1988.

Hackel, Joyce. "A Genocide Later, Rwanda Again on Edge." *Christian Science Monitor,* November 28, 1995, 6.

————. "Rwandan Troubles Made Worse by a Shake-Up." *Christian Science Monitor,* August 30, 1995, 1.

Jolly, Alison, and Richard Jolly. "Malagasy Economics and Conservation: A Tragedy Without Villains." In *Key Environments: Madagascar,* 211–217. Oxford: Pergamon Press, 1984.

Kassas, Mohamed A.F. "Ecology and Management of Desertification." *National Geographic,* 198–211.

————. *Earth '88: Changing Geographic Perspectives.* Washington, D.C.: National Geographic Society, (On reserve in Ellis Library.)

Kazaure, Z.M. "Nigerian Military Government Prepares for Return to Democratic Rule." *New York Times,* October 23, 1995, A9.

Kobani, Badey and Orage Memorial Foundation. "Until Now, Everything You've Read About the Ogoni Situation in Nigeria Has Been One-Sided." *New York Times,* December 6, 1995, A17.

Le Houerou, Henri Noel, and Hubert Gillet. "Conservation versus Desertization in African Arid Lands." In *Conservation Biology: The Science of Scarcity and Diversity,* edited by Michael E. Soule, 444–462. Sinauer Associates, 1986.

Lewis, L.A., and L. Berry. *African Environments and Resources.* Boston: Unwin Hyman Press, 1988.

Lorch, Donatella. "Even With Peace and Rain, Ethiopia Fears Famine." *New York Times,* January 3, 1996, A3.

————. "Ethiopia Deals With Legacy of Kings and Colonels." *New York Times,* December 31, 1995, 3.

————. "Trying to Avert a New Rwanda Refugee Crisis." *New York Times,* November 26, 1995, A6.

————. "Zanzibar Journal: Where Life Has No Spice, a $1 Billion Pick-Me-Up." *New York Times,* November 6, 1995, A4.

————. "A Joyful But Anxious Vote in Tanzania." *New York Times,* October 29, 1995, A8.

————. "Kenya Refuses to Hand Over Suspects in Rwanda Slayings." *New York Times,* October 6, 1995, A3.

————. "Kampala Journal: Cast Out Once, Asians Come Home." *New York Times,* March 22, 1993, A6.

MacArthur, R.H., and E.O. Wilson. *The Theory of Island Biogeography.* Princeton: Princeton University Press, 1967.

McKinley, James C. "Growing Ethnic Strife in Burundi Leads to Fears of New Civil War." *New York Times,* January 14, 1996, A1.

"Madagascar: Money Missing? Who Cares?" *Economist,* October 14, 1995, 50.

Matloff, Judith. "Southern Africa's Other Story: Economies Roar." *Christian Science Monitor,* August 7, 1997, 6.

————. "Southern Africa's Oasis May Turn to Dust." *Christian Science Monitor,* July 22, 1997, 1.

————. "W. Africans Fret Sanctions Will Stir Chaos in Nigeria." *Christian Science Monitor,* December 16, 1995, 6.

McKinley, James C. "Despite Pledge, Burundi Strife Grows." *New York Times,* September 19, 1996, A8.

McNeil, Donald G. "Air of Crisis in Zimbabwe Grows After Threat to Seize 841 Farms." *New York Times,* December 1, 1998, A15.

————. "Mozambique's Way With Austerity Puts It on Brink of an Economic Boom." *New York Times,* September 13, 1998, 10.

————. "Free Namibia Stumps the Naysayers." *New York Times,* November 16, 1997, 4.

Moose, George E. "Angola: After Decades of War, Bright Prospects." *Christian Science Monitor,* January 3, 1996, 18.

Nixon, Rob. "The Oil Weapon." *New York Times,* November 17, 1995, A23.

Noble, Kenneth B. "Zaire's Once-Wealthy Mines Now the Prey of Scavengers." *New York Times,* February 21, 1994, A1.

————. "Zaire Is in Turmoil After the Currency Collapses." *New York Times,* December 12, 1993.

————. "Glittering, New Capital Belies Turmoil in Nigeria." *New York Times,* October 31, 1993, 3.

————. "In Zaire, Starvation Is Growing in a Wealthy Land." *New York Times,* May 16, 1993.

Ntuyahaga, Monsignor. "Here Come the Cows." In *A Selection of African Prose,* edited by W.W. Whiteley, 142–145. Oxford: Clarendon, 1964.

O'Loughlin, Ed. "First Africa-Wide Effort to Restore a Democracy: So Far, So Good." *Christian Science Monitor,* March 11, 1998, 7.

————. "Now That One African Dictatorship Has Booted Another, Many Ask Why?" *Christian Science Monitor,* February 24, 1998, 7.

————. "Tribe May Be Flooded By Progress." *Christian Science Monitor,* November 12, 1997, 7.

Onishi, Norimitsu. "Nigeria Combustible As South's Oil Enriches North." *New York Times,* November 22, 1998, 1.

Perlez, Jane. "In Ethiopia, Islam's Tide Laps at the Rock of Ages." *New York Times,* January 13, 1992, A2.

Sammakia, Nejla. "Revisiting Somalia One Year Later." *Columbia Missourian,* December 12, 1993, 4.

Santoro, Lara. "Congo's 'Truce of Paris': The Promise, the Problems." *Christian Science Monitor,* December 1, 1998, 10.

————. "David and Goliath in Africa." *Christian Science Monitor,* November 18, 1998, 5.

————. "The Big Guns: Angola and Rwanda." *Christian Science Monitor,* August 27, 1998, 6.

————. "Africa's Domino Rebellion." *Christian Science Monitor,* August 21, 1998, 1.

————. "At the Heart of an African Rebellion." *Christian Science Monitor,* August 11, 1998, 1.

————. "Carving Up Congo As World Stands By." *Christian Science Monitor,* August 7, 1998, 1.

————. "At the Root of an Odd African War: Money." *Christian Science Monitor,* June 22, 1998, 1.

Schulz, William F. "The U.S. Ignores Africa at Its Own Peril." *Christian Science Monitor,* November 6, 1995, 19.

Shiner, Cindy. "Roots of Ex-U.S. Slaves Still Run Deep in Liberia." *Christian Science Monitor,* October 26, 1995, 7.

————. "Liberia Makes a Fresh Start After Six Years of Civil War." *Christian Science Monitor,* September 5, 1995, 6.

Simons, Marlise. "France Invades Comoros to Halt Coup." *New York Times,* October 5, 1995, A6.

Smith, Donald. "Horn of Africa." *St. Louis Post-Dispatch,* December 20, 1992, E1.

"South Africa: Old Wine in New Bottles" *Economist,* February 21, 1998, 45.

Stauffer, Thomas R. "West Africans Skirmish—Over Oil?" *Christian Science Monitor,* March 22, 1994, 9.

Stock, Robert. *Africa South of the Sahara: A Geographical Interpretation.* New York: Guilford, 1995.

Weber, Peter. "Neighbors Under the Gun." *Worldwatch,* July/August 1991, 35–36.

Wilson, Edward O. "Threats to Biodiversity." *Scientific American,* 261: 60–66.

"Zanzibar: Political Passion, Tourist Flesh-pots." *Economist,* March 14, 1998, 51.

"Zimbabwe: Peasants' Revolt." *Economist,* June 27, 1998, 48.

Chapter 19

Dean, Cornelia. "Burning a Landscape in Order to Save It." *New York Times,* May 26, 1998.

Kopinak, Kathryn. *Desert Capitalism: Maquiladoras in North America's Western Industrial Corridor.* Tucson: University of Arizona Press, 1966.

"The Last Communists." *Economist,* January 17, 1998, 19–21.

Loftis, Randy Lee. "Age of Fire." *Columbia Daily Tribune,* June 7, 1998, D1.

Price, Marie. "Ecopolitics and Environmental Nongovernmental Organizations in Latin America." *Geographical Review,* 84(1): 42–58, 1994.

Tedeschi, Tony. "Natural Attractions of the Caribbean and Central and South America." *Audubon,* 95(3): 113–119, 1993.

Chapter 20

"The Burning of Central America." *Economist,* May 30, 1998, 33–34.

"Fact Sheet: Coca Production and the Environment." *U.S. Department of State Dispatch,* 3(9): 165, 1992.

Manfredo, Jr., Fernando. "The Future of the Panama Canal." *Journal of Interamerican Studies & World Affairs,* 35(3): 103–128, 1993.

"Postrevolutionary Central America." *Current History,* 96(607): 49–86, 1997.

"Shipshape." *Economist,* September 13, 1997, 65.

Chapter 21

Goulding, Michael. "Flooded Forests of the Amazon." *Scientific American,* 268(3): 114–120, 1993.

Goulding, Michael et al. *Floods of Fortune: Ecology and Economy Along the Amazon.* New York: Columbia University Press. 1996.

Franco, Mario de and Ricardo Godoy. "The Economic Consequences of Cocaine Production in Bolivia: Historical, Local, and Macroeconomic Perspectives." *Journal of Latin American Studies,* 24(2): 375–406, 1992.

Marcus, Joyce. "The Amazon: Divergent Evolution and Divergent View." *National Geographic Research & Exploration,* 10(4): 384–394, 1994.

Chapter 22

"All Good Things Must Slow Down." *Economist,* March 7, 1998, 35–36.

Barkema, Alan and Mark Drabenston. "A New Agricultural Policy for a New World Market." *Economic Review,* 79(2): 59–72, 1994.

"Deadly Drug? That's Coca Leaf." *Economist,* March 7, 1998, 33–34.

Fox, Annette B. "Environment and Trade: The NAFTA Case." *Political Science Quarterly,* 110(1): 49–68, 1995.

"Guerrillas, Drugs, Votes and Clinton." *Economist,* March 7, 1998, 34.

"Latin America's Drift to the Cities." *Economist,* October 14, 1995, 36.

Zlotnik Hania. "International Migration 1965–96: An overview." *Population Development Review,* 24(3): 460–466, 1998.

Chapter 23

DePalma, Anthony. "Asian Economic Woes Hurt Canadian Tourism," *New York Times,* January 3, 1999, TR3.

————. "Shifts in Symbol and Substance Muddy Quebec Vote." *New York Times,* November 30, 1998, A4.

Hamley, Will. "Problems and Challenges in Canada's Northwest Territories." *Geography,* 78(340): 267–280, 1993.

"Hemisphere Be Damned." *Economist,* April 25, 1998, 22.

Kaplan, David H. "Population and Politics in a Plural Society: The Changing Geography of Canada's Linguistic Groups." *Annals of the Association of American Geographers,* 84: 46–67, 1994.

Parizeau, Jacques. "The Case for a Sovereign Quebec." *Foreign Policy,* 99: 69–76, 1995.

Pelly, David. "Birth of an Inuit Nation." *Geographical Magazine,* 66(4): 23–25, 1994.

Safire, William. "Name That Nation." *New York Times Magazine,* July 5, 1998, 10.

Salee, Daniel. "Identities in conflict: The Aboriginal question and the politics of recognition in Quebec." *Ethnic and Racial Studies,* 28(2): 277–314, 1995.

Chapter 24

Berle, Peter. "A Better Option for Ancient Forests." *Audubon,* 96(1): 6, 1994.

Elazar, Daniel J. *The American Mosaic: The Impact of Space, Time, and Culture on American Politics.* Boulder, CO: Westview Press, 1994.

Fost, Dan. "California Comeback." *American Demographics,* 17(7): 52–53.

Frey, William H. and Alden Speare, Jr. "America at Mid-Decade." *American Demographics,* 17(2): 23–31, 1992.

Garreau, Joel. *Edge City: Life on the New Frontier.* New York: Doubleday, 1991.

Hudson, John C. *Making the Corn Belt: A Geographical History of Middle-Western Agriculture.* Bloomington: Indiana University Press, 1994.

Jakle, John A. and David Wilson. *Derelict Landscapes: The Wasting of America's Built Environment.* Savage, MD: Rowman & Littlefield, 1992.

MacDonald, Kent. "Reston Revisited." *Landscape,* 32(2): 28–33, 1994.

Miyares, Ines M. "Changing Perceptions of Space and Place as Measures of Hmong Acculturation." *Professional Geographer,* 49(2): 214–224, 1997.

Moody, David W. "Water: Fresh Water Resources of the United States." *National Geographic Research & Exploration,* 9(1): 81–85, 1993.

Purdum, Todd S. "This Time, Los Angeles May Lose Water War." *New York Times,* June 15, 1998, A1–A14.

Uchitelle, Louis. "Now, Mid-size is Beautiful." *New York Times,* January 5, 1999, C1–C8.

"Valley of the Dammed." *Economist,* February 21, 1998, 26–27.

Zelinsky, Wilbur. "On the Superabundance of Signs in our Landscape." *Landscape,* 31(3): 30–38, 1992.

Glossary

Absolute (mathematical) location Determined by the intersection of lines such as latitude and longitude, providing an exact point expressed in degrees, minutes, and seconds.

Acid rain Precipitation that mixes with airborne industrial pollutants, causing the moisture to become highly acidic, and therefore harmful to flora and bodies of water on which it falls. Sulfuric acid is the most common component of this acid precipitation.

Adaptive reuse Finding new uses for older buildings and stores, often accompanied by a shift from decline to steady renewal in an urban neighborhood.

Age of Exploration (also called the Age of Discovery) Era of European seafaring lasting from the 15th century through the early 19th century.

Age-structure profile (population pyramid) Graphic representation of a country's population by gender and five-year age increments.

Agricultural or Neolithic (New Stone Age) Revolution The domestication of plants and animals that began about 10,000 years ago.

Altitudinal zonation In a highland area, the presence of distinctive climatic and associated biotic and economic zones at successively higher elevations. Ethiopia and Bolivia are examples of these kinds of zones. See *tierra caliente, tierra templada, tierra fria.*

Anticline *See* Folding.

Anticyclone Atmospheric high-pressure cell. In the cell, the air is descending and becomes warmer. As it warms, its capacity to hold water vapor increases, and the result is minimal precipitation.

Anti-Semitism Anti-Jewish sentiments and activities.

Apartheid The Republic of South Africa's former official policy of "separate development of the races," designed to ensure the racial integrity and political supremacy of the white minority.

Arable Suitable for cultivation.

Archipelago Chain or group of islands.

Arid China In a climatic division of China approximately along the 20° isohyet, Arid China lies to the west of the line. It characterizes more than half of the territory of the country, but accommodates less than 10 percent of China's population.

Atmospheric pollution The modification of the blanket of gases that surround the earth largely through the airborne products of industrial production and the human consumption of fossil fuels.

Atoll Low islands made of coral and usually having an irregular ring shape around a lagoon.

Balkanization Fragmentation of a political area into many smaller independent units, as in former Yugoslavia.

Barter The exchange of goods or services in the absence of cash.

Basin irrigation Ancient Egyptian system of cultivation using fields saturated by seasonal impoundment of Nile floodwaters.

Bazaar (suq) Central market of the traditional Middle Eastern city, characterized by twisting, close-set lanes, and merchant stalls.

Belief systems The set of customs that an individual or culture group has relating to religion, social contracts, and other aspects of cultural organization.

Benelux The name used to collectively refer to the countries of Belgium, Netherlands, and Luxembourg.

Biodiversity hot spots A ranked list of places scientists believe deserve immediate attention for flora and fauna study and conservation.

Biological diversity (biodiversity) The number of plant and animal species and the variety of genetic materials these organisms contain.

Biomass The collective dried weight of organisms in an ecosystem.

Biome Terrestrial ecosystem type categorized by a dominant type of natural vegetation.

Birth rate The annual number of live births per thousand people in a population.

Boat people In this text, the term given to the refugee populations who attempted to flee Cuba and Haiti in the early 1990s by launching themselves from the islands toward Florida in the hope that they would get picked up by the U.S. Coast Guard and given haven. The term had an earlier use by peoples in Southeast Asia making similar efforts to flee Vietnam.

Bourgeoisie In Marxist doctrine, the capitalist class.

Brain drain The exodus of educated or skilled persons from a poor to a rich country, or from a poor to a rich region within a country.

Break-of-bulk point A classic geographic term describing a point in transit when bulk goods must be removed from one mode of transport and installed on another. Trainloads of grain carried to a port for transhipment on cargo boats or barges is a common example.

Buffer state The term used for a generally smaller political unit adjacent to a large, or between several large, political units. Such a role often times enables the smaller state to maintain its independence because of its mutual use to the larger, proximate nations. Uruguay, between Brazil and Argentina, is an example.

Capital goods Goods used to produce other goods.

Carrying capacity The size of a population of any organism that an ecosystem can support.

Cartogram A special map in which an area's shape and size are defined by explicit characteristics of population, economy, or distribution of any stated product.

Cartography The craft of designing and making maps, the basic language of geography. In recent years, this traditional manual art has been changed profoundly through the use of computers and Geographic Information Systems (GIS), and through major improvements in machine capacity to produce detailed, colored, map products.

Cash (commercial) crops Crops produced generally for export.

Caste Hierarchy in the Hindu religion that determines a person's social rank. It is determined by birth and cannot be changed.

Chernobyl The site in the Ukraine where, in April 1986, the worst nuclear power plant accident in history occurred. It is thought that approximately 5000 people died and a zone with a twenty mile radius is still virtually uninhabitable; 116,000 people were moved from the area and cleanup continues to this day.

Chernozem A Russian term meaning "black earth." It is a grassland soil that is exceptionally thick, productive, and durable.

Chestnut soils Productive soils typical of the Russian steppe and North American Great Plains.

China proper The relatively well-watered eastern portion of China, which is the home of the great majority of the Chinese population. It is sometimes described as Humid China, as opposed to Arid China that lies to the west, and that has relatively little population.

Chokepoint A strategic narrow passageway on land or sea that may be closed off by force or threat of force.

Choropleth maps Maps that are drawn to show the differing distribution of goods or geographic characteristics (including population) across a broad area. Such maps are good for generalizations, but often mask significant local variations in the presence of the item being mapped.

Chunnel (Eurotunnel)   The 23-mile tunnel that links Britain with the European continent. It was completed in 1994 at a cost of more than 15 billion dollars, making it the single most costly project in landscape transformation ever undertaken.

Civilization   The complex culture of urban life.

Climate   The average weather conditions, including temperature, precipitation, and winds, of an area over an extended period of time.

Climatology   The scientific study of patterns and dynamics of climate.

Coal   Residue from organic material compressed for a long period under overlying layers of the earth. Coals vary in hardness and heating value. Anthracite coal burns with a hot flame and almost no smoke. Bituminous coal is used in the largest quantities; some can be used to make coke by baking out the volatile elements. Coke is largely carbon and burns with intense heat when used in blast furnaces to smelt iron ore. Peat, which is coal in the earliest stages of formation, can be burned if dried. Concern over atmospheric pollution by the sulfur in coal smoke has recently increased the demand for low-sulfur coals.

Collective farm   A large-scale farm in the Former Soviet Union that usually incorporated several villages. Workers received shares of the income after the obligations of the collective had been met.

Collectivization   The process of forming collective farms in Communist countries.

Colonization   The European pattern of establishing dependencies abroad to enhance economic development in the home country.

Colored   A South African term referring to persons of mixed racial ancestry.

Common Market   This is an earlier name given to the (current) 15 countries that make up the European Union. In 1957, an initial six countries combined to form the European Economic Community (EEC), and this supranational community has grown to have considerable economic and political importance in Europe. *See* European Union.

Compressional fault   *See* Fault.

Computer cartography   Map making using sophisticated software and computer hardware. It is a new, actively growing career field in geography.

Concentric zone model   A generalized model of a city. A city becomes articulated into contrasting zones arranged as concentric rings around its central business district.

Coniferous vegetation   Needleleaf evergreen trees; most bear seed cones.

Consumer goods   Goods that individuals acquire for short-term use.

Consumer organisms   Animals that cannot produce their own food within a food chain.

Consumption overpopulation   The concept that a few persons, each using a large quantity of natural resources from ecosystems across the world, add up to too many people for the environment to support.

Containerization   Prepackaging of items into larger standardized containers for more efficient transport.

Convectional precipitation   The heavy precipitation that occurs when air is heated by intense surface radiation, then rises and cools rapidly.

Council for Mutual Economic Assistance (COMECON)   A former economic organization consisting of the Soviet Union, Poland, East Germany, Czechoslovakia, Hungary, Romania, Bulgaria, Cuba, Mongolia, and Vietnam, now disbanded.

Crop calendar   The dates by which farmers prepare, plant, and harvest their fields. Farmers who are involved in new agricultural patterns necessitated by greater use of chemical fertilizers and new plant strains are sometimes unable to adjust to the demands of a much more exact crop calendar than has been traditional.

Crop irrigation   Bringing water to the land by artificial methods.

Cultural geography   The study of the ways in which humankind has adopted, adapted to, and modified the face of the earth, with particular attention given to cultural patterns and their associated landscapes. It also includes a culture's influence on environmental perception and assessment.

Cultural landscape   The landscape modified by human transformation, thereby reflecting the cultural patterns of the resident culture at that time.

Cultural mores   The belief systems and customs of a culture group.

Cultural diffusion   One of the most important dynamics in geography, cultural diffusion is the engine of change as crops, languages, culture patterns, and ideas are diffused from one place to other places, often in the course of human migration.

Culture   The values, beliefs, aspirations, modes of behavior, social institutions, knowledge, and skills that are transmitted and learned within a group of people.

Culture hearth   An area where innovations develop, with subsequent diffusion to other areas.

Culture system   A system in which Dutch colonizers required farmers in Java to contribute land and labor for the production of export crops under Dutch supervision.

Cyclone   A low-pressure cell that composes an extensive segment of the atmosphere into which different air masses are drawn.

Cyclonic (frontal) precipitation   The precipitation generated in traveling low-pressure cells which bring different air masses into contact.

Death rate   The annual number of deaths per thousand people in a population.

Debt-for-nature swap   An arrangement in which a certain portion of international debt is forgiven in return for the borrower's pledge to invest that amount in nature conservation.

Deciduous trees   Broadleaf trees that lose their leaves and cease to grow during the dry or the cold season and resume their foliage and grow vigorously during the hot, wet season.

Deflation   A fall in prices, such as that caused by currency devaluations.

Deforestation   The removal of trees by people or their livestock.

Delta   Landform resulting from the deposition of great quantities of sediment when a stream empties into a larger body of water.

Demographic shift   The term for major population redistributions as people move from the countryside to the city, or from one region to another.

Demographic transition   A model describing population change within a country. The country initially has a high birth rate, a high death rate, and a low rate of natural increase, moves through a middle stage of high birth rate, low death rate, and high rate of natural increase, and ultimately reaches a third stage of low birth rate, low or medium death rate, and low or negative rate of population increase.

Dependency theory   A theory arguing that the world's more developed countries continue to prosper by dominating their former colonies, the now-independent less developed countries.

Desert   An area too dry to support a continuous cover of trees or grass. A desert generally receives less than 10 inches (25 cm) of precipitation per year.

Desertification   Expansion of a desert brought about by changing environmental conditions or unwise human use.

Detritus   Material shed from rock surfaces in the process of disintegration and erosion.

Development   A process of improvement in the material conditions of people often linked to the diffusion of knowledge and technology.

Diaspora   The scattering of the Jews outside Palestine beginning in the Roman Era.

Dissection   Carving of a landscape into erosional forms by running water.

Distributary   Stream that results when a river subdivides into branches in a delta.

Donor fatigue   Public or official weariness of extending aid to needy people.

Drought avoidance   Adaptations of desert plants and animals to evade dry conditions by migrating (animals) or being active only when wet conditions occur (plants and animals).

Drought endurance   Adaptations of desert plants and animals to tolerate dry conditions through water storage and heat loss mechanisms.

Dry farming   Planting and harvesting according to the seasonal rainfall cycle.

Ecologically dominant species   A species that competes more successfully than others for nutrition and other essentials of life.

Ecology   Study of the interrelationships of organisms to one another and to the environment.

Economic development A process that generally includes a major demographic shift toward a more urban population, replacement of major imports with domestically produced goods, expansion of export industries based on manufactured goods, and a general upgrading of domestic patterns of health care, energy use, and literacy.

Ecosphere (biosphere) The vast ecosystem composed of all of Earth's ecosystems.

Ecosystem System composed of interactions between living organisms and nonliving components of the environment.

Ejido An agricultural unit in Mexico characterized by communally farmed land, or common grazing land. It is of particular importance to indigenous villages.

Endemic species A species of plant or animal found exclusively in one area.

Energy crisis The petroleum shortages and price surges sparked by the 1973 oil embargo.

Environmental assessment The process of determining the condition and value of a particular environmental setting. Used both in aspects of landscape change and in evaluating environmental perception.

Environmental determinism The belief that the physical environment has played a major role in the cultural development of a people or locale. Also called environmentalism.

Environmental perception The systematic evaluation of environmental characteristics in terms of human patterns of settlement, resources, land use, and attitudes toward the earth and its settings.

Environmental possibilism A philosophy seen in contrast to *environmental determinism* that declares that although environmental conditions do have an influence on human and cultural development, people have varied possibilities in how they decide to live within a given environment.

Escarpment A steep edge marking an abrupt transition from a plateau to an area of lower elevation.

Estuary A deepened ("drowned") river mouth into which the sea has flooded.

Ethnocentrism Regarding one's own group as superior and as setting proper standards for other groups.

European Economic Community Economic organization designed to secure the benefits of large-scale production by pooling resources and markets. The name has been changed to Economic Union. *See* Common Market and Economic Union.

Evaporation The loss of moisture from the Earth's land surfaces and its water bodies to the air through the ongoing influence of solar radiation and transpiration by plants.

Exotic species A nonnative species introduced into a new area.

Exploration The human fascination with new landscapes that leads to the search for new places, new resources, new cultural patterns, and new routes of travel and trade.

Extensive land use A livelihood, such as hunting and gathering, that requires the use of large land areas.

European Free Trade Association (EFTA) Organization that maintains free trade among its members but allows each member to set its own tariffs in trading with the outside world. Members are Iceland, Norway, Sweden, Switzerland, Austria, and Finland.

European Union The current organization begun in the 1950s as the Common Market. It now is made up of fifteen nations, including France, Germany, Italy, Belgium, Luxembourg, the Netherlands, the United Kingdom, Denmark, Ireland, Greece, Portugal, Spain, Austria, Finland, and Sweden. *See* Common Market.

Fall line Zone of transition in eastern United States where rivers flow from the harder rocks of the Piedmont to the softer rocks of the Atlantic and Gulf Coastal Plain. Falls and/or rapids are characteristic features.

Fault. A break in a rock mass along which movement has occurred. A break due to rock masses being pulled apart is a tensional or "normal" fault, whereas a break due to rocks being pushed together until one mass rides over the other is a compressional fault. The processes of faulting creating these breaks.

Fertile Crescent The arc-shaped area stretching from southern Iraq through northern Iraq, southern Turkey, Syria, Lebanon, Israel and western Jordan, where plants and animals were domesticated beginning about 10,000 years ago.

Fjord Long, narrow extension of the sea into the land usually edged by steep valley walls that have been deepened by glaciation.

Flow resource A resource that can be renewed, hence that extracts a lower cost from the environment in its utilization.

Folding The process creating folds (landforms resulting from an intense bending of rock layers). Upfolds are known as anticlines, and downfolds are called synclines.

Food chain The sequence through which energy, in the form of food, passes through an ecosystem.

Formal region *See* Region.

Four Modernizations The effort of the Chinese after the death of Mao Zedong in 1976 to focus Chinese efforts at economic development in agriculture, industry, science and technology, and defense.

Front Contact zone between unlike air masses. A front is named according to the air mass that is advancing (cold front or warm front).

Frontal precipitation *See* Cyclonic precipitation.

Fuelwood crisis Deforestation in the less developed countries caused by subsistence needs.

Functional region *See* Region.

Gaia Hypothesis A hypothesis stating that the ecosphere is capable of restoring its equilibrium following any disturbance that is not too drastic.

Gentrification The social and physical process of change in an urban neighborhood by the return of young, often professional, populations to the urban core. These peoples are often attracted by the substantial nature of the original building stock of the place and the proximity to the city center, which generally continues to have major professional opportunities. While this process brings an urban landscape back into a primary role as a tax base, this change does dispossess a considerable number of minority peoples, who tended to remain in these neighborhoods as the white population moved to the suburbs during the past five or six decades.

Geographic analysis By giving attention to the spatial aspects of a distribution, geographic analysis helps to explain distribution, density, and flow of a given phenomenon.

Geographic Information Systems (GIS) The increasingly popular field of computer-assisted geographic analysis and graphic representation of spatial data. It is based on superimposing various data layers that may include everything from soils to hydrology to transportation networks to elevation. Computer software and hardware are steadily improving, enabling GIS to produce ever more detailed and exact output.

Geography Study of the spatial order and associations of things. Also defined as study of places, study of relationships between people and environment, study of spatial organization.

Geomorphology The scientific analysis of the landforms of the earth, sometimes called *physiography*.

Glacial deposition In the process of continental and valley glaciation, the deposition of moraines that become lateral or terminal—depending on where they are deposited—in the act of glacial retreat. This same process also leads to glacial scouring as moving ice picks up loose rock and reshapes the landscape as the glacier moves forward or retreats.

Glacial scouring *See* glacial deposition.

Graben (rift valley) Landform created when a segment of the Earth's crust is displaced downward between parallel tensional faults or when segments of the crust which border it ride upward along parallel compressional faults.

Gravity flow Water flow in an irrigation or a hydroelectric system that allows for the free flow of water from a source to another area without the addition of motor or animal power. In bringing water to a city, such a characteristic is of particular importance because of its relative cheapness compared to the need to pump water upslope.

Green Revolution The transfer of high-yielding seeds, mechanization, irrigation, and massive application of chemical fertilizers to areas where traditional agriculture has been practiced.

Greenhouse Effect The observation that increased concentrations of carbon dioxide and other gasses in Earth's atmosphere causes a warmer atmosphere.

Gross domestic product (GDP) Value of goods and services produced in a country in a given year. Does not include net income earned outside the country. The value is normally given in current prices for the stated year.

Growth national product (GNP) Value of goods and services produced internally in a given country during a stated year, plus the value resulting from transactions abroad. The value is normally stated in current prices of the stated year. Such data must be used with caution in regard to developing countries because of the broad variance in patterns of data collection and the fact that many people consume a large share of what they produce.

Growth pole The term for a new city, a resource, or some development in a heretofore under-settled area that begins to attract population growth and economic development.

Guano Seabird excrement that is also found in phosphate deposits.

Gulf Stream Strong ocean current originating in the tropical Atlantic Ocean that skirts the eastern shore of the United States, curves eastward, and reaches Europe as a part of the broader current called the North Atlantic Drift.

Hacienda A Spanish term for large rural estates owned by the aristocracy in Latin America.

Haj Pilgrimage to Mecca, the principal holy city in the Islamic religion. Every Muslim is required to make this pilgrimage at least once in a lifetime, if possible.

Han Chinese The term for the original Chinese peoples who settled in North China on the margins of the Yellow River (Huang He) and who were central to the development of Chinese culture. Han Chinese make up approximately 94 percent of China's current population.

Hierarchichal rule A system by which leadership for a culture group is defined by traditional and sometimes frustrating patterns of age–sex distinction. Evident in most cultures and slowly modified by an increasing role of democratic elections and the establishment of a broader opportunity base for individual advancement.

Historical geography Concern with the historical patterns of human settlement, migration, town building, and the human use of the earth. Often the subdiscipline that best blends geography and history as a perspective on human activity.

Holocaust Nazi Germany's attempted extermination of Jews, Gypsies, homosexuals, and other minorities during World War II.

Homelands Ten former territorial units in South Africa reserved for native Africans (blacks). They had elected African governments, and some were designated as "independent" republics, although they were not recognized outside of South Africa. Formerly called Bantustans. They were abolished in 1994.

Horizontal migration The movements of pastoral nomads over relatively flat areas to reach areas of pasture and water.

Horst Block mountain formed between two roughly parallel faults; has a steep face on two sides.

Hot spot Small area of the Earth's mantle where molten magma is relatively close to the crust. Hot spots are associated with island chains and thermal features.

Household Responsibility System The Chinese system devised in 1978 that enabled their rural population to begin to free itself from the communal structures that had characterized the years of development under Chairman Mao Zedong. This innovation returned much agricultural decision-making to the farm household, although there was a required portion of the major crops that had to be sold to the government.

Humboldt (or Peru) Current The cold ocean current that flows northward along the west coast of South and North America. It plays a significant role in regional fish resources.

Humid China The eastern portion of China that is relatively well-watered and where the great majority of the Chinese population has settled. An arc from Kunming in southwest China to Beijing in North China describes the approximate western margin of this zone.

Humid continental Climate with cold winters, warm to hot summers, and sufficient rainfall for agriculture, with the greater part of the precipitation in the summer half-year.

Humid pampa A level to gently rolling area of grassland centered in Argentina with a humid subtropical climate.

Humid subtropical Climate that characteristically occupies the southeastern margins of continents with hot summers, mild to cool winters, and ample precipitation for agriculture.

Humus Decomposed organic soil material. Grasslands characteristically provide more humus than forests do.

Hunting and gathering A mode of livelihood, based on collection of wild plants and hunting of wild animals, characterizing preagricultural peoples, generally.

Hydraulic control The ability to control water in irrigation systems, rivers, urban settlements, and for the generation of electricity has been central to the development of major civilizations, and to culture groups throughout human history.

Ideology A system of political and/or economic beliefs such as communism, capitalism, autocracy, or democracy.

Igneous (volcanic) rock Formed by the cooling and solidification of molten materials. Granite and basalt are common types. Such rocks tend to form uplands in areas where sedimentary rocks have weathered into lowlands. Certain types break down into extremely fertile soils. These rocks are often associated with metal-bearing ores.

Industrial Revolution A period beginning in mid-18th century Britain that saw rapid advances in technology and the use of inanimate power.

Inflation A rise in prices.

Intensive land use A livelihood requiring use of small land areas, such as farming.

Intertillage The growing of two or more crops simultaneously in alternate rows. Also called interplanting.

Irredentism The demand for an international transfer of territory in order to place a minority population in its alleged homeland.

Irrigation Being able to overcome rainfall deficiencies by the control and transport of water to farmland has historically been central to centers of major agricultural production. This act requires engineering skills, political organization all along a water source, and considerable labor investment in the initial creation and the continuing maintenance of such a system.

Island chain Series of islands formed by ocean crust sliding over a stationary "hot spot" in the Earth's mantle.

Karst A landscape feature of an area built on limestone or dolomite, which are susceptible to differential erosion, with resultant sinks and mountains that seem to shoot straight up from limestone plains. Although the term comes from the Adriatic Coast in Europe, some of the world's most dramatic karst landscapes occur in southern China.

Keystone species A species which affects many other organisms in an ecosystem.

Kyoto Protocol A treaty on climate change signed by 160 countries in Kyoto, Japan, in 1997. It requires MDCs to reduce their greenhouse gas emissions by more than 5 percent below their 1990 levels by the year 2012; the U.S. target is 6 percent.

Lacustrine plain Floor of a former lake where glacial meltwater accumulated and sediments washed in and settled. An extremely flat surface is characteristic.

Law of Return An Israeli law permitting all Jews living in Israel to have Israeli citizenship.

Law of the Sea A United Nations treaty or convention permitting coastal nations to have greater access to marine resources.

Landscape   Originally, a term for an artist's view of a setting for human activity but now much more broadly used to define the scene that can be observed from any given point and prospect.

Landscape transformation   The human process of making over the earth's surface into a setting seen as more productive, more convenient, and more aesthetically pleasing. From initial human shelters to contemporary massive urban centers, humans have been driven to change their environmental settings, causing negative as well as positive environmental outcomes.

Large-scale map   A map constructed to show considerable detail in a small area.

Laterite   A material found in tropical regions with highly leached soils; composed mostly of iron and aluminum oxides which harden when exposed and make cultivation difficult.

Latifundia   (*sing.* latifundio)   Large agricultural Latin American estates with strong commercial orientations.

Latitude   Measurement that denotes position with respect to the equator and the poles. Latitude is measured in degrees, minutes, and seconds, which are described as parallels. The low latitudes lie between the Tropics of Cancer and Capricorn; the middle latitudes lie between the Tropics and the Arctic and Antarctic Circles; the high latitudes lie between the Polar Circles and the Poles.

Less developed countries (LDCs)   The world's poorer countries.

Lifeboat Ethics   Ecologist Garrett Hardin's argument that, for ecological reasons, rich countries should not assist poor countries.

Lithospheric plates   The top layer of sediments on the earth and the plates on which human settlement and transport takes place. The -lithic of the Neolithic, for example, comes from the same foot meaning *stone.* Above the lithosphere is the biosphere (layer of life) and above that is the atmosphere (blanket of air).

Location   Central to all geographic analysis is the concept of location. Where something "is" relates to all manner of influences, from climate to migration routes. The classic use of this term comes in the response to the question, "What is most important in real estate?" The answer: "Location, location, location." It is a crucial component in trying to understand patterns of historic and economic development.

Loess   Fine-grained material that has been picked up, transported, and deposited in its present location by wind; it forms an unusually productive soil.

Long March   The 1935–36 flight of approximately 100,000 Communist troops under Mao Zedong from south-central China to the mountains of North China. Mao was fleeing attack by General Chiang Kai-shek and the success of the Long March led to the establishment of Mao's troops as a force with the capacity to overcome the much larger and better armed troops of Chiang.

Longitude   Measurement that denotes a position east or west of the prime meridian (Greenwich, England). Longitude is measured in degrees, minutes, and seconds, and meridians of longitude extend from Pole to Pole and intersect parallels of latitude.

Malthusian scenario   The model forecasting that human population growth will outpace growth in food and other resources, with a resulting population die-off.

Map projection   A way to minimize distortion in one or more properties of a map (direction, distance, shape, or area).

Map scale   The actual distance on the earth that is represented by a given linear unit on a map.

Maquiladoras   Operations dedicated to the assembly of manufactured goods generally, in Mexico and Central America, from components initially produced in the United States or other places. With the enactment of NAFTA in 1994, there has been a massive expansion of *maquiladora* operations because of the economic energy generated by this regional economic alliance.

Marginalization   A process by which poor subsistence farmers are pushed onto fragile, inferior or marginal lands which cannot support crops for long and which are degraded by cultivation.

Marine west coast   Climate occupying the western sides of continents in the higher middle latitudes; greatly moderated by the effects of ocean currents that are warm in winter and cool in summer relative to the land.

Marshall Plan   The plan designed largely by the United States after the conclusion of World War II by which U.S. aid was focused on the rebuilding of the very Germany that had been its enemy in the war just concluded. Secretary of State George Marshall (who had been Chief of Staff of U.S. Army from 1939–45), was central to the plan's design and implementation.

Material culture   The items that can be seen and associated with cultural development, such as house architecture, musical instruments, and tools. In study of the cultural landscape, items of material culture add considerable personality to cultural identity.

Medical geography   The study of patterns of disease diffusion, environmental impact on public health, and the interplay of geographic factors, migration, and population. With the increasing ease of international movement, medical geography is becoming more important as the potential for disease diffusion increases.

Medina   An urban pattern typical of the Middle Eastern city before the 20th century.

Mediterranean (dry-summer subtropical)   Climate that occupies an intermediate location between a marine west coast climate on the poleward side and a steppe or desert climate on the equatorward side. During the high-sun period it is rainless; in the low-sun period it receives precipitation of cyclonic or orographic origin.

Meiji Restoration   The Japanese revolution of 1868 that restored the legitimate sovereignty of the emperor. Meiji means "Enlightened Rule." This event led to a transformation of Japan's society and economy.

Melanesia (Black Islands)   A group of relatively large islands in the Pacific Ocean bordering Australia.

Mental maps   In every individual's mind is a series of locations, access routes, physical and cultural characteristics of places, and often a general sense of the good or bad of locales. The term mental map is used to define such geographies. Education is often aimed at replacing subjective impressions of places and peoples with more objective information in such personal geographies.

Mercantile colonialism   The historical pattern by which Europeans extracted primary products from colonies abroad, particularly in the tropics.

Meridian   *See* Longitude.

Mestizo   In Latin America, a person of mixed European and native Indian ancestry.

Metamorphic rock   Formed from igneous or sedimentary rock through changes occurring in the rock structure as a result of heat, pressure, or the chemical action of infiltrating water. Marble, formed of pure limestone, is a common example. Such rocks are generally harder and more closely knit together than sedimentary rocks, are usually more resistant to wearing down by weathering and erosion, and tend to form uplands in areas where the sedimentary rocks have weathered into lowlands. Metal-bearing ores often are found in areas of metamorphic rock.

Metropolitan area   A city together with suburbs, satellites, and adjacent territory with which the city is functionally interlocked.

Micronesia (Tiny islands)   Thousands of small and scattered islands in the central and western Pacific Ocean, mainly north of the Equator.

Microstate   A political entity that is tiny in area and population and is independent or semi-independent.

Middle Eastern ecological trilogy   The model of mostly symbiotic relations among villagers, pastoral nomads and urbanites in the Middle East.

Migration   A temporary, periodic, or permanent move to a new location.

Milpero   The term for the Latin American farmer who engages in swidden or shifting cultivation in forest lands. The plot is often called the *milpa.*

Minifundia (*sing.* minifundio)   Small Latin American agricultural landholdings, usually with a strong subsistence component.

Mixed forest   Transitional area where both needleleaf and broadleaf trees are present and compete with each other.

Mobility   The pattern of human movement between work and residence, changing because of continual improvement in means of transportation. Increasing mobility has increased the impact that urban space has on adjacent farmlands in most developed countries.

Monsoon   A current of air blowing fairly steadily from a given direction for several weeks or months at a time. Characteristics of a monsoonal climate are a seasonal reversal of wind direction, a strong summer maximum of rainfall, and a long dry season lasting for most or all of the winter months.

Montreal Protocol   A 1989 international treaty to ban chlorofluorocarbons.

Moor   Rainy, deforested upland, covered with grass or heather and often underlain by water-soaked peat.

Moraine   Unsorted material deposited by a glacier during its retreat. Terminal moraines are ridges formed by long-continued deposition at the front of a stationary ice sheet.

More Developed Countries (MDCs)   The world's wealthier countries.

Mulatto   A person of mixed European and black ancestry.

Multiple use   A water system that enables human use of it a number of times, as in stream flow from a river into a turbine and then back into a stream or another, lower, turbine system. Because there is no waste in such a system it is highly desired and leads to sometimes complex engineering in order to achieve multiple use of this resource.

Multiple-cropping   Farming patterns in which several crops are raised on the same plot of land in the course of a year or even a season. Common examples of this are winter wheat and summer corn or soybeans in North America, or several crops of rice in the same plot in Monsoon Asia.

Nation   Commonly, a nation is a term describing the citizens of a state—or that state—but it has additional usage as a collective term for peoples of high cultural homogeneity, existing as a separate political entity. *See* Nation-state.

Nation-state   A political situation in which high cultural and ethnic homogeneity characterizes the political unit in which such people live. The term *ethnic cleansing* has been used to define some of the bloody efforts to achieve nation-state status in the collapse of the former Yugoslavia in the 1990s.

Nationalism   The drive to expand the identity and strength of a political unit which serves as home to a population interested in greater cohesion, and often expanded political power.

Natural levee   Strip of land immediately alongside a stream that has been built up by deposition of sediments during flooding.

Natural resource   Product of the natural environment that can be used to benefit people. Resources are human appraisals.

Near abroad   Russia's name for the now-independent former republics, other than Russia, of the USSR.

Neo-Malthusians   Supporters of forecasts that resources will not be able to keep pace with the needs of growing human populations.

Newly industrializing countries (NICs)   The more prosperous of the world's less developed countries.

Non-black soil zone   Areas of poor soil in cool, humid portions of the Slavic coreland, suitable for cultivation of rye.

North Atlantic Drift   A warm current, originating in tropical parts of the Atlantic Ocean, that drifts north and east, moderating temperatures of western Europe. Also called Gulf Stream.

North Atlantic Treaty Organization (NATO)   Military alliance formed in 1949 which included the United States, Canada, many European nations, and Turkey.

Oil embargo   The 1973 embargo on oil exports imposed by Arab members of the Organization of Petroleum Exporting Countries (OPEC) against the United States and the Netherlands.

Organization for Economic Cooperation and Development (OECD). *See* Organization for European Economic Cooperation.

Organization for European Economic Cooperation (OEEC)   The Organization for European Economic Cooperation was created in 1948 to organize and facilitate the European response to the 1947 Marshall Plan. In the early 1960s, the OEEC became the Organization for Economic Cooperation and Development (OECD) and this served as a multinational base for continued planning in economic and social development. The Common Market came to overshadow this organization and finally the European Union, in 1993, became the most powerful and influential multinational organization in Europe.

Orographic precipitation   The precipitation that results when moving air strikes a topographic barrier, such as a mountain, and is forced upward.

Outwash   Glacial material carried by sheets of meltwater underneath a glacier and subsequently deposited.

Ozone hole   Areas of depletion of ozone in the Earth's stratosphere, caused by chlorofluorocarbons.

Parallel   Latitude line running parallel to the equator.

Patrilineal descent system   Kinship naming system based on descent through the male line.

People overpopulation   The concept that many persons, each using a small quantity of natural resources to sustain life, add up to too many people for the environment to support.

Per capita   For every people or per people.

Perennial irrigation   Year-round irrigated cultivation of crops, as in the Nile Valley following construction of barrages and dams.

Permafrost   Permanently frozen subsoil.

Photosynthesis   The process by which green plants use the energy of the sun to combine carbon dioxide with water to give off oxygen and produce their own food supply.

Physical geography   The subdiscipline in the field of geography most concerned with the climate, landforms, soils, and physiography of the earth's surface.

Piedmont   A belt of country at an intermediate elevation along the base of a mountain range.

Pillars of Islam   The five fundamental tenets of the faith of Islam.

Place identity   In geographic analysis of a given locale, the nature of place identity becomes a means of understanding people's response to that particular place. Determination of the environmental and cultural characteristics that are most frequently associated with a certain place helps to establish that "sense of place" for a given location.

Plain   A flat to moderately sloping area, generally of slight elevation.

Planned economy   Government determination of where industrial plants should be located, what products should be produced, and how the labor force should be organized are all aspects of the planned economies found in the Soviet Union, the People's Republic of China, and Eastern Europe from the end of WWII to the early 1990s. From the early 1990s to the present time, this pattern of organization has been replaced with a market economy, or a system that lets markets—both domestic and foreign—determine the organization of resources and industrial activity.

Plantation   Large commercial farming enterprise, generally emphasizing one or two crops and utilizing hired labor. Plantations are commonly found in tropical regions.

Plate tectonics   The dominant force in the creation of the continents, mountain systems, and ocean deeps. The steady, but slow, movement of these massive plates of the earth's mantle and crust has created the positions of the continents that we have, and major patterns of volcanic and seismic activity. Areas where plates are being pulled under other plates are called subduction zones.

Plateau   An elevated plain, usually lying above 2000 feet (610 m) above sea level. Some are known as tablelands. A dissected plateau is a hilly or mountainous area resulting from the erosion of an upraised surface.

Pleistocene Overkill   A hypothesis stating that hunters and gatherers of the Pleistocene Era hunted many species to extinction.

Podzol   Soil with a grayish, bleached appearance when plowed, lacking in well-decomposed organic matter, poorly structured, and very low in natural fertility. Podzols are the dominant soils of the taiga.

Polder An area reclaimed from the sea and enclosed within dikes in the Netherlands and other countries. Polder soils tend to be very fertile.

Polynesia (Many islands) Pacific island region that roughly resembles a triangle with its corners at New Zealand, the Hawaiian Islands, and Easter Island.

Population change rate The birth rate minus the death rate in a population.

Population density The average number of people living in a square mile or square kilometer. It is a very handy statistic for generalized comparisons but often fails to provide a detailed sense of the real distribution of people.

Population explosion The surge in Earth's human population which has occurred since the beginning of the Industrial Revolution.

Postindustrial Just as the Industrial Revolution of the mid-18th century was a monumental time of change in geographic patterns, the current shifts in employment and trade that have taken place in the past two decades are also seen as significant. New importance in postindustrial society is given to information management, financial services, and the service sector.

Prairie Area of tall grass in the middle latitudes composed of rich soils that have been cleared for agriculture. Original lack of trees may have been due to repeated burnings or periodic drought conditions.

Primary consumers (herbivores) Consumers of green plants.

Primate city A city strongly dominant within its country in population, government, and economic activity. Primate cities are generally found in developing countries, although some already developed countries, such as France, have them.

Prime meridian *See* Latitude.

Producers (Self-feeders) In a food chain, organisms that produce their own food (mostly, green plants).

Push-pull forces of migration In non-forced human migration, there is generally a series of influences on the migrant that tend to push him or her away from a place. Economics and social conditions are common push factors, but images of other places (often unsubstantiated) also serve to attract the migrants and influence the future use of migration destinations.

Pyramid of biomass A diagrammatic representation of decreasing biomass in a food chain.

Pyramid of energy flow A diagrammatic representation of the loss of high quality, concentrated energy as it passes through the food chain.

Qanat Tunnel used to carry irrigation and drinking water from an underground source by gravity flow; also called foggara or karez.

Rain shadow Condition creating dryness in an area located on the lee side of a topographic barrier such as a mountain range.

Region A "human construct" that is often of considerable size and that has substantial internal unity or homogeneity and that differs in significant respects from adjoining areas. Regions can be classed as formal (homogeneous), functional, or vernacular. The *formal* region, also known as a uniform region, has a unitary quality which derives from a homogeneous characteristic. The United States of America is an example of a formal region. The *functional* region, also called the nodal region, is a coherent structure of areal units organized into a functioning system by lines of movement or influence that converge on a central node or trunk. A major example would be the trading territory served by a large city and bound together by the flow of people, goods, and information over an organized network of transportation and communication lines. *Vernacular* regions are areas that possess regional identity, such as "The Sun Belt," but share less objective criteria in the use of this regional name. *General* regions, such as the major world regions in this text, are recognized on the basis of overall distinctiveness.

Remote sensing Through the use of aerial and satellite imagery, geographers and other scientists have been able to get vast amounts of data describing places all over the face of the Earth. Remote sensing is the science of acquiring and analyzing data without being in contact with the subject. It is used in the study of patterns of land use, seasonal change, agricultural activity, and even human movement along transport lines. This process relates closely to GIS, described above.

Renewable resource A resource, such as timber, that is grown or renewed so that a continual supply of it is always available. A finite resource is one that, once consumed, cannot be easily used again. Petroleum products are a good example of such a resource—and because it takes too much time to go through the process of creation, they are not seen as renewable.

Resources Resources become valuable through human appraisal, and their utilization reflects levels of technology, location, and economic ambition. There are few resources that have been universally esteemed (water, land, defensible locales) throughout history, so patterns of resource utilization serve as indices of other levels of cultural development. *See* Natural resource.

Rift valley. *See* Graben.

Ring of Fire The long horseshoe-like chain of volcanoes that go from the southern Andes Mountains in South America up the west coast of North America and arch over into the northeast Asian island chains of Japan, the Philippines, and Southeast Asia. This zone of seismic instability and erratic vulcanism is caused by the tension built up in plate tectonics. *See* Plate tectonics.

Riparian A state containing or bordering a river.

Russification The effort, particularly under the Soviets, to implant Russian culture in non-Russian regions of the former Soviet Union and its Eastern European neighbors.

Sacred space or sacred place Any locale which people hold in reverence, such as places of worship, cemeteries, and battlefields.

Salinization The deposition of salts on, and subsequent fertility loss in, soils experiencing a combination of overwatering and high evaporation.

Savanna Low-latitude grassland in an area with marked wet and dry seasons.

Scorched earth The wartime practice of destroying one's own assets to prevent them from falling into enemy hands.

Seamount An underwater volcanic mountain.

Second law of thermodynamics A natural law stating that high quality, concentrated energy is increasingly degraded as it passes through the food chain.

Secondary consumers (carnivores) Consumers of primary consumers.

Sedentarization Voluntary or coerced settling down, particularly pastoral nomads in the Middle East.

Sedimentary rock Formed from sediments deposited by running water, wind, or wave action either in bodies of water or on land and which have been consolidated into rock. The main classes are sandstone, shale (formed principally of clay), and limestone, including the pure limestone called chalk. They are extensively associated with lowlands such as the Great Plains of North America, but they are also common in highlands. In general, they weather into more fertile soils than igneous or metamorphic rocks and are associated with most deposits of coal and petroleum.

Segmentary kinship system Organization of kinship groups in concentric units of membership, as of lineage, clan and tribe among Middle Eastern pastoral nomads.

Service sector The labor sector made up of employees in retail trade and personal services; the sector most likely to increase in employment significance in *postindustrial* society.

Settler colonization The historical pattern by which Europeans sought to create new or "neo-Europes" abroad.

Shatter belt A region having a fragmented and unstable pattern of nationalities and political units.

Shi'a Islam The branch of Islam regarding male descendants of Ali, the cousin and son-in-law of the Prophet Muhammad, as the only rightful successors to the Prophet Muhammad.

Site and situation Site is defined by the specific geographic location of a given place, while situation is defined as the accessibility of that site, and the nature of the economic and population characteristics of that locale. The combination term deals broadly with the blend of influences of setting and historical development of a place.

Slavic coreland (Fertile Triangle or Agricultural Triangle) The large area of the western Former Soviet Region containing most of the region's cities, industries and cultivated lands.

Small-scale map A map constructed to give a highly generalized view of a large area.

Soil Earth mantle made of decomposed rock and decayed organic material.

Sovereign state A political unit that has achieved political independence and maintains itself as a separate unit.

Soybean A bean crop that made its way from Asia to the New World during the Age of Discovery and has now become a major crop in the agricultural Midwest of the U.S. It is of particular significance because the plant affixes nitrogen as it grows—making it less demanding on the land—and produces a vegetable crop that is high in protein as well as an oil that can be used as a fuel and as a chemical element in a variety of paint, ink, and other products. It is bound to be a crop of growing significance in this next century as both food and petroleum products become ever more costly.

Spatial analysis *See* Spatial.

Spatial Geography is concerned at all levels with spatial organization—that is, the way in which space has been compartmentalized, modified, utilized. The patterns that are associated with the distribution of both physical and human features of the landscape are central to all geographic analysis—and such analysis is spatial analysis.

Special Economic Zones (SEZs) In China, specific urban areas were given special status in the 1980s to enable Chinese planners and entrepreneurs to begin to anticipate economic changes produced by the 1997 return of Hong Kong from Britain. The most important SEZ is Shenzhen, just adjacent to Hong Kong itself, but all of the SEZs have undergone rapid economic and demographic growth since China's current rapid pace of economic growth began in the early 1980s. These locales have served to give Chinese businesspeople a little more latitude in experimentation with free market and joint-venture activities.

St. Lawrence Seaway The St. Lawrence Seaway has a total length of 2342 mi (3796 km) between the Atlantic Ocean and its western terminus in Lake Superior. It allows major oceanic vessels to reach the Great Lakes, thus enabling ports as far inland as the west side of Lake Superior to serve as international ports. The seaway was completed in 1959 although Canada and the United States began to anticipate such a waterway project in the last years of the 19th century.

State A political unit over which an established government maintains sovereign control.

State farm (sovkhoz) A type of collectivized state-owned agricultural unit in the former Soviet Union; workers receive cash wages in the same manner as industrial workers.

Steppe Semiarid transition zone between humid and arid areas. Grass is the dominant vegetation.

Subarctic High-latitude climate characterized by short, mild summers and long, severe winters.

Subduction zone *See* Plate tectonics.

Subsidence The settling of land by the compression of lower layers of rock and soil, usually accelerated by the removal of water from below the surface. Of particular importance in urban centers, with Mexico City having one of the most persistent problems.

Sun Belt An area of indefinite extent, encompassing most of the South, plus the West at least as far north as Denver, Salt Lake City, and San Francisco. This region has shown a faster growth in population and jobs than the nation as a whole during the past several decades.

Sunni Islam The branch of Islam regarding successorship to the Prophet Muhammad as a matter of consensus among religious elders.

Sustainable development (ecodevelopment) Concepts and efforts to improve the quality of human life while living within the carrying capacity of supporting ecosystems.

Sustainable yield (natural replacement rate) The highest rate at which a renewable resource can be used without decreasing its potential for renewal.

Sweat equity The term used for the actual physical labor expended in remaking some aspect of the cultural landscape. Although it is difficult to give an economic value to this process, the term often reflects the patterns of neighborhood change associated with *gentrification.*

Swidden One of the terms used for slash-and-burn agriculture, or shifting cultivation. In this process, landless peasant farmers move into remote forest lands and cut down forest growth. After a period of drying out, the forest floor is burned and then varied crops are planted in the ash of the burned forest material. Crops may be raised as little as one year, or perhaps several years, and then the farming group moves to adjacent lands in order to take advantage of the fertilizer afforded by the burned tree growth. While swidden farmers do not make up a large percentage of the world's farm peoples, they have an enormous impact on remaining primary forest growth, especially in tropical rainforests.

Syncline *See* Folding.

Teaching water The Chinese phrase used to define the labors that must be accomplished in order to establish a productive irrigation system.

Taiga Northern coniferous (needleleaf) forest.

Technocentrists (Cornucopians) Supporters of forecasts that resources will keep pace with or exceed the needs of growing human populations.

Technological unemployment Unemployment caused by the replacement of human labor with highly sophisticated technology, even robots. *See* Postindustrialism.

Tectonic processes Processes that derive their energy from within the Earth's crust and serve to create landforms by elevating, disrupting, and roughening the Earth's surface.

Tensional fault *See* Fault.

Tertiary consumers Consumers of secondary consumers.

Theory of Island Biogeography A theoretical calculation of the relationship between habitat loss and natural species loss, in which a 90 percent loss of natural forest cover results in a loss of half of the resident species.

Tierra caliente Latin American climatic zone reaching from sea level upward to approximately 3000 feet (914 m). In this hot, wet environment are grown such crops as rice, sugarcane, and cacao.

Tierra fria Latin American climatic zone found higher than 6000 feet (1829 m) above sea level. Frost occurs and the upper limit of agriculture and tree growth is reached.

Tierra templada Climatic zone extending from approximately 3000 to 6000 feet (914–1829 m) above sea level. It is a prominent zone of European-induced settlement and commercial agriculture such as coffee-growing.

Togugawa Shogunate The period 1600 to 1868 in Japan, characterized by a feudal hierarchy and Japan's isolation from the outside world.

Trade winds Streams of air that originate in semipermanent high-pressure cells on the margins of the tropics and are attracted equatorward by a semipermanent low-pressure cell.

Transportation infrastructure In efforts of economic development it is always seen that the network of transportation facilities—railroads, highways, ports, airports—has a major impact on a government's or business's efforts to achieve higher levels of productivity.

Triangular trade The 16th- to 19th-century trading links between West Africa, Europe and the Americas, involving guns, alcohol and manufactured goods from Europe to West Africa exchanged for slaves. Slaves to the Americans were exchanged for the gold, silver, tobacco, sugar and rum carried back to Europe.

Trophic (feeding) levels Stages in the food chain.

Tropical rain forest Low-latitude broadleaf evergreen forest found where heat and moisture are continuously, or almost continuously, available.

Tundra A region with a long, cold winter when moisture is unobtainable because it is frozen, and a very short, cool summer. Vegetation includes mosses, lichens, shrubs, dwarf trees, and some grass.

Underground economy The "black market" typical of the Former Soviet Region and many LDCs.

Undifferentiated highland   Climate that varies with latitude, altitude, and exposure to the sun and moisture-bearing winds.

Vernacular region   *See* Region.

Vertical migration   Movements of pastoral nomads between low winter and high summer pastures.

Virgin and idle lands (New lands)   Steppe areas of Kazakstan and Siberia brought into grain production in the 1950s.

Warsaw Pact   Former military alliance consisting of the Soviet Union and the European countries of Poland, East Germany, Czechoslovakia, Hungary, Romania, and Bulgaria (now dissolved).

Water head   In attempting to control water for the generation of electricity or a gravity flow irrigation project, it is essential to pond up water in relative quantity so that there is an adequate force—water head—to achieve the desired goal. Dam structures are the most common engineering responses to this requirement.

Weather   The atmospheric conditions prevailing at one time and place.

Weathering   The natural process that disintegrates rocks by mechanical or physical means, making soil formation possible.

Westerly winds   Airstream located in the middle latitudes that blows from west to east. Also known as westerlies.

Westernization   The process whereby non-Western societies acquire Western traits, which are adopted with varying degrees of thoroughness.

Zero population growth (ZPG)   The condition of equal birth rates and death rates in a population.

Zionist movement   The political effort, beginning in the late 19th century, to establish a Jewish homeland in Palestine.

Zone of transition   Area where the characteristics of one region change gradually to those of another.

Place-Name Pronunciation Guide

The authors acknowledge with thanks the assistance of David Allen, Eastern Michigan University, in preparing an earlier pronunciation guide that the authors expanded and modified for this book. In general, names are those appearing in the text matter; pronunciations of others on maps can be found in Goode's World Atlas (Rand McNally) or standard dictionaries and encyclopedias. Vowel sounds are coded as: ă (uh), ā (ay), ĕ (eh), ē (ee), ī (eye), ĭ (ih), ōō (oo).

Afghanistan (af-*gann'*-ih-stann')
   Helmand R. (*hell'*-mund)
   Hindu Kush Mts. (*hinn'*-doo *koosh'*)
   Kabul (*kah'*-b'l)
   Khyber Pass (*keye'*-ber)

Albania (al-*bay'*-nee-a)
   Tirana (tih-*rahn'*-a)

Algeria (al-*jeer'*-ee-a)
   Ahaggar Mts. (uh-*hahg'*-er)
   Sahara Desert (suh-*hahr'*-a)
   Tanezrouft (*tahn'*-ez-rooft')

American Samoa
   Pago Pago (*pahng'*-oh *pahng'*-oh)
   Tutuila I. (too-too-*ee'*-lah)

Andorra (an-*dohr'*-a)

Angola (ang-*goh'*-luh)
   Benguela R. R. (ben-*gehl'*-a)
   Cabinda (exclave) (kah-*binn'*-duh)
   Luanda (loo-*an'*-duh)
   Lobito (luh-*bee'*-toh)

Anguilla (an-*gwill'*-a)

Antigua and Barbuda (an-*teeg'*-wah, bahr-*bood'*-a)

Argentina (ahr'-jen-*teen'*-a)
   Buenos Aires (*bwane'*-uhs *eye'*-ress)
   Córdoba (*kawrd'*-oh-bah)
   Patagonia (region) (patt'-hu-*goh'*-nee-a)
   Rio de la Plata (*ree'*-oh duh lah *plah'*-tah)
   Rosario (roh-*zahr'*-ee-oh)
   Tierra del Fuego (region) (tee-ehr'-uh dell *fway'*-goh)
   Tucumán (too'-koo-*mahn'*)

Armenia (ahr'-*meeny'*-uh)
   Yerevan (yair'-uh-*vahn'*)

Aruba (ah-*roob'*-a)

Australia (aw-*strayl'*-yuh)
   Adelaide (*add'*-el-ade)
   Brisbane (*briz'*-bun)
   Canberra (*can'*-bur-a)

Melbourne (*mell'*-bern)
Sydney (*sid'*-nih)
Tasmania (state) (tazz-*may'*-nih-uh)

Austria (*aw'*-stree-a)
   Graz (*grahts'*)
   Linz (*lints'*)
   Salzburg (*sawlz'*-burg)
   Vienna (vee-*enn'*-a)

Azerbaijan (ah'-zer-*beye'*-jahn)
   Baku (bah-*koo'*)

Bahamas (buh-*hahm'*-uz)
   Nassau (*nass'*-aw)

Bahrain (bah-*rayn'*)

Bangladesh (bahng'-lah-*desh'*)
   Dacca (Dhaka) (*dahk'*-a)

Barbados (bar-*bay'*-dohss)

Belarus (*bell'*-uh-roos)
   Minsk (*mensk'*)

Belgium
   Antwerp (*ant'*-werp)
   Ardennes Upland (ahr-*den'*)
   Brussels (*brus'*-elz)
   Charleroi (*shahr*-luh-*rwah'*)
   Ghent (*gent'*)
   Liège (lee-*ezh'*)
   Meuse R. (*merz'*)
   Sambre R. (*sohm*-bruh)
   Scheldt R. (*skelt'*)

Belize (buh-*leez'*)

Benin (beh-*neen'*)
   Cotonou (koh-toh-*noo'*)

Bhutan (boo-*tahn'*)
   Thimphu (*thim'*-poo')

Bolivia (boh-*liv'*-ee-uh)
   Altiplano (region) (al-tih-*plahn'*-oh)
   La Paz (lah *pahz'*)
   Santa Cruz (sahnta *krooz'*)
   Sucre (*soo'*-kray)
   Lake Titicaca (tee'-tee-*kah'*-kah)

Bosnia-Hercegovina (*boz'*-nee-uh,-herts-uh-goh-*veen'*-uh)
   Sarajevo (sahr-uh-*yay'*-voh)

Botswana (baht-*swahn'*-a)
   Gaborone (gahb'-uh-*roh'*-nee)

Brazil (bra-*zill*)
   Amazonia (region) (am-uh-*zohn'*-ih-a)
   Belo Horizonte (*bay'*-loh haw-ruh-*zonn'*-tee)
   Brasília (bruh-*zeel'*-yuh)
   Caatinga (region) (kay-*teeng'*uh)

Manaus (mah-*noose'*)
Minas Gerais (*meen'*-us zhuh-*rice*)
Natal (nah-*tal'*)
Paraíba Valley (pah-rah-*ee'*-bah)
Paraná R. (pah-rah-*nah'*)
Pôrto Alegre (por-too' al-*egg'*-ruh)
Recife (ruh-*see'*-fee)
Rio de Janeiro (*ree'*-oh dih juh-*nair'*-oh)
Salvador (sal'-vah-*dor'*)
São Paulo (sau *pau'*-loh)

Brunei (*broo'*-neye)

Bulgaria
   Sofia (*soh'*-fee-a)

Burkina Faso (bur-*keena' fah'*-soh)
   Ouagadougou (wah'-guh-*doo'*-goo)

Burundi (buh-*roon'*-dee) ("oo" as in "wood")
   Bujumbura (boo-jem-*boor'*-a) (second "oo" as in "wood")

Cambodia
   Phnom Penh (*nom' pen'*)

Cameroon (kam'-uh-*roon'*)
   Douala (doo-*ahl'*-a)
   Yaoundé (yah-onn-*day'*)

Canada
   Nova Scotia (*noh'*-vuh *scoh'*-shuh)
   Montreal (mon'-tree-*awl'*)
   Ottawa (*aht'*-ah-wah)
   Quebec (kwih-*beck'*; Fr., kay-*beck'*)
   Saguenay R. (*sag'*-uh-nay)

Cape Verde (*verd'*)

Central African Republic
   Bangui (*bahng'*-ee)

Chile (*chill'*-ee)
   Atacama Desert (ah-tah-*kay'*-mah)
   Chuquicamata (choo-kee-*kah'*-mah-tah)
   Concepción (con-sep'-*syohn'*)
   Santiago (sahn'-tih-*ah'*-goh)
   Valparaíso (vahl'-pah-rah-*ee'*-soh)

China
   Altai Mts. (al-*teye*)
   Amur R. (ah-*moor'*) ("oo" as in "wood")
   Anshan (ahn-*shahn'*)
   Baotou (*bau'*-toh')
   Beijing (bay-*zhing'*)
   Canton (kan'-*tahn'*)
   Chang Jiang (chung jee-*ahng'*)
   Chengdu (chung-*doo'*)
   Chongqing (chong-*ching'*)
   Dzungarian Basin (zun-*gair'*-ree-uhn)
   Guangdong (gwahng-*dung'*)

PG–0

Guangzhou (gwahng-*joh'*)
Hainan I. (*heye'*-*nahn'*)
Harbin (*hahr'*-ben)
Huang He (*hwahng' huh'*)
Lanzhou (lahn-*joh'*)
Lhasa (*lah'*-suh)
Li R. (*lee'*)
Liao (lih-ow')
Lüda (*loo'*-dah)
Nanjing (nahn-*zhing'*)
Peking (pea-*king'*)
Shandong (shahn-*dung'*)
Shanghai (shang-*heye'*)
Shantou (shahn-*toh'*)
Shenyang (shun-*yahng'*)
Shenzhen (shun-*zún*)
Sichuan (zehch-*wahn'*)
Taklamakan Desert (*tah'*-kla-mah-*kahn'*)
Tarim Basin (*tah'*-*reem'*)
Tiananmen (*tyahn'*-an-men)
Tianjin (tyahn-*jeen'*)
Tien Shan (*tih'*-en *shahn'*)
Tientsin (*tinn'*-*tsinn'*)
Tsinling Shan (*chinn'*-leeng *shahn'*)
Ürümqi (oo-*room'*-chee)
Wuhan (*woo'*-*hahn'*)
Xian (shee-*ahn'*)
Xinjiang (shin-jee-*ahng'*)
Xizang (shee-*dzahng'*)
Yangtze R. (*Yang'*-see)
Yunnan (yoo-*nahn'*) ("oo" as in "wood")

Colombia (koh-*lomm'*-bih-uh)
Barranquilla (bah-rahn-*keel'*-yah)
Bogotá (boh'-ga-*tah'*)
Cali (*kah'*-lee)
Cartagena (kahrt'-a-*hay'*-na)
Cauca R. (*kow'*-kah)
Magdalena R. (mahg'-thah-*lay'*-nah)
Medellín (med-deh-*yeen'*)

Comoros (*kahm'*-o-rohs')

Congo
Brazzaville (*brazz'*-uh-vill'; Fr.: brah-zah-*veel'*)
Pointe Noire (pwahnt *nwahr'*)

Costa Rica (*kohs'*-tuh *ree'*-kuh)
San José (sahn hoh-*zay'*)

Croatia (kroh-*ay'*-shuh)
Zagreb (*zah'*-grebb)

Cyprus (*seye'*-pruhs)
Nicosia (nik'-o-*see'*-uh)

Czechoslovakia (check'-oh-sloh-*vah'*-kee-a)
Bratislava (bratt'-ih-*slahv'*-a, braht-)
Brno (*ber'*-noh)
Moravia (muh-*rayve'*-ih-uh)
Ostrava (*aw'*-strah-vah)
Prague (*prahg'*)

Denmark
Copenhagen (*koh'*-pen-hahg'-gen)

Djibouti (jih-*boot'*-ee)

Dominica (dahm'-i-*nee'*-ka)

Dominican Republic (duh-*min'*-i-kan)
Santo Domingo (sant'-oh d-*min'*-goh)

Ecuador (*ec'*-wa-dawr)
Guayaquil (gweye'-uh-*keel'*)
Quito (*kee'*-toh)

Egypt
Aswan (*ahs'*-wahn)
Cairo (*keye'*-roh)
Port Said (sah-*eed'*)

El Salvador (el-*sal'*-va-dor)

Equatorial Guinea (*gin'*-ee)

Estonia (ess-*toh'*-nee-a)
Tallin (*tahl'*-in)

Ethiopia (ee'-thee-*oh'*-pea-a)
Addis Ababa (*ad'*-iss *ab'*-a-ba)
Asmara (az-*mahr'*-a)

Fiji (*fee'*-jee)
Suva (*soo'*-va)

Finland
Helsinki (*hel'*-sing-kee)
Karelian Isthmus (kuh-*ree'*-lih-uhn)

France
Alsace (region) (al-*sass'*)
Ardennes (region) (ahr-*den'*)
Bordeaux (bawr-*doh'*)
Boulogne (boo-*lohn'*-yuh)
Calais (kah-*lay'*)
Carcassonne (kahr-kuh-*suhn'*)
Champagne (sham-*pahn*-yuh)
Garonne R. (guh-*rahn'*)
Jura Mts. (*joo'*-ruh)
Le Havre (luh *ah'*-vruh)
Lille (*leel'*)
Loire R. (*lwahr'*)
Lorraine (loh-*rane'*)
Lyons (Fr., Lyon), both (*lee*-aw)
Marseilles (mahr-*say'*)
Massif Central (upland) (mah-*seef'* saw-*trahl*)
Meuse R. (*merz'*)
Moselle R. (moh-*zell'*)
Nancy (naw-*see'*)
Nantes (*nawnt'*)
Nice (*neece'*)
Oise r. (*wahz'*)
Pyrenees (peer'-uh-*neez'*)
Riviera (riv'-ih-*ehr'*-a)
Rouen (roo-*aw'*)
Saône R. (*sohn'*)
Seine R. (*senn'*)
Strasbourg (strahz-*boor'*)
Toulon (too-*law'*)
Toulouse (too-*looz'*)
Vosges Mts. (*vohzh'*)

French Guiana (gee'-*ahn*-a)
Cayenne (keye'-*enn*)

Gabon (gah-*baw'*)

Gambia (*gamm'*-bih-uh)
Banjul (*bahn*-jool)

Georgia
Abkhazia (ahb-*kahz'*-zee-uh)

Adjaria (ah-*jahr'*-ree-uh)
Caucasus Mts. (*kaw'*-kuh-suhs)
South Ossetia (oh-see'-*shee*-uh)
Tbilisi (tuh-*blee'*-see)

Germany
Aachen (*ah'*-ken)
Berlin (ber-*lin'*)
Bremen (*bray'*-men)
Chemnitz (*kemm'*-nits)
Cologne (kuh-*lohn'*)
Dresden (*drez'*-den)
Düsseldorf (*doo'*-sel-dorf)
Elbe R. (*ell'*-buh)
Erzgebirge (*ehrts'*-guh-beer-guh)
Frankfurt (*frahngk'*-foort')
Karlsruhe (*kahrls'*-roo-uh)
Leipzig (*leyep'*-sig)
Main R. (*mane'*; Ger., *mine'*)
Mannheim-Ludwïgshafen
   (*man'*-hime, *loot'*-vikhs-hah-fen)
Munich (*myoo'*-nik)
Neisse R. (*nice'*-uh)
Oder R. (*oh'*-der)
Ruhr (region) (*roor'*)
Stuttgart (*stuht'*-gahrt)
Weser R. (*vay'*-zer)
Wiesbaden-Mainz (*vees'*-bahd-en *meyents*)
Wuppertal (*voop'*-er-tahl) ("oo" as in "wood")

Ghana (*gahn'*-a)
Accra (a-*krah'*)
Kumasi (koo-*mahs'*-ee)
Tema (*tay'*-muh)
Volta R. (*vohl'*-tuh)

Greece
Aegean Sea (uh-*jee'*-un)
Piraeus (peye-*ree'*-us)
Thessaloniki (thess'-uh-loh-*nee'*-kih)

Grenada (gruh-*nayd'*-a)

Guadeloupe (gwah'-duh-*loop'*)

Guatemala (gwah'-tuh-*mah'*-luh)

Guinea (*gin'*-ee)
Conakry (*kahn'*-a-kree)

Guinea-Bissau (biw-*ow'*)

Guyana (geye-*an'*-a; geye-*ahn*-a')

Haiti (*hayt'*-ee)
Port-au-Prince (pohrt'-oh-*prants'*)

Honduras (hahn-*dur'*-as)
Tegucigalpa (tuh-goo'-sih-*gahl'*-pah)

Hungary (*hun'*-guh-rih)
Budapest (boo-duh-*pesht'*)

Iceland
Akureyri (ah'-koor-*ray'*-ree)
Reykjavik (*ray'*-kyuh-vik')

India
Agra (*ah'*-gruh)
Ahmedabad (ah'-mud-uh-*bahd'*)
Assam (state) (uh-*sahm'*)
Bengal (region) (benn-*gahl'*)
Bihar (state) (bih-*hahr'*)

Brahmaputra R. (brah′-muh-*poo*′-truh)
Deccan (region) (*deck*′-un)
Delhi (*deh*′-lih)
Ganges R. (*gan*′-jeez)
Eastern, Western Ghats (mountains) (*gahts*′)
Himalaya Mts. (himm-*ah*′-luh-yuh, -ah-*lay*-yuh)
Jaipur (*jeye*′-poor)
Jamshedpur (juhm-*shayd*′-poor)
Kashmir (region) (*cash*′-meer)
Kerala (state) (*kay*′-ruh-luh)
Madras (muh-*drass*′, -*drahs*′)
Punjab (region)
  (*pun*′-jahb, -jab; pun-*jahb*′, -*jab*′)
Rajasthan (state) (*rah*′-juh-stahn)
Uttar Pradesh (state) (*oo*′-tahr pruh-*desh*′)

Indonesia
  Bali (*bah*′-lih)
  Bandung (*bahn*′-doong)
  Celebes (Sulawesi) I.
    (*sell*′-uh-beez; sool-uh-*way*′-see)
  Irian Jaya (ihr′-ee-ahn-*jeye*′-uh)
  Jakarta (yah-*kahr*′-tah)
  Java (*jah*′-vuh)
  Kalimantan (kal-uh-*mann*′-tann)
  Medan (muh-*dahn*′)
  Palembang (pah-lemm-*bahng*′)
  Sumatra (suh-*mah*′-truh)
  Surabaya (soo′-ruh-*bah*′-yah)
  Ujung Pandang (oo′-jung pahn-*dahng*′)

Iran (ih-*rahn*′)
  Abadan (ah-buh-*dahn*′)
  Elburz Mts. (ell-*boorz*′)
  Isfahan (izz-fah-*hahn*′)
  Meshed (muh-*shed*′)
  Tabriz (tah-*breez*′)
  Tehran (teh-huh-*rahn*′)
  Zagros Mts. (*zah*′-gruhs)

Iraq (i-*rahk*′)
  Baghdad (*bag*′-dad)
  Basra (*bahs*′-ra)
  Euphrates R. (yoo-frate′-eez)
  Kirkuk (Kihr-*kook*′) ("oo" as in "wood")
  Mosul (moh-*sool*′)
  Shatt al Arab (R.) (*shaht*′ ahl ah-*rahb*′)
  Tigris R. (*teye*′-griss)

Italy
  Adriatic Sea (ay-drih-*at*′-ik)
  Apennines (Mts.) (*app*′-uh-nines)
  Bologna (boh-*lohn*′-ya)
  Genoa (*jen*′-o-a)
  Milan (mih-*lahn*′)
  Naples (*nay*′-pels)
  Palermo (pah-*ler*′-moh)
  Turin (*toor*′-in)

Ivory Coast (Côte d'Ivoire) (koht-d′-vwahr)
  Abidjan (abb′-ih-*jahn*′)

Israel
  Gaza Strip (*gahz*′-uh)
  Eilat (*ay*′-laht)
  Haifa (*heye*′-fa)
  Tel Aviv (tel′-a-*veev*′)

Jamaica (ja-*may*′-ka)
Japan
  Fukuoka (foo′-koo-*oh*′-kah)
  Hiroshima (heer′-a-*shee*′-mah)
  Hokkaido (island) (hah-*keye*′-doh)
  Honshu (island) (*honn*′-shoo)
  Kitakyushu (kee-*tah*′-*kyoo*′-shoo)
  Kobe (*koh*′-bay)
  Kyoto (kee-*oht*′-oh)
  Kyushu (island) (kee-*yoo*′-shoo)
  Kyoto (kee-*oht*′-oh′)
  Nagoya (nay-*goy*′-ya)
  Osaka (oh-*sahk*′-a)
  Sapporo (sahp-*pohr*′-oh′)
  Shikoku (island) (shih-*koh*′-koo)
  Tokyo (*toh*′-kee-oh)
  Yokohama (yoh′-ko-*hahm*′-a)

Jordan (*jor*′-dan)
  Amman (a-*mahn*′)

Kazakhstan (kuh-*zahk*′-stahn)
  Alma Ata (*al*′-mah ah-tah′)
  Aral Sea (uh-*rahl*′)
  Karaganda (kahr′-rah-*gahn*′-dah)
  Lake Balkhash (bul-*kahsh*′)
  Syr Darya (river) (*seer*′ dahr-*yah*′)

Kenya (*ken*′-ya)
  Mombasa (mahm-*bahs*-a)
  Nairobi (neye-*roh*′-bee)

Kiribati (keer′-ih-*bah*′-tih)

Korea, North
  Pyongyang (pea-ong-*yahng*′)
  Yalu River (*yah*′-loo)

Korea, South
  Inchon (*inn*′-chonn)
  Pusan (*poo*′-sahn)
  Seoul (*sohl*′)
  Taegu (teye-*goo*′)

Kuwait (koo-*wayt*′)

Kyrgyzstan (keer′-geez-*stahn*′)

Laos (*lah*′-ohs)
  Vientiane (vyen-*tyawn*′)

Latvia
  Riga (*reeg*′-a)

Lebanon (*leb*′-a-non)
  Beirut (bay-*root*′)

Lesotho (luh-*soh*′-toh)

Libya (*lib*′-yah)
  Benghazi (benn-*gah*′-zee)
  Tripoli (*tripp*′-uh-lee)

Liechtenstein (*lick*′-tun-stine′)

Lithuania (lith′-oo-*ane*′-ee-uh)
  Vilnyus (*vill*′-nee-us)

Luxembourg (*luk*′-sem-burg)

Madagascar
  Antananarivo (an-ta-nan′-a-*ree*′-voh)

Malawi (muh-*lah*′-wee)
  Lilongwe (lih-*lawng*′-way)

Malaysia

Kuala Lumpur (*kwah*′-luh *loom*′-poor)
  ("oo" as in "wood")
Sabah (*sah*′-bah)
Sarawak (suh-*rah*′-wahk)
Strait of Malacca (muh-*lahk*′-uh)

Maldives (*mawl*′-deevz)

Mali (*mah*′-lee)
  Bamako (*bam*′-ah-koh′)

Malta (*mawl*′-tuh)

Martinique (mahr′-tih-*neek*′)

Mauritania (mahr-ih-*tay*′-nee-a)

Mauritius (muh-*rish*′-us)

Mexico
  Acapulco (a-ca-*pool*-ko)
  Ciudad Juárez (see-yoo-*dahd*′ *wahr*′-ez)
  Guadalajara (gwad′-a-la-*har*′-a)
  Monterrey (mahnt-uh-*ray*′)
  Puebla (poo-*eb*′-la)
  Tijuana (tee-*hwah*′-nah)

Moldova (mohl-*doh*′-vah)
  Kishinev (kish′-in-*yeff*′)

Monaco (*mon*′-a-koh′)

Mongolia (mon-*gohl*′-yuh)
  Ulan Bator (oo-*lahn*′ *bah*′tor)

Montserrat (mont′-suh-*rat*′)

Morocco (moh-*rah*′-koh)
  Casablanca (kas-a-*blang*′-ka)
  Rabat (rah-*baht*′)

Mozambique (moh′-zamm-*beek*′)
  Beira (*bay*′-rah)
  Cabora Bassa Dam (kah-*bore*′-ah *bah*′-sah)
  Maputo (mah-*poot*′-oh′)

Namibia (nah-*mib*′-ee-a)

Nauru (nah-*oo*-roo)

Nepal (nuh-*pawl*′)
  Kathmandu (kath′-man-*doo*′)

Netherlands
  Ijsselmeer (*eye*′-sul-mahr′)
  The Hague (*hayg*′)
  Utrecht (*yoo*′-trekt′)
  Zuider Zee (*zeye*′-der-zay′)

Netherlands Antilles
  Curaçao (*koo*′-rahs-ow′)

New Zealand
  Auckland (*aw*′-klund)

Nicaragua (nik′-a-*rahg*′-wah)

Niger (*neye*′-jer; Fr., nee′-*zhere*′)
  Niamey (nee-*ah*′-may)

Nigeria (neye-*jeer*-ee-a)
  Abuja (a-*boo*′-ja)
  Ibadan (ee-*bahd*′-n)
  Kano (*kahn*′-oh)
  Lagos (*lahg*′-us)

Norway
  Bergen (*berg*′-n)
  Oslo (*ahz*′-loh)

Stavanger (stah-*vahng'*-er)
Trondheim (*trahn'*-haym')

Oman (oh-*mahn'*)
Muscat (*mus'*-kat)

Pakistan (pahk-i-*stahn'*)
Hyderabad (*heyed'*-er-a-bahd')
Islamabad (is-*lahm*-a-bahd')
Karachi (kah-*rahch'*-ee')
Karakoram (mountains) (kahr'-a-*kohr*-um)
Lahore (lah-*hohr'*)
Peshawar (puh-*shah'*-wahr)
Punjab (see India)

Papua New Guinea (*pap'*-yoo-uh new *gin'*-ee)
Port Moresby (*morz'*-bee)

Paraguay (*pair'*-uh-gweye')
Asuncíon (a-soon'-see-*ohn'*)

Peru (puh-*roo'*)
Callao (kah-*yow'*)
Chimbote (chimm-*boh'*-te)
Cerro de Pasco (*sehr'*-roh day *pahs'*-koh)
Cuzco (*koos'*-koh)
Iquitos (ee-*kee'*-tohs)
Lima (*lee'*-ma)
Machu Picchu (ma-chu-*pee*-chu)

Philippines (*fil*-i-peenz')
Luzon (island) (loo'-*zahn'*)
Mindanao (island) (minn'-dah-*nah'*-oh)
Mount Pinatubo (pea'-nah-*too'*-buh)
Visayan Islands (vih-*sah'*-y'n)

Poland
Gdansk (ga-*dahntsk'*)
Katowice (kaht-uh-*veet'*-sih)
Krakow (*krahk'*-ow)
Lodz (*looj'*)
Posnan (*pohz'*-nan-ya)
Silesia (region) (sih-*lee'*-shuh)
Szczecin (*shchet'*-seen')
Wroclaw (*vrawt'*-slahf)

Portugal (*pohr'*-chi-gal)
Lisbon (*liz'*-bon)
Oporto (oh-*pohrt'*-oh)

Qatar (*kaht'*-ar)

Romania (roh-*main'*-yuh)
Bucharest (boo'-kuh-*rest'*)
Carpathian Mts. (kahr-*pay'*-thih-un)

Russia
Altai Mts. (*al'*-teye)
Amur R. (ah-*moor'*) ("oo" as in "wood")
Angara R. (ahn'-guh-*rah'*)
Arkhangelsk (ahr-*kann'*-jelsk)
Astrakhan (as-trah-*khahn'*)
Baikal, Lake (beye-*kahl'*)
Bashkiria (bahsh-*keer'*-ee-uh)
Bratsk (*brahtsk'*)
Buryatia (boor-*yaht'*-ee-uh)
Caucasus (kaw'-*kah*-suss)
Chechen-Ingushia
(*chehch'*-yenn, enn-*gooshia'*)
Chelyabinsk (*chell'*-yah-binnsk)

Cherepovets (*chehr'*-yuh-puh-vyetz')
Chuvashia (choo-*vahsh'*-i-a)
Dagestan Rep. (dag'-uh-*stahn'*)
Irkutsk (ihr-*kootsk'*)
Ivanovo (ih-*vahn'*-uh-voh)
Izhevsk (ih-*zhehvsk'*)
Kama R. (*kah'*-mah)
Karelia (kah-*reel'*-yuh)
Kazan (kuh-*zahn'*)
Khabarovsk (kah-*bah'*-rawfsk)
Kola Pen. (*koh'*-lah)
Kolyma R. (kuh-*lee'*-muh)
Krasnoyarsk (kras'-nuh-*yahrsk'*)
Kuril Is. (*koo'*-rill)
Kuznetsk Basin (kooz'-*netsk'*)
Ladoga, Lake (*lad'*-uh-guh)
Lena R. (*lee*-nah, *lay*-nah)
Lipetsk (lyee'-*petsk'*)
Magnitogorsk (mag-*nee'*-toh-gorsk')
Moscow (*mahs'*-koh or *mahs'*-kow)
Murmansk (moor'-*mahntsk'*)
Nakhodka (nuh-*kawt*-kah)
Nizhniy Novgorod (*nizh'*-nee nahv-guh-*rahd'*)
Nizhniy Tagil (tah-*geel*)
Novokuznetsk (naw'-voh-*kooznetsk'*)
Novaya Zemlya (*naw'*-vuh-yuh zem-*lyah'*)
Novosibirsk (naw-vuh-suh-*beersk'*)
Oka R. (oh-*kah'*)
Okhotsk, Sea of (oh-*kahtsk'*)
Omsk (*ahmsk'*)
Pechora R. (peh-*chore'*-a)
Rostov (rahs-*tawf'*)
Sakhalin (sahk'-uh-*leen'*)
Samara (suh-*mahr'*-a)
Saratov (suh-*raht'*-uff)
Sayan Mts. (sah-*yahn'*)
Tatarstan (Tatar Rep.) (Tah-*tahr'*-stahn')
Togliatti (tawl-*yah'*-tee)
Tuva Rep. (*too'*-va)
Ufa (oo-*fah'*)
Ussuri R. (oo-*soor'*-i)
Vladivostok (vlah'-dih-vahs-*tawk'*)
Volga R. (*vahl*-guh, *vohl'*-guh)
Volgograd (vahl-guh-*grahd'*, vohl'-)
Voronezh (vah-*raw'*-nyesh)
Yakutsk (yah-*kootsk'*)
Yekaterinburg (yeh-*kahta'*-rinn-berg)

Rwanda (roo-*ahn*-da)

St. Kitts and Nevis (*nee'*-vis)

St. Lucia (*loo'*-sha)

St. Vincent-Grenadines (gren'-a-*deenz'*)

São Tomé and Príncipe
(sau too-*meh'*, pren-*see'*-puh)

Saudi Arabia (*saud'*-ee)
Jiddah (*jid'*-a)
Riyadh (ree-*adh'*)

Senegal (sen'-i-*gahl'*)
Dakar (da-*kahr'*)

Seychelles (say'-*shelz'*)

Sierra Leone (see-*ehr'*-a lee-*ohn'*)

Slovenia (sloh-*veen'*-i-a)
Ljubljana (*lyoo'*-b'l-yah-nah)

Somalia (so-mahl'-ee-a)
Mogadishu (mahg'-uh-*dish'*-oo)

South Africa
Bloemfontein (*bloom'*-fonn-*tayn'*)
Bophuthatswana (homeland)
(boh-poot'-aht-*swahna'*)
Drakensberg (escarpment) (*drahk*-n's-burg')
Durban (*durr'*-bun)
Johannesburg (joh-*hann'*-iss-burg)
Kwazulu (homeland) (kwah-zoo'-loo)
Natal (province) (nuh-*tahl'*)
Pretoria (prih-*tohr'*-ee-uh)
Transkei (homeland) (trans-*keye'*)
Transvaal (province) (trans-*vahl'*)

Spain
Barcelona (bahr'-suh-*lohn*-uh)
Bilbao (bill-*bah'*-oh)
Catalonia (region) (katt'-uh-*lohn'*-ee-a)
Madrid (muh-*dridd'*)
Meseta (plateau) (muh-*say'*-tuh)
Seville (suh-*vill'*)
Valencia (vull-*enn'*-shee-uh)

Sri Lanka (sree *lahnkah'*)

Sudan (*soo'*-dann)
Gezira (guh-*zeer'*-uh)
Juba (*joo'*-bah)
Khartoum (kahr'-*toom'*)
Omdurman (ahm-*durr'*-man)
Wadi Halfa (*wahdi' hawl'*-fah)

Suriname (*soor'*-i-nahm')

Swaziland (*swah'*-zih-land')

Sweden
Göteborg (yort'-e-*bor'*)
Skåne (*skohn'*)
Småland (*smoh'*-lund')

Switzerland
Basel (*bah'*-z'l)
Zürich (*zoor'*-ick)

Syria (*seer'*-i-uh)
Aleppo (a-*lep*-oh')
Damascus (da-*mas'*-kus)

Tajikistan (tah-*jick'*-ih-stahn')

Tanzania (tan'-za-*nee'*-a)
Dar es Salaam (dahr' es sa-*lahm'*)

Thailand (*teye'*-land')
Bangkok (*bang'*-kok')

Togo (*toh'*-goh)
Lomé (loh-*may'*)

Tonga (*tawng'*-a)

Trinidad and Tobago (to-*bay'*-goh)

Tunisia (too-*nee'*-zhee-a)
Tunis (*too'*-nis)

Turkey
Anatolian Peninsula (ann'-ah-*tohl'*-yun)
Ankara (*ang'*-kah-rah)

Bosporus (strait) (*bahs'*-puhr-us)
Dardanelles (strait) (dahr'-duh-*nelz'*)
Istanbul (iss'-tann-*bool'*)
Izmir (izz'-*meer'*)
Sea of Marmara (*mahr'*-muh-ruh)

Turkmenistan (turk'-menn-ih-*stahn'*)
Ashkhabad (*ash'*-kuh-bahd')

Ukraine (yoo'-*krane'*)
Crimea (cry'-*mee'*-uh)
Dnepropetrovsk (d'nyepp-pruh-pay-*trawfsk'*)
Dnieper R. (duh-*nyepp'*-er)
Donetsk (duhn-*yehtsk'*)
Kharkov (*kahr'*-kawf)
Kiev (*kee'*-yeff)
Krivoy Rog (*kree'*-voi *rohg'*)
Sevastopol (sih-*vass'*-tuh-*pohl'*)
Yalta (*yahl'*-tuh)
Zaporozhye (zah-puh-*rawzh'*-yuh)

United Arab Emirates
Abu Dhabi (*ah'*-boo dah'-bee)
Dubai (doo-*beye'*)

United Kingdom
Edinburgh (*ed'*-in-buh-ruh)
Glasgow (*glass'*-goh)
Thames R. (*timms'*)

United States
Albuquerque (*al'*-buh-ker'-kih)
Aleutian Is. (uh-*loo'*-sh'n)
Allegheny R. (*al'*-ih-gay'-nih)
Appalachian Highlands (ap'-uh-*lay'*-chih-un)
Coeur d'Alene (*kurr'* duh-*layn'*)
Des Moines (duh *moin'*)
Grand Coulee Dam (*koo'*-lee)

Juneau (*joo'*-noh)
Mesabi Range (muh-*sahb'*-ih)
Monongahela R. (muh-*nahng'*-guh-*hee'*-luh)
Omaha (oh'-mah-*hah'*)
Ouachita Mts. (*wosh'*-ih-taw')
Palouse (region) (pah-*loose'*)
Phoenix (*fee'*-nix)
Provo (*proh'*-voh)
Rio Grande (*ree'*-oh *grahn'*-day)
San Joaquin R. (sann wah-*keen'*)
Spokane (spoh'-*kann'*)
Tucson (*too'*-sonn)
Valdez (val-*deez'*)

Uruguay (*yoor'*-uh-gweye')
Montevideo (mahnt'-e-vi-*day'*-oh)

Uzbekistan (ooz-*beck'*-ih-stahn') ("oo" as in "wood")
Amu Darya (river) (ah-moo' *dahr'*-yuh, dahr-*yah'*)
Bukhara (buh-*kah'*-ruh)
Fergana Valley (fair-*gahn'*-uh)
Samarkand (*samm'*-ur-kand'; Rus., suh-mur-*kahnt'*)
Tashkent (tash-*kent'*, tahsh-)

Vanuatu (van'-oo-*ay'*-too)
Port Vila (*vee'*-la)

Venezuela (ven'-ez-*way'*-la)
Caracas (ka-*rak'*-as)
Maracaibo (mar'-a-*keye'*-boh)
Orinoco R. (oh-rih-*noh'*-koh)
Valencia (va-*len'*-see-a)

Vietnam (vee-et'-*nahm'*)
Annam (region) (uh-*namm'*)

Cochin China (*koh'*-chin)
Hanoi (ha-*noi'*)
Ho Chi Minh City (Saigon) (hoe chee *min'*; *seye'*-gahn)
Mekong R. (may-*kahng'*)

Western Samoa (sa-*moh'*-a)
Apia (ah-*pea'*-uh)

Yemen
Aden (*ah*-d'n)
S'ana (sah-*na'*)

Yugoslavia (yoo'-goh-*slahv'*-ee-a)
Belgrade (*bell'*-grade, -grahd)
Kosovo (*kaw'*-suh-voh')
Montenegro (republic) (mahn'-tee-*nay'*-groh)

Zaire (*zeye'*-eer')
Kananga (kay-*nahng'*-guh)
Kasai R. (kah-*seye'*)
Kinshasa (*keen'*-shah-suh)
Kisangani (kee'-sahn-*gahn'*-ih)
Lake Kivu (*kee'*-voo)
Lualaba R. (loo'-ah-*lahb'*-a)
Lubumbashi (loo-boom-*bah'*-shee)
Matadi (muh-*tah'*-dee)
Shaba (region) (*shah'*-bah)

Zambia (*zam'*-bee-a)
Kariba Dam (kah-*ree'*-buh)
Lusaka (loo-*sahk'*-a)
Zambezi R. (zamm-*bee'*-zee)

Zimbabwe (zim-*bahb'*-way)
Bulawayo (bool'-uh-*way'*-oh)
Harare (hah-*rahr'*-ee)
Que Que (*kway' kway'*)
Wankie (*wahn'*-kee)

Index

Italicized page numbers indicate a figure or illustration; "t" following a page number indicates a table.

Aachen, Germany, 116
Abacha, Sani, 450, 451
Abadan, Iran, 268
Abdel-Rahman, Shaykh Umar, 233
Abidjan, 452
Abkhazia, 211–212
Aboriginal peoples
    Africa, 431
    Australia, 385, *396*, 397, 398–399, 400–401, 407, 411
    autonomous okrugs, 208
    Japanese Ainu, 363, 370
    Maori, 397–400, 413
    See also Native Americans
Aborigines, Australian, 385, *396*, 397, 398–399, 400–401, 407, 411
Abortion, 305, 340
Abraham, 231, 239
Abrego, Juan Garcia, 526
Absolute location, 8, 12
Abu Dhabi, 230, 263, *263*
Abubakar, General, 451
Acapulco, 508, *508*
Accra, 452
Acid rain, 84
Adaptive reuse, 599
Addis Ababa, 444, 445
Adelaide, 402, 409
Aden, 264
Adirondack Mountains, 593
Afghanistan, *224*, 225t, 272–274, *272*, 307
Africa, Northwestern, *256*, 257–259
Africa, sub-Saharan, 419–440, 442–473, *423*
    agriculture, 38, 436–439, 449, 453, *455*
    area, 422
    basic data, 14t, 420–421t
    climate, and vegetation, 425–426
    colonialism, 430–431, *431*
    countries and cities, *423*, 450–452, *452*
    cultures, 426–427
    economy, 434–436
    environment, 422–426, *423*, *424*
    GNP, 420–421t, 454
    impact of slavery, 427–430, *430*
    natural resources, *424*, 439–440, 449–450
    politics, 435
    population, 422
    poverty, 433, 462
    public health issues, 434, 440
    racism, 462–464, 466
    regional groupings, 419, *419*
    shared traits, 433–436
    tourism, 439, 467
    trade, 437
    transportation, 424, 452, 453
    See also individual countries
African Americans, 562, 602
African National Congress (ANC), 464
Africanity, 427
Afrikaners, 462
Age of Discovery, 78, 286, 298, 448, 523
Age of Exploration, 41
Age-structure profile, 50–51, *52*, *53*
Agent Orange, 328
Agricultural geography, 19, 141–142
Agricultural Revolution, 38–40
Agricultural Triangle, 196–197
Agriculture
    cash (commercial) crops, 45, 437–438
    collectivization, 154, 182–183, *182*, 196
    cooperatives, 242
    crop breeding, 352–353
    crop calendar, 292
    crop diffusion, 74t
    crop introductions, 439
    double cropping, 367, 378
    dry farming, 39, 367–368
    enclosure, 99
    garden, 289, *289*, 349
    global competition, 108
    global warming and, 56
    Green Revolution, 34, 292, 305
    intertillage, 367–368
    landscape, *13*
    mechanization and innovation, 289, 352–353, 368
    multiple cropping, 339
    oasis farming, 291
    plantation. *See* Plantation agriculture
    privatization, 196
    rainwater harvesting, 447
    share tenancy, 292, 509
    shifting cultivation, 321, 436
    slash and burn, 290, 437, 497, *497*, 512, *512*
    See also Fertilizers; Subsistence farming; *and specific crops*
Aidid, Mohamed Farrah, 445
AIDS/HIV, 434
    Africa, 434, 440, 458
    anti-AIDS campaigns, 326
    religious clerics and, 326
    Southeast Asia, 326
Ainu, 363, 370
Air photo interpretation, 19
Air pollution, 55–57, 265, 358, 375, 512
Air transportation, 230, 585, 606
Aircraft and aerospace industry, 598, 604, 617–618
Akademgorodok, 200
Akkadian empire, 266
Akosombo Dam, 425
Akron, 612
Al-Assad, Hafez, 252–253
Al-Bashir, Omar, 256
Al-Qaddafi, Muammar, 256, 257
Alabama, *590–591*, 599, 600, 601
Alaska, 177, 572, 622–624, *623*
Alaska Highway, 622
Alaska Native Claims Settlement Act (ANSCA), 624
Albania, 78, 149, *150*, 151, 153, 159
Albedo, 447
Alberta, 573t, 579, 580–581, 583
Albuquerque, 619, 620
Alcoholism, 398
Aldabra Atoll, 473, *473*
Aleppo, 249
Aleutian Islands, 623, 624
Aleuts, 624
Alexander the Great, 214
Alexandria, 226, 255
Algeria, *224*, 225t, 233, *256*, 257–259
Algiers, 258
Ali, 232
Alice Springs, 404, *404*
Alleghany Front, 594
Allen, Paul, 619
Alliance of Small Island States, 391
Almaty, 216
Alps, the, 119, 127, *127*, *129*, 130, 145, *145*
Altitudinal zonation, 484, 488–489
Amazon Basin, 45, 481, 496
Amazon River, 529
Amazon River Lowlands, 532–532
Amboseli National Park, 457
American Civil War, 562, 564, 578, 600, 603
American Indians, 624. *See also* Native Americans
American Revolution, 564, 577, 594
American Samoa, 395
Amerindians, 490
Amin, Idi, 459
Amman, 247
Amritsar, 312, *312*
Amsterdam, 123
Amsterdam Treaty, 121
Amu Darya (Oxus) River, 215–216
Anchorage, 622

California (*continued*)
  Central Valley, 553, 614, 615
  Los Angeles Basin, 614, 615, 616–617
  population, 63, 602, 616
  Proposition 187, 507
  redwoods, 557, 614, 618
  San Francisco metropolis, 618
  Southern California, 616–618, 621
  topography, 553
  water issues, 616, *617*, 618, 621
California Air Resources Board, 621
Calvinism, 151
Cambodia, *289*, 317–318, *318*, 327
  ASEAN membership, 329
  basic data, 281t
  conflicts with Vietnam, 328
  killing fields, 329
  land mines, 329
  refugees, 326
  rice production, 323
Camel, 406
Cameroon, 420t, 427, 452, 453
Cameroon, Mt., 453
Camp David Accords, 246
CAMPFIRE program, 439
Canada, 567–585
  agriculture, 573–574, 580
  area, 549, 549t
  Atlantic Provinces, 570–575, 573t
  basic data, 549t, 573t
  climate, *554*
  Core Provinces, 573t, 575–579
  economy, 556–560, 563
  fishing industry, 573, 574
  GNP, 556, 557t
  major regions, *551*, *569*, 570
  Maritime Provinces, 570–575
  NAFTA. *See* North American Free Trade
    Agreement
  natural resources, 578–579, 580–581, 583
  the North, 582–585
  Pacific Province, 573t
  population, 549, *550*, 560–562, 561t, 567, 570
  Prairie Provinces, 579–581, 573t, 583
  Quebec separatism, 561–562, *577*, 580
  Territories, 573t
  U.S. and, 564, 567–570, *570*, 572
  Vancouver region, 581–582
Canadian North, 582–585
Canadian Pacific Railroad, 581
Canadian Shield, 550, 570, *571*, 581, 583, 584
Canals
  Bangkok, 325, *325*
  Benelux, 123
  British, 74
  "Dead-Red," 244
  Dutch, 74
  Erie, 594, 597
  French, 113
  Grand Canal, 344–345
  Jonglei, 255
  Leeds-Liverpool, 74
  North Sea, 123
  Panama, 510–515, *510*, *514*

"Soo," 578
  Volga-Don, 175
Canary Islands, 143
Canberra, 409
Cancún, 496
Cape Breton Island, 574
Cape Canaveral, 604
Cape Cod, 592
Cape Colony, 462
Cape Provinces, 465–466, *465*
Cape Town, 462, 463, 466
Cape Verde Islands, 420t, 446
Capitaincy-General of Guatemala, 509
Capitalism, 80, 156, 375
Capri, 146
Caprivi Strip, 466
Caracas, 485, 488, 525, 536
Caribbean Middle America. *See* Middle America
Caribbean Islands, 490, 496, *502*, 515–518
Carpathian Mountains, 153
Carrying capacity, 39
Carson, Rachel, 18
Cartagena Agreement, 521
Carter, Jimmy, 269, 377
Cartogram, 10, *11*
Cartography, 6, 19
Cascade Mountains, 614, 622
Cash crops, 45, 437–438, 453
Caspian Sea, 212–213
Caste system, 309, 311
Castro, Fidel, 515, 516
Catalans, 116
Catherine the Great, 177, 193
Catholicism
  anticontraception clergy, 326, 334
  Canada, 579
  England-Ireland conflict, 109
  Europe, 77–78
  Latin America, 479, *479*, 503
  Monsoon Asia, 288
  Philippines, 334
  Slavic, 151
  Southeast Asia, 326, 327
  U.S., 602
Cattle, 79, 307, 603, *609*. *See also* Cattle
    ranching; Livestock; Pastoral nomadism
Cattle ranching, 216, *405*, 407, 411, 503
Caucasus, 210–214
Cayman Islands, 517
Ceausescu, Nicolae, 158
Celebes, *289*, 317, *318*, 332
Celtic languages, 76
Center-pivot irrigation, 257
Central African Federation, 469
Central African Republic, 420t, 436, 452
Central America, 501, *502*, 508–515. *See also*
    individual countries
Central American Federation, 509
Central Arizona Project, *617*
Central Asia, *168*, 170t, 171, 185, 214–218, *215*
Central Intelligence Agency, U.S., 47
Cerro de Pasco silver, 529
Ceylon, 287, 317, 322. *See also* Sri Lanka
Chad, 420t, 446, 447, 457

Chaebols, 380
Chang Jiang (Yangtze) River, 341, 345–346, *347*,
    353
Chang Jiang Water Transfer Project, 343,
    345–346, *347*
Chao Praya River, 323, 325
Chaparral, 31, 73, 614
Charles XII, 178
Charleston, 606
Charlotte, 606
Charney Effect, 447
Chechnya, 209–210
Chelyabinsk, 182, 200
Chemical industries, 308, 598
Chengdu, 355
Chennois, 299, 302
Chernobyl nuclear accident, 84, *84*, 192, 204
Chernomyrdin, Viktor, 187
Chernozems, 172, 173, 196
Cherrapunji, India, 301
Chiang Kai-shek, 351
Chiapas, 494
Chicago, 611, 612
Chile, 481t, 493, 494, 494t, 495, 496t, 521, *522*,
    537, 540–542
Chimbote, 529
China, 337–359, *338*
  agriculture, 352–355
  Albania and, 159
  archaeological sites, *347*
  area, 337–340, *337*
  Arid China, 340–343
  basic data, 281t
  Cambodia and, 329
  cities, *336*, *338*, 355–356, *355*
  climate, 283, *345*
  Cultural Revolution, 350–351
  communization, 352–355
  compared with U.S., *337*
  future management, 356–358
  Hong Kong reversion, 288, 356, 582
  Humid China, 343–349
  industrialization, 349–351
  Japan and, 348
  Kashmir and, 311
  landscape and environment, 340–349
  one child campaign, 399–400, *399*
  Opium Wars, 177
  politics, 349
  population, 337–340
  Spratly Islands claim, 334
  steel production, 99
  Taiwan issue, 358–359
  trade, 351–352, 516
  transportation, 342, 344, 353–355, *354*
  urban migration, 290–291
  Vietnam War, 327–328
  water conservancy, 352–353
China proper, 343
Chinese Communist Party (CCP), 351
Chinese Eastern Railway, 348
Chisinau, Moldava, 195
Chittagong, Bangladesh, 308
Chokepoints, 245, 246

Java, *289*, 317, *318*, 331
Jebel Ali, 263, *263*
Jerusalem, 236, 241, 243, 426
Jewish diaspora, 240–241
Jews, 78, 151, 239–242, 247
Jidda, Saudi Arabia, 262
Jimmu, 370
Jinja, 459
Johannesburg, 469
Jokhar-Gala, 210
Jonglei Canal, 255
Jordan, *224*, 225t, 227, 244, 247–248, 252
Jordan River, 241, 243, 244
Juárez, Benito, 504
Jubail, *236*, 237
Judah, 239–240
Judaism, 223, 239–242
Jumna (Yamuna) River, 301
Juneau, 622
Jura Mountains, 109, 111, 127
Jute, 306, 308

Ka'aba, 231
Kababish, 235
Kabbah, Ahmen Tejan, 450
Kabila, Laurent, 455, 456–457, 460
Kabul, 272–273
Kabul River, 272
Kainji Dam, 425
Kalahari Desert, 425
Kalimantan, 332
Kaliningrad, 195
Kama River, 201
Kamchatka, 202, *202*
Kampala, 459
Kanaks, 390
Kananga, 454
Kangaroos, 406
Kano, Nigeria, 452
Kansas, *590–591*, 607
Kansas City, 611, 613
Karachi, 299, 308, 310
Karaganda, Kazakstan, 216–217
Karakoram mountains, 299
Karanga kingdom, 426
Karelia, 203
Karens, 325
Kariba Dam, 425, *425*, 470
Karst, 348, *348*
Karun River, 269
Kasai River, 454
Kashmir (Jammu), 311–312, 315
Katanga (Shaba), 454
Kathmandu, 300, *300*
Kauai, 624
Kazakstan, 167, *168*, 170t, 173, 183, 191, 214, 216–217, 218
Kazan, Tatarstan, 175, 177, 200, 208
Kazbek, Mt., *210*
Keating, Paul, 398, 410
Kenai Peninsula, 623
Kennedy, John F., 516
Kentucky, *590–591*, 599–600, 603, 610
Kentucky Bluegrass Region, 601, 603

Kenya, 421t, 422, 445, 458–459
Kenya-Uganda Railway, 458
Keystone species, 437
Khabarovsk, 202
Khalistan, 312
Kharkov, Ukraine, 182, 192
Khartoum, 256
Khmer empire, 317–318
Khmer Rouge, 329
Khoi people, 462, 467
Khomeini, Ayatollah Ruhollah, 26
Khrushchev, Nikita, 193
Khuzistan, 269
Khyber Pass, 299
Kibbutz, 242
Kiev, 176, 192, *194*
Kigali, 459, 460
Kikuya, 459
Kilimanjaro, Mount, 422, 455
Kim Il Sung, 377, 378
Kim Jong Il, 378
Kimberley, South Africa, 465
Kinshasa, 425, 453, 454, 453, 456
Kiribati, 391
Kirinyaga (Mt. Kenya), 422, *425*, 455
Kiriyenko, Sergei, 187
Kitakyushu, 369
Kitimat, 582
Kiwi, 413
Klamath Mountains, 614
Knoxville, 604, 606
Kobe, Japan, 366, 369, 624
Kodiak Island, 623
Kokura, Japan, 370
Kola Peninsula, 203
Komsomolskaya, 213
Kongo kingdom, 426
Korea, *364*, 376–380
    basic data, 281t
    China relations, 376, 378
    climate, 283
    Demilitarized Zone, 377–378, *378*
    isolation policy, 287
    Japan relations, 371, 376–377, 378
    nuclear weapons, 377
    See also North Korea; South Korea
Korean Americans, 562
Korean War, 348, 373, 377
Kosciusko, Mount, 402
Kosovo, 78, 89, 116, 159, 160, *161*
Kosovo Liberation Army (KLA), 159, 160
Krasnoyarsk, 205
Krays, 178
Krivoy Rog, 192
Kruger National Park, 426
Kuala Lumpur, 318, *319*, 330
Kuchma, Leonid, 193, 194
Kulaks, 182
Kumasi, 452
Kurdistan, 267
Kurds, 231, 235, 265, 266–267, 268, 271
Kuril Islands, 170, 202, 371, 377
Kursk Magnetic Anomaly (KMA), 200
Kuwait, *224*, 225t, 264–266

Gulf War, 229, 232, 250, 262, 374
    land mines, 274
    mineral resources, 259
    oil fires, 265, *265*
Kuwait City, 264
K2, Mount, 300
KwaZulu/Natal, 46
Kyoto, 369, *370*
Kyoto Climate Change Conference, 391
Kyoto Protocol, 56–57
Kyrgyzstan, *168*, 170t, 214, 217, 218
Kyushu, 363, *364*

La Paz, 487, 530
Labor
    Africa, 471
    Australia, 407
    forced. *See* Slavery
    guest workers, 85, 120
    illegal immigrant, 507
    indentured, 518
    Kanaka, 407
    Latin America, 508, 524
    Middle East, 254
    migrant contract workers, 322
    Monsoon Asia, 291
    Poland, 156, *156*, 157
    South Korea, 379
    Southeast Asia, 321–322
    unionized, 103
    See also Slavery
Labrador, 570–571, 573t
LaFranchi, Howard, 526
Lagos, 450, 451
Lahore, Pakistan, 308
Lake Baikal, 176
Lake Chad, 447
Lake Huron, 612
Lake Kinneret, 243, 248
Lake Erie, 612
Lake Malawi, 423
Lake Mead, 621
Lake Nasser, 254
Lake Powell, *561*
Lake St. Clair, 612
Lake Superior, 612
Lake Tanganyika, 422, 423
Lake Titicaca, 530
Lake Victoria, 255, 422, 423
Lalibela, 443, *444*
Lamaism, 342
L'Amour, Louis, 144
Land empire, 179
Land mines, 274, 329
Land reform
    ejido, 493, 494, 504
    Latin America, 494t
Land rights
    Australian, 398–399
    Native American, 557
Landforms, 34–35, *36–37*
Landscape, 4
    agricultural, *13*
    analysis, 12–13